Advanced Functional Materials

Yafang Han
Editor

Advanced Functional Materials

Proceedings of Chinese Materials Conference 2017

Springer

Editor
Yafang Han
Chinese Materials Research Society
Beijing
China

ISBN 978-981-13-0109-4 ISBN 978-981-13-0110-0 (eBook)
https://doi.org/10.1007/978-981-13-0110-0

Library of Congress Control Number: 2018938380

Printed on acid-free paper

This Springer imprint is published by the registered company Springer Nature Singapore Pte Ltd.
part of Springer Nature
The registered company address is: 152 Beach Road, #21-01/04 Gateway East, Singapore 189721, Singapore

Preface

This is the proceedings of the selected papers presented at Chinese Materials Conference 2017 (CMC2017) held in Yinchuan City, Ningxia, China, July 06–12, 2017.

The Chinese Materials Conference (CMC) is the most important serial conference of Chinese Materials Research Society (C-MRS) and is held each year since the early 1990s. Chinese Materials Conference 2017 had 37 symposia covering four fields of energy and environmental materials, *advanced functional materials*, high-performance structural materials, and design, preparation, and characterization of materials. More than 5500 participants attended the conference, and the organizers received more than 700 technical papers. By the recommendation of symposium organizers and after peer reviewing, 434 papers are published in the present proceedings, which are divided into four volumes as follows:

Part 1: Advances in Energy and Environmental Materials
Part 2: High-Performance Structural Materials
Part 3: Advances in Materials Processing
Part 4: Advanced Functional Materials

This is the volume for Part 4 including 104 papers selected from 11 symposia of C09: Advanced ceramic materials; D01: Metamaterials and multifunctional materials; D02: Multiferroic materials; D03: Amorphous and high entropy materials; D05: Advanced glass materials and their application; D06: Advanced optoelectronic and microelectronic materials; D07: Biomaterials; D08: Deformation behavior and flow units in metastable materials; D09: Advanced fibers and nanocomposites; D10: Polymers materials; D11: Nanoporous metallic materials.

The editors would like to give the thanks to the symposium chairs and all the paper reviewers of this volume, and the thanks should be also given to Mrs. Yiming Jia, Mr. Yongzheng Zhang for their hard work of compilation and typesetting for each paper in this volume.

Beijing, China
March 2018

Yafang Han

Contents

The Stress and Morphology Evolution of CZO Films Under Different Growth Temperature

Dong Zhao, Yunyan Liu, Shutao Li, Rendong Wang, Junshan Xiu, Gongxiang Wei, Meiling Sun, Mengzhen Song and Shenggui Fu

Abstract $Zn_{1-x}Co_xO$ (x = 0.05, CZO) thin films are deposited on Si(111) and glass substrates at different substrate temperatures from 350 to 500 °C. All the CZO films on glass substrates exhibit compressive stress, but for the films deposited on Si substrates, the change from compressive to tensile one is observed. The increase of Ts is favorable for the reduction of stress in the films. Roughness exponent α, RMS roughness w and autocorrelation length ξ are calculated. At lower temperature, for both substrates, α is less than 0.65, indicating lower temperature could not provide sufficient energy for the free migration and diffusion of the deposited atoms. The ratio of w/ξ increases with the increase of Ts from 350 to 450 °C, suggesting that the upward growth of the CZO film depends much more on growth temperature than the lateral one. When the temperature increases further, w/ξ decreases, but the morphology of CZO films shows the same evolution feature in spite of the type of substrate, indicating that Ts plays a dominate role rather than substrates in the ratio of w/ξ CZO films.

Keywords CZO · LMBE · Substrate temperature · Stress · Surface roughness

Introduction

Due to their excellent characteristics in magnetism, magneto-optics and magneto-electrics, diluted magnetic semiconductors (DMS) have drawn wide attention for their practical application like spin field-effect-transistor, high density memory, ultra-fast optical switches, semiconductor lasers and quantum spin

D. Zhao · Y. Liu (✉) · S. Li · R. Wang · J. Xiu · G. Wei · M. Sun
M. Song · S. Fu (✉)
School of Science, Shandong University of Technology, Zibo 255049
Shandong, People's Republic of China
e-mail: liuyunyan@sdut.edu.cn

S. Fu
e-mail: fushenggui@163.com

Y. Han (ed.), *Advanced Functional Materials*,
https://doi.org/10.1007/978-981-13-0110-0_1

computer [1, 2]. For its wide band gap (3.37 eV) and high excitation energy (60 meV), transition-metal (TM)-doped ZnO has been considered as a promising DMS for implementing spinelectronics device concepts [3]. In particular, ZnO doped with Co has been the subject of intense research [4], and the prominence is given to Co-doped ZnO as one of the possible candidates for room temperature ferromagnetism.

It is well known that residual stress and morphologies of growth fronts as well as their evolutions play crucial roles in determining a number of physical and chemical characteristics of deposited thin films [5, 6]. Stresses in thin films will deform the unit cell and bring in undesired anisotropy that will have much influence on physical properties of the films such as mechanical stability of structures and electrical parameters. The energy gap, effective masses and exciton resonance energy are also strongly dependent upon the strain of the films [7]. The microscopic growth dynamics can be reflected by the morphologies of growth fronts and their evolutions. Meanwhile, the studies of morphologies of growth fronts are of great importance to both basic researches and actual applications [8–10]. The remaining stress and the morphologies of thin films may vary with different processing situations, for example, deposition methods, working pressure, the substrate and substrate temperature (Ts), etc. The substrate temperature plays an important part in determining the preferred orientation, film stress, crystalline and the morphologies of growth fronts, by influencing the energy of depositing atoms [11, 12].

We have fabricated $Zn_{1-x}Co_xO$ (x = 0.05) thin films on glass and Si substrates by the LMBE technique with variation of Ts from 350 to 500 °C. In this paper, the detailed studies of substrate temperature on the structure, residual stress and morphology properties of Co-doped ZnO thin films prepared by LMBE are discussed.

Experimental Details

The Co doped ZnO films were prepared on glass and Si (111) substrates at various substrate temperatures (350–500 °C) by LMBE. The detailed procedure of the target of CZO was provided elsewhere [13]. Ablation of the target was achieved using UV KrF excimer laser at a wavelength of 248 nm (120 mJ per pulse, 10 ns and 5 Hz). The deposition system' base pressure was 5×10^{-5} Pa. In the process of deposition, the pressure of oxygen was kept at 2×10^{-3} Pa. The Si substrates were cleaned routinely by RCA. The glass substrates were cleaned firstly in boiled alcohol and in acetone for 5 min, respectively, and then were ultrasonically cleaned in alcohol for 20 min. Subsequently, the substrates were rinsed in deionized water before being dried at room temperature. The substrate holder was rotated at 5 rpm using a stepping motor during deposition to obtain uniformly films. The distance of 4 cm was maintained between the substrate and the target.

The X-ray diffraction (XRD) patterns of the CZO films were performed by using Cu kα1 radiation (λ = 0.15406 nm), D8 ADVANCE x-ray diffraction. The surface morphology was observed using atomic force microscope (AFM, Nanoscope III) in tapping mode with a UL06 probe. The typical radius and side angle of the tip were about 10 nm and 10°, seperately. The scanning area in the micrographs was 2 μm × 2 μm and the image resolution was 256 × 256.

Results and Discussion

Curves in Fig. 1a, b show the XRD spectra of the CZO films deposited on Si(111) and glass substrates at different substrate temperatures. It reveals that all the CZO films exhibit a strong peak at 2θ value around 34.2°, corresponding to the (002) preferred orientation of ZnO (JCPDS 36-1451). For the films deposited on glass substrates, the films show other diffraction peaks associated with (102) and (110) reflections of hexagonal ZnO, indicating the polycrystalline microstructure of ZnO films with dominated (002) orientation.

The position of (002) peak, and the calculated grain sizes of all the samples are listed in Table 1. Table 1 shows that the position of (002) peak of all the samples deviates from that of ZnO powder (2θ = 34.421°). It can be attribute to the residual stain in the films [14]. The strain $\varepsilon = (C_o - C)/C_o$ in the direction of the c-axis can be measured by XRD. Where C_o is the unstrained lattice constant of bulk ZnO and C is the lattice constants of the ZnO films. To derive the film stress σ_{film} parallel to the film surface, the following formula has been applied, and it is effective for a hexagonal lattice:

$$\sigma = [2C_{13} - (C_{11} + C_{12})C_{33}/C_{13}] \times \varepsilon \quad (1)$$

As to the elastic constants C_{ij}, datas of single crystalline ZnO have been used: $C_{11} = 208.8$, $C_{33} = 213.8$, $C_{12} = 119.7$, $C_{13} = 104.2$ Gpa [15], which results in the following numerical relation: $\sigma_{\text{film}} = -233\varepsilon$ (Gpa) for the stress derived from XRD [16].

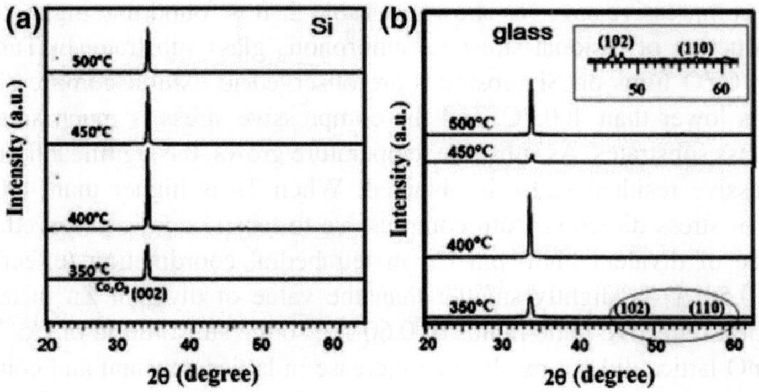

Fig. 1 The XRD pattern of ZnO:Co films on **a** Si and **b** glass substrate at different Ts from 350 to 500 °C

Table 1 The grain size, C, peak position, stress, α, w and ξ of ZnO:Co films on Si substrates at different temperature

Temperature (°C)	Grain size (nm)	C (Å) (±0.001)	2θ (deg.)	Stress (Gpa)	α	w (nm)	ξ (μm)
350	16.97	5.221	34.325	−0.64	0.65	0.8	0.019
400	30.80	5.210	34.399	−0.15	0.75	1.5	0.028
450	28.40	5.203	34.446	0.16	0.75	1.8	0.031
500	28.20	5.203	34.450	0.17	0.75	1.3	0.024

Table 2 The grain size, C, peak position, stress, α, w and ξ of ZnO:Co films on glass substrates at different temperature

Temperature (°C)	Grain size (nm)	C (Å) (±0.001)	2θ (deg.)	Stress (Gpa)	α	w (nm)	ξ (μm)
350	14.79	5.267	34.017	−2.70	0.62	0.9	0.026
400	31.13	5.241	34.192	−1.52	0.72	1.6	0.029
450	25.66	5.221	34.326	−0.63	0.80	2.0	0.030
500	34.21	5.235	34.230	−1.27	0.80	1.7	0.030

The crystal lattice constant C together with the calculated residual stress σ_{film} are also listed in Tables 1 and 2. Generally, there are two components of stress existing in the deposited films, the thermal stress and intrinsic stress. The thermal stress is because of the difference between thermal expansion coefficients of the films and the substrates induced by temperature changes. The intrinsic stress is attributed to the growth process and independent on the extrinsic substrate. The films also undergo residual stress induced by the lattice mismatch between the films and the substrates.

The negative sign in Tables 1 and 2 means that the stress is in compression condition, while the positive sign shows the tensile stress state of the films.

As for the films on glass substrates, the stresses in all the CZO films are observed to exhibit compressive stress as shown in Table 2. It is found that higher Ts profits for the reduction of residual stress on amorphous glass substrate. In Table 1, the stresses in CZO films on Si substrates are observed to exhibit compressive stress when Ts is lower than 400 °C, and the compressive stress is much smaller than those of glass substrates. As substrate temperature grows, the significantly reduction of compressive residual stress is obtained. When Ts is higher than 400 °C, the alteration of stress direction from compressive to tensile one is observed.

The size of divalent high-spin Co in tetrahedral coordination (effective ionic radius of 0.58 Å) is slightly smaller than the value of divalent Zn in tetrahedral configuration (effective ionic radius of 0.60 Å), so the substitution of Co^{2+} ions for Zn^{2+} in ZnO lattice did not result in an increase in lattice constant and compressive stress of the samples. Many reporters attributed the compressive stress in ZnO films to the defects of the film [17]. The mobility of atoms in the films increased significantly along with the increase of Ts. From non-equilibrium position, the atoms can easily diffuse to a better equilibrium position, which will result in the decrease of compressive stress and the defects of the film [17]. Also, the interstitial oxygen may be another reason for compressive stress [17, 18]. Even though the oxygen interstitials are less desired in ZnO films due to the higher formation energy, the oxygen atmosphere during our films growth would ascribe to the probability of the form of oxygen interstitials. Increasing substrate temperature, the compressive stress would be relieved for the interstitial oxygen would be out-diffused because of the thermal expansion. Moreover, as shown in Tables 1 and 2, the stress in the films deposited at 350 °C is much larger than that in other temperatures. It is noticed that the grain sizes in the films deposited at Ts of 350 °C on both Si and glass substrates are much smaller than other ones. The larger grain sizes in the films deposited at Ts higher than 400 °C may contribute to the stress variation of the films, because larger grain size may cause the decrease of the compressive stress and even the stress direction change [19].

Much smaller stress in Si substrates may due to the tensile factor as following. ① The thermal expansion coefficient of ZnO thin film and is 4.8×10^{-6}/K and the thermal expansion coefficient of Si(111) plane and ordinary glass is 2.6×10^{-6}/K and 9×10^{-6}/K, separately. The stress caused by thermal expansion of Si and CZO film should be tensile stress and those of glass and CZO films should be compressive. ② The lattice constant a of ZnO is 0.325 nm and the lattice constant of the Si(111) plane is 0.384 nm, and the stress derived from the lattice mismatch in the ZnO thin film is also the tensile stress, while the grain boundary between the glass and the ZnO thin film is non-coherent because of the amorphous substrate of glass, so the elastic distortion can be very small.

Figures 2 and 3 are the representative AFM images and one-dimensional cross section scan of surface profile of CZO films on the Si and glass substrates with different substrate temperatures from 350 to 500 °C, respectively. From Figs. 2 and 3 we can observe that substrate temperature has influence on the CZO films surface morphology. Figures 2 and 3 show that grain features of the film surface on both substrates are not obvious and the surface is relatively flat when Ts is 350 °C. As Ts is higher than 400 °C, round grain features are found and the surface roughness tends to increase, and the grains present uniform distribution in the film surface.

Under normal circumstances, the growth of the thin films can be described as a process of far-from-equilibrium, and the complex dynamics of the growth causes film morphology to evolve in the rough surface structure of random self-affine fractality. Height–height correlation function $H(r, t)$ can be effectively applied to describe the statistical evolution of the surface morphology. In describing the rough surface morphology and finding out the mechanism governing the thin film growth, three important parameters for $H(r, t)$ are the roughness exponent α, the RMS roughness w, the lateral correlation length ξ. Both w and ξ are statistical description for the global surface morphology. w shows the roughness of surface along the vertical direction in the the process of growing. ξ is a parameter which measures averagely the lateral coarsening size, and it is the distance within which surface variations are correlated, but beyond which surface fluctuations spread and are

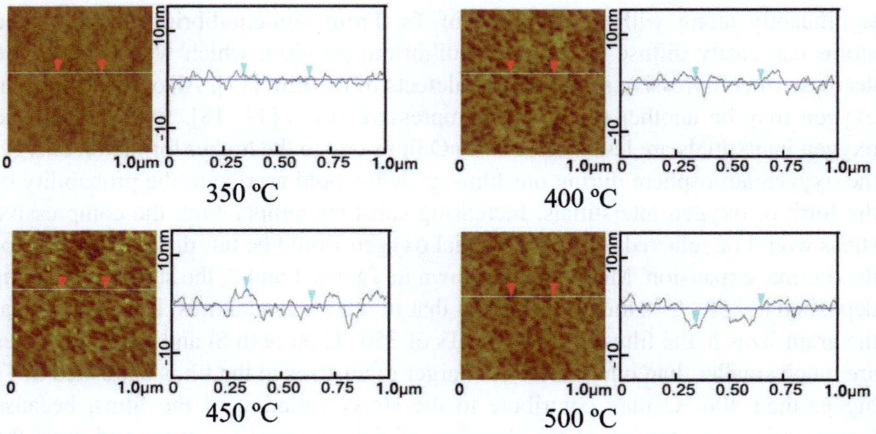

Fig. 2 The AFM images and one-dimensional cross section scan of surface profile of ZnO:Co thin films on Si substrates at 350, 400, 450 and 500 °C

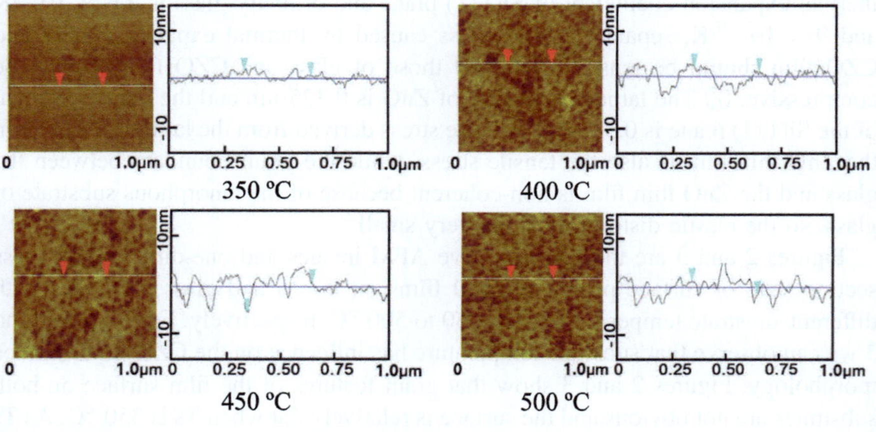

Fig. 3 The AFM images and one-dimensional cross section scan of surface profile of ZnO:Co thin films on glass substrates at 350, 400, 450 and 500 °C

uncorrelated. The roughness exponent α is a significant parameter that depicts the self-affine fractal surface. Meanwhile, it is a description of local surface roughness [5]. The detailed calculation procedure of the surface parameters was depicted in our previous work [20]. The calculation results are listed in Tables 1 and 2.

Table 1 shows that for the films on Si substrates at Ts of 350 °C, α takes the values of 0.65. Further increase of Ts from 400 to 500 °C, Ts has little effect on α values and it maintains at 0.75, which indicates a smaller fluctuation obtainable in short-range surface at higher Ts. For the Co doped ZnO films on glass substrates, α take values of 0.75 $\pm$ 0.5, except the value of 0.62 at Ts of 350 °C as revealed in Table 2. Previous reports revealed that for rounded grains or shape approximately conic and pyramidal grains, the exponents α value is from 0.7 to 0.95 [21–23]. Thus the calculated α values in Table 1 match well with the AFM images in Figs. 2 and 3. For the surface with no obvious grain feature in Figs. 2 and 3 is corresponding to α value smaller than 0.7 in Tables 1 and 2, and the films with obvious grain feature observed in Figs. 2 and 3 is corresponding to α values larger than 0.7 as shown in Tables 1 and 2.

In scaling researches for nonequilibrium films growth by deposition [24, 25], a value of $\alpha = 0.35$ is desired when surface mobility of deposited particles is not permitted. However, $\alpha = 0.66$ is wanted when surface mobility is allowed [26]. At Ts of 350 °C, the α values of 0. 65 and 0.62 in our films on Si and glass substrate, respectively, indicate that the lower Ts could not provide enough energy for the surface atoms mobility on the substrates surface. The disorder accumulation of atoms in the non-equilibrium position results in a violent height fluctuation of the local surface topography, and a large number of defects are generated, which is considered to be an important source of higher compressive stress in the CZO film at lower growth temperature. As the deposition temperature is increased further, atoms are more likely to diffuse from one place to another, which results in an improvement in crystalline of the films and strain release in the films, for the trapped atoms in the non-equilibrium position can move to a more equilibrium position. This result is coincident with the structural and stress results as mentioned above.

For the films both on glass and substrates, when Ts increase from 350 to 450 °C, the roughness w and the autocorrelation length ξ increase monotonously, then w decreases when Ts is further increased to 500 °C, and ξ decrease when Ts is 500 °C for Si substrates and unchanged for glass substrates.

To observe the impacts of temperature on CZO films' growth behavior, we calculated the ratio of w/ξ. Figure 4 shows the curve of w/ξ ratio versus the temperature Ts. The ratios of w/ξ can reflect the competition between the lateral growth and the vertical growth of the film. It is noted that the morphology of CZO films shows the same evolution feature despite the type of substrate, indicating that Ts plays a dominate role rather than substrates in the morphology evolution of CZO films. For both substrates, when the growth temperature rises from 350 to 450 °C, w/ξ becomes larger and larger, and this clearly shows the vertical growth is much more sensitive to temperature, the higher the temperature, the faster of the vertical growth than the lateral one. The mass transmission achieved by the diffusion of deposited atoms, is the dynamic process for formation of different film surface morphology. The diffusion energy of the local surface and the possible potential barrier (such as the ES barrier) in the film determine the mass transfer of the particles. In addition to the diffusion on the surface, there is a mass transfer in the vertical direction of the step barrier control [27].

If the step barrier is relatively large, the layer to layer transmission of atoms is suppressed, the adsorption atoms are not easy to jump down from the higher atomic layer, even transition from the lower atomic layer to the higher atomic layer may occur. This is easy to form a 3D island or mound mode growth and the global surface roughness increases [28]. In the range from 350 to 450 °C, the temperature

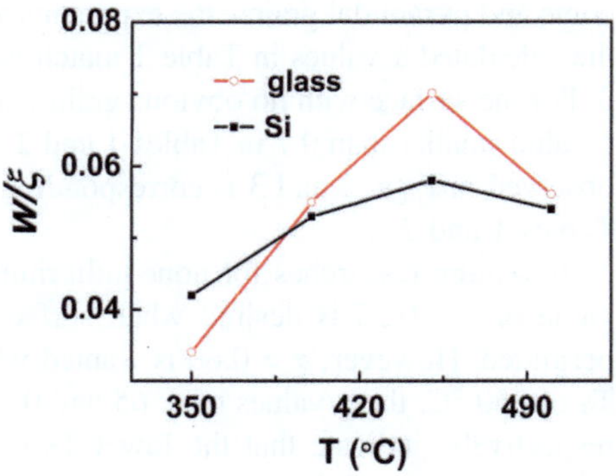

Fig. 4 The curve of w/ξ versus the temperature Ts of ZnO:Co films on Si and glass substrate

does not reach the critical temperature for the atom to overcome the step potential to jump to the lower layer resulting in the vertical migration increases with temperature. At the same time, the diffusion of the lateral atoms increases with increase of Ts, but the larger barrier of the island leads to the suppression of the lateral diffusion of the atoms. The competition of stronger vertical growth with weaker lateral one results in the increase of w/ξ when the temperature increases.

Again an increase in Ts of 500 °C might provide high surface energy and induces the agglomeration of grain structures, which results in a formation of bigger structures and it should be a diffusion process and may be favourable for the decrease in surface roughness and the formation of flat surface. At higher Ts, the desorption of deposition particles arose by the incident ions with high energy may be another cause for the reduction of RMS roughness and w/ξ.

Conclusion

In summary, we have synthesized $Zn_{1-x}Co_xO$ (x = 0.05) thin films on Si(111) and glass substrates by LMBE at various temperatures from 350 to 500 °C. The structure, stress and morphology evolution characteristics are discussed by analyzing the XRD and AFM results. We find that all the CZO films exhibit (002) preferred orientation of ZnO. In contrast to the amorphous structure of glass, Si substrate is more fitting for the formation of low stress films. All the CZO films on glass substrates exhibit compressive stress, while for the films on Si substrates, the compressive stress is only obtained at lower Ts. As substrate temperature increases, the enhanced mobility of atoms and the out-diffused interstitial oxygen caused by thermal expansion would lead to the decrease of compressive residual stress. Surface parameter α, w and ξ are preliminarily calculated. The effect of Ts on the diffusion ability of absorbed atoms or molecules could be indicated by the variation values of α, which could interpret the results as revealed in structural and stress discussion. The morphology evolution of CZO films indicates that Ts plays a dominate role rather than substrates in the morphology feature of CZO films. When the temperature increases, the growth of vertical increase is faster than the lateral one, resulting in the increase of w/ξ. The random fluctuations induced by coalesce of small grains and agglomeration of larger grains is considered to be the influence on the decrease of w/ξ when Ts increases further.

Acknowledgements The authors acknowledge the financial support of the National Science Foundation of China under Grant No. 11404191, and Shandong Provincial Natural Science Foundation (ZR2016AQ22). We are also grateful for the financial support of SDUT & Zibo City Integration Development Project (No. 2016ZBXC205) and Top Young project of SDUT.

References

1. S. Pearton, W. Heo, M. Ivill, D. Norton, Semicond. Sci. Technol. **19**, 59–74 (2004)
2. C. Liu, F. Yun, H. Morkoc, J. Mater. Sci.: Mater. Electron. **16**, 555–597 (2005)
3. V.A.L. Roy, A.B. Djurišić, H. Liu, X.X. Zhang, Y.H. Leung, M.H. Xie, J. Gao, H.F. Lui, C. Surya, Appl. Phys. Lett. **84**, 756–758 (2004)
4. W. Prellier, A. Fouchet, B. Mercey, J. Phys.: Condens. Matter **15**, 1583–1601 (2003)
5. H.N. Yang, Y.P. Zhao, G.C. Wang, T.M. Lu, Phys. Rev. Lett. **76**, 3774–3777 (1996)
6. A. Shikanai, T. Azuhata, T. Sota, S. Chichibu, A. Kuramata, K. Horino, S. Nakamura, J. Appl. Phys. **81**, 417–424 (1997)
7. R. Triboulet, J. Perrière, Prog. Cryst. Growth Charact. Mater. **47**, 65–138 (2003)
8. J.H. Kim, K.W. Chung, J. Appl. Phys. **83**, 5831–5839 (1998)
9. H.N. Yang, G.C. Wang, T.M. Lu, *Diffraction from Rough Surfaces and Dynamic Growth Fronts* (World Scientific, Singapore, 1993)
10. H.J. Qi, L.H. Huang, Z.S. Tang, C.F. Cheng, J.D. Shao, Z.X. Fan, Thin Solid Films **444**, 146–152 (2003)
11. T.B. Hur, Y.H. Hwang, H.K. Kim, I.J. Lee, J. Appl. Phys. **99**, 064308 (2006)
12. D. Zhang, C.Z. Wang. Y.L. Liu, Q. Shi, W.J. Wang, Opt. Laser. Tech. **44**, 1136–1140 (2012)
13. A. Dinia, J.P. Ayoub, G. Schmerber, Phys. Lett. A **333**, 152–156 (2004)
14. M. Chen, Z.L. Pei, C. Sun, L.S. Wen, X. Wang, J. Cryst. Growth **220**, 254–262 (2000)
15. E. Mollwo, in *Landoldt-börnstein*, ed. by O. Madelung, M, Schulz, H. Weiss. Zahlenwerte und funktionen aus Naturwiss.u.Technik. Neue Serie (Springer, Berlin, 1982), pp. 35
16. R. Cebulla, R. Wendt, K. Ellmer, J. Appl. Phys. **83**, 1087–1095 (1998)
17. W. Water, S.Y. Chu, Mater. Lett. **55**, 67–72 (2002)
18. K. Suqiyama, K. Taniquchi, K. Kuwabara, J. Mater. Sci. Lett. **9**, 489–492 (1990)
19. Y.C. Lee, S.Y. Hu, W. Water, K.K. Tiong, Z.C. Feng, J. Lumin. **129**, 148–152 (2009)
20. Y.Y. Liu, C.F. Cheng, S.Y. Yang, H.S. Song, G.X. Wei, C.S. Xue, Y.Z. Wang, Thin Solid Films **519**, 5444–5449 (2011)
21. T.J. Oliveira, F.D.A. Aarão Reis, Phys. Rev. E. **83,** 041608 (2011)
22. T.J. Oliveira, F.D.A. Aarão Reis, J. Appl. Phys. **101,** 063507 (2007)
23. J. Ebothé, A. El Hichou, P. Vautrot, M. Addou, J. Appl. Phys. **93**, 632–640 (2003)
24. M. Kardar, G. Parisi, Y. Zhang, Phys. Rev. Lett. **66**, 2348–2350 (1991)
25. D.E. Wolf, J. Villain, Europhys. Lett. **13**, 389–394 (1990)
26. D.C. Agarwal, R.S. Chauhan, D.K. Avasthi, S.A. Khan, Kabiraj, I. Sulania, J. Appl. Phys. **104**, 024304-024304-8 (2008)
27. M. Gedda, N.V.V. Subbarao, D.K. Goswami, Langmuir **30**, 8735–8740 (2014)
28. R. Kunkel, B. Poelsema, L.K. Verheij et al., Phys. Rev. Lett. **65**(6), 733–736 (1990)

Tunable Dielectric Property of Fe_3O_4/SiO_2 Meta-composite Fabricated by Magnetic-Field Assisted Self-assembly

Muyu Wang, Wenjin Zhang, Yang Huang, Chongpu Wang, Ruoyu Shi, Zhetong Liu, Yangfan Su and Zidong Zhang

Abstract In this paper, SiO_2 microsphere self-assembly process and heat treatment were combined to prepare Fe_3O_4/SiO_2 meta-composites. XRD, SEM and Precision Impedance Analyzer were used to study the phase composition, microstructure, and dielectric properties of the composites at MHz frequency range. The results show that SiO_2 microsphere matrix forms a hexagonal close packing structure and nano-Fe_3O_4 magnetic particles fill the gap between the close-packed SiO_2 microspheres. The results indicate that, in a certain frequency range (1–3 MHz), the dielectric properties of the composites can be tuned by using applied magnetic field

M. Wang · W. Zhang · Y. Huang · C. Wang · R. Shi · Z. Liu
Y. Su · Z. Zhang (✉)
Key Laboratory for Liquid-Solid Structural Evolution & Processing of Materials Ministry of Education, Shandong University, Jinan 250061, Shandong, China
e-mail: 1986zzd@163.com

M. Wang
e-mail: wmypdd@163.com

W. Zhang
e-mail: 1264180097@qq.com

Y. Huang
e-mail: qq328486124@126.com

C. Wang
e-mail: 17853140530@163.com

R. Shi
e-mail: suleimanusshi@gmail.com

Z. Liu
e-mail: lztwxhlfg2009@163.com

Y. Su
e-mail: 18661378638@163.com

Z. Zhang
Dezhou META Research Center for Innovative Materials, Dezhou 253000, China

Y. Han (ed.), *Advanced Functional Materials*,
https://doi.org/10.1007/978-981-13-0110-0_2

to change the distribution of the magnetic particles. But in the other test frequency range, the dielectric properties of the composites show no difference, whether to use magnetic field to control the distribution of the magnetic particles or not.

Keywords Permittivity · Dielectric properties · Meta-composite
Self-assembly

Introduction

Metamaterial, in which various special properties can be obtained by design a periodical structure, offers a new way to design material's properties for certain applications. Since the first experimental demonstration of double negative property in metamaterial, several interesting applications have been demonstrated, such as perfect lens, [1, 2] invisible cloaking, [3] etc. [4, 5] which can be widely applied in the communication, radar stealth technology [6] and wireless energy transmission [7, 8].

Among the various kinds of metamaterial, there exists a basic rule that is the unit size should be comparable to the wavelength [9]. This rule indicates that the fabrication cost cannot be controlled within a reasonable range for mass production in THz or visible light frequency range application, as the high-cost advanced nanofabrication technology should be chose to get a good geometric control of the nano-sized unit. Therefore, the self-assembly process has been paid more and more attention in the fabrication of metamaterial due to its low-cost, which shows great potential in practical application [10, 11].

In this paper, a modified self-assembly process is used for the fabrication of metamaterial. The SiO_2 microspheres were used to form a matrix with hexagonal close packing structure by self-assembly process. Then, followed by a typical impregnation process, in which nano-Fe_3O_4 magnetic particles will fill the gap between the close-packed SiO_2 microspheres. By using applied magnetic field, the distribution of the magnetic particles can be changed in the SiO_2 matrix. XRD, SEM and Precision Impedance Analyzer are used to study the phase composition, microstructure, and dielectric properties of the composites at MHz frequency range. The results indicate that, the distribution of the magnetic particles will influence the dielectric properties of the Fe_3O_4/SiO_2 composites, which indicate a new way for the control of the periodical structure in metamaterial.

Experiment

The SiO_2 microsphere powder has been prepared by Stöber method by using tetraethoxysilane (TEOS) as raw material. The SiO_2 matrix with hcp structure was prepared by self-assembly process. After that, the bulk matrix was put into a magnetic fluid with Fe_3O_4 nanoparticles, known as impregnation process. Two different kinds of samples were prepared by put a magnet under the beaker or not. The samples were firstly dried at 80 °C in air atmosphere to remove water and then a heat treatment under an argon atmosphere at 950 °C. The field emission scanning electron microscope (FESEM, SU-70) and LCR meter are used to investigate the morphology and the dielectric property of the samples.

Results and Discussion

The SEM images of Fe_3O_4/SiO_2 composites are shown in Fig. 1. It can be found that the SiO_2 micro-spheres tightly pack in hexagonal stacking to form a periodic structure. But the distribution of the Fe_3O_4 magnetic nanoparticles in the matrix is quite different depending on using magnetic field or not during the fabrication process.

As shown in Fig. 1a, the Fe_3O_4 nanoparticles are uniformly random distribution in the sample prepared without magnetic field. When we applied an outer magnetic field during the fabrication process, the distribution of Fe_3O_4 nanoparticles shows clearly segregation in the matrix, as shown in Fig. 1b, c. This inhomogeneous distribution of the Fe_3O_4 nanoparticles will influence the dielectric properties of the Fe_3O_4/SiO_2 composites, which will be discussed in the following part. As showed in Fig. 2, the real part of permittivity for the sample fabricated under magnetic field shows different dispersion curve, compare to the sample without magnetic field, from 1 to 3 MHz. All the parameters we use during the fabrication process are the same, except the magnetic field which will only influence the distribution of the Fe_3O_4 nanoparticles (shown in Fig. 1).

The loss tangent of the sample also shows the same tendency, as shown in Fig. 2b. This phenomenon indicates that the distribution of magnetic particles will influence the polarization behavior in the composite.

In order to get a better understanding of the polarization behavior in the composite, we use LCR meter to investigate the impedance of the sample. As show in Fig. 3, there is no obvious difference between the sample fabricated under magnetic field and without magnetic field. The impendence curve of the two kinds of sample

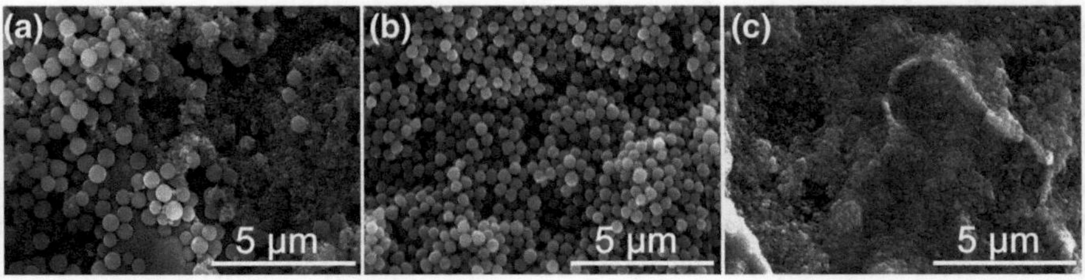

Fig. 1 SEM images of **a** sample fabricated without magnetic field, sample fabricated with magnetic field (**b**) magnetic nanoparticles poor-area and **c** magnetic nanoparticles rich-area

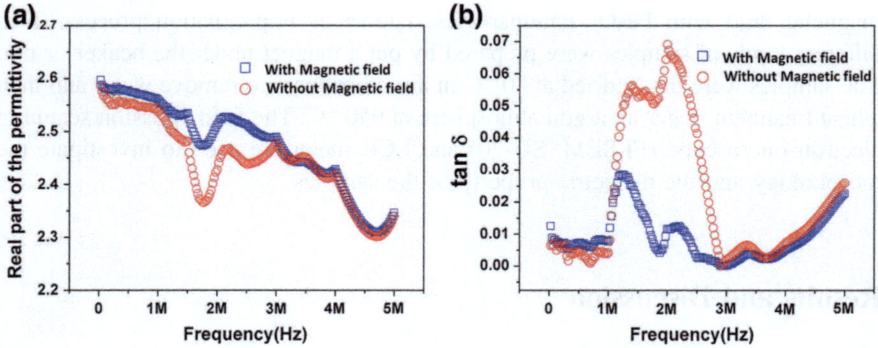

Fig. 2 The real part of the permittivity (**a**) and loss tangent (**b**) for sample fabricated with magnetic field (blue data) and without magnetic field (red data)

Fig. 3 The impedance for sample fabricated with magnetic field (blue data) and without magnetic field (red data)

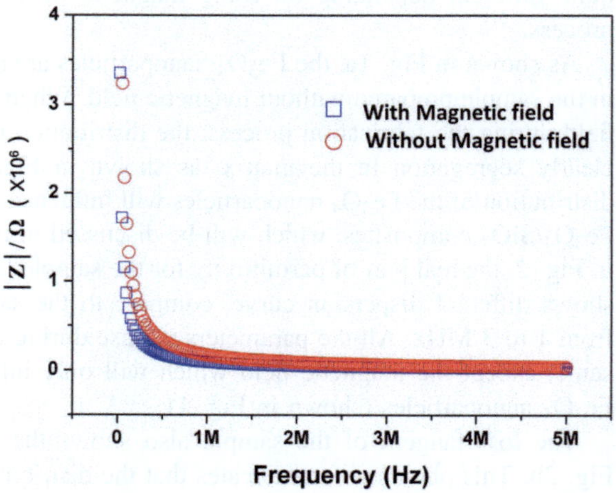

overlaps over the whole test frequency range, which indicates that the concentration of the carrier keeps the same in the samples.

But if we come to Fig. 4, in which the real part (Fig. 4a) and imaginary part (Fig. 4b) of impedance shown separately, some interesting results can be found. The imaginary part of the impedance shows no difference between the two kinds of samples. And it takes a negative value over the whole frequency range, which indicates a capacitor-like behavior. The real part of impedance shows the same tendency with the permittivity and loss tangent from 1 to 3 MHz, which indicate the difference from 1 to 3 MHz in the permittivity and loss tangent has a close relationship with the contact resistance in the sample. As shown in Fig. 1, the distribution of the magnetic particles is quite different depending on using magnetic field or not. The Fe_3O_4 magnetic nanoparticles will be aggregative along the magnetic line, leading to the changing of the contact resistance. Therefore, the pathway for

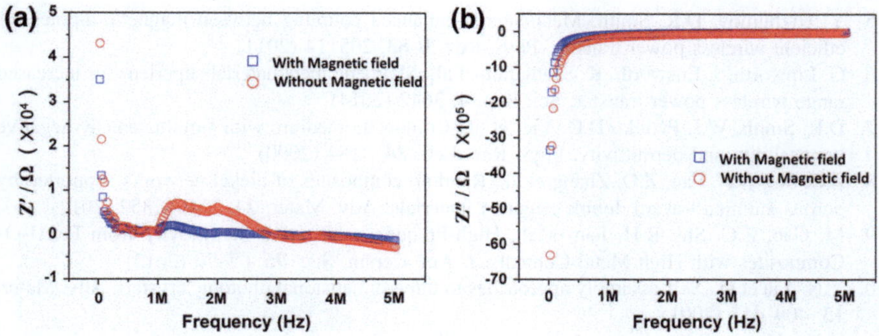

Fig. 4 The real part and imaginary part of impedance for sample fabricated with magnetic field (blue data) and without magnetic field (red data)

the carriers in the composite will be different due to this aggregation, leading to the interesting phenomenon from 1 to 3 MHz.

Conclusion

In this paper, Fe_3O_4/SiO_2 composites were prepared through self-assembly process. The distribution of the Fe_3O_4 nanoparticles can be controlled by applying an outer magnetic field. As the magnetic nanoparticles will be aggregative along the magnetic line, the permittivity and loss tangent shows different dispersion behavior from 1 to 3 MHz range. This phenomenon has a close relationship with the contact resistance in the sample. The way we used in this paper gives a new way for the control of the composites' properties, which shows great potential for the multifunctional composites' design.

Acknowledgements This study was financially supported by the National Natural Science Foundation of China (No. 51601105, No. 51402170) and the Natural Science Foundation of Shandong Province (BS2015CL020, ZR2016EMM09).

References

1. P.H. Tichit, S.N. Burokur, D. Germain et al., Design and experimental demonstration of a high-directive emission with transformation optics. Phys. Rev. B **83**, 155108 (2011)
2. F. Ding, Y. Cui, X. Ge et al., Ultra-broadband microwave metamaterial absorber. Appl. Phys. Lett. **100**, 103506 (2012)
3. Y. Pang, H. Cheng, Y. Zhou et al., Double-corrugated metamaterial surfaces for broadband microwave absorption. J. Appl. Phys. **113**, 084907 (2013)
4. N. Landy, D.R. Smith, A full-parameter unidirectional metamaterial cloak for microwaves. Nat. Mater. **12**, 25–28 (2013)

5. Y. Urzhumov, D.R. Smith, Metamaterial-enhanced coupling between magnetic dipoles for efficient wireless power transfer. Phys. Rev. B **83**, 205114 (2011)
6. G. Lipworth, J. Ensworth, K. Seetharam et al., Magnetic metamaterial superlens for increased range wireless power transfer. Sci. Rep. **4**, 3642 (2014)
7. D.R. Smith, W.J. Padilla, D.C. Vier et al., Composite medium with simultaneously negative permeability and permittivity. Phys. Rev. Lett. **84**, 4184 (2000)
8. Z.C. Shi, R.H. Fan, Z.D. Zhang et al., Random composites of nickel networks supported by porous alumina toward double negative materials. Adv. Mater. **24**, 2349–2352 (2012)
9. M. Gao, Z.C. Shi, R.H. Fan et al., High-Frequency Negative Permittivity from Fe/Al_2O_3 Composites with High Metal Contents. J. Am. Ceram. Soc. **95**, 67–70 (2012)
10. Y.N. Xia et al., Self-assembly approaches to three-dimensional photonic crystals. Adv. Mater. **13**, 409–413 (2001)
11. X.Y. Huan, P.K. Jiang, C.U. Kim, Electrical properties of polyethylene/aluminum nanocomposites. J. Appl. Phys. **102**, 124103 (2007)

Synthesis, Magnetic and Electrochemical Properties of $NiFe_2O_4$-rGO Nanohybrids

Gongqin Yan, Qiang He and Dongyan Yu

Abstract In this paper, nickel ferrite ($NiFe_2O_4$)-reduced graphene oxide (rGO) nanohybrids were synthesized via a hydrothermal strategy successfully. The morphology and phase composition of as-prepared products were analyzed by field-emission scanning electron microscopy (FESEM) and X-ray diffraction (XRD). The results show that the well-crystalized $NiFe_2O_4$ nano octahedrons with edge of 100 nm grew on the surface of rGO successfully. The magnetic property of the synthesized nanohybrids at room temperature was characterized by a vibrating sample magnetometer (VSM). The result shows that the nanohybrids have high saturation magnetization of 76.8 $emu \cdot g^{-1}$ and present a superparamagnetic property. The electrochemical property of the obtained nanohybrids was characterized by an Eight-channel battery tester through using the nanohybrids as the lithium-ion battery cathode material. It is found that the obtained nanohybrids have an initial discharge capacity and charge capacity of 1469 and 652 $mA\ h \cdot g^{-1}$ respectively when the discharge—charge rate is 0.1 C.

Keywords Nickel ferrite-reduced graphene oxide nanohybrids · Hydrothermal synthesis · Magnetic property · Electrochemistry property

Introduction

As one kind of electromagnetic materials, spinel ferrites have received highly attention for its wide application prospects in fields of high-density magnetic recording media [1], Li ion batteries [2], microwave absorption [3], fuel cells [4], gas sensors [5], photocatalyst [6]. As one of the very typical spinel ferrites, nickel ferrite has excellent electrochemical [7] and magnetic properties [8], which make

G. Yan (✉) · Q. He · D. Yu
School of Mechanical Engineering, Guangxi University of Science & Technology, Liuzhou 545006, Guangxi, People's Republic of China
e-mail: ygq@gxust.edu.cn

Y. Han (ed.), *Advanced Functional Materials*,
https://doi.org/10.1007/978-981-13-0110-0_3

nickel ferrite has a promising future in magnetic materials and lithium-ion batteries (LIBs) field.

The performance of nickel ferrites in lithium-ion batteries field can be enhanced if they were composed with nanocarbon material substrates because the electrical conductivity of nickel ferrites is poor. The nanocarbon substrates provide a 3D conductive network which improves the rate capability of the electrodes by offering a better electronic conductivity. At the same time, the substrates also served as a buffer to relieve the stress result from the volumetric expansion [9, 10]. Li et al. [11] synthesized $NiFe_2O_4$-graphene nanohybrids with the present of Polyvinylpyrrolidone (PVP) as surfactant. It is found that the composite owns great efficiency as an electrocatalyst. Li et al. [4] composited $NiFe_2O_4$ with multiwalled carbon nanotubes as a bi-efficient electrocatalyst for oxygen reduction and oxygen evolution. The multiwalled carbon nanotubes strongly cross-linked in the nanocomposite and formed conductive net which improves its electrical conductivity greatly. Yu et al. [12] successfully synthesized $NiFe_2O_4$-carbon cloth (CC) nanohybrids. The specific capacitances of CC-$NiFe_2O_4$ were as high as 1135.5 $F \cdot g^{-1}$ (in H_2SO_4) and 922.6 $F \cdot g^{-1}$ (in KOH) at a current density of 2 $mA \cdot cm^{-2}$.

Among various nanocarbon materials, graphene has been proven to be the most desirable one for metal oxides. Owing to its unique structure, graphene acquires some attractive properties, including excellent chemical/thermal stability [13], high electrical conductivity [14], superior mechanical strength [15] and large specific surface area [16]. It can be a good choice to compose nickel ferrite with graphene for its better performance in lithium-ion batteries field. Fu et al. [17] composed the $NiFe_2O_4$ particles with graphene and served the composites as anode materials for lithium-ion batteries. It is reported that the capacity of $NiFe_2O_4$-graphene composite reached 1200 $mAhg^{-1}$.

In this paper, well-crystalized $NiFe_2O_4$ nano octahedrons—reduced graphene oxide (rGO) nanohybrids were synthesized via a hydrothermal strategy successfully. The magnetic and electrochemical properties of the as-obtained nanohybrids were investigated. The results indicate that the nanohybrids have high saturation magnetization and show a superparamagnetic property, and their also exhibit a good electrochemical property. It is suggested that the as-obtained $NiFe_2O_4$-rGO nanohybrids have a potential application in the field of Li ion batteries.

Experimental Section

Synthesis of Graphene Oxide (GO) All reagents were of analytic grade and were used as received without further purification. Firstly, 0.6 g Graphite powder and 3 g $KMnO_4$ were mixed up and transferred into a Teflon-lined stainless-steel autoclave, and then 30 mL H_2SO_4 was add into the mixture. The autoclave was kept in 0 °C for 2.5 h, then heated to 70 °C and kept for 1.5 h. After the reaction system was cooled to room temperature, GO was obtained.

Reduction of Graphene oxide (GO) for Reduced Graphene Oxide (rGO) 0.6 g GO was added into 30 mL ethylene glycol (EG) followed by sonicating for 15 min and vigorous stirring for 15 min. Then NaOH was added into the mixture to keep pH value above 7 in order to prevent aggregation of GO. Then the mixture was sonicated for 30 min and vigorous stirred for 30 min. The above mixture was then transferred into a Teflon-lined stainless-steel autoclave, heated to 180 °C for 6 h. After cooled to room temperature, rGO was obtained.

Synthesis of Nickel Ferrite ($NiFe_2O_4$)-Reduced Graphene Oxide (rGO) Nanohybrids 0.035 g rGO was added into 25 mL deionized water, sonicated for 15 min and vigorous stirred for 15 min. Then 0.146 g $Ni(NO_3)_2{\cdot}6H_2O$ and 0.404 g $Fe(NO_3)_3{\cdot}9H_2O$ were added into the rGO/deionized water mixture under vigorous stirring. 5 mL NaOH aqueous solution with the concentration of 2.5 mM was added into the above mixture and stirred for 15 min. The mixture was transferred into a Teflon-lined stainless-steel autoclave, heated to 180 °C for 18 h. After cooled to room temperature, the obtained dark brown precipitant was washed with deionized water and ethanol, dried at 60 °C, then the $NiFe_2O_4$-rGO nanohybrids were obtained.

Characterization The phase composition of as-prepared product was analyzed by X-ray diffraction (XRD) with DX2700 powder diffractometer operated at 35 kV, 30 mA, using a Cu Kα. The morphology of as-prepared products were examined by field-emission scanning electron microscopy (FESEM) with a Zeiss SIGMA microscope at 10 kV. The magnetic property of those nanohybrids were measured with a BKT-4500 vibrating-sample magnetometer (VSM), at room temperature. The electrochemical property of those nanohybrids were tested by a Neware eight-channel battery tester.

Results and Discussion

Phase Composition and Morphology Fig. 1 displays the XRD pattern of the as-obtained $NiFe_2O_4$-rGO nanohybrids. It can be seen that the as-obtained product is composed of well-crystalline spinal nickel ferrite. The diffraction peaks correspond to the (220), (311), (222), (400), (422), (511), (440), (531), (620), (553), (622) and (444) crystal planes of spinal $NiFe_2O_4$ (JCPDS NO. 10-0325) respectively. And the strong diffraction peaks indicated the good crystallinity and purification of obtained nickel ferrite.

The SEM images of the as-obtained $NiFe_2O_4$-rGO nanohybrids are shown in Fig. 2. As it is shown in Fig. 2a, the $NiFe_2O_4$ nanoparticles have a narrow size distribution and are not agglomerated. From Fig. 2b–d, it can be easily seen that the $NiFe_2O_4$ nanoparticles were grown on the surface of rGO flake. The morphology of the $NiFe_2O_4$ nanoparticles could be seen in Fig. 2e. It is found that the $NiFe_2O_4$ nanoparticles have a shape of octahedron with an edge size of about 100 nm.

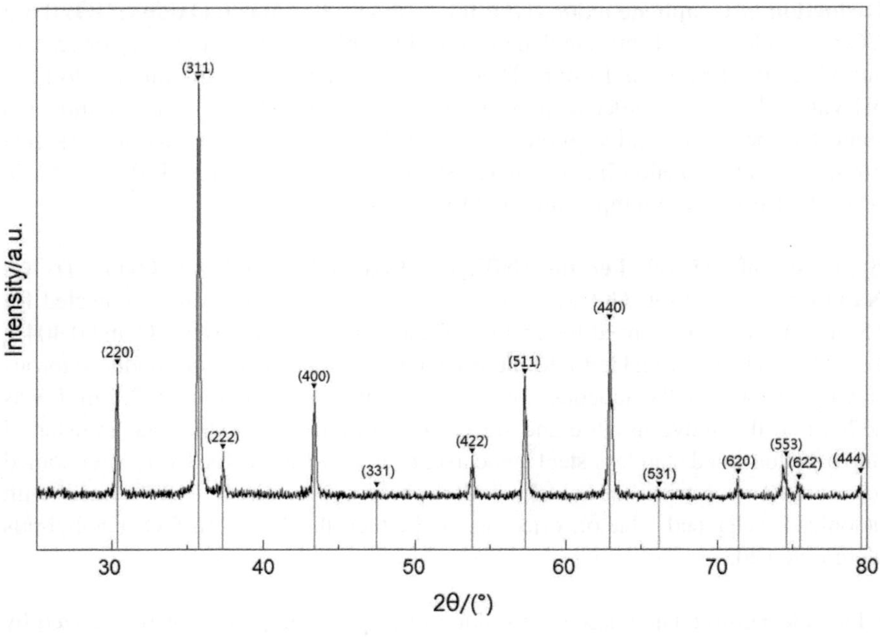

Fig. 1 XRD pattern of the as-obtained $NiFe_2O_4$-rGO nanohybrids

Magnetic Property The magnetic property of the $NiFe_2O_4$-rGO nanohybrid is carried out in the applied magnetic field of −5.2 to 5.2 kOe at room temperature. From the obtained magnetic hysteresis loop (Fig. 3), it could be found that the $NiFe_2O_4$-rGO nanohybrid is ferromagnetic at room temperature. It's saturation magnetization (M_s) is 76.8 emu·g^{-1}, which is similar to the bulk nickel ferrite's saturation magnetization. The high saturation magnetization may be attributed to the well-crystalized of the nickel ferrite nano octahedron on the surface of the graphene flake. From the magnified hysteresis loop of the inset figure in Fig. 3, it also could be seen that the coercivity (H_c) and remnant magnetization (M_r) of the obtained $NiFe_2O_4$-rGO nanohybrid are 20 Oe and 8 emu·g^{-1} respectively. The low coercivity and remnant magnetization may be attributed to the small crystal size of the nickel ferrite, which indicted that the as-obtained nanohybrid has a property of superparamagnetism.

It has reported that nickel ferrite has a high saturation magnetization and present a superparamagnetic property when the particles size is smaller than 16 nm [18]. Superparamagnetism is one of the magnetism forms which would appeared when the diameter of ferromagnetic or ferrimagnetic nanoparticles is smaller than a certain value. In the superparamagnetic state, an external magnetic field is able to magnetize the nanoparticles similarly to a paramagnet, but the magnetic susceptibility is larger than that of the paramagnet [19]. The superparamagnetic property of the $NiFe_2O_4$-rGO nanohybrid may be attributed to the small crystal size of the nickel ferrite, which causes the magnetic orientation force produced by an external

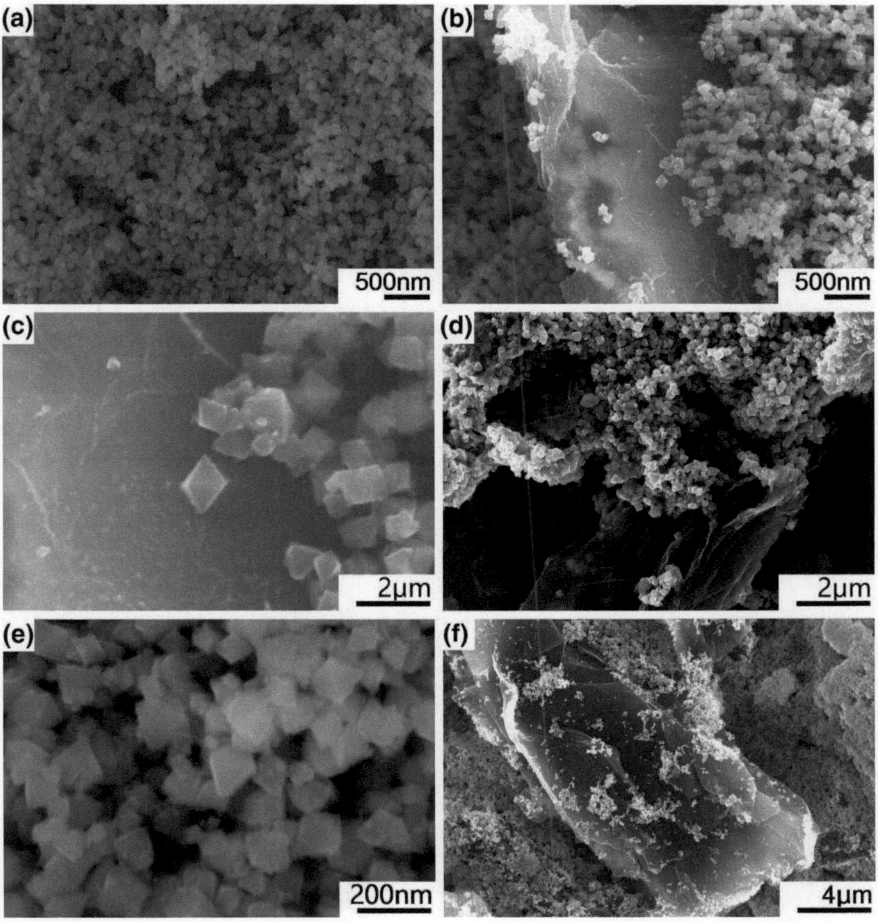

Fig. 2 SEM images of $NiFe_2O_4$-rGO nanohybrids

field is not sufficient to resist the thermal disturbances, and their magnetization is similar to that of a paramagnetic.

Electrochemical Property The electrochemical property of the as-obtained $NiFe_2O_4$-rGO nanohybrid was measured by a coin-type half-cell assembled in an argon-filled glovebox. In a typical process, the negative electrode of the coin-type half-cell was prepared by mixing of the active $NiFe_2O_4$-rGO nanohybrid, polyvinylidene fluoride (PVDF) and carbon black with a weight ratio of 7:2:1 in N-methyl-2-pyrrolidone (NMP). Then the obtained slurry coated on a circular Cu foil and dried at 60 °C for 12 h in a vacuum drying oven. PP microporous membrane, pure Li foil and $LiPF_6$ dissolved in ethyl carbonate and diethyl carbonate (a weight ratio of 1:1) with a concentration of 1 mol·L^{-1} were used as the separator, the counter electrode and the electrolyte respectively. The cells were charged and discharged on a

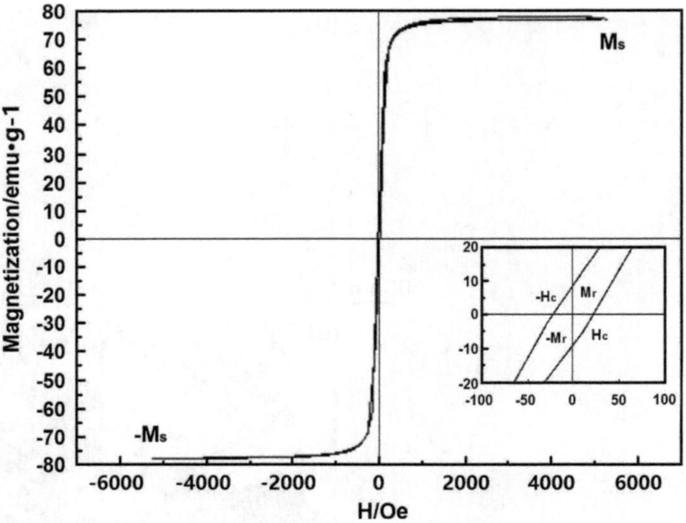

Fig. 3 The magnetic hysteresis loop of $NiFe_2O_4$-rGO nanohybrids at room temperature

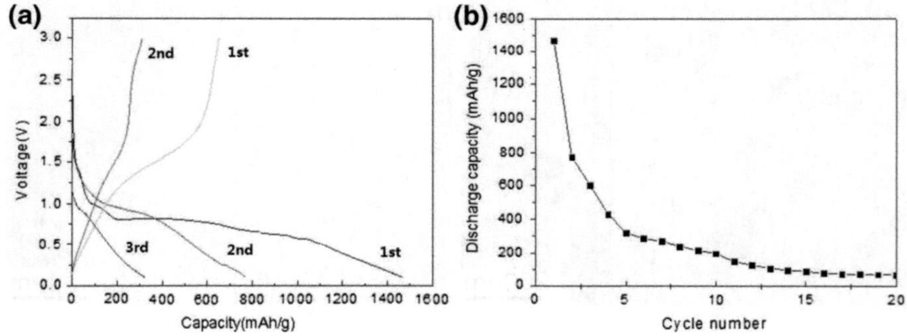

Fig. 4 The electrochemical properties of $NiFe_2O_4$-rGO nanohybrids: **a** the curves of discharge-charge at 0.1 C; **b** the cycle performance

Neware battery tester in a voltage range of 0.1–3.0 V. The curves of discharge—charge at 0.1 C and the curves of cycle number at 0.1 C are presented in Fig. 4.

As shown in Fig. 4a, the first discharge capacity and charge capacity of $NiFe_2O_4$-rGO nanohybrid (15 wt% rGO) was 1469 and 652 mA $h{\cdot}g^{-1}$ respectively, which is higher than the theoretical capacity of nickel ferrite and graphene. The large discharge and charge capacity may owe to the synergistic effect among $NiFe_2O_4$ and rGO. However, the cycle performance of the as-obtained nanohybrids is poor, there is an obvious decrease in the charge-discharge capacity in the following cycles, which indicated that the Li-ion battery anode of $NiFe_2O_4$-rGO nanohybrid have a poor cycle stability during the discharge—charge processes. This may be caused by the unstability of the nanohybrids during the discharge—

charge processes, and the nickel ferrite nano octahedrons grown on the surface of rGO may easily fall off and separate with rGO during Li-ion insertion and extraction.

Conclusion

$NiFe_2O_4$-rGO nanohybrids were synthesized by a facile hydrothermal strategy successfully. $NiFe_2O_4$ nano octahedrons with good crystallinity and an edge size of 100 nm were growth on the surface of rGO. The nanohybrids have high saturation magnetization of 76.8 emu·g^{-1}, and show a superparamagnetic property. The nanohybrids exhibit a good electrochemical property, they have an initial discharge capacity and charge capacity of 1469 and 652 mA h·g^{-1} when the discharge—charge rate is 0.1 C.

Acknowledgements This work is funded by Guangxi Natural Science Foundation (2014GXNSFBA118247).

References

1. E. Umeshbabu, G. Rajeshkhanna, G.R. Rao, Urchin and sheaf-like $NiCo_2O_4$ nanostructures: Synthesis and electrochemical energy storage application. Int. J. Hydrogen Energ. **39**, 15627–15638 (2014)
2. B. Jang, M. Park, O.B. Chae, S. Park, Y. Kim, S.M. Oh, Y. Piao, T. Hyeon, Direct synthesis of self-assembled ferrite/carbon hybrid nanosheets for high performance lithium-ion battery anodes. J. Am. Chem. Soc. **134**, 15010–15015 (2012)
3. Q.C. Liu, Z.F. Zi, M. Zhang, P. Zhang, A. Pang, J.M. Dai, Y.P. Sun, Solvothermal synthesis of hollow glass microspheres/Fe_3O_4 composites as a lightweight microwave absorber. J. Mater. Sci. **48**, 6048–6055 (2013)
4. P.X. Li, R.G. Ma, Y. Zhou, Y.F. Chen, Q. Liu, G.H. Peng, Z.H. Liang, J.C. Wang, Spinel nickel ferrite nanoparticles strongly cross-linked with multiwalled carbon nanotubes as a bi-efficient electrocatalyst for oxygen reduction and oxygen evolution. RSC Adv. **5**, 73834–73841 (2015)
5. R. Hajihashemi, A.M. Rashidi, M. Alaie, R. Mohammadzadeh, N. Izadi, The study of structural properties of carbon nanotubes decorated with $NiFe_2O_4$ nanoparticles and application of nano-composite thin film as H_2S gas sensor. Mater. Sci. Eng. C **44**, 417–421 (2014)
6. J. Li, Z. Liu, Z. Zhu, Magnetically separable ZnFe2O4, Fe2O3/ZnFe2O4 and ZnO/ZnFe2O4 hollow nanos- pheres with enhanced visible photocatalytic properties. RSC Adv. **4**, 51302–51308 (2014)
7. E.K. Heidari, B. Zhang, M.H. Sohi, A. Ataie, J.K. Kim, Sandwich-structured graphene-$NiFe_2O_4$- carbon nanocomposite anodes with exceptional electrochemical performance for Li ion batteries. J. Mater. Chem. A **2**, 8314–8322 (2014)
8. Y. Kinemuchi, K. Ishizaka, H. Suematus, W. Jiang, K. Yatsui, Magnetic properties of nanosize $NiFe_2O_4$ particles synthesized by pulsed wire discharge. Thin Solid Films **407**, 109–113 (2002)

9. Y. Yu, Y.C. Zhu, H.X. Gong, X.M. Ma, X. Zhang, N. Li, Y.T. Qian, Fe_3O_4 nanoparticles embedded in carbon-framework as anode material for high performance lithium-ion batteries. Electrochim. Acta **83**, 53–58 (2012)
10. X.M. Liu, Z.D. Huang, S.W. Oh, B. Zhang, P.C. Ma, M. Yuen, J.K. Kim, Carbon nanotube (CNT)-based composites as electrode material for rechargeable Li-ion batteries: a review. Compos. Sci. Technol. **72**, 121–144 (2012)
11. P.X. Li, R.G. Ma, Y. Zhou, Y.F. Chen, Z.Z. Zhou, G.H. Liu, Q. Liu, G.H. Peng, J.C. Wang, Solvothermally synthesized graphene nanosheets supporting spinel $NiFe_2O_4$ nanoparticles as an efficient electrocatalyst for the oxygen reduction reaction. RSC Adv. **5**, 44476–44482 (2015)
12. Z.Y. Yu, L.F. Chen, S.H. Yu, Growth of $NiFe_2O_4$ nanoparticles on carbon cloth for high performance flexible supercapacitors. J. Mater. Chem. A **2**, 10889–10894 (2014)
13. A.A. Balandin, S. Ghosh, W.Z. Bao, I. Calizo, D. Teweldebrhan, F. Miao, C.N. Lau, Superior thermal conductivity of single-layer graphene. Nano Lett. **8**, 902–907 (2008)
14. S.J. Park, J.H. An, I. Jung, R.D. Piner, S.J. An, X.S. Li, A. Velamakanni, R.S. Ruoff, Colloidal suspensions of highly reduced graphene oxide in a wide variety of organic solvents. Nano Lett. **9**, 1593–1597 (2009)
15. C. Lee, X.D. Wei, J.W. Kysar, J. Hone, Measurement of the elastic properties and intrinsic strength of monolayer graphene. Science **321**, 385–388 (2008)
16. M.D. Stoller, S.J. Park, Y.W. Zhu, J. An, R.S. Ruoff, Graphene-based ultracapacitors. Nano Lett. **8**, 3498–3502 (2008)
17. Y.S. Fu, Y.H. Wan, H. Xia, X. Wang, Nickel ferrite graphene heteroarchitectures: toward high-performance anode materials for lithium-ion batteries. J. Power Sour. **213**, 338–342 (2012)
18. J.H. Liu, L.F. Zhang, G.F. Tian, J.C. Li, F.S. Li, Fa, Structure and magnetic properties of $NiFe_2O_4$ nanoparticles prepared by low-temperature solid-state reaction. Acta Phys. Sin. **56**, 6050–6055 (2007)
19. D. Weller, A. Moser, Thermal effect limits in ultrahigh density magnetic recording. IEEE Trans. Magn. **11**, 4423–4439 (1999)

Insulating Technology Control of Soft Magnetic Composites

Jiaqi Chen, Yunpeng Qu, Ciqun Xu, Huan Ren and Yao Liu

Abstract In this paper, fabrication procedures were adjusted and investigated in soft magnetic composite (SMCs) when the micro-scale iron powder was served as matrix with amorphous silica as the insulating coating layer. The coating situations were alternated by transforming the TEOS (tetraethoxysilane) hydrolysis parameters such as temperature, PH value, and TEOS reagent dosage. It was found that the silica deposited on raw iron particles which is produced by the hydrolysis reaction from the increasing addition of TEOS reagent would make the coating layer too thick. Over-high pH value would has a great negative influence on the silica coating layers, over-high temperature would speed the hydrolysis reaction up, while silica derived from hydrolysis reaction were self-assembled to silica particles and the deposition amounts on iron particles were less. Moreover, the circumstances of insulation coating layer were characterized by introducing low frequency complex dielectric parameters as indirect assessments. The magnetic properties including saturation flux density (Bs), hysteresis loss were investigated. SEM images were presented to directly exhibit the morphology of coated iron particles. The thorough

J. Chen · Y. Qu · C. Xu · H. Ren · Y. Liu (✉)
Key Laboratory for Liquid-Solid Structural Evolution and Processing of Materials (Ministry of Education), Shandong University, Jinan 250061, China
e-mail: liuyao@sdu.edu.cn

J. Chen
e-mail: 1204638875@qq.com

Y. Qu
e-mail: 1018531959@qq.com

C. Xu
e-mail: xuciqun@126.com

H. Ren
e-mail: 843715347@qq.com

J. Chen · Y. Qu · C. Xu · H. Ren · Y. Liu
Dezhou META Research Center for Innovative Materials, Dezhou 253000, China

Y. Han (ed.), *Advanced Functional Materials*,
https://doi.org/10.1007/978-981-13-0110-0_4

investigation of coated particles' fabrication procedures greatly facilitates the applications of soft magnetic materials.

Keywords Soft magnetic composite material · Insulating · Silica

Introduction

Composites, combined by multicomponent materials, in which varieties of properties (i.e. photocatalytic performance [1], magnetic properties [2], dielectric properties [3–5]) could be constructed [6]. At present, soft magnetic composite materials are widely used in power components, computer and communications [7]. Recently, the low cost and pollution-free characteristics of the surface coated silica are popular in the research. Fe/SiO_2 soft magnetic composite materials can be used in complex forming, low dissipation and good magnetic energy, which can be widely used in electronic devices, electric vehicles, aerospace and other fields. The thickness of SiO_2 insulation coating has a great influence on the comprehensive performance of the materials, and the optimization of insulation coating process has become the focus of exploring the properties of soft magnetic materials.

The magnetic properties of soft magnetic composites are easy to be found in magnetic fields or current, and are easy to demagnetize [7]. When the current and magnetic field is removed, the magnetism will disappear. Having low loss in high frequency working environment, the performance and shape of magnetic core can be designed according to practical application and requirement [8–11]. Insulation coating process is the key to improve the magnetic performance of soft magnetic composites. With the development of electronic components, soft magnetic composites have become a hot topic. At present, the common coating of iron based soft magnetic composite materials is classified as organic coating and inorganic coating [12]. The insulation layer is easy to decompose when the stress is released at high-temperature annealing for organic coating. The inorganic coating materials include phosphate and oxides [12, 13] and its advantages are high melting point and good thermal stability.

In this paper, we provide a way to control the experimental parameters so that the results are different. TEOS addition, pH, temperature are controlled, and then different experimental samples are obtained. Introducing the dielectric constant as an indirect reference to evaluate the effect of insulating coating, a method for testing the performance of soft magnetic composites is supplemented.

Experiment

Using the control variable method, the experiment is divided into three groups, each group of experimental variables were TEOS reagent addition, temperature, pH value (Table 1).

Table 1 Each set of sample ingredients

Variables	Number	TEOS dosage (ml)	Ammonia consumption (ml)	Temperature (°C)	Stirring rate (r/min)	Reaction time (h)	Drying time (h)
TEOS	1	5	25	25	800	6	6
	2	10	25	25	800	6	6
	3	15	25	25	800	6	6
Temperature	4	10	25	20	800	6	6
	5	10	25	30	800	6	6
	6	10	25	40	800	6	6
pH value	7	10	10	20	800	6	6
	8	10	20	20	800	6	6
	9	10	30	20	800	6	6

Take experiment 2 as an example. Detailed experimental procedures to get the SiO_2-coated iron particles are as follows [14]: (i) 20 g of iron particles were washed with ethanol to clean the surface of the iron particles; (ii) the cleaned iron particles were mixed with 200 ml of ethanol and 1.5 g of oleic acid in a beaker, and the mixture was dispersed in an ultrasonic dispersion system for 30 min to activate the surface of the iron powder; (iii) after dispersing, 25 ml of aqueous ammonia (25 wt%) and 200 ml of ethanol were added to the mixture, then the mixture was stirred with a high-speed electrical stirrer with a speed of 800 rpm for 3 h at room temperature and 10 ml of TEOS was added to the mixture every hour while stirring; (iv) 3 h later, another 25 ml of aqueous ammonia (25 wt%) and 10 ml of TEOS were added to the mixture, and the suspension was stirred for another 3 h at room temperature; (v) after stirring, the suspension was filtered, washed with distilled water and ethanol several times, and then dried at 50 °C for 6 h to obtain SiO_2-coated iron particles. The sample was made into a sheet and a ring of epoxy resin as a substrate. The images and spectra of the samples were obtained by scanning a scanning electron microscope. The LCR digital bridge was tested for the sample and the dielectric constant was calculated to measure the hysteresis loop of the sample.

Results and Discussion

According to Fig. 1, the diameter of iron powder is about 100 μm and the morphology is not uniform. From the whole point of view, iron particles appear different shapes and rough surface with a great deal of uneven protrusions.

From Figs. 1 and 2, it can be found that a uniform film is wrapped on the surface of iron powder with uneven surface morphology. Coated iron powder was brighter

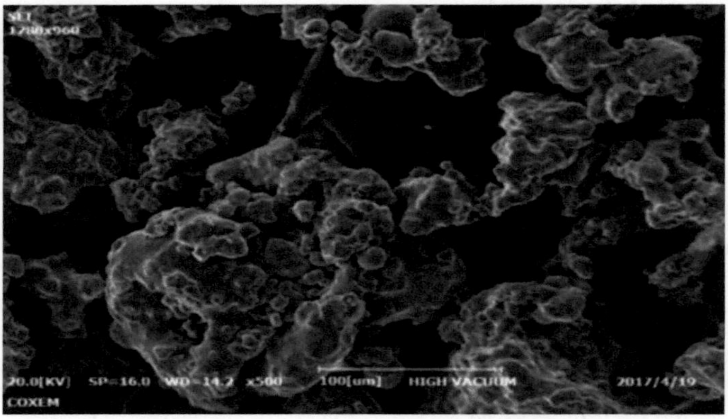

Fig. 1 Electron mirror image of iron powder

in the electron microscope images. The fundamental hydrolysation process of TEOS can be expressed using the following reactions [15]:

$$(C_2H_5O)_4Si + 4H_2O \rightarrow Si(OH)_4 + 4C_2H_5OH, \quad (1)$$

$$Si(OH)_4 \rightarrow SiO_2 + 2H_2O. \quad (2)$$

Accordingly, we were convinced that the coating layer of iron particles surface is SiO_2.

According to Fig. 3, it is found that the dielectric constant of the flake increases obviously when the coated iron powder is added. This is because the fully coated iron powder is insulated, and the electrons on the iron atoms cannot jump out. When hydrolysis parameters including temperature, PH value, and TEOS reagent dosage were alternated the dielectric constant of the flake were changed similarly which indicates that coating thickness changed sensitively with the parameters.

In Fig. 4a, when the ferromagnetic material is in an alternating magnetic field, there is an additional energy dissipation, which is a hysteresis loss, which is proportional to the area enclosed by the hysteresis loop [19]. The HD point is called coercive force, the coercivity of material hysteresis loop is large, and the loss is high [20].

The slope of the magnetization curve is the permeability of the material [14]. The magnetization curve of the material is related to the shape of the hysteresis loop in the hysteresis loop of the material.

The experimental results show that due to the introduction of silica, a non-magnetic material, the coercivity force of the coated iron powder increases. Naturally, the smaller the coercivity force, the lower the material loss according to Fig. 5. On this basis, the greater the slope, the better the magnetic properties of the permeability material [14, 21].

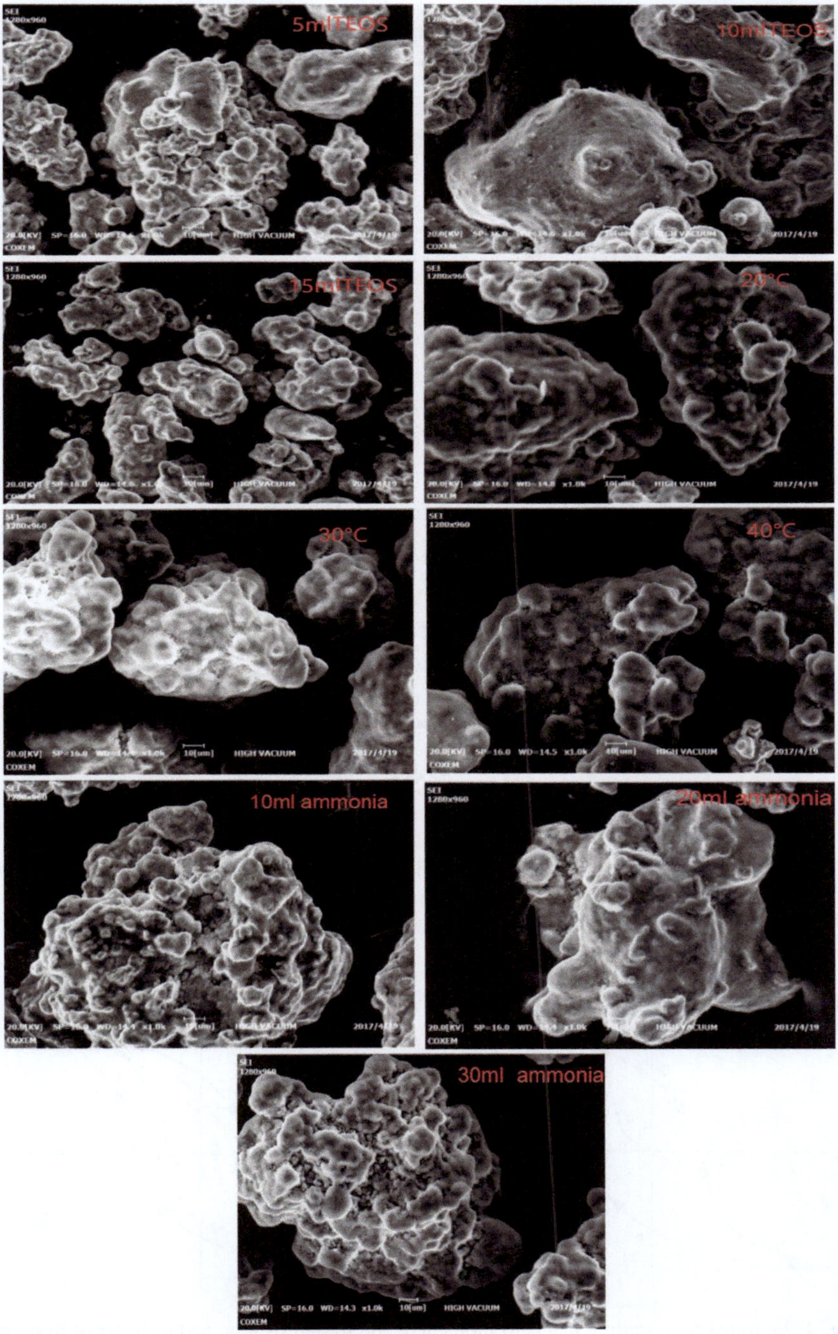

Fig. 2 Image of the coated iron powder in different experimental conditions

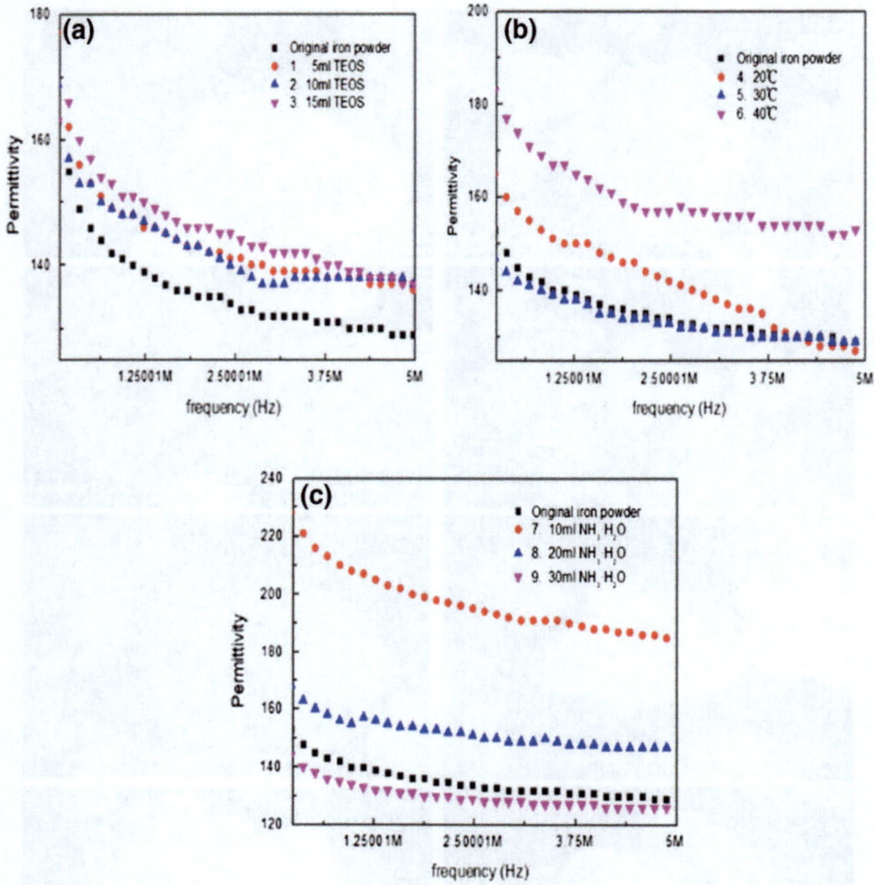

Fig. 3 Dielectric constant curve

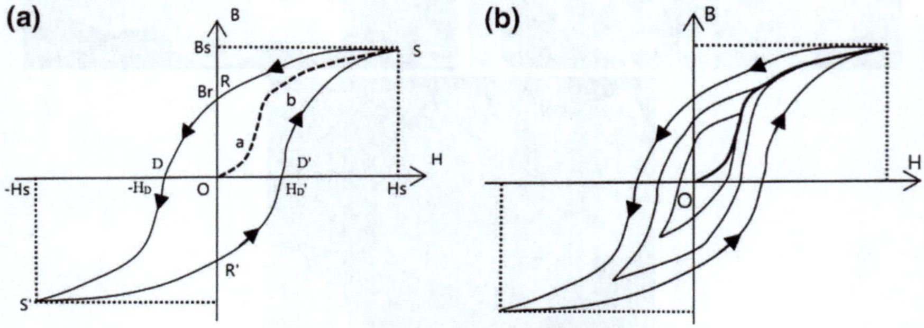

Fig. 4 Ferromagnetic starting magnetization curve and hysteresis loop and a cluster of hysteresis loops of the same ferromagnetic material [16–18]

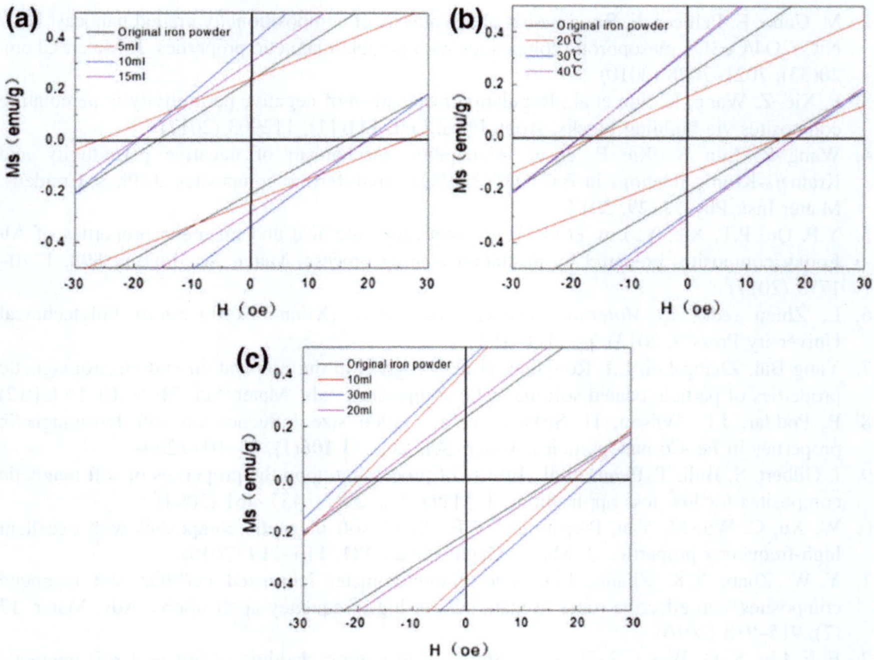

Fig. 5 Partial magnifying hysteresis loop of samples under different hydrolysis parameters

Conclusion

Obviously, if the coating is not enough to completely coat the iron particles, the insulating effect of the coated surface of the powder coating will don't work, and electron conducting behavior easily occurs when the iron particles are in contact with each other. If the insulating coating is too thick, the magnetic properties of the material will be reduced. According to the above discussion, we come to the conclusion that when the amount of TEOS reagent is 10 ml, the reaction condition is 20 °C, and when the amount of ammonia is 20 ml, the sample has the smallest coercive force, the lowest loss and the ideal coating effect.

Acknowledgements The work was supported by the National Natural Science Foundation of China [grant No. 51402170] and the Fundamental Research Funds of Shandong University [grant No. 2016WLJH40].

References

1. X. Yang, H. Cui, Y. Li et al., Fabrication of Ag3PO4-Graphene composites with highly efficient and stable visible light photocatalytic performance. Acs Catal. **3**(3), 363–369 (2013)

2. M. Cabo, E. Pellicer, E. Rossinyol et al., Synthesis of compositionally graded nanocast NiO/NiCo2O4/Co3O4 mesoporous composites with tunable magnetic properties. J. Mater. Chem. **20**(33), 7021–7028 (2010)
3. P. Xie, Z. Wang, K. Sun et al., Regulation mechanism of negative permittivity in percolating composites via building blocks. Appl. Phys. Lett. **111**(11), 112903 (2017)
4. Wang Z, Sun K, Xie P, et al., Generation mechanism of negative permittivity and Kramers-Kronig relations in BaTiO3/Y3Fe5O12 multiferroic composites. J. Phys. Condens. Matter Inst. Phys. J. **29**, 2017
5. Y.P. Qu, P.T. Xie, Y. Liu, et al., High frequency electrical and magnetic properties of Al/Epoxy composites prepared by insulation coating process. Mater. Sci. Forum. **898**, 1770–1775 (2017)
6. L. Zhien (ed.), in *Materials Science Foundation* (Xi'an: Northwestern Polytechnical University Press 9, 2013), pp. 311–317
7. Yang Bai, Zhang Lei, L.I. Rongfeng et al., Progress in the preparation and electromagnetic properties of particle-coated soft magnetic composites. Adv. Mater. Sci. **31**(7), 10–15 (2012)
8. P. Poddar, J.L. Wilson, H. Srikanth et al., Grain size influence on soft ferromagnetic properties in Fe–Co nanoparticles. Mater. Sci. Eng., B **106**(1), 95–100 (2004)
9. I. Gilbert, S. Bull, T. Evans et al., Effects of processing upon the properties of soft magnetic composites for low loss applications. J. Mater. Sci. **39**(2), 457–461 (2004)
10. W. Xu, C. Wu, M. Yan, Preparation of Fe–Si–Ni soft magnetic composites with excellent high-frequency properties. J. Magn. Magn. Mater. **381**, 116–119 (2015)
11. Y.-W. Zhao, X.K. Zhang, J.Q. Xiao, Submicrometer laminated Fe/SiO2 soft magnetic composites—an effective route to materials for high-frequency applications. Adv. Mater. **17** (7), 915–918 (2010)
12. F.-F. Liu, S.-Q. Weng, S.-H. Song, Study on insulating cladding of fe-based soft magnetic composite. Mater. Dev. Appl. **22**(5), 11–15 (2007)
13. A.H. Taghvaei, H. Shokrollahi, K. Janghorban, Properties of iron-based soft magnetic composite with iron phosphate–silane insulation coating. J. Alloys Compd. **481**(1–2), 681–686 (2009)
14. B. Yang, Z. Wu, Z. Zou, et al., High-performance Fe/SiO2 soft magnetic composites for low-loss and high-power applications. J. Phys. D Appl. Phys. **43**(36), 365003–365008(6) (2010)
15. M.Y. Lou, D.P. Wang, W.H. Huang et al., Effect of silane-coupling agents on synthesis and character of core-shell SiO_2, magnetic microspheres. J. Magn. Magn. Mater. **305**(1), 83–90 (2006)
16. V. Ginzburg, The calculation of magnetization curves and magnetic hysteresis loops for a simplified model of a ferromagnetic body. IEEE Trans. Magn. **12**(2), 119–126 (2003)
17. F. Fiorillo, L.R. Dupre, C. Appino et al., Comprehensive model of magnetization curve, hysteresis loops, and losses in any direction in grain-oriented Fe-Si. Magn. IEEE Trans. **38**(3), 1467–1476 (2002)
18. M.A. Escobar, R. Valenzuela, L.F. Magaña, Analytical prediction of the magnetization curve and the ferromagnetic hysteresis loop. J. Appl. Phys. **54**(10), 5935–5940 (1983)
19. Z. Li, J. Qiu, Study on high precision measurement method of coercive force of soft magnetic material. Magn. Mater. Devices **32**(2), 57–60 (2001)
20. B.D. Cullity, C.D. Graham, *Magnetostriction and the Effects of Stress* (Second Edition, Introduction to Magnetic Materials, 2009), pp. 241–273
21. Z. Li, D.-M. Zhang, Study on hysteresis curve measurement method for nonuniform magnetic field scanning. Power Acoust. **23**(1), 77–79 (2001)

Dielectric Properties of Cu/Epoxy Random Composites at Radio-Frequency Range

Ciqun Xu, Peitao Xie, Yunpeng Qu, Yulin Wu, Jiaqi Chen, Huan Ren and Yao Liu

Abstract Cu/epoxy composites with the different volume fraction of Cu were prepared through hot press process. The conductivity and dielectric properties of the Cu/epoxy composite were investigated. The analyze of the results indicated that the real part of permittivity ε_r' of Cu/epoxy composites increased when copper content increased, which is caused by the increasing interface connect between Cu fillers and epoxy partials. The frequency dispersions of ac conductivity follow the power law and thus indicates the hopping conduction behavior in composites.

Keywords Hot compression · Percolating composites · Dielectric property

Introduction

More and more attention has been paid to the electrically conductive composites made of an insulating polymer matrix and conductive filler due to their potential applications [1–6]. The effective concentration of electrons in the insulating polymer is limited, therefore polymers are often used as a matrix of composite [6]. Epoxy resin is a kind of highly crosslinked and an amorphous polymer which used for insulation in power transformer field [7]. At the same time, metallic material which has high electrical conductivity can act as functional filler and be added to the insulating polymeric matrix. Then the composites with a conductor/insulator structure have been produced. The conductivity and dielectric properties of the

C. Xu · P. Xie · Y. Qu · Y. Wu · J. Chen · H. Ren · Y. Liu (✉)
Key Laboratory for Liquid-Solid Structural Evolution and Processing of Materials (Ministry of Education), Shandong University, Jinan 250061, China
e-mail: liuyao@sdu.edu.cn

C. Xu · P. Xie · Y. Qu · J. Chen · H. Ren · Y. Liu
Dezhou META Research Center for Innovative Materials, Dezhou 253000, China

Y. Han (ed.), *Advanced Functional Materials*,
https://doi.org/10.1007/978-981-13-0110-0_5

composite can be adjusted by adjusting the fraction ratio between the conductor and the insulator.

It is well known that the electrical properties of composites change dramatically when the conductor content reaches a certain critical value in the conductor/insulator system. The phenomenon that the physical properties of a composite dramatically changed when the filler concentration approaches the critical value is explained by the percolation [2]. Previous research shows that the permittivity of the conductor/insulator composites changes sharply when approaching the percolation threshold [3–9]. This sudden change in conductivity and dielectric properties is due to the electrons that span across the matrix. Cu/epoxy composites are the typical conductor/insulator systems. However, the dielectric properties of Cu/epoxy composites have sacredly been reported and deserved to be investigated.

Experimental

In this paper, the Cu/epoxy composites were produced by mixing epoxy resin with the different volume content of Cu using high energy ball milling. After milling, the powder mixture was compressed at 20 MPa and 80 °C using a hot press forming method to produce a circular sample with 22 mm in diameter and around 2 mm thick. The composites with Cu mass fraction of 10, 20, 30, 40, 50, 60, 70 vol.% were prepared and referred to as Cu10, Cu20, Cu30, Cu40, Cu50, Cu60 and Cu70, respectively. The morphology of the samples was investigated by field emission scanning electron microscopy (FESEM; SU-70). The effects of Cu loading levels on electrical properties of Cu/epoxy composites were discussed. The dielectric properties of the Cu/epoxy composites at the frequency range from 10 MHz to 1 GHz were tested by Agilent E4991A precision impedance analyzer equipped with 16453A test fixture. The relationship between permittivity and frequency or dielectric parameters was measured at an AC voltage of 100 MV. Moreover, the real part (ε_r') and imaginary part ((ε_r'')) of permittivity can be described by the following formula [5, 10],

$$\varepsilon_r' = \frac{CT}{\varepsilon_0 S} \tag{1}$$

$$\varepsilon_r'' = \frac{T}{RS2\pi f \varepsilon_0} \tag{2}$$

where f is the frequency, T is the thickness of samples, C represents capacitance, R is the resistance, S is the area of the electrode plate, and ε_0 represents the absolute permittivity of free space (8.85×10^{-12} F/m).

Results and Discussions

Microstructure. FESEM images of the Cu/epoxy composites are shown in Fig. 1. We can see that the copper particles show the spherical shape and distributed homogeneously in the matrix.

For 10 and 30 vol.% samples, Cu particles are isolated in the epoxy matrix (Fig. 1a, b). With the volume fraction of copper increasing, Cu particles are close to each other. For 60 and 70 vol.% samples, Cu particles even contact and form network (Fig. 1c, d).

Conductivity. As shown in Fig. 2, the frequency dispersions of conductivities σ_{ac} for Cu/epoxy composites various from different Cu content. And conductivities were calculated by the following formula [5],

$$\sigma_{ac} = \frac{T}{RS} \tag{3}$$

where S is an area of the electrode plate, R represents the resistance, and T is the sample thickness. We can see σ_{ac} increases with higher Cu content. In other words, the ac conductivity can be largely enhanced by the Cu content increase. Besides, the σ_{ac} of Cu10, Cu20 increase with the increase of frequency. At high frequency, the

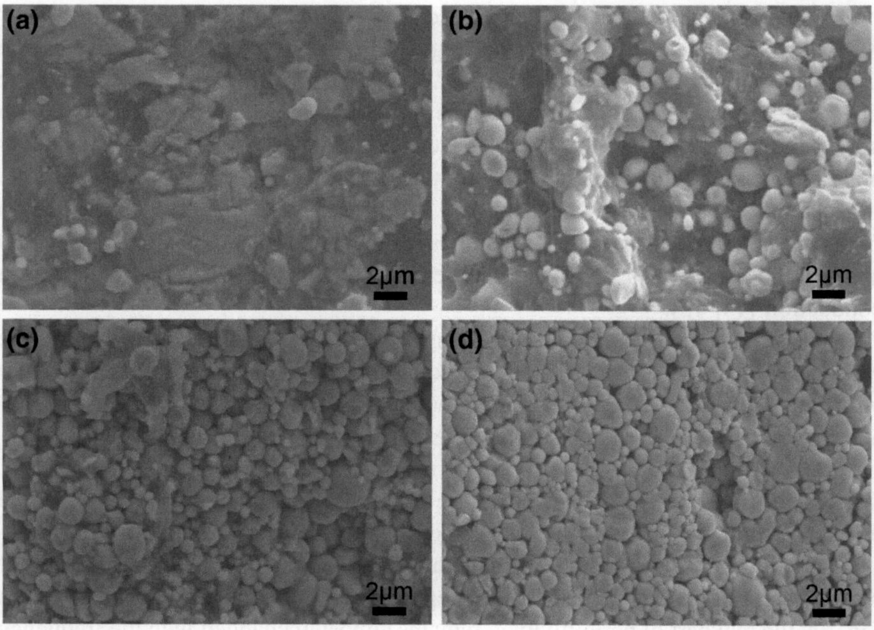

Fig. 1 FESEM images of Cu/epoxy composites with the different volume fraction of Cu. **a** Cu10, **b** Cu30, **c** Cu60, **d** Cu70

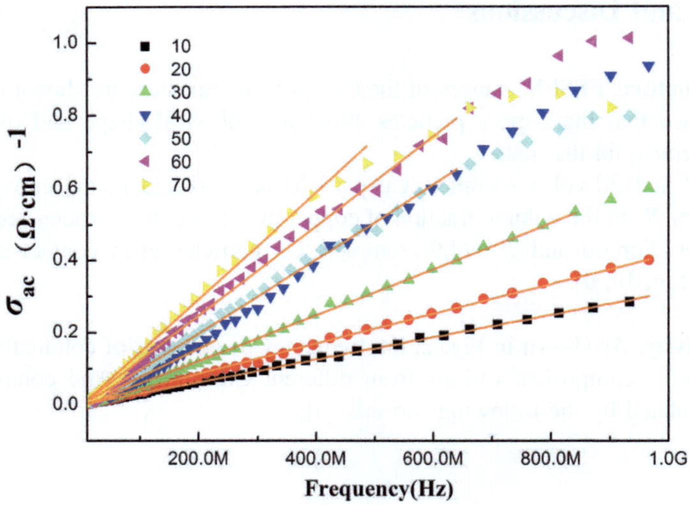

Fig. 2 The frequency dispersion of AC conductivity of Cu/epoxy composites with the different volume fraction of Cu. The solid line shows the fitted results based on the power law $\sigma_{ac} \propto \omega^n$. The conductivity curves of Cu10 and Cu20 are perfectly fitted, and these of Cu30, Cu40, Cu50, Cu60, and Cu70 can only be fitted at low frequencies

carrier motion leads to this phenomenon. Localized charge carrier movement makes it possible to fully utilize well-conducting areas [11].

The high-frequency parts of the σ_{ac}–f curves followed the power law $\sigma_{ac} \propto \omega^n$ ($0 < n < 1$), where ω represents the angular frequency ($\omega = 2\pi f$), and n is the fractional exponent between 0 and 1 [12, 13]. The solid line in Fig. 2 shows the fitted results based on the power law. This shows that the volume contents of Cu do not reach the percolation threshold. The ac conductivity of Cu/epoxy composites is hopping conductivity when the Cu content is below the percolation threshold. Interestingly, at high frequency, the σ_{ac} of Cu70 have a negative correlation to frequency. Since the skin effect in composites the σ_{ac} of Cu30, Cu40, Cu50, Cu60 and Cu70 decrease with increasing frequency, which is similar to the free electron conduction [14, 15].

The value of the exponent n at different Cu content is calculated by fitting the curves of Fig. 2 and listed in Table 1. It found that n changes in a different manner with Cu content. It is necessary to point out that the σ_{ac} of Cu30, Cu40, Cu50, Cu60 and Cu70 at low frequency also obey the power low, with abnormal $n > 1$ [11].

Dielectric property. Figure 3 shows the permittivity of the composites with the different volume fraction of Cu measured over the frequency range of 10 MHz–1 GHz. The permittivity of all the samples are positive, which is due to the electrons of the metallic region are localized by the epoxy resin isolation. As demonstrated in Fig. 3a, the real part of permittivity ε_r' remains almost unchanged with the frequency increase. Furthermore, as shown in Fig. 3b, the imaginary part of permittivity $\varepsilon_r''\varepsilon_r''$, which indicates the losses of Cu/epoxy composites, increases with increasing volume

Table 1 Illustrate the values of n for Cu/epoxy composites

Cu mass fraction (vol.%)	n
10	0.99114
20	0.98201
30	1.03017
40	1.00911
50	1.02913
60	1.02858
70	1.00136

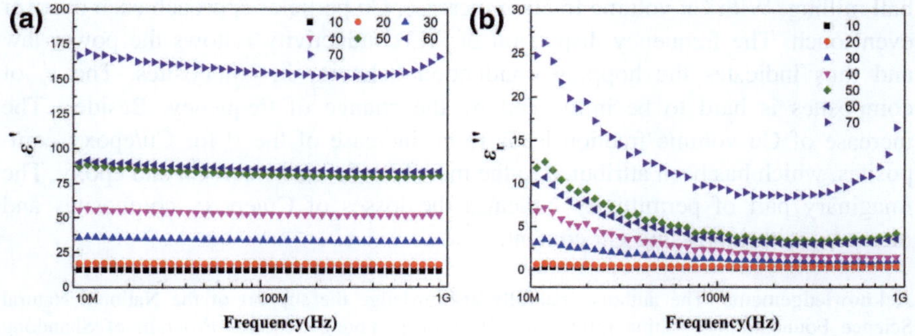

Fig. 3 Frequency dependence of real part ε_r' (**a**) and imaginary part ε_r'' (**b**) of permittivity for Cu/epoxy composites with different Cu volume fraction

Fig. 4 Variation of real permittivity with different Cu content at 10 MHz

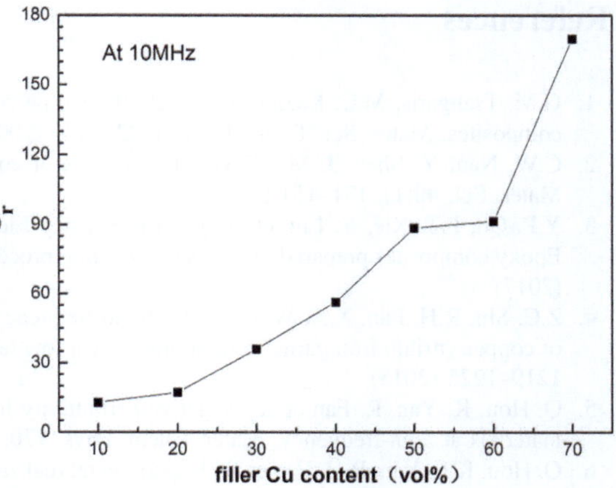

fraction of Cu. Moreover, the ε_r'' of Cu/epoxy composites decreased slightly in the frequency range from 10 MHz to 1 GHz. There are many factors cause losses, such as dielectric losses, ohmic losses, and eddy current losses [16].

Figure 4 shows the variation of ε_r' with different Cu content at 10 MHz. And there is a similar trend at any frequency. As the interface area between the copper particles and the epoxy particles increases, the phenomenon that ε_r' increases with the increase of the copper content could be observed. It should be noticed that $\varepsilon_r'\varepsilon_r'$ increased dramatically from Cu60 to Cu70 compared with other Cu content. It indicates that the electrical properties of the composites will eventually have a qualitative change when the copper content is gradually increased to a certain ratio.

Conclusions

Cu/epoxy composites were prepared through hot press process after high energy ball milling. With Cu volume fraction increase, Cu particles approach each other or even touch. The frequency dispersion of AC conductivity follows the power law and thus indicates the hopping conduction behavior in composites. The ε_r' of composites is hard to be influenced by the change of frequency. Besides, The increase of Cu volume fraction leads to an increase of the ε' for Cu/epoxy composites, which has been attributed to the more interface between Cu and epoxy. The imaginary part of permittivity indicates the losses of Cu/epoxy composites and increases with increasing Cu content.

Acknowledgements The authors gratefully acknowledge the support of the National Natural Science Foundation of China [grant No. 51402170], Young Scholars Program of Shandong University [grant No. 2016WLJH40].

References

1. G.M. Tsangaris, M.C. Kazilas, Conductivity and percolation in epoxy resin/conductive filler composites. Mater. Sci. Technol. **18**(2), 226–230 (2002)
2. C.W. Nan, Y. Shen, J. Ma, Physical properties of composites near percolation. Ann. Rev. Mater. Sci. **40**(1), 131–151 (2010)
3. Y.P. Qu, P.T. Xie, Y. Liu, et al., High Frequency Electrical and magnetic properties of Al/Epoxy composites prepared by insulation coating process. Mater. Sci. Forum **898**, 1770–1775 (2017)
4. Z.C. Shi, R.H. Fan, X.A. Wang et al., Radio-frequency permeability and permittivity spectra of copper/yttrium iron garnet cermet prepared at low temperatures. J. Eur. Ceram. Soc. **35**(4), 1219–1225 (2015)
5. Q. Hou, K. Yan, R. Fan et al., Negative permittivity in Fe–Si–Ni/epoxy magnetic composite materials at high-frequency. Mater. Chem. Phys. **170**, 113–117 (2016)
6. Q. Hou, K.L. Yan, R.H. Fan et al., Experimental realization of tunable negative permittivity in percolative Fe78Si9B13/epoxy composites. Rsc Adv. **5**(13), 9472–9475 (2015)
7. Z.M. Elimat, A.M. Zihlif, G. Ragosta, Study of ac electrical properties of aluminium epoxy composites. J. Phys. D Appl. Phys. **41**(41), 165408 (2008)
8. P.T. Xie, Z.Y. Wang, R.H. Fan et al., High frequency electrical properties of Fe/SiO2 composites by pressureless sinter. Key Eng. Mater. **727**, 257–261 (2017)
9. A.A. Hasan, Dielectric properties of epoxy/Cu, Al composites. J. Nano. Adv. Mat. **2**(2), 65–76 (2014)

10. K.L. Yan, R.H. Fan, X.A. Wang et al., Microstructure and dielectric properties of ion-doped La0.7Sr0.3MnO3 lossy ceramics at radio frequencies. Rsc Adv. **4**(49), 25804–25808 (2014)
11. P.T. Xie, R.H. Fan, Q. Hou et al., Radio-frequency electromagnetic properties of FeSiCr–epoxy composites prepared by cryomilling. Mater. Sci. Forum **848**, 18–22 (2016)
12. Z.C. Shi, S.G. Chen, R.H. Fan et al., Ultra low percolation threshold and significantly enhanced permittivity in porous metal-ceramic composites. J. Mater. Chem. C **2**(33), 6752–6757 (2014)
13. J.C. Dyre, T.B. Schrøder, Universality of ac conduction in disordered solids. Rev. Mod. Phys. **72**(3), 873–892 (1999)
14. Z.C. Shi, R.H. Fan, K.L. Yan et al., Preparation of iron networks hosted in porous alumina with tunable negative permittivity and permeability. Adv. Funct. Mater. **23**(33), 4123–4132 (2013)
15. K.L. Yan, R.H. Fan, M. Chen et al., Perovskite (La, Sr)MnO 3 with tunable electrical properties by the Sr-doping effect. J. Alloys Compd. **628**, 429–432 (2015)
16. M. Gao, Z.C. Shi, R.H. Fan et al., High-frequency negative permittivity from Fe/Al_2O_3 composites with high metal contents. J. Am. Ceram. Soc. **95**(1), 67–70 (2012)

Preparation and Photoelectric Property of Poly-3-Methylthiophene/Nano-TiO_2 Composite

Li Gao, Xiaowei Yan, Xin Yang, Wulin Ma and Haiying Du

Abstract A poly 3-methylthiophene/nano-TiO_2 composite was prepared by in situ oxidative polymerization technique. We explored properties of the composite when nano-TiO_2 was modified and unmodified, researched the representation of the composite with different molar ratios of 3-methylthiophene and anhydrous $FeCl_3$, different mass ratios of 3-methylthiophene and nano-TiO_2. The results show that the photoelectric property of poly-3-methylthiophene/modified nano-TiO_2 composite is better than that of unmodified nano-TiO_2 composite. When the molar ratio of 3-methylthiophene and anhydrous $FeCl_3$ is 1:2, the mass ratio of 3-methylthiophene and nano-TiO_2 is 2:1, the composite material has good optical properties.

Keywords Poly-3-methylthiophene · Nano-TiO_2 · Composite material
Photoelectric property

Introduction

Conductive polymers/inorganic nano-composite has characteristics of both conductive polymers and inorganic nanomaterials, which is a hot topic in the field of materials in recent years [1–5]. The polythiophene has good thermal conductivity, chemical stability and photoelectric property in the conductive polymer materials [6–8]. The composite material prepared by the combination of polythiophene and nano-inorganic compound has synergistic effect between polythiophene and inorganic substances due to the addition of nano-inorganic substances, and show the performance of photoelectric properties [9, 10]. Recently, polythiophene and its derivatives/inorganic composite materials prepare by electrochemical method, chemical oxidation polymerization, sol-gel method, blending method and so on. The inorganic nano-materials include nanao-MnO_2, SnO_2, TeO_2, TiO_2, γ-Fe_2O_3,

L. Gao · X. Yan · X. Yang · W. Ma · H. Du (✉)
College of Materials, Chemistry & Chemical Engineering, Chengdu University of Technology, Dongsanlu, Erxianqiao, Chengdu 610059, Sichuan, China
e-mail: diane201109@126.com

Y. Han (ed.), *Advanced Functional Materials*,
https://doi.org/10.1007/978-981-13-0110-0_6

ZnO, Ag, Si, carbon nanotubes, graphene oxide (GO), graphene (GNP), molecular sieve, etc. However, nano-TiO_2 explains the advantages in the chemical stability, environmental friendliness, corrosion resistance and non-toxic [11]. Preparation and application of polythiophene and its derivatives/TiO_2 nano-composite are the hot research in the composite material.

In 2003, Watanabe group [12] first prepared poly-3-hexylthiophene/TiO_2 composite material applied to solar cells, with a conversion efficiency of 0.06%. Preparation of the poly-3-octylthiophene/TiO_2 composite was also used in solar cells by Huisman group, with a conversion efficiency of 2.5% [13]. Haeldermans group [14] prepared the poly 3-thiophene sodium caproate/TiO_2 composite, which was used in dye-sensitized solar cells, with a conversion efficiency of 2.7%. In addition, Li group [15] reported that the PTh/TiO_2 composite material prepared through solid phase method, which was used in the photodegradable methyl orange reaction and shown good catalytic performance in 2009. In 2016, Xue group [16] reported the polythiophene/polypyrrole/TiO_2 composite membrane, synthesized via solid phase method and was used in electrochemical corrosion. Up to now, the photoelectric properties of polythiophene and its derivatives/TiO_2 composite have not been reported by the modification of nano-TiO_2 and the introduction of methyl in the thiophene ring. In this paper, composite material of poly-3-methylthiophene/nano-TiO_2 was prepared via in situ oxidative polymerization technique, investigated the ratio of 3-methylthiophene monomer and nano-TiO_2, 3-methyl-thiophene monomer and anhydrous $FeCl_3$, the photoelectric properties of the composite material.

Experimental

Materials and Physical Measurements. Fourier transform infrared (FTIR) spectra were recorded on German BRUKER TENSON 27 infrared spectrometer. X-ray powder diffraction (XRD) was conducted using a Dandong DX-2700 X-ray diffractometer. Fluorescence spectra were done by Japan's Hitachi (HITACHI)850 Fluorescence spectrophotometer. The resistivity was measured via ST2263 Double electric digital four probe tester. All chemicals were used of analytic reagent (AR) grade and deionized water was used in all reactions.

Preparation of Modified Nano-TiO_2. According to the literature [17, 18], the surface of nano-TiO_2 is modified by sodium dodecylbenzenesulfonate. Firstly, the pH of sodium dodecylbenzenesulfonate was adjusted to 5.0 by dropping hydrochloric acid. Then, nano-TiO_2 was added and put in the ultrasonic oscillator.

Preparation of Composite Material. CH_2Cl_2 (20 ml) and anhydrous $FeCl_3$ are added to a flask, Next, nano-TiO_2 was added to the flask and stirring for 5 min before 3-methylthiophene monomer was dripped to the flask. The reaction mixture was stirred in N_2 at r.t. for 16 h. And then the reactants were washed with absolute ethanol and distilled water several times until the filtrate was colorless. Finally, composite material was dried at 80 °C for 12 h.

Results and Discussion

FTIR Characterization

Nano-TiO_2 Modified. FT-IR of unmodified and modified nano-TiO_2 are shown in Fig. 1. The absorption peaks of modified TiO_2 appear at 1041, 1132 and 1190 cm^{-1}, which are belonged to the absorption peaks of sulfonic acid [19]. The peaks at 1400–1600 cm^{-1} correspond to benzene ring C=C scaling vibration, the peaks at 2853 and 2936 cm^{-1} correspond to C–H stretching vibration. Thus, the surface of nano-TiO_2 is successfully modified.

Composite Material. As shown in Fig. 2, the strong interaction between the carbonyl and the nano-TiO_2 lead to the carbonyl absorption peak blue-shift from 1655 to 1626 cm^{-1}. Two broad peaks at 686 and 532 cm^{-1} correspond to deformation vibrations of CH of thiophene ring, the absorption peak of 1400 cm^{-1} corresponds to symmetry deformation vibrations of CH_3, the absorption peak of 1126 cm^{-1} corresponds to the C_α–C_α single resonance absorption [20, 21]. Thus, the poly-3-methylthiophene/nano-TiO_2 composite has been successfully obtained.

X-ray Diffraction Characterization. Figure 3 presents the XRD diffraction pattern of modified, unmodified nano-TiO_2 and PDF card. All the peaks of modified and unmodified nano-TiO_2 are observed as compared to the PDF-TiO_2 card. It means that the modification process has no effect on the crystal structure of nano-TiO_2.

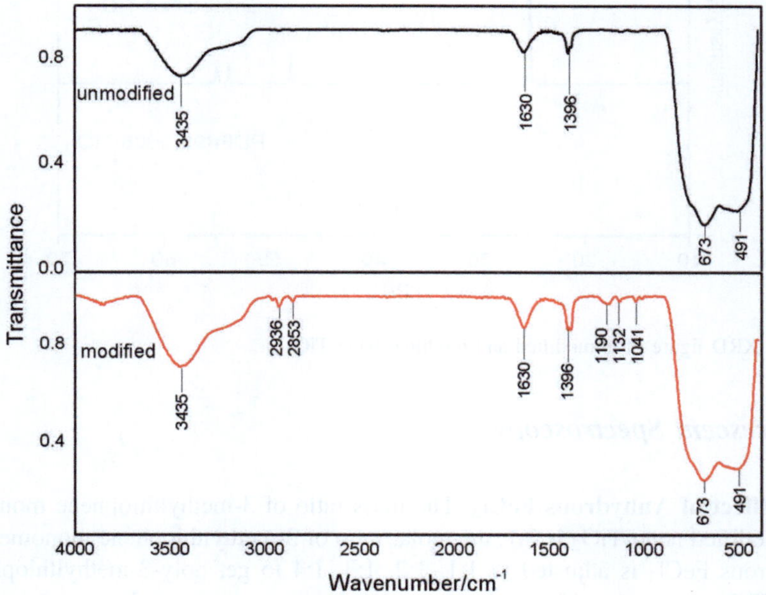

Fig. 1 FT-IR of unmodified and modified nano-TiO_2

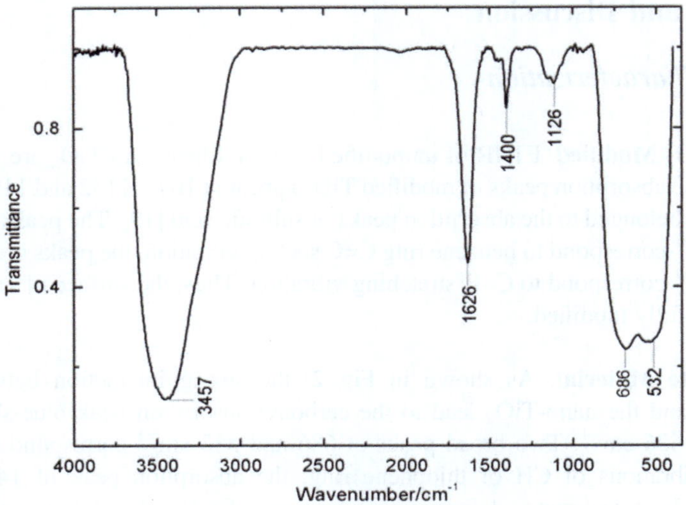

Fig. 2 FT-IR of poly-3-methylthiophene/nano-TiO_2 composite

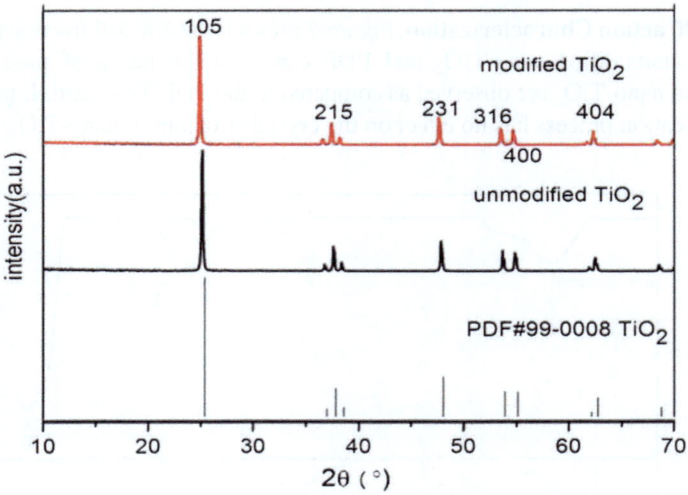

Fig. 3 XRD figure of unmodified and modified nano-TiO_2

Fluorescent Spectroscopy

The Effect of Anhydrous $FeCl_3$. The mass ratio of 3-methylthiophene monomer and modified nano-TiO_2 is 2:1, the molar ratio of 3-methylthiophene monomer and anhydrous $FeCl_3$ is adjusted to 1:1, 1:2, 1:3, 1:4 to get poly-3-methylthiophene/ nano-TiO_2 composite. Figure 4 shows the fluorescence excitation spectra under λ_{ex} = 500 nm.

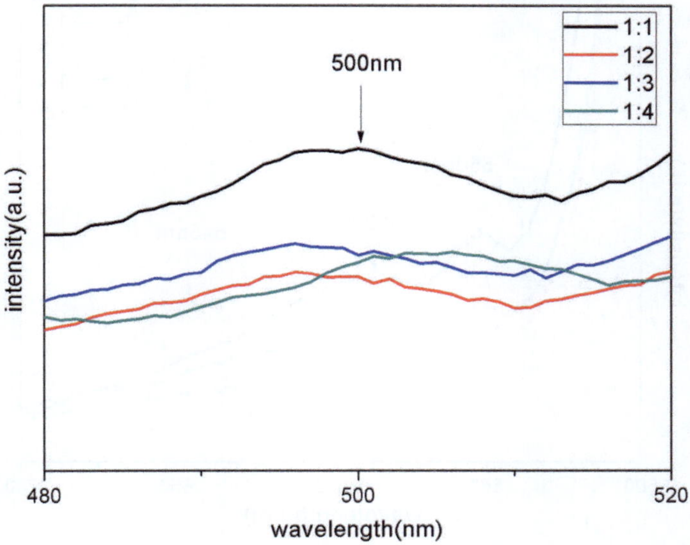

Fig. 4 Excitation spectrum at λ_{em} = 550 nm composite material of with different molar ratios of 3-methylthiophene and $FeCl_3$

Figure 5 shows all of composites have different emission intensities at 550 nm. When the molar ratio of anhydrous $FeCl_3$ and 3-methylthiophene monomer is 1:1, the composite material is superior to the other groups, and the maximum emission wavelength is 586 nm with the excitation wavelength of 550 nm. When the molar ratio is 1:2, the excitation wavelength of composite material is shifted to 645 nm. CH_3 is introduced in the 3-position of thiophene monomer to increase the conjugate degree of π electrons. At the same time, the expansion of π-electron delocalization makes the composite material more susceptible to excitation and emission, and the fluorescence emission wavelength of the composite with a molar ratio of 1:2 is red-shift [22]. The appropriate amount of oxidant $FeCl_3$ can effectively improve the polymerization degree of 3-methylthiophene monomer on the surface of nano-TiO_2, and make the composite with the molar ratio of 1:2 give the perfect fluorescence emission peak in the red range.

The Quantity of Nano-TiO_2. The molar ratio of 3-methylthiophene monomer and anhydrous $FeCl_3$ remains 1:1. The mass ratio of modified TiO_2 and 3-methylthiophene monomer is 1:1, 1:2, 1:3 to get the composite. Figure 6 shows that composite material has strong emission peaks at 551 and 582 nm, and the emission peak intensity is the largest when the mass ratio of nano-TiO_2 and 3-methylthiophene monomer is 1:2. There are two reasons to explain. Firstly, the mass of 3-methylthiophene is more than nano-TiO_2, which can be completely wrapped, to improve the degree of the two complex. Secondly, because of the low content of poly-3-methylthiophene, thiophene ring is mainly connected by α–α, the maximum overlap of the Pz orbitals connected to the C–C of thiophene. So that poly-3-methyl thiophene band gap is low, the absorption peak band appears at low frequency.

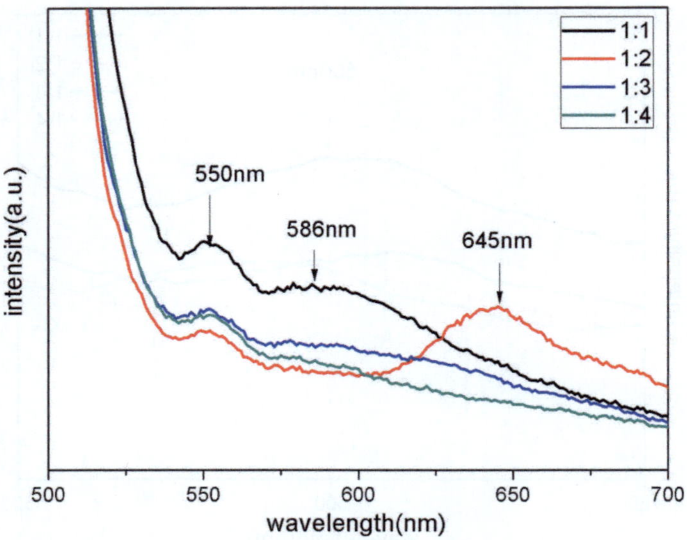

Fig. 5 Emission spectrum at λ_{ex} = 500 nm of composite material with different molar ratios of 3-methylthiophene and $FeCl_3$

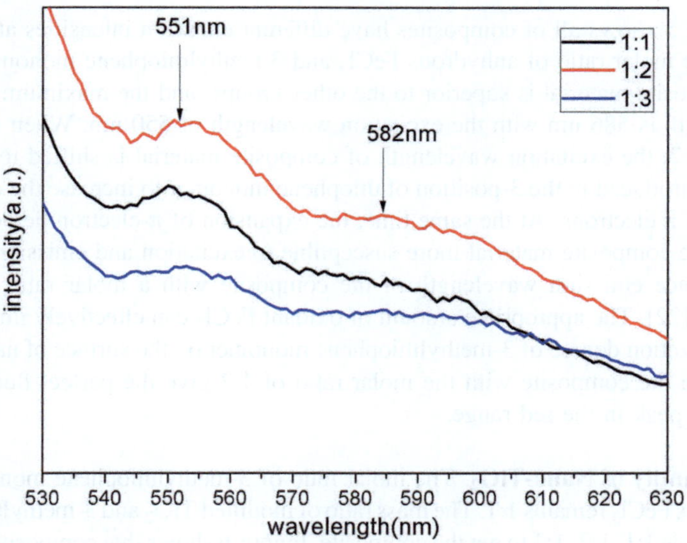

Fig. 6 Emission spectrum of the composite material is composed of different mass ratios of 3-methylthiophene and nano-TiO_2 at λ_{ex} = 500 nm

The Effect of Modified or Unmodified Nano-TiO_2. The molar ratio of 3-methylthiophene monomer, anhydrous $FeCl_3$ and nano-TiO_2 is not changed to prepare the composite, investigated the effect of modified or unmodified nano-TiO_2.

Figure 7 shows that both of composites have the emission peak at 551 and 585 nm, but the emission peak of modified nano-TiO_2 composite material is stronger. Because the modified nano-TiO_2 improve its hydrophilicity in the surface, the composite material is difficult to reunite in CH_2Cl_2. Thus, 3-methylthiophene monomer polymerization degree is better and π electron conjugated degree is higher, which lead to the fluorescence wavelength of composite material is red-shifted and fluorescence intensity is better.

Electrochemical Performance Analysis. The resistivity measurement of composite material is measured by digital four-probe measuring instrument. Once the mass ratio of nano-TiO_2 and 3-methylthiophene is fixed, the resistivity of composite will decrease with the increase of the amount of $FeCl_3$. Once the molar ratio of 3-methylthiophene monomer and oxidant $FeCl_3$ is fixed, the resistivity of the composite material will decrease with the increase of the amount of nano-TiO_2.

Because the methyl has the stronger electron-withdrawing inductive effect and the weaker for electronic conjugation effect to make the better conjugate degree of polymer, structure tend to be more complanation and better order [22]. With the increase of the proportion of nano-TiO_2 in the composite, the order of the 3-methylthiophene monomer on the surface of nano-TiO_2 is more obvious, the tightness of the poly-3-methylthiophene increases, and the conductivity increases obviously [19, 23]. Figure 8 shows that the conductivity of composite material prepared by modified nano-TiO_2 is better than that of the composite material prepared by unmodified nano-TiO_2. Because the modified nano-TiO_2 can increase the lipophilicity of its surface, combine with poly-3-methylthiophene more evenly lead

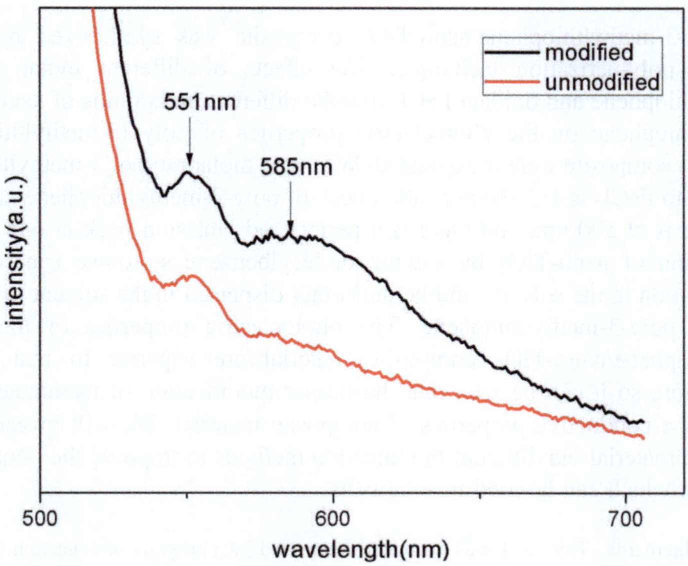

Fig. 7 Emission spectrum of composite material at $\lambda_{ex} = 500$ nm (blank: modified nano-TiO_2)

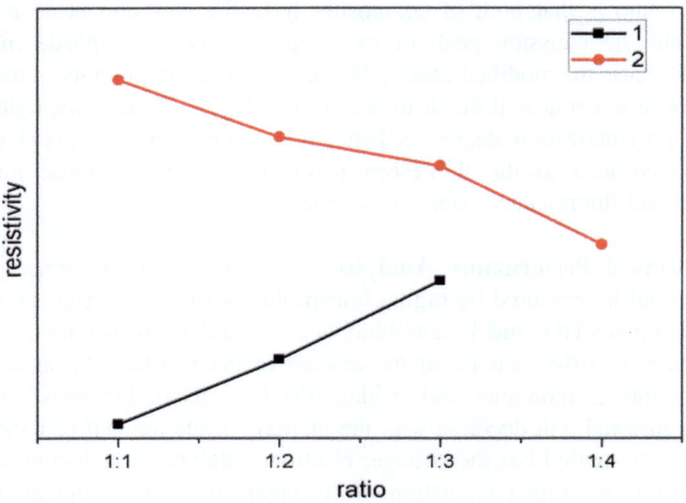

Fig. 8 Resistivity of 3-methylthiophene monomer and $FeCl_3$ different molar ratio (blank), different mass ratios of nano-TiO_2 (red)

to better conductivity. The results indicate the modified nano-TiO_2 can improve the conductivity of poly-3-methylthiophene/nano-TiO_2 composite.

Conclusion

The poly-3-methylthiophene/nano-TiO_2 composite was synthesized by in situ oxidative polymerization technique. The effects of different molar ratios of 3-methylthiophene and oxidant $FeCl_3$ and the different mass ratios of nano-TiO_2 to 3-methylthiophene on the photoelectric properties of poly (3-methylthiophene)/nano-TiO_2 composite were investigated. When the molar ratio of 3-methylthiophene monomer to $FeCl_3$ is 1:2, the excitation peak of poly-3-methylthiophene/nano-TiO_2 composite is at 500 nm, and there is a perfect red emission peak at 645 nm. The modification of nano-TiO_2 by sodium dodecylbenzene sulfonate can reduce its agglomeration in the solvent, and be uniformly dispersed in the solvent to be better form the poly-3-methylthiophene. The photoelectric properties of the poly-3-methylthiophene/nano-TiO_2 composite material are superior to that of non-modification, so it can be seen that the proper modification of nanomaterials can improve the conductive properties of composite material. We will investigate the composite material via different modification methods to improve the photoelectric properties, which can be used in solar cells.

Acknowledgements This work was financially supported by a program of education department (No. 17ZB0049) of Sichuan province, and the program of innovative entrepreneurship training (No. 201610616063) of provincial college students.

References

1. H. Shirakawa, E. Louis, A. MacDiarmid, C.K. Chiang, A.J. Heeger, Synthesis of electrically conducting organic polymers: halogen derivatives of polyacetylene, $(CH)_X$. J. Chem. Soc. Chemi. Commun. **10**, 578–580 (1977)
2. H. Naarmann, Synthesis of electrically conducting polymers. Angew. Makromol. Chem. **109**, 295–338 (1982)
3. C.O. Yoon, H.K. Sung, J.H. Kim, E. Barsoukov, J.H. Kim, The effect of low-temperature conditions on the electrochemical polymerization of polypyrrole films with high density, high electrical conductivity and high stability. Synth. Met. **99**, 201–212 (1999)
4. J. Koh, J. Kim, B. Kim, J. Kim, E. Kim, Highly efficient, iodine-free dye-sensitized solar cells with solid-state synthesis of conducting polymers. Adv. Mater. **23**, 1641–1646 (2011)
5. N. Hebestreit, J. Hofmann, U. Rammelt, W. Plieth, Physical and electrochemical characterization of nanocomposites formed from polythiophene and titaniumdioxide. Electrochim. Acta **48**, 1779–1788 (2003)
6. G.D. Sharma, P. Suresh, S. Kumar Sharma, M.S. Roy, Optical and electrical properties of hybrid photovoltaic devices from poly(3-phenyl hydrazone thiophene) (PPHT) and TiO_2 blend films. Sol. Energy Mater. Sol. Cells **92**, 1–70 (2008)
7. Y. Li, Conducting polymers. Prog. Chem. **14**, 207–211 (2002)
8. Y. Du, K. Cai, Research progress in electrical properties of polythiophene, polythiophene derivatives and polythiophene based composites. Mater. Rev. **24**, 69–73 (2010)
9. X. Li, M. Huang, H. Shen, Preparation and properties of polythiophene nanocomposites, chemical industry and engineering progress. Chem. Ind. Eng. Prog. **26**, 1243–1245 (2007)
10. T. Hai, R. Sugimoto, Effect of molar ratio of oxidizer/3-hexylthiophene monomer in chemical oxidative polymerization of poly(3-hexylthiophene). J. Mol. Struct. **1146**, 660–668 (2017)
11. A.G.R. Morales, M.O.C. Guzmán, C.C. Arteaga, A brief review on fabrication and applications of auto-organized TiO_2 nanotube arrays. Corros. Rev. **29**, 105–121 (2011)
12. H. Irie, Y. Watanabe, K. Hashimoto, Nitrogen-concentration dependence on photocatalytic activity of $TiO_{2-X}N_X$ powders. J. Phys. Chem. B. **107**, 5483–5486 (2003)
13. O. Carp, C. Huisman, A. Reller, Photoinduced reactivity of titanium dioxide. Prog. Solid State Chem. **32**, 33–177 (2004)
14. P. Vanlaeke, A. Swinnen, I. Haeldermans, G. Vanhoyland, T. Aernouts, D. Cheyns, C. Deibel, J. D'Haen, P. Heremans, J. Poortmans, J.V. Manca, P_3HT/PCBM bulk heterojunction solar cells: relation between morphology and electro-optical characteristics. Sol. Energy Mater. Sol. Cells **90**, 2150–2158 (2006)
15. S. Li, M. Ma, Y. Wei, X.S. Hong, Y.H. Wang, Preparation of PTh/TiO_2 nanocomposite by solid-state method and its photocatalytic activity under visible light irradiation. Chem. Res. Appl. **21**, 1627–1631 (2009)
16. S. Xue, Q. Liu, Preparation of Polythiophene/Polypyrrole/TiO_2 composite conductive polymers by solid-state method and its anti-corrosion properties for stainless steel. Chin. J. Inorg. Chem. **33**, 98–102 (2016)
17. J. Yu, H. Jiang, C. Shi, Study on surface modification of TiO_2 via in situ polymerization. Chem. Produc. Tech. **11**, 21–23 (2004)
18. K. Wang, J. Niu, K. Chen, Surface modification of TiO_2 and its influence on thermal insulation property of coatings. Dye Fini. **35**, 1–5 (2009)
19. X.H. Peng, Organic surface modification of titanium dioxide and electrophoretic properties of the resulting materials. J. B Univ. Chem. Technol. (Nat. Sci. Ed.) **33**, 11–14 (2006)
20. H. Yin, J. Jiang, Studies of preparation and conductivity of γ-Fe_2O_3/polythiophene nanocomposites. J. Funct. Mater. **36**, 1524–1527 (2005)

21. W. Zhang, Y. He, Q. Qi, Preparation of porous TiO_2 film of photocatalyst by microemulsion templating. J. Funct. Mater. **36**, 1590–1593 (2005)
22. S. Yu, J. Yang, J. Gu, Study on the inductive effect conjugative and orientation effect of methyl substituent. J. B Norm. Univ. (Nat. Sci.) **42**, 506–509 (2006)
23. Y. Wang, J. Wang, J. Yue, Q. Luo, D. Wang, Recent development of conducting polymer nanocomposites. Chem. Ind. Tim. **21**, 73–77 (2007)

Preparation of Nickel and Manganese Oxides Composite via Co-precipitation Method and Its Electrochemical Properties

Huan Ren, Zhong-yang Wang, Qian-qian Li and Yao Liu

Abstract A simple co-precipitation using Na_2CO_3 as precipitation was used to get Nickel and manganese oxides (Ni–Mn oxides). After that process, there is a mixture of nickel oxide, manganese oxide and its solid solution. XRD and SEM were used to get the phase and morphology of the products. In 6 mol/L KOH aqueous electrolyte, electrochemical capacitance characterization were tested by cyclic voltammetry (CV), galvanostatic charge/discharge (GCD) and electrical impedance spectroscopy (EIS) measurements. The results show that When the Ni/Mn ratio is 3/1, Mn_2O_3 appears, and at this point the compounds are NiO and Mn_2O_3 complexes. And the electrochemical performance of this composite is better than that of single-phase NiO.

Keywords Co-precipitation method · Nickel and manganese oxides
Electrochemical properties

Introduction

In the twenty-first Century, environmental problems gradually became one of the themes of the times, and the dependence of fossil fuels on the national economy led to a series of problems such as emission of automobile exhaust, and thus exacerbated the pollution. Therefore, the development of new energy to meet the production needs of all walks of life is the only way for sustainable development. As a new generation of supercapacitor elements, high power density, high rate discharge

H. Ren · Q. Li · Y. Liu (✉)
Key Laboratory for Liquid-Solid Structural Evolution and Processing of Materials (Ministry of Education), Shandong University, Jinan 250061, China
e-mail: liuyao@sdu.edu.cn

H. Ren
e-mail: rhx5200@qq.com

H. Ren · Z. Wang
Dezhou META Research Center for Innovative Materials, Dezhou 253000, China

Y. Han (ed.), *Advanced Functional Materials*,
https://doi.org/10.1007/978-981-13-0110-0_7

performance and long life, so it has great potential application value, gradually attracted people's attention [1–3]. There are two kinds of supercapacitor, one is electric double layer capacitor (EDLC) which depend on the layers' ion adsorption, another is pseudo-capacitor which depend on the electrodes' fast and fast reversible oxidation reduction reaction. Compared with the battery system, the major limitation of supercapacitors is their low energy density. In order to improve its energy density, increasing specific capacity and operating voltage become two important research directions [4]. In general, the electrode materials of supercapacitors are classified into three kinds: (1) carbon based materials with high specific surface area (2) conducting polymers (3) transition metal oxides. Transition metal oxides, especially in the presence of nanostructures, serving as supercapacitor electrode materials have higher energy density than carbon materials and better stability than conductive polymer materials [5–7]. For pseudo-capacitor's electrode material, transition metal oxides such as RuO_2, IrO_2(noble metal oxide), NiO, MnO, Co_3O_4 et al., were proved to have higher specific capacitance because of its property. Because of its high theoretical capacity (2573 F/g), environment friendly, good thermal and chemical stability, rapid redox reaction, short diffusion path and high surface area, micro nano structure of NiO has been widely used in super capacitor [8–10]. For MnO_2, although its has a theoretical specific capacity of 1370 F/g, each of the Mn also stores one electron, so its actual capacitance is only 100–400 F/g., it is important to increase the capacitance of MnO_2 base supercapacitor [11, 12].

In this article, nickel and manganese oxides have been prepared by co-precipitation at different Ni/Mn molar ratios. The morphology of the composite was characterized by SEM and its electrochemical properties were measured. The experimental results are discussed.

Experimental

Preparation of Composite. First, by using a simple co-precipitation process with Na_2CO_3 as precipitant., the precursor was synthesized. The stoichiometric amounts of $NiCl_2$ and $MnCl_2$ in different ratio (Ni:Mn = 3:1 and 4:1) were scattered in 200 ml water to form mixed solution. Then under magnetic stirring, we put moderate Na_2CO_3 solution with the concentration of 0.02 M into the mixed solution by one drop. We get the precipitated precursors which were formed and collected by centrifugation and the precipitated precursors were washed several times with deionized water. Then dry them at 120 °C for 12 h. The final process is calcining them at 500 °C for 5 h with 3 °C/min in the furnace.

Characterization of Phase and Morphology. The date of powder X-ray diffraction of prepared samples were gathered on a Rigaku D/MAX-rA diffractometer with Cu Kα radiation being operated at 40 kV and 100 mA. We use the Field Emission Scanning Electron Microscope (FESEM, SU-70, Japan) to collect morphology and microstructure.

Electrode Preparation and Electrochemical Characterization. To get the working electrodes of electrochemical capacitors, we mixed samples powder using acetylene black (AB) and Poly vinylidene fluoride (PVDF) binder with the mass ratio of 8:1:1. Adding several drops of N-Methyl pyrrolidone (NMP) into the mixture to make them homogeneous slurry. Then paint circular nickel foam with the diameter of 4 cm with the slurry of the mixture. Dry them in vacuum at 120 °C for 12 h and weigh the electrodes. Finally, we obtain the working electrode. All electrochemical measurements were done in a three-electrode experimental setup carried out in 6 M KOH aqueous electrolyte by a CHI660D (Shanghai Chenhua Apparatus, China) electrochemical workstation. The counter and reference electrodes were made of platinum wire electrode and Ag/AgCl electrode, respectively.

Result and Discussion

Figure 1 shows the crystalline structures of the samples with different Ni/Mn molar ratios using X-ray diffractometer. When the content of Mn increases to Ni/Mn ratio is 3/1, the peak of manganese oxide Mn_2O_3 appears. The phase ratio of the two groups of Ni/Mn to 3/1 is NiO and Mn_2O_3 complexes. When the Ni content is larger than Mn, the main phase of the obtained sample is NiO, and the NiO peak decreases with the decrease of Ni/Mn. At the Ni/Mn ratio of 3/1, two phases of the phase is Ni–Mn double oxide NiO and Mn_2O_3 complex.

Figure 2 is a SEM diagram of a sample of different Ni/Mn ratios at Ni/Mn > 1. As can be seen from the diagram, when the content of Ni is greater than Mn, the samples are nanoparticles and the dispersion is good. When the Ni content is larger than Mn, the main phase of the obtained sample is NiO. With the decrease of Ni/Mn, the microstructure of the sample is finely dispersed.

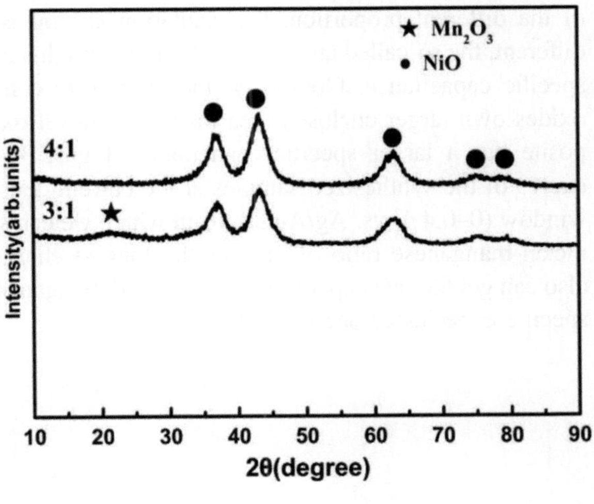

Fig. 1 XRD pattern of prepared products with different Ni/Mn molar ratio

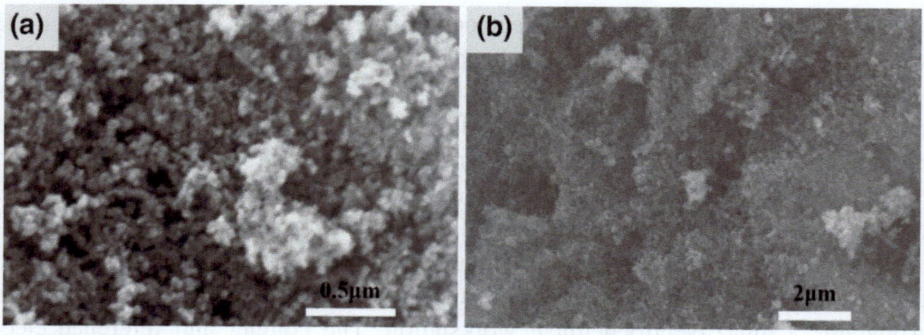

Fig. 2 SEM images of products with different Ni/Mn molar ratio (a: 3:1, b: 4:1)

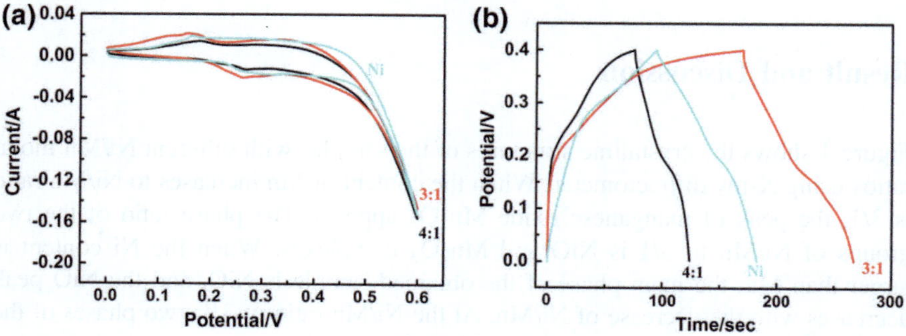

Fig. 3 CV curves (**a**) and Charge–discharge curves (**b**) of products with different Ni/Mn molar ratio

Figure 3a shows the CV curves of the electrode within the potential range from 0 to 0.5 V (vs. Ag/AgCl) of different Ni/Mn molar ratios at the scan rate 5 mV/s. As can be seen from the diagram, the curve is not a double layer capacitor rectangle, indicating that it belongs to the pseudo-capacitor capacitor and the CV curve has an obvious peak at 0.15 and 0.25 V potential. But at the same scanning speed, because of the different proportion, the excitation current is different, the peak height is different, the so-called area is also different, and this corresponds to the difference of specific capacitance. Obviously, the composite contained nickel and manganese oxides own larger enclosed area than pure nickel oxides, indicating that the composite has a larger specific capacitance. Figure 3b shows the charge–discharge curves of the synthesized samples at the current density of 1A/g within a potential window (0–0.4 V vs. Ag/AgCl), from which we can see that the we can see that the nickel-manganese ratio of 3:1 has the longest charge and discharge time and we also can get the rate capability of nickel and manganese oxides electrode and get the specific capacitance according to Eq. 1.

$$C_s = It/\Delta Vm \tag{1}$$

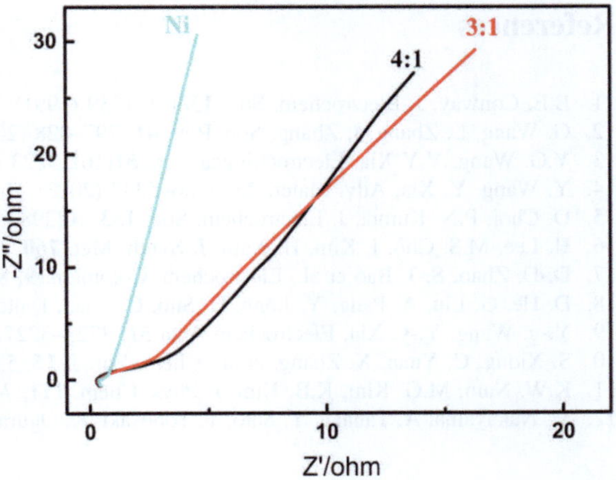

Fig. 4 Nyquist plots products with different Ni/Mn molar ratio

In Eq. 1, I is the discharge current, t is the discharge time, ΔV is the voltage range and m is the mass of the active material within the electrode.

We also obtained Nyquist plots products for samples with different Ni/Mn ratios, as shown in Fig. 4. In low frequency area, the Warburg impedance is represented by the slope of the curve. It is easier for transporting ions in the electrolyte when the Warburg impedance is small. More vertical the curve is, the smaller Warburg impedance is. The slopes of the nickel manganese oxides electrode smaller than that of pure nickel. Oxides at low frequency area lead to the slight increasing of diffusive resistance between the electrode and the electrolyte. EIS curve High-frequency curve radius represents a small charge transfer resistance, good electrochemical performance, especially 3/1 samples have the best electrochemical performance.

Conclusions

In general, the Ni–Mn oxides were prepared by a simple coprecipitation process with Na_2CO_3 as precipitant. We observed in the experiment that when Ni/Mn is 3/1, the Mn_2O_3 phase appears, and the product is Nickel and Manganese oxides Composite formed by the combination of NiO and Mn_2O_3. The compound is in nanoscale structure and the nanoparticles has good dispersibility. Its electrochemical performance is better than that of a single oxide component sample. The composite effect of nickel manganese oxides is explained, and the utilization ratio of the material is improved, and the electrochemical performance of the oxide is optimized.

Acknowledgements This study was financially supported by the National Natural Science Foundation of China [grant No. 51402170], Young Scholars Program of Shandong University [grant No. 2016WLJH40].

References

1. B.E. Conway, J. Electrochem. Soc. **138**(6), 1539 (1991)
2. G. Wang, L. Zhang, J. Zhang, Soc. Rev. **41**, 797–828 (2012)
3. Y.G. Wang, Y.Y Xia, Electrochimica Acta **51**(16), 3223 (2006)
4. Y. Wang, Y. Xia, Adv. Mater. **25**, 5336–5342 (2013)
5. D. Choi, P.N. Kumta, J. Electrochem. Soc. **153,** A2298–A2303 (2006)
6. H. Lee, M.S. Cho, I. Kim, H. Nam, J. Synth. Met. **760**, 1055–1059 (2010)
7. D.-D. Zhao, S.-J. Bao et al., Electrochem. Commum. **9**, 869–874 (2007)
8. D. He, G. Liu, A. Pang, Y. Jiang, H. Suo, C. Zhao, Dalton Trans. **46**, 1857–1863 (2017)
9. Y.-g. Wang, Y.-y. Xia, Electrochem Acta **51**, 3223–3227 (2006)
10. S. Xiong, C. Yuan, X. Zhang, et al., Chem. Eur. J. **15**, 5320–5326 (2009)
11. K.W. Nam, M.G. Kim, K.B. Kim, J. Phys. Chem. **111**, 749–758 (2007)
12. M. Nakayama, A. Tanaka, Y. Sato, T. Tonosaki, K. Ogura, Langmuir **21**, 5907–5913 (2005)

Synthesis and Properties Research of Coin-Like α-Fe_2O_3 Nanoparticles

Gongqin Yan, Qiang He and Dongyan Yu

Abstract In this paper, coin-like α-Fe_2O_3 nanoparticles were synthesized successfully through a hydrothermal method in EG/DEG mixed solution with PVP as surfactant and NaOAc as alkali source. The phase composition and morphology of as-prepared products were analyzed. It was found that the EG/DEG ratio in the mixed solution plays a key role in the formation of coin-like α-Fe_2O_3 nanoparticles. The magnetic property of the synthesized coin-like α-Fe_2O_3 nanoparticles at room temperature was also characterized. The result shows that the coin-like α-Fe_2O_3 nanoparticles have a saturation magnetization of 7.678 $emug^{-1}$ and present a superparamagnetic property. The electrochemical property of the obtained coin-like α-Fe_2O_3 nanoparticles was characterized through using the coin-like α-Fe_2O_3 nanoparticles as the cathode material of a lithium-ion battery. It is found that the obtained coin-like nanoparticles have an initial discharge capacity of 393 $mAh \cdot g^{-1}$ and a charge capacity of 236 $mAh \cdot g^{-1}$ when the charge-discharge rate is 0.1 C.

Keywords Coin-like α-Fe_2O_3 nanoparticles · Hydrothermal synthesis
Magnetic property · Electrochemistry property

Introduction

As one of the low-cost and environmental friendly semiconductor materials, α-Fe_2O_3 has been widely applied in gas sensors [1], biomolecules sensing [2], lithium ion batteries [3] and so on. Because the particle size and morphology of nanomaterial have great effect on their properties, so the synthesis of nanomaterial with expected particle size and morphology has attracted scientists's attention. Until recently, various methods for the synthesis α-Fe_2O_3 nanostructures have been exploited, such as co-precipitation [4], micro-mulsion [5], hydrothermal strategy

G. Yan (✉) · Q. He · D. Yu
School of Mechanical Engineering, Guangxi University of Science and Technology,
Liuzhou 545006, Guangxi, China
e-mail: ygq@gxust.edu.cn

Y. Han (ed.), *Advanced Functional Materials*,
https://doi.org/10.1007/978-981-13-0110-0_8

[6], chemical vapordeposition [7]. Among these methods, hydrothermal strategy has been widely used because of its low costing, easy operating, environmentally friendliness. Through hydrothermal strategy, α-Fe_2O_3 nanomaterials with various morphologies such as flake [8], line [9], threadiness [10], rod [11], tube [12], flower-like [13], ring [14], cube [15], sphere [16] have been synthesized. Lian [17] successfully synthesized α-Fe_2O_3 with various morphologies via an ionic liquid-assisted hydrothermal synthetic method. Mesoporous hollow microspheres, microcubes, and porous nanorods α-Fe_2O_3 are synthesized and the formation processes of the α-Fe_2O_3 with various morphologies have been investigated systematically. This method is expected to be a useful technique for controlling the shapes of crystallized inorganic materials. Subarna et al. [18] synthesized α-Fe_2O_3 nanostructures by a solvothermal strategy in an mixed solution of ethylenediamine (EN) and ethanol (EtOH) or water with different volume ratio. Their results showed that EN functioned as a ligand and facilitated the growth of nanostructures during the synthesis process. Dinesh et al. [19] synthesized α-Fe_2O_3 nanocups via a button-up approach in which tuning the morphology of single-crystalline hematite from hollow spheres to nanocups. Compared with α-Fe_2O_3 hollow spheres, α-Fe_2O_3 nanocups have a particular magnetic property, which would be related to their particular morphology. Zhang et al. [20] prepared α-Fe_2O_3 micro-pines though a hydrothermal reaction in $K_3[Fe(CN)_6]$ solution with different pH values at 140 °C. With the change of pH values, α-Fe_2O_3 microstructures with different morphologies including snowflake-like, paired microplates, dumbbell, column, and spindle were obtained.

In this paper, coin-like α-Fe_2O_3 nanoparticles were synthesized by a hydrothermal method. The magnetic and electrochemical properties of the as-obtained coin-like α-Fe_2O_3 nanoparticles were characterized. The results indicate that the coin-like nanoparticles show a superparamagnetism property, and their also exhibit a good electrochemical property. It is suggested that the as-obtained coin-like α-Fe_2O_3 nanoparticles have a potential application in the field of Li ion batteries.

Experimental Section

All reagents were of analytic grade and were used as received without further purification. In a typical process, EG and DEG were mixed in the proportion of 1:3 and obtain a homogeneous solution. Then 1.212 g $Fe(NO_3)_3\cdot 9H_2O$ was added to the above solution and followed by stirring for 10 min. The solution was heated to 100 °C and 3 g PVP was added to the hot solution under vigorous stirring. After cooling to room temperature, 3.5 g $NaOAc\cdot 3H_2O$ was added to the mixture under stirring. Finally, the mixture was transferred into a Teflon-lined stainless-steel autoclave and heated to 200 °C for 24 h. The obtained red precipitation was washed with deionized water and ethanol and dried at 60 °C, then the coin-like α-Fe_2O_3 nanoparticles were obtained.

The phase composition of as-prepared product was analyzed by X-ray diffraction (XRD) with DX2700 powder diffractometer operated at 35 kV, 30 mA, using a Cu Kα. The morphology of as-prepared products were examined by field-emission scanning electron microscopy (FESEM) with a Zeiss SIGMA microscope at 10 kV. The magnetic property of those nanohybrids were measured with a BKT-4500 vibrating-sample magnetometer (VSM), at room temperature. The electrochemical property of those nanohybrids were tested by a Neware eight-channel battery tester.

Results and Discussion

Phase Composition and Morphology. Figure 1 displays the XRD pattern of the as-obtained coin-like α-Fe_2O_3 nanoparticles. It can be seen that the as-obtained product is composed of well-crystalline hematite. The diffraction peaks correspond to the (012), (104), (110), (006), (113), (202), (024), (116), (211), (122), (018), (214), (300), (125) and (208) crystal planes of hematite (JCPDS No. 33-0664) respectively. And the strong diffraction peaks indicated the good crystallinity and purification of obtained hematite.

The SEM images of the as-obtained coin-like α-Fe_2O_3 nanoparticles are shown in Fig. 2. As it is shown in Fig. 2b, the coin-like α-Fe_2O_3 nanoparticles have a

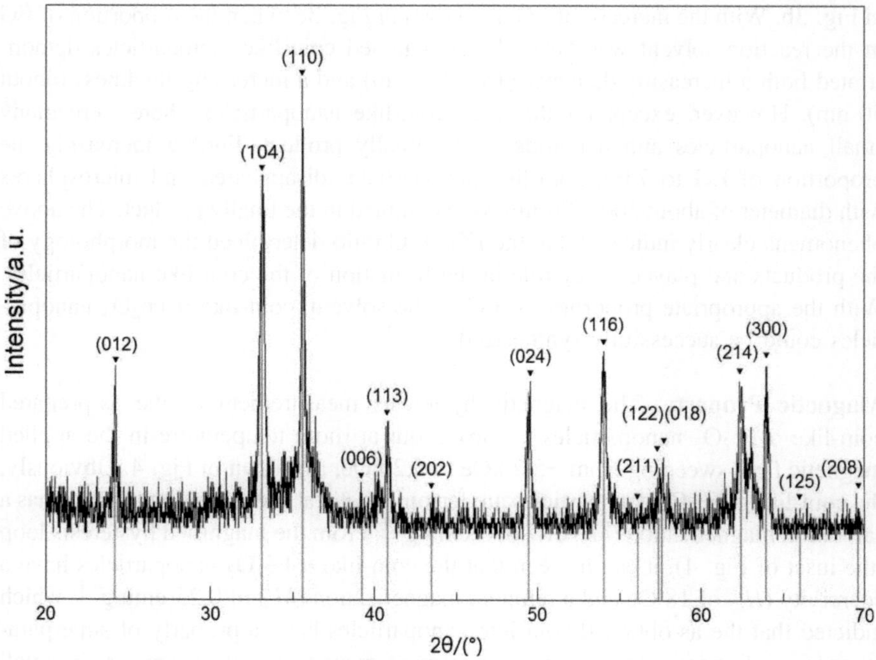

Fig. 1 XRD pattern of the as-obtained coin-like α-Fe_2O_3 nanoparticles

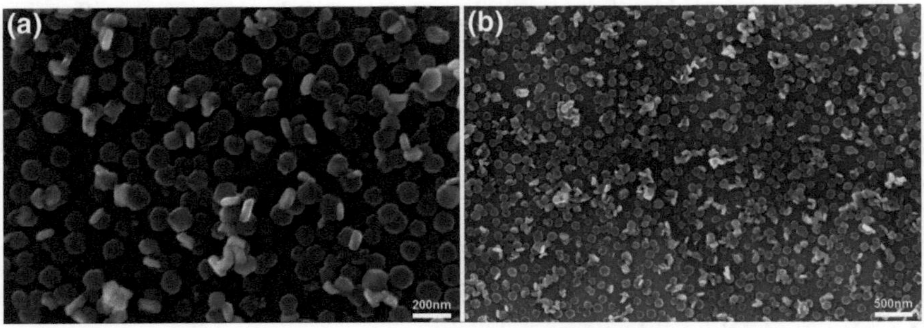

Fig. 2 SEM images of the as-obtained coin-like α-Fe_2O_3 nanoparticles

narrow size distribution and are not agglomerated. It is found that the α-Fe_2O_3 nanoparticles have an obvious coin-like shape with a diameter of about 80–100 nm and a thickness of about 30 nm.

In our research, we found that the ratio of EG and DEG plays a key role in the formation of the coin-like nanoparticles. It was found that the morphology of the obtained products was influenced by the EG/DEG ratio distinctly. Figure 3 show the SEM images of product with different EG/DEG ratio. As shown in Fig. 3a, there were no coin-like particles obtained after the hydrothermal reaction for 24 h when there was no EG in the reaction solvent. When the proportion of EG in the reaction solvent was 25%, coin-like α-Fe_2O_3 nanoparticles were obtained, as shown in Fig. 3b. With the increase of EG, as shown in Fig. 3c, when the proportion of EG in the reaction solvent was 50%, the as-obtained coin-like nanoparticles demonstrated both a increasing diameter (150–200 nm) and a increasing thickness (about 50 nm). However, except for the large coin-like nanoparticles, there were many small nanoparticles and nanorods in the finally product. Further increasing the proportion of EG to 75%, coin-like nanoparticles disappeared and microspheres with diameter of about 300–400 nm were obtained in the finally product. The above phenomena clearly indicated that the EG/DEG ratio determined the morphology of the products and played a key role in the formation of the coin-like nanoparticles. With the appropriate proportion of EG in the solvent, coin-like α-Fe_2O_3 nanoparticles could be successfully synthesized.

Magnetic Property. The magnetic hysteresis measurement of the as-prepared coin-like α-Fe_2O_3 nanoparticles is carried out at room temperature in the applied magnetic field sweeping from −5.2 kOe to 5.2 kOe, as shown in Fig. 4. Obviously, the coin-like α-Fe_2O_3 nanoparticles are ferromagnetic at room temperature and has a saturation magnetization (M_s) of 7.678 emu·g^{-1}. From the magnified hysteresis loop (the inset of Fig. 4), it can be seen that the coin-like α-Fe_2O_3 nanoparticles have a coercivity (H_c) of 18 Oe and a remnant magnetization (M_r) of 0.25 emu·g^{-1}, which indicted that the as-obtained coin-like nanoparticles have a property of superparamagnetism. Superparamagnetism is a form of magnetism which appears in small ferromagnetic or ferrimagnetic nanoparticles. In sufficiently small nanoparticles,

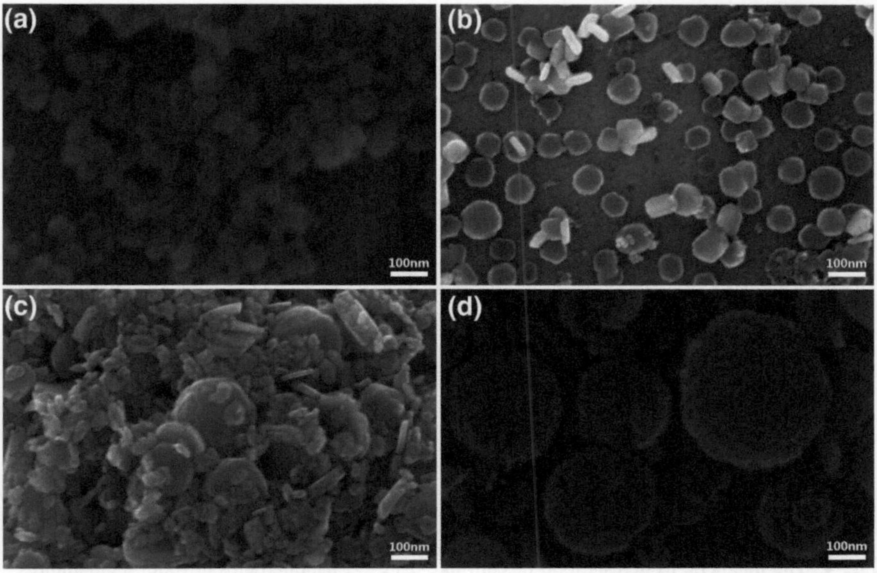

Fig. 3 The SEM images of the obtained product with different EG/DEG ratio: **a** 0:20, **b** 5:15, **c** 10:10, **d** 15:5

magnetization can randomly flip under the influence of temperature. The typical time between two flips is called the Néel relaxation time. In the absence of an external magnetic field, when the time used to measure the magnetization of the nanoparticles is much longer than the Néel relaxation time, their magnetization appears to be in average zero, which is the superparamagnetic state. In this state, an external magnetic field is able to magnetize the nanoparticles, similarly to a paramagnet. However, their magnetic susceptibility is much larger than that of paramagnets [21]. The superparamagnetic property of the coin-like α-Fe_2O_3 nanoparticles may be attributed to the small crystal size of the hematite, which causes the magnetic orientation force produced by an external field is not sufficient to resist the thermal disturbances, and their magnetization is similar to that of a paramagnetic.

Electrochemical Property. The electrochemical property of the as-obtained coin-like α-Fe_2O_3 nanoparticles was measured using a coin-type half-cell assembled in an argon-filled glovebox. The working electrode was prepared by thoroughly mixing of the active materials, carbon black and poly(vinylidene fluoride) with a weight ratio of 7:1:2 in N-methyl-2-pyrrolidone as solvent. The result slurry was then coated onto a copper foil and dried in a vacuum oven at 60 °C for 12 h. A pure lithium foil, PP microporous membrane, 1 M LiPF6 dissolved in ethyl carbonate and diethyl carbonate (1:1, wt/wt) were used as the counter electrode, the separator and the electrolyte, respectively. The cells were charged and discharged on a NEWARE battery tester in a voltage range of 0.1–3.2 V. The curves of

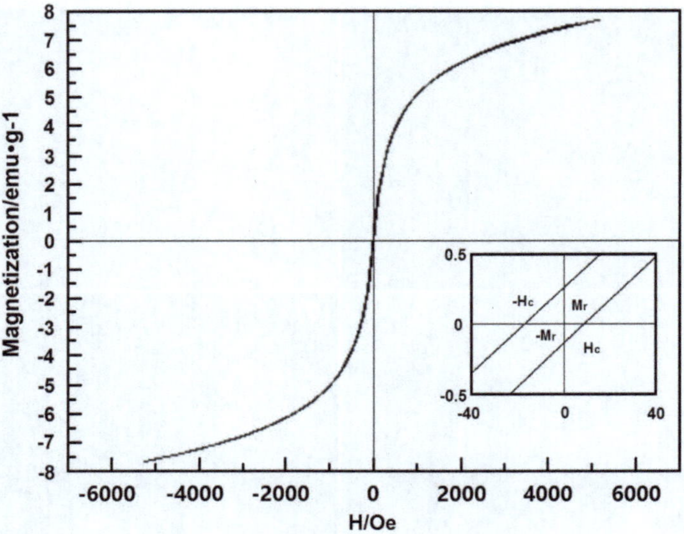

Fig. 4 The magnetic hysteresis loop of coin-like α-Fe_2O_3 nanoparticles at room temperature

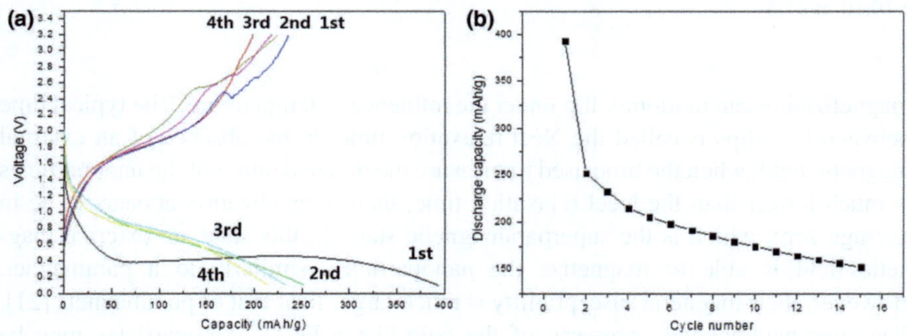

Fig. 5 The electrochemical properties of the as-obtained coin-like α-Fe_2O_3 nanoparticles: **a** the curves of discharge-charge at 0.1 C; **b** the cycle performance

discharge-charge at 0.1 C and the curves of cycle number at 0.1 C are presented in Fig. 5.

As it is shown in Fig. 5a, the first discharge and charge capacity of the as-obtained coin-like α-Fe_2O_3 nanoparticles was 393 and 236 mAh·g^{-1} respectively. However, the cycle performance of the as-obtained coin-like nanoparticles is poor, there is an obvious decrease in the charge-discharge capacity in the following cycles, which indicated that the Li-ion battery anode of coin-like α-Fe_2O_3 nanoparticles have a poor cycle stability during the charge-discharge process. This may be caused by the unstability of the coin-like nanoparticles during the charge-discharge process.

Conclusion

Coin-like α-Fe_2O_3 nanoparticles were successfully synthesized by a hydrothermal reaction in EG/DEG mixed solution with PVP as the surfactant, NaOAc as alkali source. It is found that the ratio of EG and DEG plays a key role in the formation of the coin-like nanoparticles. The as-obtained coin-like α-Fe_2O_3 nanoparticles have a saturation magnetization of 7.687 emu·g^{-1}, and show a superparamagnetic property. The as-obtained coin-like α-Fe_2O_3 nanoparticles have an initial discharge capacity of 393 mAh·g^{-1} and a charge capacity of 236 mAh·g^{-1} when the charge-discharge rate is 0.1 C.

Acknowledgements This work was supported by Guangxi Natural Science Foundation (2014GXNSFBA118247).

References

1. B. Zhang, J. Liu, X.B. Cui, Y.L. Wang, Y. Gao, P. Sun, F.M. Liu, K. Shimanoe, N. Yamazoe, G.Y. Lu, Enhanced gas sensing properties to acetone vapor achieved by α-Fe_2O_3 particles ameliorated with reduced graphene oxide sheets. Sensor. Actuat. B-Chem. **241**, 904–914 (2017)
2. S. Sundar, R. Mariappan, K. Min, S. Piraman, Facile biosurfactant assisted biocompatible α-Fe_2O_3 nanorods and nanospheres synthesis, magneto physicochemical characteristics and their enhanced biomolecules sensing ability. RSC Adv. **6**, 77133–77142 (2016)
3. H. Dong, H. Zhang, Y.L. Xu, C.J. Zhao, Facile synthesis of α-Fe_2O_3 nanoparticles on porous human hair-derived carbon as improved anode materials for lithium ion batteries. J. Power Sour. **300**, 104–111 (2015)
4. Y. Wang, S.R. Wang, Y.Q. Zhao, B.L. Zhu, F.H. Kong, D. Wang, S.H. Wu, W.P. Huang, S.M. Zhang, H_2S sensing characteristics of Pt-doped α-Fe_2O_3 thick film sensors. Sens. Actuat. B-Chem. **12**, 79–84 (2007)
5. H.S. Fei, X.C. Ai, M.Y. Gao, Y.Q. Yang, T.Q. Zhang, J.C. Shen, Luminescence of coated α-Fe_2O_3 nanoparticles. J. Lumin. **66**, 345–348 (1995)
6. J.S. Xu, Y.J. Zhu, α-Fe_2O_3 hierarchically nanostructured mesoporous microspheres: surfactant-free solvothermal combined with heat treatment synthesis, photocatalytic activity and magnetic property. Cryst. Eng. Comm. **14**, 2702–2710 (2012)
7. Q. Fu, S.M. Huang, J. Liu, Chemical vapor depositions of single-walled carbon nanotubes catalyzed by uniform Fe_2O_3 nanoclusters synthesized using diblock copolymer micelles. J. Phys. Chem. B **108**, 6124–6129 (2004)
8. L. Wang, J.M. Ma, L.B. Chen, T.H. Wang, Tailoring the subunits of α-Fe_2O_3 nanoplates for optimizing electrochemical performance. Electrochim. Acta **113**, 194–199 (2013)
9. N.M. Shen, L. Yan, X.J. Jia, Y.W. Li, Inorganic Mineralizer-Assisted Hydrothermal Synthesis of Porous α-Fe_2O_3 Nanowires with High Aspect Ratio. Chin. J. Inorg. Chem. **26**, 846–852 (2010)
10. Q. Gao, J. Luo, X.Y. Wang, M.Q. Ge, Novel hollow α-Fe_2O_3 nanofibers bia electrospinning for dye adsorption. Nanoscale Res. Lett. **10**, 176–184 (2015)
11. T.P. Almeida, M.W. Fay, Y. Zhu, P.D. Brown, Hydrothermal growth mechanism of α-Fe_2O_3 nanorods derived by near in situ analysis. Nanoscale **2**, 2390–2399 (2010)
12. P.S. Veluri, S. Mitra, Enhanced high rate performance of α-Fe_2O_3 nanotubes with alginate binder as a conversion anode. RSC Adv. **3**, 15132–15138 (2013)

13. H. Liang, B. Xu, Z. Wang, Self-assembled 3D flower-like α-Fe_2O_3 microstructures and their superior capability for heavy metal ion removal. Mater. Chem. Phys. **141**, 727–734 (2013)
14. M. Gotic, G. Drazic, S. Music, Hydrothermal synthesis of α-Fe_2O_3 nanorings with the help of divalent metal cations, Mn^{2+}, Cu^{2+}, Zn^{2+} and Ni^{2+}. J. Mol. Struct. **993**, 167–176 (2011)
15. H. Xiao, Y. Xia, W.K. Zhang, H. Huang, Y.P. Gan, X.Y. Tao, Template-free synthesis of hollow α-Fe_2O_3 microcubes for advanced lithium-ion batteries. J. Mater. Chem. A **1**, 2307–2312 (2013)
16. G. Gao, Q. Zhang, X.B. Cheng, P.Y. Qiu, R.J. Sun, T. Yin, D.X. Cui, CNTs in situ attached to α-Fe_2O_3 submicron spheres for enhancing lithium storage capacity. ACS Appl. Mater. Inter. **7**, 340–350 (2015)
17. J.B. Lian, X.C. Duan, J.M. Ma, P. Peng, T. Kim, W.J. Zheng, Hematite (α-Fe_2O_3) with various morphologies: ionic liquid-assisted synthesis, formation mechanism, and properties. ACS Nano **3**, 3749–3761 (2009)
18. M. Subarna, D. Soumen, M. Kalyan, S. Chaudhuri, Synthesis of a α-Fe_2O_3 nanocrystal in its different morphological attributes: growth mechanism, optical and magnetic properties. Nanotechnology **18**, 275608 (2007)
19. J. Dinesh, M. Uzma, M. Pranab, A. Sundaresan, M. Eswaramoorthy, Hollow spheres to nanocups: Tuning the morphology and magnetic properties of single-crystalline α-Fe_2O_3 nanostructures. Angew. Chem. **47**, 7685–7688 (2008)
20. X.L. Zhang, C.H. Sui, J. Gong, Z.M. Su, L.Y. Qu, Preparation and formation mechanism of different α-Fe_2O_3 morphologies from snowflake to paired microplates, dumbbell, and spindle microstructures. J. Phys. Chem. C **111**, 9049–9054 (2007)
21. D. Weller, A. Moser, Thermal Effect Limits in Ultrahigh Density Magnetic Recording. IEEE Trans. Magn. **11**, 4423–4439 (1999)

Preparation and Optical Properties of Self-assembled ZnO Lyche-like Aggregates

Xiaoyun Ye, Wei Li, Yongnan Lin, Shuguang Cai and Yan Lin

Abstract The self-assembled lyche-like aggregates with ZnO nanoparticles were prepared via the hydrothermal method assisted by solvent (isopropanol)-evaporation in the present of the additives. The evaporation of the solvent was demonstrated to play a critical role on the formation of lyche-like aggregates of ZnO. The structures, morphologies and optical properties of as-prepared products were investigated by X-ray diffraction (XRD), Transmission Electron Microscope (TEM), Field Emission Scanning Electron Microscope (FESEM), Fourier Transformation Infrared spectra (FT-IR), Ultraviolet visible absorption spectra (UV-vis) and Photoluminescence spectra (PL). The results show that the increased steric effect among the ZnO nanoparticles constructs the lyche-like aggregates ultimately with the decreases of the solvent in the reaction system. The infrared emissivity value in 8–14 μm of ZnO nanoparticles is 0.922. Contrastively the emissivity value of the lyche-like ZnO aggregates slightly decreases to 0.915. It can be expected as an effective method for the formation of near-spherical structure of metal oxides nanoparticles aggregates. It may be used as an unique pigment in the field of low infrared emissivity coatings or other functional materials.

X. Ye (✉) · W. Li · Y. Lin · S. Cai · Y. Lin
School of Materials Science and Engineering, Fujian University of Technology,
3 Xueyuan Road, Fuzhou 350118, China
e-mail: creekye@163.com

W. Li
e-mail: fjutlw@163.com

Y. Lin
e-mail: lyn932@163.com

S. Cai
e-mail: s.g.cai@163.com

Y. Lin
e-mail: 875917848@qq.com

X. Ye · W. Li · Y. Lin · S. Cai · Y. Lin
Fujian Provincial Key Laboratory of Advanced Materials Processing
and Application, 3 Xueyuan Road, Fuzhou 350118, China

Y. Han (ed.), *Advanced Functional Materials*,
https://doi.org/10.1007/978-981-13-0110-0_9

Keywords ZnO · Self-assembly · Infrared emissivity · Luminescence Hydrothermal

Introduction

ZnO is a kind of well-known and widely used semiconductor material which has the wide and direct band gap of 3.37 eV at room temperature. Nanostructured ZnO with varied and controlled shapes has attracted urgent concern due to the excellent physical and chemical properties as well as key applications in field emission devices, gas sensors and solar cells [1–4]. Various nanostructured ZnO, like nanosheets [5], nanowires [6], and flower-like nanorods [7] are fabricated by different techniques such as hydrothermal treatment, vapor-liquid-solid growth, sol-gel process and so on.

Nowadays, low infrared emissivity materials have been widely studied for the urgent need in the range of civil and military field with the development of infrared detector technologies. The emission performance of a material is the relatively ability of emitting energy from the surface of the material. The value for emission ability evaluation is a ratio of energy radiated from the body material and the black-body radiation respectively [8]. The materials act as pigment in the coating of low infrared emissivity and are the decisive factors on the emission performance of the coatings. Currently, many reports have focused on exploring the effect of the components [9], contents [10], calcination temperature [11] and particle morphologies [12] on the infrared emissivity aiming at the composites. However, few reports has discussed the infrared radiation property of assembled structure unit of one-component material.

In this study, ZnO particles with different morphologies have been generated via a hydrothermal method assisted by solvent evaporation. The optical properties such as PL, UV-vis and infrared emissivity are discussed. The relationships between the microstructures of ZnO nanomaterial and its infrared radiation properties are studied respectively.

Experiments

Firstly, a mixed EG/IPA solvent of 30 mL was prepared with the volume ratio of 2:1 followed by amounts of 2.634 g Zinc acetate and 0.106 g PVP. The mixtures were measured by mechanical stirring for over 2 h until the formation of a homogeneous mixture. The transparent mixture was transferred into a 100 mL of Teflon-sealed autoclave and placed at 140 °C for 2 h. One part of the resulting precipitate was recovered by centrifugation at 4000 rpm and then thoroughly washed by acetone, water and pure ethanol respectively for several times to remove the residue, and then fully dried at 80 °C. The other part of transparent mixture was treated by evaporation to remove the solvent at 140 °C for 2 h. Finally, the dry samples were calcined and exposed in air for 4 h at 550 °C. The samples with non-evaporation and evaporation were signed as Snv and Sv respectively.

The crystal structures of the samples were studied by X-ray diffraction (XRD) on a Bruker D8 advanced diffractometer under Cu K*a* (λ = 1.54056 Å). The size and morphological analysis were performed with NANO NOVA SEM450 field emission scanning electron microscope (FESEM) and JEM-2100 transmission electron microscope (TEM) using the accelerating voltage of 15 kV and 200 respectively. Fourier Transformation Infrared spectra (FT-IR) spectra were obtained using a 6700 spectrometer of American Thermo Nicolet. The photoluminescence spectra (PL) were obtained from a LS55 fluorescence spectrophotometer of Perkin Elmer with excitation wavelength of 325 nm. Ultraviolet visible absorption spectra (UV-vis) were obtained using a UV-2201 spectrophotometer of Japanese SHI-MADZU. The measurement of infrared emissivity values were performed on IRE-I Infrared emissivity measurement instrument of Shanghai Institute of Technology and Physics, China.

Results and Discussion

The morphology and microstructures of ZnO nanoparticles can be demonstrated by TEM and SEM respectively. As presented in Fig. 1, different morphologies of ZnO are clearly exhibited before and after solvent-evaporation. The homogeneous spherical particles of ~70 nm in diameter are generated without solvent evaporation (Fig. 1a, e). Assisted by HRTEM, the microstructures of the ZnO nanoparticles were further illustrated (Fig. 1b). It is confirmed that the interplanar distance of the sample is 0.278 nm, which can be identified as a wurtzite ZnO corresponding to *d* spacing of the (100) crystal plane.

When the obtained sample was treated by evaporation to remove the solvent at certain temperature, self-assembled lyche-like aggregates with ZnO nanoparticles are shown with ~580 nm in diameter (Fig. 1 c, f, g). As can be seen from the images, it is certain that there are no cores in the lyche-like aggregates. Furthermore, the inner parts of the aggregates are partially filled instead of completely hollow (Fig. 1h). Similarly, it is proved by High Resolution Transmission Electron Microscopy (HRTEM) that the lyche-like aggregates is also wurtzite ZnO identified from the fringe spacing data of 0.280 nm, which corresponds to the (100) reflection (Fig. 1d). Obviously, novel nanostructures of lyche-like ZnO aggregates are fabricated after solvent-evaporation, which is diverse from the synthesis of ZnO spherical nanoparticles by traditional hydrothermal reaction in the presence of the same additives (PVP, EG and IPA). Thus, it can be concluded that the evaporation of IPA has great effect on constructing the aggregates. At the initial reaction stage, PVP molecules, used as soft templates, are aggregated to a cluster to reduce the surface energy. The fast solvent-evaporation above the boiling points of the IPA (bp = 82.5 °C) accelerates the nucleation of Zn composite.

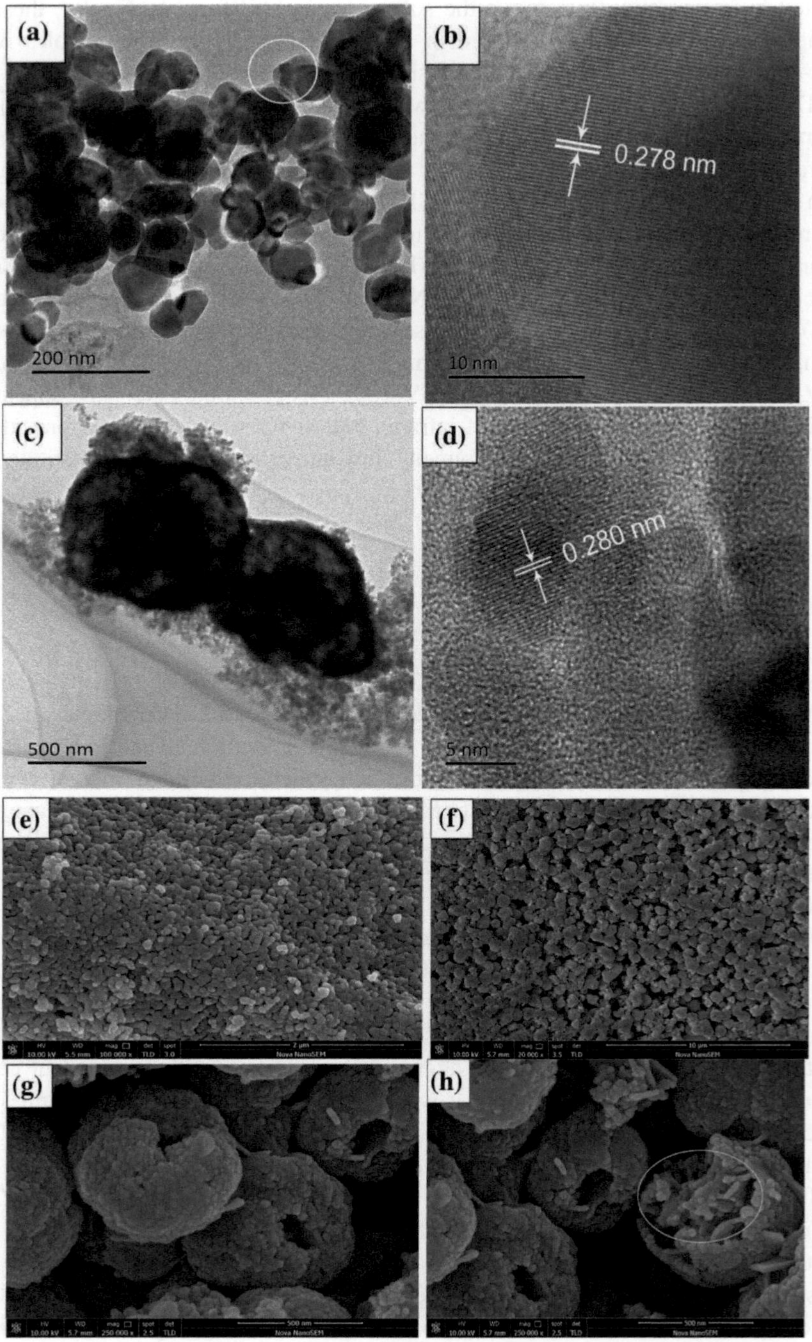

Fig. 1 TEM and SEM images of ZnO particles: Snv (**a**, **b**, **e**), Sv (**c**, **d**, **f**, **g**, **h**)

Simultaneously, the increase of the steric effect is unavoidable among the particles with the decrease of the solvent in the reaction system. Eventually, ZnO nuclei should spontaneously configurate the minimum energy, resulting in the final aggregates in the existence of EG. The schematic drawing of the grown process is shown in Fig. 2.

In FT-IR spectra (Fig. 3), the broad vibrational bands in the range of 3417–3450 cm^{-1} represent the O–H stretching which was expected owing to the existence of hydroxyl groups of ZnO with hydrophilic nature. The peaks at ~ 1620 cm^{-1} can also be seen from the curves. During the synthesis of ZnO, PVP is used as soft template, for whom the stretching vibration of C=O is usually appeared at ~ 1650 cm^{-1}. It is inferred that decomposition of PVP is probably occurred after roasting the samples at 550 °C. In consequence, the peak at ~ 1620 cm^{-1} is corresponded to the stretching vibration of –OH of adsorbed water which is in the form of hydration water in the samples. Simultaneously, the characteristic stretching mode of Zn–O vibration bands are normally seen at low absorption peaks in the range of 426–436 cm^{-1} [13].

The crystallinitity of the ZnO particles with or without solvent-evaporation are determined by XRD as shown in Fig. 4. All the identified diffraction peaks correspond to the planes of wurtzite ZnO at (100), (002), (101), (102), (110), (103), (200), (112), and (201) (JCPDS file no. 36–1451). The sharp peaks in the curves are also obvious, indicating good crystallinity of the products. No peaks of impurities are found in the pattern. The results is consist with the data from HRTEM.

The photoluminescence (PL) spectra under ambient temperature of the ZnO nanoparticles and the self-assembled lyche-like aggregates by using a 325 nm excitation are exhibited in Fig. 5. The corresponding UV-vis spectra are also inserted shown in the same figure. The PL spectra of the two samples indicate that the emission is across from a broad UV to visible along with weak blue green emission. The UV emission is generated from the emission of near band edge. The broad band in the visible region as well as two weak peaks at approximately 480

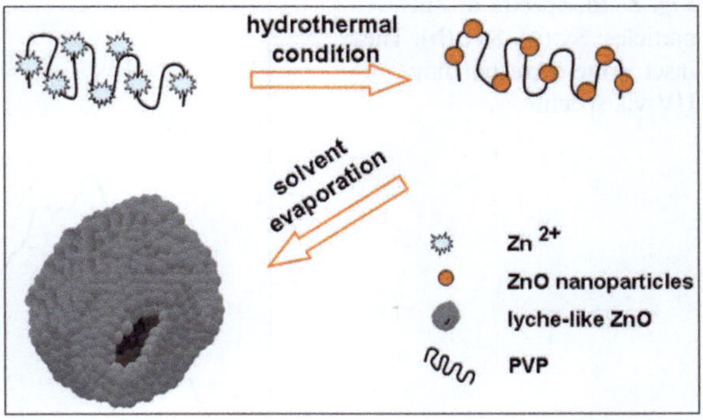

Fig. 2 Schematic illustration of the formation mechanism of lyche-like aggregates with ZnO nanoparticles

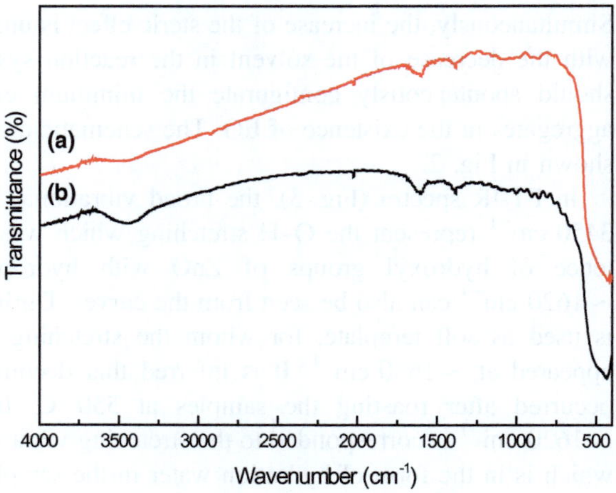

Fig. 3 FT-IR spectra of ZnO particles: Sv (**a**), Snv (**b**)

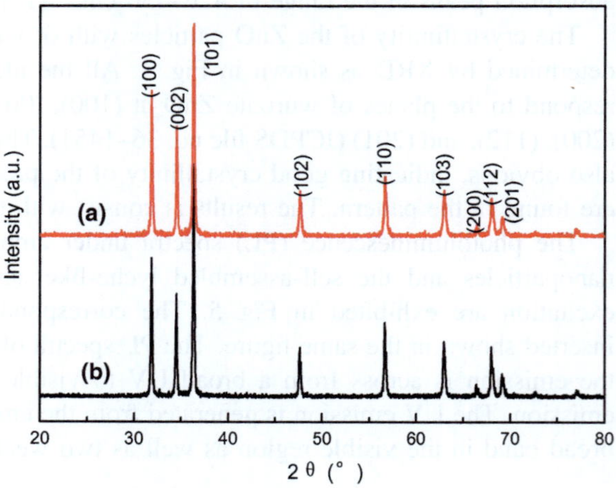

Fig. 4 XRD spectra of ZnO particles: Sv (**a**), Snv (**b**)

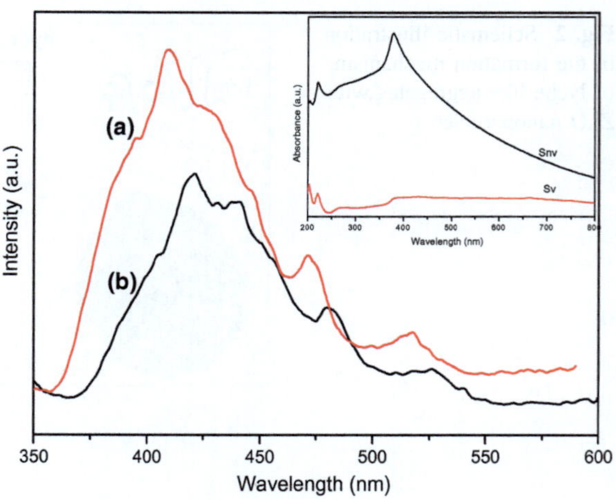

Fig. 5 PL spectra of ZnO particles: Sv (**a**), Snv (**b**). The inset is the corresponding UV-vis spectra

and 520 nm are attributed to lattice detects from internal structure of ZnO such as oxygen vacancies, zinc vacancies, or oxide antisite defect [14]. It was named the deep-level emission (DLE). Normally, a green emission at around 520 nm and a near-yellow emission at around 640 nm compose the deep-seated emission of ZnO. The green and yellow emissions can be attributed to the oxygen vacancies and oxygen interstitials respectively [15]. Thus it is indicated from the broad band that the obtained particles are crystallized imperfectly on account of small size of the nanoparticles and lyche-like aggregates. Moreover, all emission peaks of lyche-like ZnO aggregates have blue-shifted than those of nanoparticles ZnO. Simultaneously, the intensity of the peaks is also increased toward the same trend. It can be explained that the size and morphology of the samples play the key roles. During the self-assembly process of the ZnO aggregates, more zinc vacancies or oxygen vacancies give rise to the relatively high intensity of the emission. In addition, the shift of the wavelength between the nanoparticles and lyche-like aggregates could be ascribed to the competitive development of green and yellow bands [16]. In addition, a small and sharp peak at ~380 nm is observed from the UV-vis spectrum of sample Snv, which indicates the monodispersed nature of the ZnO nanoparticles [17]. It was consistent with the TEM images where highly dispersed nanoparticles were displayed. The absorption peak disappears after the formation of ZnO aggregates, owing to the scattering effect with increased dimension of samples.

The infrared emissivity data of the ZnO nanoparticles and nanoparticles-based aggregates are shown in Table 1. Results show that two types of ZnO particles present relatively high emissivity value of 0.922 and 0.915 for ZnO nanoparticles and lyche-like ZnO respectively. As an essential pigment of infrared absorbing coatings, infrared emissivity value of ZnO nanoparticles is relatively high resulted from decreased size of particles with imperfect crystallization and lattice distortion. By comparison, the infrared emissivity value is slightly decreased after assembly. In theory, the strength of the scattering ability is in accord with proper structure of the samples, which could generate the attenuation effect on the emissivity of infrared light or reduce the lattice distortion during lattice formation, resulting in the decrease of infrared emissivity [11]. In our investigation, lyche-like ZnO experiences the self-assembly process from a state of dispersed particles to a aggregation state, showing a delicate difference in infrared emissivity values probably on account of unaltered scattering ability or lattice distortion. In despite of relatively high infrared emissivity values, ZnO, one of the basic pigments on the infrared emissivity coatings, can be used as base materials to fabricate other functional composites with improved performance.

Table 1 Infrared emissivity values of samples

Samples	Infrared emissivity (εTIR at 8–14 μm)
ZnO nanoparticles	0.922
Lyche-like ZnO	0.915

Conclusion

The self-assembled lyche-like aggregates with ZnO nanoparticles were synthesized by the hydrothermal process in the presence of the additives such as ethylene glycol, poly(vinylpyrrolidone) and isopropanol (IPA). The evaporation of the solvent IPA are proven to be vital to construct lyche-like structure by means of other additives. The broad band in the visible-light region shown in PL spectra of the samples indicates the existence of ZnO lattice detects. The infrared emissivities of self-assembled lyche-like ZnO and ZnO nanoparticles are 0.915 and 0.922 in the range of 8–14 μm respectively, which indicates that the construction of the lyche-like structure has no obvious effect on the infrared emissivity.

Acknowledgements This work was financially supported by the National Natural Science Foundation of China (Grant No. 61605027), the Natural Science Foundation of Fujian Province (Grant No. 2017H6001), the Natural Science Foundation of Fujian Province (Grant No. 2015H6006), the Science and Technology Bureau of Fuzhou (Grant No. 2015-G-57), the Fujian University of Technology Foundation (Grant No. GY-Z15100).

References

1. D. Gedamu, I. Paulowicz, S. Kaps, O. Lupan, S. Wille, G. Haidarschin, Y.K. Mishra, R. Adelung, Rapid fabrication technique for interpenetrated ZnO nanotetrapod networks for fast UV sensors. Adv. Mater. **26**, 1541–1550 (2014)
2. S.N. Bai, S.C. Wu, Synthesis of ZnO nanowires by the hydrothermal method, using sol-gel prepared ZnO seed films. J. Mater. Sci. **22**, 339–344 (2011)
3. J. Kim, J.H. Yun, Y.C. Park, W.A. Anderson, Transparent and crystalline Al-doped ZnO film-embedded heterojunction Si solar cell. Mater. Lett. **75**, 99–101 (2012)
4. X.W. Sun, J.Z. Huang, J.X. Wang, Z. Xu, A ZnO nanorod inorganic/organic heterostructure light-emitting diode emitting at 342 nm. Nano Lett. **8**, 1219–1223 (2008)
5. R. Khan, P. Uthirakumar, K.B. Bae, S.J. Leem, I.H. Lee, Localized surface plasmon enhanced photoluminescence of ZnO nanosheets by Au nanoparticles. Mater. Lett. **163**, 8–11 (2016)
6. A. Reiser, A. Ladenburger, G.M. Prinz, M. Schirra, M. Feneberg, A. Langlois, R. Enchelmaier, Y. Li, R. Sauer, K. Thonke, J. Appl. Phys. **101**, 054307–054319 (2007)
7. B.W. Zhang, W.Y. Fu, H.Y. Li, X.L. Fu, Y. Wang, H. Bala, X.D. Wang, G. Sun, J.L. Cao, Z.Y. Zhang, Synthesis and enhanced gas sensing properties of flower-like ZnO nanorods decorated with discrete CuO nanoparticles. Mater. Lett. **176**, 13–16 (2016)
8. M. Mauer, P. Kalenda, M. Honner, P. Vaclkova, Composite fillers and their influence on emissivity. Phys. Procedia. **44**, 262–269 (2013)
9. Y. Yang, Y.M. Zhou, T.H. Wang, Preparation of optically active polyurethane/TiO_2/MnO_2 multilayered nanorods for low infrared emissivity. Mater. Lett. **133**, 269–273 (2014)
10. W.G. Zhang, G.Y. Xu, R.Y. Ding, J.L. Qiao, K.G. Duan, Microstructure, optimum pigment content and low infrared emissivity of polyurethane/Ag composite coatings. Phys. B **422**, 36–39 (2013)
11. X.Y. Ye, C. Zheng, L.A. Ma, X.F. Huang, Microemulsion-assisted hydrothermal preparation and infrared radiation property of TiO_2 nanomaterials with tunable morphologies and crystal form. Mater. Sci. Semicon. Proc. **31**, 295–301 (2015)

12. Y.J. Wang, Y.M. Zhou, T. Zhang, M. He, X.H. Bu, X.M. Yang, Acetate-intercalated Ni–In layered double hydroxides with low infrared emissivity: synthesis, delamination and restacked to form the multilayer films. Appl. Surf. Sci. **288**, 710–717 (2014)
13. Y. Yang, L. Ren, C. Zhang, S. Huang, T. Liu, A.C.S. Appl, Mater. Interfaces **3**, 2779–2785 (2011)
14. S.M. Zhang, C.L. Yan, H.X. Zhang, G.Y. Lu, Effects of bath temperature on the morphology of ZnO nano-rods and its optical properties. Mater. Lett. **148**, 1–4 (2015)
15. X.L. Wu, G.G. Siu, C.L. Fu, H.C. Ong, Photoluminescence and cathodoluminescence studies of stoichiometric and oxygen-deficient ZnO films. Appl. Phys. Lett. **78**, 2285–2287 (2001)
16. U. Paul, P. Santiago, Controlling the morphology of ZnO nanostructures in a low-temperature hydrothermal process. J. Phys. Chem. B **109**, 15317–15321 (2005)
17. C.L. Pueyo, M.W.E. Berg, A.D. Toni, T. Goes, S. Polarz, Nucleation and growth of ZnO in organic solvents-an in situ study. J. Am. Chem. Soc. **130**, 16601–16610 (2008)

Effects of Microwave Irradiation on Electricity of Graphene Oxide Films

Shaofeng Lin, Sai Yuan, Jianwei Zhang and Dazhi Jiang

Abstract In this study, thermally pre-reduced graphene oxide films were prepared, with respect to different microwave irradiation time and the electrical properties of those films were investigated. The interaction of GO films under intense microwave irradiation was also examined. It was found that the response of GO films to microwave irradiation was not obvious, while significant reduction could be observed for thermally pre-reduced GO films during microwave treatment. The reduction by microwave irradiation of pre-reduced GO films was confirmed through X-ray photoelectron spectroscopy (XPS) and raman spectroscopy. Compared with that of GO films, electrical conductivity of thermally pre-reduced GO films, under microwave irradiation for 5 min, was increased by three orders of magnitude approximately. The results indicated that pre-reduced GO films were effective microwave absorbent, with conversing microwave energy to heat, thus contributing to the restoration of graphite lattice and enhancing connectivity of existing graphite domains.

Keywords Graphene oxide film · Microwave irradiation · Electrical property Reduction

---

S. Lin · S. Yuan · J. Zhang (✉) · D. Jiang
Department of Materials Science and Engineering, National University of Defense Technology, Changsha, Hunan 410073, China
e-mail: jwzhang.nudt@gmail.com

S. Lin
e-mail: linsfbgg@163.com

S. Yuan
e-mail: 351109469@qq.com

D. Jiang
e-mail: jiangdz@nudt.edu.cn

Y. Han (ed.), *Advanced Functional Materials*,
https://doi.org/10.1007/978-981-13-0110-0_10

Introduction

Owing to its superior mechanical properties, excellent thermal and electrical conductivity, graphene has attracted much attention in the fields of drug delivery, electronic devices, and polymer composites [1–3]. To expand its use in above applications, large-scale production of graphene is required. The most economic and effective method for the large-scale production of graphene is chemical oxidation of graphite, conversing graphite oxide to graphene oxide (GO), then reducing GO to graphene [4, 5].

GO has numerous oxygen-containing functional groups: hydroxyl and carboxyl groups are located around the edges, whereas carbonyl and epoxide groups are in the center, resulting in intrinsic insulation of GO for the break of graphite lattice. As a result, the reduction of GO is essential for fabricating highly conductive graphene through removal of oxygen functional groups and recovery of structure defects. Generally, the reduction of GO is carried out by chemical methods, ultraviolet-assisted methods and thermal methods [6–9]. The demerit of these reduction process is that chemical reductants are environmentally unfriendly or time and energy consuming. Besides, it is really hard to remove these oxygen functional groups completely so that reduced GO is still highly disordered and obtains far poorer properties than pristine graphene. Therefore, rapid reduction of GO under mild condition seems extremely difficult.

It has been reported that graphite and amorphous carbon materials are well microwave absorbent, converting microwave energy into heat effectively and efficiently, contributing to reduction of GO [10–14]. After exposing to microwave irradiation for only a few seconds, GO powders could be reduced into pristine graphene with excellent electrical property [15]. In the present work, GO films were thermally pre-reduced, followed by microwave treatment, in order to obtain highly conductive films. The as-prepared reduced GO films were characterized by a variety of complementary methods to confirm the structure and properties. The electrical properties of these films before and after microwave treatment were also investigated.

Experiment

Materials. The GO water dispersions used in the experiment were purchased from C6G6, with size of GO ranging from 10 to 50 μm (as shown in Fig. 1).

Sample Preparation. Firstly, GO dispersion was diluted to 6 mg/ml with deionized water, and further dispersed by ultrasonic cleaner for 20 min. The obtaining dispersion was degassing at room temperature for 20 min. Then the above dispersion was casting on polyethylene terephthalate substrate, forming GO wet film with a uniform thickness of 1.2 mm. The wet film was dried at 30 °C for more than 2 days. Next, the GO film was peeled from the PET substrate and dried at 60 °C in a vacuum oven for a week approximately. GO films were cut into 10 mm × 10

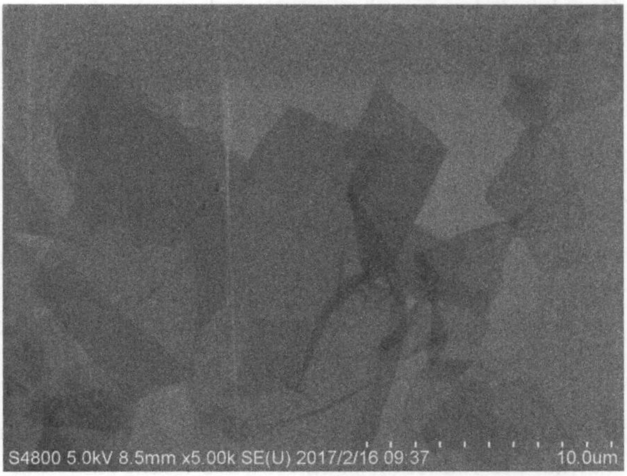

Fig. 1 SEM image of graphene oxide dispersion

mm pieces, followed by being thermally pre-reduced at 150 °C in a quartz vial under the protection of nitrogen for 1, 2, 5 h, denoted as GO-1 h, GO-2 h, GO-5 h respectively. These above films in the quartz vial were exposed to the microwave irradiation (800 W) for 1, 3 and 5 min respectively, denoted as GO-W1, GO-W3, GO-W5 for GO films.

Electrical Conductivity Measurements. The measurement of electrical conductivity of films was carried out on a Keithley 2450 Source Meter, under small currents, in order to avoid significant self-heating. The sample dimensions were 10 mm × 10 mm × 3 μm, and measured by a four-probe method. The sheet resistance (Rs, Ω/sq) of the films was obtained from current–voltage curves.

Characterizations. Scanning electron microscopy (SEM) images were taken on a Hitachi S4800 field emission system. X-ray photoelectron spectroscopy (XPS) was performed with a Escalab 250Xi System. Raman spectroscopy was obtained by LabRAM HR with a wavelength of 532 nm.

Results and Discussion

XPS Analysis. XPS was used to characterize the removal of oxygen-containing functional groups on the surface of GO after microwave treatment. As could be seen in Fig. 2a, there are four different peaks centered at 284.5, 286.7, 287.2, and 288.9 eV, corresponding to C=C in un-oxidized graphite carbon skeleton, C–OH in hydroxyl group, C–O–C in epoxide group, and O–C=O in carboxyl group, respectively. As shown from Fig. 2a–d, the peak intensity of C=C increased with thermal pre-reduction time, whereas that of oxygen-containing functional groups decreased. After microwave treatment, the hydroxyl and epoxy groups, that are the majority of oxygen-containing groups in GO film, are almost all removed and the

C=C bonds become dominant, as shown by the one single peak with a small tail in the higher binding energy region in from Fig. 2e–g.

Meanwhile, the atomic ratio of carbon and oxygen (C/O) improves from 2.13 (GO) to 2.85 for GO-5 h, indicating that graphite lattice of GO was partially restored. Furthermore, compared with that of GO-5 h films, C/O ratio of the films were significantly improved to 8.86, ascribed to efficient deoxygenation of GO and the formation of π–π conjugated structure. Also, C/O ratio of GO-5 h films increased with microwave irradiation time, while that of GO-1 h or GO-2 h films kept approximately unchanged after microwave treatment. The degree of the microwave thermal reduction is similar to the typical chemical reduction process by alkaline or hydroiodic acid [6, 8] (Table 1).

Raman Analysis. Figure 3 shows the Raman spectra of GO films before and after thermal pre-reduction or microwave treatment. Both spectra show the existence of the D band and G band. For GO films, the G band is located at 1594 cm^{-1}, while for that of GO-5 h-W5 films is located at 1589 cm^{-1}, which is close to the value of the pristine graphene, indicating the reduction of GO films during microwave treatment. Meanwhile, the existence of D band shows a red shift for GO-5 h-W5 films at 1349 cm^{-1}, comparing with that of GO films at 1354 cm^{-1}. The intensity

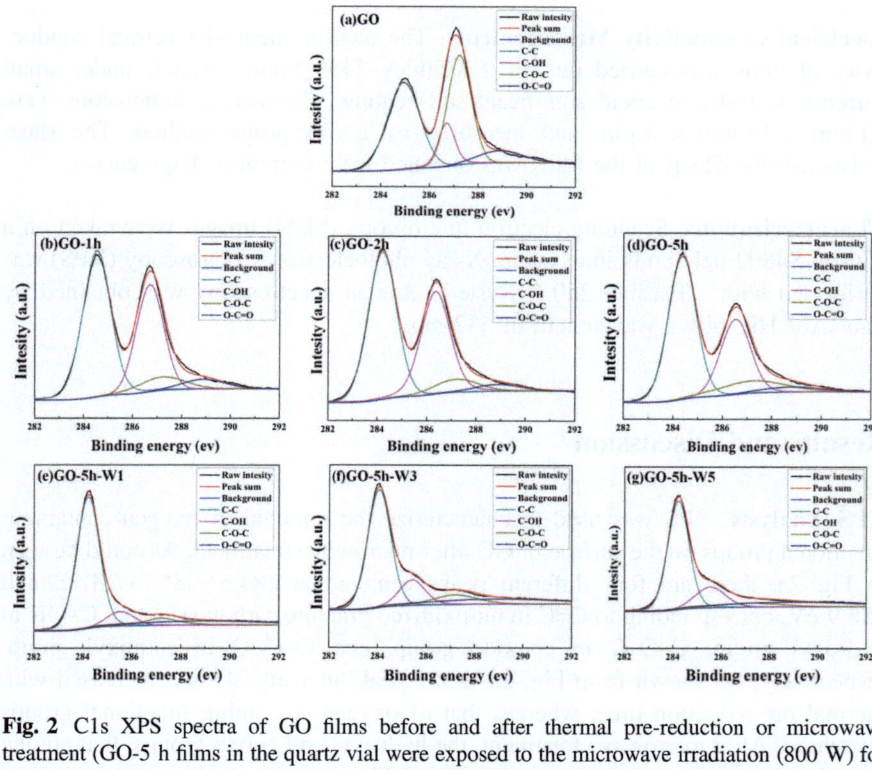

Fig. 2 C1s XPS spectra of GO films before and after thermal pre-reduction or microwave treatment (GO-5 h films in the quartz vial were exposed to the microwave irradiation (800 W) for 1, 3 and 5 min respectively, denoted as GO-5 h-W1, GO-5 h -W3, GO-5 h-W5 for GO-5 h films)

Table 1 XPS results of GO films before and after thermal pre-reduction or microwave treatment

C/O	150 °C—0 h	150 °C—1 h	150 °C—2 h	150 °C—5 h
W-0 min	2.13	2.54	2.69	2.85
W-1 min	–	2.53	2.63	6.75
W-3 min	–	2.40	2.55	8.10
W-5 min	–	2.41	2.65	8.86

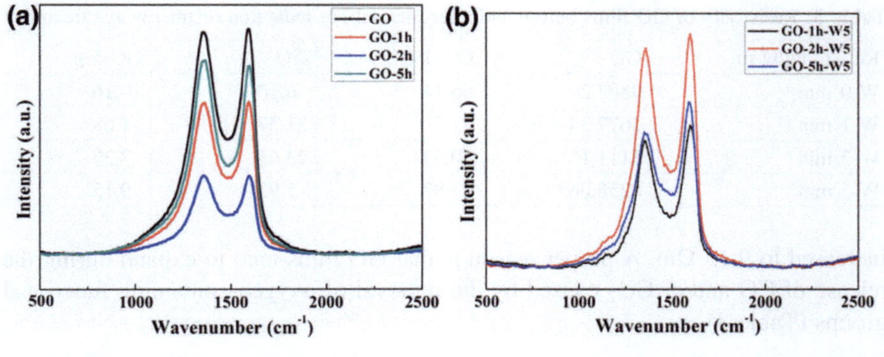

Fig. 3 Raman spectra of GO films before and after thermal pre-reduction or microwave treatment

ratio of the D band and G band, (I_D/I_G) increased slightly from 0.987 for GO films to 1.032 after thermal reduction at 150 °C, owing to a obvious increase of structural defects which is attributable to the evolution of CO and/or CO_2 species (specially from epoxy groups). However, the removal of oxygen functional groups constitutes an important step in the reduction, increasing connectivity among the existing graphite domains that were isolated by oxygen functional groups. After microwave treatment, I_D/I_G decrease significantly to 0.822, ascribed to the restore of C=C bonds of the graphitic lattice and enhancing connectivity of new C=C clusters in the films (Table 2).

Electrical Conductivity Analysis. As expected, GO film was nonconductive, with resistivity values as high as 10^4 Ωm approximately. After microwave treatment, electrical property of GO films kept nearly unchanged. On the contrary, a significant decrease of resistivity about three orders of magnitude could be observed for GO films with thermal pre-reduction, compared with that of pristine GO films. After microwave irradiation, the resistivity of GO-5 h-W1 was decreased to 1.68 Ωm, indicating well recovery of graphite lattice, verified by the XPS and Raman results. Thermal pre-reduction GO films were fine microwave absorber, turning microwave energy into heat. Therefore, oxygen functional groups were removed from GO basal and new C=C clusters were formed, enhancing the connectivity among the existing graphite domains, thus improving electrical property of these films. However, with further microwave treatment, the resistivity of GO-5 h-W5 was

Table 2 Raman results of GO films before and after thermal pre-reduction or microwave treatment

I_D/I_G	150 °C—0 h	150 °C—1 h	150 °C—2 h	150 °C—5 h
W-0 min	0.987	0.997	1.004	1.032
W-1 min	0.826	0.893	0.829	0.871
W-3 min	0.837	0.999	1.026	1.021
W-5 min	0.838	0.896	0.939	0.822

Table 3 Resistivity of GO films before and after thermal pre-reduction or microwave treatment

Resistivity/Ω m	GO	GO-1 h	GO-2 h	GO-5 h
W-0 min	9807.29	66.14	6.10	10.16
W-1 min	7677.34	47.21	21.57	1.68
W-3 min	8111.16	30.53	23.68	3.39
W-5 min	8258.06	7.87	15.92	9.15

increased to 9.15 Ωm. A proper reason is that GO films tend to expand during the release of CO and/or CO_2 caused by the removal of oxygen-containing functional groups (Table 3).

Summary

In summary, GO films with thermal pre-reduction are excellent microwave absorber, owing to partially recovery of graphite lattice and the connection improved among the existing graphite domains. After microwave irradiation for only several minutes, pre-reduced GO films are successfully further reduced, which can be confirmed by XPS and XRD characterizations. Compared with that of GO films, the resistivity of GO-5 h films is significantly decreased to three orders of magnitude. Moreover, with microwave treatment for 1 min, the resistivity of GO-5 h-W1 is declined to 1.68 Ωm, indicating that the removal of oxygen functional groups and formation of more electron transportation channels.

Funding The present work is supported by the Research Project (No. JC 15-01-03) of National University of Defense Technology.

References

1. S. Yongchao, E.T. Samulski, Synthesis of water soluble graphene. Nano Lett. **8**(6), 1679–1682 (2008)
2. P.W. Liu, Z. Jin, G. Kstsukis, L.W. Drahushuk, M.S. Strano, Layered and scrolled nanocomposites with aligned semi-infinite graphene inclusions at the platelet limit. Science **353**(6297), 364–368 (2016)

3. H. Chen, F. Guo, Y. Liu, T. Huang, B. Zheng, N. Ananth, et al., A defect-free principle for advanced graphene cathode of aluminum-ion battery. Adv. Mater. **29** (2017)
4. Z. Liu, Z. Li, Z. Xu, Z. Xia, X. Hu, L. Kou et al., Wet-Spun Continuous Graphene Films. Chem. Mater. **26**(23), 6786–6795 (2014)
5. D.A. Dikin, S. Stankovich, E.J. Zimney, R.D. Piner, G.H.B. Dommett, G. Evmenenko et al., Preparation and characterization of graphene oxide paper. Nature **448**(7152), 457–460 (2007)
6. S. Pei, J. Zhao, J. Du, W. Ren, H.-M. Cheng, Direct reduction of graphene oxide films into highly conductive and flexible graphene films by hydrohalic acids. Carbon **48**(15), 4466–4474 (2010)
7. S. Obata, H. Tanaka, K. Saiki, Electrical and spectroscopic investigations on the reduction mechanism of graphene oxide. Carbon **55**, 126–132 (2013)
8. D. Chen, L. Li, L. Guo, An environment-friendly preparation of reduced graphene oxide nanosheets via amino acid. Nanotechnology **22**(32), 325601 (2011)
9. M. Zhang, L. Huang, J. Chen, C. Li, G. Shi, Ultratough, ultrastrong, and highly conductive graphene films with arbitrary sizes. Adv. Mater. **26**(45), 7588–7592 (2014)
10. W. Chen, L. Yan, P.R. Bangal, Preparation of graphene by the rapid and mild thermal reduction of graphene oxide induced by microwaves. Carbon **48**, 1146–1152 (2010)
11. A.M. Schwenke, S. Hoeppener, U.S. Schuber, Synthesis and modification of carbon nanomaterials utilizing microwave heating. Adv. Mater. **27**, 4113–4141 (2015)
12. C. Wen, N. Zhao, D.W. Zhang et al., Efficient reduction and exfoliation of graphite oxide by sequential chemical reduction and microwave irradiation. Synth. Met. **194**, 71–76 (2014)
13. Y. Zhu, S. Murali, M.D. Stoller et al., Microwave assisted exfoliation and reduction of graphite oxide for ultracapacitors. Carbon **48**, 2106–2122 (2010)
14. M.A. Hassan, V. Abdelsayed, A.E.R.S. Khder et al., Microwave synthesis of graphene sheets supporting metal nanocrystals in aqueous and organic media. J. Mater. Chem. **19**, 3832–3837 (2009)
15. V. Damien, Y. Jieun, K. Jacob et al., High-quality graphene via microwave reduction of solution exfoliated graphene oxide. Science **353**(6306), 1413–1416 (2016)

Microwave Absorbing Performance on Polymer-Derived SiCN Ceramics Doped with Rare Earth Oxides

Yu Liu, Yurun Feng, Hongyu Gong, Xue Guo, Adil Saleem, Xiao Lin, Bingying Xie and Yujun Zhang

Abstract This study investigated the compositions, dielectric properties, magnetic properties and microwave absorbing properties of polymer-derived SiCN ceramics doped with Dy_2O_3, Eu_2O_3, Sm_2O_3, Y_2O_3 and Gd_2O_3 respectively. The complex permittivity and complex permeability were measured at the frequency range of 2–18 GHz. The reflectivity of polymer-derived SiCN ceramics doped with Eu_2O_3 reaches to the lowest value of −35 dB at 16.4 GHz. While the electromagnetic

Y. Liu · Y. Feng · H. Gong (✉) · X. Guo · A. Saleem · X. Lin
B. Xie · Y. Zhang (✉)
Key Laboratory for Liquid-Solid Structural Evolution & Processing of Materials of Ministry of Education, Shandong University, 17923 Jingshi Road, Jinan 250061, Shandong, China
e-mail: gong_hongyu@163.com

Y. Zhang
e-mail: yujunzhangcn@sdu.edu.cn

Y. Liu
e-mail: liuyu8045@126.com

Y. Feng
e-mail: tvxqfenga@163.com

X. Guo
e-mail: gx198816@163.com

A. Saleem
e-mail: adilsaleemkhan@hotmail.com

X. Lin
e-mail: 1158450341@qq.com

B. Xie
e-mail: 1390830827@qq.com

Y. Liu · Y. Feng · H. Gong · X. Guo · A. Saleem · X. Lin
B. Xie · Y. Zhang
Key Laboratory of Special Functional Aggregated Materials, Ministry of Education, Shandong University, 17923 Jingshi Road, Jinan 250061, Shandong, China

Y. Han (ed.), *Advanced Functional Materials*,
https://doi.org/10.1007/978-981-13-0110-0_11

damping coefficient reaches the maximum value of 653 at 16.1 GHz, which clearly demonstrates that by doping Eu_2O_3 the wave absorbing performance of polymer-derived SiCN can be enhanced more pointedly than the other samples.

Keywords Polymer-derived SiCN ceramics · Rare earth oxides
Microwave absorbing properties

Introduction

Due to widespread of the electromagnetic (EM) pollution, extensive studies have been carried out to exploit microwave absorbing materials [1]. In recent years, microwave absorbing materials have widely been used in stealth technology, prevention of electromagnetic wave interference, and protection of the environments from the radiation [2–4]. The traditional absorbing materials such as ferrites, graphite and carbon fiber possess strong absorptive capacity, low cost and wide bandwidth [5]. On the other hand, they could not meet some requirements like thinness, width, lightness as well as strength, which will limit further development [6]. Novel microwave absorbing materials such as SiCN, mullite and carbon nanotubes exhibit better microwave-absorbing properties and can be used to consume the energy of EM waves effectively [7, 8].

According to the lossy characteristic, microwave absorbing materials can be categorized into magnetic and dielectric absorbing materials [9]. Polymer-derived SiCN ceramics as dielectric-loss absorbing materials [10], have been attracting greater attention owing to their excellent mechanical properties, creep resistance, temperature resistance [11, 12] and oxidation resistance. It has been reported that, in order to improve the absorbing performance, the absorbing materials should satisfy the two important requirements as far as possible; First is the impedance matching characteristics (the wave enters the materials without reflection) and second is the attenuation characteristics (absorbing materials have enough loss to the entering wave) [13]. Most of researchers attempted to enhance the microwave absorbing performance by the addition of materials with magnetic lossy capability to polymer-derived SiCN ceramics [14]. Through the synergistic effect of the components, it might be also favorable to adjust the EM parameters to meet the matching condition, and allow the incident waves enter, and then attenuated [15].

As doping of Fe_2O_3 results in the significant increase of dielectric loss and magnetic loss, this paper used the polymer-derived SiCN ceramics doped with nano-sized Fe_2O_3 [hereafter referred to as polymer-derived SiCN (Fe) ceramics] as matrix [16–18]. Rare-earth elements take on unpaired 4f electrons shielded by the outermost electrons and the strong spin-orbit coupling of their angular momentum improves the overall electrical and magnetic properties [19]. Some efforts were made to dope rare earth oxides into other absorbing materials, based on the special electronic structure and EM properties of rare earth elements to adjust and optimize EM parameters and enhance microwave absorbing properties [20].

In this study, Dy_2O_3, Eu_2O_3, Sm_2O_3, Y_2O_3 and Gd_2O_3 were doped in the polymer-derived SiCN ceramics matrix. The phase composition, microstructure dielectric properties and magnetic properties were investigated. The effect of different kinds of rare earth oxides on the properties of polymer-derived SiCN ceramics was also discussed.

Experimental

The polymer-derived SiCN powders were synthesized by the polymer-derived method from a commercially available liquid-phase polysilazane, which was mixed with α-methacrylic acid and dicumyl peroxide under nitrogenous atmosphere. Then this solution was stirred until it became a transparent liquid. The mixed solution was cross-linked for 2 h in a N_2 atmosphere. The cross-linked polymer was then ball milled and passed through a 100 mesh sieve. Nano-sized ferric oxide powders and rare earth oxide powders (Dy_2O_3, Eu_2O_3, Sm_2O_3, Y_2O_3 and Gd_2O_3) were both doped into the as-received powders above with the mass ratio 5% and mixed evenly. These mixed powders were then pressed into green bodies under the pressure of 20 MPa for further pyrolysis. The final pyrolysis was carried out at the chosen temperature between 1000 and 1400 °C in a purity N_2 environment to yield samples. The prepared samples are designated as sample 1–5 doped with Dy_2O_3, Eu_2O_3, Sm_2O_3, Y_2O_3 and Gd_2O_3 respectively.

The X-ray diffraction was performed on an X-ray diffraction (XRD, D/MAX-Ultima IV, Rigaku, Japan) to study the phase composition of samples. The microstructure was investigated by scanning electron microscope (SEM, EVO18, Zeiss, German). The EM parameters of the samples can be measured by coaxial transmission and reflection method with the vector network analyzer (5244A, Agilent-N, USA) having measuring range 2–18 GHz.

Results and Discussion

Phase Compositions. The XRD spectrums of samples are shown in Fig. 1. The main diffraction peaks of the SiCN (Fe) ceramics belong to α-Fe and the main crystal plane is (110). All samples exhibit the crystal structures of α-Fe even though the rare earth oxides are different. The new crystal phases for rare earth metals are detected in the corresponding XRD patterns respectively, implying that the rare earth metals are successfully obtained through reduction reactions. However the intensity of these peaks is weak and no obvious peaks can be detected.

Microstructures. SEM analysis was used to study the surface morphology and the images of SiCN (Fe) doped with Dy_2O_3, Eu_2O_3, Gd_2O_3, Y_2O_3 and Sm_2O_3 are shown in Fig. 2a, b, c, d, e respectively. The results reveal that almost no obvious grain can be observed. According to the XRD analysis (Fig. 1) the samples are

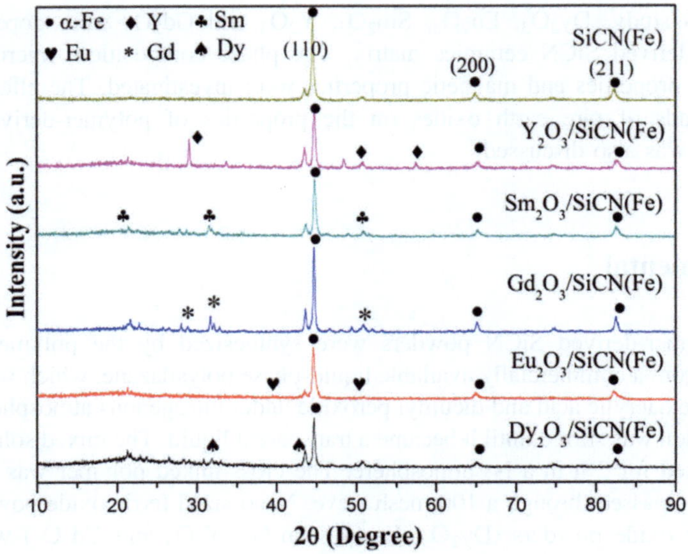

Fig. 1 XRD of samples doped with different rare earth oxides

basically amorphous and only have a few grains confirmed to be Fe phase. It can be seen from energy spectrum Fig. 2f, g, the obtained samples contain not only C, O, Si and Fe, but also the element Y and Dy respectively, which might originate from the Y_2O_3 and Dy_2O_3 doped. This result is also consistent with the result of XRD analysis that the structure of matrix is not affected by rare earth oxides doping and still amorphous without a large area of crystallization.

EM Properties. The relative complex permittivity is the key parameter to characterize dielectric properties. The real part of permittivity (ε') represents the polarization relaxation and the imaginary part of permittivity (ε'') is relative to the ability of dielectric loss. Thus the dielectric loss ($\tan\delta_\varepsilon = \varepsilon''/\varepsilon'$) is a way to predict microwave absorption properties of the materials [21]. Figure 3a, b, c show the variations of the real and the imaginary parts of permittivity (ε' and ε'') and dielectric loss tangent $\tan\varepsilon$ with frequency in the range of 2–18 GHz. It can be observed that both the ε' and ε'' of the rare earth-doped SiCN (Fe) ceramics are mostly larger than those of SiCN ceramics. Among them the sample 1 doped with Eu_2O_3 changes significantly. The real part of complex permittivity first increases to the maximum at about 15 GHz and then decrease to a minimum at about 16 GHz. There are two peaks in the ε'' spectra of the Eu_2O_3-doped SiCN ceramics (near 12.5 GHz and near 16 GHz), which might be derived from ferromagnetic resonances. As Fig. 3c shows, the $\tan\varepsilon$ spectra of sample 2, 3, 4 and 5 show insignificant variation in the whole frequency range. It can be seen from the comparison of the samples that, the value of sample 1 (doped with Eu_2O_3) declines within a small range from 2 to 12 GHz, and then has a significant fluctuation, reaching the maximum value 8 at 16 GHz. The dielectric loss angle tangent of other

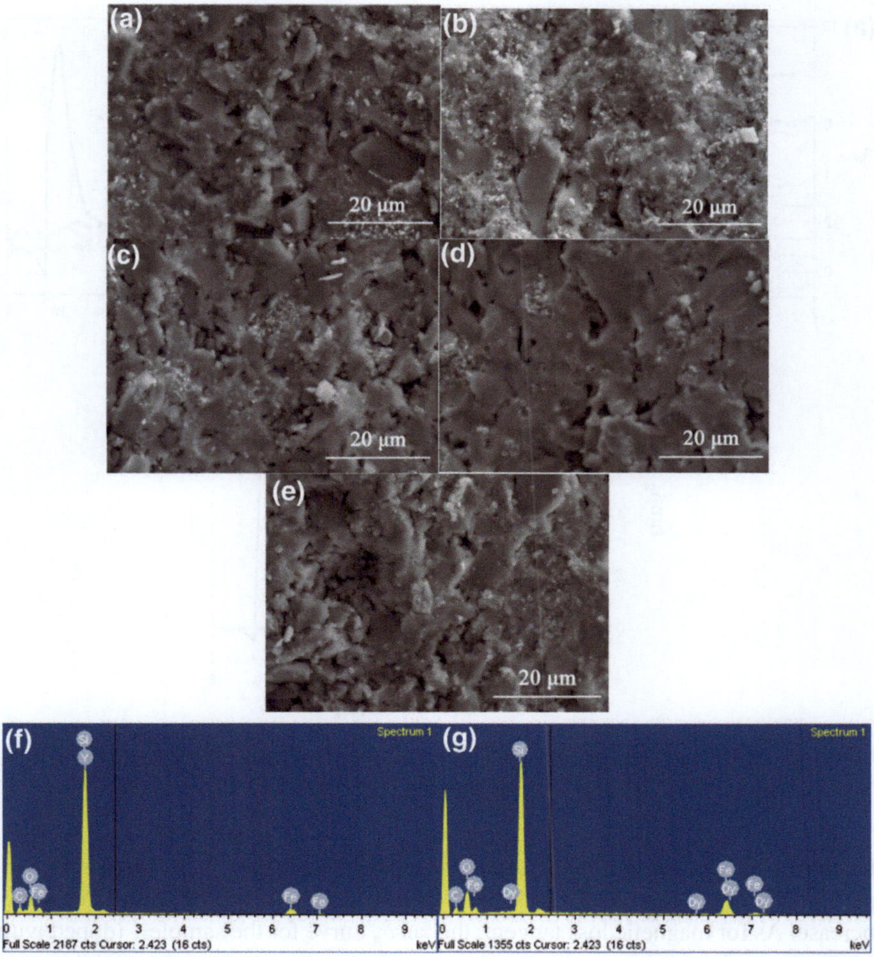

Fig. 2 The fracture morphology and energy spectrum diagram of SiCN doped different rare earth oxides pyrolysed at 1100 °C **a** Dy_2O_3/SiCN. **b** Eu_2O_3/SiCN. **c** Y_2O_3/SiCN. **d** Gd_2O_3/SiCN. **e** Sm_2O_3/SiCN. **f** EDS of Y_2O_3/SiCN. **g** EDS of Dy_2O_3/SiCN

samples keeps approximately maintained within a small range with the increasing frequency in 2–18 GHz.

Figure 4a, b, c show the frequency dependence of the real parts of permeability (μ'), the imaginary parts of permeability (μ'') and magnetic loss tangent ($\tan \delta_\mu$) of all the samples in the frequency range of 2–18 GHz. The μ' and μ'' of SiCN ceramics exhibit a trend of decrease. The Eu_2O_3-doped sample shows the visible frequency dependence of the real and imaginary parts of permeability. The μ' and μ'' values possess slight fluctuation amplitude in frequency 2–16 GHz and then increase dramatically in the following frequency band. The μ' of the samples 2, 3, 4 and 5 which were doped with Y_2O_3, Dy_2O_3, Sm_2O_3 and Gd_2O_3 exhibit frequency

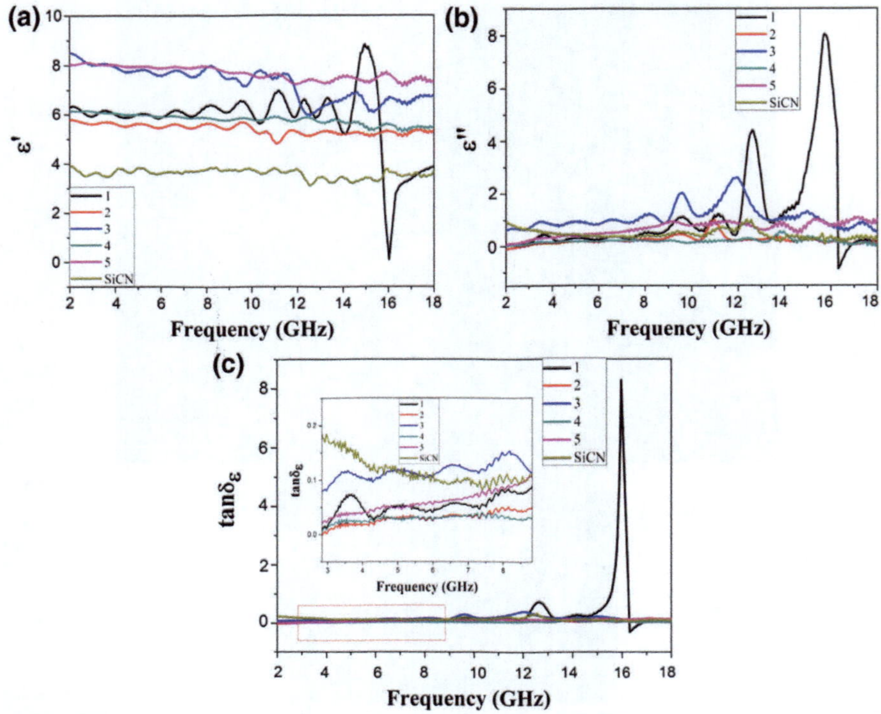

Fig. 3 Influence of rare earth oxides doping on the dielectric constant of SiCN ceramics **a** ε′. **b** ε″. **c** tan δ_ε. 1 Eu_2O_3/SiCN. 2 Y_2O_3/SiCN. 3 Dy_2O_3/SiCN. 4 Sm_2O_3/SiCN. 5 Gd_2O_3/SiCN

dependent effect, and the magnitudes of quantities decrease with increasing frequency. The μ'' of samples 2, 3, 4 and 5 tend to decrease slightly as the frequency increase. As for magnetic loss tangent, the tan δ_μ curve for the sample 1 (doped with Eu_2O_3) is nearly consistent with the changes of other samples and then exhibits two peaks with a maximum value of 13 at 17 GHz. The values of other samples along with their variation are rather similar, which illustrates their similar magnetic loss. The enhancement of the magnetic loss of Eu_2O_3-doped SiCN ceramics is most likely due to Eu nanoparticles distributed in the matrix, which function as magnetic absorbing materials. The negative ε″ and μ″ values within low frequency ranges might be caused by the measurement errors [22].

The excellent microwave absorbing properties result from the efficient complementation between permittivity and permeability, while both magnetic loss and dielectric loss contribute to the microwave absorption [6]. Furthermore, the collaboration between the dielectric loss and magnetic loss could much enhance the absorbing performance. Combining with above analysis, the polymer-derived SiCN doped with Eu_2O_3 has optimized microwave absorption properties.

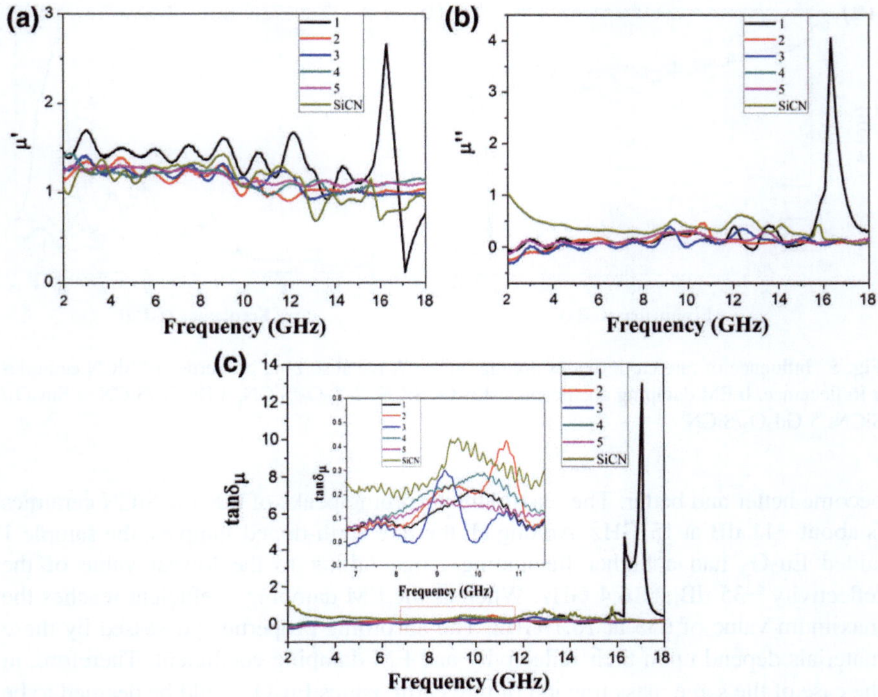

Fig. 4 Influence of rare earth oxides doping on the magnetic properties of SiCN ceramics **a** μ'. **b** μ''. **c** tan δ_μ. 1 Eu_2O_3/SiCN. 2 Y_2O_3/SiCN. 3 Dy_2O_3/SiCN. 4 Sm_2O_3/SiCN. 5 Gd_2O_3/SiCN

To reveal the microwave absorption properties of the samples, the reflectivity (R) and EM damping coefficient (α) of were calculated as the following equations: [23]

$$R_L = 20\log_0|(Z_{in} - 1)/(Z_{in} + 1)|. \tag{1}$$

$$Z_{in} = \sqrt{\mu_r/\varepsilon_r}\tan h\left[j(2\pi fd/c)\sqrt{\mu_r\varepsilon_r}\right]. \tag{2}$$

$$\alpha = (\pi f/c)\sqrt{2\mu'\varepsilon'}\sqrt{\tan\delta_\varepsilon\tan\delta_\mu - 1 + \sqrt{(\tan^2\delta_\varepsilon + 1)(\tan^2\delta_\mu + 1)}}. \tag{3}$$

where Z_{in} is the input characteristic impedance, *f* is the frequency, *c* is the velocity of light, and *d* is the thickness. Figure 5 shows the reflectivity (R) curves and the damping coefficient (α) curves for samples. To obtain materials with good EM absorption properties, the EM wave must penetrate into absorbing materials without reflection as far as possible and at the same time should be attenuated by materials as much as possible [24]. It is clearly obtained that, as the frequency increased from 2 to 18 GHz the reflectivity shows a decreasing trend. While the EM attenuation coefficient increases, indicating the microwave absorption properties of the samples

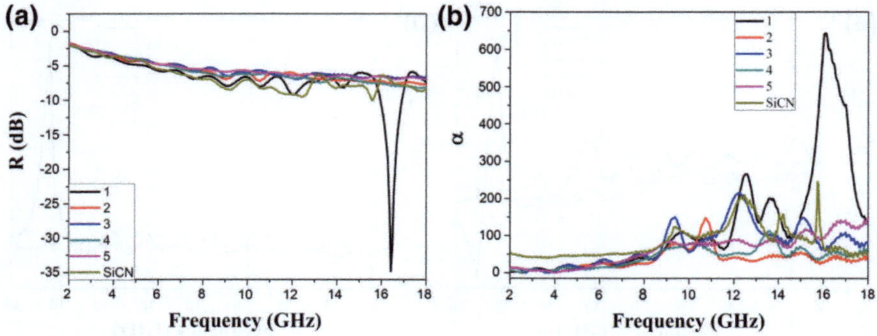

Fig. 5 Influence of rare earth oxides doping on the Wave absorbing properties of SiCN ceramics **a** Reflectance. **b** EM damping coefficient. 1 Eu_2O_3/SiCN. 2 Y_2O_3/SiCN. 3 Dy_2O_3/SiCN. 4 Sm_2O_3/SiCN. 5 Gd_2O_3/SiCN

become better and better. The maximum absorbing peaks of the raw SiCN ceramics is about −11 dB at 15 GHz. Among all the rare earth-doped samples, the sample 1 added Eu_2O_3 had a higher fluctuating range falling to the lowest value of the reflectivity −35 dB at 16.4 GHz. Whereas the EM damping coefficient reaches the maximum value of 653 at 16.1 GHz. The absorbing properties possessed by these materials depend upon their reflectivity and EM damping coefficient. Therefore, in the case of the same mass fraction of rare earth oxides Eu_2O_3 could be deemed to be a worthy dopant to increase the absorption property of polymer-derived SiCN ceramics, which is consistent with the analysis of EM properties above.

Conclusions

The polymer-derived SiCN powders were successfully synthesized by the polymer-derived method and were utilized to prepare the polymer-derived SiCN ceramics. Different kinds of rare earth oxides were doped into the matrix of polymer-derived SiCN (Fe) ceramics. The as-received samples obtained the rare earth oxides doped respectively in addition to Fe phase, and the amorphous phase was dominated. The analysis shows that in the case of the same mass fraction of rare earth oxides, the absorbing performance of polymer-derived SiCN ceramics is optimal when Eu_2O_3 is added. The value of reflectivity at 16.4 GHz is lower than that of other samples, reaching −35 dB, and the EM damping coefficient reaches the maximum value of 653 at 16.1 GHz. Generally, the microwave-absorbing performance of the polymer derived SiCN ceramics doped with rare earth oxides is quite good, which have potential applications in the microwave absorbing field.

Acknowledgements The authors would like to extend their sincere appreciation to the Deanship of The National Nature Science Foundation of China (No. 51572154) and Natural Science Foundation of Shandong Province (2015ZRE27348) for its funding.

References

1. K. Khan, S. Rehman, Microwave absorbance properties of zirconium–manganese substituted cobalt nanoferrite as electromagnetic (EM) wave absorbers. J. Mater. Res. Bull. **50**, 454–461 (2014)
2. Q. Qiu, Y.Q. Zhang, X. Zhang, Research progress of microwave absorption materials. J. Electron. Compon. Mater. **28**, 78–81 (2009)
3. X.F. Liu, L.T. Zhang, X.W. Yin, F. Ye, Y.S. Liu, L.F. Cheng, The microstructure of SiCN ceramics and their excellent electromagnetic wave absorbing properties. J. Ceram. Int. **41**, 11372–11378 (2015)
4. H.Y. Lin, H. Zhu, H.F. Guo, L.F. Yu, Microwave-absorbing properties of Co-filled carbon nanotubes. J. Mater. Res. Bull. **43**, 2697–2702 (2008)
5. L. Shi, Research progress on wave absorbing materials. J. Packag. J. **5**, 25–29 (2013)
6. H.B. Yang, T. Ye, Y. Lin, J.F. Zhu, F. Wang, Microwave absorbing properties of the ferrite composites based on graphene. J. Alloys Compd. **683**, 567–574 (2016)
7. Y.N. Fan, R.Z. Gong, Research status of electromagnetic wave absorbing ceramics materials. J. Ceram. **31**, 538–541 (2010)
8. J. Zhang, W.Y. Zhang, Z.P. Xi, Development of stealth radar wave absorbing materials. J. Rare Met. Mater. Eng. **37**, 504–508 (2008)
9. Y. Qing, W. Zhou, F. Luo, D.M. Zhu, Microwave electromagnetic properties of carbonyl iron particles and Si/C/N nano-powder filled epoxy-silicone coating. J. Physica B: Condens. Matter **405**, 1181–1184 (2010)
10. H.J. Seifrt, J.Q. Peng, H.L. Lukas, F. Aldinger, Phase equilibria and thermal analysis of Si–C–N ceramics. J. Alloys Compd. **320**, 241–261 (2001)
11. Q. Li, X.W. Yin, W.Y. Duan, B.L. Hao, L. Kong, X.M. Liu, Dielectric and microwave absorption properties of polymer derived SiCN ceramics annealed in N_2 atmosphere. J. Eur. Ceram. Soc. **34**, 589–598 (2014)
12. V.L. Nguyen, E. Zera, A. Perolo, R. Campostrini, W.J. Li, G.D. Soraru, Synthesis and characterization of polymer-derived SiCN aerogel. J. Eur. Ceram. Soc. **35**, 3295–3302 (2015)
13. H.Y. Wang, D.M. Zhu, X.F. Wang, F. Luo, Influence of silicon carbide fiber (SiCf) type on the electromagnetic microwave absorbing properties of SiCf/epoxy composites. J. Compos.: Part A **93**, 10–17 (2017)
14. F.J. Meng, M.Y. Ru, A.X. Liu, Z.L. Liu, X. Meng, Application of polysilazane-derived Si–C–N ceramics in radar absorption. J. Polym. Mater. Sci. Eng. **20**, 144–147 (2004)
15. Y. Liu, F. Luo, J.B. Su, W.C. Zhou, D.M. Zhu, Electromagnetic and microwave absorption properties of the Nickel/ Ti_3SiC_2 hybrid powders in X-band. J. Magn. Magn. Mater. **365**, 126–131 (2014)
16. H. Jiao, F. Luo, W.C. Zhou, Dielectric properties of Si/C/N nano powders. J. Inorg. Mater. **3**, 33 (2002)
17. Y.M. Li, Z.M. Zheng, C.Y. Reng, Z.J. Zhang, W. Gao, S.Y. Yang, Z.M. Xie, Preparation of Si–C–N magnetic ceramics from iron containing polysilazane. J. Appl. Organomet Chem. **17**, 120–126 (2003)
18. S.I. Andronenko, I. Stiharu, D. Menard, C. Lacroix, S.K. Misra, EPR/FMR, FTIR, X-ray and Raman investigations of Fe-doped SiCN ceramics. J. Appl. Magn. Reson. **38**, 385–402 (2010)
19. L.W. Deng, L. Ding, K.S. Zhou, S.X. Huang, Z.W. Hu, B.C. Yang, Electromagnetic properties and microwave absorption of W-type hexagonal ferrites doped with La^{3+}. J. Magn. Magn. Mater. **323**, 1895–1898 (2011)
20. J. Xu, W.H. Xiong, A.X. Zeng, C.F. Wang, Application of rare earth in the electromagnetic wave absorbing materials. J. Chin. Rare Earths **25**, 68–73 (2004)
21. F. Ye, L.T. Zhang, X.W. Yin, Y.J. Zhang, L. Kong, Q. Li, Y.S. Liu, L.F. Cheng, Dielectric and EMW absorbing properties of PDCs-SiBCN annealed at different temperatures. J. Eur. Ceram. Soc. **33**, 1469–1477 (2013)

22. D. Li, Y. Feng, D.S. Pan, L.W. Jiang, Z.M. Dai, S.J. Li, Y. Wang, J. He, W. Liu, Z.D. Zhang, Negative imaginary parts of complex permeability and microwave absorption performance of core double-shelled FeCo/C/$Fe_{2.5}Cr_{0.5}Se_4$ nanocomposites. J. RSC Adv. **6**, 73020–73027 (2016)
23. B. Zhao, G. Shao, B.B. Fan, W.Y. Zhao, Y.J. Xie, R. Zhang, Facile preparation and enhanced microwave absorption properties of core–shell composite spheres composited of Ni cores and TiO_2 shells. J. Phys. Chem. Chem. Phys. **17**, 8802 (2015)
24. Q. Li, X.W. Yin, W.Y. Duan, L.F. Cheng, L.T. Zhang, Improved dielectric properties of PDCs-SiCN by in-situ fabricated nano-structured carbons. J. Eur. Ceram. Soc. **37**, 1243–1251 (2017)

Preparation and Properties of Pressureless-Sintered Porous Si_3N_4 Ceramics

Xiao Lin, Jinming Shi, Hongyu Gong, Yujun Zhang, Yurun Feng, Xue Guo, Bingying Xie, Yu Liu and Jiajun Wei

Abstract In this paper, the effect of benzoic acid on the porosity, dielectrical and mechanical properties of porous Si_3N_4 ceramics prepared by pressureless sintering was investigated. The results showed that the addition of benzoic acid improved the porosity and dielectrical properties of Si_3N_4 remarkably by pressureless sintering at 1700 °C for 60 min. The porosity of sample reached 40.79% with 50 wt% benzoic acid addition and the reflectivity of the sample reached −45 dB at around 12 GHz.

X. Lin · H. Gong (✉) · Y. Zhang · Y. Feng · X. Guo · B. Xie
Y. Liu · J. Wei
Key Laboratory for Liquid-Solid Structural Evolution & Processing
of Materials of Ministry of Education, Shandong University, Jinan 250061, China
e-mail: hygong@sdu.edu.cn

X. Lin
e-mail: linxiao050994@sina.com

Y. Zhang
e-mail: yujunzhangcn@sdu.edu.cn

Y. Feng
e-mail: 765132878@qq.com

X. Guo
e-mail: 389702129@qq.com

B. Xie
e-mail: 1390830827@qq.com

Y. Liu
e-mail: 804512856@qq.com

J. Wei
e-mail: 1558724515@qq.com

X. Lin · H. Gong · Y. Zhang · Y. Feng · X. Guo · B. Xie · Y. Liu · J. Wei
Key Laboratory of Special Functional Aggregated Materials, Ministry of Education,
Shandong University, Jinan 250061, China

J. Shi
College of Engineering, Peking University, Beijing 100871, China
e-mail: 413735337@qq.com

Y. Han (ed.), *Advanced Functional Materials*,
https://doi.org/10.1007/978-981-13-0110-0_12

Keywords Si_3N_4 · Pressureless sintering · Benzoic acid · Dielectric properties

Introduction

Si_3N_4 ceramics attract a lot of attentions for its outstanding properties such as low dielectric constant, high hardness, wear resistance, corrosion resistant and thermal-shock resistance under room temperatures and elevated temperatures [1, 2]. Porous Si_3N_4 ceramics combine the advantages of porous ceramics and Si_3N_4 ceramics with low density, high specific surface area, heat resistance, high fracture toughness and high strength [3]. Porous Si_3N_4 ceramics have great application potential in wave-transmitting materials and wave absorbing matrix [4–6] for its excellent dielectric properties, and the changeable dielectric constant. However, porous Si_3N_4 ceramics with good properties are hard to prepared successfully because of its porosity and high sintering temperature. We decided to use pore forming agent [7] and pressureless sintering to prepare porous Si_3N_4 ceramics, this access has simple process, great controllability and the convenience for large-scale production.

Zhang et al. [8] has prepared porous Si_3N_4 ceramics with high porosity and high strength by adding nano carbon powder as pore forming agent to Si_3N_4 matrix, whose porosity was over 40%, bending strength was more than 100 MPa with addition of 5 wt% carbon. Wang et al. [9] has synthesized BN/Si_3N_4 composite wave-transmitting ceramics by pressureless sintering with porosity 40.8%, dielectric constant and dielectric loss 4.0 and 3.3×10^{-3} respectively with addition of 5 wt% PMMA.

In this study, porous Si_3N_4 ceramics was prepared by pressureless sintering with α-Si_3N_4 powders, Y_2O_3 and Al_2O_3 as sintering aids and benzoic acid as pore forming. The effect of benzoic acid on the dielectrical and mechanical properties of porous Si_3N_4 ceramics was investigated.

Experimental Procedure

α-Si_3N_4 powder (0.5 μm particle size, purity >93%, Beijing Ziguang Co. PR China), Y_2O_3 (purity >99%, Hangzhou Wanjing Co. PR China), Al_2O_3 (purity >99%, Shanghai Macklin Co. PR China) and benzoic acid (purity >99%, Guoyao Co. PR China) were used as received. α-Si_3N_4 (90 wt%), Al_2O_3 (1.5 wt%), Y_2O_3 (8.5 wt%), PVP (1 wt%) and benzoic acid (0–50 wt%) were ball milled in ethanol for 15 min with Si_3N_4 balls as media. After milling, the mixed slurries were dried in a drying oven and then pass through a 60-mesh screen. The green specimens were formed by dry pressing and then removed pore forming agent under 200 °C. These porous compacts were sintered at 1700 °C for 60 min under a nitrogen atmosphere by pressureless sintering method.

We used X-Ray diffraction (XRD, D/max-RC) to identify the phase composition. The morphology and the microstructures were characterized by scanning electron microscope (SEM, JSM-6380LA). The porosity of the sintered body was calculated by the Archimedes' method. Dielectric constant and dielectric loss were tested through coaxial transmission reflection method (Agilent-N 5244A VNA), the band range is 2–18 GHz. Flexural strength and fracture toughness were measured by three-point bending test and SENB technique respectively.

Results and Discussion

Figure 1 shows the XRD spectra of porous Si_3N_4 ceramics by adding different contents of benzoic acid. The main crystal phases of all samples are β-Si_3N_4 phase and no α-Si_3N_4 phase, indicated that the crystalline phase transition of Si_3N_4 was complete during the sintering process, no matter what amount of pore forming add in the ceramics. Besides β-Si_3N_4, there is $Y_5Si_3O_{12}N$ in the specimen. It is due to Y–Si–Al–O liquid phase [10] was formed at around 1200 °C by the reaction with sintering aids and Si_3N_4 and SiO_2 on its surface, then α-Si_3N_4 separated out β-Si_3N_4 grain by dissolution precipitation. The liquid phase promoted the phase transition and densification process of Si_3N_4 at high temperatures, and the liquid phase finally crystallized into $Y_5Si_3O_{12}N$ phase.

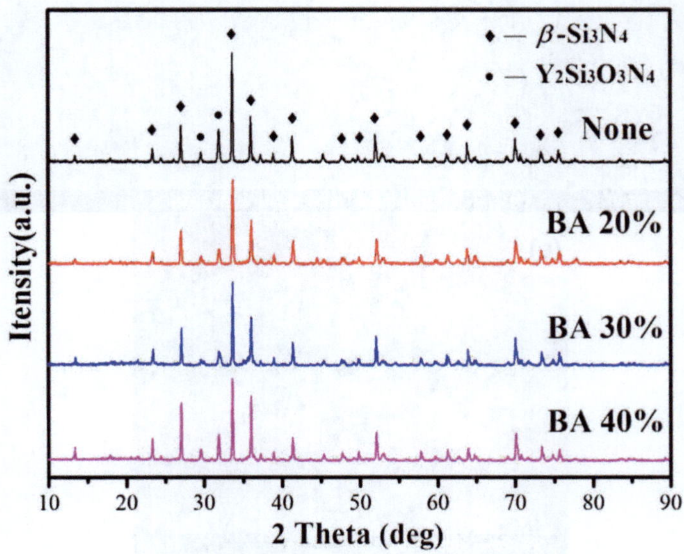

Fig. 1 XRD patterns of porous Si_3N_4 ceramics adding different contents of pore-forming agent (benzoic acid)

Figure 2 presents the SEM images of porous Si_3N_4 on fracture surfaces adding different contents of benzoic acid. We can see the equiaxed α-Si_3N_4 has been transformed into columnar β-Si_3N_4 during the sintering process. Figure 2a is Si_3N_4 without pore formers, so there are little pores inside and the grains are small and arranged closely. The benzoic acid initial particle size is small, the decomposition temperature is low, so the pore sizes are small, and shrinkage is large at high temperatures during the densification process. Therefore, the sintered body has closely spaced grains, small pore sizes and uniform distribution. As the BA content

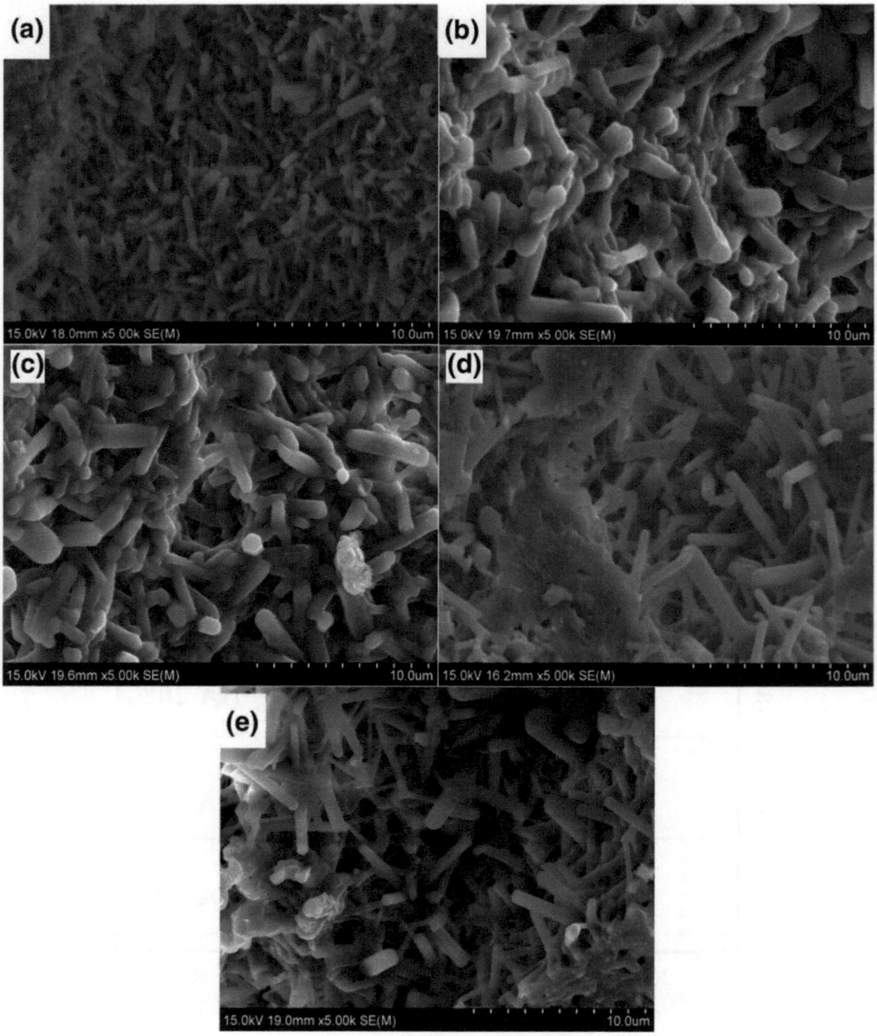

Fig. 2 SEM images of porous Si_3N_4 on fracture surfaces adding different contents of benzoic acid: **a** 0% **b** 20%; **c** 30%; **d** 40%; **e** 50%

was increased from 20 to 50 wt%, the pores sizes and quantity in the samples was increased. But the addition amount has no obvious size or size distribution influence on pores of the porous Si_3N_4 ceramics. There are some grains bonded into block areas [11], it is because the amount of β-phase has limited the growth space among grains, restrained the growth of long columnar grains, then the grains become short and thick and bonded together.

The relationship between the content of benzoic acid and porosity of Si_3N_4 ceramics has been shown in Fig. 3. With the increase of pore addition agent mass fraction, the apparent porosity of materials increases, and the amplitude decreases, this is because pores arranged more closely, and it is more difficult to exclude. At the meantime, the pores will produce a certain shrinkage during the pressureless sintering process, and the rate of grain boundary movement will be faster with the increase of the number of pores, that is to say, the porosity will be reduced more and faster during the process, and the mechanical properties will be reduced too. The porosity of specimens reached 40.79% at most with 50 wt% benzoic acid added.

The dielectric properties of porous Si_3N_4 ceramics with different contents of benzoic acid has been presented in Fig. 4. The complex dielectric constant and dielectric loss are key parameters to characterize the dielectric properties of materials, the real part (dielectric constant) is relative to polarization, the imaginary part represents dielectric loss [12]. Although the high imaginary part means good microwave absorption performance, but too high values are unfavorable to impedance matching and may lead to strong reflection and weak absorption [13].

According to the previous analysis, the affecting factors on the dielectric properties of the sample are mostly porosity. From Fig. 4a, we can see the real permittivity is low in the specimen, and the value becomes lower with the porosity goes up except with the 40 wt% benzoic acid, it is probably because the block areas in the material. The reason for some negative dielectric loss values in the images is that the testing device is unstable at the beginning of the test. The sharp peaks in the

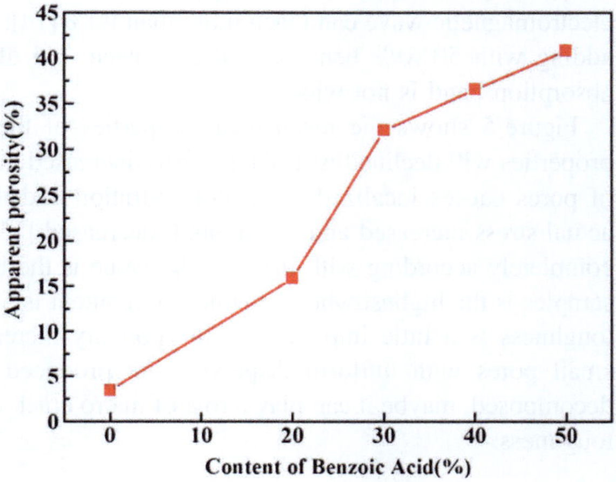

Fig. 3 Effect of adding different benzoic acid on apparent porosity of porous Si_3N_4 ceramics

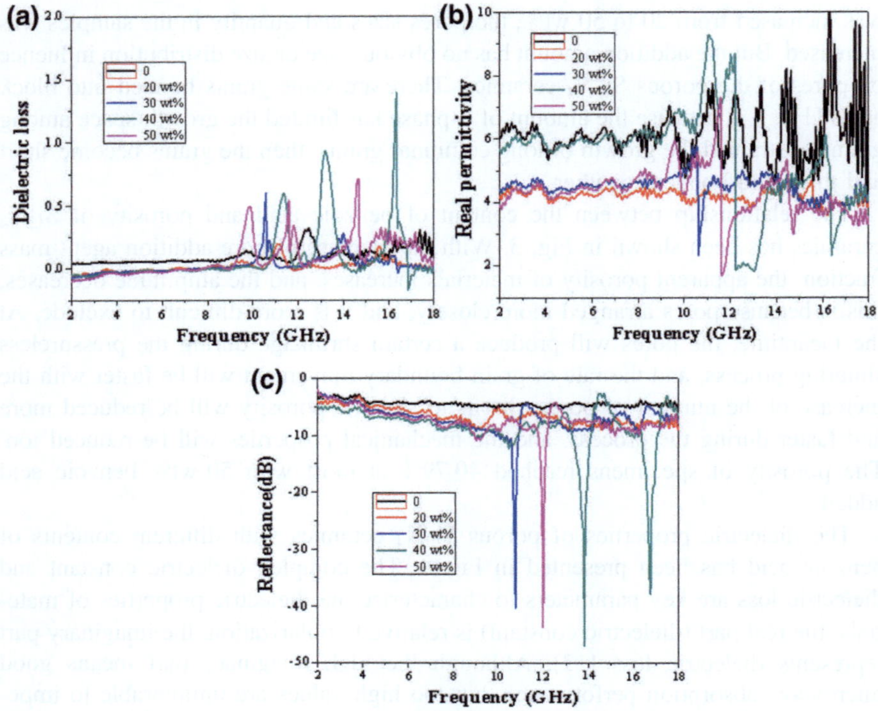

Fig. 4 Dielectric properties of porous Si_3N_4 ceramics with different contents of benzoic acid **a** real permittivity; **b** dielectric loss; **c** reflectance

picture are due to resonance absorption, and the position of the occurrence coincides with the location of the dielectric loss and reflectance peaks. The Si_3N_4 ceramics of 0 and 20 wt% benzoic acid have low dielectric loss, and there are almost no peak in the whole range. The peaks moves toward higher frequencies as the porosity increases. The reflectivity under −10 dB means the absorption rate of electromagnetic wave can reach more than 90% [14]. The reflectivity of the sample adding with 50 wt% benzoic acid can reach −45 dB at around 12 GHz, but the absorption band is not wide.

Figure 5 shows the mechanical properties of the specimens. The mechanical properties will declined with the porosity increased generally, because the presence of pores causes localized stress concentration and the reduced solid section, the actual stress increased and the strength decreased [15, 16]. The flexural strength is completely according with the law, the value is the lowest and the porosity of the samples is the highest when benzoic acid content is 50 wt%. However, the fracture toughness is a little improved as the porosity increased to 20 wt%, because the small pores with uniform dispersion are produced after pore forming agent is decomposed, maybe it can play a role of micro crack toughening and strengthen the toughness.

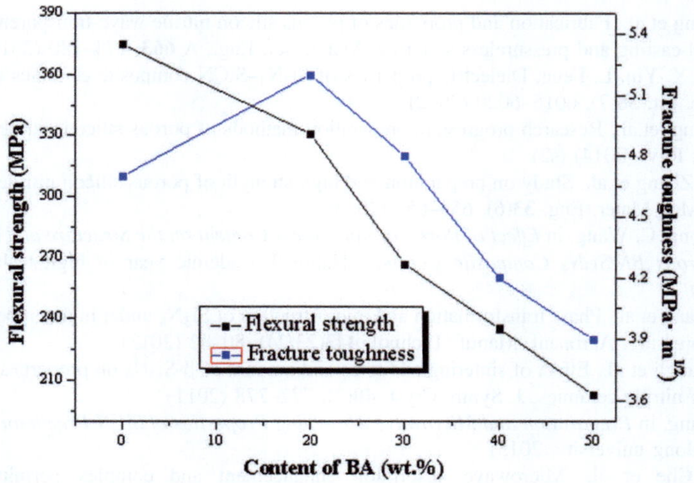

Fig. 5 Influence of the benzoic acid on the flexural strength and fracture toughness of porous Si_3N_4 ceramics

Conclusions

(1) The addition of benzoic acid improves the porosity and dielectrical properties of Si_3N_4 remarkably by pressureless sintering at 1700 °C for 60 min. The porosity of specimen reaches 40.79% with 50 wt% benzoic acid addition and the reflectivity of the sample can reach −45 dB at around 12 GHz, but the mechanical properties are reduced a little bit compared with Si_3N_4 ceramic without pore former.

(2) The specimen adding with 50 wt% benzoic acid has small and evenly distributed pores, the reflectivity of the sample is low and there are some resonance absorption peaks in the whole range, so it can be a good electromagnetic wave transmitting material if it is further improved.

Acknowledgements This work was supported by the Natural Science Foundation of China (51572154) and the Natural Science Foundation of Shandong Province (2015ZRE27348).

References

1. O.A. Lukianova, V.V. Sirota, Dielectric properties of silicon nitride ceramics produced by free sintering. Ceram. Int. **43** (2017)
2. X. Hou et al., Corrosion behavior of porous silicon nitride ceramics in different atmospheres. Ceram. Int. **43**(5), 4344–4352 (2017)
3. S. Shan, et al., Research progress of porous silicon nitride ceramics. Bull. Chin. Ceram. Soc. **28**(1), 138–142 (2009)
4. R. Raj, et al., Oxidation kinetics of an amorphous silicon carbonitride ceramic. J. Am. Ceram. Soc. **84**(8), 1803–1810 (2001)

5. X. Yang et al., Fabrication and properties of porous silicon nitride wave-transparent ceramics via gel-casting and pressureless sintering. Mater. Sci. Eng., A **663**, 174–180 (2016)
6. Q. Li, X. Yin, L. Feng, Dielectric properties of Si_3N_4–SiCN composite ceramics in X-band. Ceram. Int. **38**(7), 6015–6020 (2012)
7. P. Wang, et.al., Research progress in preparation methods of porous silicon nitride ceramics. Mater. Rev. (2014) (s2)
8. Yong Zhang et al., Study on preparation and high strength of porous silicon nitride ceramics. Rare Met. Mater. Eng. **33**(6), 655–658 (2004)
9. W. Dong, C. Wang, in *Effect of Pore Forming Agent Content on the Structure and Properties of Porous BN/Si_3N_4 Composite Ceramic*. National academic year of high tech ceramics (2012)
10. Gu Qiang et al., Phase transformation and microstructure of Si_3N_4 under high temperature and high pressure. Aeronaut. Manuf. Technol. **443**(23/24), 80–82 (2013)
11. Yu. Fangli et al., Effect of sintering additives and amount of β-Si_3N_4 on properties of porous silicon nitride ceramics. J. Synth. Cryst. **40**(3), 772–778 (2011)
12. X. Wang, in *Preparation and Microwave Absorbing Properties of SiCN Precursor Ceramics* (Shandong university, 2015)
13. R.C. Che et al., Microwave absorption enhancement and complex permittivity and permeability of Fe encapsulated within carbon nanotubes. Adv. Mater. **16**(5), 401–405 (2004)
14. L. Gao, X.H. Jin, J.G. Li et al., BN/Si_3N_4 nanocomposite with high strength and good machinability. Mater. Sci. Eng., A **415**(1), 145–148 (2006)
15. E. Rysyshkewitch, Compression strength of porous sintered alumina and zirconia[J]. J. Am. Ceram. Soc. **36**(2), 65–68 (1953)
16. S.J. Penn, N.M. Alford, A. Templeton et al., Effect of porosity and grain size on the microwave dielectric properties of sintered alumina. J. Am. Ceram. Soc. **80**(7), 1885–1888 (1997)

In Situ Synthesis of ZrB_2–SiC Composite Powders by Carbothermal Reduction Method

Bingying Xie, Jincheng Yu, Yujun Zhang, Hongyu Gong, Xiao Lin and Yu Liu

Abstract ZrB_2–SiC composite powders were synthesized in situ via carbothermal reduction using ZrO_2, HBO_2 and carbon black as the starting materials. The influences of HBO_2 contents (40.6–45.5 wt%), boron sources (HBO_2, H_3BO_3 and B_2O_3) and carbon sources (black carbon, activated carbon, graphite and petroleum coke) on phase composition and morphology of ZrB_2–SiC composite powders were analyzed. The obtained samples were characterized by X-ray diffraction (XRD) and scanning electron microscope (SEM). Results showed that when the HBO_2 content was 43.2 wt%, impurity-free ZrB_2–SiC composite powders could be successfully produced at 1600 °C for 90 min in argon atmosphere. Meanwhile, columnar ZrB_2 and granular SiC particles were combined interactively. Compared with control groups, products fabricated via black carbon revealed the better crystallinity and finer particle size.

Keywords ZrB_2, SiC · Powder · Carbothermal reduction

B. Xie · J. Yu · Y. Zhang (✉) · H. Gong · X. Lin · Y. Liu
Key Laboratory for Liquid-Solid Structural Evolution & Processing of Materials of Ministry of Education, Shandong University, Jinan 250061, China
e-mail: yujunzhangcn@163.com

B. Xie
e-mail: xby19950224@163.com

J. Yu
e-mail: 1499896714@qq.com

H. Gong
e-mail: hygong@sdu.edu.cn

X. Lin
e-mail: 1158450341@qq.com

Y. Liu
e-mail: 804512856@qq.com

B. Xie · J. Yu · Y. Zhang · H. Gong · X. Lin · Y. Liu
Key Laboratory of Special Functional Aggregated Materials, Ministry of Education, Shandong University, Jinan 250061, China

Y. Han (ed.), *Advanced Functional Materials*,
https://doi.org/10.1007/978-981-13-0110-0_13

Introduction

Carbon/carbon (C/C) composites have incomparable advantages over other materials in aerospace and other high-tech fields due to their excellent properties, such as low density, low coefficient of thermal expansion, high specific strength, and high specific modulus. Because of the internal lattice defects, C/C composites are easily oxidized at high temperature in oxidizing atmosphere, which strongly restricted their applications [1, 2]. At present, ZrB_2–SiC ultra-high temperature ceramic coatings are actively used to protect C/C composites owing to the good oxidation resistance [3–5], and spraying is widely applied to prepare the coatings for its flexible operation [6, 7]. However, due to the different densities between ZrB_2 and SiC, the components of coatings are easy to demix during spraying, if the two powders are mixed mechanically. Therefore, the synthesis of ZrB_2–SiC composite powders has become a hot issue.

There are various methods to prepare ZrB_2–SiC composite powders, including self-propagation high-temperature synthesis method, sol–gel method, carbon thermal reduction method and so on [8–11]. Among different synthesis routes, carbon thermal reduction method has the advantages of simple process, short product cycle and low production costs. In this study, ZrB_2–SiC composite powders were synthesized in situ by carbothermal reduction at 1600 °C for 90 min in Ar, using ZrO_2, HBO_2, black carbon and SiC as the starting materials. The in situ reaction could make the solidification mass transfer more fully and impede delamination effectively in spraying. The effects of HBO_2 content, different boron and carbon sources on phase composition and morphology of ZrB_2–SiC composite powders were discussed.

Experiment

Zirconia powders (ZrO_2, ≤ 10 μm), boric acid (H_3BO_3, 96%, ≤ 10 μm), boron oxide (B_2O_3, 98%, ≤ 20 μm), carbon black (C, ≤ 10 μm), activated carbon (C, ≤ 20 μm), graphite (C, ≤ 20 μm), petroleum coke (C, ≤ 20 μm) and silicon carbide (SiC, ≤ 0.5 μm) were provided as the raw materials. Since H_3BO_3 could volatilize easily which would do harm to the furnace owing to acid corrosion, H_3BO_3 was put into a vacuum drying oven at 120 °C for 12 h to get HBO_2, according to Eq. 1 [12]. In theory, ZrB_2–SiC composite powders can be produced by Eq. 2. Because of the volatilization of HBO_2 during calcination, the amount of HBO_2 was adjusted to obtain the highly pure powders.

$$H_3BO_3(s) \rightarrow HBO_2(s) + H_2O(g). \quad (1)$$

$$ZrO_2(s) + 2HBO_2(s) + 5C(s) + xSiC(s) \rightarrow ZrB_2(s) + 5CO(g) + H_2O(g) + xSiC(s). \quad (2)$$

Table 1 Content of the ingredients of different groups

Group	Composition (wt%)			
	ZrO_2	HBO_2	C	SiC
ZS01	31.8	40.6	18.2	9.4
ZS02	30.4	43.2	17.4	9.0
ZS03	29.1	45.5	16.7	8.7

Three groups of composite powders were prepared according to the ratio of Eq. (2), referred to as ZS01, ZS02, ZS03, in which the mass fraction of SiC was 20 wt%. In order to study the impact of HBO_2 content, extra HBO_2 of 80, 100, 120 wt% were added to ZS01, ZS02, ZS03 respectively. Therefore, the content of ZrO_2, HBO_2, carbon black and SiC was shown as Table 1.

The three groups of powders were mixed by ball mill at 300 rpm for 20 min with ZrO_2 balls as the grinding media. Then, the mixed powders were put into different graphite crucibles and sealed in a graphite resistance furnace. The samples were heated to 1600 °C at a rate of 10 °C/min in argon atmosphere and then held for 90 min. After cooling to room temperature, gray powders were obtained. In addition, different boron sources were added to composite powders to compare the phase composition and morphology of the products while the mass ratio of Zr and B was constant. Further, the black carbon was replaced by activated carbon, graphite and petroleum coke to study the effect of different carbon sources. XRD (D/max-RA, Japan) was used to analyze the phase composition of ZrB_2–SiC composite powders. The morphology of the products was characterized by SEM (D/max-RA, Germany).

Results and Discussion

Effect of HBO_2 Content. Figure 1 showed the XRD patterns of ZrB_2–SiC composite powders prepared with different contents of HBO_2. It was obvious that impurity-free ZrB_2–SiC powders were synthesized. When the amount of HBO_2 decreased from 43.2 to 40.6 wt%, the narrowed peaks and reduced intensities of ZrB_2 implied the increment of grain size of the products. With extra 120 wt% HBO_2, diffraction peak of ZrB_2 decreased while the HBO_2 phase was not detected in the Fig. 1c. The reason was that HBO_2 was converted to B_2O_3 during heating and a part of B_2O_3 was involved in the synthesis of ZrB_2 grains while the excess was present in the furnace as B_2O_3 vapor. The partial pressure of B_2O_3 vapor led to the change in Gibbs free energy (ΔG), which resulted in a slight decrease in the intensity of ZrB_2 phase. Meanwhile, the strongest peak of SiC appeared with 43.2 wt% HBO_2. In summary, the amount of HBO_2 had a certain influence on the crystallinity of ZrB_2–SiC composite powders.

Effect of Different Boron Sources. XRD patterns of ZrB_2–SiC composite powders prepared with different boron sources were given in Fig. 2. As Fig. 2a, b showed,

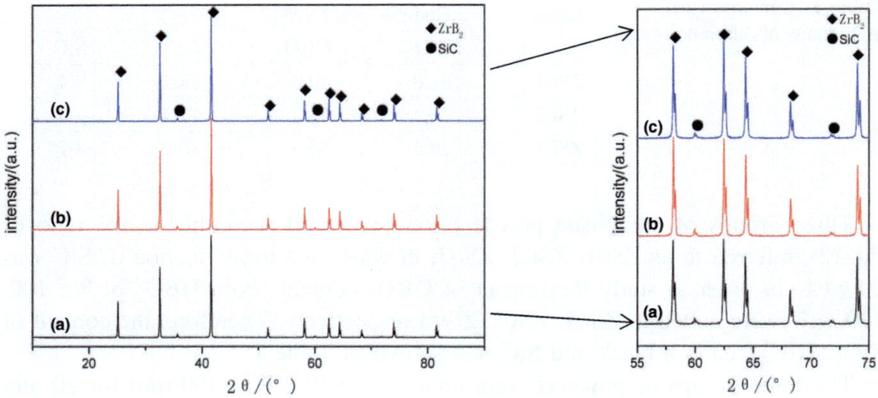

Fig. 1 XRD patterns of final products prepared with different contents of HBO_2. **a** ZS01 **b** ZS02 **c** ZS03

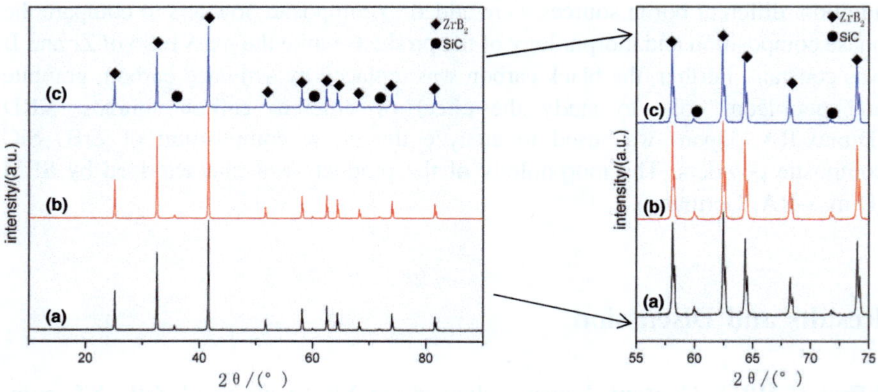

Fig. 2 XRD patterns of final products prepared with different boron source. **a** HBO_2 **b** H_3BO_3 **c** B_2O_3

powders with good crystallinity were successfully synthesized when HBO_2 or H_3BO_3 were chosen as the boron source. However, the intensity of ZrB_2 peak decreased when B_2O_3 was used instead of HBO_2 or H_3BO_3.

The resultant samples were also characterized by SEM to investigate the morphology, shown in Fig. 3. From Fig. 3a, powders synthesized by HBO_2 demonstrated uniform size of 1.5 μm. And columnar ZrB_2 and granular SiC particles were combined interactively. Grains in Fig. 3b grew larger in comparison to those in Fig. 3a, implying size increase took place with H_3BO_3. When the boron source was replaced by B_2O_3, ZrB_2 grains revealed poor crystallinity and irregular shapes.

Effect of Different Carbon Sources. XRD patterns of ZrB_2–SiC composite powders produced with different carbon sources were presented in Fig. 4. As

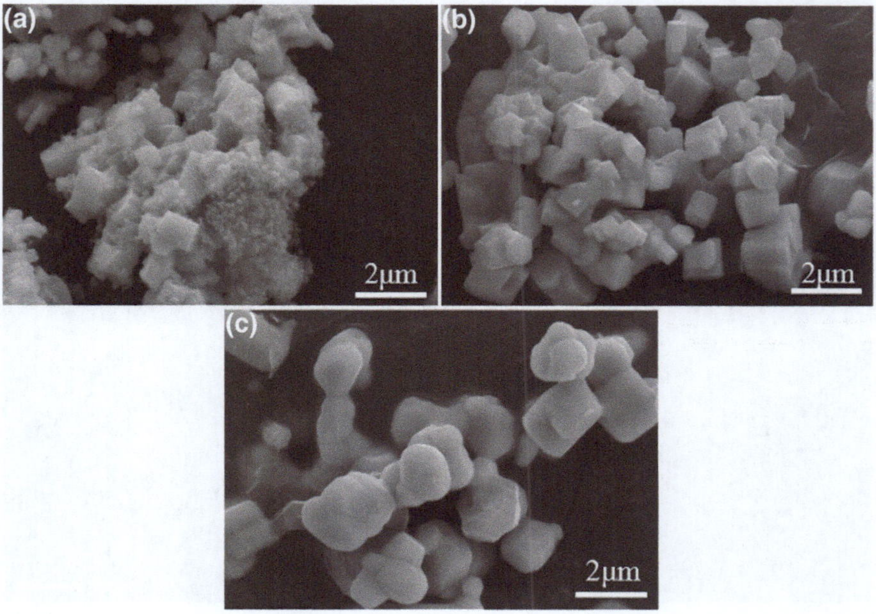

Fig. 3 SEM images of final products prepared with different boron source. **a** HBO_2 **b** H_3BO_3 **c** B_2O_3

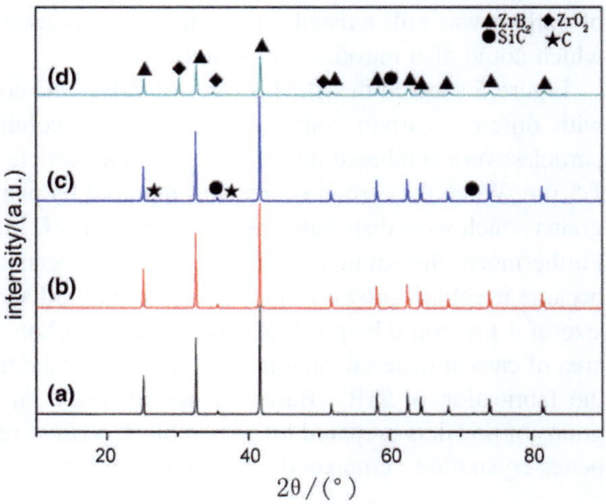

Fig. 4 XRD patterns of final products produced with different carbon source. **a** Carbon black **b** Petroleum coke **c** Graphite **d** Activated carbon

Fig. 4a, b showed, impurity-free ZrB_2–SiC composite powders could be obtained with carbon black or petroleum coke. However, the faint peaks of carbon and zirconia were detected when the carbon source was replaced by activated carbon, which was probably because that activated carbon with a small specific surface area could not play a sufficient reduction effect in the reaction. Furthermore, compared

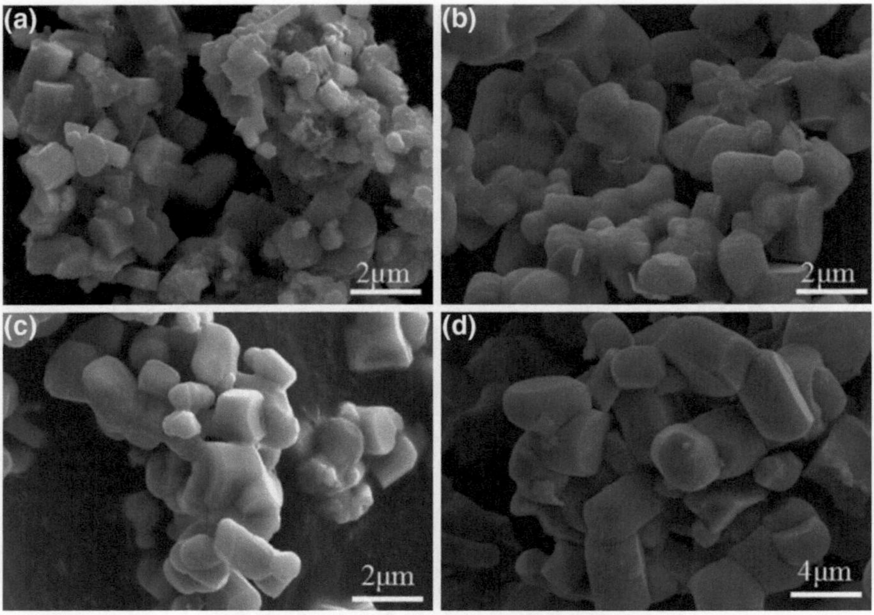

Fig. 5 SEM images of final products synthesized with different carbon source. **a** Carbon black. **b** Petroleum coke. **c** Graphite. **d** Activated carbon

with Fig. 4a, b, the intensity of ZrB_2 and SiC was reduced. From Fig. 4c, the peak of carbon was still existed when graphite was used instead of activated carbon, which could also introduce impurities.

Figure 5 showed the SEM images of ZrB_2–SiC composite powders synthesized with different carbon sources. From Fig. 5a, columnar ZrB_2 and granular SiC particles were combined interactively and the particle size of ZrB_2 was estimated to 1.5 μm. When the carbon source was replaced by petroleum coke or graphite, ZrB_2 grains which were distributed on the surface of SiC increased to about 2 μm in size. Furthermore, the grain boundaries of ZrB_2 began to melt and the morphology became irregular, shown in Fig. 5b, c. As Fig. 5d showed, ZrB_2 particles with the size of 4 μm could be produced by activated carbon. The different specific surface area of carbon materials might be the reason for the influences of carbon source on the fabrication of ZrB_2. Based on above results, it could be concluded that the grains of powders prepared by carbon black possess relatively finer particle size and better crystallinity compared with control groups.

Conclusion

Impurity-free ZrB_2–SiC composite powders were successfully produced from ZrO_2, HBO_2 SiC and carbon black by in situ carbothermal reduction method. Effects of HBO_2 content, various carbon and boron sources on the phase composition and morphology of ZrB_2–SiC composite powders were investigated. When the HBO_2 content was 43.2 wt%, the highly pure composite powders with regular shapes could be synthesized at 1600 °C for 90 min in Ar and ZrB_2 was estimated to 1.5 μm in particle size. Compared with samples prepared by petroleum coke, graphite or activated carbon, the composite powders fabricated by black carbon possessed relatively better crystallinity and finer particle size.

Acknowledgements The work is financially supported by inorganic nonmetal institute of Shandong University.

References

1. J. Wang, K. Li, W. Li et al., The preparation and mechanical properties of carbon/carbon composite joints using Ti-Si-SiC-C filler as interlayer. Mater. Sci. Eng., A **574**, 37–45 (2013)
2. M.P. Bacos, J.M. Dorvaux, O. Lavigne et al., C/C composite oxidation model: I. Morphol. Exp. Invest. Carbon **38**, 77–92 (2000)
3. Y. Zhang, H. Hu, J. Ren et al., Effect of the surface microstructure of SiC inner coating on the bonding strength and ablation resistance of ZrB_2-SiC coating for C/C composites. Ceram. Int. **42**, 18657–18665 (2016)
4. X. Zou, Q. Fu, L. Liu et al., ZrB_2-SiC coating to protect carbon/carbon composites against ablation. Surf. Coat. Technol. **226**, 17–21 (2013)
5. D. Wang, Y. Zeng, X. Xiong et al., Ablation behavior of ZrB_2–SiC protective coating for carbon/carbon composites. Ceram. Int. **41**, 7677–7686 (2015)
6. M. Tului, G. Marino, T. Valente, Plasma spray deposition of ultra high temperature ceramics. Surf. Coat. Technol. **201**, 2103–2108 (2006)
7. C. Bartuli, T. Valente, M. Tului, Plasma spray deposition and high temperature characterization of ZrB_2-SiC protective coatings. Surf. Coat. Technol. **155**, 260–273 (2002)
8. Y. Yan, Z. Huang, S. Dong et al., New route to synthesize ultra-fine zirconium diboride powders using inorganic-organic hybrid precursors. J. Am. Ceram. Soc. **89**, 3585–3588 (2006)
9. Y. Li, W. Han, H. Li et al., Synthesis of nano-crystalline ZrB_2/ZrC/SiC ceramics by liquid precursors. Mater. Lett. **68**, 101–103 (2012)
10. R.V. Krishnarao, Preparation of ZrB_2 and ZrB_2-SiC powders in a single step reduction of zircon ($ZrSiO_4$) with B_4C. Ceram. Int. **43**, 1205–1209 (2017)
11. R.V. Krishnarao, MdZ Alam, D.K. Das et al., Synthesis of ZrB_2-SiC composite powder in air furnace. Ceram. Int. **40**, 15647–15653 (2014)
12. J. Yu, L. Ma, Y. Zhang et al., Synthesis of TiB_2 powders via carbothermal reduction of TiO_2, HBO_2 and carbon black. Ceram. Int. **42**, 5512–5516 (2016)

Facile Synthesis of Mn_3O_4 Nanoparticles Decorated Graphene as Enhanced Performance Electrode for Supercapacitor

Hui Sun, Chengbao Liu, Junchao Qian, Feng Chen, Zhengying Wu and Zhigang Chen

Abstract Graphene was prepared by using the plant stem as a biotemplate through high temperature carbonization and Mn_3O_4/graphene composites were then prepared via a simple hydrothermal process. The surface morphology and structure of the Mn_3O_4/graphene hybrids were characterized by X-ray diffraction spectroscopy (XRD), Raman spectra, field emission scanning electron microscopy (FESEM), high resolution transmission electron microscopy (HRTEM), X-ray photoelectron spectrogram (XPS), N_2 adsorption–desorption isotherms and electrochemical test. The results showed that the metal oxide nanoparticles were uniformly supported on the surface of graphene. Judging from the XRD, XPS and Raman, we could acquire that the Mn_3O_4 nanoparticle was in the crystal form while the average crystallite size was about 21 nm. The composites exhibited excellent specific capacitance as high as 196 F/g and the capacitance of the composites was 88% retained after 1000 cycles in 1 M Na_2SO_4 electrolyte at a charging rate of 2 A/g. The superior electrochemical properties could be due to the improved accessible area for ions in electrolytes and enhanced conductivities. The present study provides a facile way to design a high-performance Mn_3O_4/graphene based supercapacitor electrode and the concept is extendable to other pseudo capacitive electrodes.

Keywords Graphene · Mn_3o_4 nanoparticles · Hydrothermal method
Electrochemical properties · Supercapacitor

H. Sun · C. Liu (✉) · J. Qian · F. Chen · Z. Wu · Z. Chen
Jiangsu Key Laboratory for Environment Functional Materials,
Suzhou University of Science and Technology, Suzhou 215009, China
e-mail: lcb@mail.usts.edu.cn

H. Sun
e-mail: 18862133711@163.com

J. Qian
e-mail: ziyou1900@gmail.com

H. Sun · C. Liu · J. Qian · F. Chen · Z. Wu · Z. Chen
School of Chemistry, Biology and Materials Engineering,
Suzhou University of Science and Technology, Suzhou 215009, China

Y. Han (ed.), *Advanced Functional Materials*,
https://doi.org/10.1007/978-981-13-0110-0_14

Introduction

People have a tremendous demand for energy consumption, so the energy generation and storage are the most significant issues in our modern daily life [1]. Among the most promising alternative energy storage devices, supercapacitors have attracted considerable attention because of the high specific power density, fast charge-discharge capability, high cycle life and superior reversibility [2, 3]. Based on the energy storage mechanism, supercapacitors can be classified into two types: electrical double layer capacitors (EDLCs) and pseudo-supercapacitors (otherwise known as redox capacitors) [4, 5]. For EDLCs, the energy charge is stored by the adsorption/desorption of electrolyte ions on porous carbon materials. In pseudo-supercapacitors, the energy charge is stored and released in the Faradaic electron–transfer processes of transition metal oxides (e.g., Co_3O_4, MnO_2, and Mn_3O_4) or conducting polymers.

Due to the high theoretical capacitance of $\sim$1370 F/g, Mn-based oxides have been attracting considerable research interest in the field of electrode materials for supercapacitors [6, 7]. Manganese has different oxides such as MnO_2, Mn_2O_3 and Mn_3O_4. Mixed-valent spinel Mn_3O_4 has drawn distinctive research attention owing to its particular structure and physicochemical properties, which are of great interest in energy conversion, magnetics, and catalysis, etc. In particular, the tunable capacitive behavior of Mn_3O_4 nanostructures has been investigated [8–10]. However, the poor electrical intrinsic conductivity of Mn_3O_4 materials also seriously affects the electric capacitance. A common strategy with poor electronic conductors such as transition-metal oxides is to combine them into composites with carbon, which is both lightweight and electronically conducting. As compared to other carbon matrixes such as graphite, carbon black, and carbon nanotubes, graphene is emerging as one of the most appealing carbon materials [2, 11]. The presence of graphene with high electrical conductivity is favorable to improve the dispersion of nano-particles and the electrical conductivity of the composite electrode [2, 12, 13]. In addition, nano-particles could provide extra-pseudo-capacitance, leading to a high specific capacitance.

In this work, we provided a facile biomimetic synthesis of Mn_3O_4 nanoparticles on graphene sheets by using low-cost stems as a template. Our solution-based method was prepared in aqueous state at room temperature, making it amenable for large scale production. Furthermore, the novel nanoparticles displayed an amazing capacitance of 196 F/g and 88% maintenance of capacitance after 1000 cycles even at a high current density of 2 A/g.

Experimental

Materials. All reagents used were analytically pure and purchased from China National Medicines Corporation Ltd. And the plant stems were collected from campus in autumn.

Synthesis. Typically, the plant stems were washed with deionized water and then pretreated by immersing them in ethanol/water mixtures. The stems were later added into 0.1 mol/L of dilute hydrochloric acid solution to remove dirt and P, Ca, K ions, etc. The stems were then calcinated in the tubular furnace under the N_2 atmosphere. 1.5 g $Mn(CH_3COO)_2 \cdot 4H_2O$ was added into 50 mL deionized water and stirred intensely for 12 h. The precursor with solution was then transferred to the reaction vessel and 20 mL ammonia was added. Then the reaction vessel was put into an oven (150 °C) for 12 h. Afterwards the samples were filtered to obtain the final materials.

Characterizations. The structural phases of the products were investigated by X-ray diffraction patterns (Bruker D8 ADVANCE diffractometer with a nickel-filtered Cu Kα radiation source). Raman spectra were measured on a Laser Raman spectroscopy (Renishaw In Via) with 532 nm laser as an excitation source. A scanning electron microcopy system (SEM, S-4800) was used to observe the morphologies of the Mn_3O_4/graphene composites. The distribution of Mn_3O_4/graphene was investigated by using a JEM 2100F High Resolution Transmission Electron Microscopy (HRTEM). The adsorption and desorption isotherm studies with nitrogen were performed using a Micromeritics ASAP 2020C system. XPS analysis was collected by using an X-ray photoelectron spectroscopy ESCA (Thermo Fisher Scientific, ESCALAB 250). Supercapacitor performance was obtained by using a CHI660D electrochemical workstation with a three-electrode setup.

Electrochemical Experiments. The electrochemical properties were tested with a CHI660D electrochemical workstation. A three-electrode system, with the electroactive composite materials, platinum wire, and saturated calomel electrode, were performed as the working electrode, counter electrode, reference electrode, respectively. To fabricate the electrochemical materials, Mn_3O_4/graphene was blent with acetylene black (10 wt%), and polytetrafluoroethylene emulsion (PTFE, 10 wt%) as binder, then the compound was pasted evenly onto anickel foam (1 cm × 1 cm), dried for 10 min at a room temperature. A 1 M Na_2SO_4 aqueous solution was used as the electrolyte. The fine capacitance was calculated from the galvanostatic charge–discharge curve by using the equation:

$$C = I \times \Delta t/(\Delta V \times m) \tag{1}$$

where I is the current applied (A), Δt is the discharge time (s), m is the mass of active material deposited (g), and ΔV is the potential window (V).

Results and Discussion

X-ray Diffraction Analysis. Figure 1 is the XRD pattern obtained for Mn_3O_4/ graphene composites. The diffraction profile shows characteristic signature for tetragonal hausmannite crystal structure of Mn_3O_4. All the peaks were indexed

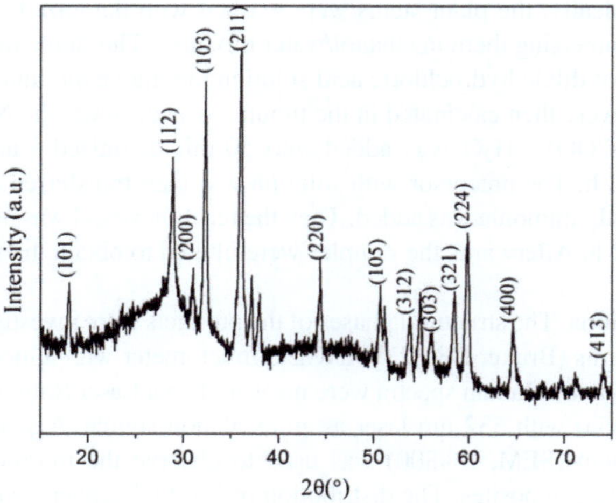

Fig. 1 XRD pattern of Mn_3O_4/graphene composites

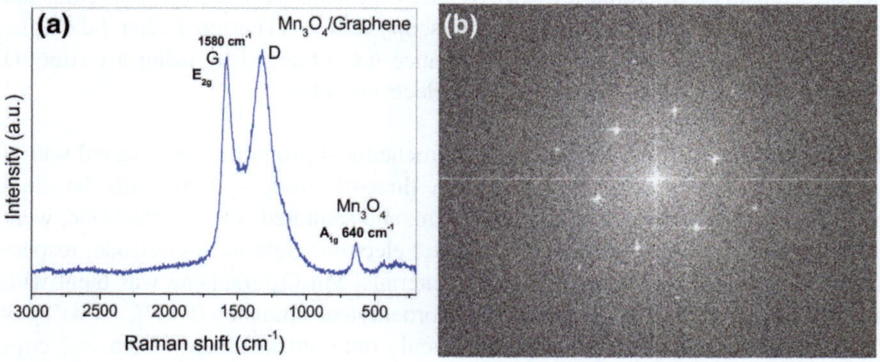

Fig. 2 **a** Raman spectra of the Mn_3O_4/graphene composite and **b** the correspond reduced FFT image of graphene

according to the Joint Committee Powder Diffraction Standards (JCPDS) file no. 80-0382. No additional peaks were detected, confirming the absence of other manganese oxides or impurity. It can be seen that the diffraction peaks of as-synthesized products at 2θ of 18.01°, 28.92°, 30.99°, 32.38°, 36.08°, 44.41°, 50.84°, 53.87°, 56.03°, 58.5°, 59.91°, 64.62° and 74.15° were respectively indexed as (101), (112), (200), (103), (211), (220), (105), (312), (303), (321), (224), (400) and (413) planes of Mn_3O_4. The average crystallite size can be estimated by using the Scherrer formula. The average crystallite size of Mn_3O_4 is about 21 nm, which is consistent with the TEM observations (see Fig. 3).

Raman Spectra Analysis. Raman spectroscopy can be used to acquire the structure of the graphene sheets in the Mn_3O_4/graphene composite. The Raman spectral

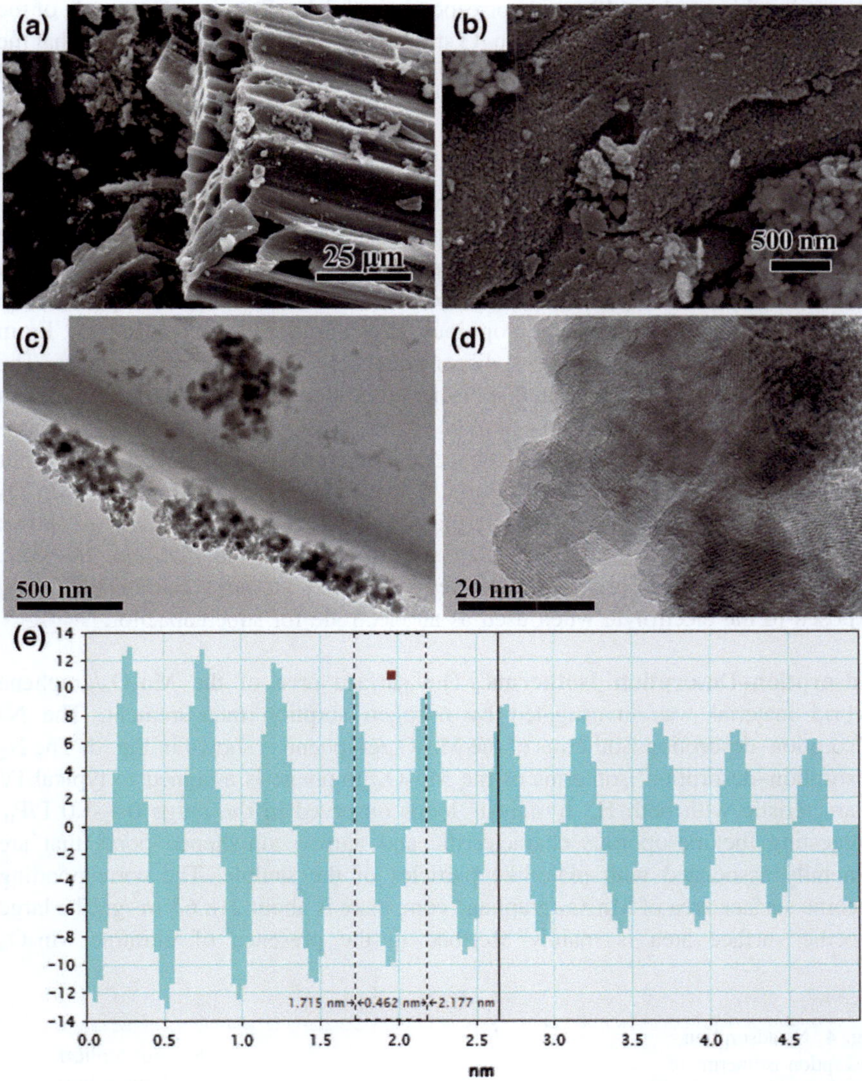

Fig. 3 SEM micrographs of **a**, **b** hollow porous structure observed under low magnifications after treatment, **c**, **d** HRTEM images of Mn_3O_4 nanoparticles/graphene after hydrothermal treatment, **e** the image of the interplanar spacing of Mn_3O_4

profile for Mn_3O_4/graphene is shown in Fig. 2a. It can be seen that two significant peaks at 1335 cm^{-1} (D-band) and 1580 cm^{-1} (G-band) are ascribed to the breathing modes of six-atom rings and the high frequency E_{2g} phonon, respectively. Consistent with the XRD data, there are no peaks assignable to this phase in the Mn_3O_4/graphene composite. The band at 640 cm^{-1} was observed in the Mn_3O_4/graphene composite, which can be attributed to the Mn_3O_4 phase and the Mn_3O_4

Raman band located at 640 cm^{-1} corresponds to the A_{1g} Raman active-mode of the octahedral site, which is due to a Mn–O stretching mode. Figure 2b showed that the graphene was further confirmed from a FFT image processed by using Digital Micrograph software.

Microstructure Characterizations. The morphology of as-prepared samples was studied by SEM. Figure 3a, b clearly illustrate that the calcined sample at 800 °C had many ordered pores and the diameter of these pores are about 15 μm. The porous structures lay the solid foundation for the circulation and impregnation of electrolyte and the morphological features of the initial biotemplate imply that the biomimetic materials are very propitious to scaffold of supercapacitors. From Fig. 3b, the Mn_3O_4 nanoparticles dispersed evenly on the graphene sheet. The porous structure of the as-prepared is further confirmed by HRTEM (Fig. 3c). It can be seen that the deposited manganese oxide closely attach to the graphene nanosheets, which may provide rich conducting channels for electron transfer between the electrolytes and electrode. The HRTEM image (Fig. 3d), Mn_3O_4 nanowires can be identified by the interplanar spacing of about 0.462 nm (as shown in Fig. 3e) indexed to the (101) planes on the basis of JCPDS no. 80-0382. Accordingly, the nanoparticles decorated nanosheet surfaces can be efficiently exposed to the electrolyte when used as an electrode for supercapacitor.

Adsorption–Desorption Isotherms. The surface area of the Mn_3O_4/graphene hybrid material was investigated by nitrogen sorption measurement. The N_2 adsorption–desorption isotherms of the Mn_3O_4/graphene is shown in Fig. 4. The N_2 adsorption–desorption isotherms of the Mn_3O_4/graphene is assigned to typical IV characteristic with both H3 hysteresis loops observed in the range 0.4–1.0 P/P_0, suggesting the mesoporous characteristic and narrow slit-shaped pores that are generally associated with plate-like particles of the sample. The corresponding specific surface area of Mn_3O_4/graphene composite is about 218.67 m^2/g. The large specific surface area is mainly ascribed to the presence of ultrafine Mn_3O_4

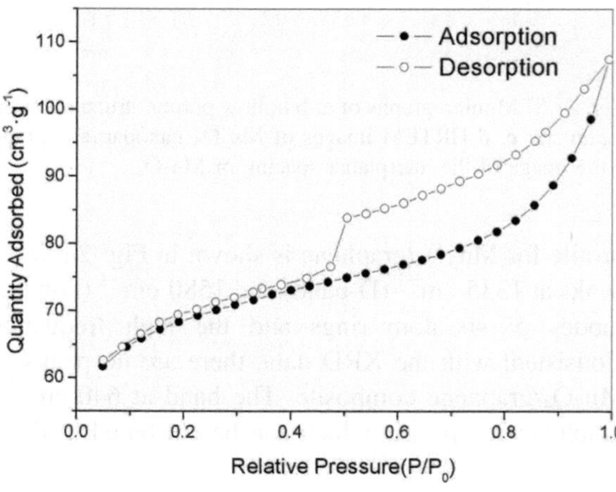

Fig. 4 N_2 adsorption–desorption isotherms of Mn_3O_4/graphene hybrid material

nanoparticles grown on graphene sheet, and is beneficial for Mn_3O_4/graphene hybrid material to accelerate the fast surface adsorption–desorption of the electrolyte ions on the electrode to contribute to a high specific capacitance.

XPS Analysis. The chemical bonding states in Mn_3O_4/graphene were further analyzed by XPS (Fig. 5). As shown in the Mn 2p region (Fig. 5a), a $2p_{3/2}$–$2p_{1/2}$ doublet at 652.9 and 641.5 eV is observed, and the splitting width (11.4 eV) is in accordance with an earlier report on Mn_3O_4. In addition, the binding energies of Mn $2p_{3/2}$ have a 0.2 eV (641.7 eV) increase after carbon doping. The C 1 s peak located at 284.8 eV is related to the C–C bond in graphene (Fig. 5b). These results are in good agreement with the XRD and Raman results.

Electrochemical Performances. The electrochemical studies for the graphene and Mn_3O_4/graphene electrodes were conducted in a three-electrode configuration. Figure 6a and b show the cyclic voltammograms (CVs) of the two electrodes over a range of scan rates from 10 to 150 mV/s. The area under the CV curve of Mn_3O_4/graphene was larger than that of graphene demonstrating that Mn_3O_4-doping sample had better capacitive properties, which was also confirmed by the charge-discharge experiments. Figure 6c and d illustrate the charge–discharge curves of the two electrodes at current density of 2 A/g. The charge-discharge curves exhibit the pseudocapacitance nature of the two electrodes. According to the discharge curves, the specific capacitance is estimated to be 196 F/g for the Mn_3O_4/graphene electrode, which is remarkably higher than the graphene electrode (25.5 F/g). Because a long cycle life is among the most important criteria for a supercapacitor, an endurance test was conducted using galvanostatic charging–discharging cycles at 2 A/g (Fig. 6e). The capacitance remained at 88% of the initial capacitance at a current density of 2 A/g after 1000 cycles, further demonstrating exceptional electrochemical stability of the Mn_3O_4 nanoparticles decorated graphene electrode.

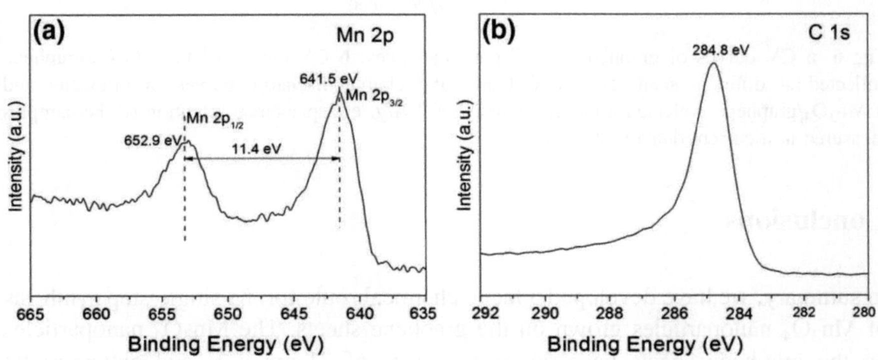

Fig. 5 XPS spectra of Mn_3O_4/graphene: **a** Mn 2p region; **b** C 1s region

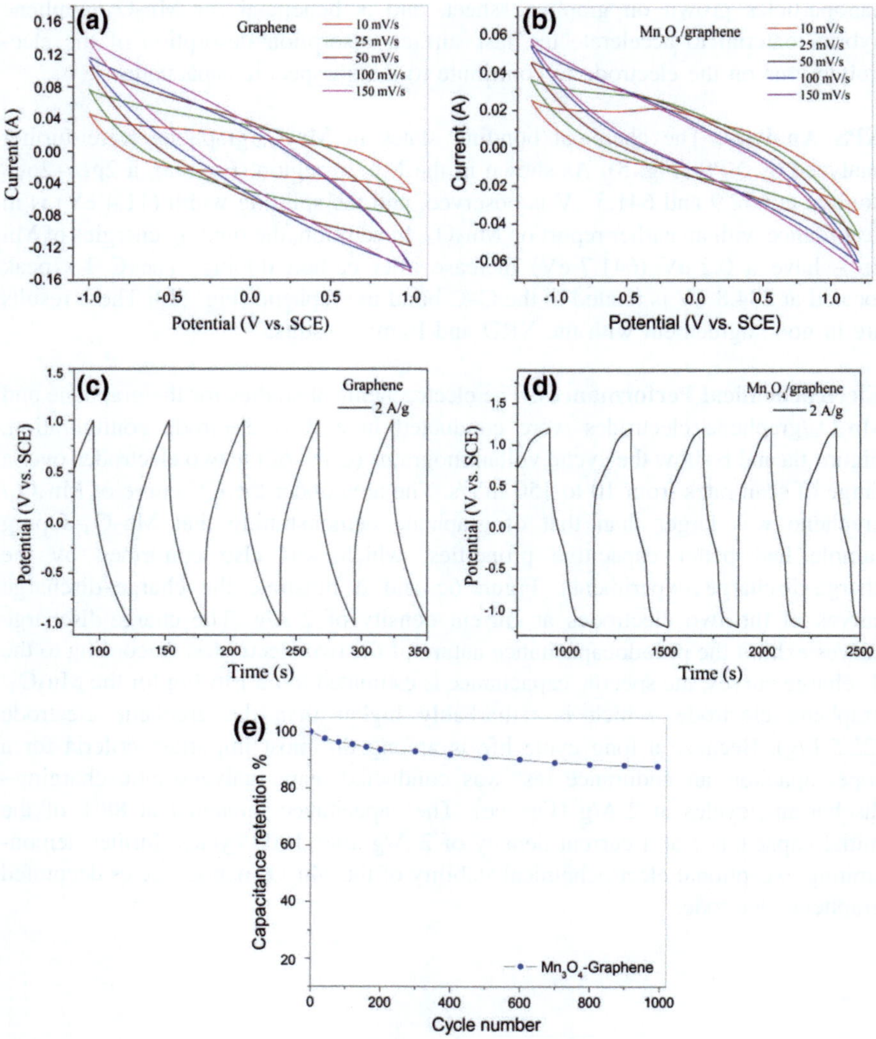

Fig. 6 **a** CV curves of graphene at different scan rates. **b** CV curves of the Mn_3O_4/graphene collected at different scan rates. **c** Galvanostatic charge–discharge curves of graphene and **d** Mn_3O_4/graphene collected a current density of 2 A/g. **e** Capacitance retention of the samples measured at a current density of 2 A/g

Conclusions

In summary, we have developed a facile chemical route for the single-step synthesis of Mn_3O_4 nanoparticles grown on the graphene sheets. The Mn_3O_4 nanoparticles on the graphene sheets have an average size of 21 nm. CV and galvanostatic charging–discharging measurements were performed to characterize these materials as supercapacitor electrodes. Compared with the pristine graphene, the Mn_3O_4/graphene composite shows exhibited excellent specific capacitance as high as 196 F/g and this electrode has good capacitance retention (88%) of its initial capacitance. The superior capacitive performance of Mn_3O_4/graphene is attributed to its unique structure, which provides good electronic conductivity, fast electron and ion transport. Such asymmetric supercapacitor offers great promise in the application of high-performance energy storage systems because of its advantages of low-cost, facile fabrication process, high energy and power density, and eco-friendly nature.

Acknowledgements We are thankful for financial support from National Natural Science Foundation of China (51478285, 21407111, 21576175); Natural Science Foundation of Jiangsu Province (BK20140280, BK20151198); Collegiate Natural Science Fund of Jiangsu Province (14KJA430004, 16KJA430008); A Project Funded by the Priority Academic Program Development of Jiangsu Higher Education Institutions (PAPD); Jiangsu Collaborative Innovation Center of Technology and Material for Water Treatment; Excellent Innovation Team in Science and Technology of Education Department of Jiangsu Province; Jiangsu Key Laboratory for Environment Functional Materials (SJHG1304, SJHG1310, SJHG1404). Creative Project of Postgraduate of Jiangsu Province (KYLX16_1356).

References

1. M.D. Stoller, S. Park, Y. Zhu, J. An, R.S. Ruoff, Graphene-based ultracapacitors. Nano Lett. **8** (10), 3498–3502 (2008)
2. L. Deng, G. Zhu, J. Wang, L. Kang, Z.-H. Liu, Z. Yang, Z. Wang, Graphene–MnO_2 and graphene asymmetrical electrochemical capacitor with a high energy density in aqueous electrolyte. J. Power Sour. **196**(24), 10782–10787 (2011)
3. J. Park, B. Kim, Y.-E. Yoo, H. Chung, W. Kim, Energy-density enhancement of carbon-nanotube-based supercapacitors with redox couple in organic electrolyte. ACS Appl. Mater. Interfaces **6**(22), 19499–19503 (2014)
4. M. Zhi, C. Xiang, J. Li, M. Li, N. Wu, Nanostructured carbon-metal oxide composite electrodes for supercapacitors: a review. Nanoscale **5**(1), 72–88 (2013)
5. K.N. Wood, R. O'Hayre, S. Pylypenko, Recent progress on nitrogen/carbon structures designed for use in energy and sustainability applications. Energy Environ. Sci. **7** (2014)
6. H. Yu, Y. Li, Hu Zuoqi, G. Wei, J. Guo, J. Liu, A carbon modified MnO_2 nanosheet array as a stable high-capacitance supercapacitor electrode. J. Mater. Chem. A **1**(34), 9809–9813 (2013)
7. Y. Liu, D. Yan, R. Zhuo, S. Li, Z. Wu, J. Wang, P. Ren, P. Yan, Z. Geng, Design, hydrothermal synthesis and electrochemical properties of porous birnessite-type manganese dioxide nanosheets on graphene as a hybrid material for supercapacitors. J. Power Sour. **242**, 78–85 (2013)
8. S. Chen, J. Zhu, H. Huang, G. Zeng, F. Nie, X. Wang, Facile solvothermal synthesis of graphene–MnOOH nanocomposites. J. Solid State Chem. **183**(11), 2552–2557 (2010)
9. J.W. Lee, A.S. Hall, J.-D. Kim, T.E. Mallouk, A Facile and Template-Free Hydrothermal Synthesis of Mn_3O_4 Nanorods on Graphene Sheets for Supercapacitor Electrodes with Long Cycle Stability. Chem. Mater. **24**(6), 1158–1164 (2012)
10. Y. Yao, C. Xu, S. Yu, D. Zhang, S. Wang, Facile synthesis of Mn_3O_4–reduced graphene oxide hybrids for catalytic decomposition of aqueous organics. Ind. Eng. Chem. Res. **52**(10), 3637–3645 (2013)

11. Y. Zhu, S. Murali, M.D. Stoller, K.J. Ganesh, W. Cai, P.J. Ferreira, A. Pirkle, R.M. Wallace, K.A. Cychosz, M. Thommes, D. Su, E.A. Stach, R.S. Ruoff, Carbon-Based Supercapacitors Produced by Activation of Graphene. Science **332**(6037), 1537–1541 (2011)
12. Z. Zhang, F. Xiao, L. Qian, J. Xiao, S. Wang, Y. Liu, Facile synthesis of 3D MnO_2–Graphene and carbon nanotube–graphene composite networks for high-performance, flexible, all-solid-state asymmetric supercapacitors. Adv. Energy Mater. **4**(10), n/a-n/a (2014)
13. P. Yang, Y. Li, Z. Lin, Y. Ding, S. Yue, C.P. Wong, X. Cai, S. Tan, W. Mai, Worm-like amorphous MnO_2 nanowires grown on textiles for high-performance flexible supercapacitors. J. Mater. Chem. A **2**(2014)

Preparation and Properties of Interconnected NiS Nanoparticle Network with Amphiphilic Polymers

Shizhen Wang, Lin Bian, Jianxun Qiu, Wei Du, Xintao Zhang, Xiaochun He, Mingjun Gao, Xiangming Li, Xingfa Ma and Guang Li

Abstract Low-dimensional functional nanocomposite with visible light activity is one of hot-spots fields in chemistry, physics and materials according to the demands of new energy and environmental fields. To simulate the process of photosynthesis of nanomaterials, in this study, NiS nanomaterials were synthesized with amphiphilic polymers as soft template. The nanocomposite with integrated multi-functionalities were characterized with several approaches. The photoconductive responses to weak visible light and 808 nm NIR (near infrared) were studied based on interdigital electrodes of Au on flexible PET (polyethylene terephthalate) film substrate with casting method. The results indicated that the resulting NiS nanocomposite is core/shell structure, showing interconnected network, looks like honeycomb, and the nanocomposite with integrated multi-functionalities exhibited photo-switching behaviors, the ratio of On/Off was 2–3 orders of magnitudes, the response and recovery were very rapidly. It would be developed the intelligent nanocomposites with external stimuli responses, light detector to NIR, biomimetic materials, nano-machine controlled by light, environmental fields, and so on.

S. Wang · L. Bian · J. Qiu · W. Du · X. Zhang · X. He · M. Gao · X. Li · X. Ma (✉)
School of Environmental and Material Engineering, Center of Advanced Functional Materials, Yantai University, Yantai 264005, China
e-mail: xingfazju@163.com; xingfamazju@aliyun.com

S. Wang
e-mail: 1355664505@qq.com

L. Bian
e-mail: 962251823@qq.com

J. Qiu
e-mail: qjxun@163.com

W. Du
e-mail: dw311@yeah.net

X. Zhang
e-mail: zxtyt@163.com

Y. Han (ed.), *Advanced Functional Materials*,
https://doi.org/10.1007/978-981-13-0110-0_15

Keywords Amphiphilic polymers · Soft template · Organic/inorganic nanocomposite · External stimuli response · Photoconductive response

Introduction

Recent two decades, the demands of new energy and clean environmental caused the great progresses of advanced functional materials. These materials includes in metal oxides [1, 2], metal sulfides [3], carbon nanomaterials [4–6], noble metals [7–14], functional polymers [15–20], organic small molecules [21, 22], et al. The emphasis of most of materials design is how to effectively utilize the visible light for sustainable development. Therefore, some functional materials with wide band gap are often modified with low band gap semiconductors.

Otherwise, in nature, photosynthesis provided a large number of energy and clean environment for human sustainably and continually. How to simulate the process of photosynthesis, there presents great challenges in materials. The photosynthesis process includes several key components: (1) antenna for light absorbance; (2) centers of photosynthesis; (3) transferring channel for charge; (4) some protein with different chain segment, and so on. In current simulating photosynthesis process, a large number of materials with visible light activity were used, such as porphyrin compounds, C_{60}, FeS, NiS, organic molecules with A/D (donor/acceptor) effects. Therefore, metal sulfide as one of the key components is good candidate for simulating the process of photosynthesis.

Among transition metal chalcogenides, nickel sulfide (NiS) is an attractive and promising semiconductor material responding in the whole solar energy spectra, due to its small direct band gap energy of about 0.9 eV. Nickel sulfides nanomaterials exist various phases, including NiS, NiS_2, Ni_3S_2, Ni_7S_6 and Ni_9S_8, and NiS form receives more attention in scientific and industrial fields owing to its excellent capacitance performance, high redox activity, which are expected to the increasing needs of energy storage devices. Currently there are many reports involving in NiS and its nanocomposites, including lithium-ion batteries [23–26], photoreduction of carbon dioxide to methane [27], rechargeable batteries [28], sensitized solar cells

X. He
e-mail: hexiaochun76@126.com

M. Gao
e-mail: gmj2188@126.com

X. Li
e-mail: li_xiangming1981@126.com

G. Li
National Laboratory of Industrial Control Technology, Institute of Cyber-Systems and Control, Zhejiang University, Hangzhou 310027, China
e-mail: guangli@zju.edu.cn

[29–31], supercapacitor [32–36], photocatalytic hydrogen evolution [37–40], and photocatalysts [41–43]. These nanocomposites on NiS are mainly based metal oxides, carbon nanomaterials, polymers composites in which NiS as co-catalyst for utilizing the visible light.

NiS/polymers nanocomposites have more advantages for treatment the organic pollutants in environmental fields. The reasons are as follow: one is design diversity of polymer materials; the other is many polymers holding high adsorption properties to organic pollutants, heavy metal ions, and low-cost. Among them, chitosan is a good candidate for adsorption to many organic pollutants. Chitosan (CS), a natural cationic polysaccharide composed of randomly distributed β-1,4-linked D-glucosamine and N-acetyl-D-glucosamine, is produced by alkaline deacetylation of chitin extracted from the shells of shrimp, lobster, and crabs, and regarded as abundant and low-cost resource. The structure of chitosan with hydroxyl and amine groups allows the uptake of adsorbates by ion-exchange or electrostatic attraction. These abundant –OH and $–NH_2$ groups of chitosan endowed excellent adsorption properties to heavy metal ions and organic pollutants, and also led to modification of chitosan easily. At same time, chitosan possesses excellent film-forming property, good biocompatibility, which is suitable to the applications in biomedical fields. The applications of chitosan and its nanocomposites are mainly including in adsorbents [44–50], sensors construction and detecting analysts [51, 52], gene and drug delivery [53–55], biomedical fields [56], antimicrobial films [57], et al.

For many years, the authors are very interested in polymer functional materials or organic-inorganic functional nanocomposites and their properties [58–62]. To develop NiS/polymer containing chitosan nanocomposite with controlled morphology and properties, in this paper, amphiphilic grafted polymers containing chitosan was synthesis, and was used as soft template for preparation of NiS nanomaterials using hydrothermal method. In simulating photosynthesis process, NiS nanoparticle as antenna for light absorbance and centers of photosynthesis, nanocomposite was integrated multi-functionality with addition of a small amount of GO (graphene oxide) as transferring channel for charge produced by light. Examination of properties was focused on photoconductive response to visible light and NIR. Excellent results were obtained.

Experimental Details

Materials. $NiSO_4{\cdot}6H_2O$ (AR), Na_2S (AR) were purchased from Sinopharm Chemical Reagent Co., Ltd. Acrylic acid (AR), ammonium persulfate (AR), styrene (AR), were purchased from Tianjin Bodi Chemical Co., Ltd, et al. Di-benzoyl peroxide (AR) et al. were commercially available.

Synthesis of Chitosan Grafted Polymer. The synthesis of chitosan grafted polymer was utilizing radial polymerization. The details is as follow: 1 g chitosan was added in three bottles of glass vessel, 200 mL H_2O and 20 mL acrylic acid were added, stirred for 1 h, then the chitosan was dissolved. The mixture was heated to 60–80 °C, 0.1–0.5 g ammonium persulfate was added for 0.5 h, 20–30 mL styrene monomer was added for 1.5–2 h. The hydrophillic chain segment is polyacrylic acid, the hydrophobic chain segment is polystyrene.

Synthesis of Nis Nanomaterials with Amphiphilic Polymers as Soft Template. 1 g $NiSO_4 \cdot 6H_2O$ was mixed with 60, 55, 50, 40 mL H_2O by ultrasound respectively, then was transferred into a 100 mL Teflon-lined stainless steel autoclave, 1, 5, 10, 20 mL above-prepared chitosan grafted polymer as soft template were added, stirred for 10 min, 1–2 g Na_2S was added by ultrasound. The autoclave was sealed tightly, and the hydrothermal reactions were carried out at 120–150 °C for 10–12 h. The resulted product was washed with deionized water repeatedly 5–6 times. The resulting product was adjusted about 300 mL solution for uses.

Synthesis of Graphene Oxide. The preparation of graphene oxide was using chemical method, which is according to the previous report utilizing carbon nanotube, concentrated sulfuric acid, concentrated nitric acid, potassium permanganate, et al. [63]. About 1 g carbon nanotube was transferred into 1000 mL GO (graphene oxide) solution.

Construction of NiS/Chitosan Grafted Polymer/GO Nanocomposite. To pave the transferring channel for charge in NiS/chitosan grafted polymer nanocomposite, in the experiment, 5, 10, 15 mL NiS/chitosan grafted polymer solution were added in 300–500 mL glass vessel, 20 mL GO solution above-synthesized were added, and ultrasounded for 15–20 min.

Morphology Observation with SEM. The scanning electron microscopy (SEM) observation was carried out with Hitachi S-4800. The obtained sample was washed with deionized water, deposited on an aluminum foil substrate, dried at room temperature and then sputtered with a thin layer of Pt on the surface for the SEM observation.

Morphology Observation with TEM. The TEM observation was carried out with JEM-1011. The sample solution was casted on cupper mesh coated with carbon film, dried at room temperature.

Measurement of FTIR Spectra. FTIR (Fourier transform infrared spectroscopy) spectra were taken with KBr, and recorded on IR Prestige-21 Fourier transform infrared spectrometer. Sample and KBr powders were mixed and pressed into a small slice, and then dried at room temperature for determination.

Measurement of Raman Spectra. Raman spectra were recorded on a LabRAM HR800 (HORLBA JY) using powers sample.

Measurement of UV-Vis Spectrum. The UV-Vis was determined by a TU-1810 spectrophotometer and the samples were using suspension.

Photoresponse of Nanocomposite to Visible Light and NIR. The nanocomposite suspension was casted on the interdigital electrodes of Au on flexible PET film substrate, after drying, the photoconductive responses to weak visible light (20–25 W) and 808 nm NIR with low-power were determined with LK2000A Electrochemical Work Station from LANLIKE Chemistry and Electron High Technology Co., Ltd (China).

Results and Discussion

It is found that, in the synthesis of NiS with hydrothermal method, the solubility of NiS/chitosan grafted polymer was increased with the amount of chitosan grafted polymer increasing. When 10, 20 mL above-prepared chitosan grafted polymer as soft template were added. The NiS/chitosan grafted polymer obtained is water-soluble, it is difficult to be washed. When 1 ml above-prepared chitosan grafted polymer as soft template were added. The obtained product was nanoparticle aggregation. In this paper, NiS/polymer nanocomposite used adding 5 mL above-prepared chitosan grafted polymer as soft template. The representative TEM images of NiS/chitosan grafted polymer is shown in Fig. 1a. The representative SEM images of NiS/chitosan grafted polymer integrated multi-functionality is shown in Fig. 1b.

Figure 1a indicated that the resulting NiS nanocomposite is core/shell structure, showing interconnected network, looks like honeycomb. NiS nanoparticle was coated with a thick layer of polymer. The size of NiS nanoparticle is about 10–50 nm. NiS nanoparticles were interconnected into network with aid of polymer.

Figure 1b also supported the interconnected network structure of NiS/chitosan grafted polymer nanocomposite.

The XRD pattern and FTIR of NiS/chitosan grafted polymer were carried out. The results are shown in Fig. 2a, b.

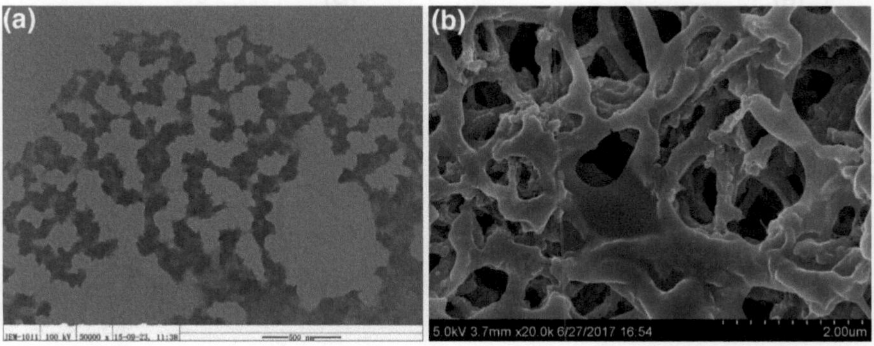

Fig. 1 TEM (**a**) and SEM (**b**) images of NiS/chitosan grafted polymer nanocomposite

Figure 2a exhibited that there are some diffraction peaks of NiS, such as 2θ in 18.53, 37.20, 40.28, 48.40, 59.16, et al. These peaks belongs to rhomb-centered lattice of NiS. There also presented some diffraction peaks of NiS_2, such as 2θ in 27.76, 31.71, 35.00, 38.74, 45.77, et al. This belongs to cubic lattice of NiS_2.

The FTIR curve of NiS/chitosan grafted polymer is very similar with that of pure chitosan. 3450 cm^{-1} is the stretching vibration of –OH, 3320 cm^{-1} belongs to the stretching vibration of N–H, 1680 cm^{-1} is the stretching vibration of –COOH group of PAA segment. 1600 cm^{-1} and 770–650 cm^{-1} is attributed to the stretching vibration of polystyrene segment of polymer soft template. Therefore, the NiS/chitosan grafted polymer nanocomposite contained some chitosan, PAA segment and polystyrene segment.

To pave the transferring channel for charge in NiS/chitosan grafted polymer nanocomposite, some GO solution were added in NiS/chitosan grafted polymer for integrated multi-functionalities. To confirm the presence of GO, Raman spectra was tested, and its UV-Vis curve was also recorded. The results are shown in Fig. 3a, b.

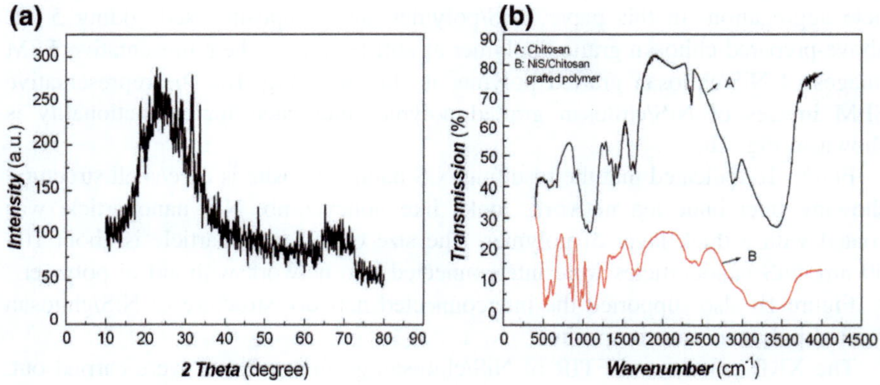

Fig. 2 XRD (**a**) and FTIR (**b**) images of NiS/chitosan grafted polymer

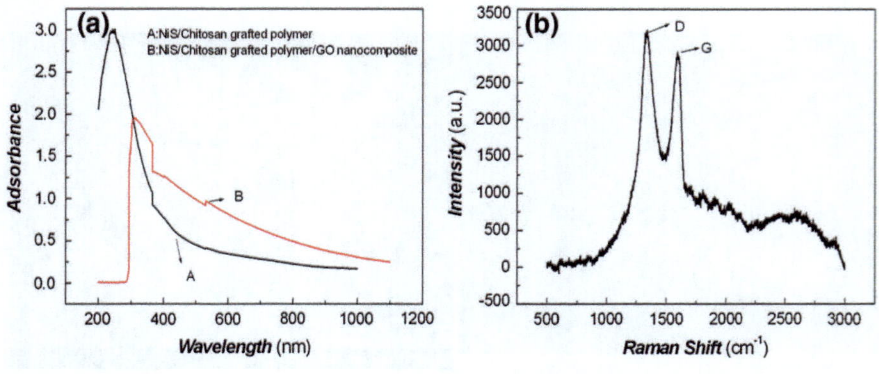

Fig. 3 UV-Vis (**a**) and Raman (**b**) spectra of NiS/chitosan grafted polymer/GO nanocomposite

Figure 3a showed that the adsorbance of NiS/chitosan grafted polymer was almost covered the whole region of visible light and extended to NIR (near-infrared), red shift was observed clearly by adding of GO.

As shown from Fig. 3b, D band (defect of C) and G band (C sp^2) were clearly observed in the nanocomposite. This illustrated that the nanocomposite had been integrated some GO.

The photoconductive properties of nanocomposite to weak visible light and 808 nm NIR were examined. The representative results are shown in Fig. 4a, b.

As shown from Fig. 4a, b, the nanocomposite integrated multi-functionalities showed good photo-switching behaviors to weak visible light and 808 nm NIR. The ratio of On/Off is two to three orders. It would be developed NIR detectors, NIR driven nano-carriers, photo-remote nanocomposites, et al.

Repeatability of photoresponse to weak visible light and 808 nm of NIR were also performed. Excellent results were obtained. The representative results are shown in Fig. 5. This illustrated that the nanocomposite integrated

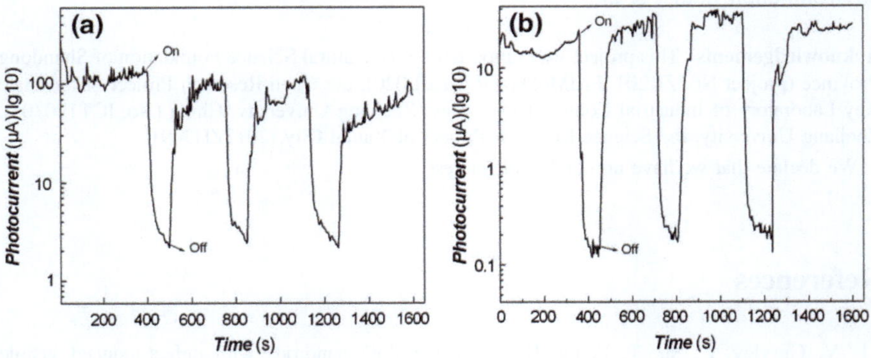

Fig. 4 Photoresponses of NiS/chitosan grafted polymer/GO nanocomposite to weak visible light (**a**) and 808 nm of 200 mW of NIR (**b**)

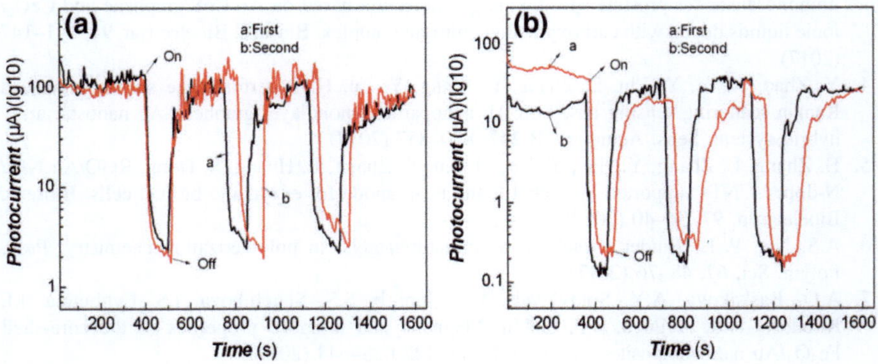

Fig. 5 Repeatability photoresponses of NiS/chitosan grafted polymer/GO nanocomposite to weak visible light (**a**) and 808 nm of 200 mW of NIR (**b**)

multi-functionalities showed good activities to visible light and NIR (near-infrared) light. Therefore, it would be developing organic-inorganic nanocomposites with external stimuli response, nanoreactors, self-cleaning films or nanocoatings, photocatalyst for treatment of organic pollutants, et al.

Conclusions

In summary, NiS nanomaterials were synthesized with amphiphilic polymers as soft template. The organic-inorganic nanocomposite with integrated multi-functionalities was prepared. The results indicated that the resulting NiS nanocomposite exhibited good photo-switching behaviors, the ratio of On/Off was 2–3 orders, the response and recovery were very rapidly. It would be developed the intelligent organic–inorganic nanocomposites with external stimuli responses, light detector to NIR, biomimetic materials, nanoreactors, self-cleaning films, coatings, environmental fields, et al.

Acknowledgements This project was supported by the Natural Science Foundation of Shandong Province (project No. ZR2013EMM008, ZR2016JL020), the Open Research Project of the State Key Laboratory of Industrial Control Technology, Zhejiang University, China (No. ICT170303), Zhejiang University, the Science Research Project of Yantai City (2013ZH349).

We declare that we have no conflict of interest.

References

1. V. Gurylev, C. Su, T. Perng, Hydrogenated ZnO nanorods with defect-induced visible light-responsive photoelectrochemical performance. Appl. Surf. Sci. **411**, 279–284 (2017)
2. I. Schmidt, A. Gad, G.R Scholz, H. Boht, M. Martens, M. Schilling, H.S. Wasisto, A. Waag, U. Schröder, Gold-modified indium tin oxide as a transparent window in photoelectronic diagnostics of electrochemically active biofilms. Biosens. Bioelectron. **94**, 74–80 (2017)
3. Y. Wei, X. Li, X. Sun, H. Ma, Y. Zhang, Q. Wei, Dual-responsive electrochemical immunosensor for prostate specific antigen detection based on Au-CoS/graphene and CeO_2/ionic liquids doped with carboxymethyl chitosan complex. Biosens. Bioelectron. **94**, 141–147 (2017)
4. Y. Zhao, X. Li, Y. Liu, L. Zhang, F. Wang, Y. Lu, High performance surface-enhanced Raman scattering sensing based on Au nanoparticle-monolayer graphene-Ag nanostar array hybrid system. Sens. Actuators, B **247**, 850–857 (2017)
5. H. Zhang, L. Zhang, Y. Han, Y. Yu, M. Xu, X. Zhang, L. Huang, S. Dong, RGO/Au NPs/N-doped CNTs supported on nickel foam as an anode for enzymatic biofuel cells. Biosens. Bioelectron. **97**, 34–40 (2017)
6. A.S. Nia, W.H. Binder, Graphene as initiator/catalyst in polymerization chemistry. Prog. Polym. Sci. **67**, 48–76 (2017)
7. A.O. Baskakova, A.Y. Solov'evab, Y.V. Ioni b, S.S. Starchikova, I.S. Lyubutina, I.I. Khodosc, A.S. Avilova, S.P. Gubin, Magnetic and interface properties of the core-shell Fe_3O_4/Au nanocomposites. Appl. Surf. Sci. **422** 638–644 (2017)

8. S. Xu, T. Gao, X. Feng, X. Fan, G. Liu, Y. Mao, X. Yu, J. Lin, X. Luo, Near infrared fluorescent dual ligand functionalized Au NCs based multidimensional sensor array for pattern recognition of multiple proteins and serum discrimination. Biosens. Bioelectron. **97**, 203–207 (2017)
9. G. Shi, Y. Li, G. Xi, Q. Xu, Z. He, Y. Liu, J. Zhang, J. Cai, Rapid green synthesis of gold nanocatalyst for high-efficiency degradation of quinclorac. J. Hazard. Mater. **335**, 170–177 (2017)
10. M. Huang, Y. Zhang, Y. Zhou, C. Zhang, S. Zhao, J. Fang, Y. Gao, X. Sheng, Synthesis and characterization of hollow ZrO_2–TiO_2/Au spheres as a highly thermal stability nanocatalyst. J. Colloid Interface Sci. **497**, 23–32 (2017)
11. J. Fang, Y. Zhang, Y. Zhou, S. Zhao, C. Zhang, M. Huang, Y. Gao, Synthesis of NiO–TiO_2 hybrids/mSiO_2 yolk-shell architectures embedded with ultrasmall gold nanoparticles for enhanced reactivity. Appl. Surf. Sci. **412**, 616–626 (2017)
12. L. Liu, P. Song, Q. Wei, X. Zhong, Z. Yang, Q. Wang, Synthesis of porous SnO_2 hexagon nanosheets loaded with Au nanoparticles for high performance gas sensors. Mater. Lett. **201**, 211–215 (2017)
13. M. Afsaria, A.A. Youzbashib, H. Nuraniana, S.M. Zahraee, Remarkable improvement of visible light photocatalytic activity of TiO_2 nanotubes doped sequentially with noble metals for removing of organic and microbial pollutants. Mater. Res. Bull. **94**, 15–21 (2017)
14. L. Huang, Y. Han, S. Dong, Highly-branched mesoporous Au–Pd–Pt trimetallic nanoflowers blooming on reduced graphene oxide as an oxygen reduction electrocatalyst. Chem. Commun. **52**, 8659–8662 (2016)
15. A.A. Oun, J. Rhim, Preparation of multifunctional chitin nanowhiskers/ZnO–Ag NPs and their effect on the properties of carboxymethyl cellulose-based nanocomposite film. Carbohydr. Polym. **169**, 467–479 (2017)
16. C. Liu, H. Tai, P. Zhang, Z. Ye, Y. Su, Y. Jiang, Enhanced ammonia-sensing properties of PANI–TiO_2–Au ternary self-assembly nanocomposite thin film at room temperature. Sens. Actuators, B **246**, 85–95 (2017)
17. M. Radoiˇci´ca, G.´Ciri´c-Marjanovi´cb, V. Spasojevi´ca, P. Ahrenkielc, M. Mitri´ca, T. Novakovi´cd, Z. Saponji´ca, Superior photocatalytic properties of carbonized PANI/ TiO_2 nanocomposites. Appl. Catal. B: Environ. **213**, 155–166 (2017)
18. Y. Li, J. Sun, L. Liu, F. Yang, A composite cathode membrane with $CoFe_2O_4$–rGO/PVDF on carbon fiber cloth: synthesis and performance in a photocatalysis-assisted MFCMBR system. Environ. Sci. Nano **4**, 335–345 (2017)
19. A.K. Solarajan, V. Murugadoss, S. Angaiah, High performance electrospun PVDF-HFP/SiO_2 nanocomposite membrane electrolyte for Li-ion capacitors. J. Appl. Polym. Sci. **134**, 45177 (2017)
20. I. Chae, S. Ahmed, H. B. Atitallah, J. Luo, Qing Wang, Zoubeida Ounaies, Seong H. Kim, Vibrational sum frequency generation (SFG) analysis of ferroelectric response of PVDF-based copolymer and terpolymer. Macromolecules **50**, 2838−2844 (2017)
21. K. Oniwa, H. Kikuchi, H. Shimotani, S. Ikeda, N. Asao, Y. Yamamoto, K. Tanigakia, T. Jin, 2-Positional pyrene end-capped oligothiophenes for high performance organic field effect transistors. Chem. Commun. **52**, 4800–4803 (2016)
22. A.S. Abd-El-Aziz, A.A. Abdelghani, B.D. Wagner, E.M. Abdelrehim, Aggregation enhanced excimer emission (AEEE) with efficient blue emission based on pyrene dendrimers. Polym. Chem. **7**, 3277–3299 (2016)
23. Y. Tan, M. Liang, P. Lou, Z. Cui, X. Guo, W. Sun, X. Yu, In situ fabrication of CoS and NiS nanomaterials anchored on reduced graphene oxide for reversible lithium storage. ACS Appl. Mater. Interfaces. **8**, 14488–14493 (2016)
24. D. Han, N. Xiao, B. Liu, G. Song, J. Ding, One-pot synthesis of core/shell-structured NiS@onion-like carbon nanocapsule as a high-performance anode material for Lithium-ion batteries. Mater. Lett. **196**, 119–122 (2017)

25. L. Zhang, Y. Huang, Y. Zhang, H. Gu, W. Fan, T. Liu, Flexible electrospun carbon Nanofi ber@NiS core/sheath hybrid membranes as binder-free anodes for highly reversible lithium storage. Adv. Mater. Interfaces **3**, 1500467 (2016)
26. X. Li, Y. Chen, J. Zou, X. Zeng, L. Zhou, H. Huang, Stable freestanding Li-ion battery cathodes by in situ conformal coating of conducting polypyrrole on NiS-carbon nanofiber films. J. Power Sour. **331**, 360–365 (2016)
27. J.H. Lee, S. Kim, S. Park, M. Kang, A p-n heterojunction NiS-sensitized TiO_2 photocatalytic system for efficient photoreduction of carbon dioxide to methane. Ceram. Int. **43**, 1768–1774 (2017)
28. H. Luo, B. Wang, Y. Li, T. Liu, W. You, D. Wang, Core-shell structured Fe_3O_4@NiS nanocomposite as high-performance anode material for alkaline nickel-iron rechargeable batteries. Electrochim. Acta **231**, 479–486 (2017)
29. Y. Lee, C.V.V.M. Gopi, M. Venkata-Haritha, S.S. Rao, H. Kim, Electrochemical growth of NiS nanoparticle thin film as counter electrode for quantum dot-sensitized solar cells. J. Photochem. Photobiol. A: Chem. **332**, 200–207 (2017)
30. H. Kim, S. Suh, S.S. Rao, D. Punnoose, C.V. Tulasivarma, C.V.V.M. Gopi, N. Kundakarla, S. Ravi, I.K. Durga, Investigation on novel CuS/NiS composite counter electrode for hindering charge recombination in quantum dot sensitized solar cells. J. Electroanal. Chem. **777**, 123–132 (2016)
31. H. Kim, H. Lee, S.S. Rao, A.E. Reddy, S. Kim, C.V. Thulasi-Varm, Well-dispersed NiS nanoparticles grown on a functionalized CoS nanosphere surface as a high performance counter electrode for quantum dot sensitized solar cells. RSC Adv. **6**, 29003–29019 (2016)
32. J. Gou, S. Xie, Z. Yang, Y. Liu, Y. Chen, Y. Liu, C. Liu, A high-performance supercapacitor electrode material based on NiS/ Ni_3S_4 composite. Electrochim. Acta **229**, 299–305 (2017)
33. B. Guan, Y. Li, B. Yin, K. Liu, D. Wang, H. Zhang, C. Cheng, Synthesis of hierarchical NiS microflowers for high performance asymmetric supercapacitor. Chem. Eng. J. **308**, 1165–1173 (2017)
34. Z. Li, A. Gu, J. Sun, Q. Zhou, Facile hydrothermal synthesis of NiS hollow microspheres with mesoporous shells for high-performance supercapacitors. New J. Chem. **40**, 663–1670 (2016)
35. W. Li, S. Wang, L. Xin, M. Wu, X. Lou, Single-crystal b-NiS nanorod arrays with a hollowstructured Ni_3S_2 framework for supercapacitor Applications. J. Mater. Chem. A **4**, 700–7709 (2016)
36. A.M. Patil, A.C. Lokhande, N.R. Chodankar, V.S. Kumbhar, C.D. Lokhande, Engineered morphologies of β-NiS thin films via anionic exchange process and their supercapacitive performance. Mater. Des. **97**, 407–416 (2016)
37. Y. Xin, Y. Lu, C. Han, L. Ge, P. Qiu, Y. Li, S. Fang, Novel NiS cocatalyst decorating ultrathin 2D TiO_2 nanosheets with enhanced photocatalytic hydrogen evolution activity. Mater. Res. Bull. **87**, 123–129 (2017)
38. J.S. Chen, J. Ren, M. Shalom, T. Fellinger, M. Antonietti, Stainless steel mesh-supported NiS nanosheet array as highly efficient catalyst for oxygen evolution reaction. ACS Appl. Mater. Interfaces. **8**, 509–5516 (2016)
39. N. Jiang, Q. Tang, M. Sheng, B. You, D. Jiang, Y. Sun, Nickel sulfides for electrocatalytic hydrogen evolution under alkaline conditions: a case study of crystalline NiS, NiS_2, and Ni_3S_2 nanoparticles. Catal. Sci. Technol. **6**, 1077–1084 (2016)
40. J. Huang, Z. Shi, X. Dong, Nickel sulfide modified TiO_2 nanotubes with highly efficient photocatalytic H_2 evolution activity. J. Energy Chem. **25**, 136–140 (2016)
41. A.L. Luna, E. Novoseltceva, E. Louarn, P. Beaunier, E. Kowalska, B. Ohtani, M.A. Valenzuela, H. Remitaa, C. Colbeau-Justin, Synergetic effect of Ni and Au nanoparticles synthesized on titania particles for efficient photocatalytic hydrogen production. Appl. Catal. B: Environ. **191**, 18–28 (2016)
42. M. Mollavali, C. Falamaki, S. Rohani, High performance NiS-nanoparticles sensitized TiO_2 nanotube arrays for water reduction. Int. J. Hydrogen Energy **41**, 5887–5901 (2016)

43. H. Derikvandi, A. Nezamzadeh-Ejhieh, Comprehensive study on enhanced photocatalytic activity of heterojunction ZnS–NiS/zeolite nanoparticles: Experimental design based on response surface methodology (RSM), impedance spectroscopy and GC-MASS studies. J. Colloid Interface Sci. **490**, 652–664 (2017)
44. X. Tang, L. Gan, Y. Duan, Y. Sun, Y. Zhang, Z. Zhang, A novel Cd^{2+} imprinted chitosan-based composite membrane for Cd^{2+} removal from aqueous solution. Mater. Lett. **198**, 121–123 (2017)
45. L. Ramos-Rivera, M. Distaso, W. Peukert, A.R. Boccaccini, Electrophoretic deposition of anisotropic a-Fe_2O_3/PVP/chitosan nanocomposites for biomedical applications. Mater. Lett. **200**, 83–86 (2017)
46. P.A. Nishad, A. Bhaskarapillai, S. Velmurugan, Enhancing the antimony sorption properties of nano titania-chitosan beads using epichlorohydrin as the crosslinker. J. Hazard. Mater. **334**, 160–167 (2017)
47. J. Li, B. Jiang, Y. Liu, C. Qiu, J. Hu, G. Qian, W. Guo, H.H. Ngo, Preparation and adsorption properties of magnetic chitosan composite adsorbent for Cu^{2+} removal. J. Cleaner Prod. **158**, 51–58 (2017)
48. V.K. Gupta, A. Fakhri, S. Agarwal, M. Azad, Synthesis and characterization of Ag_2S decorated chitosan nanocomposites and chitosan nanofibers for removal of lincosamides antibiotic. Int. J. Biol. Macromol. **103**, 1–7 (2017)
49. S. Frindy, A. Primo, H. Ennajih, A. Qaiss, R. Bouhfid, M. Lahcini, E.M. Essassi, H. Garcia, A.E. Kadib, Chitosan–graphene oxide films and CO_2-dried porous aerogel microspheres: interfacial interplay and stability. Carbohydr. Polym. **167**, 297–305 (2017)
50. T. Tolessa, X. Zhou, M. Amde, J. Liu, Development of reusable magnetic chitosan microspheres adsorbent for selective extraction of trace level silver nanoparticles in environmental waters prior to ICP-MS analysis. Talanta **169**, 91–97 (2017)
51. M. Lian, X. Chen, X. Liu, Z. Yi, W. Yang, A self-assembled peptide nanotube–chitosan composite as a novel platform for electrochemical cytosensing. Sens. Actuators, B **251**, 86–92 (2017)
52. A.R. Sadrolhosseini, M. Naseri, S.A. Rashid, Polypyrrole-chitosan/nickel-ferrite nanoparticle composite layer for detecting heavy metal ions using surface plasmon resonance technique. Opt. Laser Technol. **93**, 216–223 (2017)
53. J. Lin, Z. Liu, Q. Zhu, X. Rong, C. Liang, J. Wang, D. Ma, J. Sun, G. Wang, Redox-responsive nanocarriers for drug and gene co-delivery based on chitosan derivatives modified mesoporous silica nanoparticles. Colloids Surf. B: Biointerfaces **155**, 41–50 (2017)
54. M. Song, L. Li, Y. Zhang, K. Chen, H. Wang, R. Gong, Carboxymethyl-β-cyclodextrin grafted chitosan nanoparticles as oral delivery carrier of protein drugs. React. Funct. Polym. **117**, 10–15 (2017)
55. L. Racine, G. Costa, E. Bayma-Pecita, I. Texiera, R. Auzély-Velty, Design of interpenetrating chitosan and poly(ethylene glycol) sponges for potential drug delivery applications. Carbohydr. Polym. **170**, 166–175 (2017)
56. L. Liu, L. Mirandola, M. Chiriva-Internati, J. Chaudhuri, CdS quantum dot-chitosan-anti SP17 nanohybrid as a potential cancer biomarker. Mater. Lett. **199**, 5–8 (2017)
57. X. Zhang, G. Xiao, Y. Wang, Y. Zhao, H. Su, T. Tan, Preparation of chitosan-TiO_2 composite film with efficient antimicrobial activities under visible light for food packaging applications. Carbohydr. Polym. **169**, 101–107 (2017)
58. X. Ma, B. Zhang, Q. Cong, X. He, M. Gao, G. Li, Organic/inorganic nanocomposites of ZnO/CuO/chitosan with improved properties. Mater. Chem. Phys. **178**, 88–97 (2016)
59. Q. Cong, X. He, M. Gao, X. Ma, G. Li, ZnO/CuS heterostructured nanocomposite and its organic functionalization. Mater. Res. Innovations **18**, 740–746 (2014)
60. Q. Cong, H. Geng, X. He, M. Gao, X. Ma, G. Li, Surface modification of ZnO nanosheets with Au/polyaniline and their properties. Mater. Res. Innovations **18**, 30–36 (2014)

61. X. Ma, M. Wang, G. Li, H. Chen, R. Bai, Preparation of polyaniline-TiO_2 composite film with in-situ polymerization approach and its gas-sensitivity at room temperature. Mater. Chem. Phys. **98**, 241–247 (2006)
62. X. Ma, B. Zhang, Q. Cong, X. He, M. Gao, G. Li, Highly-enhanced performance of TiO_2 nanotube attached CdS quantum dots. Curr. Nanosci. **12**, 500–507 (2016)
63. J. Zheng, X. Ma, X. He, M. Gao, G. Li, Preparation, characterizations, and its potential applications of PANi/graphene oxide nanocomposite. Procedia Eng. **27**, 1478–1487 (2012)

Crystal Structure and Magnetic Properties of $FePd/Si_3N_4$ Composite Films

Lei Ma, Xin Zhou and Tao Liu

Abstract The ferromagnetic $FePd/Si_3N_4$ composite films were prepared by magnetron sputtering and then annealed at 500, 550, 600 and 700 °C for 3 h. The results show that the Si_3N_4 top layer can effectively inhibit the growth of FePd particles and refine the grain of FePd thin films. The crystallization of $FePd/Si_3N_4$ films increase with the rise of heat treatment temperature, and the fcc-fct phase transition process of the film increases further. When the annealing temperature rises to 700 °C, the diffraction peaks of the ordered phase are degraded instead. The reason is that the high annealing temperature disrupts the ordered arrangement of Fe and Pd, resulting in an increase in the composition of the disorder phase in the sample. The results of magnetic properties show the performance of the film was the best when the heat treatment was carried out at 550 °C for 3 h.

Keywords $FePd/Si_3N_4$ films · Magnetic properties · Annealing temperature Magnetron sputtering

Introduction

Ultra-high density magnetic recording means recording as much data as possible on a limited disk surface, and the number of recorded data is directly dependent on the size of the recording unit, so the volume of the recording unit or the grain size of the magnetic particles must be as small as possible [1]. In addition, in order to improve the signal-to-noise ratio, it is also required to reduce the magnetic coupling interaction between the grains as much as possible. However, in order to ensure the

L. Ma (✉) · X. Zhou · T. Liu
School of Materials Science and Engineering,
Guilin University of Electronic Technology, Guilin 541004, China
e-mail: malei@guet.edu.cn

L. Ma
Guangxi Key Laboratory of Information Materials,
Guilin University of Electronic Technology, Guilin 541004, China

Y. Han (ed.), *Advanced Functional Materials*,
https://doi.org/10.1007/978-981-13-0110-0_16

thermal stability and the uniform reversal of the magnetic particles, the coupling effect should be as large as possible to overcome the effects of demagnetization and thermal disturbances, so the magnetic recording material must have a moderate intergranular magnetic coupling [2]. Therefore, how to reduce the FePd grain size and the magnetic coupling between FePd particles is an important research direction in recent years. At present, the most effective method is to embed FePd ordered particles in non-magnetic metal or compound matrix to prepare nanoparticles with nano-sized FePd particles. Nanocomposite thin films are usually composed of nanoscale ferromagnetic particles which are embedded in a nonmagnetic matrix (C, Ag, Si_3N_4) [3–6]. There are two ways to prepare a granular film: one is a co-sputtering method, another method is multilayer film [7–9]. For a multilayer film on a glass substrate, the coercivity (320–1000 kA/m) can be adjust by controlling the content of the granular film and the film thickness, and the grain size is controllable (4–17 nm). This type of granular film has a weak magnetic coupling effect and a high magnetic anisotropy [6, 10–13].

In this paper, FePd/Si_3N_4 composite films were designed and prepared by multilayer films. It is found that the intergranular distance, average grain size, and particle shape of FePd particles can be affected greatly by a nonmagnetic Si_3N_4 matrix and volume fraction of matrix during heat treatment. Additionally, the FePd/Si_3N_4 composite layer shows better oxidation resistance and corrosion resistance, due to the metal particles being surrounded by an insulating Si_3N_4 matrix. In general, after annealed time at 3 h, the structure of samples could transform to an ordered face centered tetragonal (fct) structure [14]. In this study, the phase change, crystal structure and hysteresis loops were characterized when the FePd thin films is covered by an ultrathin Si_3N_4 layer.

Experimental Details

FePd thin films (47 $\pm$ 2.0 nm thickness) were prepared on quartz glass substrates by using DC magnetron sputtering. In order to avoid oxidation, an ultrathin Si_3N_4 layer (10 $\pm$ 2.0 nm thickness) was deposited on the top of the FePd films using rf magnetron sputtering. The purity of Fe and Pd were better than 99.99%. The base pressure of sputtering chamber is 6.0×10^{-5} Pa. The working gas is high-purity argon (99.99% purity). During sputtering process, the working gas pressure, the gas flow rate were maintained at 0.9 Pa, 24 SCCM, respectively. The samples were sealed in a clean quartz tube, and then annealed in vacuum of 3.0×10^{-3} Pa at 500, 550, 600, 700 °C for 3 h, respectively. The crystal structure and hysteresis loops were measured by X-ray diffraction (XRD), physical properties measurement system (PPMS-9 T), atomic force microscopy (AFM) and scanning electron microscopy (SEM). The maximum magnetic field used was $\pm$3 T. The Fe/Pd atomic ratio was 55:45 by energy dispersive X-ray spectrometer (EDS).

Results and Discussion

Figure 1 reveals XRD patterns of pure FePd and FePd/Si_3N_4 thin films with various annealing temperature for 3 h. It is observed when the annealing temperature is 500 °C, the (200) diffraction peak of the film is split into (200) and weak (002) diffraction peaks of the ordered $L1_0$ phase, indicating that the film has fcc–fct phase transition. With the increase of annealing temperature, the fcc–fct phase transition process further increased. After 600 °C, the peak intensity of the (002) diffraction peak is more obvious, indicating that the phase transition of the film has been completed. When the annealing temperature rises to 700 °C, the diffraction peaks of the ordered phase are degraded, the (002) diffraction peak becomes extremely weak, and some satellite peaks appear near the (200) diffraction peak. The reason is the high annealing temperature disrupts the ordered arrangement of Fe and Pd, resulting in an increase in the composition of the disorder phase in the sample.

Figure 2, obtained from Fig. 1 by using JADE 6.0 software, shows the grain size and c/a ratio of the FePd/Si_3N_4 thin films at different annealing temperature. As a general rule, the grain size of pure FePd thin films increases with the rise of temperature. However, for the FePd/Si_3N_4 thin films, the grain size decreased firstly and then increased with the temperature. When the temperature increases from 500 to 600 °C, the grain size of the FePd thin film is very small which is only 19.5 nm, showing that ceramic Si_3N_4 top layer helps to restrain the growth of the FePd metal particles, which directly leads to the reduction of the intergranular exchange coupling. The reason may be Si_3N_4 ceramic phase distributed between the FePd nano grains, or some Si_3N_4 matrix diffuses into the grain boundaries of FePd particles. The interfacial regulation of non-magnetic particles can effectively reduce the size of FePd magnetic grain. With the temperature rising to 700 °C, the grain size increased to 21.1 nm. This is because the higher temperature causes the growth of

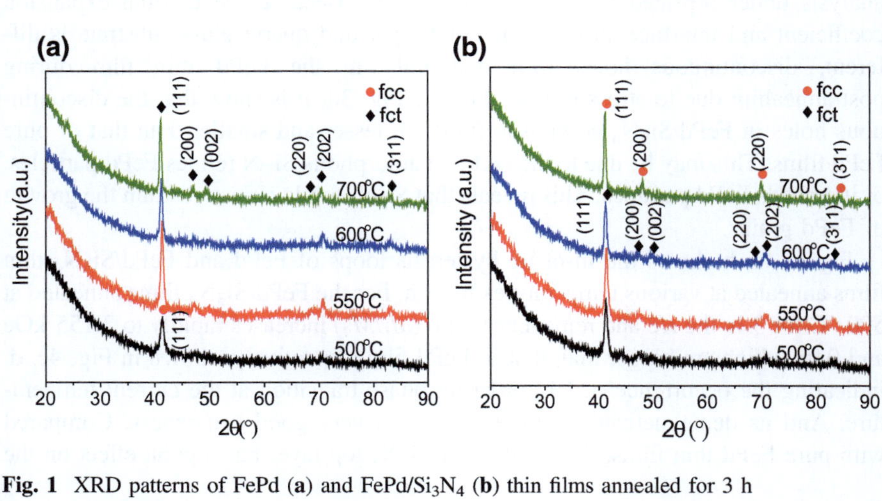

Fig. 1 XRD patterns of FePd (**a**) and FePd/Si_3N_4 (**b**) thin films annealed for 3 h

Fig. 2 Grain sizes and c/a ratio of the $FePd/Si_3N_4$ thin films versus various annealing temperature for 3 h

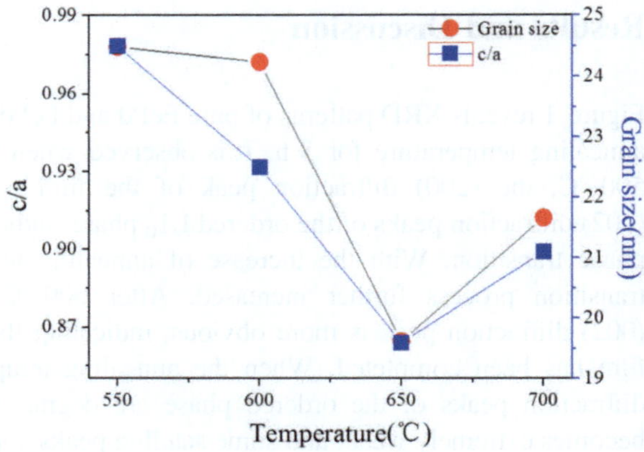

grain, which is beyond the effect of fine grains of Si_3N_4 top layer. The c/a curve shows that the c/a ratio decreases from 500 to 650 °C, but rises at 700 °C. The smaller c/a ratio indicates that the ordering degree is optimal when the sample is annealed at ~600 °C. The minimum c/a value is close to 0.861, indicating that the order degree is complete nearly. But, the temperature of 700 °C destroyed the $L1_0$ phase, and the order degree is reduced. This result reveals that Si_3N_4 top layer is beneficial for the improvement of the crystal structure of FePd thin films.

Figure 3 shows AFM and SEM images of the FePd (a, c) and $FePd/Si_3N_4$ (b, d) thin films as-prepared (a, b) and annealed (c, d) at 550 °C for 3 h. For the as-prepared films, it shows that the surface appearance is loose and homogeneous. Compared with pure FePd films, the particles of $FePd/Si_3N_4$ composite films are more obvious, subtle and homogeneous. With heating to 550 °C for 3 h, the surface morphology of $FePd/Si_3N_4$ composite films becomes compact and aggregation. In addition, discontinuous holes and stratification were clearly observed in FePd films, but the proportion of ingredients were still $Fe_{55}Pd_{45}$ determined by the EDS analysis under repeated tests, shown in Fig. 3e. Because the thermal expansion coefficient and interface structure of FePd layer and quartz glass substrate is different, discontinuous holes form randomly in the FePd thin film during post-annealing due to stress release. For the Fig. 3d, it is show that the discontinuous holes in $FePd/Si_3N_4$ composite films are lesser and smaller than that of pure FePd films. This may be due to the fact that amorphous Si–N refines FePd particles, or it embeds FePd particles. This reveals that Si_3N_4 top layer can restrain the growth of FePd grains.

Figure 4a, b shows the in-plane hysteresis loops of FePd and $FePd/Si_3N_4$ thin films annealed at various temperatures for 3 h. For the $FePd/Si_3N_4$ films annealed at 500 °C for 3 h, the *H*c and remanence ratio (*Mr*/*M*s) increases rapidly to 2.155 kOe and 0.98 which are bigger than that of FePd film respectively, shown in Fig. 4c, d, indicating the occurrence of $L1_0$ ordering phase transition at the current temperature. And its demagnetization curve shape is a very good squareness. Compared with pure FePd thin films, the addition of Si_3N_4 top layer has a great effect on the

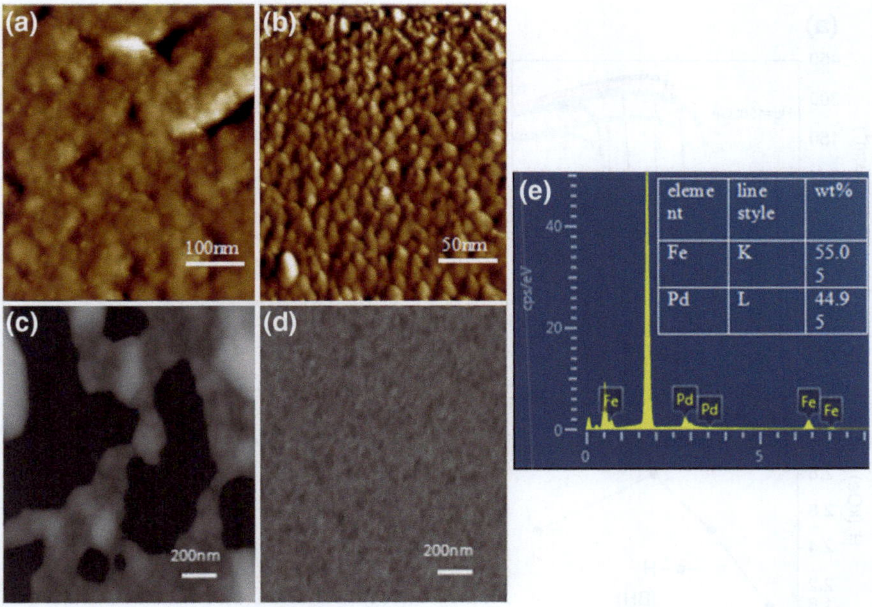

Fig. 3 AFM and SEM images of thin films: as-deposited FePd (**a**) and $FePd/Si_3N_4$ (**b**) films, FePd (**c**) and $FePd/Si_3N_4$ films (**d**) annealed at 550 °C for 3 h, and **e** EDS of annealed FePd film

shape of hysteresis loops of FePd thin film, which may be the reduction of magnetic FePd particle size with the small amount of Si_3N_4 into FePd grains. As the heat treatment temperature rises to 550 °C, the *Hc* and magnetic energy product [$(BH)_{max}$] of the $FePd/Si_3N_4$ composite film increases to 2.891 kOe and 15.5 MGOe respectively, and the shape exhibits an approximative rectangle, which is very consistent with the requirements of ultra-high density magnetic recording media. The above description shows that the addition of Si_3N_4 top layer has a positive effect on the fcc–fct phase transition process. When annealing temperature further increases to 600 °C, the *Mr*/*Ms* of the $FePd/Si_3N_4$ film decreases slightly to 0.96, and the *Hc* is further increased to 3.067 kOe, which may be due to the enhancement of the interfacial effect of Si_3N_4 and FePd layer and the defects in the crystal. However, as the annealing temperature rose to 700 °C, it was observed that the *Hc*, *Mr*/*Ms* and $(BH)_{max}$ were degraded to 2.186 kOe, 0.88 and 4.7 MGOe, respectively. This is attributed to the high annealing temperature which destroys the order degree of the film. In addition, from the hysteresis loops, it is also clearly observed that the *Ms* gradually reduced with continues improvement of annealing temperature, and the addition of Si_3N_4 top layer can reduce the annealing temperature of the fcc–fct phase transition. The results show that the addition of Si_3N_4 nonmetallic layer has a significant effect on maintaining the hard magnetic properties and improving the *Hc* and $(BH)_{max}$ of FePd films.

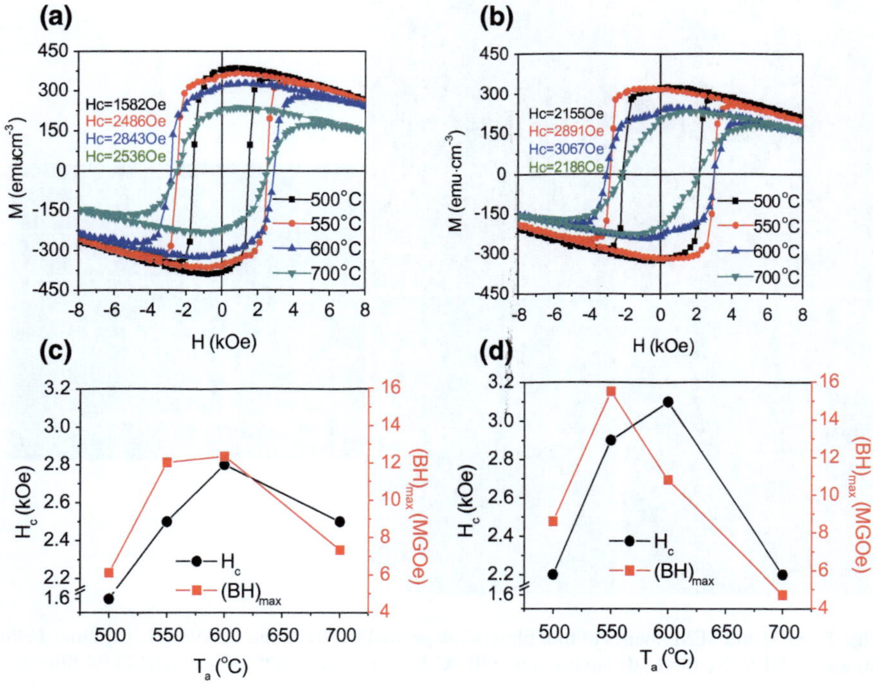

Fig. 4 In-plane hysteresis loops for FePd (**a**) and FePd/Si_3N_4 (**b**) films annealed at different temperatures for 3 h measured at 300 K, and H_c and $(BH)_{max}$ versus T_a for FePd (**c**) and FePd/Si_3N_4 (**d**) films

Conclusions

The effects of the Si_3N_4 top layer (10 ± 2.0 nm thickness) on the crystal structure and magnetic properties of FePd thin films is investigated. For FePd/Si_3N_4 composite films, the (002), (200) peaks became strong and the (001) superlattice peaks were observed. Compared with the FePd films, the lower c/a for FePd/Si_3N_4 films at the same temperature suggests that Si_3N_4 top layer promotes order phase transition. The Hc and $(BH)_{max}$ of FePd/Si_3N_4 film annealed at 550 °C increase to 2.891 kOe and 15.5 MGOe respectively, while the Mr/Ms is almost 1 and hysteresis loops show a good squareness. At 600 °C, the Hc of the FePd/Si_3N_4 film reaches a maximum of 3.067 kOe. The addition of Si_3N_4 nonmetallic layer has a significant effect on maintaining the hard magnetic properties of FePd films. The presented results indicate that Si_3N_4 top layer is beneficial for promotion of $L1_0$ ordered phase transformation, and enhanced magnetic properties of FePd thin films.

Acknowledgements This work is supported by the Guangxi Natural Science Foundation (2016GXNSFAA380030, 2016GXNSFGA380001), GUET Excellent Graduate Thesis Program (16YJPYSS32), Innovation Project of GUET Graduate Education (2018YJCX84) and the National Natural Science Foundation of China (51461012).

References

1. R. Wood, The feasibility of magnetic recording at 1 Terabit per square inch[J]. IEEE Trans. Magn. **36**, 36–42 (2000)
2. D. Weller, A. Moser, L. Folks, M.E. Best, Wen Lee, M.F. Toney, M. Schwickert, J.U. Thiele, M.F. Doerner, High K/sub u/ materials approach to 100 Gbits/in/sup 2/. IEEE Trans. Magn, **36**, 10–15 (2002)
3. J.F. Hu, J.S. Chen, Y.F. Ding, Exchange coupling assisted FePtC perpendicular recording media. Appl. Phys. Lett. **93**, 150 (2008)
4. Y. Tokuoka, Y. Seto, S. Iwata, Effect of Ag addition to $L1_0$ FePt and $L1_0$ FePd films grown by molecular beam epitaxy. J. Appl. Phys. **115**, 17 (2014)
5. T. Nagata, Y. Tokuoka, T. Kato, D. Oshima, S. Iwata, Structure and magnetic properties of (001) oriented CoPt-Ag and CoPd-Ag alloy films. J. Magn. Soc. Jpn. **40**, 17–20 (2016)
6. C.M. Kuo, P.C. Kuo, Magnetic properties and microstructure of FePt-Si_3N_4 nanocomposite thin films. J. Appl. Phys. **87**, 419–426 (2000)
7. C.P. Luo, S.H. Liou, L. Gao, Y. Liu, D.J. Sellmyer, Nanostructured FePt:B_2O_3 thin films with perpendicular magnetic anisotropy. Appl. Phys. Lett. **77**, 2225–2227 (2000)
8. Z. Qiu, L. Zhao, W. Wang, Z. Li, Microstructure and magnetic properties of FePt-MgO granular thin films fabricated by Co-sputtering method on the single crystal MgO substrate. IEEE (2015). https://doi.org/10.1109/INTMAG.2015.7157520
9. L. Zhou, T. Liu, S.Q. Zhao, L. Li, X.F. Wu, Y.L. Zhang, G. Cheng, Microstructure and magnetic properties of self-assembled FePd-SiN films. J. Mater. Sci.: Mater. Electron. **27**, 7420–7424 (2016)
10. D. Halley, A. Marty, Bayleguillemaud P. Chemical order and selection of the mechanism for strain relaxation in epitaxial FePd(Pt) thin layers. Phys. Rev. B **70**, 195–323 (2004)
11. S. Bahamida, A. Fnidiki, A. Laggoun, A comparative structural and magnetic study of $Fe_{100-x}Pd_x$ (x = 15, 20 and 36) thin films deposited on Si (100) and glass substrates. J. Magn. Magn. Mater. **392**, 139–147 (2015)
12. C. Clavero, J.M. García-Martín, J.L. Costa Krämer, G. Armelles, A. Cebollada, Y. Huttel, R. A. Lukaszew, A.J. Kellock, Temperature and thickness dependence at the onset of perpendicular magnetic anisotropy in FePd thin films sputtered on MgO(001). Phys. Rev. B **73,** 174405 (2006)
13. M. Kohda, S. Iimori, R. Ohsugi, H. Naganuma, T. Miyazaki, Y. Ando, J. Nitta, Structural and magnetic properties of $L1_0$-FePd/MgO films on GaAs and InP lattice mismatched substrates. Appl. Phys. Lett. **102**, 790 (2013)
14. T. Liu, L. Ma, S.Q. Zhao, D.D. Ma, L. Li, G. Cheng, G.H. Rao, Promotion of $L1_0$ ordered phase transformation by the Si_3N_4 TOP LAYER on FePd thin films. J. Supercond. Novel Magn. **30**, 3215–3219 (2017)

Preparation of Low-Dimensional Carbon Nanomaterials and Its Improvement of Visible Light Activity

Shisheng Lv, Lin Bian, Jianxun Qiu, Wei Du, Xintao Zhang, Xiaochun He, Mingjun Gao, Xiangming Li, Xingfa Ma and Guang Li

Abstract ZnO and its nanocomposites have been received good progresses in new energy, photocatalysts, environmental fields. However, to enhance the visible light activity of ZnO is still a great challenge. Among so many methods of modifying ZnO, carbon-doping or nanocomposite with carbon nanomaterials are simple and effective approaches. In this study, low-dimensional carbon nanomaterials were synthesized with biomass waste (such as com straw) referring to the method of chemical synthesis of graphene oxide. ZnO nanosheets/carbon nanocomposites were prepared utilizing the interaction between surface chemical groups of ZnO nanosheets and low-dimensional carbon nanomaterials produced with biomass waste. A series of characterizations were carried out. The results indicated that the band edge of nanocomposite was extended to the whole visible light and NIR. The photoconductive response to weak visible light and 808 nm NIR (near infrared) were studied based on interdigital electrodes of Au on flexible polymer film substrate with casting method. The results indicated that the ZnO/carbon

S. Lv · L. Bian · J. Qiu · W. Du · X. Zhang · X. He · M. Gao · X. Li · X. Ma (✉)
School of Environmental and Material Engineering, Center of Advanced Functional Materials, Yantai University, Yantai 264005, China
e-mail: xingfazju@163.com; xingfamazju@aliyun.com

S. Lv
e-mail: lvss19920119@qq.com

L. Bian
e-mail: 962251823@qq.com

J. Qiu
e-mail: qjxun@163.com

W. Du
e-mail: dw311@yeah.net

X. Zhang
e-mail: zxtyt@163.com

X. He
e-mail: hexiaochun76@126.com

M. Gao
e-mail: gmj2188@126.com

Y. Han (ed.), *Advanced Functional Materials*,
https://doi.org/10.1007/978-981-13-0110-0_17

nanocomposite holded photo-switching behavior, the response and recovery were very rapidly. This illustrated that the resulting nanocomposites showed good activities to weak visible light and NIR. Photocatalytic efficiency was examined by selecting typical organic pollutants, some good results were also obtained. It would be potential applications in photocatalysts, self-cleaning films, NIR detector, coatings, and organic pollutants treatment of environmental fields. It would be low-cost, effective, general method to treat the surface of several metal oxides.

Keywords Low-dimensional carbon nanomaterials · Biomass waste
Nanocomposite · Photo-induced charge generation · Photoconductive switches

Introduction

Recently, the demands of clean energy and environmental promoted the great progresses of advanced functional materials, especially carbon nanomaterials (carbon nanotube, graphene, C_{60}, carbon dots), inorganic functional materials, noble metals, functional polymers, et al. Among inorganic functional materials, ZnO is one of hot-spots materials due to its mutil-functionalities and low-cost, and has applications in chemical sensors [1–5], biosensors [6, 7], photodetectors [8–10], photocatalysts [11–16], light-responsive switches [17, 18], organic/inorganic nanocomposites with tailoring physical properties [19–24], et al. Due to the band edge of ZnO is in UV region, this is not favor to make full use of the visible light for applications in energy and environmental fields. To enhance the activity of visible light, a large number of references are focused on ZnO based nanocomposites include morphology tailoring, C, N, B doping, modification with carbon nanomaterials, metal oxides, polymers, noble metals, metal sulfides, p-n junctions, heterostructured nanocomposite, and so on.

Currently, carbon nanomaterials, especially graphene or graphene oxide are still received more attentions for various applications or modifying other materials [25–28]. In the synthesis aspects of carbon nanomaterials, one important trend is utilizing different carbon resources. For example, Chen and co-workers [25] prepared hierarchically porous carbon with cotton fabric for sustainable capacitor electrode. Xu and co-workers [26] synthesized graphene-like carbon from organic-inorganic templates for energy storage. Lim and co-workers [27] prepared structurally tunable carbon anode from sustainable source for lithium storage. Zhu and co-workers [28]

X. Li
e-mail: li_xiangming1981@126.com

G. Li
National Laboratory of Industrial Control Technology, Institute of Cyber-Systems and Control, Zhejiang University, Hangzhou 310027, China
e-mail: guangli@zju.edu.cn

synthesized highly photoluminescent carbon dots using citric acid and ethylenediamine as raw materials. *Wang and* co-workers [29] fabricated counter electrode for efficient dye-sensitized solar cells with carbonized eggshell membranes. Wu and co-workers [30] synthesized highly luminescent nitrogen-doped carbon dots for bioimaging from silk–natural proteins. Primo and co-workers [31] prepared large-area, high-quality, N-doped graphene utilizing biomass wastes. Sun and and co-workers [32] synthesized porous graphene-like nanosheets using coconut shell for high-power supercapacitors. Akhavan and co-workers [33] prepared graphene from natural and industrial carbonaceous wastes. Adolfsson and co-workers [34] valorized cellulose and waste paper to graphene oxide quantum dots. Pei and co-workers [35] synthesized carbon nanodots with highly efficient fluorescence switching by CO_2. Ortega-Liebana and co-workers [36] prepared carbon nanodots by rapid pyrolysis of organic precursors confined within nanoporous templating structures. These researches widen the ranges and applications of carbon nanomaterials. Carbon nanomaterials have widely applications of energy, environmental, biomedical fields due to their good biocompatibility and outstanding physical properties, which involving in bioimaging [37], photocatalytic energy conversion [38], resistive switching memory devices [39], flexible devices [40, 41], supercapacitors [42, 43], transparency-switchable actuator [44], Li–O_2 batteries [45], photoelectric devices and sensors [46–49], as visible light active sensitizers to ZnO for solar water splitting [50], adsorbents for pollutants [51–54], multifunctional hybrid materials [55–57], et al.

For many years, the authors are very interested in polymer functional materials or organic-inorganic functional nanocomposites and their properties studies [58–68]. The focuses are put on the sensing behaviors and photoconductive properties of nanocomposite. Materials system involves in polymers, metals oxides, metals sulfites, carbon nanomaterials (CNTs, graphene, C_{60}, nanostructured carbon materials). Synthesis of graphene oxides are used chemical oxidation utilizing graphite and carbon nanotubes (unzipping method of carbon nanotube) in our previous studies. To pursue carbon nanomaterials with low-cost, low-dimensional carbon nanomaterials were prepared with several biomass wastes (such as com straw, et al.) refer to the method of chemical synthesis of graphene oxide. Herein, carbon nanomaterials derived by com straw was introduced, and the enhancement of visible light activity to ZnO nanosheets was also reported.

Experimental Details

Materials. Zinc nitrate (AR), potassium permanganate (AR), sulfuric acid (Chemical Reagent, i.e. CP), nitric acid (CP), hydrochloric acid (AR), sodium hydroxide (AR) were purchased from Sinopharm Chemical Reagent Co., Ltd. Hexamethylenetetramine (AR), methylene blue (AR), were purchased from Tianjin Bodi Chemical Co., Ltd, et al. Com straw was available elsewhere.

Synthesis of ZnO Nanosheets. ZnO nanosheets was synthesized by a hydrothermal approach according to the previous report [64].

Syntheisis of Carbon Nanomaterials by Biomass Waste. The preparation of carbon nanostructure oxide was using chemical method, which is according to the previous report of synthesis of graphene oxide [69]. Herein, utilizing carbon derived biomass waste (carbonization of biomass waste at 400–600 °C for 3 h) instead of graphite, concentrated sulfuric acid, concentrated nitric acid, potassium permanganate, et al. [69]. About 1 g carbon derived biomass waste was transferred into 500 mL carbon nanomaterials suspension.

Assembly of Carbon Nanomaterials Containing –OH and –COOH Groups on the Surface of ZnO Nanosheets. Assembly of carbon nanomaterials containing –OH and –COOH groups on the surface of ZnO nanosheets containing –OH group is as follow: In the experiment, about 0.5–1 g ZnO nanosheets was added in 300–500 mL glass vessel, 2–5 mL carbon nanomaterials containing –OH and –COOH groups suspension and 100 mL water were added, mixed with ultrasound for 15–30 min, standing for 3–5 days, then filtrated, washed repeatedly 5–6 times, dried at room temperature.

Comparison of the Nanocomposites to the Removal of Organic Pollutants. Comparison of the nanocomposites to the removal of organic pollutants: In 50 mL transparent of vials, about 0.05 g as-prepared nanocomposite was added into 40 mL of 10^{-5} M methylene blue (methylene blue were selected as simulating organic pollutants). 10^{-5} M methylene blue solution were used as the references. The above mentioned of transparent of vials were placed with 2–3 days under exposure to natural sunlight indoor and at room temperature.

Morphology Observation with SEM. The scanning electron microscope (SEM) observation was carried out with Hitachi S-4800 and JSM-6700 (JEOL). The obtained sample was washed with deionized water, deposited on an aluminum foil substrate, dried at room temperature and then sputtered with a thin layer of Pt on the surface for the SEM observation.

Morphology Observation with TEM. The TEM observation was carried out with JEM-1011. The nanocomposite suspension were deposited on cupper mesh coated with carbon film, dried at room temperature.

Measurement of UV-Vis Spectrum. The UV-Vis was determined by a TU-1810 spectrophotometer and the samples were using suspension.

Photoresponse of Nanocomposite. The nanocomposite suspension was casted on the interdigital electrodes of Au on flexible PET (polyethylene terephthalate) film substrate, after drying, the photoconductive response to weak visible light (20–25 W) and 808 nm NIR with low-power were determined with LK2000A Electrochemical Work Station from LANLIKE Chemistry and Electron High Technology Co., Ltd (China).

Results and Discussion

First carbonization of biomass waste was carried out, then oxidation was performed according to the method of chemical synthesis of GO. The carbon nanomaterials obtained would have abundant chemical groups, such –OH, –COOH groups, et al. The representative SEM images of carbon nanomaterials prepared with biomass waste is shown in Fig. 1a. The representative TEM images of carbon nanomaterials prepared with biomass waste is shown in Fig. 1b. The representative SEM images of ZnO nanosheets modified with carbon nanomaterials prepared with biomass waste is shown in Fig. 2.

Fig. 1 SEM and TEM images of carbon nanomaterials prepared by biomass waste (**a**: SEM; **b**: TEM)

Fig. 2 SEM images of ZnO nanosheets/carbon nanocomposite

Figure 1a showed that the carbon nanomaterials prepared with biomass waste was nanorods structure, and the size is very uniform. The length of nanorods is about 40–60 nm, its diameter is about 10–20 nm.

Figure 1b indicated that the morphology of TEM is also supported the results of Fig. 1a of SEM, and its structure looks a little fluffy.

As shown form Fig. 2, it is found that the most morphologies were nanosheets structure, which belongs to ZnO [64], a small amount of carbon nanorods produced by biomass waste can also be clearly observed.

UV-Vis of ZnO nanosheets modified with carbon nanomaterials prepared with biomass waste is shown in Fig. 3a.

Figure 3a showed that the adsorption of ZnO nanosheets modified with carbon nanomaterials prepared with biomass waste was almost covered the whole region of visible light and extended to NIR (near-infrared), the edge of band is near 1000 nm. It illustrated that the ZnO nanosheets modified with carbon nanomaterials prepared with biomass waste have the activities of visible light and NIR. Its degradation to organic pollutants to 10^{-5} M methylene blue solution at weak natural light is obviously. The decolorization is completely, which is shown in Fig. 3b.

In order to explore the mechanism of ZnO nanosheets modified with carbon nanomaterials prepared with biomass waste with activity of visible light and NIR, the photoconductive response to visible light with low-power and 808 nm NIR of 200 mW were studied based on interdigital electrodes of Au on flexible PET film substrate with casting method. The representative results are shown in Fig. 4.

As shown from Fig. 4a, b, the ZnO nanosheets modified with carbon nanomaterials prepared with biomass waste showed good photo-switching behaviors to weak visible light and 808 nm NIR. The ratio of On/Off is one to two orders. It would be developed NIR detectors, NIR driven nano-carriers, nanoreactors, self-cleaning films or nanocoatings, photocatalyst, treatment of organic pollutants, photo-remote nanocomposites, et al. This is a simple and effective method to enhance the activities of visible light and 808 nm NIR for metal oxides.

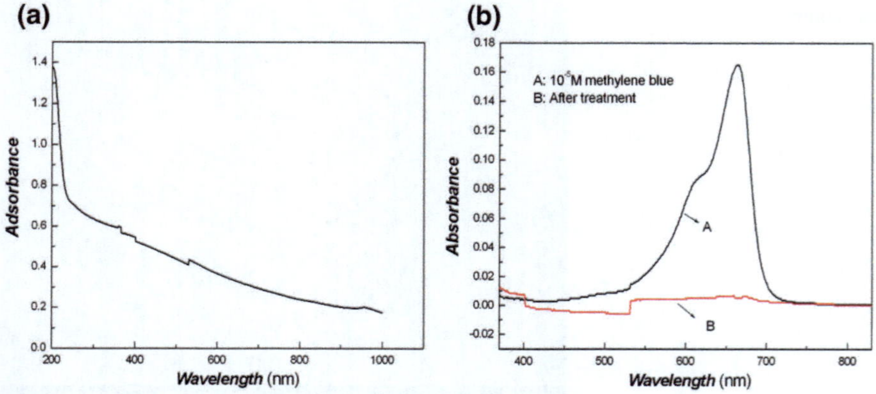

Fig. 3 UV-Vis of ZnO nanosheets/carbon nanocomposite and its degradation to organic pollutants (**a**: UV-Vis of nanocomposite; **b**: degradation of nanocomposite to organic pollutants)

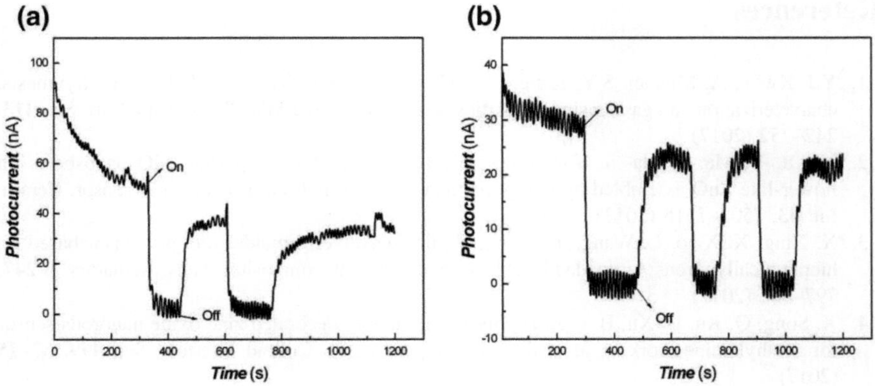

Fig. 4 Photoresponses of ZnO nanosheets/carbon nanocomposite to weak visible light and 808 nm of NIR (**a**: visible light; **b**: 808 nm NIR of 200 mW)

Conclusions

In summary, low-dimensional carbon nanomaterials were synthesized with biomass waste referring to the method of chemical synthesis of graphene oxide. Modification of ZnO nanosheets were performed utilizing the interaction between surface chemical groups of ZnO nanosheets and low-dimensional carbon nanomaterials produced with biomass waste. The photoconductive responses of the nanocomposites to visible light with low-power and 808 nm NIR were examined based on interdigital electrodes of Au on flexible PET film substrate. The results show that the ZnO nanosheets modified with the carbon nanomaterials produced with biomass waste have good photo-switching behaviors to weak visible light and 808 nm NIR. The ratio of On/Off is one to two orders, the response and recovery are also very rapidly, which is produced photo-induced charge very easily to weak visible light and NIR and possessing good visible light and NIR activities. It would be developing nanocomposites with external stimuli response, and have potential applications in photocatalysts, NIR detectors, nanoreactors, self-cleaning films, coatings, and treatment of environmental fields. This is low-cost, effective, general method to treat the surface of metal oxides.

Acknowledegements This project was supported by the Natural Science Foundation of Shandong Province (project No. ZR2013EMM008, ZR2016JL020), the Open Research Project of the State Key Laboratory of Industrial Control Technology, Zhejiang University, China (No. ICT170303), Zhejiang University, the Science Research Project of Yantai City (2013ZH349).

We declare that we have no conflict of interest.

References

1. Y.J. Kwon, A. Mirzaei, S.Y. Kang, M.S. Choi, J.H. Bang, S.S. Kim, H.W. Kim, Synthesis, characterization and gas sensing properties of ZnO-decorated MWCNTs. Appl. Surf. Sci. **413**, 242–252 (2017)
2. Y. Lu, Y. Ma, S. Ma, S. Yan, Hierarchical heterostructure of porous NiO nanosheets on flower-like ZnO assembled by hexagonal nanorods for high-performance gas sensor. Ceram. Int. **43**, 7508–7515 (2017)
3. X. Xing, X. Xiao, L. Wang, Y. Wang, Highly sensitive formaldehyde gas sensor based on hierarchicallyporous Ag-loaded ZnO heterojunction nanocomposites. Sens. Actuators, B **247**, 797–806 (2017)
4. X. Song, Q. Xu, H. Xu, B. Cao, Highly sensitive gold-decorated zinc oxide nanorods sensor for triethylamine working at near room temperature. J. Colloid Interface Sci. **499**, 67–75 (2017)
5. B. Thirumalraj, C. Rajkumar, S. Chen, K. Lin, Determination of 4-nitrophenol in water by use of a screen-printed carbon electrode modified with chitosan-crafted ZnO nanoneedles. J. Colloid Interface Sci. **499**, 83–92 (2017)
6. G. Gasparotto, J.P.C. Costa, P.I. Costa, M.A. Zaghete, T. Mazon, Electrochemical immunosensor based on ZnO nanorods-Au nanoparticles nanohybrids for ovarian cancer antigen CA-125 detection. Mater. Sci. Eng., C **76**, 1240–1247 (2017)
7. Y. Yan, H. Li, Q. Liu, N. Hao, H. Mao, K. Wang, A facile strategy to construct pure thiophene-sulfur-doped graphene/ZnO nanoplates sensitized structure for fabricating a novel "on-off-on" switch photoelectrochemical aptasensor. Sens. Actuators, B **251**, 99–107 (2017)
8. Y. Kwon, S. Kang, J. Cheon, Y. Song, J. Lee, Y. Cho, Fabrication of a Graphene/ZnO based p-n junction device and its ultraviolet photoresponse properties. Appl. Surf. Sci. **415**, 2–7 (2017)
9. J. Park, J. Lee, Y. Noh, K. Shinbc, D. Lee, Flexible ultraviolet photodetectors with ZnO nanowire networks fabricated by large area controlled roll-to-roll processing. J. Mater. Chem. C **4**, 7948–7958 (2016)
10. D. Kwon, S.J. Lee, J. Myoung, High-performance flexible ZnO nanorod UV photodetectors with a network-structured Cu nanowire electrode. Nanoscale **8**, 16677–16683 (2016)
11. L. Xu, Y. Zhou, Z. Wu, G. Zheng, J. He, Y. Zhou, Improved photocatalytic activity of nanocrystalline ZnO by coupling with CuO. J. Phys. Chem. Solids **106**, 29–36 (2017)
12. Z. Xing, Y. Chen, C. Liu, J. Yang, J. Xu, Y. Situ, H. Huang, Synthesis of core-shell ZnO/oxygen doped g-C_3N_4 visible light driven photocatalyst via hydrothermal method. J. Alloys Compd. **708**, 853–861 (2017)
13. A.N. Kadam, T.G. Kim, D.S. Shin, K.M. Garadkar, J. Park, Morphological evolution of Cu doped ZnO for enhancement of photocatalytic activity. J. Alloys Compd. **710**, 102–113 (2017)
14. R.K. Chava, M. Kang, Ag_2S quantum dot sensitized zinc oxide photoanodes for environment friendly photovoltaic devices. Mater. Lett. **199**, 188–191 (2017)
15. P. She, K. Xu, S. Zeng, Q. He, H. Sun, Z. Liu, Investigating the size effect of Au nanospheres on the photocatalytic activity of Au-modified ZnO nanorods. J. Colloid Interface Sci. **499**, 76–82 (2017)
16. A. Upneja, G. Dou, C. Gopu, C.A. Johnson, A. Newman, A. Suleimenovd, J.L. Goldfarb, Sustainable waste mitigation: biotemplated nanostructured ZnO for photocatalytic water treatment via extraction of biofuels from hydrothermal carbonization of banana stalk. RSC Adv. **6**, 92813–92823 (2016)
17. H. Yuan, J. Li, Effect of bending on resistive switching of NiO/ZnO nanocomposite thin films. J. Alloys Compd. **709**, 752–759 (2017)
18. V. Gurylev, C. Su, T. Perng, Hydrogenated ZnO nanorods with defect-induced visible light-responsive photoelectrochemical performance. Appl. Surf. Sci. **411**, 279–284 (2017)

19. J. Karbowniczek, L. Cordero-Arias, S. Virtanen, S. Misra, E. Valsami-Jones, L. Tuchscherr, Bogdan Rutkowski, Kamil Górecki, Piotr Bała, Aleksandra Czyrska-Filemonowicz, Aldo R. Boccaccini, Electrophoretic deposition of organic/inorganic composite coatings containing ZnO nanoparticles exhibiting antibacterial properties. Mater. Sci. Eng. C **77**, 780–789 (2017)
20. M.M. Ovhal, A.S. Kumar, P. Khullar, M. Kumar, A.C. Abhyankar, Photoluminescence quenching and enhanced spin relaxation in Fe doped ZnO nanoparticles. Mater. Chem. Phys. **195**, 58–66 (2017)
21. B. Chouchenea, T.B. Chaabanea, K. Mozetb, E. Girotb, S. Corbelb, L. Balanc, G. Medjahdid, R. Schneider, Porous Al-doped ZnO rods with selective adsorption properties. Appl. Surf. Sci. **409**, 102–110 (2017)
22. A.A. Oun, J. Rhim, Preparation of multifunctional chitin nanowhiskers/ZnO-Ag NPs andtheir effect on the properties of carboxymethyl cellulose-basednanocomposite film. Carbohydr. Polym. **169**, 467–479 (2017)
23. I. Khan, A.A.M. Ibrahim, M. Sohail, A. Qurashi, Sonochemical assisted synthesis of RGO/ZnO nanowire arrays for photoelectrochemical water splitting. Ultrason. Sonochem. **37**, 669–675 (2017)
24. S. Lee, J. Lee, S. Choi, J. Park, Studies of thermoelectric transport properties of atomic layer deposited gallium-doped ZnO. Ceram. Int. **43**, 7784–7788 (2017)
25. L. Chen, T. Ji, L. Mu, J. Zhu, Cotton fabric derived hierarchically porous carbon and nitrogen doping for sustainable capacitor electrode. Carbon **111**, 839–848 (2017)
26. F. Xu, T. Lin, H. Bi, F. Huang, Graphene-like carbon with three-dimensional periodicity prepared from organic-inorganic templates for energy storage application. Carbon **111**, 128–132 (2017)
27. D.G. Lim, K. Kim, M. Razdan, R. Diaz, S. Osswald, V.G. Pol, Lithium storage in structurally tunable carbon anode derived from sustainable source. Carbon **121**, 134–142 (2017)
28. S. Zhu, Q. Meng, L. Wang, J. Zhang, Y. Song, H. Jin, K. Zhang, H. Sun, H. Wang, B. Yang, Highly photoluminescent carbon dots for multicolor patterning, sensors, and bioimaging. Angew. Chem. Int. Ed. **52**, 3953–3957 (2013)
29. C. Wang, J. Liao, S. Chung, A. Manthiram, Carbonized eggshell membranes as a natural and abundant counter electrode for efficient dye-sensitized solar cells. Adv. Energy Mater. **5**, 1401524 (2015)
30. Z. Wu, P. Zhang, M. Gao, C. Liu, W. Wang, F. Leng, C. Huang, One-pot hydrothermal synthesis of highly luminescent nitrogen-doped amphoteric carbon dots for bioimaging from Bombyx mori silk—natural proteins. J. Mater. Chem. B **1**, 2868–2873 (2013)
31. A. Primo, P. Atienzar, E. Sanchez, J.M. Delgado, H. Garcı´a, From biomass wastes to large-area, high-quality, N-doped graphene: catalyst-free carbonization of chitosan coatings on arbitrary substrates. Chem. Commun. **48**, 9254–9256 (2012)
32. L. Sun, C. Tian, M. Li, X. Meng, L. Wang, R. Wang, J. Yin, H. Fu, From coconut shell to porous graphene-like nanosheets for high-power supercapacitors. J. Mater. Chem. A **1**, 6462–6470 (2013)
33. O. Akhavan, K. Bijanzad, A. Mirsepah, Synthesis of graphene from natural and industrial carbonaceous wastes. RSC Adv. **4**, 20441–20448 (2014)
34. K.H. Adolfsson, S. Hassanzadeh, M. Hakkarainen, Valorization of cellulose and waste paper to graphene oxide quantum dots. RSC Adv. **5**, 26550–26558 (2015)
35. X. Pei, D. Xiong, J. Fan, Z. Li, H. Wang, J. Wang, Highly efficient fluorescence switching of carbon nanodots by CO2. Carbon **117**, 147–153 (2017)
36. M.C. Ortega-Liebana, N.X. Chung, R. Limpens, L. Gomez, J.L. Hueso, J. Santamaria, T. Gregorkiewicz, Uniform luminescent carbon nanodots prepared by rapid pyrolysis of organic precursors confined within nanoporous templating structures. Carbon **117**, 437–446 (2017)
37. B. Chen, F. Li, S. Li, W. Weng, H. Guo, T. Guo, X. Zhang, Y. Chen, T. Huang, X. Hong, S. You, Y. Lin, K. Zeng, S. Chen, Large scale synthesis of photoluminescent carbon nanodots and their application for bioimaging. Nanoscale **5**, 1967–1971 (2013)

38. K.A.S. Fernando, S. Sahu, Y. Liu, W.K. Lewis, E. A. Guliants, A. Jafariyan, P. Wang, C.E. Bunker, Y.-P. Sun, Carbon quantum dots and applications in photocatalytic energy conversion. ACS Appl. Mater. Interfaces **7**, 8363−376 (2013)
39. P. Russo, M. Xiao, N.Y. Zhou, Carbon nanowalls: a new material for resistive switching memory devices. Carbon **120**, 54–62 (2017)
40. D. Zaharie-Butucel, L. Digianantonio, C. Leordean, L. Ressier, S. Astilean, C. Farcau, Flexible transparent sensors from reduced graphene oxide microstripes fabricated by convective self-assembly. Carbon **113**, 361–370 (2017)
41. H. Kim, J. Ahn, Graphene for flexible and wearable device applications. Carbon **120**, 244–257 (2017)
42. W. Yang, W. Yang, F. Ding, L. Sang, Z. Ma, G. Shao, Template-free synthesis of ultrathin porous carbon shell with excellent conductivity for high-rate supercapacitors. Carbon **111**, 419–427 (2017)
43. D. Schopf, M. Es-Souni, Thin film nanocarbon composites for supercapacitor applications. Carbon **115**, 449–459 (2017)
44. W. Zhang, M. Weng, P. Zhou, L. Chen, Z. Huang, L. Zhang, C. Liu, S. Fan, Transparency-switchable actuator based on aligned carbon nanotube and paraffin-polydimethylsiloxane composite. Carbon **116**, 625–632 (2017)
45. H. Lim, Y.S. Yun, Y. Ko, Y. Bae, M.Y. Song, H.J. Yoon, K. Kang, H. Jin, Three-dimensionally branched carbon nanowebs as air-cathode for redox-mediated Li-O_2 batteries. Carbon **118**, 114–119 (2017)
46. H. Yang, Y. Cao, J. He, Y. Zhang, B. Jin, J. Sun, Y. Wang, Z. Zhao, Highly conductive free-standing reduced graphene oxide thin films for fast photoelectric devices. Carbon **115**, 561–570 (2017)
47. N.A. Travlou, J. Secor, T.J. Bandosz, Highly luminescent S-doped carbon dots for the selective detection of ammonia. Carbon **114**, 544–556 (2017)
48. N. Yang, X. Jiang, Nanocarbons for DNA sequencing: a review. Carbon **115**, 293–311 (2017)
49. N.A. Travlou, T.J. Bandosz, Nanoporous carbon-composites as gas sensors: importance of the specific adsorption forces for ammonia sensing mechanism. Carbon **121**, 114–126 (2017)
50. X. Xu, Z. Bao, W. Tang, H. Wu, J. Pan, J. Hu, H. Zeng, Surface states engineering carbon dots as multi-band light active sensitizers for ZnO nanowire array photoanode to boost solar water splitting. Carbon **121**, 201–208 (2017)
51. C. Wang, S. Yang, Q. Ma, X. Jia, P. Ma, Preparation of carbon nanotubes/graphene hybrid aerogel and its application for the adsorption of organic compounds. Carbon **118**, 765–771 (2017)
52. D. Shan, S. Deng, J. Li, H. Wang, C. He, G. Cagnetta, B. Wang, Y. Wang, J. Huang, G. Yu, Carbon **119**, 101–109 (2017)
53. W. Libbrecht, A. Verberckmoes, J.W. Thybaut, P.V.D. Voort, J.D. Clercq, Soft templated mesoporous carbons: Tuning the porosity for the adsorption of large organic pollutants. Carbon **116**, 528–546 (2017)
54. H. Chen, W. Qian, Q. Xie, X. Cheng, Graphene-carbon nanotube hybrids as robust, rapid, reversible adsorbents for organics. Carbon **116**, 409–414 (2017)
55. J. Gallego, J. Tapia, M. Vargas, A. Santamaria, J. Orozco, D. Lopez, Carbon **111**, 393–401 (2017)
56. J. Wen, C. Mou, Morphology-controllable templated synthesis of three-dimensionally structured graphenic materials. Carbon **111**, 476–485 (2017)
57. J. Chen, X. Shi, L. Ren, Y. Wang, Graphene oxide/PVA inorganic/organic interpenetrating hydrogels with excellent mechanical properties and biocompatibility. Carbon **111**, 18–27 (2017)
58. X. Ma, B. Zhang, Q. Cong, X. He, M. Gao, G. Li, Mater. Chem. Phys. **178**, 88–97 (2016)
59. Q. Cong, X. He, M. Gao, X. Ma, G. Li, ZnO/CuS heterostructured nanocomposite and its organic functionalization. Mater. Res. Innovations **18**, 740–746 (2014)
60. Q. Cong, H. Geng, X. He, M. Gao, X. Ma, G. Li, Surface modification of ZnO nanosheets with Au/polyaniline and their properties. Mater. Res. Innovations **18**, 30–36 (2014)

61. X. Ma, M. Wang, G. Li, H. Chen, R. Bai, Mater. Chem. Phys. **98**, 241–247 (2006)
62. S. Lv, L. Bian, J. Qiu, X. Zhang, M. Gao, X. He, X. Ma, G. Li, Mater. Sci. Forum **898**, 1739–1748 (2017)
63. X. He, M. Wang, J. Qiu, X. Zhang, M. Gao, S. Lv, X. Ma, G. Li, Mater. Sci. Forum **898**, 2263–2271 (2017)
64. X. Ma, Y. Wang, M. Gao, H. Xu, G. Li, Catal. Today **158**, 459–463 (2010)
65. X. Ma, T. Zhu, H. Xu, G. Li, J. Zheng, A. Liu, J. Zhang, H. Du. Anal. Bioanal. Chem. **390**, 1133–1137 (2008)
66. X. Ma, G. Li, M. Wang, Y. Cheng, R. Bai, H. Chen, Chem.-A Eur. J. **12**, 3254–3260 (2006)
67. X. Ma, H. Xu, G. Li, M. Wang, H. Chen, S. Chen, Macromol. Mater. Eng. **291**, 1539 (2006)
68. X. Ma, G. Li, M. Wang, R. Bai, F. Yang, H. Chen, Green Chem. **8**, 63–69 (2006)
69. J. Zheng, X. Ma, X. He, M. Gao, G. Li, Praparation, characterizations, and its potential applications of PANi/graphene oxide nanocomposite. Procedia Eng. **27**, 1478–1487 (2012)

Corrosion Resistance of Fe-Based Stainless Steel Composite Layer Prepared by TWIAW

Dongting Wu, Yong Zou, Cong Hu and Chuanwei Shi

Abstract In this study, Fe-based stainless steel composite layer was prepared by Twin-Wire Indirect Arc Welding (TWIAW). Weld penetration increased with the increasing of the welding current and the amount of CO_2 in Ar+CO_2 protective gas. However, the CO_2 content exceeded 20% led to serious welding spatters. The phase of the surfacing layer was austenite plus ferrite. As no sensitization or slight sensitization occurred, the composite layers show better intergranular corrosion resistance, which can be attributed to the low heat input in TWIAW. The pitting corrosion resistance depends on the point defects formed on the surface of the layers, which is relevant to the stability of the process and the oxidation of the protective gas.

Keywords Twin-Wire Indirect Arc Welding · Fe-based stainless steel composite layer · Intergranular corrosion · Pitting corrosion

Introduction

Fe-based stainless steel composite materials combine the excellent corrosion resistance of stainless steel and good mechanical strength and process-ability of carbon steel, which can be prepared by the way of preparing stainless steel composite layer on the carbon steel. Surfacing of stainless steel coating on carbon steel matrix is one of the important methods for preparing stainless steel carbon steel composite material [1, 2]. Nevertheless, there are obvious differences in

D. Wu · Y. Zou (✉) · C. Hu
Key Laboratory of Liquid-Solid Structural Evolution and Processing of Materials, Ministry of Education, Shandong University, Jinan 250061, Shandong, China
e-mail: yzou@sdu.edu.cn

D. Wu
e-mail: wudongting@sdu.edu.cn

C. Shi
Shandong Jianzhu University, Jinan 250101, Shandong, China

Y. Han (ed.), *Advanced Functional Materials*,
https://doi.org/10.1007/978-981-13-0110-0_18

microstructure, thermal conductivity, thermal expansion coefficient, specific heat capacity, electromagnetic property between the stainless steel coating and the carbon steel matrix. In the process of surfacing welding, heat input should be strictly limited. Otherwise, intergranular corrosion will occur on the stainless steel side, which seriously restricts the service life of composite materials [3, 4]. Pitting corrosion is also a potential factor affecting the application of stainless steel [5].

XIE Xiang, BAO Yefeng, etc. [6] obtained stainless steel overlays on base metal Q235 by electroslag strip cladding (ESW) and submerged arc welding (SAW). They studied the electrochemical corrosion and intergranular corrosion properties of overlays in different welding process. Intergranular corrosion resistance of overlay by ESW was obviously superior to the overlay by SAW in 10% oxalic acid solution. PAN Qiang, LIU Erxi, etc. [7] adopted TIG technology to conduct surface overlaying welding on stainless steel. They found that the surface hardness of steel plate after surfacing welding was obviously increased and the corrosion resistance was better than that of Q235 steel base metal. Puli and Ram [8] used shielded-metal arc coating and friction surfacing producing martensitic stainless steel AISI 410 coatings on low-carbon steel plates. The coating produced by Friction showed much better in wear resistance and pitting corrosion resistance compared to shielded-metal arc coating.

Twin-wire indirect arc gas shielded welding (TWIAW) is a new welding technology, in which the negative and positive electrodes of a power source are connected to two wires and the work piece isn't connected to the welding circuit. The arc burning between the twin wires, and the generated heat by the arc is mainly used to melt the wires [9, 10]. This new welding method has the advantages of higher deposition efficiency and lower heat input, therefore, the TWIAW has wide application prospect in the preparation of composite layer of Fe-base stainless steel with good corrosion resistance.

Experimental Materials and Methods

The Q235 steel plate was used as the substrate, the size of which is 300 × 100 × 6 mm. Before surfacing, the surface of the testing samples is treated to remove the surface oxide and grease. The welding wire was ER308L austenitic stainless steel wire with 1.2 mm diameter. The chemical compositions of base plate Q235 and ER308L welding wire are shown in Table 1.

Table 1 The chemical composition of Q235 and ER308L (wt%)

	C	Si	Mn	P	S	Cr	Ni	Fe
Q235	0.18	0.25	0.46	0.029	0.021	–	–	99.06
ER308L	0.02	0.47	1.68	0.017	0.020	20.32	10.05	69.61

In the test, the welding current rose from 100 to 200A, and the arc voltage was adjusted from 26 to 33 V. The protective gases were 100%Ar, 93.3%Ar+6.7%CO_2, 86.7%Ar+13.3%CO_2 and 66.7%Ar+33.3%CO_2 respectively.

Microstructure The stainless steel composite layers were etched with aqua regia (V_{HCl}:V_{HNO3} = 3:1) for about 20–30 s. The samples from base metal Q235 were etched with a 3% nitric acid alcohol solution. Then, the microstructure was observed by a KEYENCE VHX-500FE digital microscope. Phases of composite layers were analyzed by a Model DX-2700(Cu Kα) X-ray diffractometer () set to working current and voltage of 30 mA and 40 kV, respectively.

Intergranular corrosion The activation current I_a and the reactivation current I_r of the Fe-based stainless steel composite layers were detected with the double loop Electrochemical potentiokinetic reactivation method (DL-EPR) application of a CS350 electrochemical workstation. The rate of reactivation $R_a = I_r/I_a$ was used to evaluate the intergranular corrosion sensitivity. Greater the R_a value is, more easily intergranular corrosion occurs. The polarization curves were measured in 0.5 mol/L H_2SO_4 + 0.01 mol/L KSCN solution at 30 °C. The platinum electrode, saturated calomel electrode (SCE) and the stainless steel composite layers were used as the counter electrode, the reference electrode and the working electrode, respectively. Before EIS tests, the samples were immersed into the electrolyte for about 5 min in order to stabilize the open-circuit potential (OCP). During potentio-dynamic polarization tests, the scan rate was 1.67 mV/s from open-circuit potential to 0.3 V. After arriving at 0.3 V, the scanning turned to opposite direction.

Pitting corrosion The pitting corrosion potential was determined by potentiodynamic scanning method. Generally, higher the potential is, better the pitting resistance is. The tests were carried out in 3.5% NaCl solution at 30 °C ± 1 °C. The ratio of the volume of the solution to the area of the sample was not less than 200 mL/cm^2. The tests were conducted at a scan rate of 20 mV/min from open-circuit potential to the moment when the anode current exceeded 1000 μA/cm^2.

SEM+EDS After electrochemical tests, the samples were ultrasonic cleaned 15 min in deionized water and absolute ethanol respectively. The corrosion morphologies and chemical elements were observed by using a JSM6610LA SEM and EDS, with the accelerating voltage of 20.0 kV.

Results

The Appearance of Composite Layer

The typical section shape of composite layer was shown in Fig. 1. When the combinations of arc voltage and welding current were 28 V-110 A, 31 V-130 A, 31 V-150 A and 33 V-180 A respectively, composite layers with better

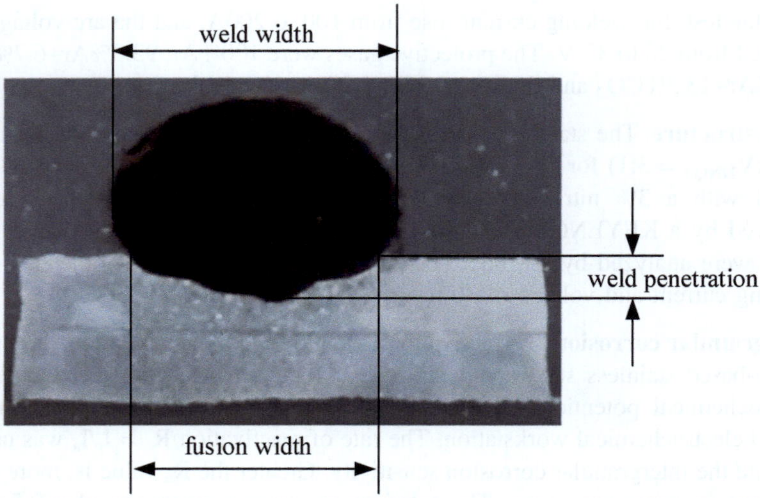

Fig. 1 The typical section shape of composite layer

Fig. 2 The shape parameters of layers in four situations. **a** 28 V-110 A. **b** 31 V-130 A. **c** 31 V-150 A. **d** 33 V-180 A

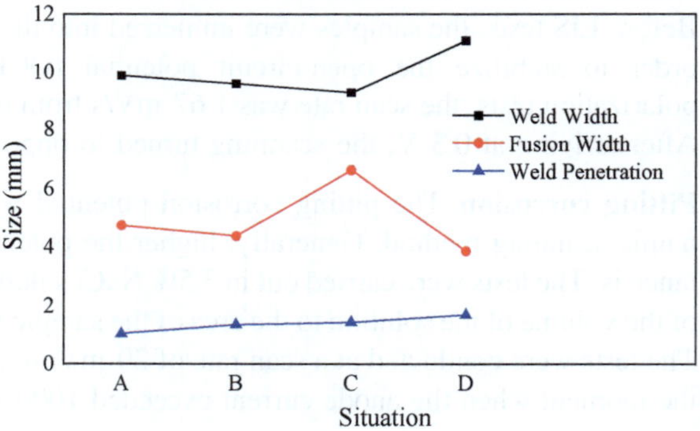

appearance were obtained. The shape parameters of layers in these four situations were summarized in Fig. 2. With the increase of heat input, the weld penetration increased gradually, and the weld width decreased firstly and then increased. When the arc voltage was 31 V and the welding current was 150 A, the weld width was the maximum, the weld width was minimum, and the appearance of weld was relatively better than the others.

The appearance of composite layers was shown in Fig. 3, in which the welding current was 150 A, the arc voltage was 31 V, and the protective gas was 100%Ar, 93.3%Ar+6.7%CO_2, 86.7%Ar+13.3%CO_2 and 66.7%Ar+33.3%CO_2 respectively. The shape parameters were shown in Fig. 4. According to Figs. 3 and 4, the layer with 86.7%Ar+13.3%CO_2 had the most uniform appearance and the best fusion effect. In situation 66.7%Ar+33.3%CO_2, spatters appeared around the weld.

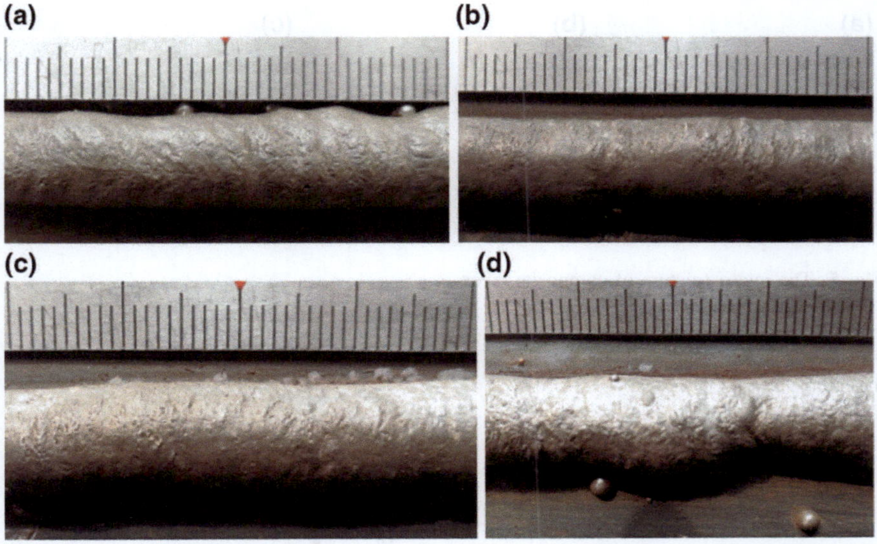

Fig. 3 The appearance of composite layers with different protective gas. **a** 100%Ar. **b** 93.3%Ar +6.7%CO_2. **c** 86.7%Ar+13.3%CO_2. **d** 66.7%Ar+33.3%CO_2

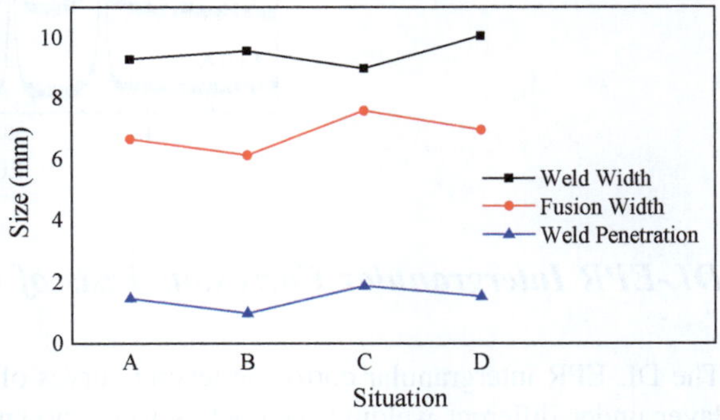

Fig. 4 The shape parameters of layers with four kinds of protective gas. **a** 100%Ar. **b** 93.3%Ar+6.7%CO_2. **c** 86.7%Ar+13.3%CO_2. **d** 66.7%Ar+33.3%CO_2

The Microstructure of Composite Layer

The typical microstructure of composite layer was shown in Fig. 5. The microstructure of stainless steel composite layer consisted of austenite and ferrite. Columnar crystals grew from the original matrix. The ferrites distributed on the austenite matrix in skeletons. There were more lath ferrites close to the surface of layer. The XRD results of the composite layer were shown in Fig. 6. As can be seen, the ratio of austenite to ferrite varied with the welding current, and with the increase of welding current, the content of ferrite increased.

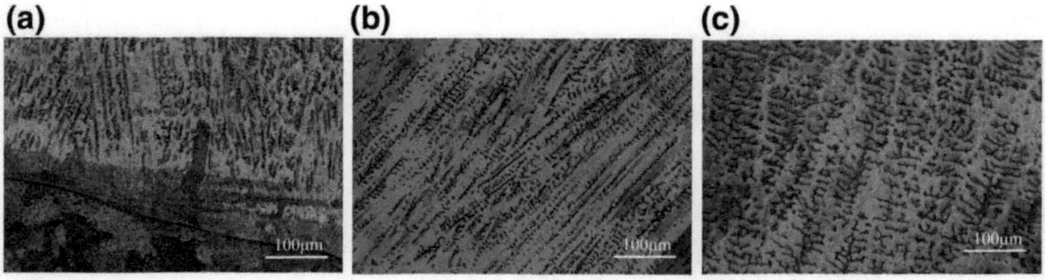

Fig. 5 The microstructure of composite layer

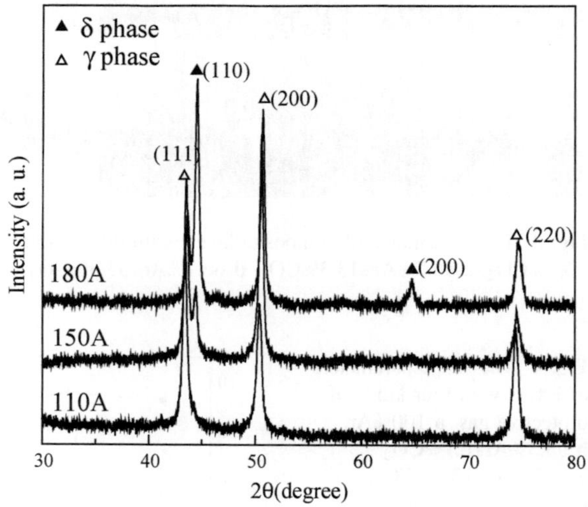

Fig. 6 The XRD results of the composite layer

DL-EPR Intergranular Corrosion Tests of Composite Layer

The DL-EPR intergranular corrosion testing curves of the stainless steel composite layer under different welding parameters were shown in Fig. 7. The parameter of the curves was shown in Table 2. When the arc voltage was 31 V and the welding current was 150 A, the minimum Ra value was close to 1%. The Ra value was relatively larger in the case of 28 V-110 A and the value was up to maximum in the case of 33 V-180 A. With the increase of CO_2 content, the Ra value increased.

The corrosion morphologies after intergranular corrosion tested were shown in Fig. 8. As the value of Ra was not high, the grain boundary corrosion ditches were not obvious when the arc voltage 31 V with the welding current from 130 to 150 A. There were a few corrosion ditches in situation 28 V-110 A, and there were clear corrosion ditches in situation 33 V-180 A. Meanwhile, there were distinct corrosion ditches when the CO_2 was mixed into the protective gas.

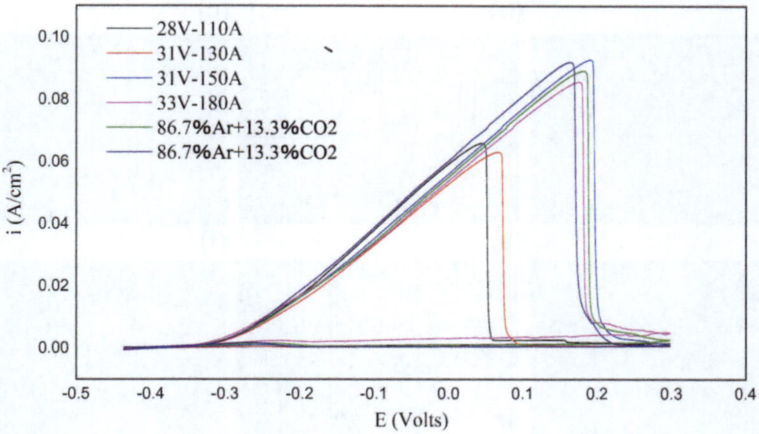

Fig. 7 The DL-EPR intergranular corrosion testing curves. **a** 28 V-110 A. **b** 31 V-130 A. **c** 31 V-150 A. **d** 33 V-180 A. **e** 31 V-150 A 86.7%Ar+13.3%CO_2. **f** 31 V-150 A 86.7%Ar +13.3%CO_2

Table 2 The parameter of DL-EPR intergranular corrosion testing curves

No	Current (A)	Voltage (V)	Protective gas	Ir (A)	Ip (A)	Ra = Ir/Ip (%)
a	110	28	Ar	0.001477	0.065859	2.24
b	130	31	Ar	0.000504	0.062923	0.80
c	150	31	Ar	0.000887	0.092746	0.96
d	180	33	Ar	0.002281	0.085614	2.66
e	150	31	93.3%Ar+6.7%CO_2	0.001429	0.089130	1.60
f	150	31	86.7%Ar+13.3% CO_2	0.001599	0.091934	1.74

Pitting Corrosion Tests of Composite Layer

Table 3 summarized the pitting potential obtained from the potentiodynamic scan of different composite layers. The pitting potential was highest in situation 31 V-150 A, and the pitting potentials were both lower when the heat input was less or more. The pitting potential turned lower with the increase of CO_2.

The pitting corrosion morphologies were shown in Fig. 9. There were corrosion pits in all situations. When the heat input is smaller or larger, pitting pits appear intensively. The resistance of pitting corrosion became worse when the protective gas mixed with CO_2.

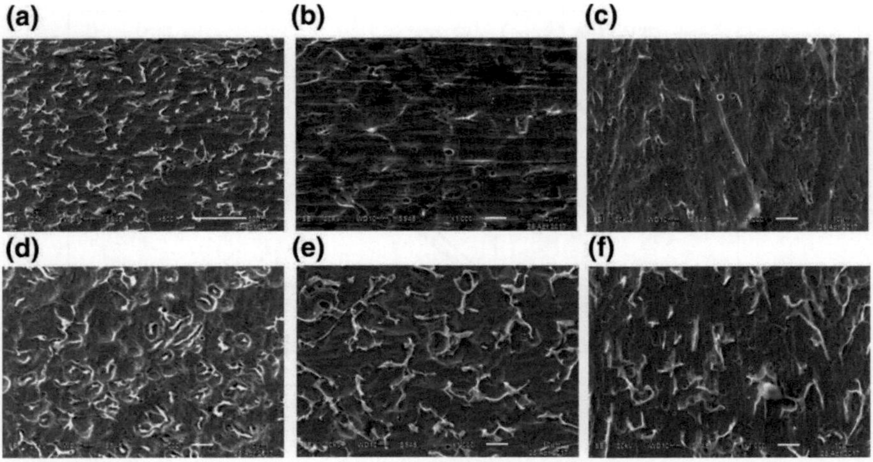

Fig. 8 The corrosion morphologies after intergranular corrosion tests. **a** 28 V-110 A. **b** 31 V-130 A. **c** 31 V-150 A. **d** 33 V-180 A. **e** 31 V-150 A 86.7%Ar+13.3%CO_2. **f** 31 V-150 A 86.7%Ar+13.3%CO_2

Table 3 The pitting potential of different composite layers

No	Current (A)	Voltage (V)	Protective gas	Pitting potential (V)
a	110	28	Ar	0.098
b	130	31	Ar	0.111
c	150	31	Ar	0.242
d	180	33	Ar	0.223
e	150	31	93.3%Ar+6.7%CO_2	0.177
f	150	31	86.7%Ar+13.3%CO_2	0.163

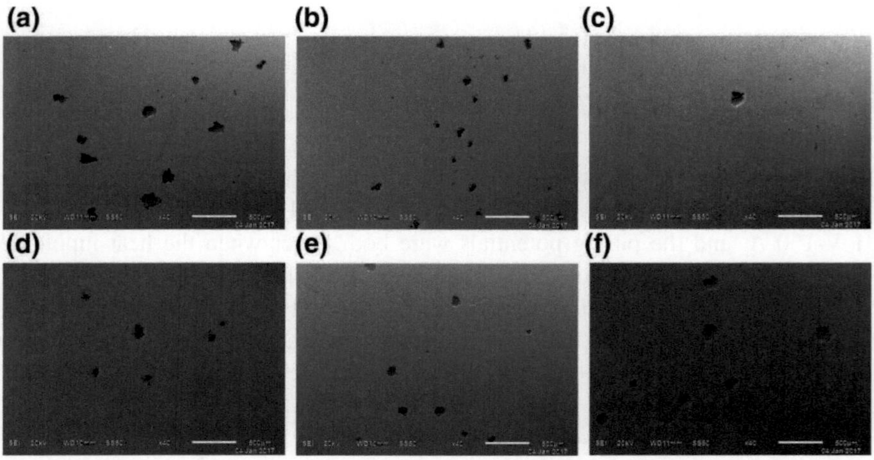

Fig. 9 The pitting corrosion morphologies. **a** 28 V-110 A. **b** 31 V-130 A. **c** 31 V-150 A. **d** 33 V-180 A. **e** 31 V-150 A 86.7%Ar+13.3%CO_2. **f** 31 V-150 A 86.7%Ar+13.3%CO_2

Discussion

In the process of TWIAW, the indirect arc burns between the twin wires and there is no active spot area on the workpiece. The energy for surfacing is mainly supplied by the indirect arc, and the heat from the droplets plays a lesser role. As the length of the arc is too short and the arc stiffness is limited, the indirect arc can't effectively heat the carbon steel and the stable welding pool cannot form. Therefore, the formation of the composite layer is poor. When the welding current is too large, the rapidly melting welding wires form a large number of droplets to the surface of the steel plate, which blocks the effective heating of the arc on the substrate. In this situation, although the total heat input is large, the melting width is still limited. When the protective 7gas mixed with CO_2, the CO_2 may break down into CO, O_2, and O under the high temperature of welding arc. The arc suffers a strong compression effect, which causes the increase of the arc stiffness and the penetrating power [11]. It can effectively heat the carbon steel plate and improve the forming of the stainless steel composite layer.

Intergranular corrosion of stainless steel is a corrosion phenomenon caused by local Cr depletion zone, which is related to the precipitation of $M_{23}C_6$. According to ISO12732:2006, no sensitization occurs when the R_a value is smaller than 1%. Slight sensitization takes place at 1–5%, and sensitization occurs at greater than 5%. For stainless steel composite layer prepared by TWIAW, the maximum value of Ra was 2.66% and the Ra values were less than 1% in situations 31 V-130 A and 31 V-150 A, which indicates that no sensitization or only slight sensitization occurred. In other words, the composite layers have good performance in resistance to intergranular corrosion. When the heat input is small, due to the poor stability of the welding process, short circuiting transition easily occurs and the weld formation is poor, resulting in inhomogeneous composition of stainless steel clad layer and poor chromium area [12]. With the increase of the welding current, the process tends to be stable, and the R_a value decreases gradually. When the arc voltage reaches 31 V and the welding current is 130 A, the welding process is very stable, the R_a value is less than 1%, which means good intergranular corrosion resistance. When the welding current is further increased, the cooling rate of the stainless steel composite layer slows down, and the high temperature residence time at 600–850 °C is prolonged, which may prompted the precipitation of $M_{23}C_6$, and the chromium deficient zone formed easily around the grain boundaries to occur intergranular corrosion [13]. With the addition of CO_2, the heat concentration of the welding arc is enhanced, the heat efficiency and the heat input are effectively improved, and the tendency of intergranular corrosion is increased. But the R_a value is still not more than 1.8% when 86.7%Ar+13.3%CO_2 is used, which means only slight sensitization and intergranular corrosion is not serious. Compared with traditional arc surfacing, TWIAW shows the advantage of suppressing effectively intergranular corrosion in surfacing layer which can be attributed to the lower heat input.

Pitting corrosion occurs on some small areas of metal surface, and it develops continuously along the hole depth as time goes by. The pitting occurs on some zone of surface defects, such as inclusions, chromium depleted zones, dislocations exposed on the surface, in which can damage to the integrity of the passive film. Pitting resistance equivalent (or pitting index) PREN is commonly used to characterize the pitting corrosion resistance [14]. For stainless steel composite layers prepared by TWIAW in various parameters, the chemical composition is similar, and the pitting resistance is also similar based on PREN. However, there are obvious differences between different samples. It is thought that this is related to the point defects formed on the surface of the layers. On the one hand, it is influenced by the welding process stability. When the heat input is low, the smaller welding current and lower arc voltage result in unstable welding process, and short circuit transition frequently occurs, which led to the formation of non-metallic inclusions in the composite layers as point defects. As evidence, EDS analysis found oxide inclusion in corrosion pit (Fig. 10 and Table 4). With the welding current and arc voltage increasing, the welding process becomes stable, the amount of inclusions and other defects decrease, which lead to much better pitting corrosion resistance. On the other hand, it is also influenced by the composition of protective gas. Comparing with 100% Ar, mixed CO_2 causes obviously larger oxidation. The oxide inclusion is easier to form in the composite layer, which leads to lower pitting potential and poorer pitting corrosion resistance [15].

Fig. 10 Oxide inclusion in corrosion pit by EDS analysis

Table 4 ZAF method standardless quantitative analysis results of oxide inclusion

Element	(keV)	Mass (%)	Atom (%)
O	0.525	28.84	54.61
Si	1.739	10.02	10.81
Cr	5.411	28.02	16.33
Mn	5.894	33.11	18.25
Total		100	100

Conclusions

Well-formed Fe-based stainless steel composite layers can be obtained by TWIAW when the arc voltage was 31 V and the welding current was 150 A. When protective gas is 86.7%Ar+13.3%CO_2, the appearance of the composite layer is relatively better. As no sensitization or slight sensitization occurred, the composite layers have good intergranular corrosion resistance, which is related to the low heat input of TWIAW. The pitting corrosion resistance is related to the point defects formed on the surface of the layers, which is depended on the stability of the process and the oxidation of the protective gas.

Acknowledgements This study has received grants from the National Natural Science Foundation of China (*No. 51171093*) and the Natural Science Foundation of Shandong Province of China (*No. ZR2016EEB31*), and the authors are deeply grateful to the support.

References

1. Z.S. Wu, H. Yun, C.R. Liu et al., Microstructure and properties of stainless steel layer on Q235. J. Weld. Technol. **03**, 13–15 (2014)
2. S. Kobayashi, R. Kobayashi, T. Watanabe, Control of grain boundary connectivity based on fractal analysis for improvement of intergranular corrosion resistance in SUS316L austenitic stainless steel. J. Acta Mater. **102**, 397–405 (2016)
3. Y. Kang, J. Choi, J. Park et al., Characterization of stainless steel surface processed using electrolytic oxidation and titanium complex ion solution. J. Appl. Surf. Sci. **415**, 161–165 (2017)
4. Y.F. Bao, Q. Ren, Z.X. Zhang et al, Study of intergranular corrosion resistance of overlay by Electroslag cladding with austenitic stainless steel. J. Trans. China Weld. Inst. **06**, 65–68+79 +132 (2016)
5. A.K. Singh, G.M. Reddy, K.S. Rao, Pitting corrosion resistance and bond strength of stainless steel overlay by friction surfacing on high strength low alloy steel. J. Defence Technol. **11**, 299–307 (2015)
6. X. Xie, Y.F. Bao, K. Yang et al, Corrosion resistance of overlay by electroslag cladding with stainless steel strip. J. Trans. China Weld. Inst. **10**, 77–80+117 (2011)
7. Q. Pan, E.X. Liu, C.Y. Feng et al., Q235 steel plate surface overlaying stainless steel modified process. J. Welded Pipe Tube **04**, 22–26 (2017)
8. R. Puli, G.D. Janaki Ram, Wear and corrosion performance of AISI 410 martensitic stainless steel coatings produced using friction surfacing and manual metal arc welding. J. Surf. Coat. Technol. **209**, 1–7 (2012)
9. C.W. Shi, Y. Zou, Z.D. Zou, D.T. Wu, Twin-wire indirect arc welding by modeling and experiment. J. Mater. Process. Technol **214**, 2292–2299 (2014)
10. C.W. Shi, Y. Zou, Z.D. Zou, H. Zhang, Physical characteristics of twin-wire indirect arc plasma. J. Vac. **107**, 41–50 (2014)
11. D. Iordachescu, L. Quintino, Steps toward a new classification of metal transfer in gas metal arc welding. J. Mater. Process. Technol. **202**, 391–397 (2008)
12. F. Iacoviello, V.D. Cocco, L. D'Agostino, Analysis of the intergranular corrosion susceptibility in stainless steel by means of potentiostatic reactivation tests. J. Procedia Struct. Integrity **3**, 269–275 (2017)

13. J. Qian, C. Chen, H. Yu et al., The influence and the mechanism of the precipitate/austenite interfacial C-enrichment on the intergranular corrosion sensitivity in 310 S stainless steel. J. Corros. Sci. **111**, 352–361 (2016)
14. D.H. Kang, H.W. Lee, Study of the correlation between pitting corrosion and the component ratio of the dual phase in duplex stainless steel welds. J. Corros. Sci. **74**, 396–407 (2013)
15. Z. Zhang, H. Jing, L. Xu et al., Influence of microstructure and elemental partitioning on pitting corrosion resistance of duplex stainless steel welding joints. J. Appl. Surf. Sci. **394**, 297–314 (2017)

Influence of Sintering Process on the Properties of Porous Ti_2AlC

Yingwen Cao, Chunsheng Guo and Yong Zou

Abstract Ti_2AlC belongs to the so-called "MAX" phases ternary carbides material group, which possess a useful combination of both metallic and ceramic properties. Its excellent oxidation and corrosion resistance indicate that porous Ti_2AlC has remarkable application prospects in LHP (loop heat pipe) wicks. In this study, Ti +Al+TiC and Ti+Al+C powder mixtures are employed as starting powders respectively and effects of different volume fraction of pore former on properties of porous Ti_2AlC were studied. The experimental results indicate that the reaction process of Ti+Al+TiC system is more stable than that of Ti+Al+C system. The porosity of the porous Ti_2AlC through sintering Ti+Al+C powder mixtures are relatively large, which is easy to form through holes, increases capillary pressure and improves the pumping capacity of the wick.

Keywords Ti_2AlC · Porous materials · Porosity · Capillary performance

Introduction

Loop heat pipe (LHP), first invented in 1976, is an efficient heat transfer device carried out via liquid-gas phase transition. It contains evaporator, condenser, vapor line and liquid line. Working liquid is pumped under capillary pressure of wick in the evaporator and evaporates to absorb heat, travels to the condenser through vapor line, and finally condenses to liquid to release heat. LHP has many advantages such as no power input, ability to operate against gravity, long transfer distance, con-

Y. Cao · Y. Zou (✉)
Key Laboratory of Liquid Structure and Heredity of Materials,
Ministry of Education, Shandong University, Jinan 250061, Shandong, China
e-mail: yzou@sdu.edu.cn

C. Guo (✉)
School of Mechanical and Electronic Engineering, Shandong University at Weihai,
Weihai 264209, Shandong, China
e-mail: guo@sdu.edu.cn

Y. Han (ed.), *Advanced Functional Materials*,
https://doi.org/10.1007/978-981-13-0110-0_19

trollable heat flow, which make it applied in space technology and electronic cooling [1–3]. In the LHP system, the capillary force supplied by wick in the evaporator is the main power source of the circulation. Therefore, manufacturing of high performance wick is the key to improve the operation properties of LHP [4]. In aspect of materials selection, sintered metals such as nickel [5] and copper [6] have been widely used due to their good pore-forming ability. On the other hand, ceramic wicks with lower thermal conductivity also had been studied, which can overcome heat leak through the metal wicks. However, most conventional ceramics present brittleness, low pore-forming ability and thermal shock resistance. Therefore, it is desirable to develop novel ceramic wicks which can meet with several constraints (e.g. good pore-forming ability, high mechanical strength and thermal shock resistance, etc.) simultaneously.

Ternary compound, MAX phase, is a new machinable ceramic material, whose general formula is $M_{n+1}AX_n$ (where n = 1, 2, 3, M = transition metal elements, A = group elements in IIIA or IVA, X = C or N). Ti_2AlC, like other MAX phases, possess a useful combination of both metallic and ceramic properties: low density [7], high elastic modulus [8], high strength and fracture toughness [9], excellent thermal and electrical conductivity [10], easy machinability, good thermal shock and oxidation resistance [11]. So, Ti_2AlC is a promising candidate material for capillary wick. There are many methods to synthesis porous Ti_2AlC: foam replication technique [12], gel-casting [13], pressureless sintering [14], self-propagating high-temperature synthesis (SHS) [15], addition of pore former [16]. However, the study focused on capillary performance of Ti_2AlC wicks is limited.

In this study, Ti+Al+TiC and Ti+Al+C powder mixtures were employed respectively as starting powders and sieved NaCl powders were used as pore former to produce porous Ti_2AlC wick. Moreover, effects of different reactants on microstructure and capillary performance of porous wicks were analyzed and discussed.

Experimental

Titanium (−40 μm, 99.9%), Aluminum (−10 μm, 99.9%), titanium carbide (-10 μm, 99%) and graphite (99%) were weighed in the stoichiometric ratio of Ti:Al: TiC = 1:1.3:1 and Ti:Al:C = 2.1:1:1. Excess aluminum addition can make up for the evaporation loss of Al, which is beneficial to obtain pure Ti_2AlC [15]. NaCl (−300 mesh) powders were used as the pore former. Powders were mixed for 2 h by ball milling and cold pressed in the graphite die (20 mm in diameter). Afterward, green bodies were sintered at 1300 °C for 1.5 h with argon atmosphere. Sintered samples were soaked in 85 °C water for 4 h to dissolve NaCl pore former and dried at 110 °C for 2 h.

The porosity of sintered samples was determined by Archimedes' principle shown in Eq. (1).

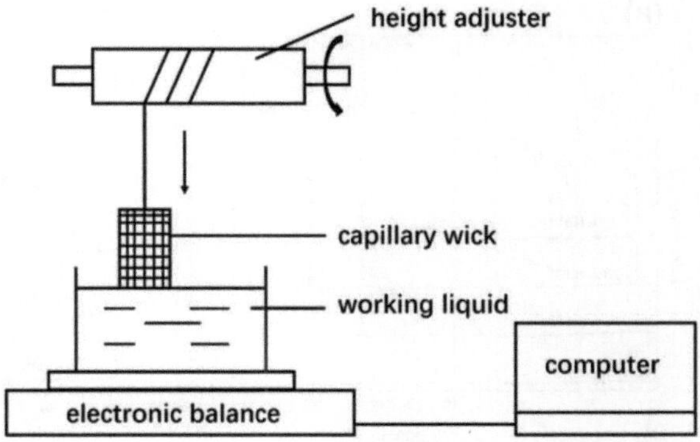

Fig. 1 Schematic diagram of capillary performance characterization

$$\varepsilon = \frac{m1 - m2 - m3}{\rho_{water} \times V} = \frac{m1 - m2 - m3}{\rho_{water} \times t \times \pi\left(\frac{d}{2}\right)^2} \tag{1}$$

where m1 is the total beaker mass when the sample is suspended in water for 2 h; m2 is the beaker mass when the saturated sample is removed from water with no water-drop outside; m3 is the mass of dry sample; ρ_{water} is the density of water; t is the thickness of the sample; d is the diameter of the sample.

The microstructure of sintered samples was characterized using SEM, which also obtained the pore distribution and pore size. The phase components were identified by XRD with Cu Kα radiation (λ radiatio) at a scanning rate of 0.02° at Capillary performance is characterized using devices shown in Fig. 1.

Results and Discussion

Phase Composition. Figure 2 shows XRD pattern of the sintered samples. It indicates that both Ti+Al+TiC and Ti+Al+C can successfully obtain Ti_2AlC phase through sintering reaction. Moreover, no NaCl peak is detected in the pattern, which suggests that NaCl did not react with the original mixture and can be removed completely during dissolving in the water. From these figures, it can be seen that the sintering products including Ti_2AlC and a litter TiC. In order to estimate the amount of TiC, the intensity of the highest diffraction peak (Ti_2AlC (103)) is defined as 100, and amount of TiC was the ratio of the integral intensities of the TiC main peaks to that of the Ti_2AlC phase. The TiC amount is about 3.1 and 6.5% for Ti+Al+TiC and Ti+Al+C, respectively. This means that sintering purity produced from Ti+Al+TiC is higher than that from Ti+Al+C. Hardness of wick is reinforced by TiC as impurity, so the existence of minor TiC is beneficial for mechanical processing of porous wick, especially when strength declines with its

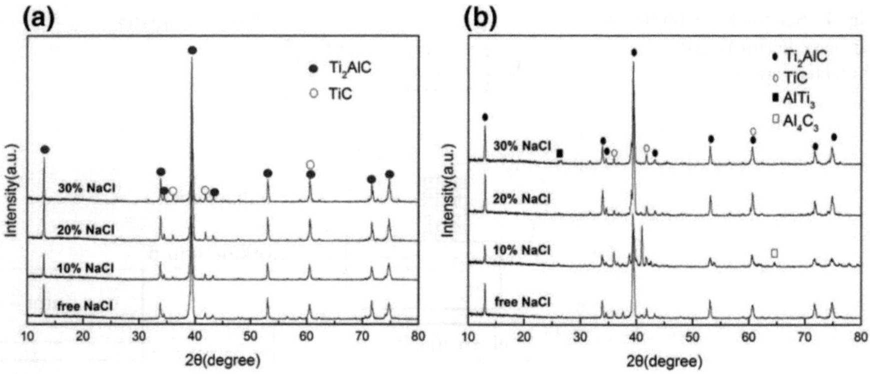

Fig. 2 XRD pattern of the end samples sintered from different reactants. **a** Ti+Al+TiC; **b** Ti+Al+C

increasing porosity. However, the rational region of TiC as impurity need discussed in next study.

Microstructure. Figure 3 shows SEM images of sintered Ti+Al+TiC samples with 0 and 30% NaCl respectively. Figure 3a indicates that porous wick can still be obtained without NaCl addition, where pores are fine and called interval pores. It is obvious that porosity obtained through loose sintering is limited and less than 10%, which is less than that of other MAX phase (such as Ti_3SiC_2) [17]. This is because the formation of liquid phase due to the addition of Al during sintering, which is beneficial to accelerate sintering reaction but deteriorated porosity due to its fluidity and filling ability. On the other side, NaCl forming pores can promote porosity by occupation. Unlike interval pores formed by the gap between the particles, the NaCl occupying pore is big pore, and the size of pores depends on the diameter of the NaCl particles. That is to say, the size of NaCl occupying pores can be controlled via regulating size of NaCl particles. Some studies [18] suggested that composite pore structure of big pores plus interval pores is more benefit for wicks, where big pores reserve more working liquid and interval pores supply stronger capillary force at the same time.

Figure 4 shows SEM images of samples obtained from different reactants with 20% NaCl addition. Compared Fig. 4a, c, it can be observed that porosity obtained from Ti+Al+C powders are relatively higher and the size of pore is bigger, but distributed non-uniformly. That's because reactions between Ti, Al and C powders are more complicated, which release more heat and melt pore formers earlier. The reaction of Ti+Al+TiC system is relatively smooth.

Figure 5a shows the relationship between porosity and volume fraction of NaCl addition. It indicates that porosity increases obviously with increasing addition of NaCl, regardless of reactants. Moreover, porosity sintered samples from Ti+Al+C and Ti+Al+TiC without NaCl addition is only 0.14 and 0.09 respectively. This

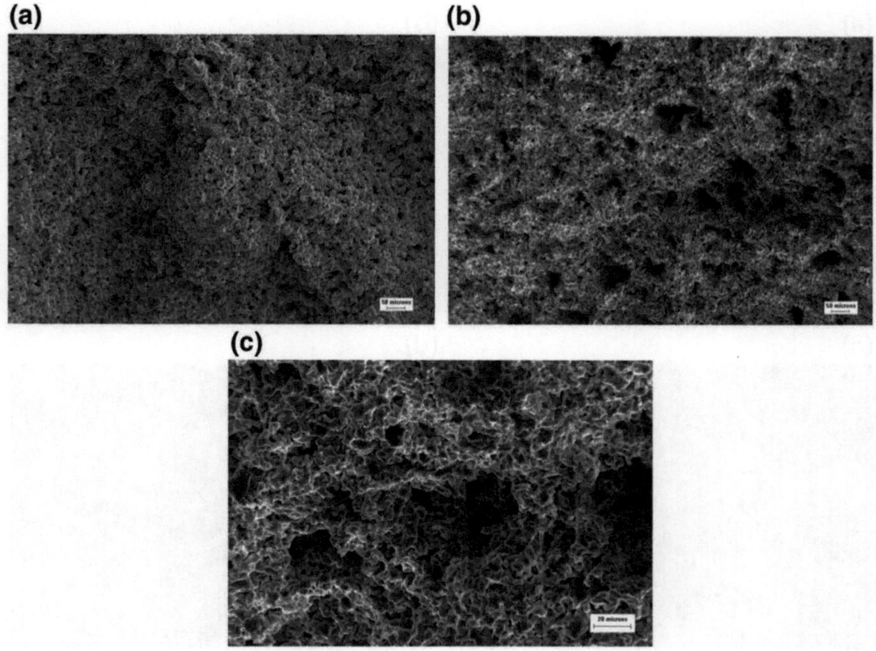

Fig. 3 SEM images of samples sintered from Ti+Al+TiC with different volume fraction of NaCl. **a** 0%, 100×; **b** 30%, 100×; **c** 30%, 500×

indicated that loose sintering could prepare porous Ti_2AlC, but the porosity is limited. Figure 5b shows the pore size distribution of samples with different NaCl additions, taking Ti+Al+TiC for example. It suggests the pore size distribution is mainly concentrated in two ranges: one is smaller than 5 μm and the other is 5–10 μm. Less than 5 μm pores are mainly from the gap between the particles; meanwhile, pores in the later size are obtained by NaCl occupying and intervals of particles. And with increasing additions of NaCl, the formation of big pores is more depending on NaCl addition. That is, both porosity and big pore size increase obviously with increasing additions of NaCl.

Capillary Performance. Capillary performance is one of the most important properties of porous wick, which depends on pore structure. Figure 6 shows the relationship between volume fractions of pore former and capillary performance, taking samples sintered from Ti+Al+TiC for example. From this curve it can be seen that the highest point of the curve represents the maximum amount of pumped liquid, which is determined by the maximum porosity; and the slope of the curve represents the pumping speed, is determined by the amount of fine pores. [19]. Sample with 30% NaCl showed the maximum slope and greater suction at the beginning of the test, the mean increase of NaCl not only improves the porosity of the wick, but also changed the amount and distribution of fine pores, improve the rate of capillary pumping.

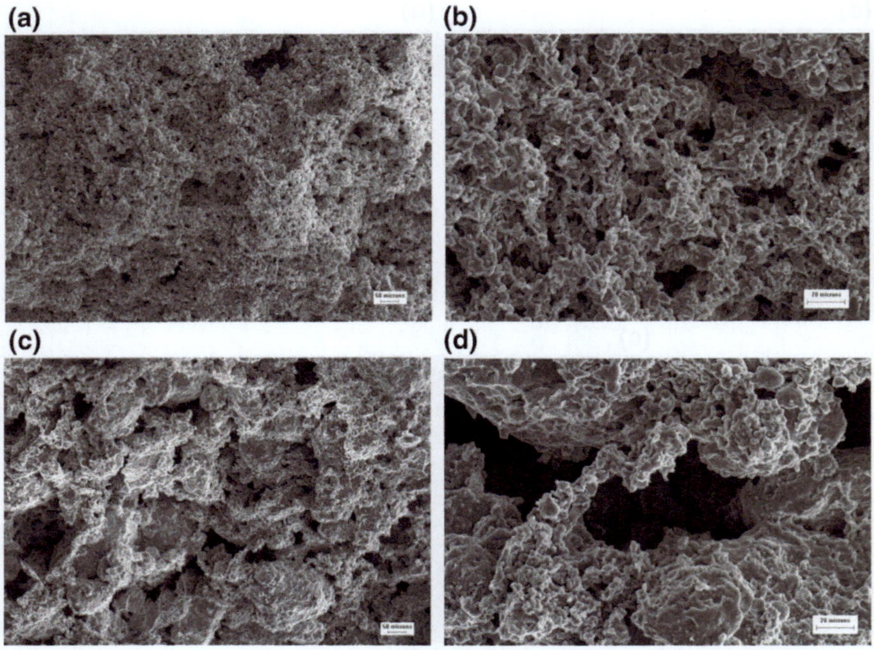

Fig. 4 SEM images sintered from different reactants with 20% NaCl. **a** Ti+Al+TiC, 100×; **b** Ti+Al+TiC, 500×; **c** Ti+Al+C, 100×; **d** Ti+Al+C, 500×

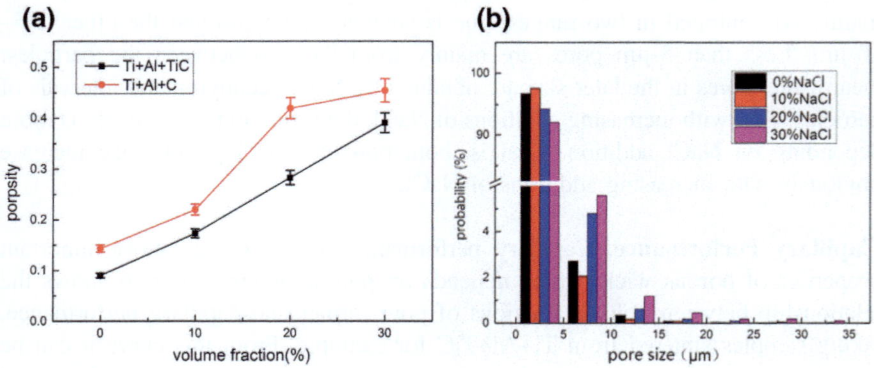

Fig. 5 Pore structure parameters of all sintered samples **a** porosity **b** pore size distribution (taking Ti+Al+TiC for example)

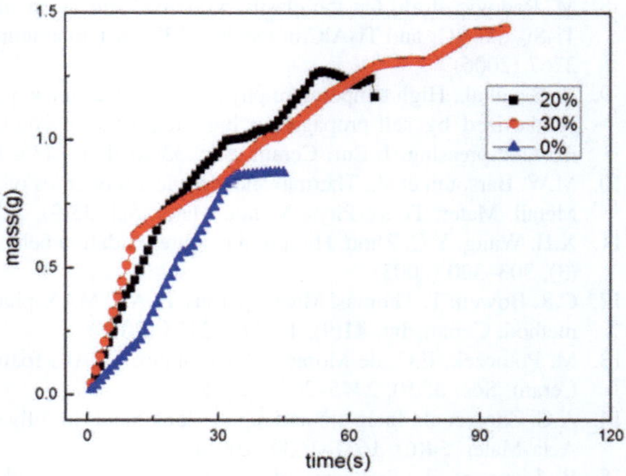

Fig. 6 Capillary curve of samples with different volume fraction of NaCl (taking Ti +Al+TiC for example)

Conclusions

In summary, the present study represents a cheap and efficient method to produce porous Ti_2AlC wick. Ti+Al+TiC and Ti+Al+C powder mixtures are employed as starting powders respectively. The experimental results indicate that the reaction process of Ti+Al+TiC system is more stable than that of Ti+Al+C system. With increasing NaCl additions, porosity increases obviously regardless of reactants and capillary performance is improved due to increase interconnected pores.

Acknowledgements The authors are greatly appreciative for supports of Natural Science Foundation of Shandong Province (Grant No. ZR2017BEE012), The Program of Science and Technology of Suzhou (Grant No. 2017GHY15122), The Open Project Program of State Key Laboratory of Petroleum Pollution Control (Grant No. PPC2017018) and CNPC Research Institute of Safety and Environmental Technology.

References

1. Y.F. Maydanik, Loop heat pipes. Appl. Therm. Eng. **25**(5–6), 635–657 (2005)
2. L.L. Vasiliev, Heat pipes in modern heat exchangers. Appl. Therm. Eng. **25**(1), 1–19 (2005)
3. T.D. Swanson, G.C. Birur, NASA thermal control technologies for robotic spacecraft. Appl. Therm. Eng. **23**(9), 1055–1065 (2003)
4. H. Li et al., Experimental investigation on the sintered wick of the anti-gravity loop-shaped heat pipe. Exp. Thermal Fluid Sci. **68**, 689–696 (2015)
5. S.-C. Wu et al., Manufacturing and testing of the double-layer wick structure in a loop heat pipe. Int. J. Heat Mass Transf. **56**(1–2), 709–714 (2013)
6. W.S. Ling et al., Thermal performance of loop heat pipe with porous copper fiber sintered sheet as wick structure. Appl. Therm. Eng. **108**, 251–260 (2016)
7. M.W. Barsoum, The MN+1AXN phases: a new class of solids. Prog. Solid State Chem. **28** (1), 201–281 (2000)

8. M. Radovic et al., On the elastic properties and mechanical damping of Ti_3SiC_2, Ti_3GeC_2, $Ti_3Si_{0.5}Al_{0.5}C_2$ and Ti_2AlC in the 300–1573 K temperature range. Acta Mater. **54**(10), 2757–2767 (2006)
9. Y. Bai et al., High temperature physical and mechanical properties of large-scale Ti_2AlC bulk synthesized by self-propagating high temperature combustion synthesis with pseudo hot isostatic pressing. J. Eur. Ceram. Soc. **33**(13–14), 2435–2445 (2013)
10. M.W. Barsoum et al., Thermal and electrical properties of Nb_2AlC, $(Ti, Nb)_2AlC$ and Ti_2AlC. Metall. Mater. Trans. Phys. Metall. Mater. Sci. **33**(9), 2775–2779 (2002)
11. X.H. Wang, Y.C. Zhou, High-temperature oxidation behavior of Ti_2AlC in air. Oxid. Met. **59** (3), 303–320 (2003)
12. C.R. Bowen, T. Thomas, Macro-porous Ti_2AlC MAX-phase ceramics by the foam replication method. Ceram. Int. **41**(9), 12178–12185 (2015)
13. M. Potoczek, E.G. de Moraes, P. Colombo, Ti2AlC foams produced by gel-casting. J. Eur. Ceram. Soc. **35**(9), 2445–2452 (2015)
14. A.G. Zhou et al., Incipient and regular kink bands in fully dense and 10 vol.% porous Ti_2AlC. Acta Mater. **54**(6), 1631–1639 (2006)
15. B. Liang et al., Synthesis of Ti_2AlC by laser-induced self-propagating high-temperature sintering. J. Alloy. Compd. **501**(1), L1–L3 (2010)
16. B. Velasco, E. Gordo, S.A. Tsipas, MAX phase Ti_2AlC foams using a leachable space-holder material. J. Alloy. Compd. **646**, 1036–1042 (2015)
17. M.J. Abu, J.J. Mohamed, Z.A. Ahmad, Synthesis of high purity titanium silicon carbide from elemental powders using arc melting method. Int. J. Refract Metal Hard Mater. **47**, 86–92 (2014)
18. D. Deng et al., Evaluation of capillary performance of sintered porous wicks for loop heat pipe. Exp. Thermal Fluid Sci. **50**, 1–9 (2013)
19. J. Li, Y. Zou, L. Cheng, Experimental study on capillary pumping performance of porous wicks for loop heat pipe. Exp. Thermal Fluid Sci. **34**(8), 1403–1408 (2010)

Influences of La on Optical and Electric Properties of $BiFeO_3$ Thin Films

Wei Cai, Kaihua Liu, Rongli Gao, Xiaoling Deng, Gang Chen and Chunlin Fu

Abstract Bismuth ferrite ($BiFeO_3$, short for BFO) is an important room-temperature single phase multiferroic materials. To further improve electric properties of BFO, La^{3+} was introduced. $Bi_{1-x}La_xFeO_3$ (x = 0–0.12) thin films were fabricated by sol-gel method. The microstructure, optical and electric properties of $Bi_{1-x}La_xFeO_3$ thin films have been investigated. XRD results show that $Bi_{1-x}La_xFeO_3$ thin films is rhombohedral distortion perovskite structure and the addition of La^{3+} can inhibit the formation of impure phase. The substitution of La^{3+} for Bi^{3+} on A sites can restrain the growth of grain and improve the roughness of thin films. The band gap of $Bi_{1-x}La_xFeO_3$ thin films is less than that of BFO thin films, and its band gap decreases first and then increases as the content of La^{3+} increases, which results from both Burstein-Moss effect and impurity level. A larger amount of La^{3+} (x = 0.12) can make leakage current of BFO thin films obviously decrease. Moreover, the remnant polarization and coercive electric field of $Bi_{1-x}La_xFeO_3$ thin films increase with the increasing of La^{3+} content. It results from the decrease of oxygen vacancy and effects of grain size.

W. Cai (✉) · K. Liu · R. Gao · X. Deng · G. Chen · C. Fu
School of Metallurgy and Materials Engineering, Chongqing University of Science and Technology, Chongqing, China
e-mail: caiwei_cqu@163.com

K. Liu
e-mail: 605149154@qq.com

R. Gao
e-mail: gaorongli2008@163.com

X. Deng
e-mail: dgxmd@163.com

G. Chen
e-mail: cgyjxy_cqust@163.com

C. Fu
e-mail: chlfu@126.com

W. Cai · R. Gao · X. Deng · G. Chen · C. Fu
Chongqing Key Laboratory of Nano-Micro Composite Materials and Devices, Chongqing, China

Y. Han (ed.), *Advanced Functional Materials*,
https://doi.org/10.1007/978-981-13-0110-0_20

Keywords Bismuth ferrite · Lanthanum · Optical · Dielectric
Ferroelectric

Introduction

Bismuth ferrite ($BiFeO_3$, short for BFO) is an important room-temperature multiferroic materials and has been paid attention extensively because of its high Curie temperature (~ 850 °C), Neel temperature (370 °C), high remnant polarization (~ 100 μC/cm^2) and narrow band gap (2.3–2.7 eV) [1–3]. But the kinetics of phase formation easily result in impure phase such as $Bi_2Fe_4O_9$, $Bi_{25}FeO_{39}$ and Bi_2O_3 due to the volatilization of some reactants (Bi^{3+}) and phase decompositions during the preparation of BFO [4, 5]. It lead to the high leakage current of BFO thin films, making it difficult to investigate its intrinsic ferroelectricity and obtain the samples with good electric properties. To further improve the electric properties of BFO thin films, some methods such as substitution [6–8], introduction of buffer layer [9] have been conducted. Among them, substitution for Bi^{3+} on A sites or Fe^{3+} on B sites is the most common method. With the partial substitution of ions on A or B sites in BFO, the crystal structure will be distorted due to the internal chemical pressure. The induced distortion can tailor electric properties of BFO thin films [10, 11]. La-doping have been investigated to improve the ferroelectric or magnetic properties of BFO thin films [12, 13]. Influences of lanthanum on microstructure, optical, dielectric and ferroelectric properties of BFO thin films have not been studied systematically. Therefore, we prepared $Bi_{1-x}La_xFeO_3$ thin films by sol-gel method and investigated the influences of La^{3+} on crystal structure, surface morphology, optical and electric properties.

Experimental

$Bi_{1-x}La_xFeO_3$ ($x = 0$, 0.06, 0.08, 0.1, 0.12) thin films were prepared by sol-gel method. Bismuth nitrate ([$Bi(NO_3)_3 \cdot 5H_2O$], 99.0%, Sinopharm Group Co. Ltd.), iron nitrate ([$Fe(NO_3)_3 \cdot 9H_2O$], 99.0%, Sinopharm Group Co. Ltd.) and lanthanum nitrate ([$La(NO_3)_3 \cdot 6H_2O$], 99.0%, Aladdin Reagent) were used as raw materials. Ethylene glycol monomethyl ether ($CH_3OCH_2CH_2OH$, 99.5%, Sinopharm Group Co. Ltd.) was used as solvent. Firstly, $Bi(NO_3)_3 \cdot 5H_2O$, $Fe(NO_3)_3 \cdot 9H_2O$ and $La(NO_3)_3 \cdot 6H_2O$ were weighed with stoichiometric ratio (the molar ratio of Bi: Fe = 1.1:1, excess of Bi to compensate its loss during heat treatment) and dissolved in ethylene glycol monomethyl ether. The solution was continuously stirred at room temperature for 30–50 min to make $Bi(NO_3)_3 \cdot 5H_2O$, $Fe(NO_3)_3 \cdot 9H_2O$ and $La(NO_3)_3 \cdot 6H_2O$ completely dissolved to form precursor solution. Secondly, acetic anhydride ($C_4H_6O_3$, 99.5%, Sinopharm Group Co. Ltd.) as dehydrating agent was added to the solution. Because the exothermal dehydration reaction can result in the rapid rise of temperature easy to cause precipitation. Citric acid ($C_6H_8O_7$, 99.5%, Sinopharm Group Co. Ltd.) as chelating agent was added until the temperature drops to room temperature. The pH value of precursor solution was adjusted to be 2–3 by adding a amount of ethanol amine (C_2H_7NO, 99.5%, Sinopharm Group Co. Ltd.) and the solution was continuously stirred for 1 h. Finally, a certain amount of ethylene glycol monomethyl ether was added to adjust the concentration of the solution to be 0.3 mol/L. The precursor solution was obtained after aging for 48 h at room temperature. The thin films were deposited on Pt(100 nm)/Ti(30 nm)/SiO_2(500 nm)/Si(500 μm) and quartz glass substrates by spin-coating method. Then, the thin films were annealed in oxygen atmosphere at 600 °C for 20 min.

The crystal structure of the thin films was determined by X-ray diffraction (DX2700, Dandong Fangyuan, China, Cu target with a working voltage of 35 kV and a current of 25 mA) in a scanning rate of 3°/min. The surface morphology of thin films were obtained by AFM (Atomic Force Microscopy, NTEGRA Prima, NT-MDT, Russia). The optical transmission characteristics of thin films have been studied using a double beam ultraviolet-visible spectrophotometer (TU1810, Persee, China) in the wavelength range of 200–1000 nm. The thickness of thin films is determined by stylus profiler (P-7, KLA-Tencor, USA). To measure electric properties, the Au top electrodes with a diameter of 200 μm and a thickness of 50 nm were deposited on the thin films using DC magnetron sputtering (JGP280, SKY, China). The room-temperature ferroelectric hysteresis loops and leakage current density were measured by ferroelectric test system (TF2000E, aixACCT, Germany).

Results and Discussion

Figure 1 shows the XRD patterns of $Bi_{1-x}La_xFeO_3$ thin films. Firstly, it is found that the main phase of $Bi_{1-x}La_xFeO_3$ (x = 0–0.12) thin films is rhombohedral distortion perovskite structure ($BiFeO_3$, *R*3*m* space group, JCPDS card no. 74-2016). There is some impure phase (indexed as $Bi_2Fe_4O_9$) in BFO thin films, and there is no impure phase observed in $Bi_{1-x}La_xFeO_3$ (x = 0.06–0.12) thin films. The result indicates that La^{3+} ions have entered into $BiFeO_3$ lattice and the addition of La^{3+} can inhibit the formation of $Bi_2Fe_4O_9$. Secondly, the diffraction peaks of the thin films slightly shift to the higher angle region as La^{3+} content (x) increases. According to Bragg equation, it is obtained that the lattice constant ($a = b = c$) of $Bi_{1-x}La_xFeO_3$ thin films (x = 0, 0.06, 0.08, 0.10, 0.12) is 0.3958, 0.3936, 0.3931, 0.3929, 0.3928 nm, respectively. It is because that the La^{3+} with smaller ionic radius (0.106 nm) substitute for Bi^{3+} with larger ionic radius (0.108 nm) on A sites so that the lattice constant decreases.

Figure 2 shows the surface morphology of $Bi_{1-x}La_xFeO_3$ thin films. Firstly, $Bi_{1-x}La_xFeO_3$ thin films show the smooth, uniform and crack-free surface.

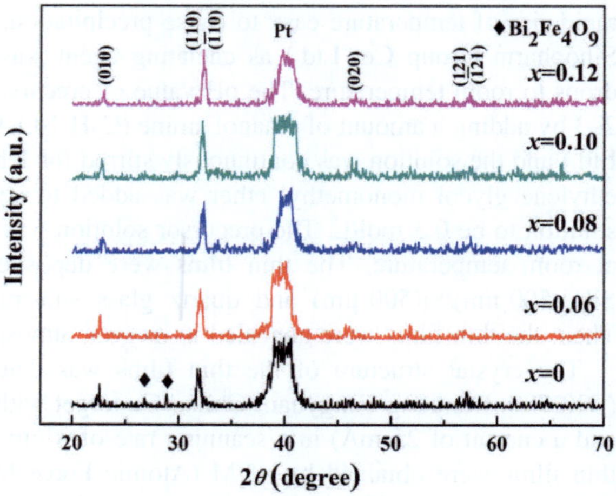

Fig. 1 XRD patterns of $Bi_{1-x}La_xFeO_3$ thin films

The average grain size of $Bi_{1-x}La_xFeO_3$ thin films ($x = 0, 0.06, 0.08, 0.10, 0.12$) is 124, 116, 72, 48 and 38 nm respectively, which indicates that the growth of grain was markedly inhibited by introduction of La^{3+}. Rare earth ions are known to suppress the grain growth in perovskite, which is attributed to their lower diffusivity [14, 15]. Secondly, the root mean square roughness (RMS) of $Bi_{1-x}La_xFeO_3$ thin films ($x = 0, 0.06, 0.08, 0.10, 0.12$) is 9.374, 8.724, 6.562, 6.312 and 5.486 nm respectively, which indicates that the addition of La^{3+} can enhance the thickness uniformity of BFO thin films.

Figure 3 shows the optical transmission spectra of $Bi_{1-x}La_xFeO_3$ thin films ($x = 0$–0.12) deposited on quartz glass substrates. Firstly, it can be seen that the transmittance of $Bi_{1-x}La_xFeO_3$ ($x = 0.06$–0.12) thin films is higher than that of BFO thin films. In generally, the transmittance of thin films depends on surface roughness, the number of defects and impurity [16] which can make carrier scattered and absorbed so that the transmittance decreases. As mentioned above, BFO thin films have impure phase ($Bi_2Fe_4O_9$) and higher surface roughness so that its transmittance is the lowest. Secondly, when the wavelength is shorter than 600 nm, the transmittance of $Bi_{1-x}La_xFeO_3$ thin films decreases sharply, which shows obvious absorption edge.

The optical band gap of $Bi_{1-x}La_xFeO_3$ thin films can be obtained by Tauc's Law:

$$(\alpha h\nu)^n = A(h\nu - E_g) \tag{1}$$

where A is a constant, α is the absorption coefficient, $h\nu$ is the photon energy, n represents the type of optical transition between valence and conduction bands, E_g is the optical band gap. For $n = 2$, the transition is direct, while $n = 0.5$ indicates an indirect optical transition [17]. The absorption coefficient α is deduced by the formula:

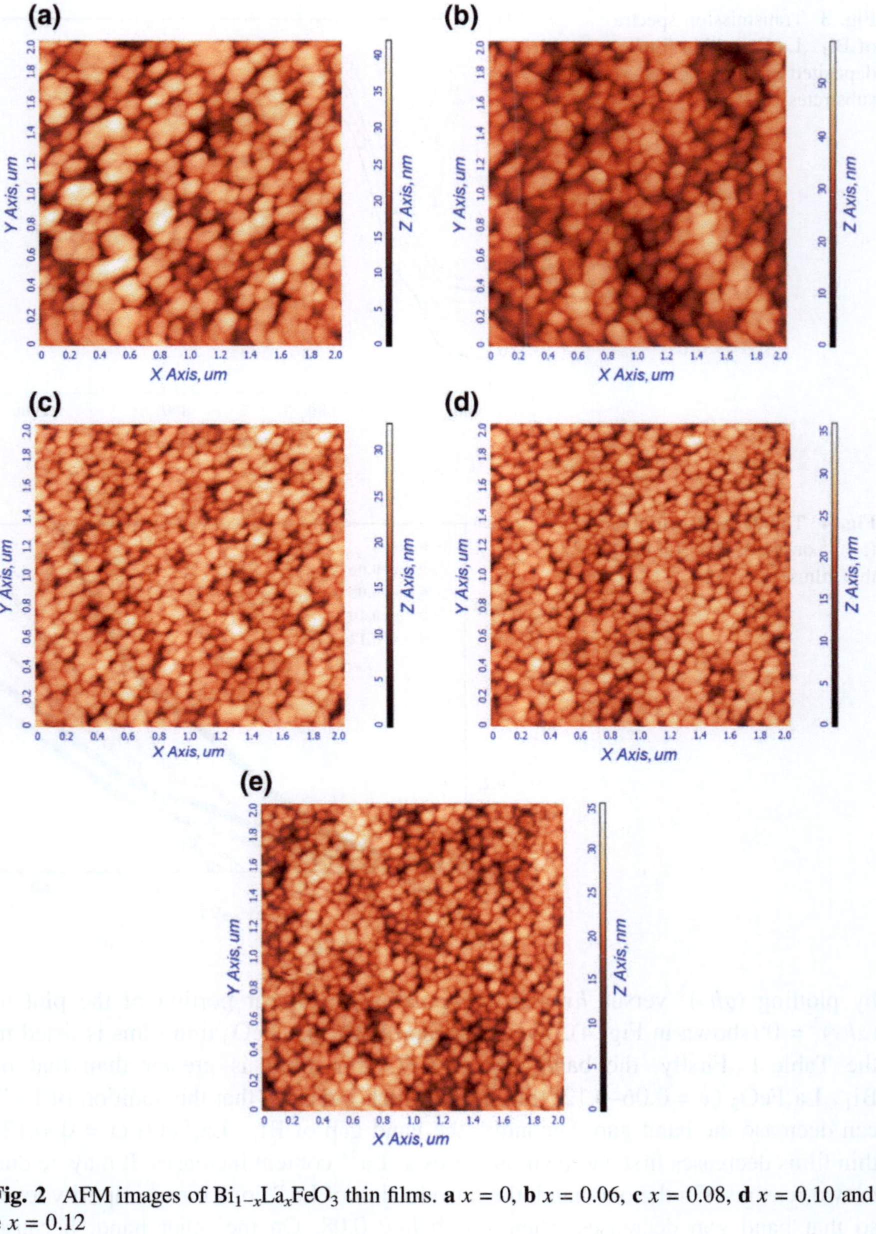

Fig. 2 AFM images of $Bi_{1-x}La_xFeO_3$ thin films. **a** $x = 0$, **b** $x = 0.06$, **c** $x = 0.08$, **d** $x = 0.10$ and **e** $x = 0.12$

$$\alpha = 1/t \cdot \ln(1/T) \qquad (2)$$

where t means the thickness of the film (t of thin films has been determined by stylus profiler) and T represents the transmittance. Because BFO thin films is direct band gap semiconductor [18, 19], n is 2. The optical band gap (E_g) can be obtained

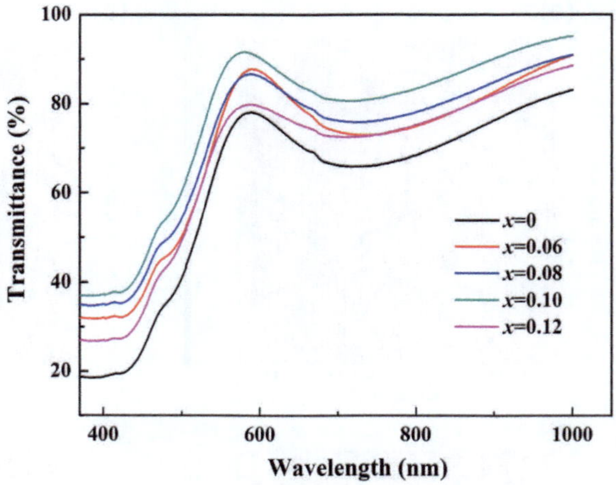

Fig. 3 Transmission spectra of $Bi_{1-x}La_xFeO_3$ thin films deposited on quartz glass substrates

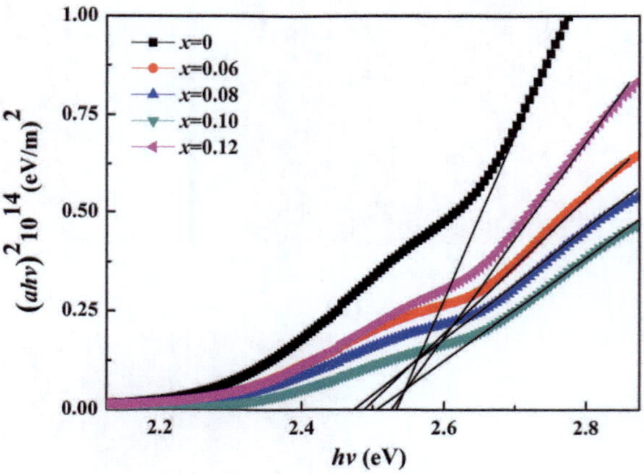

Fig. 4 The dependence of $(\alpha hv)^2$ on hv of $Bi_{1-x}La_xFeO_3$ thin films

by plotting $(\alpha hv)^2$ versus hv and extrapolating the linear portion of the plot to $(\alpha hv)^2 = 0$ (shown in Fig. 4). The band gap of $Bi_{1-x}La_xFeO_3$ thin films is listed in the Table 1. Firstly, the band gap of BFO thin films is greater than that of $Bi_{1-x}La_xFeO_3$ (x = 0.06–0.12) thin films, which indicates that the addition of La^{3+} can decrease the band gap. Secondly, the band gap of $Bi_{1-x}La_xFeO_3$ (x = 0–0.12) thin films decreases first and then increases as La^{3+} content increases. It may be due to two reasons. On the one hand, the introduction of La^{3+} ions forms impurity level so that band gap decreases when x is below 0.08. On the other hand, because $Bi_{1-x}La_xFeO_3$ is n-type conduction [20, 21] so that Burstein-Moss effect lead to the broadening of band gap in heavily-doped semiconductor ($x > 0.08$) [22].

Figure 5 shows the leakage current density-electric field (*J-E*) curves of $Bi_{1-x}La_xFeO_3$ thin films. It is found that the *J-E* curves of $Bi_{1-x}La_xFeO_3$ (x = 0, 0.06, 0.08 and 0.10) thin films at negative and positive biases are symmetric and the leakage current density change a little with the increase of La^{3+} content ($x \leq 0.10$).

Table 1 The optical band gap of $Bi_{1-x}La_xFeO_3$ thin films

x	0	0.06	0.08	0.10	0.12
Band gap (eV)	2.54	2.49	2.47	2.51	2.53

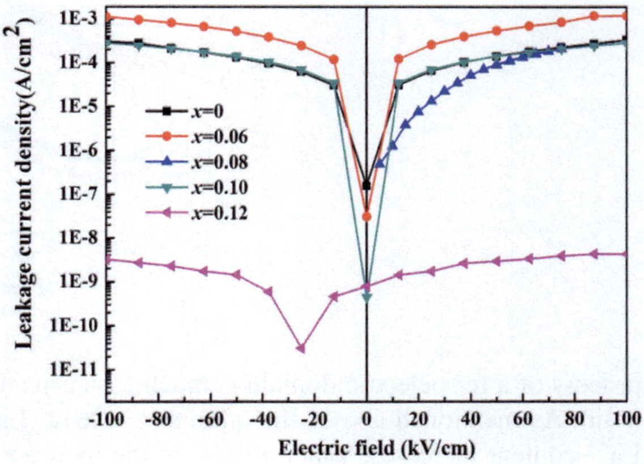

Fig. 5 Leakage current density versus electric field curves of $Bi_{1-x}La_xFeO_3$ thin films

The leakage current density of $Bi_{1-x}La_xFeO_3$ ($x = 0.12$) thin films is much lower than that of the BFO thin films, which indicates that a larger amount of La^{3+} ions can make leakage current of BFO thin films obviously decrease. It may result from two reasons. On the one hand, the high leakage current of BFO thin films results from oxygen vacancy caused by the fluctuation of Fe^{3+} to Fe^{2+} [4, 5], the introduction of La^{3+} with higher stability can reduce the concentration of oxygen vacancy. On the other hand, smaller grain size results in larger number of grain boundaries, which act as scattering center for the flow of electrons and therefore decreasing the leakage current [23]. It is noteworthy that the *J-E* curve of $Bi_{1-x}La_xFeO_3$ ($x = 0.12$) thin films is asymmetric on the electric-field axis, which may result from the defect.

Figure 6 shows the room-temperature hysteresis loops of $Bi_{1-x}La_xFeO_3$ thin films. It is seen that the remnant polarization (P_r) and coercive electric field (E_C) of $Bi_{1-x}La_xFeO_3$ thin films increase as La^{3+} content increases, which indicates the introduction of La^{3+} on A sites can improve the ferroelectricity of BFO thin films. It is well known that BFO thin films have higher leakage caused by reduction of Fe^{3+} ions ($4Fe^{3+} + 2O^{2-} = 4Fe^{2+} + O_2 \uparrow + 4V_O^{\cdot\cdot}$) and volatilization of Bi ($2Bi^{3+} + 3O^{2-} = Bi_2O_3 + 2V_{Bi}'' + 3V_O^{\cdot\cdot}$) [4–7]. The introduction of La^{3+} can inhibit the volatilization of Bi at higher temperature so that the concentration of oxygen vacancy decreases and results in the increasing of remnant polarization. Furthermore, the presence of impure phase ($Bi_2Fe_4O_9$) in BFO thin films also results in the lowest remnant polarization. The increase of coercive electric field with the increase of La^{3+} content attribute to effect of grain size. It is known that energy barrier for switching ferroelectric domain must be broken through and energy barrier increases as grain size decreases [24, 25]. So reversal polarization

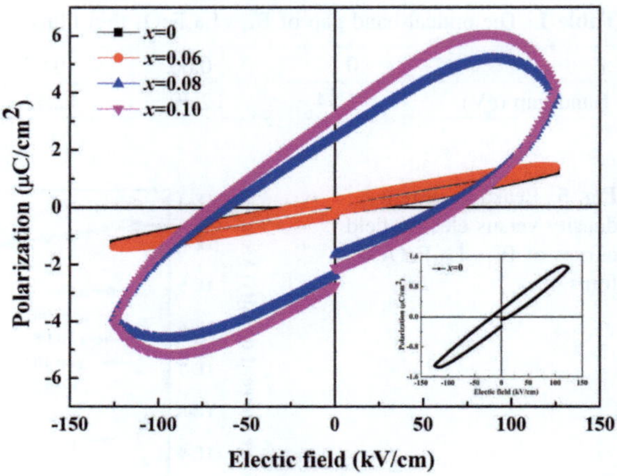

Fig. 6 Room-temperature hysteresis loops of $Bi_{1-x}La_xFeO_3$ thin films measured at 1 kHz

process of a ferroelectric domain is much easier inside a large grain than in a small grain. As mentioned above, the grain size of $Bi_{1-x}La_xFeO_3$ thin films decreases as La^{3+} content increases, which results in the increase of coercive electric field.

Summary

$Bi_{1-x}La_xFeO_3$ thin films were prepared by sol-gel method. $Bi_{1-x}La_xFeO_3$ (x = 0–0.12) thin films is rhombohedral distortion perovskite structure. and the introduction of La^{3+} can inhibit the formation of impurity ($Bi_2Fe_4O_9$). The lattice constant of $Bi_{1-x}La_xFeO_3$ thin films decreases as La^{3+} content increases. The average grain size and roughness of $Bi_{1-x}La_xFeO_3$ thin films decrease with the increasing of La^{3+} content. The substitution of La^{3+} for Bi^{3+} on A sites can decrease the band gap of BFO thin films and the optical band gap of $Bi_{1-x}La_xFeO_3$ (x = 0–0.12) thin films decreases first and then increases as La^{3+} content increases, which results from both Burstein-Moss effect and impurity level caused by La^{3+} doping. A larger amount of La^{3+} (x = 0.12) can make leakage current of BFO thin films obviously decrease. Moreover, the remnant polarization and coercive electric field of $Bi_{1-x}La_xFeO_3$ thin films increase as La^{3+} content increases, which results from the decrease of oxygen vacancy and effects of grain size.

Acknowledgements This work was supported by the National Natural Science Foundation of China (Grant No. 51372283, 51402031, 61404018), the Program for Innovation Teams in University of Chongqing, China (Grant No. CXTDX201601032), the Science and Technology Innovation Project of Social Undertakings and People's Livelihood Guarantee of Chongqing, China (Grant No. CSTC2017SHMSA90015).

References

1. S.M. Lam, J.C. Sin, A.R. Mohamed, A newly emerging visible light-responsive $BiFeO_3$, perovskite for photocatalytic applications: a mini review. Mater. Res. Bull. **90**, 15–30 (2017)
2. S. Nandy, P.S.V. Mocherla, C. Sudakar, Photoconductivity induced by nanoparticle segregated grain-boundary in spark plasma sintered $BiFeO_3$. J. Appl. Phys. **121**, 203102 (2017)
3. M.M. Yang, Z.D. Luo, D.J. Kim, M. Alexe, Bulk photovoltaic effect in monodomain $BiFeO_3$ thin films. Appl. Phys. Lett. **110**, 183902 (2017)
4. H. Tao, J. Lv, R. Zhang, R.P. Xiang, J.G. Wu, Lead-free rare earth-modified $BiFeO_3$ ceramics: phase structure and electrical properties. Mater. Design **120**, 83–89 (2017)
5. Y. Heo, S.B. Hu, P. Sharma, K.E. Kim, B.K. Jang, C. Cazorla, C.H. Yang, J. Seidel, Impact of isovalent and aliovalent doping on mechanical properties of mixed phase $BiFeO_3$. ACS Nano **11**, 2805–2813 (2017)
6. H. Lzumi, T. Yoshimura, N. Fujimura, Growth and ferroelectric properties of La and Al codoped $BiFeO_3$ epitaxial films. J. Appl. Phys. **121**, 174102 (2017)
7. F. Burkert, M. Janowski, X. Zhang, I. Takeuchi, C.A. Kuntscher, Chemical pressure effect in Sm and La substituted ferroelectric $BiFeO_3$ thin films: Insights from infrared spectroscopy. J. Appl. Phys. **121**, 144103 (2017)
8. W. Sun, Z. Zhou, J. Luo, K. Wang, J.F. Li, Leakage current characteristics and Sm/Ti doping effect in $BiFeO_3$ thin films on silicon wafers. J. Appl. Phys. **121**, 064101 (2017)
9. Y. Feng, C. Wang, S.L. Tian, Y. Zhou, C. Ge, H.Z. Guo, M. He, K.J. Jin, G.Z. Yang, Effects of $BaTiO_3$ and $SrTiO_3$ as the buffer layers of epitaxial $BiFeO_3$ thin films. Sci. China. Phys. Mech. **60**, 067711 (2017)
10. P.P. Lv, X.M. Jiang, J. Yan, G.D. Hu, Stable self-polarization in Nd and Ti codoped $BiFeO_3$ films. J. Mater. Sci. Mater. Electron. **28**, 2233–2240 (2017)
11. J. Kolte, A.S. Daryapurkar, M. Agarwal, D.D. Gulwade, P. Gopalan, Effect of substrate temperature on the structural and electrical properties of La and Mn co-doped $BiFeO_3$ thin films. Thin Solid Films **619**, 308–316 (2016)
12. S.J. Guo, C.H. Yang, X.M. Jiang, P.P. Lv, G.D. Hu, High ferroelectric performance of $Bi_{0.9}La_{0.1}FeO_3$ thick film by optimizing preparation precursor solution. J. Sol-Gel. Sci. Techn. **80,** 174–179 (2016)
13. H.P. Guo, J.J. Tian, H.W. Zheng, F.R. Tan, W.F. Zhang, Magnetic and optical properties of La-doped $BiFeO_3$ films prepared by sol–gel route. J. Mater. Sci. Mater. El. **26**, 700–704 (2014)
14. X.J. Chou, J.W. Zhai, H.T. Jiang, X. Yao, Dielectric properties and relaxor behavior of rare-earth (La, Sm, Eu, Dy, Y) substituted barium zirconium titanate ceramics. J. Appl. Phys. **102**, 084106 (2007)
15. P. Pandit, S. Satapathy, P.K. Gupta, Effect of La substitution on conductivity and dielectric properties of $Bi_{1-x}La_xFeO_3$ ceramics: an impedance spectroscopy analysis. Phys. B **406**, 2669–2677 (2011)
16. A. Gaur, P. Singh, N. Choudhary, D. Kumar, M. Shariq, K. Singh, N. Kaur, D. Kaur, Structural, optical and magnetic properties of Nd-doped $BiFeO_3$ thin films prepared by pulsed laser deposition. Phys. B **406**, 1877–1882 (2011)
17. J. Tauc, *Amorphous and liquid semiconductors* (Academic Press, New York, 1974), p. 516
18. J.F. Ihlefeld, N.J. Podraza, Z.K. Liu, R.C. Rai, X. Xu, T. Heeg, Y.B. Chen, J. Li, R.W. Collins, J.L. Musfeldt, X.Q. Pan, J. Schubert, R. Ramesh, D.G. Schlom, Optical band gap of $BiFeO_3$ grown by molecular-beam epitaxy. Appl. Phys. Lett. **92**, 142908 (2008)
19. F. Yan, T.J. Zhu, M.O. Lai, L. Lai, Enhanced multiferroic properties and domain structure of La-doped $BiFeO_3$ thin films. Scr. Mater. **63**, 780–783 (2010)
20. Y.P. Guo, B. Guo, W. Dong, H. Li, H.Z. Liu, Evidence for oxygen vacancy or ferroelectric polarization induced switchable diode and photovoltaic effects in $BiFeO_3$ based thin films. Nanotechnology **24**, 275201 (2013)

21. Z.K. Liu, F. Yan, The application of bismuth-based oxides in organic-inorganic hybrid photovoltaic devices. J. Am. Ceram. Soc. **95**, 1944–1948 (2012)
22. S.H. Deng, M.Y. Duan, M. Xu, L. He, Effect of La doping on the electronic structure and optical properties of ZnO. Phys. B **406**, 2314–2318 (2011)
23. K. Sen, K. Singh, A. Gautam, M. Singh, Dispersion studies of La substitution on dielectric and ferroelectric properties of multiferroic $BiFeO_3$ ceramic. Ceram. Int. **38**, 243–249 (2012)
24. W. Cai, C.L. Fu, Z.B. Lin, X.L. Deng, Vanadium doping effects on microstructure and dielectric properties of barium titanate ceramics. Ceram. Int. **37**, 3643–3650 (2011)
25. W. Cai, C.L. Fu, J.C. Gao, X.L. Deng, Dielectric properties, microstructure and diffuse transition of Al-doped $Ba(Zr_{0.2}Ti_{0.8})O_3$. J. Mater. Sci. Mater. Electron. **21**, 796–803 (2010)

Effect of Mn-Doping on the Structure, Magnetic Properties of $Fe_{17}Gd_2$ Compound

Yongbin Guo, Mufen He, Lei Ma, Dao Wang and Xin Zhou

Abstract The $(Fe_{1-x}Mn_x)_{17}Gd_2$ (x = 0.0–0.4) compounds were prepared by arc melting method, then annealed at 700 °C for two weeks. The effect of Mn-doping on the structure and magentic properties were studied by X-ray diffraction (XRD) and vibrating sample magnetometer (VSM). The crystal structure of $(Fe_{1-x}Mn_x)_{17}Gd_2$ maintains the Th_2Zn_{17}-type rhombohedral structure, and the unit cell volume and lattice parameter a, b and c of $(Fe_{1-x}Mn_x)_{17}Gd_2$ increase with the addition of Mn content. With the increase of Mn-doping, the magnetic-type of $(Fe_{1-x}Mn_x)_{17}Gd_2$ compounds obviously transform from ferromagnetism to antiferromagnetism, and the saturation magnetization (M_s) can be tuned continuously from 80 Am^2/kg at x = 0.0 to ~0 Am^2/kg at x = 0.4. At the same time, the Curie temperature (T_C) of compounds reduced to room temperature. It indicates that Mn doping can effectively tune the structure, magnetic-type and T_C of $Fe_{17}Gd_2$ compound.

Keywords Magnetic properties · $Fe_{17}Gd_2$ compound · Rhombohedral structure

Introduction

The properties of the R_2Fe_{17} compounds (R represents the rare-earth) have attracted interest for many years [1–5]. The crystallographic structure, the magnetization, the magnetostriction and the thermodynamic properties of the compounds have been intensively studied [6–8]. The study found that the Fe-rich intermetallic compound R_2Fe_{17} showed a moderate Curie temperature (Tc) at around room temperature [9]. In addition, three different magnetic ordering types and a field-induced magne-

Y. Guo · M. He (✉) · L. Ma · D. Wang · X. Zhou
School of Materials Science and Engineering, Guilin University of Electronic Technology, Guilin 541004, China
e-mail: hemf815@guet.edu.cn

M. He · L. Ma
Guangxi Key Laboratory of Information Materials, Guilin University of Electronic Technology, Guilin 541004, China

Y. Han (ed.), *Advanced Functional Materials*,
https://doi.org/10.1007/978-981-13-0110-0_21

toresistance effect have been observed in compounds with the $ThCr_2Si_2$-type structure [10–12]. Compared with $R_2Fe_{17-x}T_x$ compound (T=Al, Ga and Si) in which the lattice parameters exhibit monotonic behaviour with increasing x, $R_2Fe_{17-x}Mn_x$ (R=Nd [13], Pr [14], Tb [15], Dy [16], Er [17]) compounds show unexpected behaviour in that the unit-cell volume first exhibit a slight maximum around x = 0.5–2 with the increase of x. Following previous studies that Mn doping can effectively adjust the T_C in $Fe_{17}Dy$ [16] and $Er_2Fe_{17-x}Mn_x$ [17], we report our experimental investigation on $(Fe_{1-x}Mn_x)_{17}Gd_2$ compounds.

In this work, we focus our attention on the effect of the substitution of Mn for Fe in Gd_2Fe_{17} compound on the crystal structure and magnetic properties, such as the saturation magnetization (M_s), the Tc and magnetic transition with the doping of Mn content. The results would help us have a better understanding of the interrelation between structure and magnetic properties.

Experiment

The $(Fe_{1-x}Mn_x)_{17}Gd_2$ (x = 0.0–0.4) compounds were prepared by arc melting method. The purities of Fe, Mn and Gd are better than 99.99%. In order to ensure a homogeneity of the composition, the samples were repeatedly melted 3–4 times, and the weight loss for all the samples was less than 0.5 wt%. Approximately 3 wt % excess Mn was added during melting. The ingots were wrapped in a tantalum foil, sealed in a quartz tube filled argon, annealed at 700 °C for two weeks and then cooled in ice-water. The X-ray diffraction (XRD) measurements were carried out on a PANalytical PIXcel^{3D} X-ray powder diffractometer using Cu Kα_1 radiation. The data were collected between 20° and 120° in 2θ with a step of 0.026°. Analysis of XRD patterns was carried out using the Fullprof suite package. The magnetic measurements were performed using the vibrating sample magnetometer (VSM, LakeShore, Inc.). The T_C of compounds were performed using the vibrating sample magnetometer option of the physical property measurement system (PPMS-9 manufactured by Quantum Design, Inc.).

Results and Discussion

Crystal Structure. The $Fe_{17}Gd_2$ compound crystallized in the typical Th_2Zn_{17} rhombohedral structure with $R\bar{3}m$ space group. Figure 1 shows the crystal structure of $Fe_{17}Gd_2$. Figure 2 depicts the XRD patterns recorded for powder specimens of $(Fe_{1-x}Mn_x)_{17}Gd_2$ (x = 0.0–0.4) compounds. All the diffraction peaks can be indexed as the Th_2Zn_{17}-type structure (rhombohedral). With the addition of Mn content, the positions of diffraction peaks shift to a small angle slightly, which is due to the radius of Mn is greater than that of Fe. When the replacement occurs, a

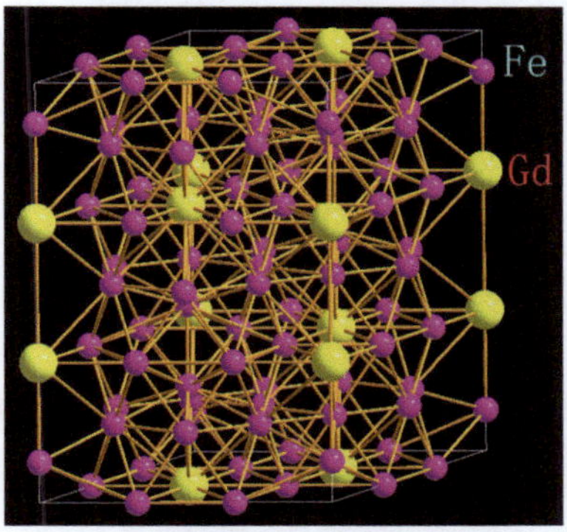

Fig. 1 The crystal structure of $Fe_{17}Gd_2$ compound

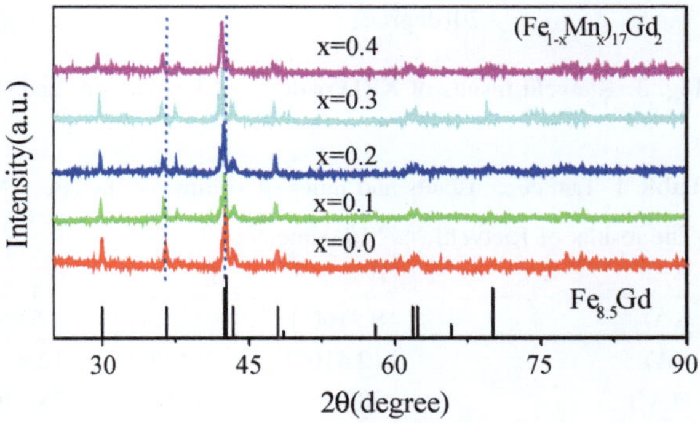

Fig. 2 XRD patterns of $(Fe_{1-x}Mn_x)_{17}Gd_2$ (x = 0.0–0.4) compounds

small amount of Mn atoms does not cause large distortion of the lattice, and lattice structure remains the same. In order to further study the change in the structure of the $(Fe_{1-x}Mn_x)_{17}Gd_2$ (x = 0.0–0.4) with increasing Mn content, Rietveld refinement was carried out. The results show a good fitting between experimental and calculated profiles for x = 0.0 and 0.3 compounds in Fig. 3, which is based on XRD data and the structural model of $Fe_{17}Gd_2$ (Th_2Zn_{17}-structure) [18]. The results of Rietveld, lattice constants and unit-cell volumes of the $(Fe_{1-x}Mn_x)_{17}Gd_2$ compounds were calculated respectively, as shown in Table 1. The unit-cell volume of the $(Fe_{1-x}Mn_x)_{17}Gd_2$ compounds increases with the Mn content, which could be caused by replacement of the larger Mn atom for Fe atom. When $x \leq 0.2$, Mn atoms substituted firstly Fe atoms on the 6c-site with the largest Wigner-Seitz unit-cell volume and bond length [19].

Magnetic Properties. Figure 4a shows the hysteresis loops for $(Fe_{1-x}Mn_x)_{17}Gd_2$ (x = 0.0–0.4) compounds measured at 300 K. It indicates the compounds are soft magnetic materials. There is a significant transition of ferromagnetism (FM) to antiferromagnetism (AFM) when the Mn content x exceeds 0.2. The magnetic moment of the $(Fe_{1-x}Mn_x)_{17}Gd_2$ compounds except for x = 0.3, 0.4 are close to saturation at 0.5T. The saturation magnetization (M_s) of x = 0.0, 0.1, 0.2 compounds are 77.13, 53.82 and 22.29 Am^2/kg respectively, which decreases

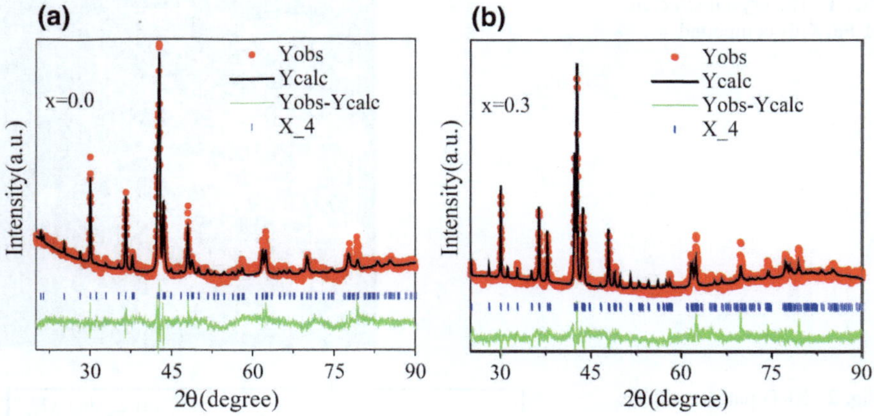

Fig. 3 Rietveld results of XRD patterns of $Fe_{17}Gd_2$ (**a**) and $(Fe_{0.7}Mn_{0.3})_{17}Gd_2$ (**b**) compounds

Table 1 Lattice constants and unit-cell volume of the $(Fe_{1-x}Mn_x)_{17}Gd_2$ compounds

The results of Rietveld	Mn content/x				
	0	0.1	0.2	0.3	0.4
a(Å)	8.516(8)	8.525(4)	8.532(1)	8.539(1)	8.545(5)
c(Å)	12.416(9)	12.423(2)	12.435(3)	12.445(7)	12.465(6)
V(Å^3)	782.73(7)	783.18(7)	783.70(3)	784.09(4)	784.50(8)
GOF	3.5%	2.1%	1.8%	2.7%	2.6%
R_{wp}	15.6%	12.4%	14.6%	11.2%	10.3%
R_{exp}	8.3%	9.1%	7.8%	5.6%	9.7%

drastically with the increase of Mn-doping amount. This is due to the fact that the total magnetization in this kind of compounds mainly comes from the magnetization of the sublattice composed of rare earth and 3d metal, while the 3d electron of Mn is one electron less than Fe and usually exhibits antiferromagnetism. With the addition of Mn content, more Mn atoms substitute for the Fe atoms, which weaken the ferromagnetic interaction between Fe and Fe. The antiferromagnetic interaction between Fe and Mn is enhanced accordingly, which results in a reduction of total magnetization. Figure 4b shows the magnetic phase transition interval of $(Fe_{1-x}Mn_x)_{17}Gd_2$ (x = 0.0–0.4) compounds with the change of Mn content. No. 1 shows the region of FM, and No. 2 means the region of AFM. When $x < 0.2$ the compounds are obviously FM, and the compounds are AFM when $x > 0.3$. In the interval area of x = 0.2–0.3, it is the mixed region of AFM to FM transition. The temperature dependencies of magnetization curves for $(Fe_{1-x}Mn_x)_{17}Gd_2$ (x = 0.1–0.4) compounds taken in applied magnetic field $\mu_0 H = 0.02$ T were presented in Fig. 5. The experimental results show the T_C of the $Fe_{17}Gd_2$ is 472 K, which is consistent with previous report [1]. It also can be seen that this compounds exhibits FM in low temperature and paramagnetic at room temperature,

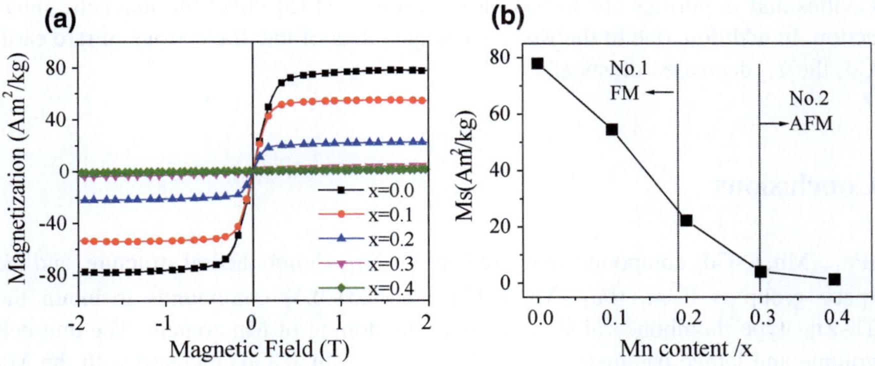

Fig. 4 Magnetic hysteresis loops (**a**) and magnetization dependence on Mn concentration (**b**) of $(Fe_{1-x}Mn_x)_{17}Gd_2$ (x = 0.0–0.4) compounds

Fig. 5 Temperature dependence of magnetization of $(Fe_{1-x}Mn_x)_{17}Gd_2$ (x = 0.1–0.4)

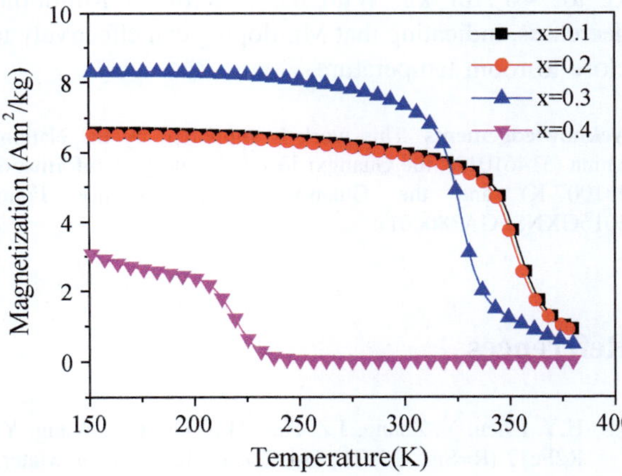

Table 2 The T_c of alloys components

Alloys components	T_c (K)	References
x = 0.0	472 (±2)	[1]
x = 0.1	354 (±2)	This work
x = 0.2	352 (±2)	This work
x = 0.3	326 (±2)	This work
x = 0.4	220 (±2)	This work

respectively. The T_C dependence on Mn-doping of the $(Fe_{1-x}Mn_x)_{17}Gd_2$ compounds are showed in Table 2. It indicates that the T_C of $(Fe_{1-x}Mn_x)_{17}Gd_2$ compounds decrease with increasing Mn content, especially when x is more than 0.3. This phenomenon may be attributed to the weak anti-ferromagnetic coupling interaction between Mn atoms and Fe atoms. With the increasing of Mn content, the lattice distortion and anti-ferromagnetic coupling are further enhanced, and the cavities and impurities are formed at same time, which effect the magnetic interaction. In addition, due to the weak exchange effect of the 4f electrons of rare earth Gd, the T_C decreases drastically.

Conclusions

$(Fe_{1-x}Mn_x)_{17}Gd_2$ compounds are a typical Th_2Ni_{17} rhombohedral structure, and the space group is R$\bar{3}$m. $(Fe_{1-x}Mn_x)_{17}Gd_2$ (x = 0.0–0.4) compounds maintain the Th_2Zn_{17}-type rhombohedral structure with the doping of Mn content. The unit-cell volume and lattice parameter a, b and c of $(Fe_{1-x}Mn_x)_{17}Gd_2$ increase with the Mn content. The $(Fe_{1-x}Mn_x)_{17}Gd_2$ compounds obviously transform from ferromagnetism to antiferromagnetism when the Mn content x exceeds 0.2. The Mn-doping reduces drastically the saturation magnetization of the compounds from 77.13 Am^2/kg to $\sim$0 Am^2/kg. With the increase of Mn-doping, the T_C of the compound decreases, indicating that Mn-doping can effectively tuned the T_C of the compounds close to room temperature.

Acknowledgements This work is supported by the National Natural Science Foundation of China (51461012), the Guangxi Key Laboratory of Information Materials (161011-Z, 161005-K, 151007-K) and the Guangxi Natural Science Foundation (2016GXNSFAA380030, 2016GXNSFGA380001).

References

1. H.Y. Chen, Y. Zhang, J.Z. Han, H.L. Du, C.S. Wang, Y.C. Yang, Magnetocaloric effect in R2Fe17 (R=Sm, Gd, Tb, Dy, Er). J. Magn. Magn. Mater. **320**, 1382–1384 (2008)
2. K. Takeda, Magneto-optical Kerr rotation spectra in R2Fe17 single crystals. J. Magn. Magn. Mater. **272–276**, 1863–1864 (2004)
3. T. Kamimori, K. Kinoshita, J. Mochimaru, K. Konishi, H. Tange, Magnetovolume effect of Fe sublattice in R2 Fe17. J. Magn. Magn. Mater. **230**, 993–995 (2001)
4. Y.G. Wang, F. Yang, N. Tang, Changpin Chen, Qidong Wang, Structure and magnetic properties of R2Fe17−xMnx compounds (R=Tb, Y). J. Magn. Magn. Mater. **167**, 237–240 (1997)
5. X.C. Kou, F.R. de Boer, R. Grossinger, G. Wiesinger, H. Suzuki, H. Kitazawa, T. Takamasu, G. Kido, Magnetic anisotropy and magnetic phase transitions in R2Fe17 with R = Y, Ce, Pr, Nd, Sm, Gd, Tb, Dy, Ho, Er, Tm and Lu. J. Magn. Magn. Mater. **177**, 1002–1007 (1998)
6. H.R. Kirchmayr, C.A. Poldy, in *Handbook on Physics and Chemistry of Rare Earths*, vol. 4, ed. by K.A. Gschneidner, J. Eyring (North-Holland, Amsterdam, 1979), p. 55
7. K.H.J. Buschow, in *Handbook of Ferromagnetic Materials*, vol. 7, ed. by E.P. Wohlfarth (North-Holland, Amsteram, 1993), p. 307
8. J.J.M. Franse, R.J. Radwanski, K.H.J. Buschow (ed.), *Handbook of Magnetic Materials* (Amsterdam, 1993), pp. 307–501
9. X.C. Zhong, Z.W. Liu, D.C. Zeng, K.A. Gschneidner Jr., V.K. Pecharsky, Magnetocaloric effect of $Pr_2Fe_{17-x}Mn_x$ alloys. Rare Met. **33**, 552–555 (2014)

10. J.H.V.J. Brabers, K. Bakker, H. Nakotte, F.R. de Boer, S.K.J. Lenczowskiand, K.H. J. Buschow, Giant magnetoresistance in polycrystalline $SmMn_2Ge_2$. J. Alloys Comp. **199**, L1–L3 (1993)
11. J.H.V.J. Brabers, A.J. Nolten, F. Kayzel, S.H.J. Lenczowski, K.H.J. Buschow, F.R. de Boer, Phys. Rev. B **50**, 16410–16417 (1994)
12. G. Ventufini, R. Welter, E. Ressouche, B. Malaman, Neutron diffraction study of $Nd_{0.35}La_{0.65}Mn_2Si_2$: a $SmMn_2Ge_2$-like magnetic behaviour compound. J. Magn. Magn. Mater. **150**, 197–212 (1995)
13. W.B. Yelon, Z. Hu, M. Chen, H. Luo, P.C. Ezekwenna, G.K. Marasinghe, W.J. James, K.H. J. Buschow, D.P. Middleton, F. Pourarian, Neutron diffraction and magnetic studies of $Nd_2Fe_{17-x}Tx$ (T=Si, Mn) alloys. IEEE Trans. Magn. **32**, 4431 (1996)
14. Z.G. Sun, S.Y. Zhang, H.W. Zhang, B.G. Shen, Magnetohistory effect of $Pr_2Fe_{17-x}Mn_x$ compounds. J. Alloys Compd. **349**, 1–5 (2003)
15. Y.G. Wang, F.M. Yang, C.P. Chen, N. Tang, P.H. Lin, Q.D. Wang, Investigation of magnetic properties of $Tb_2Fe_{17-x}Mn_x$ compounds. J. Appl. Phys. **84**, 6229–6232 (1998)
16. Y.B. Guo, L. Ma, S.Q. Zhao, D. Wang, X. Zhou, L. Li, Y.S. Du, G.H. Rao, J. Supercond. Nov. Magn. https://doi.org/10.1007/s10948-017-4210-6
17. J.L. Wang, M.R. Ibarra, C. Marquina, B. García-Landa, W.X. Li, N. Tang, W.Q. Wang, F.M. Yang, G.H. Wu, J. Appl. Phys. **92**, 1453 (2002)
18. Information on http://mits.nims.go.jp/index.en.html
19. Z.G. Sun, H.W. Zhang, J.Y. Wang, B.G. Shen, Structure and magnetic properties of $Pr_2Fe_{17-x}Mn_x$ (x=0–9) compounds. J. Appl. Phys. **86**, 5152–5156 (1999)

Microstructure and Electric Properties of $(Sr_{1-x}Ca_x)_3Sn_2O_7$ Ceramics with Ruddlesden-Popper Structure

Fengqi Wang, Hai Zhang, Wei Cai, Rongli Gao and Chunlin Fu

Abstract $(Sr_{1-x}Ca_x)_3Sn_2O_7$ (x = 0, 0.1 and 0.2) ceramics were prepared by conventional solid solution method. The microstructure, dielectric and ferroelectric properties of $(Sr_{1-x}Ca_x)_3Sn_2O_7$ ceramics have been investigated systematically. The XRD results show that $(Sr_{1-x}Ca_x)_3Sn_2O_7$ ceramics is orthorhombic structure and the lattice constants decrease as Ca^{2+} content increases. The grain size of $(Sr_{1-x}Ca_x)_3Sn_2O_7$ ceramics decreases with the increasing of Ca^{2+} content, which results in the decrease of dielectric constant and dielectric loss. The compactness of $(Sr_{1-x}Ca_x)_3Sn_2O_7$ ceramics deteriorates as Ca^{2+} content increases, which causes the increase of leakage current density. Moreover, the introduction of Ca^{2+} in $(Sr_{1-x}Ca_x)_3Sn_2O_7$ ceramics leads to the increase of the remnant polarization. But the remnant polarization changes a little with Ca^{2+} content when x is above 0.1. The coercive field increases with the increasing of Ca^{2+} content.

Keywords Ruddlesden-Popper · $(Sr_{1-x}Ca_x)_3Sn_2O_7$ · Microstructure Ferroelectric

---

F. Wang · H. Zhang · W. Cai (✉) · R. Gao · C. Fu
School of Metallurgy and Materials Engineering, Chongqing University of Science and Technology, Chongqing 401331, China
e-mail: caiwei_cqu@163.com

F. Wang
e-mail: wangle_17@126.com

H. Zhang
e-mail: 958728340@qq.com

R. Gao
e-mail: gaorongli2008@163.com

C. Fu
e-mail: chlfu@126.com

W. Cai · R. Gao · C. Fu
Chongqing Key Laboratory of Nano-Micro Composite Materials and Devices, Chongqing 401331, China

Y. Han (ed.), *Advanced Functional Materials*,
https://doi.org/10.1007/978-981-13-0110-0_22

Introduction

In recent years, ferroelectrics have attracted a considerable number of attentions because of its spontaneous polarization and potential applications in novel devices, such as nonvolatile memory [1], solar cell [2] and variable capacitor [3]. However, there are still fewer ferroelectrics that can be applied widely in practice. Recently, some new ferroelectrics have been found by theoretical calculations [4] and parts of the new ferroelectric materials have been confirmed experimentally. The hybrid improper ferroelectricity (HIF) caused by two different octahedral rotation has been proposed in Ruddlesden-Popper (R-P) structure ($A_{n+1}B_nO_{3n+1}$) by the first-principle calculations [5–7]. The ferroelectricity of $Ca_3Ti_2O_7$ [8], $Ca_3(Ti,Mn)_2O_7$ [5] and (Ca, Sr)$_3$Ti$_2$O$_7$ [9] with $A_3B_2O_7$ ($n = 2$) R-P structure results from octahedral tilting ($a^-a^-c^0$) and octahedral rotation ($a^0a^0c^+$) and have been demonstrated in experiment.

The optical properties of $Sr_3Sn_2O_7$ compounds belong to $A_3B_2O_7$ ($n = 2$) R-P structure have been investigated [10, 11]. Mulder et al. have predicted that $Sr_3Sn_2O_7$ has hybrid improper ferroelectricity, which results from the condensation of complex lattice distortions [6] and the HIF of $Sr_3Sn_2O_7$ ceramics has been experimentally confirmed by Wang et al. [12] The chemical property of calcium is very similar to that of strontium, therefore the partial substitution of Sr^{2+} on A sites in $Sr_3Sn_2O_7$ by Ca^{2+} might provide an alternative to effectively improve the electric properties of $Sr_3Sn_2O_7$. However, little information on the dielectric and ferroelectric properties of $(Sr_{1-x}Ca_x)_3Sn_2O_7$ ceramics is available. In this paper, the microstructure and electric properties of $(Sr_{1-x}Ca_x)_3Sn_2O_7$ ($x = 0$, 0.1, and 0.2) ceramics were systematically studied.

Experimental Details

$(Sr_{1-x}Ca_x)_3Sn_2O_7$ ($x = 0$, 0.1 and 0.2) ceramics were synthesized by solid-state reaction method. High purity raw materials $SrCO_3$ (99.0%), $CaCO_3$ (99.0%), SnO_2 (99.5%) were stoichiometrically mixed and milled 8 h in distilled water and zirconia media. After the slurry was dried, the mixture was calcined at 1300 °C for 10 h in air. The obtained powder was milled for 8 h, dried and then calcined at 1400 °C for 10 h. The homogenous powder added with 7 wt% binder (polyvinyl alcohol, PVA) was pressed into disk-shaped pellets with a diameter of 10 mm and thickness of 1 mm at 10 MPa pressure. Finally, the pellets were sintered at 1450 °C for 10 h in air.

Crystal structure of the $(Sr_{1-x}Ca_x)_3Sn_2O_7$ ceramics was confirmed by X-ray diffraction (XRD, Smartlab, Rigaku, Japan) with Cu $K\alpha$ radiation ($\lambda = 0.154$ nm). The surface morphology of ceramics was characterized by scanning electron microscopy (SEM, S-3700N, Hitachi, Japan). To measure the electric properties, $(Sr_{1-x}Ca_x)_3Sn_2O_7$ ceramics were polished and coated with silver electrodes. The dielectric properties of the samples were measured by an impedance analyzer (LCR, HP 4980A, Agilent, USA). The ferroelectric hysteresis loop and leakage current density were obtained by ferroelectric test system (TF2000E, aixACCT, Germany) at room temperature.

Results and Discussion

Figure 1 shows the XRD patterns of $(Sr_{1-x}Ca_x)_3Sn_2O_7$ ceramics at room temperature. Firstly, it is found that the diffraction patterns of $(Sr_{1-x}Ca_x)_3Sn_2O_7$ (x = 0, 0.1 and 0.2) ceramics are indexed as orthorhombic structure ($Sr_3Sn_2O_7$, *CmCa* space group, JCPDS card no. 25-0914). No secondary phase is observed in all the samples, which indicates that Ca^{2+} ions have entered into $Sr_3Sn_2O_7$ lattice. The lattice constants can be calculated by Bragg Eq. 1 and orthorhombic *d*-spacing Eq. 2:

$$2\mathrm{d}sin\theta = \mathrm{n}\lambda \tag{1}$$

$$\frac{1}{\mathrm{d}^2} = \frac{\mathrm{h}^2}{\mathrm{a}^2} + \frac{\mathrm{k}^2}{\mathrm{b}^2} + \frac{\mathrm{l}^2}{\mathrm{c}^2} \tag{2}$$

where d is the interplanar spacing, θ is the diffraction angle, n is the number of interplanar spacing, λ is the wavelength of X-ray (0.154 nm), h, k, l are the indices of crystallographic plane, and a, b, c are the lattice constants. The dependences of the lattice constants and unit-cell volume on Ca^{2+} content are plotted in Fig. 2. It shows that the lattice constants (a, b and c) and unit-cell volume decrease with the increasing of Ca^{2+} content, which may be due to that Ca^{2+} with smaller ionic radius (0.106 nm) substitutes for Sr^{2+} with larger ionic radius (0.118 nm) on A sites.

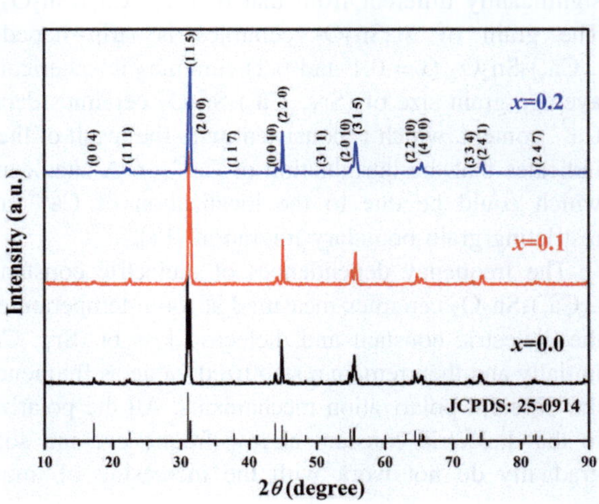

Fig. 1 XRD patterns of $(Sr_{1-x}Ca_x)_3Sn_2O_7$ ceramics

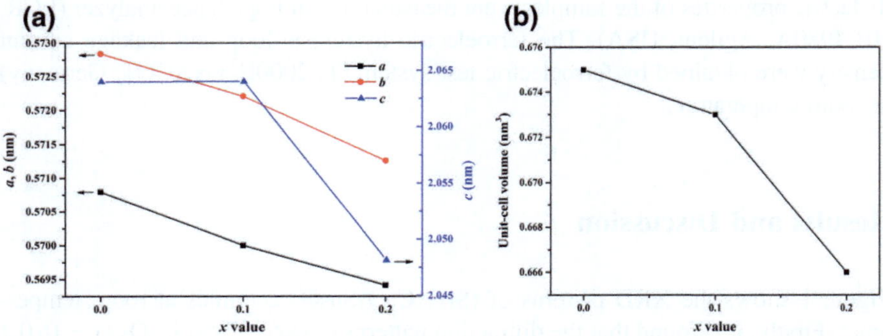

Fig. 2 Dependences of lattice constants (**a**) and unit-cell volume (**b**) on Ca^{2+} content (x) in $(Sr_{1-x}Ca_x)_3Sn_2O_7$ ceramics

Secondly, it is well known that the crystalline grain size can be obtained according to the full width at half maximum (*FWHM*) by the Scherrer formula:

$$D = \frac{0.89\lambda}{\beta cos\theta} \quad (3)$$

where D is the crystalline grain size (nm), λ is the wavelength of X-rays (0.154 nm), β is the full width at half maximum reflection height and θ is the diffraction angle. It is obvious that the *FWHM* increases as Ca^{2+} content increases, which indicates that the crystalline grain size of $(Sr_{1-x}Ca_x)_3Sn_2O_7$ ceramics decreases with the increasing of Ca^{2+} content according to Eq. 3.

Figure 3 shows the surface morphology of $(Sr_{1-x}Ca_x)_3Sn_2O_7$ ceramics. Firstly, it is seen that the compactness of $(Sr_{1-x}Ca_x)_3Sn_2O_7$ ceramics deteriorates to a certain extent as Ca^{2+} content increases. There are some pores in $(Sr_{1-x}Ca_x)_3Sn_2O_7$ (x = 0.1 and 0.2) ceramics. Secondly, the shape of grain in $Sr_3Sn_2O_7$ ceramics is significantly different from that of $(Sr_{1-x}Ca_x)_3Sn_2O_7$ (x = 0.1 and 0.2) ceramics. The grain of $Sr_3Sn_2O_7$ ceramics is strip-shaped, while the grain of $(Sr_{1-x}Ca_x)_3Sn_2O_7$ (x = 0.1 and 0.2) ceramics is spherical. Finally, it is found that the average grain size of $(Sr_{1-x}Ca_x)_3Sn_2O_7$ ceramics decreases with the increasing of Ca^{2+} content, which is consistent with the result of the Scherrer formula. The result indicates that the introduction of Ca^{2+} on A sites can inhibit the growth of grain, which could be due to the localization of Ca^{2+} at the grain boundaries thus restricting grain boundary migration [13].

The frequency dependences of dielectric constant and dielectric loss of $(Sr_{1-x}Ca_x)_3Sn_2O_7$ ceramics measured at room-temperature are shown in Fig. 4. Firstly, the dielectric constant and dielectric loss of $(Sr_{1-x}Ca_x)_3Sn_2O_7$ ceramics decrease initially and then remain a stabilized value as frequency increases. It may be due to the different polarization mechanisms. All the polarization mechanisms contribute to the dielectric constant at low frequency, and some polarization mechanisms gradually do not work with the increasing of frequency so that the dielectric

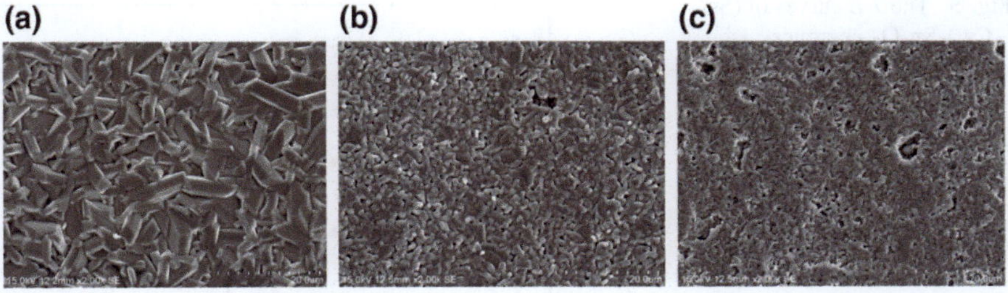

Fig. 3 SEM images of $(Sr_{1-x}Ca_x)_3Sn_2O_7$ ceramics: **a** $x = 0$; **b** $x = 0.1$; **c** $x = 0.2$

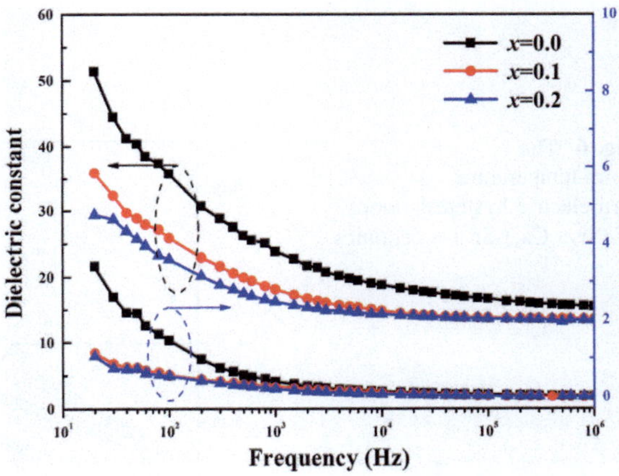

Fig. 4 The frequency dependences of dielectric constant and dielectric loss of $(Sr_{1-x}Ca_x)_3Sn_2O_7$ ceramics

constant decreases [14]. Secondly, the dielectric constant of $(Sr_{1-x}Ca_x)_3Sn_2O_7$ ceramics decreases as Ca^{2+} content increases, which results from the effects of grain size. It is well known that the larger grain size, the higher dielectric constant [15]. As mentioned above, the grain size of $(Sr_{1-x}Ca_x)_3Sn_2O_7$ ceramics decreases with the increasing of Ca^{2+} content. Thirdly, the dielectric loss of $(Sr_{1-x}Ca_x)_3Sn_2O_7$ ceramics decreases with the increasing of Ca^{2+} content when frequency is lower than 10^4 Hz, and the dielectric loss change a little with frequency when frequency is above 10^4 Hz.

Figure 5 shows the leakage current density-electric field (*J-E*) curves of $(Sr_{1-x}Ca_x)_3Sn_2O_7$ ceramics. Firstly, the leakage current density of $(Sr_{1-x}Ca_x)_3Sn_2O_7$ ceramics increases as Ca^{2+} content increases, which results from the deterioration of compactness with the increase of Ca^{2+} content. The variation of conductivity in $(Sr_{1-x}Ca_x)_3Sn_2O_7$ ceramic is mainly due to the lattice distortion caused by the introduction of Ca^{2+} with smaller ionic radius [16]. Secondly, it is found that the *J-E* curve of $Sr_3Sn_2O_7$ ceramics is asymmetric, and *J-E* curves of $(Sr_{1-x}Ca_x)_3Sn_2O_7$ ($x = 0.1$ and 0.2) is symmetric in the electric-field axis, which indicates that there is internal bias field in $Sr_3Sn_2O_7$. The internal bias field may result from the asymmetric distribution of defect such as space charges [17]. On the one hand, the

Fig. 5 The *J-E* curves of $(Sr_{1-x}Ca_x)_3Sn_2O_7$ ceramics

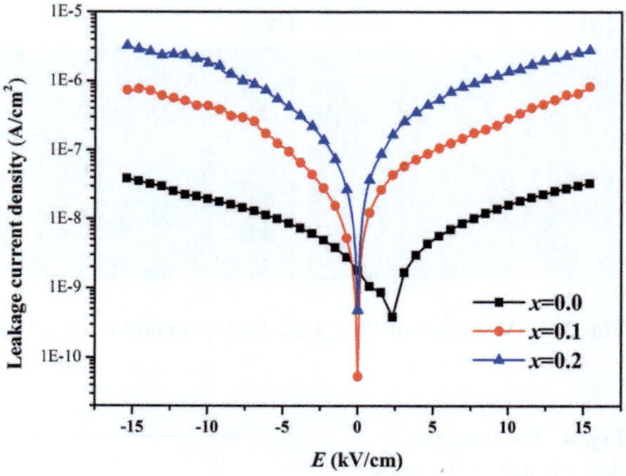

Fig. 6 The room-temperature ferroelectric hysteresis loops of $(Sr_{1-x}Ca_x)_3Sn_2O_7$ ceramics

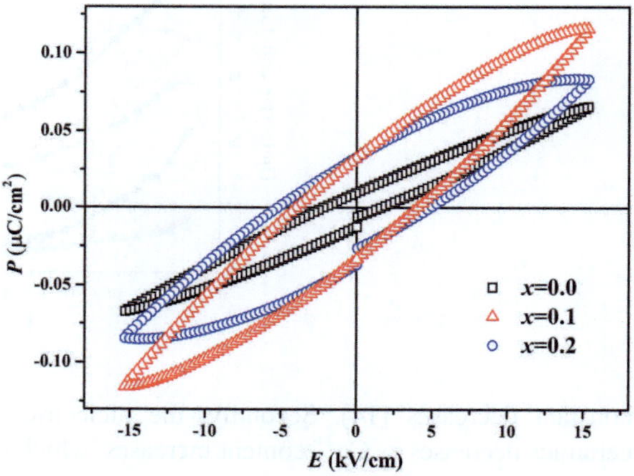

contributions of internal bias field in crystal lattice results in the change of tunneling current in the pure samples. On the other hand, the introduction of Ca^{2+} can restrict the generation of defect such as oxygen vacancy and reduce the internal bias field, which make *J-E* curves of $(Sr_{1-x}Ca_x)_3Sn_2O_7$ (x = 0.1 and 0.2) ceramics symmetric.

Figure 6 shows the room-temperature ferroelectric hysteresis loops (*P-E*) of $(Sr_{1-x}Ca_x)_3Sn_2O_7$ ceramics measured at 10 Hz. Firstly, it is found that *P-E* curves of $(Sr_{1-x}Ca_x)_3Sn_2O_7$ ceramics show non-linear characteristics and the polarization is lower, which indicates poor ferroelectricity. Recent theoretical results have confirmed that a noncentrosymmetric structure originates from HIF, which results from the condensation of complex lattice distortion [6]. Wang et al. have prepared the $Sr_3Sn_2O_7$ ceramics and found that there is robust ferroelectricity at room temperature [12]. Secondly, the remnant polarization (P_r) of $(Sr_{1-x}Ca_x)_3Sn_2O_7$ (x = 0.1 and 0.2) ceramics is obviously higher than that of $Sr_3Sn_2O_7$ ceramics. It may be due to that Ca^{2+} enlarges the space for oxygen octahedron rotation. But the remnant

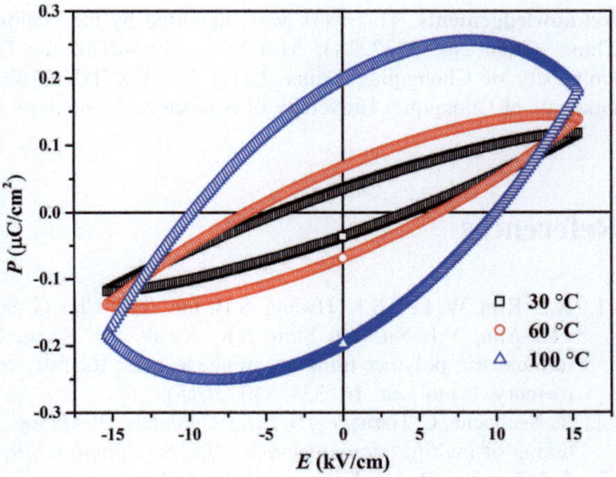

Fig. 7 Hysteresis loops of $(Sr_{0.9}Ca_{0.1})_3Sn_2O_7$ ceramics measured at different temperature and 10 Hz

polarization of $(Sr_{1-x}Ca_x)_3Sn_2O_7$ ceramics changes a little as Ca^{2+} content increases. Moreover, the coercive field increases with the increasing of Ca^{2+} content. These results illustrate that a certain amount of Ca^{2+} on A sites will improve the ferroelectricity of $Sr_3Sn_2O_7$ ceramics.

As mentioned above, $(Sr_{0.9}Ca_{0.1})_3Sn_2O_7$ ceramics presents the lower dielectric loss, the lower leakage current density, and the higher remnant polarization. The temperature dependence of hysteresis loops of $(Sr_{0.9}Ca_{0.1})_3Sn_2O_7$ ceramics are shown in Fig. 7. It is seen that the remnant polarization increases with the rise of temperature, which is different from normal ferroelectrics. It may be due to that the defects such as oxygen vacancy are induced at higher temperature and leads to space charge polarization.

Conclusion

$(Sr_{1-x}Ca_x)_3Sn_2O_7$ ($x = 0$, 0.1 and 0.2) ceramics were fabricated by conventional solid solution method. It is found that $(Sr_{1-x}Ca_x)_3Sn_2O_7$ ceramics is orthorhombic structure. The lattice constants decrease with the increasing of Ca^{2+} content. The average grain size of $(Sr_{1-x}Ca_x)_3Sn_2O_7$ ceramics decreases as Ca^{2+} content increases. And the reduction of grain size causes the decrease of dielectric constant and dielectric loss. The compactness of $(Sr_{1-x}Ca_x)_3Sn_2O_7$ ceramics deteriorates as Ca^{2+} content increases, which results in the increase of leakage current density. Moreover, the substitution of Ca^{2+} for Sr^{2+} on A sites of $(Sr_{1-x}Ca_x)_3Sn_2O_7$ ceramics leads to the increase of remnant polarization. But the remnant polarization has a little change with Ca^{2+} content when x is above 0.1. And the coercive field increases as Ca^{2+} content increases.

Acknowledgements This work was supported by the National Natural Science Foundation of China (Grant No. 51372283, 51402031, 61404018), the Program for Innovation Teams in University of Chongqing, China (Grant No. CXTDX201601032), the Innovation Project for Graduate of Chongqing University of Science & Technology (Grant No. YKJCX1620207).

References

1. K.L. Kim, W. Lee, S.K. Hwang, S.H. Joo, S.M. Cho, G. Song, S.H. Cho, B. Jeong, I. Hwang, J.H. Ahn, Y.J. Yu, T.J. Shin, S.K. Kwak, S.J. Kang, C. Park, Epitaxial growth of thin ferroelectric polymer films on graphene layer for fully transparent and flexible nonvolatile memory. Nano Lett. **16**, 334–340 (2015)
2. R. Nechache, C. Harnagea, S. Li, L. Cardenas, W. Huang, J. Chakrabartty, F. Rosei, Bandgap tuning of multiferroic oxide solar cells. Nat. Photonics **9**, 61–67 (2015)
3. J. Acharya, C.R. Ma, E. Brown, J. Li, J. Wu, Probing effect of temperature on energy storage properties of relaxor-ferroelectric epitaxial $Pb_{0.92}La_{0.08}Zr_{0.52}Ti_{0.48}O_3$ thin film capacitors. Thin Solid Films **616**, 711–716 (2016)
4. N.A. Benedek, Origin of ferroelectricity in a family of polar oxides: the Dion-Jacobson phases. Inorg. Chem. **53**, 3769–3777 (2014)
5. X.Q. Liu, J.W. Wu, X.X. Shi, H.J. Zhao, H.Y. Zhou, R.H. Qiu, W.Q. Zhang, X.M. Chen, Hybrid improper ferroelectricity in Ruddlesden-Popper $Ca_3(Ti, Mn)_2O_7$ ceramics. Appl. Phys. Lett. **106**, 202903 (2015)
6. A.T. Mulder, N.A. Benedek, J.M. Rondinelli, C.J. Fennie, Turning ABO_3, antiferroelectrics into ferroelectrics: design rules for practical rotation-driven ferroelectricity in double perovskites and $A_3B_2O_7$, Ruddlesden-Popper compounds. Adv. Funct. Mater. **23**, 4810–4820 (2013)
7. N.A. Benedek, C.J. Fennie, Hybrid improper ferroelectricity: a mechanism for controllable polarization-magnetization coupling. Phys. Rev. Lett. **106**, 107204 (2011)
8. X. Li, L. Yang, C.F. Li, M.F. Liu, Z. Fan, Y.L. Xie, C.L. Lu, L. Lin, Z.B. Yan, Z. Zhang, J.Y. Dai, J.M. Liu, S.W. Cheong, Ultra-low coercive field of improper ferroelectric $Ca_3Ti_2O_7$ epitaxial thin films. Appl. Phys. Lett. **110**, 042901 (2017)
9. Y.S. Oh, X. Luo, F.T. Huang, Y.Z. Wang, S.W. Cheong, Experimental demonstration of hybrid improper ferroelectricity and the presence of abundant charged walls in $(Ca, Sr)_3Ti_2O_7$ crystals. Nat. Mater. **14**, 407–413 (2015)
10. B.F. Lei, S.Q. Man, Y.L. Liu, S. Yue, Luminescence properties of Sm^{3+}-doped $Sr_3Sn_2O_7$ phosphor. Mater. Chem. Phys. **124**, 912–915 (2010)
11. S. Kamimura, H. Yamada, C.N. Xu, Strong reddish-orange light emission from stress-activated $Sr_{n+1}Sn_nO_{3n+1}$:Sm^{3+} (n = 1, 2, ∞) with perovskite-related structures. Appl. Phys. Lett. **101**, 091113 (2012)
12. Y.Z. Wang, F.T. Huang, X. Luo, B. Gao, S.W. Cheong, The first room temperature ferroelectric Sn insulator and its polarization switching kinetics. Adv. Mater. **29**, 1601288 (2017)
13. R. Maleque, M.D. Rahaman, A.K.M.A. Hossain, Influence of Ca^{2+} ions substitution on structural, microstructural, electrical and magnetic properties of $Mg_{0.2-x}Ca_xMn_{0.5}Zn_{0.3}Fe_2O_4$ ferrites. J. Mater. Sci. Mater. Electron. **28**, 13185–13200 (2017)
14. F.C. Chiu, A review on conduction mechanisms in dielectric films. Adv. Mater. Sci. Eng. **2014**, 1–18 (2014)
15. B. Liu, L. Li, X.Q. Liu, X.M. Chen, $Sr_{n+1}Ti_nO_{3n+1}$ (n = 1, 2) microwave dielectric ceramics with medium dielectric constant and ultra-low dielectric loss. J. Am. Ceram. Soc. **100**, 496–500 (2017)

16. S. Tsuda, N. Kikugawa, K. Sugii, S. Uji, S. Ueda, M. Nishio, Y. Maenoet, Mott transition extremely sensitive to impurities in $Ca_3Ru_2O_7$ revealed by hard x-ray photoemission studies. Phys. Rev. B. **87**, 241107 (2013)
17. P. Gao, C.T. Nelson, J.R. Jokisaari, S.H. Baek, C.W. Bark, Y. Zhang, E. Wang, D.G. Schlom, C.B. Eom, X.Q. Pan, Revealing the role of defects in ferroelectric switching with atomic resolution. Nat. Commun. **2**, 591–596 (2011)

Microstructural Regulation and Optical Performance of Bismuth Ferrite Nanowires by Precipitant

Jing Zhang, Xinlong Fang, Shenglan Wu, Chunlin Fu, Wei Cai, Rongli Gao and Jinyi Ma

Abstract Bismuth ferrite ($BiFeO_3$, short for BFO) is a single-phase and multi-ferrous material with ferroelectricity and weak ferromagnetism at room temperature. One-dimensional nanomaterials have specific physical properties due to anisotropy and unique size effects. In this paper, bismuth ferrite nanowires were prepared by hydrothermal method. The effects of precipitant ($NH_3 \cdot H_2O$, NaOH) on the microstructures of bismuth ferrite nanowires were studied. The results show that NaOH is a precipitant to promote the formation of nanowires, and the single crystal nanowires can be obtained using NaOH precipitation agent. The nanowires are arranged in the same direction. The diameter is less than 45 nm and the length

---

J. Zhang (✉) · X. Fang · S. Wu · C. Fu · W. Cai · R. Gao · J. Ma
School of Metallurgical and Materials Engineering,
Chongqing University of Science and Technology, Chongqing 401331, China
e-mail: 1181807264@qq.com

X. Fang
e-mail: 1019732856@qq.com

S. Wu
e-mail: 986117822@qq.com

C. Fu
e-mail: chlfu@126.com

W. Cai
e-mail: caiwei_cqu@163.com

R. Gao
e-mail: gaorongli2008@163.com

J. Ma
e-mail: ma_jinyi163@163.com

X. Fang · C. Fu · W. Cai · R. Gao
Chongqing Key Laboratory of Nano-Micro Composite Materials and Devices,
Chongqing 401331, China

J. Ma
26th Institute of China Electronics Group Corporation, Chongqing 400060, China

Y. Han (ed.), *Advanced Functional Materials*,
https://doi.org/10.1007/978-981-13-0110-0_23

changes from several tens nm to several microns. The absorption of nanowires is stronger than that of nanoparticles at wavelength of 230–400 nm.

Keywords Bismuth ferrite · Nanowire · Precipitant · Regulation

Introduction

$BiFeO_3$ (BFO) is one of several rare single-phase multiferroic materials, which are both ferroelectric and weakly ferromagnetic at room temperature [1]. It have high Curie temperature ($T_C \sim 1103$ K) [2] and high Neel temperature ($T_N \sim 643$ K) [3]. Owing to its excellent performance, BFO systems could be used to develop novel applications in the field of radio, television, solar battery and so on [4–6]. At present, various kinds of fabrication methods, such as solid-state [7], sonochemical and microemulsion [8], sol-gel [9], solution-combustion [10], the soft chemical route [11], co-precipitation [12] and hydrothermal [13], have been applied to obtain BFO with excellent performance. Among them, the hydrothermal method has been widely used to fabricate fine powders with narrow particle size and controlled stoichiometry. Some researchers have applied mineralizer [14], polymer [15], surfactant-assisted [16] hydrothermal and mechanochemical [17] methods to obtain nano-scaled BFO with special excellent performances [17].

Hydrothermal method is widely used in the preparation of bismuth ferrite. The type of precipitant is the key factor. In this report, we research the effects of precipitant species ($NH_3 \cdot H_2O$, NaOH) on the microstructures of BFO nanowires.

Experimental

The chemical regents used in the work were bismuth nitrate ($Bi(NO_3)_3 \cdot 5H_2O$), iron nitrate ($FeCl_3 \cdot 9H_2O$), ammonia ($NH_3 \cdot H_2O$), acetone(CH_3COCH_3) and sodium hydroxide (NaOH). All the chemicals were analytical grade purity and were used as received without further purification.

Firstly, 10 mmol $Bi(NO_3)_3 \cdot 5H_2O$ and 10 mmol $FeCl_3 \cdot 9H_2O$ were dissolved in 50 mL CH_3COCH_3 to form aqueous solutions. Secondly, the NaOH (or $NH_3 \cdot H_2O$) solution was slowly added to the above solution to coprecipitate $Fe^{3+}$and Bi^{3+} ions by constant stirring, and then a brown precipitate was formed. The sediment was centrifuged out and washed with distilled water several times until the pH value was neutral. Thirdly, 5 mol/L NaOH was added with stirring. Then the solution was transferred to a sealed, teflon-lined steel autoclave and heated at 200 °C for 48 h. The black powder obtained after filtration was washed with distilled water and ethanol and dried at 60 °C for characterization. Fourthly, 0.0020 g BFO samples were dissolved in 20 mL alcohol, before ultrasonic dispersion for 30 min at room temperature.

X-ray diffraction was performed on a Rigaku X-ray diffractometer (XRD DX-2700) with high-intensity Cu Kα radiation. The morphology of BFO was observed by scanning electron microscopy (SEM S-3700N). The diffraction spots and lattice phases were observed by transmission electron microscopy (TEM JSM-7800F). The transmittance was observed by UV-visible spectrophotometer (TU-1810).

Results and Discussion

X-ray Diffraction. Figure 1 shows the XRD pattern of the samples synthesized under different hydrothermal conditions. The BFO samples are in accordance with the standard XRD diffraction peaks (JCPDS 74-2016) and belong to a rhombohedral perovskite structure (R3c group). It can be seen that the two samples do not have impurity phase basically. When $NH_3{\cdot}H_2O$ is as precipitant, the crystallinity of the powder sample is higher. The lattice parameters of the product prepared by NaOH and $NH_3{\cdot}H_2O$ were calculated by the data of Jade 6 software on XRD data. From the data (in Table 1), it can be seen that the precipitant effect on the purity of bismuth ferrite is weak.

Scanning Electron Microscopy. Figure 2 shows SEM photographs of BFO samples prepared at different hydrothermal conditions. As shown in Fig. 2a, b, when $NH_3{\cdot}H_2O$ is used as precipitant, there are a little of nanowires and a large number of round cake, cubic and some irregular nanoparticles, besides, the length of the line is short. However, as shown in Fig. 2c, d, when NaOH is used as precipitant, there are many nanowires and other round cake and some irregular nanoparticles. Although the nanowires are agglomerated with the rest of the particles, we found that the nanowires are more uniform and slender rod-like.

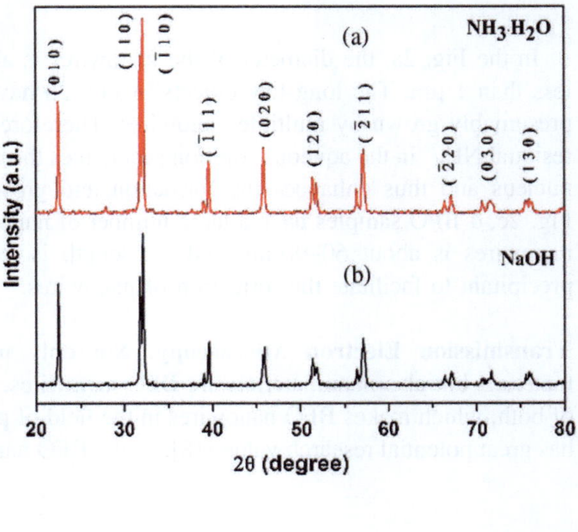

Fig. 1 XRD patterns of the as-prepared samples synthesized by different precipitants: **a** $NH_3{\cdot}H_2O$, **b** NaOH

Table 1 Unit cell parameters of the product

Alkali solution	NaOH	$NH_3{\cdot}H_2O$
$a = b = c$ (Å)	3.9495	3.9468
$\alpha = \beta = \gamma$ (°)	89.91	89.95

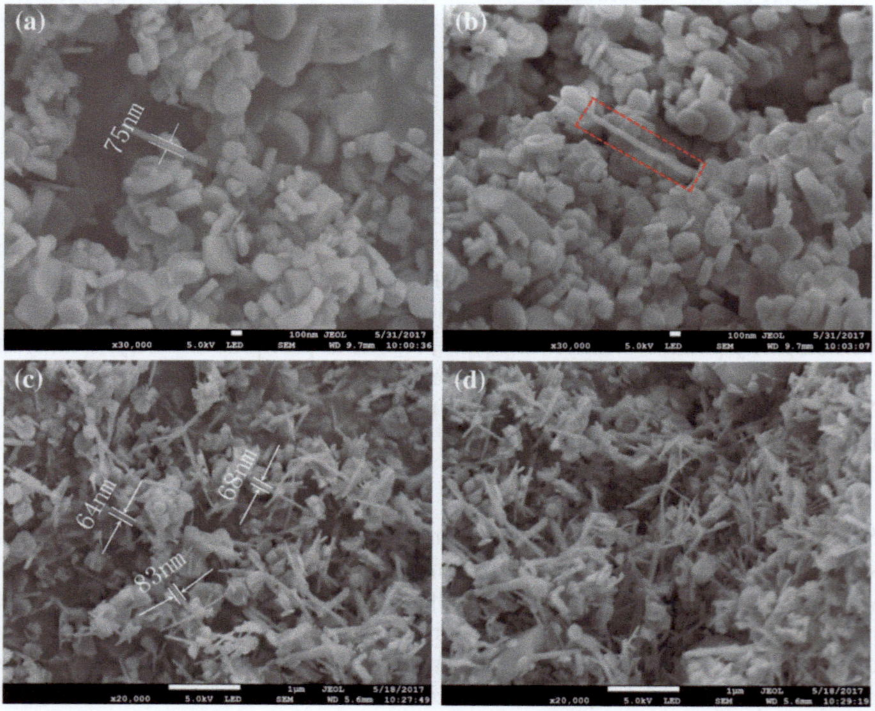

Fig. 2 SEM images of the as-prepared samples synthesized by different precipitants: **a**, **b** $NH_3{\cdot}H_2O$ and **c**, **d** NaOH

In the Fig. 2a, the diameter of the nanowires is about 75 nm and the length is less than 1 μm. The long-line objects in Fig. 2b have a layered structure and are presumably grown by multiple nanowires. Therefore it can be speculated that the residual NH_4^+ in the aqueous solution accelerates the crystallization rate of the BFO nucleus and thus enhances the formation and growth of the nanowires. In the Fig. 2c, d BFO samples have a large number of nanowires and the diameter of the nanowires is about 60–90 nm and the length is less than 1.5 μm. NaOH is a precipitant to facilitate the formation of nanowires.

Transmission Electron Microscopy. Not only narrow bandgap ferroelectric materials but also nanomaterials are BFO nanowires, it have some unique features of both, which makes BFO nanowires in the field of photovoltaic power generation has great potential research value [18], so the BFO nanowires need further research.

The TEM image (Fig. 3) shows that the as-prepared sample consists of $BiFeO_3$ nanowires, and the diameter of a single $BiFeO_3$ nanowire is less than 45 nm and the length changes from several tens nm to several microns. Figure 3b is a high-resolution TEM image recorded in the rectangular area, displaying an intact, orderly structure. Figure 3c indicates that the nanowire shows an excellent single-crystalline structure. The planes with interplanar spacings of 0.245, 0.412 and 0.271 nm are corresponding to the (2 1 1), (0 0 1) and (2 1 2) crystal faces.

UV-visible Spectrophotometer. Due to the potential application of bismuth ferrite in optical detectors and solar cells, the optical transmittance of BFO is also very important [19–21]. The variation of transmittance as a function of wavelength for BFO samples with different precipitant is plotted in Fig. 4. When NaOH is used as the precipitant, the BFO sample exhibits strong absorption peak at wavelength of 230–400 nm. When $NH_3{\cdot}H_2O$ is used as the precipitant, the BFO sample exhibits strong absorption peak at wavelength of 230–600 nm and the strongest absorption is at the wavelength of 600 nm. The absorption peaks of nanoparticles and nanowires are different, it indicates that the bandgaps of $BiFeO_3$ may change with the surface morphology. Nanowires having stronger absorption than nanoparticles at wavelength of 230–400 nm, it is possible due to that the nanowires have a larger surface area, resulting in stronger absorption of light.

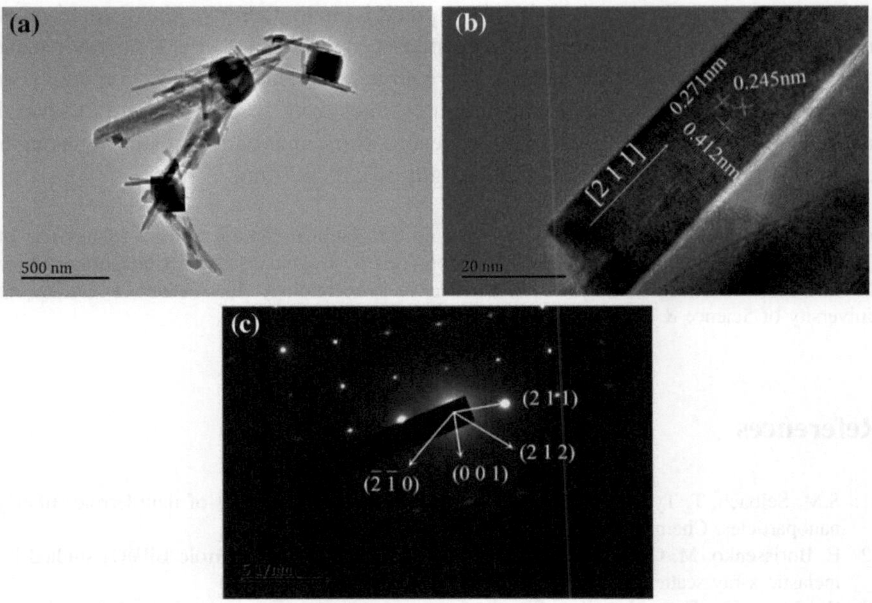

Fig. 3 **a** TEM image of $BiFeO_3$ nanowires, **b** HRTEM image of a typical portion of a corresponding $BiFeO_3$ nanowire, **c** SAED pattern of an individual $BiFeO_3$ nanowire

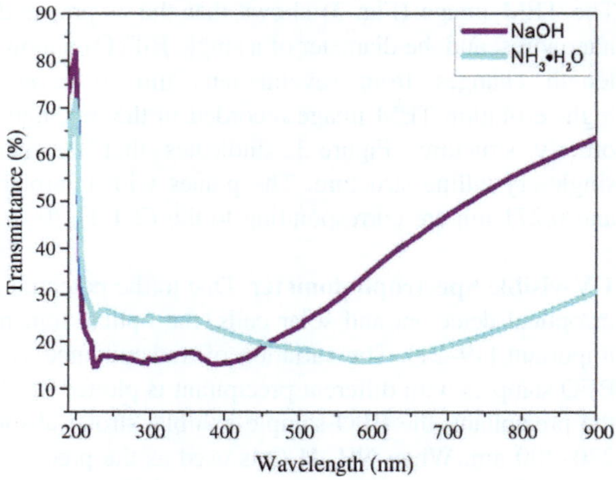

Fig. 4 Transmittance of the as-prepared samples synthesized by different precipitant

Conclusions

The bismuth ferrite nanowires were prepared by hydrothermal method, and the morphology of bismuth ferrite nanowires was systematically analyzed. XRD results show that when $NH_3{\cdot}H_2O$ is as a precipitant, there is a higher degree of crystallization. The SEM results show that the nanowires prepared by $NH_3{\cdot}H_2O$ as precipitants are very few, and NaOH is a precipitant to facilitate the formation of nanowires. From the results of TEM, it can be seen the whole $BiFeO_3$ nanowires are single crystal structures. The nanowires are arranged in the same direction and have a diameter less than 45 nm and the length changes from several tens nm to several microns. UV-visible spectrophotometer results show that nanowires have stronger absorption than nanoparticles at wavelength of 230–400 nm.

Acknowledgements This work was supported by the National Natural Science Foundation of China (51372283), the Program for Innovation Teams in University of Chongqing, China (CXTDX201601032), the Program of Science and Technology Innovation of Chongqing University of Science & Technology (YKJCX1620210).

References

1. S.M. Selbach, T. Tybell, M.A. Einarsrud, Size-dependent properties of multiferroic $BiFeO_3$ nanoparticles. Chem. Mater. **26**, 6478 (2007)
2. E. Borissenko, M. Goffinet, A. Bosak, Lattice dynamics of multiferroic $BiFeO_3$ studied by inelastic x-ray scattering. J Phys. Condens. Mat. **25**, 102201 (2013)
3. W. Mao, X. Wang, L. Chu, Simultaneous enhancement of magnetic and ferroelectric properties in Dy and Cr co-doped $BiFeO_3$ nanoparticles. Phys. Chem. Chem. Phys. **9**, 6399–6405 (2016)

4. F. Xue, L. Liang, Y. Gu, Erratum: "Composition- and pressure-induced ferroelectric to antiferroelectric phase transitions in Sm-doped $BiFeO_3$ system". Appl. Phys. Lett. **108**, 012903 (2016)
5. P. Sharma, Y. Heo, B.K. Jang, Morphotropic phase elasticity of strained $BiFeO_3$. Adv. Mater. Interfaces. **12** (2016)
6. M. Hasan, M.A. Hakim, M.A. Basith, Size dependent magnetic and electrical properties of Ba-doped nanocrystalline $BiFeO_3$. AIP Adv. **3**, 13–20 (2016)
7. M. Hasan, M.F. Islam, R. Mahbub, A soft chemical route to the synthesis of $BiFeO_3$ nanoparticles with enhanced magnetization. Mater. Res. Bull. **73**, 179–186 (2016)
8. P.R. Vanga, R.V. Mangalaraja, M. Ashok, Effect of co-doping on the optical, magnetic and photocatalytic properties of the Gd modified $BiFeO_3$. J. Mater. Sci. Mater. Electron. **27**, 5699–5706 (2016)
9. P. Saxena, A. Kumar, P. Sharma, Improved dielectric and ferroelectric properties of dual-site substituted rhombohedral structured $BiFeO_3$ multiferroics. J. Alloy. Compd. **682**, 418–423 (2016)
10. A. Azam, A. Jawad, A.S. Ahmed, Structural, optical and transport properties of Al^{3+} doped $BiFeO_3$ nanopowder synthesized by solution combustion method. J. Alloy. Compd. **509**, 2909–2913 (2011)
11. A.H.M. Gonzalez, A.Z. Simões, L.S. Cavalcante, Soft chemical deposition of $BiFeO_3$ multiferroic thin films. Appl. Phys. Lett. **90**, 819 (2007)
12. H. Ke, W. Wang, Y. Wang, Factors controlling pure-phase multiferroic $BiFeO_3$ powders synthesized by chemical co-precipitation. J. Alloy. Compd. **509**, 2192–2197 (2011)
13. C. Fu, F. Sun, J. Hao, The growth, enhanced optical and magnetic response of $BiFeO_3$ nanorods synthesized by hydrothermal method. J. Mater. Sci. Mater. Electron. **27**, 8242–8246 (2016)
14. M. Mahesh Kumar, V.R. Palkar, K. Srinivas, Ferroelectricity in a pure $BiFeO_3$ ceramic. Appl. Phys. Lett. **76**, 2764–2766 (2000)
15. J. Silva, A. Reyes, H. Esparza, $BiFeO_3$: a review on synthesis, doping and crystal structure. Integr. Ferroelec. **126**, 47–59 (2011)
16. Y. Lv, J. Xing, C. Zhao, The effect of solvents and surfactants on morphology and visible-light photocatalytic activity of $BiFeO_3$ microcrystals. J. Mater. Sci. Mater. Electron. **26**, 1525–1532 (2015)
17. A.A. Cristóbal, P.M. Botta, Mechanochemically assisted synthesis of nanocrystalline $BiFeO_3$. Mater. Chem. Phys. **139**, 931–935 (2013)
18. J. Yang, S. Wang, C. Zhang. Effect of surfactant on the morphology and optical performance of $BiFeO_3$ single crystal nanowires. J. Tianjin Normal Univ. (2017)
19. C. Ederer, N.A. Spaldin, Influence of strain and oxygen vacancies on the magnetoelectric properties of multiferroic bismuth ferrite. Phys. Rev. B **71**, 4103 (2005)
20. G.S. Lotey, N.K. Verma, Multiferroic properties of Tb-doped $BiFeO_3$ nanowires. J. Nanopart. Res. **15**, 1553 (2013)
21. S. Li, R. Nechache, C. Harnagea, Single-crystalline $BiFeO_3$ nanowires and their ferroelectric behavior. Appl. Phys. Lett. **101**, 737 (2012)

Improvement of the Thermal Stability of Laminated Magnetoelectric (ME) Composites

Zhan Shi, Mingyue Xu, Qun Huang, Shuiyuan Yang, Yong Lu, Jiajia Han, Cuiping Wang and Xingjun Liu

Abstract In order to search for ME laminated composites with a good thermal stability, the temperature dependent ME coefficient in two kinds of laminates bonded by welding, namely, Ni/Pb(Zr,Ti)O_3/Ni and Metal-glass/Pb(Zr,Ti)O_3/Metal-glass were investigated. The ME effect of all laminated composites at temperature range from −20 to 100 °C was measured. Results showed that both two laminates bonded by welding exhibited very good thermal stability. A new phenomenon was found that the ME charge coefficient slowly increased with increasing temperature, which resulted from the increase of piezoelectric layer's capacitance. These ME laminated composites bonded by welding supposed a good performance at high temperature, which were suitable for high temperature applications.

Keywords Laminated ME composites · ME effect · Thermal stability

Introduction

ME effect is the effect that an applied magnetic field induces an electric polarization or vice versa. ME effect was first observed in single-phased materials (Cr_2O_3) in 1961 [1]. However, until now, very few single-phase multiferroic materials are discovered. Their ME effect is relatively weak and usually occurs at very low temperature. Although the novel physics in single-phase ME materials implies many possible applications, low ME effect and low working temperature still limit their practical device application. Recently, ME laminates, which consist of magnetostrictive materials and piezoelectric materials, have attracted increasing interest of the researchers due to the giant ME effect above room temperature. Many studies on the giant ME coefficients in layered composites have been made, and many potential applications were proposed, such as magnetic field sensors, transformers,

Z. Shi (✉) · M. Xu · Q. Huang · S. Yang · Y. Lu · J. Han · C. Wang · X. Liu
College of Materials, Xiamen University, Xiamen 361005, China
e-mail: shizhan@xmu.edu.cn

Y. Han (ed.), *Advanced Functional Materials*,
https://doi.org/10.1007/978-981-13-0110-0_24

gyrators, memory, and microwave devices [2–9]. To meet the requirement of ME devices used in extreme environment, a broadband temperature dependent analysis on ME effect of these ME devices is much required.

Recently, some theoretical studies on the influence of temperature on ME effect have been carried out. For example, Zhou et al. [10–13] established theoretical models to predict ME coefficients under different temperature for the Terfenol-D/Pb(Zr,Ti)O_3(PZT)/Terfenol-D, Terfenol-D/Pb(Mg,Nb)O_3-PbTiO_3/Terfenol-D and Terfenol-D/BiTiO_3/Terfenol-D, respectively. In which, a common conclusion was draw that ME coefficient decreased with increasing temperature. In addition, Yao et al. [14] built a mechanical-thermo-magneto coupling model for FeCuNbSiB/Terfenol-D/PZT/Terfenol-D/FeCuNbSiB laminated composite. All of these results indicated that as the temperature rose, ME voltage coefficient decreased. In the experimental research on the temperature dependent ME effect, Fang et al. [15] measured the ME coefficients from 20 to 80 °C for three kinds of ME laminates, namely, Terfenol-D/PZT/Terfenol-D, Ni/PbZT/Ni and Metal-glass/PZT/Metal-glass, which showed that both the ME coefficients and the resonant frequency decreased with increasing temperature. Meanwhile Gutiérrez et al. [16] fabricated V4040/poli2,6/V4040 laminates using a new high temperature piezoelectric polymer poli2,6 polyimide. And the ME coefficient was found keeping constant under test temperature range. Cótica et al. [17] studied the ME effect of $NiFe_2O_4$/PZT/$NiFe_2O_4$ at low temperature. The ME coefficients decreased very fast as the temperature decreased from 300 to 100 K.

In general, these theoretical or experimental results demonstrated that the ME effect in laminates had a bad thermal stability in test temperature range. Hence, in this work, we try to optimize the combination method of ME laminates. Two kinds of tri-layered laminates (Ni/PZT/Ni and Metal-glass/PZT/Metal-glass) bonded by welding were prepared. For comparison, we also designed the same tri-layered laminates bonded by means of epoxy. Variation of ME coefficient was measured at a temperature range from −20 to 100 °C. We aimed at finding a kind of laminated ME composites which possess good thermal stability. This study would be of importance for the practical device applications for ME laminates.

Experimental Details

The Preparation of Laminated ME Composites. The L-T (a longitudinal magnetization and transverse polarization) mode tri-layer ME composites with two magnetostrictive layers and one piezoelectric layer, which we studied in this paper was showed in Fig. 1. The external magnetic field (H_{AC} or H_{DC}) was parallel to the longitudinal direction of the ME composites, and the magnetostrictive layers were magnetized along the longitudinal direction and the piezoelectric layer was polarized along its thickness direction. All laminates had the same piezoelectric phase of PZT-5 (Acoustics, Chinese Academy of Sciences, Beijing, China) with the dimension of $10 \times 3 \times 0.5$ mm^3. Ni and Metal-glass were used as the magnetic

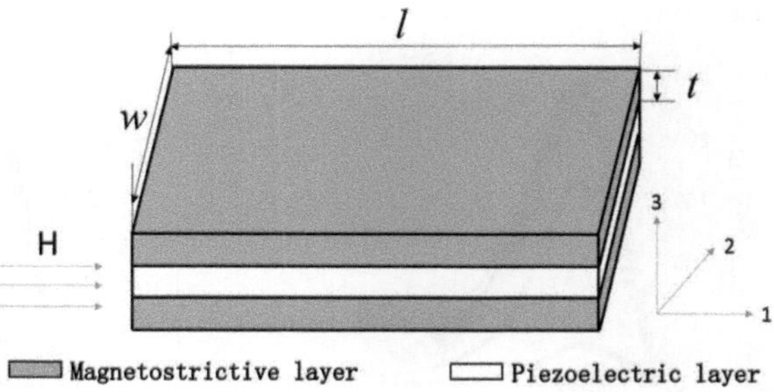

Fig. 1 L-T mode ME laminated composites

phase. And they had the same lateral dimension as the PZT, but the thicknesses were 0.5 and 0.08 mm respectively. The L-T type ME composites were fabricated by bonding two magnetostrictive layers and one piezoelectric layer by means of welding as shown in Fig. 1. The combination by epoxy is the common way to prepare ME laminates in previous researches. For comparison, ME laminates bonded by heat-resistant epoxy were also prepared.

Measurement Method of ME Coefficient. The ME effect is a coupled (or cross) two-field effect, in which either a magnetic field or an electrical field induces an electrical polarization as well as a magnetization [5]. In this paper, the ME coefficient was measured by a dynamic method. First, a perturbed H_{AC} field of 1 kHz was generated via a coil driven by a signal generator. And the static H_{bias} field was generated by an electromagnet. The H_{bias} and H_{AC} were applied on the ME composites and the induced ME voltage or ME charge was measured. The voltage δV generated across the piezoelectric layer was amplified by a lock-in-amplifier. The charge δQ was amplified by a charge amplifier. The ME voltage coefficient is estimated by $\alpha_{ME} = \delta E/\delta H = \frac{\delta V}{t}/\delta H$, where t is the thickness of PZT. The ME charge coefficient is estimated by $\alpha_{QME} = \delta Q/\delta H$.

Results and Discussion

The variation of ME voltage coefficient α_{ME} with H_{bias} for four laminates was shown in Fig. 2. From Fig. 2, it can be found that temperature variations do not affect the shape of $\alpha_{ME} \sim H_{bias}$ curves, which is that the curves first increase and then decrease with increasing bias magnetic field. However, different combination method results in quite different thermal stability. Figure 3 shows the maximum ME voltage coefficient α_{ME} of four laminates at different temperatures. From Fig. 3, it can be clearly seen that the maximum ME voltage coefficients α_{ME} of Ni/PZT/Ni and Metal-glass/PZT/Metal-glass bonded by an epoxy begin to decrease as temperature increased to 40 °C. By contrast, the maximum ME voltage coefficients α_{ME} of Ni/PZT/Ni and Metal-glass/PZT/Metal-glass bonded by welding exhibit a good thermal stability. Zhou et al. [10] demonstrated that the influence of temperature on the ME effect was mainly caused by the magnetostrictive layers. In their

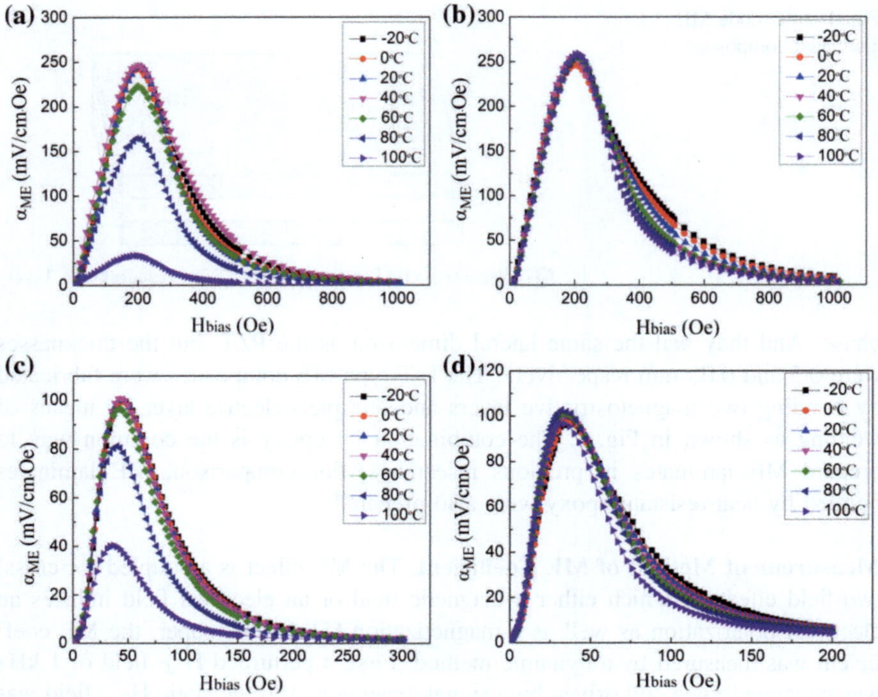

Fig. 2 Variation of ME voltage coefficient α_{ME} with H_{bias} for **a** Ni/PZT/Ni (bonded by an epoxy), **b** Ni/PZT/Ni (bonded by welding), **c** Metal-glass/PZT/Metal-glass (bonded by an epoxy), and **d** Metal-glass/PZT/Metal-glass (bonded by welding) tri-layered composites at different temperatures

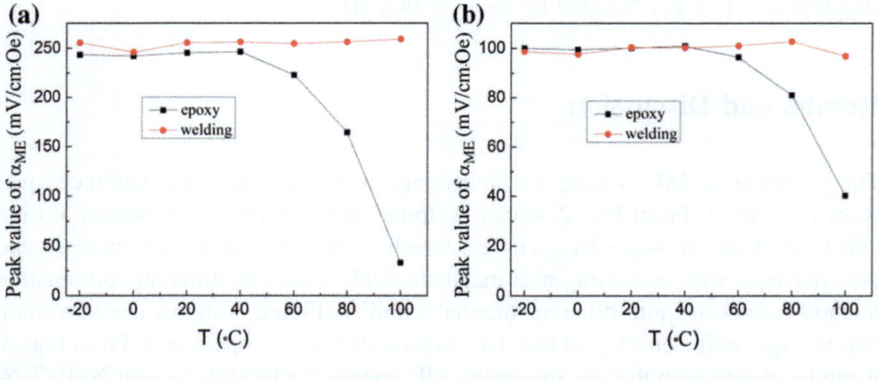

Fig. 3 Peak value of ME voltage coefficient α_{ME} with temperature T for **a** Ni/PZT/Ni (bonded by an epoxy or welding), **b** Metal-glass/PZT/Metal-glass (bonded by an epoxy or welding) tri-layered composites

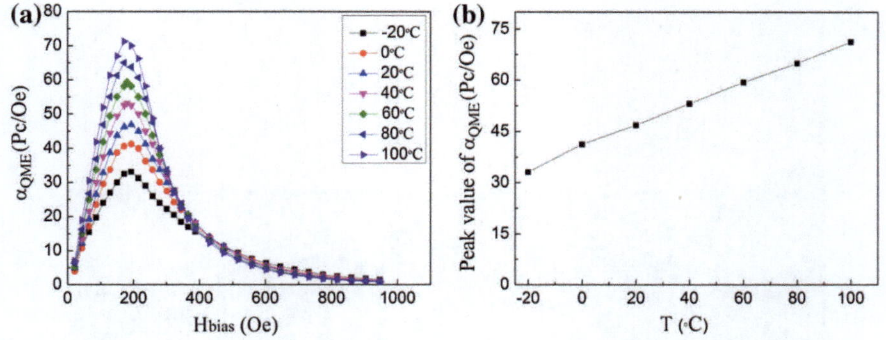

Fig. 4 **a** Variation of ME charge coefficient α_{QME} with H_{bias} for Ni/PZT/Ni (bonded by welding) tri-layered composite at different temperatures. **b** Peak value of ME charge coefficient α_{QME} with temperature T for Ni/PZT/Ni (bonded by welding) tri-layered composite

opinion, the magnetization and output strain in magnetostrictive layers decreased with the increasing temperature, so the strain transferred to the piezoelectric layer became smaller correspondingly. In Fig. 3, the decrease of ME effect in laminates with epoxy may result from the decrease of stiffness of epoxy. By welding, the measurement temperature range is far below the melting point of the solder, so the ME coupling seems stable under the test temperature range. It can be concluded that the reduction of the strain transfer is the main reason of the decrease of ME coefficient. Here, the new combination method by welding shows good performance in prepared ME laminates in the whole test temperature range.

Figure 4 shows the variation of ME charge coefficient α_{QME} with H_{bias} for Ni/PZT/Ni (bonded by welding) tri-layered composite at different temperatures and the peak value of ME charge coefficient α_{QME} with various temperature. As shown in Fig. 4, the ME charge coefficient α_{QME} of Ni/PZT/Ni (bonded by welding) increases with increasing temperature. This phenomenon is quite different with previous researches. In general, the ME coupling effect will decrease with increasing temperature. To investigate the reason of this abnormal phenomenon, comprehensive analysis was carried out. After investigating the capacitance of the PZT layer under various temperature, it was found that the capacitance of the PZT layer increased with increasing temperature. According to the ME voltage coefficient mentioned in Figs. 2 and 3, the ME voltage kept almost constant under various temperature. Because the charge is the product of voltage and capacitance, i.e. $Q = V \times C$, where Q, V, C is the charge, the voltage and the capacitance. The increase of the capacitance will result in the increase of the charge under the same voltage. This phenomenon can be used for the compensation of the ME coefficient under high temperature range.

Figure 5 shows the interface optical microscope diagram of Ni/PZT/Ni bonded by welding. As shown in Fig. 5, the Ni layer and the PZT layer combine well, and the solder layer is clearly visible too. This good interface combination provides a basis for the good temperature stability of the Ni/PZT/Ni bonded by welding.

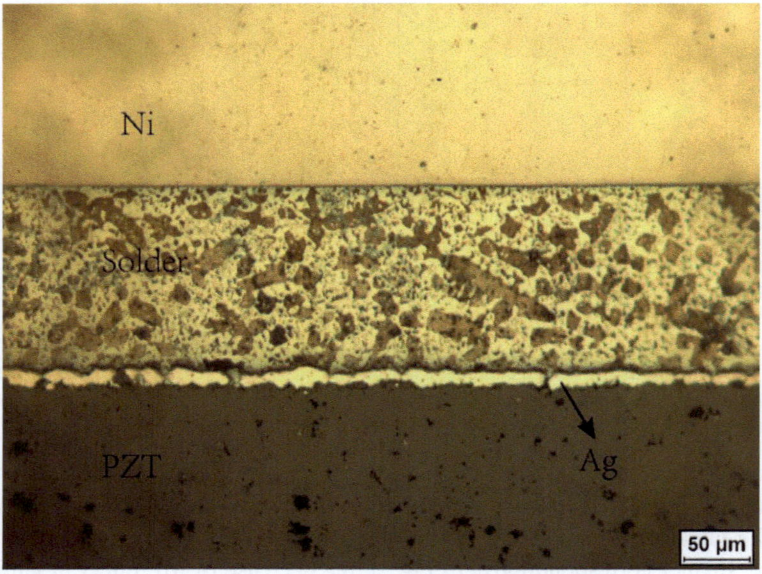

Fig. 5 Interface optical microscope diagram of Ni/PZT/Ni bonded by welding

Conclusions

In clear contrast with the temperature behavior of four kinds of tri-layered laminates (Ni/PZT/Ni and Metal-glass/PZT/Metal-glass) bonded by means of welding or epoxy, it could be found that the ME laminated composites bonded by means of welding showed very good thermal stability of ME voltage coefficient from −20 to 100 °C, and the ME voltage coefficients kept almost constant at the whole test temperature range. It can be concluded that high temperature solder is helpful to improve thermal stability. This research is beneficial for ME devices operating in extreme environments. An abnormal increase of ME charge coefficient with increasing temperature is observed as well. It can provide a new compensation method for ME coefficient when ME devices are used in high temperature.

Acknowledgements This work was supported by the Natural Science Foundation of Fujian Province of China (Grant No. 2016J01256).

References

1. D.N. Astrov, Magnetoelectric effect in chromium oxide. Sov. Phys. JETP **13**, 729–733 (1961)
2. H. Zhang, C.J. Lu, C.B. Xu, Y.J. Xiao, J.G. Gui, C.H. Lin, Y. Xiao, Improved magnetoelectric effect in magnetostrictive/piezoelectric composite with flux concentration effect for sensitive magnetic sensor. AIP Adv. **5**, 047114 (2015)

3. L. Zeng, M. Zhou, K. Bi, M. Lei, Giant magnetoelectric effect in negative magnetostrictive/ piezoelectric/positive magnetostrictive semiring structure. J. Appl. Phys. **119**, 034102 (2016)
4. J. Ma, J. Jiao, C. Fang, H. Deng, S. Wang, D. Lin, H. Xu, X. Zhao, H. Luo, Improvement of magnetoelectric properties in metglas/$Pb(Mg_{1/3}Nb_{2/3})O_3$–$PbTiO_3$/metglas laminates with screen-printed ID-electrodes by poling optimization. J. Alloy. Compd. **656**, 793–797 (2016)
5. C.W. Nan, Magnetoelectric effect in composites of piezoelectric and piezomagnetic phases. Phys. Rev. B **50**, 6082–6088 (1994)
6. N.A. Spaldin, M. Fiebig, The renaissance of magnetoelectric multiferroics. Science **309**, 391–392 (2005)
7. W. Eerenstein, N.D. Mathur, J.F. Scott, Multiferroic and magnetoelectric materials. Nature **442**, 759–765 (2006)
8. C.W. Nan, M.I. Bichurin, S.X. Dong, D. Viehland, G. Srinivasan, Multiferroic magnetoelectric composites: historical perspective, status, and future directions. J. Appl. Phys. **103**, 031101 (2008)
9. N. Ortega, A. Kumar, J.F. Scott, R.S. Katiyar, Multifunctional magnetoelectric materials for device applications. J. Phys. Condens. Matter **27**, 504002 (2015)
10. H.M. Zhou, X.L. Cui, A theoretical study of the nonlinear thermo-magneto-electric coupling effect in magnetoelectric laminates. Smart Mater. Struct. **23**, 105014 (2014)
11. H.M. Zhou, X.L. Cui, Static magnetoelectric coupling of magnetoelectric laminated composites under combined temperature and stress loadings. J. Appl. Phys. **115**, 083905 (2014)
12. Y. Xiao, H.M. Zhou, X.L. Cui, Nonlinear resonant magnetoelectric coupling effect with thermal, stress and magnetic loadings in laminated composites. Compos. Struct. **128**, 35–41 (2015)
13. H.M. Zhou, X.L. Cui, Numerical simulation of nonlinear thermo-magneto-electric coupling effect in ME laminates. AIP Conf. Proc. **1648**, 490006 (2015)
14. H. Yao, Y. Shi, Y.W. Gao, A mechanical-thermo-magneto model for self-biased magnetoelectric effect in laminated composite. J. Magn. Magn. Mater. **401**, 1046–1053 (2016)
15. F. Fang, Y.T. Xu, W. Yang, Magnetoelectric coupling of laminated composites under combined thermal and magnetic loadings. J. Appl. Phys. **111**, 023906 (2012)
16. J. Gutiérrez, A. Lasheras, J.M. Barandiarán, J.L. Vilas, M. San Sebastián, L.M. León, Temperature response of magnetostrictive/piezoelectric polymer magnetoelectric laminates. Mater. Appl. Sens. Transducers **495**, 351–354 (2012)
17. L.F. Cótica, S. Betal, C.T. Morrow, S. Priya, R. Guo, A.S. Bhalla, Thermal effects in magnetoelectric properties of $NiFe_2O_4$/$Pb(Zr_{0.52}Ti_{0.48})O_3$/$NiFe_2O_4$ tri-layered composite. Integr. Ferroelectr. **174**, 203–209 (2016)

Mechanism of Preparing Laminar Composite Using Ti-Based Amorphous Alloy

Rongxia Zhang, Wei Wu and Yuansong Zeng

Abstract In this article amorphous alloy was used in preparing Ti-Al system laminar composite. Amorphous alloy strips and TC4 were stacked interactively. After thermoforming process a new structure laminar composite was prepared by using the activity performance of amorphous alloy. Every layer phase composition was determined after TEM and energy dispersive spectrum analysis. The special diffusion mechanics of this process was disclosed, which including two stages. The first was the plastic deformation stage before amorphous alloy crystallization and interface atoms bonded together under van der Waals force in the first stage. The second was crystal plane diffusion. The result of test displayed that the conversion zone was much thicker because of using the amorphous alloy. With tension testing and fracture analysis, the laminar composite expressed well crack extension property. The conversion zone had a strengthening faction and it could consume some damage energy from outer force. TC4 had toughening faction which could prevent the material breaking crisply.

Keywords Amorphous alloy · Preparing · Laminated composites
Mechanism

R. Zhang (✉) · W. Wu · Y. Zeng
Beijing Aeronautical Manufacturing Technology Institute, Beijing, China
e-mail: zrxia1979@163.com

W. Wu
e-mail: wuweifiles@163.com

Y. Zeng
e-mail: zengys@163.com

R. Zhang · W. Wu · Y. Zeng
Aeronautical Key Laboratory for Plastic Forming Technologies, Beijing, China

R. Zhang · W. Wu · Y. Zeng
Beijing Key Laboratory of Digital Plasticity Forming Technology
and Equipment, Beijing, China

Y. Han (ed.), *Advanced Functional Materials*,
https://doi.org/10.1007/978-981-13-0110-0_25

Introduction

The metal-based composite comes from the shell nacre structure. Among intermetallic compounds, titanium-aluminum intermetallics have become a hotspot in recent years because of their low density, high specific strength, high temperature mechanical properties and excellent oxidation resistance [1]. The amorphous alloy is a kind of novel material similar to glass which has the characteristics of short-range order and long-range disordered structure, the highly unsaturated surface and high surface energy. So it has a strong activation ability and high activity center density for the reaction molecules, functionally similar to the shell pearl layer of organic self-assembly [2]. It has been proved that the thickness of diffusion layer can increase by using amorphous as welding active agent for TiAl base alloy and TC4 [3, 4]. In this paper, referring to nacre structure and growth mode, we got the Ti/Ti-Al laminated composites by the characteristic of high energy activation of amorphous materials. The paper focuses on the mechanism of atomic diffusion and strengthening and toughening of amorphous self-assembled laminated composite materials.

Preparation of Amorphous Laminated Composites

Pure titanium, pure aluminum and titanium based amorphous alloys were used as raw materials, and the layers are overlapped in sequence according to the Fig. 1. We got the laminar composite material under hot-pressing technology in the thermal-mechanical simulation test machine Gleeble3500. The process parameters were 873 K, 10 MPa, 8 h. The Al foil was completely reacted and the Ti-based amorphous was allowed to remain. The materials used in the test are shown in Table 1.

Phase Composition Analysis of Amorphous Self-assembled Laminated Composites

SEM, TEM and energy dispersive spectrum analysis were carried out to study the element diffusion law and the composition of each phase. Figure 2a is the whole interface microstructure of layered structure after hot pressing in 873 K, 8 h,

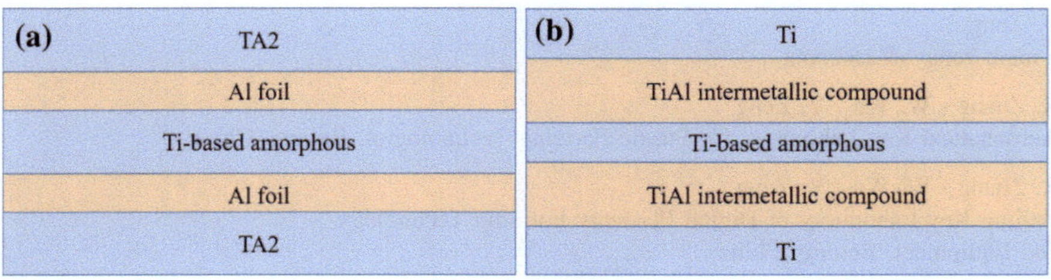

Fig. 1 Principle of material preparation **a** before the reaction; **b** after the reaction

Table 1 Composition of raw material

Grades	Alloy								
	Thickness (mm)	Chemical composition, %							
		Main ingredient (%)	Impurity content (%), ≤						
TA2	0.8	Ti	Fe	C	N	H	O	–	
		Rest	0.30	0.08	0.05	0.015	0.35	–	
$Ti_{46}Zr_{26}Cu_{17}Ni_{11}$ amorphous (17 mm width)	0.03	Ti	Zr	Cu	Ni				
		Rest	26	17	11				
1050	0.02	Al	Fe	Si	Mn	Zn	Ti	Mg	
		Rest	0.40	0.25	0.05	0.05	0.03	0.05	
TC4	0.8	Al	V	Fe	Si	C	N	H	O
		5.5–6.75	3.5–4.5	0.30	0.15	0.08	0.05	0.015	0.20

10 MPa. It can be seen that the laminated material consists of 5 layers, layer A is the unreacted pure titanium plate, layer B, C and D is the reaction layer, E is the amorphous layer. We analyzed the composition of the reaction layer by EDS of TEM, and obtained the phase composition as shown in Table 2. The red circles in Fig. 3 were the analyzed areas.

From Table 2, it can be seen that the atomic ratio of Ti and Al of the B layer is about 1: 3, judged as Al_3Ti, as shown in Fig. 3a. Al_3Ti is spherical grains with an average diameter of about 1.83 ± 0.24 μm. Results can be found by line scan that the content of the Ni element in the thin layer C is increased. The atomic ratio observed by transmission electron microscopy (TEM) accords with Al_3Ni. The grain morphology is shown in Fig. 3b Al, Ti and Zr in the layer D appear plat forms in the vicinity of the interface, the proportion of elements is relatively constant, and there is no obvious element segregation phenomenon. The structure of the D layer, which is based on the spectrum analysis of Table 2, is in accordance with $Al_{75}Ti_{15}Zr_{10}$ and belongs to $Al_3(Ti_{0.6}Zr_{0.4})$structure. The grain morphology is shown in Fig. 3c. In the three phase system of Al-Ti-Zr, the elements Ti and Zr can be replaced with each other. It is known from the literature that the formation free energy of Al_3Ti is smaller than that of Al_3Zr, so the Al_3Ti is preferentially formed at thermodynamic state [5–11]. With the increase of Zr concentration in Al_3Ti, partial Ti is replaced by Zr, and finally the Zr atom is saturated to form a stable substitutional solid solution $Al_3(Ti_{0.6}Zr_{0.4})$. As the radius of the Zr atom is larger than Ti, there is lattice distortion of solid solution, which is changed from spherical to the long strip [12].

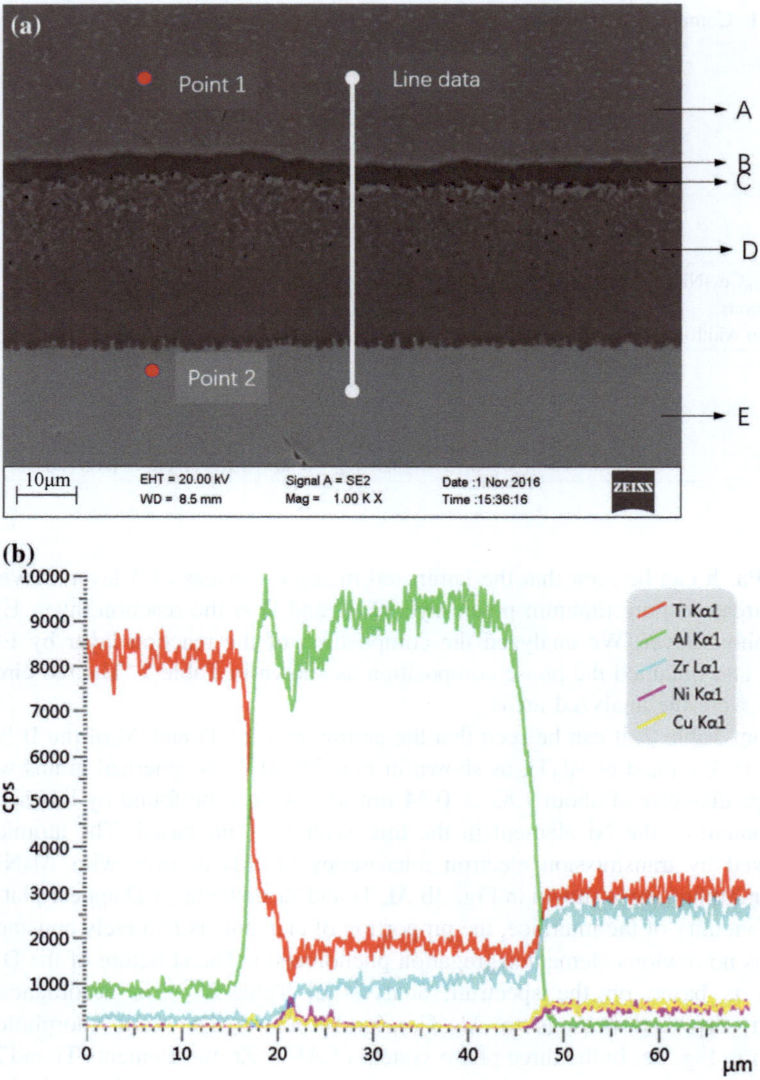

Fig. 2 Micrograph of reaction interface **a** micrograph of section, **b** line scanning of section

Table 2 EDS element composition analysis results of different interface layer

Test position	Chemical element (at.%)			
	Al	Ti	Zr	Ni
Interface layer B	73.22	26.78		
Interface layer C	72.88			27.12
Interface layer D	72.98	16.38	10.64	

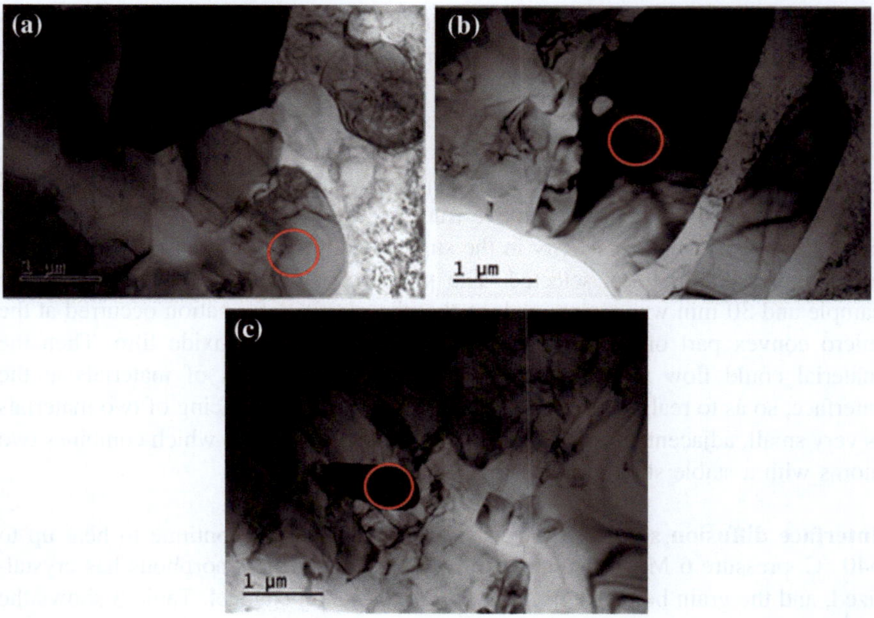

Fig. 3 TEM bright field images of the interface layer **a** layer B; **b** layer C; **c** layer D

Analysis of Diffusion Mechanism

At present, the understanding of the diffusion process of Ti-Al system is still not complete, through the labeling test van Loo and Rieck found that: in the Ti/Al interdiffusion process, Al is the only diffusion component. [13] In some subsequent studies, this result was also used to explain the growth behavior of $TiAl_3$. In addition, Luo and Acoff argued that the diffusion of Ti in Al is much more significant than that of Al in Ti [14].

The mechanism of diffusion bonding of amorphous alloys is also very different from that of crystalline alloys because of their different structures from crystalline alloys. Since the amorphous alloy has no grain boundary, the diffusion of atoms only includes volume diffusion and surface diffusion. Lin etc. proposed a theoretical model for the diffusion bonding of amorphous alloys. It is considered that the diffusion bonding process of amorphous alloys is composed of two stages, that is, the plastic deformation stage and the void contraction stage. According to this theory, combined with the experimental results. The diffusion bonding of amorphous materials for preparing laminated composites is also divided into two processes. The first process is the plastic deformation stage. The plastic deformation of amorphous occurs before crystallization under pressure. The second process is the interface diffusion stage. At this point, the amorphous has crystallized and diffusion is between the crystals.

Plastic deformation stage. During the first stage of plastic deformation, the surface of amorphous, Ti and Al foil should be treated by sandpaper grinding, ultrasonic degreasing, and removing oxide film by acid solution. But after processing the material surface with many micro-bump is still rough at the micro level. Because the surface of the aluminum foil is easily oxidized and covered with an oxide film, only contact is present at the protruding spikes without any pressure in such solid surfaces, and it is difficult to achieve true contact between metals. By using the characteristics of super plasticity in the supercooled liquid region of amorphous, a temperature of 490 °C is selected. The pressure of 15 MPa was applied to the sample and 30 min was maintained, so that the plastic deformation occurred at the micro convex part of the amorphous surface to break the oxide film. Then the material could flow into the depression area of two kinds of materials at the interface, so as to realize a close contact. When the atomic spacing of two materials is very small, adjacent atoms will produce van der Waals force which combines two atoms with a stable structure, as shown in Fig. 4.

Interface diffusion stage. On the basis of the first stage, continue to heat up to 640 °C, pressure 6 MPa, insulation 6 h. At this stage, the amorphous has crystallized, and the grain boundary is an important diffusion channel. Table 3 shows the specific contents of the elements of point 1 and 2 in Fig. 2a. It can be seen that the Zr, Cu and Ni elements in the amorphous alloy reach the Ti layer through the original Al foil layer, but no Al atoms are found in the amorphous layer. Indicating that amorphous has been crystallized, but the atoms still have a strong ability to diffuse. In the reaction with Al foil, diffusion of amorphous atoms into Al is

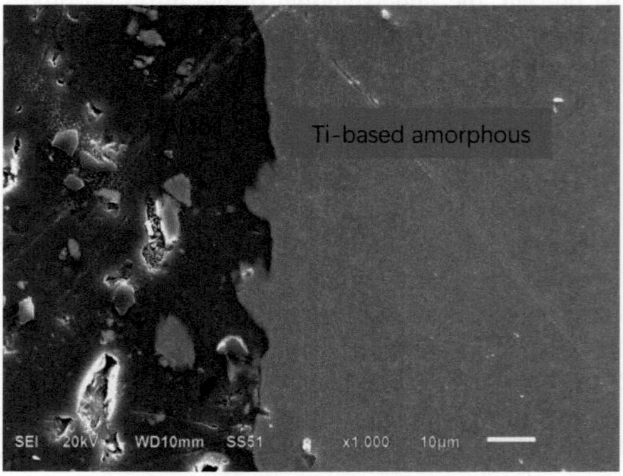

Fig. 4 Specimen of the plastic deformation stage

Table 3 Element contents in different layers

Contents	Elements				
	Al/%	Ti/%	Ni/%	Cu/%	Zr/%
1	57.98	16.35	10.04	5.37	10.27
2		43.09	10.34	12.6	33.97

predominant. As shown in Fig. 2a, the layer B is a reaction layer of Ti and Al foil with a thickness of about 6 μm. The layer C and D are reaction layers of Ti-based amorphous and Al foil with a thickness of about 29 μm, which is about 5 times thickness of layer B, and it is also proved that the amorphous layer has obvious activity in the preparation of the laminated composites, which can promote the interface diffusion and enhance the interfacial bonding degree.

After crystallization of amorphous, Ti, Zr, Cu and Ni elements quickly spread to the Al layer, part of the atoms reach the Ti layer. With the increase of temperature and time, Al_3Ti is generated firstly between the reaction interfaces of Ti and Al. The Al3Ti layer prevents the elements in the amorphous layer from further diffusing into the Ti plate. And the elements such as Ni appear to be enriched on the Al_3Ti side. The supersaturated Ni atoms react chemically with Al to form Ni_2Al_3 and grow in a single-phase layered manner, after Ni is saturated with Al. The growth of Ni_2Al_3 is accompanied with the dissolution process, and $NiAl_3$ is precipitated during the cooling process. In this process, a large number Ti and Zr elements of amorphous diffuse into the Al layer, react with each other to produce $Al_{75}Ti_{15}Zr_{10}$. The diffusion of Ni atoms is becoming more and more difficult. There will not be a continuous $NiAl_3$ single phase layer in the subsequent cooling process, which leads to the formation of a continuous thin layer of $NiAl_3$ precipitates near the interface.

Strengthening and toughening mechanism of composite materials. In order to study the strengthening and toughening mechanism of laminated composites, the tensile test and tensile fracture analysis were carried out to reveal the strengthening and toughening mechanism of the composites. Refer to the GB/T 228-2002 "tensile test method of metallic materials at room temperature", the tensile stress-strain curve of 10^{-3}/s is shown in Fig. 5.

The tensile curve of Fig. 5 shows a broken line shape, which is a special deformation form of laminated composites, and the fracture points on the curve correspond to the brittle fracture of the reaction layer. The continuous rise stage is the plastic tensile deformation of the TC4 plate.

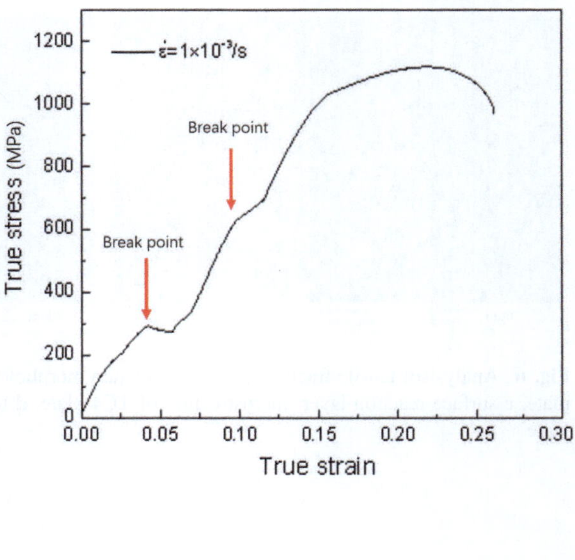

Fig. 5 Tensile stress-strain curve at room temperature

Figure 6 is the tensile fracture morphology and Fig. 6a is the picture of the whole fracture. It can be seen that the TC4 sheet is peeled off due to the brittle fracture of the reaction layer. Figure 6b shows the fracture surface of the middle TC4 plate. It can be seen that the upper and lower edges are relatively smooth, which is the brittle section of the typical cleavage fracture river pattern, and the middle is the plastic ductile fracture dimple of TC4 plate. Figure 6c shows the micro morphology of the surface reaction layer on the TC4 plate, which shows that there is a uniform distribution of cracks in the reaction layer, and the angle between the crack direction and tensile direction is 45 degrees. Compared with the surface morphology of the tensile fracture surface of the pure TC4 plate (Fig. 6d), it only shows the unstable orange peel shape. It can be seen from the fracture morphology that the fracture is composed of fracture along the layer and fracture through the layer. The fracture surface of the reaction zone is a smooth section, which belongs to the extension along the layer. The fracture process is mainly the main crack beginning, and then extending, and then connecting and last forming a cleavage fracture initiation source. If it can meet the critical crack size required by Griffiths brittle cleavage fracture, it will cause the brittle cleavage fracture of the whole reaction layer. As the load increases, the crack penetrates and extends through the TC4 layer, the transgranular ductile fracture occurs, eventually leading to the fracture of the entire specimen [15–19].

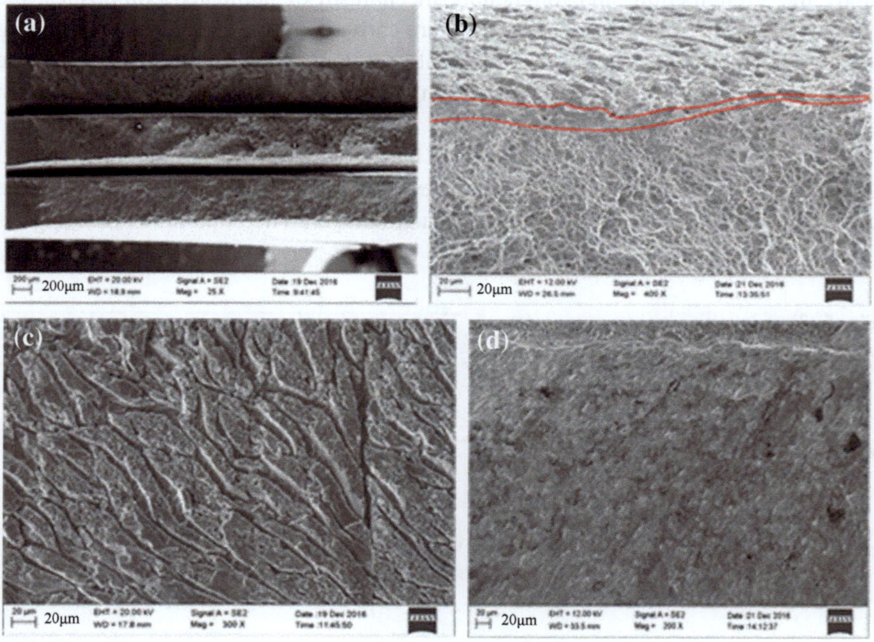

Fig. 6 Analysisof tensile fracture. **a** Overall fracture morphology, **b** fracture morphology of TC4 plate, **c** surface reaction layer microstructure of TC4 plate, **d** tensile surface morphology of TC4 plate

The interface of the layered material acts as a deflection to the crack. The frequent deflection of the crack not only causes the extension of the crack propagation path, but also causes the crack to shift from the favorable direction of the stress to the unfavorable direction. The crack propagation resistance increases and the matrix is thus strengthened. At the same time, when the reaction layer is deformed and broken, the TC4 plate layer also undergoes plastic deformation, which reduces the stress intensity factor of the crack tip and increases the crack resistance. Impact of external forces on the material could be damped, so that the specimen can implement good elongation before yielding. The above analysis shows that the reaction layer plays a role of pinning reinforcement, which can consume some of the external forces on the damage energy of the material, while the TC4 plate layer toughening effect to prevent the overall brittle fracture.

Conclusion

In this paper, the amorphous was used in the preparation of laminated composites. It is a new technology to enhance the reaction interfacial bonding degree by using the activity of amorphous. Through the analysis of this paper, the laminated composites have a multi-layer structure with a certain crack resistance, which have broad application prospects such as used as a new armor protection material to enhance the resilience of armor weapons. In the future, the following two aspects should also be carried out in-depth study: First, the impact of blank ratio on the performance of composite materials; Second, the fatigue properties and failure mechanism of composites. If the composites can be integrated design and control on shape and performance, it will promote the engineering application process of laminated composites.

References

1. H. Zhang, Preparation and properties of amorphous composite materials. Mater. China **29** (11), 8–15 (2010)
2. W.H. Wang, The nature and properties of amorphous matter. Prog. Phys. **33**(5), 177–351 (2013)
3. S. Xu, Vacuum brazing TC4 titanium alloy with Ti-based amorphous filler, Master Thesis, Zhengzhou University, 2014.4
4. S. Chen, Research on technology and mechanism of laser welding-brazing for Ti/Al dissimilar alloys, Ph.D. Dissertation, Harbin Institute of Technology, 2009, 3
5. J. Shuying, Study on diffusion-solution zones in Al/Fe, Al/Ni and Al/Ti liquid/solid interfaces, Master Thesis, China University of Petroleum (East China), 2010.10
6. U.R. Kattner, J.C. Lin, Y.A. Chang, Thermodynamic assessment and calculation of the Ti-Al system. Metall. Mater. Trans. A **23**(8), 2081–2090 (1992)

7. E. Fischer, C. Colinet, Study on diffusion-solution zones in Al/Fe, Al/Ni and Al/Ti liquid/solid interfaces. An updated thermodynamic modeling of the Al-Zr system. J. Phase Equilib. Diffus. **36**(5), 404–413 (2015)
8. G. Herzer, Modern soft magnets: amorphous and nanocrystalline materials. Acta Mater. **61**(3), 718–734 (2013)
9. G. Lian, Y.J. Dong, *Manual of Abiothermo Dynamics* (Northeastern University Press, Shenyang, 1993)
10. H.L. Chen, E. Doernberg, P. Svoboda et al., Thermodynamics of the Al_3Ni phase and revision of the Al-Ni system. Thermochim. Acta **512**(1), 189–195 (2011)
11. F. Yang, Phase relations of Al-Ti-Zr ternary system, Master Thesis, Central South University, 2014.4
12. L. Xu, Synthesis of Ti/$TiAl_3$ laminate composites and γ-TiAl alloys by reactive diffusion routes, Ph.D. Dissertation, Institute of Metal Research, Chines Academy of Sciences, 2005.12
13. F.J.J. van Loo, G.D. Rieck, Acta Metall. **21**, 73–84 (1973)
14. H. Sun, The research on the process and mechanism of the AI based amorphous alloys during diffusion bonding, Master Thesis, Lanzhou University of Technology, 2010.4
15. Y. Cao, C. Guo, S. Zhu et al., Fracture behavior of Ti/Al3Ti metal-intermetallic laminate (MIL) composite under dynamic loading. Mater. Sci. Eng., A **637**, 235–242 (2015)
16. X. Li, Deformation behavior of amorphous Fe78Si9B13 alloy and Fe78Si9B13/Ni laminated composite, Harbin Institute of Technology, 2007
17. M. Konieczny, Microstructural characterization and mechanical response of laminated Ni-intermetallic composites synthesized using Ni sheets and Al foils. Mater. Charact. **70**, 117 (2012)
18. Q.W. Wang, G.H. Fan, L. Geng et al., Formation of intermetallic compound layer in multi-laminated Ni-(TiB_2/Al) composite sheets during annealing treatment. Micron **45**, 150–154 (2013)
19. Y. Cao, C. Guo, S. Zhu et al., Fracture behavior of Ti/Al_3Ti metal-intermetallic laminate (MIL) composite under dynamic loading. Mater. Sci. Eng. A **637**, 235–242 (2015)

Effect of Zr and Cu on Mechanical Properties, Glass Forming Ability and Thermal Stability of $Zr_{63.68}Cu_{16.2}$ $Ni_{10.12}$ Al_{10} Amorphous Alloy

Chunyan Li, Jinfeng Yin, Juanqiang Ding, Fuping Zhu, Yanchun Zhao and Shengzhong Kou

Abstract The effects of Zr and Cu content on mechanical properties, glass forming ability and thermal stability of $Zr_{63.68-x}Cu_{16.2+x}$ $Ni_{10.12}$ Al_{10} (x = 0, 0.45, 0.9, 1.35, 1.8, 2.25, 2.7) bulk metallic glasses (BMGs) has been studied. The specimens with diameters of 2, 4 and 6 mm cut from the same ladder-shaped BMG rod by copper mold suction casting method were used in the present investigation. The results show that with the increase of Cu content and the decrease of Zr content the prepared alloy specimens with diameter of 2 mm are completely amorphous. The plastic strain shows the trend of increasing at first and then decreasing. At x = 1.8, the plastic strain reaches the maximum and its value is 38.74%, about 4.3 times of $Zr_{63.68}Cu_{16.2}Ni_{10.12}Al_{10}$ (BMG) 9.10%. The elastic modulus is 51,757 MPa, the yield strength is 2079 MPa, the compressive strength and the fracture strength are 3861 MPa. With the increase of specimens size, there is obvious scale effect in plastic strain. It is also found that the plastic strain is up to 30.94% for specimens with diameter of 4 mm, showing a great plasticity. With the increase of Cu atoms and the decrease of Zr atoms, the width of the super-cooled liquid region ($\Delta T_x = T_x - T_g$) and parameter ($\gamma = T_x/(T_g + T_l)$) both increase first and then decrease, the maximum values can reach 98 and 0.395 respectively at x = 1.35. The parameter γ tends to be constant at x = 1.8–2.7.

C. Li (✉) · J. Yin · J. Ding · F. Zhu · Y. Zhao · S. Kou
College of Material Science and Technology,
Lanzhou University of Technology, Lanzhou 730050, China
e-mail: licywz@163.com

J. Yin
e-mail: 2386331413@qq.com

C. Li · Y. Zhao · S. Kou
State Key Laboratory of Advanced Processing and Recycling of Non-Ferrous Metals,
Lanzhou University of Technology, Lanzhou 730050, China

C. Li
Wenzhou Engineering Institute of Pump and Valve,
Lanzhou University of Technology, Wenzhou 325000, Zhejiang, China

Y. Han (ed.), *Advanced Functional Materials*,
https://doi.org/10.1007/978-981-13-0110-0_26

Keywords Mechanical properties · Serrated flow · Work hardening
Scale effect

Introduction

Metallic glass is formed by rapid solidification under large cooling rate, getting across the crystallization process and occurring below the glass transition temperature. Compared with crystalline alloys, bulk metallic glasses (BMGs) have some special properties, such as their strength is much higher than that of crystalline alloys and almost close to the theoretical strength, but the BMGs have relatively low room temperature plasticity. Therefore, no symptoms of brittle fracture will be happened, which, lead to disaster easily. If the materials are applied to practice. As a kind of promising structural material, the wide application has been limited due to the poor plasticity of BMGs [1, 2]. Because Zr-based BMGs have many advantages, such as high strength, high hardness, high wear resistance and low elastic modulus, they are one of the most advantageous structural materials [3]. It has been found that the room temperature plasticity of metallic glasses could be improving by compositing [4], microalloying [5, 6] and prefabricatiing defects [7]. Therefore, how to optimize the mechanical properties of bulk metallic glasses has become the focus of research in this field for many years. In recent years, scholars have done many researches on the mechanical properties of BMGs [8–13]. Based on the study of bulk metallic glasses $Zr_{63.68}Cu_{16.2}Ni_{10.12}Al_{10,}$ the effects of the content of Zr, Cu elements on mechanical properties, thermal stability and glass forming ability of $Zr_{63.68}Cu_{16.2}Ni_{10.12}Al_{10}$ alloys have been studied in this paper.

Experimental Raw Materials and Methods

The experimental materials were composed of high-purity zirconium (Zr 99.99%), copper (Cu 99.99%), aluminum (Al 99.98%), nickel (Ni 99.99%). According to the percentage of atoms, the alloy composition were designed as $Zr_{63.68-x}Cu_{16.2+x}$ $Ni_{10.12}$ Al_{10} (x = 0, 0.45, 0.9, 1.35, 1.8, 2.25, 2.7). Using magnetic suspension smelting process to prepare alloy ingot under high pure argon atmosphere. In order to ensure the alloy composition uniform, each ingot was remelted 3 or 4 times. The alloy bars with diameters of 2 mm were fabricated by copper mold suction casting method under argon atmosphere. The amorphous structure of specimens was identified by Bruker Axs D8 Advance X-ray diffraction (XRD, Cu Ka radiation). The glass transition and crystallization behavior of the specimens were evaluated by STA409 thermal differential calorimeter (DSC) at a heating rate of 20 K/min. Uniaxial compression tests of the specimens were carried out by WDW-100D universal testing machine at room temperature with the strain rate of $2.7 \times 10^{-4}\ s^{-1}$, and the aspect ratio of all specimens was about 2:1. The fracture surface was observed by Quanta Feg450 field emission scanning electron microscope (SEM).

Results and Analysis

XRD result analysis. Figure 1 shows the XRD diffraction pattern of $Zr_{63.68-x}Cu_{16.2+x}Ni_{10.12}Al_{10}$ (x = 0, 0.45, 0.9, 1.35, 1.8, 2.25, 2.7) BMG specimens with diameters of 2 mm. It can be seen from Fig. 1, the diffraction peaks of all specimens show a unique diffuse peaks of amorphous materials, indicating that with he increase of the content of Cu elements and the decrease of Zr elements, the specimens with diameters of 2 mm are completely amorphous.

Mechanical Properties Analysis. The room temperature compression stress-strain curve for $Zr_{63.68-x}Cu_{16.2+x}Ni_{10.12}Al_{10}$ (x = 0, 0.45, 0.9, 1.35, 1.8, 2.25, 2.7) BMGs with diameters of 2 mm as shown in Fig. 2. All specimens have undergone elastic deformation, yield, plastic deformation, fracture and so on. Serrated flow behavior is produced in the stage of plastic deformation for all specimens, there have been obvious process hardening when Cu content is x = 0.45, 0.9, 1.35, 1.8. As shown in Fig. 3, the plastic strain ε_p increases first and then decreases with the increase of the content of Cu elements and the decrease of Zr elements, it increases from the base of 9.10% suddenly increased to 29.38%, then slowly increased to 38.74%, and finally suddenly dropped to 12.72%, indicating that the plasticity of the BMGs system is very sensitive to the change of composition. Although there are some fluctuations in the change of compressive strength σ_{bc} and fracture strength σ_c, the whole show a tendency to increasing firstly and then decreasing, when x = 1.8, both values are 3861 Mpa. The yield strength σ_s tends to level with the increase of Cu content only

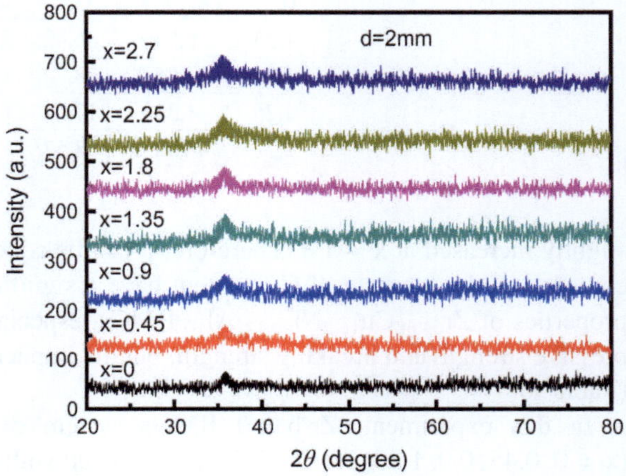

Fig. 1 XRD diffraction pattern of alloy specimens

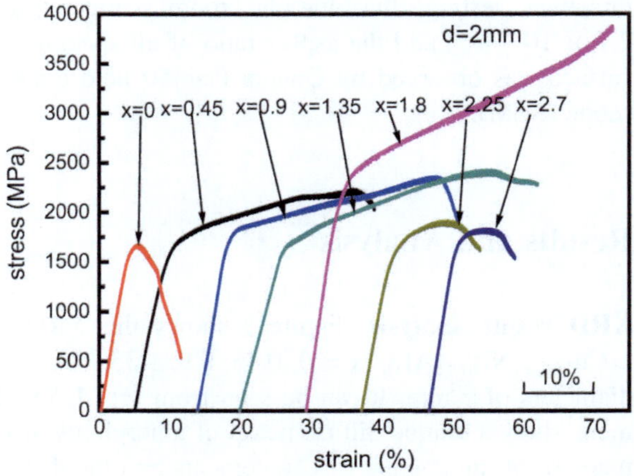

Fig. 2 Stress-strain curves of BMG specimens

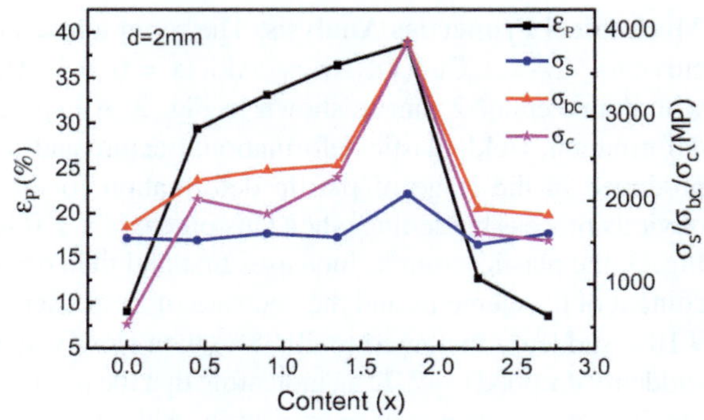

Fig. 3 ε_p-x, σ_s-x, σ_{bc}-x, σ_c-x graph

Table 1 Mechanical parameters of alloy specimens

x	E/Mpa	ε_p/%	σ_s/Mpa	σ_{bc}/Mpa	σ_c/Mpa
0	41,997	9.10	1539	1675	527
0.45	36,194	29.38	1517	2236	2016
0.9	42,904	33.14	1601	2358	1840
1.35	26,840	36.42	1556	2421	2277
1.8	51,757	38.74	2079	3861	3861
2.25	41,390	12.72	1470	1908	1614
2.7	46,485	8.57	1605	1828	1519

slightly increased at x = 1.8. Therefore, it can be concluded that the decrease of Zr content and the increase of Cu content have a significant effect on the mechanical properties of $Zr_{63.68}Cu_{16.2}Ni_{10.12}$ Al_{10} BMGs, especially to the plastic strain, compressive strength and breaking strength, but the impact on the yield strength is weak (Table 1).

In this experiment, Zr-based BMGs system of $Zr_{63.68-x}Cu_{16.2+x}Ni_{10.12}Al_{10}$ (x = 0, 0.45, 0.9, 1.35, 1.8, 2.25, 2.7) is prepared with high-purity Zr, Cu, Ni and Al as raw materials. The lattice types of them show that Zr is BCC, but Cu, Ni, Al are FCC. Their respective atomic radius are R (Zr) = 0.16 nm, R (Cu) = 0.128 nm, R (Ni) = 0.125 nm, and R(Al) = 0.143 nm, it is obvious that the atomic radius of Zr is larger than Cu atomic radius. As we all know that the larger the radius, the worse the fluidity. The content of Cu increase and the content of Zr decrease in the alloy system not only mean the increase of FCC atom, and the same time the decrease of BCC atom, which will improve the density of alloy system, but also means that small atoms replace big atoms, will help to get more compact accumulation structure to reducing the free volume, improving the solid-liquid interface energy, hindering long-range diffusion, and increasing the structural heterogeneity of the alloy in the process of rapid cooling [14, 15] (soft area, hard area). Because the soft region in the bulk metallic glasses is more easily activated under the stress condition to becoming the nucleation of the shear band, this structural inhomogeneity plays an important role in strengthening the plasticity of bulk metallic glasses [16–20]. In addition, it is found that the Zr based BMGs is one of the most stable BMGs, the local structure is mainly composed of twenty faces cluster [21], the structure is characterized by local five-fold symmetry (LFFS) which was defined as the percentage of the area of the Pentagon in the Tyson (Voronoi) polyhedron. Plastic deformation of BMG was reacted by LFFS structural parameter, it has been shown that the plastic deformation of BMGs tends to occurring in regions with higher locally translational symmetry. In other words, a region with a lower five-fold symmetry, mainly because these regions have higher translational symmetry and lower energy barrier so that they can accommodate a relatively large number of plastic shape variables. On the contrary, it is difficult to producing plastic deformation in the higher area of LFFS owing to its high stability. It will expand to the higher LFFS region when the plastic deformation reaches saturation in the lower region of LFFS, and the total energy released by the energy of sound and light causes the final rupture. In this experiment, which are beneficial to generate more numbers of lower local five-fold symmetry regions with the increase of the number of Cu atoms and the decrease of the number of Zr atoms. But superfluous of the amount of Cu atoms increase and Zr atoms decrease are bad for generating more numbers of lower local five-fold symmetry regions.

Shear Band Analysis. Brittle fracture has always been the focus of research in the research field of amorphous materials, the fracture surface of the BMGs specimens were obtained by compression test, which can be used to observing and analyzing the deformation mechanism. Figure 4 shows the shear bands morphology of $Zr_{63.68-x}Cu_{16.2+x}Ni1_{0.12}Al_{10}$ (x = 0, 0.45, 0.9, 1.35, 1.8, 2.25, 2.7) alloy specimens with diameter of 2 mm after the compression fracture at room temperature.

The seven different composition of the alloy specimens near the fracture surface basically have different numbers of shear bands, (a), (f), (g) specimens show that numbers of shear bands at near the fracture surface are few, especially the (a), (f) the sample that without the main shear bands, there are only a small amount of the secondary shear bands and the roots type shear bands, The crack is mainly extended along the direction of a single main shear band until fracture; There are

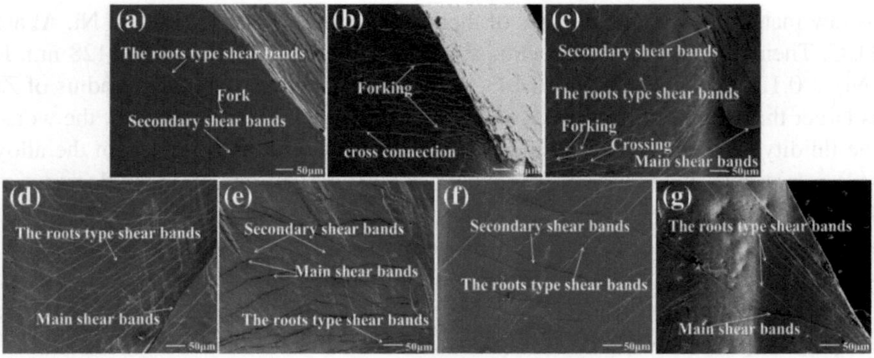

Fig. 4 Shear band morphology of $Zr_{63.68\text{-}x}Cu_{16.2+x}$ $Ni_{10.12}$ Al_{10} (x = 0, 0.45, 0.9, 1.35, 1.8, 2.25, 2.7) amorphous alloy sample. **a** x = 0; **b** x = 0.45; **c** x = 0.9; **d** x = 1.35; **e** x = 1.8; **f** x = 2.25; **g** x = 3.15

high density roots type shear bands and a small amount of the main shear bands at near the fracture surfaces of (b), (d) specimens, the high density the roots type shear bands have both horizontal distribution and vertical distribution, They interact with each other to forming multiple shear bands; The sample (c) has a high-density horizontal wavy shear bands and a small number of main shear bands near the fracture surface, and a large number of vertical roots type shear bands were distributed away from the fracture surface. Owing to the plasticity of (e) specimen is excellent, the compressed sample is a pie shape. Compared to the previous four specimens, on the lateral surface of (e) specimen coexist high-density main secondary and roots type shear bands, and the same time the main shear bands have branching and rebranching behavior to forming the shape of the dendritic shear bands [22] and a fractal distribution is presented [23]. From the main shear bands split into secondary shear bands or root type shear bands. they interact with each other become the main carrier of metallic glasses plastic deformation. Above analysis found that the greater the density of the main shear bands, the better the plasticity, and the existence of roots type shear bands can increase the strength; the more intensely their interaction, the faster the intensity increases, this is consistent with the stress-strain curve shown in Fig. 2.

Scale effect analysis. The XRD diffraction pattern of the $Zr_{61.88}Cu_{18}Ni_{10.12}Al_{10}$ BMGs specimens with 2, 4, 6 mm diameter as shown in Fig. 5. It can be seen that amorphous material show characteristic diffuse diffraction peaks at diffraction angles of $2\theta = 30°–50°$ with the increase of specimens size, there are no sharp diffraction peaks. It shows that the preparation of the samples are completely amorphous with the increase of sample size.

The compressive stress-strain curves of different diameters of $Zr_{61.88}Cu_{18}Ni_{10.12}Al_{10}$ BMGs as shown in Fig. 6. Table 2 shows corresponding mechanical parameter. Figure 7 shows the bar chart of ε_p, σ_{bc}, σ_c with the change of dimension. It can be seen from Fig. 6 that the plastic strain decreases gradually

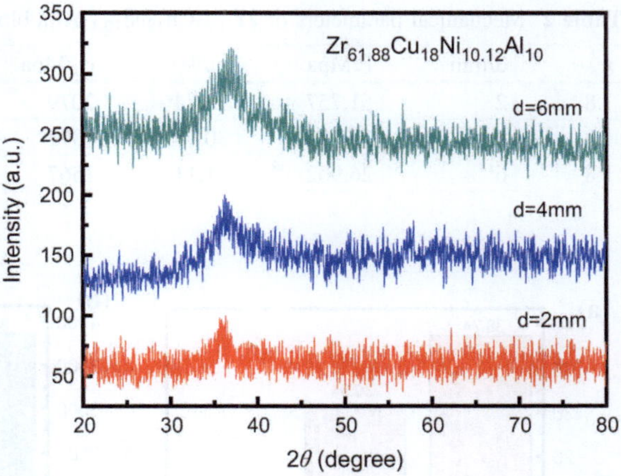

Fig. 5 XRD diffraction pattern of $Zr_{61.88}Cu_{18}Ni_{10.12}Al_{10}$ BMGs with different diameters

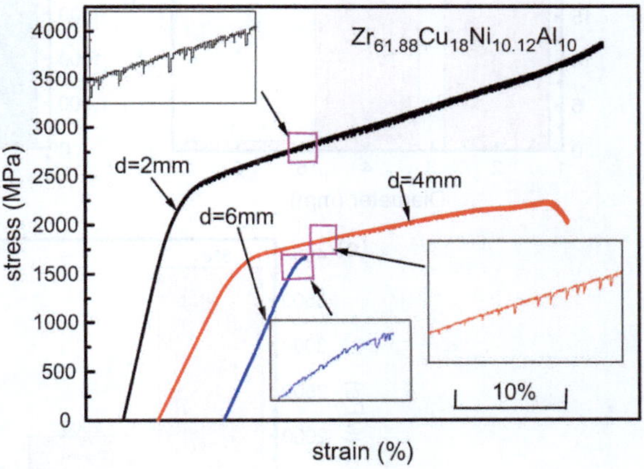

Fig. 6 Compressive stress-strain curve of $Zr_{61.88}Cu_{18}Ni_{10.12}Al_{10}$ BMGs with different diameters

with the increase of the size of the alloy samples, occurring brittle fracture when the specimen size increases from 2 mm to 6 mm in diameter, that is, fracture occurs at the initial stage of plastic deformation after yielding. As shown in Fig. 7a, b, c, the plastic strain, compressive strength and fracture strength decrease gradually, there is obvious scale effect with the increase of alloy specimen size. However, it can be seen from (a) that the plastic strain decreases from 38.74 to 30.94% when the sample size increases from 2 to 4 mm. The plastic strain suddenly decreases from 30.94 to 1.11% when the sample size increases from 4 to 6 mm, indicating that plastic strain is very sensitive to the change of dimension. At the same time, it is concluded that the alloy has a very large room temperature plasticity. The general BMGs has no plasticity when the size is 4 mm, but the alloy has a plastic strain of 30.94%, showing excellent mechanical properties.

The fracture and shear bands morphology of $Zr_{61.88}Cu_{18}Ni_{10.12}Al_{10}$ BMGs at different sizes as shown in Fig. 8. (a), (b), (c) are the fracture morphologies of specimens with 4 mm diameter in different regions, it can be seen that most areas

Table 2 Mechanical parameters of $Zr_{61.88}Cu_{18}Ni_{10.12}Al_{10}$ block BMGs with different diameters

x	d/mm	E/Mpa	ε_p/%	σ_s/Mpa	σ_{bc}/Mpa	σ_c/Mpa
1.8	2	51,757	38.74	2079	3861	3861
1.8	4	23,251	30.94	1392	2231	2000
1.8	6	26,902	1.11	1567	1670	1670

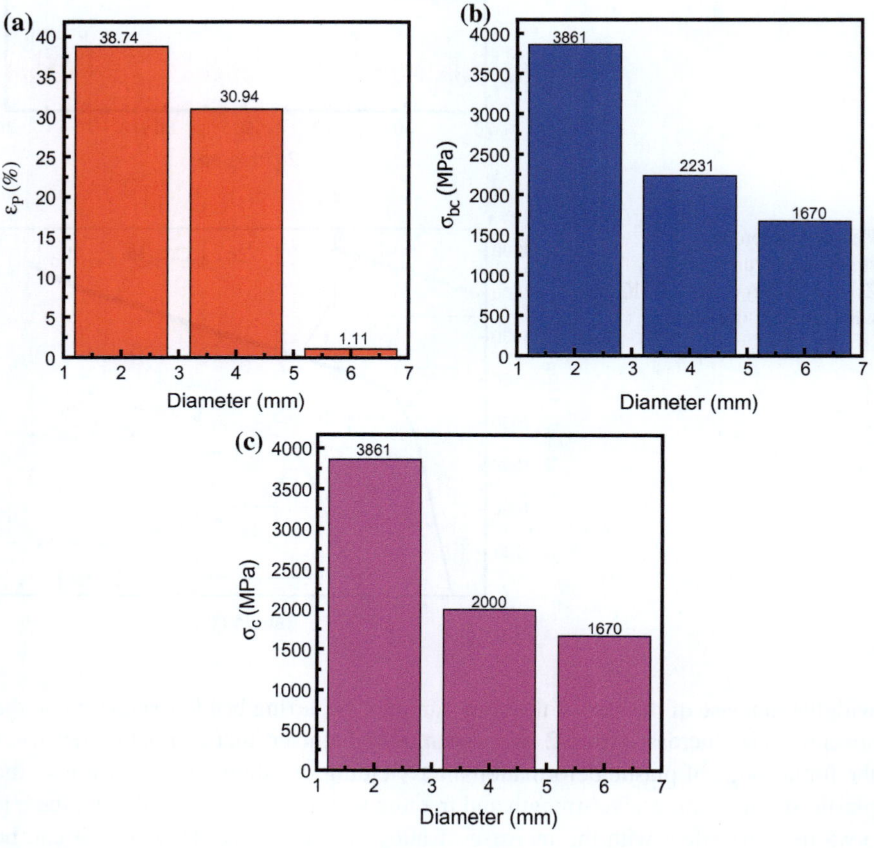

Fig. 7 Histogram of εp, σbc, σc varying with different dimension

show dense venation, the distribution is more uniform. Because of them great plasticity make the dimples deeper, as shown in (a), (b), (c) the thin wire arrow, tiny streaks can also be seen in certain dimples. In the square area of (a), (c), we can see some of the small stripes also have branching behavior, which is related to the improvement of plasticity. There are a few small molten droplets and large molten droplet as shown in pachytene arrows and circular areas. The reason for the occurrence of the droplet is that it is locally melted in some very few local regions,

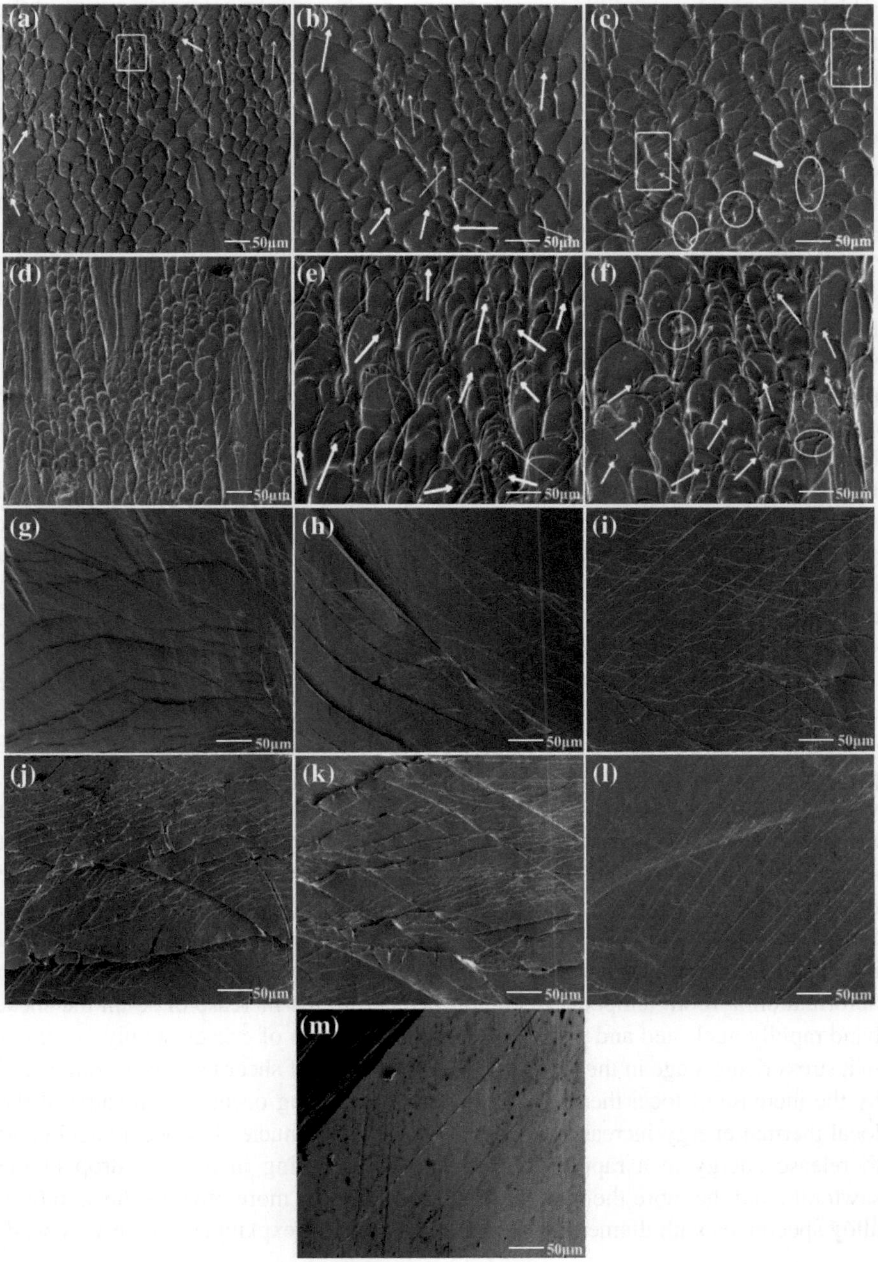

Fig. 8 Fracture morphology and shear bands morphology of $Zr_{61.88}Cu_{18}Ni_{10.12}Al_{10}$ BMGs at different sizes. **a**, **b**, **c** The fracture morphology of d = 4 mm; **d**, **e**, **f** The fracture morphology of d = 6 mm; **g**, **h**, **i** The shear bands morphology of d = 2 mm; **j**, **k**, **l** The shear bands morphology of d = 4 mm; **m** The shear bands morphology of d = 6 mm

which is caused by a rapid increase in the temperature of the amorphous zone at the moment of the high degree of localized fracture of the amorphous alloy [24, 25]. (d), (e), (f) are the fracture morphologies of specimens with 6 mm diameter in different regions, also show venation but uneven distribution, the density are small, dimple are shallow. There are plenty of small molten droplets and large molten droplet, some dimples also exist in tiny streaks. (g), (h), (I) are the shear band morphologies of specimens with 2 mm diameter in different regions, (g), (h) region are mainly a high density primary shear bands, and with the increase of the stress. The shear bands forking-again forking along the stress direction to form dendritic shear band morphology. (i) region is intensive roots type shear zone generated in the main shear bands and secondary shear bands extended front end, they interweave each other like electric shock pattern. (j), (k), (l) are the shear band morphologies of samples with 4 mm diameter in different regions, (j), (k) region are dominated by a large number of wavy shear bands, and between the wavy shear bands exist a high density of secondary shear bands and the root type shear bands intense interaction. This kind of high density corrugated shear band is a special shear band shape in compression fracture of large size and large plastic metallic glasses. (l) region is mutually parallel root type shear bands produced in the tip of a corrugated shear bands. (m) is the shear band morphology of specimens with 6 mm diameter, it can be seen that there are very few shear bands near the fracture surface, which is a typical brittle fracture.

In summary, combined with mechanical analysis, it can be concluded that the denser the choroid, the more uniform the distribution and the better the plasticity; The denser the shear bands, the more intense the interaction and the better the plasticity. Regardless of the fracture morphology or the shear bands morphology, with the sample size increases, its density gradually reduced, and the corresponding mechanical properties become worse. In addition, from Fig. 6, it is apparent that serrated flow phenomenon, the specimens with 2 mm and 4 mm in diameter have the trend that the amplitude of serrated flow increases with the increase of strain. Specimen of 6 mm diameter appears big serration after yielding, fracture occurred in the early stage of plastic deformation. The reason is that in the process of plastic deformation at room temperature compression, with the increase of strain the shear band rapidly nucleated and grew up under the condition of energy supply, resulting in a stress rising stage in the serration. But the growth of shear band is accompanied by the increase of local thermal energy so that softening occurs when some of the local thermal energy increases to a certain extent, and nucleation occurs again and to release energy in a rapidly expanding way, resulting in a stress drop in the sawtooth, and the more the number of shear bands, the more obvious the serration, alloy specimens with diameter of 2 mm and 4 mm can explain this point very well.

DSC analysis. In order to gain a better understand of the effect of Cu, Zr on the thermal stability and glass forming ability of $Zr_{63.68}Cu_{16.2}Ni_{10.12}Al_{10}$ alloys. Thermal analysis of $Zr_{63.68-x}Cu_{16.2+x}Ni_{10.12}Al_{10}$ (x = 0, 0.45, 0.9, 1.35, 1.8, 2.25, 2.7) alloys are carried out by differential scanning calorimetry to obtain the DSC curve of each sample as shown in Fig. 9, the heating rate is 20 k/min. Table 3 is

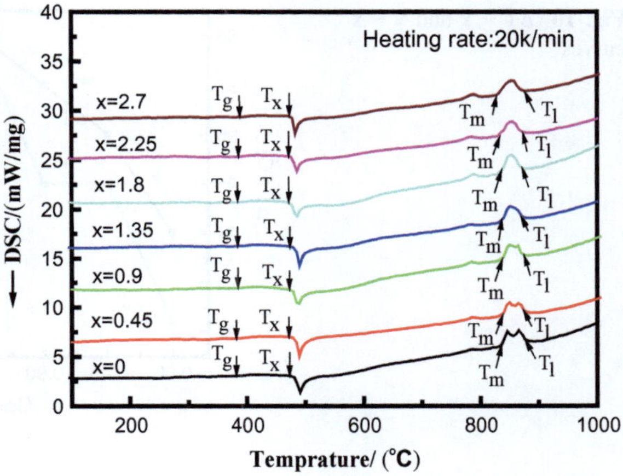

Fig. 9 DSC curve for $Zr_{63.68-x}Cu_{16.2+x}Ni_{10.12}Al_{10}$ (x = 0, 0.45, 0.9, 1.35, 1.8, 2.25, 2.7) alloys

Table 3 Thermodynamic parameters of different DSC curves

x	T_g/°C	T_x/°C	T_m/°C	T_l/°C	ΔT_x/°C	
0	389	483	836	874	94	0.383
0.45	388	484	837	874	96	0.384
0.9	385	482	836	870	97	0.384
1.35	385	483	837	869	98	0.385
1.8	383	477	835	867	94	0.383
2.25	387	480	832	866	93	0.383
2.7	388	480	831	865	92	0.383

corresponding characteristic thermodynamic parameters, including glass transition temperature T_g, crystallization start temperature T_x, solid phase temperature Tm, liquid phase temperature T_l, the width of super-cooled liquid region $\Delta T_x(T_x - T_g)$, parameter gamma $\gamma = T_x/(T_g + T_l)$. Figure 10 shows the ΔT_x and γ curve with the change of Cu content. The thermal stability is characterized by the width of super-cooled liquid zone, the greater the width of super-cooled liquid phase, the better the thermal stability of the alloy. The formation ability of amorphous is characterized by parameter γ, and the greater the γ, the stronger the glass formation ability. It can be clearly seen from Fig. 10 that with the increase of Cu atoms and the reduction of Zr atoms, the width of super-cooled liquid region ΔT_x and parameter γ tend to increase first and then decrease, and reach the maximum at Cu content x = 1.35. The values are ΔTx = 98 C and $\gamma = 0.385$, respectively, with high thermal stability and amorphous forming ability. As can be seen from Fig. 10, the parameter γ remains horizontal after x = 1.8, indicating that they glass forming ability remains constant at Cu content x = 1.8–2.7. Primarily because the big Zr atoms are replaced by the small Cu atoms, the thermal stability and amorphous forming ability are improved. When the big Zr atoms are replaced by the small Cu atoms, a large number of dense packing structures are formed, the number of free

Fig. 10 ΔT − x and γ − x curves

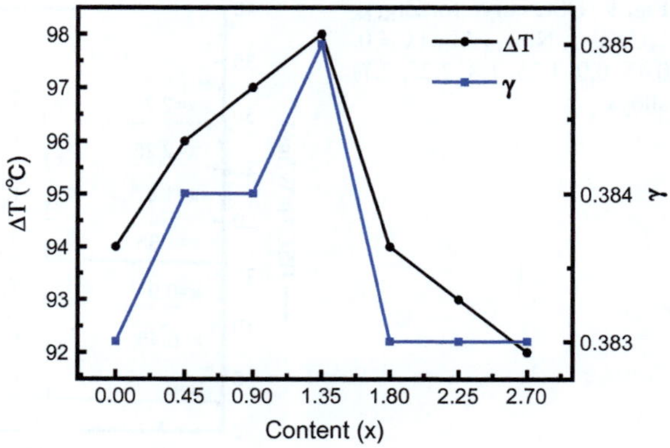

volumes are reduced, the solid-liquid interface energy and viscosity coefficient of amorphous liquid are increased. The changes above in the system improve the viscosity of super-cooled liquid, and result in the difficulties of atomic long-range diffusion and nucleation, so the amorphous structure is easily formed.

In this alloy system, the thermal stability and glass formation ability increase first and then decrease with the increase of Cu atoms and the decrease of Zr atoms, and reach the maximum at x = 1.35. It is proved that highly stable metallic glasses tend to have strong glass forming ability [26, 27]. It can also be concluded that the increase in the amount of Cu atoms and the reduction of Zr atoms can improve thermal stability and glass formation ability of $Zr_{63.68}Cu_{16.2}Ni_{10.12}$ Al_{10} BMGs.

Conclusion

(1) The XRD analysis show that all specimens with a diameter of 2 mm are completely amorphous with the increase of Cu content and the decrease of Zr content. The mechanical properties of alloys with diameter of 2 mm are analyzed that Zr, Cu elements have a significant influence on the mechanical properties of $Zr_{63.68}Cu_{16.2}Ni_{10.12}Al_{10}$ alloys, especially the influence on plasticity and compressive strength. When the x = 1.8, plastic strain is the largest, reaching 38.74%, which is about 9.10% of 4.3 times of $Zr_{63.68}Cu_{16.2}Ni_{10.12}Al_{10}$ alloy. The modulus of elasticity is 51,757 MPa, the yield strength is 2079 MP, the compressive strength and fracture strength are 3861 MPa. Obvious serrated flow behavior and work hardening phenomenon are observed in the process of plastic deformation, indicating that the increase in the amount of Cu atoms and the reduction of Zr atoms can increase the plastic and improve the intensity of the alloys.

(2) The mechanical properties of $Zr_{61.88}Cu_{18}Ni_{10.12}Al_{10}$ alloys with diameter of 2, 4, 6 mm are analyzed, it is found that the plastic strain, compressive strength and breaking strength are all decrease with the increase of the size of the

sample, especially, the change of plastic strain is more obvious. Therefore, it is concluded that the plastic strain has obvious scale effect. It is also found that the plastic strain at the diameter of 4 mm is as high as 30.94% and has large plasticity, showing they have excellent mechanical properties.

(3) The thermal analysis shows that the width of the super-cooled liquid region ΔT_x and parameter γ first increase and then decrease with the increase of Cu atoms and the decrease of Zr atoms, the maximum values reach 98 and 0.395 respectively at x = 1.35. It can improve the thermal stability and glass formation ability to a certain extent with the increase of Cu atoms and the decrease of Zr atoms.

Acknowledgements This work was supported by the National Natural Science Foundation of China (Grant No. 51661016, 51571105, 51551101, 51661017); Wenzhou city public welfare science and technology project (2011CB612203); The National Basic Research Program of China (973 Program) (Grant No. 2011CB612203).

References

1. C.T. Liu, L. Heatherly, D.S. Eaton, Test environments and mechanical properties of Zr-base bulk amorphous alloys. Metall. Mater. Trans. A **29**, 1811–1820 (1998)
2. X.L. Guo, D.J. Li, Y.M. Wang, J.B. Qiang, C. Dong, Fracture behavior of bulk amorphous alloy $Zr_{65}A1_{7.5}Ni_{10}Cu_{17.5}$ under uniaxial compression at room temperature. Acta. Metall. Sin. **39,** 1089–1093 (2003)
3. J.S. Huang, Y. Liu, S.Q. Chen, Z.M. Liu, B.Y. Huang. Research progress and application of Zr based amorphous alloys. Chin. J. Nonferr. Met. **13**, 1321–1332 (2003)
4. D.C. Hofmann, J.Y. Suh, A. Wiest, G. Duan, M.L. Lind, Designing metallic glass matrix composites with high toughness and tensile ductility. Nature **451**, 1085–1089 (2008)
5. L.Y. Chen, Z.D. Fu, G.Q. Zhang, X.P. Hao, Q.K. Jiang, New class of plastic bulk metallic glas. Phys. Rev. Lett. **100**, 652–663 (2008)
6. Z. Tan, Y.F. Xue, X.W. Cheng, L. Zhang, W.W. Chen, L. Wang, H.F. Zhang, H.M. Fu, Effect of element fitting on compositions optimization of Al-Cu-Ti amorphous alloy by mechanical alloying. Trans. Nonferr. Met. Soc. China **25**, 3348–3353 (2015)
7. F. Jiang, L. He, M.B. Zhong, Z.H. Han, Q. Zhao, Orientation effect of pre-introduced shear bands in a bulk-metallic glass on its “work-ductilising”. Mater. Sci. Eng. A **496**, 285–290 (2008)
8. M.Q. Jiang, Research progress in plasticity theory of amorphous alloys. Mater. China **33**, 257–264 (2014)
9. C. Sun, H.W. Chen, Y.F. Xue, C. Chen, J. Liu, J. Teng, Study on low temperature dynamic compressive properties of $Ti_{40}Zr_{25}Ni_8Cu_9Be_{18}$ bulk amorphous alloys. Rare Met. Mater. Eng. **43**, 960–963 (2014)
10. L. Liu, H. Song, X.J. Zhao, W. Liu, Effect of sample size on Pd based bulk amorphous alloy internal shear bands. Rare Met. Mater. Eng. **44**, 1522–1526 (2015)
11. N.Q. Pan, Z.H. Du, Z.W. Zhu, X.Y. Lei, L.Z. Xu, Temperature rise and fracture temperature rise in the shear band of Zr-Ti-Ni-Cu-Be bulk amorphous alloy. Chin. J. Nonferr. Met. **26**, 973–979 (2016)
12. Z.F. Zhang, R.T. Qu, Z.Q. Liu, Research progress in fracture behavior and strength theory of metallic glass. Acta Metall. Sin. **52**, 1171–1182 (2016)

13. Y.J. Sun, Y.H. Zhai, Low temperature compressive deformation behavior of ZrCuNiAl bulk amorphous alloys. Heat Treat. Met. **41**, 25–28 (2016)
14. H.B. Yu, X. Shen, Z. Wang, L. Gu, W.H. Wang, H.Y. Bai, Tensile plasticity in metallic glasses with pronounced β relaxations. Phys. Rev. Lett. **108**, 015504 (2012)
15. Y.H. Liu, G. Wang, M.X. Pan, W.H. Wang, Super plastic bulk metallic glasses at room temperature. Science **315**, 1385–1388 (2007)
16. H.B. Yu, W.H. Wang, H.Y. Bai, Y. Wu, M.W. Chen, Phys. Rev. B **81**, 220201 (2010)
17. H.L. Peng, M.Z. Li, W.H. Wang, Structural signature of plastic deformation in metallic glasses. Phys. Rev. Lett. **106**, 135503 (2011)
18. M.H. Lee et al., Phys. Status Solidi RRL **3**, 46–49 (2009)
19. J.M. Park, D.H. Kim, J. Eckert, Internal state modulation-mediated plasticity enhancement in monolithic Ti-based bulk metallic glass. Intermetallics **29**, 70–74 (2012)
20. Z.W. Zhu, L. Gu, G.Q. Xie, W. Zhang, A. Inoue, H.F. Zhang, Z.Q. Hu, Relation between icosahedral short-range ordering and plastic deformation in Zr-Nb-Cu-Ni-Al bulk metallic glasses. Acta Mater. **559**, 2814–2822 (2011)
21. Q. Wang, C.T. Lin, Y. Yang, Y.D. Dong, Phys. Lett. 106 (2011) 215505
22. F.L. Liu, Y.J. Guo, J.H. Yan, G.N. Tang, Formation mechanism of compressibility related to shear band of bulk metallic glass. Mater. Herald B. **28**, 117–121 (2014)
23. D. Ma, A.D. Stoica, X.L. Wang, Power-law scaling and fractal nature of medium-range order in metallic glasses. Nat. Mater. **8**, 30–34 (2009)
24. C.T. Liu, L. Heatherly, D.S. Easton, Test environments and mechanical properties of Zr-base bulk amorphous alloys. Metall. Mater. Trans. A **29**, 1811–1820 (1998)
25. W.J. Wright, R. Saha, W.D. Nix, Deformation mechanisms of the $Zr_{40}Ti_{14}Ni_{10}Cu_{12}Be_{24}$ bulk metallic glass. Mater. Trans. **42**, 642–649 (2001)
26. W.H. Wang, C. Dong, C.H. Shek, Bulk metallic glasses. Mater. Sci. Eng. R **44**, 45–89 (2004)
27. W.L. Johnson, Bulk glass-forming metallic alloys: science and technology. MRS Bull. **24**, 42–56 (1999)

Manifestation of Dynamic Anomaly in Metallic Glass-Forming Liquids by Their Crystallization Processes

Nannan Ren, Chao Zhou, Lina Hu, Fan Zhou, Jilin Tang, Xiaokun Song, Yining Liu, Yijing Zang and Yuzhong Chen

Abstract More and more evidence has indicated the general existence of fragile-to-strong transition (FST) in metallic glass-forming liquids (MGFLs), but the thermodynamic properties corresponding to this dynamic anomaly remain unclear. For rare earth-based metallic glasses, the crystallization process is of importance to understand the dynamic anomaly of their supercooled liquids. In the present work, we studied the effects of thermal history (including cooling rate of fabrication, aging temperature and DSC scanning rate) on the crystallization behaviors of Ce-based, Pr-based and La-based hyperquenched glasses. Abnormal changes of crystallization processes were found in all these three rare-earth-based metallic glass ribbons, despite the distinct difference in their respective specific behaviors. The general non-monotonic change of crystallization characteristics corresponds to the non-monotonic structural evolution of supercooled liquid at a certain fictive temperature, which just falls in the temperature range where the fragile-to-strong transition occurs. Based on the observed abnormal crystallization behavior and its relation to the dynamic anomaly in supercooled liquids, as well as the convenience and reliability of measuring thermodynamic properties, we expect to provide a deep insight into the evolution of the supercooled metallic liquids.

Keywords Metallic glass-forming liquid · Dynamic anomaly · Thermal history Crystallization behavior · Non-monotonic change

---

N. Ren · L. Hu (✉)
Shenzhen Research Institute of Shandong University, Shenzhen 518057, China
e-mail: hulina0850@sina.com

N. Ren · C. Zhou · L. Hu · F. Zhou · J. Tang · X. Song · Y. Liu · Y. Zang · Y. Chen
Key Laboratory for Liquid-Solid Structural Evolution and Processing of Materials (Ministry of Education), Shandong University, Jinan 250061, China

Y. Han (ed.), *Advanced Functional Materials*,
https://doi.org/10.1007/978-981-13-0110-0_27

Introduction

As an abnormal dynamic phenomenon, the fragile-to-strong transition (FST) which occurs in supercooled liquids has aroused a lot of interests since it was firstly reported in water in 1999 [1]. Although the initial studies focused on non-metallic glass-forming liquids (such as water, SiO_2 and BF_2) [2–7], more and more evidence has indicated the general existence of such dynamic anomaly in metallic glass-forming liquids (MGFLs) [8–15]. By systematically measuring the viscosity data in Cu-, La-, Pr- and Al-based MGFLs [10–12], our previous work has found the fragile nature of liquids at the liquidus temperature and the strong nature approaching T_g. Similarly, Busch et al. also observed an abrupt viscosity change in the supercooled region in Zr-based MGFLs [8]. Johnson et al. verified the existence of the fragile-to-strong transition in CuZr-based alloys by detecting the crossover of the specific volume change with temperature [9]. Further evidence of the fragile to strong transition was also given by simulational and experimental work from the aspect of structural characteristics, as well as by the measurements of coefficient of thermal expansion of MGFLs [16–28].

However, the thermodynamic aspect corresponding to this dynamic anomaly remains unclear. Due to the difficulty of directly detecting the thermodynamic properties in the supercooled liquids, i.e., "no-man's land" of MGFLs, the hyperquenching-annealing-thermal scanning strategy has been widely used to investigate how the thermodynamic properties evolve in such region of liquids [29–32]. By survey on more than twenty metallic and non-metallic glasses, it has been found that the fragile-to-strong transition temperature, Tf-s, is generally about 1.36Tg [33]. For CuZrAl hyperquenched glass ribbons, an abnormal three-step relaxation pattern has been observed around the Tf-s [29, 30]. It hints that this abnormal sub-Tg relaxation pattern is a thermodynamic evidence of the FST phenomenon. However, such sub-Tg thermodynamic anomaly is only observed in several metallic glass systems. In rare earth-based metallic glasses, even changing the cooling rate of fabrication in a large range such abnormal sub-Tg relaxation pattern can not be observed, although the FST also exists in liquids of them [10, 11, 30, 34]. In their case, there is a hint that the crystallization processes of them are important, because different crystallizations also reflect different supercooled liquids.

Bear it in mind, we studied the effects of different thermal history (such as cooling rate of fabrication, aging temperature and DSC rates) on the crystallization behavior of Ce-based, Pr-based, and La-based metallic glasses. Through studying the influence of the above factors on the crystallization behavior, we hope to discover the general scenario concerning how the dynamic FST phenomenon is reflected by their thermodynamic properties.

Experimental Procedures

In this work, a series of hyperquenching–annealing–calorimetric experiments have been done to investigate the characteristics of crystallization processes in $Ce_{60}Al_{20}Cu_{20}$, $Pr_{55}Ni_{25}Al_{20}$ and $La_{55}Al_{25}Ni_{5}Cu_{15}$ metallic glasses. The studied alloys were prepared by arc-melting the elemental metals with purities ranging from 99.9 to 99.999%. Then we got different fresh glass ribbons by single roller spinning method by changing the fabrication rate (i.e., cooling at 6–30 m/s) or a 2 mm round bar prepared by vacuum copper mold casting method. The amorphous nature of all the fresh hyperquenched (HQ) samples was confirmed by X-ray diffraction experiments (not shown here). The thermal responses of the 20 ± 0.5 mg fresh HQ glass ribbons (GRs) were detected at different heating rates (2–40 K/min) using a Netzsch DSC 404 calorimeter with highpurity standard Al_2O_3 crucibles in argon. The heat flow (DSC) signals were then obtained. The largest error range of temperature for our calorimeter is 1 K. The characteristic temperatures T_g of $Ce_{60}Al_{20}Cu_{20}$, $Pr_{55}Ni_{25}Al_{20}$ and $La_{55}Al_{25}Ni_{5}Cu_{15}$ HQ GRs are 396, 475 and 446 K respectively obtained from the DSC curves at 20 K/min and the detailed procedure for the determination of T_g is described elsewhere [34, 35]. To determine the heat capacity (C_p) curve of the GRs, both the baseline and the reference sample (sapphire) were measured. The heat capacity curve is only used for the calculation of T_f. The detailed experimental procedures on how to determine the fictive temperature of the ribbons are referred to Refs. [36, 37].

Results

The Crystallization Behavior of Ce-Based Metallic Glasses. First, we compared the crystallization behaviors of Ce-based metallic glasses with different cooling rates at 10 K/min scanning heating rate. As shown in Fig. 1, the patterns and positions of the crystallization peak are strongly affected by the cooling rate. Although the positions of crystallization peak changes distinctly with the cooling rate, the general tendency of such change is lacking. However, for the peak patterns, the 6 m/s ribbon shows three special crystallization peaks. Especially around 470 K (shown in the inset), there is a distinct exothermic peak which is not observed in other samples. This indicates that the 6 m/s samples have different structural heterogeneity from other samples, leading to different nuclei formed and grow in succession during crystallization process.

Considering the aging annealing treatment can also change the structural state of metallic glass and its thermal history, we treated metallic glass ribbons under different spinning rates with isothermal annealing at different temperatures for 30 min. Figure 2 shows that the 6 and 20 m/s amorphous ribbons exhibit different crystallization behaviors under different annealing temperatures. Given the area of the crystallization peaks can reflect the exothermic enthalpy, we show the area

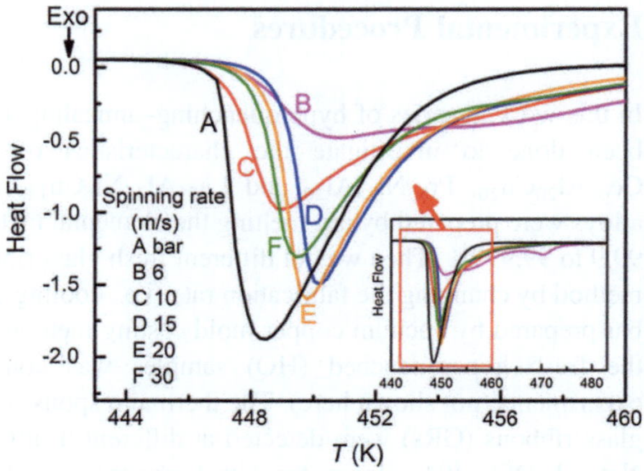

Fig. 1 Crystalline parts on DSC curves of Ce-based metallic glasses in different cooling rates. Upscan rate: 10 K/min

information for different annealing temperatures in the insets. In Fig. 2a, no obvious change of the exothermic ratio of the two crystallization exothermic peaks can be observed when annealed below 353 K. However, when the annealing temperature is 363 K, the area of the first crystallization peak decreases, but that of the second peak increases. The ratio of the two crystallization exothermic peaks displays an abrupt decrease (see the blue line in the inset of Fig. 2a). When the annealing temperature continues to rise to 373 K, the area of the first crystallization peak increases distinctly, and that of the second peak decreases. The exothermic proportion of the first crystallization exothermic peak exhibits an increase. Similar non-monotonic changing tendency is also observed in Fig. 2b. Although there is only one main exothermic peak, the area of this peak decreases at 363 K, and also reaches the maximum at 373 K. Similar non-monotonic behaviors are also observed in the ribbons fabricated by 10, 15, and 30 m/s. According to Refs. [30, 34], the non-monotonic changing tendencies of crystallizations in Figs. 1 and 2 both indicate that the structural evolution of the supercooled liquid is non-monotonic.

Figure 3 shows the change in crystallization behavior with the upscanning rate in Ce-based metallic glass at spinning rates of 6 and 15 m/s. In Fig. 3a, below the upscanning rate of 20 K/min, the main crystallization peaks are split into a peak and a shoulder, illustrating that the main crystallization process of $Ce_{60}Al_{20}Cu_{20}$ metallic glass is divided into two stages. When the heating rate reaches above 20 K/min, the shoulder disappears. This indicates that the crystallization phase corresponding to the shoulder is very unstable, so it can only be observed at certain conditions. Whereas, for the ribbons with a rapid cooling rate (Fig. 3b), its fictive temperature T_f and the energy state are higher, so that they crystallize as a whole in the thermal activation of crystallization, showing only one main crystallization peak even at the upscanning rate of 5 K/min. The above results show that the crystallization process of $Ce_{60}Al_{20}Cu_{20}$ metallic glass is closely related to the temperature T_f where the supercooled liquids are frozen in. When the fictive temperature T_f is low, various types of clusters in supercooled liquids have enough time to relax themselves to get low energy states, and then crystallize independently when the

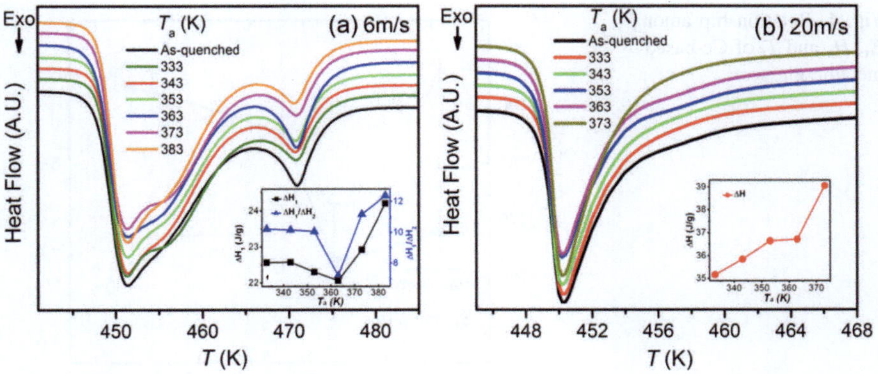

Fig. 2 Crystalline characteristics of Ce-based metallic glasses after isothermally annealed at various temperatures. Spinning rates: **a** 6 m/s; **b** 20 m/s. Insets **a** The area of the first crystallization peak and the ratio of area for the first to the second crystallization peak; **b** the area of the crystallization peak

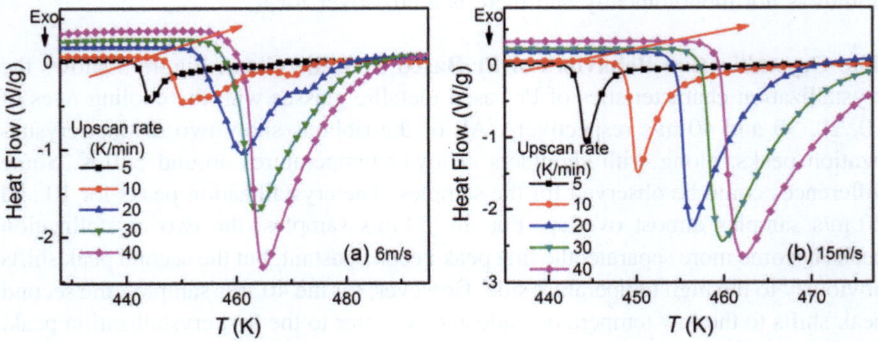

Fig. 3 Influence of upscan rate on crystalline behavior of Ce-based metallic glasses in different spinning rates: **a** 6 m/s; **b** 15 m/s

reheating rate is lower enough. Thus, Fig. 3 suggests that there are competitions between different types of local clusters in supercooled liquids of Ce-based metallic glass. Such competitions between clusters have been believed to be the origin of the FST phenomenon [12, 20, 29, 30].

By using the Kissinger formula [38] to fit and calculate crystallization activation energy E_x, the changing trend of E_x with the fictive temperature, T_f, has been shown in Fig. 4, as well as the crystallization enthalpy. As generally believed, the higher the cooling rate, the higher the fictive temperature, the more the exothermic enthalpy corresponding to the crystallization processes and the lower the E_x. Therefore, the variation trend in Fig. 4 as a whole should be linear. However, we notice that the 6 m/s ribbon shows abnormal crystallization characteristics, i.e., its activation energy and exothermic enthalpy are both smaller than ribbons fabricated by other cooling rates. This further verifies the abnormal properties of supercooled

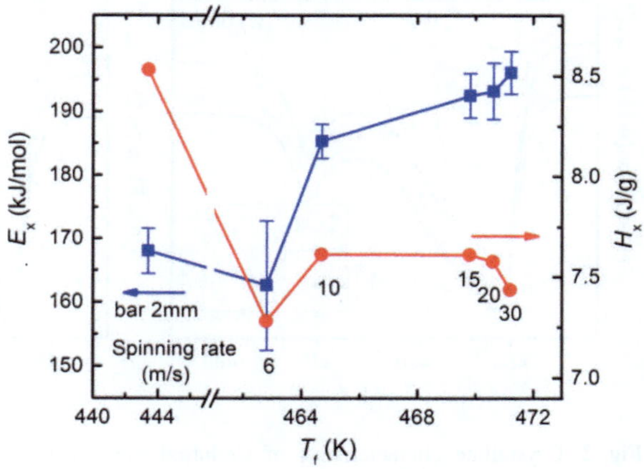

Fig. 4 Relationship among E_x, H_x and T_f of Ce-based metallic glasses

liquids at around 463 K. The structures in supercooled liquid at 463 K are quite different from those at other temperatures. The structural evolution of supercooled liquids is not monotonically and there is a crossover for it.

The Crystallization Behaviors of Pr-Based Metallic Glass. Figure 5 shows the crystallization characteristics of Pr-based metallic glasses with the cooling rates of 10, 20, 30 and 40 m/s respectively. All of the ribbons show two distinct crystallization peaks, along with shoulders at lower temperatures around 540 K. Some differences could be observed for the samples. The crystallization peaks for 10 and 20 m/s samples almost overlap. For the 30 m/s samples, the two crystallization peaks become more separate: the first peak keeps constant, but the second peak shifts obviously to the high temperature side. However, for the 40 m/s samples, the second peak shifts to the low temperature side and is nearer to the first crystallization peak. Besides, compared with 10 and 20 m/s samples, the 30 m/s samples have the largest

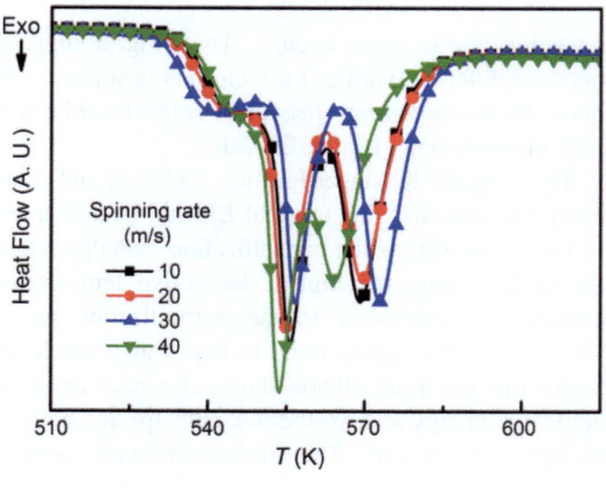

Fig. 5 Crystalline parts on DSC curves of Pr-based metallic glasses in different cooling rates. Spinning rate: 20 m/s

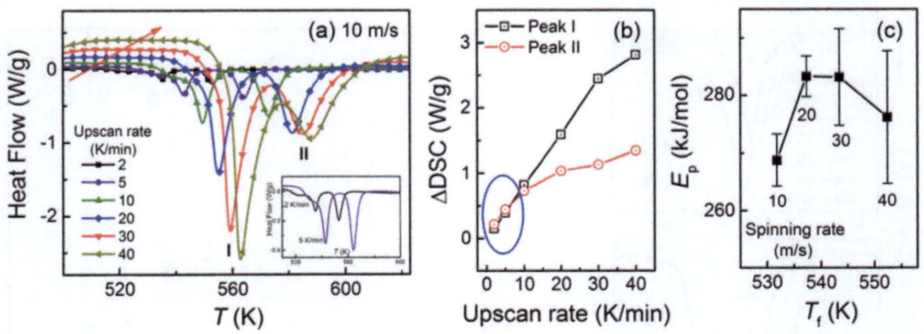

Fig. 6 **a** Influence of upscan rate on crystalline behaviors of Pr-based metallic glasses in the spinning rate of 10 m/s. **b** Upscan rate dependence of DSC differences between the heat flow of two crystalline peaks and that of super-cooled liquids. **c** Relationship between E_p and T_f of Pr-based metallic glasses

area of the second peak and the smallest area of the first peak. Nevertheless, the 40 m/s samples have the largest area of the first crystallization peak and the smallest area of the second crystallization peak among the investigated samples.

Figure 6 shows the crystallization behaviors of Pr-based metallic glass under the influences of thermal upscanning rate. As shown in Fig. 6a, with the thermal upscanning rate increasing, the onset temperature and the peak temperature of crystallization peak I and II both gradually shift towards high temperature region. However, when the thermal upscanning rate is lower than 10 K/min, the exothermic extent of crystallization peak I is evidently smaller than that of crystallization peak II, as shown in Fig. 6a. When the upscanning heating rate is relatively fast, the exothermic extent of crystallization peak I increases dramatically while that of crystallization peak II increases slowly. This changing tendency could be observed in Fig. 6b. The difference between the heat flow value of the two crystallization peaks and that of supercooled liquid is shown in Fig. 6b respectively, with the increase in upscanning rates. Given that there are unnoticeable shoulders in crystallization peaks of Pr-based metallic glass, we take the first crystallization peak temperature as characteristic temperature T_P to research variation on activation energy of crystallization as the cooling rate changes, as shown in Fig. 6c. With the cooling rate decreasing, the activation energy of crystallization E_P gradually increases in the cooling rate ranging from 20 to 40 m/s while decreases for ribbons with the cooling rate of 10 m/s. Since the onset temperature of crystallization corresponds to embryo nucleation, while the peak temperature of crystallization corresponds to the growth rate after nucleation, we think that the crystal nuclei growth of amorphous ribbons with the spinning rate of 10 m/s required less energy, even less than that of ribbons with the spinning rate of 40 m/s. Figure 6c hints the non-monotonic evolution of amorphous structure with the fictive temperature during the cooling process.

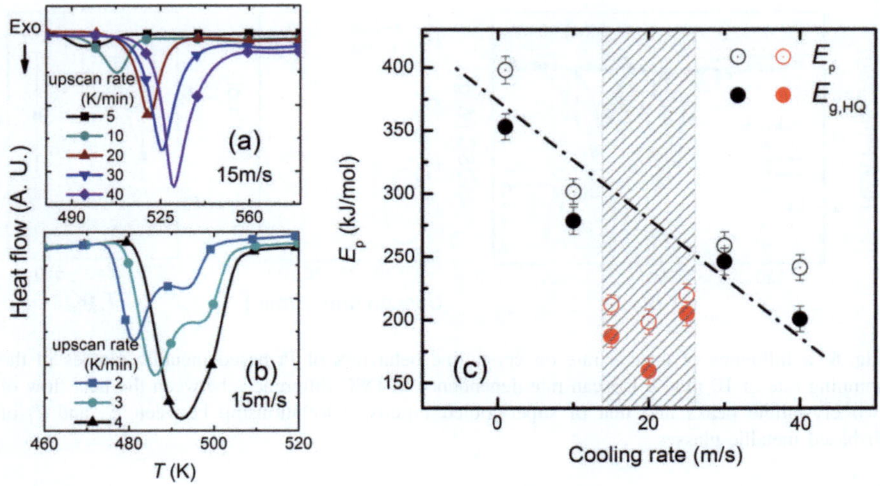

Fig. 7 **a**, **b** Influence of the DSC upscan rate on the crystallization peak in the as-quenched $La_{55}Al_{25}Ni_5Cu_{15}$ GRs in 15 m/s spinning rate; **c** the spinning rate dependence of activation energy of crystallization (E_p) and glass transition (E_g, HQ) for the as-quenched $La_{55}Al_{25}Ni_5Cu_{15}$ GRs (Reprinted with permission from Ref. [34]. Copyright 2014 American Chemical Society.)

Crystallization Behaviors of La-Based Metallic Glasses. In our previous work [34], the abnormal crystallization behavior of $La_{55}Al_{25}Ni_5Cu_{15}$ metallic glasses has been observed. As shown in Fig. 7a, the crystallization behaviors of $La_{55}Al_{25}Ni_5Cu_{15}$, which was cooled with 15 m/s, manifest noticeable differences between the upscanning rates of 5–40 and 2–4 K/min. When the thermal upscanning rate is reduced below 4 K/min, the primary crystallization peak of the metallic glass will split into one crystallization peak and one shoulder. Similar phenomena have been observed in ribbons with the cooling rate of 20 and 25 m/s [34]. These phenomena are similar to the responds of Ce-based metallic glass ribbons to different thermal upscanning rates in Fig. 3. We fit the crystallization peak temperature T_P at different thermal scanning rates by Kissinger formula in Fig. 7b. It has been found that the crystallization peak activation energy E_p does not increase monotonically with the decrease of the cooling rate, and there is an obvious abnormal decrease of E_p at the cooling rate of 15–25 m/s. This phenomenon is consistent with the non-monotonic change of the crystallization activation energy of the Ce- and Pr-based metallic glasses in Figs. 4 and 6.

At the same time, we found that the crystallization behaviors of the metallic glass ribbons frozen at different fictive temperatures exhibit different annealing responses, as shown in Fig. 8. In Fig. 8b, the crystallization behavior of the 25 m/s ribbons is almost completely independent of the annealing temperature. The crystallization behavior of the La-based amorphous ribbons with 15 m/s is very similar to that of the 25 m/s samples. However, when the ribbons of 20 m/s were annealed at 368 and 388 K, the position of the crystallization peak shifts towards the lower temperature and the main peak is separated into two peaks. Besides, the peak

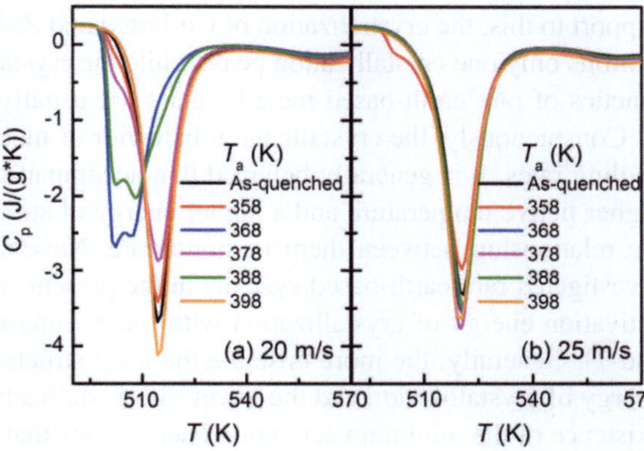

Fig. 8 Influence of annealing temperatures on the crystallization peak of the La-based metallic glass ribbons. Spinning rates: **a** 20 m/s; **b** 25 m/s

position of the first peak is similar to that of the crystallization peak shoulder when the ribbons are annealed at 358 and 378 K. The difference between Fig. 8a, b indicates that the microstructures of the 20 m/s ribbons are more complex than those of the 25 m/s ribbons. It is directly related to the frozen states of the initial topology and the frozen cluster configuration at different fictive temperatures. In this point, such difference is consistent with the abnormal drop of E_p in Fig. 7b.

Discussion

According to the above experimental results, the change of annealing temperature has little effect on the crystallization behavior of metallic glass ribbons, except for La-based and the previously reported Cu-based glasses [30]. Our previous work has shown that the sub-T_g annealing has a significant effect on the sub-T_g enthalpy relaxation pattern [29–31]. This kind of effect is a three-stage change in some metallic glasses, which is considered to be evidence of the fragile-to-strong transition in supercooled region [29, 30]. It indicates that the sub-T_g annealing process affects mainly the atom motion prior to α relaxation (i.e., glass transition), especially in the slow β relaxation, and whether such annealing affects the nucleation and growth of crystal nuclei during crystallization is system related.

We notice that the decrease in the heating rate leads to the splitting of crystallization peak of rare earth-based metallic glasses, whereas such phenomenon is not observed in Cu- and Zr-based glasses [30]. We suspect that this is due to structural inhomogeneity caused by differences in atomic radius and interactions of the constituent elements. The maximum atomic radius of Cu-based and Zr-based metallic glass is Zr (159 pm) and the atomic radius difference is generally less than 27.6%, while the atomic radius of rare earth elements is very large (>182 pm), resulting in a difference in atomic radius greater than 42.8%. That is, from the topological order of view, the heterogeneity of rare earth-based metallic glasses is clearly much larger than that of Cu-based and Zr-based metallic glasses. As a support to this, the crystallization of Cu-based and Zr-based metallic glasses usually exhibits only one crystallization peak, while the crystalline phase and crystallization kinetics of rare earth-based metallic glass are usually more complex.

Conspicuously, the crystallization behavior of metallic glass is in the grip of the cooling rates. It is generally believed that accompanied by a faster cooling rate is a higher fictive temperature and a higher energy of structures of metallic glasses, and the relationship between them is monotonic. Nevertheless, we notice that for the investigated rare-earth-based systems in the present work, there exist minimums of activation energy of crystallization with the cooling rate increasing (see Figs. 4, 6 and 7). Generally, the more unstable the local structure is, the lower the activation energy of crystallization and the lower the crystallization temperature are. Thus, the existence of the minimum activation energy hints that the structure of metallic glass within that range of cooling rate (i.e., within the fictive temperature T_f range) evolves non-monotonically. Based on the area-matching method of enthalpy [37], we have figured out the crossover temperature T_f' of such transition is around $1.2T_g \sim 1.3T_g$ for rare-earth-based alloys. This finding agrees well with what has been reported in Cu- and Zr-based hyperquenched ribbons [12, 30]. As a manifestation of the dynamic fragile-to-strong transition, similar crystallization anomaly of the $Cu_{46}Zr_{46}Al_8$ metallic glasses was also observed [30]. Figure 9 clearly demonstrated that the structures of 12 m/s ribbons are different from those of 25 m/s ribbons. The two crystallization peaks come from the metastable structures and are only present at a certain fictive temperature range. We have concluded that, during the cooling process there is a competition between high temperature clusters and low temperature clusters in Cu-based metallic liquids [30]. For large cooling rates, the liquid is frozen at a high temperature, and the high temperature clusters exert a dominant influence on the following crystallization. When the cooling rate is low, the low temperature clusters become dominant.

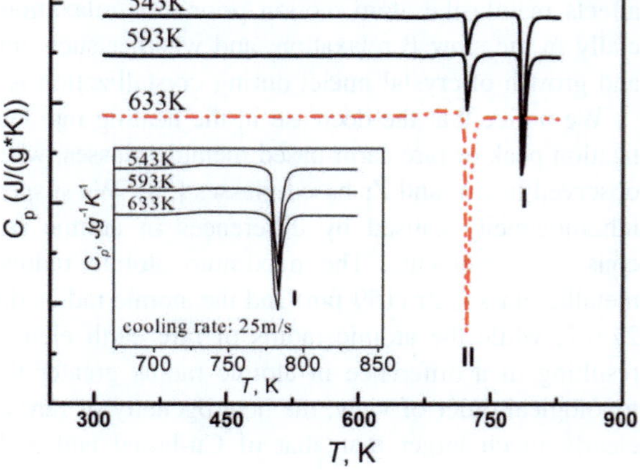

Fig. 9 Influence of annealing temperatures (T_a) of the as-quenched Cu-based metallic glass ribbons spun at 12 m/s on the crystallization peaks (on the DSC output ~ T curves). Inset crystallization behavior of the samples quenched at 25 m/s for a comparison. T_g = 695 K (Reproduced from Ref. [30], with the permission of AIP Publishing.)

Summary

In summary, by regulating various heat history and cooling/heating conditions, we can observe and characterize the dynamic anomaly of metallic glass liquid indirectly from thermodynamically abnormal crystallization. We find that, for rare-earth-based metallic glasses, although the sub-T_g three-stage relaxation pattern prior to the crystallization is not observed, the abnormal change of crystallization behavior with the thermal history seems general. Especially with changing the cooling rate of fabrications, all the rare-earth-based ribbons exhibit a non-monotonic evolution of crystallization characteristics. The crossover temperature for this evolution is around 1.2–1.3T_g, which just falls in the temperature range where the fragile-to-strong transition occurs [30]. The present work hopes to use a convenient way to help understanding how supercooled metallic liquids evolve during cooling thermodynamically.

Acknowledgements This work was supported by National Key Research Program of China [Grant No. 2016YFB0300500] and Shenzhen Science and Technology Research and Development Funds [Grant No. JCYJ20160429183214979].

References

1. K. Ito, C.T. Moynihan, C.A. Angell, Thermodynamic determination of fragility in liquids and a fragile-to-strong liquid transition in water. Nature **398**, 492–495 (1999)
2. S.H. Chen, F. Mallamace, C.Y. Mou, M. Broccio, C. Corsaro, The violation of the Stokes-Einstein relation in supercooled water. Proc. Natl. Acad. Sci. U. S. A. **103**, 12974–12978 (2006)
3. J. Barrat, J. Badro, P. Gillet, A strong to fragile transition in a model of liquid silica. Mol. Simul. **20**, 17–25 (1996)
4. I. Saika-Voivod, P.H. Poole, F. Sciortino, Fragile-to-strong transition and polyamorphism in the energy landscape of liquid silica. Nature **412**, 514–517 (2001)
5. B.W.H. van Beest, G.J. Kramer, R.A. van Santen, Force fields for silicas and aluminophosphates based on ab initio calculations. Phys. Rev. Lett. **64**, 1955–1958 (1990)
6. I. Saika-Voivod, F. Sciortino, P.H. Poole, Free energy and configurational entropy of liquid silica: fragile-to-strong crossover and polyamorphism. Phys. Rev. E **69**, 041503 (2004)
7. M. Hemmati, C.T. Moynihan, C.A. Angell, Interpretation of the molten BeF_2 viscosity anomaly in terms of a high temperature density maximum, and other waterlike features. J. Chem. Phys. **115**, 6663–6671 (2001)
8. C. Way, P. Wadhwa, R. Busch, The influence of shear rate and temperature on the viscosity and fragility of the $Zr_{41.2}Ti_{13.8}Cu_{12.5}Ni_{10.0}Be_{22.5}$ metallic-glass-forming liquid. Acta Mater. **55**, 2977–2983 (2007)
9. J.J.Z. Li, W.K. Rhim, C.P. Kim, K. Samwer, W.L. Johnson, Evidence for a liquid–liquid phase transition in metallic fluids observed by electrostatic levitation. Acta Mater. **59**, 2166–2171 (2011)
10. C. Zhang, L. Hu, X. Bian, Y. Yue, Fragile-to-strong transition in Al-Ni-M (M=La, Pr, Nd) metallic glasses. Chin. Phys. Lett. **27**, 116401 (2010)
11. C. Zhang, L. Hu, Y. Yue, J.C. Mauro, Fragile-to-strong transition in metallic glass-forming liquids. J. Chem. Phys. **133**, 14508 (2010)

12. C. Zhou, L. Hu, Q. Sun, H. Zheng, C. Zhang, Y. Yue, Structural evolution during fragile-to-strong transition in CuZr(Al) glass-forming liquids. J. Chem. Phys. **142**, 64508 (2015)
13. K. Georgarakis, D.V. Louzguine-Luzgin, J. Antonowicz, G. Vaughan, A.R. Yavari, T. Egami, A. Inoue, Variations in atomic structural features of a supercooled Pd-Ni-Cu-P glass forming liquid during in situ vitrification. Acta Mater. **59**, 708–716 (2011)
14. J. Orava, H. Weber, I. Kaban, A.L. Greer, Viscosity of liquid Ag–In–Sb–Te: evidence of a fragile-to-strong crossover. J. Chem. Phys. **144**, 194503 (2016)
15. S. Wei, P. Lucas, C.A. Angell, Phase change alloy viscosities down to T_g using Adam-Gibbs-equation fittings to excess entropy data: a fragile-to-strong transition. J. Appl. Phys. **118**, 034903 (2015)
16. N.A. Mauro, M. Blodgett, M.L. Johnson, A.J. Vogt, K.F. Kelton, A structural signature of liquid fragility. Nat. Commun. **5**, 4616 (2014)
17. K. Georgarakis, L. Hennet, G.A. Evangelakis, J. Antonowicz, G.B. Bokas, V. Honkimaki, A. Bytchkov, M.W. Chen, A.R. Yavari, Probing the structure of a liquid metal during vitrification. Acta Mater. **87**, 174–186 (2015)
18. S. Wei, M. Stolpe, O. Gross, W. Hembree, S. Hechler, J. Bednarcik, R. Busch, P. Lucas, Structural evolution on medium-range-order during the fragile-strong transition in $Ge_{15}Te_{85}$. Acta Mater. **129**, 259–267 (2017)
19. Z. Evenson, T. Schmitt, M. Nicola, I. Gallino, R. Busch, High temperature melt viscosity and fragile to strong transition in Zr-Cu-Ni-Al-Nb(Ti) and $Cu_{47}Ti_{34}Zr_{11}Ni_8$ bulk metallic glasses. Acta Mater. **60**, 4712–4719 (2012)
20. K.N. Lad, N. Jakse, A. Pasturel, Signatures of fragile-to-strong transition in a binary metallic glass-forming liquid. J. Chem. Phys. **136**, 104509 (2012)
21. J.C. Bendert, A.K. Gangopadhyay, N.A. Mauro, K.F. Kelton, Volume expansion measurements in metallic liquids and their relation to fragility and glass forming ability: an energy landscape interpretation. Phys. Rev. Lett. **109**, 185901 (2012)
22. S. Lan, Y. Ren, X.Y. Wei, B. Wang, E.P. Gilbert, T. Shibayama, S. Watanabe, M. Ohnuma, X.L. Wang, Hidden amorphous phase and reentrant supercooled liquid in Pd-Ni-P metallic glasses. Nat. Commun. **8**, 14679 (2017)
23. J.H. Na, S.W. Sohn, W.T. Kim, D.H. Kima, Two-step-like anomalous glass transition behavior in Ni–Zr–Nb–Al–Ta metallic glass alloys. Scr. Mater. **57**, 225–228 (2007)
24. S. Küchemann, K. Samwer, Ultrafast heating of metallic glasses reveals disordering of the amorphous structure. Acta Mater. **104**, 119–124 (2016)
25. H.W. Sheng, H.Z. Liu, Y.Q. Cheng, J. Wen, P.L. Lee, W.K. Luo, S.D. Shastri, E. Ma, Polyamorphism in a metallic glass. Nature **6**, 192–197 (2007)
26. W. Xu, M.T. Sandor, Y. Yu, H. Ke, H. Zhang, M.Z. Li, W.H. Wang, L. Liu, Y. Wu, Evidence of liquid–liquid transition in glass-forming $La_{50}Al_{35}Ni_{15}$ melt above liquidus temperature. Nat. Commun. **6**, 7696 (2015)
27. S. Wei, F. Yang, J. Bednarcik, I. Kaban, O. Shuleshova, A. Meyer, R. Busch, Liquid–liquid transition in a strong bulk metallic glass-forming liquid. Nat. Commun. **4**, 2083 (2013)
28. G. Li, Y.Y. Wang, P.K. Liaw, Y.C. Li, R.P. Liu, Electronic structure inheritance and pressure-induced polyamorphism in lanthanide-based metallic glasses. Phys. Rev. Lett. **109**, 125501 (2012)
29. L. Hu, Y. Yue, C. Zhang, Abnormal sub-T_g enthalpy relaxation in the CuZrAl metallic glasses far from equilibrium. Appl. Phys. Lett. **98**, 81904 (2011)
30. L. Hu, C. Zhou, C. Zhang, Y. Yue, Thermodynamic anomaly of the sub-T_g relaxation in hyperquenched metallic glasses. J. Chem. Phys. **138**, 174508 (2013)
31. C. Wang, L. Hu, C. Wei, X. Tong, C. Zhou, Sub-T_g relaxation patterns in Cu-based metallic glasses far from equilibrium. J. Chem. Phys. **141**, 164507 (2014)
32. C.A. Angell, Y.Z. Yue, L.M. Wang, J. Copley, S. Borick, S. Mossa, Potential energy, relaxation, vibrational dynamics and the boson peak of hyperquenched glasses. J. Phys.-Condes. Matter. **15**, S1051–S1068 (2003)

33. C. Zhou, L. Hu, Q. Sun, H. Zheng, C. Zhang, Y. Yue, Structural evolution during fragile-to-strong transition in CuZr(Al) glass-forming liquids. J. Chem. Phys. **142**, 64508 (2015)
34. X. Yang, C. Zhou, Q. Sun, L. Hu, J.C. Mauro, C. Wang, Y. Yue, Anomalous crystallization as a signature of the fragile-to-strong transition in metallic glass-forming liquids. J. Phys. Chem. B **118**, 10258–10265 (2014)
35. B. Zhang, C. Tang, W. Xu, W. Pan, J. Wang, H. Zhou, Kinetics of glass transition of $Ce_{65}Al_{20}Co_{15}$ metallic glass. Mater. Chem. Phys. **142**, 707–711 (2013)
36. L.N. Hu, Y.Z. Yue, Secondary relaxation behavior in a strong glass. J. Phys. Chem. B **112**, 9053–9057 (2008)
37. Y.Z. Yue, J.D. Christiansen, S.L. Jensen, Determination of the fictive temperature for a hyperquenched glass. Chem. Phys. Lett. **357**, 20–24 (2002)
38. H.E. Kissinger, Variation of peak temperature with heating rate in differential thermal analysis. J. Res. Natl. Bur. Stan. **57**, 217–221 (1956)

Effects of Iron on Microstructure and Properties of $CoCrFe_xNi$ Multi-principal Element Alloys

Linge Han, Hui Jiang, Dongxu Qiao, Yiping Lu and Tongmin Wang

Abstract In this work, $CoCrFe_xNi$ (x = 2, 3, 4) multi-principal element alloys were designed and synthesized by vacuum arc melting technique, and effects of Fe on microstructure and properties of $CoCrFe_xNi$ alloys were investigated. The crystal phase and microstructure of these alloys were studied by X-ray diffraction and scanning electron microscopy, while mechanical properties were measured by Vickers hardness and compressive tests, respectively. The $CoCrFe_xNi$ alloys presented a single face-centered cubic (FCC) structure with columnar crystal morphology, indicating that the Fe content doesn't change the microstructure of $CoCrFe_xNi$ alloys. However, the lattice parameters of FCC phases decreased with the addition of Fe element. It was found that all the alloys exhibited excellent ductility without fracture. In addition, the yield strength decreased, and the Vickers hardness varied from 124 HV for $CoCrFe_2Ni$ to 109 HV for $CoCrFe_4Ni$.

Keywords Multi-principal element alloys · Microstructure · Crystal phase
Mechanical properties

L. Han · H. Jiang · D. Qiao · Y. Lu (✉) · T. Wang
Key Laboratory of Solidification Control and Digital Preparation Technology (Liaoning Province), School of Materials Science and Engineering, University of Technology, Dalian, China
e-mail: luyiping@dlut.edu.cn

L. Han
e-mail: hanlinge1993@mail.dlut.edu.cn

H. Jiang
e-mail: jianghui2013@mail.dlut.edu.cn

D. Qiao
e-mail: qiaodongxu@mail.dlut.edu.cn

T. Wang
e-mail: tmwang@dlut.edu.cn

Y. Han (ed.), *Advanced Functional Materials*,
https://doi.org/10.1007/978-981-13-0110-0_28

Introduction

Traditional alloys have been developing for thousands years, which usually have one principal element as the matrix. A new sort of alloys were researched, called multi-principal element alloys (MPEAs) or high entropy alloys (HEAs) [1, 2], which broke through the traditional alloy design concept. In traditional research of physical metallurgy and phase diagrams, intermetallic compounds which caused brittleness would be generated in multiple principal alloys. However, these MPEAs could simplify the microstructure due to the high entropy effect, and form solid solution such as face-centered cubic (FCC), body-centered cubic (BCC), or hexagonal close-packed (HCP) phases [3–7]. These alloys have good comprehensive properties, such as excellent corrosion resistance, wear resistance, thermal stability, etc. [8–10].

The typical CoCrFeNi MPEA has attracted extensive research attention because of its good ductility with a single FCC structure. Many CoCrFeNi-based MPEAs were investigated, such as $CoCrFeNiTi_x$ alloys, $CoCrFeMo_xNi$ alloys, $CoCrFeNb_xNi$ alloys, and so on [11–13]. Above referred research is an efficient strengthening method through addition of larger-atomic size elements (Ti, Mo and Nb). However, the effects of inherent elements on CoCrFeNi alloy also have been rarely researched. In addition, the research has important significance for practical industrial applications. Therefore, effects of Fe element on the microstructure and properties of $CoCrFe_xNi$ MPEAs were investigated in this study.

Experimental Procedures

The purity of master elements is higher than 99.9 wt%. Table 1 shows the nominal atomic percent of $CoCrFe_xNi$ alloys. Alloys ingots were prepared by vacuum arc furnace with re-melting 4–5 times in a Ti-gettered high purity argon atmosphere with the pressure of 0.05 MPa and casted in a water-cooling copper crucible. Alloys ingots were the fastener-like shape with 30 mm in diameter and 10 mm in length. The crystal structures of alloys were analyzed by X-ray diffractometer (Empyrean) with CuKα radiation and a 2θ range from 20° to 100°. Microstructures of alloys were measured through Scanning Electron Microscope (SEM). Moreover, chemical compositions were determined by Energy dispersive spectrometry (EDS). And pure aqua regia was chose (HCL: HNO3 = 3:1) as the corrosive liquid to etch samples before observing microstructures. Universal testing machine tested compressive

Table 1 Nominal composition of $CoCrFe_xNi$ alloys

Alloys	Co (at.%)	Cr (at.%)	Fe (at.%)	Ni (at.%)
$CoCrFe_2Ni$	20	20	40	20
$CoCrFe_3Ni$	16.7	16.7	50	16.7
$CoCrFe_4Ni$	14.3	14.3	57.1	14.3

yield strength and ductility at a strain rate of 1×10^{-3} s^{-1}. The experimental samples of compressive test were cut from ingots with a size of $\Phi 5 \times 10$ mm. Every alloy prepared two samples to assure the accuracy of experimental results. The hardness was carried out by Vickers hardness tester under the load of 500 g applied for 15 s. Every ingot selected at least 5 points to take experiment and calculated the average value as the final results.

Results and Discussion

Crystal Structure. Table 2 shows the values of valence electron concentration (VEC), electro-negativity difference (χ), atomic radius difference (δ), and mixing enthalpy (ΔH). For $CoCrFe_xNi$ alloys, the values of VEC > 8.0, $\delta < 4.72\%$, and 7.27 kJ mol^{-1} < ΔH < 4 kJ mol^{-1} indicate that these alloys should be possessed a simple FCC solid solution structure according to the reports in Refs. [4, 14]. Moreover, the value of χ is below 0.117 in $CoCrFe_xNi$ alloys indicating inexistence of topological close-packed (TCP) phase [15]. The XRD patterns of $CoCrFe_xNi$ alloys are presented in Fig. 1. The XRD peaks are indexed as a single FCC solid solution structure, which consisted with above theoretical prediction. Figure 1b is the magnifying pattern of $(111)_{FCC}$ peak. The peak of $(111)_{FCC}$ moves right gradually with the addition of Fe element. The lattice parameters of CoCrFeNi alloy [16] and $CoCrFe_xNi$ are 3.5752, 3.5709, 3.5543, 3.5473 Å, respectively. The results are in agree with the change of $(111)_{FCC}$ peak in the Fig. 1b.

Table 2 Values of VEC, χ, δ and ΔH for $CoCrFe_xNi$ (x = 2, 3, 4) alloys

Alloys	VEC	χ	δ (%)	ΔH (kJ mol^{-1})
Fe_2	8.200	0.107	1.055	−3.040
Fe_3	8.170	0.079	0.964	−2.560
Fe_4	8.143	0.073	0.894	−2.204

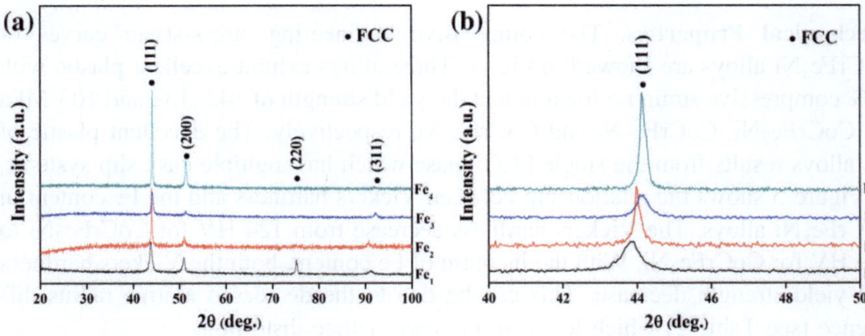

Fig. 1 XRD patterns of $CoCrFe_xNi$ alloys, **a** XRD pattern; **b** magnifying peak of $(111)_{FCC}$

Fig. 2 Microstructure of $CoCrFe_xNi$ alloys, **a** x = 2, **b** x = 3, **c** x = 4

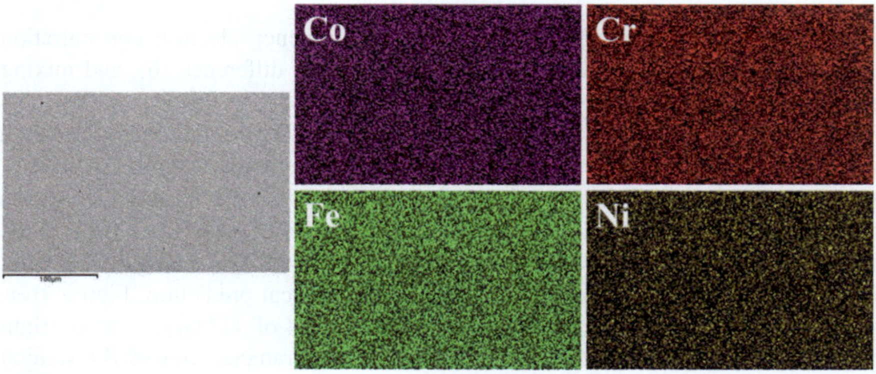

Fig. 3 EDS mappings of $CoCrFe_2Ni$ alloys

Microstructure. Figure 2 exhibits the low magnification SEM images of $CoCrFe_xNi$ alloys. Because water-cooling copper crucible had rather fast cooling rate and ingots weren't thick enough. Grains grew up from bottom to top figuring columnar crystal morphology with a size of 400–500 μm. From EDS mappings of $CoCrFe_2Ni$ alloy shown in Fig. 3, the elements distribution were uniform which was in accord with the microstructure identified the single phase structure. Other alloys were same as $CoCrFe_2Ni$ alloy, and the results weren't put on the article to avoid repetition.

Mechanical Properties. The compressive engineering stress-strain curves of $CoCrFe_xNi$ alloys are showed in Fig. 4. Three alloys exhibit excellent plastic with 50% compressive strain no fracture and the yield strength of 142, 132 and 103 MPa for $CoCrFe_2Ni$, $CoCrFe_3Ni$ and $CoCrFe_4Ni$, respectively. The excellent plastic of the alloys results from the single FCC phase which has multiple easy slip systems.

Figure 5 shows the relationship between Vickers hardness and the Fe content in $CoCrFe_xNi$ alloys. The Vickers hardness decrease from 124 HV for $CoCrFe_2Ni$ to 109 HV for $CoCrFe_4Ni$. With the increase of Fe content, both the Vickers hardness and yield strength decrease. This can be due to the decreased atomic radius difference (see Table 2) which leads to the lower lattice distortion.

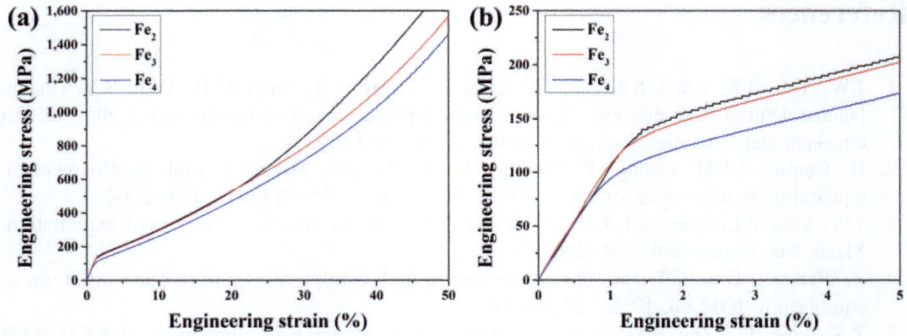

Fig. 4 Compressive engineering stress-strain curves of CoCrFe$_x$Ni alloys, **a** the integral map, **b** the local zoom picture

Fig. 5 Vickers hardness of CoCrFe$_x$Ni alloys

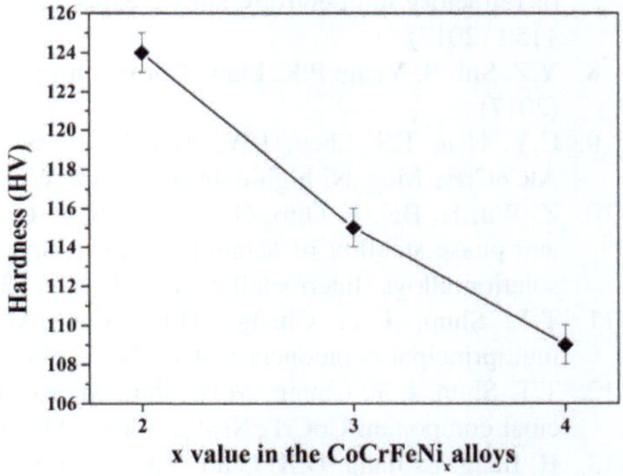

Conclusions

In this study, effects of Fe element on microstructure and mechanical properties of CoCrFe$_x$Ni MPEAs were investigated. And conclusions can be drawn as follows:

(1) CoCrFe$_x$Ni alloys present a single FCC structure and the lattice parameters decrease with the addition of Fe element. All the alloys exhibit large columnar crystal morphology with a size of 400–500 μm.

(2) CoCrFe$_x$Ni alloys possessed good ductility without fracture in compressive test. With Fe content increased, both the Vickers hardness and yield strength decrease which can be due to the decreased atomic radius difference.

Acknowledgements This work was supported by the National Natural Science Foundation of China (Nos. 51671044, 51471044 and 51525401), Dalian Support Plan for Innovation of High-level Talents (Top and Leading Talents, 2015R013), the Fundamental Research Funds for the Central Universities, and the National Key Research and Development Program of China (No. 2016YB0701203).

References

1. J.W. Yeh, S.K. Chen, S.J. Lin, J.Y. Gan, T.S. Chin, T.T. Shun, C.H. Tsau, S.Y. Chang, Nanostructured high-entropy alloys with multiple principal elements: novel alloy design concepts and outcomes. Adv. Eng. Mater. **6**, 299–303 (2004)
2. B. Cantor, I.T.H. Chang, P. Knight, A.J.B. Vincent, Microstructural development in equiatomic multicomponent alloys. Mater. Sci. Eng. A **375–377**, 213–218 (2004)
3. J.W. Yeh, Y.L. Chen, S.J. Lin, S.K. Chen, High-entropy alloys—a new era of exploitation. Mater. Sci. Forum **560**, 1–9 (2007)
4. Z. Wang, S. Guo, C.T. Liu, Phase selection in high-entropy alloys: from nonequilibrium to equilibrium. JOM **66**, 1966–1972 (2014)
5. Z.S. Nong, J.C. Zhu, R.D. Zhao, Prediction of structure and elastic properties of AlCrFeNiTi system high entropy alloys. Intermetallics **86**, 134–146 (2017)
6. W. Huo, H. Zhou, F. Fang, X. Hu, Z. Xie, J. Jiang, Strain-rate effect upon the tensile behavior of CoCrFeNi high-entropy alloys. Mater. Sci. Eng. A **689**, 366–369 (2017)
7. H.W. Yao, J.W. Qiao, J.A. Hawk, H.F. Zhou, M.W. Chen, M.C. Gao, Mechanical properties of refractory high-entropy alloys: experiments and modeling. J. Alloy. Compd. **696**, 1139–1150 (2017)
8. Y.Z. Shi, B. Yang, P.K. Liaw, Corrosion-resistant high-entropy alloys: a review. Metals. **7**, 43 (2017)
9. C.Y. Hsu, T.S. Sheu, J.W. Yeh, S.K. Chen, Effect of iron content on wear behavior of $AlCoCrFe_xMo_{0.5}Ni$ high-entropy alloys. Wear **268**, 653–659 (2010)
10. Z. Wu, H. Bei, F. Otto, G.M. Pharr, E.P. George, Recovery, recrystallization, grain growth and phase stability of a family of FCC-structured multi-principal element equiatomic solid solution alloys. Intermetallics **46**, 131–140 (2014)
11. T.T. Shun, L.Y. Chang, M.H. Shiu, Microstructures and mechanical properties of multiprincipal component $CoCrFeNiTi_x$ alloys. Mater. Sci. Eng. A **556**, 170–174 (2012)
12. T.T. Shun, L.Y. Chang, M.H. Shiu, Microstructure and mechanical properties of multiprincipal component $CoCrFeNiMo_x$ alloys. Mater. Charact. **70**, 63–67 (2012)
13. H. Jiang, L. Jiang, D.X. Qiao, Y.P. Lu, T.M. Wang, Z.Q. Cao, T.J. Li, Effect of Niobium on microstructure and properties of the $CoCrFeNb_xNi$ high entropy alloys. J. Mater. Sci. Technol. (2016)
14. L. Jiang, Y.P. Lu, H. Jiang, T.M. Wang, B.N. Wei, Z.Q. Cao, T.J. Li, Formation rules of single phase solid solution in high entropy alloys. Mater. Sci. Technol. 1–5 (2016)
15. Y. Dong, Y.P. Lu, L. Jiang, T.M. Wang, T.J. Li, Effects of electro-negativity on the stability of topologically close-packed phase in high entropy alloys. Intermetallics **52**, 105–109 (2014)
16. H. Jiang, K.M. Han, D.X. Qiao, Y.P. Lu, Z.Q. Cao, T.J. Li, Effects of Ta addition on the microstructures and mechanical properties of CoCrFeNi high entropy alloy. Mater. Chem. Phys. 1–6 (2017)

Effects of La and Be on Glass-Forming Ability of $Al_{84}Co_8Y_8$ Metallic Glass

Qiang Yang, Yonghao Yu, Zhiping Sun and Zhiming Wang

Abstract The influences of Lanthanum (La) and Beryllium (Be) on glass forming ability (GFA) of amorphous $Al_{84}Co_8Y_8$ were studied through replacing Y with 2, 4, 6 and 8 at.% La and Be, respectively. To directly compare the GFA of $Al_{84}Co_8Y_8$ with different additions of La and Be, the melt-quenched alloys were fabricated by melt-spinning method under different rotating speeds. The amorphous nature of melt-quenched alloys was investigated by X-ray diffraction (XRD) and differential scanning calorimeter (DSC). The results indicate that certain amount of La can promote the GFA of $Al_{84}Co_8Y_8$, and the best composition is $Al_{84}Co_8Y_4La_4$. However, it is found that Be has not distinct influence on the GFA of $Al_{84}Co_8Y_8$. In addition, the investigation on thermal stability of amorphous $Al_{84}Co_8Y_{8-x}La_x$ alloys shows that the thermal stability of amorphous $Al_{84}Co_8Y_8$ alloy can be improved by La addition with a peak at the composition $Al_{84}Co_8Y_6La_2$. It is also found that the crystallization behavior of amorphous $Al_{84}Co_8Y_8$ alloy will be changed by the addition of La, in the form of transforming from two-step to three-step crystallization process.

Keywords Glass forming ability · Amorphous $Al_{84}Co_8Y_8$ · Lanthanum Beryllium

Q. Yang · Y. Yu · Z. Sun · Z. Wang (✉)
School of Mechanical and Automotive Engineering,
Qilu University of Technology, Jinan 250353, China
e-mail: zhi820426@163.com

Q. Yang
e-mail: qlgdnb@163.com

Y. Yu
e-mail: 1692733759@qq.com

Z. Sun
e-mail: 18353108673@163.com

Y. Han (ed.), *Advanced Functional Materials*,
https://doi.org/10.1007/978-981-13-0110-0_29

Introduction

Aluminum-based metallic glasses have drawn more and more attentions for their high hardness, high strength, and superior corrosion resistance [1–5]. It could be traced back to 1988 when Aluminum-based amorphous alloys were obtained by rapid solidification [6–8]. Aluminum alloy with amorphous state is better than its normal crystalline state in terms of mechanical properties. Notably, Inoue has summed up three empirical rules for amorphous formation in the multi-component alloys, which indicates that the alloys satisfying the three empirical rules have special atomic configuration [9].

In present, the practical application of aluminum-based glass is limited by its weak GFA [10]. Therefore, the enhancement of GFA of aluminum-based glass is still the primary task to be done. Al-TM-RE (TM: transition metal, RE: rare earth metal) is the most popular system of Al-based amorphous. Many researches have been done on Al-Ni system [11, 12], but few reports on Al-Co-Y system. However, Co significantly improves the strength of Al-based amorphous without decreasing its ductility [13], and Y can enhance the GFA. Therefore, Al-Co-Y system is of interesting and needs further investigation.

Generally, the forming ability of amorphous alloy can be improved by increasing components of the alloy. This is called the "confusion principle" [14], that is, the introduction of new components can further improve the glass forming ability (GFA). In the present work, the effects of element Lanthanum with larger atomic radius and Beryllium with smaller atomic radius on the GFA of Al-Co-Y metallic glass alloy were studied in the viewpoint of atomic packing efficiency. Meanwhile, the influences of La on thermal stability and crystallization behavior of Al-Co-Y metallic glass alloy were also investigated.

Experiments

Alloys were pre-alloyed by arc-melting the mixture of pure ingots under Ar gas protection. The melt-quenched ribbons were fabricated from the pre-alloyed samples by single-roller melt-spinning method under Ar gas protection. To compare the glass forming ability, different rotating speeds of copper roller (80.59, 34.54, 23.03, 18.42 m/s in linear velocity) were adopted in the melt-spinning process. The amorphous nature of melt-quenched ribbons was investigated by X-ray diffraction (D/MAX-RA with Cu-Kα radiation). Thermal stability and crystallization behavior of the ribbons were performed using a Netzsch DSC404 calorimeter at a heating rate of 10 K/min.

Results and Discussion

Effect of La And Be on GFA of $Al_{84}Co_8Y_8$. Figure 1 shows the XRD patterns of as-quenched $Al_{84}Co_8Y_{8-x}La_x$ (x = 0, 2, 4, 6, 8) ribbons quenched at the cooling rates of (a) 80.59 m/s, (b) 34.54 m/s, (c) 23.03 m/s, (d) 18.42 m/s. Typical broad peaks appear at 2θ between 30–45° and no distinct diffraction peaks corresponding to a crystalline phase are observed for all the ribbons quenched at the rotating speed of 80.59 m/s, indicating fully amorphous structures are obtained. With the decreasing of rotating speed of copper roller to 23.03 m/s, crystalline phase firstly appears in ribbons with x = 6 and 8. The crystalline phase in ribbon with x = 6 is α-Al. However, in ribbon with x = 8, not only α-Al but also Al_4La phase is found. Only when the rotating speed decreases to 18.42 m/s, do the ribbons with x = 0, 2 and 4 appear crystalline phase. In these ribbons, α-Al, Al_9Co_2, Al_5Co_2 and some unknown phases are found. Comparing with the intensity and numbers of diffraction peaks in Fig. 1d, it can be known that the ribbon with x = 4 has the best GFA. From the results above, it is known that the GFA of $Al_{84}Co_8Y_8$ can be enhanced by partial substitution of La for Y. However, large amount of La leads to the decrease of GFA of $Al_{84}Co_8Y_8$, such as when x > 6.

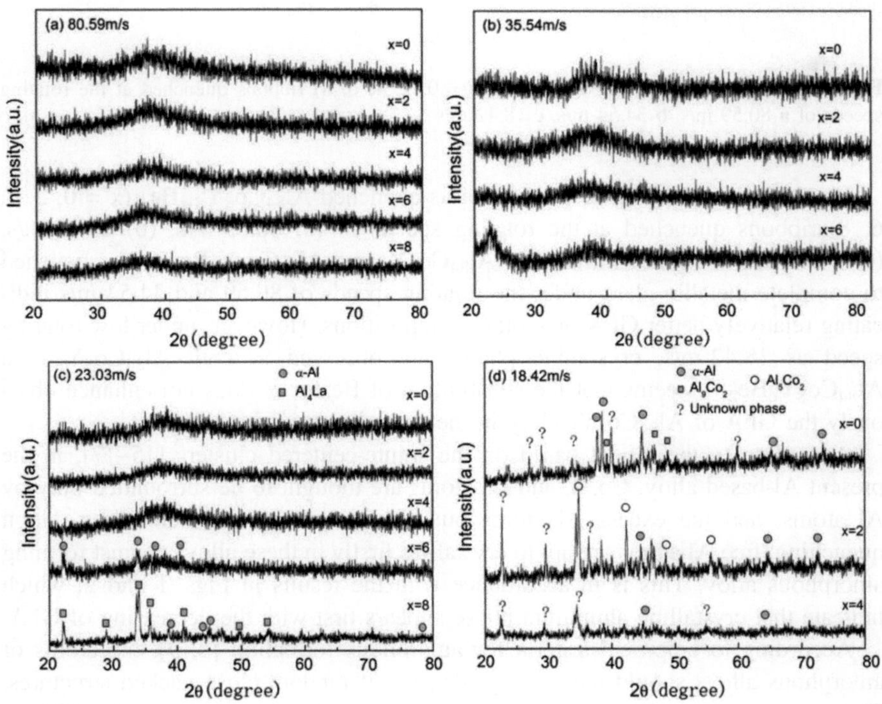

Fig. 1 XRD patterns of $Al_{84}Co_8Y_{8-x}La_x$ (x = 0, 2, 4, 6, 8) ribbons quenched at the rotating speeds of **a** 80.59 m/s, **b** 34.54 m/s, **c** 23.03 m/s, **d** 18.42 m/s

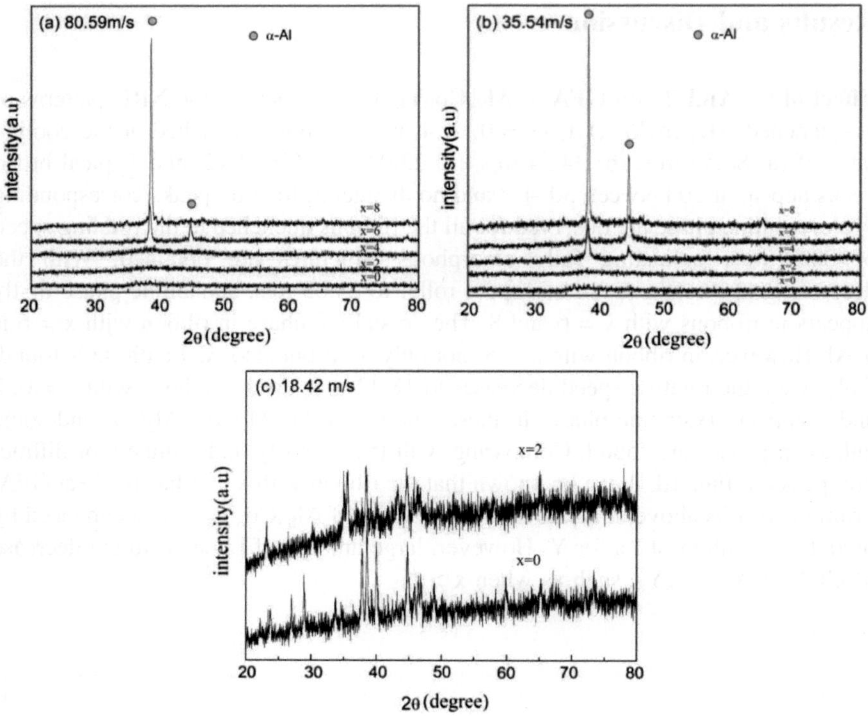

Fig. 2 XRD patterns of $Al_{84}Co_8Y_{8-x}Be_x$ (x = 0, 2, 4, 6, 8) ribbons quenched at the rotating speeds of **a** 80.59 m/s, **b** 34.54 m/s, **c** 18.42 m/s

Figure 2 shows the XRD patterns of as-quenched $Al_{84}Co_8Y_{8-x}Be_x$ (x = 0, 2, 4, 6, 8) ribbons quenched at the rotating speeds of (a) 80.59 m/s, (b) 23.03 m/s, (c) 18.42 m/s. It is shown that only $Al_{84}Co_8Y_8$ and $Al_{84}Co_8Y_6Be_2$ were quenched to complete metallic glass under the rotating speeds of 80.59 and 34.54 m/s indicating relatively better GFA than other compositions. However, under low rotating speed as 18.42 m/s, crystalline phases are observed in both $Al_{84}Co_8Y_8$ and $Al_{84}Co_8Y_6Be_2$. It seems that the substitution of Be for Y does not enhance obviously the GFA of $Al_{84}Co_8Y_8$ alloy in the present composition range.

According to the model based on the solute-centered clusters [15–17], in the present Al-based alloy, Co, Y and La atoms are thought to be surrounded only by Al atoms, and the excess Al atoms out of clusters are considered free. Upon quenching, free Al atoms are apt to crystallize firstly in these alloys against forming amorphous alloy. This is in accordance with the results in Figs. 1 and 2, which indicate that crystalline aluminum phase appears first with the decreasing of GFA.

According to empirical criteria for amorphous formation [1, 9], the atoms of amorphous alloys should have a high density of random close packed structures. These structures can depress the form of crystalline phase and improve the GFA by increasing the liquid-solid interfacial energy and viscosity which hinders the long range diffusion of alloy elements. Efficient atomic packing (EAP) is a fundamental consideration in investigating the GFA. Referring to the efficient atomic packing model proposed by Miracle [15–17], the theoretical coordination number of a solute-centered cluster (N^T) has a close relationship with the atomic radius ratio of the solute to solvent elements (R), as shown below [18]:

$$N^T = \begin{cases} \dfrac{4\pi}{6\arccos\left\{\sin\left(\frac{\pi}{3}\right)\left[1-\frac{1}{(R+1)^2}\right]^{\frac{1}{2}}\right\}-\pi} & for\ 0.225 \le R < 0.414 \\ \dfrac{4\pi}{8\arccos\left\{\sin\left(\frac{\pi}{4}\right)\left[1-\frac{1}{(R+1)^2}\right]^{\frac{1}{2}}\right\}-2\pi} & for\ 0.414 \le R < 0.902 \\ \dfrac{4\pi}{10\arccos\left\{\sin\left(\frac{\pi}{5}\right)\left[1-\frac{1}{(R+1)^2}\right]^{\frac{1}{2}}\right\}-3\pi} & for\ R \ge 0.902 \end{cases} \quad (1)$$

The packing efficiency (η) of solute centered clusters can be obtained by a simple representation as follows [19, 20]:

$$\eta = \begin{cases} N/N^T & for\ N < N^T \\ r_{solute}/\left(R_N^* \times r_{Al}\right) & for\ N > N^T \end{cases} \quad (2)$$

where N is the actual coordination number of the given solute element, R_N^* is the specific value of atomic radius ratio of the solute to solvent elements derived from Eq. 1 by letting $N^T = N$. r_{solute} and r_{Al} are the atomic radii of solute and aluminum, respectively.

Based on the above model, the packing efficiencies of various solute centered clusters in present study are calculated and listed in Table 1. It shows that, among Co, Y, La and Be centered clusters, La centered cluster has the highest atomic packing efficiency, however Be centered clusters has the lowest. That is to say, the addition of large atomic La may reduce the number of free Al in present $Al_{84}Co_8Y_{8-x}La_x$ alloy, and consequently promotes its GFA. However, Be do an opposite effect. It should be note that too much La as x > 4 will decrease the GFA, and this maybe is attributed to the change of configuration of La centered cluster. Such as that in Fig. 1c, when x = 8, not only α-Al but also Al_4La precipitated in the quenched ribbons.

Table 1 Atomic radius (r), theoretical coordination number N^T around solute elements, most possible coordination number N, efficient atomic packing (EAP) clusters and atomic packing efficiency η

	r (nm)	N^T	N	EAP cluster	η (%)
Al	0.143	–	–	–	–
Co	0.125	10.88	11	$Al_{11}Co$	98.9
Y	0.181	17.28	17	$Al_{17}Y$	98.4
La	0.187	17.99	18	$Al_{18}La$	99.7
Be	0.111	9.741	10	$Al_{10}Be$	97.2

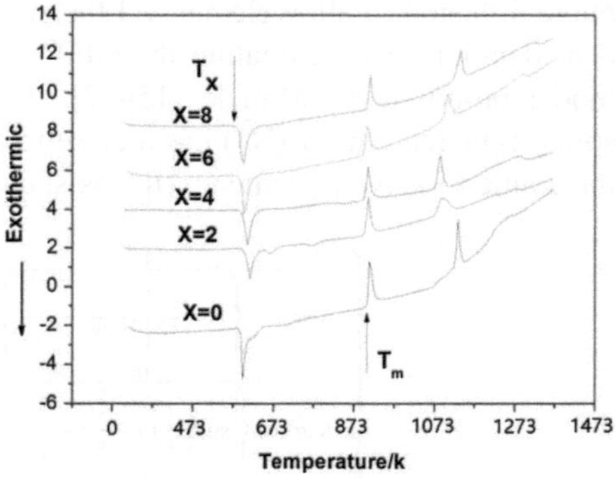

Fig. 3 Differential scanning calorimetry curves of melt-spun $Al_{84}Co_8Y_{8-X}La_x$ (x = 0, 2, 4, 6, 8) ribbons at a heating rate of 10 K/min

Effect of La on Thermal Stability and Crystallization Behavior of Amorphous $Al_{84}Co_8Y_8$. Figure 3 shows the differential scanning calorimetric (DSC) curves of $Al_{84}Co_8Y_{8-X}La_x$ (x = 0, 2, 4, 6, 8) amorphous alloys at a heating rate of 10 K/min. The onset crystallization temperature (T_x) and the onset melting temperature (T_m) are marked in the DSC curves by arrows. There is no distinct glass transition temperature (T_g) before crystallization of these amorphous alloys. This may be attributed to the small fraction of crystallites formed initially during the rapid cooling process, which could influence or cover T_g. The similar phenomenon was also found in other metallic glass forming systems such as Al-Fe-La [21]. The related parameters on curves are listed in Table 2.

From Table 2, it can be seen that $Al_{84}Co_8Y_6La_2$ has the highest T_x, indicating the best thermal stability of amorphous. It is considered to be attributed the formation of La-centered clusters. These clusters reduce the amount of free Al atoms and alternate permutation, resulting in the suppression of diffusion rate of Al atoms and the free energy of the system. Moreover, as is seen in Table 2, $Al_{84}Co_8Y_6La_2$ amorphous alloy has the lowest value of ΔT_m as 302.4 K and highest value of T_{rx} as 0.6642. Larger T_x/T_m and lower ΔT_m in some amorphous alloy systems usually correspond to better GFA [1, 12]. However, in present work, $Al_{84}Co_8Y_6La_2$ has better GFA than other compositions except $Al_{84}Co_8Y_6La_4$.

It also can be seen from Fig. 3 that the crystallization of amorphous $Al_{84}Co_8Y_8$ is a two-step process (T_{x1} = 569 K, T_{x2} = 605 K), however, it is a three-step

Table 2 The related parameters of the $Al_{84}Co_8Y_{8-X}La_x$ (x = 0, 2, 4, 6, 8) amorphous alloys

Alloys	T_x (K)	T_m (K)	$T_{rx} = T_x/T_m$	$\Delta T_m = T_m - T_x$ (K)
$Al_{84}Co_8Y_8$	592.3	908.1	0.6522	315.8
$Al_{84}Co_8Y_6La_2$	598.3	900.7	0.6642	302.4
$Al_{84}Co_8Y_4La_4$	597.5	900.5	0.6635	303
$Al_{84}Co_8Y_2La_6$	593.7	899.4	0.6601	305.7
$Al_{84}Co_8La_8$	586.9	906.7	0.6473	319.8

process (T_{x1} = 578 K, T_{x2} = 623 K, T_{x3} = 706 K) for amorphous $Al_{84}Co_8Y_6La_2$. In other words, the addition of La changes the crystallization behavior of $Al_{84}Co_8Y_8$ amorphous alloy in the form of transforming from two-step to three-step crystallization process.

Summary

The influences of Lanthanum (La) and Beryllium (Be) on glass forming ability (GFA) of amorphous Al84Co8Y8 were studied through replacing Y with 2, 4, 6, 8at.% La and Be, respectively. The results indicate that appropriate amount of La can promote the GFA of Al84Co8Y8, and the best composition is Al84Co8Y4La4. However, it is found that Be has not distinct influence on the GFA of Al84Co8Y8. This phenomenon is attributed to that the addition of La can effectively improve the liquid bulk density coefficient, but Be does an opposite effect.

The thermal stability of amorphous Al84Co8Y8 alloy can be improved by La addition with a peak at the composition Al84Co8Y6La2, meanwhile, the crystallization behavior of this amorphous is changed by the addition of La in the form of transforming from two-step crystallization process to three-step crystallization process.

Acknowledgements This work was supported by the National Natural Science Foundation of China (grant number 51601102); Natural Science Foundation of Shandong Province (grant number ZR2014EEQ012).

References

1. A. Inoue, Amorphous, nanoquasicrystalline and nanocrystalline alloys in Al-based systems. Prog. Mater. Sci. **43**, 365–520 (1998)
2. Y. He, S.J. Poon, G.J. Shiflet, Synthesis and properties of metallic glasses that contain aluminum. Science **241**, 1640–1642 (1988)
3. J.H. Perepezko, Nucleation-controlled reaction sand metastable structures. Prog. Mater. Sci. **49**, 263–284 (2004)
4. B.J. Yang, J.H. Yao, Y.S. Chao, J.Q. Wang, E. Ma, Developing aluminum-based bulk metallic glasses. Phil. Mag. **90**, 3215–3231 (2010)
5. B.J. Yang, J.H. Yao, J. Zhang, H.W. Yang, J.Q. Wang, E. Ma, Al-rich bulk metallic glass with plasticity and high specific strength. Scripta Mater. **61**, 423–426 (2009)
6. A. Inoue, K. Ohtera, A.P. Tsai, T. Masumoto, Aluminum-based amorphous alloys with tensile strength above 980 MPa (100 kg/mm^2). Jpn. J. Appl. Phys. **27**, L479–L482 (1988)
7. G.J. Shiflet, Y. He, S.J. Poon, Mechanical properties of aluminum-rich Al-Fe-Gd metallic glass. Scr. Metall. **22**, 1661–1664 (1988)
8. Y. He, H. Chen, G.J. Shiflet, S.J. Poon, On the structural nature of aluminum-based metallic glasses. Philos. Mag. Lett. **61**, 297–303 (1990)
9. A. Inoue, Stabilization of metallic supercooled liquid and bulk amorphous alloys. Acta Mater. **48**, 279–306 (2000)

10. F. Li, X.J. Liu, H.Y. Hou, G. Chen, G.L. Chen, Structural origin underlying poor glass forming ability of Al metallic glass. J. Appl. Phys. **110**, 013519 (2011)
11. Y. He, G.J. Shiflet, S.J. Poon, Synthesis and properties of aluminum-based metallic glasses containing rare earths. J. Alloy. Compd. **207–208**, 349–354 (1994)
12. A. Inoue, N. Matsumoto, T. Masumoto, Al-Ni-Y-Co amorphous alloys with high mechanical strengths, wide supercooled liquid region and large glass forming capacity. Mater. Trans. JIM **31**, 493–500 (1990)
13. D.V. Louzguine, A. Inoue, Influence of a supercooled liquid on crystallization behaviour of Al–Y–Ni–Co metallic glass. Mater. Lett. **54**, 75–80 (2002)
14. A.L. Greer, Confusion by design. Nature **366**, 303–304 (1993)
15. D.B. Miracle, A structural model for metallic glasses. Nat. Mater. **3**, 697–702 (2004)
16. D.B. Miracle, The efficient cluster packing model—an atomic structural model for metallic glasses. Acta Mater. **54**, 4317–4336 (2006)
17. D.B. Miracle, O.N. Senkov, A geometric model for atomic configurations in amorphous Al alloys. J. Non-Cryst. Solids **319**, 174–191 (2003)
18. M. Widom, I. Al-Lehyani, J.A. Moriarty, First-principles interatomic potentials for transition-metal aluminides. III. Extension to ternary phase diagrams. Phys. Rev. B **62**, 3648 (2000)
19. Z. Zhang, X.Z. Xiong, J.J. Yi, J.F. Li, Effects of substitution of La by other rare-earth elements on the glass forming ability of $Al_{86}Ni_9La_5$alloy. J. Non-Cryst. Solids **369**, 1–4 (2013)
20. J.J. Yi, X.Z. Xiong, A. Inoue, L.T. Kong, J.F. Li, Glass forming ability of Al–Ni–La alloys with Si addition. J. Alloy. Compd. **650**, 578–583 (2015)
21. Guihua Li, Xiufang Bian, Kaikai Song, Jing Guo, Xuelian Li, Caidong Wang, Effect of Si addition on glass forming ability and thermal stability of Al–Fe–La alloys. J. Alloy. Compd. **471**, L47–L50 (2009)

Effect of Si and Y Addition on Glass Transition Kinetics of ZrCu Based Bulk Metallic Glass

Baoqing Zhang, Xinhang Lu, Kun Liu, Zhi Zhang and Guohua Cao

Abstract $Zr_{47}Cu_{44}Al_9$, $(Zr_{47}Cu_{44}Al_9)_{98.5}Si_{1.5}$ and $[(Zr_{47}Cu_{44}Al_9)_{98.5}Si_{1.5}]_{97.5}Y_{2.5}$ bulk metallic glass (BMG) were prepared by copper mould casting and studied using X-ray diffraction and differential scanning calorimetry. The influence of addition elements on the glass transition kinetics of the alloys were analyzed by the VFT equation. It is revealed that the glass transition behaviours of ZrCu based BMG were changed obviously by Si and Y addition, and the sensitivity of the glass transition temperature (T_g) to the heating rate is significantly increased. The fragility parameters m were reduced from 35.4 of $Zr_{47}Cu_{44}Al_9$ to 31.5 of $(Zr_{47}Cu_{44}Al_9)_{98.5}Si_{1.5}$, and that of alloy reached 30.7 after adding Y. The viscosity versus temperature relationship of alloy changes from characteristic of brittle fluid to strong fluid. The normalized glass transition temperature and heating rate show a clear VFT relationship. With the addition of Si and Y elements, the melt viscosity is close to that of the strong fluid.

Keywords Bulk metallic glass · Element addition · Glass transition kinetics VFT relation · Fragility parameter

Introduction

Bulk metallic glass (BMG) has attracted more attention for their scientific purposes and technological applications due to high strength, large plasticity, low elastic modulus and high corrosion resistance [1–3]. The development of BMG with excellent glass forming ability (GFA) has been a long-standing issue. An enormous

B. Zhang · X. Lu · Z. Zhang
School of Materials Science and Engineering,
Henan Polytechnic University, Jiaozuo 454000, China

K. Liu · G. Cao (✉)
School of Physics and Electronic Information Engineering,
Henan Polytechnic University, Jiaozuo 454000, China
e-mail: ghcao@hpu.edu.cn

Y. Han (ed.), *Advanced Functional Materials*,
https://doi.org/10.1007/978-981-13-0110-0_30

amount of research has centered on preparing large BMG, and various theories have been proposed, such as confusion principle [4] and three empirical rules [5]. Introducing new elements to the alloy can result in more complex structure and enhance the GFA, which is favor to forming metallic glass. Xu et al. prepared Cu-based BMG rods with 10 mm in diameter by adding of 5 at.% yttrium [6]. Liang et al. found that NiTi-based alloys with silicon addition display excellent GFA [7]. The researches reveal that the GFA of an alloy is closely associated with its liquid behavior [8–10]. The viscosity of glass-forming systems undergoes great changes near the glass transition. Therefore, it would be investigating to clarify if the influence of the microalloying is related to the liquid behavior for BMG.

ZrCu based BMG possesses high tensile strength, good plasticity, free from toxic elements, without noble metals, possess strong GFA and wide range of amorphous forming components, so it is regarded as a promising structural material [11–13]. A series of ZrCu based BMG were prepared by micro-alloying method, and multicomponent alloys were obtained. In this paper, BMGs of $Zr_{47}Cu_{44}Al_9$, $(Zr_{47}Cu_{44}Al_9)_{98.5}Si_{1.5}$, and $[(Zr_{47}Cu_{44}Al_9)_{98.5}Si_{1.5}]_{97.5}Y_{2.5}$ has been prepared. After addition alloying elements, the critical size of BMG has been increased from 4 mm to over 8 mm [14–16]. In order to understand the influence of alloying elements on the amorphous forming ability, the glass transition kinetics of different alloys was investigated by means of thermal analysis.

Experimental Details

The purity of starting materials was better than 99.9%. $Zr_{47}Cu_{44}Al_9$, $(Zr_{47}Cu_{44}Al_9)_{98.5}Si_{1.5}$, and $[(Zr_{47}Cu_{44}Al_9)_{98.5}Si_{1.5}]_{97.5}Y_{2.5}$ alloy samples with a diameter of 2–8 mm were prepared by copper mould casting according to the component ratio. In order to ensure uniform composition, each sample is smelted 3–4 times under the condition of argon protection.

The structure of the samples were measured by D/max-2500/PC X ray diffraction (XRD), and the diffract meter uses Cu target and K alpha radiation. Thermal analysis was characterized by STA449C differential scanning calorimeter (DSC) under the argon atmosphere. Heating rates is in the range of 1–40 K/min. Thermal analysis samples were intercepted from the diameter 2 mm samples.

Results and Discussion

Figure 1 displays the XRD patterns of $Zr_{47}Cu_{44}Al_9$, $(Zr_{47}Cu_{44}Al_9)_{98.5}Si_{1.5}$ and $[(Zr_{47}Cu_{44}Al_9)_{98.5}Si_{1.5}]_{97.5}Y_{2.5}$ samples related with their critical diameter. Each of these alloys possesses a single broad diffraction peaks without an obvious crystalline peak. It have been found that the maximum size of amorphous sample is

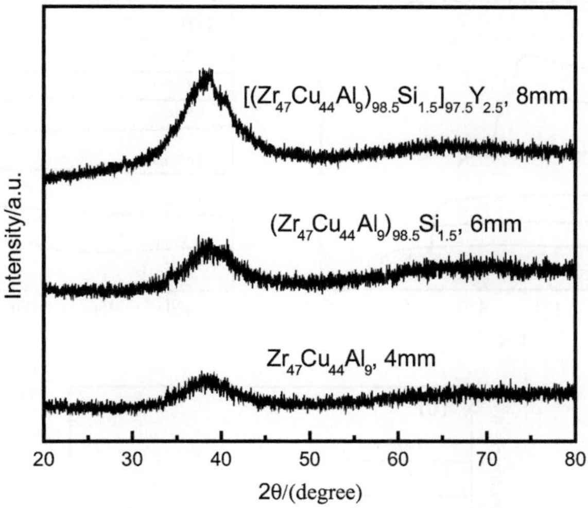

Fig. 1 XRD patterns of the $Zr_{47}Cu_{44}Al_9$, $(Zr_{47}Cu_{44}Al_9)_{98.5}Si_{1.5}$ and $[(Zr_{47}Cu_{44}Al_9)_{98.5}Si_{1.5}]_{97.5}Y_{2.5}$ samples related with their critical diameter

4 mm for $Zr_{47}Cu_{44}Al_9$, 6 mm for $(Zr_{47}Cu_{44}Al_9)_{98.5}Si_{1.5}$, and 8 mm for $[(Zr_{47}Cu_{44}Al_9)_{98.5}Si_{1.5}]_{97.5}Y_{2.5}$, respectively, indicating the addition of Si and Y can obviously improve the amorphous forming ability of the alloy.

Figure 2 shows the DSC curves of $Zr_{47}Cu_{44}Al_9$, $(Zr_{47}Cu_{44}Al_9)_{98.5}Si_{1.5}$, and $[(Zr_{47}Cu_{44}Al_9)_{98.5}Si_{1.5}]_{97.5}Y_{2.5}$ glassy alloys obtained at various heating rates of 1, 2, 5, 10, 20 and 40 K/min. It is clearly seen that the glass transition temperature (T_g) moves gradually to the high temperature region as the heating rate increases. With the increase of heating rate, the relaxation time of amorphous alloy at various temperatures according to the corresponding structural state is obviously insufficient, and the hysteresis becomes increasingly evident.

The remarkable characteristic of glass transition is that the kinetic properties of supercooled liquid vary with the temperature. Viscosity η is one of the important properties of liquid transport dynamics. As the temperature decreases, η increases dramatically. Usually, when the viscosity is $10^{13.5}$ poise, the liquid is converted to glass, and the corresponding temperature is called the glass transition temperature T_g [17]. The greater the viscosity of the liquid, the lower the nucleation rate and growth rate in cooling process, the stronger the tendency to form glass. The viscosity of a liquid depends on the nature of the liquid, and is also a function of temperature. The viscosity can be directly studied by the method of direct measurement when the liquid is cooled. Moreover, for many amorphous alloy systems, the characteristic relation between the liquid viscosity and the temperature can be reflected by investigating the glass transition during heating or cooling condition. Therefore, the viscosity of liquid is determined by the fitting relation between the T_g and the heating rate [18].

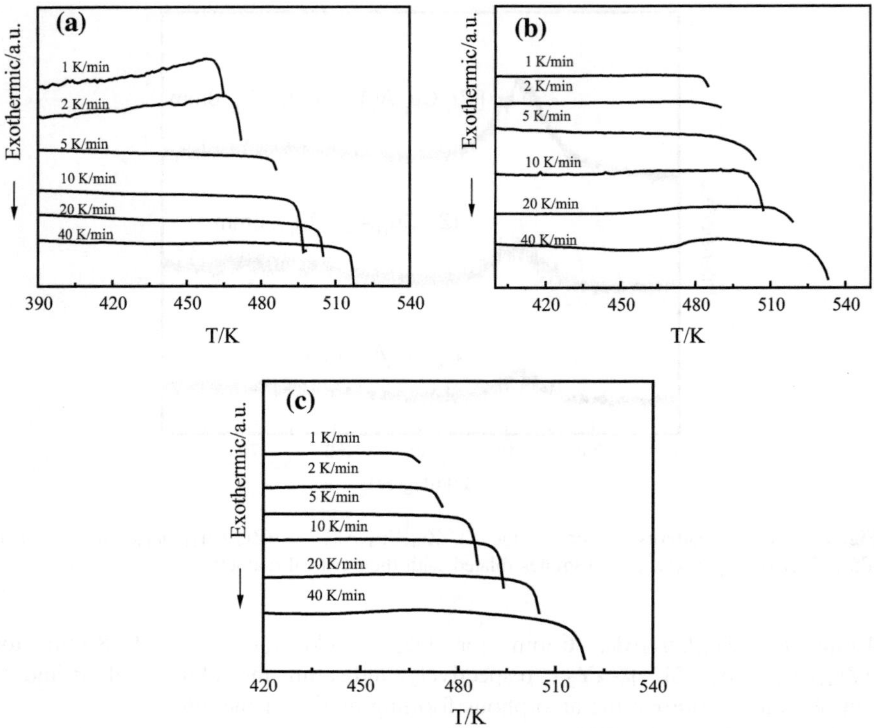

Fig. 2 DSC curves of $Zr_{47}Cu_{44}Al_9$ (**a**), $(Zr_{47}Cu_{44}Al_9)_{98.5}Si_{1.5}$ (**b**), and $[(Zr_{47}Cu_{44}Al_9)_{98.5}Si_{1.5}]_{97.5}Y_{2.5}$ (**c**) at various heating rates

Bruning and Samwer [19] studied the glass transformation behavior of some glass forming liquid in the larger cooling speed range. The relationship between the glass transition behavior and the cooling rate is subject to the Vogel-Fulcher-Tammann (VFT) equation [19, 20]:

$$\beta = B\,\exp[DT_0/(T - T_0)] \tag{1}$$

where β is heating or cooling rate, B has the dimension of a heating rate, D is strength parameter, and T_0 is VFT temperature. Under heating conditions, T_g is generally taken as the starting temperature of glass transition instead of the turning point temperature.

In order to deeply reveal the influence of the addition of trace elements on the liquid properties and GFA of base alloy, it is necessary to discuss the glass transition of BMG with the increase of heating rate. Therefore, the VFT relations of three amorphous alloys have been analyzed. Figure 3 displays the variations of T_g with the heating rates for $Zr_{47}Cu_{44}Al_9$, $(Zr_{47}Cu_{44}Al_9)_{98.5}Si_{1.5}$ and

Fig. 3 Variations of T_g with the heating rates for $Zr_{47}Cu_{44}Al_9$, $(Zr_{47}Cu_{44}Al_9)_{98.5}Si_{1.5}$, and $[(Zr_{47}Cu_{44}Al_9)_{98.5}Si_{1.5}]_{97.5}Y_{2.5}$

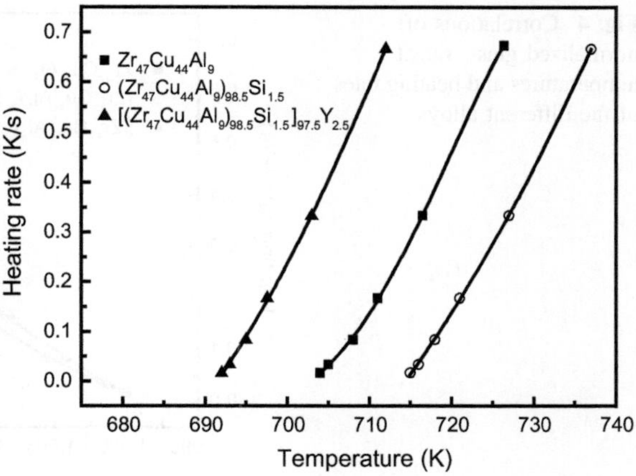

$[(Zr_{47}Cu_{44}Al_9)_{98.5}Si_{1.5}]_{97.5}Y_{2.5}$. Fitting of the measured data was implemented by the equation

$$\ln \beta = \ln B - DT_0/(T - T_0) \quad (2)$$

The adjustable parameters of such a fit are B, D and T_0. The calculated values are listed in Table 1.

From Table 1, it is clearly seen that the glass transition behavior of the ZrCuAl BMG after adding Si and Y has changed significantly, and the sensitivity of T_g to heating rate is obviously enhanced. The structure of supercooled liquid can be quantitatively assessed by the fragility index m. Basing the VFT fits, the m can be calculated as:

$$m = \frac{DT_0T}{\ln 10(T - T_0)^2} \quad (3)$$

The m [21, 22] of the alloy was obtained by taking the parameters of 20 K/min heating rate. The m values of $Zr_{47}Cu_{44}Al_9$, $(Zr_{47}Cu_{44}Al_9)_{98.5}Si_{1.5}$ and $[(Zr_{47}Cu_{44}Al_9)_{98.5}Si_{1.5}]_{97.5}Y_{2.5}$ are 35.4, 31.5 and 30.7, respectively. With the increase of element components, the fragility parameter of three alloys decreases. The viscosity versus temperature relationship of alloys indicates that the alloys change from characteristic of brittle fluid to strong fluid. The results also display that the Si and Y addition can increase the GFA of the base alloy, namely,

Table 1 VFT parameters of three alloys derived from the best fit of the experimental data

BMGs	ln*B*(K/s)	*D*	T_0 (K)
$Zr_{47}Cu_{44}Al_9$	2.29	0.141	687.4
$(Zr_{47}Cu_{44}Al_9)_{98.5}Si_{1.5}$	2.15	0.162	692.5
$[(Zr_{47}Cu_{44}Al_9)_{98.5}Si_{1.5}]_{97.5}Y_{2.5}$	2.94	0.251	662.4

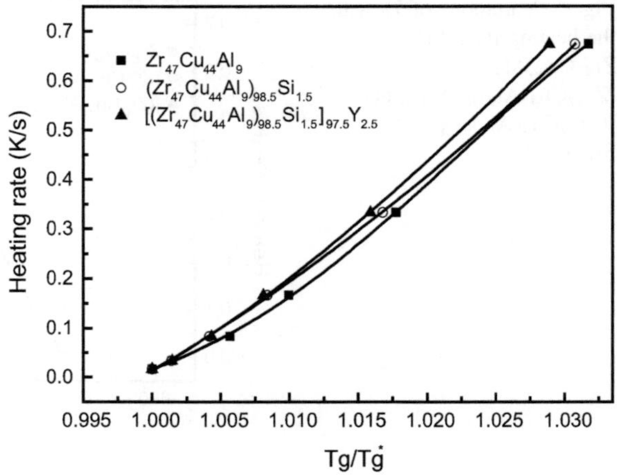

Fig. 4 Correlations of normalized glass transit temperatures and heating rates of the different alloys

obviously enhance the viscosity of the alloy. The viscosity data obtained in this paper are different from those obtained by direct measurement methods, which is due to the difference between the two research methods. This phenomenon is found in the other studies [18]. Although there are differences in data, the two methods can be well represented for the characterization of the VFT relations.

Figure 4 shows how the fragility of three alloys is deduced from the dependence of the heating rate on T_g normalized to T_g^*, where T_g^* is the glass transition temperature at the heating rate of 1 K/min. From the figure, the normalized glass transition temperature and heating rate show a clear VFT relationship. With the increase of heating rate, the dimensionless normalized glass transition temperature increases gradually. In addition, the dynamic characteristics of glass transition with increasing heating rate can also reflect the fragility of alloys [23]. The results indicate that the viscosity characteristics of the melt are closer to the strong fluid when the Si and Y elements are added to the alloys.

Conclusions

(1) For $Zr_{47}Cu_{44}Al_9$, $(Zr_{47}Cu_{44}Al_9)_{98.5}Si_{1.5}$, and $[(Zr_{47}Cu_{44}Al_9)_{98.5}Si_{1.5}]_{97.5}Y_{2.5}$, the critical diameter enhances gradually with the increase of adding element. The addition of Si and Y improves the GFA of ZrCuAl based BMG.

(2) The glass transition behaviours of ZrCu based BMG were changed obviously by Si and Y addition, and the sensitivity of the glass transition temperature to the heating rate is significantly increased. The fragility parameters were reduced from 35.4 of $Zr_{47}Cu_{44}Al_9$ to 31.5 of $(Zr_{47}Cu_{44}Al_9)_{98.5}Si_{1.5}$, and the fragility parameter of alloy reached 30.7 after adding Y.

(3) The normalized glass transition temperature and heating rate show a clear VFT relationship. With the addition of Si and Y, the viscosity characteristics of the melt are closer to the strong fluid.

Acknowledgements Financial supports from the National Natural Science Foundation of China (Grant Nos. 51571085 and 51301062), the Foundations of Henan Educational Committee under Grant No. 13B430019, and the Doctoral fund project of Henan Polytechnic University under Grant No. B2012-007.

References

1. W.L. Johnson, Bulk amorphous metal an emerging engineering material. JOM **54**, 40–43 (2002)
2. X.J. Liu, G.L. Chen, H.Y. Hou, X. Hui, K.F. Yao, Z.P. Lu, C.T. Liu, Atomistic mechanism for nanocrystallization of metallic glasses. Acta Mater. **56**, 2760–2769 (2008)
3. J. Das, M.B. Tang, K.B. Kim, R. Theissmann, F. Baier, W.H. Wang, J. Eckert, Work-hardenable ductile bulk metallic glass. Phys. Rev. Lett. **94**, 205501 (2005)
4. A.L. Greer, Confusion by design. Nature **366**, 303–304 (1993)
5. A. Inoue, Stabilization of metallic supercooled liquid and bulk amorphous alloys. Acta Mater. **48**, 279–306 (2000)
6. D. Xu, G. Duan, W.L. Johnson, Unusual glass-forming ability of bulk amorphous alloys based on ordinary metal copper. Phys. Rev. Lett. **92**, 245504 (2004)
7. W.Z. Liang, Z.L. Ning, F.F. Sun, J. Shen, Formation and compressive properties of a Ni-Ti-Zr-Al-Cu-Si amorphous alloy. Rare Met. Mater. Eng. **35**, 1441–1444 (2006)
8. I. Gallino, O. Gross, G.D. Fontana, Z. Evenson, R. Busch, On the kinetic and thermodynamic fragility of the $Pt_{60}Cu_{16}Co_2P_{22}$ and $Pt_{57.3}Cu_{14.6}$Ni5.3$P_{22.8}$ bulk metallic glasses. J. Alloy. Compd. **615**, 535–539 (2014)
9. R. Busch, The thermophysical properties of bulk metallic glass-forming liquids. JOM **52**, 39–42 (2000)
10. J.C. Qiao, J.M. Pelletier, Dynamic mechanical relaxation in bulk metallic glasses: a review. J. Mater. Sci. Technol. **30**, 523–545 (2014)
11. Y.C. Zhao, S.Z. Kou, H.L. Suo, R.J. Wang, Y.T. Ding, Overheating effects on thermal stability and mechanical properties of $Cu_{36}Zr_{48}Ag_8Al_8$ bulk metallic glass. Mater. Des. **31**, 1029–1032 (2010)
12. Y. Wu, H. Wang, H.H. Wu, Z.Y. Zhang, X.D. Hui, G.L. Chen, D. Ma, X.L. Wang, Z.P. Lu, Formation of Cu-Zr-Al bulk metallic glass composites with improved tensile properties. Acta Mater. **59**, 2928–2936 (2011)
13. S.Z. Kou, Y.C. Zhao, S.W. Xue, H. Suo, R. Wang, Effect of supercooling treatment on thermal stability and mechanical properties of $Cu_{45}Zr_{42}Al_8Ag_5$ bulk metallic glasses. Chin. J. Rare Met. **35**, 213–217 (2011)
14. B.Q. Zhang, C.Z. Fan, H. Tian, S.F. Shan, Z.J. Zhan, R.P. Liu, W.K. Wang, ZrCuAlSi bulk amorphous alloy formation and mechanical properties. Rare Met. Mat. Eng. **38**, 331–334 (2009)
15. Y. Zhang, J. Chen, G.L. Chen, X.J. Liu, Influence of yttrium addition on the glass forming ability in Cu-Zr-Al alloys. Mater. Sci. Eng. A **483**(484), 235–238 (2008)
16. B.Q. Zhang, X.Y. Zhang, G. Li, B.R. Sun, C.Z. Fan, Z.J. Zhan, R.P. Liu, W.K. Wang, Effect of yttrium addition on glass forming ability of ZrCuAlSi Alloy. Chin. Phys. Lett. **6**, 219–221 (2009)

17. P. Wen, Progress in the theoretical investigations of glass transitions. Physics **33**, 587–592 (2004)
18. Y.Z. Jia, Study on forming ability and properties of FeCoNbSiBCu alloy, Doctoral Dissertation of Harbin Institute of Technology, 2008, pp. 65–67
19. R. Brüning, K. Samwer, Glass transition on long time scales. J. Phys. Rev. B. **46**, 11318–11322 (1992)
20. N. Mitrovic, S. Roth, J. Eckert, Kinetics of the glass-transition and crystallization process of $Fe_{72-x}Nb_xAl_5Ga_2P_{11}C_6B_4$(x=0,2) metallic glasses. Appl. Phys. Lett. **8**, 2145–2147 (2001)
21. L.M. Martinez, C.A. Angell, A thermodynamic connection to the fragility of glass-forming liquids. Nature **410**, 663–667 (2001)
22. T. Hecksher, A.I. Nielsen, N.B. Olsen, J.C. Dyre, Little evidence for dynamic divergences in ultraviscous molecular liquids. Nat. Phys. **4**, 737–741 (2008)
23. R. Rusch, Y.J. Kin, W.L. Johnson, Thermodynamics and kinetics of the undercooled liquid and the glass transition of the $Zr_{41.2}Ti_{13.8}Cu_{12.5}Ni_{10.0}Be_{22.5}$ alloy. Appl. Phys. Lett. **77**, 4039–4043 (1995)

Effects of Pre-compression on the Microstructure and Corrosion Resistance of $Cu_{50}Zr_{50}$ Bulk Metallic Glass Matrix Composites

Sicheng Zhai, Wen Wang and Yan Wang

Abstract The effects of the pre-compression and cyclic compression treatment at room temperature on the microstructures, microhardness, and corrosion resistance in seawater solution of the $Cu_{50}Zr_{50}$ bulk metallic glass composites (BMGCs) have been investigated. Our work shows that pre-compression treatment suppressed the precipitation of B_2-CuZr and $Cu_{10}Zr_7$ nano-crystals from amorphous matrix and induced phase transformations, $B_2 - CuZr + Cu_{10}Zr_7 \rightarrow B'_{19} - CuZr$. The compressive fracture strength of the as-cast sample reaches up to 1566 MPa while preserving a certain degree of plasticity with a larger microhardness of 392 HV. Moreover, appropriate pre-compression and cyclic compression treatment can enhance the microhardness of the $Cu_{50}Zr_{50}$ BMGCs. The 5 cycles compression treatment under the 1273 MPa shows the enhanced corrosion resistance in the seawater solution, while the as-cast sample exhibits the excellent pitting corrosion resistance.

Keywords Bulk metallic glass composites · Pre-compression · Microhardness
Corrosion resistance

Introduction

Due to the disordered structures, bulk metallic glasses (BMGs) exhibit very different properties from crystalline materials, such as high strength, high hardness, large elastic strain limit, good corrosion resistance and good magnetic properties [1–4], attracting much attention in both academic research and industrial applications. However, the bottleneck issues that limit the engineering applications of BMGs are the room temperature brittleness and strain softening behaviors [5, 6]. Recently, BMG-matrix composites (BMGCs), especially those with in situ

S. Zhai · W. Wang · Y. Wang (✉)
School of Materials Science and Engineering, University of Jinan,
No. 336, West Road of Nan Xinzhuang, Jinan 250022, China
e-mail: mse_wangy@ujn.edu.cn

Y. Han (ed.), *Advanced Functional Materials*,
https://doi.org/10.1007/978-981-13-0110-0_31

crystalline phases, have been prepared to overcome some of the weaknesses of monolithic BMGs. In situ BMGCs included the nanocrystalline ductilization, dendrite or solid-solution ductilization and transformation-mediated ductilization [7]. Several types of BMGCs, such as Ti-based [8], CuZr-based [9, 10] and Ni-based [11] with excellent properties have been developed successfully by elements addition, heat treatment and pre-compressed treatment [12]. Minor addition of Ni (1 at.%) can change the glass forming ability (GFA) of the CuZr-based BMGCs, exhibiting the work-hardening ability [13]. The pre-compression treatment at room temperature for the rare-earth-based BMGs induced the precipitation of nano-crystals, which can significantly improve the thermal stability and the corrosion resistance. Moreover, cyclic compression process has been studied for Zr-based MGMCs, resulting in plastic softening, which could be attributed to the increased free volume induced by the cyclic compression [14].

In recent years, CuZr-based BMGCs have been widely studied due to the good glass forming ability, high thermal stability, excellent corrosion resistance and work-hardening ability. Compared with other types of BMGCs, the CuZr-based BMGCs shape-memory B_2-CuZr crystalline phases, which could undergo martensitic transformation during deformation and turn into the B19′ phase [15]. It has been reported that the precipitation of the B_2-CuZr crystalline phase in the glassy matrix in these composites could display pronounced work-hardening ability by controlling the size and distribution of the B_2-CuZr phases [16, 17].

In the present work, we choose different pre-compression stress and times of cyclic pre-compression on $Cu_{50}Zr_{50}$ BMGCs. Microstructure, thermal stability, mechanical properties and corrosion resistance in seawater solution for $Cu_{50}Zr_{50}$ BMGCs were investigated.

Experimental

A master alloy ingot of $Cu_{50}Zr_{50}$ was fabricated by arc melting mixtures of the composing element (99.9% purity) under a Ti-gettered argon atmosphere. The ingot was repeated remelted at least four times to ensure its compositional homogeneity. Cylindrical rods with 3 mm in diameter and 90 mm long were produced by copper-mold suction casting method. Microstructures of the as-cast and pre-compressed $Cu_{50}Zr_{50}$ BMGMCs were determined by X-ray diffraction (XRD, Rigaku D8 Advance) using Cu Kα radiation. The thermal stability of the tested samples were analyzed by differential scanning calorimetry (DSC, TGA/DSC 1, Mettler-Toledo) at a heating rate of 10 K/min. The compression experiment and pre-compression treatment were carried out by using an MTS Criterion testing machine. The cylindrical rods with a dimension of $\phi 3 \times 6$ mm were prepared for compression tests. It has been found that the as-cast $Cu_{50}Zr_{50}$ BMGMC failed when the stress reached 1566 MPa with a compression speed of 0.02 mm/min and a holding time of 60 s. Therefore, the stresses for the pre-compression treatment were chosen to be 990, 1273 and 1415 MPa, respectively. Subsequently, 3, 5 and 7 cycles of 1273 MPa pre-compressed sample were carried out. Microhardness of the $Cu_{50}Zr_{50}$ BMGMCs with the different pre-compression treatments was measured by a Vickers hardness tester (HV-10B) with a load of 1000 g and a dwell time of 15 s. The HV measurement for the each sample was repeated ten times in order to give the averaged values and error bars.

Potentiodynamic polarization measurements of the as-cast and pre-compressed $Cu_{50}Zr_{50}$ BMGMCs was carried out by the electrochemical workstation (LK2005A) in a seawater solution. The transverse cross sections of the tested samples should be grinded smoothly by the 2000-grit SiC papers, followed by ultrasonic cleaning. The seawater solution was prepared by artificial sea salt and distilled water with the mass ratio of 1:30 and the salt-ion concentration of about 3.3%, which is closer to the concentration of the salt in sea. The counter electrode and reference lelctrode were the platinum plate and Ag-AgCl (in saturated KCl), respectively. Prior to the potentiodynamic polarization curves, the samples were immersed in the fresh seawater solution for about 10 min when the open circuit potentials became almost steady. The measurements have been repeated three times at a potential sweep rate of 3.0 mV/s. The fracture surfaces of the $Cu_{50}Zr_{50}$ sample was observed by field emission scanning electron microscope (FESEM, QUANTA FEG 250, FEI).

Results and Discussion

Figure 1a displays the XRD patterns of the as-cast and pre-compressed $Cu_{50}Zr_{50}$ BMGMCs. The as-cast ϕ3 mm sample presents crystalline peaks corresponding to the B_2-CuZr, $Cu_{10}Zr_7$ and B_{19}'-CuZr phases on the top of bulk amorphous, which is consistent with the related reports [18, 19]. The $Cu_{10}Zr_7$ phase may be harmful to the mechanical properties and result in the catastrophically brittle fracture under uniaxial stress [20]. The curve of 990 MPa pre-compressed sample is similar to the as-cast. When the stress increases to 1273 MPa, The $Cu_{10}Zr_7$ phase slightly reduced, indicating that the appropriate pre-compression induced phase transformations, $\mathrm{B_2} - \mathrm{CuZr} + Cu_{10}\mathrm{Zr}_7 \rightarrow \mathrm{B}'_{19} - \mathrm{CuZr}$. Therefore, the cyclic compression test of the 1273 MPa pre-compressed sample was carried out and the XRD curves were shown in Fig. 1b. It was clearly seen that with the increasing of cycles, the precipitation of B_2-CuZr phase has been gradually reduced.

To understand the effect of pre-compression and cyclic compression on the phase structure evolution of the CuZr-based BMGCs, DSC analysis was carried out during continuously heating process. The DSC curves of as-cast and 1273 MPa pre-compressed $Cu_{50}Zr_{50}$ BMGCs are shown in Fig. 2a. One endothermic peak corresponding to the melt state are observed for each tested sample. It indicates that the melting point (T_m) and liquid point (T_l) becomes slightly reduced with increasing the pre-compression stress. It can be seen that the transformation in tested samples contains a glass transition process followed by an exothermic heat

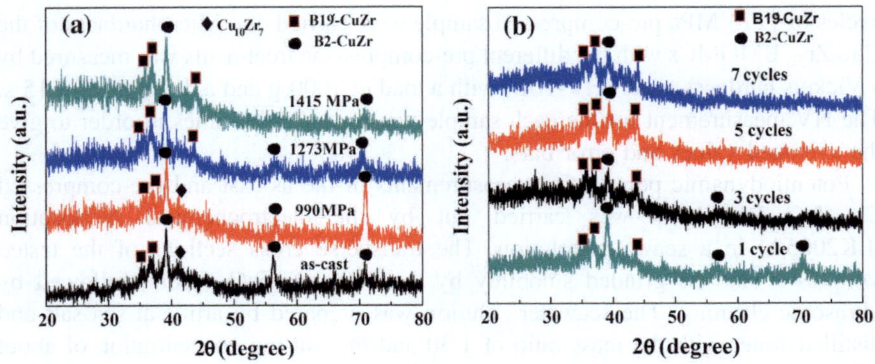

Fig. 1 XRD patterns of the $Cu_{50}Zr_{50}$ BMGCs under different pre-compression stress (**a**) and under the cyclic pre-compression of 1273 MPa (**b**)

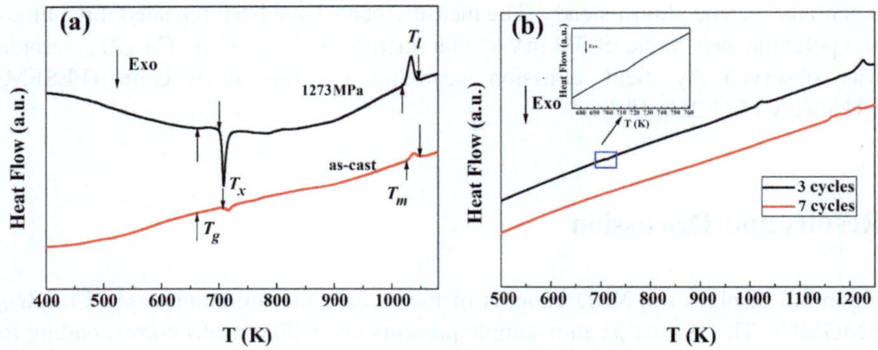

Fig. 2 DSC curves of the $Cu_{50}Zr_{50}$ BMGCs under different pre-compression stress (**a**) and under the cyclic pre-compression of 1273 MPa (**b**) at the heating rate of 10 K/min

characteristic for crystallization. The glass transition temperature (T_g), crystallization onset temperature (T_x), and the super-cooled liquid region ($\Delta T_x = T_x - T_g$) are obtained from DSC curves. As the stress increases from 0 to 1273 MPa, T_g is almost constant, but T_x monotonously decreases from 707 to 701 K. Correspondingly, ΔT_x that reflects the resistance to crystallization decreases from 46 to 39 K, indicating that pre-compression treatment reduces the thermal stability of the glassy alloy, which could be attributed to the structural heterogeneity serving as nucleation site for the precipitation of crystalline phase [21]. The DSC curves of 3 and 7 cycles under 1273 MPa pre-compressed sample are shown in Fig. 2b. Compared to the 7 cycles sample, 3 cycles sample exhibits an obvious exothermic peak corresponding to the crystallization state of amorphous structure.

Figure 3a shows the variation of the Vickers hardness (HV) of the as-cast and pre-compressed Cu_{50}Zr50 BMGCs. The HV values of the as-cast sample is higher than the pre-compressed samples, resulting from decreasing of the atomic packing

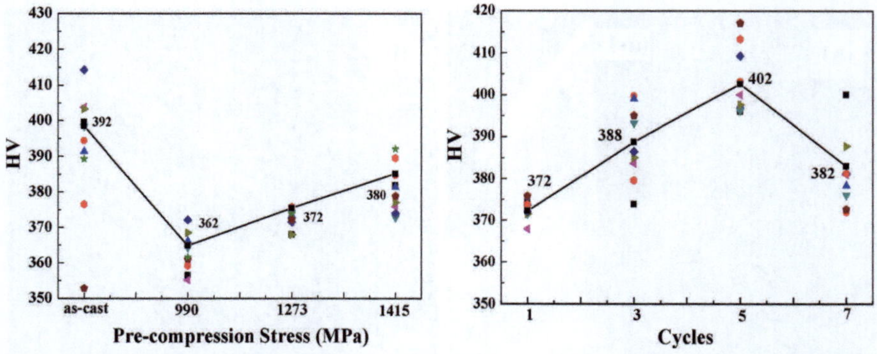

Fig. 3 Variation of HV values of $Cu_{50}Zr_{50}$ BMGCs under the different pre-compression stress (**a**) and cyclic compression (**b**)

density [22]. It can be found that the values of HV of the pre-compressed samples shows the increased trend with increasing stresses. The values of the HV of the cyclic pre-compressed samples under 1273 MPa are shown in the Fig. 3b. It can be found that the values of HV reveal the variation trend of first increasing and then decreasing. The sample after 5 cycles exhibits the excellent HV among the tested samples, which is due to the precipitations of fewer B_2-CuZr and B_{19}'-CuZr nano-crystals may be able to effectively hinder the propagation of shear bands and enhance the microhardness. Compared to the as-cast one, the sample after 5 cycles increased by more than 10 HV. Appropriate pre-compression stress and cycles can increase the microhardness of $Cu_{50}Zr_{50}$ BMGCs.

The uniaxial compression stress-strain curves measured from the $Cu_{50}Zr_{50}$ BMGC are shown in Fig. 4. It can be observed that the fracture strength (σ_f) of the as-cast sample reaches up to 1566 MPa. The inset of Fig. 4 shows the serration-like pattern of the as-cast sample, indicating that the $Cu_{50}Zr_{50}$ BMGC possesses a certain plasticity [23].

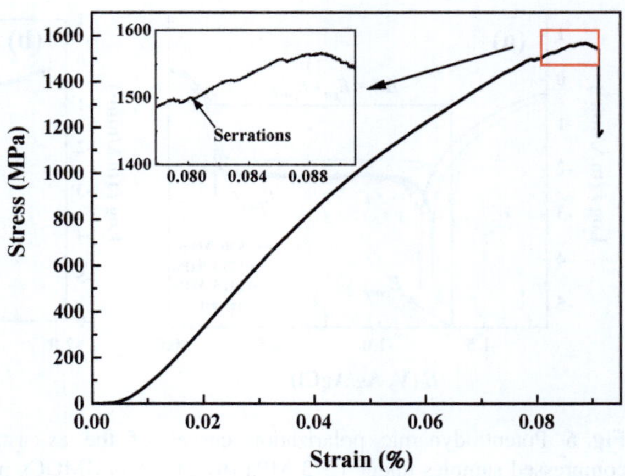

Fig. 4 Engineering stress-strain curves of the as-cast $Cu_{50}Zr_{50}$ BMGCs under quasi-static compression test

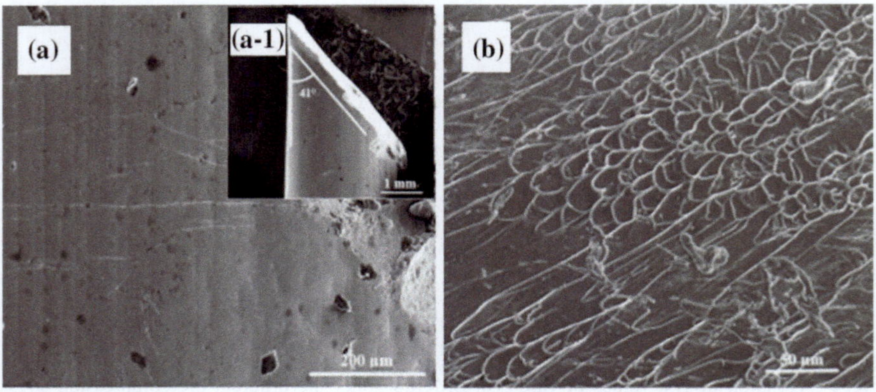

Fig. 5 FESEM micrographs recorded from the side surfaces (**a**) and sections (**b**) coupled with fractured angle (**a-1**) of the fractured as-cast $Cu_{50}Zr_{50}$ BMGCs

Figure 5 presents the FESEM micrographs from the side surfaces and sections of the fractured as-cast sample. From the Fig. 5a, it can be seen that the fracture angle between major fracture surface and compression axis of the as-cast sample is approximately 41°. Moreover, there are slender and shallow shear bands on the side surface of fractured as-cast sample. The fractography of the fractured as-cast sample exhibits the typical characteristic of vein-like and river-like patterns caused by local viscous flow during shear-band deformation [24], as shown in Fig. 5b. The primary and secondary shear bands can be easily distinguished, and the secondary shear bands intersect with the primary shear bands, meaning that the propagation of shear bands has been constrained [25].

Figure 6 shows the electrochemical polarization curves of the $Cu_{50}Zr_{50}$ BMGCs in the seawater solution after pre-compression and cyclic compression. The values of corrosion potential (E_{corr}), pitting potential (E_{pit}), passive current density (I_{pass})

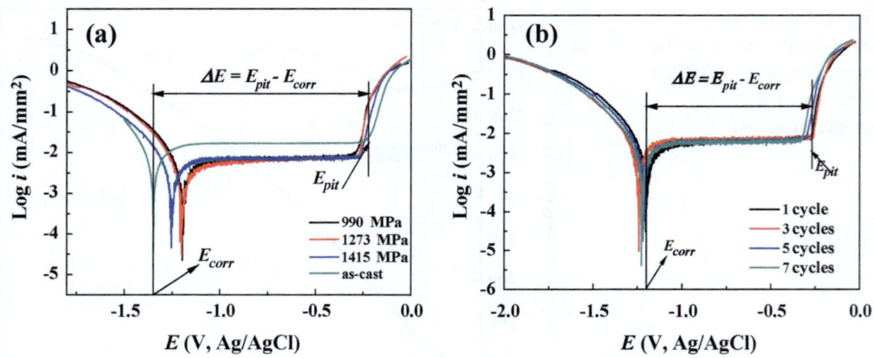

Fig. 6 Potentiodynamic polarization curves of the as-cast, pre-compressed (**a**) and cyclic compressed samples under 1273 MPa (**b**) $Cu_{50}Zr_{50}$ BMGCs in seawater solution

Table 1 Corrosion parameters derived from the polarization tests in seawater solution of the as-cast, pre-compressed and cyclic compressed $Cu_{50}Zr_{50}$ BMGCs

Samples	E_{corr}	E_{pit}	ΔE	I_{pass}
as-cast	−1.349	−0.222	1.127	16.67×10^{-3}
990 MPa	−1.197	−0.279	0.918	6.9×10^{-3}
1273 MPa	−1.202	−0.270	0.932	6.7×10^{-3}
1415 MPa	−1.252	−0.263	0.989	6.8×10^{-3}
3 cycles	−1.243	−0.258	0.985	7.1×10^{-3}
5 cycles	−1.219	−0.299	0.920	6.3×10^{-3}
7 cycles	−1.228	−0.322	0.906	6.1×10^{-3}

and passive region ($E_{pit} - E_{corr}$) are identified in Fig. 6 and summarized in Table 1. Overall, the profiles of the polarization curves with different stresses and cycles are almost the same. Only the values of the corrosion potential has different shifts. All tested samples probably undergo a general corrosion process and tend to form an oxide film on the alloy anode surface, developing into a stable protective layer in the seawater solution. It is obvious that the E_{corr} values change to more positive potentials with increasing the stresses. However, with the increasing cycles, E_{corr} increases firstly and then decreases. The values of i_{pass} display an decreasing tendency. Generally, the higher the E_{corr} is, the lower the I_{pass} is, indicating significant improvement of the corrosion resistance in the seawater solution [26]. The values of E_{corr} and I_{pass} of the 5 cycles under the 1273 MPa are −1.219 V and 6.3×10^{-3} mA/mm^2, respectively, showing the enhanced corrosion resistance in the seawater solution. Moreover, the as-cast $Cu_{50}Zr_{50}$ BMGC has the higher value of ΔE (1.127) and E_{pit} (−0.222 V), indicating that the as-cast sample possess the excellent pitting corrosion resistance against localized corrosion in the seawater solution [27].

Conclusions

In present study, the effects of the precompression and cyclic compression treatment on the microstructure, microhardness and corrosion resistance in the seawater solution of the $Cu_{50}Zr_{50}$ BMGCs have been investigated. Pre-compression treatment induced phase transformations, resulting in the microstructural changes. The compressive fracture strength of the as-cast sample reaches up to 1566 MPa and a certain degree of plasticity, as well as the larger microhardness of 392 HV and the excellent pitting corrosion resistance. The 5 cycles compression treatment under the 1273 MPa shows the enhanced corrosion resistance in the seawater solution.

Acknowledgements The authors acknowledge financial support from the National Natural Science Foundation of China (No. 51671095), and Key Research Development Program of Shandong Province of China (No. 2015GGx102016).

References

1. Y.H. Li, W. Zhang, T.L. Qi, New soft magnetic $Fe_{25}Co_{25}Ni_{25}(P, C, B)_{25}$ high entropy bulk metallic glasses with large supercooled liquid region. J. Alloy. Compd. **693**, 25–31 (2017)
2. Y.D. Gu, Z. Zheng, S.Z. Niu, W.J. Ge, Y. Wang, The seawater corrosion resistance and mechanical properties of $Cu_{47.5}Zr_{47.5}Al_5$ bulk metallic glass and its composites. J. Non-Cryst. Solids **380**, 135–140 (2013)
3. Z.L. Ning, W.Z. Liang, M.X. Zhang, Z.Z. Li, H.C. Sun, A.L. Liu, J.F. Sun, High tensile plasticity and strength of a CuZr-based bulk metallic glass composite. Mater. Des. **90**, 145–150 (2016)
4. D.C. Hofmann, Materials science. Shape memory bulk metallic glass composites. Science **329**, 1294–1295 (2010)
5. Z.F. Zhang, J. Eckert, L. Schultz, Difference in compressive and tensile fracture mechanisms of $Zr_{59}Cu_{20}Al_{10}Ni_8Ti_3$ bulk metallic glass. Acta Mater. **51**, 1167–1179 (2003)
6. P. Gargarella, S. Pauly, M. Samadi Khoshkhoo, U. Kuhn, J. Eckert, Phase formation and mechanical properties of Ti-Cu-Ni-Zr bulk metallic glass composites. Acta Mater. **65**, 259–269 (2014)
7. S. Pauly, S. Gorantla, G. Wang, U. Kühn, J. Eckert, Transformation-mediated ductility in CuZr-based bulk metallic glasses. Nat. Mater. **9**, 473–477 (2010)
8. L.Y. Li, J.S. Li, J. Wang, H.C. Kou, Tune the mechanical properties of Ti-based metallic glass composites by additions of nitrogen. Mater. Sci. Eng. A **694**, 93–97 (2017)
9. R. Wei, L.B. Chen, H.P. Li, F.S. Li, Compression-compression fatigue behavior of CuZr-based bulk metallic glass composite containing B2 phase. Intermetallics **85**, 54–58 (2017)
10. Z.Q. Liu, R. Li, G. Liu, W.H. Su, H. Wang, Y. Li, M.J. Shi, X.K. Luo, G.J. Wu, T. Zhang, Microstructural tailoring and improvement of mechanical properties in CuZr-based bulk metallic glass composites. Acta Mater. **60**, 3128–3139 (2012)
11. Farzaneh Bahrami, Rasool Amini, Amir Hossein Taghvaei, Microstructure and corrosion behavior of electrodeposited Ni-based nanocomposite coatings reinforced with $Ni_{60}Cr_{10}Ta_{10}P_{16}B_4$ metallic glass particles. J. Alloy. Compd. **714**, 530–536 (2017)
12. J.W. Qiao, H.L. Jia, P.K. Liaw, Metallic glass matrix composites. Mater. Sci. Eng. R **100**, 1–69 (2016)
13. Y.J. Liu, H.W. Yao, T.W. Zhang, Z. Wang, Y.S. Wang, J.W. Qiao, H.J. Yang, Z.H. Wang, Designing ductile CuZr-based metallic glass matrix composites. Mater. Sci. Eng. A **682**, 542–549 (2017)
14. X.D. Wang, R.T. Qu, Z.Q. Liu, Z.F. Zhang, Shear band propagation and plastic softening of metallic glass under cyclic compression. J. Alloy. Compd. **695**, 2016–2022 (2017)
15. B.A. Sun, K.K. Song, S. Pauly, P. Gargarella, J. Yi, G. Wang, C.T. Liu, J. Eckert, Y. Yang, Transformation-mediated plasticity in CuZr based metallic glass composites: a quantitative mechanistic understanding. Int. J. Plast. **85**, 34–51 (2016)
16. Z.Q. Liu, G. Liu, R.T. Qu, Z.F. Zhang, S.J. Wu, T. Zhang, Microstructural percolation assisted breakthrough of trade-off between strength and ductility in CuZr-based metallic glass composites **4**, 4167–4172 (2014)
17. Y. Wu, Y.H. Xiao, G.L. Chen, C.T. Liu, Z.P. Lu, Bulk metallic glass composites with transformation-mediated work-hardening and ductility. Adv. Mater. **22**, 2770–2773 (2010)

18. R. Wei, Y. Chang, S. Yang, C.J. Zhang, L. He, Strain rate sensitivity variation in CuZr-based bulk metallic glass composites containing B_2-CuZr phase. Rare Met. Mater. Eng. **45**, 542–547 (2016)
19. K.K. Song, S. Pauly, Y. Zhang, P. Gargarella, R. Li, N.S. Barekar, U. Kühn, M. Stoica, J. Eckert, Strategy for pinpointing the formation of B_2 CuZr in metastable CuZr-based shape memory alloys. Acta Mater. **59**, 6620–6630 (2011)
20. J.F. Ding, Z.Q. Liu, H. Wang, T. Zhang, Large-sized CuZr-based bulk metallic glass composite with enhanced mechanical properties. J. Mater. Sci. Technol. **30**, 590–594 (2014)
21. W. Zhou, J.X. Hou, W.P. Weng, Microstructure, thermal stability and mechanical properties of Zr-Cu-Al-Sn bulk metallic glass. J. Non-Cryst. Solids **429**, 208–212 (2015)
22. J. Gu, L.X. Zhang, Y.H. Wang, S. Ni, S.F. Guo, M. Song, Combined effect of isothermal annealing and pre-compression on mechanical properties of $Cu_{36}Zr_{48}Al_8Ag_8$ bulk metallic glass. Trans. Nonferrous Met. Soc. **26**, 1620–1628 (2016)
23. J. Gu, M. Song, S. Ni, X.Z. Liao, S.F. Guo, Improving the plasticity of bulk metallic glasses via pre-compression below the yield stress. Mater. Sci. Eng. A **602**, 68–76 (2014)
24. Z.F. Zhang, G. He, J. Eckert, L. Schultz, Fracture mechanisms in bulk metallic glassy materials. Phys. Rev. Lett. **91**, 045505 (2003)
25. J.J. Wu, W.H. Li, Y. Pan, J.Y. Qi, J.G. Wang, Microalloying and microstructures of Cu-based bulk metallic glasses & composites and relevant mechanical properties. Mater. Des. **89**, 1130–1136 (2016)
26. Q. Xing, K. Zhang, Y.Z. Wang, J.F. Leng, H.L. Jia, P.K. Liaw, Y. Wang, Effects of pre-compression on the microstructure, mechanical properties and corrosion resistance of $RE_{65}Co_{25}Al_{10}$ (RE = Ce, La, Pr, Sm and Gd) bulk metallic glasses. Intermetallics **67**, 94–101 (2015)
27. K. Zhang, X.L. Gao, Y.G. Dong, Q. Xing, Y. Wang, Effect of annealing on the microstructure, microhardness, and corrosion resistance of $Ni_{62}Nb_{33}Zr_5$ metallic glass and its composites. J. Non-Cryst. Solids **425**, 46–51 (2015)

Effect of Annealing Temperature of $Fe_{78}Si_9B_{13}$ Amorphous Ribbons for Activation of Persulfate on Azo Dye Degradation

Junjun Li, Lin Wu, Pengfei Liu, Lanjun Liu and Bo Zhang

Abstract The effect of annealing temperature of $Fe_{78}Si_9B_{13}$ amorphous ribbons for activation of persulfate on azo dye Orange G degradation was investigated. The ribbons annealed below 623 K have very rapid degradation rate, and the degradation efficiency within 10 min was 97%. The kinetic analysis elucidated that the degradation process could be described by pseudo-first-order kinetic model. The reaction rate was not strictly decreased with the increasing annealing temperature. In particular, the ribbons with different structures by annealing at different temperatures, after the reaction, show obviously discrepant surface corrosion morphology which also plays a role in the reaction rate.

Keywords Amorphous ribbon · Annealing temperature · Persulfate
Azo dye

Introduction

Azo dye is an important part of synthetic textile dyes, representing the most widely used dyes in textile processing. Azo dyes have aroused serious environmental problems because of poorly biodegradable contaminants with toxicity, mutagenicity

J. Li · L. Wu · P. Liu · L. Liu · B. Zhang (✉)
Institute of Amorphous Matter Science, School of Materials Science and Engineering, Hefei University of Technology, Hefei 230009, China
e-mail: bo.zhang@hfut.edu.cn

J. Li
e-mail: lijunjun@mail.hfut.edu.cn

L. Wu
e-mail: 18555508579@163.com

P. Liu
e-mail: pfliu@mail.hfut.edu.cn

L. Liu
e-mail: liulanjun@hfut.edu.cn

Y. Han (ed.), *Advanced Functional Materials*,
https://doi.org/10.1007/978-981-13-0110-0_32

and carcinogenicity. The traditional treatment techniques such as adsorption [1], coagulation [2], photo-catalysis [3], and biosorption [4], are usually time-consuming, high costs, and limited applicability [5]. Therefore, it is important to explore an effective technique to remove dyes from wastewater [6].

Advanced oxidation process (AOP) are possibly the best technologies to totally eliminate organic carbons in wastewater, which has been extensively studied as a promising technique [7]. Most of the non-biodegradable compounds are easily mineralized into CO_2, small molecules or other harmless products by AOP [8]. It is recently reported that sulfate radicals ($SO_4\cdot^-$) with a high redox potential are easy to be produced by Fe^{2+} by persulfate. However, the secondary pollution with the high ferrous irons leaching rate, required additional pressure to eliminate redundant ferrous ions, is inevitable [9]. Owing to the superior properties of non-toxicity, effectiveness and low price [10], many studies extract the zero valent iron (ZVI) as a persulfate activation catalyst [11]. The ZVI as a source to slow releasing Fe^{2+} and provides a novel method to generate sulfate radicals without involving iron ions. Thus, an appropriate Fe-based catalyst with sustainable development is great needed for wastewater remediation.

Fe-based amorphous alloys are a special type of ZVI materials [12]. There are some Fe-based amorphous alloys, such as Fe-B [13], Fe–Si-B [14], Fe-Mo-Si-B [15], and Fe–Si-B-Cu-Nb amorphous alloys [16], have been reported that possess the ultrafast degradation and mineralization efficiency in the wastewater. Z. Jia et al. introduced ultrafast $Fe_{78}Si_9B_{13}$ metallic glass as a catalyst for activation of persulfate on methylene blue degradation [16]. Based on Z. Jia's work, we extract $Fe_{78}Si_9B_{13}$ amorphous ribbons, which isothermal annealed at different temperatures (from 573 to 823 K, interval of 50 K), as catalysts, an alternative Fe^{2+} releasing source for activating persulfate, to investigate the Orange G degradation. To explore the relationship between the degradation property and the structure, kinetic analysis of the degradation process by as-spun ribbons and the ribbons annealed at different temperatures were also carried out. The corrosion morphology of the samples after the reaction was reported.

Experimental

Alloy ingots with compositions of $Fe_{78}Si_9B_{13}$ were prepared by arc-melting mixtures of high purity Fe (99.95%), Si (99.99%), B (99.99%) under a Ti-gettered argon atmosphere. All ingots were remelted five times to properly mix the elements of the materials. Amorphous alloys were subsequently produced in the form of ribbons, typically 26–28 μm thickness and 2 mm wide, by using melt spinning under an argon atmosphere. Annealing treatments of as-spun samples were performed at the temperature of 573, 623, 673, 723, 773 and 823 K for 10 min, and specimens were protected from oxidation by the argon atmosphere inside the silica tube. The structures and phase contents of as-spun and annealed samples were examined by X-ray diffraction (XRD) with Cu-Kα radiation. The samples were heated to 1673 K in 20 K/min to obtain the thermodynamic parameters by a differential scanning calorimeter (DSC, Nestal STA449F3, Germany) under high purity dry nitrogen flow. The Orange G (OG) dye used throughout this study is supplied by aladdin Chemical Co., Ltd (China). The chemicals such as $Na_2S_2O_8$ and methanol are all of the analytical grades.

In the degradation procedure, OG was dissolved in deionized water to get the solution. 250 mL stock solutions, quantitative $Na_2S_2O_8$ and pretreated ribbons were added in the beaker. The mass of the ribbons was 0.5 g to make sure the same surface area contacting with the solutions. The breakers containing the solutions and ribbons were agitated at 100 r/min in a temperature-controlled (298 K) water-bath trough. About 3 mL solution was taken from the beaker at intervals, adding 50 μL methanol immediately to stop the reaction and subjected to UV-vis spectrum scanning from 200 to 700 nm using a UV-visible spectrophotometer (CARY 5000, Agilent, America). The scanning electron microscope (SEM, SU8020, Hitachi, Japan) was used for the investigation of surface morphology of the ribbons after the reaction. All the initial dye solutions were freshly prepared before each experiment in order to minimize variance in concentration due to hydrolysis and direct photolysis of OG.

Results and Discussions

In order to explore the role of the annealing temperature of samples in the degradation of OG, the structure of the samples and corresponding reaction rate were tested. Figure 1 presents the XRD curves of the as-spun ribbons as well as the ribbons annealed at 573, 623, 673, 723, 773 and 823 K. The as-spun ribbons and the ribbons annealed at 573, 623, 673 K exhibit a broad hump in the range of $2\theta = 40–50°$, indicating that the structures of the ribbons are amorphous state. It shows that first traces of crystallization were detected after annealing at 723 K. When the ribbons annealed temperature ramped to 773 and 823 K, greater amounts of crystallization precipitate, with the only phase detected being bcc-Fe. Figure 2 shows the DSC scans for the as-spun and annealed ribbons. The as-spun ribbons and annealed ribbons below 773 K have the obvious exothermic peaks, but the ribbons annealed at 823 K have no exothermic peak. Combining with Figs. 1 and 2, the ribbons annealed at 723 and 773 K were amorphous/nano-Fe (bcc) composite with the exothermic peaks of DSC curve, and the ribbons annealed at 823 K were fully crystallized.

The quantitative evaluation of the degradation rate was further conducted in this study. Figure 3 shows the absorbance at 480 nm dependency on reaction time of OG solutions reacting with the ribbons. By non-linear curve fit, it can be seen the

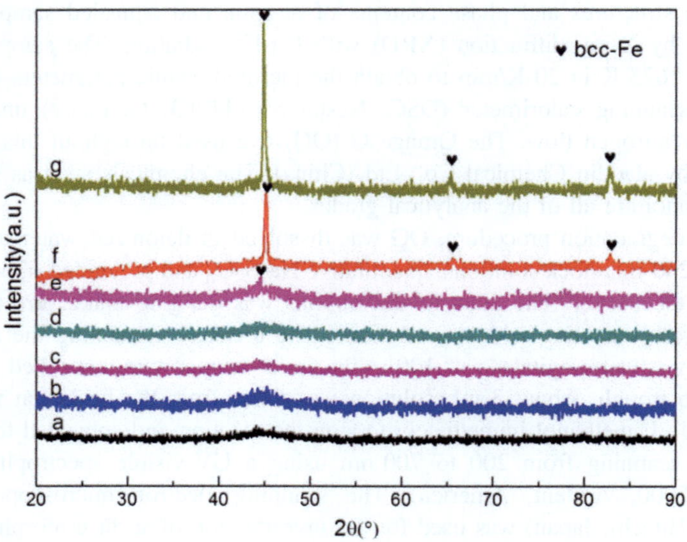

Fig. 1 Room temperature XRD patterns of **a** as-spun $Fe_{78}Si_9B_{13}$ amorphous ribbons and the ribbons annealed at **b** 573 K, **c** 623 K, **d** 673 K, **e** 723 K, **f** 773 K, **g** 823 K

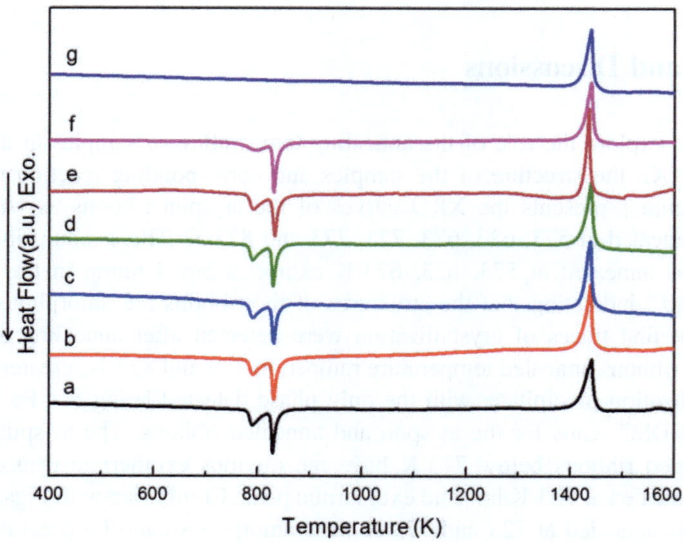

Fig. 2 DSC curves of **a** as-spun $Fe_{78}Si_9B_{13}$ amorphous ribbons and the ribbons annealed at **b** 573 K, **c** 623 K, **d** 673 K, **e** 723 K, **f** 773 K, **g** 823 K

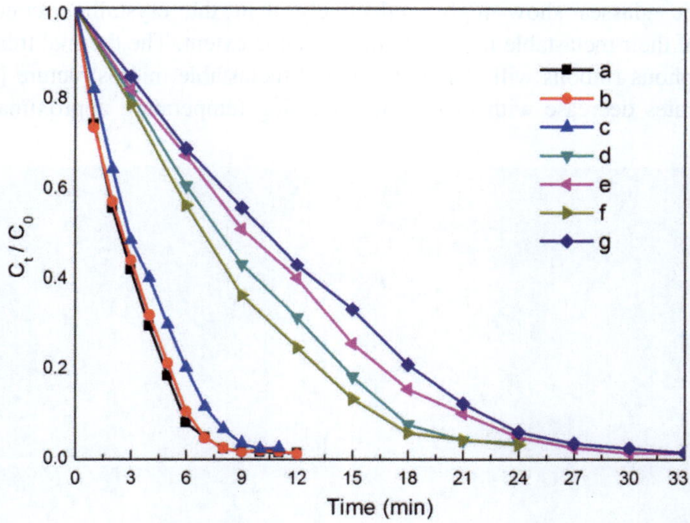

Fig. 3 C_t normalized by C_0 dependence on the reaction time at 298 K for the degradation process of **a** as-spun $Fe_{78}Si_9B_{13}$amorphous ribbons and the ribbons annealed at **b** 573 K, **c** 623 K, **d** 673 K, **e** 723 K, **f** 773 K, **g** 823 K

Table 1 Results of nonlinear regression of the experimental data

T (K)	k (min^{-1})	Goodness of fit (R^2)
as-spun	0.314	0.988
573	0.297	0.990
623	0.236	0.993
673	0.099	0.973
723	0.081	0.975
773	0.113	0.977
823	0.071	0.981

degradation process fits well with the first-order exponential decay kinetics proposed by Nam and Tratnyek shown in Eq. (1), which can be testified by the correlation coefficient R^2 (Table 1).

$$C_t/C_0 = \exp(-k \cdot t). \quad (1)$$

where C_0 is the initial dye concentration; C_t is the dye concentration at time t (min) and k (min^{-1}) is the observed first-order rate constant, which is an indication of the reaction rate. The experiments were investigated when the initial OG concentration is 20 ppm, persulfate dosage is 1.00 mmol/L, and ribbons dosage is 0.500 g/L at 298 K. As shown in Fig. 3, the as-spun ribbons show the highest reaction rate of 0.314 min^{-1}. The ribbons annealed below 623 K also have a very rapid degradation rate with higher than 0.236 min^{-1}, the degradation efficiency within 10 min is 97%.

Metallic glasses show higher effectively than the crystalline counterparts, because of their metastable microstructure to some extent. The thermal treatment of the amorphous ribbons will change the initial metastable microstructure [17]. The reaction rates decrease with increasing annealing temperature approximately, but

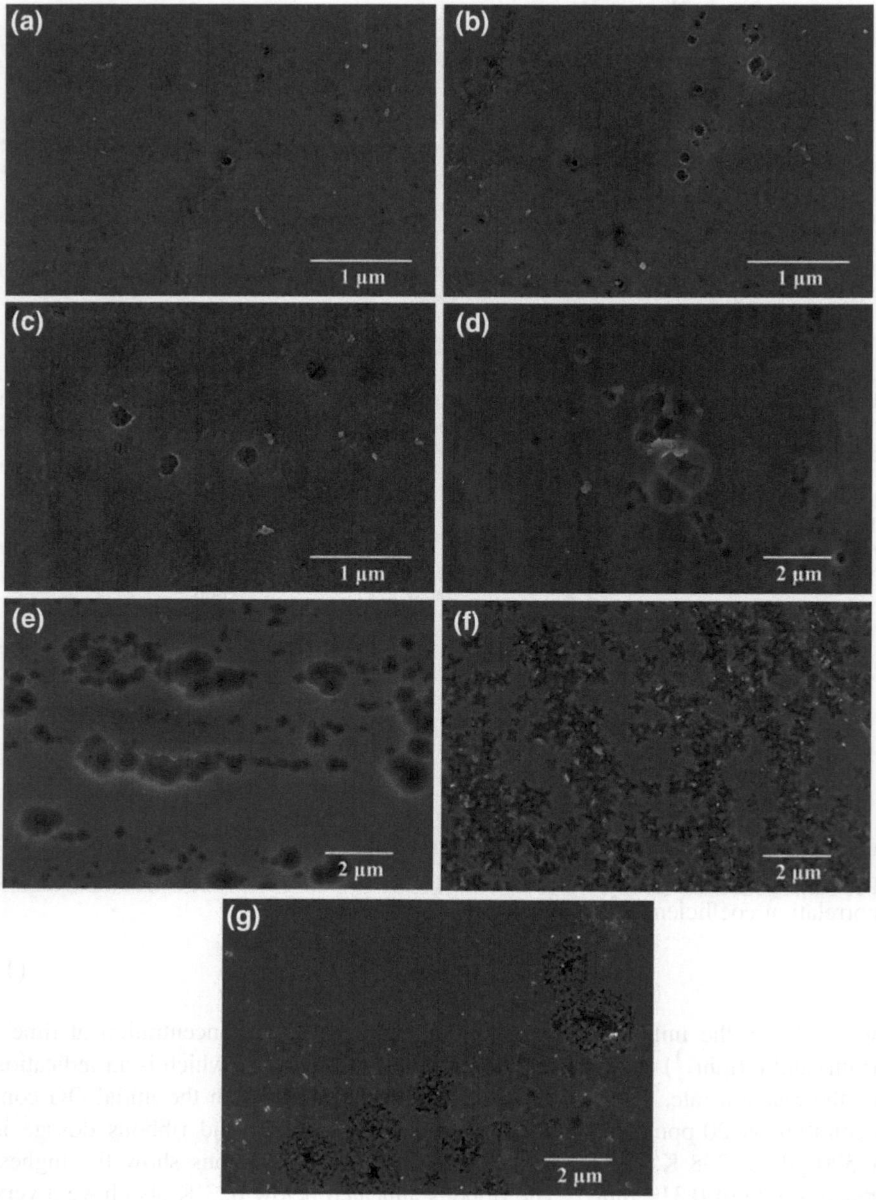

Fig. 4 SEM micrographs of free surfaces of **a** as-spun $Fe_{78}Si_9B_{13}$ amorphous ribbons and the ribbons annealed at **b** 573 K, **c** 623 K, **d** 673 K, **e** 723 K, **f** 773 K, **g** 823 K after the reaction

the reaction rate of the ribbon annealed at 773 K is higher than the ribbon annealed at 673, 723 and 823 K, which is possibly attributed to the specific corrosion morphology (show in Fig. 4f). As Table 1 shows, the degradation rate of as-spun ribbons is four times larger than that of the ribbons annealed at 823 K with crystallization at the same reactive temperature. On annealing the ribbons, nanoscale bcc-Fe is formed in the amorphous matrix, introducing grain boundaries, dislocations or segregations to amorphous ribbons, thus decreasing the chemical uniformity, bringing side effects on surface catalysis. Figure 4 shows the SEM images of the as-spun amorphous ribbons and the ribbons annealed at 573, 623, 673, 723, 773, 823 K after reaction. The ribbons free surfaces have obviously different characteristics after the reaction. As seen from Fig. 4a, circular pits with the diameter of 200–300 nm appear on the free surface of the as-spun ribbons. The ribbons annealed at 573 K (Fig. 4b) and 623 K (Fig. 4c) have the same appearance as the as-spun ribbons (Fig. 4a). There are micro-cracks in the center of the circular pits on the surface of the ribbons annealed at 673 K (Fig. 4d) and 723 K (Fig. 4e), and the diameter of the pits are absolutely bigger than the above in Fig. 4a, b, c). The star-shaped pits on the surface of the ribbons annealed at 773 K (Fig. 4f) cause the greater possibility to the following corrosion on the surface, which means the higher reaction rate. From the Fig. 4g, the circular corrosion morphology is formed by dendritic expansion.

Conclusion

The investigation of OG dye degradation indicates that persulfate can be effectively activated by the as-spun and the annealed ribbons. The OG solutions with concentration of 20 ppm can be degraded by the as-spun ribbons and the ribbons annealed below 623 K nearly completely in 10 min under the conditions of persulfate dosage of 1.00 mmol/L, ribbons dosage of 0.500 g/L. The ribbons annealed above 673 K have obviously lower reaction rates than the ribbons annealed below 623 K, though both of them possess amorphous structure. All the dye degradation process can be described by a pseudo-first-order kinetic model. The surface corrosion morphology after the reaction has an effect on reaction rate, the star-shaped pits on the surface of the ribbons annealed at 773 K can make it possible to the following corrosion on the surface, which results in the rate of the ribbons annealed at 773 K is higher than annealed at 673 and 723 K.

Acknowledgements The authors are grateful for the financial support from the National Key Research and Development Project (No. 2016YFB0300500) and the State Key Development Program for Basic Research of China (No. 2015CB856800).

References

1. S. Papic, Removal of some reactive dyes from synthetic wastewater by combined Al(III) coagulation/carbon adsorption process. Dyes Pigm. **62**, 291–298 (2004)
2. E. Ellouze, D. Ellouze, A. Jrad, R.B. Amar, Treatment of synthetic textile wastewater by combined chemical coagulation/membrane processes. Desalination **33**, 118–124 (2011)
3. T.A. Saleh, V.K. Gupta, Photo-catalyzed degradation of hazardous dye methyl orange by use of a composite catalyst consisting of multi-walled carbon nanotubes and titanium dioxide. J. Colloid Interface Sci. **371**, 101–106 (2012)
4. A.H. Chen, S.M. Chen, Biosorption of azo dyes from aqueous solution by glutaraldehyde-crosslinked chitosans. J. Hazard. Mater. **172**, 1111–1121 (2009)
5. M.Q. Cai, X.Q. Wei, Z.J. Song, M.C. Jin, Decolorization of azo dye Orange G by aluminum powder enhanced by ultrasonic irradiation. Ultrason. Sonochem. **22**, 167–173 (2015)
6. S. Liang, Z. Jia, W. Zhang, W. Wang, L. Zhang, Rapid malachite green degradation using Fe 73.5 Si 13.5 B 9 Cu 1 Nb 3 metallic glass for activation of persulfate under UV–Vis light, Mater. Des. **119**, 244–253 (2017)
7. Z. Jia, J. Miao, H.B. Lu, D. Habibi, W.C. Zhang, L.C. Zhang, Photocatalytic degradation and absorption kinetics of cibacron brilliant yellow 3G-P by nanosized ZnO catalyst under simulated solar light. J. Taiwan Inst. Chem. Eng. **60**, 267–274 (2016)
8. M. Pera-Titus, V. García-Molina, M.A. Baños, J. Giménez, S. Esplugas, Degradation of chlorophenols by means of advanced oxidation processes: a general review. Appl. Catal. B **47**, 219–256 (2004)
9. H. Li, J. Wan, Y. Ma, Y. Wang, M. Huang, Influence of particle size of zero-valent iron and dissolved silica on the reactivity of activated persulfate for degradation of acid orange 7. Chem. Eng. J. **237**, 487–496 (2014)
10. J. Cao, W.X. Zhang, D.G. Brown, D. Sethi, Oxidation of lindane with Fe(II)-activated sodium persulfate. Environ. Eng. Sci. **25**, 221–228 (2008)
11. X. Jiang, Y. Wu, P. Wang, H. Li, W. Dong, Degradation of bisphenol A in aqueous solution by persulfate activated with ferrous ion. Environ. Sci. Poll. Res. **20**, 4947–4953 (2013)
12. B. Lin, X. Bian, P. Wang, G. Luo, Application of Fe-based metallic glasses in wastewater treatment. Mater. Sci. Eng. B **177**, 92–95 (2012)
13. Y. Tang, Y. Shao, N. Chen, K.F. Yao, Rapid decomposition of Direct Blue 6 in neutral solution by Fe–B amorphous alloys. RSC Adv. **5**, 6215–6221 (2015)
14. Z. Jia, W.C. Zhang, W.M. Wang, D. Habibi, L.C. Zhang, Amorphous Fe 78 Si 9 B 13 alloy: an efficient and reusable photo-enhanced Fenton-like catalyst in degradation of cibacron brilliant red 3B-A dye under UV–vis light. Appl. Catal. B **192**, 46–56 (2016)
15. C. Zhang, Z. Zhu, H. Zhang, Z. Hu, On the decolorization property of Fe–Mo–Si–B alloys with different structures. J. NonCryst. Solids **358**, 61–64 (2012)
16. Z. Jia, X. Duan, W. Zhang, W. Wang, H. Sun, S. Wang, L.C. Zhang, Ultra-sustainable $Fe_{78}Si_9B_{13}$ metallic glass as a catalyst for activation of persulfate on methylene blue degradation under UV-Vis light. Sci. Rep. **6** (2016)
17. Z. Jia, X.G. Duan, P. Qin, W.C. Zhang, W.M. Wang, C. Yang, H.Q. Sun, S.B. Wang, L.C. Zhang, Disordered atomic packing structure of metallic glass: toward ultrafast hydroxyl radicals production rate and strong electron transfer ability in catalytic performance. Adv. Funct. Mater. **27**, 1702258 (2017)

Microstructures and Mechanical Properties of $Ni_xCoCrFeMo_{0.1}$ Multi-component Alloys

Xiaoxue Chang, Hui Jiang, Yiping Lu and Tingju Li

Abstract Five multi-principal component $Ni_xCoCrFeMo_{0.1}$ (x = 1, 2, and 3, in molar ratio) alloys were prepared by vacuum arc-melting in a water-cooled copper crucible. The effects of Ni on the microstructure and mechanical properties were studied. Results indicated the CoCrFeNi alloy exhibits a single face-centered cubic (FCC) solid-solution structure. The hardness of all these alloys have slightly different after annealing, which indicates the series of alloys have high thermal stability and good high temperature application.

Keywords High entropy alloy · Morphology · Organizational characteristics
Performance characteristics

Introduction

High entropy alloys or multi-component alloys are a kind of new material which contains more than three basic element components. The mixture of multiple principle elements in an equimolar or near-equimolar ratio gives a huge space to design new alloys [1–3]. These alloys have high mixing entropy, which tend to form simple solution phase [2–7] and exhibit outstanding properties such as excellent plastic, high strength and et al. [8–10].

X. Chang · H. Jiang · Y. Lu (✉) · T. Li
School of Materials Science and Engineering, Dalian University of Technology, 2 Linggong Street, Dalian 116024, China
e-mail: luyiping@dlut.edu.cn

X. Chang
e-mail: cxx0777@dlut.edu.cn

H. Jiang
e-mail: jianghui2013@mail.dlut.edu.cn

Y. Han (ed.), *Advanced Functional Materials*,
https://doi.org/10.1007/978-981-13-0110-0_33

Recently, many researchers have investigated the microstructure and properties of $CoCrFeNiMo_x$ alloys. The addition of Mo contributed to the formation of a hard second phase in the FCC matrix, which resulted in the high strength/hardness [11]. Specially, the $CoCrFeNiMo_{0.1}$ alloy exhibited an excellent comprehensive properties. However, the effect of Ni on the microstructure and properties of the $CoCrFeNiMo_{0.1}$ alloys has seldom reported. Thus, in this paper, $Ni_xCoCrFeMo_{0.1}$ alloys were prepared and the microstructure and properties of the alloys were investigated in detail.

Experimental Procedures

Chemical compositions of the $Ni_xCoCrFeMo_{0.1}$ (x = 1, 2 and 3, denoted as Ni_1, Ni_2 and Ni_3, the corner marks represent the mole ratio of the elements) alloy system used in this investigation are listed in Table 1. Alloys were prepared through vac melting under Ar atmosphere in a water-cooled crucible with a purity above 99.95%. In order to enhance homogeneity, alloy ingots re-melted 6 times resulting in an ingot size of 20 mm (length) × 15 mm (width).

X-ray diffractometer(XRD) analysis was performed using an EMPUREAN diffractometer with Co Ka radiation scanning from 30 to 130° in 2θ by scanning step of 0.033. Compress tests was performed on 5 × 10 mm samples using universal testing machine with a strain rate of 1×10^{-3}/s. The microstructures were characterized using an olympic GX51 optical metallographic microscope (OM), a Zeiss Supra55 scanning electron microscope (SEM) equipped with the energy-dispersive spectrometer (EDS) and a Tecnai G220 S-Twin transmission electron microscope (TEM). OM and SEM specimens were polished to a mirror surface without scratches after grinding to 2000-grit SiC paper, subsequently surface corrosion with chloroazotic acid. TEM samples were primarily ground below 80 μm and then punched to Φ3 mm circle sheets, followed by twin-jet electro-polishing using a mixed solution of HNO_3:C_2H_5OH = 1:9. The hardness tests were performed by using EVERONE Vickers hardness tester with parameters of 500 g and 15 s. Each specimen was chosen from five different points to test for average. The ingot was annealed at 1200° h for 4 h and then cooled along with the GSL-1500X-OTF vacuum tube furnace (denoted as Ni_1-h, Ni_2-h and Ni_3-h alloy).

Table 1 The parameters of the $Ni_xCoCrFeMo_{0.1}$ system

No.	Composition	VEC	δ	ΔH	χ	ΔS_{mix}
1	$NiCoCrFeMo_{0.1}$	8.20	1.47	−3.90	0.109	12.20
2	$Ni_2CoCrFeMo_{0.1}$	8.55	1.32	−4.01	0.103	11.66
3	$Ni_3CoCrFeMo_{0.1}$	8.79	1.21	−3.85	0.097	10.85

Results and Discussion

Crystal Structure. There are many empirical parameters used to predict a single phase solid-solution structured in high entropy alloys. Such as atomic size difference (δ), mixing enthalpy (ΔH_{mix}), mixing entropy (ΔS_{mix}), electro-negativity difference ($\Delta\chi$) and valence electron concentration (VEC) [12]. It is tending to form the face-centered cubic(FCC) solid-solution if $\delta < 4.27\%$, $\chi < 1.1$ [13], VEC > 8 [14], -7.27 kJ/mol < ΔH_{mix} < 4 kJ/mol [15].

From the above parameters of the $Ni_xCoCrFeMo_{0.1}$ system shown in Table 1, the values of all these parameters (δ, χ, VEC, ΔH_{mix}) of $Ni_xCoCrFeMo_{0.1}$ alloys are all in FCC range, which tend to form FCC solid-solution structure. Based on the previous prediction, $Ni_xCoCrFeMo_{0.1}$ alloys should be possessed FCC structure.

Figure 1 shows in the XRD patterns of the casting and annealing $Ni_xCoCrFeNiMo_{0.1}$ alloys. As we can see, all alloys display a single FCC solid-solution structure without other minor phase. The results also fit our previous predictions.

Microstructure. The low magnification morphology of the casting and annealing $Ni_xCoCrFeNiMo_{0.1}$ alloys are shown in Fig. 2. Figure 2a, b, c show the Ni_1, Ni_2 and Ni_3 organization under the metallographic microscope. The organizations of Ni_1, Ni_2 and Ni_3 alloys are the similar typical dendritic crystal with grains size above 500 μm, and no precipitated phase can be found. Figure 2d, e, f show the Ni_1-h, Ni_2-h and Ni_3-h organization under the metallographic microscope. After annealing, the composition segregation was eliminated.

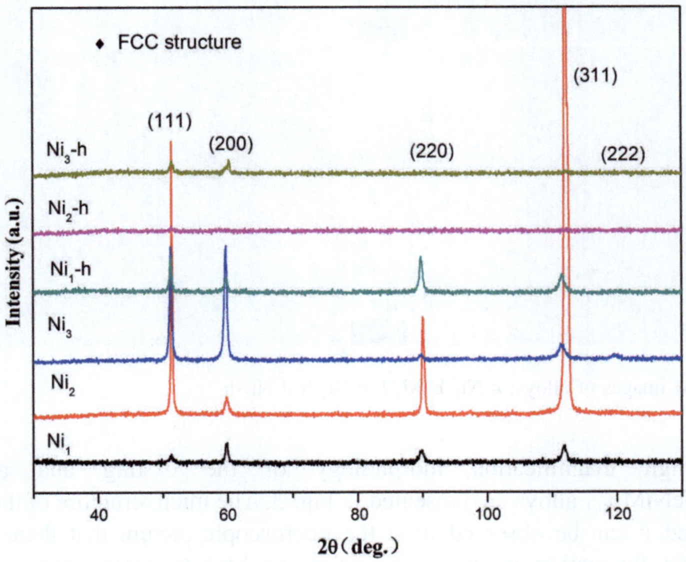

Fig. 1 XRD patterns of the casting and annealing $Ni_xCoCrFeNiMo_{0.1}$ alloys

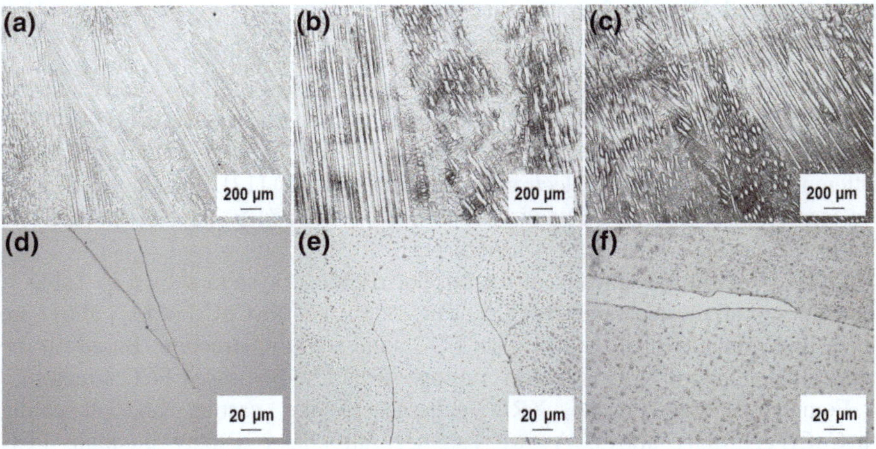

Fig. 2 OM images of alloys: **a** Ni_1 **b** Ni_2 **c** Ni_3 and **d** Ni_1-h **e** Ni_2-h **f** Ni_3-h

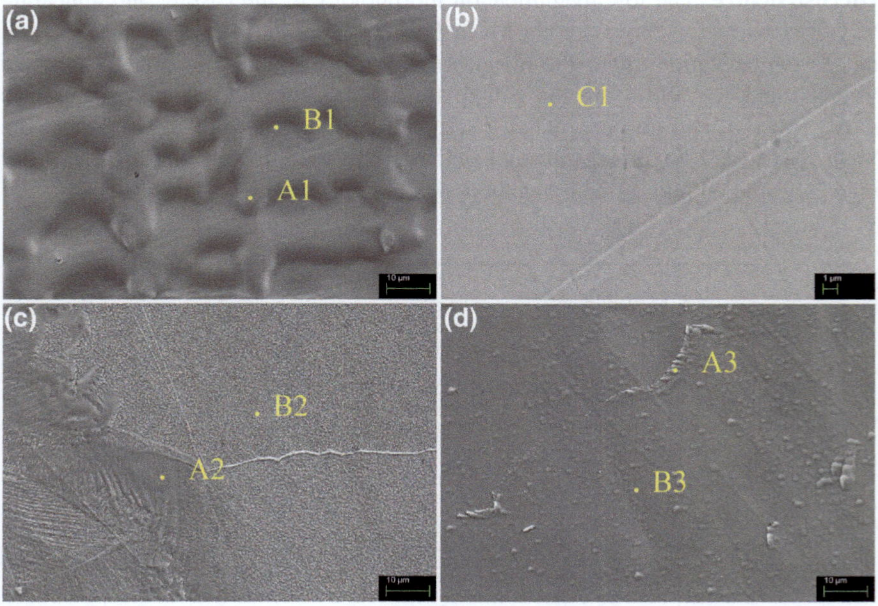

Fig. 3 SEM images of alloys: **a** Ni_1 **b** Ni_1-h **c** Ni_2-h **d** Ni_3-h

The high magnification morphology of the casting and annealing $Ni_xCoCrFeNiMo_{0.1}$ alloys are presented in Fig. 3. The microstructure of the alloy is similar, and it can be observed from the microscopic picture that there is a Mo composition fluctuation in the grain. After the heat treatment, the ingredients fluctuate disappear. Composition distribution of sample after heat treatment is

Table 2 The EDS results in different regions (at.%)

Alloy	Region	Ni	Co	Cr	Fe	Mo
Ni_1	A1	23.38	24.14	24.97	24.09	3.43
	B1	25.55	24.47	23.09	24.56	2.34
Ni_1-h	C1	20.97	27.81	21.58	26.78	2.86
Ni_2-h	A2	38.01	19.25	19.29	21.13	2.31
	B2	38.96	19.67	20.00	18.83	2.54
Ni_3-h	A3	49.62	15.47	16.12	16.12	2.68
	B3	47.13	16.57	17.35	16.64	2.31

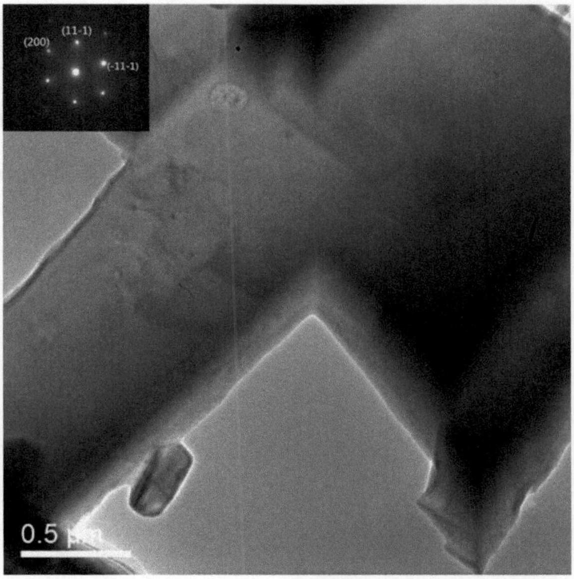

Fig. 4 TEM images and SEAD patterns of annealed at 1473 K for 4 h Ni_2-h alloy

similar to casting samples. The results of EDS map scanning in different regions shown in Table 2.

Figure 4 shows the annealed Ni_2-h alloy TEM images and SEAD patterns. The transmission zone axis is [011]. It can be seen, microstructure of alloy is uniform and the alloy is a single FCC solid-solution.

Mechanical Properties. Figure 5 shows the Vickers hardness curves of the $Ni_xCoCrFeMo_{0.1}$ alloys. The hardness of Ni_1 alloy is the highest among the three alloys while the hardness of Ni_2 alloy is similar to Ni_3 alloy. The hardness curves of material after heat treatment have the same trend. This can be explained that Ni element is soft than other element except Fe, but the proportion of Fe in all alloys is almost unchanged. The increase of the Ni content makes the hardness decrease.

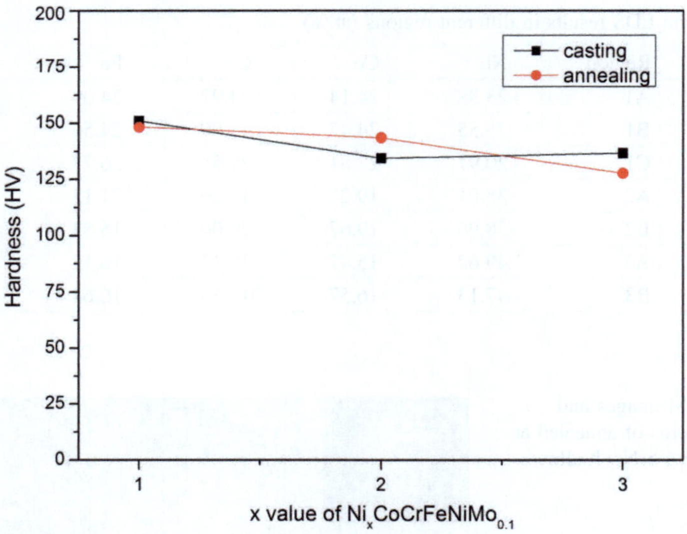

Fig. 5 Variations in the alloy hardness of $Ni_xCoCrFeMo_{0.1}$ (x = 1, 2 and 3) alloys

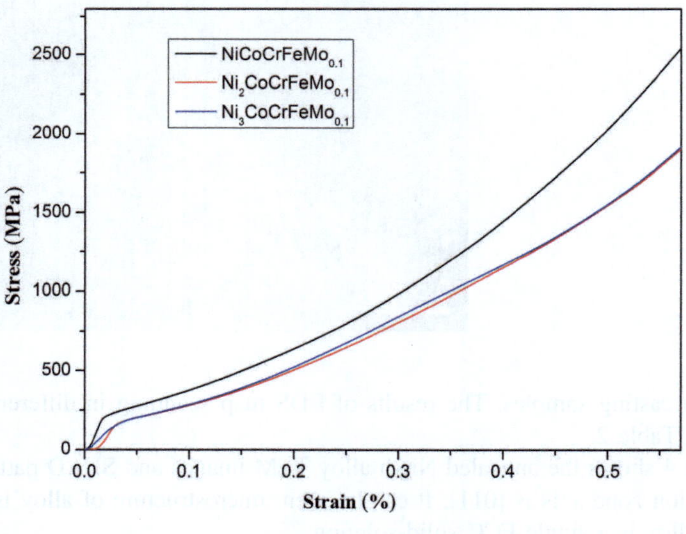

Fig. 6 Compressive engineering stress–strain curves for $Ni_xCoCrFeMo_{0.1}$ alloys

Figure 6 displays the compressive engineering stress–strain curves of $NixCoCrFeMo_{0.1}$ alloys and their specific properties are listed in Table 3. The compressive yield strength and fracture strength of the alloys decrease as the Ni content increased. But the ductility is still good, which all of the compression test samples can be compressed above 60% of the original height without cracking.

Table 3 The parameters of specific properties for the $Ni_xCoCrFeMo_{0.1}$ system

No.	Composition	Plasticity index (%)	Intensity index (Mpa)
1	$NiCoCrFeMo_{0.1}$	Above 60	260
2	$Ni_2CoCrFeMo_{0.1}$	Above 60	215
3	$Ni_3CoCrFeMo_{0.1}$	Above 60	192

With the increase of Ni content from 1 to 3, the yield strength decrease from 260 to 192 MPa. It is found that the single phase FCC solid-solution alloy has excellent plasticity due to various slip systems. It is the major factor contributing to the ductility of the alloy.

Conclusions

In this study, the effect of Ni element on the microstructure and mechanical properties of $Ni_xCoCrFeMo_{0.1}$ ($x = 1$, 2, and 3) alloys were investigated. From XRD, SEM, and EDS analyses consequences, the microstructures of the Ni_1, Ni_2 and Ni_3 alloys are determined to be a single FCC solid-solution phase.

The alloy exhibits excellent plasticity with the fracture compression rate is over 60%. And the hardness and yield stress decrease with the increased Ni content. The highest yield strength of 260 MPa is found in the Ni_1 alloy.

The heat treatment can eliminate composition segregation and has no obvious impact on the hardness, which illustrate the series of alloys have a very good thermal stability.

Acknowledgements This work was supported by the National Natural Science Foundation of China (Nos. 51671044, 51471044 and 51525401), Dalian Support Plan for Innovation of High-level Talents (Top and Leading Talents, 2015R013), the Fundamental Research Funds for the Central Universities, and the National Key Research and Development Program of China (No. 2016YB0701203).

References

1. J.W. Yeh, S.K. Chen, S.J. Lin, J.Y. Gan, T.S. Chin, T.T. Shun, C.H. Tsau, S.Y. Chang, Nanostructured high-entropy alloys with multiple principal elements: novel alloy design concepts and outcomes. Adv. Eng. Mater. **6**, 299–303 (2004)
2. B. Cantor, I. Chang, P. Knight, A. Vincent, Microstructural development in equiatomic multicomponent alloys. Mater. Sci. Eng. A Struct. **375**, 213–218 (2004)
3. W.H. Liu, Z.P. Lu, J.Y. He, J.H. Luan, Z.J. Wang, B. Liu, Y. Liu, M.W. Chen, C.T. Liu, Ductile CoCrFeNiMox high entropy alloys strengthened by hard intermetallic phases. Acta Mater. **116**, 332–342 (2016)

4. M.S. Lucas, G.B. Wilks, L. Mauger, J.A. Munoz, O.N. Senkov, E. Michel, J. Horwath, S.L. Semiatin, M.B. Stone, D.L. Abernathy, E. Karapetrova, Absence of long-range chemical ordering in equimolar FeCoCrNi. Appl. Phys. Lett. **100**, 1–4 (2012)
5. Z. Wang, Y. Huang, Y. Yang, J. Wang, C.T. Liu, Atomic-size effect and solid solubility of multicomponent alloys. Scr. Mater. **94**, 28–31 (2015)
6. W.H. Liu, Y. Wu, J.Y. He, T.G. Nieh, Z.P. Lu, Grain growth and the Hall-Petch relationship in a high-entropy FeCrNiCoMn alloy. Scr. Mater. **68**, 526–529 (2013)
7. A. Takeuchi, K. Amiya, T. Wada, K. Yubuta, W. Zhang, High-entropy alloys with a hexagonal close-packed structure designed by equi-atomic alloy strategy and binary phase diagrams. JOM-US **66**, 1984–1992 (2014)
8. O.N. Senkov, S.V. Senkova, C. Woodward, Effect of aluminum on the microstructure and properties of two refractory high-entropy alloys. Acta Mater. **68**, 214–228 (2014)
9. K.M. Youssef, A.J. Zaddach, C. Niu, D.L. Irving, C.C. Koch, A novel low-density, high-hardness, high-entropy alloy with close-packed single-phase nanocrystalline structures. Mater. Res. Lett. **3**, 95–99 (2015)
10. B. Gludovatz, A. Hohenwarter, D. Catoor, E.H. Chang, E.P. George, R.O. Ritchie, A fracture-resistant high-entropy alloy for cryogenic applications. Science **345**, 1153–1158 (2014)
11. L. Jiang, Z.Q. Cao, J.C. Jie, J.J. Zhang, Y.P. Lu, T.M. Wang, T.J. Li, Effect of Mo and Ni elements on microstructure evolution and mechanical properties of the CoFeNixVMoy high entropy alloys. J. Alloy. Compd. **649**, 585–590 (2015)
12. H. Jiang, H. Zhang, T. Huang, Y. Lu, T. Wang, Microstructures and mechanical properties of $Co_2Mo_xNi_2VW_x$ eutectic high entropy alloys. Mater. Des. **109**, 539–546 (2016)
13. Y. Dong, Y. Lu, L. Jiang, T. Wang, T. Li, Effects of electro-negativity on the stability of topologically close-packed phase in high entropy alloys. Intermetallics **52**, 105–109 (2014)
14. H. Jiang, L. Jiang, Y.P. Lu, T.M. Wang, Z.Q. Cao, Microstructure and mechanical properties of the W-Ni-Co system refractory high-entropy alloys. Mater. Sci. Forum **816**, 324–329 (2015)
15. L. Jiang, Y.P. Lu, H. Jiang, T.M. Wang, B.N. Wei, Formation rules of single phase solid solution in high entropy alloys. Mater. Sci. Technol. Lond. **32**, 588–592 (2016)

Influences of Impurity Alloying on Thermal Stability and Mechanical Properties of $Zr_{50}Ti_4Y_1Al_{10}Cu_{25}Ni_7Co_2Fe_1$ Bulk Metallic Glass

Chen Chen, Guowen Sun, Hang Zhang, Xiaodong Jia, Ran Wei and Fushan Li

Abstract In this study, influences of impurity concentration tuned by mixing sponge Zr and high-purity Zr on thermal stability and mechanical properties of a super-multicomponent $Zr_{50}Ti_4Y_1Al_{10}Cu_{25}Ni_7Co_2Fe_1$ BMG were investigated. The results show that, different concentrations of impurities in Zr raw materials have inconspicuous influences on glass transition temperature (T_g) and crystallization temperature (T_x), but can affect the crystallization process of this BMG. Moreover, it is interesting that, the plasticity of this alloy reaches a maximum value when using 80% sponge Zr + 20% high-purity Zr, indicating that proper amount of impurities have positive effects on mechanical properties of this BMG, and such phenomenon may be caused by the formation of nano-crystals in the glassy matrix. This study provides a guideline for the exploration of Zr-based BMGs which are applicable for industrial production.

Keywords Bulk metallic glass · Super-multicomponent · Impurity
Thermal stability · Mechanical properties

C. Chen · G. Sun · H. Zhang · X. Jia · R. Wei · F. Li (✉)
School of Materials Science and Engineering,
Zhengzhou University, Zhengzhou 450001, China
e-mail: fsli@zzu.edu.cn

C. Chen
e-mail: chenchenmse@zzu.edu.cn

G. Sun
e-mail: 1176096785@qq.com

H. Zhang
e-mail: 1179977412@qq.com

X. Jia
e-mail: xdjia703@qq.com

R. Wei
e-mail: weiranmse@zzu.edu.cn

Y. Han (ed.), *Advanced Functional Materials*,
https://doi.org/10.1007/978-981-13-0110-0_34

Introduction

Zr-based bulk metallic glasses (BMGs) possess excellent glass forming ability (GFA), good thermal stability and attractive mechanical properties, which are very promising in various applications as engineering materials [1–4]. However, there still exist some challenges for industrial production of this series of materials, and it is one of the most important issues to explore the compositions that could form BMGs and have high mechanical properties under high levels of impurities in raw materials because oxygen impurity in the low-purity Zr raw materials (e.g. sponge Zr) generally deteriorates the mechanical properties for Zr-based BMGs [5–8]. Recently, a super-multicomponent $Zr_{50}Ti_4Y_1Al_{10}Cu_{25}Ni_7Co_2Fe_1$ BMG fabricated using sponge Zr and Ti with high GFA and good mechanical properties was reported, which shows intriguing properties that this BMG produced by low-purity raw materials could have similar thermal stability and higher plasticity compared to its high-purity counterpart [9]. As impurities in raw materials could also be regarded as alloying elements, it is expected that tuning the concentration of impurities would lead to further enhancement of the mechanical properties of this super-multicomponent Zr-based BMGs. Moreover, the investigation focusing on the influences of impurity concentration on thermal stability and mechanical properties would be conductive to better understanding of the role of impurities on these two aspects, which is beneficial for the design of Zr-based BMGs with high performance and low cost.

Therefore, in this study, the effects of impurity concentration on thermal stability and mechanical properties of $Zr_{50}Ti_4Y_1Al_{10}Cu_{25}Ni_7Co_2Fe_1$ were systematically investigated. In order to realize the modification of impurity concentration in this BMG, both high-purity Zr and sponge Zr were used and various proportion of these two raw materials were adopted in the fabrication of different alloy ingots. This work is beneficial for the exploration of high-performance Zr-based BMGs containing impurities, which is of significance to the industrial application of Zr-based BMGs.

Experimental

Alloy ingots with nominal composition of $Zr_{50}Ti_4Y_1Al_{10}Cu_{25}Ni_7Co_2Fe_1$ (at.%) were fabricated by arc melting the mixtures of the raw materials under a high-purity argon atmosphere, and all the purities of the raw materials, except sponge Zr, are over 99.9 wt%. The composition of sponge Zr used in this work is listed in Table 1. All the ingots were re-melted four times to ensure compositional homogeneity. Based on the proportion of sponge Zr and high-purity Zr, the alloy ingots were denoted as Z1–Z6, and the ratios of different Zr raw materials in these ingots are listed in Table 2. Ribbon specimens and rod specimens with 2 mm in diameter were prepared by melt-spinning method and copper mold injection casting method,

Table 1 The composition of sponge Zr (wt%)

Zr + Hf	Cr	Cl	Ni	C	Mg	N	H	O	Bal.
≥99.4	0.020	0.130	0.010	0.050	0.060	0.025	0.005	0.140	≤0.16

Table 2 Proportion of sponge Zr and high-purity Zr utilized in fabrication of Z1–Z6 specimens

	Z1 (%)	Z2 (%)	Z3 (%)	Z4 (%)	Z5 (%)	Z6 (%)
Sponge Zr	0	20	40	60	80	100
High-purity Zr	100	80	60	40	20	0

respectively, and the fabrication processes were under high-purity argon atmosphere. All the Z1–Z6 specimens were examined by a UltimaIV X-ray diffractometer with Cu K_α radiation, which were proved to have fully glassy structures under the resolution limit of XRD (not shown in this paper). A NETZSCH DSC 404C differential scanning calorimeter (DSC) was employed to test the thermal behaviors of Z1–Z6 ribbon specimens at a heating rate from 0.17 to 0.66 K s^{-1}. A Shimadzu AG-IC testing machine was utilized to examine the mechanical properties of Z1–Z6 rod specimens. The specimens for the compressive tests at room temperature were with 2 mm in diameter and 4 mm in height, and the strain rate was 4.17×10^{-4} s^{-1}. The side and fractured morphologies of specimens were observed by FEI QUANTA 200 scanning electron microscope (SEM). A JEM 2100 transmission electron microscope (TEM) was employed for further observation of nano-scale microstructure in specimens, and the specimens were cut from the rods using electron-discharging machine and ground to <50 μm in thickness and then etched through ion milling.

Results and Discussion

Figure 1 shows the DSC curves of Z1–Z6 specimens at a heating rate of 0.33 K s^{-1}. The glass transition temperature (T_g), onset crystallization temperature (T_x), crystallization peak temperatures (T_{p1} and T_{p2}) and the end of crystallization temperature (T_O) are marked in this figure and listed in Table 3. It could be observed that, all the DSC curves exhibit a distinct glass transition process followed by a supercooled liquid region and exothermic process, and the specimens containing different concentrations of impurities exhibit slight differences in the shapes and peak positions of DSC curves. Moreover, it is found that Z1–Z6 specimens possess almost identical T_g and T_x, as well as the width of supercooled liquid region (ΔT_x), indicating that the thermal stability of this alloy is insensitive to the change of casting conditions and atmosphere in this work. However, the crystallization behaviors of this super-multicomponent Zr-based BMG are changed by impurities, which could be deduced by the positive correlation with the enhancement of

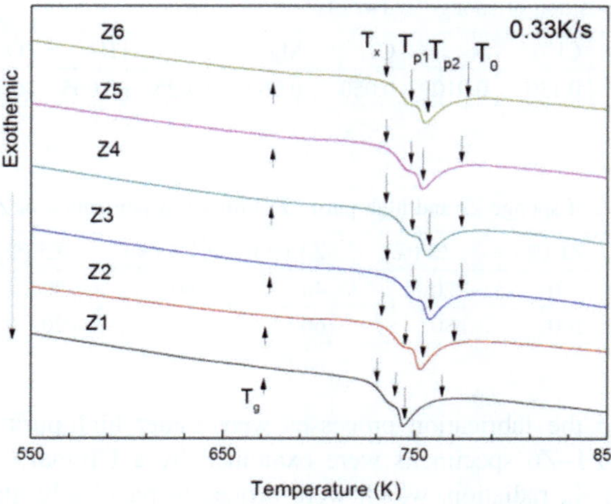

Fig. 1 DSC curves of Z1–Z6 specimens

Table 3 Thermal parameters and mechanical properties of Z1–Z6 specimens

	T_g (K)	T_x (K)	ΔT_x (K)	T_{p1} (K)	T_{p2} (K)	T_O (K)	$(T_O - T_x)$ (K)	σ_f (MPa)	ε_p (%)
Z1	667	727	60	732	741	760	33	1830	2.2
Z2	668	730	62	737	750	766	36	1850	2.1
Z3	669	731	62	740	755	770	39	1840	1.9
Z4	669	730	61	739	754	770	40	1840	2.4
Z5	669	729	60	739	752	771	42	1885	6.7
Z6	668	727	59	738	752	773	46	1865	4.5

impurity concentration values and the increase of T_{p1}, T_{p2} and T_O (also listed in Table 3). This result indicates that, impurity content would not significantly influence the thermal stability of this super-multicomponent Zr-based BMG, but could affect its crystallization behavior to some extent.

Typical stress-strain curves of Z1–Z6 specimens are shown in Fig. 2. All the specimens show an elastic deformation process, and then exhibit a yielding process followed by serrated flow during deformation. The fracture strength (σ_f) and plasticity (ε_p) are also listed in Table 3. It is found that, the Z1–Z6 specimens exhibit similar fracture strength ranging from 1830 to 1885 MPa, while the plasticity shows a maximum value at Z5. Interestingly, the plasticity values of Z1–Z4 have no obvious differences ranging from 1.9 to 2.4%, but the specimens produced using 80% sponge Zr and 20% high-purity Zr exhibit the increase of plasticity for almost two times before fracture, while the specimens prepared without using high-purity Zr display slight decrease in plasticity, indicating that proper amount of impurities are beneficial for the enhancement of plasticity in $Zr_{50}Ti_4Y_1Al_{10}Cu_{25}Ni_7Co_2Fe_1$ BMG.

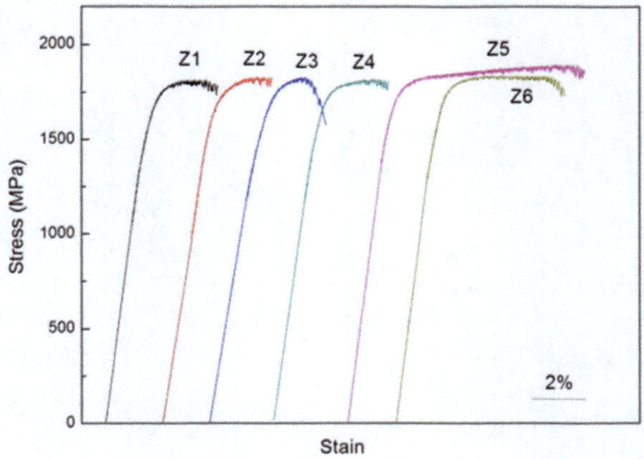

Fig. 2 Compressive stress-strain curves of Z1–Z6 specimens

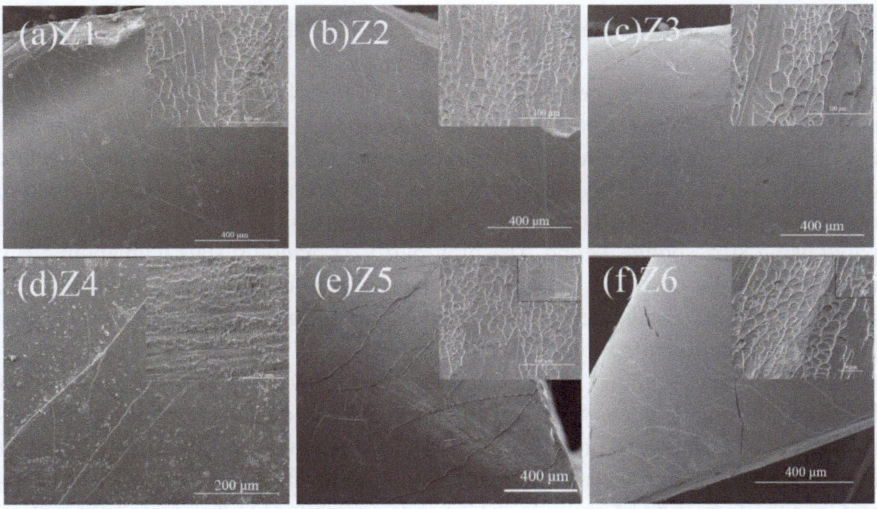

Fig. 3 Side surfaces and fractured surfaces (insets) of **a** Z1, **b** Z2, **c** Z3, **d** Z4, **e** Z5 and **f** Z6

The fracture behaviors of Z1–Z6 were further investigated. The SEM images for morphologies of side surfaces of Z1–Z6 specimens are shown in Fig. 3a–f, and the insets are their corresponding fractured surfaces. It could be observed that Z5 and Z6 specimen has higher density of multiple shear bands compared to Z1–Z4 specimens, implying superior ability of plastic deformation. Moreover, it is obvious that a higher fraction of slip morphology and fishbone-like patterns emerges in the fractured surface of Z5 and Z6 specimen, indicating high plasticity of BMGs due to steady flow in the main shear band before fracture [10].

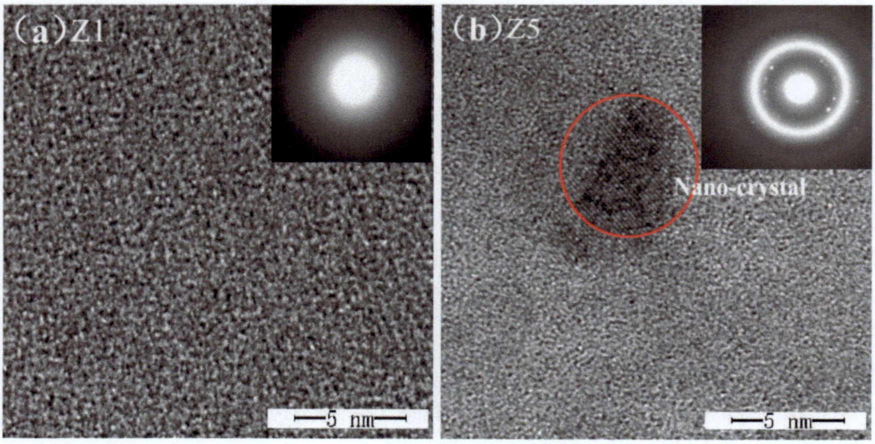

Fig. 4 HRTEM images of **a** Z1 and **b** Z5, insets are their corresponding SAED patterns

In order to further understand the origin of this phenomenon, the nano-scale structure of Z1 and Z5 were observed by TEM. The high-resolution TEM (HRTEM) images of the specimens are shown in Fig. 4a, b, and the insets are their corresponding selected area electron diffraction (SAED) patterns. As shown in Fig. 4a, the Z1 specimen prepared without sponge Zr has fully amorphous structure due to the disordered atomic arrangement in HRTEM image and typical diffraction ring for amorphous structure in the SAED pattern. As for the Z5 specimens prepared using 80% sponge Zr and 20% high-purity Zr, nano-crystals with about 5 nm in scale could be observed in the glassy matrix, as marked with circle in Fig. 4b, and the ring-like arrangement of diffraction spots in the SAED pattern could also prove the existence of nano-crystals.

From the investigations above, impurity contents would affect the plasticity of BMGs to varying degrees. As the precipitation of nano-crystals in Z5 is the obvious structure feature different from that of Z1, and the previous studies demonstrate that the embedded nano-crystals in glassy matrix promote the nucleation of shear bands and thus the formation of multiple shear bands, which lead to the stable propagation of shear bands and homogeneous deformation in BMGs [11–13], it indicates that nano-crystallization might have a positive effect on the enhancement of plasticity in this super-multicomponent Zr-based BMG.

Generally, impurities in sponge Zr, especially oxygen, would lead to significant decrease in plasticity of Zr-based BMGs due to brittleness of glassy matrix and the formation of brittle crystalline phases [5–8]. However, the super-multicomponent Zr-based BMG selected in this work shows special properties that could increase plasticity with the existence of proper amount of impurities in Zr raw materials, indicating that impurity elements have positive influences on mechanical properties of this BMG. Therefore, the impurities in sponge Zr could be regarded as alloying elements, and based on this analysis we could better understand the change of properties in this BMG after tuning impurity concentration. With low concentration of impurity elements alloyed in $Zr_{50}Ti_4Y_1Al_{10}Cu_{25}Ni_7Co_2Fe_1$ BMG, the impurities might be well dissolved in the glassy matrix, which might have inconspicuous influences on mechanical properties. When more impurity elements added in this BMG, these elements could not be fully dissolved in the glassy matrix, thus they precipitate as the form of nano-crystals and then enhance the plasticity of this BMG.

In order to discuss the factors that result in the easy formation of nano-crystals in the as-cast state of impurity-containing $Zr_{50}Ti_4Y_1Al_{10}Cu_{25}Ni_7Co_2Fe_1$ BMG, the Kissinger equation ($\ln(T^2/\beta) = -(E/RT) + C$) for calculating crystallization activation energy (E_x) [14] of Z1 and Z5 was utilized, and the calculation result shows that, Z5 exhibits lower E_x (362.8 kJ/mol) compared to Z1 (395.3 kJ/mol), indicating that crystalline phases would be easier to be formed with the addition of impurities (the DSC results for calculation are not shown in this paper). Such phenomenon is possibly resulted from the supersaturation of impurity elements in the glassy matrix of BMG, which may have stronger tendency to separate out compared to its impurity-free counterpart. Based on the analysis above, it could be deduced that such impurities might also have strong trend to precipitate at the casting state, which could be regarded as a possible reason for the easy formation of nano-crystals. In addition, it is noticed that, the formation of nano-crystals in the impurity-containing Zr-based BMGs is rarely reported in previous studies, which indicates the particularity of this super-multicomponent Zr-based BMG and deserves further investigations.

Summary

In this work, thermal stability and mechanical properties of $Zr_{50}Ti_4Y_1Al_{10}Cu_{25}Ni_7Co_2Fe_1$ BMG specimens containing different concentrations of impurities tuned by using the mixture of sponge Zr and high-purity Zr were systematically investigated. It is found that, the thermal stability of this alloy is insensitive to the impurity content. Moreover, it is intriguing that, the glassy specimens possess maximum plasticity value using proper proportion of sponge Zr and high-purity Zr. Nano-crystals formed in the glassy matrix during casting is a possible reason that results in the enhancement of plasticity of this super-multicomponent Zr-based BMG. This work is beneficial for further composition design and commercial applications of Zr-based BMGs.

References

1. A. Peker, W.L. Johnson, A highly processable metallic glass: $Zr_{41.2}Ti_{13.8}Cu_{12.5}Ni_{10.0}Be_{22.5}$. Appl. Phys. Lett. **63**(17), 2342–2344 (1993)
2. A. Inoue, T. Zhang, Fabrication of bulk glassy $Zr_{55}Al_{10}Ni_5Cu_{30}$ alloy of 30 mm in diameter by a suction casting method. Mater. Trans. **37**(2), 185–187 (1996)

3. Y.H. Liu, G. Wang, R.J. Wang et al., Super plastic bulk metallic glasses at room temperature. Science **315**(5817), 1385–1388 (2007)
4. Q. He, J.K. Shang, E. Ma et al., Crack-resistance curve of a Zr-Ti-Cu-Al bulk metallic glass with extraordinary fracture toughness. Acta Mater. **60**(12), 4940–4949 (2012)
5. R.D. Conner, R.E. Maire, W.L. Johnson, Effect of oxygen concentration upon the ductility of amorphous $Zr_{57}Nb_5Al_{10}Cu_{15.4}Ni_{12.6}$. Mater. Sci. Eng. A **V419**(1–2), 148–152 (2006)
6. Z.H. Han, L. He, Y.L. Hou et al., Understanding the mechanism for the embrittlement of a monolithic Zr-based bulk metallic glass by oxygen. Intermetallics **17**(7), 553–561 (2009)
7. V. Keryvin, C. Bernard, J.C. Sanglebœuf et al., Toughness of $Zr_{55}Cu_{30}Al_{10}Ni_5$ bulk metallic glass for two oxygen levels. J. Non-Cryst. Solids **352**(26–27), 2863–2868 (2006)
8. Z.P. Lu, H. Bei, Y. Wu, G.L. Chen et al., Oxygen effects on plastic deformation of a Zr-based bulk metallic glass. Appl. Phys. Lett. **92**(1), 148 (2008)
9. C. Chen, Y.Y. Cheng, T. Zhang, Synthesis of impurity-insensitive Zr-based bulk metallic glass. J. Non-Cryst. Solids **439**, 1–5 (2016)
10. B.A. Sun, J. Tan, S. Pauly et al., Stable fracture of a malleable Zr-based bulk metallic glass. J. Appl. Phys. **112**(10), 219 (2012)
11. T. Moriya, L. Hiraishi, N. Horie et al., Ductilization of BMGs by optimization of nanoparticle dispersion. J. Alloy. Compd. **434**(1), 6–9 (2007)
12. L. Zhang, F. Jiang, D. Zhang et al., In-situ precipitated nanocrystal beneficial to enhanced plasticity of Cu-Zr based bulk metallic glasses. Adv. Eng. Mater. **10**(10), 943–950 (2008)
13. K. Hajlaoui, A.R. Yavari, A. Lemoulec et al., Plasticity induced by nanoparticle dispersions in bulk metallic glasses. J. Non-Cryst. Solids **353**(3), 327–331 (2007)
14. F. Liu, H.F. Wang, Z. Chen et al., Determination of activation energy for crystallization in amorphous alloys. Mater. Lett. **60**(29–30), 3916–3921 (2006)

Corrosion Resistance of Fe-Based Bulk Amorphous Alloy with Sulfide Inclusion

Y. Zuo, S. L. Wang, Y. Huang and S. X. Wang

Abstract Corrosion resistance of Fe-based bulk metallic glass with sulfide inclusion was investigated in HCl, H_2SO_4, NaCl and NaOH solutions at different ambient temperature. The results indicate that Fe-based amorphous alloy generally exhibits more excellent corrosion resistance in alkaline solutions than that in acidic solutions, and presents high spontaneous passivated ability and evident passivation region in all solutions. Electrochemical impedance spectroscopy results display that all of the Nyquist curves are composed of just a single capacitive loop, and the electrode reaction is mainly controlled by the electrode potential. The mass loss exceeds 0.67 mg/cm^2 in 0.5 M H_2SO_4 solution, and rapider than that in another solutions. With increase of ambient temperature, the corrosion resistance decreases in all solutions. No pitting corrosion occurs in acidic solutions, though the inclusion particle is dissolved firstly. Seemly, some pit can be formed in NaCl solution, when ambient temperature exceeds 60 °C.

Keywords Fe-based bulk metallic glass · Sulfide · Pitting corrosion
Ambient temperature

Introduction

Due to excellent corrosion resistance and high microhardness, Fe-based amorphous alloys present an extensive potential application as anti-corrosive or anti-wearing materials, such as nuclear power, thermal power, ship, automobile, sports equipment, and so on. So many Fe-base amorphous alloy systems as Fe–Co–B–Si–Nb–Cr [1], Fe–Co–Cr–Mo–C–B–Y [2], Fe–Cr–Mo–C–B [3], Fe–Si–B–Nb–Cu [4], have been developed in the last decades. Since the high brittleness, Farmaer et al. [5] has attempted the applications of Fe-based amorphous coatings on the

Y. Zuo · S. L. Wang (✉) · Y. Huang · S. X. Wang
National Defence Key Disciplines Laboratory of Light Alloy Processing Science
and Technology, Nanchang Hangkong University, Nanchang 330063, Jiangxi, China
e-mail: slwang70518@nchu.edu.cn

Y. Han (ed.), *Advanced Functional Materials*,
https://doi.org/10.1007/978-981-13-0110-0_35

containers for the transportation, aging and disposal of spent nuclear fuel. The neutron absorption cross section in transmission for thermal neutrons of $Fe_{49.7}Cr_{17.7}Mn_{1.9}Mo_{7.4}W_{1.6}B_{15.2}C_{3.8}Si_{2.4}$ (SAM2X5) with high boron content is four times greater than that of borated stainless steel, and twice as good as nickel-based alloy (C-4) with additions of Gd. The Fe-based amorphous coating with the composition of $Fe_{54.2}Cr_{18.3}Mo_{13.7}Mn_2W_6B_{3.3}C_{1.1}Si_{1.4}$(wt%) on the marine pump impellers exhibits 2–3 times erosion rate than that of SUS304, believed to enhance effectively the lifetime of pump impellers in sand-containing seawater [6]. The corrosion rate of Fe-based amorphous coating of $Fe_{48}Cr_{16}Mo_{16}C_{14}B_2Mn_2Y_2$ is 1/1000 of that of 20G steel in the artificial simulation environment of power plant Boiler [7]. However, for all Fe-based amorphous alloys above mentioned, the produce cost is very high since high pure elements, which limits the extensive application in industries.

In our group, the Fe–C–Si–B–P–Cr–(Mo, Al, Co) amorphous alloy was developed using cast iron, industrial ferroalloys and commercial grade elements [8, 9]. Effects of elements, pH, ambient temperature on corrosion resistance were investigated in the published literatures [10, 11]. Since the impurity elements of oxygen and sulphur in raw materials, some oxide and sulphide can be detected in some bulk amorphous alloy. The effect of sulphur on the glass forming ability and corrosion resistance was investigated [12, 13]. However, effect of sulphide on corrosion behaviors of Fe-based amorphous alloy is unknown. In this paper, the corrosion resistance of Fe-based bulk amorphous alloy with sulphide inclusion will be discussed in difference solutions.

Experimental

The Fe-based amorphous alloy rods with the nominal composition of $Fe_{66.6}C_{7.1}Si_{3.3}B_{5.5}P_{8.7}Cr_{2.3}Mo_{2.5}Al_{2.0}Co_{1.0}S_{1.0}$ (at.%) were fabricated by the water cooling copper mold suction-casting, and the detailed processing and the composition of raw materials were shown in the literature [9]. Prior to immersion tests and electrochemical measurements, the specimens were degreased in acetone, washed in distilled water and dried in air. Corrosion rates were estimated from the weight loss after immersed in 0.5 M HCl, H_2SO_4, NaCl and NaOH solutions, respectively, open to air at room temperature for 30 days. The electrochemical measurements were conducted by a potentiostat (CHI 660E PARSTAT) and a three-electrode cell. The counter electrode was platinum and the reference electrode was a saturated calomel electrode (SCE). The potentiodynamic polarizations were measured in 1 M HCl, H_2SO_4, NaCl and NaOH solutions with 60 mV/min sweep rate at 20, 40 and 60 °C, respectively. The potentiostatic polarization were conducted by applying a potential of 0.5 V in acidic solutions or 0 V (SCE) in alkaline solutions for 2 h at different ambient temperatures. The electrochemical impedance was measured from 0.01 Hz to 100 kHz with the AC mode after the specimens were immersed for

20 min to obtain a stable open circuit potential. After immersion, the surface morphologies were observed by scanning electron microscope (SEM).

Results and Discussions

Figure 1 shows the potentiodynamic polarization curves in 1 M HCl, H_2SO_4, NaCl and NaOH solutions at different ambient temperature. In HCl solution, the specimen exhibits spontaneous passivation when scanning voltage exceeds 0.3 V, and presents transpassivation as scanning voltage is larger than 1.2 V in different ambient temperature. As the ambient temperature increases, corrosion current density and passive current density also increases, while corrosion potential and passive potential region seem no variation. The all related electrochemical results are shown in Table 1. Compared with HCl solution, a similar tendency can been attained in 1 M H_2SO_4, NaCl and NaOH solutions.

The current density versus passivation time during the potentiostatic polarization in four solutions is shown in Fig. 2. The applied potential of 0.5 V was selected

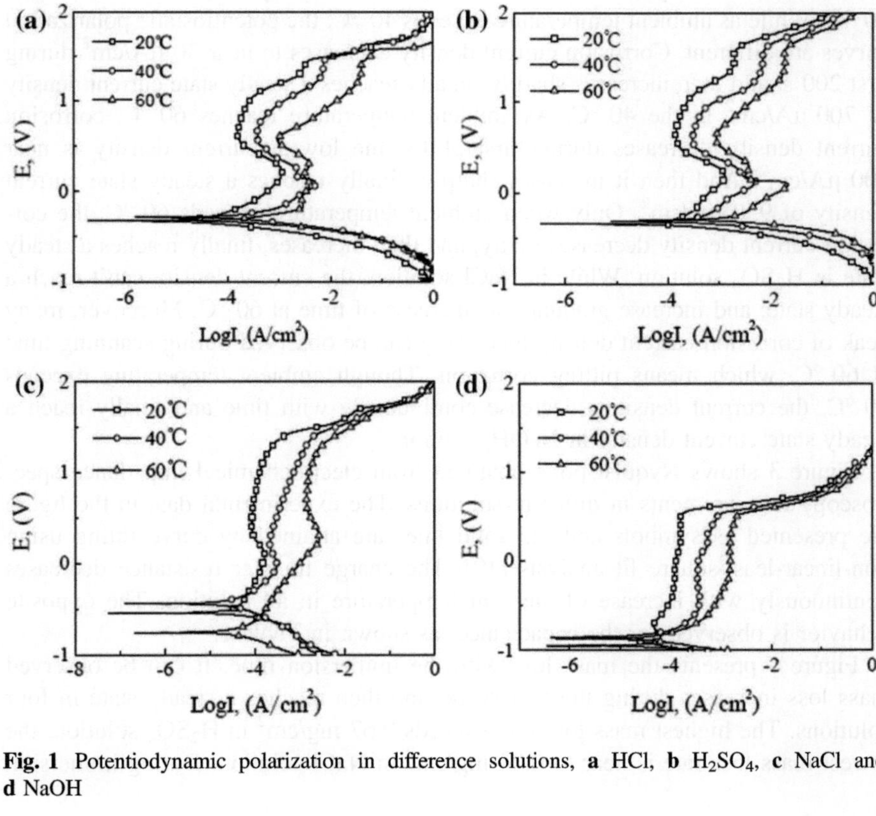

Fig. 1 Potentiodynamic polarization in difference solutions, **a** HCl, **b** H_2SO_4, **c** NaCl and **d** NaOH

Table 1 The results obtained from electrochemical tests

Solutions	Temperature (°C)	I_{corr} (μA/cm^2)	I_{pass} (μA/cm^2)	E_{corr} (−V)	$E_{pit}-E_{pass}$	CPE (μA/cm^2)	R_P (Ω/cm^2)
HCl	20	33.9	154.9	0.32	1.24	25.ro2	28,209
	40	111.2	382.8	0.28	1.21	108.3	6169
	60	125.3	955.0	0.32	1.13	499.4	734
H_2SO_4	20	66.1	135.8	0.33	1.44	9.8	115,690
	40	232.0	300.1	0.31	1.51	51.6	9925
	60	461.6	598.4	0.29	1.56	315.1	1104
NaCl	20	46.8	104.7	0.54	1.30	2.3	12,365
	40	50.1	489.8	0.43	1.43	2.9	30,673
	60	79.4	1013.6	0.75	1.0	5.3	10,290
NaOH	20	12.2	131.8	0.94	1.41	0.7	8589
	40	16.5	182.0	0.90	1.03	1.0	2613
	60	29.0	208.9	0.91	1.01	7.43	1293

from the potentiodynamic polarization curves. In HCl solution, the current densities decrease continuously with time and finally reach a steady state current density at 20 °C. While as ambient temperature exceeds 40 °C, the potentiostatic polarization curves are different. Corrosion current density decreases to near 30 μA/cm^2 during first 200 s, and then increases sharply, finally reaches a steady state current density of 700 μA/cm^2 in the 40 °C. As ambient temperature reaches 60 °C, corrosion current density decreases during first 500 s, the lowest current density is near 200 μA/cm^2. And then it increases sharply, finally reaches a steady state current density of 950 μA/cm^2. Only when ambient temperature exceeds 60 °C, the corrosion current density decreases firstly, and then increases, finally reaches a steady state in H_2SO_4 solution. While in NaCl solution, the current density can't reach a steady state, and increase gradually as increase of time at 60 °C. Moreover, many peak of corrosion current density increasing can be observed during scanning time at 60 °C, which means pitting corrosion. Though ambient temperature exceeds 60 °C, the current densities decrease continuously with time and finally reach a steady state current density in NaOH solution.

Figure 3 shows Nyquist polts obtained from electrochemical impedance spectroscopy measurements in different solutions. The experimental data in the figure are presented as symbols and the solid lines are attained by curve fitting using non-linear-least-square fit analysis [10]. The charge transfer resistance decreases continuously with increase of ambient temperature in all solution. The opposite behavior is observed for the capacitance, as shown in Table 1.

Figure 4 presents the mass loss with the immersion time. It can be observed mass loss increases during the first stage, and then reaches a steady state in four solutions. The highest mass loss rate exceeds 0.67 mg/cm^2 in H_2SO_4 solution, the lowest mass loss rate is near to 0.04 mg/cm^2 in NaCl solution. During immersion

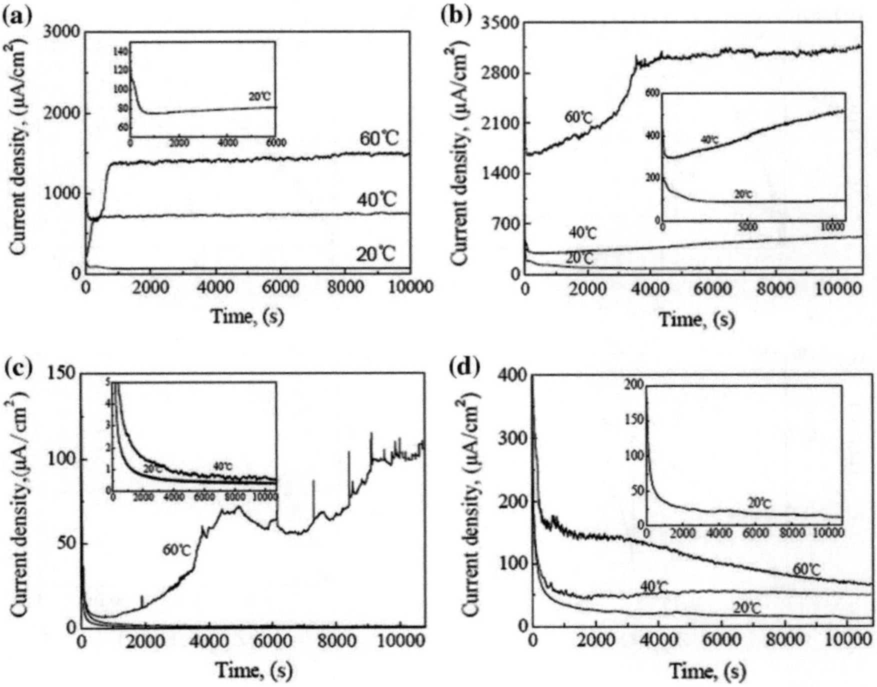

Fig. 2 Potentiostatic polarization in difference solutions, **a** HCl, **b** H_2SO_4, **c** NaCl and **d** NaOH

test, as no passive film is formed on the surface, the mass loss increases with time at first stage. The mass loss is stable gradually, since the formation of passive film and the precipitation of corrosion products on the surface with immersion time increase.

The surface morphologies immersed in four solutions are shown in Figs. 5, 6, 7 and 8. The 'dried riverbed' cracked morphologies are clearly observed on the surface immersed in acidic solutions, as shown in Fig. 5 (HCl solution) and Fig. 6 (H_2SO_4 solution). While no crack, just smoothing surfaces are remained in NaCl (Fig. 7) and NaOH (Fig. 8) solutions. In acidic solutions, firstly, the groove is observed near the inclusion particle in Fig. 5b, and then some cracks originate from the inclusion particles, even pass through them, as shown in Figs. 5c and 6b. Finally, more and more cracks are formed and propagated along all direction on the surface, thus the 'dried riverbed' cracked morphologies are formed, as shown in Figs. 5d and 6d. According to the variety of corrosion morphology characteristics, it may be deduced that the inclusion particles are dissolved firstly, and some corrosion products are formed on the position of inclusion particles. With acceleration of corrosion products on the surface, the microstructure in the position of inclusion particles becomes more loosing, compared to other. Therefore, the cracks will initiate and propagate with the great internal stress in the cast specimen.

In alkaline solutions, it is well know that Fe_2O_3 is main corrosion product and it is stable. So a greater amount of Fe_2O_3 will precipitated on the surface and this was

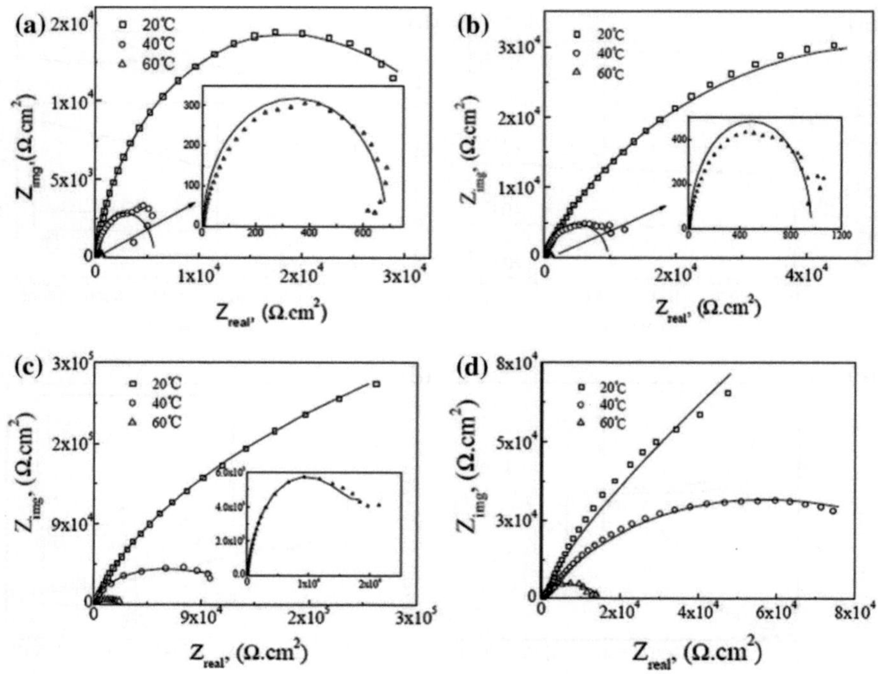

Fig. 3 Nyquist plots n in difference solutions, **a** HCl, **b** H_2SO_4, **c** NaCl and **d** NaOH

Fig. 4 Corrosion rate in difference solution at room temperature

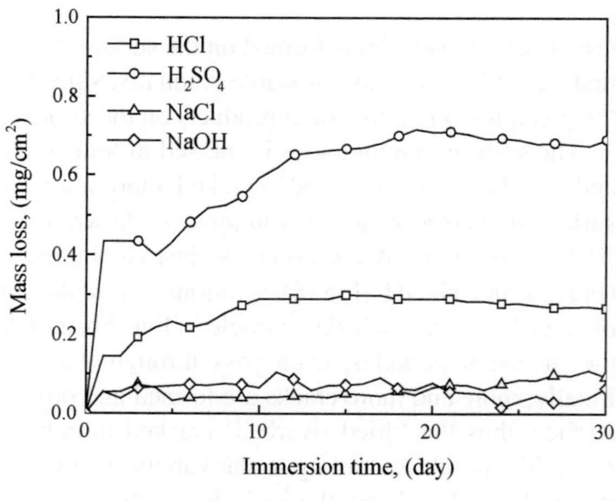

indentified in the literature [10]. The acicular shape Fe_2O_3 layer is porosity and loosing, which is benefit to the release of internal stress in cast specimen. Therefore, no crack is formed on the surface.

It is well known that the reaction rate is accelerated generally since the intensifying activity of ions or elements with increase of ambient temperature. Though

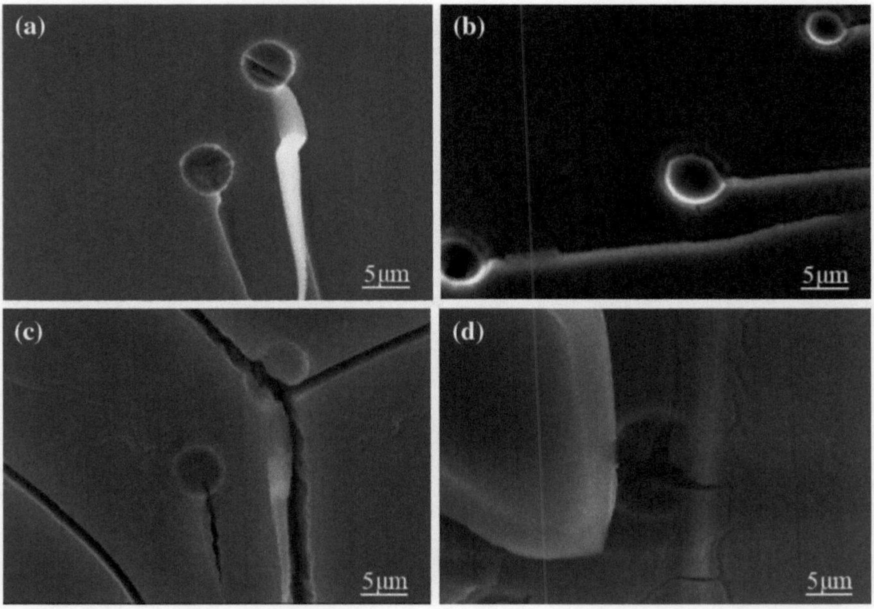

Fig. 5 Corrosive surface morphology in 0.5 M HCl solution, **a** as-cast, **b** 1 day, **c** 3 days and **d** 7 days

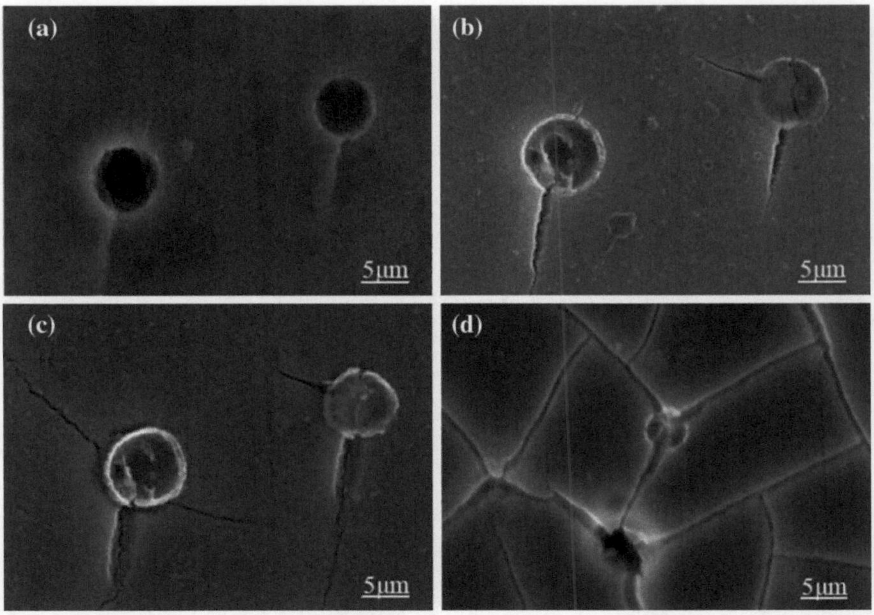

Fig. 6 Corrosive surface morphology in 0.5 M H_2SO_4 solution, **a** as-cast, **b** 1 day, **c** 3 days and **d** 7 days

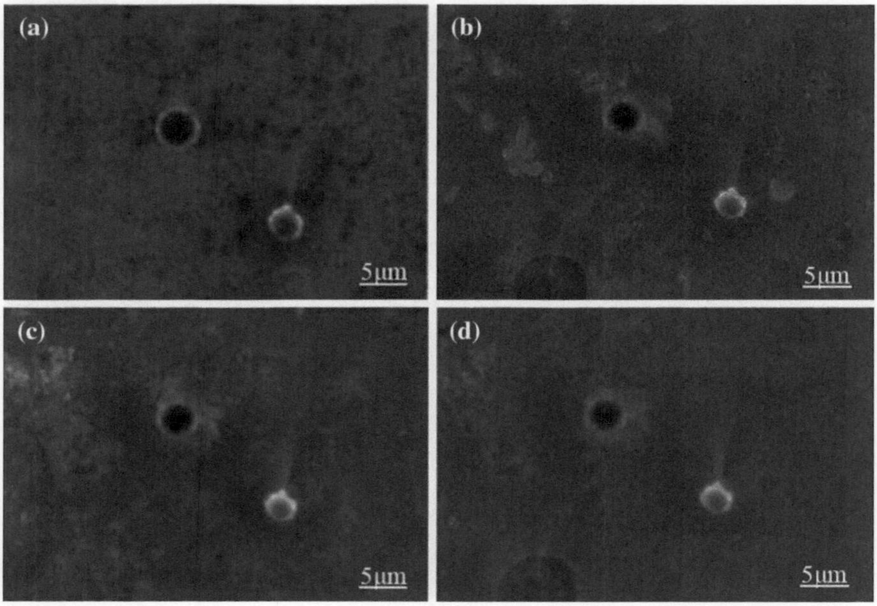

Fig. 7 Corrosive surface morphology in 0.5 M NaCl solution, **a** as-cast, **b** 1 day, **c** 3 days and **d** 7 days

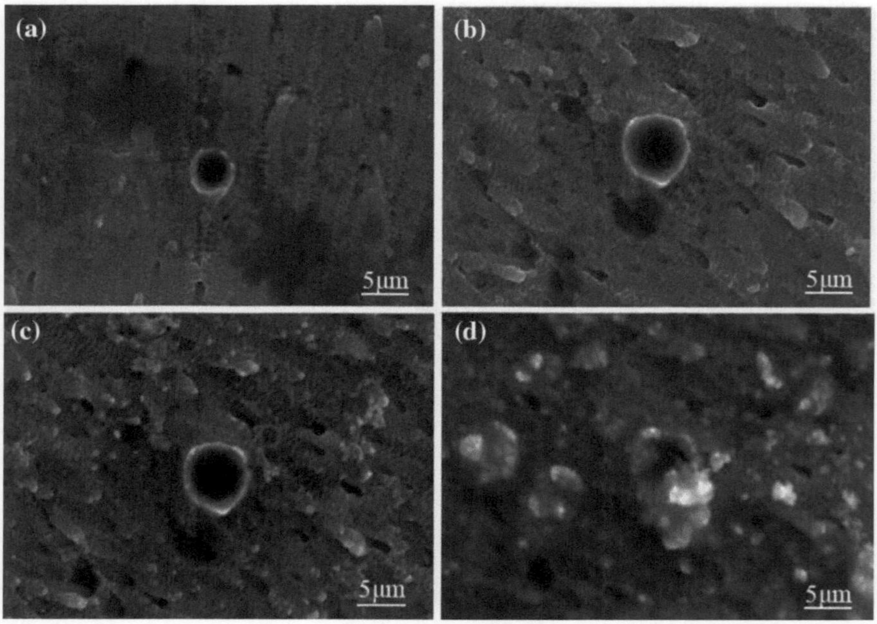

Fig. 8 Corrosive surface morphology in 0.5 M NaOH solution, **a** as-cast, **b** 1 day, **c** 3 days and **d** 7 days

$Fe_{48.7}Cr_{18}Mn_{1.9}Mo_{7.4}W_{1.6}V_{15.2}C_{3.8}Si_{2.4}$ amorphous alloy coating is spontaneously passivated with a low passive current density in different ambient temperatures, the passive current density progressively increases and the pitting potential decreases as temperature increases [14]. While the erosion rates of FeBSiNb amorphous coating decrease with increase of test temperature, since the formation of a thick oxide, and surface characteristic modification during erosion testing [15]. Liu [16] found that the Fe-based amorphous composite exhibits an excellent hot corrosion resistance and excellent thermal stability in molten Na_2SO_4 + K_2SO_4 salt at 923 K, due to the formation of a continuous and protective Cr_2O_3 oxide which acts as an effective barrier to the inward diffusion of corrosive species during the hot corrosion process.

In general, the CPE is related to the layer capacitance of the electrode. The thicker the corrosion product is, the higher the CPE is. With increase of ambient temperature, the corrosion reaction is also accelerating, thus the corrosion product is accumulating. The R_{ct} is charge transfer resistance. The increase of R_{ct} indicates an increase in the insulating character of the oxide film [17].

Conclusions

(1) With increase of ambient temperature, the corrosion rate of Fe-based amorphous alloy will accelerate in all solutions.
(2) No pitting corrosion occurs in acidic solutions, though the inclusion particle is dissolved firstly. Seemly, some pit can be formed in NaCl solution, when ambient temperature exceeds 60 °C.
(3) Fe-based amorphous alloy generally exhibits more excellent corrosion resistance in alkaline solutions than that in acidic solutions.

Acknowledgements The work was supported by the National Natural Science Foundation of China (No. 51461031), Department of Education Fund of Jiangxi Province (GJJ150733), Natural Science Foundation of Jiangxi Province (20161BBH80031) and Power Construction Corporation of China (DJ-ZDXM-2014-34).

References

1. Z.L. Long, Y. Shao, A. Inoue, Cr effects on magnetic and corrosion properties of Fe–Co–Si–B–Nb–Cr bulk glassy alloys with high glass-forming ability. Intermetallics **15**, 1453–1458 (2007)
2. H.B. Fan, W. Zhang, J. Shen, Corrosion behavior of $Fe_{41}Co_7Cr_{15}Mo_{14}C_{15}B_6Y_2$ bulk metallic glass in sulfuric acid solutions. Metallur. Mater. Trans. A **42**, 1524–1533 (2011)
3. S.J. Pang, T. Zhang, A. Inoue, Bulk glassy Fe–Cr–Mo–C–B alloys with high corrosion resistance. Corr. Sci. **44**, 1847–1856 (2002)

4. A. Pardo, E. Otero, M.C. Merino, Influence of Cr addition on the corrosion resistance and magnetic properties of amorphous $Fe_{73.5}Si_{13.5}B_9Nb_3Cu_1$ in simulated industrial environments. J. Non-Cryst. Solids **287**, 421–427 (2001)
5. J.C. Farmer, J. Haslam, S. Day, The corrosion resistance of Fe-based amorphous metals: $Fe_{49.7}Cr_{17.7}Mn_{1.9}Mo_{7.4}W_{1.6}B_{15.2}C_{3.8}Si_{2.4}$ and other compositions. Mater. Sci. Tech. 318–392 (2007)
6. Y. Wang, Y.G. Zheng, Z.M. Yao, Investigation of HVOF-sprayed Fe-based amorphous metallic coatings with high erosion resistance and corrosion resistance, in *The 6th China Corrosion Conference*, pp. 1030–1036 (2011)
7. J. Lu, C. Liu, Y. Han, Analysis of utilizing anti-corrosion Fe-based amorphous coating material with the flue gas waste heat in the power plant boiler. Electr. Power Sci. Eng. **30**, 13–18 (2014)
8. H.X. Li, J.E. Gao, S.L. Wang, Formation, crystallization behavior, and soft magnetic properties of FeCSiBP bulk metallic glass fabricated using industrial raw materials. Metall. Mater. Trans. A **43A**, 2615–2619 (2012)
9. S.L. Wang, H.X. Li, S. Yi, Effects of Cr contents in Fe-based bulk metallic glasses on the glass forming ability and the corrosion resistance. Mater. Chem. Phys. **113**, 878–883 (2009)
10. S.L. Wang, H.X. Li, S. Yi, Effects of electrolyte pH on the electrochemical behavior of Fe-based bulk metallic glass. Met. Mater. Inter. **18**, 791–797 (2012)
11. S.L. Wang, S. Yi, The corrosion behaviors of Fe-based bulk metallic glasses in a sulfuric solution at 70 & #xB0;C. Intermetallics **18**, 1950–1953 (2010)
12. S.L. Wang, J.C. Cheng, S. Yi, Precipitation of sulfide particle in situ formed in Fe-based bulk metallic glass. Foundy **62**, 491–495 (2013)
13. S.L. Wang, S.B. Chen, S. Yi, Effects of sulphide on pit corrosion in metallic glass steel. Mater. Res. Innov. **18**, 638–641 (2014)
14. R.Q. Guo, C. Zhang, L. Liu, Corrosion and wear resistance of a Fe-based amorphous coating in underground environment. Intermetallics **30**, 94–99 (2011)
15. J.B. Cheng, X.B. Liang, Y.X. Chen, High-temperature erosion resistance of FeBSiNb amorphous coatings deposited by arc spraying for boiler applications. J. Therm. Spray. Techn. 820–827 (2013)
16. W.Y. Liu, Y. Hou, C. Liu, Hot corrosion behavior of a centimeter Fe-based amorphous composite coating prepared by laser cladding in molten Na_2SO_4 + K_2SO_4 salts. Surf. Coat. Techn. **270**, 33–38 (2015)
17. F. Rosalbino, S. Delsante, G. Borzone, Influence of rare earth metals on the characteristics of anodic oxide films on aluminium and their dissolution behaviour in NaOH solution. Corr. Sci. **52**, 322–328 (2010)

Synthesis of In-Suit $Ti_{46.8}Zr_{19.5}Nb_{11.7}Cu_5Be_{14.5}Ag_{2.5}$ Amorphous Composite Materials with High Tensile Ductility

Yonghua Shen, Huanwu Cheng, Weiwei Chen and Yunfei Xue

Abstract A high tensile ductility in-suit $Ti_{46.8}Zr_{19.5}Nb_{11.7}Cu_5Be_{14.5}Ag_{2.5}$ amorphous composite materials was prepared successfully by arc melting and injection casting into a water cooling copper mold. The composition and microstructure were characterized by X-ray diffraction (XRD) with Cu Kα radiation and scanning electron microscope (SEM). The results showed the fully developed dendritic structure of the β phase with green color was uniformly distributed in the amorphous matrix, while, the size and proportion of second phase were 30 μm and 78% respectively. Standard size of tensile specimen was designed to test its tensile mechanical properties. The occurrence of rare necking deformation in amorphous materials showed the $Ti_{46.8}Zr_{19.5}Nb_{11.7}Cu_5Be_{14.5}Ag_{2.5}$ amorphous composite materials possessed a good ductility ability. The fracture contained the fiber area, radiation area and shear lips. Large proportion dimple area showed the dendritic phase distributed uniformly played a great role in the process of deformation.

Keywords Amorphous alloy · Necking · In situ dendritic phase
Dimple · Tensile ductility

Y. Shen · H. Cheng (✉) · W. Chen · Y. Xue
School of Materials Science and Engineering,
Beijing Institute of Technology, Beijing 100081, China
e-mail: chenghuanwu@bit.edu.cn

Y. Shen
e-mail: shenyh1990@163.com

W. Chen
e-mail: wwchen@bit.edu.cn

Y. Xue
e-mail: xueyunfei@bit.edu.cn

Y. Shen · H. Cheng · W. Chen · Y. Xue
National Key Laboratory of Science and Technology
on Materials Under Shock and Impact, Beijing 100081, China

Y. Han (ed.), *Advanced Functional Materials*,
https://doi.org/10.1007/978-981-13-0110-0_36

Introduction

As amorphous metals do not have long-range structural order like crystalline metals, they exhibit unique properties such as ultrahigh strength, high elastic strain limit, excellent abradability and corrosion resistance, which enable their potential application in structural and functional materials field [1–4]. Unfortunately, their processing and application are severely limited by the poor plasticity and subsequent catastrophic fracture. Upon yielding, the plastic deformation of amorphous metals are restricted to highly localized into narrow shear bands. These shear bands can rapidly propagate across the whole sample after their initiation and initiate cracking, leading to macroscopic catastrophic fracture. The deformation is highly localized into narrow shear bands, followed by the rapid propagation of these shear bands across the sample and subsequently premature catastrophic fracture [4–7].

The attainment of both strength and toughness is a big challenge for amorphous metals. The most efficient way to improve plasticity is inhibiting the rapidly propagate of shear bands. The deformation occurred through the development of highly organized patterns of regularly spaced shear bands distributed uniformly throughout the sample. By adjusting the composition and preparation technology, the introduction of second phase can effectively hinder the propagation of individual shear band within materials and seed the initiation of multiple shear bands throughout the sample, which greatly improves the deformation ability [8, 9].

Compared to previous researches [10, 11], we adjust the alloy composition by introducing the element Ag. Lots of study showed that the introduction of Ag could improve the thermal stability, the glass forming ability and the mechanical properties [12, 13]. In present work, we are committed to prepare amorphous composite materials with high ductility. $Ti_{46.8}Zr_{19.5}Nb_{11.7}Cu_5Be_{14.5}Ag_{2.5}$ amorphous composite materials with high tensile ductility were successfully prepared. Then, the tensile properties were performed and the mechanism of deformation was revealed.

Experiment

Alloy ingots with composition of $Ti_{46.8}Zr_{19.5}Nb_{11.7}Cu_5Be_{14.5}Ag_{2.5}$ were prepared by arc melting under Ti-gettered argon atmosphere. The alloy elements with a purity of 99.9% or better were used. The ingot was remelted at least four times to ensure its composition homogeneity. Then specimens with diameter of 5 mm were produced by injection casting into a water cooling copper mold. The size, morphology and distribution of in-suit crystal phase were observed by optical microscope (OM). Structural investigation of the samples was carried out using XRD with Cu Kα radiation and transmission electron microscopy (TEM). Glass transition and crystallization behavior were examined using differential scanning calorimeter (DSC) under a continuous argon flow under a heating rate of 0.33 K/s. Tensile properties were tested at room temperature under a strain rate of 1e-3/s using the Instron5985 electronic universal material testing machine. The structure and the fracture surfaces of the specimens were studied using a SEM (S4800).

Results and Discussions

Microstructure Characteristics. Figure 1 was XRD patterns of $Ti_{46.8}Zr_{19.5}Nb_{11.7}Cu_5Be_{14.5}Ag_{2.5}$ (at.%) amorphous composite materials. Both the diffuse amorphous peaks between 30° and 50° and relatively sharp peaks with structure of β dendritic crystal phase were discovered in XRD. The microstructure of amorphous composites in Fig. 2, showed that the fully developed dendritic structure of the β phase with green color was uniformly distributed in the amorphous matrix. The dendritic crystal was body-centered cubic (b.c.c.) solid solution containing primarily Ti, Zr, and Nb, as verified by XRD and, Energy Dispersive Spectrometer (EDS). The size of the dendritic phase (L) characterized by the span distance of the single dendrite was about 30 μm. This gave an estimated fraction of 22% β phase by volume (78% amorphous phase). Hence, in the process of material preparation,the amorphous composites produced two-phase microstructure, consisting of β phase and amorphous phase.

Figure 3 showed DSC curves of melt-spun $Ti_{46.8}Zr_{19.5}Nb_{11.7}Cu_5Be_{14.5}Ag_{2.5}$ amorphous composite materials. It could be observed an endothermic peak at about 660 K, proving that the sample occurred clear glass transition in the process of heating. The supercooled liquid region ΔTx = Tx − Tg, was about 68 K, followed by a sharp exothermic crystallization peak, which showed the samples transition from the supercooled liquid region to the crystallization phase stability. The occurrence of typical glass transformation endothermic peak and a larger degree of

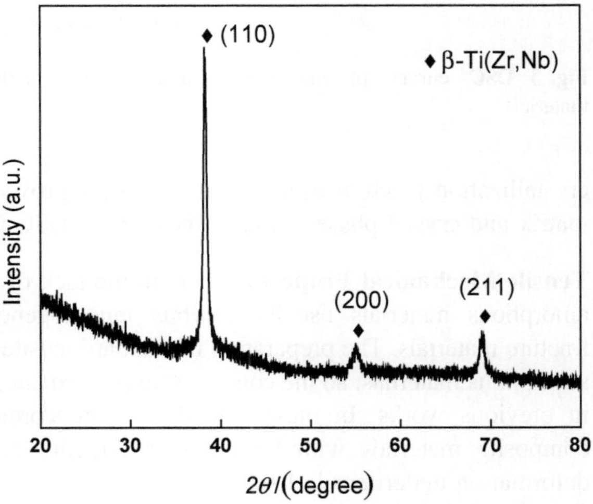

Fig. 1 XRD patterns of amorphous composite materials

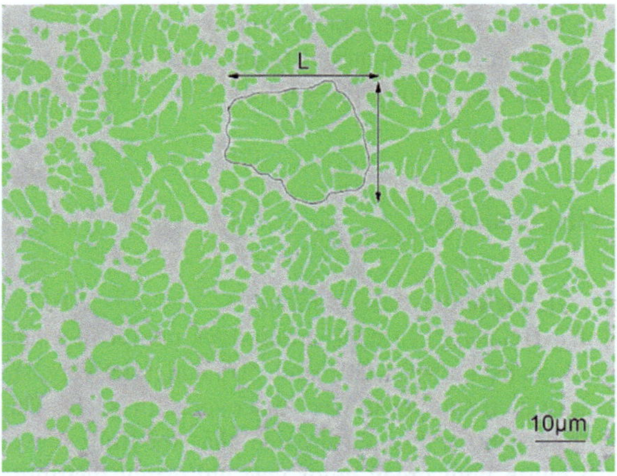

Fig. 2 Microstructure morphology of amorphous composite materials

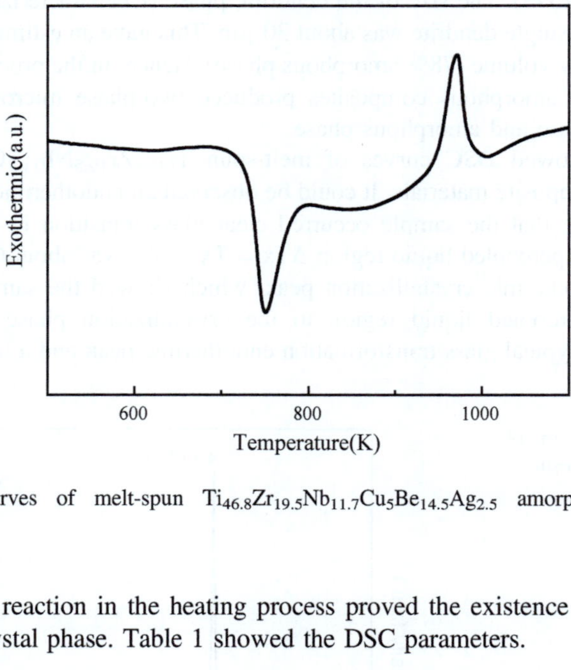

Fig. 3 DSC curves of melt-spun $Ti_{46.8}Zr_{19.5}Nb_{11.7}Cu_5Be_{14.5}Ag_{2.5}$ amorphous composite materials

crystallization reaction in the heating process proved the existence of amorphous matrix and crystal phase. Table 1 showed the DSC parameters.

Tensile Mechanical Properties. Due to the lack of dislocation and other defects, amorphous materials usually exhibits inhomogeneous deformation like brittle fracture materials. The preparation of standard tensile specimen was difficult for the amorphous materials, so the compressive performance was the main research object in previous works. In present work, we are committed to prepare amorphous composite materials with high tensile ductility and reveal the mechanism of deformation under tensile loading.

Table 1 DSC parameters: the glass transition temperature Tg, crystallization temperature Tx, the alloy melting point Tm and supercooled liquid region ΔTx = Tx − Tg

Composition	Tg/K	Tx/K	Tm/K	ΔTx/K
$(Ti_{48}Zr_{20}Nb_{12}Cu_5Be_{15})_{97.5}Ag_{2.5}$	665	733	949	68

Figure 4 was macroscopic fracture of amorphous composite materials under tensile state. The size of standard tensile specimen was shown in Fig. 4a. Figure 4b was the fracture morphology after the tensile mechanical properties test. The necking deformation was occurred during the tests, which was typical deformation characteristics in plastic metal. It indicated that the tensile specimen experienced a greater degree of deformation.

Figure 5 showed the tensile stress–strain curve of amorphous composite materials. It could be seen that it experienced the elastic deformation about 2.4%, and the amorphous composite materials yield about 1209 MPa, followed the large plastic deformation about 15.8% with a high compressive strength of 1343 MPa.

The occurrence of rare necking deformation in amorphous materials showed the $Ti_{46.8}Zr_{19.5}Nb_{11.7}Cu_5Be_{14.5}Ag_{2.5}$ amorphous composite materials possessed a good ductility ability. Figure 6 showed tensile fracture of $Ti_{46.8}Zr_{19.5}Nb_{11.7}Cu_5Be_{14.5}Ag_{2.5}$ amorphous composite materials. It could be seen that the fracture morphology was typical ductile cup cone fracture characteristics, and the tensile fracture was divided into fiber area, radiation area and shear lips. Researches had shown that fiber area was a signal of slow crack extension, while characters of radiation area indicated the materials experienced rapidly expanding of crack. Under certain conditions, according to the area and the proportion of three zone in the

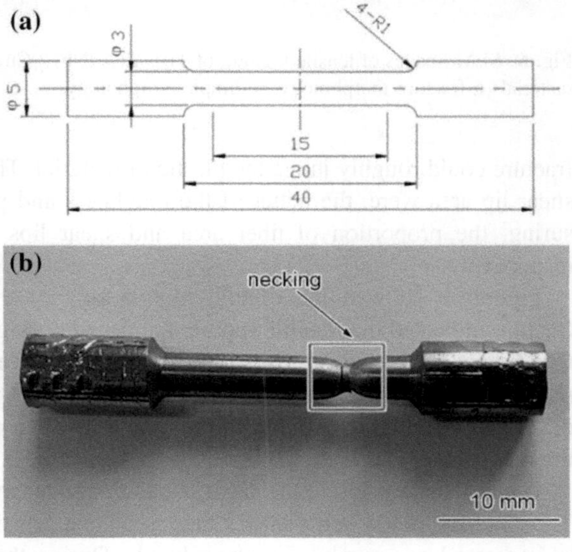

Fig. 4 Macroscopic fracture of amorphous composite materials under tensile state **a** tensile samples; **b** fracture morphology

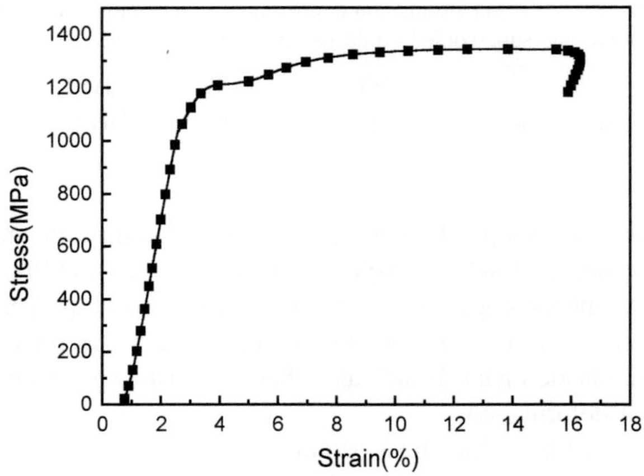

Fig. 5 Tensile stress-strain curves of $Ti_{46.8}Zr_{19.5}$-$Nb_{11.7}Cu_5Be_{14.5}Ag_{2.5}$ amorphous composite materials

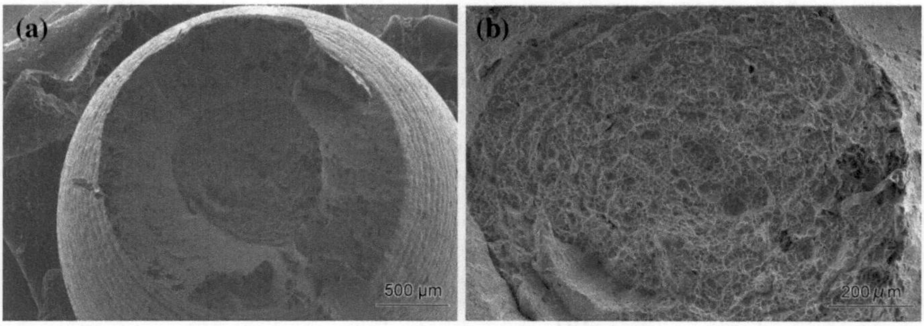

Fig. 6 SEM images of tensile fracture of $Ti_{46.8}Zr_{19.5}Nb_{11.7}Cu_5Be_{14.5}Ag_{2.5}$ amorphous composite materials **a** fracture morphology; **b** dimple morphology

fracture could roughly judge the plastic of material. The bigger of fiber area and the shear lip area were, the better of the toughness and plastic material had. By measuring, the proportion of fiber area and shear lips area were 15.6, and 46.7% respectively.

Figure 6b showed the dimple morphology of the fracture. Dimples on the fracture indicated that ductile rupture happened during fracture process. The size of the dimple was uniformly distributed, which showed the amorphous composite materials experienced an uniform deformation. This was consistent with the fully developed dendritic structure of the β phase in green color uniformly distributed in the amorphous matrix.

Under tensile loading, the shear bands firstly generated in the amorphous phase. The dendritic phase in the matrix hindered the expansion of the shear bans, leading to generate large numbers of shear bands. During the movement, the shear bands cut across or got around the dendritic phase, finally the movement direction of the shear bands were changed. All these movement consumed more energy, increasing the strength of the material and the plastic deformation.

Conclusions

In-suit $Ti_{46.8}Zr_{19.5}Nb_{11.7}Cu_5Be_{14.5}Ag_{2.5}$ amorphous composite materials with high tensile ductility were successfully prepared by arc melting and injection casting into a water cooling copper mold. XRD and microscope analyses were used to observe the composition and microstructure. The results showed the fully developed dendritic structure of the β phase with green color was uniformly distributed in the amorphous matrix, while the size and proportion of second phase was 30 μm and 78% respectively. Standard size of tensile specimen was designed to test its tensile mechanical properties. The occurrence of rare necking deformation in amorphous materials shows that the $Ti_{46.8}Zr_{19.5}Nb_{11.7}Cu_5Be_{14.5}Ag_{2.5}$ amorphous composite materials prepared by in-suit second crystal phase possess a good ductility ability. The fracture contains the fiber area, radiation area and shear lips. Large proportion dimple area shows the dendritic phase distributed uniformly plays a great role in the process of deformation. The shear bands cut across or get around the dendritic phase, finally the movement direction of the shear bands are changed. All these movements can consume more energy, increasing the strength of the material and the plastic deformation.

Acknowledgements This work was supported by Beijing Materials Analysis & Testing Union (MATU) and Testing center of USTB (china).

References

1. Z.Q. Hu, Q.H. Song, H.F. Zhang et al., *Metastable Metal Materials* (Science Press, Beijing, 2006)
2. W.L. Johnson, M.D. Demetriou, J.S. Harmon et al., Rheology and ultrasonic properties of metallic glass-forming liquids: a potential energy landscape perspective. MRS Bull. **32**, 644–650 (2007)
3. M. Martin, N.N. Thadhani, Mechanical properties of bulk metallic glasses. Prog. Mater Sci. **55**, 759–839 (2010)
4. A. Inoue, Stabilization of metallic supercooled liquid and bulk amorphous alloys. Acta Mater. **48**, 279–306 (2000)
5. F. Spaepen, A microscopic mechanism for steady state inhomogeneous flow in metallic glasses. Acta Mater. **25**, 407–415 (1977)
6. C.A. Schuh, T.C. Hufnagel, U. Ramamurty, Mechanical behavior of amorphous alloys. Acta Mater. **55**, 4067–4109 (2007)
7. D.C. Hofmann, J.Y. Suh, A. Wiest, G. Duan, M.L. Lind, M.D. Demetriou, W.L. Johnson, Designing metallic glass matrix composites with high toughness and tensile ductility. Nature **451**, 1085–1090 (2008)

8. D.C. Hofmann, J.Y. Suh, A. Wiest, M.L. Lind, M.D. Demetriou, W.L. Johnson, Development of tough, low-density titanium-based bulk metallic glass matrix composites with tensile ductility. Proc. Natl. Acad. Sci. U. S. A. **105**, 20136–20140 (2008)
9. P. Wang, H.C. Kou, J. Bai, J. Wang, R. Hu, J.S. Li, Preparation and properties of Ti-based metallic glass composites. Acta Metall. Sin. **29**, 120–123 (2011)
10. Y.H. Shen, H.W. Cheng, Y.F. Xue, L. Wang, W.J. Chen, B.P. Wang, H. Liu, C.K. Zhu, Effect of sample size upon quasi-static compressive loading of $Ti_{48}Zr_{20}Cu_5Nb_{12}Be_{15}$ amorphous composites. Mater. Sci. Forum **816**, 96–101 (2015)
11. Y.H. Shen, H.W. Cheng, Y.F. Xue, B.P. Wang, W.W. Chen, W.J. Chen, L. Wang, J. Liu, Compression properties of $Ti_{48}Zr_{20}Nb_{12}Cu_5Be_{15}$ in-situ crystal phase amorphous alloy. Rare Metal Mat. Eng. **45**, 92–96 (2016)
12. Y. Li, L.J. Yue, K. Xie, W.J. Ying, Effect of micro-compositions Y and Ag on the glass-forming ability of Cu-based amorphous alloys. Chin. Rare Earths **32**, 30–34 (2011)
13. S.T. Wang, *Glass Forming, Crystallization and Mechanical Property of CuZrAl and CuZrAlAg BMG* (Yanshan University, 2007)

Effect of Argon Filling Ratio on Heat Transfer Coefficient of Double-Glazing

Junjie Li, Yingliang Tian, Shibing Sun, Jinwei Li, Lu Zhang and Keyu Chen

Abstract In this work the effect of argon filling ratio and air space thickness on the heat transfer coefficient of double-glazing was studied. The heat transfer coefficient of double-glazing was measured by heat flow meter method (based on standards ISO10291, ISO 10292, ISO8301 and DIN EN 12939), during this process the testings of argon filling ratio and emissivity of coated glass surface were involved. For the thermal performance of double-glazing, the influence of various factors on heat transfer coefficient was analyzed. The comparison was conducted between coated double-glazing (one glass pane coated with low-emissivity) and non-coated double-glazing. Within the thickness of air space in the range of 9–18 mm, the higher the air space thickness, the higher the filling ratio of argon gas, the better the thermal insulation performance achieved. Under the same gas filling ratio, the heat transfer coefficient of coated double-glazing (emissivity is 0.13) which is much lower than that of non-coated double-glazing, can effectively meet the requirement of building energy efficiency in China.

Keywords Argon · Air space thickness · Double-glazing · Coated glass
Heat transfer coefficient · Building energy efficiency

Introduction

Building energy consumption is an important part of the total social energy consumption. According to the statistics, buildings contribute to 28.6% of energy consumption in China [1–3] and to about 40% in Western countries [4–6]. With the development of China's urbanization, urban population will greatly increase, and

J. Li (✉) · Y. Tian · S. Sun · J. Li · L. Zhang · K. Chen
School of Materials Science and Engineering,
Beijing University of Technology, Beijing 100124, China
e-mail: lijunjie518@163.com

Y. Tian
e-mail: boli106@126.com

Y. Han (ed.), *Advanced Functional Materials*,
https://doi.org/10.1007/978-981-13-0110-0_37

the proportion of building energy consumption will continue to raise. It is expected that China's building energy consumption will be more than industrial, transportation, and eventually live in the first place of social energy consumption, reaching 33% and even more. Furthermore, China has become the world's largest energy consumer according to the Statistical Review of World Energy released by British Petroleum (BP) June 2011 [7]. Therefore, some measures should be taken to reduce building energy consumption in China.

To reduce building energy consumption, the thermal insulation performance of building envelope should be improved. Recently, the trends in the design of buildings, mainly relating to office building, have radically changed and a more frequent use of large glass wall is observed [8]. Glass envelopes, such as windows, doors as well as facades, are important parts of building envelope which provide interior with plenty of light, air ventilation and also the ability to view the outside [9]. Besides, the use of a large number of glasses has brought a better modern fashion aesthetic effect to the buildings [10]. However, they have, in general, poor thermal performance characteristics among the building components, and hence, they play a significant role in the heating and cooling demand of buildings in winter and summer [11, 12]. In China, for example, the heat transfer coefficient of most building glass windows is above 2.8 W/(m^2 k) [13], which account for more than 50% of the building energy consumption [14].

In order to improve the insulation performance of glass windows and curtain walls, many construction engineers and scientific research workers have progressed. Multi-pane glass is an effective means to achieve energy efficiency [15]. With the increase of the number of layers of glass, the thermal insulation performance is increased significantly. However, too much glass layer can increase the cost of manufacturing process, so the double layer glass is commonly used. Generally, the heat transfer coefficient of non-coated double-glazing is about 2.9 W/(m^2 k) (12 mm air space thickness), which obviously has more effective thermal performance compared with single layer glass, but it can not meet the new energy saving requirements. In the UK, for example, the new energy saving requirement of glass windows is lower than 1.4 W/(m^2 k) [16]. Accordingly, China has also adjusted the energy efficiency standards for the windows in some areas that to be required in the region of or lower than 1.5–2.0 W/(m^2 k) since 2013 [17]. Obviously, to increase the number of layers of glass is not enough. In recent years, there have been many ways to further improve the thermal performance of double-layer glazing glass. Vacuum glazing, for example, is a good thermal insulator, but the technology faces some challenges in maintaining vacuum for long period [18, 19]. Similarly, aerogel glazing PCM glazing, TIM glazing are very efficient for thermal regulation of building envelop, but the commercialization is still a challenging due to their high cost in China [20–22].

Nowadays, the coated double-glazed filled with argon is widely being used among European countries. Argon is a colorless, tasteless, non-toxic, stable, low thermal conductivity of inert gas, and the content in the air is relatively high, about 0.934%. Argon is not sensitive to the solar energy, and it is widely used in double-glazing to improve its thermal insulation performance [23]. At present, the use of argon filled double-glazing in China is very low, so it is suitable for promotion of the glass to meet the requirements of China's new building energy efficiency standards. However, the effect of the filling ratio and the thickness of argon gas on the heat transfer performance of different double-glazing are still not clear. In this paper, the influence of different argon filling ratio and filling thickness and coating on heat transfer coefficient of double-glazed is studied. This research has very important practical application value for building energy saving.

Double-Glazed Sample Preparation

Structure of Double-glazing. Double-glazed sample is composed of two pieces of glass through the aluminum spacer support, surrounded by a rubber seal showing structure of glass + air space + glass (sample dimension: 300 mm × 300 mm). There are two kinds of glass panes (thickness 6 mm) which can be used in the sample: non-coated glass pane and coated glass pane. The air space thickness is 9, 12, 15, 18 mm. All of the double-glaze samples was produced by Beijing LAN-STAR Hualin Curtain Wall Glass Co. Ltd. (Fig. 1).

Argon Filling. In order to effectively control and to ensure the content of argon in the air space of the double-glazed sample, the air tightness of the insulating glass was detected. Specifically, the double-glazed sample was placed in the water firstly, and filled with 50% excess air, then it was observed whether the rubber sealing edge has gas overflow. After air tightness detection, the pure argon was filled with pressure-relief mode to control the argon filling ratio. The argon filling ratio of double-glazed sample was controlled at 0, 10, 20, 30, 40, 50, 60, 70, 80, 85, 90, 95 and 97%. In order to ensure the accuracy of the argon filling ratio, the Gas glass Laser was used to detect the argon content. The instrument, provided by Finland sparklike Co. Ltd., is capable to accurately measure the content of argon gas and other insert gases [24] (Fig. 2).

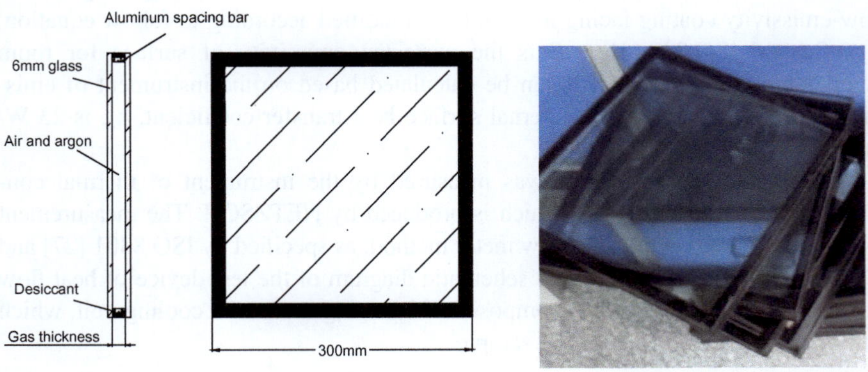

Fig. 1 Structure of double-glazing

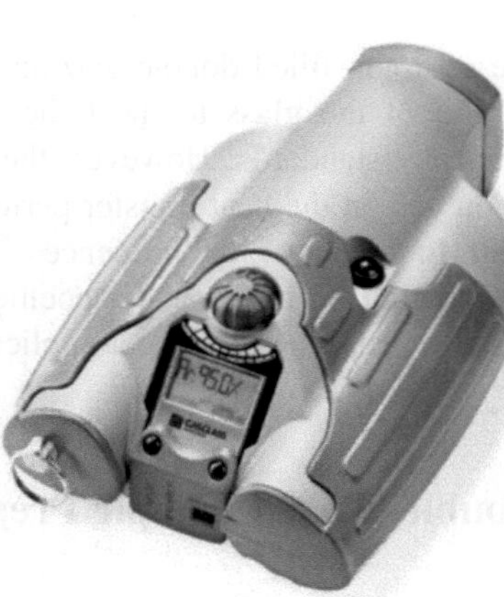

Fig. 2 Gasglass Handheld v2

Testing and Performance Evaluation of Double-glazed. The coefficient of heat transfer, the U-value, is an important index to reflect the insulation performance of double-glazing. The calculation of U-value depends on the thermal resistance of the multiple glazing as well as the heat transfer coefficient of both the external surface and internal surface according to the following equation [ISO 10291] [25]:

$$\frac{1}{U} = R + \frac{1}{h_e} + \frac{1}{h_i} \tag{1}$$

where

R is the thermal resistance of the multiple glazing, in square metres Kelvin per watt (m^2 k/W);

h_e is the external surface heat transfer coefficient, in watt per square metre Kelvin W/(m^2 k);

h_i is the internal surface heat transfer coefficient, in watt per square metre kelvin W/(m^2 k).

For non-coated double glazing, $h_i = 8$ W/(m^2 k); For double glazing with a low-emissivity coating facing inward, h_i, is modified according to follow equation: $h_i = 3.6 + 4.4\varepsilon/0.837$. Here ε is the corrected emissivity of surface for room temperature radiation [26], it can be calculated based on the instrument of emissivity testing in Fig. 3. The external surface heat transfer coefficient, h_e, is 23 W/(m^2k)[ISO 10291].

The thermal resistance, R, was measured by the instrument of thermal conductivity detector HFM 436, which is produced by NETZSCH. The measurement principle is based on the heat flow meter method, as specified by ISO 8301 [27] and DIN EN 12939 [28] of the. The schematic diagram of the test device of heat flow meter is shown in Fig. 4. It is composed of a heating unit and a cooling unit, which are located on both sides of the sample.

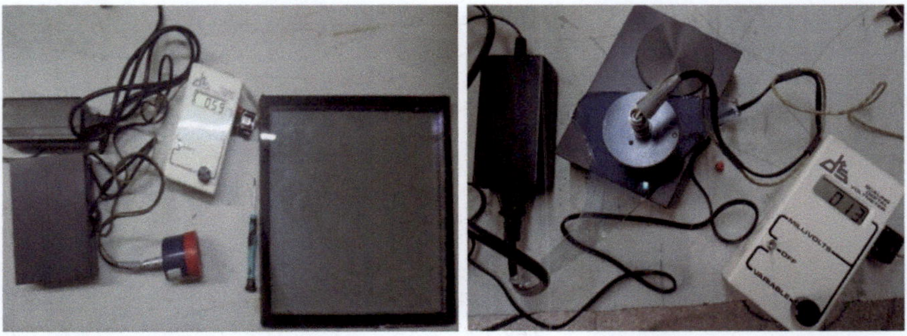

Fig. 3 Emissivity testing instrument

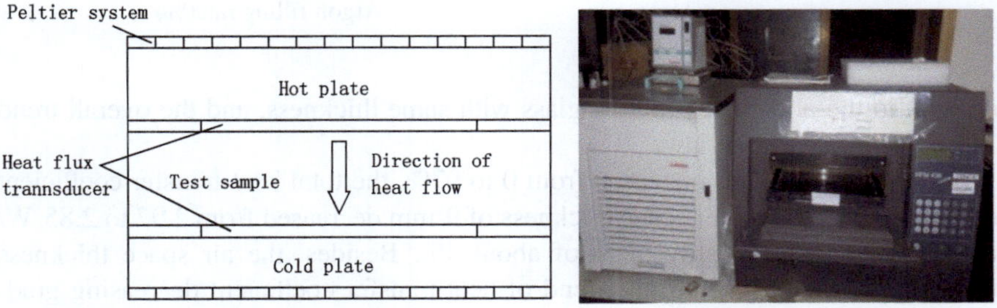

Fig. 4 Schematic diagram of measuring device of heat flow meter method

The thermal resistance, R, is calculated with the following equation:

$$R = \frac{2(T_1 - T_2)}{(q_1 + q_2)N(t)} \tag{2}$$

where

T_1 is the average specimen hot side temperature, in kelvins;

T_2 is the average specimen cold side temperature, in kelvins;

q_1 when measuring the balance of the instrument, the hot side heat flow parameters of the sample;

q_2 when measuring the balance of the instrument, the cold side heat flow parameters of the sample;

N(t) correction factor, instrument measurement parameters.

Results and Discussion

The Effect of Argon Filling Ratio. The effect of argon filling ratio on heat transfer coefficient of non-coated (without low-e coating) double-glazing were carried out (Fig. 5). The heat transfer coefficient decreases with the increase of the filling ratio

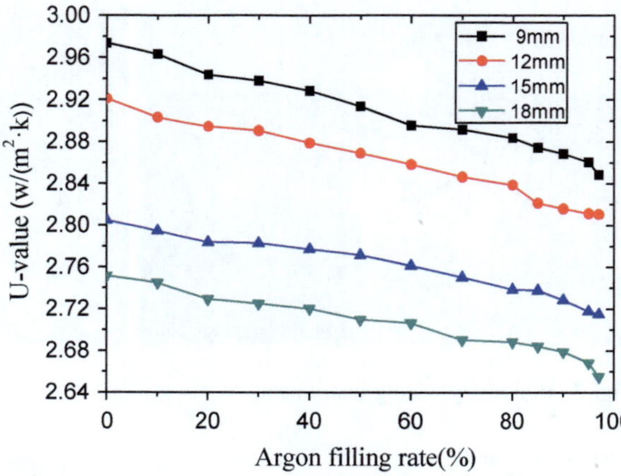

Fig. 5 Effect of argon filling ratio on overall heat transfer coefficient of non-coated double-glazed windows

of argon to the non-coated hollow glass with same thickness, and the overall trend showed a linear decline law.

As argon filling ratio increased from 0 to 97%, the total heat transfer coefficient of double-glazing with air space thickness of 9 mm decreased from 2.97 to 2.85 W/(m^2 k) which is an improvement of about 4%. Besides, the air space thickness increased from 9 to 18 mm, the trend of heat transfer coefficient decreasing gradually weakened. As for thickness of 12 mm, the heat transfer coefficient decreased from 2.92 to 2.81 W/(m^2 k).As for thickness of 15 mm, the heat transfer coefficient decreased 2.9% (from 2.80 to 2.72 W/(m^2 k)).As for thickness of 18 mm, the heat transfer coefficient decreased also 2.9% from 2.75 to 2.66 W/(m^2 k). It indicated that argon filling ratio can improve the thermal insulation performance of non-coated double-glazing, but the effect is not obvious, only at the range of 2.9–4%. So it is not enough by the means of filling argon to improve the thermal performance of non-coated double-glazing.

Then some tests were also done in the case of coated double-glazing (has one glass pane coating with low emissivity layer). The reduction trend could also be seen as a result of argon filling on the heat transfer coefficient of coated double-glazing. As it is shown in the Fig. 6, the descent velocity of heat transfer coefficient is greater.

As the argon filling ratio increased from 0 to 97%, the heat transfer coefficient of coated double-glazing with air space thickness of 9 mm decreased 11.0% from 1.63 to 1.45 W/(m^2 k) The similar trend can also be seen when the gas thickness increased from 9 to 18 mm. As for air space thickness of 12 mm, the heat transfer coefficient decreased 12.7% from 1.50 to 1.32 W/(m^2 k). As for the glass with a thickness of 15 mm, it decreased 13.8% from 1.41 to 1.24 W/(m^2 k). Furthermore, As for the glass with a thickness of 18 mm, the heat transfer coefficient decreased 12.8% from 1.33 to 1.17 W/(m^2 k). Obviously, the overall effect of filling of argon on the heat transfer coefficients for the coated double-glazed are more stronger than non-coated double-glazed. It can be explained by that the radiation rate of

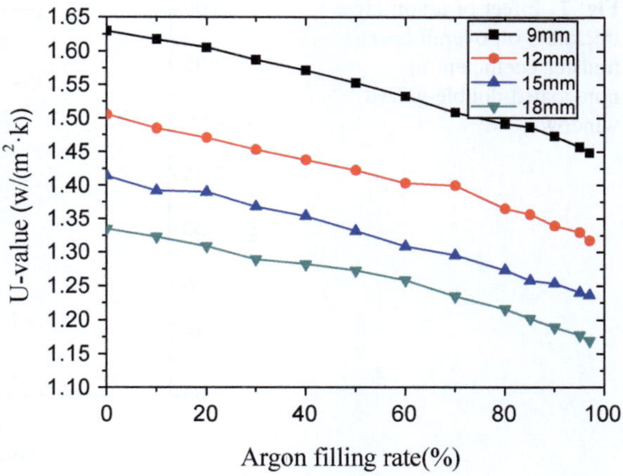

Fig. 6 Effect of argon filling ratio on overall heat transfer coefficient of coated double-glazed windows

non-coated double-glazed is relatively high, so the hot side of surface would radiate a large amount of energy to the inner layer of gas, resulting in a air space with elevated temperature of air space, and then the heat is transferred to another layer of glass, and finally leads to the integrated increase of heat transfer coefficient, but once the coated glass pane was used in double-glazed, the thermal transfer coefficient will be significantly reduced.

The Effect of Air Space Thickness. When the filling ratio of argon gas is constant, the heat transfer coefficient of double-glazing is gradually decreased with the increase of the air space thickness. In the case of non-coated double-glazing, as it is shown in the Fig. 7, as the air space thickness increased from 9 to 18 mm, the overall trend of U-value decreased obviously. More specifically, when the argon filling rate is zero, as air space thickness increased from 9 mm to 18 mm, the U-value decreased from 2.97 to 2.75 W/(m^2 k), decrease by 7.4%; when the argon filling ratio is 50%, the U-value down from 2.91 to 2.71 W/(m^2 k), decrease by 6.9%; when the argon filling ratio is 97%, the U-value down from 2.85 to 2.67 W/(m^2 k), decreases by 6.3%.

In the case of coated double-glazed, as it is shown in the Fig. 7, as the air space thickness increased from 9 to 18 mm, the overall decrease trend of U-value become more obviously. when the argon filling ratio is zero, the U-value decreased from 1.63 to 1.33 W/(m^2 k), decrease by 18.4%; when the argon filling ratio is 50%, the U-value down from 1.55 to 1.27 W/(m^2 k), decrease by 18.1%; when the argon filling ratio is 97%, the U-value down from 1.45 to 1.17 W/(m^2 k), decreases by 19.3%.

In addition, the heat transfer coefficient of two kinds of double-glazing with same argon filling ratio and same thickness was also compared. For the same argon ratio filling in the same thickness of two kinds of double-glazing, the U-value of coated double-glazing is much lower than non-coated one. As for the gas thickness 9 mm, when argon filling amount is 0, 50 and 97%, the U-value of non-coated double-glazed is 2.97, 2.91, 2.85 W/(m^2 k), respectively, while coated double-glazing accordingly 1.63, 1.55 and 1.45 W/(m^2 k), which provides about

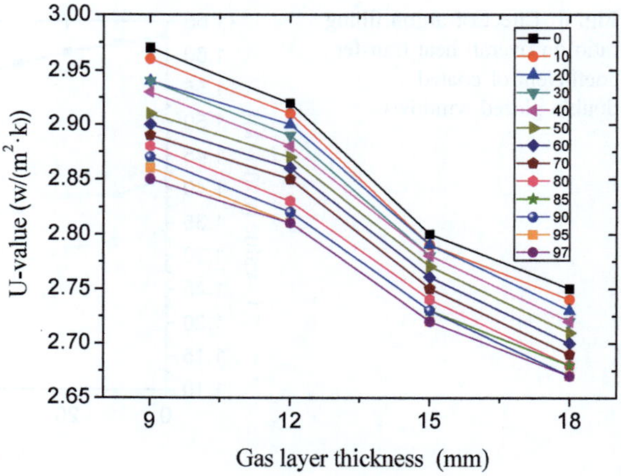

Fig. 7 Effect of argon layer thickness on overall heat transfer coefficient of non-coated double-glazed windows

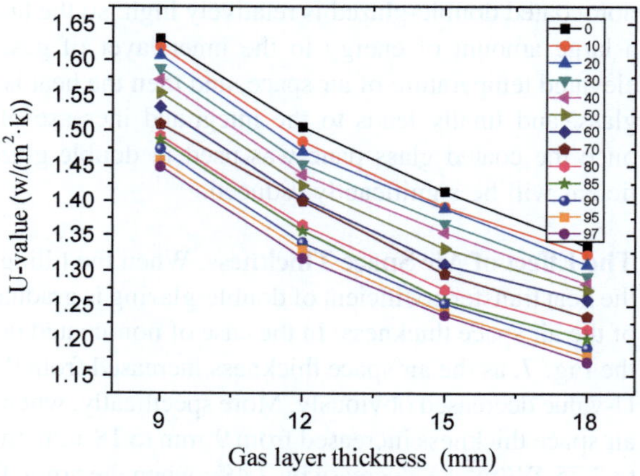

Fig. 8 Effect of argon layer thickness on overall heat transfer coefficient of coated double-glazed windows

47% of the impact. Furthermore, it can be find in the Fig. 8 that more the thickness of air space is, the more important effect on U-value induced by the coating. As for the glass with a gas thickness 18 mm, when argon filling amount is 0, 50 and 97%, the U-value of 2.95, 2.71 and 2.67 W/(m^2 k) was tested in non-coated double-glazing and 1.33, 1.27 and 1.17 W/(m^2 k) in coated double-glazing, here the coated glass pane provide about 55% impart. Such phenomenon can be explained as follows. The structure of the double-glazed makes the inner gas in a closed state, which reduced heat transfer accounted by the rapid gas flowing. The air space thickness increase is helpful to improve the partial gas thermal resistance, reduce the energy through heat conduction heat transfer; despite the air space thickness increases will encourage the convective heat transfer, the effect of heat convection on heat transfer coefficient is far from heat conduction.

Conclusion

In this paper the effect of argon filling ratio and air space thickness on the windows insulating performance was analyzed for the case of non-coated (without low-e coating) double-glazing and coated double-glazing. The effect of argon filling ratio can reduce about 3.7% of U-value on non-coated double-glazing and around 13% on coated double-glazing. The effect of air space thickness from 9 to 18 mm can reduce around 7% of U-value on non-coated double-glazing and around 19% on coated double-glazing. The more relative energy savings obtained for double-glazing having low emissivity coatings can save about 51%.

In order to improve the thermal performance of double-glazed windows, coated glass surface with low-emissivity is of great useful, besides the effect of argon filling can not be ignored. Coated double-glazing, filling with more argon, is of great choice for china to reduce the high building energy consumption.

References

1. J. He, Z.F. Yan, H.J. Liu, On the current energy consumption and countermeasures of large public buildings in China. Adv. Mater. Res. 374–377 (2011)
2. S.G. Li, H. Wang, Summarization of present building energy consumption and corresponding strategies in China. Environ. Sci. Manag. **33**, 6–9 (2008)
3. Y. Zhang, C.Q. He, B.J. Tang, Y.M. Wei, China's energy consumption in the building sector: a life cycle approach. Energy Build. **94**, 240–251 (2015)
4. L. Navarro, A.D. Gracia, A. Castell, L.F. Cabeza, Experimental study of an active slab with PCM coupled to a solar air collector for heating purposes. Energy Build. (2016)
5. L. Zhao, J.L. Zhang, R.B. Liang, Development of an energy monitoring system for large public buildings. Energy Build. **66**, 41–48 (2013)
6. S. Prívara, J. Cigler et al., Building modeling as a crucial part for building predictive control ☆. Energy Build. **56**, 8–22 (2013)
7. V. Buyanov, BP: statistical review of world energy 2011. Econ. Policy **4** (2011)
8. I. Pérez-Grande, J. Meseguer, G. Alonso, Influence of glass properties on the performance of double-glazed facades. Appl. Therm. Eng. **25**, 3163–3175 (2005)
9. E. Cuce, C.H. Young, S.B. Riffat, Performance investigation of heat insulation solar glass for low-carbon buildings. Energy Convers. Manag. **88**, 834–841 (2014)
10. G.I. Kiani et al., Cross-dipole band pass frequency selective surface for energy-saving glass used in buildings. IEEE Trans. Antennas Propag. **59**, 520–525 (2011)
11. A. Jonsson, A. Roos, Visual and energy performance of switchable windows with antireflection coatings. Sol. Energy **84**, 1370–1375 (2010)
12. C. Oleskowicz-Popiel, M. Sobczak, Effect of the roller blinds on heat losses through a double-glazing window during heating season in Central Europe. Energy Build. **73**, 48–58 (2014)
13. J. Yi, China building energy conservation stratagems study. Eng. Sci. (2011)
14. R. Yin, P. Xu, P. Shen, Case study: energy savings from solar window film in two commercial buildings in Shanghai. Energy Build. **45**, 132–140 (2012)
15. M. Calero, F. Mathieux, C. Baldassarri, Y. Roderick, K. Allacket, Using life cycle based environmental assessment in developing innovative multi-functional glass-polymer windows. SB13 Graz (2013)

16. E. Cuce, C.H. Young, S.B. Riffat, Thermal performance investigation of heat insulation solar glass: a comparative experimental study. Energy Build. **86**, 595–600 (2015)
17. DB11/891, Design standard for energy efficiency of residential buildings (2012)
18. S. Sadineni, B.S. Madala, R.F. Boehm, Passive building energy savings: a review of building envelope components. Renew. Sustain. Energy Rev. **15**, 3617–3631 (2011)
19. R.E. Collins, T.M. Simko, Current status of the science and technology of vacuum glazing. Sol. Energy **62**, 189–213 (1998)
20. C. Buratti, E. Moretti, Glazing systems with silica aerogel for energy savings in buildings. Appl. Energy **98**(5), 396–403 (2012)
21. C. Buratti, E. Moretti, Glazing systems with silica aerogel for energy savings in buildings. Appl. Energy **98**, 396–403 (2012)
22. Y. Sun et al., Thermal evaluation of a double glazing façade system with integrated Parallel Slat Transparent Insulation Material (PS-TIM). Build. Environ. **105**, 69–81 (2016)
23. M. Harvey et al., Insulating glass units, EP, US 6238755 B1 (2001)
24. H. Nakano, T. Goto, Argon gas laser device, US, US4615033 (1986)
25. ISO 10291, Glass in building, determination of steady-state u values (thermal transmittance) of multiple glazing, Guarded hot plate method first edition (1994)
26. ISO 10292, Glass in building, calculation of steady-state u values (thermal transmittance) of multiple glazing (1994)
27. ISO 8301, Thermal insulation, determination of steady-state thermal resistance and related properties, Heat flow meter apparatus (1991)
28. DIN EN 12939, Thermal performance of building materials and products-determination of thermal resistance by means of guarded hot plate and heat flow meter methods-thick products of high and medium thermal resistance, German version (2001)

Hydrothermal Synthesis and Characterization of $Cs_{0.32}WO_3$ Rod-like Nanoparticles with Excellent Near-Infrared Shielding Property

Shuying Lin, Xiaoqiang Zhang, Yue Huang, Zhonghua Zhou and Shirley Shen

Abstract Cesium tungsten bronze (Cs_xWO_3) nanoparticles were successfully synthesized via a hydrothermal method using sodium tungstate and cesium carbonate as raw materials, and citric acid and D-malic acid as mixed reducing agents. The results of XRD and SEM show that the sample with Cs/W (molar ratio) = 0.5 promotes formation of hexagonal $Cs_{0.32}WO_3$ with high crystallinity during hydrothermal synthesis. Moreover, with increasing reaction time, the morphology of the sample changes from irregular-shaped nanoparticles to rod-like nanoparticles. $Cs_{0.32}WO_3$ powders with high crystallinity and rod-like morphology have high near-infrared absorption. Additionally, transparent thermal insulation coatings and NIR shielding glass samples were prepared by coating the mixed dispersed solution of $Cs_{0.32}WO_3$ with organic silicon resin on glass substrate. The results of thermal insulation characterized by UV-Vis-NIR spectra show that NIR transmittance of glass substrate is 82% and that of the as-prepared NIR shielding glass is only around 12%, which is reduced dramatically, illustrating that the synthesized rod-like $Cs_{0.32}WO_3$ nanoparticles with high NIR absorption have promising prospects in the fields of automotive and architectural glass.

Keywords Cesium tungsten bronze · Rod-like nanoparticles · Hydrothermal synthesis · Near-infrared shielding properties

---

S. Lin · X. Zhang · Y. Huang · Z. Zhou (✉)
College of Materials, Xiamen University, Xiamen 361005, China
e-mail: zzh@xmu.edu.cn

Y. Huang · Z. Zhou
Fujian Key Laboratory of Advanced Materials, Xiamen 361005, Fujian, China

S. Shen
CSIRO Manufacturing, Clayton, VIC 3168, Australia

Y. Han (ed.), *Advanced Functional Materials*,
https://doi.org/10.1007/978-981-13-0110-0_38

Introduction

Transparent heat-shielding coatings with a high transmittance of visible light as well as with a high absorption of near-infrared light (NIR) is of great significance for energy-saving and environmental protection, especially for application of automotive and architectural windows. Because human skin is sensitive and has an uncomfortable burning feeling when exposed to NIR light of sunlight, it is necessary to develop transparent heat-shielding coatings or heat-shielding glass, which can be able to efficiently shield NIR (wavelength between 780 and 2500 nm) heat rays [1].

Transparent conductive films such as indium tin oxides (ITOs) and antimony-doped tin oxides (ATOs) films have been applied in transparent heat-insulating materials. However, ITOs and ATOs only shield NIR light with wavelength longer than 1500 nm [2]. Recently, Pen et al. [3] reported that Cs_xWO_3 ($0 < x < 1$) belonging to tungsten bronzes is a kind of non-stoichiometric semiconductor material with narrow band gap and exhibits high transparence of visible light and intense absorption for NIR light with wavelength longer than 1100 nm. Due to the blueshift of NIR absorption peak comparing with ITOs and ATOs, Cs_xWO_3 attracts more and more scientific interesting. Exploring environment-friendly preparation method and improving heat-shielding properties are hotspots for Cs_xWO_3 materials. As is known that tungsten (W) in the tungsten oxide (WO_3) is hexavalent (+6), while the valence state of W in cesium tungsten bronze (Cs_xWO_3) is lower than hexavalent, thus reducing atmosphere is necessary for synthesis of Cs_xWO_3. Takede et al. [4] synthesized $Cs_{0.32}WO_3$ nanoparticles by drying the mixed aqueous solution of cesium salt and ammonium tungstate, and then heating at 800 °C under hydrogen atmosphere. On the contrary, Guo et al. [5] synthesized $Cs_{0.32}WO_3$ nanoparticles by solvothermal reaction at 180–200 °C by using WCl_6 and CsOH as raw materials, acetic acid as reducing agent without using a heat treatment under reducing atmosphere. However these raw materials of WCl_6 and CsOH are neither economical nor environmental friendly. Liu et al. [6] reported the $Cs_{0.32}WO_3$ powders synthesized by hydrothermal reaction at 190 °C with citric acid as reducing agent, and with sodium tungstate and cesium carbonate as raw materials. However, the morphology of the obtained $Cs_{0.32}WO_3$ powders was irregular agglomerated particles, and the transparency and heat-shielding properties need to be improved. To improve heat shielding property of Cs_xWO_3, Mattox et al. [7] pointed out that localized surface plasmon resonance (LSPR) is main contribution for NIR absorption of Cs_xWO_3 and morphologies of tungsten bronzes Cs_xWO_3 have a remarkably effect on LSPR, therefore, have important effect on NIR heat-shielding property. Rod-like F-doped cesium tungsten bronzes ($Cs_{0.33}WO_{x\text{-}3}F_x$) nanoparticles, was then synthesized by a hydrothermal method with hydrofluoric acid (HF) as fluorine source, which achieved higher NIR shielding ability than irregular agglomerated nanoparticles [8]. Liu et al. [8] indicated that HF addition in hydrothermal synthesis process promotes formation of rod-like structure. However, due to the fact that HF is harmful to environment, there is still a need to explore a new process for preparing rod-like $Cs_{0.32}WO_3$ nanoparticles.

In this paper, $Cs_{0.32}WO_3$ rod-like nanoparticles have been synthesized via a hydrothermal reaction using sodium tungstate and cesium carbonate as raw materials, and citric acid and D-malic acid as mixed reducing agents. The influence of crystal structure, crystallinity and morphology on absorbance of the powders have also been investigated. A transparent heat shielding coating and heat shielding glass covered by the coating have also been prepared by adding inorganic NIR absorbent $Cs_{0.32}WO_3$ into organic silicon resin and compared with the glass substrate without the coating in terms of the heating shielding property.

Experimental

Preparation of Cesium Tungsten Bronze ($Cs_{0.32}WO_3$) Nanoparticles. $Cs_{0.32}WO_3$ powders were synthesized via a hydrothermal reaction by using sodium tungstate and cesium carbonate as raw materials, and citric acid and D-malic acid as mixed reducing agents. Firstly, Na_2WO_4 was dissolved in 20 ml distilled water under stirring and ion-exchanged with strong-acid-styrene type cation-exchange resin. Secondly, the resin column was flashed with 30 ml of distilled water to obtain yellowish tungstic acid solution with a concentration of 0.53 ml/L. Thirdly, Cs_2CO_3 was added according to different predesigned Cs/W molar ratios. After that, the citric acid and D-malic acid with weight ratio of 1.41:1 were added into the solution and pH value was adjusted to 1. The solution was then poured into Teflon-lined autoclave to undergo a hydrothermal reaction at 175 °C. The period varied from 24 to 72 h according to Liu et al. [6]. Finally, the obtained precipitate was washed several times with deionized water and absolute ethanol, followed by drying at 80 °C for 2 h, then grinded in an agate mortar to obtain $Cs_{0.32}WO_3$ powders. As a comparison, the same experiments were carried out by using citric acid or D-malic acid as a sole reducing agent.

Preparation of $Cs_{0.32}WO_3$ Coatings and Heat Shielding Glass. In order to evaluate NIR-shielding effect, $Cs_{0.32}WO_3$ coatings and heat shielding glass samples were prepared. 2 ml ethanol dispersion solution (solid concentration 30.8 wt%) of $Cs_{0.32}WO_3$ powders and 8 ml organic silicon resin (film forming agent) were mixed and stirred for 30 min to obtain a coating liquid, then the $Cs_{0.32}WO_3$ coating liquid was spin-coated on soda lime glass substrate by 250 rpm and 30 s, after drying for 30 min at 200 °C, $Cs_{0.32}WO_3$ coating on glass samples were obtained.

Testing and Characterizations. X-ray diffraction (XRD, Bruker AXS (D8 advance)) was used to analyze phase structure of the samples and the spectra was collected under the following conditions: graphite monochromatic copper λ (Cu, Kα) = 0.154 nm, at 40 kV, 40 mA over a range of 10–80° with a step of 0.0167° for 0.1 s. Scanning electron microscope (SEM, Hitachi SU-70, 10 kV) was used for morphology characterization. The NIR absorbance of $Cs_{0.32}WO_3$ powders were examined by diffuse reflectance spectroscopy (Varian, Cary 5000). The absorption spectra were scanned in the range of 380–2500 nm with a step of 600 nm/min. The optical properties of the as-prepared heat shielding glass with $Cs_{0.32}WO_3$ coatings were also characterized by UV-Vis-NIR spectrometer (Lambda 950, Perkin Elmer).

Results and Discussion

Influence of Different Predesigned Cs/W (Molar Ratio). Figure 1 shows XRD patterns of the as-prepared $Cs_{0.32}WO_3$ powders by using citric acid and D-malic acid (weight ratio of 1.41:1) as mixed reducing agents with different predesigned Cs/W molar ratios. Figure 1a, b present XRD profiles of the samples synthesized with Cs/W = 0.1 and 0.2 respectively. The mixed phases of hexagonal $Cs_{0.32}WO_3$ (JCPDS No. 83-1334) and WO_3 (JCPDS No. 72-2141) are clearly observed for these two samples. The XRD diffraction patterns of Cs/W = 0.3, 0.4, 0.5 and 0.6 (Fig. 1c–f, respectively) agree well with hexagonal cesium tungsten bronze $Cs_{0.32}WO_3$ without any impurity peak, which illustrates that free Cs^{+1} has been inserted into the crystal lattice of tungsten bronze. Meanwhile, for the samples with molar ratio of Cs/W in the range of 0.1–0.5, the intensity of the main diffraction peak (2θ = 27.77°) of XRD increases with the increase of Cs concentration (XRD intensity has been calibrated), which indicates that the crystallinity of $Cs_{0.32}WO_3$ increases with the increase of Cs concentration. Obviously, the main diffraction peak of the sample with Cs/W = 0.5 (Fig. 1e) shows the sharpest and highest intensity, suggesting it has the highest crystallinity of $Cs_{0.32}WO_3$ among these

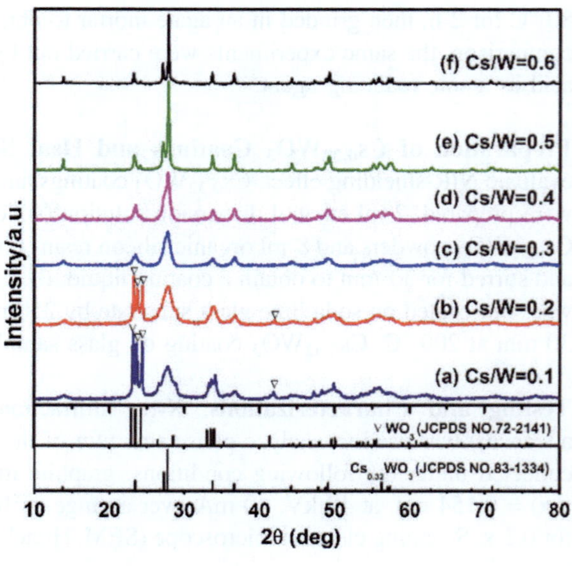

Fig. 1 XRD patterns of as-prepared $Cs_{0.32}WO_3$ powders with different predesigned Cs/W molar ratio

samples. Subsequently, the intensity of the main diffraction peak for the sample with Cs/W = 0.6 (Fig. 1f) decreases, indicating that the crystallinity of $Cs_{0.32}WO_3$ decreases.

The results of absorbance spectra of the $Cs_{0.32}WO_3$ powders synthesized with Cs/W (molar ratio) from 0.1 to 0.6 (Fig. 2) shows that NIR absorbance of C $Cs_{0.32}WO_3$ powders increases with the increase of Cs concentration when Cs/W mole ratio ≤ 0.5, but decreases at the high Cs/W mole ratio of 0.6. The variation trend of NIR absorbance with Cs/W is coincident with that of the crystallinity variation trend derived from XRD patterns (Fig. 1). NIR absorption is strongly related to plasmon resonance of free carriers [9]. In $Cs_{0.32}WO_3$ tungsten bronze structure, Cs^{+1} ion can be stably embedded into the hexagonal tungsten bronze structure without losing easily [10], considering the size of Cs^{+1} ion (0.17 nm) is similar to the size of hexagonal tunnels (0.163 nm). The crystallinity and NIR absorbance of $Cs_{0.32}WO_3$ raise with the increase of Cs doping amount, which is probably attributed to the increasing charge carriers' concentration. The sample of Cs/W = 0.5 with the highest crystallinity (Fig. 1e), exhibits the best NIR absorbance (Fig. 2e). The sample with Cs/W = 0.6 has a lower NIR absorbance and lower crystallinity than that of Cs/W = 0.5, probably because that excessive addition of Cs+ reduces its reactivity and generates W-defects [8].

Influence of Reaction Time on Morphology and Properties. Figure 3a shows XRD patterns of $Cs_{0.32}WO_3$ powders synthesized via a hydrothermal reaction by using citric acid and D-malic acid (weight ratio of 1.41:1) as a mixed reducing agent at 175 °C for 24, 48 and 72 h, respectively. All samples can be indexed as hexagonal cesium tungsten bronze $Cs_{0.32}WO_3$ crystal without any impurity peak. From SEM images of $Cs_{0.32}WO_3$ for different times, irregularly agglomerated particles forms after 24 h (Fig. 3c); Rod-like particles forms and are coexistent with irregularly agglomerated particles in the products after 48 h (Fig. 3d) and rod-like particles dominates after 72 h (Fig. 3e). The irregularly agglomerated particles transform into rod-like nanoparticles with about 50 nm in diameter and 1 μm in length. According to Fig. 3a, the intensity of main XRD diffraction peak

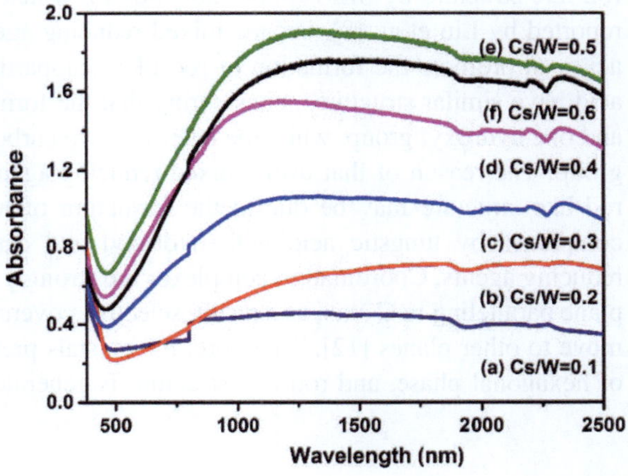

Fig. 2 Diffuse reflectance UV-Vis-NIR absorption spectra of $Cs_{0.32}WO_3$ with different predesigned Cs/W molar ratios

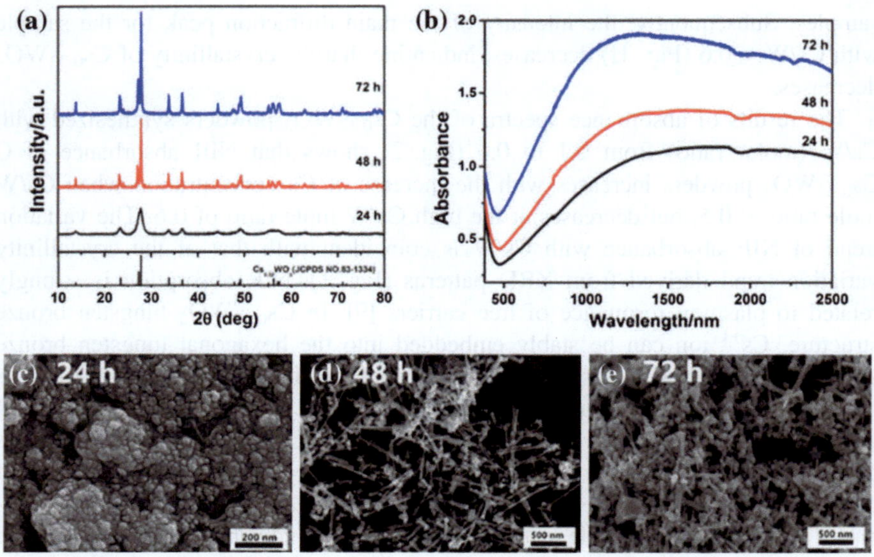

Fig. 3 XRD patterns (**a**), diffuse reflectance UV-Vis-NIR absorption spectra of $Cs_{0.32}WO_3$ powders prepared according to different reaction durations (**b**), SEM images of $Cs_{0.32}WO_3$ synthesized at 175 °C for different time of 24 h (**c**), 48 h (**d**) and 72 h (**e**)

($2\theta = 27.77°$) of $Cs_{0.32}WO_3$ enhances with the increases of hydrothermal time. With the formation of rod-like nanoparticles, the intensity of main XRD diffraction peak increases and the crystallinity increases. The $Cs_{0.32}WO_3$ with a high aspect ratio often shows stronger intensities of specific reflection in XRD pattern, which is consistent with the work reported by Guo et al. [11].

In this experiment, samples synthesized by a sole reducing agent were also prepared. The sample of Cs/W (molar ratio) = 0.5 with the hydrothermal reaction for 72 h, using either citric acid (0.55 mol/L) or D-malic acid (0.55 mol/L) as a sole reducing agent, being detected to be $Cs_{0.32}WO_3$ by XRD (pattern not shown), presents morphology of irregularly agglomerated nanoparticles but no formation of rod-like structure by SEM (photo not shown), which is consistent with the results reported by Liu et al. [8]. So, the mixed reducing agent of citric acid and D-malic acid can promote the formation of rod-like nanoparticles. Citric acid and D-malic acid have similar structures, considering that the former has three carboxyl groups and one hydroxyl group, while the latter has two carboxyl groups and one hydroxyl group. The reason of that using mixed reducing agents promotes the formation of rod-like structure may be due to the formation of tungsten oxygen coordination complexes by tungstic acid with hydroxyl and carboxyl groups of the mixed reducing agents. Coordination complexes are strongly linked to the crystallographic plane paralleling to C axis, as a result, selective coverage would drive the nucleus to move to other planes [12]. Therefore, the crystals preferentially grow along C axis of hexagonal phase, and rod-like structure is generated.

UV-Vis-NIR absorption spectra (Fig. 3b) of $Cs_{0.32}WO_3$ powders, synthesized via the hydrothermal reaction by using citric acid and D-malic acid (weight ratio of 1.41:1) as the mixed reducing agent at 175 °C for 24, 48 and 72 h, show that $Cs_{0.32}WO_3$ powders exhibit strong absorption of NIR in the range of 1000–2500 nm. The NIR absorption of $Cs_{0.32}WO_3$ improves with the increase of reaction time. The $Cs_{0.32}WO_3$ powders obtained by reaction time of 72 h has the highest absorption at 1314 nm, and the absorbance is up to 1.92. Based on above results, great crystallinity and rod-like morphology are main factors for achieving higher NIR absorption property of $Cs_{0.32}WO_3$ powders. The formation of $Cs_{0.32}WO_3$ rod-like structure can improve NIR absorbance, which is consistent with the reported by Guo et al. [13]. The localized surface plasmon resonance (LSPR) is closely related to morphology of nanoparticles and free carrier concentration. The nano rod-like morphology may also be helpful for NIR shielding ability due to LSPR is generated along both the length of the nanorod (the longitudinal plasmon band) and the width of the nanorod (the transverse plasmon band). It suggests that a greater aspect ratio leads to a higher light absorption [14]. According to the above discussion, it concludes that the NIR absorption ability of $Cs_{0.32}WO_3$ powders strongly relates to the high crystallinity and rod-like structure.

NIR Shielding Properties of $Cs_{0.32}WO_3$ Coatings and Heat Shielding Glass. Transparent NIR shielding coatings were prepared by mixing an organic silicon resin and the as-prepared $Cs_{0.32}WO_3$ nanoparticles, which synthesized via the hydrothermal reaction by using citric acid and D-malic acid (weight ratio of 1.41:1) as the reducing agent with Cs/W (molar ratios) = 0.5 and with a heat treatment of 175 °C for 72 h. The transparent NIR shielding coatings were coated on soda lime glass substrate to obtain a heat shielding glass sample (coating thickness 4.8 μm from SEM, herein SEM image not shown). Figure 4a, b show UV-Vis-NIR spectrum of glass substrate and heat shielding glass sample with the $Cs_{0.32}WO_3$ coating, respectively. Obviously, the $Cs_{0.32}WO_3$ coating maintains a high transmittance in visible light region. Moreover, its shielding ability in NIR region improves significantly comparing with controlled glass substrate. Table 1 lists data of UV transmittance (Tuv), visible light transmittance (Tl) and NIR transmittance (Tir). Tl of the glass sample with $Cs_{0.32}WO_3$ coating is up to 75.2%, while its Tir is only 11.7%. Comparing with glass substrate, the Tir decreases by 70.3%. The visible light transmittance of $Cs_{0.32}WO_3$ coating ($Tl_{coating}$) is calculated to be 85.1% by a formula (($Tl_{coating} = 100\% - (Tl_{glass} - Tl_{glass\ with\ coatings})$), without considering the loss caused by reflection of the film). Similarly, the transmittance of NIR and UV of the $Cs_{0.32}WO_3$ coating can be calculated, and the results are listed in Table 1. Tl of $Cs_{0.32}WO_3$ coating (thickness 4.8 μm) is as high as 85.1%, while its Tir is only about 29.7% and NIR shielding rate is up to 70.3%. The results show that the as-prepared $Cs_{0.32}WO_3$ coating has excellent transparency and NIR shielding ability. The reflectance of $Cs_{0.32}WO_3$ coating (Fig. 4c) occurs also low at NIR wavelength range, which indicates that NIR shielding ability of the $Cs_{0.32}WO_3$ nanoparticles is mainly caused by absorption, rather than by reflection.

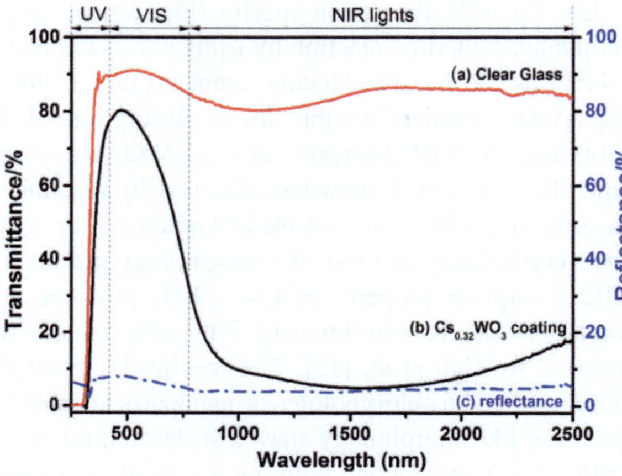

Fig. 4 Transmittance spectra of **a** glass substrate; **b** $Cs_{0.32}WO_3$ coating on glass substrate; and **c** reflectance spectra of $Cs_{0.32}WO_3$ coating on glass

Table 1 UV, visible light and NIR transmission data of glass substrate, $Cs_{0.32}WO_3$ coatings on glass substrate and $Cs_{0.32}WO_3$ coating

Transmittance (%)	Glass substrate (%)	$Cs_{0.32}WO_3$ coatings on glass substrate (%)	$Cs_{0.32}WO_3$ coating (%)
T_{UV} (250–380 nm)	82.4	41.7	59.3
T_L (380–780 nm)	90.1	75.2	85.1
T_{NIR} (780–2500 nm)	82	11.7	29.7

Conclusion

Rod-like $Cs_{0.32}WO_3$ nanoparticles with excellent NIR absorption properties have been successfully synthesized via the hydrothermal reaction by using citric acid and D-malic acid as the mixed reducing agent. UV-Vis-NIR spectra reveal an eminent optical property of $Cs_{0.32}WO_3$ coating with the transmittance of visible light reaching about 85%, while its NIR transmittance is only 29.7%, having NIR cut down by 70.3%. The improvement of NIR shielding ability of $Cs_{0.32}WO_3$ is related to the increase of crystallinity and the formation of rod-like structure, which results from the increase of free carrier concentration. Further studies on formation mechanism of rod-like structure are in progress.

Acknowledgements This work is supported by the Science and Technology Major Program of Fujian Province (2014HZ0005), China.

References

1. F. Shi, J.X. Liu et al., Hydrothermal synthesis of Cs_xWO_3 and the effects of N_2 annealing on its microstructure and heat shielding properties. J. Mater. Chem. Technol. **30**(4), 342–346 (2014)

2. C. Guo, S. Yin, Q. Dong et al., Near-infrared absorption properties of Rb_xWO_3 nanoparticles. J. CrystEngComm **14**(22), 7727–7732 (2012)
3. Z.J. Pen, J.X. Liu, F. Shi et al., Synthesis and near-infrared shielding property of cesium tungsten bron. J. Chin. Ceram. Soc. **40**(6), 0454–5648 (2012)
4. H. Takeda, K. Adachi, Near infrared absorption of tungsten oxide nanoparticle dispersions. J. Am. Ceram. Soc. **90**(12), 4059–4061 (2007)
5. C. Guo, S. Yin, P.L. Zhang et al., Novel synthesis of homogenous Cs_xWO_3 nanorods with excellent NIR shielding propertiesby a water controlled-release solvothermal process. J. Chem. Mater. **20**, 8227–8229 (2010)
6. J.X. Liu, Q. Xu, F. Shi et al., Dispersion of $Cs_{0.33}WO_3$ particles for preparing its coatings with higher near infrared shielding properties. J Appl. Surf. Sci. **309**, 175–180 (2014)
7. T.M. Mattox, A. Bergerud, A. Agrawal et al., Influence of shape on the surface plasmon resonance of tungsten bronze nanocrystals. J. Chem. Mater. **26**(5), 1779–1784 (2014)
8. J. Liu, J. Luo, F. Shi et al., Synthesis and characterization of F-doped $Cs_{0.33}WO_{3-x}F_x$ particles with improved near infrared shielding ability. J. Solid State Chem. **221**, 255–262 (2015)
9. J.X. Liu, F. Shi, X.L. Dong et al., Morphology and phase controlled synthesis of Cs_xWO_3 powders by solvothermal method and their optical properties. J. Powder Technol. **270**, 329–336 (2015)
10. C. Guo, S. Yin, Q. Dong et al., Facile synthesis of homogeneous Cs_xWO_3 nanorods with excellent low-emissivity and NIR shielding property by a water controlled-release process. J. Mater. Chem. **21**, 5099–5105 (2011)
11. C. Guo, S. Yin, L. Huang et al., Synthesis of one-dimensional potassium tungsten bronze with excellent near-infrared absorption property. J. ACS Appl. Mater. Interfaces **3**(7), 2794–2799 (2011)
12. J.Y. Luo, *Study on preparation and properties of M_xWO_3 tungsten bronzes* (D. Dalian Polytechnic University, Dalian, 2014)
13. C. Guo, S. Yin, L. Huang et al., Discovery of an excellent IR absorbent with a broad working waveband: Cs_xWO_3 nanorods. J. Chem. Commun. **47**(31), 8853–8855 (2011)
14. W.H. Ni, X.S. Kou, Z. Yang et al., Tailoring longitudinal surface plasmon wavelengths, scattering and absorption cross sections of gold nanorods. J. Acs Nano **2**(4), 677–686 (2008)

High Order Laterally-Coupled Distributed-Feedback GaSb-Based Diode Lasers at 1.9 μm Wavelength

Huan Li, Sheng-wen Xie, Yu Zhang, Shu-Shan Huang, Jin-Liang Wang and Zhi-Chuan Niu

Abstract We reported on GaSb-based laterally coupled distributed-feedback (LC-DFB) quantum well lasers. The lateral grating of the LC-DFB lasers were patterned along both sidewalls of the ridge waveguide. Gratings with higher order could enhance the lithographic tolerance for lower resolution patterning, yielding lasers more amenable to mass-manufacturing. The lasers emitting wavelength was at 1.9 μm in a single longitudinal mode at room temperature. A low wavelength dependent current tuning coefficient of 0.13 nm/mA was measured for the laser output. We obtained high output power of 13.6 mW, which were suitable for gas analysis.

Keywords Distributed feedback lasers · Laterally-corrugated ridge waveguide Higher-order grating · Ultraviolet exposure lithographic

H. Li (✉) · J.-L. Wang
School of Physics and Nuclear Energy Engineering, Beihang University,
Beijing 100191, China
e-mail: huanzi@buaa.edu.cn

J.-L. Wang
e-mail: wangjinliang@buaa.edu.cn

H. Li · S. Xie · Y. Zhang · S.-S. Huang · Z.-C. Niu
State Key Laboratory for Supperlattices and Microstructures,
Institute of Semiconductors, Chinese Academy of Sciences,
Beijing 100083, China
e-mail: owen-xie@163.com

Y. Zhang
e-mail: zhangyu@semi.ac.cn

S.-S. Huang
e-mail: shshuang@semi.ac.cn

Z.-C. Niu
e-mail: zcniu@semi.ac.cn

S. Xie · Y. Zhang · S.-S. Huang · Z.-C. Niu
College of Materials Science and Opto-Electronic Technology,
University of Chinese Academy of Sciences, Beijing 101408, China

Y. Han (ed.), *Advanced Functional Materials*,
https://doi.org/10.1007/978-981-13-0110-0_39

Introduction

There is a wide variety of applications for mid-infrared laser sources with emission wavelength at 1.5–3 μm. In particular, distribute feedback (DFB) GaSb-based diode lasers are suitable for sensing of atmospheric greenhouse gasses [1], since molecules such as CH_4 and CO_2 have strong absorption lines in this wavelength region [2, 3].

For accurate sensing, the tunable lasers with single mode emission wavelength are needed. The traditional DFB grating is introduce a buried grating deep into the epilayer structure so as to realize the good coupling of the optical field and grating [4, 5]. However, this method requires the regrowth of epitaxial and increase the manufacturing difficulty. Another drawback present in the regrowth approach is the rapid oxidation of Al-containing layers in normal atmosphere. AlGaAsSb layers were used wildly for GaSb based laser, making the regrowth approach very challenging [6]. We mitigate this problem by using surface gratings which placed on the sides of the ridge waveguide [7]. This technique requires only one epitaxial growth step, thus simplifying the growth process and preventing unwanted oxidation of AlGaAsSb layers [8].

In this paper we report the growth of GaSb-based type-I quantum well diode lasers and the fabrication process of laterally coupled DFB Bragg gratings. Gratings with twenty-one order were fabricated by ultraviolet exposure. These devices can produce single-mode emission near 1.9 μm and generate more than 13 mW of CW emission at room temperature. A wavelength dependent current tuning coefficient of 0.13 nm/mA was measured for the laser output.

LC-DFB Device Fabrication

The lasers were fabricated by molecular beam epitaxy (MBE) on an n-GaSb (100) substrate. Active region of the laser structure comprised of one $In_{0.2}GaSb$ quantum well (10 nm) with $Al_{0.35}GaAsSb$ barrier layers (20 nm) embedded in $Al_{0.35}GaAsSb$ waveguide layer (270 nm). The undoped region was surrounded by lattice matched $Al_{0.55}GaAsSb$ cladding layers (1500 nm). The structure was capped by a 250 nm highly p-doped GaSb contact layer for achieving a good Ohm contact with the p-side electrode mental. Typical epitaxial structure of the devices is shown in Fig. 1.

We designed our Bragg gratings by using the modified coupled mode theory [9]. 2D finite elements numerical calculations [10] were used to determine the electric field distribution of the fundamental guided mode and evaluate the modal effective index n_{eff}. Assuming an effective-medium index for the grating region, the n_{eff} was calculated to be 3.45 [11]. The Bragg condition are related to the laser mode wavelength λ-DFB and the grating period Λ. The coupling coefficient, κ [12], is determined by a perturbation analysis, based on coupled waves theory [13, 14]. The

Fig. 1 Epitaxial structure used to fabricate the LC-DFB laser

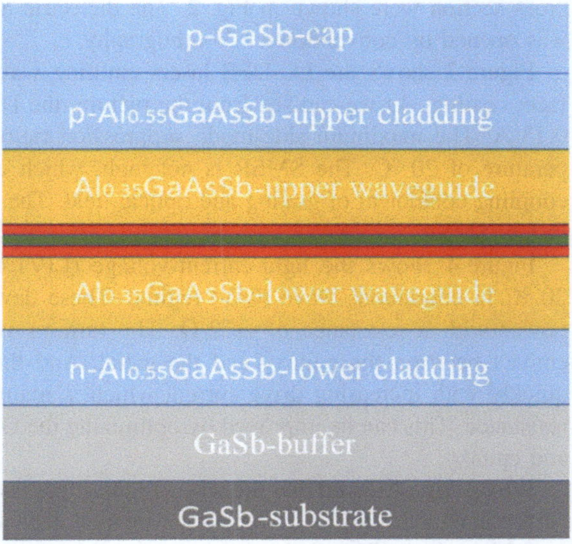

duty cycle of grating is defined by the ratio of grating length and the period. Based on the above theoretical analysis, the cavity length was designed to be 1500 μm. The twenty-one higher-order grating has a period of 6 μm with a duty cycle of 0.6. The ridge waveguide widths were 6 μm. The ridge and lateral grating were etched using induction-coupled plasma (ICP) with a depth of 2.2 μm. After the etching process, the wafer was passivated with 200 nm SiO_2 as insulating layers, followed by via etching into the areas where electrical injection will be provided. Finally, ohmic of p-contacts were formed by TiPtAu. The NiGeAu/Au was deposited on thinned GaSb to form a N surface electrode. Lasers were cleaved to the length of 1500 μm and width of 500 μm.

Measurements and Analysis

Laser characterization was measured by current-light-voltage (ILV) curves of the devices with a thermal power head. The morphology of grating was measured by SEM. Optical spectra were measured with Vertex70 Fourier infrared spectrometer and infrared fluorescence test system. All measurements were done for a component with as-cleaved facets.

The DFB Grating Was Etched by Inductively Coupled Plasma. We using photoresist and silicon dioxide layers served as the double mask for etching of the semiconductor cladding layer. The grating pattern was etched until to the waveguide layers by using inductively coupled plasma with a mixture of $SiCl_4$ and Ar_2 as the process gas. The scanning electron microscope image of the grating and the cross section were shown in Fig. 2. The dielectric SiO_2 layer on top of the ridge was opened by conventional UV lithography.

Figure 3 shows the LC-DFB lasers emitting wavelength near 1.9 μm without facet coating. As we can see from the picture, the DFB lasers threshold current is 0.13 A. The maximum side-mode suppression ratio (SMSR) is 13.4 dB in temperature of 20 °C. The SMSR is not high, which could be account for the low coupling coefficient of grating and optical field. The etching depth may also result in low SMSR.

Figure 4 shows the light-current-voltage (LIV) characteristics of the laser at 20 °C in continuous wave (CW) regime. These device can emit over 14 mW of power with a resistance of 6.72 Ω. The turn-on voltage of 1.3 V shows Ohm contact between the epitaxy and the metal. And the operation voltage increases quickly with increasing drive current which is mainly caused by the large series resistance. This can be improved by optimizing the Ohm contact between the metal and epitaxy.

We clearly observed that the wavelength of the laser gradually red shifted as the current increased (Fig. 5). And the wavelength dependent current tuning coefficient is 0.13 nm/mA. That behavior can probably be explained that the band gap becomes narrower when the concentration of the conduction band increases as the injection current increases. Another reason may be that the non radiative loss causes the lasers temperature rise, which affects the refractive index and band gap of the material, that causes the wavelength to drift. But the laser output spectrum remained locked to the DFB grating over a current range from 130 to 157 mA at 20 °C. At low current levels, the intensity of gain spectrum is low. At higher current levels, we observed that the intensity of gain spectrum gradually increase and shift toward longer wavelengths with single longitudinal mode. As shown in Fig. 6, the wavelength increases linearly with the increase of current. When the drive current

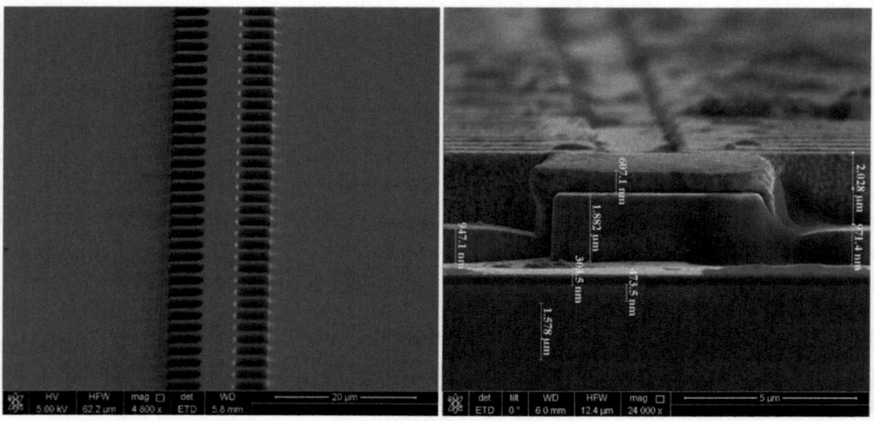

Fig. 2 SEM picture of the feedback grating (left) and the cross-section of a 5 μm-wide laser ridge topped with a titanium-platinum-gold contact layer (right)

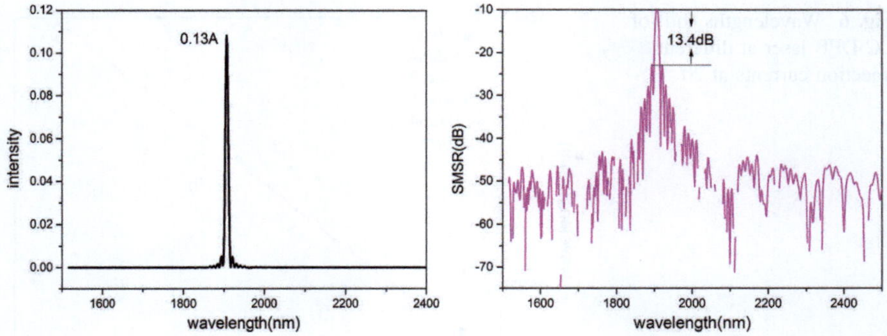

Fig. 3 The left picture shows the emission spectrum at a wavelength of 1.91 μm under an injected current of 130 mA in temperature of 20 °C. The right picture shows a single longitudinal mode operation with an SMSR of 13.4 Db

Fig. 4 Light-current-voltage and characteristic curve of 21 order with the duty ratio of 0.6 of LC-DFB CW laser at room temperature

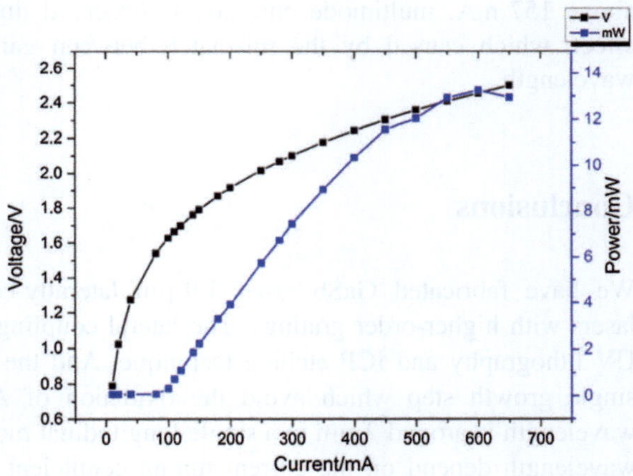

Fig. 5 Emission wavelengths of LC-DFB laser at different injection currents

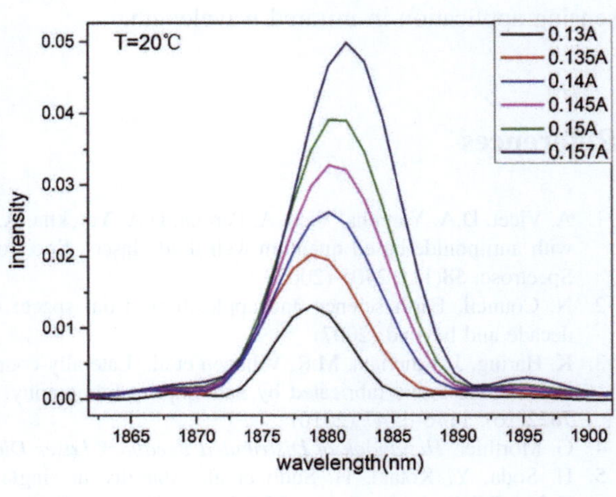

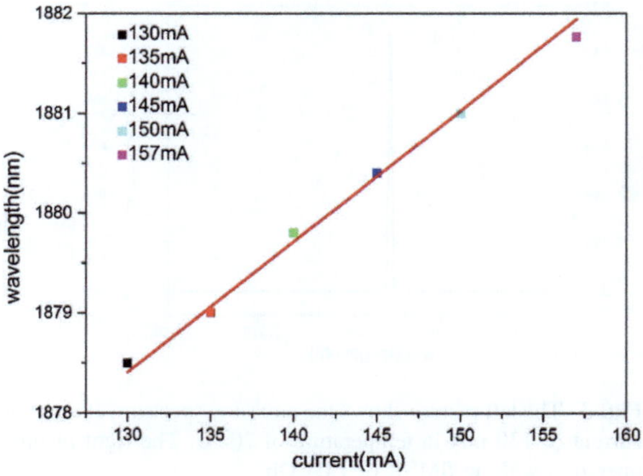

Fig. 6 Wavelengths shift of LC-DFB laser at different injection currents at 20 °C

above 157 mA, multimode emission is observed due to the aggravated thermal effect, which caused by the mismatch between gain spectrum and the grating wavelength.

Conclusions

We have fabricated GaSb-based 1.9 μm laterally-coupled distributed feedback lasers with higher-order gratings. The lateral coupling gratings were fabricated by UV lithography and ICP etching technique. And the epitaxy was completed in a single growth step which avoid the oxidation of Al. The CW laser emission wavelength is around 2 μm in a single longitudinal mode at room temperature. The wavelength depend on the current tuning coefficient is 0.13 nm/mA. The device emits an output power over 14 mW at room temperature which is enough for gas sensing application in infrared wavelength.

References

1. A. Vicet, D.A. Yarekha, Vicet A. Perona, D.A. Yarekha, A. Pérona et al., Trace gas detection with antimonide-based quantum-well diode lasers. Spectrochim. Acta Part A Mol. Biomol. Spectrosc. **58**(11), 2405 (2002)
2. N. Council, Earth science and applications from space: national imperatives for the next decade and beyond (2007)
3. K. Haring, J. Viheriala, M.R. Viljanen et al., Laterally-coupled distributed feedback InGaSb/GaSb diode lasers fabricated by nanoimprint lithography. Proc. SPIE—Int. Soc. Opt. Eng. **7822**(16), 1146–1147 (2010)
4. G. Morthier, *Handbook of Distributed Feedback Laser Diodes* (2013)
5. H. Soda, Y. Kotaki, H. Sudo et al., Stability in single longitudinal mode operation in GaInAsP/InP phase-adjusted DFB lasers. Quantum Electron. IEEE J. **23**(6), 804–814 (1987)

6. P. Apiratikul, L. He, C.J.K. Richardson, 2 μm laterally coupled distributed-feedback GaSb-based metamorphic laser grown on a GaAs substrate. Appl. Phys. Lett. **102**(23), 031107 (2013)
7. M. Kamp, J. Hofmann, F. Schäfer et al., Lateral coupling—a material independent way to complex coupled DFB lasers. Opt. Mater. **17**(1–2), 19–25 (2001)
8. R.M. Briggs, C. Frez, M. Bagheri et al., Single-mode 2.65 μm InGaAsSb/AlInGaAsSb laterally coupled distributed-feedback diode lasers for atmospheric gas detection. Opt. Express **21**(1), 1317 (2013)
9. W. Streifer, D. Scifres, R. Burnham, Analysis of grating-coupled radiation in GaAs:GaAlAs lasers and waveguides. IEEE J. Quantum Electron. **12**(8), 494–499 (1976)
10. R. Millett, K. Hinzer, H. Schriemer, Fabrication-tolerant higher order laterally coupled distributed feedback lasers. Proc. SPIE—Int. Soc. Opt. Eng. (2008) 7099:70991P-70991P-12
11. C.A. Yang, Y. Zhang, Y.P. Liao et al., 2-μm single longitudinal mode GaSb-based laterally coupled distributed feedback laser with regrowth-free shallow-etched gratings by interference lithography. Chin. Phys. B **25**(2), 181–185 (2016)
12. C.H. Chen, L.H. Chen, Q.M. Wang, Coupling coefficients of gain-coupled distributed feedback lasers with absorptive grating. Electron. Lett. **32**(14), 1288–1290 (1996)
13. R.R. Millett, K. Hinzer, T.J. Hall et al., Simulation analysis of higher order laterally-coupled distributed feedback lasers. IEEE J. Quantum Electron. **44**(12), 1145–1151 (2008)
14. W. Streifer, D.R. Scifres, R. Burnham, Coupled wave analysis of DFB and DBR lasers. Quantum Electron. IEEE J. **13**(4), 134–141 (1977)

GaSb-Based QWs 2 μm High Power Laser Diode

Kelu Zhang, Shengwen Xie, Yu Zhang, Yingqiang Xu, Jinliang Wang and Zhichuan Niu

Abstract 2 μm high-power GaSb-based type-I quantum well diode lasers were fabricated in this study. Under direct current, the output power of the lasers with uncoated cavity length of 2 mm and 100-μm-wide ridge is about 0.533 W with injection current of 3 A. The maximum conversion efficiency of single emitter is 12.67%. Under pulse current, the output power is 1.946 W with injection current of 14 A of 1 k Hz and 5% duty cycle. The output power of the laser with coated cavity is increased to 2.466 W with injection pulsed current of 16 A.

Keywords Compressive strain quantum well · Mid-infrared · Fabry-Perot cavity

---

K. Zhang (✉) · J. Wang
School of Physics and Nuclear Energy Engineering, Beijing University of Aeronautics and Astronautics, Beijing 100191, China
e-mail: klzhang@buaa.edu.cn

J. Wang
e-mail: wangjinliang@buaa.edu.cn

K. Zhang · S. Xie · Y. Zhang · Y. Xu · J. Wang · Z. Niu
State Key Laboratory for Super Lattices and Microstructures, Institute of Semiconductors, Chinese Academy of Science, Beijing 100083, China
e-mail: xsw@semi.ac.cn

Y. Zhang
e-mail: zhangyu@semi.ac.cn

Y. Xu
e-mail: yingqxu@semi.ac.cn

Z. Niu
e-mail: zcniu@semi.ac.cn

S. Xie · Y. Zhang · Y. Xu · Z. Niu
College of Materials Science and Optic-Electronic Technology, University of Chinese Academy of Sciences, Beijing 101408, China

Y. Han (ed.), *Advanced Functional Materials*,
https://doi.org/10.1007/978-981-13-0110-0_40

Introduction

High power semiconductor lasers emitting near 2 μm can be used in pumping solid state laser, metal cutting [1], medical treatment, free space applications and electrical confrontation for their low power consuming, high working temperature, superior efficiency, good temperature stability and easily assembly, thus we have done a lot of research to improve the output power of mid-infrared laser [2]. Considering that the energy band corresponding to 1.7–3.5 μm, GaSb is the ideal material for this wavelength range. Great efforts have been made to improve the performance on the lasers made from this material system. In 2008, at Fraunhofer Laboratory in Germany, the electro-optical characteristic of different emitter was researched. The single emitter at 1908 nm the operation current has been ramped up to a current of 4 A resulting in an output power of 1.17 W. Slope efficiency of 0.33 W/A and a maximum wall-plug efficiency of 26.5% are remarkable results for the GaSb based material system [3]. In 2015, Institute of Semiconductors, Chinese Academy of Sciences, the single uncoated LD device with the maximum output power was 1.058 W under continuous wave (CW) operation were fabricated by the GaSb-based AlGaIn/AsSb material [4]. In 2011, the state university of New York at stony brook, the wavelength of the quantum well laser was extended to 3.44 μm [5]. In 2012 Germany Walter Schottky Institute, the laser wavelength was extended to 3.73 μm in pulse condition at room temperature [6]. In 2016, High-power two-stage cascade GaSb-based type-I quantum well diode lasers emitting near 2 μm with cavity length of 3 mm generated about 2 W at the current of 6 A [12]. At present, Institute of Semiconductors, Chinese Academy of Sciences, Changchun Institute of Optics, Fine Mechanics and Physics, Chinese Academy of Sciences [7] and China Electronics Technology Group Corporation also conduct research on this kind of quantum well laser [8, 9], and obtain the corresponding results.

Experiment

It is found that the resistance of the un-doped waveguide layer is a key factor affecting the output power of semiconductor laser, and the thickness of the un-doped waveguide layer is the decisive factor affecting the resistance. For reach the higher power, we studied the relationship between the thickness of different un-doped waveguide layer and the output power.

The compressive strain quantum well AlGaInAsSb laser structure used here was grown on (100)-oriented 2-in. GaSb substrates by solid source Gen-IIMBE system. The active region is 10 nm wide $In_{0.2}Ga_{0.8}Sb$ QWs. The active region is sandwiched between $Al_{0.25}Ga_{0.75}As_{0.02}Sb_{0.98}$ waveguide layer, and the different waveguide thickness is 130, 180 and 270 nm respectively. The waveguide core is embedded between 2 μm wide n-doped and p-doped $Al_{0.5}Ga_{0.5}As_{0.02}Sb_{0.98}$ cladding layer which is lattice-matched to GaSb substrate. The cap layer is 260 nm

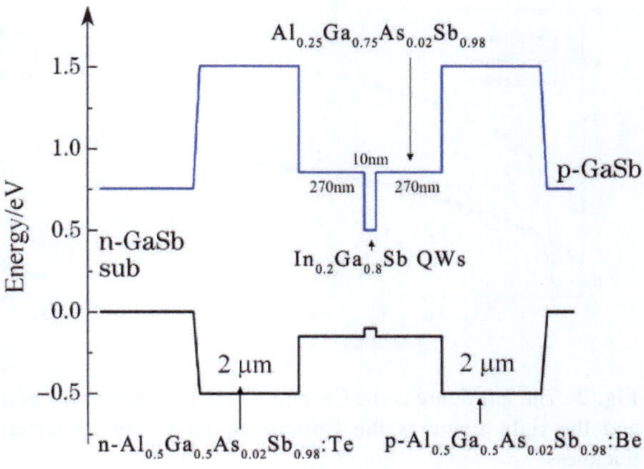

Fig. 1 Energy structure of the QWs laser

which is heavily p-doped GaSb for low resistance. The doping concentration of p-GaSb layer and p-doped Cladding layer are 2×10^{19} cm^{-3} and 5×10^{18} cm^{-3} respectively. The doping concentration of the n-GaSb layer and the n-doped Cladding layer are 1.2×10^{18} cm^{-3} and 4×10^{17} cm^{-3} respectively. The optical field is limited by the waveguide layer and the barrier layer [10]. The energy structure of the QWs laser is shown in Fig. 1.

After the MBE growth, 2000 μm cavity length, 100 μm wide and 2.1 μm deep ridges were fabricated using the contact standard optical lithography technique and the inductively coupled plasma etching technology. A 250 nm thick SiO_2 insulation layer was deposited using PECVD. Then a 1900 μm length and 90 μm wide injection window was opened with the contact standard optical lithography technique and the dry-etching. After that, we deposited 500/500/10,000 Å Ti/Pt/Au as p-contact electrode. The N-side substrate was thinned to 130 μm with physical polishing, and 500/3000 Å AuGeNi/Au was sputtered with thermal evaporation, then Ohm contact is obtained by rapid thermal annealing. Finally the N-side was splashed 500/4000 Å Ti/Au as N-side electrode. The finished epitaxial wafer was chipped into single laser bars with 2000 μm long and 500 μm wide. Then the single devices were mounted junction side down by Indium solder on copper heat sinks (C-mount).

Results and Analysis

Based on the experiments of different thickness of un-doped waveguide layer, we analyzed the results. The characteristic of current-voltage and current-output power are shown in Fig. 2.

As shown in Fig. 2, we found that the opening voltage of 270, 180 and 130 nm waveguide layer are 1.2, 0.75 and 0.72 V respectively, and the output optical power are 299, 350 and 533 mW respectively. The threshold current decreases with the

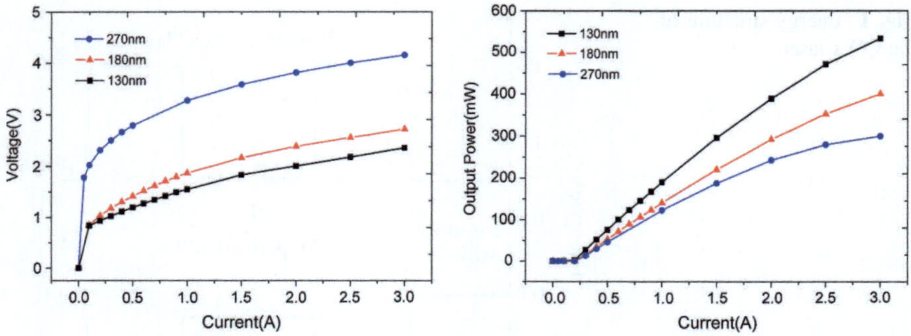

Fig. 2 The left figure is the Current-Voltage characteristic of different waveguide layer thickness, and the right figure is the Current-Output power characteristic of different waveguide layer thickness

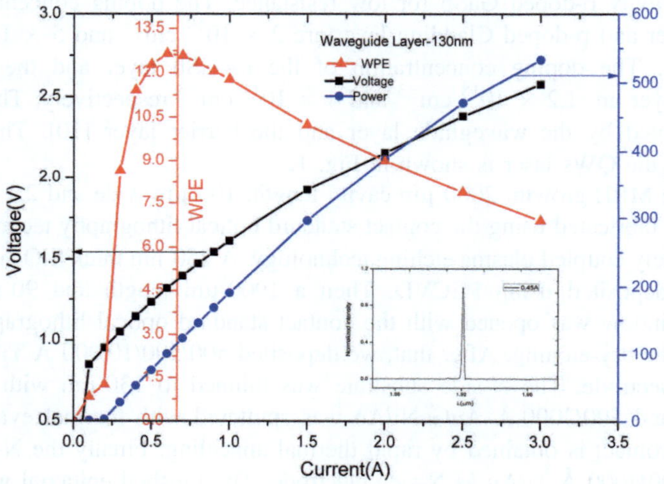

Fig. 3 I-V-P and WPE feature

thickness of the waveguide layer. From the Current-Voltage curve, we found that the greater the thickness of the waveguide layer, the greater the series resistance. And the series resistance of the device caused a serious heat that will limit the device's output optical power. By comparison experiments, we found that the device of 130 nm waveguide layer had the lowest working voltage and the highest output optical power. So we focus on research the characteristic of laser with 130 nm un-doped waveguide layer in detail. The fitting result is shown in Fig. 3.

From Fig. 3 we found that the slope efficiency of the uncoated single laser is 244 mW/A, the laser threshold current is 0.2 A, and the threshold current density is 116.96 A/cm^2. The maximum power of the device is 533 mW, the maximum power efficiency is 6.9%, and the maximum power conversion efficiency is 12.67%.

We also compared the Power-Current characteristic of the laser on the working condition of continuous current and pulse current. The result is shown in Fig. 4.

The Power-Current characteristic of coated device with pulse condition is shown in red line in Fig. 4. The Power-Current characteristic of uncoated device with pulse condition is shown in black line. The Power-Current characteristic of uncoated device with continuous current is shown in blue line. Under the driving current of 3 A, the output power of uncoated laser is 533 mW. The power increased up to 1.946 W with pulse current of 14 A (1000 Hz pulse frequency, 50 μm pulse width, 5% duty ratio). The output power of the coated laser is 2.466 W at 16 A pulse current, and it is still in increasing with the current increase. It found that as the injected current increases, the device produces more heat, which decrease the slope efficiency, and the optical power tend to be saturated. When the injection current increased to a certain extent, the laser will be deteriorated due the COMD effect.

The drifting of the peak wavelength is obtained at room temperature which is shown in Fig. 5. (The intensity is normalized.) The range of the injection current is 0.35–0.55 A.

At 0.45 A injecting current, the peak wavelength of the device is 1928.0 nm. With the increase of the injection current, the peak wavelength moves to the long wavelength (red shift), and the peak wavelength is 1929.5 nm when the injection current is 0.65 mA. As injecting current is higher, the redshift phenomenon is more obvious, which can be seen in the built-in small map clearly. The red shift of the wavelength was caused mainly by heat effects [11] and band gap decrease [12]. As the temperature of the device increases, the refractive index and the band gap of the material are changed, and the band gap becomes narrow, causing the output wavelength becomes larger. Considering wavelength drift limits its application at specific wavelength, we should avoid the drift.

Fig. 4 The P-I features of two operating models

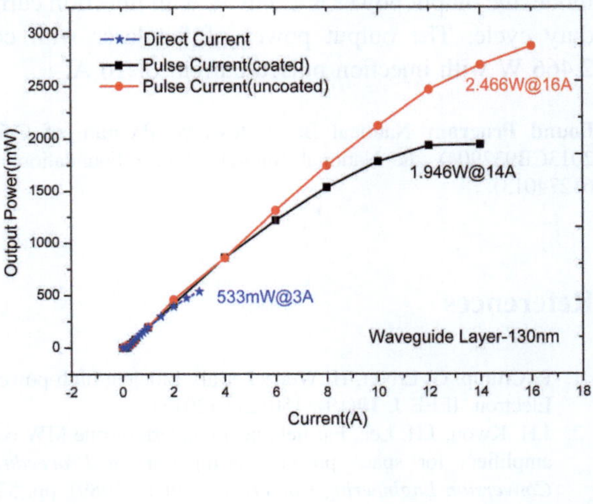

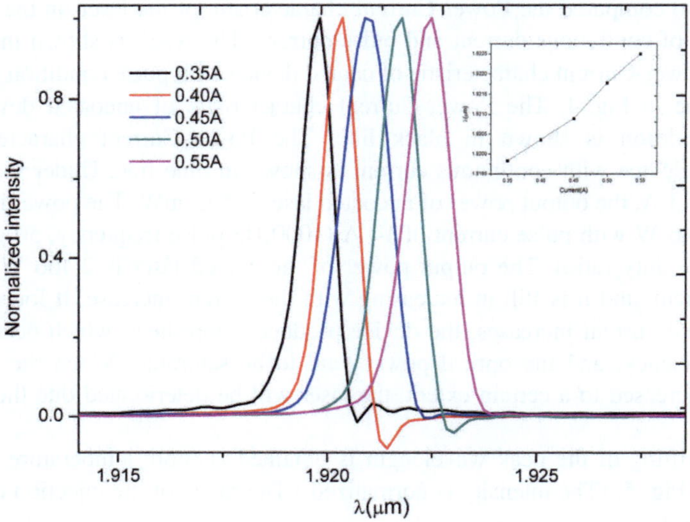

Fig. 5 The peak wavelength with injection current ranging from 0.35 to 0.55 A

Summary

In conclusion, by experimental optimize the epitaxial growth conditions and optimize the thickness of the symmetric un-doped waveguide layer, combined with the standard laser technology of Fabry-Perot cavity, the 2 μm high-power GaSb-based type-I quantum well lasers have been fabricated. Under continuous wave operation mode, the output power of the lasers with uncoated cavity length of 2 mm and 100-μm-wide ridge is about 0.533 W with injection current of 3 A. The maximum conversion efficiency of single device is 12.67%. Under pulsed power operation mode, the output power is 1.946 W with injection current of 14 A of 1 kHz and 5% duty cycle. The output power of the laser with coated cavity is increased to 2.466 W with injection pulsed current of 16 A.

Found Program National Basic Research Program of China (Grant Nos. 2014CB643903, 2013CB932904), the National Natural Science Foundation of China (Grant Nos. 61435012, 61274013).

References

1. P. Crump, G. Erbert, H. Wenzel et al., Efficient high-power laser diodes. Sel. Top. Quantum Electron. IEEE J. **19**(19), 1501211 (2013)
2. J.H. Kwon, J.H. Lee, Far field beam pattern of one MW combined beam of laser diode array amplifiers for space power transmission, in *Proceedings of 24th Intersociety Energy Conversion Engineering Conference*, vol 1, (1989), pp. 521–526

3. Y. Zhang, G.W. Wang, B. Tang et al., Molecular beam epitaxy growth of InGaSb/AlGaSb strained quantum well diode lasers. J. Semiconductors **32**, 103002 (2011)
4. Y. Liao, Y. Zhang, C. Yang et al., High-power, high-efficient GaSb-based quantum well diodes emitting at 2 μm, in *National Conference on Molecular Beam Epitaxy* (2015)
5. M.T. Kelemen, J. Gilly, M. Rattunade et al., Mid-infrared high-power diode lasers and modules, in *Proceedings of SPIE* (2010), 7583
6. R. Liang, G. Kipshidze, T. Hosoda et al., 3.3–3.4 μm diode lasers based on triple-layer GaInAsSb quantum wells. IEEE **26**, 664–666 (2014)
7. L.-j. Wang, Y.-q. Ning, L. Qin, etc., Development of high power diode laser. Chin. J. Lumin., **36**(1), 1–19 (2015)
8. L. Lin, H. Chen, etc., High efficiency semiconductor laser with 976 nm asymmetrical waveguide structure. Devices Technol. **5**, 281–285 (2013)
9. H. Chen, X. Che, L. Lin, etc., 808 nm High efficiency laser diodes. Devices Technol. **48**(7), 418–421 (2011)
10. J. del Amol, S. Swirhum, R.M. Swanson et al., Simultaneous measurement of hole lifetime, hole mobility and bandgap narrowing in heavily doped n-type Silicon. IEDM Tech. Dig., 290–293 (1985)
11. C.r. Li, C.l. Song, S.f. Li et al., Experimental investigation of the change of semiconductor laser output wavelength corresponding to operation current. Infrared Laser Eng. **32**(2), 144–147 (2003)
12. T. Hosoda, T. Feng, L. Shterengas et al., High power cascade diode lasers emitting near 2 μm. Appl. Phys. Lett. **108**(13), 1089 (2016)

Thermal Stability of Er_2O_3–Al_2O_3 Thin Films Grown on Si Substrates

Xiaojie Pan, Zhifang Zhang, Yanyan Zhu, Zebo Fang and Haijing Cao

Abstract The thermal stability of Er_2O_3 and Al_2O_3 doped Er_2O_3 thin films deposited on Si substrates has been investigated by x-ray diffraction and x-ray photoelectron spectroscopy. The structures for the as-grown Er_2O_3 and Al_2O_3 doped Er_2O_3 films on Si substrates are found to convert from amorphous to polycrystalline at the annealing temperatures above 450 °C in O_2 ambience. The crystallinity and the surfaces roughness of Er_2O_3 thin films on p-type Si (001) substrates are decreased if Al_2O_3 is doped in them. However, the result is complicated if these Er_2O_3 and Al_2O_3 doped Er_2O_3 thin films deposited on n-type Si(001) substrates.

Keywords Rare earth oxides · Thermal stability · X-ray photoelectron spectroscopy

Introduction

Er_2O_3 films have been studied because of their different applications in future microelectronic and optical devices in these years. For example, it is an alternate for SiO_2 high κ dielectric in the complementary metal-oxide-semiconductor (CMOS) device because of its high dielectric constant (κ = 10–14) [1–4]. It is also an selective emitters used in thermophotovoltaic (TPV) cell which can directly convert infrared light or thermal energy into electrical energy. The valence 4f electrons in erbium ions are very active, which lie in the 5s/5p electron orbits, which result in a very interesting variety of emission and absorption [5, 6].

X. Pan · Z. Zhang · Y. Zhu (✉) · H. Cao
Department of Physics, Shanghai University of Electric Power, Shanghai 200090, China
e-mail: yyzhu@shiep.edu.cn

Z. Fang (✉)
Department of Physics, Shaoxing University, Shaoxing 312000, China
e-mail: csfzb@usx.edu.cn

Y. Han (ed.), *Advanced Functional Materials*,
https://doi.org/10.1007/978-981-13-0110-0_41

For all the applications, the thermal stability is a very important parameter independent of the common electrical and optical demand. It is not only the requirement of the device work environment but also the requirement of the manufactural process. For example, a specifically fastidious step in the conventional CMOS process is the 900–1000 °C dopant drive-in annealing. Meanwhile, the thermophotovoltaic cell is working at the temperatures above 500 °C. The stability of this material is crucial to guarantee compatibility with the working environment and the synthesis step [1]. It is investigated that most oxide films do not exhibit sufficient thermal stability because the construction of the oxide films is easily transformed from amorphous to polycrystalline and it reacts with the Si substrates. Some researchers have focused on silicate and oxide laminates (alloys) in an attempt to get over the poor thermal stability [6]. Especially, a structure made up of two oxides is a promising solution for enhancing thermal stability to maintain their electric and optical properties. It has been observed that the mechanical properties can be improved and further properties can be tuned when Al_2O_3 is added to silica [6]. On the other hand, the optical and electrical properties such as refractive index and reflectivity can be tuned by alloying of Er_2O_3 and Al_2O_3 to form amorphous Er_2O_3–Al_2O_3 (ErAlO) films [7]. However, their thermal properties have not been reported. In this paper, we studied the thermal stability of Er_2O_3 and Al_2O_3 doped Er_2O_3 (ErAlO) films after they were annealed at different temperatures.

Experiment

The Er_2O_3 films were deposited by radio-frequency magnetron sputtering using a Er_2O_3 composite ceramic target with a pressure of about 3.0×10^{-4} Pa, while ErAlO films were deposited using a $(Er_2O_3)_{0.9}(Al_2O_3)_{0.1}$ target. Both the Er_2O_3 and ErAlO films were deposited at room temperatures. The Si (001) substrates with a resistivity of 2–10 Ω cm were cleaned in HF (1%) solution for approximately 30 s (s) to remove the native oxide. The deposition was carried out on unheated substrates in the mixed gas of O_2 and Ar, where oxygen partial pressure $R = P_{O_2}/(P_{O_2} + P_{Ar})$ is 1%. There are other procedures, such as the mixed gas pressure was kept at 1.0 Pa, the Er_2O_3 and ErAlO films were then annealed in O_2 ambient for 30 s at different temperatures. The films thickness is about 30 nm as measured by spectroscopic ellipsometer. The structure and composition of Er_2O_3 and ErAlO films are checked by X-ray diffraction (XRD), X-ray photoelectron spectroscopy (XPS) and atomic force microscopy (AFM) respectively.

Results and Discussion

From Fig. 1, the θ-2θ x-ray diffraction curves for the as-grown Er_2O_3 and ErAlO films. Clearly, the as-deposited sample is amorphous because no signal was detected.

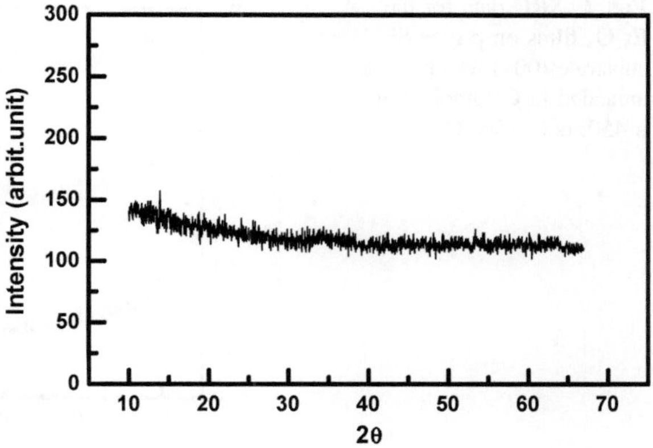

Fig. 1 XRD pattern of the as-grown Er_2O_3 and ErAlO films

Figures 2 and 3 show the XRD results for the Er_2O_3 and ErAlO films deposited on Si substrates(001) which were annealed at different temperatures in O_2 atmosphere. It is observed that after these two films were annealed at 450 °C, one peak around 44.53° appear which are attributed to the reflection of (134) series planes of the Er_2O_3 cubic bixbyte phase, as shown in Figs. 2a and 3a. And the intensity(area) of the peak at 44.53° for the Er_2O_3 film is much larger than that one for ErAlO films indicating that the thermal stability of ErAlO films is better than that of Er_2O_3 film after annealed in O_2 ambient at 450 °C. After the Er_2O_3 films were annealed at 600 °C, another two strong peaks appear, as shown in Fig. 2b. The peak at 29.12° is Er_2O_3 (222) signal. Another peak at 58.91° comes from substrate Si (004) planes. The substrate Si (004) signal here may attribute to the large grains and then the pinpoints or pinholes after the it was annealed at 600 °C. For the ErAlO films, after they were annealed at 600 °C, only one weak Er_2O_3 (222) peak at 29.12° is observed except for the Er_2O_3 (134) peak at 44.53°, as it is shown in Fig. 3b. Therefore, the two kinds of films after annealed in O_2 ambient at 600 °C are both poly-crystallized. The crystallinity for the Er_2O_3 must be increased compared to that of ErAlO film because of its stronger XRD peaks. After the ErAlO film was annealed at 750 °C, only one Er_2O_3 (134) peak at 44.53° is observed, as shown in Fig. 3c. For the Er_2O_3 film annealed at 750 °C, a broad weak peak around 29.12° appear which comes from Er_2O_3 (222) planes, as shown in Fig. 2c. These results imply that when a proper annealing temperature is chosen, Er_2O_3 or the ErAlO film can preferential growth. And all the Er_2O_3 signal from ErAlO film are much smaller than that from undopped Er_2O_3 film which indicates that this is a promising solution for enhancing the thermal stability of Er_2O_3 film on p-type Si (001) substrates.

In Figs. 2 and 3, for the ErAlO film, no signal is detected from any compound containing Al, revealing that the dopant content of Al_2O_3 is small and out of the measuring accuracy of XRD. And Al_2O_3 do not react with Er_2O_3 or Si substrates. In order to further confirm the chemical composition of ErAlO film, the XPS measurement was applied. Figure 4a, b show the Er 4d and Al 2p XPS spectra for the as-grown ErAlO film and the film annealed at 750 °C. Notice that the Er 4d peak

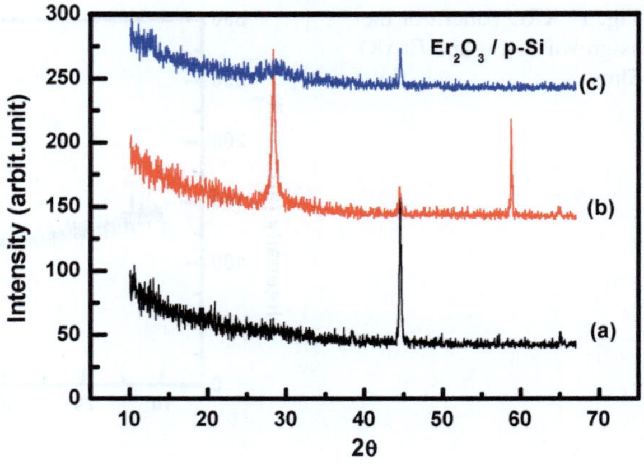

Fig. 2 XRD data for the Er_2O_3 films on p-type Si substrates(001) which were annealed in O_2 ambient at **a** 450, 600, 750 °C

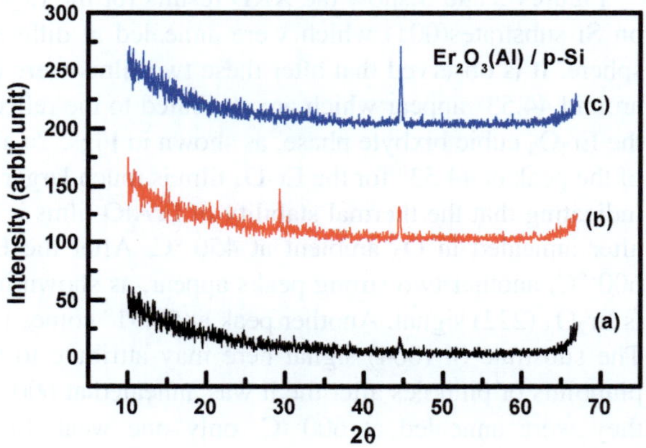

Fig. 3 XRD data for the ErAlO films on p-type Si(001) substrates which were annealed in O_2 ambient at **a** 450 °C, **b** 600 °C, and **c** 750 °C, respectively

located at 168.5 eV for Er_2O_3 accorded with Pan et al. report [8, 9]. We should point out that Al_2O_3 do not react with Er_2O_3, in which Er has a higher binding energy than that in Er_2O_3. In the Al 2p spectra as illustrated by Fig. 2b, the peak at 74.5 eV meet with the Al 2p signal from Al_2O_3 which further confirm Al_2O_3 do not react with Er_2O_3. In comparison, the theory of Pan et al. [8, 9] suggested that for as-grown Er_2O_3 films prepared by reactive magnetron sputtering, a large Er silicate is shaped after annealing at a temperature above 800 °C. The chemical stability of our samples found here is obviously superior to theirs.

Figure 5 reports the AFM images of the Er_2O_3 films on p-type Si substrates(001) annealed at different temperatures. The root-mean-square surface roughness value are 18, 21, and 26 nm for the films annealed at 450, 600, and 750 °C, respectively. For the annealed sample, the roughness increases and pinpoints is observed. Such a rough surface may led to the appearance of the signal from substrates in XRD results which is consistent with the previous XRD data.

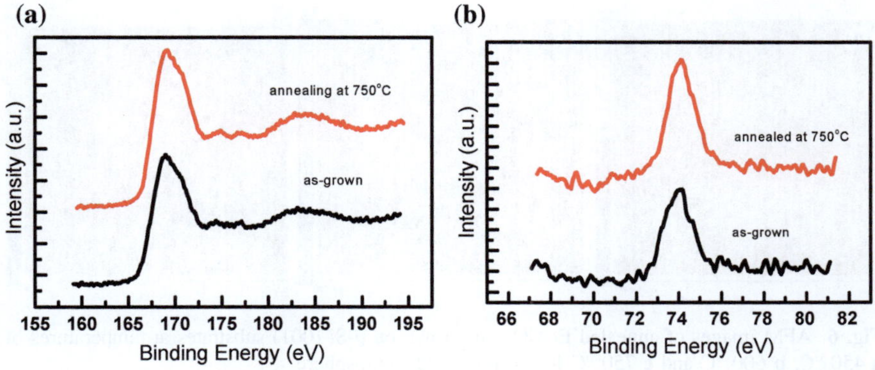

Fig. 4 XPS spectra of **a** Er 4d core level and **b** Al 2p core level derived from the as-deposited ErAlO films and annealed samples at 750 °C in O_2 atmosphere

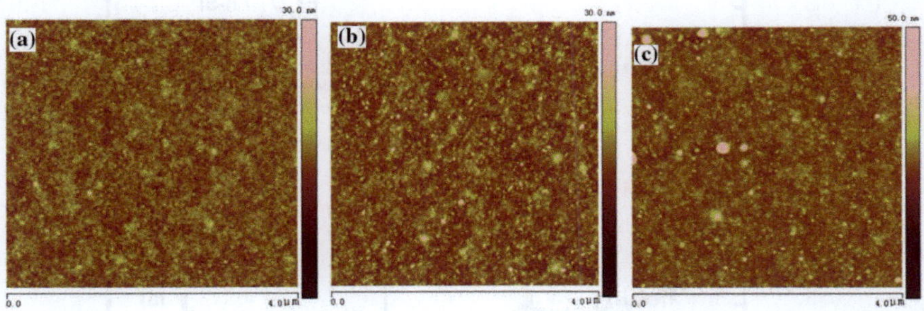

Fig. 5 AFM images of annealed Er2O3 film grown on Si substrates(001) at temperatures of: **a** 450 °C, **b** 600 °C and **c** 750 °C for 30 min in O_2 atmosphere

Figure 6 shows the AFM images of the ErAlO films on p-type Si(001) substrates annealed at different temperatures. The corresponding root-mean-square surface roughness values are 1.8, 2.0 and 2.1 nm, respectively. This value is much smaller than that of the Er_2O_3 films above and the results in other work [10]. For the annealed sample, the roughness increases and no pits or voids appears. It is obvious that the surface roughness for the Er_2O_3 films on p-type Si(001) substrates annealed at different temperatures is much large as the corresponding ErAlO films.

However, the fact that doping Al_2O_3 can enhance the thermal stability of Er_2O_3 film is not always correct when these film are deposited on n-type Si(001) substrates. Figures 7 and 8 show the XRD results for the Er_2O_3 and ErAlO films deposited on n-type Si substrates(001) which were annealed in O_2 ambient at 450, 600, and 750 °C, respectively. The crystalline phase around 29.12°, 44.53°, 48.78°, and 52.46° are caused by the reflection of (222), (134), (440), and (423) series

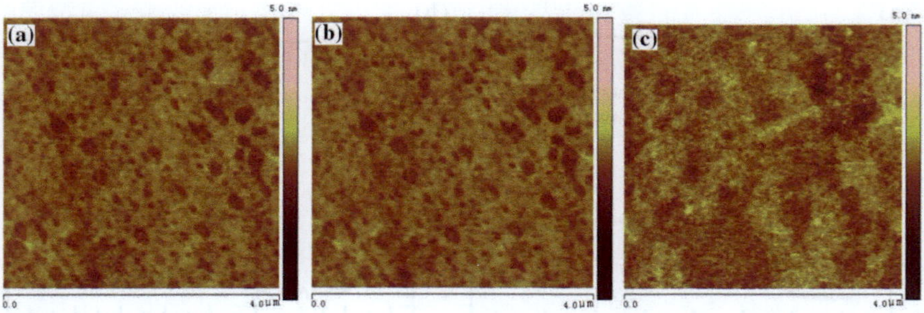

Fig. 6 AFM images of annealed ErAlO film grown on p-Si (001) substrates at temperatures of: **a** 450 °C, **b** 600 °C and **c** 750 °C for 30 min in O_2 atmosphere

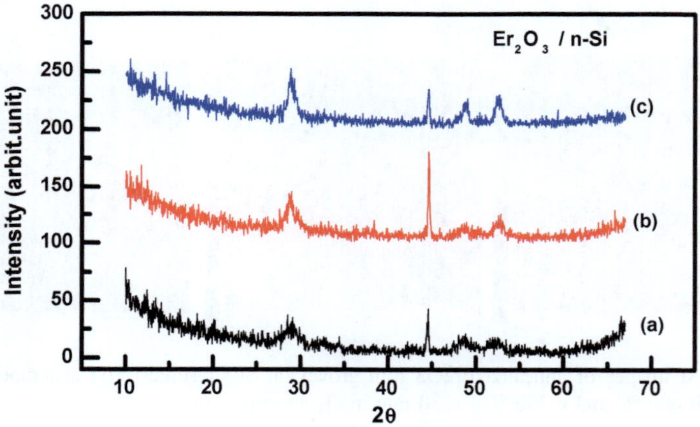

Fig. 7 XRD data for the Er_2O_3 films on n-type Si(001) substrates which were annealed in O_2 ambient at **a** 450, 600, 750 °C, respectively

planes of the Er_2O_3 cubic bixbyte phase, revealing that these films are polycrystalline in structure. When the sample annealed in O_2 atmosphere at 750 °C, these four peaks become sharper, indicating the crystallinity is increased with increasing the annealing temperatures. In Fig. 4c, after the Er_2O_3 films on n-type Si(001) substrates annealed at 750 °C in O_2 ambient, a wide weak peak around 28.01° appears, which is the signal from the reflection of (211) series planes of the Er_2O_3 bixbyte phase. From the XRD results, it may conclude that the thermal stability of ErAlO films on n-type Si(001) substrates are better than that of pure Er_2O_3 films when they are annealed at the temperatures lower than 600 °C while the results will be complex for the films annealed at high temperatures.

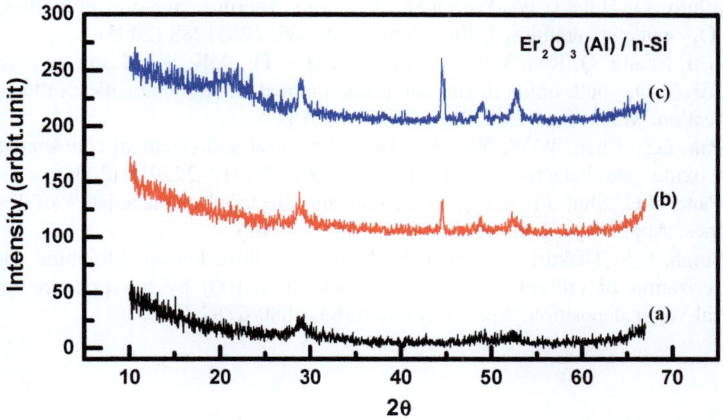

Fig. 8 XRD data for the ErAlO films on n-type Si(001) substrates which were annealed in O_2 ambient at **a** 450, 600, 750 °C, respectively

Summary

In summary, annealing experiments have been done to investigate the thermal stability of Er_2O_3 thin films deposited on Si substrates. The as-grown amorphous films are found to become polycrystalline at temperatures above 450 °C in O_2 ambience. The thermo stability of ErAlO films on p-type Si(001) substrates is found to be increased compared that of Er_2O_3 films. However, the stability of ErAlO films on n-type Si(001) substrates is not increased obviously.

Acknowledgements This work was supported by National Natural Science Foundation of China (Grant No. 51672172), and by the Natural Science Foundation of Shanghai (Grant No. 15ZR1418700).

References

1. S. Chen, Y.Y. Zhu, R. Wu, Y. Wu, Y.L. Fan, Z.M. Jiang, Thermal stability of Er_2O_3 thin films grown epitaxially on Si substrates. J. Appl. Phys. **101**, 064106–064111 (2007)
2. Z.B. Fang, Y.Y. Zhu, J.L. Wang, Z.M. Jiang, Annealing effects on the structure and electrical characteristics of amorphous Er_2O_3 films. Chin. Phys. B **18**, 3542 (2009); W.J. Tobler, W. Durisch, Appl. Energy **85**, 483–485 (2008)
3. Q.Y. Liu, Z.B. Fang, T. Ji, S.Y. Liu, Y.S. Tan, J.J. Chen, Y.Y. Zhu, Band alignment and band gap characterization of La_2O_3 Films on Si substrates grown by radio frequency Magnetron sputtering. Chin. Phys. Lett. **31**, 027702–027705 (2014)
4. B. Yao, Z.B. Fang, Y.Y. Zhu, T. Ji, G. He, A model for the frequency dispersion of the high-k metal-oxide-semiconductor capacitance in accumulation. Appl. Phys. Lett. **100**, 222903–222905 (2012)
5. C. Ferrari, F. Melino, M. Pinelli, P.R. Spina, M. Venturini, Overview and status of thermophotovoltaic systems. Energy Procedia. **45**, 160–163 (2014)

6. N. Narihito, O. Hideki, W. Yoshiharu, Y. Hiroo, Thermal emission properties of Al_2O_3/$Er_3Al_5O_{12}$ eutectic ceramics. J. Eur. Ceram. Soc. **25**, 12851288 (2005)
7. H.J. Su, J. Zhang, Q. Ren, Y.F. Deng, L. Liu, H.Z. Fu, A.K. Soh, Laser zone remelting of Al_2O_3/$Er_3Al_5O_{12}$ bulk oxide in situ composite thermal emission ceramics: influence of rapid solidification. Mater. Res. Bull. **48**, 544–849 (2013)
8. T.M. Pan, C.L. Chen, W.W. Yeh, S.J. Hou, Structural and electrical characteristics of thin erbium oxide gate dielectrics. Appl. Phys. Lett. **89**, 222912–222915 (2006)
9. T.M. Pan, W.H. Shu, J.L. Hong, Structural and electrical characteristics of Er_2TiO_5 gate dielectrics. Appl. Phys. Lett. **90**, 222906–222911 (2007)
10. M.P. Singh, C.S. Thakur, K. Shalini, N. Bhat, S.A. Shivashankar, Structural and electrical characterization of erbium oxide films grown on Si(100) by low-pressure metalorganic chemical vapor deposition. Appl. Phys. Lett. **83**, 2889–2892 (2003)

$(GaSb)_{0.5}$–$Ge_{1.6}Te$ Alloys for High-Temperature Phase Change Memory Applications

Yuan Xue, Sannian Song, Shuai Yan, Tianqi Guo, Lanlan Shen, Liangcai Wu, Zhitang Song and Songlin Feng

Abstract In this paper, phase change characteristics of $(GaSb)_{0.5}$–$Ge_{1.6}Te$ alloy were investigated for long data retention phase change memory application. $(GaSb)_{0.5}$–$Ge_{1.6}Te$ film has high crystallization temperature (357 °C) and large crystallization activation energy (4.57 eV), resulting in a good data retention ability (251 °C for 10 years). The prominent advantages can be seen in comparison with those of pure GeTe and $Ge_2Sb_2Te_5$. The fine crystal grain size is smaller than GeTe owing to GaSb doping, which is contributed to operation speed. Furthermore, as short as 20 ns electrical pulse can achieve Reset operation. The pulse width of 200 ns requires only a Reset voltage of 1.5 V, which is much lower than that of GeTe-based cells.

Keywords GaSb · Doping · Data retention · Phase change memory

Introduction

As a new generation nonvolatile memory, Phase Change Memory (PCM) has made great achievements over the past decades. Compared with other competitors, PCM may not excel in any category but still has an excellent all-round performance, such as high density, low power consumption, good endurance, fast programming

Y. Xue (✉) · S. Song (✉) · S. Yan · T. Guo · L. Shen · L. Wu
Z. Song · S. Feng
State Key Laboratory of Functional Materials for Informatics,
Shanghai Institute of Microsystem and Information Technology,
Chinese Academy of Sciences, Shanghai 200050, China
e-mail: xueyuan@mail.sim.ac.cn

S. Song
e-mail: songsannian@mail.sim.ac.cn

Y. Xue · S. Yan · T. Guo · L. Shen
University of the Chinese Academy of Sciences, Beijing 100049, China

Y. Han (ed.), *Advanced Functional Materials*,
https://doi.org/10.1007/978-981-13-0110-0_42

capability, and fabrication compatibility with complementary metal-oxide-semiconductor (COMS) [1, 2]. Providing that fast enough switching materials can be found and very large cycle numbers can be reached, PCM is replacement of DRAM in the future [3]. Chalcogenide materials are the heart of PCM technology, stability and operation speed are two points to be improved in PCM technology. Except for traditional phase change material $Ge_2Sb_2Te_5$, some other systems have been developed to improve the material properties. GeTe has been investigated as a potential material due to its higher crystallization temperature (T_c) and larger resistance contrast [4–6]. But the data retention ability (117 °C) badly limits further to meet high temperature applications. Thus GaSb is introduced into the GeTe film as the dopant owing to high thermal stability of amorphous state and fast crystallization speed. This work explored the electrical performance and physical properties of $(GaSb)_{0.5}$–$Ge_{1.6}Te$ film, and meanwhile, adopted the microscopic analysis to verify the test results.

Experiments

100 nm-thick $Ge_{1.6}Te$ and $(GaSb)_{0.5}$–$Ge_{1.6}Te$ films were deposited on SiO_2/Si substrates by co-sputtering GeTe, GaSb and Sb targets at room temperature with a base pressure of 2.0×10^{-4} Pa. The concentration of GaSb in GeTe was evaluated by energy dispersive spectroscopy (EDS) through averaging three measurements and the resistance was in situ measured at the heating rate of 20 °C/min from room temperature to 450 °C. X-ray diffraction measurement was employed to characterize the structure of films in the 2θ range of 20–60°. The microstructure was investigated by transmission electron microscopy (TEM), and the samples were fabricated by depositing 40 nm films on carbon supporting membranes and annealed by in situ heating experiments. In order to test the performance of device, a 190 nm heating electrode was deposited using 0.13 μm CMOS technology. Then, 50 nm thick $(GaSb)_{0.5}$–$Ge_{1.6}Te$ film was fabricated as phase change layer. Finally, 10 nm TiN and 300 nm Al was deposited as adhesion layer and top electrode, respectively.

Results and Discussion

Figure 1a shows the curves of resistance of $Ge_{1.6}Te$ (GeTe) and $(GaSb)_{0.5}$–$Ge_{1.6}Te$ films as a function of temperature at different heating rate from 10 to 40 °C/min. The resistance value decreases gradually until the crystallization temperature (T_c) determined by the minimum of the first derivative of the R-T curve, displaying a sharp drop. As the heating rate increases, T_c shifts to higher temperature from 351 to 365 °C, which is because atoms cannot spread in time during crystalline process with higher heating rate. And T_c increases specifically from 203 °C of GeTe to

358 °C (20 °C/min) of $(GaSb)_{0.5}$–$Ge_{1.6}Te$. It indicated the addition of GaSb being an effective method to enhance the thermal stability of amorphous GeTe material and thus improves the data retention of the PCM cell. The sheet resistance almost keeps constant as the temperature exceeds 400 °C, indicating a complete crystallization. The sheet resistance ratio of amorphous and crystalline phase is about two orders of magnitude. The data retention ability can be characterized by the isothermal change in resistance on the basis of Arrhenius equation:

$$t = \tau \exp(E_a/(K_B T)) \tag{1}$$

where τ is a proportional time constant, E_a is activation energy of crystallization and k_B is Boltzamann's constant [7]. The failure time is defined as the time when the resistance decreases to half of its initial value at the specific isothermal temperature. A fitting line based on the failure time versus $1/k_BT$ was plotted in Fig. 1b. The E_a and $T_{10\text{-year}}$ of GeTe and $(GaSb)_{0.5}$–$Ge_{1.6}Te$ are 3.15 eV/117 °C and 4.57 eV/251 °C, respectively. Therefore, this alloy can store information for longer and safer than GeTe does, which is sufficient for automotive systems.

Figure 2 shows the XRD patterns of GeTe and $(GaSb)_{0.5}$–$Ge_{1.6}Te$ thin films annealed at the different temperature for 3 min. In the sample of GeTe film, there are some crystalline peaks at 320 °C. And with the increasing of annealing temperature, the crystal grains further grow with the width of peak (202) and (220) decreasing. Being different from pure GeTe material, no diffraction peaks were seen in $(GaSb)_{0.5}$–$Ge_{1.6}Te$ thin films at 320 °C, which demonstrates its better amorphous thermal stability than pure GeTe. The comparison between the pure GeTe film and $(GaSb)_{0.5}$–$Ge_{1.6}Te$ film shows a decrease in the intensity of the (202) and (220) peaks at 360 °C. The diffraction peaks of $(GaSb)_{0.5}$–$Ge_{1.6}Te$ film become broaden, which illustrated the grain size is effectively restrained by incorporating GaSb. The small grain size may induce more grain boundaries and

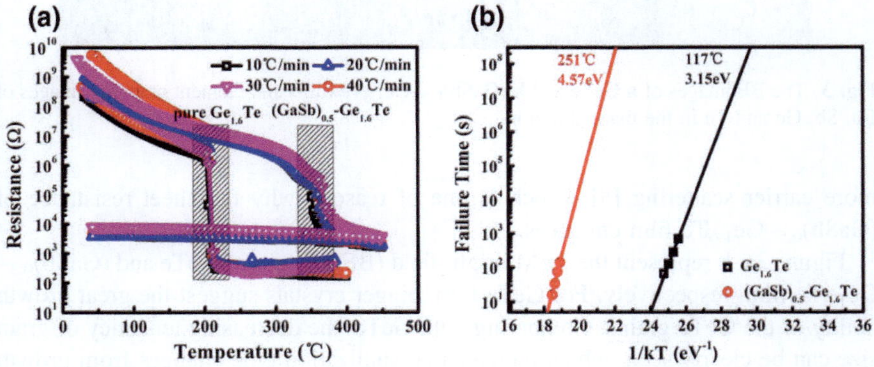

Fig. 1 **a** Temperature dependence of the resistance measured of $Ge_{1.6}Te$ and $(GaSb)_{0.5}$–$Ge_{1.6}Te$ films with various heating rates. **b** The corresponding Arrhenius extrapolation plots of failure time versus $1/k_BT$

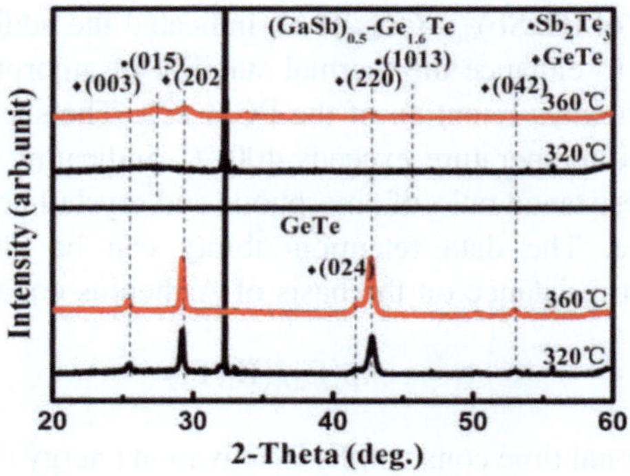

Fig. 2 XRD measurements of GeTe and $(GaSb)_{0.5}$–$Ge_{1.6}Te$ films annealed at the different temperature

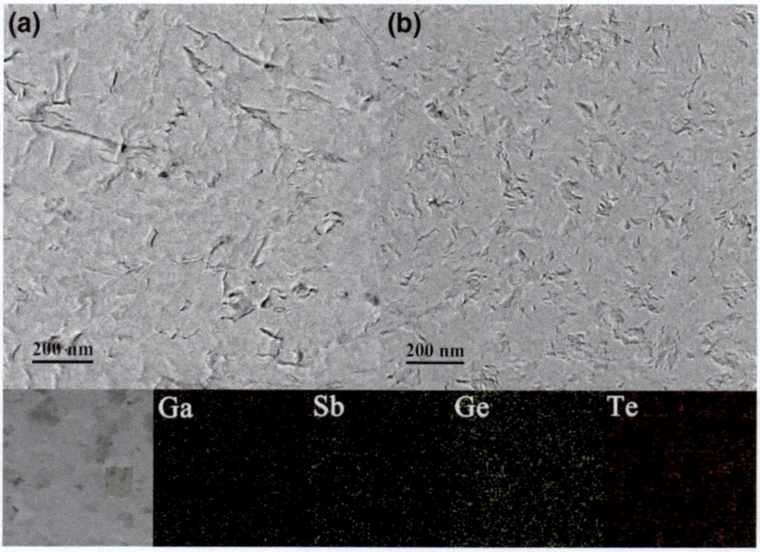

Fig. 3 The BF images of **a** GeTe and **b** $(GaSb)_{0.5}$–$Ge_{1.6}Te$ films and element mapping images of Ga, Sb, Ge and Te in the marked region

more carrier scattering [8], which is one of reasons why the sheet resistance of $(GaSb)_{0.5}$–$Ge_{1.6}Te$ film enhances.

Figure 3a, b represent the TEM bright-field (BF) images of GeTe and $(GaSb)_{0.5}$–$Ge_{1.6}Te$ film, respectively. For GeTe film, bigger crystals suggest the great growth ability of the GeTe grains. Comparing with GeTe, the decreasing tendency of grain size can be clearly seen, which indicated crystallization type changes from growth domination to nucleation domination. A smaller crystal grain size will lead to the increase of the ratio of boundary interfaces for a same large volume area [9]. This phenomenon may promote the reliability of memory devices. Moreover, from the distribution of elements after heating, it can be observed that there is almost no Ga element in the grain, which is consistent with the SEAD patterns. On the other hand, the element mapping images show that the distribution of other three elements is uniform. This analysis is also from the viewpoint about the literature of GaSb doping $Ge_2Sb_2Te_5$ alloys [10]. In a word, GaSb doping is a potential solution.

The typical plan-view TEM images of GeTe and $(GaSb)_{0.5}$–$Ge_{1.6}Te$ films in situ annealed during observation are presented in Fig. 4a–g along with the associated diffraction patterns in the insets. The heating rate is fixed at 10 °C/min. As the annealing temperature arising, many small crystal grains began to appear in the $(GaSb)_{0.5}$–$Ge_{1.6}Te$ film, and diffraction rings related to the polycrystalline diffraction can be observed in the corresponding SAED patterns, which is identical as the rhombohedral GeTe and the rhombohedral Sb_2Te_3. It can be seen obviously from Fig. 4j that the nucleation regions of thin film distribute widely and uniformly. When the temperature increase to 350 °C, the grain are ready to rapid grow as shown in Fig. 4h, indicating a transformation of the crystallization mechanism from nucleation to crystal growth [11]. As the annealing temperature goes up to higher temperature, crystallization of all regions is more completed. It is indicated that, with the annealing temperature, the probability for crystal growth in the film is dominant over nucleation as the activation barrier for growth is surmounted [12].

Low Reset power consumption is important for PCRAM application [13]. Figure 5a compares the current-voltage (I-V) curves for GeTe and $(GaSb)_{0.5}$–$Ge_{1.6}Te$ memory devices. Initially, for the high resistance of amorphous state, the voltage increases sharply under the applied current. Until reaching the threshold voltage and current of 2.2 V/2.8 μA, the low conductive state is changed. Obviously, these are markedly less than those of GeTe, 3.6 V/4 μA. Then Fig. 5b shows the R-V curves of the cells with different voltage pulse widths. When the applied voltage is lower than certain voltage, the resistance of the cell almost keeps

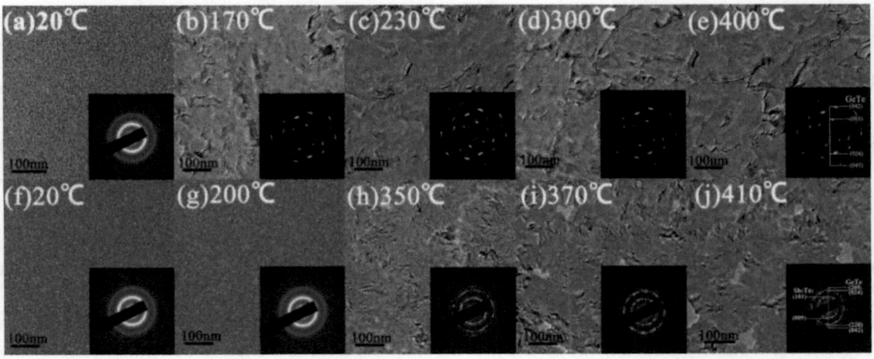

Fig. 4 TEM BF images and the corresponding SAED patterns of **a–e** GeTe and **f–j** $(GaSb)_{0.5}$–$Ge_{1.6}Te$ films

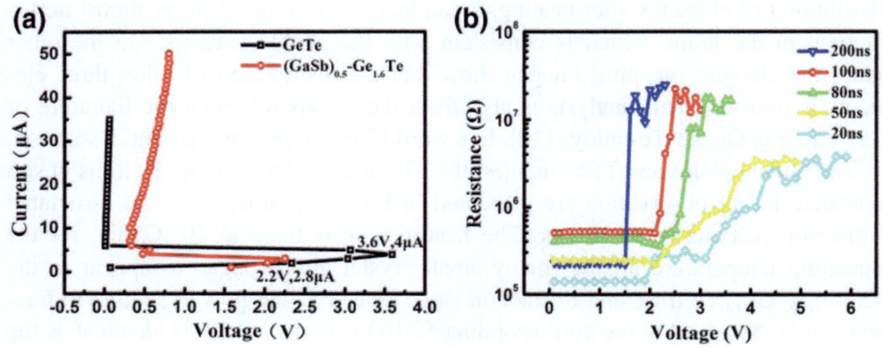

Fig. 5 **a** Current-Voltage (I-V) curves for GeTe and $(GaSb)_{0.5}$–$Ge_{1.6}Te$- based PCM cells. **b** Resistance-voltage characteristics with different voltage pulse widths

constant ($\sim 10^5$) with the increase of pulse voltage. With wider reset pulse width, smaller voltage is required to trigger the resistance switching. This kind of cell can readily achieve Reset operation at 20 ns pulse along with the resistivity change of nearly two orders of magnitude. Noticeably, a Reset voltage of 1.5 V can complete Reset operation under 200 ns pulse width. It is worthy mention that the lower Reset voltage is beneficial to the required dissipated energy being proportional to square of the Reset voltage. Thus, from the above comparisons, PCM devices based on $(GaSb)_{0.5}$–$Ge_{1.6}Te$ may be possessed a lower Reset power consumption.

Conclusions

In summary, phase change abilities of $(GaSb)_{0.5}$–$Ge_{1.6}Te$ phase change film was systemically investigated. The crystallization temperature and crystallization activation of $(GaSb)_{0.5}$–$Ge_{1.6}Te$ film are much higher than those of GeTe film, which is attributed to its good thermal stability. Possessing a 10 years' temperature of 251 °C, $(GaSb)_{0.5}$–$Ge_{1.6}Te$ based PCRAM is sufficient for applications used in aerospace. Short Reset time of 20 ns can be easily achieved due to the formation of uniform smaller grains by the GaSb dopant. Expectedly, the new phase change material gives rise to mitigating the energy of write operation.

Acknowledgements This work was supported by the National Key Research and Development Program of China (2017YFA0206101, 2017YFB0701703), "Strategic Priority Research Program" of the Chinese Academy of Sciences (XDA09020402), National Integrate Circuit Research Program of China (2009ZX02023-003), National Natural Science Foundation of China (61376006, 61401444, 61504157, 61622408), Science and Technology Council of Shanghai (17DZ2291300).

References

1. Y. Hu, H. Zou, L. Yuan, J. Xue, Y. Sui, W. Wu, J. Zhang, X. Zhu, S. Song, Z. Song, Scripta Mater. **115**, 19–23 (2016)
2. X. Ji, L. Wu, L. Cao, M. Zhu, F. Rao, Y. Zheng, W. Zhou, Z. Song, S. Feng, Appl. Phys. Lett. **106**, 243103 (2015)
3. S. Raoux, Annu. Rev. Mater. Res. **39**, 25–48 (2009)
4. E. Gourvest, S. Lhostis, J. Kreisel, M. Armand, S. Maitrejean, A. Roule, C. Vallée, Appl. Phys. Lett. **95**, 031908 (2009)
5. N.N. Yu, H. Tong, X.S. Miao, Appl. Phys. Lett. **105**, 121902 (2014)
6. A.V. Kolobov, P. Fons, J. Tominaga, M. Hase, J. Phys. Chem. C **118**, 10248–10253 (2014)
7. M.H.R. Lankhorst, L. van Pieterson, M. van Schijndel, B.A. Jacobs, J.C.N. Rijpers, IEEE, 419–421 (2002)
8. M. Zhu, L. Wu, F. Rao, Z. Song, X. Li, C. Peng, X. Zhou, K. Ren, D. Yao, S. Feng, J. Alloy. Compd. **509**, 10105–10109 (2011)
9. R.L. Penn, J.F. Banfield, Science, **281**, 969–971 (1998)
10. Z. Zhang, S. Song, Z. Song, Y. Cheng, Y. Gu, L. Wu, B. Liu, S. Feng, J. Non-Cryst. Solids **381**, 54–57 (2013)
11. F. Gan, Z. Song, T. Guo, S. Song, L. Li, L. Shen, B. Liu, Z. Song, M. Qi, S. Feng **9818**, 98180U (2016)
12. X. Zhou, L. Wu, Z. Song, F. Rao, Y. Cheng, C. Peng, D. Yao, S. Song, B. Liu, S. Feng, B. Chen, Appl. Phys. Lett. **99**, 032105 (2011)
13. L. Wu, M. Zhu, Z. Song, S. Lv, X. Zhou, C. Peng, F. Rao, S. Song, B. Liu, S. Feng, J. Non-Cryst. Solids **358**, 2409–2411 (2012)

A Stretchable and Flexible Strain Sensor Based on Graphene Sponge

Fei Han, Jinhui Li, Yuan Zhang, Guoping Zhang, Rong Sun and Chingping Wong

Abstract Flexible strain sensors have attracted extensive research all over the world for their great potential applications in wearable electronics, man-machine interaction system, and flexible human health monitoring system. Nevertheless, there still remains some problems such as low stretchability and poor durability. In this study, a highly flexible, stretchable and sensitive strain sensor has been fabricated with a low-cost strategy. The strain sensor was manufactured by graphene-coated polyurethane sponge and packaged by PDMS. The obtained strain sensor could be stretched up to 100% of its original length, showing excellent flexibility and stretchability. In addition, the gauge factor of 2 indicates high sensitivity of the sensor. Besides, the strain sensor exhibited high durability and reliability after 300 stretching-releasing cycles. The facile fabrication process, low-cost materials, and good performance of graphene-sponge based flexible strain sensor shows a good prospect for the application to flexible wearable devices.

F. Han · J. Li · Y. Zhang · G. Zhang (✉) · R. Sun (✉) · C. Wong
Guangdong Provincial Key Laboratory of Materials for High Density Electrical Packaging, Shenzhen Institutes of Advanced Technology, Chinese Academy of Science, Shenzhen, China
e-mail: gp.zhang@siat.ac.cn

R. Sun
e-mail: rong.sun@siat.ac.cn

F. Han
e-mail: fei.han@siat.ac.cn

J. Li
e-mail: li.jh@siat.ac.cn

Y. Zhang
e-mail: yuan.zhang1@siat.ac.cn

F. Han · Y. Zhang
Nano Science and Technology Institute, University of Science and Technology of China, Suzhou, China

C. Wong
Department of Electronic Engineering, The Chinese University of Hong Kong, Hong Kong, China

Y. Han (ed.), *Advanced Functional Materials*,
https://doi.org/10.1007/978-981-13-0110-0_43

Keywords Graphene sponge · Stretchable · Flexible · Strain sensor

Introduction

Recently, with the rapid development of the wearable electronics, strain sensors have drawn a great deal of attentions for their important function in the operation system [1–4]. But the conventional semiconductor based and metal foils based strain gauges are not suitable for application of flexible electronics because of their low strain range (<5%) and low resolution [5, 6]. With the purpose of develop flexible strain sensors with high stretchability and sensitivity, different types strain sensors have been fabricated, such as piezoelectric sensing, [7, 8] transistor sensing, [9, 10] piezoresistive sensing, [11, 12] and capacitive sensing mechanisms [13, 14]. By means of special advantages, piezoresistive sensors have already been further researched [11, 12, 15]. The above-mentioned advantages, such as simple design, low-cost fabrication, and easy signal collection, make it output resistance change signals instantly when external mechanical stimuli imposed on. Commonly, stretchable polymer substrates (e.g., polyurethane, [16] polydimethylsiloxane (PDMS), [17] and natural rubber [18]) coupled with conductive materials (e.g., carbon blacks, [18] carbon nanotubes, [11] graphene, [19] and metal nanoparticles/nanowires [20, 21]) constitute piezoresistive sensors. Among them, graphene has been widely used in various aspects, such as flexible display, [22] transistor, [23] and biological materials [24]. Importantly, the graphene based stretchable strain sensors as one of the important smart materials have been investigated widely. For example, a graphene coated nanotube aerogel has been reported which could be compressed to 60% of its original states [25]. And the gauge factor value of the graphene coated nanotube aerogel was 0.5. Gauge factor (GF), an indicator of sensitivity, defined as $((R - R_0)/R_0)/\Delta\varepsilon$, where R, R_0, ε represented the resistances at strain, relaxed states, and the strain, respectively. In addition, a fractured microstructure designed graphene-based piezoresistive pressure sensor with high sensitivity has been prepared which was a promising material for artificial skin [16]. Besides, a reduced graphene oxide (RGO) based composites which provided the fragile RGO aerogel with super-flexible 3D architectures were fabricated and exhibited potential applications in multifunctional strain sensors [1].

The graphene-based piezoresistive sensors mentioned above are most pressure sensors which used the polyurethane sponge as flexible carrier. Polyurethane sponge (PUS), a porous flexible substrate, has been widely researched in the application of flexible sensors [16]. When PUS coated by conductive materials, both electronic conductivities and mechanical flexibilities make it having a great application potential in the fabrication of flexible strain sensors.

In this work, a highly stretchable and flexible strain sensor based on graphene coated PUS [graphene sponge (GS)] has been prepared. The obtained GS strain sensor could be stretched up to 100% of its original length, showing excellent flexibility and stretchability. In addition, the GF of 2 indicates high sensitivity of the sensor. Besides, 300 stretching-releasing cycles exhibited high durability and reliability of the sensor. The facile fabrication process, low-cost materials, and good performance of GS based flexible strain sensor shows a good prospect for the application to flexible wearable electronics.

Experimental Section

Materials. Commercial polyurethane sponges (PUS, one millimetre in thickness, the pore size ranges from 100 to 300 μm) were purchased from Chengxing Hong Co., Shenzhen, China. Graphene oxide (GO) were purchased from Sinopharm Chemical Keagent co., Ltd., China. Hydroiodic acid (HI) was purchased from Aladdin Co., Shanghai, China. PDMS prepolymer and the curing agent were supplied as two-part liquid component kits from Dow Corning (Sylgard 184).

Preparation of GS. PUS was cut into cuboids with 30 mm in length and 20 mm in width, then cleaned by deionized water (DIW) and ethanol for several times and dried at 90 °C for 2 h. After that, PUS pieces were completely dipped into the GO solution (2 mg/mL) under the condition of vacuum oven for 20 min and got the graphene oxide coated PUS (GOS). Then, PUS pieces were dipped in the hot HI (80 °C) for 15 min and washed by adequate DIW and dried in 70 °C. Repeated the above step for four cycles and resulted in the graphene sponges.

Preparation of Graphene Sponge Based Strain Sensor. For the substrate fabrication, 6 g PDMS and 0.6 g curing agent was poured into vial bottle and sufficient mixed via vacuum mixer. After that, the PDMS was poured into polytetrafluoroethylene (PTFE) mold and baked at 70 °C for 15 min and elastomer mixtures were half cured. Then the GS was put between two half-cured PDMS then baked them together for 4 h to make the GS tightly adhered on the PDMS. Finally got the GS strain sensor.

Characterization. Scanning electronic micrographs (SEM) were recorded with Nova Nano SEM 450. Powder X-ray diffraction (XRD) patterns were recorded on an X-ray diffractometer (D8 Advance/Bruker) with monochromated Cu radiation. The tensile test was carried out on an electronic universal testing machine (RGM-4000, REGER Co. Ltd., China). The electrical properties of tensile test were recorded by a digital source meter (Keithly 2410 Source Meter instrument).

Results and Discussion

The fabrication process of strain sensor is illustrated in Fig. 1, consists of two steps: (1) Preparation of GS; (2) packaged by PDMS. Firstly, a piece of commercial PUS (Fig. 2a) was fully immersed in GO solution and reduced by hot HI. Repeated the immersion and reduction process for four cycles got the GS, as shown in Fig. 2b. Secondly, the as-prepared GS was put between two half-cured PDMS then cured

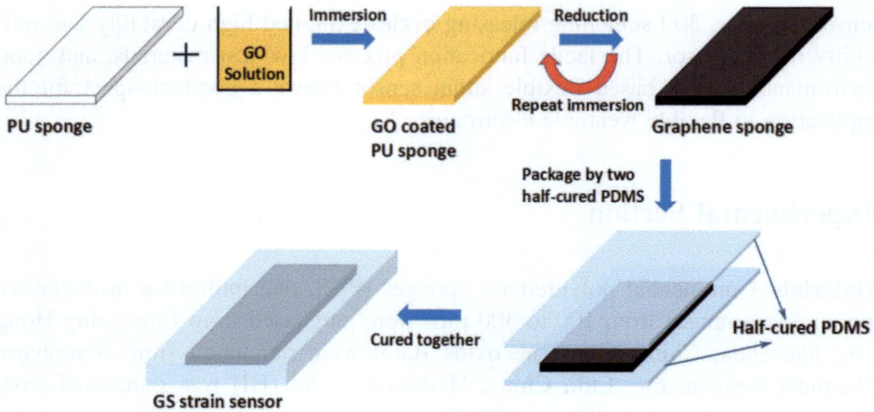

Fig. 1 **a** Schematic illustration of the fabrication process

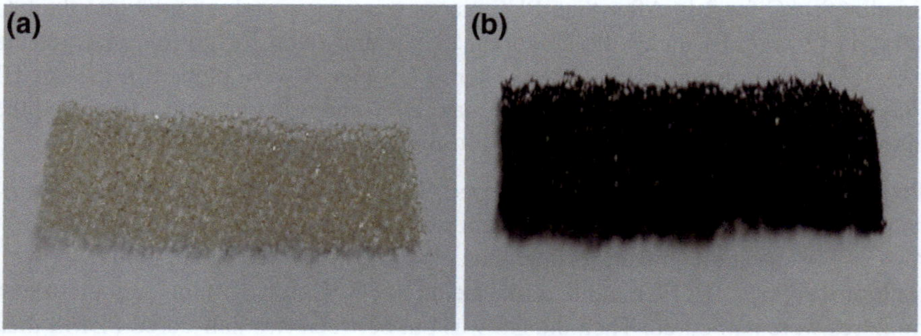

Fig. 2 **a** The photograph of PU sponge. **b** The photograph of graphene sponge

together and got the GS strain sensor (the optical photograph of GS strain sensor exhibits in Fig. 4b). During the fabrication process of GS, repeated immersion and reduction was an important method which could increase the coated graphene layer on PUS and the conductivity of the GS. Besides, the half-cured PDMS, as a vital stretchable substrate, provided the strain sensor with good packaging. The reason for the selected of half-cured PDMS was that half-cured PDMS possess strong adhesive while with poor fluidity [26]. After the GS and half-cured PDMS cured together, the GS strain sensor was endowed with high stretchability and rebound elasticity because of the fully cured PDMS.

Figure 3 exhibits the SEM of original PUS, and GS with different immersion-reduction cycles. As shown in Fig. 3a, the original PUS exhibits high porosity, and the size of the holes ranges from 100 to 300 μm. Different with other commercial PUS, the thickness of PUS in this experiment was only 1 mm. The reason for the chosen of this kind of PUS was that ultrathin PUS possess the good

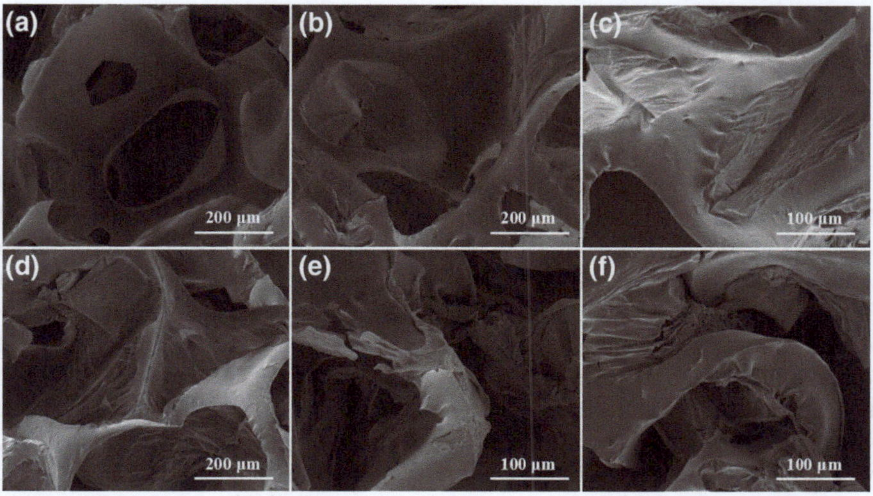

Fig. 3 **a** SEM images of original PUS. **b–f** SEM images of GS with different absorption-reduction cycles, one, two, three, four, five, respectively

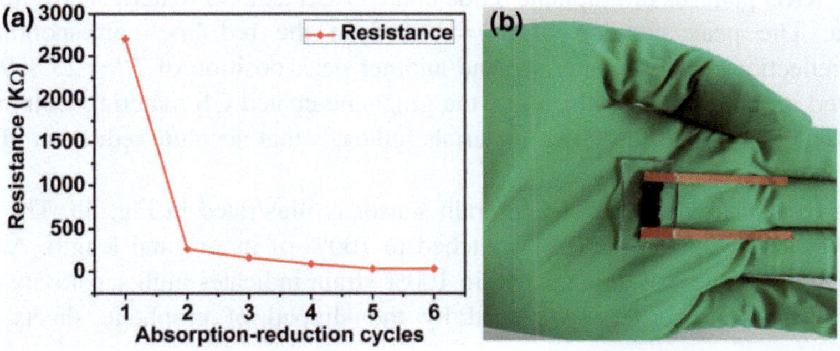

Fig. 4 **a** Variation of resistances with the increased absorption-reduction cycles. **b** Optical photograph of GS strain sensor

stretchability and excellent adsorptive property for the GO solution. The variation of resistance with the increased absorption-reduction cycles as shown in Fig. 4a. The 2.7 MΩ of the GS after first reduction process could not meet the requirement of high-performance strain sensor. As the absorption-reduction cycles increased, the resistance of GS decreased rapidly to 30 KΩ after five cycles. Although the obtained GS possess good conductivity of 30 KΩ, compared with the reported lumpy graphene sponge, the resistance of our GS was much higher because of no redundant graphene which filled the hole of PUS. But this kind of parameters were suitable for the application in stretchable electronics. The SEM of GS in different cycles are depicted from Fig. 3b–f. In Fig. 3b, the relative smooth surface of one cycled GS indicated that a thin layer of graphene coated on PUS without overlapped. As the absorption-reduction cycles increased, multilayer graphene coated on the PUS which exhibit the wrinkle and overlap morphology in Fig. 3c, d. When the cycles increased to five (Fig. 3f), the PUS completely coated by overlapped and wrinkled graphene, and the conductivity of formed GS tended to be stable.

In addition, during the fabrication process, there were some factors that effected the successful preparation of GS strain sensor, such as the concentration of HI solution (the reduction time of every cycle was the same), the concentration of GO solutions, and the form of PDMS. In this experiment, we fixed the concentration of GO solution and adjusted the other factors. In detail, we could not get the absolutely reduced GS if the concentration of HI solution was lower (the volume ratio of HI solution and deionized water was below 1:4). But if the concentration of HI was higher (the volume ratio of HI solution and deionized water was above 1:1), the structure of PUS would be destroyed because of the strong causticity of the HI. So, we chosen the proper volume ratio of HI solution and deionized water of 1:2. Besides, the half-cured PDMS played an important role in the system of the strain sensor which has reported [26]. The porosity of the GS would not be retained if the PDMS was liquid. On the contrary, the completely cured PDMS would lost the viscidity which as the substrate for the GS attached to.

The XRD patterns of graphene oxide coated PUS and GS materials are shown in Fig. 5a. The peak position of $2\theta = 10.98°$ in the red line corresponding to (002) reflection of GOS material. And another peak position of $2\theta = 23.859°$ was observed in the black line illustrated the graphene-coated GS material. Comparison with XRD patterns of these two materials indicates that absolute reduction of GOS [27].

The resistance variation of GS strain sensor is illustrated in Fig. 5b. The fabricated GS strain sensor could be stretched to 100% of its original length. And the averaged GF value was about 2 within 100% strain indicates high sensitivity of the sensor. The high sensitivity caused by the slipped of graphene sheets when

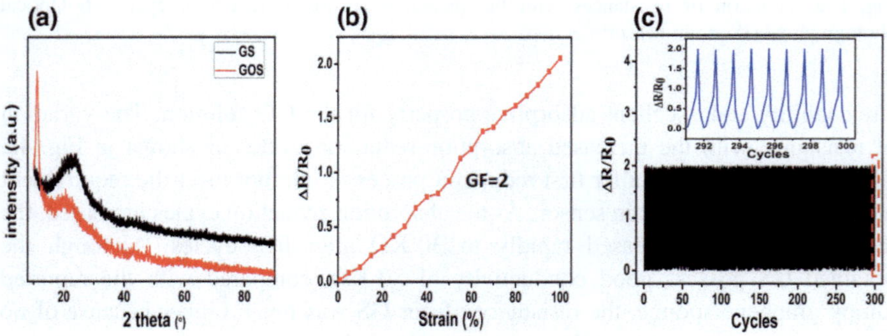

Fig. 5 **a** XRD patterns of graphene oxide coated PUS and GS. **b** Relative resistance changes of GS and NGS strain sensor when stretched. **c** Relative change in resistance under repeated stretching-relaxing of 100% strain for 300 cycles, the inset graph is the enlargement region of the red part

stretched and finally cracks occurred among the graphene layer [4]. Specifically, the GS material kept its continuous structure which provided the electrically conductive path in the initial state. When the sensor was stretched by external force, the graphene sheets slipped and gradually results in the microcracks arose due to the separation of the graphene sheets. And those microcracks decreased the connection area of graphene sheets and the resistance of the sensor.

In addition, 300 cycles, stretched to 100% and backed to relaxed states, demonstrates the stability and reproducibility of the sensor in Fig. 5c. The inset of Fig. 5c is the enlargement graph of the last nine cycles in the 300 cycles. The high sensitivity, stretchability, and good stability brought the GS strain sensor with great application potential in wearable electronics.

Summary

Graphene based PUS stretchable strain sensor was fabricated by repeated absorption-reduction process and packaged by PDMS. The as-prepared GS strain sensor can be stretched up to 100% of its original states and GF value of is 2. The sensor maintains good stability after 300 stretching-releasing cycles. The facile preparation method, low-cost materials, high sensitivity, and good stability make it a promising candidate for flexible wearable devices.

Acknowledgements This work was financially supported by Guangdong and Shenzhen Innovative Research Team Program (No.2011D052, KYPT20121228160843692), NSFC-Guangdong Jointed Funding (U1601202), NSFC-Shenzhen Robot Jointed Funding (U1613215), Natural Science Foundation of China (21601065), R&D Funds for basic Research Program of Shenzhen (Grant No. JCYJ20150401145529012), Key Deployment Project of Chinese Academy of Sciences (Grant No. KFZD-SW-202) and the National Key R&D Project from Minister of Science and Technology of China (2016YFA0202702).

References

1. R. Xu, Y. Lu, C. Jiang, J. Chen, P. Mao, G. Gao, L. Zhang, S. Wu, Facile fabrication of three-dimensional graphene foam/poly(dimethylsiloxane) composites and their potential application as strain sensor. ACS Appl. Mater. Interfaces **6**, 13455–13460 (2014)
2. X. Wu, Y. Han, X. Zhang, C. Lu, Highly sensitive, stretchable, and wash-durable strain sensor based on ultrathin conductive layer@Polyurethane Yarn for tiny motion monitoring. ACS Appl. Mater. Interfaces **8**, 9936–9945 (2016)
3. S.C. Mannsfeld, B.C. Tee, R.M. Stoltenberg, C.V. Chen, S. Barman, B.V. Muir, A.N. Sokolov, C. Reese, Z. Bao, Highly sensitive flexible pressure sensors with microstructured rubber dielectric layers. Nat. Mater. **9**, 859–864 (2010)
4. J. Li, S. Zhao, X. Zeng, W. Huang, Z. Gong, G. Zhang, R. Sun, C.P. Wong, Highly stretchable and sensitive strain sensor based on facilely prepared three-dimensional graphene foam composite. ACS Appl. Mater. Interfaces **8**, 18954–18961 (2016)

5. A.A. Barlian, W.-T. Park, J.R. Mallon, A.J. Rastegar, B.L. Pruitt, Semiconductor piezoresistance for microsystems. Proc. IEEE **97**, 513–552 (2009)
6. A. Ajovalasit, B. Zuccarello, Local reinforcement effect of a strain gauge installation on low modulus materials. J. Strain Anal. Eng. Des. **40**, 643–653 (2005)
7. I. Lee, H. Sung, Development of an array of pressure sensors with PVDF film. Exp. Fluids **26**, 27–35 (1999)
8. A. Shirinov, W. Schomburg, Pressure sensor from a PVDF film. Sens. Actuat. A. **142**, 48–55 (2008)
9. T. Someya, T. Sekitani, S. Iba, Y. Kato, H. Kawaguchi, T. Sakurai, A large-area, flexible pressure sensor matrix with organic field-effect transistors for artificial skin applications. Proc. Natl. Acad. Sci. U.S.A. **101**, 9966–9970 (2004)
10. T. Someya, Y. Kato, T. Sekitani, S. Iba, Y. Noguchi, Y. Murase, H. Kawaguchi, T. Sakurai, Conformable, flexible, large-area networks of pressure and thermal sensors with organic transistor active matrixes. Proc. Natl. Acad. Sci. U.S.A. **102**, 12321–12325 (2005)
11. T. Yamada, Y. Hayamizu, Y. Yamamoto, Y. Yomogida, A. Izadi-Najafabadi, D.N. Futaba, K. Hata, A stretchable carbon nanotube strain sensor for human-motion detection. Nat. Nanotechnol. **6**, 296–301 (2011)
12. M. King, A. Baragwanath, M. Rosamond, D. Wood, A. Gallant, Porous PDMS force sensitive resistors. Procedia Chem. **1**, 568–571 (2009)
13. C. Metzger, E. Fleisch, J. Meyer, M. Dansachmüller, I. Graz, M. Kaltenbrunner, C. Keplinger, R. Schwödiauer, S. Bauer, Flexible-foam-based capacitive sensor arrays for object detection at low cost. Appl. Phys. Lett. **92**, 013506 (2008)
14. C.H. Mastrangelo, X. Zhang, W.C. Tang, Surface-micromachined capacitive differential pressure sensor with lithographically defined silicon diaphragm. J. Microelectromech. Syst. **5**, 98–105 (1996)
15. M. Hussain, Y.-H. Choa, K. Niihara, Conductive rubber materials for pressure sensors. J. Mater. Sci. Lett. **20**, 525–527 (2001)
16. H.B. Yao, J. Ge, C.F. Wang, X. Wang, W. Hu, Z.J. Zheng, Y. Ni, S.H. Yu, A flexible and highly pressure-sensitive graphene-polyurethane sponge based on fractured microstructure design. Adv. Mater. **25**, 6692–6698 (2013)
17. Y. Li, Y.A. Samad, T. Taha, G. Cai, S.-Y. Fu, K. Liao, Highly flexible strain sensor from tissue paper for wearable electronics. ACS Sustain. Chem. Eng. **4**, 4288–4295 (2016)
18. S. Tadakaluru, W. Thongsuwan, P. Singjai, Stretchable and flexible high-strain sensors made using carbon nanotubes and graphite films on natural rubber. Sensors **14**, 868–876 (2014)
19. J.J. Park, W.J. Hyun, S.C. Mun, Y.T. Park, O.O. Park, Highly stretchable and wearable graphene strain sensors with controllable sensitivity for human motion monitoring. ACS Appl. Mater. Interfaces **7**, 6317–6324 (2015)
20. S. Zhang, H. Zhang, G. Yao, F. Liao, M. Gao, Z. Huang, K. Li, Y. Lin, Highly stretchable, sensitive, and flexible strain sensors based on silver nanoparticles/carbon nanotubes composites. J. Alloys Compd. **652**, 48–54 (2015)
21. S. Gong, W. Schwalb, Y. Wang, Y. Chen, Y. Tang, J. Si, B. Shirinzadeh, W. Cheng, A wearable and highly sensitive pressure sensor with ultrathin gold nanowires. Nat Commun. **5**, 3132 (2014)
22. K.-Y. Shin, J.-Y. Hong, J. Jang, Flexible and transparent graphene films as acoustic actuator electrodes using inkjet printing. Chem. Commun. **47**, 8527–8529 (2011)
23. X. Yang, G. Liu, A.A. Balandin, K. Mohanram, Triple-mode single-transistor graphene amplifier and its applications. ACS Nano **4**, 5532–5538 (2010)
24. Y. Liu, X. Dong, P. Chen, Biological and chemical sensors based on graphene materials. Chem. Soc. Rev. **41**, 2283–2307 (2012)
25. F.-Y. Chang, R.-H. Wang, H. Yang, Y.-H. Lin, T.-M. Chen, S.-J. Huang, Flexible strain sensors fabricated with carbon nano-tube and carbon nano-fiber composite thin films. Thin Solid Films **518**, 7343–7347 (2010)

26. H. Chen, Z. Su, Y. Song, X. Cheng, X. Chen, B. Meng, Z. Song, D. Chen, H. Zhang, Omnidirectional bending and pressure sensor based on stretchable CNT-PU sponge. Adv. Funct. Mater. **27**, 1604434 (2017)
27. I.K. Moon, J. Lee, R.S. Ruoff, H. Lee, Reduced graphene oxide by chemical graphitization. Nat. Commun. **1**, 73 (2010)

Self-healable Silver Nanowire-Based Composite for Elastic Strain Sensor

Feng Liu, Jinhui Li, Fei Han, Lei Ling, Xinxiu Wu, Guoping Zhang, Rong Sun and Ching Ping Wong

Abstract Stretchable and flexible sensors attract increasing interest due to their potential applications such as healthcare monitoring, wearable displays, and electronic skins. In these advanced application, high sensitivity, low detection limit, and conductive stability are essential features. Besides, the service life is also one of the important indicators, prolonged service life can greatly cut the cost. Self-healing can help materials recover its integrity in time after damage, which is an effective

F. Liu · J. Li · F. Han · L. Ling · X. Wu · G. Zhang (✉) · R. Sun (✉)
Shenzhen Institutes of Advanced Technology,
Chinese Academy of Sciences, Shenzhen, Guangdong, China
e-mail: gp.zhang@siat.ac.cn

R. Sun
e-mail: rong.sun@siat.ac.cn

F. Liu
e-mail: feng.liu2@siat.ac.cn

J. Li
e-mail: li.jh@siat.ac.cn

F. Han
e-mail: fei.han@siat.ac.cn

L. Ling
e-mail: lei.ling@siat.ac.cn

X. Wu
e-mail: xx.wu1@siat.ac.cn

F. Liu
College of Materials Science and Engineering, Shenzhen University,
Shenzhen, Guangdong, China

C. P. Wong
School of Materials Science and Engineering, Georgia Institute of Technology,
Atlanta, GA, USA

C. P. Wong
Department of Electronic Engineering, Faculty of Engineering,
The Chinese University of Hong Kong, Hong Kong, China

Y. Han (ed.), *Advanced Functional Materials*,
https://doi.org/10.1007/978-981-13-0110-0_44

way to extend service life, thereby reducing costs and improving real engineering applications. In this work, a self-healable, stretchable and sensitive strain sensor based on the nanocomposite of AgNWs and polyurethane (PU) in the configuration of sandwich structure (AgNWs thin film was embedded between two layers of PU) was fabricated via vacuum filtration and transferring process. The as-prepared PU–AgNWs–PU sandwich structural composite showed a high gauge factors of 15 at the strain range of more than 11%. More importantly, the composite could be efficiently repaired by simply thermal treatment after damaged. All the results indicated that the as-fabricated PU–AgNWs–PU sandwich structural composites exhibited great potential applications as smart material for elastic strain sensor.

Keywords Strain sensor · Self-healable · AgNWs · High sensitivity

Introduction

Stretchable, wearable, flexible and low-cost strain sensors are highly desirable in next generation of portable and foldable devices. Strain sensors can transfer mechanical deformations into electrical change, which have found broad applications in rehabilitation and health care monitor [1–3], sports [4, 5] and entertainment [6–8], etc. So far, many strain sensors have been reported combining elastic substrate with carbon nanotubes [9, 10], semiconductor [11, 12], graphene [13, 14], conductive polymer [15]. However, strain sensors responding to mechanical deformation by the change of resistance or capacitance would easily results in microcracks or mechanical fractures under cyclic deformation. The gradually expanding microcracks will finally undermine the strain sensor's integrity, which may lead to the failure of functionality. Therefore, strain sensor is expected to be capable to repair themselves spontaneously to improve its reliability, while maintaining its sensitivity after damage.

To realize the properties mentioned above, few works have been reports in self-healing sensors or conductors [16, 17]. Bao and co-workers developed a highly stretchable and autonomous self-healing material using the dynamic nature of the metal-ligand bond, which enabled the elastomer to restore high dielectric strength after the self-healing process [18]. Lee et al. have reported a self-healing piezoresistive strain sensor using different electronic conductors of hydrogel which was cross-linked by hydrogen bonding [19]. Zhu and co-workers reported a recoverable motion sensor based on flexible self-healing nanocomposite, which involved incorporating modified nanoparticles in self-healing polymer matrix based on dynamic Diels-Alder (DA) adducts [20]. Though these reported sensors and conductors showed excellent self-healing properties, and could restore their functionalities after a cutting/healing process, most of them were based on hydrogels and their mechanical properties were poor. Besides, the complicated polymerization and fabrication process hindered their large-scale application. Therefore, strain sensors with the feature of repeatable self-healing, considerable mechanical strength and simple preparation process simultaneously are highly desirable for the next-generation devices.

Herein, a self-healable strain sensor based on the AgNWs network and Polyurethane (PU) is fabricated using facile vacuum filtration and water-assisted transferring process. The AgNWs network layer is sandwiched between two layers of PU, which is based on the dynamic Diels-Alders adducts [21] (Hereinafter referred to as DAPU). This configuration was constructed to improve the mechanical properties and sensing stability in terms of conductivity and recovery. The as-prepared PU–AgNWs–PU composite exhibits strong piezoresistivity with high gauge factor (GF) and self-healing properties, the healed PU–AgNWs–PU composite could maintain similar sensitivity. These flexible, healable strain sensors with facile fabrication process would provide a new design and fabrication options for next generation flexible electronics.

Experiment

Materials. Silver nanowire suspension (2 mg/mL in ethanol) were supplied by XFNANO Materials, 4,4-Diphenylmethane diisocyanate, (MDI, 99%), furfury alcohol (FA) and N,N′-(4,4′-diphenylmethane)bismaleimide (BMI) were supplied by Aladdin and used as received. Poly(tetramethylene glycol) (PTMEG, Mn = 2000 g mol^{-1}) was purchased from Aladdin and used after 2 h of drying under vacuum at 110 °C. Dimethyl formamide (DMF) was dried with molecular sieves for more than 24 h and freshly distilled before use.

Preparation of PU–AgNWs–PU Composites. In order to prepare the AgNWs-PU composite, 1.0 g of MDI was dispersed in 10.0 g of anhydrous DMF and stirred at room temperature for 30 min, Then 4.0 g of PTMEG 2000 in 15.0 g of anhydrous DMF was added slowly to the above solution and reacted at 80 °C for 2 h in a N_2 atmosphere. Afterwards, 0.4 g FA and 0.754 g BMI were dispersed in 9.6 g of anhydrous DMF before dropping in the solution and reacted at 80 °C for 3 h in a N_2 atmosphere which resulted in a solution of DAPU. The resultant DAPU solution was pour into a Teflon model (with size of 50 × 15 × 2.5 mm), and then placed into an oven with given temperature of 70 °C for 30 min.

The silver nanowire percolation network was prepared by vacuum filtration method. Specifically, 2 mL of AgNWs suspension (2 mg/mL) was diluted with 500 mL water and dispersed adequately using ultrasonic setup for 3 min, and then the diluted suspension was vacuum filtrated to obtain the silver nanowire network. The film was cut into (with size of 5 × 30 mm) strip. Afterward, the AgNWs strip was stacked on the DAPU film with the AgNWs surface downward against the surface of DAPU, a light force was applied to press the strip for a while and peel the membrane off, thanks to the adhesion of the not-completely-dry DAPU, the AgNWs network can fully transferred from membrane to the DAPU surface. And then, DAPU solution was pour onto the AgNWs surface to make the sandwich structure, after 24 h of drying the PU–AgNWs–PU composite was obtained.

Characterization. Scanning electronic micrographs (SEM) were recorded with a Nova NanoSEM 450. Thermogravimetry analysis (TGA) was made on a TA SDTQ600 thermogravimetric analyzer. Differential scanning calorimetry (DSC) was performed on a Q20-1173 DSC thermal system (TA instruments, New Castle, USA) in the temperature range of 50–200 °C at a heating rate of 10 °C/min. Nitrogen gas was purged at a flow rate of about 50 mL/min. The tensile test was carried out on a stretching machine (AG-X Plus 100N). The electrical properties were recorded by a digital source meter (Keithley 2000).

Results and Discussions

The schematic of fabrication process is shown in Fig. 1 (the fabrication procedures are described in detail in the Experimental Section). Compared with most reported strain sensor fabricated by depositing or embedding the sensing layer on or into the flexible matrix in which the structure of the strain sensor can be easily damaged even after mild touching, the sandwich configuration strain sensor here exhibited higher reliability due to the fully encapsulated in the DAPU substrate.

Due to the adhesion between the DAPU substrate and AgNWs network, the AgNWs could be totally transferred onto the DAPU surface as shown in Fig. 2a, and the AgNWs will not be washed away and maintain its connection when the liquid DAPU solution cast onto the AgNWs film. The liquid PU solution then penetrates into the interconnected pore of the AgNWs network, owing to the low viscosity and low surface energy of the PU solution. After drying, all AgNWs are buried into the DAPU substrate without any voids.

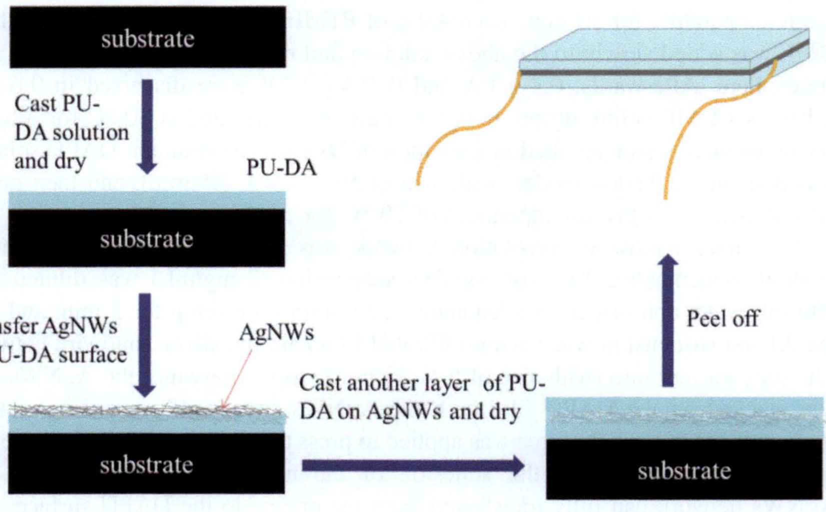

Fig. 1 Schematic diagram of the fabrication procedure of the PU–AgNWs–PU composite

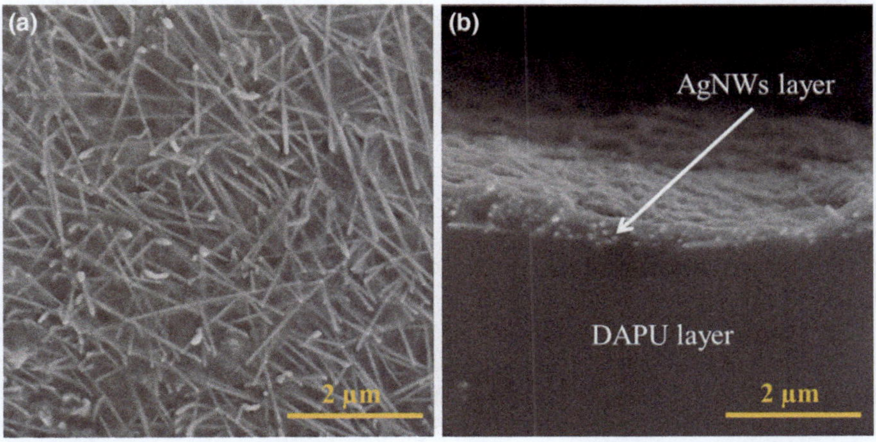

Fig. 2 **a** The SEM image of the surface of the AgNWs when transferred onto DAPU. **b** The cross-sectional SEM image of the sandwich structured sensor

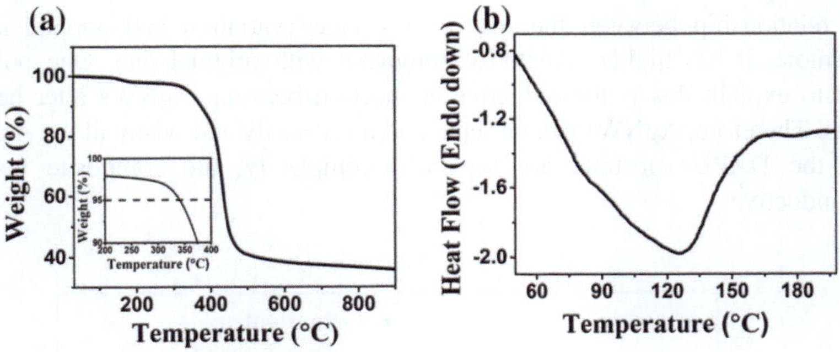

Fig. 3 **a** TGA curves of the DAPU. **b** DSC curve of the DAPU

The cross-sectional optical image of the sample are shown in Fig. 2b, which indicating the silver nanowire was perfect encapsulated between two layers of DAPU films, showing relatively uniform thickness of 400 nm.

In order to endow the strain sensor with self-healing property, we employ the polyurethane cross-linked by using bifunctional maleimide via DA chemistry. The thermal stability of the as-prepared PU was evaluate by TGA under nitrogen atmosphere between 30 and 900 °C as shown in Fig. 3a. As we can see from the insert curve in Fig. 3a, the temperature at 5% weight loss (T_5) is about 340 °C which is considered a high thermal stability. Besides, the DA chemistry was applied to the PU successfully which can be confirmed by the peak of the retro-DA reaction as shown in Fig. 3b. The DSC curve of as-prepared PU shows a significant endothermic peak at 120–130 °C which should be owed to the retro-DA reaction.

The response of the sandwich-structured strain sensor before and after healing is demonstrated under strain test, which is shown in Fig. 4. The relative resistance variations ($\Delta R/R_0 = (R - R_0)/R_0$, where R denote the resistance of the strain sensor with applied strain) increased as function of the increasing strain. Herein, the slope of the curve represents the sensitivity (GF $= \delta(\Delta R/R_0)/\delta\varepsilon$). According to the slope, the relative resistance versus strain curve can be divided into two sections, under lower strain ($\leq 11\%$), the sandwich structured strain sensor has lower sensitivity with gauge factor of 1, and has high sensitivity with gauge factor of 15 (which is much higher than those reported AgNWs based strain sensor [22]) under higher strain ($\geq 11\%$).

To investigate the healing property of the sandwich-structured strain sensor, the strain sensor was cut into two part with a razor blade as shown in Fig. 5a, put them together and applied a light force to ensure the contact between each other (resistance in this moment can be observed in Fig. 5c) and then heated for 2 h at 130 °C, allowed to cool down slowly to 60 °C, and place in 60 °C for 24 h. After the retro-DA reaction, the scratch was almost disappeared as shown in Fig. 5b, and the resistance restored simultaneously as can be seen in Fig. 5d.

The response of healed sandwich-structured strain sensor is shown in the black curve in Fig. 4. Though the healed strain sensor can sustain lower strain, show linear relationship between the relative resistance variation and applied strain, furthermore, it has higher sensitivity compared with original one. One possible reason to explain this is the reductive connection between AgNWs after healing process. Therefore, AgNWs can be separated more easily and when all the AgNWs inside the DAPU substrate are separated completely, the composite become non-conductive.

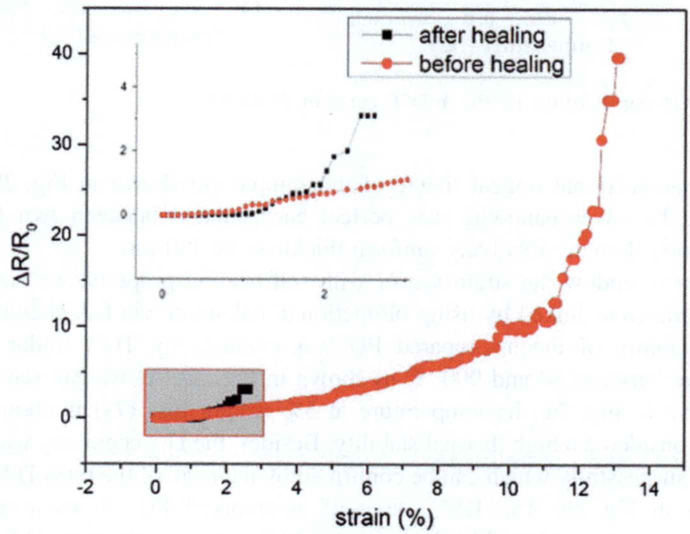

Fig. 4 Response of the sandwich-structured strain sensor to the strain from 0 to 14%

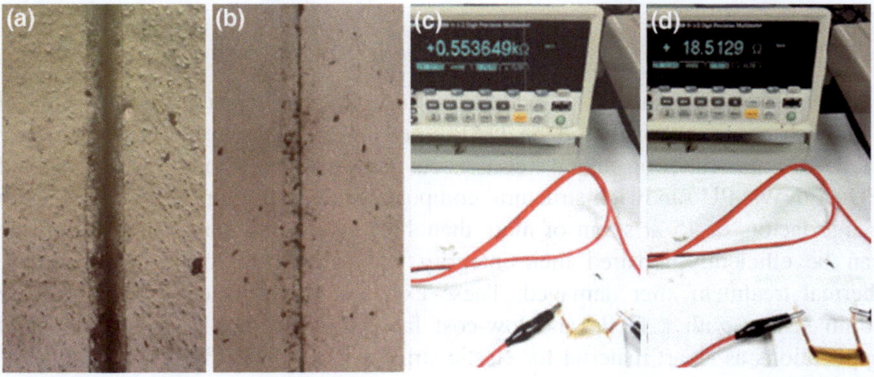

Fig. 5 Optical image of sample **a** before and **b** after healed, and the resistance measurement **c** before and **d** after healed

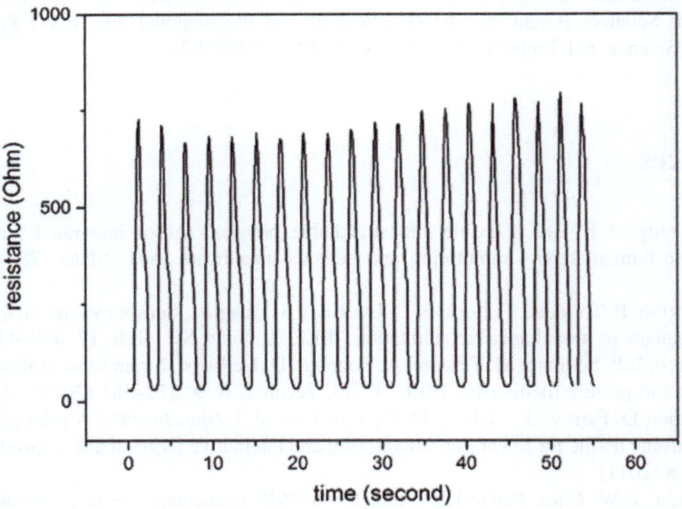

Fig. 6 Electrical resistance change of strain sensor in the first 20 stretch-release cycles under the strain of 10%

We further investigate the stability of the strain sensor. In the cyclic stretch and release test, the samples was stretched to a given strain of 10%, and released to its original state at room temperature. As shown in Fig. 6, the strain sensor exhibits outstanding sensing stability. The resistance of stretched sensor shows very small fluctuations. The change in the resistance is reproducible where the resistance is only dependent on the tensile strain. In other words, the strain sensor has repeatable deformation and stable relationship between deformation and electrical resistance.

Conclusions

In conclusion, we successfully prepared a strain sensor with the structure of sandwich configuration based on DAPU and silver nanowires, by using a very facile vacuum filtration and water-assisted transferring method. The as-prepared PU–AgNWs–PU sandwich structural composite shows strong piezoresistivity with gauge factors of 15 at strain of more than 11%. More importantly, the composite can be efficiently repaired their integrity and sensing performance by simply thermal treatment after damaged. These excellent performances of the healable strain sensor with a facile and low-cost fabrication process show the potential applications as smart material for elastic strain sensor.

Acknowledgements This work was financially supported by Guangdong and Shenzhen Innovative Research Team Program (No.2011D052, KYPT20121228160843692), NSFC-Guangdong Jointed Funding (U1601202), NSFC-Shenzhen Robot Jointed Funding (U1613215), Natural Science Foundation of China (21601065), R&D Funds for basic Research Program of Shenzhen (Grant No. JCYJ20150401145529012), Key Deployment Project of Chinese Academy of Sciences (Grant No. KFZD-SW-202) and the National Key R&D Project from Minister of Science and Technology of China (2016YFA0202702).

References

1. T.Q. Trung, N.E. Lee, Flexible and stretchable physical sensor integrated platforms for wearable human-activity monitoring and personal healthcare. Adv. Mater. **28**, 4338–4372 (2016)
2. T. Giorgino, P. Tormene, F. Lorussi, D. De Rossi, S. Quaglini, Sensor evaluation for wearable strain gauges in neurological rehabilitation. IEEE. T. Neur. Sys. Reh. **17**, 409–415 (2009)
3. F. Lorussi, E.P. Scilingo, M. Tesconi, A. Tognetti, D. De Rossi, Strain sensing fabric for hand posture and gesture monitoring. IEEE. T. Inf. Technol. B. **9**, 372–381 (2005)
4. R. Helmer, D. Farrow, K. Ball, E. Phillips, A. Farouil, I. Blanchonette, A pilot evaluation of an electronic textile for lower limb monitoring and interactive biofeedback. Procedia Eng. **13**, 513–518 (2011)
5. C.-X. Liu, J.-W. Choi, Patterning conductive PDMS nanocomposite in an elastomer using microcontact printing. J. Micromech. Microeng. **19**, 085019 (2009)
6. J. Lee, S. Kim, J. Lee, D. Yang, B.C. Park, S. Ryu, I. Park, A stretchable strain sensor based on a metal nanoparticle thin film for human motion detection. Nanoscale **6**, 11932–11939 (2014)
7. N. Lu, C. Lu, S. Yang, J. Rogers, Highly sensitive skin-mountable strain gauges based entirely on elastomers. Adv. Funct. Mater. **22**, 4044–4050 (2012)
8. X. Xiao, L. Yuan, J. Zhong, T. Ding, Y. Liu, Z. Cai, Y. Rong, H. Han, J. Zhou, Z.L. Wang, High-strain sensors based on ZnO nanowire/Polystyrene hybridized flexible films. Adv. Mater. **23**, 5440–5444 (2011)
9. Y. Zhang, L. Yang, Y. Ge, Transparent strain sensors through randomly self-assembly of single walled carbon nanotubes at the air/water interface. J. Nanosci. Nanotechnol. **17**, 4931–4935 (2017)
10. T. Giffney, E. Bejanin, A.S. Kurian, J. Travas-Sejdic, K. Aw, Highly stretchable printed strain sensors using multi-walled carbon nanotube/silicone rubber composites. Sensor. Actuat. A-Phys. **259**, 44–49 (2017)

11. M. Amjadi, A. Pichitpajongkit, S. Lee, S. Ryu, I. Park, Highly stretchable and sensitive strain sensor based on silver nanowire-elastomer nanocomposite. ACS Nano **8**, 5154–5163 (2014)
12. S. Gong, D.T. Lai, B. Su, K.J. Si, Z. Ma, L.W. Yap, P. Guo, W. Cheng, Highly stretchy black gold E-skin nanopatches as highly sensitive wearable biomedical sensors. Adv. Electron. Mater. **1**, 1400063 (2015)
13. M. Chen, L. Zhang, S. Duan, S. Jing, H. Jiang, C. Li, Highly stretchable conductors integrated with a conductive carbon nanotube/graphene network and 3D porous poly (dimethylsiloxane). Adv. Funct. Mater. **24**, 7548–7556 (2014)
14. Y.R. Jeong, H. Park, S.W. Jin, S.Y. Hong, S.S. Lee, J.S. Ha, Highly stretchable and sensitive strain sensors using fragmentized graphene foam. Adv. Funct. Mater. **25**, 4228–4236 (2015)
15. B. Sun, Y.-Z. Long, S.-L. Liu, Y.-Y. Huang, J. Ma, H.-D. Zhang, G. Shen, S. Xu, Fabrication of curled conducting polymer microfibrous arrays via a novel electrospinning method for stretchable strain sensors. Nanoscale **5**, 7041–7045 (2013)
16. X. Liu, C. Lu, X. Wu, X. Zhang, Self-healing strain sensors based on nanostructured supramolecular conductive elastomers. J. Mater. Chem. A. **5**, 9824–9832 (2017)
17. S. Liu, Y. Lin, Y. Wei, S. Chen, J. Zhu, L. Liu, A high performance self-healing strain sensor with synergetic networks of poly (ε-caprolactone) microspheres, graphene and silver nanowires. Compos. Sci. Technol. **146**, 110–118 (2017)
18. C.-H. Li, C. Wang, C. Keplinger, J.-L. Zuo, L. Jin, Y. Sun, P. Zheng, Y. Cao, F. Lissel, C. Linder, A highly stretchable autonomous self-healing elastomer. Nat. Chem. **8**, 618–624 (2016)
19. G. Cai, J. Wang, K. Qian, J. Chen, S. Li, P.S. Lee, Extremely stretchable strain sensors based on conductive self-healing dynamic cross-links hydrogels for human-motion detection. Adv. Sci. **4**, 1600190 (2017)
20. Y. Yang, B. Zhu, D. Yin, J. Wei, Z. Wang, R. Xiong, J. Shi, Z. Liu, Q. Lei, Flexible self-healing nanocomposites for recoverable motion sensor. Nano. Energy. **17**, 1–9 (2015)
21. J. Li, G. Zhang, L. Deng, S. Zhao, Y. Gao, K. Jiang, R. Sun, C. Wong, In situ polymerization of mechanically reinforced, thermally healable graphene oxide/polyurethane composites based on Diels-Alder chemistry. J. Mater. Chem. A. **2**, 20642–20649 (2014)
22. S. Zhao, J. Li, D. Cao, G. Zhang, J. Li, K. Li, Y. Yang, W. Wang, Y. Jin, R. Sun, C.-P. Wong, Recent advancements in flexible and stretchable electrodes for electromechanical sensors: strategies, materials, and features. Acs Appl. Mater Inter. 12147–12164 (2017)

Morphology Evolution of Monolayer MoS_2 Flakes with Seed Promotor Grown by CVD

Xin Zhang, HongBin Zhao, QingZhu Zhang and Feng Wei

Abstract We report a developed preparation process on the monolayer MoS_2 which was grown on SiO_2/Si substrates with seed promotor by atmospheric-pressure chemical vapor deposition (CVD) method. It is indicated that growth temperature and proportion of precursors play significant roles on the morphology of the monolayer MoS_2 which can change from three-point star to triangle and hexagon. The dimension of the MoS_2 flake is mainly dependent on the growth temperature, while its morphology is mainly influenced by the amount of the loaded MoO_3. Raman spectra and AFM images show that the MoS_2 flakes of the three morphologies are all monolayer.

Keywords Molybdenum disulfide · CVD growth · Seed promotor
Monolayer

X. Zhang (✉) · H. Zhao · Q. Zhang · F. Wei (✉)
State Key Laboratory of Advanced Materials for Smart Sensing,
General Research Institute for Nonferrous Metals, Beijing, China
e-mail: zhangxin_grinm@126.com

F. Wei
e-mail: weifeng@grinm.com

H. Zhao
e-mail: zhaohongbin@grinm.com

Q. Zhang
e-mail: zhangqingzhu@ime.ac.cn

Q. Zhang
Key Laboratory of Microelectronic Devices Integrated Technology,
Institute of Microelectronics of Chinese Academy of Sciences, Beijing, China

Y. Han (ed.), *Advanced Functional Materials*,
https://doi.org/10.1007/978-981-13-0110-0_45

Introduction

In recent years, two-dimensional transition metal dichalcogenides (TMDs) have received extensive attention by their graphene-like structures and remarkable physical and chemical properties [1, 2]. Among all the TMDs, molybdenum disulfide (MoS_2) receives much concerns due to the direct band gap and strong photoluminescent effects, which enable its application for logical and optoelectronic devices [3, 4]. Several methods have been developed in the last few years to obtain atomically thin MoS_2, including exfoliation, hydrothermal synthesis, and chemical vapor deposition (CVD) [5, 6]. Among these methods, CVD is an effective growth method to obtain large-area and atomically thin MoS_2 flakes [7–9]. In this method, various domain shapes have been obtained, such as triangles, hexagons, three-point stars and so on [8, 9]. It has also been reported that different shape of MoS_2 flakes can be obtained by using different amount of loaded S powder in CVD [9, 10]. Besides, the shape of MoS_2 flakes are usually influenced by several parameters, such as the growth temperature, the growth pressure, and the flow rate of the carrier gas. However, only few research studied the effects of those parameters in detail.

In this work, we studied the morphology evolution of monolayer MoS_2 flakes grown by CVD under atmospheric pressure using perylene-3,4,9,10-tetracarboxylic acid tetrapotassium salt (PTAS) as seed promotor. The growth temperature and the Mo: S ratio which significantly affect the morphology of monolayer MoS_2 flakes were investigated. Raman spectroscopy and atomic force microscopy (AFM) techniques are employed to study the crystal structure of MoS_2. Finally, we propose a controllable process to achieve the monolayer MoS_2 with large area.

Experimental

CVD was used to grow MoS_2 under atmospheric pressure by a double-zone furnace. Si with a 300 nm layer of SiO_2 used as the growth substrates. After cleaned by piranha solution to achieve a hydrophilic surface, a small droplet (~ 2 μL) of PTAS solution (10^{-4} M) was dropped and spread out on the SiO_2 substrate. Then the substrates were placed face down above a quartz boat containing x mg (x = 1, 5 or 10) of MoO_3 powder. The quartz boat and the other one containing 100 mg of S powder were loaded into 2-in. quartz tube at a low-temperature and high-temperature zone respectively, where the boat of S was placed 18 cm from the boat of MoO_3 at the upstream of the tube. The flowrate was 10 sccm with Ar carrier gas. Both of the two zones were heated to their target temperature at the rate of 15 °C/min and the growth time was 10 min. Figure 1 illustrates the sketch of MoS_2 growth system.

The morphology of the MoS_2 flakes were characterized by AFM (Bruker Dimension Icon) and optical microscopy (Nikon Eclipse LV-N). The structure of

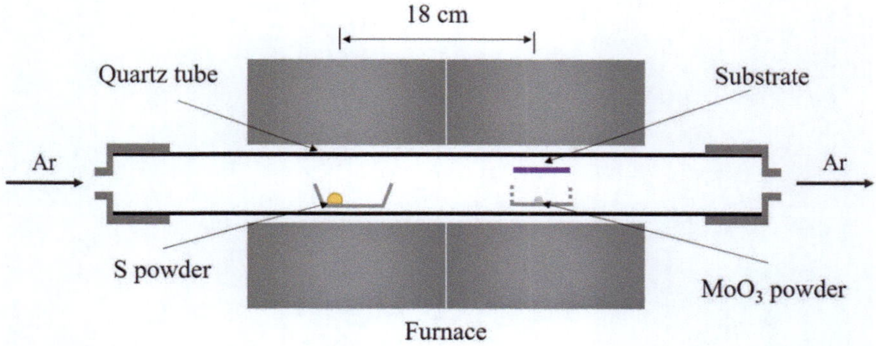

Fig. 1 Schematic illustration of the MoS_2 CVD system

the MoS_2 flakes was determined by Raman (Horiba LabRAM ARAMIS) analyzer. The excitation wavelength for Raman measurement was 532 nm.

Results and Discussion

To make sure the influence of growth temperature, we kept the loaded MoO_3 powder (5 mg) as constant and changed the temperature of MoO_3 in the range of 620–750 °C. The optical images of MoS_2 flakes grown at different temperatures were shown in Fig. 2. Figure 2a–f shows MoS_2 flakes grown at 620, 650, 680, 700, 720 and 750 °C, respectively. Triangular flakes were formed under the temperature of 650–720 °C (Fig. 2b–e). With the increase of temperature, lateral dimension of MoS_2 flakes became larger, even up to 190 μm. However, when it came to 720 °C (Fig. 2e), tetragonal flakes were observed as well and we could not even find triangular flakes at 750 °C (Fig. 2f). MoO_3 could be reduced to tetragonal MoO_2 at a high temperature and Mo-rich condition, while high temperature largely increased the evaporation amount of MoO_3 powder [11–13].

Since Mo: S ratio is crucial to the quality of the 2D MoS_2 flakes, it is necessary to make explicit in the growth process. To obtain different Mo: S ratio in the growth region, we changed the amount of MoO_3 powder (1, 5, 25 mg), while kept the growth temperature (720 °C) as constant. Figure 3a–c shows the optical images of MoS_2 flakes grown with 1, 5 and 25 mg of MoO_3 powder at 720 °C respectively. Because that a small amount of MoO_3 (1 mg) can lead to a S-rich situation and higher growth temperature can contributed to larger side length, flakes with the shape of three-point star appears with a lateral dimension of 159 μm (shown in Fig. 3a). When the amount of MoO_3 powder was increased to 25 mg, the shape of the film became hexagonal. In this situation, the growth parameter might change from S-rich condition to a balance condition (Mo: S $\approx$ 1:2) where the Mo: S ratio corresponded to the stoichiometric ratio of MoS_2, and the growth of S edges was almost the same as Mo edges, resulting in the formation of hexagonal MoS_2 flakes [14].

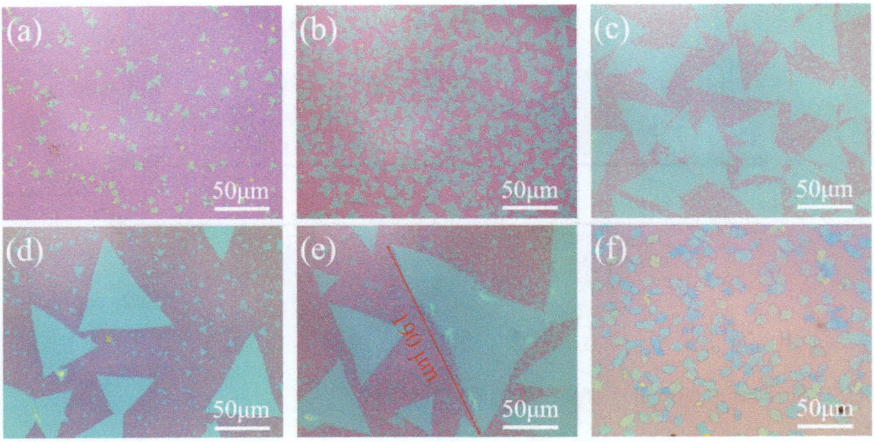

Fig. 2 Optical microscope images of MoS_2 flakes grown with 5 mg of MoO_3 powder at different temperatures of **a** 620, **b** 650, **c** 680, **d** 700, **e** 720 and **f** 750 °C

Fig. 3 Optical microscope images of MoS_2 flakes grown at 720 °C with **a** 1 mg, **b** 5 mg and **c** 25 mg of MoO_3 powder

To confirm the number of layers of MoS_2 flakes, AFM and Raman spectroscopy were used to characterize the structure of the flakes. Figure 4a–c show AFM results of a typical MoS_2 flake. Cross-sectional height was 0.8 nm (Fig. 4c), consistent with the thickness of monolayer MoS_2 (0.6–0.9 nm) [15]. The Raman spectra of the three-point star, hexagonal and triangular flakes were similar, showing two characteristic Raman peak, corresponding to in-plane E^1_{2g} and out-of-plane Raman modes (A_{1g}) [15]. The difference between the two modes was 20.5, 20.7, 20.2 cm^{-1} respectively, exactly coincides with the monolayer MoS_2 flakes reported before [8, 16–18]. The results of AFM and Raman spectroscopy fully illustrate that the MoS_2 flakes of the three morphologies were all monolayer. Figure 4e shows the Raman spectroscopy of tetragonal flakes in Fig. 2f. Six main peaks were observed in the range of 300–800 cm^{-1}, which corresponded to Si ($\sim$520 cm^{-1}) and MoO_2 ($\sim$349, 457, 494, 569, 736 cm^{-1}), respectively. As a result, the tetragonal flake is proved as MoO_2.

We fabricated a backgate device and simply recorded the dependence of photocurrent generation on gate voltage at V_{ds} = 2 V, showed in Fig. 5a, b. Under both

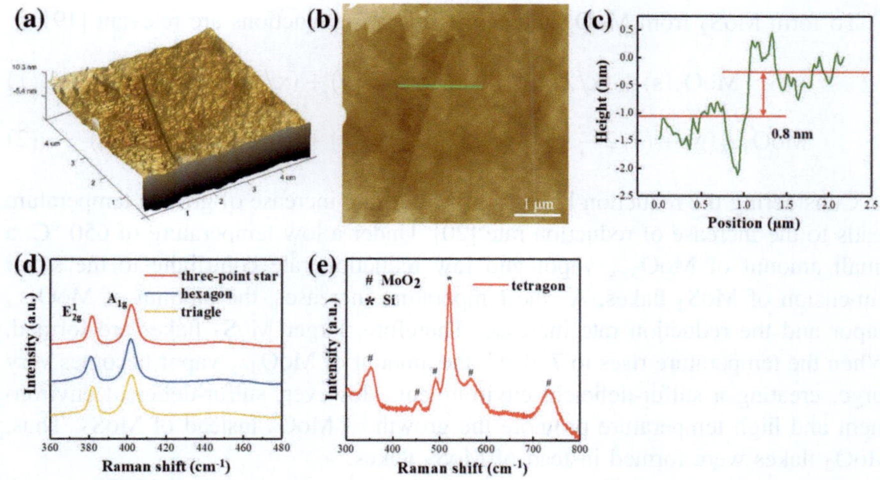

Fig. 4 **a–c** AFM images of a MoS_2 flake and height profile taken across the green line in panel b. Raman spectra: **d** for three morphologies of MoS_2 and **e** for tetragonal MoO_2

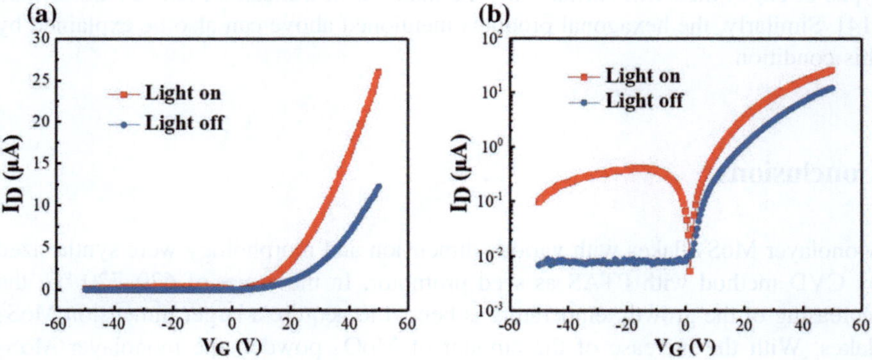

Fig. 5 Light response (I_D–V_G) of the device in bright (visible light) and dark states at V_{ds} = 2 V

visible light and dark conditions, the device showed the characteristics of field-effect transistors with n-channel. In the case of light-on, both the ON and OFF current were improved and the OFF current rose by about an order of magnitude over the dark condition. These indicated that our device had an obvious light response characteristic.

Under a relatively low temperature (620 °C), MoS_2 flakes with the shape of three-point star were obtained. Lower temperature in growth region means a smaller amount of MoO_3 vapor and S-rich situation in the growth region. The growth of S edges cannot catch up with that of Mo edges in the initial hexagonal nuclei, so that S edge is able to form only a curved side instead of a straight one, which looks like a three-point star [14].

To form MoS_2 from MoO_3 and S, the following reactions are relevant [19]:

$$MoO_3(s) + (x/2)\,S\,(g) \rightarrow MoO_{3-x}(g) + (x/2)\,SO_2(g) \tag{1}$$

$$MoO_{3-x}(g) + (7/2 - x/2)\,S\,(g) \rightarrow MoS_2(s) + (3/2 - x/2)SO_2(g) \tag{2}$$

Considering the reduction kinetics of MoO_3, the increase of growth temperature leads to the increase of reduction rate [20]. Under a low temperature of 650 °C, a small amount of MoO_{3-x} vapor and low reduction rate contribute to the small dimension of MoS_2 flakes. As the temperature increases, the amount of MoO_{3-x} vapor and the reduction rate increase. Therefore, larger MoS_2 flakes are formed. When the temperature rises to 750 °C, the amount of MoO_{3-x} vapor becomes very large, creating a sulfur-deficient environment. However, sulfur-deficient environment and high temperature promote the growth of MoO_2 instead of MoS_2. Thus, MoO_2 flakes were formed instead of MoS_2 flakes.

Under the fixed growth temperature (720 °C), small amount of MoO_3 powder create a S-rich situation. Consequently, magnified triangular flakes are formed because of the high reduction rate at a high temperature. When the Mo: S ratio in the growth region is close to 2:1, the difference in the rate of growth between two types of edges then will shrink, thus the flakes with truncated triangles are formed [14]. Similarly, the hexagonal products mentioned above can also be explained by this condition.

Conclusions

Monolayer MoS_2 flakes with various dimension and morphology were synthesized by CVD method with PTAS as seed promotor. In the range of 620–720 °C, the increasing of the growth temperature is benefit to acquire a larger dimension MoS_2 flakes. With the increase of the amount of MoO_3 powder, the monolayer MoS_2 flakes evolve from three-point star to triangle and hexagon. Besides, the as-prepared monolayer MoS_2 revealed significant photoresponse property. It is a promising way to obtain homogenous and high-quality monolayer MoS_2 flakes by optimizing the growth temperature and the proportion of the precursors.

Acknowledgements This work was supported by GRINM Innovation Fund.

References

1. Q.H. Wang, K. Kalantar-Zadeh, A. Kis, J.N. Coleman, M.S. Strano, Electronics and optoelectronics of two-dimensional transition metal dichalcogenides. J. Nat. Nanotechnol. **7**, 699–712 (2012)

2. K. Kalantar-Zadeh, J.Z. Ou, T. Daeneke, M.S. Strano, M. Pumera, S.L. Gras, Two-dimensional transition metal dichalcogenides in biosystems. J. Adv. Funct. Mater. **25**, 5086–5099 (2015)
3. A. Kuc, N. Zibouche, T. Heine, Influence of quantum confinement on the electronic structure of the transition metal sulfide TS_2. J. Phys. Rev. B **83**, 245213 (2011)
4. B. Radisavljevic, A. Radenovic, J. Brivio, V. Giacometti, A. Kis, Single-layer MoS_2 transistors. J. Nat. Nanotechnol. **6**, 147–150 (2011)
5. J.N. Coleman, M. Lotya, A. O'neill, S.D. Bergin, P.J. King, U. Khan, K. Young, A. Gaucher, S. De, R.J. Smith, I.V. Shvets, S.K. Arora, G. Stanton, H.Y. Kim, K. Lee, G.T. Kim, G.S. Duesberg, T. Hallam, J.J. Boland, J.J. Wang, J.F. Donegan, J.C. Grunlan, G. Moriarty, A. Shmeliov, R.J. Nicholls, J.M. Perkins, E.M. Grieveson, K. Theuwissen, D.W. Mccomb, P.D. Nellist, V. Nicolosi, Two-dimensional nanosheets produced by liquid exfoliation of layered materials. J. Sci. **331**, 568–571 (2011)
6. M.M. Benameur, B. Radisavljevic, J.S. Heron, S. Sahoo, H. Berger, A. Kis, Visibility of dichalcogenide nanolayers. J. Nat. Nanotechnol. **22**, 125706 (2011)
7. Y.H. Lee, X.Q. Zhang, W. Zhang, M.T. Chang, C.T. Lin, K.D. Chang, Y.C. Yu, J.T. Wang, C.S. Chang, L.J. Li, T.W. Lin, Synthesis of large-area MoS_2 atomic layers with chemical vapor deposition. J. Adv. Mater. **24**, 2320–2325 (2012)
8. A.M. Van Der Zande, P.Y. Huang, D.A. Chenet, T.C. Berkelbach, Y. You, G.H. Lee, T.F. Heinz, D.R. Reichman, D.A. Muller, J.C. Hone, Grains and grain boundaries in highly crystalline monolayer molybdenum disulphide. J. Nat. Mater. **12**, 554–561 (2013)
9. S. Najmaei, Z. Liu, W. Zhou, X. Zou, G. Shi, S. Lei, B.I. Yakobson, J.C. Idrobo, P.M. Ajayan, J. Lou, Vapour phase growth and grain boundary structure of molybdenum disulphide atomic layers. J. Nat. Mater. **12**, 754–759 (2013)
10. S.Y. Yang, G.W. Shim, S.-B. Seo, S.-Y. Choi, Effective shape-controlled growth of monolayer MoS_2 flakes by powder-based chemical vapor deposition. J. Nano Res. **10**, 255–262 (2016)
11. X.S. Wang, H.B. Feng, Y.M. Wu, L.Y. Jiao, Controlled synthesis of highly crystalline MoS_2 flakes by chemical vapor deposition. J. J. Am. Chem. Soc. **135**, 5304–5307 (2013)
12. B. Chen, Q. Yu, Q. Yang, P. Bao, W. Zhang, L. Lou, W. Zhu, G. Wang, Large-area high quality MoS_2 monolayers grown by sulfur vapor counter flow diffusion. J. RSC Adv. **6**, 50306–50314 (2016)
13. J. Li, C. Hu, H. Wu, Z. Liu, S. Cheng, W. Zhang, H. Shu, H. Chang, Facile preparation of single MoS_2 atomic crystals with highly tunable photoluminescence by morphology and atomic structure. J. Cryst. Growth Des. **16**, 7094–7101 (2016)
14. S. Wang, Y. Rong, Y. Fan, M. Pacios, H. Bhaskaran, K. He, J.H. Warner, Shape evolution of monolayer MoS_2 crystals grown by chemical vapor deposition. J. Chem. Mater. **26**, 6371–6379 (2014)
15. C. Lee, H. Yan, L.E. Brus, T.F. Heinz, J. Hone, S. Ryu, Anomalous lattice vibrations of single- and few-layer MoS_2. J. ACS Nano **4**, 2695–2700 (2010)
16. H. Schmidt, S. Wang, L. Chu, M. Toh, R. Kumar, W. Zhao, A.H. Neto, J. Martin, S. Adam, B. Ozyilmaz, G. Eda, Transport properties of monolayer MoS_2 grown by chemical vapor deposition. J. Nano Lett. **14**, 1909–1913 (2014)
17. Y. Gong, Z. Lin, G. Ye, G. Shi, S. Feng, Y. Lei, A.L. Elias, N. Perea-Lopez, R. Vajtai, H. Terrones, Z. Liu, M. Terrones, P.M. Ajayan, Tellurium-assisted low-temperature synthesis of MoS_2 and WS_2 monolayers. J. ACS Nano **9**, 11658–11666 (2015)
18. X. Ling, Y. Lin, Q. Ma, Z. Wang, Y. Song, L. Yu, S. Huang, W. Fang, X. Zhang, A.L. Hsu, Y. Bie, Y.H. Lee, Y. Zhu, L. Wu, J. Li, P. Jarillo-Herrero, M. Dresselhaus, T. Palacios, J. Kong, Parallel stitching of 2D materials. J. Adv. Mater. **28**, 2322–2329 (2016)
19. Y. Feldman, E. Wasserman, D.J. Srolovitz, R. Tenne, High-rate, gas-phase growth of MoS_2 nested inorganic fullerenes and nanotubes. J. Sci. **267**, 222–225 (1995)
20. J. Dang, G.H. Zhang, K.-C. Chou, R.G. Reddy, Y. He, Y. Sun, Kinetics and mechanism of hydrogen reduction of MoO_3 to MoO_2. J. Int. J. Refract. Met. Hard Mater. **41**, 216–223 (2013)

Automatic Switching System of Voltage Pulse and Current Pulse Test in Phase Change Memory Devices

Shuai Yan, Daolin Cai, Yuan Xue, Zhitang Song, Yifeng Chen and Yaoyao Lu

Abstract Voltage pulse and current pulse test of phase change memory (PCM) device is realized by computer controlling pulse generator, digital source meter, current pulse generating circuit, switching circuit and other hardware devices. The PCM device test includes several modules of the relationship between current and voltage, voltage and current, resistance and voltage pulse width or height, resistance and current pulse width or height, and fatigue characteristics. In these test modules, the automatic switching system achieves the automatic switching between voltage pulse and current pulse test. The automatic switching system of voltage pulse and current pulse test can not only achieve the automation of engineering testing, but also can reduce the noise interference.

Keywords Phase change memory · Automatic switching system
Remote controlling · Voltage test · Current test

Introduction

Phase change memory (PCM) based on the chalcogenide materials holds promise as the next generation of nonvolatile memory, due to its low power consumption, high operating speed and high storage density [1]. The data storage capability of PCM is based on the property of phase change material which is capable of fast and reversible switch between the amorphous phase (high-resistance or RESET state)

S. Yan (✉) · D. Cai (✉) · Y. Xue · Z. Song · Y. Chen · Y. Lu
State Key Laboratory of Functional Materials for Informatics,
Shanghai Institute of Microsystem and Information Technology,
Chinese Academy of Sciences, Shanghai 200050, China
e-mail: yanshuai@mail.sim.ac.cn

D. Cai
e-mail: caidl@mail.sim.ac.cn

S. Yan · Y. Xue · Y. Lu
University of the Chinese Academy of Science, Beijing 100049, China

Y. Han (ed.), *Advanced Functional Materials*,
https://doi.org/10.1007/978-981-13-0110-0_46

and the crystalline phase (low-resistance or SET state) [2, 3]. In order to convert the PCM cell from crystal to amorphous, the RESET pulse should be a high peak electric pulse with a fast falling edge which ensures that the phase change material is melted and rapidly quenched. Adding a low peak electric pulse with a longer width to the PCM device converts the phase change material from crystal to amorphous [4].

At present, the PCM device test is divided into two kinds of method, which include voltage pulse test and current pulse test. The voltage pulse test is simple and easy to implement, but it has some flaws. The voltage pulse test is very damaging to the device for the reset operation, is difficult to achieve the set operation. Besides, the voltage pulse test can't accurately measure the transient current across the device and accurately calculate the power consumption [5]. The current pulse test can accurately measure the transient current across the device and accurately calculate the device power consumption [6]. So a good test system requires both a voltage pulse test system and a current pulse test system. In the traditional test scheme, the two test systems are independent, but interrelated, not only increase the complexity of the operation, but also introduce noise interference [7]. The introduction of automatic switching system for voltage and current testing can solve these problems.

Experimental Design and Setup

Automatic switching system of voltage pulse and current pulse test consists of the remote computer controlling circuit, solid state optronics relays drive circuit, solid state optronics relays switching circuit and the current source circuit [8]. The hardware schematic of the system is shown in Fig. 1.

(1) The current source circuit uses a self-developed by the group of a high-speed programmable constant current driver chip, which has two identical constant current output channels I_{data1} and I_{data2} and two constant current input channels I_{in1} and I_{in2}. The constant current output I_{data1} and I_{data2} are controlled by the V_{data} input for turning on or turning off, and the output frequency can be as high as 10 MHz. The ratio of the input negative DC currents I_{in1} and I_{in2} to the output currents I_{data1} and I_{data2} is 1:4. The current pulse width of the chip is 100–2000 ns, the output current amplitude can be up to 20 mA and the output current change is less than 5%.
(2) Voltage pulse, current pulse and resistance measurement switching circuit is the core of the entire switching system. The solid state optronics relays switching circuit consists of six solid state optronics relays R0, R1, R2, R3, R4 and R5. The solid state optronics relays switching circuit is shown in Fig. 2. The 1 port of the solid state optronics relays is connected to a pull-up resistor and an external solid state optronics relays drive circuit output, and The 2 and 3 ports provide a path for signal switching. The opening and closing of the 2 and 3

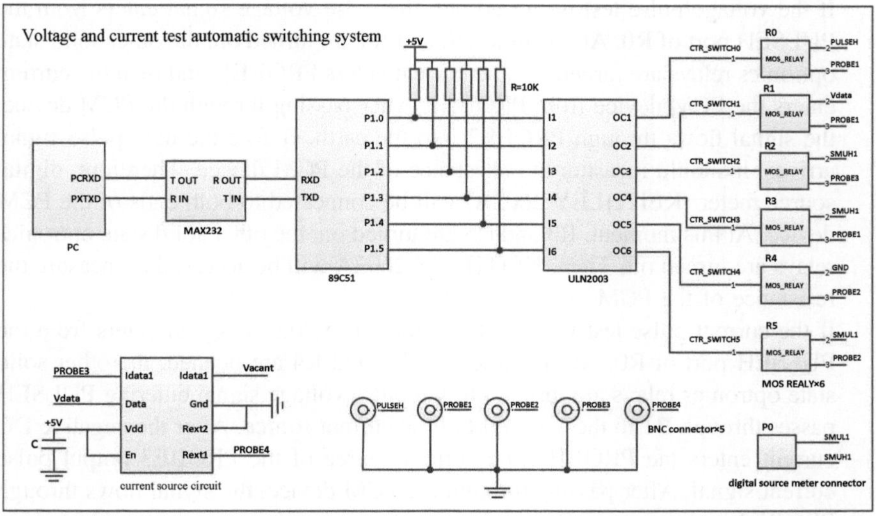

Fig. 1 Hardware schematic of the system

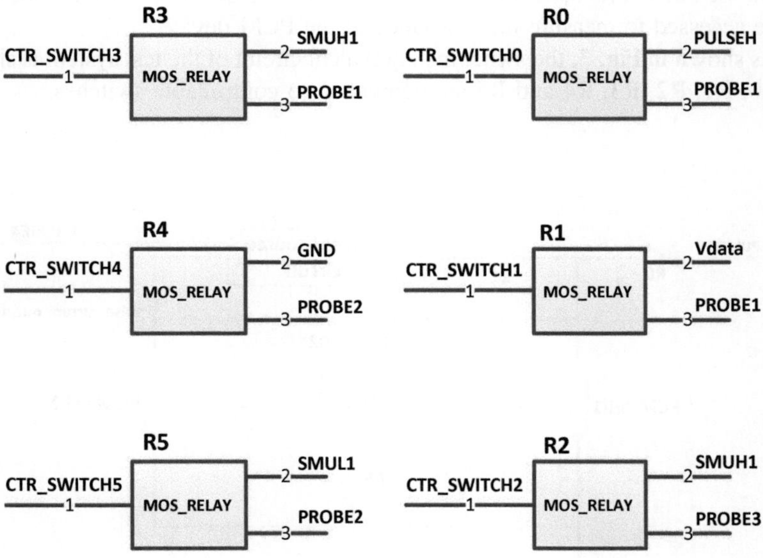

Fig. 2 Solid state optronics relay switching circuit

ports are determined by the output of the solid state optronics relays circuit. If the 1-port input (the output of the solid state optronics relays driver circuit) is low voltage and the 2 and 3 ports are turned off, The port input (the output of the solid state optronics relays drive circuit) is high and the 2 and 3 ports are closed to turn on.

If the voltage pulse test is carried out, the pulse voltage signal enters from the PULSEH port of R0. At this time, R0 and R4 are turned on, the other solid state optronics relays are turned off, the current enters PROBE1, and then the current enters the PCM device from PROBE1. After passing through the PCM device, the signal flows through PROBE2 into the earth. Before the next pulse signal arrives, it should measure the resistance of the PCM device. Therefore, digital source meter (KEITHLEY 2602A) will be connected to both ends of the PCM device. At this moment, R3 and R5 are turned on, the other solid state optronics relays are turned off. Then, KEITHLEY 2602A will be accessed to measure the resistance of the PCM device.
If the current pulse test is carried out, the pulse voltage signal enters from the PULSEH port of R0. At this time, R0, R1 and R4 are opened, the other solid state optronics relays are turned off, the pulse voltage signal entering PULSEH passes through R1 to the V_{data} port of the current source. After the negative DC current enters the PROBE4, the current source of the PROBE3 output pulse current signal. After passing through the PCM device, the signal flows through PROBE2 into the earth. Before the next pulse signal arrives, it should measure the resistance of the PCM device. Therefore, KEITHLEY 2602A will be connected to both ends of the PCM device. At this time, R2 and R5 are turned on, the other solid state optronics relays are turned off, KEITHLEY 2602A will be accessed to measure the resistance of the PCM device.
As shown in Fig. 3, the switching equivalent circuit of the test system is shown. R0, R1, R2, R3, R4 and R5 are equivalent to controllable switches.

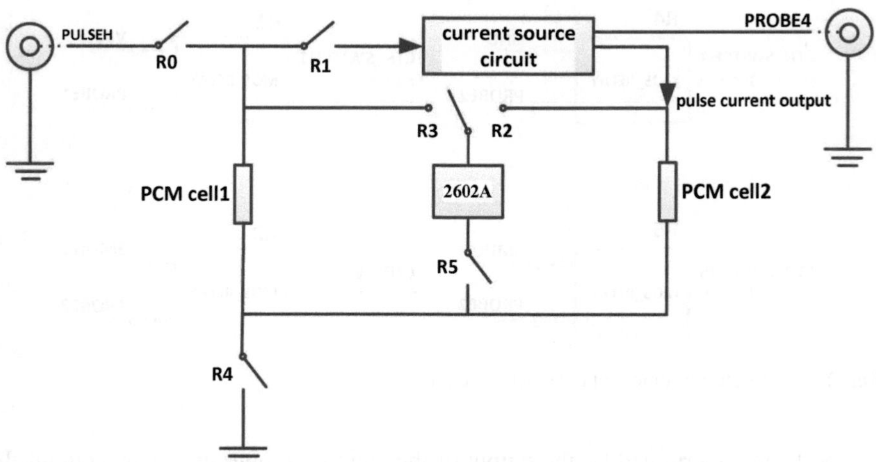

Fig. 3 Solid state optronics relay switching circuit

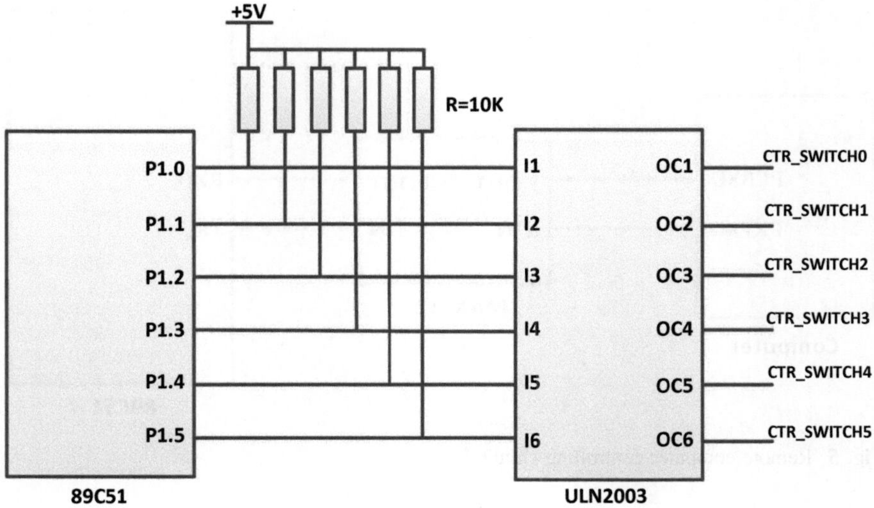

Fig. 4 Solid state optronics relay drive circuit of the system

(3) The solid state optronics relays drive circuit consists of ULN2003 and 89C51 microcontroller. The solid state optronics relays drive circuit of the system is shown in Fig. 4. Because of 89C51 microcontroller with a low load capacity, the switching of solid state optronics relays can be controlled by ULN2003. In addition, UL2003 voltage input port need an external 10 KΩ pull-up resistor.
(4) The remote computer controlling circuit is composed of 89C51 microcontroller and MAX232 serial chip. The main role of the circuit is 89C51 microcontroller to accept the computer's instructions to control the driver chip UL2003 output high and low voltage signals, and ultimately control the opening and closing of the solid state optronics relays. Figure 5 shows the remote computer controlling circuit.

The core of the software design is AT89C51 microcontroller program preparation. The AT89C51 is a low-power, high-performance CMOS 8-bit microcomputer with 4 K bytes of Flash Programmable and Erasable Read Only Memory (PEROM). The device is manufactured using Atmel's high density nonvolatile memory technology and is compatible with the industry standard MCS-51 instruction set and pinout. The chip combines a versatile 8-bit CPU with Flash on a monolithic chip, the Atmel AT89C51 is a powerful microcomputer which provides a highly flexible and cost effective solution to many embedded control applications.

The AT89C51 provides the following standard features: 4 K bytes of Flash, 128 bytes of RAM, 32 I/O lines, two 16-bit timer/counters, a five vector two-level interrupt architecture, a full duplex serial port, on-chip oscillator and clock circuitry. In addition, the AT89C51 is designed with static logic for operation down to zero

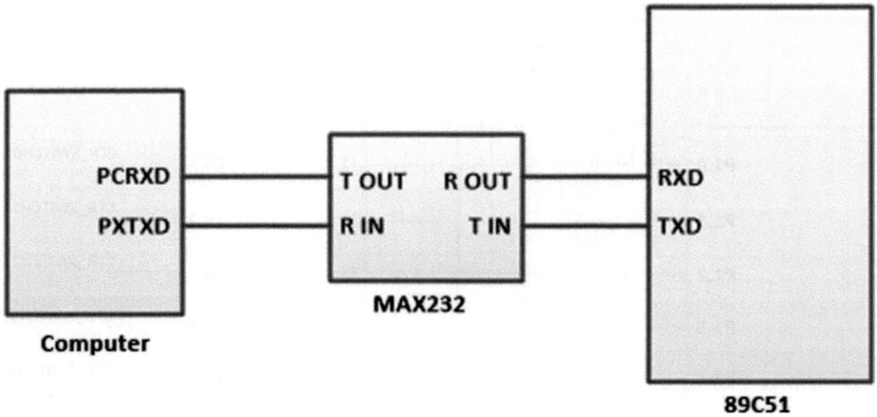

Fig. 5 Remote computer controlling circuit

frequency and supports two software selectable power saving modes. The Idle Mode stops the CPU while allowing the RAM, timer/counters, serial port and interrupt system to continue functioning. The Power Down Mode saves the RAM contents but freezes the oscillator disabling all other chip functions until the next hardware reset.

According to the hardware software design, software structure is shown in Fig. 6. Firstly, AT89C51 receive the data from the PC. Then according to the instructions of the PC, the port P1.0–P1.5 output the high and low levels. As the port P1.0–P1.5 drive capability is too weak, the microcontroller uses ULN2003 to indirectly control the opening and closing of the solid state optronics relays.

Results and Discussion

In the PCM device test [9], the measurement of the resistance occupies an important position. High-precision resistance measurements are more accurate to reflect the nature of the device after the effect of electrical signal. The resistance values of the experiment were 10 and 20 MΩ, and each of the resistors was tested for fifty times in succession. As a contrast, the experiment tested the two different resistors before and after the automatic switching system was used.

As shown in Fig. 7a, the measured resistance of the 10 MΩ resistor fluctuates between 7 and 24 MΩ. But After using the automatic switching system, the measured resistance of the 10 MΩ resistor fluctuates between 10.5 and 13.5 MΩ from the Fig. 7b.

The measured resistance of the 20 MΩ resistor before and after the automatic switching system was used is shown in Fig. 8. As shown in Fig. 8a, the experiment

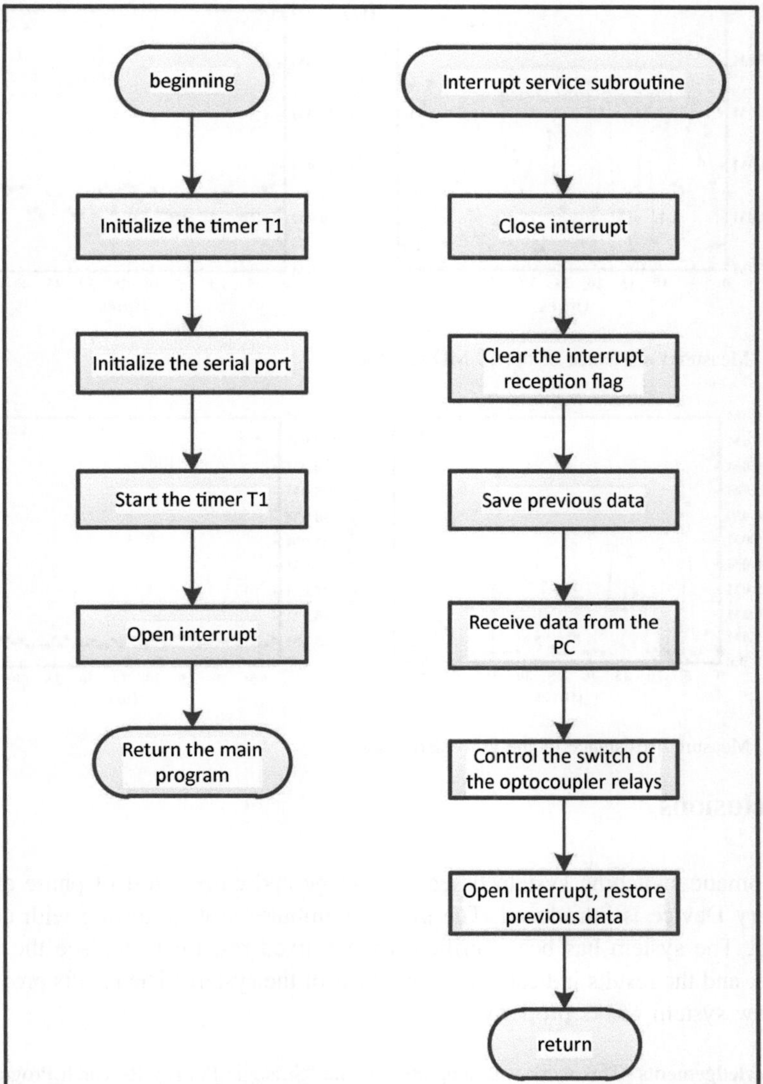

Fig. 6 Software structure according to the hardware software design

can't normally measure the resistance of 20 MΩ resistor. But from the Fig. 8b, the measured resistance of the 10 MΩ resistor fluctuates between 20 and 30 MΩ. Therefore, after using automatic switching system, the resistance measurement accuracy has been significantly improved.

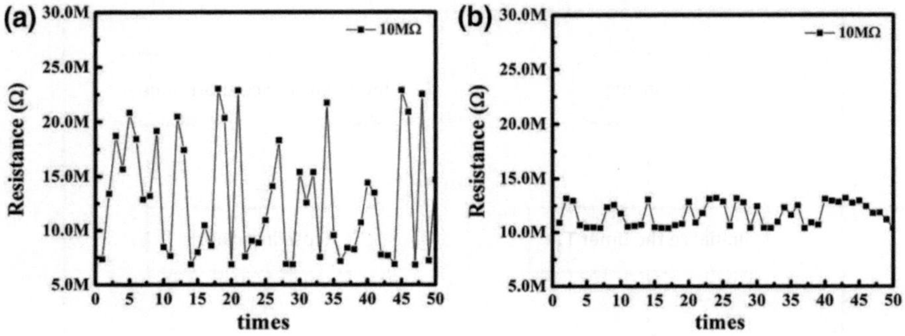

Fig. 7 Measured resistance of the 10 MΩ resistor

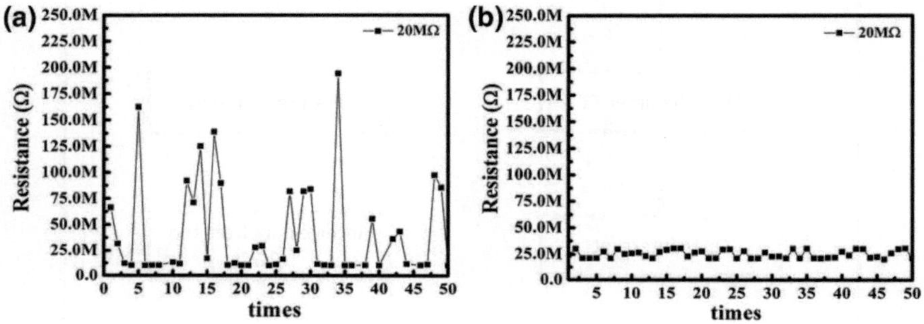

Fig. 8 Measured resistance of the 20 MΩ resistor

Conclusions

A automatic switching system based on voltage and current test of phase change memory Device is introduced. The system combines voltage testing with current testing. The system has been verified using a fixed resistor to replace the actual device, and the results indicate the correctness of the system. The results prove that our new system works properly.

Acknowledgements This work was supported by the "Strategic Priority Research Program" of the Chinese Academy of Sciences (XDA09020402), National Integrate Circuit Research Program of China (2009ZX02023-003), National Natural Science Foundation of China (61376006, 61401444, 61504157,61622408), Science and Technology Council of Shanghai (14ZR1447500, 15DZ2270900).

References

1. S. Lai, Current status of the phase change memory and its future. IEEE, 255–258 (2003)
2. X. Chen, Y. Wang, Y. Cheng, X. Zhou, S. Lv, Y. Chen, Y. Wang, M. Zhou, H. Chen, Y. Zhang, Z. Song, G. Feng, RESET distribution improvement of phase change memory: the impact of pre-programming. IEEE Electron Dev. Lett. **35**, 536–538 (2014)

3. Y. Wang, D. Cai, Y. Chen, Y. Wang, H. Wei, R. Huo et al., Optimizing set performance for phase change memory with dual pulses set method. ECS Solid State Lett. **4**, Q32–Q35 (2015)
4. Y. Wang, D. Cai, Y. Chen, Y. Wang, H. Wei, R. Huo et al., Reduction of reset current in phase change memory by pre-programming. ECS J. Solid State Sci. Technol. **5**, Q13–Q16 (2016)
5. X. Chen, W. Yuchan, S. Zhitang, C. Yifeng, W. Yueqing, C. Hongpeng, R. Feng, Test system of current pluses based on phase change mmeory device, Semicond. Optoelectron. **35**, 136–143 (2014)
6. X. Chen, Y. Wang, S. Li, Y. Chen, L. Xu, Y. Wang, M. Zhou, G. Li, Z. Song, Test system of current pulse in phase change memory devices, in *Proceeding of 2012 International Workshop on Information Storage and Ninth International Symposium on Optical Storage*, vol. 8782, p. 878214 (2013)
7. L.P.S.E.G. Yeo, R. Zhao, K.G. Lim, T.C. Chong, I. Adesida, Parasitic capacitance effect on programming performance of phase change random access memory devices. Appl. Phys. Lett. **96**, 043506 (2010)
8. S.Z.-T. Liang Shuang, L.I.U. Bo, C.H.E.N. Xiao-gang, F.E.N.G. Song-lin, Device unit test system on phase change memory. Semicond. Technol. **31**, 614–617 (2006)
9. G.-S. Park, J.-B. Park, H.-S. Baik, J.-H. Lee, H. Jeong, K. Kim, Phase-change behavior of stoichiometric $Ge_2Sb_2Te_5$ in phase-change random access memory. J. Electrochem. Soc. **154**, H139 (2007)

Synthesis and Characterization of New Block Copolymers for Direct Self-assembly

Haibo Li, Bing Li, Dejun Liu, Mark Neisser, Caleb L. Breaux and Clifford L. Henderson

Abstract Due to its high resolution, easy process, and independence to complex optical exposure systems, Directed Self Assembly (DSA) is a promising alternative patterning technology in microlithography area. In this paper, a new block copolymer was designed and synthesized through anionic polymerization. It has large molecular weight and low PDI. The formation of lamellae was confirmed and the L_0 of the polymer was analysed by using Small Angle X-ray Scattering (SAXS) and 2D-Fourier transform of SEM images. ICP dry etch data showed that the two blocks had a good etch rate ratio, which was good for pattern transfer. A neutral underlayer was crafted for the BCP to phase separate on and form perpendicular features. Annealing temperature and time were investigated for the phase separation of this BCP/Neutral layer system. It could be found this system shows phase separation after spin coating, without any anneal step, by analysis of finger print patterns. 50 nm lines could be found after spin coating. The dependence of L_0 on MW is described along with approaches and results of directed self-assembly using graphoepitaxy and this polymer and neutral layer.

Keywords DSA · Anionic polymerization

H. Li · B. Li (✉) · D. Liu · M. Neisser
Kempur Microelectronics, Inc., Beijing 101312, China
e-mail: libing@kempur.com

C. L. Breaux
Department of Chemical and Biomolecular Engineering,
Georgia Institute of Technology, Atlanta, GA 30332, USA

C. L. Henderson
Department of Chemistry and Biomolecular Engineering,
University of South Florida, Tampa, FL 33620, USA

Y. Han (ed.), *Advanced Functional Materials*,
https://doi.org/10.1007/978-981-13-0110-0_47

Introduction

Due to its high resolution, easy process, and independence to complex optical exposure systems, Direct self-assembly (DSA) has become a promising alternative patterning technology in microlithography [1]. The main component in a DSA process is the block copolymer (BCP). Generally, it has two blocks. Under certain conditions, phase separation will take place and ordered structures will be formed. Depending on the volume ratio of the blocks, it can form a cubic lattice of spheres, a hexagonal lattice of cylinders, and one-dimensional lamellar structures. The last two are attractive to the lithography industry. The ability of BCP to form ordered structures through phase separation depends on χN, where χ is the Flory parameter characterizing the interaction between the two sequences, and N is the degree of polymerization. N determines the pitch (L_o) that is formed; the smaller the N, the smaller the feature size. Thus to form a small feature size and maintain the phase separation, one need to have a high χ polymer, which means to increase the incompatibility of the two units in the BCP. Correspondingly, large pitch structures might be better made with lower χ polymers.

As an enabling technology for Moore's law, lots of work focuses on extending current leading edge optical lithography (ArFi), using high χ polymers, new processes, and simulation works [2, 3]. DSA materials and process also could be used to extend other optical lithographies, like I line and KrF exposure tools. Generally the features made by older exposure tool are larger compared to those made with current lithography tools, for instance the resolution limitation of KrF tool is around 130 nm. To use DSA in such technical node with a proper density multiplication, the BCP needs to have a larger N, this means a small χ polymer can be used, which will also lower the phase separation barrier, leading to easy phase separation and less defects [4].

In this paper we reported a new block copolymer poly (4-tertbutylstyrene)-block-poly(propylmethacrylate) (PtBS-b-PPMA) (Fig. 1). The polymer was synthesized by anionic polymerization, with high molecular weight and a low polydispersity index. Small angle X-ray scattering was used to characterize the ordered structure of the polymer after annealing. The dependence of L_o on N was

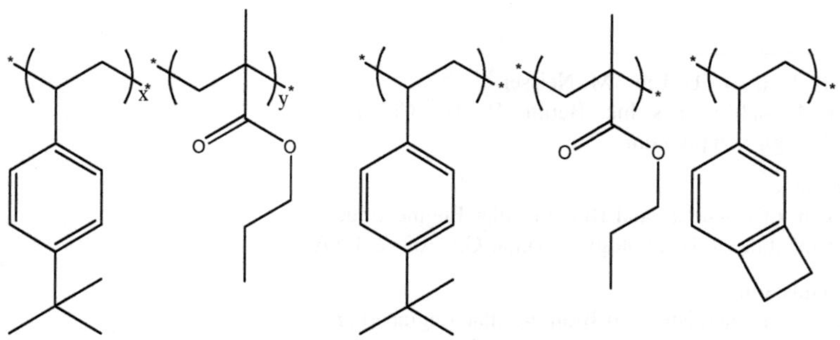

Fig. 1 Structure of BCP (left) and Neutral layer polymer (right)

investigated, a good relationship between pitch and polymer molecular weight was found, and this could be use as guideline to control the CD at the lithography process. Thermal annealing was then applied to the polymers. Ordered structures were formed with a pitch of 50–80 nm, suitable for graphoepitaxy with KrF lithography. Finally etch resistance was tested to check the feasibility of pattern transfer through dry etch.

Experimental

Reagents. 4-tert-butylstyrene (tBS), 1,1-diphenylethylene (DPE), sec-butyllithium (s-BuLi) (1.4 M in hexanes), dibutylmagnesium (DBMg) (1 M in heptane), n-butyllithium (n-BuLi) (2.5 M in hexanes), trioctylaluminum (TOA) (25 wt% in hexanes), and calcium hydride powder were purchased from Sigma-Aldrich. Propyl methacrylate (PMA) was purchased from Alfa Aesar. Tethrahydrofuran (THF) was purchased from VWR and purified by solvent purification system to remove trace water.

The monomers need well purified before anionic polymerization. The tBS was added into a flame-dried flask and went thought freeze-pump- thaw cycle three time, then vacuum distilled at 80 °C into another flask with dry DBMg. This solution was stirred at room temperature for 20 min, and vacuum distilled again into a new flame-dried flask. The purified tBS was packed with Ar and stored at −16 °C.

PMA was stirred with CaH overnight after three FPT cycles and refilled with Ar. The PMA was vacuum distilled into a flame-dried flask, titrated with TOA under Ar protection until the solution turned into bright yellow color and color didn't disappear in 20 min. Then the purified PMA was obtained by another vacuum distillation, packed with Ar and stored at −16 °C.

DPE was added into a flame-dried flask and went through three FPT cycles with a refill of Ar. The DPE was vacuum distilled into a flame-dried flask with dry n-BuLi at 110 °C. The solution turned into a dark red color and was stirred for 20 min. the purified DPE was obtained by another vacuum distillation, packed with Ar and stored at −16 °C.

Anhydrous methanol was degassed by three FPT cycle before termination of anionic polymerization.

Polymerization. The LiCl was added into flame-fried flask in the glovebox and THF was dispensed into flask followed by one FPT cycle. The whole reaction was maintain at −78 °C by dry ice/acetone bath. The initiator sec-BuLi solution was added into THF by airtight syringe and stirred for 20 min. Then, tBS monomer was added by airtight syringe to start polymerization. The solution turned bright orange. After 60 min stirring, DPE was added, and the solution turned red. After 60 more minutes, the PMA was injected into the reactor, turning the solution colourless. After 180 min, the reaction was terminated by degassed methanol. The reacted solution was concentrated by rotavapor and added into cold methanol dropwise. White precipitate was collected and washed by decane to remove the dead homopolymer PtBS. The block copolymer PtBS-b-PPMA was obtained after vacuum dry overnight.

The random polymer for neutral layer was synthesized by free radical polymerization using tBS, PMA and 4-Vinylbenzocyclobutene (VBCB) as crosslinking unit. All the monomer and AIBN as initiator was added into flask with toluene. The solution was stirred at 70 °C for 8 h. Then, the solution was add into cold methanol dropwise and white precipitated polymer was collected. The random copolymer PtBS-r-PPMA was obtained after vacuum dry overnight.

Small Angle X-ray Scattering Instrument. Xeuss 2.0, Detector: Pilatus 300 k, Distance between sample and detector 2.5 m.

Sample preparation: Put 0.2 g of PtBS-b-PPMA powder on an aluminium foil mold with 5 mm diameter and 1 mm thickness. 0.1 g of PGMEA was dropped into the mold. After 30 min, the viscous solution was formed in the mold, then put it on the hotplate at 90 °C for 30 min to remove most solvent. The mold with polymer was baked under vacuum to anneal at 150 °C for 18 h. The annealed BCP disk was obtained after peeling off the foil mold.

Results and Discussion

Synthesize of Block Copolymer (BCP). By controlling synthesize process, block copolymers with different molecular weight and different molar ratio were prepared, the PDIs were all within 1.05 which is good for LER control for guide direct self-assembly. The following Table shows the summary of BCPs we made: (Table 1).

Small Angle X-ray Scatter (SAXS). Block copolymers with different Mw were annealed and sent for SAXS test, SAXS was used to calculate the L_o, following the equation below:

$$L_o = \frac{2\pi}{q^*} \quad (1)$$

From the following plot, we can found that as the molecular weight increases, the peak from SAXS gets sharper, indicating a lamella pattern formation, and the primary peak q^* gets smaller, indicate a larger L_o. For sample A, there is no q^* observed means this sample is below the ODT region, the N is too small to get

Table 1 Property summary of synthesized BCP

BCP	Mw (K/mol)	PDI	Molar ratio (tBS:PPMA)
A	23	1.03	46:54
B	52	1.02	45:55
C	145	1.02	49:51
D	218	1.03	49:51
E	268	1.05	45:55

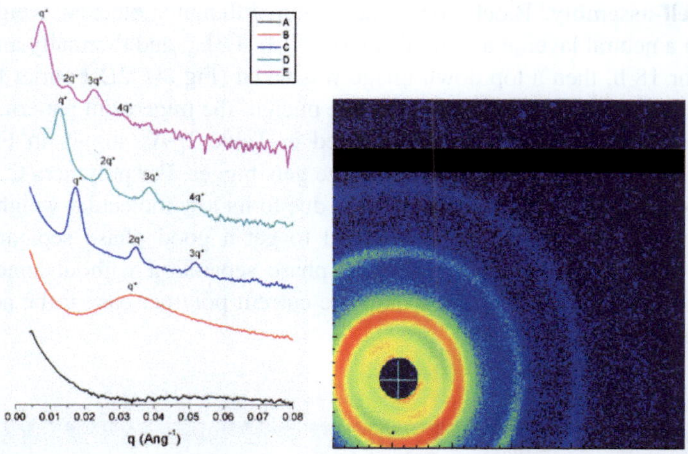

Fig. 2 SAXS data. Left: SAXS curve of different BCP, right: diffraction pattern of polymer E

proper phase separation, and for sample B; there is a broad q^* which means a semi-ordered pattern, and the N is at the edge of phase separation (Fig. 2).

Leibler theory [5] was used to fit the SAXS data to get χ, a_{BCP}, and K, from this, χ and a_{BCP} were fitted to be 0.027 and 0.75 nm. Then these parameters were applied to all the other samples, the relationship between L_o and N was established from the experimental data. From the least squares fit shown in the Fig. 3, for this BCP:

$$L_o = 0.082N^{0.92} \tag{2}$$

This can be used for future design of BCP based on the guide pattern size and the pitch requirement of the process.

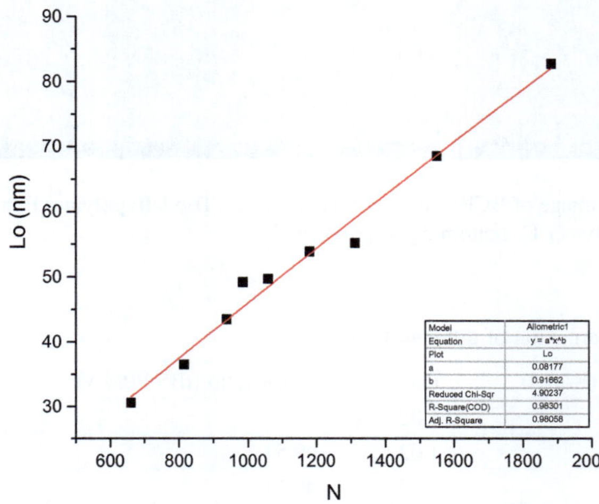

Fig. 3 L_o dependence on N (black dot: experiment data, red line: fitted)

Direct Self-assembly. Block copolymers with different molecular weights were coated on a neutral layer at a film thickness of 1.0–1.3L_o, and thermally annealed at 180 °C for 18 h, then a top down image was taken (Fig. 4). 2D-Fourier transform of the SEM image was used to calculate the pitch of the fingerprint pattern. Polymer information and pitch size are summarized in Table 2. As shown in Fig. 4 and Table 2, with increase of N, the feature size gets bigger. For polymers C, D, E, all form good finger print, but for polymer B, due to its low molecular weight and the low χ property of this polymer, it's hard to get a good phase separation. One property of this polymer is it undergoes phase separation without annealing, as shown in Fig. 5. This advantage makes the current polymer easy to be adapted in the lithography process [6, 7].

Fig. 4 Top down image of BCP after thermal annealing. (Top left: polymer E, top right: polymer D, bottom left: polymer C, bottom right: polymer B)

Table 2 Pitch information of different BCP

BCP	Mw (K/mol)	PDI	Molar ratio (tBS:PPMA)	Pitch (nm)
B	52	1.02	45:55	NG
C	145	1.02	49:51	50
D	218	1.03	49:51	82
E	268	1.05	45:55	91

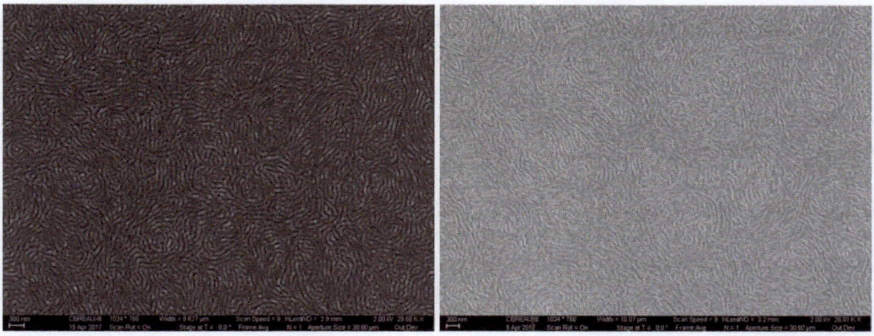

Fig. 5 Top down image of BCP after spin coating (left) and after thermal annealing (right)

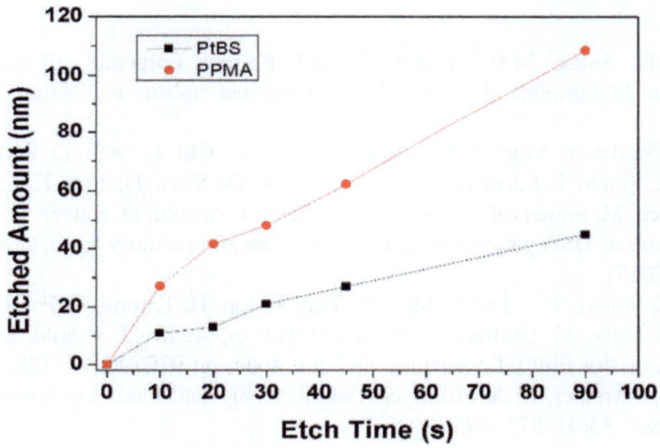

Fig. 6 Etch rate compare of two blocks

Etch Test. Two homopolymers, PtBS and PPMA, were coated on bare silicon separately and etched by standard etch gas (Ar/O_2), then film thickness at different etch time were collected, from these data the etch rate of each unit in the BCP can be calculated, from the Fig. 6, PtBS and PPMA has big etch rate bias (PPMA 2.3X of PtBS), that means dry etch can be used to transfer the pattern to substrate smoothly.

Conclusion

A new block copolymer with high molecular weight and narrow polydispersity was synthesized through anionic polymerization. SAXS was used to characterize the Lo of polymers with different molecular weight, and a good correlation between L_o and N was found. The correlation can be a guide for future polymer design for different applications. Thermal annealing shows good finger print patterns, and pitches from 50 to 90 can be achieved with different BCPs. Etch rates are studied, and a big etch rate bias is found, indicating a good match to current lithography pattern transfer requirements. Guide pattern direct self-assembly is under research, and more work will be reported later.

Acknowledgements This research was part of Research on Directed Self-assembly materials for advanced lithography Project, supported by Beijing Municipal Science & Technology Commission No. Z161100002116001.

Special thanks to Stephen Sirard (Lam Research Corporation) for providing etching data and Dr. Tingting's group at Beijing University of Chemical Technology for providing SAXS data.

References

1. S.O. Kim, H.H. Solak, M.P. Stoykovich, N.J. Ferrier, Epitaxial self-assembly of block copolymers on lithographically defined nanopatterned substrates. Nature **424**(6947), 411 (2003)
2. C.-C. Liu, E. Franke, Y. Mignot, S. LeFevre, S. Sieg, C. Chi, L. Meli, D. Parnell, K. Schmidt, M. Sanchez, L. Singh, T. Furukawa, I. Seshadri, E.A. De Silva, H. Tsai, K. Lai, H. Truong, R. Farrell, R. Bruce, M. Somervell, D. Sanders, N. Felix, J. Arnold, D. Hetzer, A. Ko, A. Metz, M. Colburn, D. Corliss, DSA patterning options for logics and memory applications, pp 1014603–1014603-13 (2017)
3. C.-C.C. Liu, E. Franke, F.L. Lie, S. Sieg, H. Tsai, K. Lai, H. Truong, R. Farrell, M. Somervell, D. Sanders, N. Felix, M. Guillorn, S. Burns, D. Hetzer, A. Ko, J. Arnold, M. Colburn, DSA patterning options for FinFET formation at 7 nm node, pp 97770R-97770R-15 (2016)
4. N.A. Lynd, A.J. Meuler, M.A. Hillmyer, Polydispersity and block copolymer self-assembly. Prog. Polym. Sci. **33**(9), 875–893 (2008)
5. L. Leibler, Theory of microphase separation in block copolymers. Macromolecules **13**(6), 1602–1617 (1980)
6. U. Nagpal, M. Müller, P.F. Nealey, J.J. De Pablo, Free energy of defects in ordered assemblies of block copolymer domains. ACS Macro Lett. **1**(3), 418–422 (2012)
7. R. Ruiz, H. Kang, F.A. Detcheverry, E. Dobisz, D.S. Kercher, T.R. Albrecht, J.J. de Pablo, P.F. Nealey, Density multiplication and improved lithography by directed block copolymer assembly. Science **321**(5891), 936–939 (2008)

In Vitro Cytocompatibility and Osteogenic Potential of Biodegradable Mg–Sr Alloys

Chen Liu, Lili Tan, Ying Zhao, Peng Wan, Haobo Pan, Xinbing Zhao, Yongdong Xu and Ke Yang

Abstract The cytocompatibility and osteogenic ability of novel as-extruded Mg–xSr (x = 0.25, 1.0, 1.5, 2.5 wt%) alloys are systematically investigated by in vitro cell adhesion and proliferation, alkaline phosphatase (ALP) activity, real-time RT-PCR evaluation, and mineralization tests. The results indicate that in addition to no cytotoxicity towards MC3T3-E1 cells, the Mg–Sr alloys, particularly Mg–1.5Sr, obviously promote the adhesion, proliferation, alkaline phosphatase activity, matrix mineralization and collagen secretion of MC3T3-E1 cells compared to pure Mg. Nearly all the osteogenesis-related genes, namely Runx2, Opn, Sp7, and Bmp2, are up-regulated for Mg–xSr, compared to pure Mg. In particular, Mg–1.5Sr reveals the best osteogenic ability. Our work suggest that novel as-extruded Mg–1.5Sr alloy, which exhibits excellent in vitro cytocompatibility and osteogenic ability, is promising in expediting clinical acceptance of biodegradable magnesium alloys.

Keywords Magnesium alloys · Cytocompatibility · Osteogenesis Strontium

C Liu and L. Tan these two authors contributed equally to this paper.

C. Liu · X. Zhao
Department of Materials Science and Engineering,
Zhejiang University, Hangzhou, China
e-mail: chenliu52s@163.com

C. Liu · Y. Xu
Ningbo Branch of Chinese Academy of Ordnance Science, Ningbo, China

L. Tan · P. Wan · K. Yang (✉)
Institute of Metal Research, Chinese Academy of Sciences, Shenyang, China
e-mail: kyang@imr.ac.cn

L. Tan
e-mail: lltan@imr.ac.cn

Y. Zhao (✉) · H. Pan
Shenzhen Institutes of Advanced Technology,
Chinese Academy of Sciences, Shenzhen, China
e-mail: ying.zhao@siat.ac.cn

Y. Han (ed.), *Advanced Functional Materials*,
https://doi.org/10.1007/978-981-13-0110-0_48

Introduction

Due to the specific physical properties and chemical structure, magnesium-based metals possess significant advantages compared with the traditional metallic implant materials [1–5]. They have been demonstrated with extraordinary biological properties, including (1) excellent biocompatibility: magnesium is an indispensable element in the body. Mg^{2+} is involved in many metabolic reactions; (2) unique degradability: magnesium and its alloys can easily corrode in physiological environment so that repeated surgery for implant removal is not needed; (3) appropriate mechanical properties: the elastic moduli of magnesium alloys are similar to those of cortical bones, which can effectively avoid the stress shielding effect. Based on above advantages, Mg-based metals have been widely studied in recent years as potentially biodegradable metallic materials for implantation applications, especially in orthopedic fields [6–8].

An ideal magnesium alloy implant would maintain mechanical integrity in the human body for at least 12 weeks. While the bone tissue heals, it will eventually be replaced by new bone tissue [9, 10]. It is also important that the corrosion rate and degradation products, such as subcutaneous gas bubbles, should be within the body's acceptable ranges [11]. However, magnesium is highly liable to corrode in physiological conditions that contains chloride ion, which accordingly results in a dramatically decrease in its mechanical properties prior to the tissue healing [12–14]. From the material perspective, alloying may bean effective approach to diminish the degradation rates of magnesium alloys. Alloying elements such as aluminum, manganese, zinc, calcium, and rare earth metals have been used to improve the mechanical properties and reduce the corrosion rate of Mg [15–18]. However, from the perspective of in vivo application, not all the alloying elements are suitable. Aluminum is well known as a neurotoxicant and its accumulation has been suggested to be an associated phenomenon in various neurological disorders [19]. Severe hepatotoxicity has been detected after the administration of those rare earth metals of cerium, praseodymium and yttrium [20]. Comparatively, calcium, manganese, zinc and perhaps a very small amount of low toxicity rare earth elements are optional as suitable alloying elements for biodegradable Mg alloys.

Strontium (Sr) has been recently reported as a potential alloying element for medical applications [13]. From the standpoint of corrosion and mechanical properties, Sr can improve the mechanical properties and corrosion resistance of certain Mg alloys. Zeng et al. reported that Sr had a significant grain-refinement effect in AZ31 magnesium alloy and could improve both yield strength and elongation of AZ31 alloy [21]. Fan et al. indicated that Sr could improve both the mechanical properties and corrosion resistance of AZ91D magnesium alloy in NaCl solution due to the formation of Al_4Sr phase [22]. From the viewpoint of biocompatibility, Sr is a natural component in human bones and about 98% of Sr in human body is localized in bone tissues [12, 23]. The role of Sr in bone is known to promote the growth of osteoblasts and prevent bone resorption [24]. With aid of the bone formation stimulation effect by Sr, Sr salts at low dose were put forward as a therapeutic agent to treat osteoporosis through increasing the bone mass and reducing the incidence of fractures [25, 26]. Sr-substituted hydroxyapatite (SrHA) was demonstrated to exert a positive effect on osteogenesis around the bone cement interface [27, 28].

Recently, novel as-rolled Mg–Sr alloys have been developed for skeletal applications. It is reported that as-rolled Mg–2Sr alloy possessed the best combination of corrosion resistance, high strength and in vivo biocompatibility among various Mg–Sr alloys (with Sr contents ranging from 1 to 4 wt%) [13]. Brar et al. studied as-cast Mg–xSr (x = 0.5, 1.0, 1.5 wt%) alloys and found the lowest degradation rate from Mg–0.5Sr [29]. Bornapour et al. indicated that the as-cast Mg–0.5Sr with a slow degradation rate could not lead to thrombosis during three weeks of implantation and a Sr-substituted HA layer was formed on Mg–0.5Sr which is known to enhance bone cell growth and proliferation around bone implants [12]. Additionally, our previous work has confirmed that the as-extruded Mg–xSr (x = 0.25, 1.0, 1.5, 2.5 wt%) alloys possessed similar mechanical properties to cortical bone, controlled corrosion rates, good cytocompatibility as well as excellent antibacterial properties [30]. The above results show that superior Mg–Sr alloys can be used as potential candidate for future orthopedic applications. However, up till now a comprehensive evaluation of osteogenic ability of as-extruded Mg–Sr alloy is not reported. The aim of this study was to examine the in vitro cytocompatibility and osteogenic ability of biodegradable Mg–Sr alloys.

Materials and Methods

Materials preparation. The experimental binary Mg–xSr (x = 0.5, 1.5, 2.5, 4.0 wt %) alloys were fabricated using pure Mg (99.9%) and a Mg-30wt% Sr master alloy in a high-purity graphite crucible under the protection of a mixed gas atmosphere of SF_6 (1 vol%) and carbon dioxide (CO_2). The melting was held at 730 °C for 30 min and stirred with a graphite rod. The melt was then poured into a steel mold preheated to 300 °C. The binary Mg–Sr alloys were encapsulated in quartz tubes under vacuum and then homogenization treated at 500 °C for 20 h followed by quenching in water. The as-cast ingots were hot extruded into bars at 390 °C with extrusion ratio of 64:1. The as-extruded pure Mg was prepared following the aforementioned procedure as the control group. It was subsequently machined into samples with dimensions of Φ10 × 3 mm^3. All the samples were ground with SiC papers up to 2000 grit, ultrasonically cleaned in acetone, absolute ethanol and distilled water in turn, and finally dried. The compositions of the Mg–Sr alloys measured by inductively coupled plasma atomic emission spectrometry (ICP-AES, Optima 7300DV, USA) are listed in Table 1.

Extracts preparation. Extracts were prepared using serum-free modified Eagle's medium alpha (α-MEM, Hyclone, USA) as the extraction medium with a ratio of extraction medium/sample surface area of about 1.25 cm^2/mL in a humidified atmosphere containing 5% CO_2 at 37 °C for 24 h according to the ISO 10993-5.

Table 1 Chemical composition of the Mg–Sr alloys (wt%)

Nominal composition		Mg–0.5Sr	Mg–1.5Sr	Mg–2.5Sr	Mg–4.0Sr
Actual composition	Sr	0.25	1.0	1.5	2.5
	Mg	Balance			

Mg magnesium; *Sr* strontium

Prior to immersion, the samples were sterilized by ultraviolent irradiation for 30 min. After immersion, the supernatant was withdrawn, centrifuged, and filtered to prepare the extraction medium. It was remained at 4 °C prior to cell tests and the extract from pure Mg served as the control.

Cytotoxicity and cell proliferation assay. The cytotoxicity and cell proliferation were evaluated by an indirect contact assay according to ISO 10993-5. Murine calvarial preosteoblasts (MC3T3-E1, ATCC CRL-2594) were seeded on 96-well cell culture plates at an initial density of 5×10^3 cells/100 mL and incubated for 24 h to allow cell attachment. The medium was then replaced with 100 μL of the extracts. After culturing for 1, 3, and 5 days, MTT solutions were added to each well and incubated at 37 °C for 4 h to form formazen which was then dissolved using dimethyl sulfoxide (DMSO). The optical density (OD) was determined on a microplate reader (Thermo Scientific Multiskan GO, USA) at 490 nm with a reference wavelength of 570 nm to determine the cell viability in comparison with the control.

Live/dead staining. The MC3T3-E1 cells were seeded on 12-well cell culture plates at a density of 2×10^4cells/well for 24 h to allow cell attachment. The medium except the control group was then replaced by 500 μL of the extracts. After incubation for another 24 h, the cells adhered to the plates and cultured in extracts were stained by the Live/Dead staining. The living and dead cells were, respectively, stained with Calcein-AM and iodide pyridine (Sigma) for 30 min at 37 °C, and then visualized using a fluorescence microscope (Olympus IX71, Japan).

Cell morphology. The MC3T3-E1 cells were seeded on 12-well cell culture plates at a density of 2×10^4 cells/well for 24 h to allow cell attachment. The medium except the control group was then replaced by 500 μL of the extracts. After incubation for 12 h, the unattached cells were removed by rinsing with phosphate buffered saline (PBS) solution. The cells were fixed with 4% paraformaldehyde at 4 °C for 30 min and permeabilized with 0.5% Triton X-100 in PBS solution for 5 min. The cells were subsequently rinsed with PBS solution and the F-actin stress fibers and nuclei were stained with rhodamine phalloidin (Cytoskeleton, USA) and 40, 6-diamidino-2- phenylindole (DAPI, Dojindo, Japan), respectively. The cytoskeleton and cell nuclei were observed by fluorescence microscopy (Olympus IX71, Japan).

Alkaline phosphatase (ALP) activity. The MC3T3-E1 cells were seeded on 12-well cell culture plates at a density of 3×10^4 cells/well and incubated in different extracts supplemented with 100 nM dexamethasone, 0.2 mM ascorbic acid, and 10 mM β-glycerophosphate. The extracts were changed every 3 days. After 4, 7 and 14 days, the cells were washed three times with PBS and lysed in 0.2 vol% Triton X-100. The alkaline phosphatase (ALP) activity was determined by a colorimetric assay using an ALP reagent containing p-nitrophenyl phosphate (p-NPP) as the substrate (Beyotime, China). The absorbance of p-nitrophenol was monitored at 405 nm. The intracellular total protein content was determined using the MicroBCA protein assay kit (Thermo Pierce, USA) and the ALP activity was normalized to the total protein content.

Extracellular matrix (ECM) mineralization and collagen secretion. The extracellular matrix (ECM) mineralization and collagen secretion by the MC3T3-E1 cells in the Mg–Sr alloys extracts were evaluated by the Alizarin Red and Sirius Red staining, respectively. After incubating for 15 days, the cells with an initial concentration of 3×10^4/well were washed and fixed. Afterwards, they were stained using 40 mM Alizarin Red (Sigma) at a pH of 4.2 to show the mineralization or 0.1% Sirius Red (Sigma) to reveal the collagen. The unbound stain was washed with distilled water or 0.1 M acetic acid prior to photographing by a digital camera (Nikon D3200, Japan). In the quantitative analysis, the Alizarin Red or Sirius Red stain was dissolved in 10% cetylpyridinum chloride in 10 mM sodium phosphate (pH 7) or 0.2 M NaOH/methanol (1:1). The absorbance was measured at 620 or 540 nm.

Real-time polymerase reaction (PT-PCR). RT-PCR was used to assess the expression of the osteogenesis-related genes. The MC3T3-E1 cells were seeded on 12-well cell culture plates at a cell density of 3×10^4 cells/well and after 24 h, the culture medium was replaced by different material extracts supplemented with fresh serum. The MC3T3-E1 cells were incubated for 7, 10 and 14 days. The total RNA was isolated using a Trizol reagent (Invitrogen, USA) and the concentration of RNA was detected by measuring the optical absorbance at 260 nm on the Thermo 2000c (USA). 1 μg of the RNA was reversely transcribed into complementary DNA (cDNA) using Superscript III (Invitrogen, USA) in a volume of 20 μl. The forward and reverse primers of the selected genes are listed in Table 2. The expressions of the osteogenesis-related genes, including runt-related transcription factor 2 (Runx2), osteopontin (Opn), osterix (Sp7) and bone morphogeneticprotein-2 (Bmp2) were quantified by Real-time PCR (Biorad

Table 2 Primer pairs used in real-time PCR analysis

Gene	Forward primer	Reverse primer
Mouse β-actin	5′-GGCTGTATTCCCCTCCATCG-3′	5′-CCAGTTGGTAACAATGCCATGT-3′
Runx2	5′-AGAGTCAGATTACAGATCCC AGG-3′	5′-TGGCTCTTCTTACTGAGAGAGG - 3′
Opn	5′-TCTGATGAGACCGTCACTGC-3′	5′-AGGTCCTCATCTGTGGCATC-3′
Sp7	5′-ATGGCGTCCTCTCTGCTTG-3′	5′-TGAAAGGTCAGCGTATGGCTT-3′
Bmp2	5′-GGGACCCGCTGTCTTCTAGT-3′	5′-TCAACTCAAATTCGCTGAGGAC-3′

CFX96, USA) on the SYBR Green PCR Master Mix (Applied Biosystems, USA). The relative mRNA expression level of each gene was normalized to the housekeeping gene β-actin (mouse β-actin) and determined by the Ct values.

Statistical analysis. The cell tests were performed in triplicate. The experimental results were expressed as mean ± standard deviations and the data were analyzed using SPSS 13.0 software. A two-way analysis of variance (ANOVA) followed by a Student-Newman-Keuls post hoc test was used to determine the level of significance in cytotoxicity, cell proliferation, ALP activity and osteogenesis-related gene expressions assay, and an one-way ANOVA in extracellular matrix mineralization and collagen secretion test. $p < 0.05$ was considered to be statistically significant, and $p < 0.01$ was considered to be highly statistically significant.

Results and Discussions

Cytotoxicity. Figure 1 illustrates the relative cell growth rates of murine calvarial preosteoblasts MC3T3-E1 after 1, 3, and 5 days of incubation in the extracts of as-extruded Mg–xSr alloy (x = 0.25, 1.0, 1.5, 2.5 wt%) and pure Mg, respectively. It can be seen that no cytotoxicity is found for all the as-extruded Mg–Sr alloys and their cell viabilities are shown over 100% and higher than that of pure Mg, especially after 3 and 5 days culture (#$p < 0.05$, ##$p < 0.01$). The extract of the as-extruded Mg–1.5Sr alloy leads to the highest cell viability. All Mg–Sr alloys reveal Grade 0 cytotoxicity according to ISO10993-5, suggesting good cytocompatibility and acceptable biosafety for in vivo applications.

Cell proliferation. Figure 2 shows MC3T3-E1 cells proliferation of as-extruded Mg–Sr alloys and pure Mg extraction mediums after 1, 3, and 5 days of incubation. It can be found that the cells number for all the as-extruded Mg–Sr alloy extracts

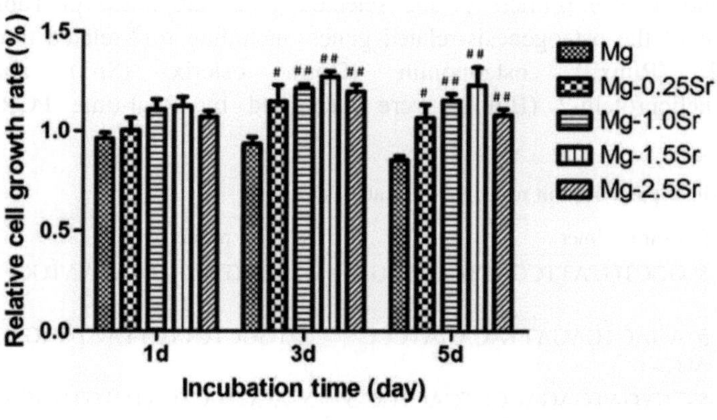

Fig. 1 Relative cell growth rates after 1, 3, and 5 days of incubation in the extracts of as-extruded Mg–Sr alloys and pure Mg. #$p < 0.05$ and ##$p < 0.01$ compared with pure Mg

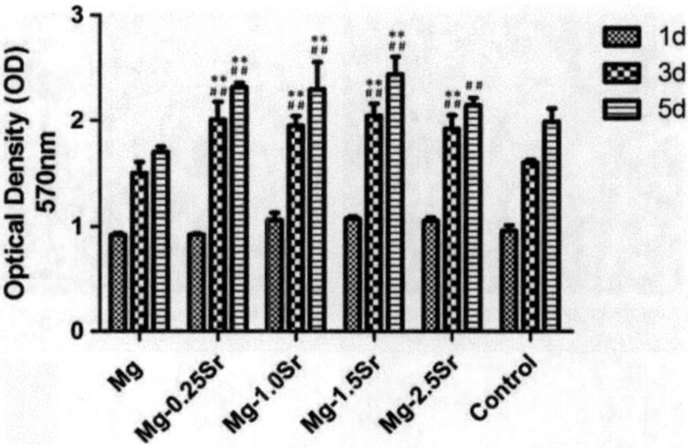

Fig. 2 Comparison of proliferation of MC3T3-E1 cells cultured in as-extruded Mg–Sr alloy and pure Mg extraction mediums for 1, 3, and 5 days. $^{**}p < 0.01$ compared to the Control; $^{\#\#}p < 0.01$ compared to pure Mg

shows no significant difference when compared with pure Mg and control group on Day 1. However, after 3 days incubation, the cells number of as-extruded Mg–Sr alloys rise up dramatically and exceed pure Mg and control group, while pure Mg group shows slightly lower than the control group. After 5 days, the MC3T3-E1 cells incubated in all Mg–Sr alloys group show further increase. In particular, Mg–1.5Sr group rises significantly. These results indicate Mg–Sr alloys effectively promote MC3T3-E1 cells proliferation compared to pure Mg and control group, suggesting that Sr addition significantly improves in vitro cytocompatibility of magnesium metals.

Cell adhesion. Figure 3 depicts the Live/dead staining of MC3T3-E1 cells after 24 h of incubation in the extracts of as-extruded Mg–Sr alloys and pure Mg, respectively. Typical microscope fields are selected to show viable (green) and non-viable (red) cells. It is observed that most cells in all groups are stained green with occasionally distributed red-stained non-viable cells. The numbers of the viable cells in pure Mg and Mg–0.25Sr groups are similar to that of the control. With increasing Sr content in the alloys, more viable cells are observed, and the cells almost reach to 100% confluence in Mg–1.5Sr group. In comparison, the viable cells in the Mg–2.5Sr group are slightly decreased, but still higher than that of the control. The results indicate Mg–Sr alloy with appropriate Sr amount is favorable to enhance cytocompatibility of magnesium metals.

Cell morphology. Figure 4 reveals the cytoskeletons of the MC3T3-E1 cells after incubation for 12 h in extracts of as-extruded Mg–Sr alloys and pure Mg, respectively. Compared to the control, the attached cells for the pure Mg and as-extruded Mg–Sr alloys extracts show more spreading and superior filopodia extension. For the as-extruded Mg–Sr alloys, the cells are polygonal, elongated and thicker, with fine pseudopodia extension. The MC3T3-E1 cells cultured in the Mg–1.0Sr and Mg–1.5Sr extracts show more plump focal adhesion via well-organized F-actin stress fibers (red filaments) compared to Mg–0.25Sr and Mg–2.5Sr.

Osteogenic differentiation. Figure 5 shows the ALP activity of the MC3T3-E1 cells after incubation for 4, 7 and 14 days in extracts of as-extruded Mg–Sr alloys

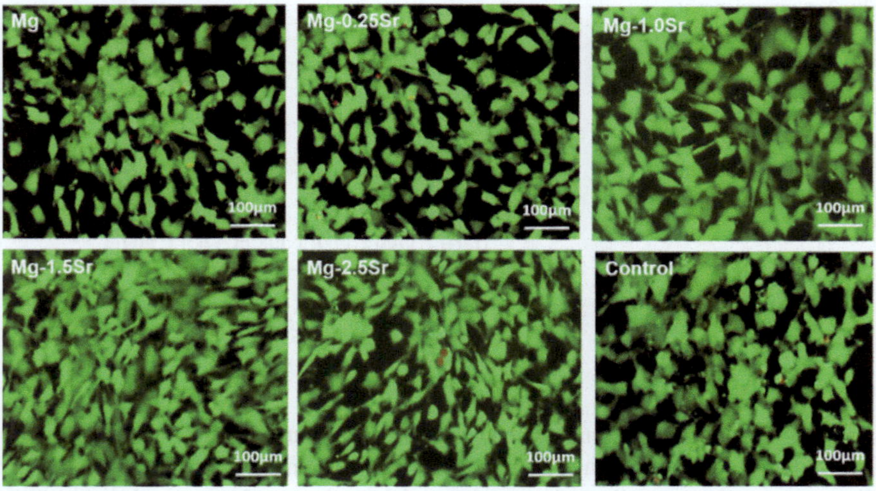

Fig. 3 Live/dead staining of MC3T3-E1 cells after 24 h of incubation in the extracts of as-extruded Mg–Sr alloys and pure Mg. Calcein-AM for viable cells (green) and iodide pyridine for non-viable cells (red)

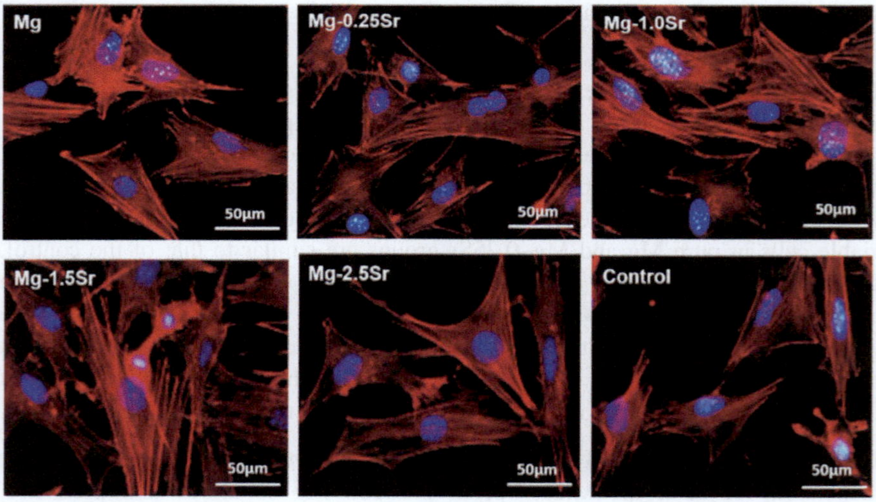

Fig. 4 Cytoskeleton staining of MC3T3-E1 cells after 12 h of incubation in the extracts of as-extruded Mg–Sr alloys and pure Mg. DAPI for nuclei (blue) and rhodamine phalloidin for F-actin stress fibers (red)

and pure Mg. There is no significant difference among all the extracts on day 4. After incubation for 7 days, ALP levels of the Mg–Sr alloys are significantly enhanced compared to those of pure Mg and control. The distinct ALP activities are emerged after 14 days incubation. Mg–1.5Sr shows the highest ALP activity which is

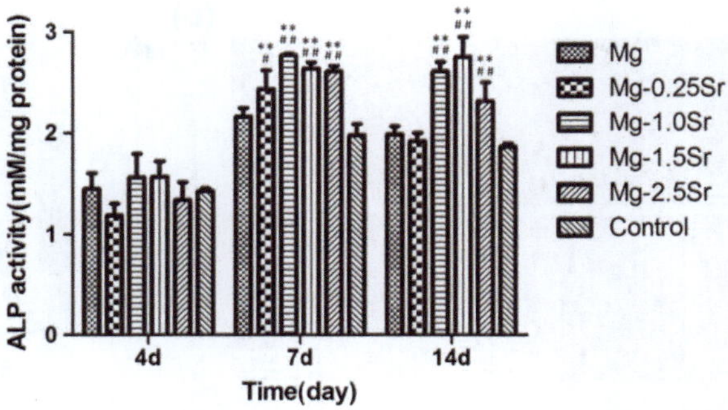

Fig. 5 ALP activity of MC3T3-E1 cells after incubation for 4, 7, and 14 days in the extracts of as-extruded Mg–Sr alloys and pure Mg. $**p < 0.01$ compared to the control; $\#p < 0.05$ and $\#\#p < 0.01$ compared to pure Mg

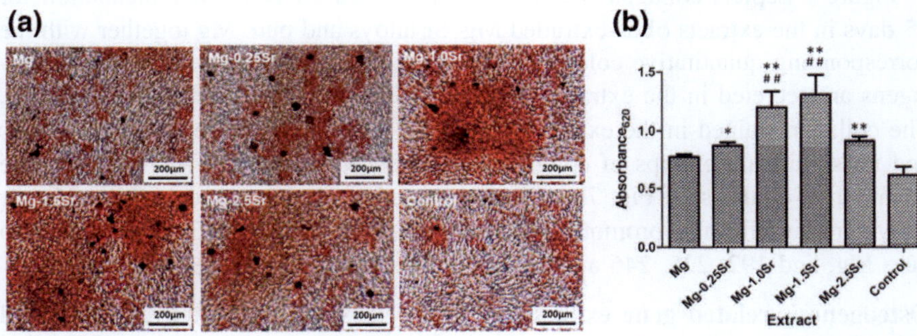

Fig. 6 **a** Extracellular matrix mineralization of MC3T3-E1 cells after incubation for 15 days in the extracts of as-extruded Mg–Sr alloys and pure Mg and **b** colorimetrically quantitative analysis. $*p < 0.05$ and $**p < 0.01$ compared to the control; $\#\#p < 0.01$ compared to pure Mg

followed by Mg–1.0Sr, Mg–2.5Sr, pure Mg and Mg–0.25Sr in sequence. The ALP activity of Mg–1.0Sr and Mg–1.5Sr are higher than those of pure Mg and control throughout the experimental period, showing the best osteogenic differentiation.

Figure 6 displays the ECM mineralization of MC3T3-E1 cells after incubation for 15 days in various Mg–Sr alloys and pure Mg extracts together with the corresponding quantitative colorimetric analysis. Figure 6a reveals mineralized calcium nodules formed in various extracts. A small portion of the stained calcium nodules in the pure Mg extract suggests a relatively weak extracellular calcium deposition. Comparatively, more depositions are observed in the Mg–Sr alloys extracts, especially for Mg–1.0Sr and Mg–1.5Sr. Correspondingly, the quantitative results as shown in Fig. 6b indicate that significant enhanced ECM mineralizations are detected in the Mg–1.0Sr and Mg–1.5Sr extracts compared to other groups. The ECM mineralization levels in Mg–0.25Sr and Mg–2.5Sr extracts are lower than those in Mg–1.0Sr and Mg–1.5Sr ones, but still higher than the pure Mg and control.

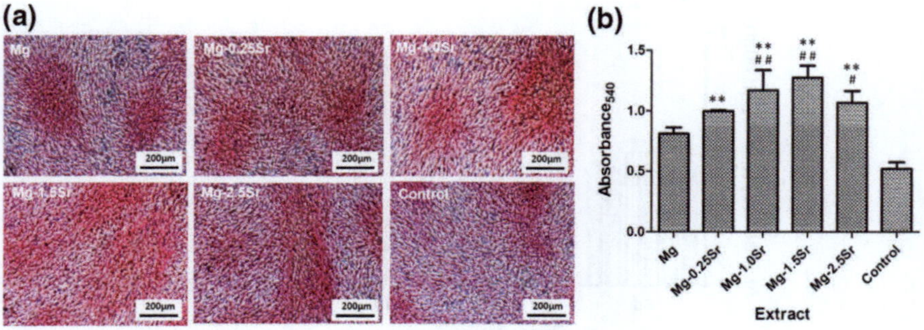

Fig. 7 **a** Collagen secretion from MC3T3-E1 cells after incubation for 15 days in the extracts of as-extruded Mg–Sr alloys and pure Mg and **b** colorimetrically quantitative analysis. $^{**}p < 0.01$ compared to the control; $\#p < 0.05$ and $\#\#p < 0.01$ compared to pure Mg

Figure 7 depicts collagen secretion from MC3T3-E1 cells after incubation for 15 days in the extracts of as-extruded Mg–Sr alloys and pure Mg together with the corresponding quantitative colorimetric analysis. Figure 7a shows that denser collagens are secreted in the extracts of the Mg–Sr alloys than pure Mg and control. The collagen stained in the extract of pure Mg is less than those of Mg–Sr alloys and no significant clumps of collagen are deposited in control. According to the quantitative analysis in Fig. 7b, collagen secretion in the extracts of the Mg–Sr alloys are significantly promoted to about 123, 144, 157 and 132% compared to pure Mg, and 192, 225, 246 and 205% relative to the control, respectively.

Osteogenesis-related gene expressions. Figure 8 shows the osteogenesis-related gene expressions of Runx2, Opn, Sp7 and Bmp2 after culture for 7, 10, and 14 days with MC3T3-E1 cells in the extracts of as-extruded Mg–Sr alloys and pure Mg. In general, the gene expressions are time dependent. After 7 days of incubation, the extract of Mg–1.5Sr induces the highest Runx2 expression followed by the Mg–1.0Sr, Mg–0.25Sr, pure Mg and Mg–2.5Sr groups in sequence. For genes Opn, Sp7 and Bmp2, although no statistically significant differences are found among the various groups, Mg–Sr alloys groups still show higher expression levels, especially for Mg–1.0Sr and Mg–1.5Sr groups, than that of pure Mg. After culturing for 10 days, the Runx2 expressions of Mg–1.0Sr, Mg–1.5Sr and Mg–2.5Sr are much higher than that of pure Mg. The highest expressions of genes Opn and Sp7 are induced by the extracts of Mg–1.0Sr and Mg–1.5Sr, respectively. There is no statistical difference in the Bmp2 expression among the various groups. Then on day 14, Mg–Sr alloys, especially Mg–1.0Sr and Mg–1.5Sr, yield higher Runx2, Sp7 and Bmp2 expressions compared to the pure Mg group even though no statistical difference can be found for gene Opn.Generally, the Runx2, Opn, Sp7 and Bmp2 expressions of all the Mg–Sr samples increase at first and then decrease gradually with Sr contents increase. Mg–1.0Sr and Mg–1.5Sr show the most significant up-regulation of osteogenesis-related gene expressions.

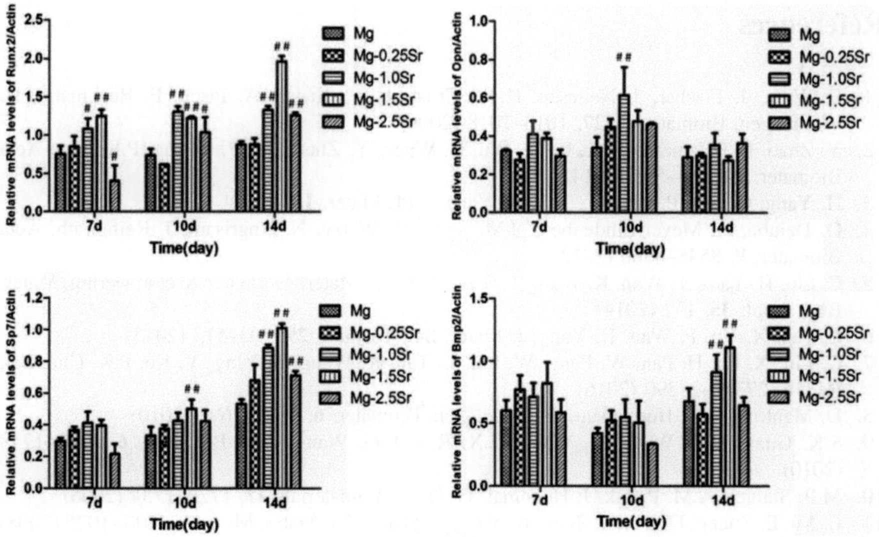

Fig. 8 Osteogenesis-related gene expressions by MC3T3-E1 cells after incubation for 7, 10 and 14 days in the extracts of as-extruded Mg–Sr alloys and pure Mg. #$p < 0.05$ and ##$p < 0.01$ compared to pure Mg

Conclusion

The present study systematically evaluates the cytocompatibility and osteogenic ability of the as-extruded Mg–xwt%Sr (x = 0.25, 1.0, 1.5, 2.5) alloys. No cytotoxicity towards MC3T3-E1 cells is observed for the Mg–Sr alloys and proper Sr alloying to Mg metal is favorable to enhance the adhesion, proliferation, alkaline phosphatase activity, matrix mineralization and collagen secretion of MC3T3-E1 cells. In particular, the optimal as-extruded Mg–1.5Sr alloy shows the best osteogenic differentiation inducibility. Our study suggests that degradable as-extruded Mg–Sr alloys with proper Sr content show good potential as orthopedic implants.

Acknowledgements This work was financially supported by the National Key Research on Development Program of China (No. 2016YFC1101804), National Natural Science Foundation of China (Nos.81401773, 51501218,81572113), the funding from Institute of Metal Research, Chinese Academy of Sciences (No.2015-ZD01), Guangdong Provincial Science and technology projects (2014A010105033), Guangdong Natural Science Foundation (2014A030310129), Shenzhen Peacock Programs (KQCX20140521115045444 and 110811003586331), Shenzhen-Hong Kong Technology Cooperation Funding Scheme (SGLH20150213143207910), Shenzhen Science and Technology Research Funding (JCYJ20160229195249481, JCYJ20160429185449249, JCYJ20160608153641 020) and the Public Technology Research Project of Zhejiang Province (Grant No.2016C31030).

References

1. F. Witte, J. Fischer, J. Nellesen, H.-A. Crostack, V. Kaese, A. Pisch, F. Beckmann, H. Windhagen, Biomaterials **27**, 1013–1018 (2006)
2. Y. Zhao, M.I. Jamesh, W.K. Li, G. Wu, C. Wang, Y. Zheng, K.W. Yeung, P.K. Chu, Acta Biomater. **10**, 544–556 (2014)
3. H. Yang, C. Liu, P. Wan, L. Tan, K. Yang, APL Mater. **1**, 052104 (2013)
4. D. Dziuba, A. Meyer-Lindenberg, J.M. Seitz, H. Waizy, N. Angrisani, J. Reifenrath, Acta Biomater. **9**, 8548–8560 (2012)
5. C. Liu, H. Yang, P. Wan, K. Wang, L. Tan, K. Yang, Materials science & engineering. Mater. Biol. Appl. **35**, 1–7 (2014)
6. L. Tan, X. Yu, P. Wan, K. Yang, J. Mater. Sci. Technol. **29**, 503–513 (2013)
7. C. Liu, X. Fu, H. Pan, W. Peng, W. Lei, L. Tan, K. Wang, Z. Ying, Y. Ke, P.K. Chu, Sci. Rep. **6**, 27374–27390 (2016)
8. D. Mantovani, H. Hermawan, D. Dube, Acta Biomater. **6**, 1693–1697 (2010)
9. S.K. Guan, H.X. Wang, X. Wang, C.X. Ren, L.G. Wang, Acta Biomater. **6**, 1743–1748 (2010)
10. M.P. Staiger, A.M. Pietak, J. Huadmai, G. Dias, Biomaterials **27**, 1728–1734 (2006)
11. L. Xu, E. Zhang, D. Yin, S. Zeng, K. Yang, J. Mater. Sci. Mater. Med. **19**, 1017–1025 (2008)
12. M. Bornapour, N. Muja, D. Shum-Tim, M. Cerruti, M. Pekguleryuz, Acta Biomater. **9**, 5319–5330 (2013)
13. X.N. Gu, X.H. Xie, N. Li, Y.F. Zheng, L. Qin, Acta Biomater. **8**, 2360–2374 (2012)
14. X.N. Zhang, J.N. Li, Y. Song, S.X. Zhang, C.L. Zhao, F. Zhang, L. Cao, Q.M. Fan, T.T. Tang, Biomaterials **31**, 5782–5788 (2010)
15. L. Xu, G. Yu, E. Zhang, F. Pan, K. Yang, J. Biomed. Mater. Res. Part A **83**, 703–711 (2007)
16. Y.F. Zheng, Z.J. Li, X.N. Gu, S.Q. Lou, Biomaterials **29**, 1329–1344 (2008)
17. N. Hort, Y. Huang, D. Fechner, M. Stormer, C. Blawert, F. Witte, C. Vogt, H. Drucker, R. Willumeit, K.U. Kainer, F. Feyerabend, Acta Biomater. **6**, 1714–1725 (2010)
18. R. Walter, M.B. Kannan, Mater. Lett. **65**, 748–750 (2011)
19. S. El-Rahman, Pharmacol. Res. **47**, 189–194 (2003)
20. Y. Nakamura, Y. Tsumura, Y. Tonogai, T. Shibata, Y. Ito, Toxicol. Sci. **37**, 106–116 (1997)
21. X. Zeng, Y. Wang, W. Ding, A.A. Luo, K. Sachdev Anil, Metall. Mater. Trans. A **37A**, 1333–1341 (2006)
22. Y. Fan, G.H. Wu, C.Q. Zhai, Mater. Sci. Forum **5**, 567–570 (2007)
23. Y. Li, Q. Li, S. Zhu, E. Luo, J. Li, G. Feng, Y. Liao, J. Hu, Biomaterials **31**, 9006–9014 (2010)
24. P.J. Marie, Osteoporos. Int. : J. Established Result Cooperation Between Eur. Foundation Osteoporos. Nat. Osteoporos. Foundation USA **16**(Suppl 1), S7–10 (2005)
25. J.Y. Reginster, O. Bruyere, A. Sawicki, A. Roces-Varela, P. Fardellone, A. Roberts, J. P. Devogelaer, Bone **45**, 1059–1064 (2009)
26. R. Rizzoli, R.D. Chapurlat, J.M. Laroche, M.A. Krieg, T. Thomas, I. Frieling, S. Boutroy, A. Laib, O. Bock, D. Felsenberg, Osteoporos. Int.: J. Established Result Cooperation Between Eur. Foundation Osteoporos. Nat. Osteoporos. Foundation USA **23**, 305–315 (2012)
27. J. Yan, J.F. Sun, P.K. Chu, Y. Han, Y.M. Zhang, J. Biomed. Mater. Res. Part A **101**, 2465–2480 (2013)
28. G.X. Ni, Z.P. Yao, G.T. Huang, W.G. Liu, W.W. Lu, J. Mater. Sci. Mater. Med. **22**, 961–967 (2011)
29. H.S. Brar, J. Wong, M.V. Manuel, J. Mech. Behav. Biomed. Mater. **7**, 87–95 (2012)
30. C. Liu, P. Wan, L.L. Tan, K. Wang, K. Yang, J. Orthop. Transl. **2**, 139–148 (2014)

Fabrication of Urethral Tissue Engineering Scaffolds Based on Multi-scale Structure of Bacterial Cellulose Matrix Materials: A Preliminary Study

Zhe Li, Xiangguo Lv, Zhiyong Yan, Yongbo Yao, Chao Feng, Huaping Wang and Shiyan Chen

Abstract Tissue engineering (TE) approaches provide an effective strategy for developing functional substitutes to repair or replace failing urethra. In urethral TE, one of the most important challenges is design and fabrication the urethral scaffolds that possess biomimetic multi-scale structures composed of urethral shape and micropore microporous structure with nanofibers for more biomimetic the native urethral ECM. In this study, ultrasonography was used to reconstruct a 3D digital model of human urethra. Then, the micro- and nano- structural characteristics of urethral ECM were studied by using FF-SEM. The indirect 3D printing technology and template biosynthesis method were used to fabricate a 3D gelatin/silk-bacterial cellulose (Gel/silk-BC) urethral scaffold, with defined macro-, micro- and nanostructure. The Gel/silk-BC scaffold has a urethral shape in macro-size, and highly interconnected micropore (140 ± 34 μm) with surface decorated of BC nanofibers (24.6 ± 5.0 nm), which serve as mimics for the native urethra ECM.

Keywords Urethral tissue engineering · Bacterial cellulose · Multi-scale structure Microporous structure · Nanofibrous structure

---

Xiangguo Lv: these authors contributed equally to the work.

Z. Li (✉) · Z. Yan · Y. Yao
College of Materials and Textile Engineering, Jiaxing University, Zhejiang 314001, China
e-mail: lizhe830817@163.com

X. Lv · C. Feng (✉)
Department of Urology, Affiliated Sixth People's Hospital, Shanghai Jiaotong University, Shanghai 201620, China
e-mail: fengchaomed@sina.com

H. Wang · S. Chen (✉)
College of Materials Science and Engineering, Donghua University, Shanghai 201620, China
e-mail: chensy@dhu.edu.cn

Y. Han (ed.), *Advanced Functional Materials*,
https://doi.org/10.1007/978-981-13-0110-0_49

Introduction

Urinary tract defects are common clinical problem that caused by inflammation, trauma, malignancy and congenital diseases [1]. Many attempts have been made to develop the urethra rebuilding. The rapidly expanding of tissue engineered urinary tract grafts as an alternative to autologous grafts has attracted much attention in recent years. Tissue engineering (TE) approaches provide an effective strategy for developing functional substitutes to repair or replace failing urethra [2].

In TE technique, biomaterials play a critical role as scaffolds to provide three-dimensional microenvironment for support cell spreading, proliferation, migration, infiltration, vascularization and tissue ingrowth [3]. Over the past several decades, various natural and synthetic biomaterials, including collagen, small intestinal submucosa (SIS) [4, 5], bladder acellular matrix (BAM) [6], silk fibroin [7], polylactic acid (PLA) and polyglycolic acid (PGA) have been used to reconstruct the urinary tract [8]. However, some limitations still exist in used urethral TE strategies, such as stricture recurrence, fistula formation and foreign body responses [9–11].

Nowadays, much interest has been given on the new generation of 3D fibrous-based scaffolds with multi-scale structure to biomimetic the native urethral tissue [12–14]. The design of 3D scaffolds is still far from fulfilling all requirements. It demands the understanding and regulation of multiple variables that ranging from the macroscale to the micro- and nano-scales. Those structural features can have significant effects on cell behaviors in normal and regenerative tissue. It has been proved the nanofibrous materials with high porosity and large surface area can influence cell-material interaction facilitate mass transport, and promote cell behaviors such as cell attachment, proliferation and expression of matrix components [15]. Furthermore, the individual 3D scaffold with urethral shape might shorten the operative time and enhance the regeneration of urethral tissue.

The purpose of this work is to demonstrate the feasibility of constructing the urethral TE scaffolds that possess multi-scale structures features composed of macroscopic morphology and microporous structure with nanofibers for more biomimetic the native urethral extracellular matrix (ECM).

Materials and Methods

Reconstruction of 3D Digital Model of Urethral Tissue. A 3D digital model of a human urethra reconstructed from the medical imaging technology of ultrasonography. One healthy male 34 year-old volunteer was examined by an anterior urethral ultrasonography (Aplio 500, Toshiba, 7–14 MHz). The hand-held transducer was placed on the ventral surface of the penis and the perineum. The transverse multi-section scans were then performed with DICOM file format. Those 2D images were uploaded onto a digital imaging management system, and added together to create the 3D model reconstruction (WRML file).

Analysis of the Micro/Nano Structure of Urethral Acellular Matrix. Three human urethras were obtained from 23 to 62 years old (mean 33) male patient, accept the radical resection of penis carcinoma. Those urethral tissues were treated by the decellularized procedure. Briefly, tissues were cut into sections with a thickness of approximately 10 mm, treated with 0.1% (v/v) ammonium hydroxide and 0.5 wt% Triton X-100 in ultrapure water for 14 days at 37 °C, and the medium was changed every 3 days. To remove the excess chemical, the samples were rinsed with a lot of ultrapure water for 3 times. After that, the treated tissues were frozen at −20 °C for 12 h, and lyophilized for 24 h. The micro- and nano- structural characteristics of the urethral acellular matrix were studied by using FF-SEM.

Preparation of Multi-scale Galetin/Silk-BC Scaffold for Urethral Tissue Engineering. By using the indirect 3D printing technique, a negative silicone mold of urethra was constructed. A urethra replica was produced by the fused deposition modelling (FDM) printing of polylactic acid (PLA) using a desktop FDM 3D printer (Flashforge Creator, China). The stereolithography (STL) file of the 3D digital urethral model was processed in software and the G-code file for the FDM 3D printer was generated. Then, the negative urethra mold was manufactured by casting silicone solution around the PLA urethra model. The silicone mold comprise of two parts for relatively simple retrieval of the urethra model.

The preparation of silk fibroin and gelatin solutions have been introduced in our recent works [7]. Briefly, the extracted silk was dissolved in 9.3 M LiBr solution at 60 °C. The silk solution was diluted with ultrapure water and filtered. Then, the solution was dialyzed against ultrapure water using dialysis tube. (molecular weight cutoff of 1.2–1.4 × 10^4 Da). Finally, the dialysis tube was immerse into a 20 wt% PEG (molecular weight: 2 × 10^4 Da) solution for 6 h to concentrate. After that, the silk fibroin solution with concentration of about 10 wt% was stored at 4 °C. 2 wt% Gelatin was dissolved in ultrapure water at 37 °C, with constant stirring for 1 h.

After that, gelatin and silk fibroin solutions were blended at pH 7.0 with 70/30 ratio, subsequently, the mixed solution was poured into the negative silicone mold and frozen at −20 °C for 12 h. The frozen sample with the negative mold was freeze-dried for 24 h to manufacture a 3D urethral-shaped microporous gelatin/silk scaffold (Gel/silk scaffold) and stored at 4 °C until use.

The template biosynthesis method was used to fabricate the multi-scale urethral scaffold by cultivation *Gluconacetobacter xylinus* using the Gel/silk scaffold as a template [16]. About 2 mL bacterial suspensions with the culture medium at the initial bacterial density of 3.8 × 10^5 cells/mL were seeded onto the Gel/silk scaffold. The mixed complex was incubated statically at 30 °C, for 7 days to prepare gelain/silk-BC composite scaffolds (Gel/silk-BC scaffold). The multi-scale urethral scaffold was lyophilized to obtain 3D urethral-shaped microporous nanofibrous scaffolds and stored at 4 °C before further usage.

Morphology and Structure Characterization. The morphology of the samples were characterized by field emission scanning electron microscopy (FE-SEM). The urethral acellular matrix and Gel/silk-BC scaffold were lyophilized for 24 h, cutted into small pieces. All samples were coated with an evaporated gold thin film and observed with a Hitachi S-4800 (Hitachi, Japan). From the FE-SEM images, the micropore size, nanofiber diameter and distributions were obtained by ImageJ software for at least 200 individual data points.

Results and Discussion

The transverse multi-section ultrasound images of normal penile urethra were obtained by using the ultrasonography (Fig. 1a). Those section images with the interval of 100 μm thickness were exported to the DICOM file format. Then, the contour of urethral tissue was extracted from the ultrasound image (Fig. 1b). The 3D digital model of normal urethra with length of 10 mm was reconstructed based on 101 images of urethral ultrasonography (Fig. 1c).

The gross view of the native urethra shown in Fig. 2a, it has an asymmetric tubular cylinder with the wall thickness gradually thickened. Meanwhile, urethral acellular matrix is a white porous sponge with structural features of three-dimensionally open and interconnected micropore (Fig. 2b). Furthermore, the micro- and nano- structural characteristics of the urethral acellular matrix were observed by using FE-SEM. The cross-section morphology of the urethral acellular matrix with different magnifications was revealed and the distributions of micropore and nanofiber size were analyzed by by ImageJ software (Fig. 3). Actually, the urethal acellular matrix with 3D microporous nanofibrous structure consists mainly of collagen fibrils in nanocale. The nanofiber of average diameter is 30.2 ± 6.6 nm (Fig. 3c, d), which are assembled into highly interconnected micropore (146 ± 30 μm).

From those results, we know that the native urethral extracellular matrix exhibits multi-scale structures for an asymmetric tubular cylinder with microporous nanofibrous structure. In order to fabricate the multi-scale structure of tissue engineered urethral scaffolds, the indirect 3D printing technology and template biosynthesis method were used to control the multi-scale structures with defined macro-, micro- and nanostructure.

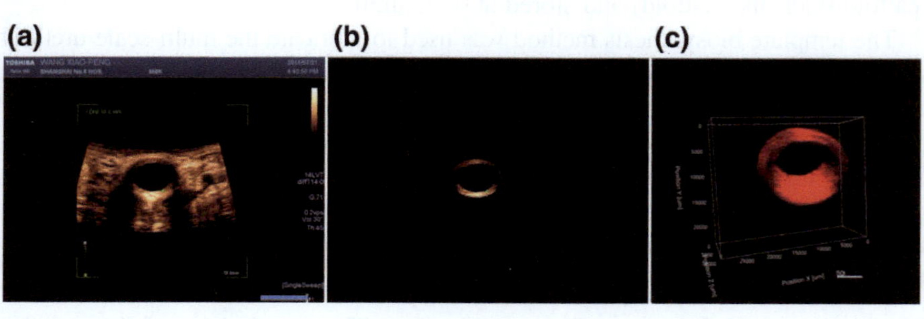

Fig. 1 **a** Urethral transverse image from ultrasound, **b** the contour of urethral tissue was extracted from the ultrasound image, **c** the reconstructed 3D digital urethra

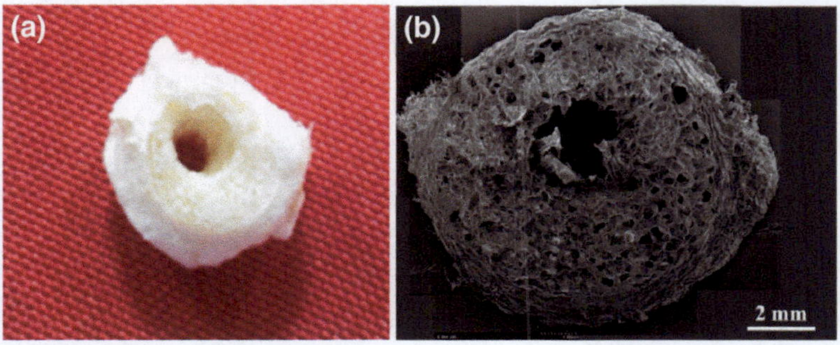

Fig. 2 Digital photograph (**a**) and SEM image (**b**) of urethral acellular matrix

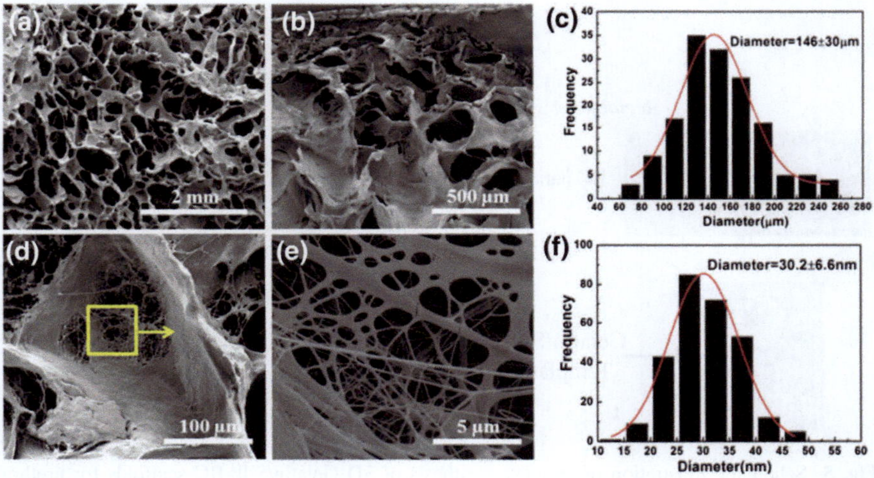

Fig. 3 FE-SEM images of urethral acellular matrix with different magnifications (**a**, **b**) and the micropore diameter distribution (**c**); higher magnification of the micropore wall in urethral acellular matrix (**d**, **e**) and the nanofiber diameter distribution (**f**)

Using the indirect 3D printing technique, the urethral negative silicone mold was constructed (Fig. 4). Form the 3D digital model and PLA urethra model, the macro-size was measured. The urethra segment, with an outer diameter of 13.4 ± 0.6 mm, inner diameter of 8.9 ± 0.7 mm, wall thickness of 1.4 ± 0.4 to 3.1 ± 0.4 mm, and cylinder length of 10 mm, is an asymmetric tubular cylinder.

Schematic illustration of template synthesis of 3D multi-scale urethral composite scaffolds is outlined in Fig. 5. In our previous studies, both silk fibrin and gelatin exhibited excellent biocompatibility and biodegradability [7]. It has been proved to be useful for urinary tract reconstruction. Furthermore, 3D microporous nanofibrous Gel/BC composite scaffolds fabricated by the template biosynthesis method have

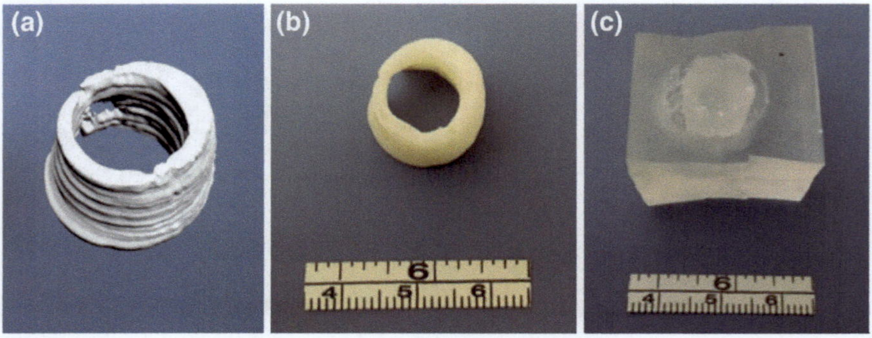

Fig. 4 **a** 3D digital model of the urethra; **b** 3D printed urethral model by PLA, **c** Silicone urethral mould

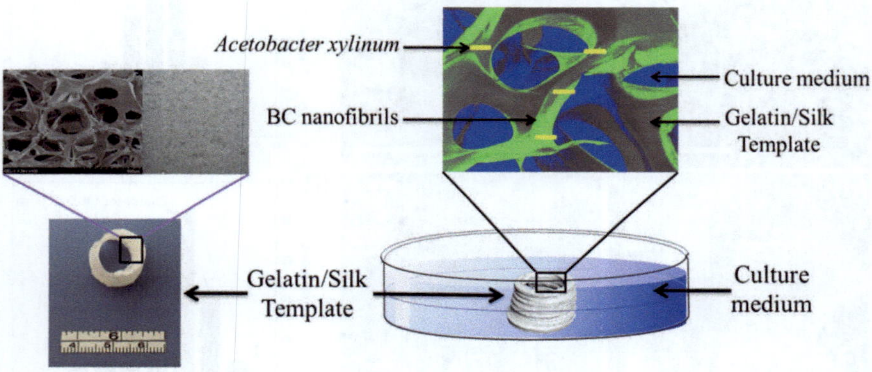

Fig. 5 Schematic illustration of template synthesis of 3D Gelatin/Silk-BC scaffolds for urethral tissue engineering

been investigated in our previous research. The results shown that the Gel/BC scaffolds exhibit good biocompatibility, promoting cell attachment, infiltration and vascularization [16].

The microporous Gel/silk composite scaffolds can be obtained by lyophilization, and the pore architecture can be regulated by the freezing temperature, concentration, and mixture ratio. In this study, we chosen the mixture ration of gelatin/silk with 70/30, and the freezing temperature is established at −20 °C. As shown in Fig. 5, gelatin/silk template is a tubular cylinder with the highly interconnected micropore. When the bacterial suspensions of *Gluconacetobacter xylinus* were seeded onto the gelatin/silk microporous scaffold, bacteria subsequently adhered onto the inwall of the template scaffold, and then nanofibers of cellulose were biosynthesized for 7 day.

Finally, a 3D microporous nanofibrous scaffold of Gel/silk-BC with urethral shape, high microporous and nanofibers was manufacture by template biosynthesis method (Fig. 6a inset). The cross profile morphology of Gel/silk-BC scaffold were observed by FE-SEM. Meanwhile, the micro-pore size and nanofiber diameter were analyzed by ImageJ analysis. As shown in Fig. 6a, the multi-scale scaffold of Gel/silk-BC is comprised by open and interconnected pore in 3D with micropore size (140 ± 34 μm), which is close to the native urethra (146 ± 30 μm).

The Gel/silk-BC urethral scaffold with 3D microporous nanofibrous structure was manufactured by forming BC nanofibrous network on the surface of micropore wall throughout the multi-scale scaffold. Under higher magnification, the nanoscale roughness is proved to be caused by BC nanofibrous network (Fig. 6c) and this 3D non-woven nanofibrous matrix architecture of BC showed close resemblance to the native urethra ECM. The average nanofiber diameter of BC in the Gel/silk-BC is 24.6 ± 5.0 nm, which has no statistically significant differences with the native urethra (30.2 ± 6.6 nm). The mechanical performance of Gel/silk-BC urethral scaffold was determined at 37 °C in distilled water, and the preliminary result of the tensile stress can reach to 0.54 ± 0.04 MPa.

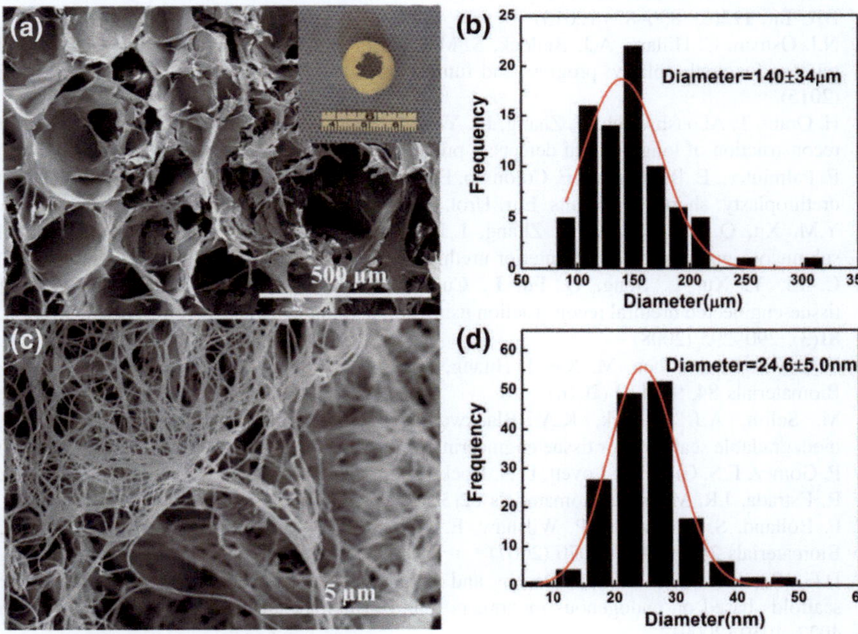

Fig. 6 FE-SEM images of 3D Gelatin/Silk-BC scaffolds with high porous structure (**a** ×100 magnification, Inset: digital photograph) and the micropore diameter distribution (**b**); higher magnification of the micropore wall in Gelatin/Silk-BC scaffolds (**c**) and the nanofiber diameter distribution (**d**)

Conclusions

Our study indicate that the feasibility of fabrication multi-scale 3D urethral scaffold, which could structures similar to the native urethra. Furthermore, the multi-scale structure of this scaffold could be controlled by adjusting the fabrication process according to the demanding of multi-scale hierarchical structural design of native tissue. It provides a novel method in engineering the 3D cell microenvironment for urethral tissue. By this method, it is possible to design and manufacture 3D nanofibrous-based scaffolds with multi-scale structure to biomimetic the native tissue or organ architecture.

Acknowledgements The authors acknowledge financial support from the National Natural Science Foundation of China (51703078 and 51573024) and the Youth Foundation of Jiaxing University.

References

1. D.M. Stein, D.J. Thum, G. Barbagli, S. Kulkarni, S. Sansalone, A. Pardeshi, C.M. Gonzalez, BJU Int. **112**(6), 830–834 (2013)
2. N.I. Osman, C. Hillary, A.J. Bullock, S. MacNeil, C.R. Chapple, Tissue engineered buccal mucosa for urethroplasty: progress and future directions. Adv. Drug Deliv. Rev. **82**, 69–76 (2015)
3. H. Orabi, T. Aboushwareb, Y. Zhang, J.J. Yoo, A. Atala, Cell-seeded tubularized scaffolds for reconstruction of long urethral defects: a preclinical study. Eur. Urol. **63**(3), 531–538 (2013)
4. E. Palminteri, E. Berdondini, F. Colombo, E. Austoni, Small intestinal submucosa (SIS) graft urethroplasty: short-term results. Eur. Urol. **51**(6), 1695–1701 (2007)
5. Y.M. Xu, Q. Fu, Y.L. Sa, J. Zhang, L.J. Song, C. Feng, Outcome of small intestinal submucosa graft for repair of anterior urethral strictures. Int. J. Urol. **20**(6), 622–629 (2013)
6. C. Li, Y. Xu, L. Song, Q. Fu, L. Cui, S. Yin, Preliminary experimental study of tissue-engineered urethral reconstruction using oral keratinocytes seeded on BAMG. Urol. Int. **81**(3), 290–295 (2008)
7. X. Lv, Z. Li, S. Chen, M. Xie, J. Huang, X. Peng, R. Yang, H. Wang, Y. Xu, C. Feng, Biomaterials **84**, 99–110 (2016)
8. M. Selim, A.J. Bullock, K.A. Blackwood, C.R. Chapple, S. MacNeil, Developing biodegradable scaffolds for tissue engineering of the urethra. BJU Int. **107**(2), 296–302 (2011)
9. P. Gomez, E.S. Gil, M.L. Lovett, D.N. Rockwood, D. Di Vizio, D.L. Kaplan, R.M. Adam, C. R. Estrada, J.R. Mauney, Biomaterials **32**(30), 7562–7570 (2011)
10. F. Bolland, S. Korossis, S.P. Wilshaw, E. Ingham, J. Fisher, J.N. Kearney, J. Southgate, Biomaterials **28**(6), 1061–1070 (2007)
11. D.G. Barrett, M.N. Yousaf, Design and applications of biodegradable polyester tissue scaffolds based on endogenous monomers found in human metabolism. Molecules **14**(10), 4022–4050 (2009)
12. J.J. Green, J.H. Elisseeff, Mimicking biological functionality with polymers for biomedical applications. Nature **540**(7633), 386–394 (2016)

13. S.M. Oliveira, R.L. Reis, J.F. Mano, Towards the design of 3D multiscale instructive tissue engineering constructs: current approaches and trends. Biotechnol. Adv. **33**(6), 842–855 (2015)
14. K. Wang, X. Wang, C. Han, W. Hou, J. Wang, L. Chen, Y. Luo, Adv. Mater. **29**(2) (2017)
15. X. Wang, B. Ding, B. Li, Biomimetic electrospun nanofibrous structures for tissue engineering. Mater. Today **16**(6), 229–241 (2013)
16. Z. Li, X. Lv, S. Chen, B. Wang, C. Feng, Y. Xu, H. Wang, RSC Adv. **6**(48), 42229–42239 (2016)

Peripheral Nerve Repair with Electrospinning Composite Conduit

Fei Lin, Xinyu Wang, Yiyu Wang, Rong Zhu, Yuanjing Hou, Zhengwei Cai, Yi Li and Zimba Bhahat

Abstract Peripheral nerve injury is a common disabling disease, but autograft can't meet the real needs due to its limitations. As a single material nerve conduit can't repair nerve damage effectively, the study of composite nerve conduit is significantly important. In this article, a biomimetic artificial nerve conduit was prepared with PDLLA/β-TCP/I-collagen(c) via electrospinning method. The material properties and biocompatibility of PDLLA/β-TCP/I-collagen composite nerve conduit were studied, and then the feasibility of PDLLA/β-TCP/I-collagen conduit in repairing 10 mm rat sciatic nerve defects for 3 months was explored. PDLLA(a), PDLLA/β-TCP(b) and autograft groups(d) were set. PDLLA/β-TCP/I-collagen

F. Lin · X. Wang (✉) · Y. Wang · R. Zhu · Y. Hou · Z. Cai · Z. Bhahat
State Key Laboratory of Advanced Technology for Materials Synthesis and Processing, Wuhan, China
e-mail: wangxinyu@whut.edu.cn

F. Lin
e-mail: 792380302@qq.com

Y. Wang
e-mail: 419459797@qq.com

R. Zhu
e-mail: 610127313@qq.com

Y. Hou
e-mail: 993294455@qq.com

Z. Cai
e-mail: 1984972906@qq.com

Z. Bhahat
e-mail: bahat.zimba@yahoo.com

F. Lin · X. Wang · Y. Wang · R. Zhu · Y. Hou · Z. Cai · Z. Bhahat
Biomedical Materials and Engineering, Research Centre of Hubei Province, Wuhan, China

Y. Li
Institute of Textiles and Clothing, The Hong Kong Polytechnic University, Kowloon, China
e-mail: tcliyi@polyu.edu.hk

Y. Han (ed.), *Advanced Functional Materials*,
https://doi.org/10.1007/978-981-13-0110-0_50

showed significant porosity and low swelling ratios. PDLLA/β-TCP/I-collagen has well biocompatibility and low antigenicity through subcutaneous test for 3 weeks from the analysis of Hematoxylin-eosin and Immunohistochemical staining analysis. The muscle recovery rate in PDLLA/β-TCP/I-collagen group is better than PDLLA and PDLLA/β-TCP groups after 3 months implantation. The myelinated nerve fiber diameter and myelin sheath thickness in PDLLA/β-TCP/I-collagen is bigger than in PDLLA and PDLLA/β-TCP groups ($p < 0.05$), which is closed to autologous transplantation groups ($p > 0.05$) through the observation of transmission electron microscopy (TEM). PDLLA/β-TCP/I-collagen composite conduits were prepared for the first time which could repair the nerve defects effectively.

Keywords Peripheral nerve · Nerve conduits · Electrospinning PDLLA/β-TCP/I-collagen

Introduction

Injury to peripheral nerve may result into impaired function of the nerve or even total loss of function of the nerve depending on the extent to which the nerve and the surrounding connective tissues have been damaged. Regeneration is usually suboptimal when an adult peripheral nerve tissue is injured and poor recovery has been associated with an increase in the length of the injury gap. Short range and small area nerve injury can be repaired with autologous transplantation method, which is considered as the gold standard for peripheral nerve repair, but long distance and large area peripheral nerve injury cannot be repaired with autologous transplantation due to insufficient donors for the transplants, donor site morbidity, the sensation loss of donor area, etc. On the other hand, allografts nerve repair is more likely to be associated with tissue rejection due to the immune system; so, the allografts have lower value in peripheral nerve regeneration compared to autografts [1, 2]. It is because of such shortcomings associated with autografts and allografts that is why the artificial nerve conduits have attracted people's interest in research as the alternative way of repairing peripheral nerve injury. The use of artificial nerve conduits for nerve injury repair is a well-established and effective method of treating peripheral nerve injury although finding the suitable materials for preparing such artificial nerve conduits is a huge challenge for researchers. This is so because when selecting the materials to be used in the preparation of the artificial nerve conduits, there has to be a thorough examination and analysis of the properties of the materials in terms of their mechanical strength, biocompatibility, hydrophilicity and they should easily be formable. The synthesized artificial nerve conduits can connect and fix both ends of the defective nerves, then the axons can grow along the lumen from the proximal to the distal. Artificial nerve conduits can prevent the invasion of fibrous scar tissue and avoid the formation of neuroma. At the same time, the internal microenvironment of the artificial nerve conduit can be artificially regulated by manipulating the materials used to produce the conduits [3].

Many materials have been reported to have been used as artificial nerve conduits. In the past, biological transplants like the venous duct, intestinal mucosa, amniotic membrane and the skeletal muscle could be used to bridge defective peripheral [1, 4–6]. The main reason for using these biological implants is that they contain the basement membrane which is rich in protein and collagen that can promote Schwann cells adhesion, reproduction and migration. The disadvantage of using biological implants is that they easily collapse during the process of nerve repair. In addition, smooth hypertrophy scar tissue proliferation occurs sometimes on the regenerating nerves leading to poor repair effect. At the same time, the material source for these biological transplants are limited hence not suitable for vast applications. Due to this, people began trying to use non-degradable materials like silicone, polyethylene and Teflon as nerve conduits to repair peripheral nerve injury [7–9]. Lundborg was the first to establish the artificial nerve conduit model which could connect the defect nerves at both ends and provide the appropriate microenvironment for repairing the 10 mm sciatic nerve defect in rats [10]. The use of non-degradable nerve conduits would cause chronic nerve compression and inflammatory response once implanted in the body [11]. Furthermore, non-degradable nerve conduits need secondary surgery to remove the conduits and because of this, that is why current research focuses on biodegradable materials. Biodegradable materials can provide a temporary environment for nerve repair and later the materials undergo self-degradation as the nerve repair is being completed without the need for surgical removal of the materials. Some of the most common biodegradable materials include: chitosan, collagen, chitin and polyglycolic acid [12–15].

There are a number of methods that can be used to make artificial nerve conduits. Some of these methods include: gas foaming [16], biopolymer replication [17], freeze-drying [18] and sol-gel [19]. In order to prepare a more efficient porous biomaterial scaffold for peripheral nerve repair, there is need for selecting an effective method for preparing the nerve conduit. Electrospinning has been reported to have many advantages when it comes to preparing such biomaterial. By controlling the electrospinning parameters, bionic and semi-permeable biomaterial structures could be prepared.

In this article, a biomimetic artificial nerve conduit was prepared with PDLLA/β-TCP/I-collagen via electrospinning method. PDLLA is highly degradable and it has good mechanical properties, however, its hydrophilicity and cell compatibility are weak [11, 20, 21] and also it lacks cell recognition signal [21, 22]. I-collagen has better hydrophilicity and cell compatibility which can promote cell adhesion, proliferation and migration effectively. The composite with PDLLA and I-collagen could improve the cell affinity of PDLLA and the mechanical properties of I-collagen. The addition of β-TCP can neutralize the acidic substances produced by PDLLA and I-collagen and the degradation product of β-TCP could promote the repair of peripheral nerves.

Experimental Section

Materials. PDLLA; β-TCP; Dichloromethane; Paraformaldehyde; Ethyl acetate; Ethanol; Pentobarbital sodium; Dimethylbenzene; Neutral balsam were purchased from Sinopharm Chemical Reagent Co. Ltd (China). Phosphate buffered saline (PBS); I-collagen were purchased from Soho Biotechnology Co. Ltd. (China). Surgical suture; centrifuge tubes; hematoxylin-eosin were purchased from Wuhan goodbio technology CO. Ltd (China). H_2O_2; TNF-α; Epon 812 resin; Lead citrate were purchased from Beyotime Institute of Biotechnology (China). All other reagents were of analytical grade.

Fabrication of the Nerve Conduits. The electrospinning solution of pure PDLLA was prepared by dissolving PDLLA (w/v% = 10%) in the mixture of dichloromethane and ethyl acetate (volume ratio = 7:3). For the PDLLA/β-TCP film, extra β-TCP (w/v% = 3%) was added into the PDLLA solution. For the PDLLA/β-TCP/collagen film, the collagen (w/v% = 3%) was added into the PDLLA/β-TCP solution. The electrospinning solution was loaded into a syringe that had a blunt-end needle attached to it. The releasing of the polymer solution was controlled by a syringe pump at a feeding rate of 0.8 mL/h and DC voltage of +3 kV was applied to the needle. The ejected fibers were collected on a collector covered with aluminum foil. Negative voltage of −3 kV was applied to the collector. The distance between the needle and the collector was 15 cm. After electrospinning, the obtained polymer films were put into the vacuum drying oven for 72 h at 40 °C to remove the residual solvent. The films of pure PDLLA, PDLLA/β-TCP and PDLLA/β-TCP/collagen were labeled as sample a, b and c, respectively. These films were pasted into conduits with dichloromethane. All these conduits were sterilized with ethylene oxide ultraviolet light irradiation before further assessments.

Swelling Ratio Test. The swelling ratio were measured according to conventional gravimetric procedure. Different materials were prepared in the same shape (10 × 10 × 5) mm, each group has five parallel samples. The samples were put in a vacuum drying oven for 10 h until their weight could no longer change. Then the samples were weighed by electronic scales and marked as m_0 at first, then they were put in centrifuge tubes with deionized water and then incubated at 37 °C. Thereafter, the samples were taken out at a set time, wiped off the excess moisture on the surface with filter paper, then weighed the mass of sample as m_t before calculating the SR (Swelling Ratio) using the formula below [23]:

$$SR = \frac{m_t - m_0}{m_0} \tag{1}$$

(m_0 indicate the mass of dry scaffolds at time 0, m_t indicate the mass of wet scaffolds at time t).

Porosity. Porosity is the percentage of material pores that account for the total material volume, which is an important assessment of porous materials. In this article, the weight spectrometry method was used in testing the porosity. Different materials were prepared having the same shape (10 × 10) mm with each material having five parallel samples. The samples were then weighed after being dried and the weight was marked as Ws. The pycnometer was then filled with ethanol and the weight was measured and recorded as W_1. The dried samples were then put in a pycnometer and the air was removed by vacuum drying the oven for 1 h. After the completion of the degassing, the pycnometer was filled with ethanol and the weight was measured and marked as W_2. The sample was then taken out and the mass of the ethanol and the pycnometer was measured and marked as W_3 [24].

Material volume (V_m):

$$V_m = \frac{W_1 - W_2 + W_s}{\rho} \tag{2}$$

Material pores volume (V_p):

$$V_p = \frac{W_2 - W_3 - W_s}{\rho} \tag{3}$$

(ρ is the density of ethanol)

Material pores Porosity(θ):

$$\theta = \frac{V_p}{V_p + V_m} = \frac{W_2 - W_3 - W_s}{W_1 - W_3}. \tag{4}$$

Experimental Animal and Surgical Procedure. The study plan was approved by the Animal Care and Use Committees of Wuhan University of Technology, China and performed in accordance with the National Institutes of Health Guide for the Care and Use of Laboratory Animals (No. 85-23, revised 1996). The materials used in this study meets the requirements of medical material through the cytotoxicity test of RSC96 cells. Having satisfied the above requirements, the nerve conduits were now subjected to in vivo experimentation. Three groups of materials were cut into "0.5 cm × 0.5 cm" sizes and sterilized with ethylene oxide. 8 weeks old, female, 36 Wistar (200–250 g) rats were supplied by Disease Control and Prevention of Hubei Province (China) (license No. SCXK (E) 2015-0018) and randomly distributed into four groups [(a) PDLLA; (b) PDLLA/β-TCP; (c) PDLLA/β-TCP/I-collagen; (d) Blank control groups], each group had 3 parallel samples. The hair on the chosen site at the back of the rats was shaved cleanly after anesthetized with 10 mL/kg pentobarbital sodium. The skin is then wiped with iodine, cut and three different materials (a; b; c) were implanted into the intended places in different rats. There were no materials implanted into rats subcutaneous in blank control groups (d), other experimental steps are the same as experimental groups, stitch the wound with surgical suture. The suture was disinfected with iodine and washed with alcohol. The wounds of rats were treated and shaved on a regular basis. At 3 days, 7 days and 14 days, the rats were sacrificed and dissected for analysis. The subcutaneous tissues (0.3–0.5 cm) around the materials were collected and used in histological research. 40 wistar rats were prepared for bridging the ends of defect nerves. There were four groups of the materials used including the autograft group and each group of the materials was allocated 10 rats. A 10 mm nerve defect in each rat was bridged with different nerve conduits made from the four materials. The regenerated nerves were collected 3 months after implantation and used for lead citrate staining which can observe the morphology of myelinated nerve fiber.

Hematoxylin-eosin Staining. The subcutaneous tissues that were in contact with the implanted materials were collected from all rats including experimental groups and blank control groups after the rats were anesthetized with 10 mL/kg pentobarbital sodium. The middle part of the subcutaneous tissues was fixed with immobilization solution (containing: 1% glutaraldehyde, 0.1 M cacodylate and 1% paraformaldehyde). Then it was dehydrated (in the rank: 25, 50, 75 and 100%) and embedded with paraffin following standard histological techniques. The samples of these subcutaneous tissues were then cut into slices (4 μm) before dewaxing it in water. The slices were used for hematoxylin-eosin staining in which the slices were stained with hematoxylin and eosin.

Immunohistochemical Staining. The slices were put into an EDTA tube containing an antigen repair solution (pH = 9) to repair antigen. From the EDTA tube, they were then placed in PBS (pH = 7.4) and shaken three times on a decolorizing shaker. The samples were then put in 3% H_2O_2 solution and incubated for 25 min. The slices were then dried and signed around in circle using a pen (to prevent the antibody flow away). Then the tissue samples were evenly covered with 3% Bovine Serum Albumin (BSA) and the container was kept closed for 30 min at room temperature. Thereafter, BSA was drained and the samples were dried before being put into the humidifying box and incubated overnight at 4 °C. Then IgG was added and the samples were incubated at 20–37 °C for 20 min then they were placed in PBS (pH = 7.4) and shaken three times on a decolorizing shaker. The slices from both hematoxylin-eosin staining and immunohistochemical staining were observed and analyzed by the optical microscope (DMLP-MP30, Leica).

Muscle Recovery Rate. After the surgical operation, the rats were kept in the animal house at 23 °C. Three months later, the skin of the right leg was opened after the rats were anesthetized with 10 mL/kg pentobarbital sodium. Then the soleus muscles around the sciatic nerves were collected and weighed with electronic scales and the weight was marked as M_1. Again, using the same method, the soleus muscles mass of the left leg was also weighed and its weight was marked as M_2. The muscle recovery rate (R) was then calculated using the formula below:

$$R(\%) = \frac{M_1}{M_2} \times 100\%. \tag{5}$$

Lead Citrate Staining. The middle sections of the regenerated nerves were collected from each group randomly for transmission electron microscopy. The rats were anesthetized with 10 ml/kg pentobarbital sodium through intraperitoneal injection before observing the regenerated nerves. The regenerated nerves were washed with 0.1 M PBS, fixed with 1% glutaraldehyde in PBS at 20 °C for 2 h and dehydrated in ethanol, then embedded in Epon 812 resin. The samples were then put into an oven at 37 °C for overnight. From the oven, the samples were cut into 60–80 nm thick slices through an ultramicrotome and later stained with lead citrate. The slices were then analysed using the scanning electron microscopy (SEM, HT7700, Hitachi, Japan) to observe the morphology of the myelinated nerve fiber.

Statistical Analysis. The experimental data was expressed as mean ± standard error. The statistical significance analysis was analyzed by software SPSS10.0 (SPSS Inc, Chicago, IL, USA). Statistical significance was shown at $P \leq 0.05$.

Results

Swelling Ratio Test. The swelling ratio of different groups of materials increased continuously within the first 2 h and then tend to balance after 2 h. The figure above shows that the swelling ratio of c is the greatest compared to a and b with a P-Value of $P < 0.05$. This can be attributed to the swelling of collagen which increase the water absorption and consequently the swelling of the porous scaffold material. I-collagen has a lot of hydroxyl and carboxyl groups which are hydrophilic groups, so the addition of collagen could improve porous material hydrophilicity and swelling ratio significantly. Different materials were prepared in the same shape (10 × 10) mm, each group had five parallel samples. Swelling ratio of c was maintained at 17.71 ± 1.34% (n = 5) and it kept the stability and maintained the shape of porous scaffold material. The low swelling ratio of porous scaffold materials can absorb a lot of water without affecting the integrity of its structure. The mesh structure of the porous scaffold material not only ensures the stability of the scaffold material, but also provide the channels to transport nutrition and waste (Fig. 1).

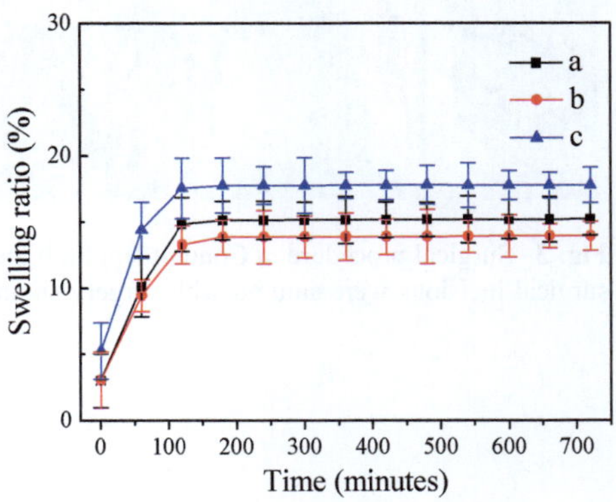

Fig. 1 Swelling ratio of different groups at 37 °C. **a** PDLLA; **b** PDLLA/β-TCP; **c** PDLLA/β-TCP/I-collagen

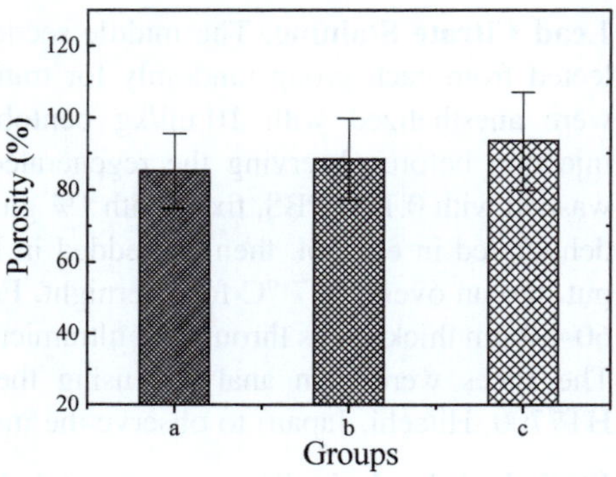

Fig. 2 Porosity of different material groups. **a** PDLLA; **b** PDLLA/β-TCP; **c** PDLLA/β-TCP/I-collagen

Porosity. As shown in Fig. 2, the porosity of a is less than b while c has the largest porosity compared to both a and b ($P < 0.05$). This can be attributed to the instability of I-collagen during the electrospinning process which causes the pore sizes to become bigger than in materials a and b. Different materials were prepared having the same shape (10 × 10) mm with each material having five parallel samples. The porosity of a, b and c are more than 80% (n = 5) while that of c was up to 90% (n = 5) and it showed better pore structure that could promote the entry of nutrients and the excretion of metabolites effectively. Besides the scaffold materials having a high porosity, most of the scaffolds appeared as irregular polygons with pores directly connected to each other and this is beneficial when it comes to the transportation of nutrients and metabolites. This would promote cell adhesion, growth and migration (Fig. 3).

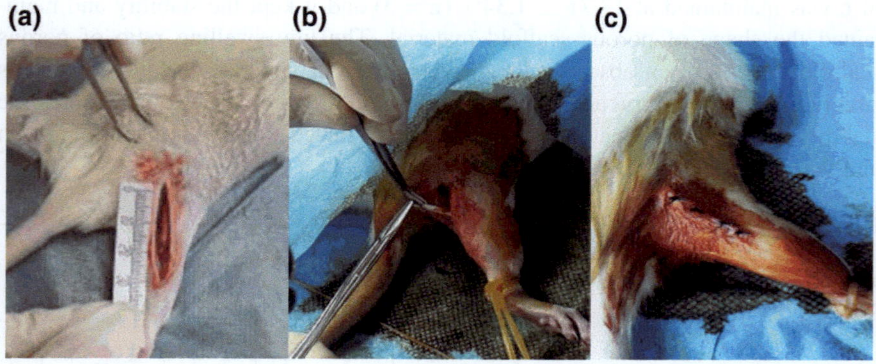

Fig. 3 Surgical procedure. **a** Conduit length; **b** the defect nerves were bridged with conduit; **c** the surgical incisions were sutured with surgery sutures

Surgical Procedure

The composite films (PDLLA, PDLLA/β-TCP, PDLLA/β-TCP/collagen) were stuck into conduits (14 mm length; 2 mm inner diameter; 0.2 mm thickness) with dichloromethane. The conduits were implanted in wistar rats that bridged the ends of the injured nerves in distal and proximal parts after the rats were anesthetized with 10 mL/kg pentobarbital sodium through intraperitoneal injection. The hair of right legs was shaved clean, daubed tincture of iodine and made skin incision before cutting off the sciatic nerve of 5 mm length until the broken ends of the remaining nerve retracts to 10 mm. The proximal and distal ends of the injured nerve were bridged by nerve conduits using the 9-0 absorbable sutures and the depth of 2 mm. Proximal and distal injured nerves were then put into the nerve conduits. In the autologous nerve groups, the defect nerves were bridged with autologous nerves and connected by 9-0 absorbable sutures (PGA, Shanghai), the surgical incisions were sutured with surgery sutures. All rats were put into animal house after disinfection and kept the room temperature at 23 °C. In this article the group of autografts were used as a control group since they are now widely used in clinics as the gold standard.

Hematoxylin-eosin. There was a series of body tissue immune response after the biomedical materials were implanted into body. Humoral immune response would occur in the early stages where B-lymphocytes would secrete products that activate neutrophils and these neutrophils infiltrates the surface of the materials. Cellular immunity become the mainly way to protect body from foreign substance after humoral immunity. T-lymphocytes identified the antigenic determinants which were exposed under the interaction of B-lymphocytes and then these antigenic determinants were cracked by the cellular immunity. At the same time, fibrocyte begin to proliferate gradually to form fibrous sac which separate the materials with body tissues. Therefore, it is necessary to evaluate the biocompatibility of the implanted biomedical material. The nucleus of cells would be dyed blue through hematoxylin-eosin staining. The stronger the antigenic nature of the implanted material, the stronger the cellular response and the more the number of cells appearing on the staining results. In hematoxylin-eosin staining, the depth of color indirectly reflects the degree of cell inflammation.

As shown in Fig. 4, after 1 week of implantation, the hematoxylin-eosin staining images of the subcutaneous tissue showed numerous cells that indicate each group of the materials had different degrees of cell response. Fibrous sac appeared in a1 and b1 which had clear demarcation with connective tissue and this is an indication of a more severe cell response. The quantity of cells in a1 and b1 is more intensive than in c1. Sample d1, which is also the blank control group, had the least quantity of cells. After 2 weeks of implantation, there was a great decrease in the quantity of cells in each group of materials. There was no typical macrophages and cysts in each group which indicates that the three groups of materials are biocompatible. By analysing the quantity of cells from the samples, it shows that there was low cell response is sample c2 which consequently implies that c2 had better biocompatibility as compared to a2.

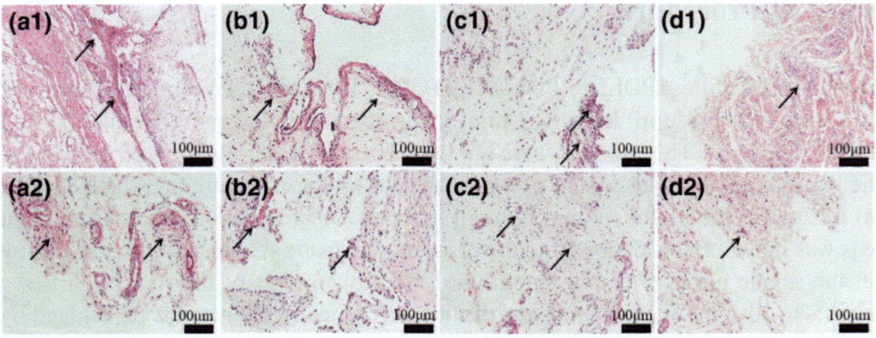

Fig. 4 Hematoxylin-eosin staining results of subcutaneous tissues within 2 weeks. **a1**, **a2** PDLLA; **b1**, **b2** PDLLA/β-TCP; **c1**, **c2** PDLLA/β-TCP/I-collagen; **d1**, **d2** autologous groups; **a1**, **b1**, **c1**, **d1** at 1 week; **a2**, **b2**, **c2**, **d2** at 2 weeks. The black arrows show various cells

Immunohistochemical Staining. Immunohistochemical staining is an effective means to detect the existence of the antigen TNF-α which is a kind of proinflammatory cytokine that plays a significant role in the body immune response. This cytokine mainly affects cell proliferation and apoptosis. The darker the color on the stained samples, then the more serious the inflammation. Higher concentrations of TNF-α is not conducive when it comes to repairing injured nerves.

Figure 5 shows immunohistochemical staining images of subcutaneous tissue samples that were collected at different times after implantation (3 days; 1 week; 2 weeks). The yellow granules represent TNF-α positive expression, with the extension of operation time. The TNF-α positive expression increase firstly and then decrease in all groups as time prolongs. On the third day after surgery, there was a significant positive expression in each group and the strength of positive expression is similar which indicate that the body began to produce obvious inflammatory responses. The seventh day after surgery, the positive expression of TNF-α in all groups was more prominent than the positive expression at 3 days which achieved the peaks. The positive expression in a2 and b2 was higher than in c2 and d2; and that of c2 did not differ very much from that of d2. Image c2 showed slight inflammatory response as compared to images a2 and b2. On the fourteenth day after surgery, the positive expression of TNF-α in all groups weakened gradually. The positive expression of c3 was the lowest when compared to that of a3 and b3 and it was closer to that of d3 which is similar to normal tissues. The positive expression in a3 and b3 still showed existence of some inflammation response which appeared as a large number of macrophages as compared to c3. So, the inflammation of c3 is minimal than that of a3 and b3 and this shows that it has good biocompatibility and security.

Muscle Recovery Rate. The condition of the soleus muscles which are found around the sciatic nerves plays a significant role in reflecting the recovery of sciatic nerve function. Soleus muscles around injured sciatic nerves were collected and the muscle recovery rate was calculated 3 months after abridgement of defect nerves.

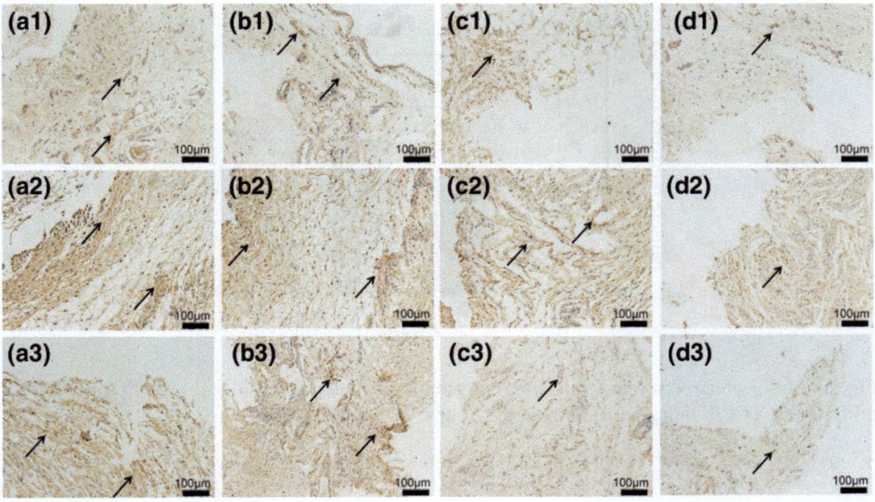

Fig. 5 Immunohistochemical staining of subcutaneous tissues in rats after sciatic nerve injury. **a1**, **a2** and **a3** are from PDLLA materials; **b1**, **b2** and **b3** are from PDLLA/β-TCP materials; **c1**, **c2** and **c3** are from PDLLA/β-TCP/I-collagen materials; **d1**, **d2** and **d3** are from autologous group f materials. Staining images **a1**, **b1**, **c1** and **d1** were taken after 3 days; **a2**, **b2**, **c2** and **d2** after 1 week; **a3**, **b3**, **c3** and **d3** after 2 weeks. Black arrows indicate the inflammatory factor

All rats had myophagism at 1 month which caused the soleus muscles to degenerate and lose weight. The recovery of the soleus muscles began from 2 months and the speed of the recovery in different conduits showed some variations. As shown in Fig. 5, the muscle recovery rate was higher in c followed by d and then a. Above all, d had the highest rate although the difference to that of c was little. It was discovered that the recovery of sciatic nerve function in c was much better as compared to that of a and b. The muscle recovery rate difference between a and c showed significance difference ($P < 0.05$); that of between b and c also showed significance difference ($P < 0.05$); while that of between c and d didn't show any significance difference ($P > 0.05$). This shows that the recovery of soleus muscles in PDLLA/β-TCP/I-collagen which was closer to that of the autologous group had significantly improved as compared to pure PDLLA and PDLLA/β-TCP (Figs. 6 and 7).

Lead Citrate Staining

Regenerated nerve microstructure was observed through the transmission electron microscopy and myelin sheath thickness and myelinated nerve fiber diameter was measured so as to aid in evaluating regenerated nerve maturation. The changes in the diameter of the myelinated nerve fiber and the thickness of the myelin sheath

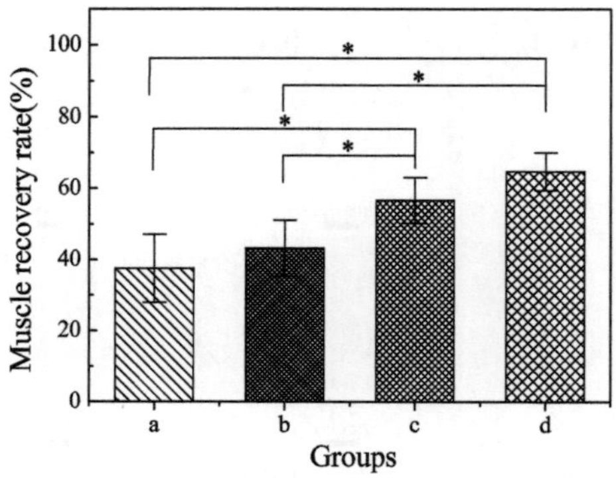

Fig. 6 Muscle recovery rate from different groups of materials. **a** PDLLA; **b** PDLLA/β-TCP; **c** PDLLA/β-TCP/I-collagen; **d** autologous groups. *$P < 0.05$

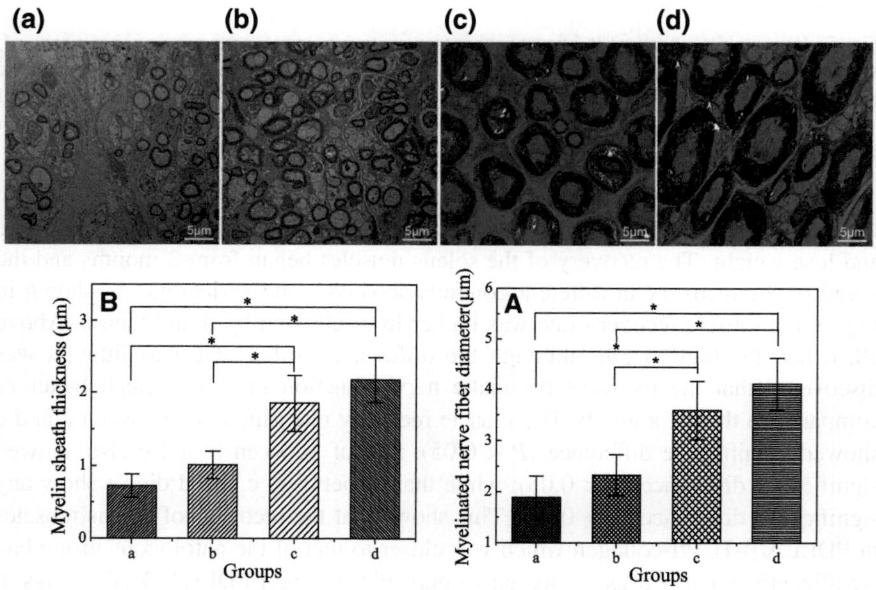

Fig. 7 Lead citrate staining of regenerate nerves at 3 months after sciatic nerve injury. **a** PDLLA; **b** PDLLA/β-TCP; **c** PDLLA/β-TCP/I-collagen; **d** autologous groups; **A** myelinated nerve fiber diameter; **B** myelin sheath thickness. *$P < 0.05$

reflect the speed of recovery in these injured sciatic nerves. It was found that samples c and d had a large number of nerve fibers, but few nerve fibers were found in a and b. Furthermore, the myelinated nerve fibers in c which were also closer to those in d appeared larger than those in a and b. The morphology of myelinated nerve fibers in c is more regular, round and their size well-proportional than a and b.

The myelinated nerve fiber diameter in c was significantly greater as compared to that of a and b ($P < 0.05$), but it was less than that of d ($P < 0.05$). Similar, then myelin sheath thickness in c was significantly greater than that from a and b ($P < 0.05$), but it was closer to d ($P > 0.05$). By analyzing both the myelinated nerve fiber diameter and the myelin sheath thickness, it indicates that the materials for sample c has a better recovery in nerve regeneration in relation to the materials for samples a and b. The results for sample c were closer to those for sample d with those from d being better than c.

Discussion

PDLLA/β-TCP/I-collagen composite conduit was made from PDLLA, β-TCP and I-collagen, which was prepared via electrospinning. PDLLA is widely used as biomaterials and tissue engineering scaffold, due to its well biocompatibility and mechanical properties. However, the drawback in cellular affinity limits its application in biomaterials and tissue engineering. Li et al. [25] electrospun P(LLA-CL) nanofibers and PLLA nanofiber yarns used in nerve repair, but the produced materials showed poor cell compatibility. So, the addition of Collagen has shown to be of importance as it improves the cell compatibility of nerve conduits. Wu et al. [26] prepared conduits (LC-YE-PLGA NGC) that promoted SCs proliferation and migration which contain laminin. Collagen has a lot of laminin that has better cellular affinity that can mediate cell adhesion, migration and proliferation. Low antigenicity and cytotoxicity of collagen plays an important role in vivo in that it had no obvious rejection reaction between tissues and nerve conduits. On the other hand, the mechanical properties of collagen have shown to be poor [27] which can't sustain the morphology and size of the material in nerve conduits. The ideal nerve conduit should also have appropriate mechanical properties to provide a suitable stress environment. In addition, such conduit should be porous and permeable to permit the ingress of cells and nutrients [28]. Since β-TCP can upregulates the mRNA expression of cytoskeletal protein and neutralize the acid substance, β-TCP composites with PDLLA and collagen can adjust the pH of microenvironment in nerve conduits [29]. Busra Mammadov et al. electrospun glycosaminoglycan and laminin mimetic peptide nanofiber gels which present a promising treatment option for peripheral nerve injuries and showed the advantages of electrospinning [30]. The conduits prepared via electrospinning method have a larger porosity than traditional method. I-collagen has a lot of hydroxyl and carboxyl groups which are hydrophilic groups, so the addition of collagen could improve porous material hydrophilicity and swelling ratio significantly. I-collagen has a better cell compatibility that is conducive to reduce immune rejection and promote the repair of nerve defects than PDLLA and PDLLA/β-TCP.

Conclusions

PDLLA/β-TCP/I-collagen composite conduit which has well biodegradability and cellular compatibility was prepared successfully via electrospinning. PDLLA/β-TCP/I-collagen exhibits long term preservation of shape during swelling ratio test. This can protect the artificial nerve conduit microenvironment and maintain the pore structure that is conducive for nerve regeneration. High porosity of PDLLA/β-TCP/I-collagen has the advantage in promoting the transportation of nutrients and metabolites and this is beneficial to cell adhesion and growth. Noticeably, PDLLA/β-TCP/I-collagen group has better biocompatibility compared to PDLLA and PDLLA/β-TCP groups when implanted into subcutaneous tissue as it shows lower inflammatory response after 3 weeks of implantation. The muscle recovery rate in PDLLA/β-TCP/I-collagen group is greater than PDLLA and PDLLA/β-TCP groups, which is close to autologous groups at 3 months. Furthermore, myelinated nerve fiber diameter and myelin sheath thickness of regenerate nerves in PDLLA/β-TCP/I-collagen are larger than PDLLA and PDLLA/β-TCP groups, which is also close to autologous groups. Experimental studies have shown that PDLLA/β-TCP/I-collagen has well neurological repair effect which is close to autologous groups. This shows that DLLA/β-TCP/I-collagen artificial nerve conduits will be a potential candidate for repairing peripheral nerve injury due to its cytocompatibility characteristics coupled with more other advantages that it has.

Acknowledgements This work was supported by the Hong Kong, Macao and Taiwan Science & Technology Cooperation Program of China (No. 2015DFH30180), the Key Technology Research Plan of Wuhan Municipality (No. 2014060202010120) and the Science and Technology Support Program of Hubei Province (No. 2015BCE022).

References

1. M.J. Moore, J.A. Friedman, E.B. Lewellyn et al., Biomaterials **27**(3), 419 (2006)
2. P. Prang, R. Müller, A. Eljaouhari et al., Biomaterials **27**(19), 3560 (2006)
3. V.B. Doolabh, M.C. Hertl, S.E. Mackinnon, Rev. Neurosci. **7**(1), 47 (1996)
4. S. Leungwattanakij, T.J Bivalacqua et al. Br. J. Urol. Int. **92**(1), 119 (2003)
5. M.F. Meek, A.S. Varejão, S. Geuna, Tissue Eng. **10**(8), 1027 (2004)
6. Y. Su, B.F. Zeng, C.Q. Zhang et al., Brain Res. **1145**(1), 41 (2007)
7. S. Saneinejad, M.S. Shoichet, J. Biomed. Mater. Res. **50**(4), 465 (2000)
8. B.L. Young, P. Begovac, D.G. Stuart et al., J. Neurosci. Methods **10**(1), 51 (1984)
9. M. Borkenhagen, R.C. Stoll, P. Neuenschwander et al., Biomaterials **19**(23), 2155 (1998)
10. G. Lundborg, F.M. Longo, S. Varon, Brain Res. **232**(1), 157 (1982)
11. C.A. Heath, G.E. Rutkowski, Trends Biotechnol. **16**(4), 163 (1998)
12. G.R. Evans, K. Brandt, A.D. Niederbichler et al., J. Biomater. Sci. Polym. Ed. **11**(8), 869 (2000)
13. N. Sinis, A. Kraus, N. Tselis et al., J. Brachial Plexus Peripheral Nerve Injury **4**(1), 19 (2009)
14. R.C. Young, M. Wiberg, G. Terenghi, Br. J. Plast. Surg. **55**(3), 235 (2002)
15. M.D. Bender, J.M. Bennett, R.L. Waddell et al., Biomaterials **25**(7), 1269 (2004)
16. J.R Jones, L.L. Hench. J. Biomed. Mater. Res. Part B Appl. Biomater. **68b**(1), 36 (2004)

17. M. Azami, F. Moztarzadeh, M. Tahriri, J. Porous Mater. **17**(3), 313 (2010)
18. H. Fu, Q. Fu, N. Zhou et al., Mater. Sci. Eng. C **29**(7), 2275 (2009)
19. J.R. Jones, S. Ahir, L.L. Hench, J. Sol-Gel. Sci. Technol. **29**(3), 179 (2004)
20. G.R. Evans, K. Brandt, M.S. Widmer et al., Biomaterials **20**(12), 1109 (1999)
21. E. Lieb, M. Hacker, J. Tessmar et al., Biomaterials **26**(15), 2333 (2005)
22. J.F. Alvarez-Barreto, M.C. Shreve et al., Tissue Eng. **13**(6), 1205 (2007)
23. B.B. Mandal, S. Kapoor, S.C. Kundu, Biomaterials **30**(14), 2826 (2009)
24. R. Zhang, P.X. Ma, J. Biomed. Mater. Res. **44**(4), 446 (1999)
25. D. Li, X. Pan et al., J. Mater. Chem. B **45**(3), 8823 (2015)
26. T. Wu, D. Li et al. J. Mater. Chem. B **5**(1), 3186 (2015)
27. C. Gardin, L. Ferroni et al., Int. J. Mol. Sci. **13**(5), 6452 (2012)
28. B. Sun, X.J. Jiang et al. J. Mater. Chem. B **27**(3), 5389 (2015)
29. T. Qiu, Y. Yin et al., J. Biomed. Mater. Res. Part A **102**(10), 3734 (2014)
30. B. Mammadov, M. Sever et al., RSC Adv. **112**(6), 110535 (2016)

Preparation and Characterization of Mg–Zn–Al LDHs Nanoplates via In Situ Growth on Mg–Zn–Zr–Sr Alloy

Wei Wang, Xiao Li, Yun Zhao and Minfang Chen

Abstract Magnesium (Mg) and its alloy have been regarded as revolutionary bioabsorbable materials. However, high corrosion activity limits their applications as biomaterials. To improve the corrosion resistance and in vitro degradation of Mg–Zn–Zr–Sr alloy (MZZS), Mg–Al hydrotalcite (LDHs) was fabricated through in situ growing approach by hydrothermal treatment. The characteristics of the coatings were investigated using SEM, XRD, FT-IR and EDS. The corrosion resistance of the LDH coatings was also studied using potentiodynamic polarization and hydrogen evolution test. The results demonstrated that the LDHs coatings exhibited a promising application of improving corrosion resistance and degradability in vitro.

Keywords Magnesium alloy · Hydrotalcite · Corrosion resistance Biodegradability

W. Wang · X. Li · Y. Zhao · M. Chen (✉)
School of Materials Science and Engineering,
Tianjin University of Technology, Tianjin 300384, China
e-mail: mfchentj@126.com

W. Wang
e-mail: wangweiyincta1@yeah.net

X. Li
e-mail: lixiaotyang@126.com

Y. Zhao
e-mail: yun_zhaotju@163.com

Y. Zhao · M. Chen
Key Laboratory of Display Materials and Photoelectric Device
(Ministry of Education), Tianjin 300384, China

Y. Han (ed.), *Advanced Functional Materials*,
https://doi.org/10.1007/978-981-13-0110-0_51

Introduction

In recent years, there has been a rising tendency to replace the conventional implant materials (e.g., titanium, stainless steel, bioceramic material) with magnesium (Mg) and its alloys in medical aspect, characterized as 'revolutionary' devices for orthopedic implants, due to their low density, high strength-to-weight ratio, capability of biodegradation and low Young modulus similar to natural human bone. Furthermore, Mg is the macroelement in human beings, which has an indispensible physiological effect [1]. However, fast corrosion rate can result in undesired loss of mechanical properties prior to predesigned service duration. Besides, the chemically active behavior could cause local accumulation of OH^- and hydrogen gas bubbles, which induce tissue inflammation and prolong the osseointegration, even second surgery [2–6].

Surface treatment has been used to tailor the corrosion rate of magnesium-based alloys [3]. Several methods have been adopted successfully, such as chemical conversion coating [7], micro-arc oxidation (MAO) [8, 9], polymer coatings [10], and sol-gel coatings [11]. Hydrothermal process is simple and comparatively cheap method to prepare magnesium hydroxide coatings on Mg alloys. Unfortunately, pitting corrosion caused by $Mg(OH)_2$ and aggressive species (such as Cl^-) deteriorates the coating integrity, thus greatly shorten the service time. Layered double hydroxides (LDHs) are a group of promising lamellar hydroxides that consist of positively charged hydromagnesite-like host layers and anions as corrosion inhibitor. LDHs can be expressed by the chemical formula: $\left[M^{2+}_{1-x}, M^{3+}_{x}(OH)_2\right]^{x+} A_{x/2^{n-}} \cdot mH_2O$, where the cations M^{2+} and M^{3+} reside in the octahedral holes in a brucite-like layer, and the anions A^{n-} is positioned in the hydrated interlayer galleries [12–15]. Recently, researchers have showed considerable interests on fabrication of the LDHs coatings for improving the corrosion resistance of degradable metals by simple hydrothermal process (e.g., magnesium and its alloy). Lin developed in situ LDHs films on AZ91 alloys in aqueous HCO_3^-/CO_3^- medium and suggested coated samples exhibit better corrosion resistance [16]. Chen adopted the in situ method to prepare Mg–Al LDH on AZ31 alloy and then modified the LDH coatings with phytic acid [17]. Zeng fabricated LDHs on AZ31 alloy by a urea hydrolysis method, and LDH film significantly improved the corrosion resistance of AZ31 alloy [18]. Ishizaki prepared $Mg(OH)_2$/Mg–Al LDHs composite coating on AMCa602 alloy by steam coating with aid of a self-designed Teflon-line autoclave. However, no interlaced lotus leaf-like structures were found in spite of characteristic peaks of (003) and (006) attributed to LDH can be observed. Their group further introduced Al source in order to accelerate LDH nucleation, and the correlation between Mg–Al LDH content, bath duration, and protection efficiency were discussed in detail [19]. Liu demonstrated a facile way to synthesize corrosion resistant LDH film on AZ91D alloy by a similar process which confers improved anticorrosion [20]. In summary, the substrate material chosen for LDH coating fabrication fall into AZ series or Al-rich alloy. In addition, it is noteworthy to realize that Zinc (Zn) content can react with Al^{3+} to grow Zn–Al LDH incorporated into conversion coating [21]. Unfortunately, although the biocompatibility of LDHs have been confirmed with respect to biological nanoparticles and drug delivery system in recent years, LDH coating on biomedical alloys, Al-free Mg alloys in particular. Moreover, zinc (Zn) can also be introduced into LDH structure, which have been neglected in low Zn content alloy.

In our previous work, a novel Mg-based alloy, Mg–Zn–Zr–Sr alloy (MZZS) was fabricated and discussed with respect to mechanical properties, corrosion resistance, and cytocompatibility [22]. Compared with Mg–Zn–Zr series alloy (MZZ) investigated by our group, it showed enhanced antibacterial properties, biocompatibility and proper mechanical properties. However, in vitro test disclosed a decline in corrosion resistance of MZZS alloy, thus proper surface modification is necessary. In this study, $Mg(OH)_2$/Mg–Al LDH film was synthesized on the surface of a biomedical Mg–Zn–Zr–Sr alloy (MZZS) via a simple hydrothermal reaction using only aluminum nitrate aqueous solution. The correlation between $Al(NO_3)_3$ concentration in aqueous solution and corrosion resistance of coated samples was also investigated.

Materials and Methods

Substrate Material and LDH Film Fabrication. The MZZS alloy coupons were cut from extruded bar designed and fabricated in our previous research [22] (Table 1) with size of 10 mm × 10 mm × 2 mm. Before coated, they were ground and polished with 800#, 1500#, 3000# SiC abrasive paper progressively to achieve a smooth surface, oxide/hydroxide layer on the surface of Mg alloy until the surface was shinning, and then degreased with acetone, then washed with acetone and deionized water in an ultrasonic bath for 5 min, and dried with hair drier.

The LDH-coated samples were prepared as follow: Aluminum nitrate aqueous solutions containing 1, 10, and 100 mM (pH = 12, adjusted by 2M NaOH) were prepared and poured to Teflon-lined stainless. MZZS alloy substrates were placed vertically to the bottom of the autoclave. The reactor was heated via electric furnace at 125 °C for 48 h (denoted as HT1, 2, 3, respectively). Finally, samples were taken out after naturally cooling and rinsed with deionized water.

Coatings Characterization. The surface morphologies and cross-sectional morphologies of the coated samples were observed using field emission scanning electron microscopy (FE-SEM; ZEISS, MERLIN, Germany) equipped with energy-dispersive X-ray spectrometer (EDX, Oxford Isis). The cross-section samples were prepared by a focus ion beam (FIB, FEI NOVA-600). Pt film was

Table 1 The chemical components of MZZS alloy

	Zn (wt%)	Zr (wt%)	Sr (wt%)	Mg
Mg–Zn–Zr–Sr (MZZS)	3.2	0.8	0.3	Balance

deposited on the area to protect the area from ion damage. The crystallographic structures that developed on the samples were characterized by an glancing-angle X-ray diffraction diffract meter (GAXRD; RigakuD/Max 2500PC) with Cu target (λ = 0.154 nm, 40 kV, 40 mA) at a scanning rate of 2°/min in the glancing angle of 2°, within the range of 5°–80°. Fourier transform infrared spectroscopy (FT-IR, TENSOR) were used to characterize the chemical bonding of the coatings.

In Vitro Corrosion Behavior Evaluation. All experiments were conducted at 37 ° C in simulated body fluid (SBF). The electrolyte used to simulate in vitro conditions was composed of: 8.0 g/l NaCl, 0.35 g/l $NaHCO_3$, 0.22 g/l Kill, 0.23 g/l $KH_2PO_4{\cdot}H_2O$, 0.31 g/l $MgCl_2{\cdot}6H_2O$, 39 ml Hal (1.0 M), 0.29 g/l $CaCl_2$, 0.07 g/ L Na_2SO_4, and 6.11 g/L $(CH_2OH)CNH_2$. 1 M Hal was used to adjust the pH value to 7.4.

Hydrogen Evolution Test. H_2 evolution test was conducted in a self-design set-up. The evolution of hydrogen was tested by placing the substrates in SBF at 37 °C under an inverted funnel for 120 h time span. Briefly, samples were sterilized under UV radiation for 1 h per side. A burette was placed over the top of the funnel above the samples to capture hydrogen released. The samples were hung by fishing line. The solution level in each burette was measured each day. The ratio of the surface area to the volume of SBF was set at 20 cm^2/ml. According to the volume of hydrogen, the corrosion rate of specimens can be calculated as follows:

$$r = PVM/RTAt \tag{1}$$

where r is the corrosion rate (mg/(cm^2d)), P is standard atmospheric pressure (Pa), V is volume of hydrogen, R is 8.314 J/(molK), T is temperature (K), M is the molar mass (g/mol), A is the surface area (cm^2), and t is immersion time (day).

Immersion Test. Each MZZS, HT1, HT2 and HT3 coated specimen was molded into epoxy resin with only one side being exposed. These samples were immersed in SBF to evaluate their in vitro degradation behavior. According to ASTM G31-72, specimens were soaked into 20 ml of SBF at temperature of 37 °C for 14 days. The medium was refreshed and collected every 24 h to monitor the pH values and Mg^{2+} concentration. After immersion, the specimens were washed with deionizer water and dried at ventilation oven for 24 h. The surface morphologies and element mapping were detected by FE-SEM equipped with EDX.

Results and Discussion

Characterization of the Coatings. The surface morphologies of HT-coated samples are depicted in Fig. 1a–c and inset images shown in top left. Nanoplate-like structures were observed on all the treated samples, but this structure on the surface of HT2 was smallest and most compact. For the samples prepared with the aqueous

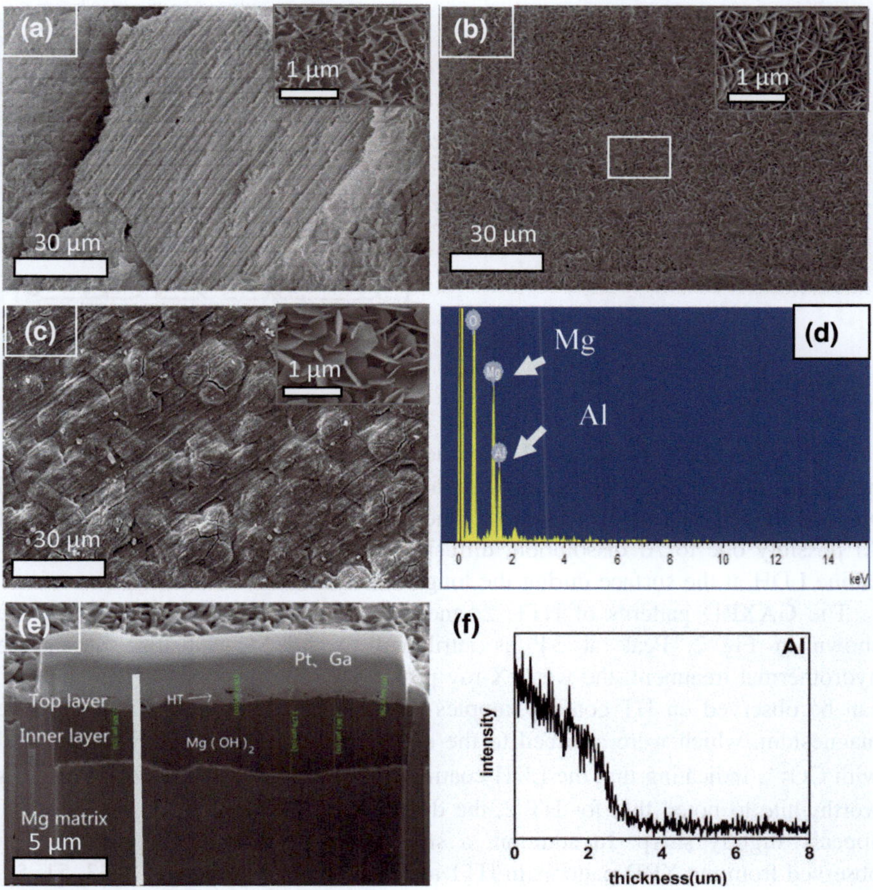

Fig. 1 Representative FE-SEM images of the HT1 (**a**), HT2 (**b**), HT3 (**c**), EDS spectrum acquired from HT2 (**d**), cross-section images of the HT2 (**e**) with the line-scan of Al (**f**)

solution containing 1 mM aluminum nitrate, a relatively rough surface with macroscopic cracks throughout the surface. Figure 1b demonstrated that the HT2 coating was compact over the entire magnesium alloy substrate compared to HT1 and 3. The small nanosheets formed nearly vertically on the substrate surface, and no trace of cracks could be found on the film. However, HT3 coating had blistered and cracked, which could peel off in corrosive ions. Figure 1e depicts the cross-sectional microstructure of the HT2 coating, demonstrating a dense, uniform coating on the substrate. The coating contains two structural layers: the dense inner thick-layer and the porous thin layer. The thickness of the coating was ca.5 µm. A distinct boundary of between the MZZS alloy and hydrothermal layer were observed from the cross-section line-scan. The elemental mapping images in

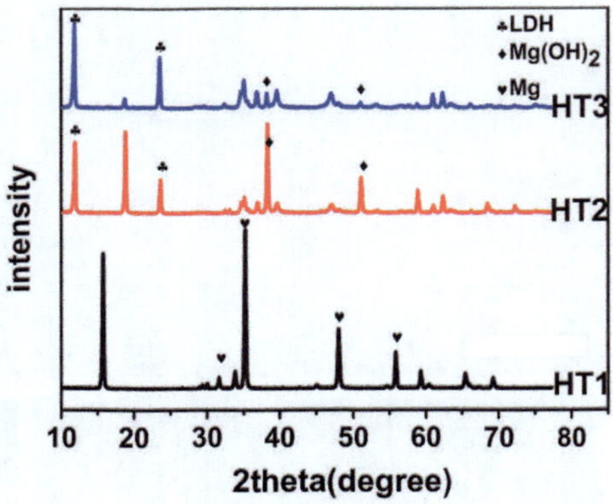

Fig. 2 GAXRD patterns of HT1-3 samples

Fig. 1d of the HT2 films reveal that the film included mostly Mg, Al and O, indicating that the films comprised primarily $Mg(OH)_2$ and Mg–Al LDH, but Al was enriched in the top layer (Fig. 1f). The inner layer had a small concentration of Al possibly due to Al dissolution, diffusion, and preferential deposition of crystalline LDH at the surface during the long time hydrothermal reaction.

The GAXRD patterns of HT1, 2, and 3 coating formed on MZZS alloy are shown in Fig. 2. Peak at 34° is attributed to the Mg substrate. After the hydrothermal treatment, the weak X-ray peaks at the 2θ positions of 12° and 23° can be observed on HT coating samples besides the strong peaks from the bare magnesium, which were ascribed to the characteristic peaks of LDH intercalated with CO_3^{2-}, indicating that the LDH coating was synthesized on MZZS alloy. It is worthwhile to noted that for HT 2, the diffraction peak of $Mg(OH)_2$ at $2\theta = 38°$ appears slightly sharp. In addition, a small amount of $Mg(OH)_2$ can also be observed from the XRD patterns in HT1 and 2, for several peaks at 18, 33, 51, 58, 62, 68, and 72° are also clearly observable in all of the GAXRD patterns. With the concentration of aluminum nitrate reaching 100 mM, the peak intensity of LDH increased gradually [16, 23–25]. These GAXRD patterns indicate that crystallization of LDH is dependent of the amount of $Al(NO_3)_3$ added to the solution.

In order to further characterize the structure of the coating, coating power was scratched from MZZS alloy. Figure 3 presents the FTIR spectra of the as-prepared Mg–Al LDH coating. As observed, for HT1, the peak at 3700 cm^{-1} is attributed to the OH stretching vibration of $Mg(OH)_2$, strong band at 500 cm^{-1} is assigned to the Mg–O stretching vibration. These characteristic adsorption signals are in consistence with $Mg(OH)_2$. However, when increasing the concentration of $Al(NO_3)_3$, more adsorption peaks are in line with Mg–Al LDH. The O–H symmetric contraction of the H_2O between layers of the LDH coating and the H_2O absorbed on the LDH surface are observed at ca.3696, 3409 and 1633 cm^{-1}. The absorption band at 1394 cm^{-1} is ascribed to the asymmetric stretching peak in C–O(CO_3^{2-}). The bands at 950, 780 and 560 cm^{-1} are assigned to Al–OH vibration mode, and

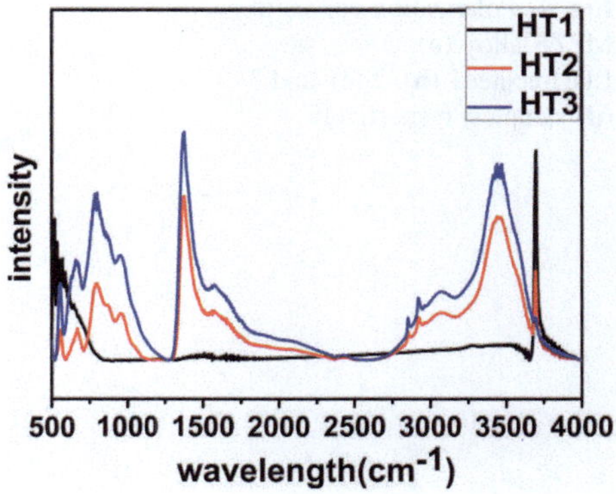

Fig. 3 ATR-FTIF spectroscopy of HT1-3 coating

peaks at 450 cm^{-1} is assigned to Mg–O vibration modes [24, 26, 27]. The FTIR results agrees well with the XRD patterns and FTIR spectra, it is concluded that the concentration of $Al(NO_3)_3$, indeed determines the phase composition, and affects the morphology and corrosion resistance. Based on the characterization results above discussed, Mg–Al LDH/$Mg(OH)_2$ bilayer structures have successfully formed on MZZS alloy by one simple process.

The Effect of Concentration of Aluminum Nitrate on Corrosion Resistance. The effect of aluminum nitrate concentration during the hydrothermal reaction was assessed. The concentration of Al^{3+} was in the range of two orders of magnitude (1, 10, 100 mM). To evaluate the corrosion resistance, electrochemical test and in vivo experiment were presented and the analysis was given below.

The potentiodynamic polarization curve test is a commonly used method employed to investigate the corrosion resistance of the conversion coating [29]. Figure 4 shows the potentiodynamic polarization curves of the prepared hydrotalcite coating in SBF. As shown in Fig. 4, the corrosion potential (E_{corr}) of the untreated MZZS alloy is −1.62 V (vs. SCE), while that of the HT1-3 samples is −1.40, −1.31 and −1.48 V (vs. SCE), respectively. According to the Tafel extrapolation method, the corrosion current density (I_{corr}) of the substrate are estimated to be 1.856×10^{-5} A cm^{-2}, whilst that of the HT1-3 sample is 1.515×10^{-6} A cm^{-2}, 1.405×10^{-7} A cm^{-2} and 1.906×10^{-6} A cm^{-2}, respectively (Table 2). It is clearly seen that the I_{corr} of the coated samples decrease by approximately one to two orders of magnitude compared to the magnesium alloy. Depending on $Al(NO_3)_3$ concentrations, E_{corr} shifts toward positive values slightly. The current densities of all the treated samples were decreased obviously compared to the treated Mg alloy, indicating that all coated samples have enhanced corrosion resistance. In this case, the film HT2 is the most corrosion resistance among all of the samples. This is consistence with the SEM images, which HT2 film is smooth, condense and no crack among three samples with appropriate thickness. However, uneven surface could be even seen in Fig. 1a, similar to image shown in Fig. 1c, a bulge and fracture happened occasionally. In this case, the HT2 and 3 had no obvious difference in XRD, FTIR outcomes besides Mg–Al LDH intensity.

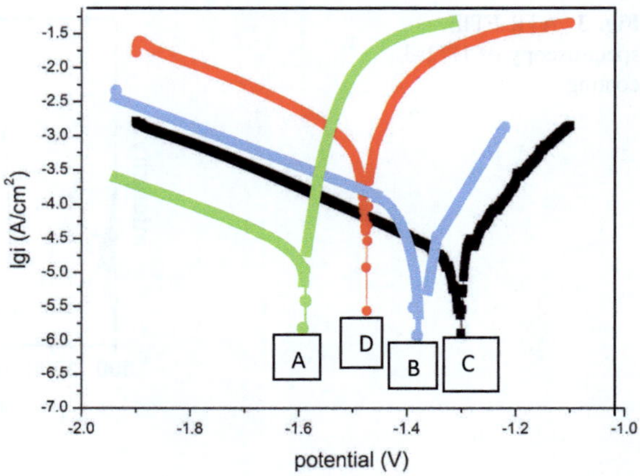

Fig. 4 Polarization curves of MZZS alloy (**a**), LDH-coated1 (**b**), 2 (**c**), and 3 (**d**) samples, respectively

Table 2 Corrosion potential (E_{corr}), corrosion current density (i_{corr}), and polarization resistance (Rp)

Samples	E_{corr} (V/SCE)	I_{corr} (A/cm^2)	R_p (ohm)
MZZS	−1.62	1.856×10^{-5}	5.47×10^5
HT1	−1.40	1.515×10^{-6}	3.72×10^7
HT2	−1.31	1.405×10^{-7}	4.54×10^7
HT3	−1.48	1.906×10^{-6}	6.74×10^8

However, in contrast, there existed in unnegligible difference on the corrosion resistant.

Further study of hydrogen evolution rate (HER) in SBF were carried on bare and coated alloy to acquire the degradability information. As illustrated in Fig. 5, at the initial stage of the immersion, coated samples show low degradation rates compared with bare samples. During the immersion test, bubbles emerged on the surfaces of the MZZS alloy, whilst no bubbles were detected on the surface of the HT treated samples for first 24 h. It can be observed that degradation rate of HT2 coated alloy (4.5 ml/(cm^{2}h), respectively), was lower than that of bare alloy (15.6 ml/(cm^2 h)) in the first 24 h, suggesting a better corrosion resistance. In subsequent time, a

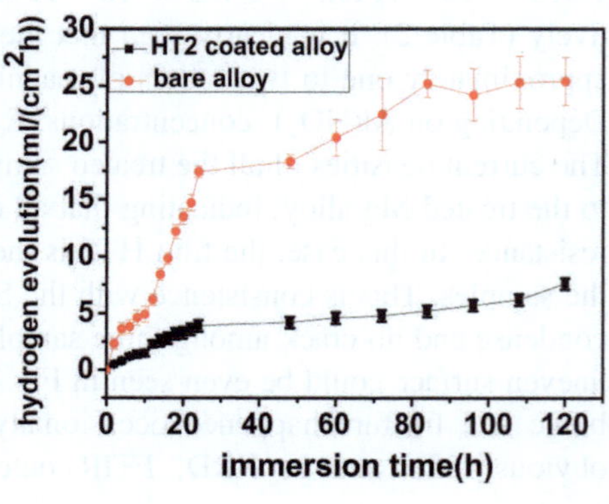

Fig. 5 Hydrogen evolution rate of the MZZS alloy and HT2 coated samples in SBF for 120 h

considerable difference maintained and gradually became larger, which corroborated the thickness and uniformity play important role in enhancing the protection efficiency.

It is generally been accepted that enhanced anticorrosion is benefited from the thickness, uniformity and compactness of the coating. According to the SEM images, EDS mappings, and corrosion tests, it can be confirmed that the LDH/Mg $(OH)_2$ coatings can be synthesized on the surface of MZZS alloy by the hydrothermal synthesis method. When the $Al(NO_3)_3$ content in electrolyte was relatively low (1–10 mM), LDH contents were higher and coatings were more compact. But excessive $Al(NO_3)_3$ source promoted the nucleation and Mg–Al LDH crystals grew up rapidly, thus corridors were blocked by LDH nucleus. Furthermore, Mg^{2+} and OH^- that released from micro galvanic corrosion reaction, could be difficult to diffuse to the surface through pores, insufficient $Mg(OH)_2$ could cause less reactant for LDH synthesizing.

Formation Mechanism. The formation mechanism of the hierarchical hydrotalcite/ $Mg(OH)_2$ coating during the one-step hydrothermal process is proposed here.

Firstly, the MZZS alloy has many β phases such as the Mg_2Zn_3 and $Mg_{17}Sr_2$ dispersed in the alloy acting as cathodes relative to the Mg matrix. The galvanic corrosion of Mg alloy can produce Mg^{2+}, resulting in a increase of the OH^- concentration and H_2 release [28].

Anodic reaction:

$$Mg \rightarrow Mg^{2+} + 2e \tag{2}$$

Cathodic reaction:

$$2H_2O + 2e \rightarrow 2OH^- + H_2 \uparrow \tag{3}$$

Total reaction:

$$Mg + 2H_2O \rightarrow Mg(OH)_2 \downarrow + H_2 \uparrow \tag{4}$$

Al^{3+} first forms $[Al(OH)_4]^-$ in alkali aqueous:

$$Al^{3+} + 4OH^- \rightarrow \left[Al(OH)_4\right]^- \tag{5}$$

The higher alkalinity means more OH^- in the solution, resulting in initial deposition and nucleation of $Mg(OH)_2$ particles on the MZZS alloy. Meantime, higher pH value initiates Zn^{2+} release:

$$Zn + 2OH^- + 2H_2O \rightarrow Zn(OH)_4^{2-} + H_2 \uparrow \tag{6}$$

Generally speaking, Zn is omitted in the forming process of Mg–Al LDH doped coating, for the content of alloying has no influence on Mg–Al LDH forming, though trace of Zn dissolve and position in the lattice of LDH.

The initial $Mg(OH)_2$ coating was porosity and thin, providing through-out pores that electrolyte could penetrate into and react with substrate. Owing to the $[Al(OH)_4]^-$ concentration gradient, Al source diffuse in pores and transport into the lattice of $Mg(OH)_2$, generating Mg–Al LDH. As $[Al(OH)_4]^-$ can also replace Mg^{2+} on the surface of $Mg(OH)_2$ film, LDH dense coating could block the permeation of $[Al(OH)_4]^-$ and, as a result, a rather loose inner layer could be formed. Meanwhile the OH^- in the interlayer is exchanged by CO_3^{2-}, thus leading to the formation of the LDH($Mg_6Al_2(OH)_{16}CO_3{\cdot}4H_2O$) coating.

Mg–Al LDH has good biocompatibility and low toxicity, in addition to commonly accepted viewpoint that Cl^- trapping behavior. Nevertheless, previous studies concentrated on industrial application, few literature discussed extended it to bio-corrosion coating and investigate comparatively of the in vitro degradability.

It is suggested that the introduction of Mg–Al LDH into hydromagnesite improve the corrosion resistance to some extent. The anticorrosive mechanism of the LDH coating involves two parts: barrier effect and ion-exchange ability [29]. On one hand, the coating can act as a barrier layer against Cl^- attack because of the high density of the LDH film, protecting the bare alloys to a certain extent. On the other hand, ion-exchange mechanism is of prominent importance for the corrosion resistance of the LDH coating. Miyata proposed the anion preference series as follow: $CO_3^{2-} > SO_4^{2-} > OH^- > F^- > Cl^- > Br^- > NO_3^-$, due to the high affinity of CO_3^{2-} to the layer, it is difficult to substitute the CO_3^{2-} with Cl^- [30]. Nevertheless, Chen introduced $CaCl_2$ to immersion test to study the ion-exchange of LDH and verified $CaCO_3$ formation [17]. The Ca^{2+}/Mg^{2+} in SBF could easily react with CO_3^{2-} in the interlayer to form a protective $CaCO_3/MgCO_3$ film. $MgCO_3$ is much more easily dissolved to form $Mg(OH)_2$ in SBF. The formation of $CaCO_3/MgCO_3$ film can inhibit the expansion and spread of the pitting corrosion. The ion exchange reaction of the LDH coating on the magnesium alloys in SBF can be written as follows:

$$\mathrm{LDH} - \mathrm{CO_3^{2-}} + \mathrm{Ca^{2+}}/\mathrm{Mg^{2+}} \rightarrow \mathrm{LDH} + \mathrm{CaCO_3}/\mathrm{MgCO_3} \tag{7}$$

$$\mathrm{MgCO_3} + 2\mathrm{OH^-} \rightarrow \mathrm{Mg(OH)_2} + \mathrm{CO_3^{2-}} \tag{8}$$

$$\mathrm{LDH} + \mathrm{Cl^-} \rightarrow \mathrm{LDH} - \mathrm{Cl^-} \tag{9}$$

Based on the experimental results, a corrosion protection mechanism model of the LDH coating is proposed, and this model is used to illustrate of ion-exchange/deposition. This would result in a lower concentration of chloride ions and protect the $Mg(OH)_2$ from degradable too fast.

Figure 6 shows the morphology and corresponding EDS spectra of the pre-immersed sample and of the coated samples immersed in SBF for different time. After immersion for 7 days (Fig. 6), most of the area of the samples still exhibits the nanoplate-like structure, similar to the sample before immersion, except for plenty of white fine particles observed on the film surface. EDS analysis of the LDH

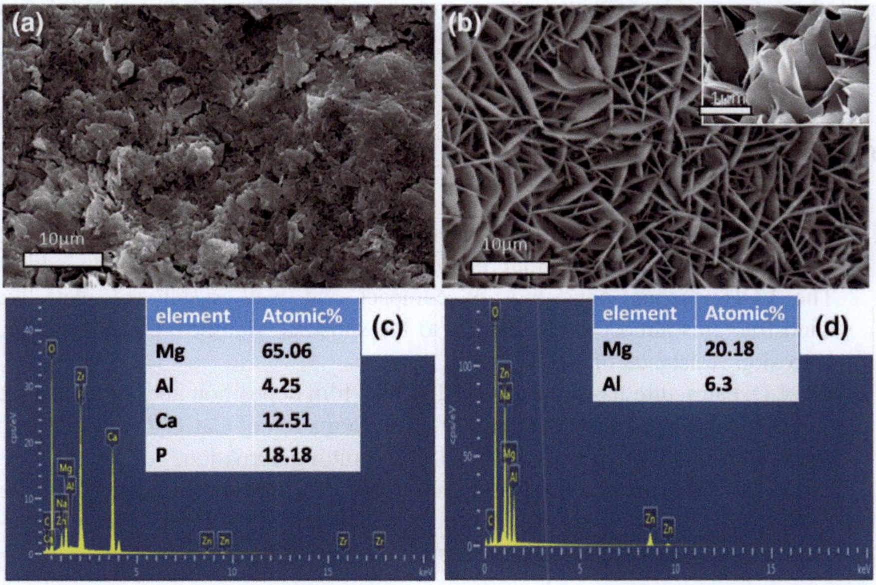

Fig. 6 SEM morphologies and EDS mappings of LDH coating (**a**, **c**) and the sample after 7 days immersion test (**b**, **d**), respectively

coating immersed for 7 days is studied for the element of the MZZS and hydrotacite film coated one. As shown in Fig. 6, apart from Mg, Al, O, Cl, C, certain amount of Ca and P is present in the coating, which demonstrates the corrosion products have not completely covered the Mg–Al LDH coating at all. Furthermore, Ca and P are essential element and a platform for new bone forming.

Cytotoxicity Evaluation. To evaluate the possible cytotoxity of the LDH coating, a preliminary cytotoxicity assay is carry out. The cell viability results from the MTT tests assessing the cytotoxicity of MZZS alloy, MgZn–Al LDH coated alloy (treated with 1 mM $Al(NO_3)_3$), are shown in Fig. 7. The MC3T3-E1 cell viability cultured

Fig. 7 Cell viabilities of MC3T3-E1 cells cultured in extraction medium of bare alloy, HT-coated alloy and control group

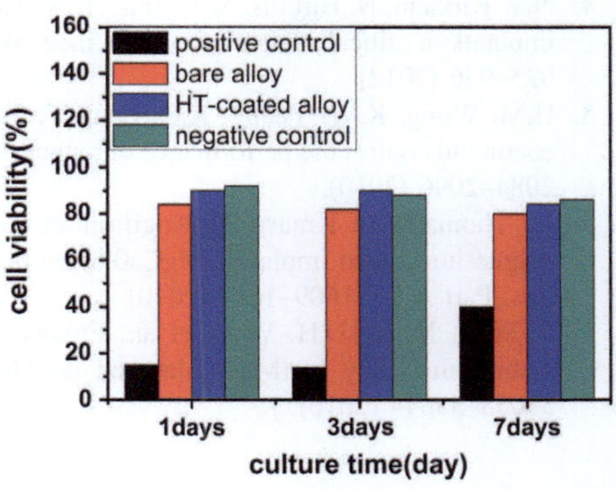

with the extracts of bare, coated and control groups for 1, 4 and 7 days are all larger than 90%, which verifies the non-toxicity of the LDH coating.

Conclusions

(1) A Mg–Al LDH coating with nanosized lamellar structures was fabricated by the simple one-step hydrothermal process with only aluminum nitrate added. The LDH coating consisting of compact and well-crystallized plate-like nanostructures almost perpendicular to the substrate significantly improve the corrosion resistance of MZZS alloy.
(2) The LDH structure has the $CaCO_3/Mg(OH)_2$ film deposition and ion-exchange ability by absorbing and retaining Cl^-. The formation of $CaCO_3/Mg(OH)_2$ film can inhibit the expansion and spread of the pitting corrosion.
(3) $CaCO_3$ particles induce Ca–P nucleation and deposition. After immersion for 7 days, CaP particles can be detected by EDX mapping. Further deep researches will be carried out and give more information in detail.

Acknowledgements The authors acknowledge the financial support for this work from the National Nature Science Foundation of China (No. 51371126 and No. 51271131), Major science and technology projects in Tianjin (No. 15ZXQXSY00080) and Science and Technology developing Foundation of Tianjin High Education (No. 20110301).

References

1. F. Witte, N. Hort, C. Vogt, S. Cohen, K.U. Kainer, R. Willumeit et al., Degradable biomaterials based on magnesium corrosion. Curr. Opin. Solid State Mater. Sci. **12**, 63–72 (2008)
2. M.P. Staiger, A.M. Pietak, J. Huadmai, G. Dias, Magnesium and its alloys as orthopedic biomaterials: a review. Biomaterials **27**, 1728–1734 (2006)
3. H. Hornberger, S. Virtanen, A.R. Boccaccini, Biomedical coatings on magnesium alloys-a review. Acta Biomater. **8**, 2442–2455 (2012)
4. N.T. Kirkand, N. Birbilis, M.P. Staiger, Assessing the corrosion of biodegradable magnesium implants: a critical review of current methodologies and their limitations. Acta Biomater. **8**, 925–936 (2012)
5. H.M. Wong, K.W. Yeung, K.O. Lam, V. Tam, P.K. Chu, A biodegradable polymer-based coating to control the performance of magnesium alloy orthopaedic implants. Biomaterials **31**, 2084–2096 (2010)
6. M. Thomann, C. Krause, N. Angrisani et al., Influence of a magnesium-fluoride coating of magnesium-based implants (MgCa0.8) on degradation in a rabbit model. J. Biomed. Mater. Res. Part A **93**, 1609–1619 (2010)
7. F. Peng, H. Li, D.H. Wang et al., Enhanced corrosion resistance and biocompatibility of magnesium alloy by Mg–Al-layered double hydroxide. ACS Appl. Mater. Interfaces. **8**, 35033–35044 (2016)

8. T. Douglas, E. Pamula, D. Hauk, Porous polymer/hydroxyapatite scaffolds: characterization and biocompatibility investigation. J. Mater. Sci. Mater. Med. **20**, 1909–1915 (2009)
9. T. Imwinkelried, S. Beck, T. Iizuka, B. Schaller, Effect of a plasma electrolytic coating on the strength retention of in vivo and in vitro degraded magnesium implants. Acta Biomater. **9**, 8643–8649 (2013)
10. Ping Shi, Bo Niu, Preparation and characterization of PLA coating and PLA/MAO composite coatings on AZ31 magnesium alloy for improvement of corrosion resistance. Surf. Coat. Technol. **262**, 26–32 (2015)
11. M.F. Montemor, M.G.S. Ferreira, Analytical characterization and corrosion behaviour of bis-aminosilance modified with carbon nanotubes activated with rare-earth salts applied on AZ31 magnesium alloy. Surf. Coat. Technol. **202**, 4766–4774 (2008)
12. P. Gunawan, R. Xu, Direct assembly of anisotropic layered double hydroxide (LDH) nanocrystals on spherical template for fabrication of drug-LDH hollow nanospheres. Chem. Mater. **21**, 781–783 (2009)
13. J.M. Oh, S.J. Choi, S.H. Lee et al., Inorganic drug-delivery nanovehicle conjugated with cancer-cell-specific ligand. Adv. Funct. Mater. **19**, 1617–1624 (2009)
14. Z. Gu, B.E. Rolfe, Z.P. Xu et al., Enhanced effects of low molecular weight heparin intercalated with layered double hydroxide nanoparticles on rat vascular smooth muscle cells. Biomaterials **31**, 5455–5462 (2010)
15. J.L. Gunjakar, T.W. Kim, H.N. Kim et al., Mesoporous layer-by-layer ordered nanohybrids of layered double hydroxide and layered metal oxide: highly active visible light photocatalysts with improved chemical stability. J. Am. Chem. Soc. **133**, 14998–15007 (2011)
16. J.Y. Uan, J.K. Lin et al., Direct growth of oriented Mg–Al layered double hydroxide film on Mg alloy in aqueous HCO_3^-/CO_3^{2-} solution. J. Mater. Chem. **20**, 761–766 (2010)
17. J. Chen, Y.W. Song, D.Y. Shan, E.H. Han, Study of the corrosion mechanism of the in situ grown Mg-Al–CO_3^{2-} hydrotalcite film on AZ31 alloy. Corros. Sci. **65**, 268–277 (2012)
18. F. Zhang, Z.G. Liu, R.H. Zeng et al., Corrosion resistance of Mg-Al-LDH coating on magnesium alloy AZ31. Surf. Coat. Technol. **258**, 1152–1158 (2014)
19. T. Ishizaki, N. Kamiyama, K. Watanabe et al., Corrosion resistance of $Mg(OH)_2$/Mg-Al layered double hydroxide composite film formed directly on combustion-resistant magnesium alloy AMCa602 by steam coating. Corros. Sci. **92**, 76–84 (2015)
20. L.D. Wang, K.Y. Zhang, G.C. Liu et al., Hydrothermal synthesis of corrosion resistant hydrotalcite conversion coating on AZ91D alloy. Mater. Lett. **106**, 111–114 (2013)
21. J.Y. Uan, J.K. Lin, K.L. Jeng, Crystallization of a chemical conversion layer that forms on AZ91D magnesium alloy in carbonic acid. Corros. Sci. **53**, 3832–3839 (2011)
22. Z. Li, M.F. Chen, W. Li et al., The synergistic effect of trace Sr and Zr on the microstructure and properties of a biodegradable Mg-Zn-Zr-Sr alloy. J. Alloys Compds. **702**, 290–302 (2017)
23. J.K. Lin, J.Y. Uan, Formation of Mg, Al-hydrotalcite conversion coating on Mg alloy in aqueous HCO_3^-/CO_3^{2-} and corresponding protection against corrosion by the coating. Corros. Sci. **51**, 1181–1188 (2009)
24. N. Kamiyama, G. Panomsuwan, E. Yamamoto et al., Effect of treatment time in the $Mg(OH)_2$/Mg–Al LDH composite film formed on Mg alloy AZ31 by steam coating on the corrosion resistance. Surf. Coat. Technol. **286**, 172–177 (2016)
25. J. Wang, D.D. Li, Q. Liu et al., Fabrication of hydrophobic surface with hierarchical structure on Mg alloy and its corrosion resistance. Electrocimica Acta **55**, 6897–6906 (2010)
26. R.C. Zeng, Z.G. Liu, F. Zhang et al., Corrosion resistance of in-situ Mg–Al hydrotalcite conversion film on AZ31 magnesium alloy by one-step formation. Trans. Nonferrous Met. Soc. China **25**, 1917–1925 (2015)
27. F. Jing, Y. Chen, X.H. Liu et al., In-situ hydrothermal crystallization $Mg(OH)_2$ films on magnesium alloy AZ91 and their corrosion resistance properties. Mater Chem. Phys. **143**, 322–329 (2013)

28. R.C. Zeng, Y. Hu, S.K. Guan et al., The influence of bicarbonate, sulphate, hydrogen phosphate and dihydrogen phosphate ions in saline solution. Corros. Sci. **86**, 171–182 (2014)
29. R.C. Zeng, Z.G. Liu, F. Zhang et al., Corrosion of molybdate intercalated hydrotalcite coating on AZ31 Mg alloy. J. Mater Chem. A **2**, 13049–13057 (2014)
30. S. Miyata, Anion-exchange properties of hydrotalcite-like compounds. Clays Clay Miner. **31**, 305–311 (1983)

Microstructure, Mechanical Properties and Corrosion Behavior of Mg–Zn–Zr Alloys with Different Compositions

Luanxiang Wang, Renbo Song, Feng Gao, Changhong Cai and Liang Huang

Abstract Mg–Zn–Zr alloys were studied as a promising biomaterial in terms of their excellent biocompatibility and mechanical properties. Cast Mg–Zn–Zr alloys with different compositions were immersed in Hank's solution at 37 °C to investigate the effect of Zn content on the microstructure, mechanical properties and corrosion behavior. The microstructure observation showed the alloys were mainly composed of primary α-Mg matrix and a small amount of MgZn second phase. Mechanical tests showed that Mg–3Zn–0.6Zr alloy had optimum mechanical properties with the tensile strength of 219 MPa, yield strength of 65 MPa, elongation of 21% and hardness of 46 HV. The corrosion morphologies and immersion test proved that Mg–2Zn–0.6Zr alloy exhibited best corrosion resistance with an average corrosion rate of 1.68 mm/year.

Keywords Mg–Zn–Zr alloys · Biomaterial · Corrosion behavior
Immersion test

L. Wang · R. Song (✉) · C. Cai · L. Huang
School of Materials Science and Engineering, University of Science and Technology Beijing, Beijing 100083, China
e-mail: songrb@mater.ustb.edu.cn

L. Wang
e-mail: wlx9306wawj@163.com

C. Cai
e-mail: cchong@126.com

L. Huang
e-mail: wlx9306wawj@163.com

F. Gao
Ansteel Mining Engineering Corporation, Anshan 114004, China
e-mail: gaofeng02@ansteel.com.cn

Y. Han (ed.), *Advanced Functional Materials*,
https://doi.org/10.1007/978-981-13-0110-0_52

Introduction

In recent years, metal materials have often been implanted into the human body as biological materials to help repair human tissue. Stainless steels and titanium alloy are employed as implant materials due to their high corrosion resistance currently. However, the facts that a second surgery is needed to remove the stainless steels or titanium alloy from human body after the tissue recovery severely restrict their applications [1]. Otherwise, the long-term existence in human body of these implant materials will lead to osteopenia for the reason of stress shielding [2]. In that case, people make use of Mg to serve as a new type implant material for its biodegradability. Mg has an elastic modulus of 42 GPa, similar to that of natural bone compared with conventional implant materials like Ti alloys and stainless steel. The matching in elastic modulus is extremely important as it helps to avoid stress shielding [3]. Nevertheless, Pure Mg can't meet the requirements of biomaterials because of its low mechanical properties.

Research findings [4] reveal that Mg alloys have shown huge potential as degradable implants because of its excellent mechanical properties and corrosion resistance after a trace addition of some alloying elements, including aluminum, manganese, zinc, rare earth (RE) and so on. However, some elements are harmful to human body after being added. For example, excessive amounts of aluminum may cause Alzheimer's disease for its neurotoxicity [5]. On the other hand, zinc and zirconium have been fully studied to have better biocompatibility, exhibiting less cytotoxicity compared to aluminum in human body. What's more, the addition of zinc is beneficial to improve the strength properties of magnesium alloy because of the solution strengthening and aging strengthening [6]. When the content of zinc in magnesium alloy is controlled within a certain range, the alloy shows good corrosion resistance [7]. In this paper, Mg alloys were fabricated with trace addition of Zn and Zr element with the aim of being suitable degradable biomaterials. A comprehensive investigation of microstructure, mechanical properties and corrosion behavior was studied.

Experiment

The Mg–Zn–Zr alloys were prepared using high purity Mg (99.99 wt%) ingots, high purity Zn (99.995 wt%) ingots and Mg–30Zr (wt%) master alloys as raw materials. Table 1 shows the chemical compositions of Mg–Zn–Zr alloys. The specimens with a gage length of 25 mm and a gage diameter of 5 mm were used for the tension tests, and the specimens with dimensions of 10 mm × 10 mm × 5 mm were used for the hardness measurement. The cast ingots were cut into cylindrical samples of ø11 mm × 5 mm in dimension for immersion test. All the samples were ground wet and stepwise up to 4000 grit, followed by ultrasonic cleaning in acetone, ethanol in turn for 5 min, respectively.

Table 1 Chemical compositions of cast alloys in wt%

Alloy	Zn	Zr	Fe	Ni	Cu	Al	Mg
Mg–1Zn–0.6Zr	0.90	0.44				0.066	Balance
Mg–2Zn–0.6Zr	1.87	0.54	<0.01	<0.01	<0.01	0.060	Balance
Mg–3Zn–0.6Zr	2.75	0.56				0.070	Balance

For the observation of cast microstructure and corrosion morphology, an optical microscope optical microscopy (OM, Axio Imager M2m) and a field emission scanning electron microscope (FE-SEM, FEI Quanta FEG 450) equipped with an energy dispersive X-ray spectroscope (EDS, EDAX) were used. Intermetallic secondary phases present in the cast alloys were analyzed by X-ray diffraction (XRD, Cu target, 20 kV).

In this study, immersion tests was carried out in Hank's solution at 37 °C. Hank's solution was prepared with distilled water and high-purity chemicals, composed of 8.0 g/L NaCl, 0.4 g/L KCl, 0.14 g/L $CaCl_2$, 0.35 g/L $NaHCO_3$, 0.1 g/L $MgCl_2 \cdot 6H_2O$, 0.06 g/L $MgSO_4 \cdot 7H_2O$, 0.06 g/L KH_2PO_4, 0.06 g/L $Na_2HPO_4 \cdot 12H_2O$ and 1.0 g/L $C_6H_{12}O_6$. The pH value was adjusted to 7.4 with NaOH solution. For immersion test, the specimens were respectively put in wide mouth jars containing Hank's solution and the immersion solution was renewed every 24 h. After the test, a mixture of CrO_3 (200 g/L) and $AgNO_3$ (10 g/L) was used to get rid of the corrosion products. The weight loss tests were repeated three times using LE104E analytical balance.

Results and Discussion

Microstructural analysis. Figure 1 shows the optical micrographs of all examined cast Mg alloys. Analyzed by Image-Pro Plus, the average grain size of Mg–1Zn–0.6Zr, Mg–2Zn–0.6Zr and Mg–3Zn–0.6Zr were about 194, 81 and 106 μm, respectively. It reveals that Zn can effectively refine grains to a certain extent, but not means that the grain size gets smaller when the Zn content keeps increasing.

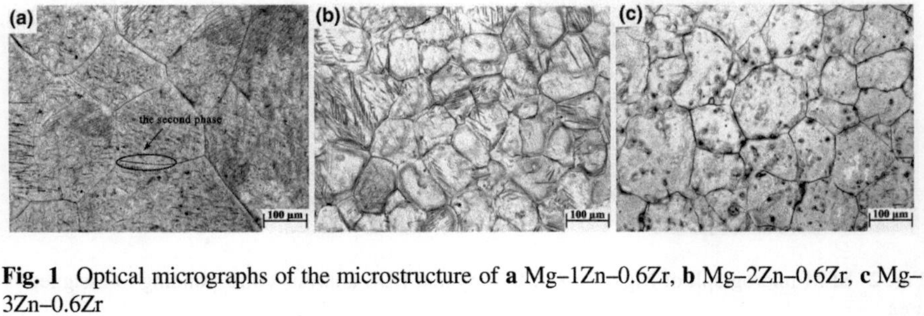

Fig. 1 Optical micrographs of the microstructure of **a** Mg–1Zn–0.6Zr, **b** Mg–2Zn–0.6Zr, **c** Mg–3Zn–0.6Zr

Based on optical micrographs, the microstructure of investigated alloys was composed of primary α-Mg matrix and a small amount of second phase. It was found that the amount of second phase increased when more Zn was added to test alloys. The scanning electron microscopy confirmed the existence of intermetallic phase which was mainly distributed along grain boundaries, exhibiting as sparse particles, as shown in Fig. 2. Indicated from EDS analysis, the granular second phase was rich in Zn and almost Zr free, suggesting it was binary intermetallic compound but not ternary intermetallic compound. Figure 3 shows the X-ray diffractograms of cast specimens of Mg–Zn–Zr alloys. A reflection peak of MgZn intermetallic phase with a pretty low intensity appeared in the test alloys apart from clear Mg reflections. As shown in the Mg–Zn phase diagram there exists a eutectic composition of Mg_7Zn_3 phase at about 325 °C, which eventually transforms into α-Mg solid solution and MgZn intermetallic during solidification [8]. As a result, considering all the results including SEM micrographs, EDS analysis and X-ray diffractograms, it could be confirmed that the second phase presented in Mg–Zn–Zr alloys was MgZn.

Mechanical properties. Table 2 displays the mechanical properties of the Mg–Zn–Zr alloys in comparison with the reported properties of natural bone and pure Mg [9]. It can be clearly seen, the tensile strength, yield strength, elongation and hardness of test alloys increased with the increase of Zn content from 1 to 3%. After the addition of Zn and Zr, the strength of test alloys had a significant improvement, helping the alloys to be outstanding substitutes of bone tissue with higher strength. Owe to the function of fine grain strengthening, the significant improvement was obtained in both strength and ductility by Mg–Zn–Zr alloys compared to pure Mg material. The mechanical properties of Mg–3Zn–0.6Zr were totally better than that of Mg–2Zn–0.6Zr though the grain size of the former was greater than that of the latter, and the reason may lies in the rich of second phase distributed in α-Mg matrix of Mg–3Zn–0.6Zr. To summarize, the comprehensive effect of fine grain strengthening, solid solution strengthening and second phase strengthening led to the optimal mechanical properties of Mg–3Zn–0.6Zr alloy.

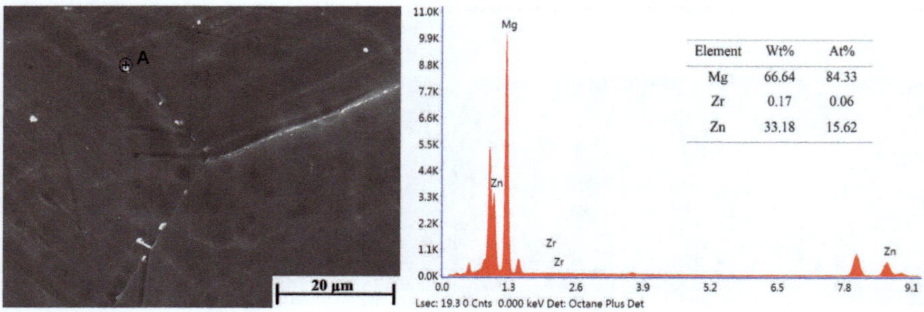

Fig. 2 SEM micrograph of Mg–2Zn–0.6Zr alloy and EDS analysis corresponding to assigned A area

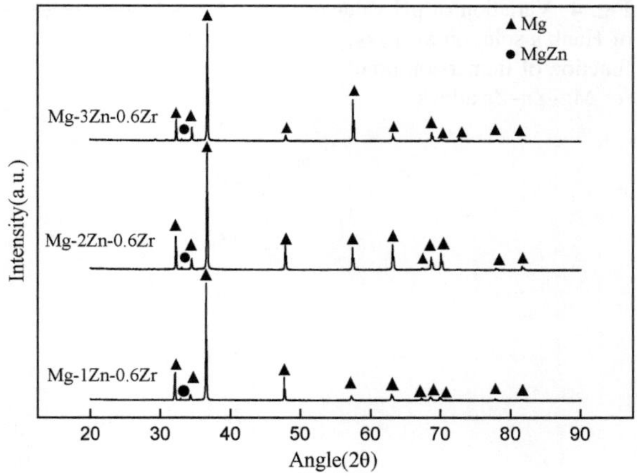

Fig. 3 X-ray diffractograms of Mg–Zn–Zr alloys

Table 2 Mechanical properties of the Mg–Zn–Zr alloys

Material	Tensile strength (MPa)	Yield strength (MPa)	Elongation (MPa)	Hardness (HV)
Natural bone	35–283	–	1–2	–
Pure Mg	66	19	7	–
Mg–1Zn–0.6Zr	168	41	13	40
Mg–2Zn–0.6Zr	196	51	18	42
Mg–3Zn–0.6Zr	219	65	21	46

Immersion Test. The variation of pH value of Hank's solution as a function of immersion time for three Mg–Zn–Zr alloys was illustrated in Fig. 4. As can be seen, the pH value increased sharply at the initial stage of the immersion process, after that, the upward trend of pH value slowed and finally reached a relatively steady stage. It is well known that the variation trend of pH value reflects the corrosion resistance of magnesium alloys [10]. The initial pH values for the three alloys were the same, and as the process went on, the increasing speed was different obviously, leading to the trend: the Mg–1Zn–0.6Zr alloy had the higher pH value than the Mg–3Zn–0.6Zr alloy throughout the immersion test period, and the Mg–2Zn–0.6Zr alloy had the lowest pH value all the time. From the variation trend, the Mg–2Zn–0.6Zr showed the best corrosion resistance, and the corrosion resistance of Mg–3Zn–0.6Zr alloy was better than that of Mg–1Zn–0.6Zr alloy. It is reported that the increase in pH value was likely to cause haemolysis, which were unfavourable for the growth and proliferation of cells [11]. Therefore, the Mg–2Zn–0.6Zr alloy exhibited better biocompatibility than the other two alloys.

Figure 5 shows the average corrosion rate of three Mg–Zn–Zr alloys after 240 h of immersion in Hank's solution. The average corrosion was calculated from mass loss using a formula [12]. The average corrosion rates of Mg–1Zn–0.6Zr, Mg–2Zn–0.6Zr and Mg–3Zn–0.6Zr were about 11.67, 1.68 and 3.56 mm/year, respectively.

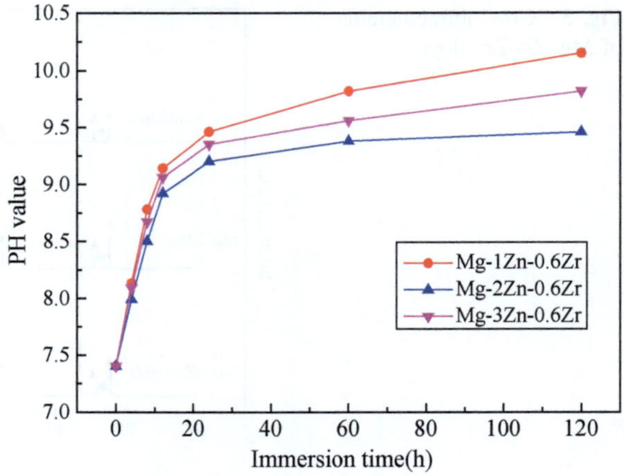

Fig. 4 Variation of pH value of Hank's solution as a function of immersion time for Mg–Zn–Zr alloys

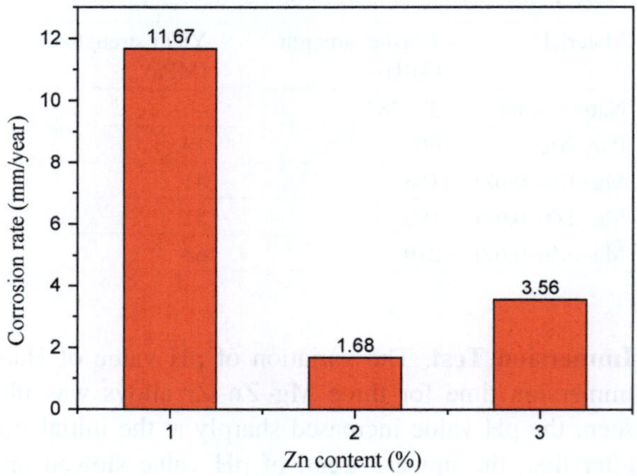

Fig. 5 Average corrosion rate calculated from mass loss for Mg–Zn–Zr alloys

The Mg–2Zn–0.6Zr alloy had a lowest average corrosion rate for 240 h in Hank's solution, demonstrating it exhibited best corrosion resistance, which was a superiority for using as biomaterials. As the same time, the Mg–3Zn–0.6Zr alloy showed better corrosion resistance than Mg–1Zn–0.6Zr alloy.

Figure 6 illustrates the representative surface morphologies of Mg–Zn–Zr alloys after immersion in Hank's solution at 37 °C for 240 h. The surface morphologies with corrosion product removed was shown in Fig. 6a–c, respectively. It can be clearly seen that all the alloys appeared corrosion pits, cracks and caves after immersion, but the severity of corrosion was quite different. Mg–2Zn–0.6Zr exhibited shallower pits compared to Mg–3Zn–0.6Zr and the latter formed a few corrosion caves, indicating Mg–2Zn–0.6Zr had better corrosion resistance than Mg–3Zn–0.6Zr. As for Mg–1Zn–0.6Zr, the corrosion caves expanded and propagated, and finally connected with each other, leading to collapse in the local region.

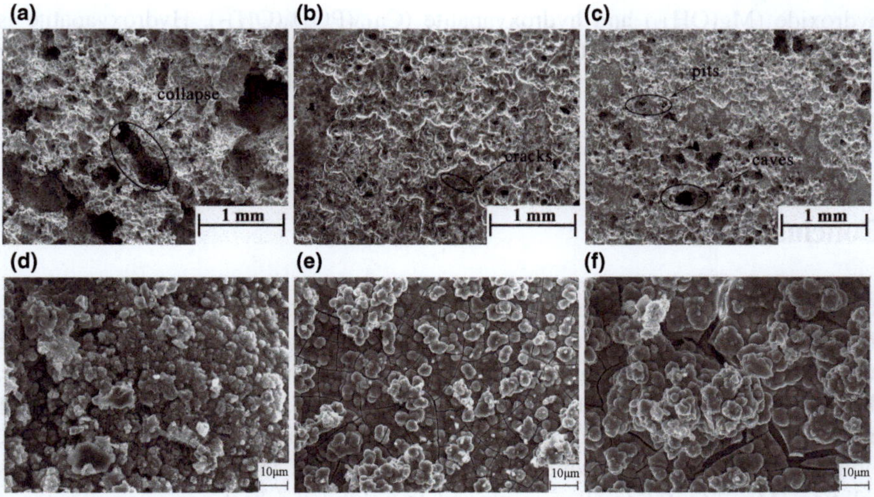

Fig. 6 SEM micrographs after immersion in Hank's solution for 240 h: **a**, **d** Mg–1Zn–0.6Zr; **b**, **e** Mg–2Zn–0.6Zr; **c**, **f** Mg–3Zn–0.6Zr

Figure 6d, e and f show the morphologies of their corrosion products after immersion. It is noteworthy that plenty of flower-like white particles conglomerating to clusters were deposited on the surface of Mg–Zn–Zr alloys, among which the products on Mg–2Zn–0.6Zr tended to be much smaller, fewer and uniformly distributed. It can be speculated that the deeper pits cause more cluster corrosion product. Cracks were visible on the naked alloys, with the finest cracks presented in Mg–2Zn–0.6Zr and cracks partly concealed by the products in Mg–1Zn–0.6Zr. Figure 7 shows the EDS results of corrosion product of test alloys after immersion in Hank's solution at 37 °C for 240 h. Analyzed by EDS, the white corrosion products contained O, Mg, P and Ca elements. Based on the previous research [13], the white products deposited on the surface were mainly composed of magnesium

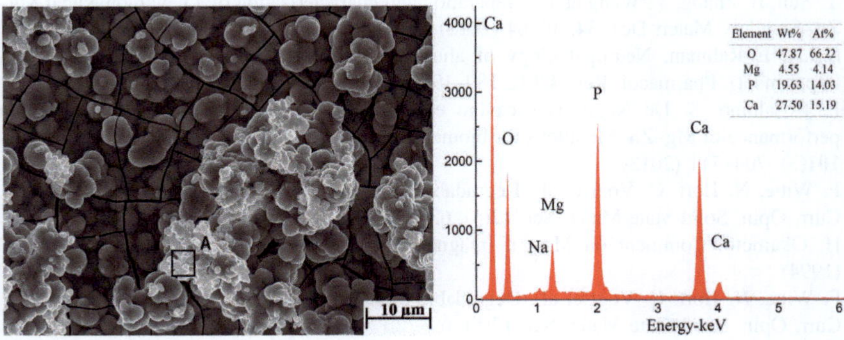

Fig. 7 EDS results of corrosion product of Mg–2Zn–0.6Zr alloy after immersion in Hank's solution for 240 h

hydroxide ($Mg(OH)_2$) and hydroxyapatite ($Ca_{10}(PO_4)_6(OH)_2$). Hydroxyapatite is the primary inorganic substance of bone, the presence of which is beneficial to accelerate the healing of bone tissue, demonstrating that Mg–Zn–Zr alloys have good biocompatibility [14].

Conclusion

The cast Mn–Zn–Zr alloys with different compositions were investigated as degradable metallic implants. The alloys are mainly composed of primary α-Mg matrix and a small amount of MgZn second phase. With the increase of Zn content, the mechanical properties are improved, and the Mg–3Zn–0.6Zr alloy show best mechanical properties with tensile strength of 219 MPa, yield strength of 65 MPa, elongation of 21% and hardness of 46 HV. Through immersion test in Hank's solution, Mg–2Zn–0.6Zr alloy is proved to exhibit the best corrosion resistance with an average corrosion rate of 1.68 mm/year. A large number of white corrosion products are adhered onto the entire surface of alloys, which are considered to be mixture consisting of magnesium hydroxide and hydroxyapatite.

Acknowledgements This work was supported by the National Key Project of Research and Development Plan (No. 2016YFB0700303).

References

1. M.P. Staiger, A.M. Pietak, J. Huadmai et al., Magnesium and its alloys as orthopedic biomaterials: a review. Biomaterials **27**(9), 1728–1734 (2006)
2. H. Hornberger, S. Virtanen, A.R. Boccaccini, Biomedical coatings on magnesium alloys—A review. Acta Biomater. **8**(7), 2442–2455 (2012)
3. F. Witte, J. Fischer, J. Nellesen et al., In vitro and in vivo corrosion measurements of magnesium alloys. Biomaterials **27**(7), 1013–1018 (2006)
4. Y. Sun, B. Zhang, Y. Wang et al., Preparation and characterization of a new biomedical Mg–Zn–Ca alloy. Mater. Des. **34**, 58–64 (2012)
5. S.S.A. El-Rahman, Neuropathology of aluminum toxicity in rats (glutamate and GABA impairment). Pharmacol. Res. **47**(3), 189–194 (2003)
6. F. Rosalbino, S. De Negri, G. Scavino et al., Microstructure and in vitro degradation performance of Mg–Zn–Mn alloys for biomedical application. J. Biomed. Mater. Res. Part A **101**(3), 704–711 (2013)
7. F. Witte, N. Hort, C. Vogt et al., Degradable biomaterials based on magnesium corrosion. Curr. Opin. Solid State Mater. Sci. **12**(5), 63–72 (2008)
8. H. Okamoto, Comment on Mg–Zn (magnesium-zinc). J. Phase Equilib. **15**(1), 129–130 (1994)
9. F. Witte, N. Hort, C. Vogt et al., Degradable biomaterials based on magnesium corrosion. Curr. Opin. Solid State Mater. Sci. **12**(5), 63–72 (2008)
10. J. Fan, X. Qiu, X. Niu et al., Microstructure, mechanical properties, in vitro degradation and cytotoxicity evaluations of Mg–1.5Y–1.2Zn–0.44Zr alloys for biodegradable metallic implants. Mater. Sci. Eng. C **33**(4), 2345–2352 (2013)

11. H.R.B. Rad, M.H. Idris, M.R.A. Kadir et al., Microstructure analysis and corrosion behavior of biodegradable Mg–Ca implant alloys. Mater. Des. **33**, 88–97 (2012)
12. Y. Zong, G. Yuan, X. Zhang et al., Comparison of biodegradable behaviors of AZ31 and Mg–Nd–Zn–Zr alloys in Hank's physiological solution. Mater. Sci. Eng., B **177**(5), 395–401 (2012)
13. X. Zhang, G. Yuan, L. Mao et al., Effects of extrusion and heat treatment on the mechanical properties and biocorrosion behaviors of a Mg–Nd–Zn–Zr alloy. J. Mech. Behav. Biomed. Mater. **7**, 77–86 (2012)
14. S. Zhang, X. Zhang, C. Zhao et al., Research on an Mg–Zn alloy as a degradable biomaterial. Acta Biomater. **6**(2), 626–640 (2010)

Preparation and Characterization of Polymer Nanoparticles with Multiple Response

Yushun Jin, Weijin Zhang, Jinhong Zhang, Tianyang He, Lujia Zhang, Yan Shang and Zimo Wang

Abstract Intelligent nanoparticles with core-shell structure have attracted much attention in the field of drug controlled release. 5-fluorouracil (5-FU) is an anti-metabolic drug that inhibits the formation of pyrimidine nucleotides in tumor cells and has a significant effect on a variety of tumors. However, 5-FU is more toxic and has more harmful reactions. In order to reduce the toxicity, the study of its controlled release system is attracting more and more attention. In this paper, PLA macromolecular monomers with double bond functional groups were prepared by ring-opening polymerization of lactide with hydroxyethyl methacrylate as an initiator. And then, N-isopropylacrylamide (NIPAm) was used as the thermosensitive monomer and methacrylic acid (AA) as the pH sensitive monomer, and the PLA macromonomer was subjected to free radical copolymerization. PH reaction, biodegradable properties of amphiphilic copolymer PLA-g-P(NIPAm-co-AA), and 5-fluorouracil are as a kind of model drug preparation of multi-functional nano-anti-tumor drugs. The results of in vitro drug release showed that PLA-g-P (NIPAm-co-AA) nanoparticles released 5-FU slowly in the buffer solution, and the cumulative drug release rate increased slowly with time and showed no bursting, good temperature and pH response. That can more accurately target controlled release drugs, reduce the toxicity of normal tissues and organs, improve its positioning and targeting ability.

Keywords Intelligent nanoparticle · Amphiphilic copolymer · Drug release

Y. Jin (✉) · W. Zhang · J. Zhang · T. He · L. Zhang · Y. Shang · Z. Wang
Department of Materials Science and Engineering,
Beijing Institute of Petrochemical Technology, Beijing 102617, China
e-mail: jinyushun@bipt.edu.cn

W. Zhang
College of Materials Science and Engineering,
Beijing University of Chemical Technology, Beijing 100029, China
e-mail: 15010401887@163.com

Y. Han (ed.), *Advanced Functional Materials*,
https://doi.org/10.1007/978-981-13-0110-0_53

Introduction

With the drawbacks of traditional drug delivery system, the multi-response amphiphilic polymer nanoparticles, as a new drug delivery system, has become one of the hotspots at home and abroad. The polymer nanoparticles have good solubilization effect on the drug, it also has the advantages of long periodicity and preferentially accumulates in the solid tumors with drug targeting, controlling release and so on [1–3]. Drugs can be encapsulated in the hydrophobic core of micelles by electrostatic, hydrophobic interactions, π-π overlap, etc. [4]. In addition, the drug can be prepared by combining the drug with a weak chemical bond on the polymer. In the case of external environment transformation, the chemical action of the bonding drug and the polymer breaks to release the drug. The 5-fluorouracil (5-FU) belongs to the anticancer antimetabolite, influences the synthesis of DNA, and that is affected by the pseudo metabolite RNA protein synthesis, thus the 5-FU is the first-line therapy of malignant tumors, but it metabolism rapidly, short half-life in vivo and systemic administration of selectivity is not high, also there are many toxic effects. Clinically in order to achieve the numerical value of vivo plasma concentration, often using large doses, sustained administration or repeated and other means, which greatly increased the side effects of 5-FU. PLA-g-P(NIPAm-co-AA) is an amphiphilic molecule consisting of hydrophilic and hydrophobic linkages. It has temperature responsiveness, pH responsiveness and biodegradability. It is used as carrier to prepare nanometer drug, which not only increases water insolubility Drug solubility, reduce liver metabolism of drugs, but also to achieve targeted therapy. In order to prolong the in vivo half-life of 5-fluorouracil and improve the efficacy of 5-fluorouracil and reduce its toxicity, we studied the amphiphilic graft copolymer of PLA-g-P(NIPAm-co-AA) and used it as a carrier to prepare 5-Fluorouracil nano-micelles, to study its release in vivo.

Experiments

Raw Materials and Reagents. Hydroxyethyl methylacrylate, purity quotient ≥99.0%, J&K Scientific; stannous octoate, analytical reagent, Johnson Matthey Chemicals Limited; anhydrous methanol, analytial reagent, Beijing Chemical Works; methylbenzene, analytial reagent, Beijing Chemical Works; chloroform, analytial reagent, Beijing Chemical Works; L-lactide (L-LA), purity quotient ≥99.5%, Ji'nan Daigang Biomaterial Co., Ltd.; N-isopropylacrylamide(NIPAm), purity quotient ≥99.0%, J&K Scientific; acrylic acid (AA), purity quotient ≥99.0%, Beijing Xudong chemical works; azodiisobutyronitrile (AIBN), purity quotient ≥99.0%; 5-fluorouracil, purity quotient ≥98.0%, Sahn Chemical Technology Co., Ltd; Polyvinyl alcohol (PVA) analytial reagent, Sinopharm Chemical Reagent Beijing Co., Ltd; Dimethyl sulfoxide (DMSO), analytial reagent, Tianjin Fu chen Chemical Reagent Factory; dipotassium hydrogen phosphate, analytical reagent, Tianjin Guangfu Science and Technology Development Co., Ltd.; monopotassium phosphate, analytial reagent, Tianjin Guangfu Science and Technology Development Co., Ltd.; dialysis bag, MWCO7000, Beijing Jingke Hongda Biological Technology Co., Ltd.

Synthesis of Multi-response Polymers

Synthesis of Polylactic Acid Macromolecules. The L-lactide, hydroxyethyl methylacrylate (HEMA) and stannous octoate were added to a three-necked round-bottle flask with a magnetic stirrer. The mixture was immersed in an oil bath and stirred at 140 °C under nitrogen. After 5 h later, end the process of this reaction. Afterward cooled it to room temperature, dissolved in methylene chloride after that poured it out. The product was precipitated by adding methanol to the dichloromethane solution and the PLA macromonomer was obtained by suction filtration, washing and vacuum drying.

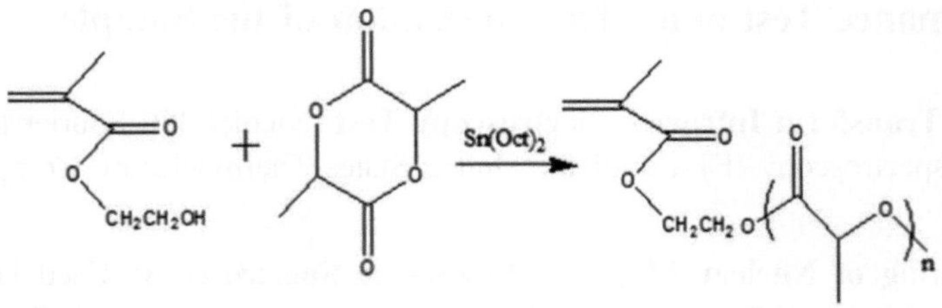

Synthesis of PLA-g-P(NIPAm-co-AA) Polymers. The NIPAm and polylactic acid were dissolved in a certain amount of toluene and chloroform. then, the two polymer solution were mixed evenly after adding a certain amount of acrylic acid and initiator AIBN, respectively. The reaction was conducted at 70 °C for 5 under nitrogen. After polymerization, the precipitate was precipitated by adding ice methanol to the polymer solution. Thereafter the product was purified by filtration, washed and dry, respectively, to obtain the final copolymer.

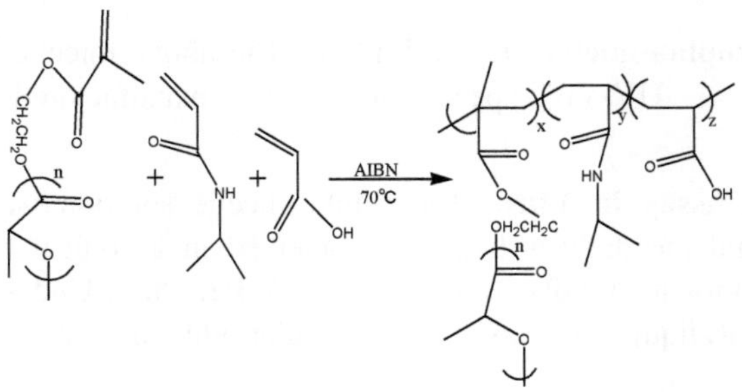

Preparation of Nanoparticles

Nanoparticles Preparation of PLA-g-P(NIPAm-co-AA). The graft copolymer PLA-g-P(NIPAm-co-AA) was dissolved in dimethyl sulfoxide. The DMSO solution of the polymer was slowly added drop by drop to the distilled water and continue stirring 1 h after dripping off. The solution of PLA-g-P(NIPAm-co-AA) micellar solution with nanostructure was prepared by dialysis bath for 48 h and the organic solvent was removed once every 6 h in the system.

Nanoparticles Preparation of PLA-g-P(NIPAm-co-AA)/5-fluorouracil. The graft copolymer PLA-g-P(NIPAm-co-AA) and 5-fluorouracil were dissolved in dimethyl sulfoxide and distilled water, add blank, respectively. The DMSO solution of the polymer was slowly added drop by drop to the 5-FU aqueous solution and the solution was continued continue stirring 1 h after drip off. The solution of PLA-g-P (NIPAm-co-AA)/5-FU micellar solution with nanostructure was prepared by dialysis bath for 48 h and remove organic solvent from the system every 6 h.

Performance Test and Characterization of the Sample

Fourier Transform Infrared Spectroscopy Test. Nicolet-380 Fourier transform infrared spectroscopy (FT-IR) of the United States Thermoelectric Company was used.

The Testing of Nuclear Magnetic Resonance Spectroscopy. Used DMSO as solvent and 0.03 vol.% tetramethylsilane (TMS) as internal standard. The material structure was determined by DRU AVANCE 400 MHz NMR spectrometer from BRUKER, Switzerland.

Particle Size Test. The particle size distribution of micelles was measured by NANOSIGHT LM20 nano-particle tracking analyzer, UK. Malvern Instruments Ltd.

SEM Determination. Using a JSM-7500F field emission scanning electron microscope from Nippon Electronics Co, Ltd. The micelles were dropped onto the coverslips and were dried enough to observe their surface morphology.

UV-Vis Spectrophotometer Determination. The above micelles were measured by UV-2450, a UV-Vis spectrophotometer, manufactured by Shimadzu Corporation.

Drug Release Assay in Vitro. The 5 mL micelle solution was placed in the dialysis bag, and the dialysis bag was immersed in a conical bottle containing 100 mL phosphoric acid buffer solution (pH = 7.4/2.5/5.8). Under the condition of 37 °C, the 10 mL liquid were taken out at regular intervals and the absorbance was measured by UV spectrophotometer and then another 10 mL buffer solution was added in time.

The cumulative release rate of drugs at different times is calculated by Eq. 1

$$E_r = \frac{V_e \sum_{i=1}^{n-1} C_i + V_0 C_n}{m_d} \tag{1}$$

V_e The displacement volume of the releasing medium solution
C_i The concentration of the release fluid at the time of the ith replacement
C_n The concentration of the release fluid at the time of the nth replacement
V_0 Release the volume of the media solution
m_d Quality of 5-fluorouracil in drug loaded micelles.

Results and Discussion

Characterization of PLA-g-P(NIPAm-co-AA). Figure 1b is the infrared spectrum of polylactic acid macromolecules. The main absorption peaks of polylactic acid appear at 3416 cm^{-1} (OH), 2975 cm^{-1}, 1384 cm^{-1} (C–H stretching vibration and in-plane bending vibration Peak of the methyl), 1758 cm^{-1} (C=O), 1184 cm^{-1}, 1090 cm^{-1} (C–O). In addition, the C=C stretching vibration peak in hydroxyethyl methacrylate appeared at 1637 cm^{-1}, indicating that hydroxyethyl methacrylate has been attached to the polylactic acid segment. The obtained product is polylactic acid macromolecular monomer containing C=C bond.

Figure 1a is an infrared spectrum of PLA-g-P(NIPAm-co-AA). The characteristic absorption peaks of lactic acid chain in the figure are 1759 cm^{-1} (C=O), 1184 cm^{-1}, 1090 cm^{-1} (C–O), 2975 cm^{-1} and 1386 cm^{-1} (CH_3). Moreover,

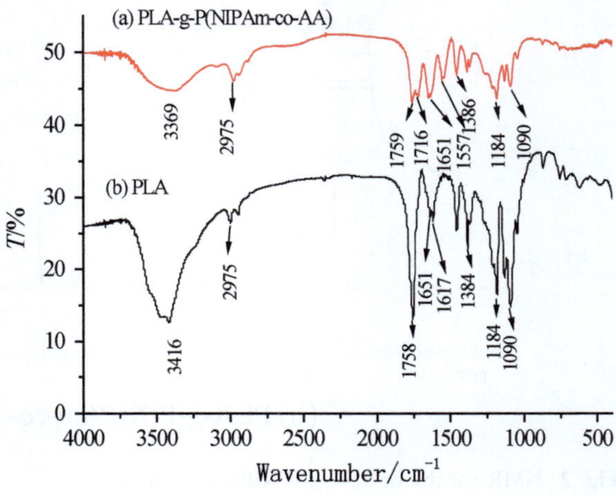

Fig. 1 FT-IR spectra of the PLA-g-P(NIPAm-AA)

absorbances at 1651 cm^{-1} and 1557 cm^{-1} were characteristic absorptios of the C=O stretching and N-H (In-plane bending vibration) due to the presence of NIPAm. The AA absorbances occurred at 1716 cm^{-1} (C=O) and 3369 cm^{-1} (O–H). The resulting products are the PLA-g-P(NIPAm-co-AA) (Fig. 2).

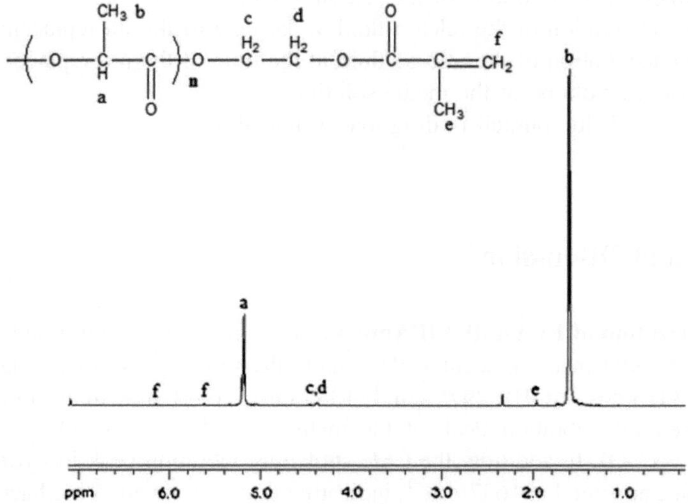

(a) PLA.

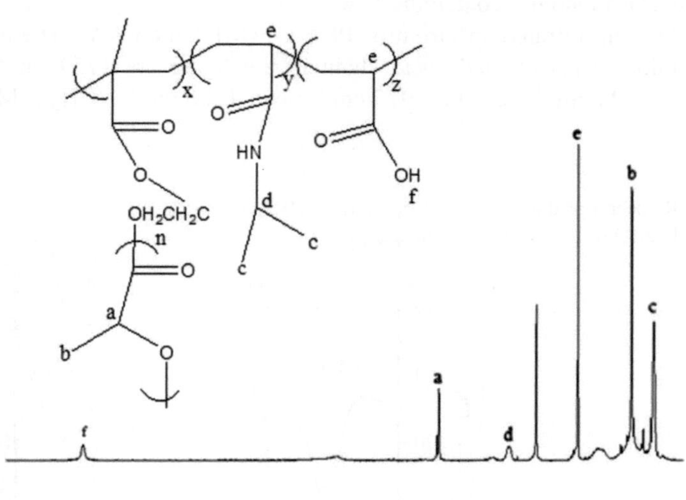

(b) PLA-g-P(NIPAm-co-AA).

Fig. 2 NMR spectra of PLA-g-P(NIPAm-co-AA)

Characterization of PLA-g-P(NIPAm-co-AA)/5-FU Drug-loaded Micelles. PLA-g-P(NIPAm-co-AA) self-assembled into micelles in aqueous solution, and hydrophobic segment PLA formed the micelle core, and hydrophilic segment P(NIPAm-co-AA) formed the micelle shell, while 5-FU entered the micelle core. It can be seen from Fig. 3 that the particle size of PLA-g-P(NIPAm-co-AA)/5-FU drug loaded micelles is 58 ± 17 nm.

Figure 4 is the SEM diagram of PLA-g-P(NIPAm-co-AA)/5-FU loaded micelles. It can be seen that the polymer nanoparticles are spherical, with no obvious defects on the surface, and the particle size distribution is uniform and the particle size is about 20 nm. This consequence indicates that the size of the micelles is contracted after drying, lest the size of the particles determined by SEM is smaller than that in the micellar solution.

When the PLA-g-P(NIPAm-co-AA)/5-FU micellar solution was diluted 10 times with deionized water and DMSO, the UV spectra showed that the absorbance at 244 nm was 0.616 and 3.720, respectively. This is because 5-FU has entered the micelles of PLA-g-P(NIPAm-co-AA), and the rate at which 5-FU diffuses from the micelle core to the solution outside the micelle is relatively slow, so that in the deionized water 5-FU absorbance is low. When PLA-g-P(NIPAm-co-AA) was

Fig. 3 Particle size distribution of PLA-g-P(NIPAm-co-AA)/5-FU loaded micelles

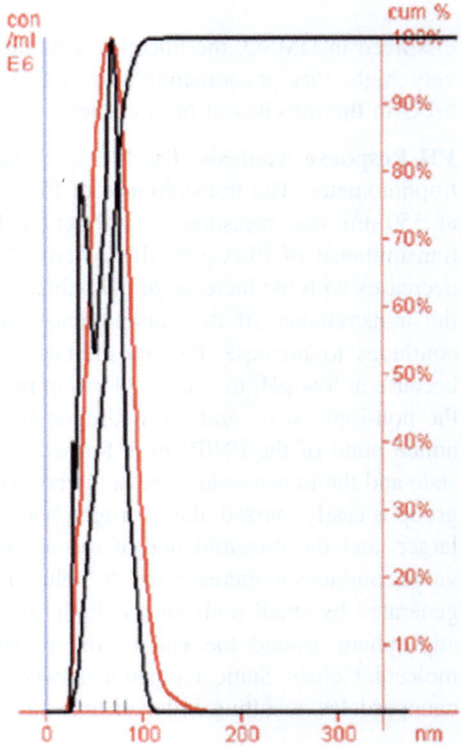

Fig. 4 SEM diagram of PLA-g-P(NIPAm-co-AA)/5-FU loaded micelles

dissolved in DMSO, the micelles were destroyed, so that the 5-FU Absorbance is very high, this phenomenon shows that PLA-g-P(NIPAm-co-AA)/5-FU system 5-FU in the micelles of the micelles.

PH Response Analysis. The PLA-g-P(NIPAm-co-MAA) was measured by spectrophotometer. The transmittance of PLA-g-P(NIPAm-co-MAA) micellar solution at 550 nm was measured at different pH. It can be seen from Fig. 5 that the transmittance of PLA-g-P(NIPAM-co-AA) nanoparticles increases first and then decreases with the increase of pH value. When the pH value increases from 1 to 8, the transmittance of the nanoparticles increases gradually. When the pH value continues to increase, the transmittance of the nanoparticles decreases. This is because at low pH, the carboxyl group in the PAA is difficult to ionize, it exists in the non-ionic state, and the hydrogen bond between the carboxyl group and the amide bond of the PNIPAm is formed so that the nanoparticle is in a contracted state and the transmission rate is lower; When the pH value increases, the carboxyl group is easily ionized, the hydrogen bond effect is weakened, the particle becomes larger, and the transmittance of the nanoparticles becomes larger. When the pH value continues to increase and the solution is alkaline, the free cation concentration generated by small ionization is high, and a large number of cations form an ion atmosphere around the end of the polymer chain, thus shielding the polymer molecular chain Static repulsion between the sections, resulting in shrinkage of nanoparticles, resulting in lower transmittance.

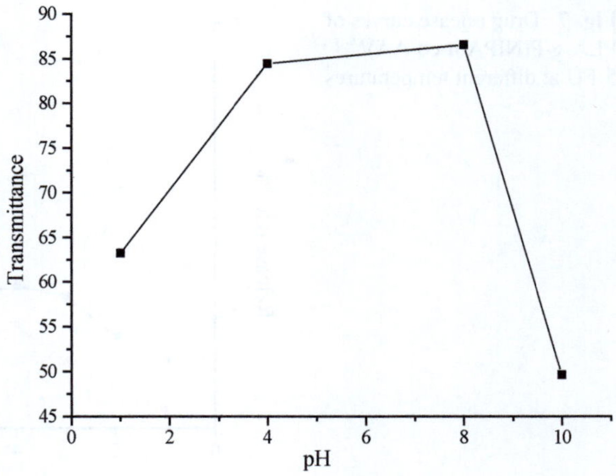

Fig. 5 Effect of pH on the transmittance of PLA-g-P (NIPAm-co-AA) nanoparticles

Drug Release Performance. The DMSO solution with a certain concentration of micelle was prepared and its absorbance was measured. According to the standard curve of DMSO solution of 5-FU, drug loading was 40.26 mg/g, and the entrapment efficiency was 12.18%.

Accurately weighing some 5-FU dissolved in pH = 7.4 buffer solution, and dilute into different concentrations of 5-FU standard solution. The ultraviolet absorption spectrum was determined at room temperature, and the maximum absorption wavelength was determined. Then the absorbance relationship curve between solution concentration and maximum absorption wavelength was plotted (see Fig. 6). The standard operating curve equation was obtained by linear regression of absorbance A to concentration C: Eq. 2

$$A = 0.0343C + 0.0113, \quad R^2 = 0.9973 \tag{2}$$

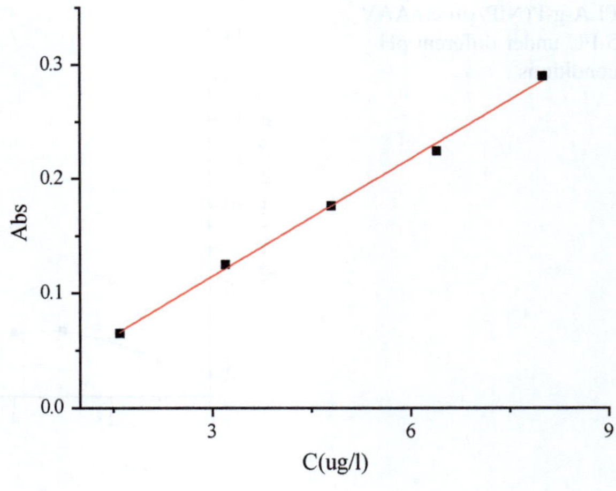

Fig. 6 5-fluorouracil standard curve of DMSO solution

Fig. 7 Drug release curves of PLA-g-P(NIPAm-co-AA)/5-FU at different temperatures

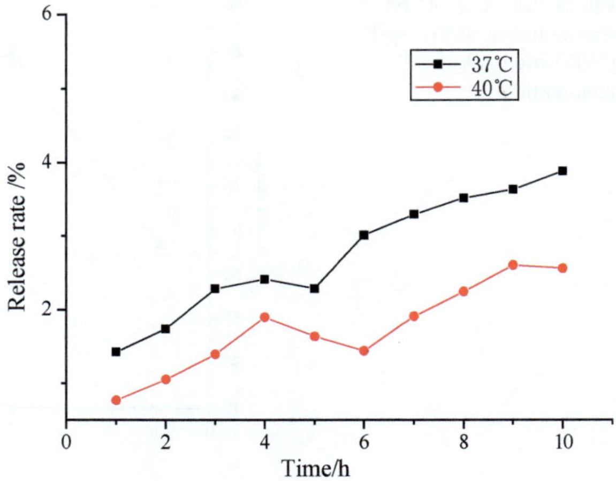

Figure 7 shows the drug release profile of PLA-g-P(NIPAm-co-AA)/5-FU at different temperature. The 5-FU is located in the core in core-corona structure of the micelles. Therefore the effect of the 5-FU controlled release mainly depends on the diffusion speed, there are other factors such as superficial area, density, surface morphology of the carrier. It can be seen that the release rate increases steadily with release time, indicating that the micelle has a control effect of drug release effect and the release rate at 37 °C were higher than 40 °C.

Figure 8 shows the drug release curves of PLA-g-P(NIPAm-co-AA)/5-FU at different pH conditions. It can be seen that PLA-g-P(NIPAm-co-AA)/5-FU nanoparticles the release rate were significantly higher than under the condition of weak base of release rate. The release rate of PLA-g-P(NIPAm-co-AA)/5-FU nanoparticles at pH5.8 were higher than that at pH2.5, and the pH value of PLA-g-P

Fig. 8 Drug release curves of PLA-g-P(NIPAm-co-AA)/5-FU under different pH conditions

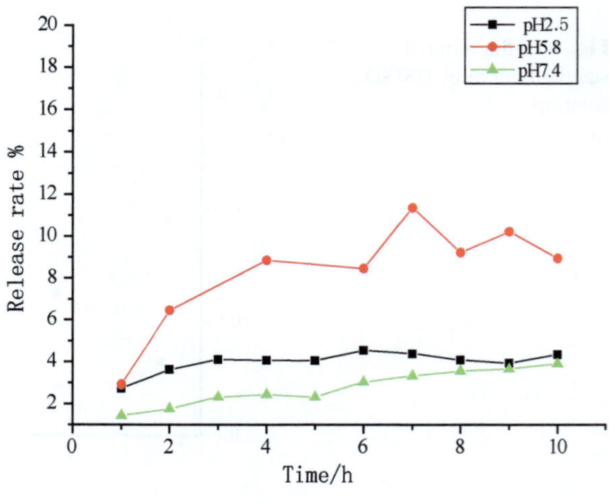

(NIPAm-co-AA)/5-FU nanoparticles were close to 5.8, so the drug in the stomach and other parts of the release were significantly less than in the tumor around the release of cells, indicating that the drug-loaded micelles have a good target therapy.

Conclusions

Poly (lactic acid) macromonomer with double bond functional groups was synthesized by ring-opening polymerization of lactide with hydroxyethyl methacrylate as initiator. Polymerization of PLA-g-P(NIPAm-co-AA) with temperature and pH dual reactivity was obtained by free radical polymerization with NIPAm, AA and PLA. The Nano-drug micelles with particle size of 58 $\pm$ 17 nm were prepared by PLA-g-P(NIPAm-co-AA) as the polymer carrier and 5-fluorouracil as the target drug. The nano-drug micelles slowly released 5-FU in buffer solution, and the cumulative drug release rate increased slowly with time and no burst phenomenon.

By adjusting the proportion of each component in the copolymer, it is possible to control and regulate the particle size of nano-drug micelles and the responsiveness to different environments, so as to achieve the purpose of preparing anti-tumor drugs with different drug release rate.

References

1. M.L. Adams, A. Lavasanifar, G.S. Kwon, J. Pharm. Sci. **92**, 1343 (2003)
2. K. Yasugi, Y. Nagasaki, M. Kato et al., Preparation and characterization of polymer micelles from poly(ethylene glycol)-poly(D, L-lactide)block copolymers as potential drug carrier. J. Controlled Release **62**(1–2), 89 (1999)
3. K. Kataoka, T. Matsumoto, M. Yokoyama et al., Doxorubicin-loaded polu (ethylene glycol)-poly (β-benzyl-L-aspartate) copolymer micelles: their pharmaceutical characteristics and biological significance . J. Controlled Release **64**(1–3), 143 (2000)
4. V.P. Torchilin, Structure and design of polymeric surfactant-based drug delivery systems. J. Controlled Release **73**(2/3), 137–172 (2001)

High-Strength $FeCrMo_{0.2}(AlNi)_{0.5}$ High Entropy Alloy Strengthened by B2 Precipitate

Yong Dong, Xingyu Ding, Wei Fu, Yongqi Cheng and Zhengrong Zhang

Abstract The purpose of this study is to investigate the effects of Al and Ni addition on the microstructure and mechanical properties of $FeCrMo_{0.2}$ alloy. The $FeCrMo_{0.2}$ alloy displayed a disordered BCC solid solution structure. With the addition to Al and Ni elements, an ordered peak around 30.6° appeared in $FeCrMo_{0.2}(AlNi)_{0.5}$ alloy. A uniform microstructure was obtained in $FeCrMo_{0.2}$ alloy while dense nano-scale or micron-scale particles precipitated in $FeCrMo_{0.2}(AlNi)_{0.5}$ alloys. Besides, with the addition to Al and Ni elements, the yield strength increased from 550 to 1600 MPa while the plastic strain ε_p decreased to 30%. And the Vickers hardness increased from HV380 to HV522. The strengthening mechanisms in the alloy system were the solid solution strengthening and the second-phase strengthening.

Keywords High entropy alloy · Precipitated-phase strengthening
Solution strengthening · Compressive properties · Hardness

Introduction

In the past decade, the high entropy alloys (HEAs), developed by Yeh et al. [1] and Cantor et al. [2], have been attracting more and more attention because of their special microstructures and excellent properties. The HEAs were defined as those alloys consisting at least five principal elements each with an atomic percentage between 5 and 35% [1] and having configurational entropy larger than 1.5 R in random solution state [3]. Previous studies have showed that the HEAs not only can form a single-phase solid solution [4–6], but also can obtain multiple-phase solid solution or intermetallics [7–9]. The single-phase HEAs with face-centered cubic (FCC) structure usually have high plasticity but low tensile strength [10, 11] while

Y. Dong (✉) · X. Ding · W. Fu · Y. Cheng · Z. Zhang
School of Materials and Energy, Guangdong University of Technology,
Guangzhou 510006, China
e-mail: dongyong5205@163.com

Y. Han (ed.), *Advanced Functional Materials*,
https://doi.org/10.1007/978-981-13-0110-0_54

the single-phase HEAs with body-centered cubic (BCC) structure exhibit low ductility [12, 13]. The HEAs with multi-phase intermetallics were almost no plastic [14]. Encouragingly, the balance of mechanical properties can be achieved for the HEAs consisting a disordered solid solution phase plus one or two ordered intermetallics phases [15–19]. Li et al. [15] reported a non-equiatomic dual-phase alloy to overcome the strength-ductility trade-off by partial martensitic transformation of FCC phase to hexagonal close-packed (HCP) phase. He et al. [16] designed a L12 phases precipitation-hardened FCC HEAs by thermomechanical processing and microstructure controlling to obtain extraordinary balanced tensile properties. Lu et al. [17] proposed a new eutectic-HEAs (EHEAs) approach to balance the strength, ductility, and castability at the same time. Liu et al. [18] presented a demonstration of using complex hard topologically close-packed (TCP) intermetallics particles to manipulate the properties of FCC-type HEAs. Above all, the studies on multi-phase HEAs have a significant research value for real engineering applications.

However, the above researches mainly focus on the HEAs with disordered FCC solid solution phase plus one or two ordered intermetallics phases. In recent years, few work on the HEAs with disordered BCC solid solution phase plus one or two ordered intermetallics phases was reported because of the brittleness [20]. For example, the previous study showed that the near-eutectic $AlCrFeNiMo_{0.2}$ alloy with disordered BCC and ordered B2 structures had excellent compressive properties [21]. Unfortunately, in tensile test the alloy presented a low tensile strength and had no plasticity (unpublished), which limited its potential application. In $AlCrFeNiMo_{0.2}$ alloy [22], the volume fraction of FeCrMo-rich disordered BCC solid solution was approximate to that of NiAl-rich ordered B2 intermetallic. The BCC solid solution cannot suppress the propagation of micro-cracks originated at the brittle B2 phases. Therefore, the volume fraction, morphology, grain size, and distribution of the B2 intermetallics need to be controlled to improve the ductility. In this paper, a $FeCrMo_{0.2}$–NiAl pseudo binary eutectic phase diagram was built to guide the design of the B2 phases' precipitation-hardened BCC matrix microstructure by careful control of alloy composition. Then, the microstructures and mechanical properties of $FeCrMo_{0.2}Al_{0.5}Ni_{0.5}$ HEA were studied systematically.

Methodology

Alloy design. According to the authors' previous study [23], in near-eutectic $AlCrFeNiMo_{0.2}$ alloy, during the solidification a irregularly-shaped second NiAl-rich phases precipitated in FeCrMo-rich phases while a needle-like second FeCrMo-rich phases precipitated in NiAl-rich phases, which indicated that there existed a limited solubility between the two phases. Therefore, a hypothetical ($FeCrMo_{0.2}$)-(NiAl) pseudo binary eutectic phase diagram can be built as seen in Fig. 1. According to the diagram, the alloys with a composition region between the line S1 and S2, a B2 precipitation-hardened BCC matrix microstructure can be

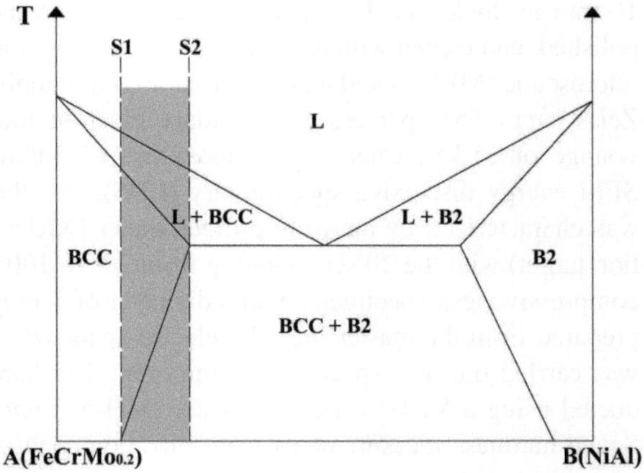

Fig. 1 $(FeCrMo_{0.2})$–(NiAl) pseudo binary eutectic phase diagram

obtained and varied by the control of the alloy composition and thermal treatment. After some exploratory experiments, it can be found that the expected microstructures can appear when the $0.15 \leq x \leq 0.5$ in $(FeCrMo_{0.2})_{1.0}(NiAl)_x$ alloy system. Then the microstructures, mechanical properties and strengthening mechanism of $FeCrMo_{0.2}(AlNi)_{0.5}$ alloy was selected as a typical case to study systematically.

Moreover, the empirical parameter method was adopted to predict the stability of the BCC phases (disordered BCC phases and ordered B2 phases). The atomic-size difference (δ), the mixing enthalpy (ΔH), the electro-negativity difference (Δχ), and valence electron concentration (VEC) of $FeCrMo_{0.2}$ and $FeCrMo_{0.2}(AlNi)_{0.5}$ alloys are calculated. The calculation formulas of δ, ΔH, Δχ and VEC can be found in [24, 25]. The Results were listed in Table 1. According to the study of Zhang et al. [24] and Guo et al. [25], the BCC solid solution phases are stable and no intermetallics form when $\delta \leq 6.6$, $\Delta H \geq -15$ kJ/mol and VEC < 6.8. From the Table 1, the prediction parameters of $FeCrMo_{0.2}$ and $FeCrMo_{0.2}(AlNi)_{0.5}$ alloys are within the region range of BCC solid solution.

Experimental Procedure. The purity of each raw materials used in this study is higher than 99.9 wt%. The $FeCrMo_{0.2}$ and $FeCrMo_{0.2}(AlNi)_{0.5}$ alloys were prepared by arc melting mixture of pure metals in a Ti-gettered high purity argon atmosphere. After the vacuum pressure reached to 3×10^{-3} Pa, melting and casting were performed with the pressure of 0.01 atm by backfilled to high-purify argon (Ar). The alloys were repeatedly melted at least five times in order to ensure chemical homogeneity. Then the alloys were directly solidified in a water-cooled copper hearth and the solidified button ingots were about 25 mm in diameter and

Table 1 The δ, ΔH, and VEC of $FeCrMo_{0.2}$ and $FeCrMo_{0.2}(AlNi)_{0.5}$ alloys

Alloys	δ (×100)	ΔH (kJ/mol)	VEC
$FeCrMo_{0.2}$	3.50	−1.16	6.9
$FeCrMo_{0.2}(AlNi)_{0.5}$	5.39	−9.02	6.78

10 mm in thickness. The specimens cutting from the button ingots were ground, polished, and etched with aqua regia for examining microstructures under an optical microscope (MEF-4) and a field-emission-gun scanning electron microscope (SEM, Zeiss supra 55) operated at secondary electron mode and with an accelerating voltage of 15 kV. Chemical compositions of different phases were analyzed by SEM energy dispersive spectrometry (EDS), and the crystal structure of samples was characterized by an x-ray diffractometer (XRD, EMPYREAN with Cu radiation target) with the 2θ scan ranging from 20° to 100° at a speed of 4°/min. For the compressive test, specimens with a diameter of 5 mm and a length of 10 mm were prepared from the master ingot by electro-spark wire-electrode cutting and the test was carried out at a speed of 0.6 mm/min. The hardness measurement was conducted using a Vickers hardness tester (MH-5L) under a load of 1000 g for 15 s. Seven hardness measurements were made on each sample to obtain the averaged experimental data.

Results and Discussion

Crystal Structure. Figure 2 showed the XRD patterns of $FeCrMo_{0.2}$ and $FeCrMo_{0.2}(AlNi)_{0.5}$ alloys. The $FeCrMo_{0.2}$ alloy displayed a disordered BCC solid solution structure. The lattice constant of the BCC phases was calculated to be 0.291 nm by the strongest peak of (110). With the addition to Al and Ni elements, an ordered peak around 30.6° appeared in $FeCrMo_{0.2}(AlNi)_{0.5}$ alloy. The constant of the ordered B2 phases was calculated to be 0.2916 nm which was similar with that of BCC solid solution, which led to the overlap of BCC and B2 peaks. The formation of the B2 phases was attributed to the strong binding forces between Al and transition element [1, 9].

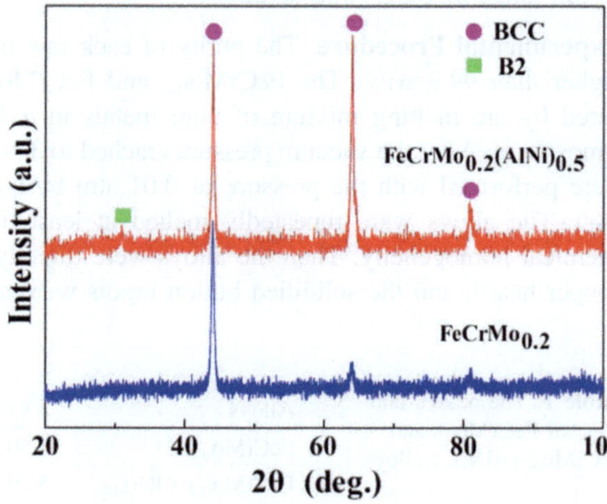

Fig. 2 XRD patterns of $FeCrMo_{0.2}$ and $FeCrMo_{0.2}(AlNi)_{0.5}$ alloys

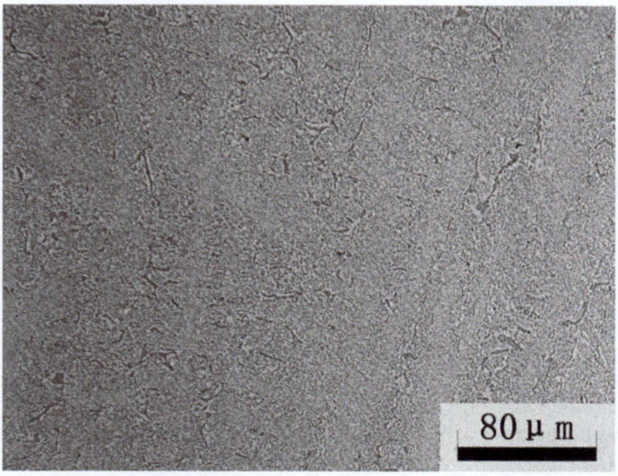

Fig. 3 Metallography microstructure of $FeCrMo_{0.2}(AlNi)_{0.5}$ alloy

Microstructures. The metallography microstructure of $FeCrMo_{0.2}(AlNi)_{0.5}$ alloy was shown in Fig. 3. It can be seen that the alloy showed a typical dendrite/interdendrite microstructure. And dense precipitates formed in dendrite. The SEM secondary electron images of $FeCrMo_{0.2}$ and $FeCrMo_{0.2}(AlNi)_{0.5}$ alloys were shown in Fig. 4. Before the SEM tests, the $FeCrMo_{0.2}$ and $FeCrMo_{0.2}(AlNi)_{0.5}$ alloys were etched with aqua regia. It can be seen that a uniform microstructure was obtained in $FeCrMo_{0.2}$ alloy while dense nano-scale or micron-scale particles precipitated in $FeCrMo_{0.2}(AlNi)_{0.5}$ alloys. The volume fraction of the precipitates was estimated to be 23% by using the Image-Pro Plus software. Table 2 listed the chemical composition of different regions in atomic percentage by EDS. From the EDS results, it can be concluded that the matrix (denoted as 1 in Fig. 4b) was enriched with Fe, Cr, and Ni elements while the precipitates (denoted as 2 in Fig. 4b) was enriched with Al and Ni elements. Therefore, it can be deduce that the

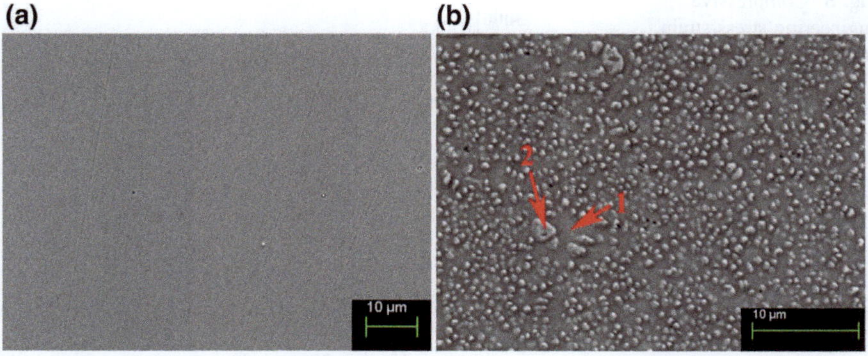

Fig. 4 SEM secondary electron images of **a** $FeCrMo_{0.2}$ and **b** $FeCrMo_{0.2}(AlNi)_{0.5}$ alloys

Table 2 Chemical composition of different regions in atomic percentage by EDS (at. %)

Alloys	Regions	Al	Cr	Fe	Ni	Mo
$FeCrMo_{0.2}$	Nominal	–	45.5	45.5	–	9.1
	Experiment	–	45.9	44.8	–	9.3
$FeCrMo_{0.2}(AlNi)_{0.5}$	Nominal	15.6	31.2	31.2	15.6	6.3
	Experiment	15.1	32.6	31.3	14.5	6.5
	1	13.2	34.7	32.7	12.2	7.2
	2	38.5	8.3	13.1	39.1	1.0

Table 3 Chemical mixing enthalpies of the atomic pairs among the alloying elements (kJ/mol) [23]

	Al	Cr	Fe	Ni	Mo
Al	–	–	–	–	–
Cr	−10	–	–	–	–
Fe	−11	−1	–	–	–
Ni	−22	−7	−2	–	–
Mo	−5	0	−2	−7	–

(FeCrMo)-rich phase and (NiAl)-rich phase were (Cr, αFe)-type BCC solid solution and NiAl-type B2 intermetallic compound, which is also accordance with the XRD results. Table 3 lists the chemical mixing enthalpies of the atomic pairs among the experiment elements [23]. The mixing enthalpy between aluminum and nickel are −22 kJ/mol, which led to the formation of the NiAl-type B2 intermetallic compound.

Mechanical Properties. The compressive engineering stress-strain curves of $FeCrMo_{0.2}$ and $FeCrMo_{0.2}(AlNi)_{0.5}$ alloys were plotted in Fig. 5. The $FeCrMo_{0.2}$ alloy had a yield strength of 550 MPa and did not fracture at a strain of 50%. With the addition to Al and Ni elements, the yield strength increased from 550 to 1600 MPa while the plastic strain ε_p decreased to 30%. Besides, the Vickers

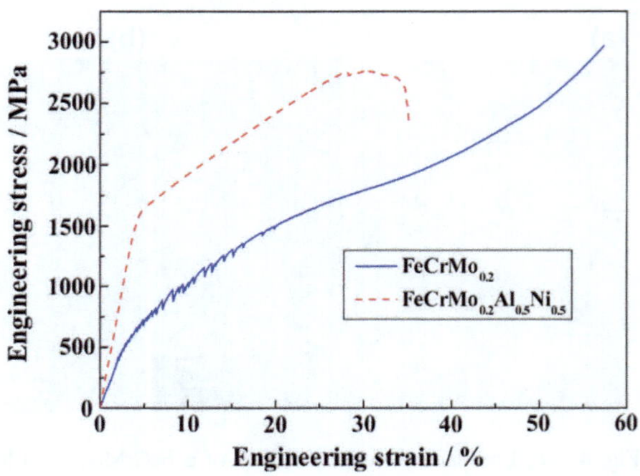

Fig. 5 Compressive engineering stress-strain curves of $FeCrMo_{0.2}$ and $FeCrMo_{0.2}(AlNi)_{0.5}$ alloys

hardness increased from HV380 to HV522. Moreover, the serration behavior happened in $FeCrMo_{0.2}$ alloy, which may caused by the accumulation and the releasing of stress due to the dislocation pin of Mo solute atoms [26].

The strengthening mechanisms in the $FeCrMo_{0.2}(AlNi)_{0.5}$ alloy were analyzed. Due to no work hardening and obvious grain size change, the contributions of the dislocation strengthening and the grain boundary strengthening can be ignored. Besides, there occurred a slight increase of lattice constant of the BCC-structure matrix and the lattice distortion effect of HEAs when Al and Ni elements were added to $FeCoMo_{0.2}$ alloy, which indicated that there was a contribution of the solid solution strengthening. From Fig. 4b, dense nano-scale or micron-scale particles precipitated in $FeCrMo_{0.2}(AlNi)_{0.5}$ alloys. The precipitated particles can hinder the dislocation movement and the effect of strengthening is more obvious with much smaller and more disperse precipitated phases. Thus, the solid solution strengthening and the second-phase strengthening can be considered as the mainly strengthening mechanisms of $FeCrMo_{0.2}(AlNi)_{0.5}$ alloy.

Summary

The $FeCrMo_{0.2}$ and $FeCrMo_{0.2}(AlNi)_{0.5}$ alloys were produced by vacuum arc melting. The $FeCrMo_{0.2}$ alloy exhibits a disordered BCC solid solution structure. With the addition to Al and Ni elements, a typical dendrite/interdendrite microstructure is obtained and dense B2 precipitates are formed in BCC dendrite. From the EDS results, the matrix is enriched with Fe, Cr, and Ni elements while the precipitates is enriched with Al and Ni elements. The yield strength increases from 550 to 1600 MPa and the Vickers hardness increases from HV380 to HV522 due to the solid solution strengthening of Al and Ni element and second-phase strengthening of B2 precipitates.

Acknowledgements This research was supported by the "one-hundred young talents" projects of Guangdong University of Technology (No. 220413575).

References

1. J.W. Yeh, S.K. Chen, S.J. Lin et al., Nanostructured high-entropy alloys with multi-principal elements: novel alloy design concepts and outcomes. Adv. Eng. Mater. **6**(5), 299–303 (2004)
2. B. Cantor, I.T.H. Chang, P. Knight et al., Microstructural development in equiatomic multicomponent alloys. Mater. Sci. Eng., A **375–377**, 213–218 (2004)
3. J.W. Yeh, Alloy design strategies and future trends in high-entropy alloys. JOM **65**(12), 1759–1771 (2013)
4. B. Gludovatz, A. Hohenwarter, D. Catoor et al., A fracture-resistent high-entropy alloy for cryogenic applications. Science **345**, 1153–1158 (2014)
5. O.N. Senkov, S.L. Semiatin, Microstructure and properties of a refractory high-entropy alloy after cold working. J Alloys Compd. **649**, 1110–1123 (2015)

6. Y. Zhang, T.T. Zuo, Z. Tang et al., Microstructures and properties of high-entropy alloys. Prog. Mater Sci. **61**, 1–93 (2014)
7. T.T. Zuo, R.B. Li, X.J. Ren et al., Effects of Al and Si addition on the structure and properties of CoFeNi equal atomic ratio alloy. J. Mag. Mag. Mater. **371**, 60–68 (2014)
8. N.D. Stepanov, N.Y. Yurchenko, D.V. Skibin et al., Structure and mechanical properties of the $AlCr_xNbTiV$ (x = 0, 0.5, 1, 1.5) high entropy alloys. J. Alloys Compd. **652**, 266–280 (2015)
9. A.M. Manzoni, H.M. Daoud, R. Voelkl et al., Influence of W, Mo and Ti trace elements on the phase separation in $Al_8Co_{17}Cr_{17}Cu_8Fe_{17}Ni_{33}$ based high entropy alloy. Ultramicroscopy **159**, 265–271 (2015)
10. M.J. Yao, K.G. Pradeep, C.C. Tasan et al., A novel, single phase, non-equiatomic FeMnNiCoCr high-entropy alloy with exceptional phase stability and tensile ductility. Scripta Mater. **72–73**, 5–8 (2014)
11. N.D. Stepanov, D.G. Shaysultanov, G.A. Salishchev et al., Effect of V content on microstructure and mechanical properties of the $CoCrFeMnNiV_x$ high entropy alloys. J. Alloys Compd. **628**, 170–185 (2015)
12. O.N. Senkov, G.B. Wilks, D.B. Miracle et al., Refractory high-entropy alloys. Intermetallics **18**(9), 1758–1765 (2010)
13. O.N. Senkov, J.M. Scott, S.V. Senkova et al., Microstructure and room temperature properties of a high-entropy TaNbHfZrTi alloy. J. Alloys Compd. **509**(20), 6043–6048 (2011)
14. T.T. Shun, L.Y. Chang, M.H. Shiu, Microstructures and mechanical properties of multiprincipal component $CoCrFeNiTi_x$ alloys. Mater. Sci. Eng. A **556**, 170–174 (2012)
15. Z.M. Li, K.G. Pradeep, Y. Deng et al., Metastable high-entropy dual-phase alloys overcome the strength-ductility trade-off. Nature **534**, 17981 (2016)
16. J.Y. He, H. Wang, H.L. Huang et al., A precipitation-hardened high-entropy alloy with outstanding tensile properties. Acta Mater. **102**, 187–196 (2016)
17. Y.P. Lu, Y. Dong, S. Guo et al., A promising new class of high-temperature alloys: eutectic high-entropy alloys. Sci. Rep. **4**, 6200 (2014)
18. W.H. Liu, Z.P. Lu, J.Y. He et al., Ductile $CoCrFeNiMo_x$ high entropy alloys strengthening by hard intermetallic phases. Acta Mater. **116**, 332–342 (2016)
19. J.Y. He, W.H. Liu, H. Wang et al., Effects of Al addition on structural evolution and tensile properties of the FeCoNiCrMn high-entropy alloy system. Acta Mater. **62**, 105–113 (2014)
20. Y.J. Zhou, Y. Zhang, Y.L. Wang et al., Solid solution alloys of $AlCoCrFeNiTi_x$ with excellent room-temperature mechanical properties. Appl. Phys. Lett. **90**(18), 181904 (2007)
21. Y. Dong, Y.P. Lu, J.R. Kong et al., Microstructure and mechanical properties of multi-component $AlCrFeNiMo_x$ high entropy alloys. J. Alloys Compd. **573**, 96–101 (2013)
22. Y. Dong, L. Jiang, Z.Y. Tang et al., Effect of electromagnetic field on microstructure and properties of bulk $AlCrFeNiMo_{0.2}$ high-entropy alloy. J. Mater. Eng. Perform. **24**, 4475–4481 (2015)
23. Y. Dong, L. Jiang, H. Jiang et al., Effects of annealing treatment on microstructure and hardness of bulk $AlCrFeNiMo_{0.2}$ eutectic high-entropy alloy. Mater. Des. **82**, 91–97 (2015)
24. Y. Zhang, Y.J. Zhou, J.P. Lin et al., Solid-solution phase formation rules for multi-component alloys. Adv. Eng. Mater. **10**(6), 534–538 (2008)
25. S. Guo, C. Ng, J. Lu et al., Effect of valence electron concentration on stability of fcc or bcc phase in high entropy alloys. J. Appl. Phys. **109**(10), 103505 (2011)
26. S. Chen, X. Xie, B. Chen et al., Effects of temperature on serrated flows of $Al_{0.5}CoCrCuFeNi$ high-entropy alloy. JOM **67**(10), 2314–2320 (2015)

Compressive Strength of Al_2O_3 Composites Reinforced with Three-Dimensional Carbon Fiber Preform

Chaoyang Fan, Qingsong Ma and Kuanhong Zeng

Abstract The static compressive strength and the impact resistance of three-dimensional carbon fiber preform reinforced Al_2O_3 (C/Al_2O_3) composites were investigated in this paper. It was found that the static compressive strength in Z direction was 422.7 MPa, which is higher than that in X direction. The results from split-hopkinson pressure bar experiment indicated that the dynamic compressive strength of C/Al_2O_3 composites enhanced when the strain rate ranged from 400 to 1600 s^{-1}, followed by decline at 2700 s^{-1}. The relationship between structure and compressive strength is discussed.

Keywords Compressive strength · Impact behavior · Alumina composites Carbon fiber reinforcement · Sol

Introduction

The intrinsically brittle fracture behavior limits its application although alumina (Al_2O_3) ceramics are well known as high performance structural materials. To improve the fracture toughness of monolithic Al_2O_3 ceramics, the second phases are introduced [1–10]. Of all the second phases, continuous fiber reinforcement has drawn much attention due to its outstanding damage tolerance [4–10]. So far, oxide fiber has been paid much more attention than SiC fiber and C fiber for reinforcing Al_2O_3 ceramics, and the well-developed fabrication technologies include slurry

---

C. Fan · Q. Ma (✉) · K. Zeng
Science and Technology on Advanced Ceramic Fibers and Composites Laboratory,
National University of Defense Technology, Changsha 410073, China
e-mail: nudtmqs1975@163.com

C. Fan
e-mail: 362676161@qq.com

K. Zeng
e-mail: zkhong147@126.com

Y. Han (ed.), *Advanced Functional Materials*,
https://doi.org/10.1007/978-981-13-0110-0_55

infiltration and heat treatment (SIH) [4], reaction-bonding aluminum oxide (RBAO) [6], sol-gel [7–9] and electrophoretic deposition (EPD) [10].

As an extensively used reinforcement for high performance composites, three-dimensional (3D) fiber preform displays better flexibility in structure design, comprehensive performance and adaptability to complex shape. For the fabrication of 3D fiber reinforced Al_2O_3 composites, especially large-size complex parts, gas infiltration and liquid impregnation are superior to the EPD and RBAO routes. However, proper gaseous raw materials for the deposition of Al_2O_3 are not available at present. Inhomogeneous distribution of Al_2O_3 matrix is likely to be obtained in SIH-derived 3D fiber reinforced Al_2O_3 composites because the large size Al_2O_3 particles in slurry are apt to sedimentation. Although the sol-gel route does not have the deficiency of SIH, the transformation efficiency from sol via gel to Al_2O_3 is very low, resulting in 3D fiber reinforced Al_2O_3 composites with high porosity and low strength [7–9].

Recently, the route of sol impregnation-drying-heat treatment (SIDH) using sols with high solid content has been adopted to fabricate 3D fiber preform reinforced oxide composites, for example, Nextel^{TM}/(Al_2O_3–SiO_2) [11], SiO_{2f}/(ZrO_2–SiO_2) [12], SiC_f/SiO_2 [13], and C_f/mullite [14–16]. These studies indicate that the SIDH route is attractive since it can improve fabrication efficiency of the sol-gel route and reserve its advantage of homogeneous distribution of Al_2O_3 matrix. In our previous study [17], 3D carbon fiber preform reinforced Al_2O_3 (C/Al_2O_3) composites were fabricated through the SIDH route, and their thermal stability and oxidation resistance of the composites were investigated. In this paper, the static compressive strength and the dynamic impact behavior of 3D C/Al_2O_3 composites were reported.

Experimental Procedure

Raw materials. Plain weave carbon fiber cloths were laminated and stitched by carbon fiber bundles to form 3D carbon fiber preform with a fiber volume fraction of 35%. The fiber was ex-PAN carbon fiber with a trademark of T300 and a density of 1.76 g/cm^3. An Al_2O_3 sol with a solid content of 30wt% and a colloid particle diameter of 20 $\sim$ 30 nm was used as raw material for matrix.

Fabrication of C/Al_2O_3 composites. Carbon fiber preform was desized by thermal treatment at 1400 °C for 1 h, followed by vacuum impregnation of the Al_2O_3 sol. After soaked in sol for 6 h, the preform was dried at 200 °C for 4 h and then heated at 1400 °C for 1 h in inert atmosphere with a heating rate of 10 °C/min. The fabrication of C/Al_2O_3 composites was accomplished after the cycle of impregnation-drying-heating was repeated 28 times. The C/Al_2O_3 composites showed an apparent density of 2.63 g/cm^3 and a total porosity of 15.5%. The flexural strength and fracture toughness of the composites were 208.5 MPa and 8.1 MPa m$^{1/2}$, respectively.

Characterization. The samples with a size of 12 mm × 4 mm × 4 mm were cut from the as-prepared C/Al_2O_3 composites for the test of static compressive strength. Split-hopkinson pressure bar (SHPB) was employed to measure dynamic impact behavior of the composites. The sample size for SHPB test was 4 mm × 4 mm 4 mm. The compressive strength tests were carried out on X and Z direction of sample, respectively. The impact behavior test was performed on Z direction of sample.

Results and Discussion

Static compressive strength. Five specimens were selected for the test of static compressive strength. Table 1 shows the compressive strengths in X and Z direction of C/Al_2O_3 composites. It is found that the static compressive strength in Z direction is 422.7 MPa which is higher than that in X direction. At the same time, the small fluctuation of compressive strength suggests the good uniformity of C/Al_2O_3 composites.

The fracture surfaces of the specimens after compressive strength test are shown in Fig. 1. When the compressive load was put on X direction, the specimen fractures along diagonal. The resulted fracture surface is relatively flat and makes an angle of ~18° with loading direction. Rough fracture surface and powder desquamation are observed when Z direction was compressed. The fracture surface makes an angle of ~45° with loading direction.

Figures 2 and 3 illustrate compressive load-displacement curves and compressive failure modes of C/Al_2O_3 composites on different directions, respectively. As shown, distinguishing compressive failure behaviors are found due to different structure in X and Z directions. In the C/Al_2O_3 composites, several large pores are found among carbon fiber cloths and the relatively uniform distribution of some micropores are observed [17]. Thus, the composites is better densified in X direction than Z direction. As a result, the slope of ascendant stage of load-displacement curve is higher in X direction than Z direction. In X direction, when compressive load reaches the maximum value, carbon fibers yield to shear

Table 1 Static compressive strengths in X and Z direction of C/Al_2O_3 composites

Sample number	Static compressive strength (MPa)	
	X direction	Z direction
1	332.4	432.5
2	354.1	453.8
3	341.3	394.6
4	326.4	397.3
5	297.3	435.4
Statistical average	330.3 ± 21.2	422.7 ± 25.8

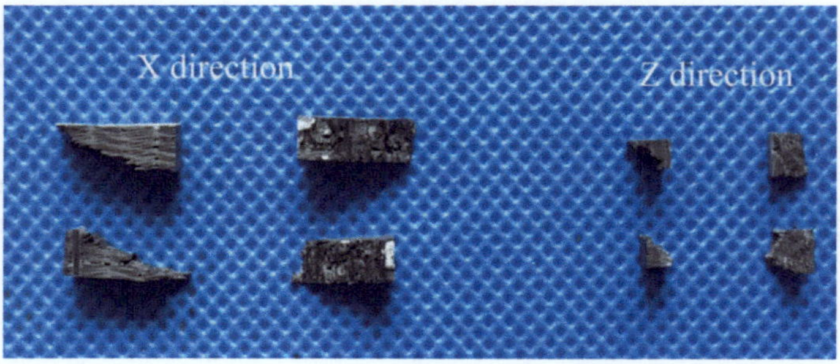

Fig. 1 Optical photos of the specimens after compressive strength test

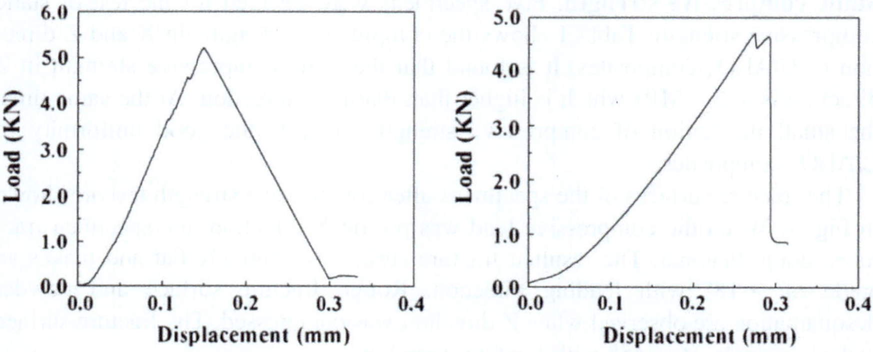

(a) X direction (b) Z direction

Fig. 2 Compressive load-displacement curves of C/Al_2O_3 composites in different directions

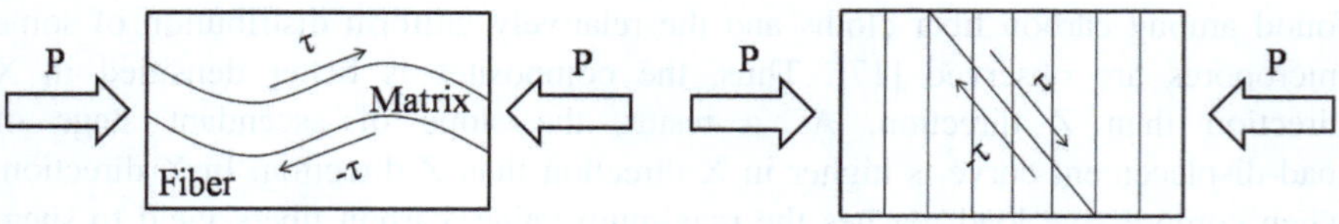

(a) X direction (b) Z direction

Fig. 3 Compressive failure modes of C/Al_2O_3 composites on different directions

stress derived from the surrounding matrix and begin to rupture. The created cracks propagate along the surface between carbon fiber cloths, leading to delamination failure of composites. During the propagation of cracks, fiber bridging and pull-out mechanisms can dissipate energy of cracks, which is reflected by the non-vertical decline of load after the maximum point.

When Z direction of the composites was compressed, loose matrix is consolidated at first. On increasing compressive load, shear stress due to slippage of matrix is gradually enhanced. When the shear stress is high enough, carbon fibers begin to fracture. The formed cracks spread throughout the thickness, resulting in rupture failure of the composites. In this direction, it is necessary for cracks to split the pressed fibers and matrix in order to destroy the composites. At the same time, the physical interfacial bonding is strengthened due to the compressive load, impairing the fiber bridging and pull-out effects. Accordingly, higher compressive strength is obtained and the abrupt decline of load after the maximum point is observed.

Dynamic compressive strength. Figure 4 is the SHPB equipment for the test of dynamic compressive strength. The C/Al_2O_3 composites were impacted on Z direction. Stress-strain curves of C/Al_2O_3 composites at various strain rates are illustrated in Fig. 5. As shown, a pre-rupture strain of ~ 0.009 and a compressive strength of ~ 500 MPa are found at a strain rate of 400 s^{-1}. With increasing strain rate to 800 s^{-1}, the strain decreases to ~ 0.008 and the compressive strength is

Fig. 4 SHPB equipment for the test of dynamic compressive strength of C/Al_2O_3 composites

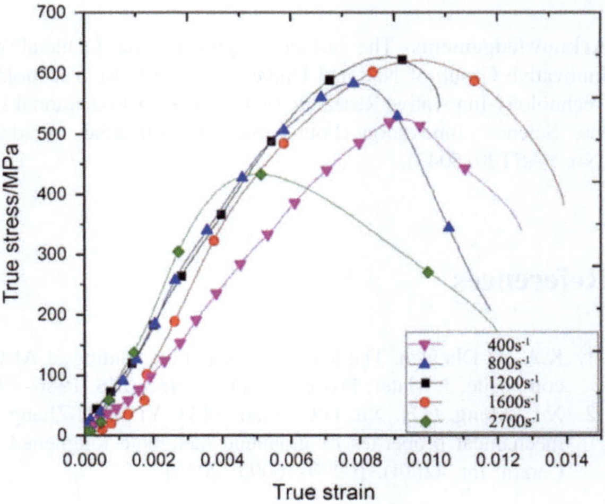

Fig. 5 Stress-strain curves of C/Al_2O_3 composites at various strain rates

elevated to ~580 MPa. Subsequently, the strain increases to ~0.009 at a strain rate of 1200 s^{-1} and ~0.10 at a strain rate of 1600 s^{-1}, respectively. Both of the compressive strengths at strain rates of 1200 and 1600 s^{-1} are ~620 MPa. At a strain rate of 2700 s^{-1}, the strain and the compressive strength decrease to ~0.005 and ~410 MPa, respectively.

Based on its special structure, the compression failure of C/Al_2O_3 composites includes three stages, namely, compaction of pores, shear damage of fibers and compressive damage of matrix. When strain rate is low (<800 s^{-1}), the plastic deformation of matrix can not only heal pores but also prevent crack propagation to a certain extent, enhancing compressive strength. On increasing strain rate to 800–1600 s^{-1}, carbon fibers can resist the shear stress originating from inner slippage. As a result, the compressive strength keeps stable. If strain rate is overlarge (2700 s^{-1}), the normal stress imposed on the composites is quickly elevated to the threshold of compressive damage. Consequently, catastrophic breakage of the composites occurs and compressive strength decreases.

It is noticed that the composites exhibit the highest compressive strength and the maximum strain when the strain rate is 1600 s^{-1}. Under the circumstance, the composites show favorable prospect for bulletproof application [18]. Of course, the compressive strength and fracture toughness should be further improved in order to meet the bulletproof demands.

Summary

The static and dynamic compressive strengths of C/Al_2O_3 composites have been characterized. The static compressive strengths in X and Z directions are 330.3 MPa and 422.7 MPa, respectively. The dynamic compressive strength increases with the elevation of strain rate from 400 to 1600 s^{-1}, then declines at strain rate of 2700 s^{-1}.

Acknowledgements The authors appreciate the financial support of the Aid Program for Innovative Group of National University of Defense Technology, Aid Program for Science and Technology Innovative Research Team in Higher Educational Institutions of Hunan Province, and the Science Innovation Foundation of Shanghai Academy of Spaceflight Technology (No. SAST2015043).

References

1. K.A. Al-Dheylan, The low velocity impact loading of Al_2O_3/SiC whisker reinforced ceramic composite. J. Mater. Process. Technol. **155–156**, 1986–1994 (2004)
2. X.L. Meng, C.H. Xu, G.C. Xiao, M.D. Yi, Y.B. Zhang, Microstructure and anisotropy of mechanical properties of graphene nanoplate toughened Al_2O_3-based ceramic composites. Ceram. Int. **42**(14), 16090–16095 (2016)

3. S.S. Huang, W.C. Zhou, F. Luo, P. Wei, D.M. Zhu, Mechanical and dielectric properties of short carbon fiber reinforced Al_2O_3 composites with MgO additive. Ceram. Int. **40**(2), 2785–2791 (2014)
4. F.W. Zok, Developments in oxide fiber composites. J. Am. Ceram. Soc. **89**(11), 3309–3324 (2006)
5. E. Volkmann, K. Tushtev, D. Koch, C. Wilhelmi, J. Goring, K. Rezwan, Assessment of three oxide/oxide ceramic matrix composites: mechanical performance and effects of heat treatments. Compos. Part A **68**, 19–28 (2015)
6. B. Wilshire, F. Carreño, Deformation and damage processes during tensile creep of ceramic-fibre-reinforced ceramic matrix composites. J. Eur. Ceram. Soc. **20**(4), 463–472 (2000)
7. Z.F. Chen, L.T. Zhang, L.F. Cheng, Y.D. Xu, G.F. Han, Preparation of C fiber reinforced alumina-mullite composite by PIP and CVI. J. Inorg. Mater. **18**(3), 638–644 (2003). (in chinese)
8. Z.F. Chen, L.T. Zhang, L.F. Cheng, Y.D. Xu, P. Xiao, Preparation of C fiber reinforced alumina-mullite composite assisted by silica sol infiltration. J. Aeronaut. Mater. **21**(4), 28–32 (2001). (in chinese)
9. Z.F. Xie, J.Y. Xiao, Z.H. Chen, X.Y. Wang, W.W. Zheng, D.Z. Jiang, Processing and mechanical properties of 3-dimensional braided carbon fiber preform reinforced alumina matrix composites prepared by sol-gel method. J. Natl. Univ. Defense Technol. **20**(5), 14–18 (1998). (in chinese)
10. E. Stoll, P. Mahr, H.G. Krüger, H. Kern, B.J.C. Thomas, A.R. Boccaccini, Fabrication technologies for oxide-oxide ceramic matrix composites based on electrophoretic deposition. J. Eur. Ceram. Soc. **26**(9), 1567–1576 (2006)
11. Y. Wang, H.T. Liu, H.F. Cheng, J. Wang, Effective fugitive carbon coatings for the strength improvement of 3D Nextel^{TM} 440/aluminosilicate composites. Mater. Lett. **126**, 236–239 (2014)
12. Y. Xiang, Q. Wang, F. Cao, Y.W. Ma, D.L. Quan, Sol-gel process and high-temperature property of SiO_2/ZrO_2–SiO_2 composites. Ceram. Int. **43**(1), 854–859 (2017)
13. Q. Wang, F. Cao, Y. Xiang, Z.H. Peng, Effects of ZrO_2 coating on the strength improvement of 2.5D SiC_f/SiO_2 composites. Ceram. Int. **43**(1), 884–889 (2017)
14. K.W. Dai, C.C. Zhang, Q.S. Ma, H.T. Liu, Fabrication and properties of 3D-C_f/mullite composites with SiC interfacial coatings. Adv. Mater. Res. **833**, 150–153 (2014)
15. H.T. Liu, Q.S. Ma, W.D. Liu, Mechanical and oxidation resistance properties of 3D carbon fiber-reinforced mullite matrix composites prepared by sol-gel process. Ceram. Int. **40**(5), 7203–7212 (2014)
16. S.L. Liang, Q.S. Ma, H.T. Liu, Microstructures and mechanical properties of three-dimensional braided carbon fiber reinforced mullite composites with different sols as raw materials. Mater. Sci. Forum **816**, 27–32 (2015)
17. C.Y. Fan, Q.S. Ma, K.H. Zeng, Thermal stability and oxidation resistance of C/Al_2O_3 composites fabricated from a sol with high solid content. Ceram. Int. **43**(14), 10983–10990 (2017)
18. B.H. Jiang, R.Q. Zhang, Dynamic compressive mechanical properties of a carbon fiber woven reinforced composite: experimental study. Acta Materiae Compositae Sinica **22**(2), 109–115 (2005)

Influence of Compaction Pressure on the Properties of Silica Ceramic Cores

Jianbo Yu, Zhigang Yang, Zongcheng Song, Zhongming Ren and Kang Deng

Abstract Based on the application background of the ceramic cores for the hollow blades, silica ceramic cores were prepared by dry-pressing and subsequent pressureless sintering. The silica powders coated by silicone resin were used as the raw materials. The influence of compaction pressure on the properties of silica ceramic cores was analyzed. The results showed that with the increasing of compaction pressure, both the shrinkage rate and apparent porosity of the samples decreased. On the contrary, the increasing compaction pressure promoted the increasing of bulk-density and bending strength of the samples. The phase composition of the sample was composed of amorphous and crystallized silica. A number of cracks formed at the crystallization front of the grain owing to the volume change of crystallized silica during the cooling.

Keywords Silica ceramic cores · Compaction pressure · Bending strength
Apparent porosity

J. Yu (✉) · Z. Song · Z. Ren · K. Deng
State Key Laboratory of Advanced Special Steels, School of Materials Science and Engineering, Shanghai University, Shanghai 200072, China
e-mail: jbyu@shu.edu.cn

Z. Song
e-mail: 862239893@qq.com

Z. Ren
e-mail: renzm2201@163.com

K. Deng
e-mail: dengk2000@163.com

Z. Yang
School of Materials Science and Engineering, Shijiazhuang Tiedao University, Shijiazhuang 050043, China
e-mail: 203yangzhigang@163.com

Y. Han (ed.), *Advanced Functional Materials*,
https://doi.org/10.1007/978-981-13-0110-0_56

Introduction

Ceramic cores are widely used to form the complex cooling passages of hollow blades [1, 2]. To improve the match the properties between ceramic cores and alloy blades, excellent properties of ceramic cores are needed. There are two requirements for the ceramic cores. On the one hand, high strength is needed to resist the erosion from high temperature molten alloy during the casting process. On the other hand, high porosity is necessary to make the acid or alkaline liquid into the inner of ceramic cores because the ceramic cores need to be removed from the hollow blades [3, 4]. However, the two requirements are contradictory because the high strength will accompany with the decreasing of porosity. Therefore, the balance between the strength and porosity is necessary to make the ceramic cores wide application in the hollow blades.

Silica ceramic cores are one of the most widely researched cores in the application of the Ni-based hollow blades owing to their high dimensional stability, thermal shock resistance and leachability [5]. Crystallized silica at high temperature undergoes the phase transformation during the cooling, which will lead to the volume change of 5vol.% [6]. This problem may go against the properties of the ceramic cores. Silicone resin as a polymer precursor can be used as a binder joining the particles together by applying external pressure and further connects the particles by forming network after the curing at the temperature of 200–250 °C. And then silicone resin can transform into the silica products during heat treatment at high temperature of >700 °C [7, 8].The use of silicone resin in the preparation of silica ceramic cores can promote the high production yield and bending strength of ceramic cores, which will further broaden the application of ceramic cores. As the silica particles coated by the silicone resin can show the good formability, the kind of powders are suitable for the fabrication of ceramic cores with good properties.

In this paper, silica ceramic cores were prepared by dry-pressing and subsequent sintering. The raw materials used were silica powders coated by silicone resin. The influence of compaction pressure on the properties of silica ceramic cores was investigated.

Experimental Procedures

The silica particles coated by silicone resin were used as the raw materials. The coating process was based on the reports of previous literature [9]. The silica ceramic core samples were prepared by dry-pressing method. The dimension of metal mold was 4 mm × 5 mm × 45 mm. The compaction pressures were 5, 10, 15 and 20 Mpa, respectively. The holding time of pressure was about 3 min. Afterward, the samples were firstly heated to 250 °C for 6 h to promote the polymerization of the silicone resin, and then sintered at 1250 °C for 2 h in air atmosphere. The heating rate is 5 °C/min.

The shrinkage rate of the samples was determined by measuring the dimension change before and after sintering. The relative density and apparent porosity of the samples were calculated by the Archimedes method in the distilled water. The fracture microstructure of the samples was scanned by SEM (SU-1500, Japan). The phase composition of the samples was analyzed by XRD (D/max-2500, Japan). The bending strength of the samples was measured by the three-point bending method (WDW-300). The span length was 30 mm. The loading rate was 0.5 mm/min. Each experimental result was given as average value of ten.

Results and Discussion

Figure 1 shows the shrinkage rate of the samples as a function of the compaction pressure. It is seen that with the increasing of pressure, the shrinkage rate of the samples shows a decreasing trend. When the pressure is 5 Mpa, the shrinkage rate is 2.8%. When the pressure reaches to 20 Mpa, the shrinkage rate decreases to 1.2%. The increasing compaction pressure can decrease the gaps between the silica particles and make the green body more densification. The gaps between the grains will further decrease after sintering. The sample with bigger gaps will produce bigger shrinkage by decreasing the distance between the grains owing to the existence of the sintering force. Therefore, the increasing compaction pressure is beneficial for the degradation of shrinkage rate in the samples.

Figure 2a and b show the bulk-density and apparent porosity of the samples as a function of the compaction pressure, respectively. With the increasing of the compaction pressure, the bulk-density of the samples increases. On the contrary, the apparent porosity decreases. When the pressure is 5 Mpa, the bulk-density is 1.53 g/cm^{-3} and the apparent porosity is 30.08%. When the pressure reaches 20 Mpa, the bulk-density reaches 1.67 g/cm^{-3} and the apparent porosity decreases

Fig. 1 Shrinkage rate of the samples as a function of the compaction pressure

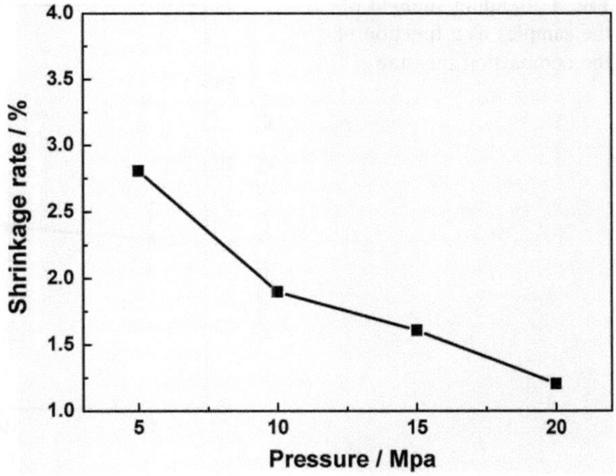

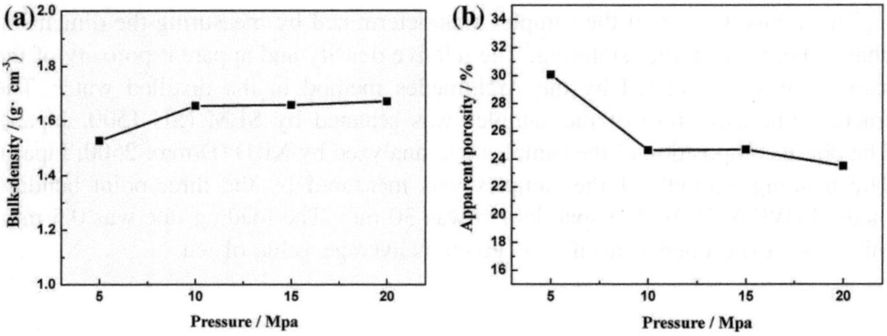

Fig. 2 Bulk-density and apparent porosity of the samples as a function of the compaction pressure **a** bulk-density; **b** apparent porosity

23.51%. As described above, the increasing compaction pressure leads to the decrease of shrinkage rate. Additionally, the increasing pressure can promote the densification of green body including the shortening of distance between the silica grains. After sintering, the densification of the sample also shows the increasing trend. As a result, the bulk-density increases and apparent porosity decreases with the compaction pressure.

Figure 3 shows the bending strength of the samples as a function of the compaction pressure. It is seen that with the increasing of the pressure, the bending strength of the samples at ambient temperature increases. When the pressure is 5Mpa, the bending strength is 15.54 ± 2.15 Mpa. When the pressure reaches 20Mpa, the bending strength reaches 18.55 ± 1.99 Mpa. The trend is caused by densification of the samples with the compaction pressure.

Figure 4 shows the fracture morphologies of the samples as a function of the compaction pressure. It is seen that when the pressure is 5 Mpa, big pores or gaps exist and the silica grains are very loose in the sample. The obtained sample shows

Fig. 3 Bending strength of the samples as a function of the compaction pressure

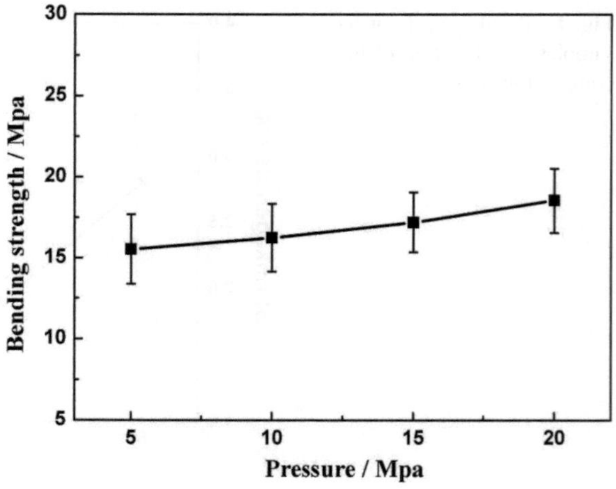

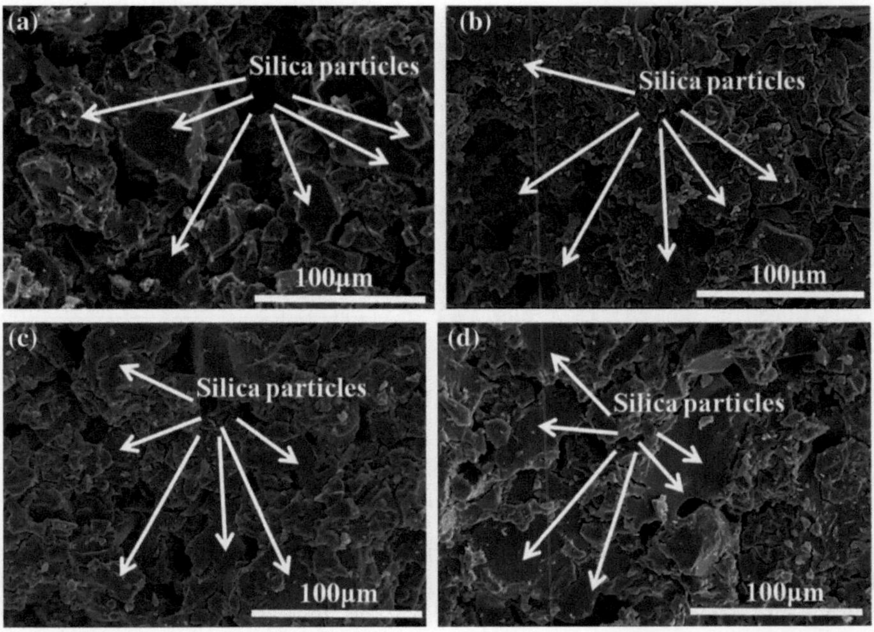

Fig. 4 Fracture morphologies of the samples as a function of pressure. **a** 5 Mpa; **b** 10 Mpa; **c** 15 Mpa; **d** 20 Mpa

high porosity. With the increasing of the compaction pressure, the size of pores or gaps decreases and the grains are increasingly tight in the samples. Correspondingly, the apparent porosity of the samples gradually decreases and the bulk-density gradually increases.

Figure 5a shows the crystallized zone morphology of the sample prepared at 20 Mpa pressure. It is seen obviously that the crystallized front exists on the surface of the silica grains. When the amorphous silica grains were used as the raw materials, the crystallization process happens from outside to inner of the grain during sintering, leading to the formation of cristobalite [2]. However, during the cooling, the cristobalite undergoes the phase transformation with volume contraction of about 5vol.% at about 200–270 °C, leading to the formation of more cracks at the crystallized front owing to the introduction of tensile stresses [6], as shown in Fig. 5a. XRD pattern of the sample prepared at 20 Mpa pressure is shown in Fig. 5b. It is seen that the phase composition of the sample is mainly composed of amorphous silica and a large number of crystallized cristobalite. During sintering, the increasing temperature gradually decreases the viscosity of glass phase, promoting the transformation of amorphous silica to cristobalite. Meanwhile, when the pressure is 20 Mpa, the distance between the amorphous silica particles is small in the sample, leading to decrease of the atom diffusion path. Thus, the cristobalite easily separates out from the amorphous silica in the sample.

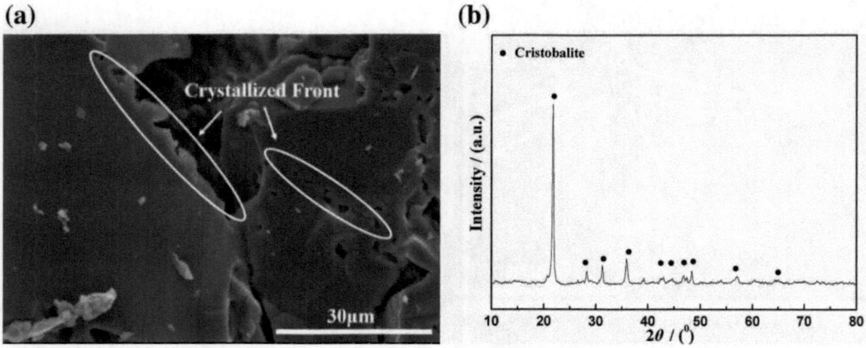

Fig. 5 **a** Crystallized zone morphology; **b** XRD pattern of the sample at 20 Mpa pressure

Conclusion

In this paper, silica ceramic cores have been prepared by dry-pressing method using the amorphous silica powders coated by silicone resin as the raw materials and subsequent sintering at 1250 °C for 2 h. The increasing pressure gradually decreases the shrinkage rate and apparent porosity of the samples. On the contrary, the bulk-density of the samples increases significantly, leading to the increasing of the bending strength at ambient temperature. Crystallization process starts from the surface of the silica grain and cracks occurs at the crystallized zone owing to the volume change during cooling.

Acknowledgements We would like to acknowledge the financial support provided by United Innovation Program of Shanghai Commercial Aircraft Engine (AR910), the National Natural Science Foundation of China (Nos. U1560202, 51690162).

References

1. J.C. Rotger, what is a good ceramic core. INCAST **6**, 21 (2008)
2. A. Kazemi, M.A. Faghihi-Sani, H.R. Alizadeh, Investigation on cristobalite crystallization in silica-based ceramic cores for investment casting. J. Eur. Ceram. Soc. **33**, 3397–3402 (2013)
3. G.H. Gu, L.M. Cao, Development of ceramic cores for investment casting hollow blades. Foundry Technol. **23**, 81–83 (2002)
4. L.M. Cao, The processing characters of ceramics core used for directional and single crystal blades. Mater. Eng. **5**, 20–21 (1995)
5. L. Huang, L. Duffrene, J. Kieffer, Structural transitions in silica glass: thermo-mechanical anomalies and polyamorphism. J. Non-Cryst. Solids **349**, 1–9 (2004)
6. S. Krug, J.R.G. Evans, J.H. Ter Maat, Residual stresses and cracking in large ceramic injection mouldings subjected to different solidification schedules. J. Eur. Ceram. Soc. **20**, 2535–2541 (2000)

7. Z.G. Yang, J.B. Yu, C.J. Li, W.D. Xuan, Z.Q. Zhang, K. Deng, Z.M. Ren, Preparing porous Si-based ceramic core using thermosetting silicon resin injection method. J. Inorg. Mater. **30**, 0147–0152 (2015)
8. Z.G. Yang, J.B. Yu, Z.M. Ren, C.J. Li, K. Deng, Fabrication and characterization of porous alumina-based ceramics using silicone resin as binder. Trans. Indian Ceram. Soc. **75**, 40–46 (2016)
9. J.B. Yu, Z.G. Yang, C.J. Li, K. Deng, Z.M. Ren, Investigation on properties of the silica ceramic cores for the hollow blades prepared by the conversion of the silicon resin. Key Eng. Mater. **680**, 267–271 (2016)

Performances of Multiphase Ceramics of SiC and Praseodymium Aluminates

Wengao Pan, Laner Wu, Yong Jiang and Zhenkun Huang

Abstract Multiphase ceramics of SiC based composites were prepared by hot pressed (HP) and pressureless sintering with Al_2O_3 and Pr_2O_3 as sintering additives. The compositions were designed according to the two binary compounds of $PrAlO_3$(PrAP) and $PrAl_{11}O_{18}$[β(Pr)] formed in Al_2O_3–Pr_2O_3 system. And the performances of the multiphase ceramics with different content of the second phases were investigated. The results showed that the weight loss, linear shrinkage and relative density of the samples in the group 1/B (SiC–$PrAl_{11}O_{18}$) containing more Al_2O_3 were higher than those of the samples in the group 1/1(SiC–$PrAlO_3$) by pressureless sintering, and its supreme relative density was 91.5%. By hot pressed sintering, the relative density of the densest sample in group 1/1 and group 1/B reached 94.8 and 95%, respectively. In group 1/1, sample with 10wt% of additives possessed the greatest hardness of 21.1 GPa, and the maximal flexure strength of 494.9 MPa was achieved when additives content increased to 15 wt%. The sample containing 10wt% of additives in group 1/B had the best fracture toughness of 7.4 MPa $m^{1/2}$.

Keywords SiC · Praseodymium aluminates · Multiphase ceramics
Performances

W. Pan · L. Wu (✉) · Y. Jiang · Z. Huang
School of Materials Science and Engineering,
Beifang University of Nationalities, Yinchuan 750021, China
e-mail: lanerwu@126.com

W. Pan
e-mail: wgao_pan@126.com

Y. Jiang
e-mail: jynxyc@126.com

Z. Huang
e-mail: zkh_08@163.com

Y. Han (ed.), *Advanced Functional Materials*,
https://doi.org/10.1007/978-981-13-0110-0_57

Introduction

SiC based ceramics are widely used owing to its super hardness, wear resistance, high temperature resistance, antioxidation as well as its better comprehensive performance. However, SiC is hard to density without additives owing to its strong covalent bond. In previous researches, additives of Al–B–C [1–3] were used for densification of SiC ceramics. Nevertheless, temperature of the solid-state sintering is high, fracture toughness is low and grains are coarse [4]. Subsequently, liquid-sintering technology mainly with oxide additives was developed. Recently, liquid phase sintering SiC ceramics with Al_2O_3–RE_2O_3 (RE-rare earth:La, Nd, Y, Yb, Dy, Er, Tm, Sc, Ho) additives have been reported, and Al_2O_3–Y_2O_3 system is most successful [5–8]. Pr_2O_3 is one of the rare earth oxides and can prevent growth of grains effectively [9], while less attentions are paid on SiC ceramics with Pr_2O_3–Al_2O_3 additives.

According to the related phase diagrams [10, 11], $PrAlO_3$(PrAP) and $PrAl_{11}O_{18}$ [β(Pr)] formed in Pr_2O_3–Al_2O_3 system, and SiC coexists with each of them. The melting point of β(Pr) is low (about 1850 °C), and samples with β(Pr) phase possess a higher hardness and shrinkage rate in previous experiment [11]. Accordingly, β(Pr) is considered to promote the densification of SiC ceramics effectively, and influences of PrAP is also worth investigating, simultaneously. In present work, multiphase ceramics of SiC–$PrAlO_3$ (Group 1/1) and SiC–$PrAl_{11}O_{18}$ (Group 1/B) were prepared and their performances were investigated.

Experimental

The starting powders used in the experiment were α-SiC (UF15, D_{50} = 0.6 μm, purity > 98% with 0.5 wt% O, H.C. Starck, Germany); Al_2O_3 (D_{50} = 0.3 μm, purity > 99.99%, Xuancheng Jingrui New Material Co., Ltd., China); Pr_6O_{11} (D_{50} = 2–3 μm, purity > 99.95%, Baotou Research Institute of Rare Earths, China); Commercial praseodymium oxide usually is Pr_6O_{11}, it can release lattice oxygen when temperature raised [12], which may oxidize SiC. Therefore, raw powder Pr_6O_{11} was calcined at 1200 °C for 2 h in a vacuum muffle furnace in argon atmosphere for getting purity Pr_2O_3 before experiment. Besides, Al_2O_3 powders were calcined at 1000 ºC for 2 h in air to remove the possible hydrates.

Eight samples using for the experiment were divided into two groups of 1/1 and 1/B(B = 11) according to the different mole rate of additive component (Pr_2O_3/Al_2O_3). Samples labeled with prefix as SPA (i.e. SiC/Pr_2O_3/Al_2O_3) and postfix as four-digit number where the front two and the rear two respectively represented the additive content and the mole ratio of Pr_2O_3/Al_2O_3, as shown in Table 1. Starting materials were manually mixed in an agate mortar for 2 h by using anhydrous ethanol as medium. After fully mixed and dried in an oven, batches of the powder mixtures were vertically axial pressed in steel dies of 20 mm × 20 mm at 80 MPa

Table 1 Sample code and its compositions

Additive contents (wt%)	Mole ratio (Pr_2O_3/Al_2O_3)	
	Group 1/1	Group 1/B
10	SPA1011	SPA101B
15	SPA1511	SPA151B
20	SPA2011	SPA201B
25	SPA2511	SPA251B

*B = 11

for 60 s and then cold isostatic pressed under 250 MPa for 60 s. Sintering were carried out in a graphite furnace (Materials Research Furnaces, Inc., USA) in Ar atmosphere at 1800–2000 °C for 1 h. Some samples were hot pressed (HP) under 30 MPa. The heating and cooling rate was 10 and 20 °C/min, respectively.

Results and Discussion

Weight Loss. The components of the samples contained with SiC, Al_2O_3, Pr_2O_3 and also SiO_2 impurity from SiC powder, and chemical reactions shown in formula (1) and (2) may take place in sintering process [13, 14]. Gas of Al_2O, CO, SiO formed and volatilized, therefore, mass of the biscuit reduced.

$$T > 1950\,K; 2SiO_2 + SiC \rightarrow 3SiO(g) + CO(g) \quad (1)$$

$$T > 2123\,K; SiC + Al_2O_3 \rightarrow Al_2O(g) + CO(g) + SiO(g) \quad (2)$$

Results shown that, weight loss of two groups with different additive mole ratio increased as the sintering temperature and additive content raised. Curves of the weight loss of group 1/1 and 1/B changing with temperature were respectively shown in Fig. 1a and b. Weight loss of group 1/B was generally higher than that of 1/1, and both of them achieved the maximum of 34% and 13% at 2000 °C, respectively. Comparing with group 1/1, group 1/B contained more Al_2O_3 and could react with more SiC substrate to form more gas, which promoted the weight loss. Besides, it can be revealed from Fig. 1b that weight loss of group 1/B was sensitive and it exceeded the total additive content at 1950 °C, indicating the occurrence of over sintering. Wight loss of sample SPA251B was 16.5% at 1900 °C and reached up to 34% at 2000 °C. It was reasonable that sintering temperature of group 1/B should be controlled at below 1900 °C considering the sintering activity of SiC, simultaneously.

Linear Shrinkage. Linear shrinkage of samples derived from the average value of radial shrinkage and altitudinal shrinkage. Both linear shrinkage percentage of two groups soared as temperature and additive content increased. Linear shrinkage of group 1/1 and 1/B was 9.5 and 20% at 2000 °C, respectively, as shown in

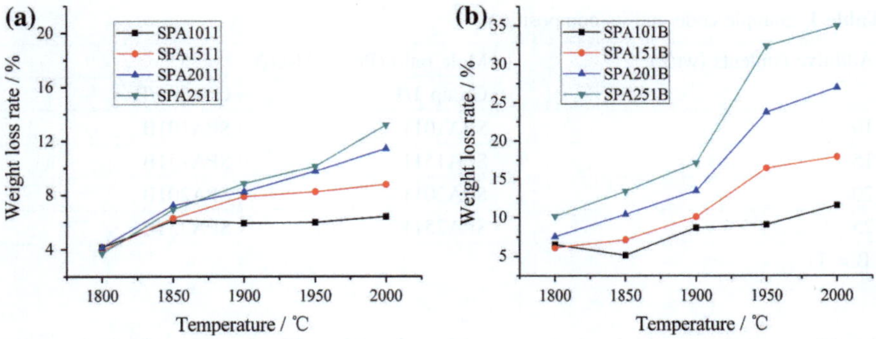

Fig. 1 Weight loss rate of two groups under different temperatures. **a** Group 1/1; **b** group 1/B

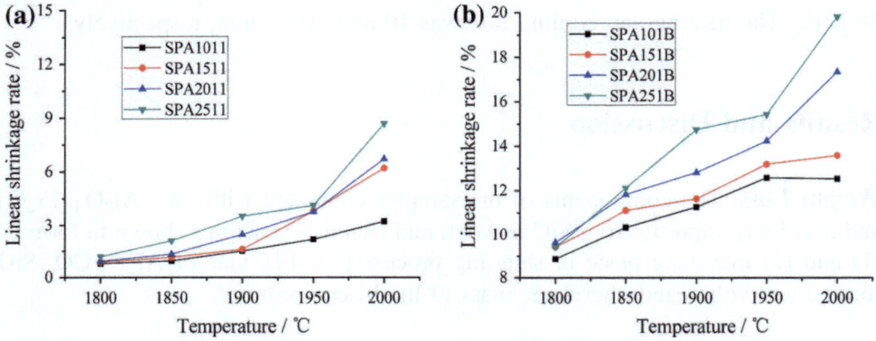

Fig. 2 Linear shrinkage percentages of two groups under different temperatures. **a** Group 1/1; **b** group 1/B

Fig. 2a and b. Linear shrinkage of group 1/1 was relatively less owing to melting point of the second phase $PrAlO_3$ was higher (about 2100 °C), and the linear shrinkage may increase as temperature further raised. Increasing content of Al_2O_3 improved the linear shrinkage of group 1/B, and SiC ceramic samples in the group should possess a higher relative density, theoretically.

Relative Density. Figure 3a and b were relative density of group 1/1 and 1/B, respectively. It can be seen that, relative density of group 1/1 generally increased with increment of the sintering temperatures although it had local fluctuation, and sample SPA2511 with the most content of additives achieved its supreme relative density of 78% at 2000 °C. While relative density of group 1/B presented a trend of first increase and then decrease along with the temperatures raising, and sample SPA101B with the least content of additives possessed the highest relative density of 91.5%. Comparing with group 1/1, both weight loss and linear shrinkage of group 1/B were obviously larger owing to the more content of Al_2O_3. Densification of SiC ceramics in group 1/B could be improved effectively if temperatures were suitable.

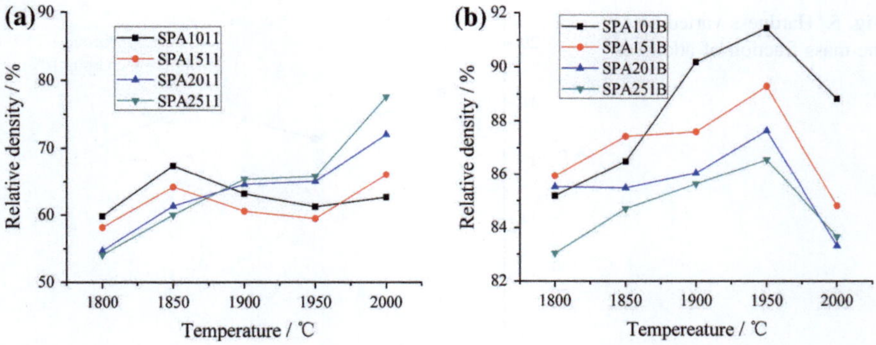

Fig. 3 Relative density of two groups under different temperatures. **a** Group 1/1; **b** group 1/B

Samples with better experimental results by pressureless sintering were further investigated by hot pressed sintering at 1850 °C, and their properties as well as microstructure were analyzed. The relative density of group 1/1 and 1/B by HP were shown in Fig. 4. It indicated that densification of SiC ceramics in two groups were achieved at 1850 °C, respectively. Relative density of the both groups reached about 95% and sintering temperatures reduced comparing with those by pressureless sintering. Influence of HP process was obvious on group 1/1 while obscure on group 1/B. Along with increment of additive content, relative density of group 1/1 decreased while that of group 1/B increased first and then decreased and achieved its maximum at 20 wt% additives, which was different from the situation by pressureless sintering that relative density decreased with increment of additive content.

Hardness. Hardness, fracture toughness, flexural strength and microstructure of samples sintered by HP at 1850 °C were investigated. Hardness was measured by micro-vickers (432SVD, Wolpert Measuring Instruments Ltd.), no less than five

Fig. 4 Relative density of two groups prepared by HP at 1850 °C

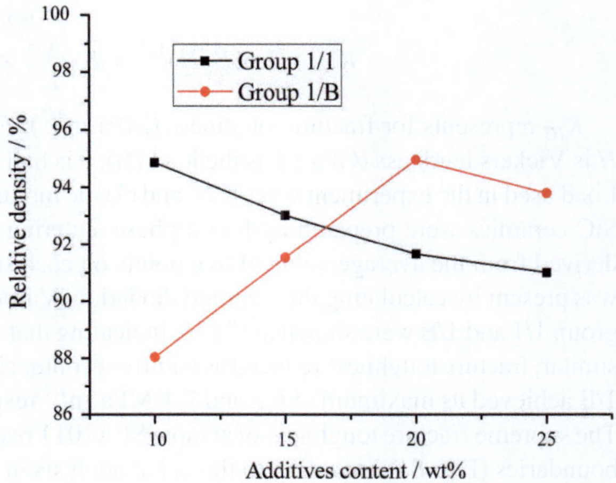

Fig. 5 Hardness varied on the mass fraction of additives

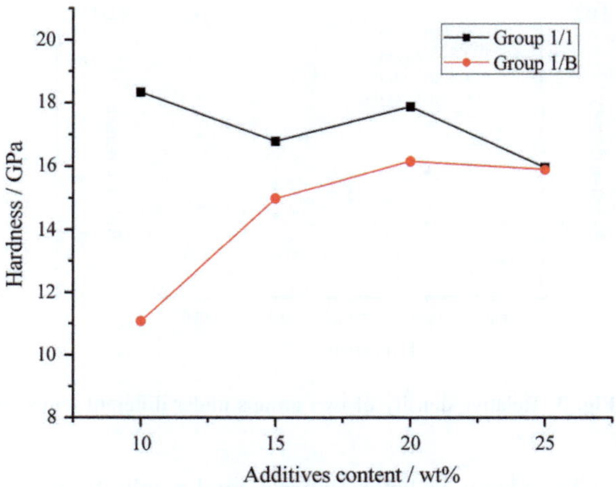

points in each sample were chosen for hardness test, and units of HV were transformed into GPa. Result shown that hardness of group 1/1 was higher than that of group 1/B, generally, as shown in Fig. 5. Along the increasement of additives content, hardness of samples in group 1/1 fluctuated at 19.5 GPa and had its peak value of 21.1 GPa at 10 wt% additives, while that of group 1/B soared first and then tend to stabilization with the maximum of 19.6 GPa at 25 wt% additives. Effective bearing area reduced owing to the presence of pores in powder metallurgy material and its performance decreased. Hardness and strength of materials increased generally as its density raised [15]. Therefore, variation trends of hardness and relative density of samples in group 1/B were coincident. The fine and equiaxed grains (Fig. 8c) in sample SPA1011 contributed to its higher hardness.

Fracture Toughness. It reflects the capability of materials in resistance to crack propagation. According to indentation method, fracture toughness was measured and calculated by formula (3).

$$K_{IC} = 1.6(E/H)^{1/2} \times F/c^{3/2} \times 10^{-8} \tag{3}$$

K_{IC} represents for fracture toughness (MPa $m^{1/2}$); E is elasticity modulus (GPa); H is Vickers hardness (GPa); F is the load (N); c is half length of crack diagonal (m). Load used in the experiment was 98 N, and elastic modulus was about 390 GPa when SiC ceramics were prepared by liquid phase sintering [16, 17]. Fracture toughness derived from the average value of five points on each sample, and dispersion degree was present by calculating the standard deviation. Variations of fracture toughness in group 1/1 and 1/B were shown in Fig. 6, indicating that variations of two groups were similar, fracture toughness reduced as additives content increased. Both group 1/1 and 1/B achieved its maximum of 6.2 and 7.4 MPa $m^{1/2}$ respectively at 10wt% additives. The supreme fracture toughness of sample SPA1011 resulted from its circuitous grain boundaries (Fig. 8a) according to the detail analysis in microstructure section.

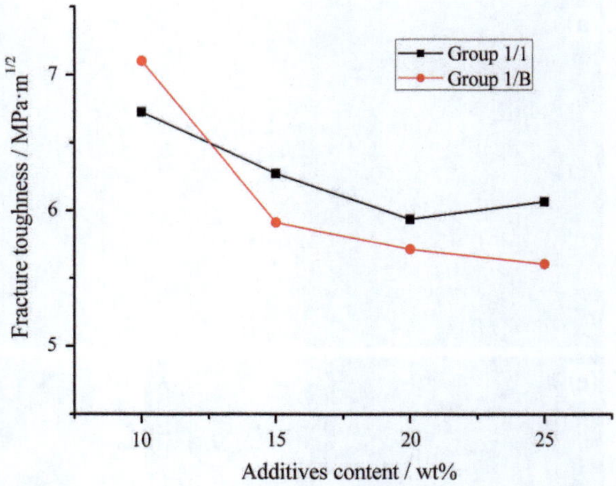

Fig. 6 Fracture toughness varied on the mass fraction of additives

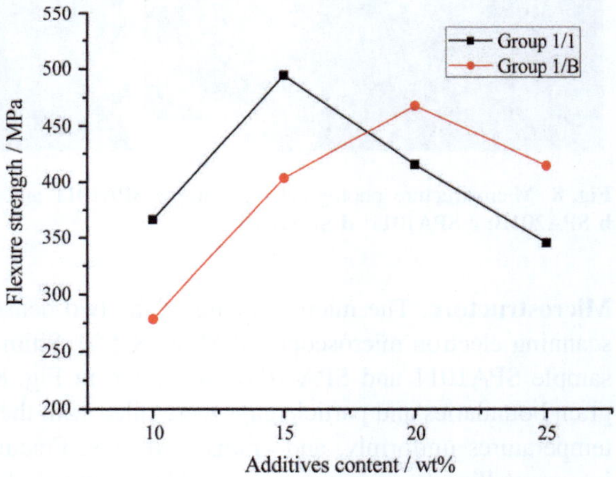

Fig. 7 Flexural strength varied on the mass fraction of additives

Flexural Strength. The flexural strength of the sample was determined by three points bending method on a universal testing machine (CMT5305, MTS Industry Co. Ltd.), and five strips in each sample were selected for test. The flexural strength of group 1/1 and 1/B are shown in Fig. 7, indicating that both of them increased first and then decreased. Besides, group 1/1 and 1/B achieved the peak flexural strength of 494.9 and 467.7 MPa at 15 and 20 wt% additives, respectively. Variation trends of flexural strength in two groups were consistent, while the maximum values and its corresponding additives content were different. Strength of material is closely related to its density as mentioned above. Accordingly, both relative density and flexural strength of group 1/1 were higher than those of group 1/B when additives content was low, while they decreased as additives content raised. Grains with a tendency of growing axially (Fig. 8d) in sample SPA201B gave rise to the situations of grains drawing and then the enhancement of its flexural strength.

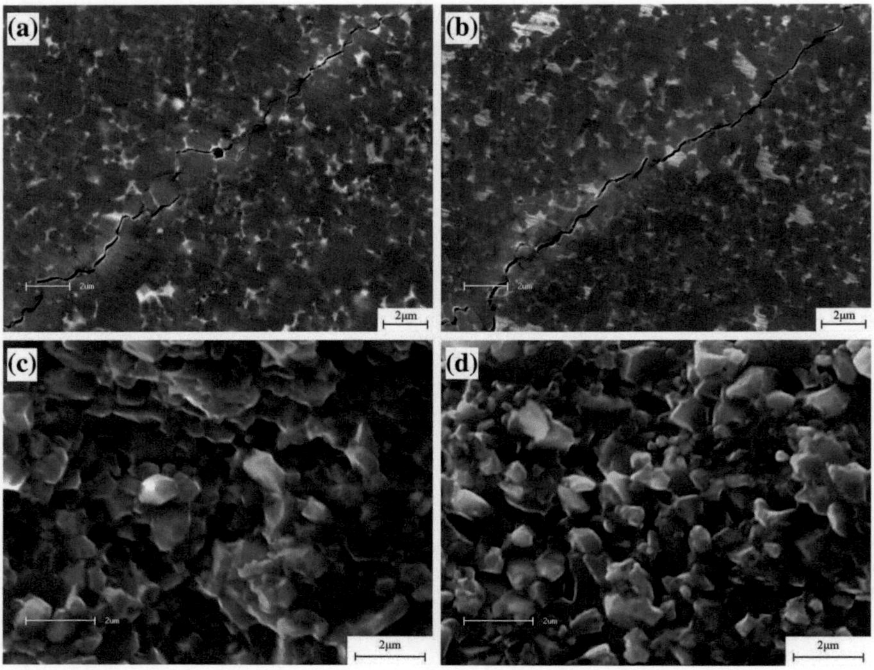

Fig. 8 Microstructure photographs of sample SPA1011 and SPA201B by SEM. **a** SPA1011; **b** SPA201B; **c** SPA1011; **d** SPA201B

Microstructure. The microstructure of the two densest samples was observed by scanning electron microscope (SEM, SSX-550, Shimadzu). Surface photograph of sample SPA1011 and SPA201B are shown in Fig. 8a and b. It can be seen that grain boundaries and particle gaps were filled with the second phase formed at high temperatures uniformly, and grains were fine. Fracture mode of two samples are intercrystalline fracture typically, cracks extended along the grain boundaries circuitously from lower left quarter to top right corner. Materials were in unstable state when applied stress exceeded the critical value, and cracks extend preferentially from source along the route with a weaker binding force. Liquid phase filling in grain boundaries weakened the binding force between interfaces, which made cracks extend along grain boundaries circuitously rather than thread grains directly. Besides, the stress concentration resulting from flaws reduced owing to the grain refinement, which decreased crack produce from the source. The grain sizes of the sample SPA201B and SPA1011 were similar (1–2 μm), while grain boundaries of sample SPA1011 were more circuitous, absorbing more energy before breakage, therefore, the sample possessed a higher fracture toughness.

Figure 8c and d were fracture photograph of sample SPA1011 and SPA201B, respectively. It showed that no abnormal growth occurred, and SiC particles were bonded by the second phase that distributed in matrix uniformly. Besides, grains drawing which can result in improvement of strength and fracture toughness happened in both of the two samples when they were bending broken. Comparing with the equiaxed grains of sample SPA1011, grains of sample SPA201B tended to grow axially and grains pull-out more obviously to improve its flexural strength.

Conclusions

Relative densities of the samples in group 1/B(SiC–$PrAl_{11}O_{18}$) are generally higher than those of the samples in group 1/1(SiC–$PrAlO_3$) by pressureless sintering. Group 1/B with more Al_2O_3 possesses the highest relative density of 91.5%, indicating its effective influence on the densification of SiC ceramics.

By hot pressed sintering at 1850 °C, the samples with 10 wt% of additives in group 1/1 possess the maximal relative density of 94.8%, the supreme hardness of 21.1 GPa and the best fracture toughness of 6.2 MPa $m^{1/2}$, while the samples containing 15 wt% of additives has the greatest flexural strength of 494.9 MPa. In group 1/B, the relative density and flexural strength is the highest (95% and 467.7 MPa, respectively), when the samples contained 20 wt% additives. The samples with 25 wt% additives achieve the super hardness of 19.6 GPa, while the samples with 10 wt% additives obtain the best fracture toughness of 7.4 MPa $m^{1/2}$. In terms of the microstructure, the grains are uniform and fine (1–2 μm), besides, the grains drawing has been observed when they are broken and the fracture mode is intercrystalline fracture typically.

Acknowledgements The present work was financially supported by National Natural Science Foundation of China (NSFC51362001) and Colleges and Universities Oversea Talents Introduce Project (X2017006).

References

1. R.A. Alligeo, L.B. Coffin, J.R. Tinklpaugh, Pressure-sintered silicon carbide. J. Am. Ceram. Soc. **39**(11), 386–389 (1956)
2. W. Bocker, H. Landfermann, H. Hausner, Sintering of alpha silicon carbide with additions of aluminum. Powder Metall. Int. **11**(2), 83–85 (1979)
3. K. Biswas, Solid state sintering of SiC-ceramics. Mater. Sci. Forum **624**, 71–89 (2009)
4. Y.W. Chang, *Researches on preparation of high performance SiC ceramic materials* (Nanjing Univ. Sci. Technol., Nanjing, 2009)
5. J.L. Hu, C.Y. Hu, X. Liu et al., Research progress on preparation methods of SiC powder and its sintering aids. Bull. Chin. Ceramic Soc. **33**(9), 2280–2284 (2014)
6. H.W. Yu, K. Raju, Y.P. Ji et al., Effects of Al_2O_3–RE_2O_3 additive for the sintering of SiC and the fabrication of SiCf/SiC composites. J. Korean Ceram. **50**(6), 364–371 (2013)
7. F. Rodríguez-Rojas, A.L. Ortiz, F. Guiberteau et al., J. Eur. Ceram. Soc. **30**(15), 3209–3217 (2010)
8. H.Q. Liang, X.M. Yao, H. Zhang et al., The effect of TiC on the liquid phase sintering of SiC ceramics with Al_2O_3 and Y_2O_3 additives. Key Eng. Mater. **602–603**, 197–201 (2014)

9. T.T. Li, C.Q. Peng, R.C. Wang et al., Effects of $Pr_6\ O_{11}$ and sintering process on relative density and thermal conductivity of BeO ceramics. Min. Metal. Eng. **30**(5), 104–107 (2010)
10. M. Mizuno, T. Yamada, T. Noguchi, Phase diagram of the system Al_2O_3-Pr_2O_3 at high temperature. J. Ceram. Soc. Jpn. **85**(1), 24–29 (1977)
11. W. Pan, L. Wu, Y. Jiang et al., J. Ceram. Sci. Technol. **7**(4), 433–440 (2016)
12. P.L. Ying, C. Li, Q. Xin, J. Catal. **14**(6), 488–492 (1993)
13. V.G. Borisov, B.F. Yudin, Reaction thermodynamics in the SiO_2–SiC system. Refract. Ind. Ceram **9**(3), 162–165 (1968)
14. T. Grande, H. Sommerset, E. Hagen et al., Effect of weight loss on liquid-phase-sintered silicon carbide. J. Am. Ceram. Soc. **80**(4), 1047–1052 (1997)
15. X.H. Qu, *Principle and Process of Powder Metallurgy* (Metallurgical Industry Press, Beijing, 2013)
16. Y.H. Chen, F. Han, Y. Jiang et al., Liquid phase sintered silicon carbide with AlN-Y_2O_3. Ordnance Mater. Sci. Eng. **32**(5), 1–3 (2009)
17. Y.H. Chen, L.E. Wu, Y. Jiang et al., Effective of AlN content on liquid sintering silicon carbide. China Ceram. **45**(7), 14–16 (2009)

Influence of SPS Sintering Temperature on Properties of ZrB_2–SiC–Cr_3C_2 Ceramic

Qi Li, Fengwei Guo, Lamei Cao and Xiaosu Yi

Abstract To investigate the influence of sintering temperature on properties of ZrB_2-16vol.%SiC-4vol.%Cr_3C_2 ceramic (ZSC), ZSC was prepared by spark plasma sintering (SPS) process with different sintering temperature (1700 and 1800 °C). The microstructures and mechanical properties were characterized, and the oxidation behaviors were mainly studied at 1500 °C in air. The results showed that ZSC sintered at 1700 °C had better mechanical properties with higher hardness (13.37 GPa), fracture toughness (5.3 MPa $m^{1/2}$), bending strength at 25 °C (529 MPa) and 1600 °C (128 MPa). However, ZSC sintered at 1800 °C had better oxidation resistance, the thickness of each oxidation layer was thinner and the oxidation depth was much shallower than ZSC sintered at 1700 °C.

Keywords ZrB_2–SiC · Cr_3c_2 · Ceramic · Spark plasma sintering
Sintering temperature

Introduction

In recent years, hypersonic vehicles have became a hotspot of new equipment system in the world. During supersonic flight and re-entry process, the hot-end components and thermal protection materials will serve in extremely harsh envi-

Q. Li (✉) · F. Guo · L. Cao · X. Yi
Science and Technology on Advanced High Temperature
Structural Materials Laboratory, Beijing Institute of Aeronautical Materials,
Beijing 100095, China
e-mail: liqi1988china@126.com

F. Guo
e-mail: pandaguofw@163.com

L. Cao
e-mail: lamei.cao@biam.ac.cn

X. Yi
e-mail: yi_xiaosu@sina.cn

Y. Han (ed.), *Advanced Functional Materials*,
https://doi.org/10.1007/978-981-13-0110-0_58

ronment at high temperature, which are supposed to have excellent comprehensive properties. ZrB_2 based ceramic has become one of the most attractive candidate materials in this field due to its high melting point, high strength, high thermal conductivity and relatively low density [1–4].

Spark plasma sintering (SPS) technique gets rapid development in preparation of ZrB_2 based ceramic with high performance recently, due to its low sintering temperature (200–300 °C lower than traditional sintering process), shorter sintering time (only 3–10 min), low energy consumption, etc [4–8]. During SPS process, sintering temperature is one of the most important factors. The effect of SPS sintering temperature on properties of ZrB_2–SiC (20–60 vol.%) ceramics was researched by Akin during 2000–2100 °C, which suggested that the sintering temperature had a significant influence on microstructures and mechanical properties [6]. ZrB_2–SiC–Cr_3C_2 was proved having better mechanical properties and oxidation resistance than ZrB_2–SiC [9], however the influence of sintering temperature on properties of ZrB_2–SiC–Cr_3C_2 haven't been reported.

In this article, effect of sintering temperature on ZrB_2-16 vol.% SiC-5 vol.% Cr_3C_2 (ZSC) is studied. Two kinds of ZSC ceramics were successfully prepared by SPS process with sintering temperature of 1700 °C and 1800 °C (abbreviated as ZSC1700 and ZSC1800, separately). The microstructures and mechanical properties of ZSC1700 and ZSC1800 were compared. The mass gains at 1500 °C in air and the cross section analysis after oxidizing for 10 h were analyzed to research the effect of sintering temperature on oxidation resistance of ZSC. The results could give technical support for preparing ZSC with excellent comprehensive properties.

Experiment

Commercial powders produced by Qinhuangdao Eno High-Tech Material Development CO., LTD. were used to prepare the ceramics. ZrB_2, particle size: 10–15 μm, purity: 99.5%; α-SiC, particle size: 2 μm, purity: 99.9%; Cr_3C_2, particle size: 1–3 μm, purity: 99.5%.

SPS Sintering to Prepare ZSC1700 and ZSC1800. The powder mixtures mixed with a volume proportion of ZrB_2:SiC:Cr_3C_2 = 80:16:4 were filled into the ZrO_2 jar and ball milled for 6 h taking ZrO_2 as media. Then, the slurries were dried in a rotary evaporator and sieved. The dried powder mixture was then filled into a graphite die in SPS furnace under argon atmosphere to sinter for 5 min at 1700 or 1800 °C with a mechanical pressure of 30 MPa and a heating rate of 100 °C/min. Taking the graphite die cooled to room temperature out to get ZSC1700 or ZSC1800 ceramic.

Measurement. Apparent density was calculated by the Archimedes drainage method. The microstructure and phase composition of ceramics are studied using the XRD method and a Nova Nano scanning electron microscopy (SEM). The chemical composition of the phases formed was inferred from energy dispersive X-ray spectroscopy (EDX). Vickers microhardness (HV 0.5) was measured with a load of 4.9 N on a hardness tester. Fracture toughness was measured on a CERA brittle bending machine with 30 mm span and 0.02 mm/min rate of head movement. Three-point bending strength was measured on a INSTRON 3365 bending tester with 30 mm span and 0.5 mm/min rate of head movement. Three-point bending strength at 1600 and 1800 °C were tested on a YKM-2200 bending tester with 30 mm span and 0.5 mm/min rate of head movement, the heating rate was 40 °C/min, holding time was 10 min. The oxidation experiment wes carried out on a tube furnace at 1500 °C in air.

Results and Discussions

Microstructures. SEM images of ZSC1700 and ZSC1800 were showed in Fig. 1. The microstructures showed that they were both formed by three phases: gray phase, black phase and dark gray phase. Three phases distributed evenly and there were very little pores. The EDX analysis indicated that gray phase was ZrB_2, black phase was SiC, and dark gray phase was Cr–B, which was possibly produced by the reaction of Cr_3C_2 and B_2O_3 during SPS sintering. Due to the existence of Cr–B, the bonding force between different phases was enhanced. By increasing the sintering temperature, the ZrB_2 grain size grew significantly and SiC grain got local aggregation in ZSC1800 which was not good for the mechanical properties [10, 11]. Besides, the new formed Cr–B phase could fill the pores in the ceramic to increase the relative density. The theoretical densities, apparent densities and relative densities are listed in Table 1. The theoretical density was calculated from the volume proportion and densities of ZrB_2, SiC and Cr_3C_2, regardless the reaction between ZrB_2 and Cr_3C_2 during sintering. Practical density divided by theoretical density to get relative density. The apparent densities were both close to theoretical densities. The relative density of ZSC1700 was 95.5%. By increasing the sintering temperature, the relative density of ZSC1800 increased to 98.9%.

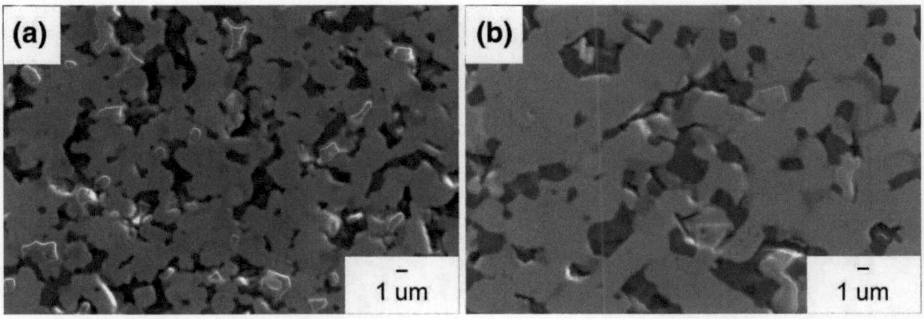

Fig. 1 SEM images of ZSC1700 (**a**) and ZSC1800 (**b**)

Table 1 Theoretical, apparent and relative densities of ZSC1700 and ZSC1800

Sample	ρ_0 (g/cm^3)	ρ (g/cm^3)	Relative density (%)
ZSC1700	5.65	5.40	95.5
ZSC1800	5.65	5.58	98.9

Table 2 Mechanical properties of ZSC1700 and ZSC1800

Sample	HV0.5 (GPa)	Fracture toughness at room temperature (MPa m$^{1/2}$)	Bending strength (MPa)		
			25 °C	1600 °C	1800 °C
ZSC1700	13.4	5.3	529	128	32
ZSC1800	11.0	3.7	457	77	Bending

Mechanical Properties. The Vicker's hardness, fracture toughness, and bending strength at 25, 1600 and 1800 °C of ZSC1700 and ZSC1800 were measured and listed in Table 2. By increasing the sintering temperature to 1800 °C, the mechanical properties got worse. The hardness decreased from 13.4 GPa to 11.0 GPa, the reason is that the increased sintering temperature is beneficial to form the soft Cr–B phase which make the hardness lower. The fracture toughness decreased from 5.3 to 3.7 MPa·m$^{1/2}$ was due to the growth of ZrB_2 grain which improved the brittleness.

The bending strength at 25 °C decreased significantly from 529 MPa of ZSC1700 to 457 MPa of ZSC1800. The bending strength at 1600 °C decreased from 128 MPa of ZSC1700 to 77 MPa of ZSC1800. At 1800 °C, the bending strength of ZSC1700 was 32 MPa, but the bending strength of ZSC1800 couldn't be measured due to the bending deformation. Because of the grain growth at higher sintering temperature, the bending strength decreased. Besides, the formation of soft Cr–B phase was improved also decreased the bending strength, the bending deformation of ZSC1800 at 1800 °C also proved this. According to the cross sectional SEM images of ZSC1700 and ZSC1800 fractured at 25 °C (Fig. 2), the grain size of ZSC1800 is larger. A large number of transgranular cleavage plane could be observed on both of ZSC1700 and ZSC1800. The fracture of ZSC1700 and ZSC1800 were both transgranular mode.

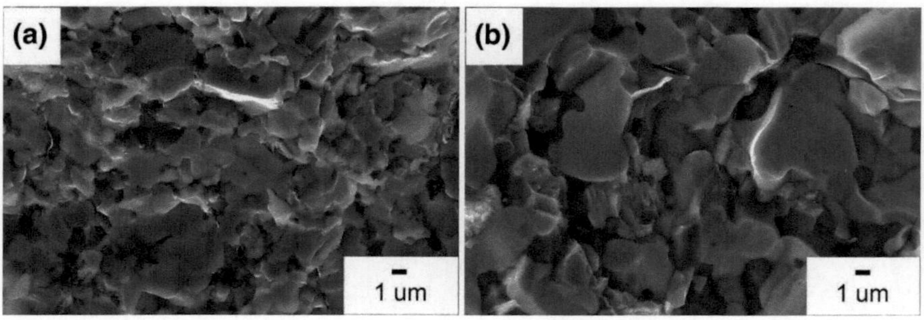

Fig. 2 Cross sectional SEM images of ZSC1700 (**a**) and ZSC1800 (**b**) fractured at 25 °C

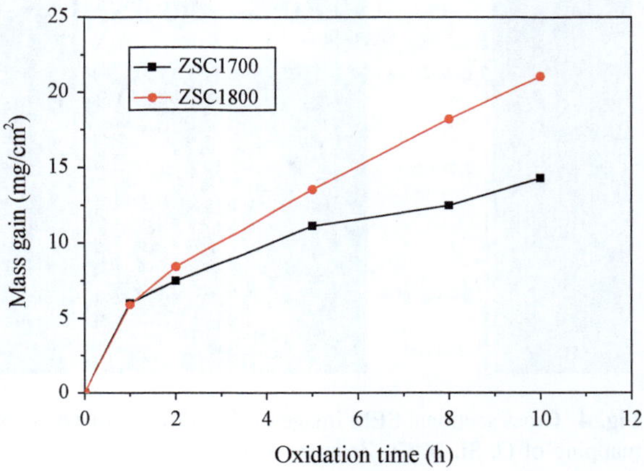

Fig. 3 Mass gain changes of ZSC1700 and ZSC1800 oxidized at 1500 °C in air as a function of time

Oxidation Behavior. Oxidation experiments were carried out at 1500 °C in air, the mass gains of ZSC1700 and ZSC1800 with different oxidation time were lined in Fig. 3. Before 1 h, the mass gain rates of ZSC1700 and ZSC1800 were fast and their mass gains were almost the same. After 1 h, The growth rates of mass gain got slow down。The gap between mass gains of ZSC1700 and ZSC1800 increased gradually. After oxidation for 10 h, the mass gain of ZSC1700 was 14.3 mg/cm^2, which was much lesser than that of ZSC1800 (21.0 mg/cm^2).

The oxide scale of ZSC could be divided into four layers: SiO_2 rich layer, Cr depletion layer, ZrO_2 rich layer and SiC depletion layer, which is different from the three layers (SiO_2 rich layer, ZrO_2 rich layer and SiC depletion layer) of oxide scale on ZrB_2–SiC ceramic [12–14]. During the oxidation process, Cr oxide (Cr_2O_3) was formed, Cr_2O_3 could evaporate under high oxygen pressure close to the surface, and was nonvolatile under low oxygen pressure far from the surface [15]. Therefore, Cr depletion layer was formed under the SiO_2 rich layer.

The cross sectional image of oxide scale on ZSC1700 oxidized at 1500 °C in air for 10 h is shown in Fig. 4. According to the elemental mapping of O, Si, B, Cr, Zr at the same area, the thickness of each layer can be calculated and are listed in Table 3. The SiO_2 rich layer has high content of Si and O, the thickness is about 80 μm. The Cr depletion layer has no Cr, about 50 μm. The ZrO_2 rich layer has high content of O, but has no Si and B, about 370 μm. The SiC depletion layer has high content of Zr and B, but has no Si, about 130 μm. The cross sectional image of oxide scale on ZSC1800 oxidized at 1500 °C in air for 10 h (Fig. 5) was almost the same, but with different thickness of each layer: the SiO_2 rich layer is about 95 μm, the Cr depletion layer is about 20 μm, the ZrO_2 rich layer is about 120 μm, the SiC depletion layer is about 95 μm.

After oxidation for 10 h, a glossy and dense SiO_2 rich layer was formed both on the surfaces of ZSC1700 and ZSC1800. The SiO_2 rich layer on ZSC1800 was thicker than ZSC1700, which can prevent infiltration of oxygen more efficient. Cr_2O_3 could evaporate under high oxygen pressure. Therefore the thickness of Cr depletion layer could indicate the difficulty degree of oxygen infiltration. The

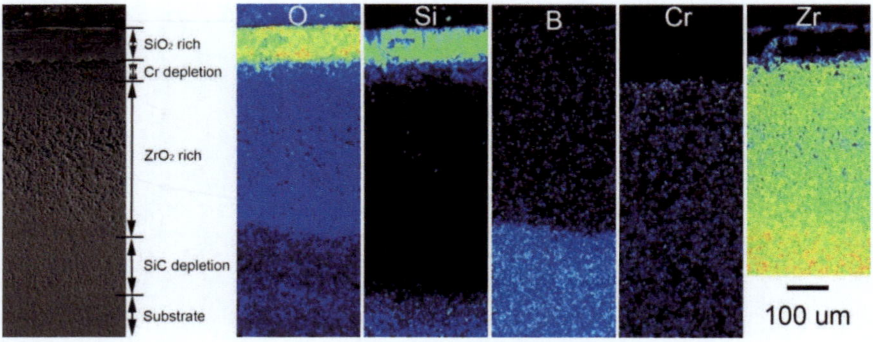

Fig. 4 Cross sectional SEM image of ZSC1700 oxidized at 1500 °C in air for 10 h and elemental mapping of O, Si, B, Cr, Zr at the same area

Table 3 Thicknesses of oxidation layers on ZSC1700 and ZSC1800

Sample	Thickness (μm)					
	Oxidation layer	SiO_2 rich layer	Cr depletion layer	ZrO_2 rich layer	SiC depletion layer	Oxidation depth
ZSC1700	630	80	50	370	130	550
ZSC1800	330	95	20	120	95	235

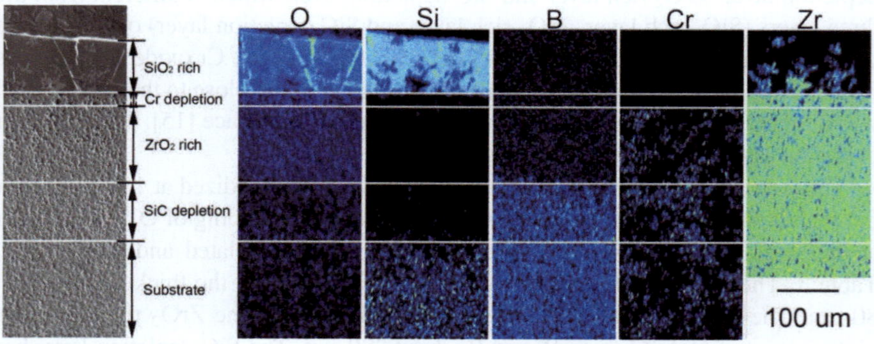

Fig. 5 Cross sectional SEM image of ZSC1800 oxidized at 1500 °C in air for 10 h and elemental mapping of O, Si, B, Cr, Zr at the same area

thinner the Cr depletion layer is, the more difficult the oxygen infiltrates. The Cr depletion layer of ZSC1800 is much thinner than ZSC1700, suggest that the oxidation resistance of ZSC1800 is better. Due to the evaporation of oxide, ZrO_2 rich layer became loose. The ZrO_2 rich layer of ZSC1800 is denser than ZSC1700, which is due to the high relative density of ZSC1800. In ZSC1700, part of the oxides evaporated and the volume expansion of oxide [15] couldn't fill the porous

ZrO_2 rich layer, therefore the infiltration rate of oxygen was faster. Thus, the ZrO_2 rich layer in ZSC1700 was much thicker. The oxygen infiltration in ZSC1800 with high relative density was difficult, therefore SiC depletion layer of ZSC1800 was thinner. According to the thicknesses of Cr depletion layer, ZrO_2 rich layer and SiC depletion layer, the oxidation depth of ZSC1700 was about 550 μm, the oxidation depth of ZSC1800 was about 235 μm which was much shallower than ZSC1700.

Although the mass gain of ZSC1700 oxidized for 10 h at 1500°C in air was lesser than ZSC1800, the Cr depletion layer, ZrO_2 rich layer and SiC depletion layer of ZSC1800 were all much thinner than ZSC1700. The oxidation depth of ZSC1800 was much shallower than ZSC1700. Therefore, the oxidation resistance of ZSC1800 was better than ZSC1700. The higher mass gain of ZSC1800 was due to the less evaporation of oxide. In a word, the oxidation resistance of ZSC was significantly enhanced by increasing the SPS sintering temperature.

Summary

ZSC ceramics have been successfully prepared by SPS process with sintering temperature of 1700 °C and 1800 °C. Cr–B phase forms during sintering due to the reaction of Cr_3C_2 and oxide on the grain surface, the bonding force between different phases enhances. By increasing the sintering temperature, the grain grew significantly, the relative density got higher. ZSC1700 have better mechanical properties with higher hardness (13.37 GPa), fracture toughness (5.3 $MPa{\cdot}m^{1/2}$), bending strength at 25 °C (529 MPa) and 1600 °C (128 MPa). But ZSC1800 has better oxidation resistance that the thickness of each oxidation layer is much thinner and the oxidation depth is much shallower. In summary, ZSC ceramic sintered at 1700 °C has better mechanical properties, but has better oxidation resistance when sintered at 1800 °C.

References

1. M.M. Opeka, I.G. Talmy, E.J. Wuchina, J.A. Zaykoski, S.J. Causey, Mechanical, thermal, and oxidation properties of refractory hafnium and zirconium compounds. J. Eur. Ceram. Soc. **13**, 2405 (1999)
2. K. Upadhya, J.M. Yang, W.P. Hoffman, Advanced materials for ultrahigh temperature structural applications above 2000 °C. Am. Ceram. Soc. Bull. **0**, 51 (1997)
3. W.G. Fahrenholtz, G.E. Hilmas, I.G. Talmy, J.A. Zaykoski, Refractory diborides of zirconium and hafnium. J. Am. Ceram. Soc. **5**, 1347 (2007)
4. S. Chakraborty, P.K. Das, D. Ghosh, Spark plasma sintering and structural properties of ZrB_2 based ceramics: a review. Rev. Adv. Mater. Sci. **0**, 182 (2016)
5. A. Bellosi, F. Monteverde, S. Diletta, Fast densification of ultra-high-temperature ceramics by spark plasma sintering. Int. J. Appl. Ceram. Technol. **1**, 32 (2006)

6. I. Akin, M. Hotta, F.C. Sahin, O. Yucel, G. Goller, T. Goto, Microstructure and densification of ZrB_2–SiC composites prepared by spark plasma sintering. J. Eur. Ceram. Soc. **11**, 2379 (2009)
7. L. Zhang, D.A. Pejakovic, J. Marschall, M. Gasch, Thermal and electrical transport properties of spark plasma-sintered HfB_2 and ZrB_2 ceramics. J. Am. Ceram. Soc. **8**, 2562 (2011)
8. Q. Li, L. Cao, X. Yi, Oxidation behavior of a SPS sintered ZrB_2–SiC–$MoSi_2$ ceramic at 1500 °C. MATEC Web Conf. **67**, 06079 (2016)
9. O.N. Grigoriev, V.B. Vinokurov, L.I. Klimenko, N.D. Bega, N.I. Danilenko, Sintering of zirconium diboride and phase transformations in the presence of Cr_3C_2. Powder Metall. Metal Ceram. **3**, 185 (2016)
10. W.G. Fahrenholtz, E.J. Wuchina, W.E. Lee, Y. Zhou, *Ultra-high Temperature Ceramics: Materials for Extreme Environment Applications* (Wiley, Hoboken, 2016), pp. 167–196
11. J. Watts, G. Hilmas, W.G. Fahrenholtz, Mechanical characterization of ZrB_2–SiC composites with varying SiC particle sizes. J. Am. Ceram. Soc. **12**, 4410 (2011)
12. E. Opila, S. Levine, J. Lorincz, Oxidation of ZrB_2- and HfB_2-based ultra-high temperature ceramics: effect of Ta additions. J. Mater. Sci. **19**, 5969 (2004)
13. O.N. Grigoriev, B.A. Galanov, V.A. Lavrenko, A.D. Panasyuk, S.M. Ivanov, A.V. Koroteev, K.G. Nickel, Oxidation of ZrB_2–SiC–$ZrSi_2$ ceramics in oxygen. J. Eur. Ceram. Soc. **11**, 2397 (2010)
14. Y. Wang, B. Ma, L. Li, L. An, Oxidation behavior of ZrB_2–SiC–TaC ceramics. J. Am. Ceram. Soc. **1**, 374 (2012)
15. M.S. Li, *High Temperature Corrosion of Metals* (Metallurgical Industry Press, Beijing, 2001), pp. 139–143

Effect of Different Crystallization Temperature on Dielectric Properties of $BaO–SrO–PbO–TiO_2–SiO_2–Nb_2O_5$ Glass-Ceramic Composites

Junyou Chen, Qingmeng Zhang, Feihu Tan, Peng Zhou and Min Zhou

Abstract Nanocomposite dielectrics in $2.5BaO–18.5SrO–8.5PbO–8TiO_2–20SiO_2–20Nb_2O_5$ system were prepared via melt-quenching followed by controlled crystallization. X-ray diffraction studies revealed that $Pb_2Nb_2O_7$, $(Ba,Sr,Pb)Nb_2O_6$ and $(Sr,Pb)TiO_3$ phases formed from the as-quenched glass annealed in the temperature range from 800 to 1000 °C. $Pb_2Nb_2O_7$, $(Ba,Sr,Pb)Nb_2O_6$ and $(Sr,Pb)TiO_3$ crystallized at 800 °C and then $Pb_2Nb_2O_7$ disappears at 900 °C, while $PbNb_2O_6$ formed at 900 °C. The microstructural observation showed that randomly oriented, the nanometer-sized crystalline grains were found with residual glass concentrated at phase boundaries. Meanwhile, the crystalline grain of the samples grew with the increase in crystallization temperature. And the dielectric constant is enhanced with the increasing crystallization temperature as well. The dielectric constant of the nanocomposites crystallized at different temperatures showed good temperature and electric field stability. In particular, the sample had a high dielectric constant of 900 after crystallized at 1000 °C, and the dielectric loss still remained at a low level of 0.0042, which is expected to be used as a promising candidate material for high voltage ceramic capacitor applications.

Keywords Glass-ceramic dielectric property · Crystallization · High dielectric constant · Low dielectric loss

---

J. Chen · Q. Zhang (✉) · F. Tan · P. Zhou · M. Zhou
Advanced Electronic Materials Institute, General Research Institute for Nonferrous Metals, Beijing, China
e-mail: zhangqm@grinm.com

J. Chen
e-mail: material_chen@163.com

F. Tan
e-mail: tanfeihu@163.com

P. Zhou
e-mail: zhoupeng534@163.com

M. Zhou
e-mail: 18810723831@sina.cn

Y. Han (ed.), *Advanced Functional Materials*,
https://doi.org/10.1007/978-981-13-0110-0_59

Introduction

In recent years, electronic devices continuously pursue lightweight and miniaturization, which requires the capacitor to have high energy storage density [1]. Dielectric material is a key component of capacitor devices, so the performance of dielectric material greatly determines whether the capacitor has a high energy storage density. Glass-ceramics as a new type of nanocomposites have become the focus of researchers. Glass-ceramic is a nano-composite material comprising the glass matrix and the ceramic phase [2]. The ceramic phase of high dielectric constant and the non-void glass matrix make the glass-ceramic dielectric with high dielectric constant and high breakdown strength, and the glass-ceramic dielectric has a good energy storage density [3]. Much work has been focused on obtaining simultaneously high dielectric constant and high breakdown strength in the nanocomposite dielectrics through composition design and process control [4].

At present, the main glass-ceramic systems for high energy density capacitor applications are titanate system and niobate system. For titanate system, the dielectric properties of manganese-doped barium strontium titanate glass-ceramics have been investigated. The results are shown that the dielectric constant and the dielectric loss measured at room temperature pass through a maximum with increasing MnO_2 concentration. When the amount of MnO_2 is 0.2 mol%, both the dielectric constant ($\sim$380) and the dielectric loss (0.02) have the maximum values [5]. For the glass-ceramic in the system of $BaO–PbO–TiO_2–B_2O_3–Al_2O_3$, by controlling the crystallization time, glass-ceramic samples realized relatively low dielectric losses (0.01–0.03) and variable permittivity in the range 30–650 [6].

For niobate system, Han et al. [7]. Investigated the effect of substitution of Pb for Sr at the A-site in $ANb_2O_6–NaNbO_3–SiO_2$ (A=[(1−x)Pb, xSr]) glass-ceramics. At the point of x = 0.6, the highest dielectric constant ($\sim$600) and relatively high breakdown strength (36.7 kV/mm) was obtained, resulting in the highest energy storage density. In the glass-ceramic system of $SrO–BaO–Nb_2O_5–SiO_2$, Wang et al. [8] investigated the effect of crystallization time on the dielectric properties of this system. The experimental results show that dielectric constant enhanced as the crystallization time is prolonged. The maximum dielectric constant ($\sim$500) is present at 5 h of crystallization time and the corresponding dielectric loss is 0.06.

In the present research of titanate system and niobate system, the energy storage density of glass-ceramic dielectric materials has been greatly improved. However, the dielectric constant of glass-ceramics is not large enough, especially compared with sintered ferroelectric ceramics. Most of the reported dielectric constant values are only several hundred, and most of the glass-ceramic nanocomposites currently research has higher dielectric loss, which is a serious threat to the lifetime of capacitors. Therefore, it is of major practical significance to develop a glass-ceramic system with high dielectric constant and low dielectric loss.

In this work, we try to fabricate a glass-ceramic dielectric containing both a titanate ceramic phase and a niobate ceramic phase by a controlled crystal technique in the system of 2.5BaO–18.5SrO–8.5PbO–8TiO_2–20SiO_2–20Nb_2O_5. And the microstructural evolution and corresponding dielectric properties of the glass-ceramic nanocomposite dielectrics with different crystallization temperatures were investigated systematically.

Experimental Procedures

The nanocomposites with nominal composition of 2.5BaO–18.5SrO–8.5PbO–8TiO_2–20SiO_2–40Nb_2O_5 in cation mole percentage were prepared using reagents of barium carbonate ($BaCO_3$, Analytical Reagent viz. A.R.), strontium carbonate ($SrCO_3$, Analytical Reagent viz. A.R.), lead(II) oxide (PbO, A.R.), niobium pentoxide (Nb_2O_5, A.R.), titanium dioxide (TiO_2, A.R.) and silicon dioxide (SiO_2, A. R.). The powder was well mixed and melted in a platinum crucible for 3 h at 1420 ° C. Then the homogeneous melt was rapidly cast into a stainless steel mold pre-heated in an oven to form the glass bulks. The as-quenched glass bulks was then placed in a crystallization oven at 600 °C for 2 h to reduce the internal stress. Approximately 0.5 g of ground glass was used for thermal analysis (Model SDT 2960, TA Instruments, New Castle, DE) up to 1100 °C using a rate of 10 °C/min. Based on the crystallization temperature determined by DTA, the controlled crystallization was carried out at temperatures of 800, 900 and 1000 °C to precipitate nanometer-sized dielectric ceramic phases from the glass matrix.

Phase identification was performed by X-ray diffraction (XRD; model MSAL-XD2, Micro-Structure Analysis Laboratory, Beijing) using Cu–Ka radiation at room temperature and over the 2 theta range from 10° to 90°. Microstructure observation was carried out using a scanning electron microscope (FE-SEM:Hitachi S4800). Temperature dependence of dielectric constant was measured by a precision LCR meter (HP-4284A, Agilent Technologies, Inc., USA). The capacitance–voltage (C–V) curves were measured at room temperature using a ferroelectric tester (RT6000HVA, Radiant Technology, Albuquerque, NM, USA). All the glass-ceramic samples were immersed in silicon oil to avoid flashover.

Results and Discussion

The crystallization temperature of glass was guided by DTA analysis. As shown in the inset of Fig. 1, an endothermic baseline appears at 710 °C corresponding to the glass transition temperature, and an exothermic peak appear at 800 °C, respectively, which is associated with the crystallization of ceramic phases from glass matrix. The crystallization behavior of BaO–SrO–PbO–TiO_2–SiO_2–Nb_2O_5 glass-ceramic dielectrics with different crystallization temperature was investigated by XRD analysis, as shown in Fig. 2. When the quenched glass is crystallized at 800 °C, $Pb_2Nb_2O_7$ phase with pyrochlore structure, tungsten bronze structured (Ba,Sr,Pb)

Fig. 1 DTA profile of as-quenched 2.5BaO–18.5SrO–8.5PbO–8TiO$_2$–20SiO$_2$–20Nb$_2$O$_5$ glass

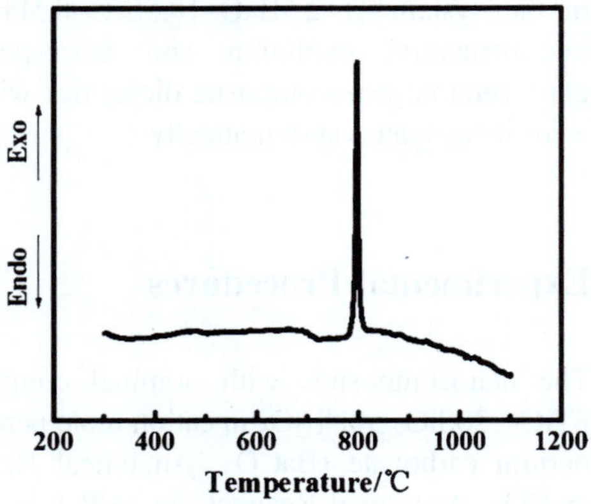

Fig. 2 XRD patterns of 2.5BaO–18.5SrO–8.5PbO–8TiO$_2$–20SiO$_2$–20Nb$_2$O$_5$

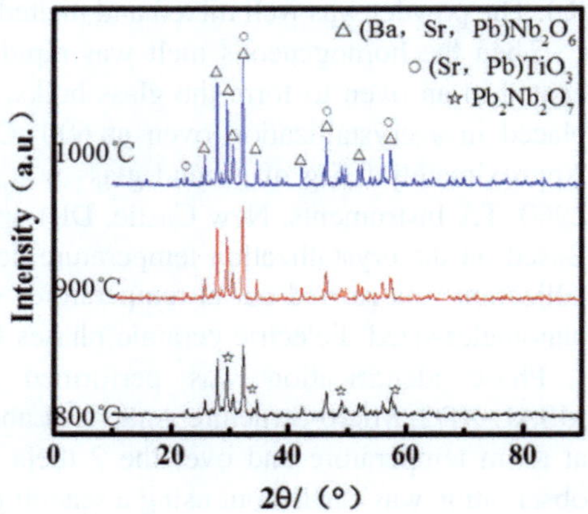

Nb$_2$O$_6$ phase and perovskite structured (Sr,Pb)TiO$_3$ phase are formed. When the temperature needs to be further increased to 900 °C, Pb$_2$Nb$_2$O$_7$ phase disappears completely. For samples crystallized at 900 and 1000 °C, (Ba,Sr,Pb)Nb$_2$O$_6$ and (Sr, Pb)TiO$_3$ phases become the dominating ones. All samples with crystallization temperatures of 800, 900 and 1000 °C were investigated by using SEM. Different stages of crystallization are evident in the micrographs. From the Fig. 3, it is obvious that the crystalline grain of the samples was grown with the increase in crystallization temperature. The average grain size was about 500 nm annealing at 1000 °C. The ceramic grains are uniformly dispersed in the glass matrix.

Figure 4 indicates the relative dielectric constant and dielectric loss tangent measured at 1 kHz at room temperature as a function of the crystallization temperature. The dielectric constant is enhanced with the increase in crystallization temperature. This is attributed to the precipitation of crystal phases from the glass matrix. The glass-ceramic samples annealed at 1000 °C shares the highest dielectric

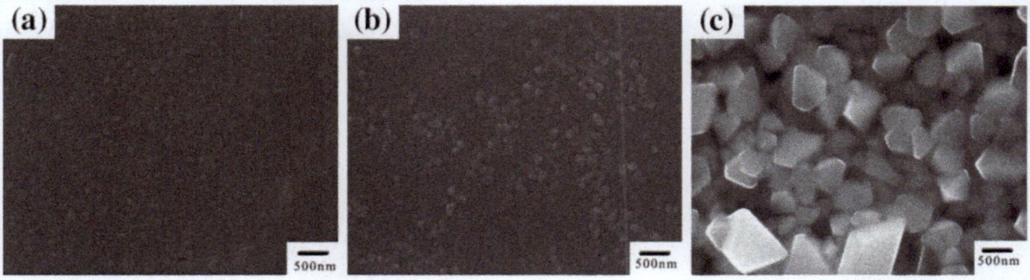

Fig. 3 SEM images of glass-ceramics crystallized at **a** 800 °C, **b** 900 °C, **c** 1000 °C

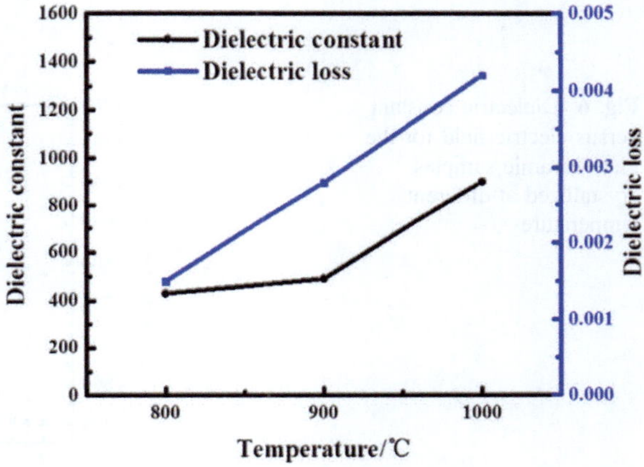

Fig. 4 Relative dielectric constant and dielectric loss tangent as a function of crystallization temperature at 1 kHz

constant($\sim$900). The reason why the dielectric constant increases significantly after annealed at 1000 °C maybe contributed to two factors, one is the formation of high dielectric constant phase (Sr,Pb)TiO_3 in the glass matrix, and the other is the growth of crystallized grain. Meanwhile, the dielectric loss of the samples gradually increases with increasing crystallization temperature. Obviously, the dielectric loss factors of the BaO–SrO–PbO–TiO_2–SiO_2–Nb_2O_5 glass-ceramics can maintain relatively low values (0.0015–0.0042), which are promising for high voltage capacitor applications. It is worth mentioning that the sample has a very high dielectric constant of 900 after crystallized at 1000 °C, but the dielectric loss remains at a very low level of 0.0042.

Figure 5 shows the temperature dependence of the dielectric constant of different samples at a frequency of 1 kHz in the temperature range of −40 to 80 °C. The dielectric constant of the nanocomposites formed at different crystallization temperatures shows good temperature stability. In the measured temperature range, the variation of dielectric constant is lower than 7% compared to that at 900 °C.

Figure 6 shows the electric field dependence of dielectric constant of the glass-ceramics crystallized at different temperatures. Samples crystallized at

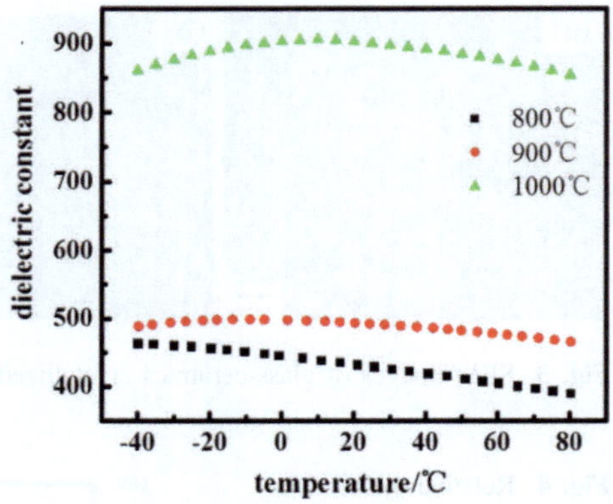

Fig. 5 Dielectric constant as a function of testing temperature for the samples at different crystallization temperature (f = 1 kHz)

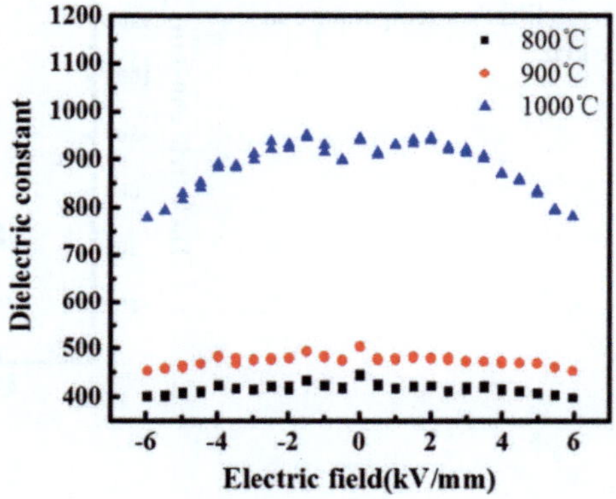

Fig. 6 Dielectric constant versus electric field for the glass-ceramic samples crystallized at different temperature

the temperatures of 800 and 900 °C. The dielectric constant varies little when the applied electric field is scanning from 0 to 6 kV/mm. For samples crystallized at 1000 °C, the electric field stability of dielectric constant is slightly deteriorated, which may attribute to the formation of high dielectric constant (Sr,Pb)TiO_3 phase precipitation [4] and growth with the increase of crystallization temperature [9].

Figure 7 indicates the leakage current of BaO–SrO–PbO–TiO_2–SiO_2–Nb_2O_5 glass-ceramic with the different crystallization temperature. It can be observed that the leakage current density of the sample exhibits a parabolic trend as the electric field intensity increases. Samples annealed at 800 °C have the lowest leakage current density. Meanwhile, the leakage current density of the glass-ceramic showed a rising trend with the increase in crystallization temperature. This is attributed to the crystalline grain of the samples was grown with the increase in crystallization temperature. The crystalline grains are in contact with each other. Under the action of the external electric field, the ions migrate and the distance between the two barriers decreases. The resistivity decreases and the leakage current density increases [10].

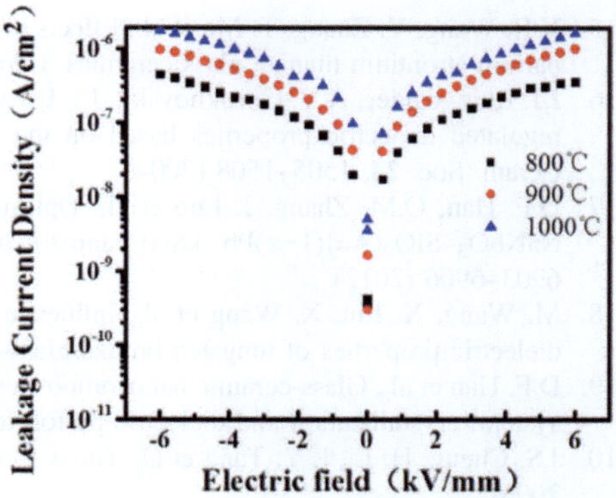

Fig. 7 Leakage current of the glass-ceramics samples crystallized at different temperature

Conclusions

Glass-ceramics in the $BaO–SrO–PbO–TiO_2–SiO_2–Nb_2O_5$ system has been fabricated successfully. Three dielectric phases, $Pb_2Nb_2O_7$, $(Ba,Sr,Pb)Nb_2O_6$ and $(Sr,Pb)TiO_3$ phases form from the as-quenched glass annealed in the temperature range from 800 to 1000 °C. $Pb_2Nb_2O_7$, $(Ba,Sr,Pb)Nb_2O_6$ and $(Sr,Pb)TiO_3$ crystallizes at 800 °C and then $Pb_2Nb_2O_7$ disappears at 900 °C, while $PbNb_2O_6$ forms at 900 °C. Microstructural observation shows that the crystalline grain of the samples grows with the increase in crystallization temperature. The average grain size is about 500 nm annealing at 1000 °C. The ceramic grains are uniformly dispersed in the glass matrix. The dielectric constant of the nanocomposites formed at different crystallization temperatures shows good temperature and electric field stability. In particular, the sample has a very high dielectric constant of 900 after crystallized at 1000 °C, but the dielectric loss remains at a very low level of 0.0042. It is a good candidate dielectric for high voltage ceramic capacitor applications.

Acknowledgements This work was supported by the National Natural Science Foundation of China (No. 51477012).

References

1. Y. Zhou et al., Structural optimization and improved discharged energy density for niobate glass-ceramics by La_2O_3 addition. J. Am. Ceram. Soc. **96**, 372–375 (2013)
2. J. Du, B. Jones, M. Lanagan, Preparation and characterization of dielectric glass-ceramics in $Na_2O–PbO–Nb_2O_5–SiO_2$ system. Mater. Lett. **59**, 2821–2826 (2005)
3. C. Li, Q.M. Zhang, Q. Tang et al., Dielectric and Energy Storage Properties of BaO–SrO–NaO–NbO–SiO Glass-ceramics with different crystallization times. J. Electron. Mater. **6**, 45 (2016)
4. Q.M. Zhang, J. Luo, Q. Tang et al., Nanocomposite dielectrics in $PbO–BaO–Na_2O–Nb_2O_5–SiO_2$ system with high breakdown strength for high voltage capacitor applications. J. Nanosci. Nanotechnol. **12**, 8832–8835 (2012)

5. X.R. Wang, Y. Zhang, T. Ma et al., Effects of MnO_2 concentration on dielectric properties of barium strontium titanate glass ceramics. Ceram. Int. **38**, S57–S60 (2012)
6. J.J. Ruiz-Valdés, A.V. Gorokhovsky, J.I. Escalante-Garcí et al., Glass–ceramic materials with regulated dielectric properties based on the system $BaO–PbO–TiO_2–B_2O_3–Al_2O_3$. J. Eur. Ceram. Soc. **24**, 1505–1508 (2004)
7. D.F. Han, Q.M. Zhang, J. Luo et al., Optimization of energy storage density in ANb_2O_6–$NaNbO_3$–SiO_2(A=[(1−x)Pb, xSr]) nanostructured glass-ceramic dielectrics. Ceram. Int. **38**, 6903–6906 (2012)
8. M. Wang, X. Liu, X. Wang et al., Influence of crystallization time on microstructures and dielectric properties of tungsten-bronze glass-ceramics. Ceram. Int. **47**, 2535–2540 (2012)
9. D.F. Han et al., Glass-ceramic nanocomposites in the [(1−x)PbO–xBaO]–Na_2O–Nb_2O_5–SiO_2 system: crystallization and dielectric performance. Solid State Sci. **14**, 661–667 (2012)
10. J.S. Cheng, H. Li, L.Y. Tang et al., *Glass Ceramics* (Chemical Industry Publishing, Beijing, 2006)

Densification of Tantalum Carbide Ceramics with 1–10 mol% LaB_6

Zhuang Lu and Limeng Liu

Abstract TaC ceramics were prepared by hot pressing at 1900 °C for 30 min under 30 MPa mechanical pressure. The effect of adding 1–10 mol% LaB_6 as a sintering aid on densification, the phase composition, microstructure and mechanical properties of the consolidated materials were investigated. TaC and LaB_6 reacted to form the solid solution. The grain average size decreased from 10 µm at 1 mol% LaB_6 addition to 4 µm at 10 mol% LaB_6 addition. The 5 mol% LaB_6 composition had good flexural strength of 312 MPa.

Keywords Ultra high-temperature ceramic · Tantalum carbide
Hot pressing

Introduction

Ultra high-temperature materials (UHM) are essential for long-time reentry flight applications in aerospace. Components of leading edge and nose cap in new conceptual space craft will expose to temperatures >2000 °C under complex stresses. Harsh thermal and mechanical conditions limit materials choice for such applications only to a few carbides, borides and nitrides of the IV_B-VI_B transition metals having a melting point in excess of 3000 °C [1]. Among carbides, borides and nitrides of the transition metals, TaC is outstanding for its high melting point up to ~3983 °C. In fact, this melting point is the second highest, only a few tens of degrees below the first which is observed on a TaC–HfC solid solution [2–4]. The TaC-based ceramics also have better mechanical properties relative to other

Z. Lu · L. Liu (✉)
School of Materials Science and Engineering,
North Minzu University, Yinchuan 750021, China
e-mail: llm681@sina.com

Z. Lu
e-mail: lu_zhuang@126.com

Y. Han (ed.), *Advanced Functional Materials*,
https://doi.org/10.1007/978-981-13-0110-0_60

UHTCs. Room temperature flexural strengths of the TaC ceramics were 400–700 MPa [5–8]. In comparison, strength of ZrC and other UHTCs fall in the range of 220–400 MPa [9, 10].

Monolithic tantalum carbide ceramics are extremely hard to densify due to strong covalent bonds. Various methods have been tried to obtain fully dense tantalum carbide ceramics. A relative density of 97.5% was obtained for pure TaC [11] after sintering at 2300 °C for 0.5 h by using a fine TaC powder an average particle size around ~250 nm [12]. Other approaches to enhance densification of tantalum carbide ceramics were designed to avoid entrapping residual pores inside the grains [13].

Experimental

TaC (purity > 99%, 0–1.5 μm provided by Ningxia Orient Tantalum industry, Yinchuan, China) and LaB_6 (purity > 99.5% provided by Aladdin Industrial Corporation, Shanghai, China) powers were used as starting materials. The compositions under investigation are listed in Table 1. The power mixtures in proper proportions were mixed in plastic bottles containing ZrO balls and ethanol as mixing media. After being dried, the powder mixtures, about 30 g each batch, were loaded into graphite dies for hot pressing (15 t Hot Press Furnace, Materials Research Furnace, Inc., USA). The hot pressing (HP) was carried out at 1900 °C under 30 MPa for 30 min in a argon flux using. The furnace was pumped to a vacuum <100 micro Torr, followed by fluxing with argon at a flowing rate of 2.0 L/min before heating. Subsequently, temperature was increased at 20 °C/min to 1900 °C under 30 MPa uniaxial pressure. After hold for 30 min, the pressure was released and the heating power was shut down to allow cooling.

Table 1 Composition, densification, XRD analysis result and average grain size of the investigated materials

Material	Staring composition (g)		Relative density (%)	Phase composition	Average grain size (μm)
	TaC	LaB_6			
TaC	32	0	85.9	TaC	18 ± 13
TaC-1mol% LaB_6	31.67	0.33	96.5	TaC	10 ± 5
TaC-2mol% LaB_6	31.32	0.68	96.4	(Ta,La)C, TaB_2	8 ± 2
TaC-5mol% LaB_6	30.32	1.68	94.5	(Ta,La)C, TaB_2	5 ± 2
TaC-10mol% LaB_6	28.95	3.05	87.3	(Ta,La)C, TaB_2	4 ± 2

About 1.0 mm material was ground off from each surface of the consolidated sample to eliminate possible contamination. The samples were sliced, ground and polished to 1 μm finish. Phase analysis was done by X-ray diffraction (XRD-6000, Shimadzu, Japan) using Cu Kα radiation.

Densities of the consolidated materials were measured by the Archimedes' principle. Relative densities were calculated as fraction of the measured densities to the theoretical. Theoretical densities of the various compositions were estimated by the law of mixtures.

Polished, thermally etched and fractured surfaces were observed by scanning electron microscopy (SEM; SSX-550, Shimadzu, Japan). Thermal etching was done by heating the polished surfaces at 1600 °C for 30 min in flowing argon. Grain sizes were measured on the SEM micrographs by the Image Pro software (Image Pro-Plus 5.0.1, Media Cybernetics, Sliver Spring, MD). Average grain size was determined according to international standard (ISO 13383-1:2012). Flexure strength was tested at room temperature by three-point bending. Averages of ten specimens for each composition were reported.

Results and Discussion

Densification. Density measurement results of the consolidated materials are listed in Table 1. Densification of the TaC ceramics was increased by addition of LaB_6. The pure TaC composition had a relatively low density of ~85.9%. A relative density value of 96.5, 96.4, 94.5 and 87.3%, was reached in the 1, 2, 5 and 10 mol % LaB_6 composition samples, respectively.

Phase compositions. XRD patterns of the obtained materials are show in Fig. 1. TaC was the only crystalline phase detected in the 1 mol% LaB_6 composition. TaB_2 was presented in the other compositions, with peaks of the TaC phase shift toward lower 2θ position. La may substitute for Ta in the TaC lattice to form a solid solution. La^{3+} ion (0.117 nm) substitution for the smaller Ta^{4+} (0.068 nm) would expand the TaC lattice, hence shift of the peaks towards lower 2θ positions.

Lattice parameter of the TaC phase and the (Ta,La)C solid solutions (ss) were calculated. The calculated a (=0.4455) for the TaC phase in the 1 mol% LaB_6 TaC composition was in consistency with stoichiometric TaC (a = 0.460 nm, PDF 65-0282). The a value of (Ta,La)Css in the consolidated 2, 5 and 10 mol% LaB_6 composition was 0.4456, 0.4457 and 0.4460 nm, respectively. Increase in a values indicated larger La content in the (Ta,La)Css with increase in the LaB_6 addition.

Microstructure. Microstructures of the consolidated materials are shown in Fig. 2. Residual pores were occasionally observed in the 1 mol% and the 2 mol% LaB_6 compositions. In the 10 mol% LaB_6 material, large pores were observed on the polished-and thermally etched surface. These large pores were the main reason for the low relative density (87.3%) of the TaC sample. Thermal etching of the polished surfaces

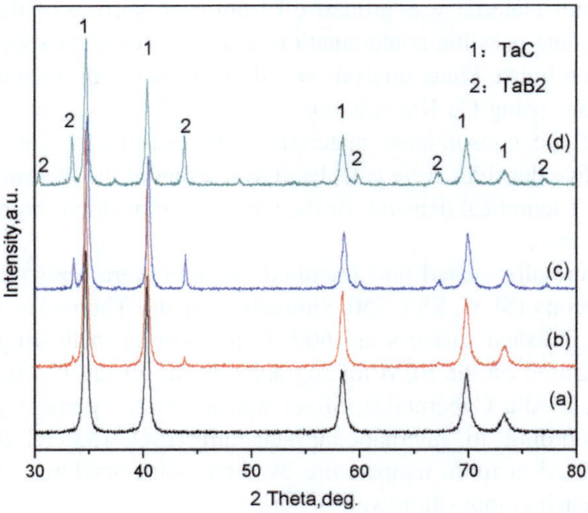

Fig. 1 XRD patterns of the TaC ceramics with **a** 1 mol%, **b** 2 mol%, **c** 5 mol% and **d** 10 mol% LaB_6

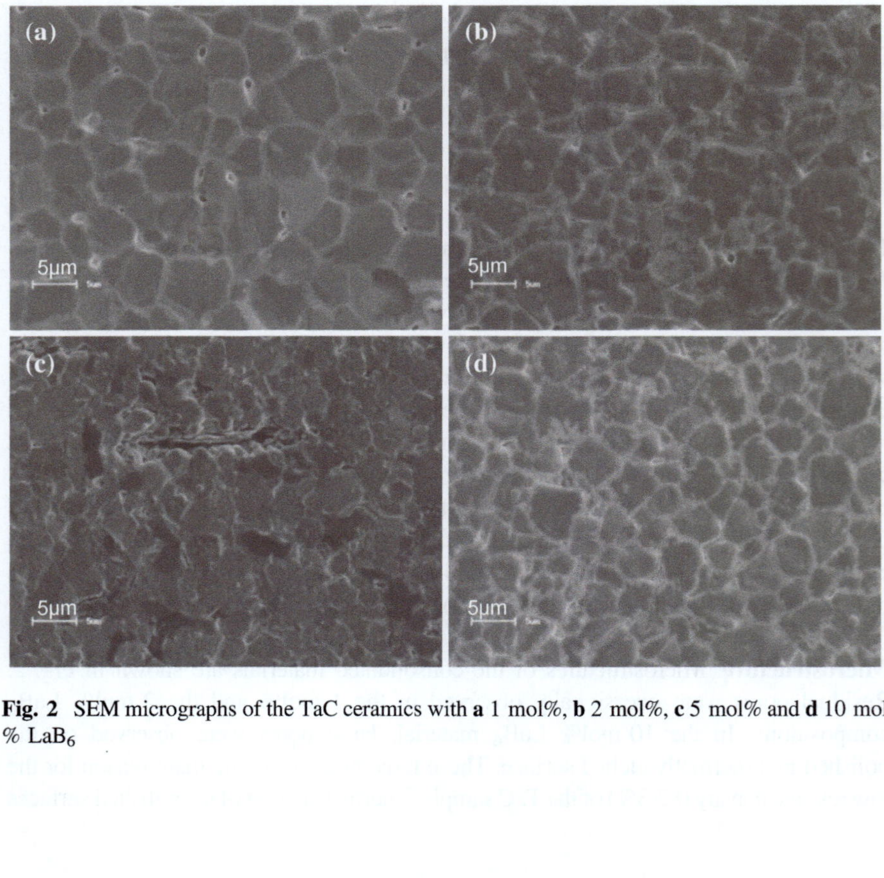

Fig. 2 SEM micrographs of the TaC ceramics with **a** 1 mol%, **b** 2 mol%, **c** 5 mol% and **d** 10 mol % LaB_6

exposed grain boundaries to allow measurement of the grain sizes. The grain size measurement results are listed in Table 1. Average diameter of the TaC grains decreased from ~ 18 μm to ~ 4 μm with increase of LaB_6 addition from 0 to 10 mol%.

The pure TaC material had an average grain diameter of ~18 μm. Relative to the 1–1.5 μm particles in the TaC starting powder, rapid grain growth took place during sintering at 1900 °C, indicating high mobility of the grain boundaries. LaB_6 particles pinned the grain boundaries to explain the significant refinement of the grains in the consolidated materials where LaB_6 was added. For instance, in Fig. 2d, the average grain size in the 10 mol% LaB_6 material was only ~4 μm, decreased by about 5 times relative to the pure TaC.

Mechanical Properities. Flexural strength of the 1, 2, 5 and 10 mol% LaB_6 composition was 270, 272, 312 and 194 MPa, respectively. The low strength of the 10 mol% LaB_6 TaC material was due to the high porosity. The other materials reached full densities. Therefore, the difference in the strength values was caused by the difference in the microstructures.

Fractographs of the consolidated materials are shown in Fig. 3. The TaC ceramics were predominated by intergranular fracture. Residual pores were frequently observed on the fracture surface of the pure TaC. No large flaws as crack origins were observed in the other materials. The intergranular fracture indicated weakened bonding between the grains.

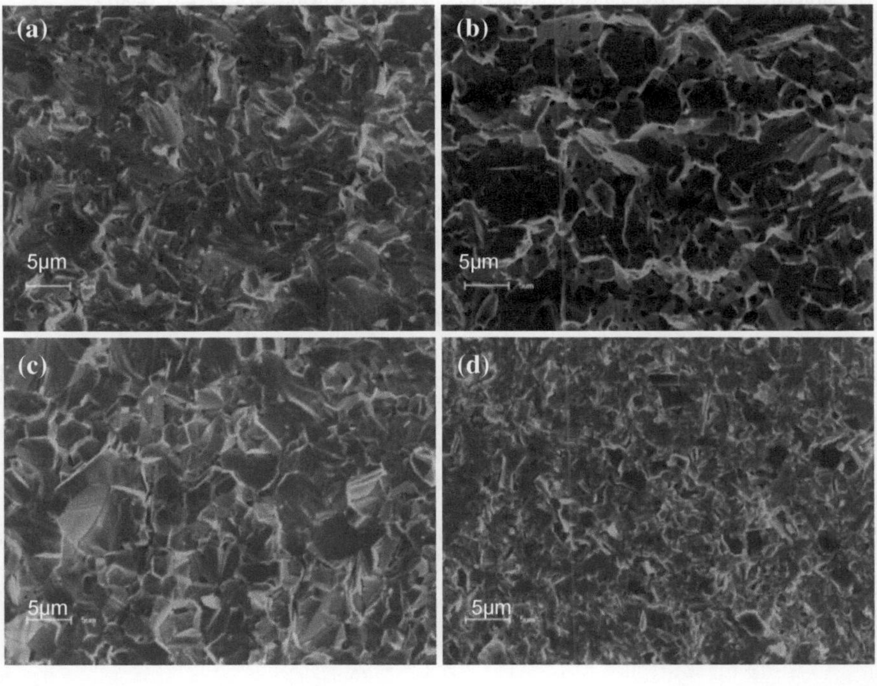

Fig. 3 SEM fractographs of the TaC ceramics with **a** 1 mol%, **b** 2 mol%, **c** 5 mol% and **d** 10 mol % LaB_6

Conclusions

(1) TaC ceramics with 1–10 mol% LaB_6 as an additive have been fabricated by hot pressing at 1900 °C for 30 min. Full density can be obtained by the addition of 1–2 mol% LaB_6.

(2) TaC and LaB_6 react to form TaB_6. TaC lattice expands due to the formation of (Ta,La)C solid solution.

(3) The consolidated TaC ceramics with 2 mol% LaB_6 has the flexural strength of 272 MPa.

Acknowledgements LML thanks financial support from the National Natural Science Foundation of China under Grant no. 51662001.

References

1. W.G. Fahrenholtz, G.E. Hilmas, Refractory diborides of zirconium and hafnium. J. Am. Ceram. Soc. **90**, 1347–1364 (2007)
2. R.A. Andrievskii, N.S. Strel'nikova, N.I. Poltoratskii, E.D. Kharkhardin, V.S. Smirnov, Melting point in systems ZrC–HfC, TaC–ZrC, TaC–HfC. Powder Metall. Met. **C6**, 65–68 (1967)
3. S.Z. Nazarova, E.Z. Kurmaev, N.I. Medvedeva, A. Mowes, Physical properties and electronic structure of TaC–HfC solid solutions. Russ. J. Inorg. Chem. **52**, 233–237 (2007)
4. S.A. Ghaffari, M.A. Faghihi-Sani, F. Golestani-Fard, H. Mandal, Spark plasma sintering of TaC–HfC UHTC via disilicides sintering aids. J. Eur. Ceram. Soc. **33**, 1479–1484 (2013)
5. K. Hackett, S. Verhoef, R.A. Cutler, D.K. Shetty, Phase constitution and mechanical properties of carbides in the Ta–C system. J. Am. Ceram. Soc. **92**, 2404–2407 (2009)
6. X.H. Zhang, G.E. Hilmas, W.G. Fahrenholtz, Densification and mechanical properties of TaC-based ceramics. Mater. Sci. Eng., A **501**, 37–43 (2009)
7. M. Sygnatowicz, R.A. Cutler, D.K. Shetty, ζ-Ta_4C_3-x: a high fracture toughness carbide with rising-crack-growth-resistance (R-curve) behavior. J. Am. Ceram. Soc. **98**, 2601–2608 (2015)
8. L.M. Liu, F. Ye, Y. Zhou, Z.G. Zhang, Microstructure and mechanical properties of spark plasma sintered $TaC_{0.7}$ ceramics. J.Am. Ceram. Soc. **93**, 2945–2947 (2010)
9. S. Ran, S. Huang, O.V. Biest, J. Vleugels, High-strength ZrB_2-based ceramics prepared by reactive pulsed electric current sintering of ZrB_2–ZrH_2 powders. J. Eur. Ceram. Soc. **32**, 2537–2543 (2012)
10. D. Sciti, S. Guicciardia, M. Nygren, Spark plasma sintering and mechanical behaviour of ZrC-based composites. Scr. Mater. **59**, 638–641 (2008)
11. J.X. Liu, Y.M. Kan, G.J. Zhang, Synthesis of Ultra-fine hafnium carbide powder and its pressureless sintering. J. Am. Ceram. Soc. **93**, 980–986 (2010)
12. G.V. Samonov, R.Y. Petrikina, Sintering of metals, carbides, and oxides by hot-pressing. Phys. Sinter. **2**(3), 1 (1997)
13. N. Alexandre, M. Desmaison, F. Valin, M. Bonocoeur, Solid state reaction between tantalum (Ta) and tantalum carbide (TaC) powders during hiping. Key engineering materials. Trans. Tech. Publications. **868**, 132–136 (1997)

Effects of Fine Composite Powders Addition on Properties of Corundum-Spinel Refractory Castables

Hai Tang, Nian Jiang, Wenjie Yuan, Chengji Deng and Hongxi Zhu

Abstract The enhanced properties of refractory castables are usually designed by the addition of multifunctional mineralizers and fine powders to satisfy the requirements of steelmaking industry. In this work, the effects of fine composite powders addition on properties of corundum-spinel refractory castables were investigated. Phase composition, microstructure, physical and mechanical properties of corundum-spinel refractory castables with different contents of fine composite powders were characterized. The results showed that the smaller spinel in the fine composite powders improved the sintering activity of the spinel particles, which had an influence on the morphology and the distribution of CA_6 for the limited space. As the consequence of this, the cracks caused by in situ CA_6 generation decreased and the firm link between particles was gained, finally the strength of castables containing fine composite powders improved.

Keywords Castables · Fine composite powders · Properties · CA_6

Introduction

Materials designed in the alumina-rich zone of the Al_2O_3–MgO–CaO system have been widely used in high temperature industry such as linings of the steel ladle [1]. As it known to all calcium aluminate cement is commonly selected as binder of castables due to its placing advantages (suitable working, setting and demoulding time, as well as mechanical strength at the green stage, price and availability) [2]. Besides, the product calcium hexaluminate (CA_6) between CaO and Al_2O_3 is a high-temperature calcium aluminate phase with the highest Al_2O_3 content in CaO–Al_2O_3 system, which can be used in high-temperature industries such as refractories field [3]. Within the past half century, magnesium aluminate spinel ($MgAl_2O_4$ or

H. Tang · N. Jiang · W. Yuan (✉) · C. Deng · H. Zhu
The State Key Laboratory of Refractories and Metallurgy,
Wuhan University of Science and Technology, Wuhan 430081, China
e-mail: yuanwenjie@wust.edu.cn

Y. Han (ed.), *Advanced Functional Materials*,
https://doi.org/10.1007/978-981-13-0110-0_61

MA) has been widely used for refractory and structural ceramic materials on account of its excellent mechanical property at high temperatures, thermal shock resistance and slag corrosion resistance [4]. These aspects above-mentioned ensure the calcium aluminate cement combined corundum-spinel castables presents excellent properties in the tough service environment.

In order to extend the service life of materials and obtain better products, the properties of corundum-spinel castables should be further improved [5]. At present, adding mineralizes such as titanium component, boron component and barium component is regarded as the mainly method to modify the properties of castables including the strength of materials [6–8]. However, the purity and particle size of raw materials can have a great influence on the properties of materials. Based on this background, the influence of fine composite powders addition on the properties of corundum-spinel refractory castables was studied in this work.

Materials and Methods

Corundum-spinel refractory castables were prepared according to the compositions listed in Table 1. Tabular alumina (d $\leq$ 6 mm) was used as aggregates. The castables' matrix comprised reactive alumina, spinel (d $\leq$ 325 mesh), calcium aluminate cement. 1 wt% of additives (accelerator and retarder) was added. The reactive alumina was replaced by fine composite powders on half and all respectively in the experiment. Based on the chemical composition (Table 2) of fine composite powders, the added composite powder's content was calculated on the standard that the quality of alumina was invariant. Except that, the content of spinel was adjusted to make sure that the ratio of spinel in castables was unchanged. The added water content for vibrocasting was about 4.3 wt%.

Table 1 Compositions of alumina-magnesia refractory castables

Raw material	Content (wt%)		
	X	Y	Z
Tabular alumina ($\leq$6 mm) (Almatis)	50	50	50
Tabular alumina ($\leq$1 mm) (Almatis)	24	24	24
Spinel ($\leq$325 mesh) (AR78)	14.5	11.9	9.3
Reactive alumina (CL370) (Almatis)	8	4	0
Calcium aluminate cement (Secar71) (Kerneos)	2.5	2.5	2.5
Fine composite powder (E-SY288)	0	6.6	13.2
Additives (Refpac 288 and 388)	1	1	1

Table 2 Chemical compositions of fine composite powders

Chemical composition	Content (wt%)
Al_2O_3	89
MgO	11

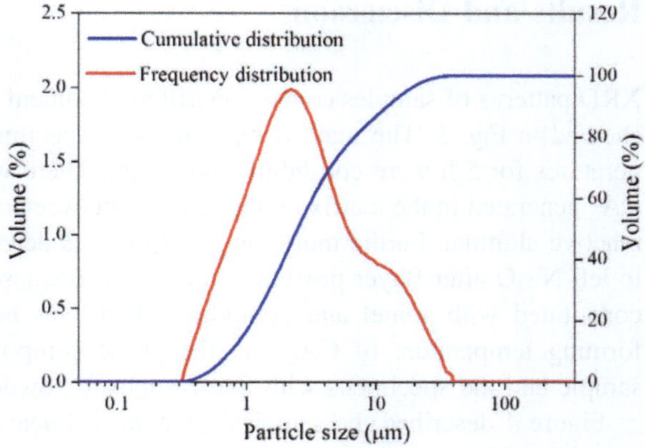

Fig. 1 Cumulative distribution and frequency distribution of fine composite powders

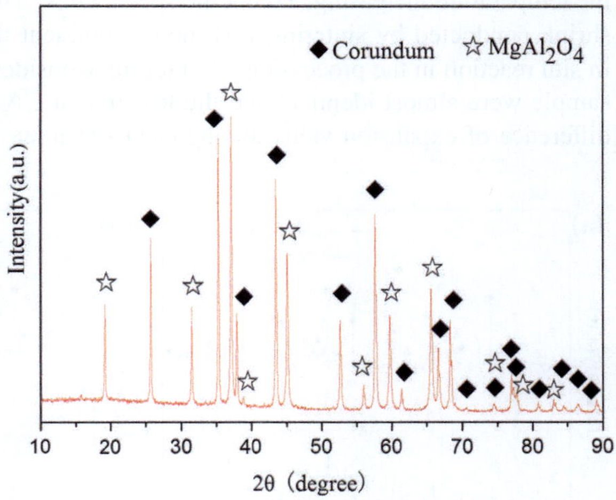

Fig. 2 XRD patterns of fine composite powders added into the corundum-spinel refractory castables

The XRD patterns and particle size distribution of fine composite powders were shown in Figs. 1 and 2 respectively. The main phases of fine composite powders were corundum and $MgAl_2O_4$. The d_{50} of fine composite powder was 2.49 μm, which was smaller than that of the spinel (≤ 325 mesh) listed in Table 1.

After casting, all the castables were cured at 25 °C for 24 h with a relative humidity of 100%, and then dried at 110 °C for 24 h. All samples were calcined at 1400–1600 °C for 5 h, respectively. The permanent linear change (PLC) was measured in compliance with GB/T 5988–2007. The apparent porosity and the bulk density of samples were tested by the Archimedes technique. The cold modulus of rupture (CMOR) was measured with three-point bending tests. The phase composition of castables was analyzed by means of X-ray diffraction (XRD, Philips, X'pert Pro MPD, Netherlands). The microstructure and the chemical composition of the specimens were observed and measured with scanning electron microscopy (SEM, JEOL JSM-6610, Japan) and an energy-dispersive spectrometer (EDS, Bruker QUANTAX200-30, Germany).

Results and Discussion

XRD patterns of samples containing different content fine composite powders were showed in Fig. 3. The main compositions of specimens calcined at different temperatures for 5 h were corundum and spinel. There was a small amount of in situ CA_6 generated in the matrix by the reaction between calcium aluminate cement and reactive alumina. Furthermore, the β-Al_2O_3 was detected by X-ray diffraction due to left Na_2O after Bayer processing of Al_2O_3. Because the fine composite powders constituted with spinel and corundum, there was no obvious distinction on the forming temperature of CA_6 and the phase composition between the reference sample and the specimens with fine composite powders addition.

Figure 4 described the samples' permanent linear change varied with temperature rising. The permanent linear change of all samples continuously decreased with the temperature increasing, except the specimen X. This tendency demonstrated the shrink conducted by sintering was more prominent than the expansion caused by in situ reaction in the processing of sintering. Considering the compositions of each sample were almost identical and the less in situ CA_6 generated in the matrix, the difference of expansion value among each specimen was small.

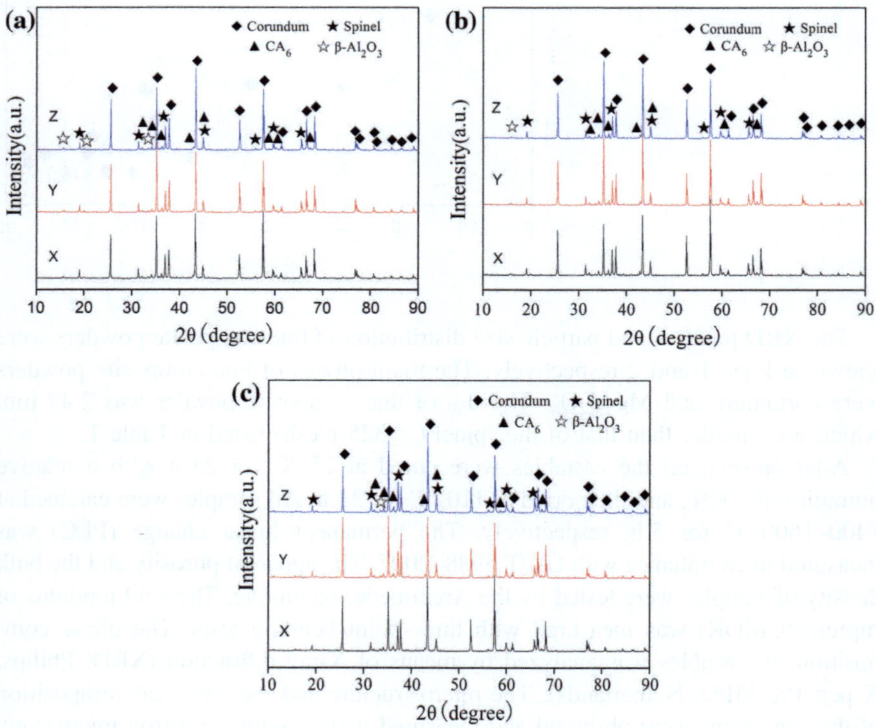

Fig. 3 XRD patterns of corundum-spinel refractory castables sintered at different temperatures: **a** 1400 °C, **b** 1500 °C and **c** 1600 °C

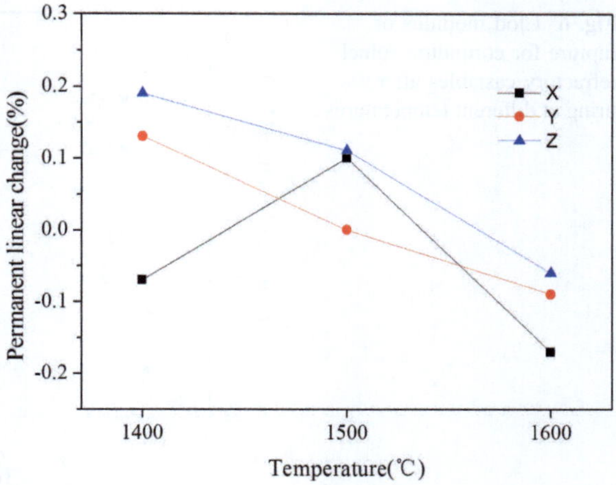

Fig. 4 Permanent linear change of samples sintered at different temperature for 5 h

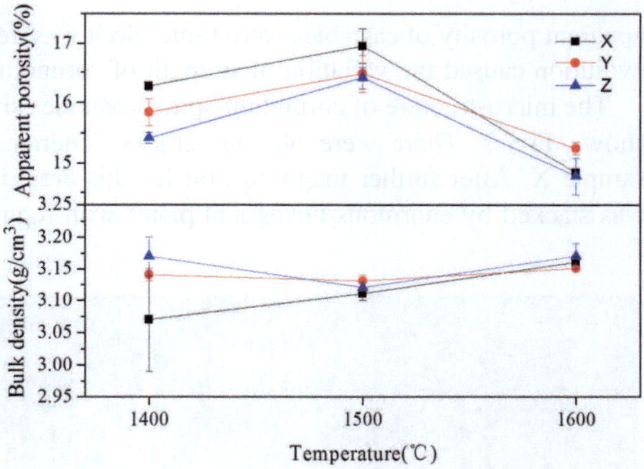

Fig. 5 Apparent porosity and bulk density of specimens calcined at different temperatures

The apparent porosity and bulk density of samples sintered at different temperatures are presented in Fig. 5. The apparent porosity of specimens rose at first and went down latter with the temperature rising. The variation tendency of corundum-spinel refractory castables' bulk density was contrary with that of the apparent porosity. In general, the distinction of the data among each sample was fairly small, which proved that the fine composite powders had limited influence on densification of corundum-spinel refractory castables.

The line chart (Fig. 6) drew the cold modulus of rupture of samples dealt with different temperatures for 5 h. On the whole, the strength of samples containing fine composite powders was higher than that of the reference sample except for the sample Y sintered at 1600 °C for 5 h. It was worthy to notice that strength of samples with fine composite powders addition was improved from 24 to 33 MPa after the materials sintered at 1500 °C for 5 h. Based on the discussion above, effects of the fine composite powders addition on phase composition, bulk density and

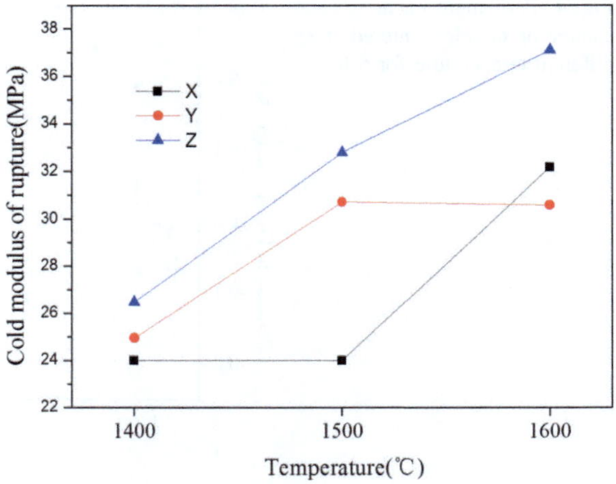

Fig. 6 Clod modulus of rupture for corundum-spinel refractory castables after firing at different temperatures

apparent porosity of castables were finite. So it was deduced that the microstructure evolution caused the variation of strength of corundum-spinel castables.

The microstructure of corundum-spinel castables sintered at 1600 °C for 5 h was shown Fig. 7. There were obvious cracks generated in the matrix of reference sample X. After further magnification for this area, it was observed that this area was stacked by enormous hexagonal plates with regular edge (CA_6). According to

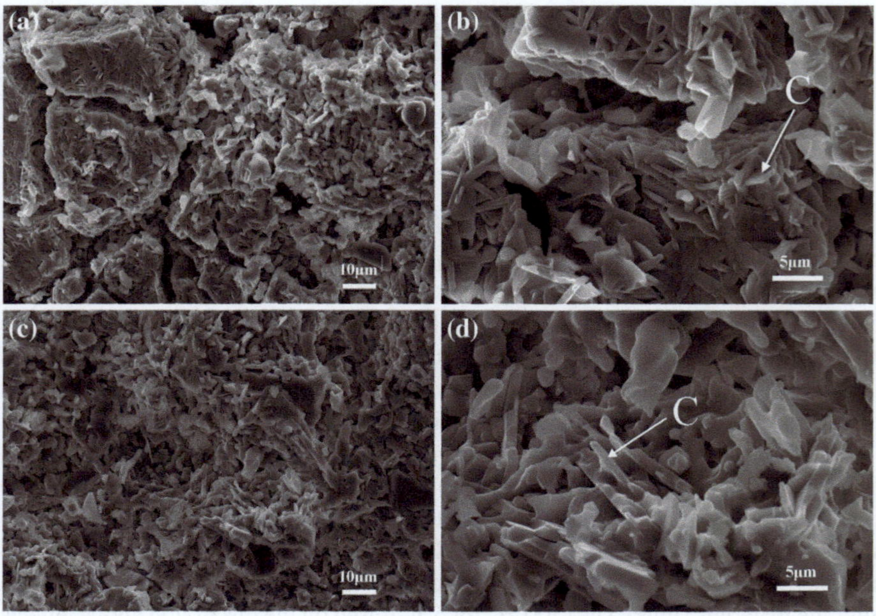

Fig. 7 SEM of corundum-spinel refractory castables (X-(**a**)(**b**) and Z-(**c**)(**d**)) after firing at 1600 °C for 5 h. (C-CA_6)

the literature [9], the flack structure of CA_6 restrains its density enhancement. Besides, the formation of CA_2 and CA_6 leads to the volume expansion of 8 and 13.6% respectively [10]. These two features of CA_6 conducted the appearance of cracks and the relative lower strength of reference sample under the same condition. While the microstructure of castables containing fine composite powders presented totally different characteristics, the better link between particles was achieved in the matrix as shown in Fig. 7c, the number of CA_6 was dramatic decreased and the morphology of CA_6 was more irregular than that of the reference sample as shown in Fig. 7d. Due to smaller size spinel (d_{50} = 2.49 μm) in the fine composite powders, the better sintering activity between spinel particles was gained, which had an influence on the distribution of CA_6 by the limited space [11], as well as decreased the cracks caused by in situ CA_6 generation, finally the strength of corundum-spinel refractory castables with the fine composite powders addition improved.

Conclusion

The effects of fine composite powders addition on the properties of corundum-spinel refractory castables are investigated in this work. Based on the discussion above, the following conclusions are drawn. The higher strength of corundum-spinel refractory castables is gained with the fine composite powders addition especially after calcined at 1500 °C for 5 h, which can be explained by the microstructure evolution of castables. With the function of smaller spinel in fine composite powders, the sintering activity of spinel is improved. As the consequence, the distribution of CA_6 is influenced by limited space, then the cracks caused by in situ CA_6 generation are decreased, and the firm link between particles is gained, finally the strength of castables containing fine composite powders improves.

Acknowledgements The authors are thankful to National Natural Science Foundation of China (NSFC No. 51502214).

References

1. A.H. De Aza, P. Pena, S. de Aza, Ternary system Al_2O_3–MgO–CaO: I, Primary phase field of crystallization of spinel in the subsystem $MgAl_2O_4$–$CaAl_4O_7$–CaO–MgO. J. Am. Ceram. Soc. **82**, 2193–2203 (1999)
2. M.A.L. Braulio, G.G. Morbioli, D.H. Milanez, V.C. Pandolfelli, Calcium aluminate cement source evaluation for Al_2O_3–MgO refractory castables. Ceram. Int. **37**, 215–221 (2011)
3. J.H. Chen, H.Y. Chen, W.J. Mi, Z. Cao, B. Li, C.J. Liang, Substitution of Ba for Ca in the structure of $CaAl_{12}O_{19}$. J. Am. Ceram. Soc. **100**, 413–418 (2017)

4. X.L. Yin, M. Chen, N. Wang, L. Xu, Improvement of densification and mechanical properties of $MgAl_2O_4$–$CaAl_4O_7$–$CaAl_{12}O_{19}$ composite by addition of MnO. Ceram. Int. **43**, 4706–4711 (2017)
5. D.L. Xu, S.Y. Wang, X.H. Zhang, Current state and prospects for 12th five-year plan of China's refractories industry, in *Proceedings of 1st China International Refractory Production and Application Conference* (Guangzhou, China, 2011), pp. 1–9
6. W.J. Yuan, H. Tang, H. Shang, J. Li, C.J. Deng, H.X. Zhu, Effects of TiO_2 addition on kinetics of in situ spinel formation and properties of alumina-magnesia refractory castables. J. Ceram. Sci. Technol. **8**, 121–128 (2017)
7. L. Liu, M. Chen, L. Xu, X.L. Yin, W.J. Sun, Effect of BaO addition on densification and mechanical properties of Al_2O_3–MgO–CaO refractories. Metals **84**, 1–9 (2016)
8. I.D. Giovannelli Maizo, A.P. Luz, C. Pagliosa, V.C. Pandolfelli, Boron sources as sintering additives for alumina-based refractory castables. Ceram. Int. **43**, 10207–10216 (2017)
9. J.H. Chen, H.Y. Chen, W.J. Mi, Z. Cao, B. Li, C.J. Liang, Substitution of Ba for Ca in the structure of $CaAl_{12}O_{19}$. J. Am. Ceram. Soc. **100**, 413–418 (2017)
10. W.J. Yuan, C.J. Deng, H.X. Zhu, Effects of TiO_2 addition on the expansion behavior of alumina-magnesia refractory castables. Mater. Chem. Phys. **162**, 724–733 (2015)
11. Y.Q. Li, Y.W. Li, S.L. Jin, N. Li, Synthesis and its microstructure of calcium hexaluminate material by reaction sintering. Refract. Naihuo Cailiao **38**, 318–323 (2004)

Preparation and Bench Test of SiC Honeycomb Ceramics with Macroporous Walls

Yuzhi Bao, Hu Li, Wei Jiang and Chunlei Duan

Abstract Honeycomb ceramics is the main part of Diesel Particulate Filter. Because of its using environment it must have excellent performance such as thermal shock resistance. Silicon carbide can exactly meet the requirements. Silicon carbide honeycomb ceramic can be obtained via the processes including mixing materials, extrusion moulding, alternative block and sintering. The results of the bench test indicate that silicon carbide honeycomb ceramic can be used in Diesel Particulate Filter successfully.

Keywords Honeycomb ceramics · Silicon carbide · DPF

Introduction

The honeycomb ceramic is a kind of ceramic which has positive hexagonal perforated holes. It first used as a heat exchanger, but with car exhaust pollution has become increasingly severe, requiring a kind of catalytic transfer with better performances including small pressure force loss, heat resistance, thermal shock resistance, good wear resistance for purify the car exhaust. Relying on the above characteristics, honeycomb ceramics become the mainstream of vehicle exhaust catalytic purification carrier. The diameter of the honeycomb ceramics is widely used. Since 1969 years Germany exhibited a company after the first honeycomb

Y. Bao (✉) · H. Li · W. Jiang · C. Duan
Department of New Materials, Ningxia Machinery Research Institute Co. Ltd., NO. 420, Baohuzhong Road, Jinfeng District, Yinchuan, Ningxia, China
e-mail: marryanna@yeah.net

H. Li
e-mail: lihu@nxjixie.com

W. Jiang
e-mail: jiangwei@nxjixie.com

C. Duan
e-mail: dcl163@163.com

Y. Han (ed.), *Advanced Functional Materials*,
https://doi.org/10.1007/978-981-13-0110-0_62

ceramic catalytic transposed, the United States and Japan for similar effect, particularly the United States, Japan and other countries made the atmosphere clean method, and made strict rules and standards for exhaust. In order to promote its development, many companies in the United States were engaged in the production of honeycomb ceramic in 1995.

DPF shows in Fig. 1 mainly consists of 3 processes [1]: particulate filter is empty, particle filter is full, and particle filter is reduction.

(1) The particulate filter is empty. The flow resistance of the exhaust gas is very low due to the absence of any stock in the DPF, which will not affect the normal work of the engine.
(2) The particle filter is full. Due to the formation of soot particles, the smoke particles in DPF gradually increased, and the exhaust and exhaust resistance increased. The fuel consumption and power generation were affected by the increase of exhaust backpressure.
(3) The reduction process of the particle filter. Because DPF gathered a lot of carbon particles which is affect exhaust backpressure of the filter, exhaust gas was difficult to discharge, which greatly restricted the power and fuel economy of the engine. According to the signal of exhaust gas pressure sensor, the engine control unit let the engine to clean and restore the carbon smoke itself and burn the smoke particles gathered inside in high temperature. However, when the engine is in the reduction process, only the coke for internal DPF reduction, and for non coke ash material cannot restore decomposition, the ash mainly comes from oil and fuel composition, because the ash does not burn, it

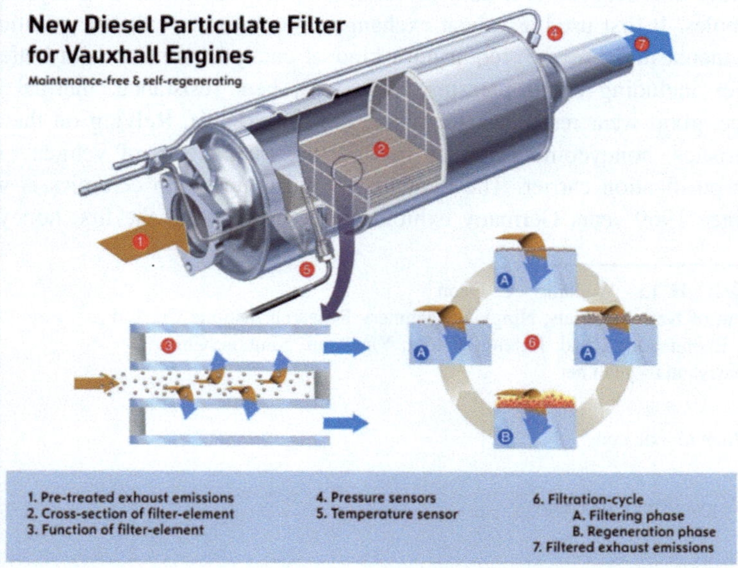

Fig. 1 How the DPF looks like

Table 1 Performance requirements of wall-flow ceramic

Testing items	Performance requirements
Bulk density (g/cm^3)	≤ 0.8
Porosity (%)	40–60
Thermal conductivity (W/M K)	>14
Thermal expansion coefficient (0–600 °C) ($\times 10-6$ °C $-$ 1) CTE	4.3–4.4
Compressive strength (MPa)	Direction A 10–18
	Direction B > 10
Pore size distribution	50–300

will reduce the effective volume of the filter, after a certain period of time gathered particles in the filter will make the filter's function deterioration. Therefore, in order to ensure the actual efficiency of DPF, reduce the emission of pollutants, prolong the service life of the DPF, reduce the maintenance cost, reduce the loss of engine power and reduce fuel consumption, diesel vehicles should choose low ash oil, and keep regular cleaning of the DPF by reduction treatment.

In this paper, the hole technological process [2] and the key technology of silicon carbide wall-flow diesel particulate filters will be presented. And the final product will be taking bench test to verify compliance with standard requirements (Table 1).

Technological Process

As you see in Fig. 2, the technological process of silicon carbide wall-flow ceramic [3] always be including the mixing materials, refining mud, souring, extrusion molding, drying, alternate block, sintering and final testing. Figure 3 shows the two most important processes, one is mixing materials, another is extrusion molding. When using powders with different grain size, the porosity of green body is different. Water content is the most important factors in extrusion molding and must be control in small fluctuation range as far as possible.

Mixing Materials. In this process, as major ingredient three kind of silicon carbide powder with different diameters (D50: A:15–18 μm, B:5–8 μm, C:less than 1.0 μm) mixed with water, lubricant and binder in mud mixer and mixing 10–15 h. Table 2 shows the porosity between the proportion of A and C when the proportion of B is 65 wt%. When added 15% starch, the porosity increase to 54.66%.

Extrusion Molding. After mixing materials, the mud must be refined 3–5 times in the pug mill to get good uniformity. During refining process the water loss is between 0.3–1.0% (wt) especially depending on the temperature of the environment

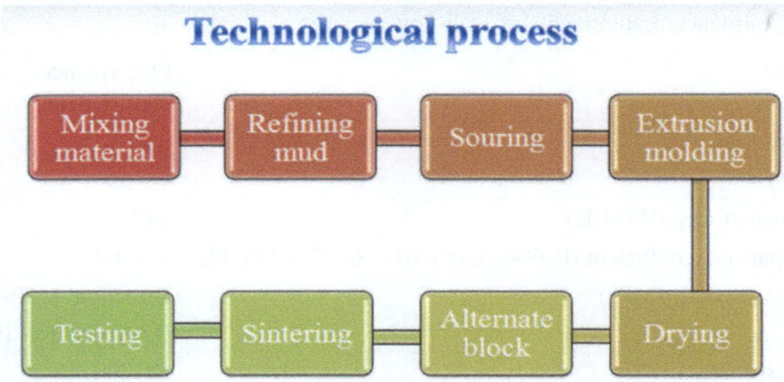

Fig. 2 Technological process of silicon carbide wall-flow ceramic

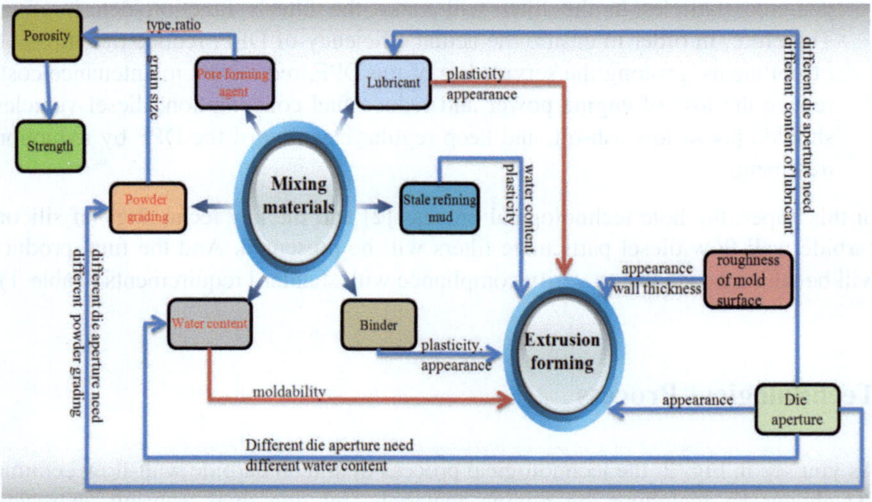

Fig. 3 Key process and influence factors

Table 2 The porosity between the proportion of A and C when the proportion of B is 65 wt%

C (wt%)					
A (wt%)	Porosity (%)				
	10	8	6	4	2
25	36.8				
27		38.7			
29			41.2		
31				43.9	
33					47.6

Fig. 4 Screw-type extruder

Fig. 5 Refining machine

and its moisture. So when mixing mud this need to be considered. In order to get green body with sleek look, suitable amount of lubricant would be added. When using 200 mesh per-square-inch to extrude and then get the green body with suitable water content and lubricant etc. Putting the green body in microwave dryer for 4–8 h. Figures 4 and 5 shows the equipments used.

Sintering. Using pressureless sintering, its temperature is between 2000–2200 °C. After the small pieces of green body sintered, they can be adhesive together to form a cylinder. The binder past must be strong enough with no cracking at all when it be sintered. Figures 6 and 7 show the sintered body and its SEM.

Alternative block is always after sintering and the block pug is soft with desirable viscos-plasticity. If the viscos-plasticity is low, the block pug can not be fastened in the hole, and non-full filled on the opposite. So control the water content of the block pug is the most important thing.

Fig. 6 The sintered body with 200 mesh PSI

Fig. 7 500 and 2000 SEM of the sintered body

Table 3 The results of the bench test of silicon carbide DPF

Testing item	Standard requirement	Testing results	Testing institutions	
Pressure drop characteristics	Pressure drop less than 8.5 kPa	7.1 kPa	TIANJIN SwARC Automotive Research Laboratory	
Filtration efficiency	Filter efficiency is high than 85%	88.2%	TIANJIN SwARC Automotive Research Laboratory	
Thermal cycle test	Carrier without crack, without leaking channels	Carrier without crack, without leaking channels	TIANJIN SwARC Automotive Research Laboratory	

Performance Testing

According to *NJ451-2008 Specifications for environmental protection product of diesel exhaust reprocessing device,* the bench test can be taken of the DPF. Table 3 shows the results of the bench test of silicon carbide DPF.

Conclusion

Using silicon carbide powder with different grain size and some pore forming material and precise control of water content, lubricant and reasonable technological process, silicon carbide honey-comb ceramic with high porosity, the value is between 50–55% can be get. The result of the bench test indicate that silicon carbide honey-comb ceramic can be used in DPF successfully.

References

1. "Study: 'Clean fuel' not always successful" (March 1, 2011) UPI NewsTrack, Vancouver, British Columbia "Canadian researchers say a program by one of the world's largest cities [New Delhi] to switch its vehicles to clean fuel has not significantly improved emission levels"
2. A. Dey, N. Kayal, O. Chakrabarti, R.F. Caldato, C.M. André, M.D.M. Innocentini, Perme-ability and nanoparticle filtration assessment of cordierite-bonded porous SiC ceramics. Ind. Eng. Chem. Res. **52**, 18362–18372 (2013)
3. J. Saggio-Woyansky, C.E. Scott, W.P. Minnear, Processing of porous ceramics. Mater. Sci. Eng. **71**(11), 1674–1682 (1992)

Effects of Annealing on Properties of PLZT (8/68/32) Electrically Controlled Light Scattering Ceramics

Feifei Song, Dazhi Sun, Xia Zeng, Wenxiu Cheng, Pingsun Qiu, Bin Xia and Xiyun He

Abstract PLZT (8/68/32) electrically controlled light scattering ceramics with fully dense microstructure and high transmittance were prepared by a hot-press sintering method. The effects of annealing on dielectric, ferroelectric, optical and electrically controlled light scattering properties were examined and analysed. The XRD patterns indicated that all sintered specimens were well crystallized and the peak of $2\theta \sim 30.8°$ gradually shifted to high degree. The temperature dielectric spectrum manifested the obvious relax or phase transition. And the double-look-like ferroelectric hysteresis loops revealed the tendency to transform a slim loop with the significant reduction in remnant polarization and coercive field via annealing treatment. Then the transmittance (from 58 to 62%, E = 0, λ = 632.8 nm) and contrast ratio (from 545:1 to 699:1) clearly increased after annealing treatment. The light scattering behavior of the PLZT samples was improved obviously.

F. Song · X. Zeng · W. Cheng · P. Qiu · B. Xia · X. He (✉)
Key Laboratory of Transparent Opto-Functional Inorganic Materials,
Shanghai Institute of Ceramics, Chinese Academy of Sciences,
1295 Dingxi Road, Shanghai 200050, China
e-mail: xyhe@mail.sic.ac.cn

F. Song
e-mail: feifeisong1991@163.com

X. Zeng
e-mail: zengxia@mail.sic.ac.cn

W. Cheng
e-mail: wxcheng@mail.sic.ac.cn

P. Qiu
e-mail: psqiu@mail.sic.ac.cn

B. Xia
e-mail: jason.xia@sjtu.edu.cn

D. Sun (✉)
Key Laboratory of Resource Chemistry of Education Ministry,
Department of Chemistry Laboratory, Shanghai Normal University,
100 Guilin Road, Shanghai 200234, China
e-mail: sundazhi@shnu.edu.cn

Y. Han (ed.), *Advanced Functional Materials*,
https://doi.org/10.1007/978-981-13-0110-0_63

Keywords PLZT ceramics · Annealing · Light scattering

Introduction

Lanthanum doped lead zirconate titanate (PLZT) ceramics were first produced by Haertling in 1972 at the Sandia National Laboratory [1], thence their excellent electro-optic properties were gradually reported. Recently, the PLZT ceramics are receiving more and more attention for their high optical transparency and unique electrically controlled light scattering performance, which promise a well application in the electro-optic modulation devices for the developing modern laser technology [2, 3].

Previous works have been carried out to study the influence of the various parameters including temperature, Zr/Ti ratio, La content, etc. on the properties of the PLZT electrically controlled light scattering ceramics. L. Nie compared the different effects between the slow annealing and the rapid annealing on the properties of PLZT (X/70/30) [4]. X. Zhang studied the temperature dependence of electric-induced scattering performance for PLZT ceramics [5]. B. Xia investigated the electrically controlled light scattering performances of PLZT (X/70/30) ceramics by changing the La content [6]. And it has been demonstrated that the domain and domain walls will influence the ferroelectric and light scattering performance [7], especially for the contrast ratio ($R_c = T_0/T_E$:1, T_0: the transmittance without electric field, T_E: the transmittance with electric field). However, the internal stress in the ceramics is a key factor to influence the domain and domain walls. So it is important to figure out the inner relationship between annealing and internal stress of the PLZT ceramics.

In this work, the PLZT (8/68/32) ceramics were prepared by hot-pressing sintering method. The aim is to diminish the effect of the internal stress on the electro-optic properties of the PLZT ceramics by a proper annealing treatment. The transmittance, ferroelectric, dielectric properties and electrically controlled light scattering performance of PLZT ceramics were examined and analyzed.

Experimental Procedure

The raw materials PbO (99.7%), La_2O_3 (99.99%), ZrO_2 (99.9%), TiO_2 (99.995%) powders based on the formula $Pb_{0.92}$ $La_{0.08}(Zr_{0.68}Ti_{0.32})_{0.98}O_3$ were selected to prepare PLZT ceramics. 10wt% PbO was added to prevent lead loss during the hot-press sintering process. First, the powders ZrO_2 and TiO_2 in stoichiometric proportions were ball milled in the ethanol medium, dried and calcined at 950 °C. Then the powders PbO and La_2O_3, the pre-synthesized ZrO_2 and TiO_2 compounds were all mixed and ball milled for 4 h, and then dried and sieved. Second, the PLZT powders mixed with a small amount of PVA as a binder were cold uniaxial pressed into cylindrically shaped specimens with the diameter of 25 mm. All disks were sintered at 1240 °C for 16 h in an O_2 atmosphere with the same axial press of ~160 kg/cm^2. The pellets were annealed at 500 or 600 °C for 10 h to reduce internal stress resulting from the sintering process. Finally, all samples were cut and polished to be carried out a series of tests.

The scanning electron micrographs (SEM) of the fracture planes of sintered samples were obtained using a Hitachi TM-3000 (Japan). The phase structures of the ceramics samples were examined by X-ray diffraction (XRD, D8ADVANCE, BRUKER/AXS, German). The polarization versus electric field (P-E) hysteresis loops were measured on a Work Station equipment (Radiant, Technologies, USA) at room temperature in a silicone oil bath. The electrically controlled light scattering performances were studied using He–Ne laser emitter by examining the changes of sample's transmittance under an applied electric field from about −15 to +15 kV/cm (λ = 632.8 nm, the test distance ~20 cm).

Results and Discussions

Microstructure and Phase. The microstructures of PLZT ceramics are shown in Fig. 1. All samples present a dense and uniform microstructure, and the grains are well-crystallized. They all exhibit an intergranular fracture with a few trans-crystalline in the materials.

The X-Ray diffraction (XRD) patterns of sintered PLZT ceramics at different annealing temperature are tested shown as Fig. 2a. The patterns indicate that the sintered PLZT ceramics all present a pure perovskite phase structure. However, the peak of 2θ–30.8° gradually shifted to high degree with the ascent of annealing temperature, as shown in Fig. 2b. This phenomenon indicates the internal stress can be cut down at some extent through annealing treatment.

Fig. 1 SEM fracture graphs of PLZT ceramics. **a** No treatment. **b** Annealing at 500 °C, **c** annealing at 600 °C

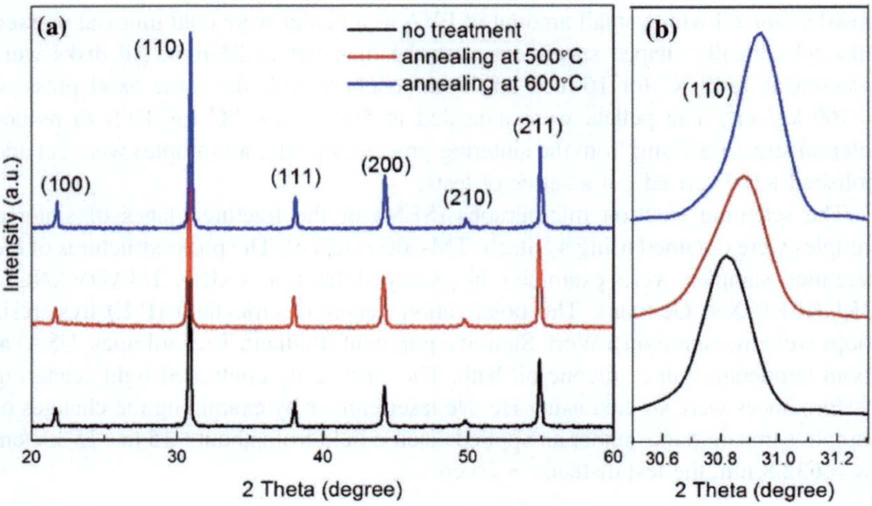

Fig. 2 XRD patterns of PLZT ceramics **a** 2θ = 60°–60°, **b** 2θ°–30.8°

Dielectric and Ferroelectric Properties. In order to study the effect of annealing treatment on the dielectric property of PLZT ceramics, the following formula is used:

$$\varepsilon_r = \frac{Ct}{A\varepsilon_0} \quad (1)$$

where ε_r is dielectric constant, t and A are the thickness and area of samples respectively, C is the capacitance value and ε_0 is the vacuum dielectric constant (8.8538×10^{-12} F/m). Then the temperature dependence of the dielectric constant is shown in Fig. 3. From Fig. 3a–c, we can see that the dielectric constants of the samples tend to elevate with the rise of temperature when the frequency from 10^2 to 10^6 Hz whether or not it has been annealed. Figure 3d shows that the dielectric constant will similarly rise through annealing treatment. And the dielectric constant increases with the rise of the temperature. As a result, the material exhibits obvious relax or phase transitions.

From the hysteresis curves at the room temperature as shown in Fig. 4, the annealing treatment can be found to influence the shape of the hysteresis loops obviously as well as the numerical value of remnant polarization (P_r) and coercive field (E_r), see Table 1. The double-look-like curves of PLZT materials with anti-ferroelectric phase become slim with the rise of the annealing temperature [8]. The plausible reason is attributed to the easier electric-field induced phase transition in the material. So it can be deduced that the internal stress in the PLZT ceramics can be reduced effectively via higher annealing temperature.

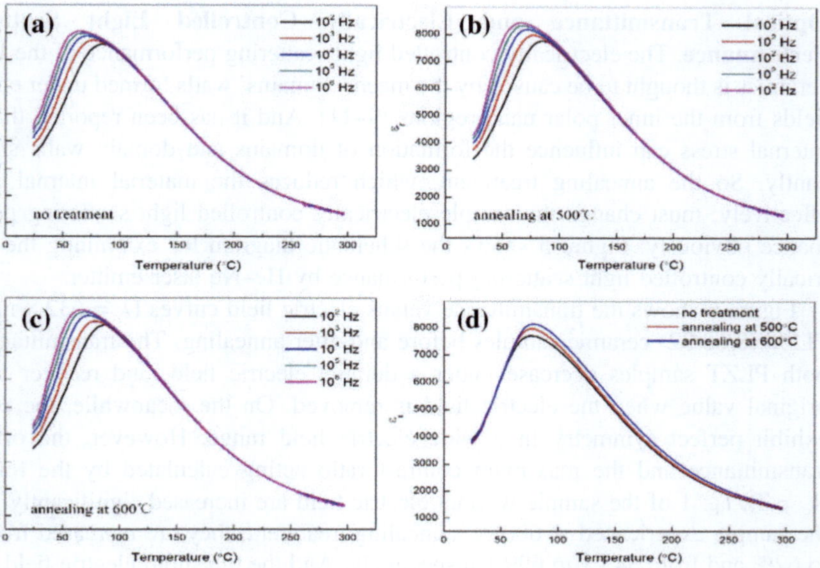

Fig. 3 **a–c** Dielectric constant (εr) for PLZT ceramics at different frequencies, **d** Dielectric constant (εr) for PLZT ceramics with different processing method at 1 kHz

Fig. 4 Ferroelectric hysteresis loops of PLZT ceramics

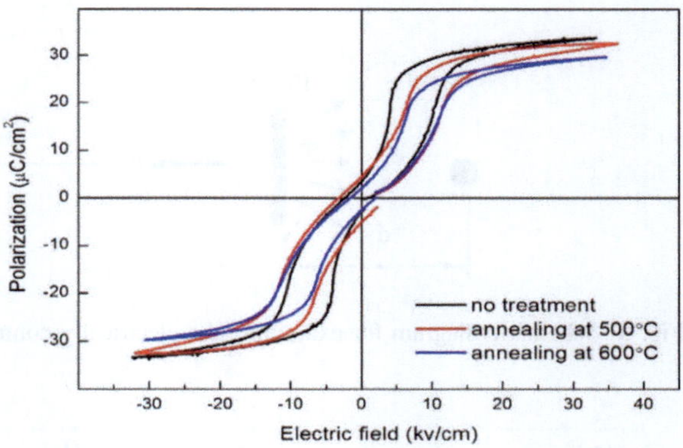

Table 1 Parameters of hysteresis

Annealing temperature (°C)	P_S (μc/cm^2)	P_r (μc/cm^2)	E_c (kv/cm)
No treatment	33.8	3.42	1.87
500	32.5	2.73	1.23
600	29.7	1.8	1.17

Optical Transmittance and Electrically Controlled Light Scattering Performance. The electrically controlled light scattering performance of the PLZT ceramics is thought to be caused by the macro-domains' walls formed under electric fields from the inner polar nano-regions [9–11]. And it has been reported that the internal stress can influence the formation of domains and domain walls significantly. So the annealing treatment, which reduces the material internal stress effectively, must change the sample electrically controlled light scattering performance obviously. Figure 5 shows the schematic diagram for examining the electrically controlled light scattering performance by He–Ne laser emitter.

Figure 6 shows the transmittance versus electric field curves (λ = 632.8 nm) of PLZT (8/68/32) ceramic samples before and after annealing. The transmittance of both PLZT samples decrease under a definite electric field, and recover to the original value when the electric field is removed. On the meanwhile, the curves exhibit perfect symmetry in a wide electric field range. However, the original transmittance and the maximum contrast ratio acting calculated by the formula $R_c = T_0/T_E$: 1 of the sample without electric field are increased significantly when the sample experienced at 600 °C annealing treatment, they are increased from 58 to 62% and from 545:1 to 699:1 respectively. And the threshold electric field, with which the sample transmittance decreases abruptly, decreases either after annealing. These improvements are also attributed to the internal stress reducing in the PLZT electrically controlled light scattering ceramics after suitable annealing treatment.

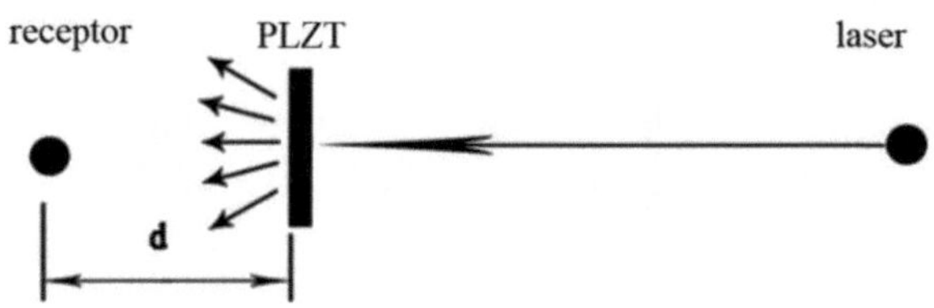

Fig. 5 Schematic diagram for examining the electrically controlled light scattering performance

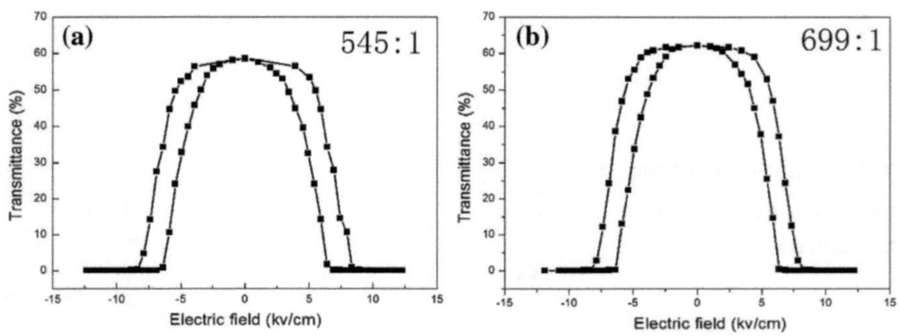

Fig. 6 Transmittance verves electric field curves for PLZT ceramics (λ = 632.8 nm, test distance = 20 cm). **a** No annealing treatment, **b** annealing at 600 °C

Summary

The PLZT (8/68/32) electrically controlled light scattering ceramics with a pure perovskite phase structure have been prepared by a hot-press sintering method. A uniform and pore-free homogeneous microstructure and high transmittance are present. The dielectric, ferroelectric and optical properties are all improved by the annealing treatment. Especially, the electrically controlled light scattering behavior has been ameliorated obviously by the annealing treatment, which may decrease the material internal stress effectively. It can be thought that a suitable annealing is an effective and economical method to improve the electro-optic properties of PLZT ceramics.

Acknowledgements The authors thank the National Natural Science Foundation of China (Grant Nos. 51602327), the National Key Research and Development Program of China (2016YFB0701004) and the Key Laboratory of Transparent Optical Inorganic Materials of Chinese Academy of Sciences (KLTOIM) for supporting the research.

References

1. G.H. Haertling, C.E. Land, Hot-Pressed (Pb, La)(Zr, Ti)O_3 ferroelectric ceramics for electrooptic applications. J. Am. Ceram. Soc. **54**, 1–11 (1971)
2. N. Balke, I. Bdikin, S.V. Kalinin, Electromechanical imaging and spectroscopy of ferroelectric and piezoelectric materials: state of the art and prospects for the future. J. Am. Ceram. Soc. **92**, 1629–1647 (2009)
3. A. Kumada, An electrooptic image storage device using field induced reversible transitions between AFE and FE phases of PLZT ceramics ferroelectric. J. Ferroelectr. **7**, 367–369 (1974)
4. L. Nie, X. He, C. Cheng, Effect of anisotropy on the ferroelectric, optical, and electro-optic properties of PLZT transparent ceramics prepared by uniaxial hot-press sintering techniques. J. Opt. Mater. Express. **6**, 3565–3574 (2016)
5. X. Zhang, B. Xia, X. He, Temperature dependence of electric-induced light scattering performance for PLZT ceramics. J. Am. Ceram. Soc. **97**, 1389–1392 (2014)
6. B. Xia, X. He, X. Zeng, Electrically induced light scattering performances of lanthanum-modified lead zirconate titanate transparent ceramics. J. Electroceram. **29**, 192–197 (2012)
7. Q. Yin, B. Zhu, *Microstructure, Property and Processing of Functional Ceramics* (Beijing Metallurgical Industry Press, Beijing, 2005)
8. G. Du, R. Liang, J. Wang, Evidence of macro-micro domain transition in poled PMN–PZT ceramics. J. Ceram. Int. **39**, 9299–9303 (2013)
9. B. Xia, X. He, X. Zeng, Light scattering properties for PLZT transparent ceramics of different Zr/Ti ratios. J. KEM **512**, 1537–1540 (2012)
10. I. Matsuyama, S. Jyomura, Grain size dependence of optical transmission in 7.6/70/30 PLZT. J. Am. Chem. Soc. **58**, 347 (1975)
11. Z. Cao, A. Ding, X. He, Preparation and characteristics of nanoporous PLZT ferroelectric thin films prepared via a one-step CSD process. J. Cryst. Growth **273**, 489–493 (2005)

Luminescence Properties of Lu^{3+}-Doped YAG Prepared by a Solid-State Reaction Method

Haili Li, Hui Chen, Mengxi Guo and Na Zhang

Abstract Lu^{3+} is commonly used as activator and sensitizer for making light materials. In the present investigation Lu^{3+}-doped YAG phosphor was synthesized by the high temperature solid-state method, whose chemical formula was $Y_{3-x-0.06}Lu_xAl_5O_{12}$:$0.06Ce^{3+}$, and x = 0–2.94, then the structure and luminescence characteristics of the sample were studied by X-ray diffraction, scanning electron microscopy and fluorescence spectroscopy. It was found that adding a small amount of Lu^{3+} to replace Y can make a significant improvement on the morphology, from rod-shape to nearly spherical. The emission spectra showed that the emission intensity at room temperature was 12% higher than that of undoped Lu^{3+}, and with the Lu^{3+} concentration increasing the peak wavelength blue shift happened. Lu^{3+}-doped YAG phosphor can reduce the drop of its emission intensity with rising temperature, which reduce the drop by about 7% compared to undoped Lu^{3+} at 185°C.

Keywords YAG phosphor · Lu^{3+}-doped · Morphology · Emission intensity

H. Li · H. Chen (✉) · M. Guo · N. Zhang
Optoelectronic Materials and Devices,
Chongqing Academy of Science and Technology, Chongqing, China
e-mail: ch@cstc.gov.cn

H. Li
e-mail: lihl8383@163.com

M. Guo
e-mail: guomx0719@sina.com

N. Zhang
e-mail: naxzhang@126.com

Y. Han (ed.), *Advanced Functional Materials*,
https://doi.org/10.1007/978-981-13-0110-0_64

Introduction

White LEDs have been considered as the most popular light source, because it has many advantages, such as longer life time, higher luminescent efficacy, lower energy consumption and so on [1–4]. The current commercial white LEDs are combined the LED chip with YAG:Ce phosphors to obtain the white light [5]. However, due to the lack of warm color components when the YAG:Ce phosphor was excited, the LEDs have high correlated color temperature (Tc) and a low color rendering index (CRI), which cannot meet the requirements of some special illumination fields.

Consequently, to improve the luminescent characteristics of the YAG:Ce phosphor is very urgent. Many papers report the YAG replace Y^{3+} with Gd^{3+} or other large trivalent ions in the main lattice [6–8], and also replace Al^{3+} with Ga^{3+} [9, 10]. In particular, due to the exceptionally high Z number (71) and high density (9.42 g/cm^3), Lu_2O_3 is regarded as a promising scintillating host, so Lu was commonly used as activator and sensitizer for making light materials. Li [11] synthesized $[(Gd_{1-x}Lu_x)_{1-y}Tb_y]_3Al_5O_{12}$ green phosphor, and found that the addition of Lu was able to fully stabilize the garnet lattice to against thermal decomposition, simplify the reaction pathway and lower the temperature of garnet formation. Shao [12] found that replacing Y with Lu can improve thermal stability of phosphor. Liu [13] also demonstrated that LuAG possesses a higher luminescent intensity, quantum yield and thermal stability than YAG phosphor. However, there are few attentions on the correlation between the structure and Lu^{3+} concentration doped in the YAG phosphor, and also few reports to research the thermal quenching, though it is very important. Phosphor powder works in a temperature more than 120 °C environment because it is directly contacted with LED chips and subjected to the heat generated by LEDs. Therefore, improving its photoluminescence efficiency at elevated temperature or thermal quenching resistance is of great importance for the applications of LED lighting.

In this paper, we synthesized Lu^{3+}-doped YAG phosphor, luminescence properties, the crystal structure, emission intensity at elevated temperature and thermal quenching resistance of the phosphor with different Lu^{3+} concentration are studied.

Experimental

Synthesis of $Y_{3-x-0.06}Lu_xAl_5O_{12}$:$0.06Ce^{3+}$. $Y_{3-x-0.06}Lu_xAl_5O_{12}$:$0.06Ce^{3+}$ were prepared by a solid-state reaction. The starting materials were Y_2O_3 (99.9%), Al_2O_3 (99%) and Ce_2O_3 (99.9%), Lu_2O_3 (99.99%), and were purchased from Beijing Youyan Rare Earth New Material Co., Ltd., Beijing Huachuang Zhongshi Science & Technology Development Co., Ltd., Desunmet ceramic material Co., Ltd. respectively. According to the chemical metering of chemical compounds, the materials were weighed, mixed, and then placed in the furnace and firing at

1500 °C for 2 h and followed by 1600 °C for 15 h under reducing atmosphere. After cooling down to room temperature naturally, the phosphors were obtained.

Characterization. The phosphor was synthesised in 1700 L high-temperature tube furnace (Luoyang city sigma furnace industry co., Ltd.). The crystal composition of phosphors was measured by X-ray diffraction (XRD), which was carried out with a DX-2500 spectrometer (Dandong, China) (Cu-Kα radiation, λ = 1.5418 Å). SEM images were carried out with Quanta-250 scanning electron microscope (American FEI Cooperation). The photoluminescence properties of the synthesized phosphors were characterized using a Hitachi F-7000 Fluorescence Spectrophotometer (Everfinephoto-e-info Co., Ltd.). The temperature-dependence PL intensity of the phosphors was characterized using an Ever-fine EX-1000 Exciting Spectra and Thermal Quenching Analyzer (EverfinePhoto-e-info Co., Ltd.).

Result and Discussions

Crystal Structure. All prepared samples were used power X-ray diffraction to verify their crystal structure and phase purity. The powder XRD patterns with different Lu^{3+} concentration were showed as Fig. 1. It is obvious that all the diffraction peaks can be indexed to the standard data of $Y_3Al_5O_{12}$ (JCPDS No. 33-0040), which illustrates that the structure of YAG doped with Lu^{3+} does not change. As the radius of Lu^{3+} and Y^{3+} are similar [14–16], so the crystal structure remains constant when Lu^{3+} substitutes for Y^{3+}, But as demonstrate by the SEM images in Fig. 2, the particle size and morphology of the powers were extremely sensitive to the addition of Lu^{3+}. The addition of small amount of Lu^{3+} to replace Y can make a significant improvement on the morphology, from rod-shape to nearly spherical. For the same type of phosphor, good crystallinity and morphology can

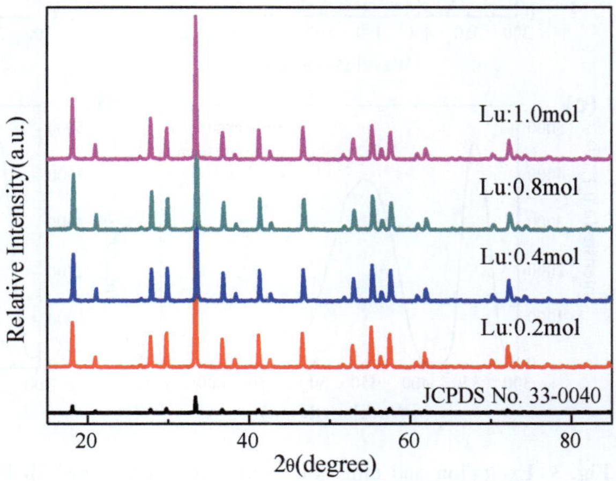

Fig. 1 X-ray diffraction patterns of the phosphor with different Lu^{3+} concentration

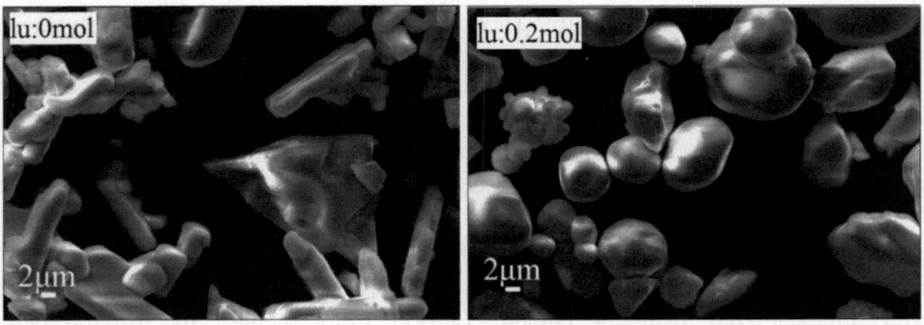

Fig. 2 SEM images of $YAG:Ce^{3+}$ with different Lu^{3+} concentration

affect the luminescence efficiency and pc-LEDs packaging significantly. The results confirm that Lu^{3+}-doped YAG phosphor is superior to pure YAG in the aspects of both crystal transformation and morphology control.

Photoluminescence Properties. The concentration of $Ce^{3+} = 0.06$ was tested by experiment. The PL strength intensity increases with the increase of Ce^{3+}, and when $Ce^{3+} = 0.06$, it is at its maximum. As the concentration of the quenching process, it decreases with the increase of Ce^{3+} concentration [17]. The excitation and emission spectra of $Y_{3-x-0.06}Lu_xAl_5O_{12}:0.06Ce^{3+}$ are shown in Fig. 3a–d. The excitation

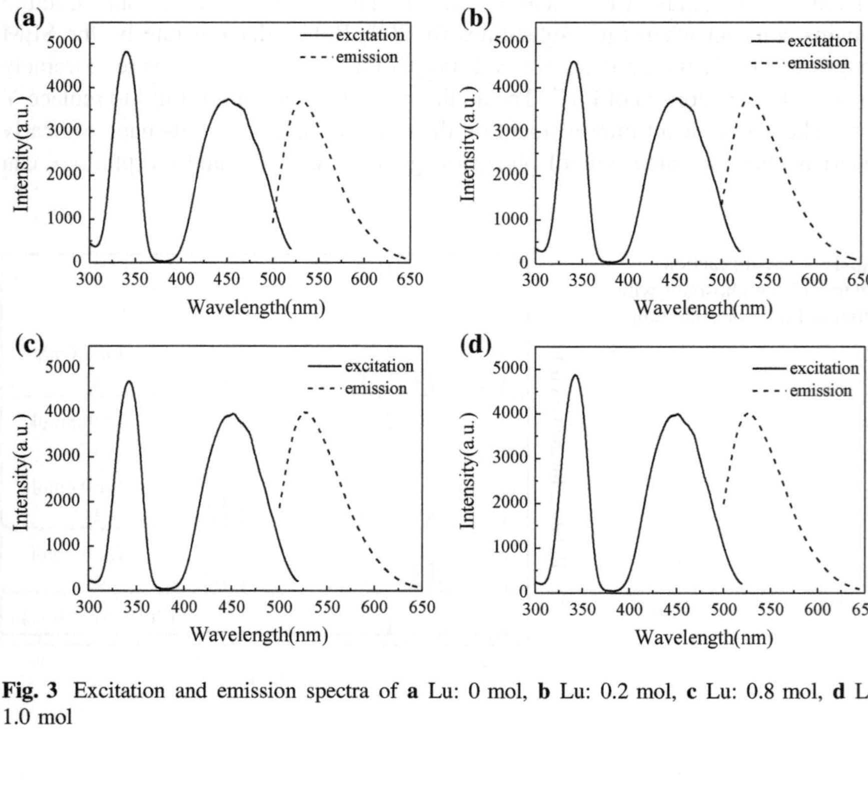

Fig. 3 Excitation and emission spectra of **a** Lu: 0 mol, **b** Lu: 0.2 mol, **c** Lu: 0.8 mol, **d** Lu: 1.0 mol

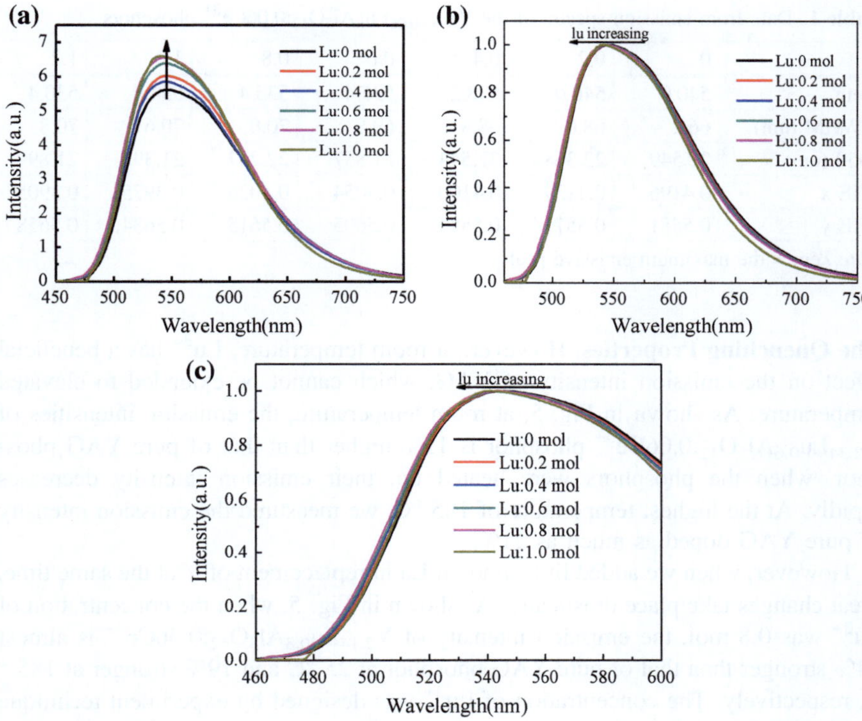

Fig. 4 Emission spectra of $Y_{3-x-0.06}Lu_xAl_5O_{12}$:$0.06Ce^{3+}$. **a** Real spectra, **b**, **c** normalized spectra (λex = 450 nm)

band around 340 nm shows a red-shift, but the amount of Lu^{3+} is so small that it caused slightly blue-shift of its emission peak, as shown in Fig. 4, and the emission intensity of the phosphor increases gradually with increasing Lu^{3+} concentration. And this result is different with the report [12] that the emission intensity slightly decreased with the increase of Lu^{3+} concentration. The emission and normalized emission spectra are shown in Fig. 4a–c, which produces a wide range of asymmetric emission bands from 450 to 750 nm. The peak of the launch was blue shifted, which was due to the chemical substitution of Lu^{3+} [18, 19].

The data of phosphors emission spectra are shown in Table 1. Furthermore, with the increase of Lu^{3+} concentration, the chromaticity coordinates (CIE) of these phosphors change from (0.4196, 0.5551) to (0.3905, 0.5658). Therefore, with the increase of Lu^{3+} concentration, the colour of $Y_{3-x-0.06}Lu_xAl_5O_{12}$:$0.06Ce^{3+}$ phosphors changes from green-yellow to green. And with the increase of Lu^{3+} concentration, the full widths at half maximum (FWHM) increases, which is consistent with the change of chromaticity coordinates (CIE). The particle size (D50) of the phosphor is located within the range of 21–23 μm, which is very uniform and beneficial for the fabrication and commercialization of these phosphors. Therefore the YAG phosphors doped with Lu^{3+} will be a promising green component on white LEDs with high Ra or cool colour temperature.

Table 1 Data from emission spectra of the $Y_{3-x-0.06}Lu_xAl_5O_{12}:0.06Ce^{3+}$ phosphors

x	0	0.2	0.4	0.6	0.8	1.0	1.2
λem	540.0	540.0	538.2	535.4	535.4	534.6	530.4
FWHM (nm)	66.2	68.0	68.8	69.8	70.0	70.6	70.8
D50 (um)	22.549	22.368	22.544	21.857	22.267	21.392	21.899
CIE x	0.4196	0.4139	0.4100	0.4054	0.4026	0.3975	0.3905
CIE y	0.5551	0.5571	0.5589	0.5605	0.5615	0.5634	0.5658

Note λem = the maximum emissive peak

The Quenching Properties. However, at room temperature, Lu^{3+} has a beneficial effect on the emission intensity of YAG, which cannot be extended to elevated temperature. As shown in Fig. 5, at room temperature, the emission intensities of $Y_{2.34}Lu_{0.6}Al_5O_{12}:0.06Ce^{3+}$ phosphor is 12% higher than that of pure YAG phosphor, when the phosphors were heated up, their emission intensity decreases rapidly. At the highest temperature of 185 °C, we measured the emission intensity of pure YAG doped as much as 21%.

However, when we added little amount Lu in replacement of Y at the same time, great changes take place drastically. As shown in Fig. 5, when the concentration of Lu^{3+} was 0.8 mol, the emission intensity of $Y_{2.14}Lu_{0.8}Al_5O_{12}:0.06Ce^{3+}$ is almost 14% stronger than that of pure YAG phosphor at 25 °C and 19% stronger at 185 ° C, respectively. The concentration of Lu^{3+} was designed by experiment technique (DOE) for the stronger emission intensity at room temperature. As the temperature increases, the emission intensity of the phosphor still decreases, but the rate of the change is obviously decreased. Specifically, the decay value of the phosphor as shown in Table 2, at 185 °C, the emission intensity of $Y_{2.34}Lu_{0.6}Al_5O_{12}:0.06Ce^{3+}$ drops 14% from the level at room temperature, representing 7% points of improvement as compared to pure YAG phosphor.

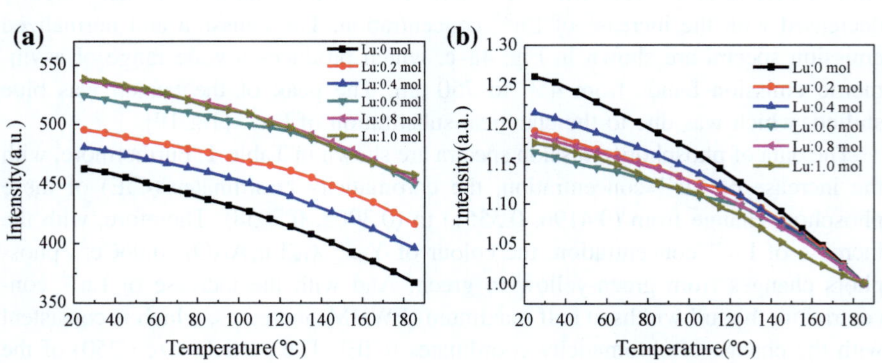

Fig. 5 Temperature dependence of the emission intensities of $Y_{3-x-0.06}Lu_xAl_5O_{12}:0.06Ce^{3+}$ phosphors with different Lu^{3+} doping concentration (**a** Real spectra) (**b** normalized)

Table 2 Thermal quenching effect of the phosphors $Y_{3-x-0.06}Lu_xAl_5O_{12}$:$0.06Ce^{3+}$

Lu^{3+} Con. (mol)	a	b	R^2	Ea (eV K^{-1})	DV* (%)
0	484.26	−0.5910	0.9837	0.1224	21
0.2	512.91	−0.4665	0.9604	0.0879	16
0.4	500.03	−0.5243	0.9757	0.1025	18
0.6	540.19	−0.4262	0.9507	0.0750	14
0.8	556.85	−0.5118	0.9673	0.0888	16
1.0	558.17	−0.5276	0.9818	0.0896	15

*decay values of the phosphors from 25°C to 180°C

To evaluate the thermal quenching thoroughly, the data of fitting curves (y = a + bx) are given in Table 2. The activation energy (Ea) for the phosphor $Y_3Al_5O_{12}$:$0.06Ce^{3+}$ are calculated to be 0.1224 eV K^{-1} and Ea decrease with the addition of Lu^{3+}, respectively, according to the following equation [20]

$$\ln\left(\frac{I_0}{I}\right) = \ln A - \frac{E_a}{kT} \quad (1)$$

where I_0 and I are the luminescence intensity of the phosphor at 25 °C and testing temperature, respectively; A is a constant; k is the Boltzmann constant (8.617×10^{-5} eV K^{-1}). Table 2 illustrates that the PL intensity changes linearly well with temperature. The activation energy of Lu^{3+}-doped YAG phosphor is lower than pure YAG phosphor, indicates that pure YAG phosphor is much more sensitive to heating energy than Lu^{3+}-doped YAG phosphor [21] and less stable for high temperature.

Conclusions

In summary, Lu^{3+}-doped YAG phosphor has been successfully prepared. The well-crystallized (Y,Lu)AG phosphor shows an efficient excitation bands around blue areas. The amount of Lu^{3+} can change the emission intensity, and the emission spectra can change from green-yellow to the green. And the sizes of $Y_{3-x-0.06}Lu_xAl_5O_{12}$:$0.06Ce^{3+}$ phosphors were about 22 μm. The effect of Lu on the temperature dependence of the emission intensity of YAG has been investigated. It is found that the addition of Lu can significantly improve the thermal quenching performance of pure YAG. The $Y_{3-x-0.06}Lu_xAl_5O_{12}$:$0.06Ce^{3+}$ phosphor shows greatly improved photoluminescence performance at both room and elevated temperatures. All the results show that Lu^{3+}-doped YAG phosphor has excellent properties than pure YAG, and can be a good phosphor for white LEDs.

Acknowledgements This work was supported by Chongqing Science and Technology Commission, China, under contract no. CSTC 2011GGB50012.

References

1. W.J. Park, M.K. Jung, T. Masaki et al., Characterization of YVO4: Eu^{3+}, Sm^{3+} red phosphor quick synthesized by microwave rapid heating method. Mater. Sci. Eng. B **146**, 95–98 (2008)
2. A. Boukerika, L. Guerbous, Investigation of the structural and photoluminescence properties of Ce^{3+}-doped LuAG nanopowders prepared via sol-gel method. Opt. Mater. **40**, 14–19 (2015)
3. T.-G. Kim, Y.S. Kim, S.-J. Im, Energy transfer and brightness saturation in $(Sr,Ca)_2P_2O_7$: Eu^{2+}, Mn^{2+} phosphor for UV-LED lighting. J. Electrochem. Soc. **156**, J203–J207 (2009)
4. C. Daqin, X. Weidong, L. Xiaojuan et al., Advances in transparent glass-ceramic phosphors for white light-emitting diodes—A review. J. Eur. Ceram. Soc. **35**, 859–869 (2015)
5. Sun Xiaoyuan, Z. Jiahua, Z. Xia et al., A white light phosphor suitable for near ultraviolet excitation. J. Lumin. **122**, 955–957 (2007)
6. Z. Antic, R. Krsmanovic, M. W et al., Preparation, structural and spectroscopic studies of $(Y_xLu_{1-x})_2O_3:Eu^{3+}$ nanopowders. Opt. Mater. **32**, 1612–1617 (2010)
7. Z. Andric, M.D. Dramicanin, M. Mitric et al., Polymer complex solution synthesis of $(Y_xGd_{1-x})_2O_3:Eu^{3+}$ nanopowders. Opt. Mater. **30**, 1023–1027 (2008)
8. L. Seijo, Z. Barandiaran, Red shifts of the yellow emission of YAG:Ce^{3+} due to tetragonal fields induced by cationic substitutions. RSC Adv. **6**, 25741–25743 (2016)
9. J.L. Jennifer, G. Gundiah, A.K. Cheetham, Structure-property correlations in Ce-doped garnet phosphors for use in solid state lighting. Chem. Phys. Lett. **441**, 250–254 (2007)
10. Z. Jun, Y. Bobo, Q. Xinglu, et al. Effect of Al/Ga substitution on photoluminescence and chromatic properties of $Y_3Al_{5-x}Ga_xO_{12}:Ce^{3+}$ phosphor. J. Mater. Sci. Mater. Electron. **27**, 8074–8079 (2016)
11. Li Jiguang, L. Jinkai, Z. Qi et al., Photoluminescent and cathodoluminescent performances of Tb^{3+} in the Lu^{3+}-stabilized gadolinium aluminate garnet solid-solutions of $[(Gd_{1-x}Lu_x)_{1-y}Tb_y]_3Al_5O_{12}$. RSC Adv. **5**, 59686–59695 (2015)
12. S. Qiyue, Dong Y, Jiang Jianqing et al., Temperature-dependent photoluminescence properties of $(Y,Lu)_3Al_5O_{12}:Ce^{3+}$ phosphors for white LEDs applications. J. Lumin. **131**, 1013–1015 (2011)
13. L. Jiaqing, W. Xiaojun, X. Tongtong et al., $Lu_3(Al,Si)_5(O,N)_{12}:Ce^{3+}$ phosphors with broad emission band and high thermal stability for white LEDs. J. Lumin. **158**, 322–327 (2015)
14. V. Babin, V. Gorbenko, A. Krasnikov et al., Photoluminescence of $Lu_3Al_5O_{12}$:Bi and $Y_3Al_5O_{12}$:Bi single crystalline films. Radiat. Meas. **45**, 331–335 (2010)
15. R.D. Shannon, Revised effective ionic radii and systematic studies of interatomic distances in halides and chalcogenides. Acta Crystallogr. **32**, 751–767 (1976)
16. D.J. Robbins, The effects of crystal field and temperature on the photoluminescence excitation efficiency of Ce^{3+} in YAG. J. Electrochem. Soc. **126**, 1550–1555 (1979)
17. T.Y. Tien, E.F. Gibbons, R.G. DeLosh et al., Ce^{3+} activated $Y_3Al_5O_{12}$ and some of its solid solutions. J. Electrochem. Soc. **120**, 278–281 (1973)
18. Y.X. Pan, W. Wang, G.K. Liu et al., Correlation between structure variation and luminescence red shift in YAG:Ce. J. Alloy. Compd. **488**, 638–642 (2009)
19. H.T. Kim, J.H. Kim, J.-K. Lee et al., Green light-emitting $Lu_3Al_5O_{12}$:Ce phosphor powders prepared by spray pyrolysis. Mater. Res. Bull. **47**, 1428–1431 (2012)
20. L. Weiren, C. Yeh, C.-H. Huang et al., $(Ba,Sr)Y_2Si_2Al_2O_2N_5:Eu^{2+}$: a novel near-ultraviolet converting green phosphor for white light-emitting diodes. J. Mater. Chem. **21**, 3740–3744 (2011)
21. Y.-C. Chiu, C.-H. Huang, T.-J. Lee et al., Eu^{2+}-activated silicon-oxynitride $Ca_3Si_2O_4N_2$: a green-emitting phosphor for white LEDs. Opt. Express **19**, A331–339 (2011)

Effects of Sintering Temperature on Microstructure, Electric Properties of $Ba_{0.7}Sr_{0.3}TiO_3$ Ceramics

Haifeng He, Wei Cai, Rongli Gao, Gang Chen, Xiaoling Deng and Chunlin Fu

Abstract Barium strontium titanate is an important electronic functional ceramics, which has a strong dielectric nonlinearity and excellent ferroelectricity. In the present study $Ba_{0.7}Sr_{0.3}TiO_3$ ceramics were synthesized by conventional solid state reaction method. The effects of sintering temperature on the microstructure, dielectric and ferroelectric properties of $Ba_{0.7}Sr_{0.3}TiO_3$ ceramics have been investigated. The XRD results show that $Ba_{0.7}Sr_{0.3}TiO_3$ ceramics sintered at 1350–1450 °C were single tetragonal phase. The lattice constant and tetragonality decreased as sintering temperature increasing. The grain size increased with the rise of sintering temperature. The dielectric properties of $Ba_{0.7}Sr_{0.3}TiO_3$ ceramics sintered at 1350 °C were superior to that of the ceramic samples sintered at 1400 and 1450 °C. The decrease of tetragonality with the rise of sintering temperature leaded to the fall of the Curie temperature of $Ba_{0.7}Sr_{0.3}TiO_3$ ceramics. $Ba_{0.7}Sr_{0.3}TiO_3$ ceramics sintered at 1350–1450 °C were relaxor ferroelectrics. The diffuseness weakened as sintering temperature increase. The remnant polarization and coercive field gradually increased with the rise of sintering temperature.

H. He · W. Cai · R. Gao · G. Chen · X. Deng · C. Fu
School of Metallurgy and Materials Engineering,
Chongqing University of Scince and Technology, Chongqing 401331, China
e-mail: 736105073@qq.com

R. Gao
e-mail: gaorongli2008@163.com

G. Chen
e-mail: cgyjxy_cqust@163.com

X. Deng
e-mail: dgxmd@163.com

C. Fu
e-mail: chlfu@126.com

W. Cai (✉) · R. Gao · G. Chen · X. Deng · C. Fu
Chongqing Key Laboratory of Nano-Micro Composite Materials
and Devices, Chongqing 401331, China
e-mail: caiwei_cqu@163.com

Y. Han (ed.), *Advanced Functional Materials*,
https://doi.org/10.1007/978-981-13-0110-0_65

Keywords $Ba_{0.7}sr_{0.3}tio_3$ · Microstructure · Electrical properties · Relaxor

Introduction

Barium titanate ($BaTiO_3$) is widely used in a variety of capacitor devices, because of its high dielectric constant [1, 2]. In order to further improve its electric properties, Sr^{2+} ions are introduced into barium titanate to substitute for Ba^{2+} ions on A sites and form $Ba_{1-x}Sr_xTiO_3$ (short for BST) [3]. BST has attracted great attention for its potential applications for the microwave technology, due to its high dielectric constant, low dielectric loss and large dielectric nonlinearity [4–6]. Sintering temperature is an important process parameter to fabricate ceramic materials by conventional solid state method. Some research shows the sintering temperature has a certain influence on microstructure, the Curie temperature and electrical properties of $BaTiO_3$-based ceramics [7–11]. The sintering kinetics of nanometer $BaTiO_3$ ceramics at different sintering temperature show that the shrinkage and density of the ceramic samples change regularly with the sintering temperature [12]. For barium strontium titanate, the experimental results show that the sintering temperature has significant effects on its electric properties [13]. However, there is a little information on effects of sintering temperature on dielectric and ferroelectric properties of $Ba_{0.7}Sr_{0.3}TiO_3$ ceramics. In this paper, $Ba_{0.7}Sr_{0.3}TiO_3$ ceramics were prepared via conventional solid state reaction method. The effects of sintering temperature on microstructure, dielectric and ferroelectric properties of $Ba_{0.7}Sr_{0.3}TiO_3$ have been investigated in detail.

Experiment Details

$Ba_{0.7}Sr_{0.3}TiO_3$ ceramics were prepared by conventional solid state reaction method. The starting raw materials were high purity $BaCO_3$ (99.9%), TiO_2 (99.9%), $SrCO_3$ (99.5%) powders, which were stoichiometrically mixed and milled in distilled water and zirconia media for 8 h (500 r/min). After drying, the mixture was calcined in an alumina crucible at 1200 °C for 5 h in air. The calcined powders were milled and dried. The powders added with 6 wt% binder (liquid paraffin) were compacted into disk-shaped pellets with a diameter of 10.0 mm and thickness of 1.0 mm with 12 MPa pressure. The green pellets were sintered at 1350, 1400 and 1450 °C for 6 h in air, respectively.

X-ray diffraction data of ceramic samples were obtained by XRD (SmartLab-9, Rigaku, Japan) with Cu $K\alpha$ radiation at 40 kV and 30 mA. The surface morphology of ceramic sample was examined by scanning electron microscope (SEM, S-3700 N, Hitachi, Japan). In order to measure electric properties, silver paste was painted on the polished samples and fired at 830 °C for 15 min as electrode. The dielectric properties of the ceramics were determined by impedance analyzer (LCR,

HP 4980A, Agilent, USA) with variation of frequency (20 Hz–2 MHz) at 1 V/mm from 20 to 120 °C. The dielectric constant was calculated from the capacitance using the following equation:

$$\varepsilon = Ch/\varepsilon_0 A \quad (1)$$

where C is the capacitance of samples (F), ε_0 is the free space dielectric constant value (8.85×10^{-12} F/m), A the capacitor area (m^2) and h the thickness (m) of the ceramics. The polarization-electric field (P–E) curve were obtained using a ferroelectric test system (TF2000E, aixACCT, Germany).

Results and Discussions

Figure 1 shows the X-ray diffraction patterns of $Ba_{0.7}Sr_{0.3}TiO_3$ ceramics sintered at different temperature. Firstly, it is seen that $Ba_{0.7}Sr_{0.3}TiO_3$ ceramics sintered at 1350–1450 °C are single phase with perovskite structure and there is no evidence of any additional phase, which implies that Sr^{2+} ions have entered the unit cell maintaining the perovskite structure of solid solution. It is found that there are two diffraction peaks about 45° diffraction angle, corresponding to (200) and (002) crystal face, which indicates that $Ba_{0.7}Sr_{0.3}TiO_3$ ceramics sintered at 1350–1450 °C belong to tetragonal perovskite structure at room temperature. Secondly, it is found that the diffraction peaks of $Ba_{0.7}Sr_{0.3}TiO_3$ ceramics shift to higher angle

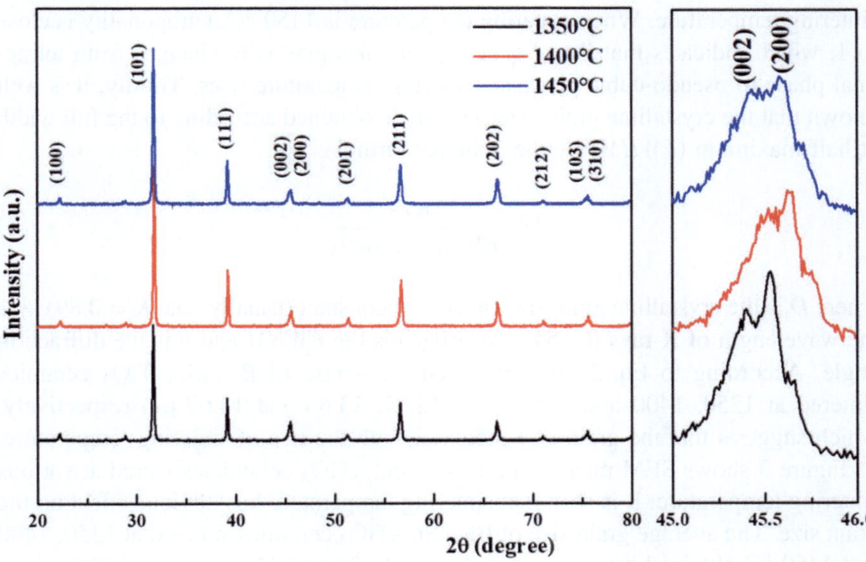

Fig. 1 XRD patterns of $Ba_{0.7}Sr_{0.3}TiO_3$ ceramics sintered at different temperature

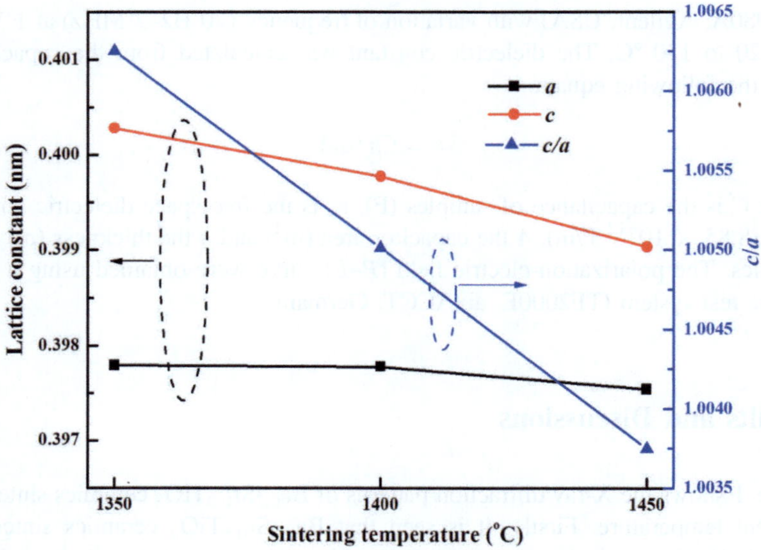

Fig. 2 Sintering temperature dependences of lattice constant and tetragonality in $Ba_{0.7}Sr_{0.3}TiO_3$ ceramics

side with the rise of sintering temperature. This shift indicates the lattice constant decreases with the increase of sintering temperature according to Bragg equation. The lattice constant (*a* and *c*) and tetragonality (*c*/*a*) of $Ba_{0.7}Sr_{0.3}TiO_3$ ceramics sintered at different temperature were calculated and plotted in Fig. 2. It can be seen that the lattice constant and tetragonality gradually decrease with the increasing of sintering temperature. When sintering temperature is 1450 °C, tetragonality is close to 1, which indicates that $Ba_{0.7}Sr_{0.3}TiO_3$ ceramics gradually changes from tetragonal phase to pseudo-cubic phase as sintering temperature rises. Thirdly, it is well known that the crystalline grain size (*D*) can be obtained according to the full width at half maximum (*FWHM*) by the Scherrer formula:

$$D = \frac{K\lambda}{FW(S) * \cos(\theta)} \qquad (2)$$

where *D* is the crystalline grain size (nm), *K* is constant (usually take $K = 0.89$), λ is the wavelength of X-rays (0.154 nm), *FW*(*S*) is the *FWHM* and θ is the diffraction angle. According to Eq. 2, the calculated grain size of $Ba_{0.7}Sr_{0.3}TiO_3$ ceramics sintered at 1350, 1400 and 1450 °C is 12.92, 13.66 and 14.67 μm respectively, which suggests that the grain size increases with the rise of sintering temperature.

Figure 3 shows SEM micrograph of $Ba_{0.7}Sr_{0.3}TiO_3$ ceramics sintered at various sintering temperature. It is seen that sintering temperature has obvious effect on the grain size. The average grain size of $Ba_{0.7}Sr_{0.3}TiO_3$ ceramics sintered at 1350, 1400 and 1450 °C obtained by intercept method is 9, 20 and 24 μm respectively, which

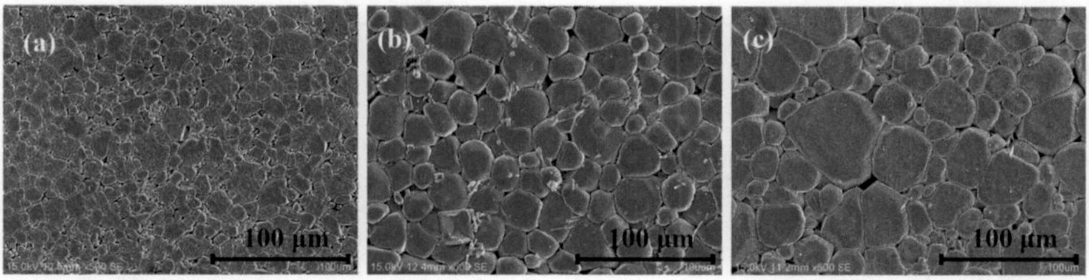

Fig. 3 SEM micrograph of $Ba_{0.7}Sr_{0.3}TiO_3$ ceramics sintered at various temperatures **a** 1350 °C, **b** 1400 °C, **c** 1450 °C

indicates that the grain size increases with the rise of sintering temperature. The change law of grain size is consistent with XRD results. Moreover, there are some small grains in the ceramic samples sintered at 1350 °C, and the number of small grain decreases with the rise of sintering temperature.

The frequency dependences of dielectric properties in $Ba_{0.7}Sr_{0.3}TiO_3$ ceramics sintered at various sintering temperature are given in Fig. 4. Firstly, it is found that the dielectric constant of $Ba_{0.7}Sr_{0.3}TiO_3$ ceramics sintered at 1350–1450 °C decrease as frequency increases. It can be explained by different polarization mechanisms. All polarization mechanisms contribute to the dielectric constant at lower frequency. But some polarization mechanism (such as space-charge polarization and dipole polarization) gradually cannot catch up with the change of electric field as frequency increases, which make dielectric constant decreases. Secondly, as frequency increases, the dielectric loss of $Ba_{0.7}Sr_{0.3}TiO_3$ ceramics sintered at 1350 °C changes little and the dielectric loss of $Ba_{0.7}Sr_{0.3}TiO_3$ ceramics sintered at 1400 and 1450 °C decreases continuously. Thirdly, the dielectric constant of $Ba_{0.7}Sr_{0.3}TiO_3$ ceramics sintered at 1350 °C is higher than that of the ceramic samples sintered at 1400 and 1450 °C, and the dielectric loss of $Ba_{0.7}Sr_{0.3}TiO_3$ ceramics sintered at 1350 °C is obviously lower than that of the other samples, which indicates that the dielectric properties of $Ba_{0.7}Sr_{0.3}TiO_3$ ceramics sintered at 1350 °C are superior.

The temperature dependences of dielectric properties in $Ba_{0.7}Sr_{0.3}TiO_3$ ceramics measured at different frequencies are shown in Fig. 5. It is seen that the dielectric peaks of $Ba_{0.7}Sr_{0.3}TiO_3$ ceramics sintered at 1350–1450 °C show obvious broadening characteristics, which suggests that there are diffuse phase transition (DPT) in $Ba_{0.7}Sr_{0.3}TiO_3$ ceramics sintered at 1350–1450 °C. The Curie temperature (T_C) corresponding to the maximum dielectric constant shifts to high temperature region with the increase of frequency, which shows frequency dispersion phenomenon. The results indicate that $Ba_{0.7}Sr_{0.3}TiO_3$ ceramics is relaxor ferroelectrics. The temperature dependences of dielectric loss in $Ba_{0.7}Sr_{0.3}TiO_3$ ceramics sintered at 1350–1450 °C is similar with that of dielectric constant.

Dielectric constant and dielectric loss versus temperature curves of $Ba_{0.7}Sr_{0.3}TiO_3$ ceramics sintered at different temperatures measured at 1 kHz are given in Fig. 6. Firstly, it is seen that the maximum dielectric constant of $Ba_{0.7}Sr_{0.3}TiO_3$ ceramics decreases as sintering temperature rises. Secondly, it is

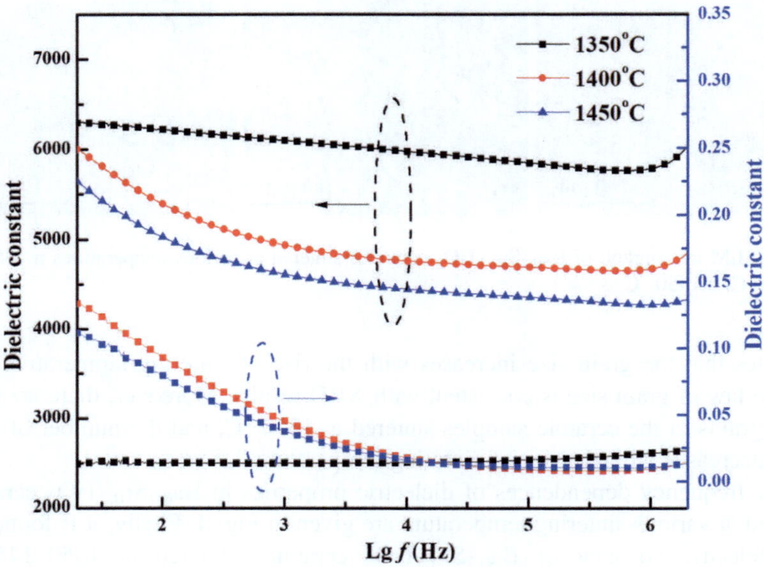

Fig. 4 Frequency dependences of room-temperature dielectric properties in $Ba_{0.7}Sr_{0.3}TiO_3$ ceramics sintered at various temperature

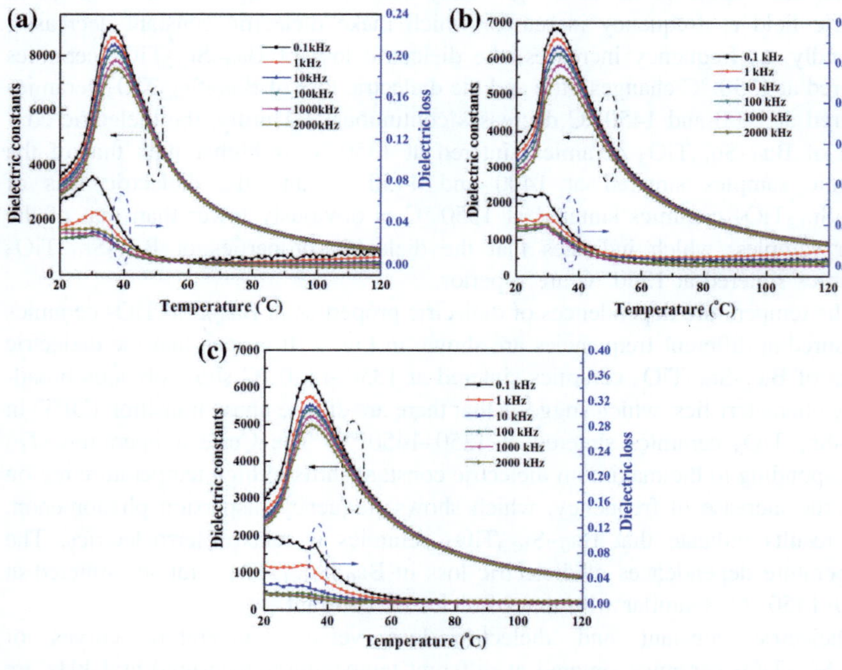

Fig. 5 Temperature dependences of dielectric properties in $Ba_{0.7}Sr_{0.3}TiO_3$ ceramics sintered at various temperature measured at different frequencies **a** 1350 °C, **b** 1400 °C, **c** 1450 °C

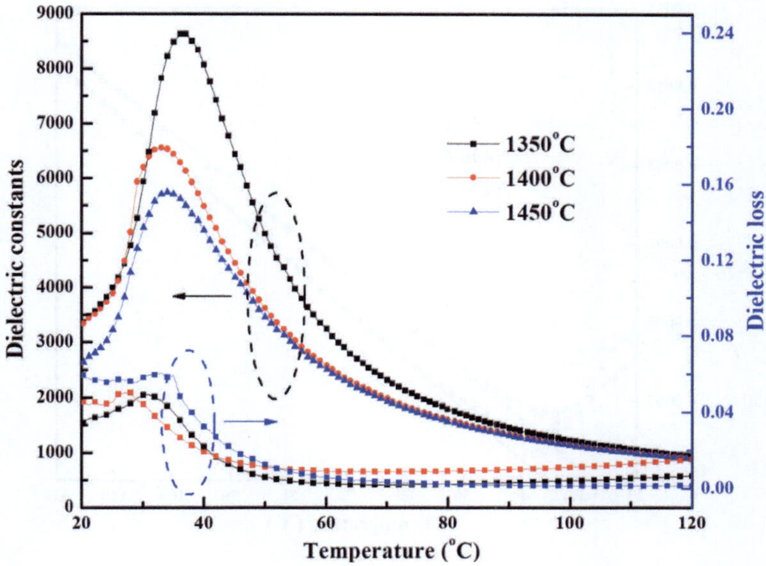

Fig. 6 Temperature dependences of dielectric properties in $Ba_{0.7}Sr_{0.3}TiO_3$ ceramics sintered at different temperature measured at 1 kHz

found that the Curie temperature (T_C) of $Ba_{0.7}Sr_{0.3}TiO_3$ ceramics sintered at 1350 °C, 1400, and 1450 is 36, 33 and 34 °C, respectively. The result suggests that the rise of sintering temperature lead to the decrease of the Curie temperature. In generally, the Curie temperature of $BaTiO_3$-based materials falls as the decreasing of tetragonality [14, 15]. As mentioned above, the tetragonality of $Ba_{0.7}Sr_{0.3}TiO_3$ ceramics decreases with the rise of sintering temperature so that the Curie temperature falls. Moreover, the broadening of dielectric peaks in $Ba_{0.7}Sr_{0.3}TiO_3$ ceramics sintered at 1350–1450 °C is different, which indicates that DPT is different.

DPT is generally characterized by the broadening of dielectric peak at the phase-transition temperature, a large separation between the maxima of dielectric constant and dielectric loss and a deviation from the Curie-Weiss law in the vicinity of T_C [16]. It is known that the dielectric constant of a normal ferroelectric above the Curie temperature follows the Curie-Weiss law described by [17, 18]

$$1/\varepsilon = (T - T_0)/C \tag{3}$$

where T_0 is the Curie-Weiss temperature and C is the Curie-Weiss constant. The parameter ΔT_m which is often used to show the degree of the deviation from the Curie-Weiss law is defined as

$$\Delta T_m = T_{CW} - T_m \tag{4}$$

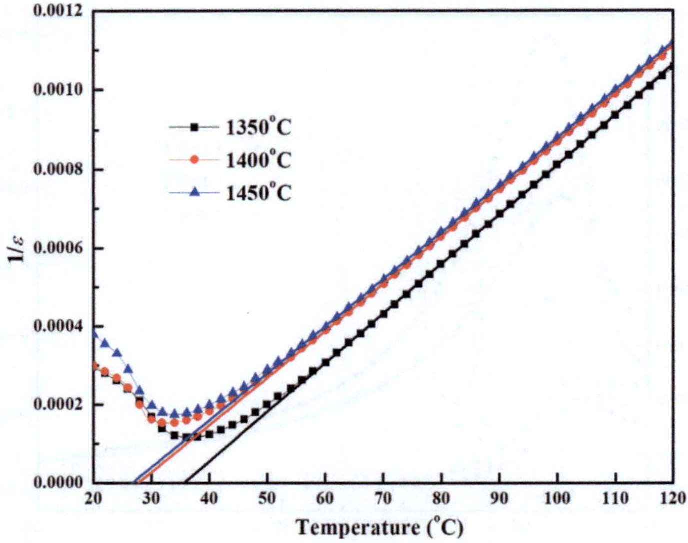

Fig. 7 Temperature dependence of the inverse dielectric constant in $Ba_{0.7}Sr_{0.3}TiO_3$ ceramics sintered at different temperature and measured at 1 kHz

where T_{CW} denotes the temperature from which the dielectric constant starts to deviate from the Curie-Weiss law and T_m represents the corresponding temperature of the maximum dielectric constant.

Figure 7 shows the inverse of dielectric constant as a function of temperature in $Ba_{0.7}Sr_{0.3}TiO_3$ ceramics. The solid lines are plotted for the ceramic samples by Eq. 3 and some parameters such as T_0, C and T_{CW} have been obtained from the fittings and listed in Table 1. It is found that the dielectric behavior of $Ba_{0.7}Sr_{0.3}TiO_3$ ceramics does not completely follow Curie-Weiss law when temperature is above the Curie temperature. A deviation from the Curie-Weiss law starting at T_{CW} can be seen and T_{CW} of $Ba_{0.7}Sr_{0.3}TiO_3$ ceramics sintered at different temperature is different. ΔT_m of $Ba_{0.7}Sr_{0.3}TiO_3$ ceramics sintered at 1350, 1400 and 1450 °C is 27, 26 and 23 °C respectively, which suggests that the diffuseness of the phase transition decreases as sintering temperature rises.

Table 1 Curie-Weiss temperature (T_0), the Curie-Weiss constant (C), T_{CW} and ΔT_m of $Ba_{0.7}Sr_{0.3}TiO_3$ ceramics sintered at different temperature and measured at 1 kHz

Sintering temperature (°C)	1350	1400	1450
T_0 (°C)	36	28	27
T_m (°C)	36	33	34
C ($\times 10^5$ K)	2.23209	1.70274	1.31836
T_{CW} (°C)	63	59	57
ΔT_m (°C)	27	26	23

A modified Curie-Weiss law was proposed to describe the diffuseness of the ferroelectric phase transition as [18]:

$$1/\varepsilon - 1/\varepsilon_m = (T - T_m)^{\gamma}/C\prime \tag{5}$$

where γ and C' are constant, ε_m and T_m represent the maximum of dielectric constant and the corresponding temperature. The diffuseness constant (γ) gives information on the character of the phase transition. The plots of $\ln(1/\varepsilon - 1/\varepsilon_m)$ as a function of $\ln(T - T_m)$ in $Ba_{0.7}Sr_{0.3}TiO_3$ ceramics are shown in Fig. 8. A linear relationship is observed in $Ba_{0.7}Sr_{0.3}TiO_3$ ceramics sintered at different temperature. The slope of the fitting curves by Eq. 5 is used to obtain γ value (listed in Table 2). The diffuseness constant γ of $Ba_{0.7}Sr_{0.3}TiO_3$ ceramics sintered at 1350, 1400, and 1450 °C is 1.45, 1.36 and 1.34. It is evident that the diffuseness of the phase transition of $Ba_{0.7}Sr_{0.3}TiO_3$ ceramics weakens with the rise of sintering temperature.

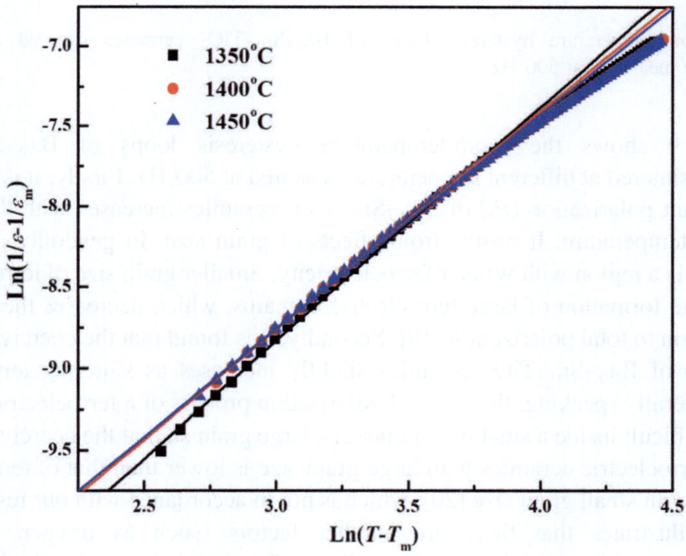

Fig. 8 Plots of $\ln(1/\varepsilon - 1/\varepsilon_m)$ as a function of $\ln(T - T_m)$ for $Ba_{0.7}Sr_{0.3}TiO_3$ ceramics sintered at different temperatures

Table 2 Corresponding temperature (T_m) of dielectric constant maximum, inverse maximum dielectric constant ($1/\varepsilon_m$) and the diffuseness constant (γ) for BST ceramics sintered at different temperatures

Sintering temperature (°C)	1350	1400	1450
T_m (°C)	36	33	34
$1/\varepsilon_m$ ($\times 10^{-4}$)	1.159	1.548	1.7446
γ	1.45	1.36	1.34

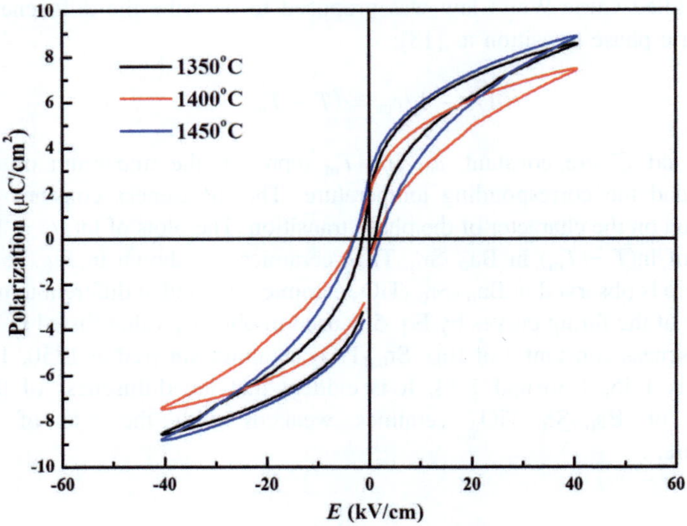

Fig. 9 Room-temperature hysteresis loops of $Ba_{0.7}Sr_{0.3}TiO_3$ ceramics sintered at various temperature measured at 500 Hz

Figure 9 shows the room-temperature hysteresis loops of $Ba_{0.7}Sr_{0.3}TiO_3$ ceramics sintered at different temperature measured at 500 Hz. Firstly, it is seen that the remnant polarization (P_r) of $Ba_{0.7}Sr_{0.3}TiO_3$ ceramics increases with the rise of sintering temperature. It results from effects of grain size. In generally, the grain boundary is a region with weaker ferroelectricity. Smaller grain size of ferroelectrics inhibits the formation of large ferroelectric domains, which decreases the effective contribution to total polarization [19]. Secondly, it is found that the coercive electric field (E_C) of $Ba_{0.7}Sr_{0.3}TiO_3$ ceramics slightly increases as sintering temperature rises. Generally speaking, the reversal polarization process of a ferroelectric domain is more difficult inside a small grain than in a large grain so that the coercive electric field of ferroelectric ceramics with large grain size is lower than that of ferroelectric ceramics with small grain size [20], which is not in accordance with our results. The conflict illustrates that there are another factors (such as oxygen vacancy) influencing coercive electric field. According to Fig. 9, it is found that the hysteresis loop of $Ba_{0.7}Sr_{0.3}TiO_3$ ceramics sintered at 1350 °C is symmetric, and the hysteresis loops of $Ba_{0.7}Sr_{0.3}TiO_3$ ceramics sintered at 1400 and 1450 °C are asymmetric on electric-field axis, which indicates that there are internal bias field caused by defect such as oxygen vacancy in the samples sintered at 1400 and 1450 °C. Therefore, the existence of oxygen vacancy in $Ba_{0.7}Sr_{0.3}TiO_3$ ceramics sintered at 1400 and 1450 °C may be the reason for unusual change of coercive electric field.

Summary

$Ba_{0.7}Sr_{0.3}TiO_3$ ceramics have been prepared by solid state reaction method. The effects of sintering temperature on microstructure, dielectric and ferroelectric properties of $Ba_{0.7}Sr_{0.3}TiO_3$ ceramics have been investigated. XRD results show that $Ba_{0.7}Sr_{0.3}TiO_3$ ceramics sintered at 1350–1450 °C belong to tetragonal perovskite structure, and the lattice constant and tetragonality gradually decreases with the rise of sintering temperature. The average grain size of $Ba_{0.7}Sr_{0.3}TiO_3$ ceramics increases with the rise of sintering temperature. $Ba_{0.7}Sr_{0.3}TiO_3$ ceramics sintered at 1350 °C have higher dielectric constant and lower dielectric loss than that of the ceramic samples sintered at 1400 and 1450 °C. The Curie temperature of $Ba_{0.7}Sr_{0.3}TiO_3$ ceramics decreases to a certain extent as sintering temperature rises, which results from the decreasing of tetragonality. There is obvious broadening dielectric peak and frequency dispersion in $Ba_{0.7}Sr_{0.3}TiO_3$ ceramics sintered at 1350–1450 °C, which indicates that $Ba_{0.7}Sr_{0.3}TiO_3$ ceramics is relaxor ferroelectrics. The diffuseness of $Ba_{0.7}Sr_{0.3}TiO_3$ ceramics weakens as sintering temperature rises. The remnant polarization and the coercive field gradually increase with the rise of sintering temperature.

Acknowledgements This research is supported by the Program for Innovation Teams in University of Chongqing (Grant No. CXTDX201601032), the Science and Technology Innovation Project of Social Undertakings and People's Livelihood Guarantee of Chongqing, China (Grant No. cstc2017shmsA0192) and the Program of Science and Technology Innovation of Chongqing University of Science & Technology (YKJCX1620206).

References

1. J.K. Lee, K.S. Hong, J.W. Jang, Roles of Ba/Ti ratios in the dielectric properties of $BaTiO_3$ ceramics. J. Am. Ceram. Soc. **84**, 2001–2006 (2010)
2. G.G. Genchi, A. Marino, A. Rocca, V. Mattoli, G. Ciofani, Barium titanate nanoparticles: promising multitasking vectors in nanomedicine. Nanotechnology **27**, 232001 (2016)
3. G.S. Jeon, S.N. Jang, D.S. Kim, S.I. Cho, Improved breakdown strength and energy density in thin-film polyimide nanocomposites with small barium strontium titanate nanocrystal fillers. J. Phys. Chem. C **117**, 6958–6965 (2016)
4. S.R. Kwon, W.B. Huang, S.J. Zhang, F.G. Yuan, X.N. Jiang, Study on a flexoelectric microphone using barium strontium titanate. J. Micromech. Microeng. **26**, 045001 (2016)
5. L.M. Garten, M. Burch, A.S. Gupta, R. Haismaier, V. Gopalan, E.C. Dickey, S. Trolier-McKinstry, Relaxor ferroelectric behavior in barium strontium titanate. J. Am. Ceram. Soc. **99**, 1645–1650 (2016)
6. Y.L. Gao, V.V. Shvartsman, D. Gautam, M. Winterer, D.C. Lupascu, Nanocrystalline barium strontium titanate ceramics synthesized via the "organosol" route and spark plasma sintering. J. Am. Ceram. Soc. **97**, 2139–2146 (2014)
7. S.H. Hu, C.X. Luo, P.W. Li, J. Hu, G. Li, H.B. Jiang, W.D. Zhang, Effect of sintered temperature on structural and piezoelectric properties of barium titanate ceramic prepared by nano-scale precursors. J. Mater. Sci. Mater. El. **28**, 9322–9327 (2017)

8. M. Alkathy, S.R. Gayam, K.C.J. Raju, Effect of sintering temperature on structural and dielectric properties of Bi and Li co-substituted barium titanate ceramic. Ceram. Int. **42**, 15432–15441 (2016)
9. D. Xu, W. Li, L. Wang, W. Wang, W.D. Fei, Effect of sintering temperature on the phase structure and electrical properties of Al–N co-doped $BaTiO_3$ piezoceramics. Rsc. Adv. **4**, 34008–34013 (2014)
10. H.L. Gong, X.H. Wang, S.P. Zhang, X.Y. Yang, L.T. Li, Influence of sintering temperature on core-shell structure evolution and reliability in Dy modified $BaTiO_3$ dielectric ceramics. Phys. Status Solidi A **211**, 1213–1218 (2014)
11. K. Mimura, K. Kato, Dielectric properties of barium titanate nanocube ordered assembly sintered at various temperatures. Jpn. J. Appl. Phys. **53**, 09PA03 (2014)
12. M.L.V. Mahesh, V.V.B. Prasad, A.R. James, Effect of sintering temperature on the microstructure and electrical properties of zirconium doped barium titanate ceramics. J. Mater. Sci. Mater. El. **24**, 4684–4692 (2013)
13. F. Chen, Q.X. Liu, X.G. Tang, Y.P. Jiang, Relaxation associated with oxygen vacancies at high temperatures and leakage current in $Ba_xSr_{1-x}TiO_3$ ceramics. J. Electron. Mater. **45**, 3174–3182 (2016)
14. K. Uchino, E. Sadanaga, T. Hirose, Dependence of the crystal structure on particle size in barium titanate. J. Am. Ceram. Soc. **72**, 1555–1558 (2010)
15. B.D. Begg, E.R. Vance, J. Nowotny, Effect of particle size on the room-temperature crystal structure of barium titanate. J. Am. Ceram. Soc. **77**, 3186–3192 (2010)
16. X. Fu, W. Cai, G. Cheng, R.L. Gao, Effects of Sn doping on the microstructure and dielectric and ferroelectric properties of $Ba(Zr_{0.2}Ti_{0.8})O_3$ ceramics. J. Mater. Sci. Mater. El. **28**, 8177–8185 (2017)
17. W. Cai, C.L. Fu, J.C. Gao, X.L. Deng, Effect of Mn doping on the dielectric properties of $BaZr_{0.2}Ti_{0.8}O_3$ ceramics, J. Mater. Sci-Mater. El. 21 (2009) 317–325
18. C.L. Fu, J.N. Liang, W. Cai, G. Chen, X.L. Deng, Effect of vanadium doping on the electric properties of barium titanate hafnate ceramics. J. Mater. Sci. Mater. El. **24**, 2438–2444 (2013)
19. X.J. Shi, J.P. Wang, Y.Y. Zhao, S.J. Liu, L.X. Zhang, Competition effects of grain boundary and aging on the hysteresis loop behavior of $(Ba_{0.8}Sr_{0.2})(Ti,Mn)O_3$ ceramics. Ceram. Int. **42**, 4734–4738 (2016)
20. V.R. Mudinepalli, L. Feng, W.C. Lin, B.S. Murty, Effect of grain size on dielectric and ferroelectric properties of nanostructured $Ba_{0.8}Sr_{0.2}TiO_3$ ceramics. J Adv. Ceram. **4**, 46–53 (2015)

Effect of Fused Silica Fiber on the Shrinkage and Properties of Silica Based Ceramic Core

Xin Li, Jiansheng Yao, Dingzhong Tang, Shuxin Niu, Junhao Yan and Chunxiao Cao

Abstract To improve the dimensional accuracy and high-temperature performance of ceramic core, the content of fused silica fiber and its length effects on mechanical and dimension behavior of silica-based ceramic cores were investigated. In order to simulate the single crystal blades casting process, the sintered samples were also heated up to 1550 °C and kept for 0.5 h. The sintering and as-cast shrinkage, creep deformation and bending strength were characterized. The result showed that fused silica short fiber could obviously reduce the shrinkage rate of ceramic core. Short fiber has an anisotropic effect on inhibition of contraction. The shrinkage in the direction of length and injection direction was less than that of the other direction. The fused silica fiber with higher length/diameter ratio maintains the creep deformation resistance at high-temperature acting as skeleton structure.

Keywords Ceramic cores · Fused silica fiber · Flexural strength
Creep deformation · Anisotropic shrinkage

Introduction

Fused silica-based ceramics cores are widely used in investment castings to produce hollow, precise, and complex-shaped internal cooling passages of superalloy gas turbine blades [1–3]. A ceramic core with precise and complex shapes must have excellent properties such as high dimensional stability, low thermal expansion, high thermal shock resistance, high thermal stability, and chemical inertness against the metal, enough mechanical weakness to prevent the generation of recrystallizations. The mineralizer is one of most critical materials in ceramic cores. Traditionally,

X. Li (✉) · J. Yao (✉) · D. Tang · S. Niu · J. Yan · C. Cao
Science and Technology on Advance High-Temperature Structural Materials Laboratory, Beijing Institute of Aeronautical Materials, Beijing 100095, China
e-mail: Lxin_86@163.com

J. Yao
e-mail: Tju_yjs@163.com

Y. Han (ed.), *Advanced Functional Materials*,
https://doi.org/10.1007/978-981-13-0110-0_66

zircon or alumina powder as mineralizer is used in most silica based ceramic cores to optimize the high-temperature performance [4–6]. However the high-temperature properties are improved,but it can't solve the problem of dimensional accuracy and introduces other issues, such as large shrinkage, the contradiction between strength and dimension, which affect the rate of qualified products and quality of turbine blade.

Forming gas turbine blades to stringent dimensional tolerances is challenging because ceramic cores shrink in the sintering and casting process, which leads to deformation defects and also cause cracks. Large shrinkage could lead to high internal stresses that make the core deform in the process of directional solidification, and even failure of the green body of ceramic core. Obviously, improving the shrinkage for better of the ceramic core is very important to avoid casting defects and improve casting dimensional accuracy. Lu [7] researched that aluminum silicate fiber dispersion in the ceramic slurry in investment casting could effectively decrease the shrinkage and resist cracking of the ceramic mold. Lu [8] report that the SiC whisker additive could adjust and control the shrinkage and bending strength of ceramic shell. Zhang et al. [9] reported that with Al_2O_3 fiber content increased,the creep deformation resistance of aluminum-based ceramic cores decreased, and the leachability improved.

In this study, low content of short-cut fused silica fiber was added into silica-based ceramic core materials. The effect of fused silica fiber on the properties and microstructure of silica-based ceramic core material, including the three-dimensional sintering shrinkage and casting shrinkage, had been investigated. The fused silica fiber was selected as the reinforcement due to the fact that its compositions (99% silica and other oxide inclusions) are similar to the silica-based ceramic core, and it has good physical and chemical compatibilities with the ceramic core.

Experiment Procedure

Fused silica powder with different particle size distributions was used as the raw materials in this study. Alumina powder was used as a mineralizer. The fused silica fiber used are 1900 type and the addition changes from 0 to 4.0 wt%. In order to obtain good dispersion of fused silica fiber in ceramic slurry, fused silica fiber was cut short into about 1 and 2 mm length. Based on the observation under a optical microscope, as Fig. 1 shows, the diameter of the fused silica fiber ranges from 5 to 10 μm, and the surface of the fiber is smooth. Characteristics of the used fused silica and alumina powder, fused silica fiber were illustrated in Table 1. Purity and PSD were measured with XRF and laser particle size analyzer, respectively.

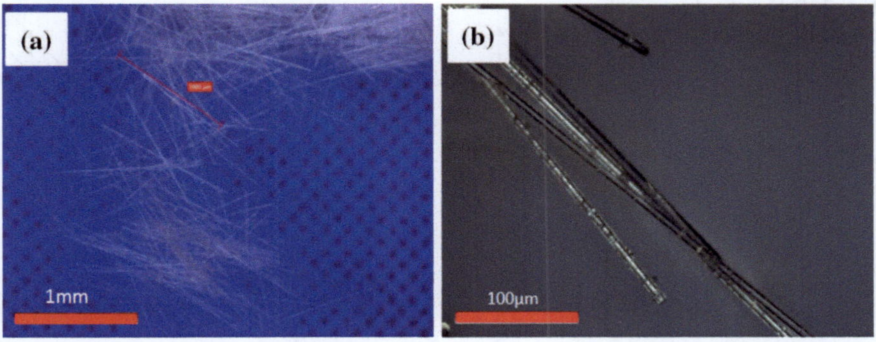

Fig. 1 Optical microscope image of fused silica fiber: **a** 1 mm; **b** 100 μm

Table 1 Characteristics of fused silica fiber, fused silica powder, and alumina powder, as raw materials

Raw material	Purity (%)	PSD (μm)			Diameter (μm)	Length (mm)
		D10	D50	D90		
Fused silica fiber	$SiO_2 \geq 99.0$	–			5–10	1.2
Fused silica powder	≥99.95	7.5	34.4	104.0		
Alumina powder	≥95	3.2	15.9	40.5		

The ceramic slurry consisted of 90 wt% fused silica powder and 10 wt% alumina powder. The different ceramic slurry containing 0, 1, 2, 3 and 4 wt% fused silica fiber with 1 and 2 mm, were mixed uniformly at the same speed and temperature using an electric mixer. The slurry was pressed into a metal mold by ceramic injection molding machine and burying pre-sintered in α-Al_2O_3 powder to 1220 °C for 2 h. Flexural strength of sintered samples was determined through three-point bending tests. The specimens with dimensions of 4 mm × 10 mm × 120 mm were measured after sintering and as cast heat treatment to calculate the shrinkage. All the three-dimensional direction shrinkages were measured and compared. For testing the creep deformation, the ceramic core sample was manufactured with dimensions of 2 mm × 4 mm × 120 mm and subjected to a simulated casting heat treatment at 1550 °C for 0.5 h by twin-cantilever beam method (Fig. 2).

After sintered and as-cast, the fused silica fiber was characterized by X'pert Pro type X-ray diffractometer (XRD) with CUKα radiation. The microstructure and fracture surface of ceramic core samples were observed using a Hitachi S-2500 scanning electron microscope (SEM).

Fig. 2 Schematic drawing of ceramic core pellets

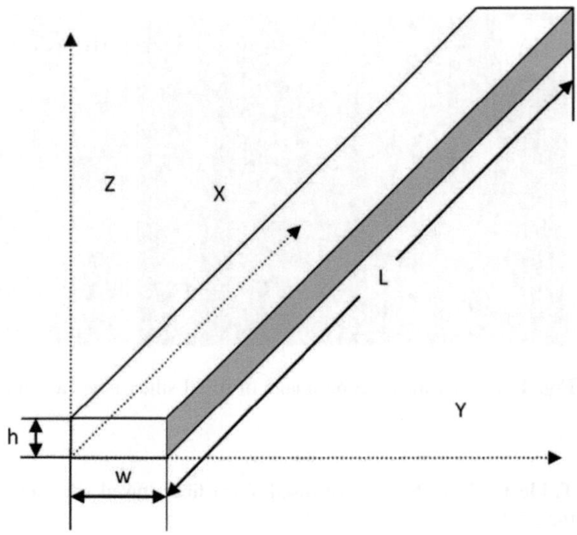

Results and Discussion

Sintering and as-Cast Shrinkage

It is well known that shrinkage determines the precision quality of ceramic core. The greater the shrinkage of the ceramic core, the lower the quality accuracy. The fused silica fiber dispersed in the ceramic core matrix inhibited the ceramic contraction. As was shown in Fig. 3, with the increase of fused silica fiber content, ceramic core with different length of fused silica fiber showed a similar decreasing trend. The arrest effect of 2 mm length silica fiber on shrinkage of ceramic core was more evident than that of 1 mm length fused silica fiber. The shrinkage reduction of the ceramic core with fiber could be interpreted in two aspects. Firstly, fused silica fiber scattered the capillary shrinkage stress, which would decrease the sintering shrinkage. Secondly, the fiber with 1–2 mm plays a similar role to the coarse particles on forming a skeleton structure in ceramic matrix.

The shrinkage of ceramic core samples was measured at different axial locations, as was shown in Fig. 4. The sintering shrinkage character showed an obvious anisotropy in different axial locations. The X-axis showed the lowest shrinkage while the Z-axis showed the largest shrinkage. In injection molding process the orientation of short fiber was different in the matrix. The orientation of most short fiber was along the X-axis direction, which was the injected direction, as was shown in Fig. 5. For the size of the sample in the Y-axis was larger than that in the Z-axis, the short fiber oriented to Y-axis or formed an acute angle with Y-axis was more likely than that of Z-axis. Fused silica fiber showed the strongest effect on inhibiting the shrinkage in the X-axis direction, and the weakest effect on inhibiting shrinkage

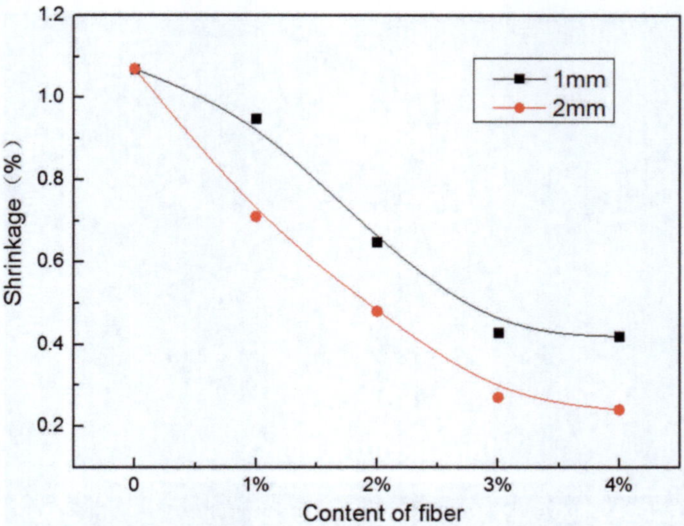

Fig. 3 Effect of fiber content on shrinkage rate of ceramic core

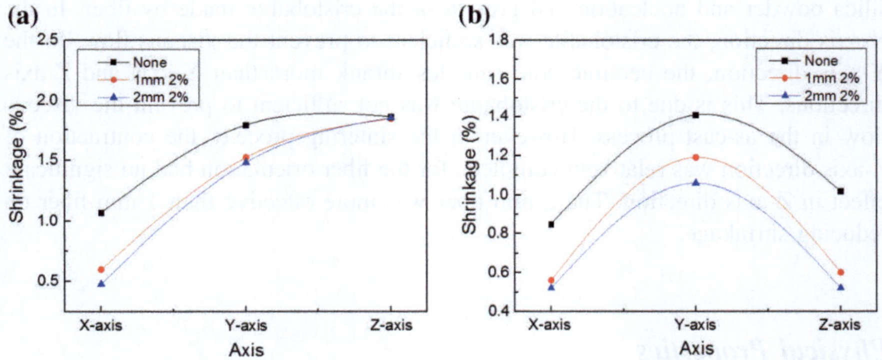

Fig. 4 Effect of fused silica fiber on shrinkage in different direction: **a** sintering process; **b** as-cast process

in the Z-axis direction. When there is no fiber, the shrinkage of the core in three directions showed a certain difference, but with the introduction of fiber, the trend of this divergence becomes more apparent.

The casting shrinkage of ceramic core appeared different tendency, compared with different directions, as indicated in Fig. 4b. In the as-cast process, the fused silica fiber was transformed into cristobalite phase, which inhibited further viscous flow of that particular fused silica powder and shrinkage decreased. The observed shrinkage changes were therefore a result of fiber orientation and secondly sintering competition between the two thermally activated process of viscous flow of fused

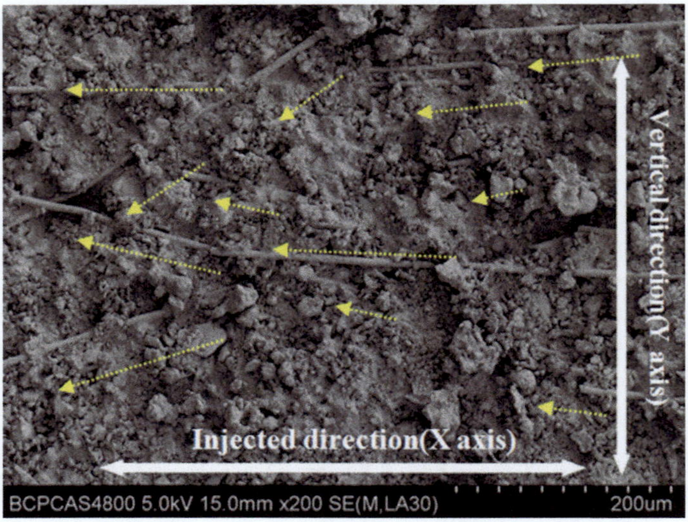

Fig. 5 Surface of ceramic core reinforce by fused silica fiber

silica powder and nucleation and growth of the cristobalite made by fiber. In the X-axis direction, the cristobalite was sufficient to prevent the viscous flow. In the Y-axis direction, the ceramic core samples shrank more than X-axis and Z-axis directions. This is due to the cristobalite was not sufficient to prevent the viscous flow in the as-cast process. However in the sintering process, the contraction of Z-axis direction was relatively complete, for the fiber orientation had no significant effect in Z-axis direction. The 2 mm fiber was more effective than 1 mm fiber on reducing shrinkage.

Physical Properties

Flexural strength of specimens with different content and length of short fused silica fiber was shown in Fig. 6. Flexural strength of ceramic core decreased with increasing the fiber content in the ceramic matrix, especially 2 mm fused silica fiber affects severity the flexural strength. The flexural strength showed the lowest value (12.5 MPa) when the 2 mm fiber content was 4%. The fused silica fiber stops the ceramic core from contracting, meaning that the porosity increased in the matrix. When beginning to act on the ceramic cores, though the external load was transferred by the fiber, the pores accelerate the expanding of the microcrack. The microstructure of ceramic core revealed that the 2 mm fused silica fiber was not dispersed in the ceramic matrix uniformly that generated large pores or microcrack that caused flexural strength significantly decreased, as was shown in Figs. 7 and 8.

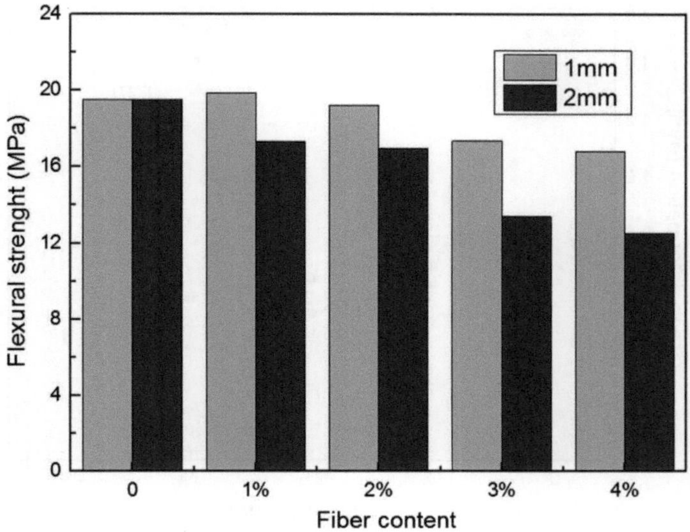

Fig. 6 Flexural strength of ceramic cores with various fused silica fiber content at room temperature

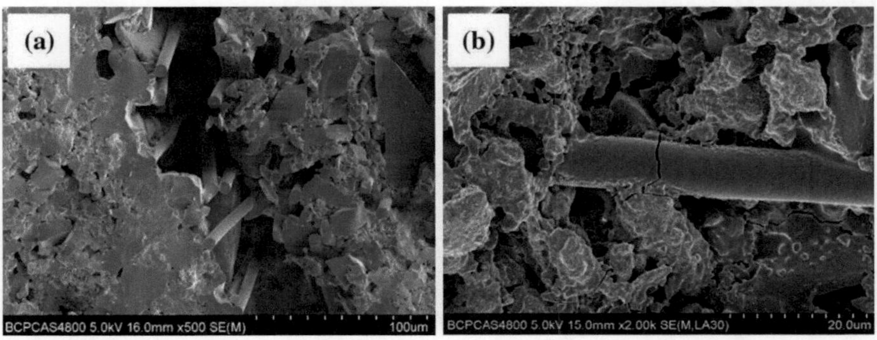

Fig. 7 SEM graphs of fracture surface of ceramic cores: **a** sintering process, ×500; **b** as-cast process, ×2000

The addition of fused silica fiber specimen had the effect of significantly reducing the creep deformation of the ceramic cores at 1550 °C (Fig. 8). This was most likely due to an increase in fiber content and crystallization. As shown in Fig. 7b, the fused silica fiber showed smooth surface while the fine silica powder become caramelized structure. The XRD result of fused silica fiber illustrates that the fused silica fiber mainly transformed into cristobalite, as shown in Fig. 9, which act as a nucleator promoting cristobalite formation more readily. The introduction of fused silica fiber with high aspect ratio and good high-temperature resistance to

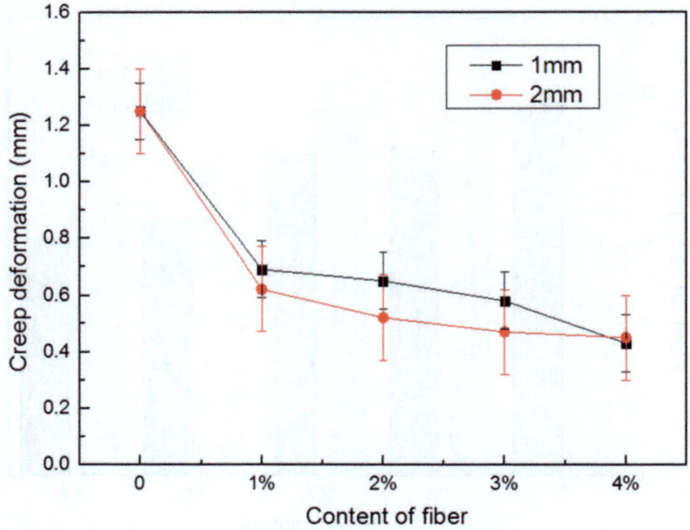

Fig. 8 Creep deformation for ceramic core samples containing different content of silica fiber

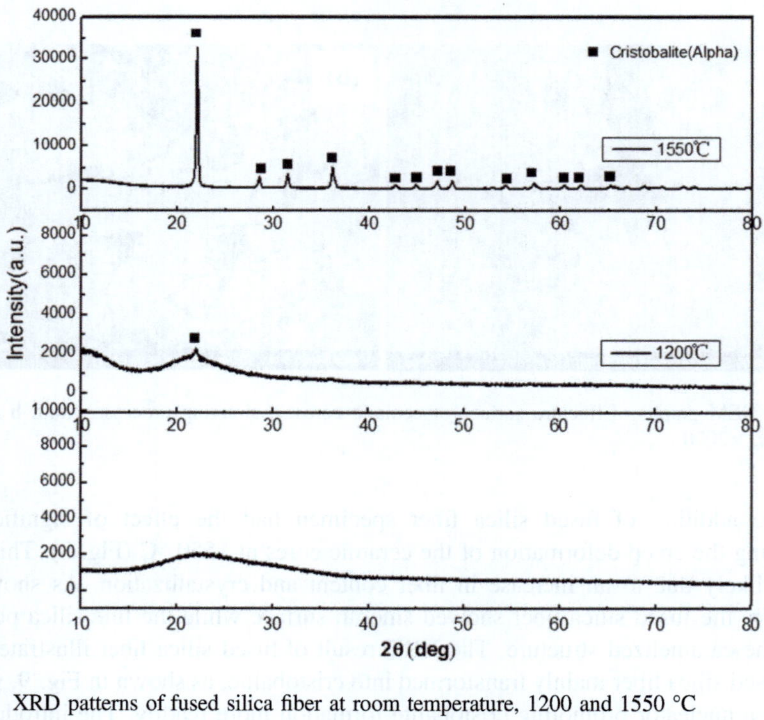

Fig. 9 XRD patterns of fused silica fiber at room temperature, 1200 and 1550 °C

creep deformation was also equivalent to increase the overall content of fused silica coarse powder, which had enhanced the skeleton structure that improved the creep deformation resistance of ceramic cores by bridging action.

Conclusions

In this work, the content of fused silica fiber and its length effects on mechanical and dimension behavior of silica-based ceramic cores were investigated. Results showed that fused silica short fiber could obviously reduce the shrinkage rate of ceramic core. Short fiber has an anisotropic effect on inhibition of contraction. The shrinkage in the direction of length and injection direction was less than that of the other direction. The 2 mm silica fiber had a better effect than 1 mm silica fiber on controlling the shrinkage.

At the same time, the fused silica fiber with higher length/diameter ratio maintains the creep deformation resistance at high-temperature acting as skeleton structure. The contradiction between the strength and sintering shrinkage of ceramic cores was resolved to a certain extent, which was helpful to reduce the microcracks.

References

1. A. Mitchell, New casting directions for gas turbine construction, in *Proceedings of the 24th BICTA Conference on Investment Casting*, Oxford (1999)
2. M. Thomas, Materials development in aero gas turbines, in *Proceedings of the 5th International Charles Parsons Turbine Conference* (2000), pp. 709–723
3. X.G. Liu, Study on high-strength ceramic core for investment casting of complex hollow turbine blade (Foundry, 2015)
4. B.Q. Wang, Y.U. Jianbo, Z.M. Ren et al., Fabrication and properties of porous alumina-based ceramic core. J. Inorg. Mater. **27**(3), 239–244 (2012)
5. P.J. Wilson, The role of zircon particle size distribution, surface area and contamination on the properties of silica-zircon ceramic materials. J. Eur. Ceram. Soc. **31**, 1849–1855 (2011)
6. Arno Kaiser, Markus Lobert, Rainer Telle, Thermal stability of Zircon ($ZrSiO_4$). J. Eur. Ceram. Soc. **28**, 2199–2211 (2008)
7. D. Lu, W. Zhao, Y. Jiang et al., Effect of aluminum silicate fiber modification on crack-resistance of a ceramic mould. China Foundry **9**(4), 322–327 (2012)
8. Z.L. Lu, Y.X. Fan, D.S. Yang et al., The effect of SiC whisker on the performances of Al_2O_3 matrixceramics mould for hollow turbine blade, in *IEEE International Symposium on Assembly and Manufacturing* (IEEE, 2013), pp. 85–87
9. R. Zhang, Y. Qin, P. Wei, Influence of Al_2O_3 fiber on the properties Al_2O_3-based ceramic core. Rare Metal Mater. Eng. **36**, 675–677 (2007)

Temperature Dependence of Ga^{3+}-Doped $CaS:Eu^{2+}$ Phosphors

Na Zhang, Suqin Luo, Renju Cheng, Haili Li, Ruoyu Zhang, Bin Jiang and Liu Yang

Abstract The temperature dependence of the photoluminescence spectra of Ga^{3+}-doped $CaS:Eu^{2+}$ phosphors were investigated. The results showed that at room temperature, the peak position was blue shifted by increasing the Ga^{3+} concentration from 0 to 0.2 mol. The peak intensity showed bandwidth broadened and emission intensity decreased with the increase of temperature. With the increasing of Ga^{3+} dopant, the declining rate and dilated trend aggravated. Increasing FWHMs and decreasing emission intensities can be explained in terms of thermal quenching in the configuration coordinate diagram. The formula explanation is due to the dominant electron-phonon interaction for thermal decay and bandwidth broadening.

Keywords Luminescence · $CaS:Eu^{2+}$ phosphor · Ceramics

N. Zhang · S. Luo (✉) · R. Cheng · H. Li · R. Zhang · L. Yang
Chongqing Academy of Science & Technology, Chongqing 401123, China
e-mail: ssuqinluo@163.com

N. Zhang
e-mail: naxzhang@126.com

R. Cheng
e-mail: chengrenju@126.com

H. Li
e-mail: lihl8383@163.com

R. Zhang
e-mail: sgzry@foxmail.com

L. Yang
e-mail: liu.yang@sytrontechnology.com

B. Jiang
College of Materials Science and Engineering,
Chongqing University, Chongqing 400045, China
e-mail: jiangbinrong@cqu.edu.cn

Y. Han (ed.), *Advanced Functional Materials*,
https://doi.org/10.1007/978-981-13-0110-0_67

Introduction

The yellow phosphor (YAG:Ce^{3+}) is excited by a blue GaN chip, and it produces a yellow emission and leads to white light that can be seen by human eyes. However, the color rending of the white LED is a lack of the red color. The red phosphor must be improvement in a white LED in order to change this situation [1–4]. Europium doped alkaline earth sulfides have been known as good red emitting phosphors and can in general be synthesized at low cost [5–7]. Use rare earth ions as a dopant have been widely studied to modify the luminescence properties of CaS:Eu^{2+} and the persistence or thermo-luminescence properties has become a hot spot of research [8–12]. A few issues, however, exist for this material system as red emitting phosphors for LED lighting applications. One of these issues is that CaS:Eu^{2+} has very poor thermal quenching resistance, that is, its photoluminescence efficiency drops quickly as the temperature increases [13, 14]. In previous works [15], the concentration quenching of CaS:Eu^{2+} can be improved by adding small amount of Mg^{2+} and Ga^{3+}.

It was reported that addition of small amount of Ga into CaS:Eu^{2+} leads to significant increase of the emission intensity at room temperature [1]. So in this article we will study that the emission intensity and bandwidth of CaS:Eu^{2+}, Ga^{3+} at elevated temperature. With the rise of the temperature, the emission peak shifts and an explanation is presented for this phenomenon.

Experimental

CaS:Eu^{2+} phosphor powders were produced using solid state synthesis approach. The starting materials included $CaCO_3$ (analytical grade), S (analytical grade), and Eu_2O_3 (99.99%). For small amount dopants of Ga^{3+}, Ga_2O_3 were used for raw materials. These starting materials were weighed by the stoichiometry of the compound and thoroughly mixed, followed by firing at 1000 °C under reducing atmosphere for extended amount of time. The firing temperature and the amount of time were optimized so that the photoluminescence intensities of the synthesized phosphors are the strongest.

The excitation and emission spectra of the synthesized phosphors were recorded by Hitachi F-7000 Fluorescence Spectrophotometer. The dependence on temperature of the photoluminescence (PL) intensity of the phosphors was measured by Thermal Quenching Analyzer and Ever-fine EX-1000 Exciting Spectra.

Results and Discussion

The Effect of Eu^{2+} Concentration. It is universally known that the effect of the concentration of activators on the emission intensity of the phosphors, Fig. 1 describes the emission spectra of CaS:x$Eu^{2+}$with x = 0.01, 0.03, 0.05, 0.07 and 0.08 mol. As shown in Fig. 1, too small or excessive Eu^{2+}-doped concentration can cause the extreme reduction in luminescence intensities of phosphors. It is obtained from the experimental results that the concentration of Eu^{2+} with the best luminescence intensity is 5 mol%, This is in consistence with what has been reported elsewhere (e.g., [12, 16]). The luminescent centers of Eu^{2+} are isolated when Eu^{2+}-doped concentration is small, which leads to lower energy transfer among Eu^{2+} ions. The luminescence intensity decreased extremely when the concentration of Eu^{2+} exceeded 5 mol%. Owing to energy migration among dopant ions, the phenomenon is known as concentration quenching [17].

The Effect of Ga^{3+}-Doped Concentration on the Emission Spectrum and the Full Widths at Half Maximum (FWHM). The emission spectra of CaS:0.05Eu^{2+}, xGa^{3+} with x = 0, 0.02, 0.06, and 0.2 mol by exciting at 460 nm are normalized in Fig. 2. As shown in Fig. 2, the effect of the Ga^{3+} concentration is on the peak position. With the rise of Ga^{3+}concentration, the emission spectrum is blue shifted. The intensity increased slightly with increasing x. The emission peak position is at about 650 nm, which is because the typical 5d–4f transition in divalent europium ions (Eu^{2+}). The peak position is blue shifted because a small amount of Ga substitute Ca, and that leads to the changes in the crystallographic environment of Eu^{2+} ion in the CaS host.

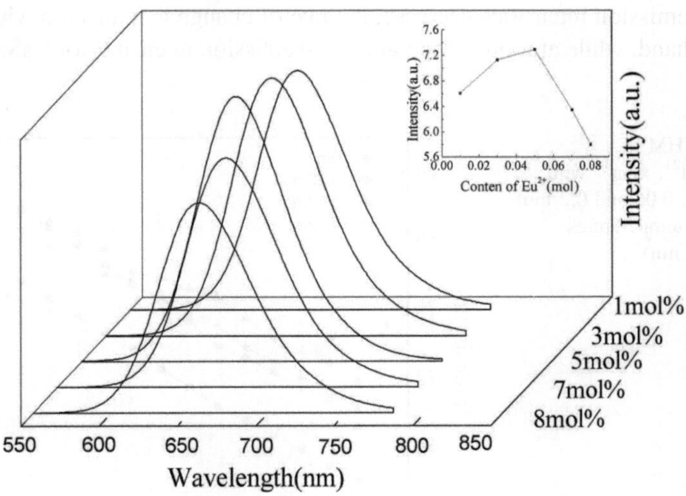

Fig. 1 Emission spectra of CaS:xEu^{2+} phosphors with different Eu^{2+} doping ratios excited at 460 nm

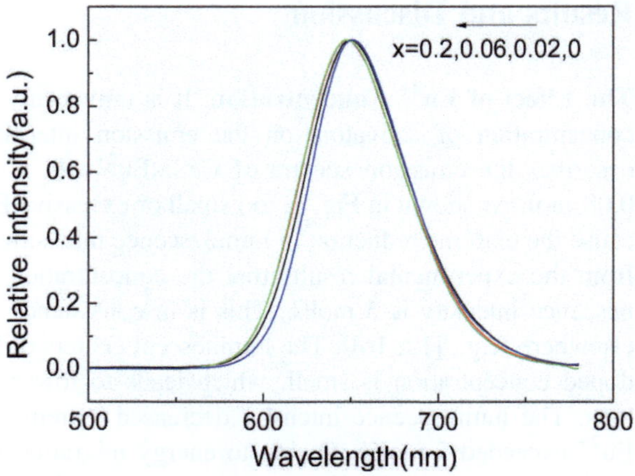

Fig. 2 Normalized emission spectra of CaS:0.05Eu^{2+}, xGa^{3+} with x = 0,0 .02,0 .06 and 0.2 mol (λex = 460 nm)

Figure 3 shows the full widths at half maximum (FWHM) of CaS:0.05Eu^{2+}, xGa^{3+} with x = 0, 0.02, 0.06 and 0.2 mol for various temperatures (λex = 460 nm). As can be seen from Fig. 3, the emission intensities of FWHM increase as an increase in temperature, the rate of change is increased significantly. On the other hand, the emission band is broadened with increasing Ga^{3+} concentration. This is because the symmetry of the crystal field changes with the increase Ga^{3+} concentration. So the environment of Eu^{2+} becomes inhomogeneous resulting in bandwidth broaden.

Temperature Dependence of the Phosphors. Figure 4 shows photoluminescence intensities of CaS:0.05Eu^{2+},xGa^{3+} with x = 0, 0.02, 0.06 and 0.2 mol for various temperatures (λex = 460 nm). As can be seen from Fig. 4, with increasing temperature, emission intensities decrease, the rate of change is reduced obviously. On the other hand, while at room temperature, the emission intensities of CaS:0.05Eu^{2+}

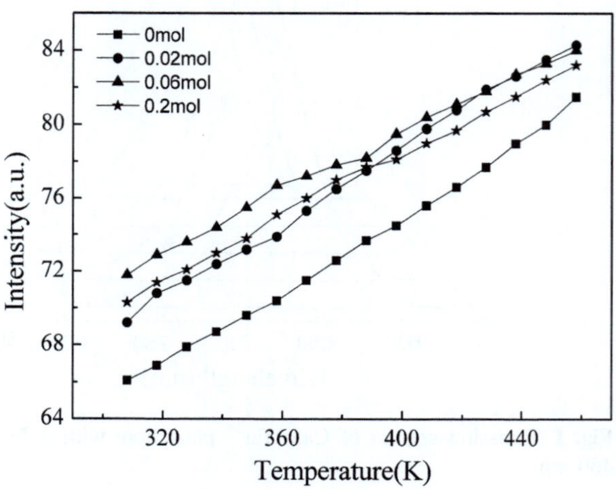

Fig. 3 FWHM of CaS:0.05Eu^{2+}, xGa^{3+} with x = 0, 0.02, 0.06 and 0.2 mol for various temperatures (λex = 460 nm)

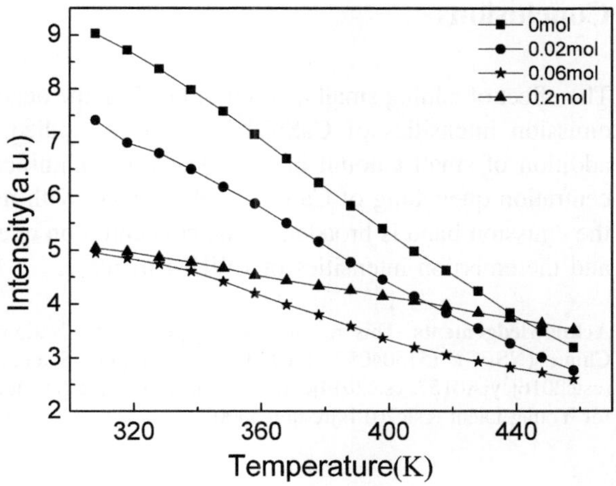

Fig. 4 Photoluminescence intensities of CaS:0.05Eu^{2+}, xGa^{3+} with x = 0, 0.02, 0.06 and 0.2 mol for various temperatures (λex = 460 nm)

with Ga^{3+} at 0 and 0.2 mol are 80.6 and 48.2% higher than that of CaS:0.05Eu^{2+} with Ga^{3+} at 0.06 and 0.2 mol respectively. As the phosphors were heated up, their emission intensities all drop quickly. At the highest temperature we measured, or at 185 °C, the emission intensities of them drop as much as 65%.

In general, the dependence on temperature of the emission intensity of a phosphor can be explained by the equation [18]:

$$I(T) = \frac{I(0)}{1 + A\ \exp\left(-\Delta E/_{KT}\right)} \tag{1}$$

where I(T) is the photoluminescence intensity at a temperature T, I(0) is the photoluminescence intensity at 0 K, A is the constant, k is the Boltzmann's constant, and ΔE is the energy barrier for thermal quenching. Fitting our experiment data to Eq. 1 gives the energy barriers for thermal quenching of CaS:0.05Eu^{2+}, xGa^{3+} with x = 0, 0.02, 0.06 and 0.2 mol are 0.084, 0.079, 0.056 and 0.288 eV, respectively.

The thermal quenching of the photoluminescence of CaS:0.05Eu^{2+}, xGa^{3+} with x = 0, 0.02, 0.06 and 0.2 mol are believed to be determined by several non-radiative decay paths. In particular, in the configurational coordinate diagram, the quenching of 5d–4f emission in Eu^{2+} can be described by non-radiative decay, which was brought about by thermally assisted crossing between the excite state and the ground state [19, 20]. Thermal activation can probably lead to this non-radiative transition, and it strongly relies on temperature causing the decrease of thermal quenching. The non-radiative transition is stronger with increasing Ga^{3+} concentration. So the increasing Ga^{3+} concentration leads to the decrease of thermal quenching.

Conclusions

The effect of adding small amount of Ga^{3+} on the dependence on temperature of the emission intensities of CaS:Eu^{2+} has been studied. It has been found that the addition of small amount of Ga^{3+} leads to a significant improvement of the concentration quenching of CaS:Eu^{2+} phosphors. With increasing Ga^{3+} concentration, the emission band is broadened, the concentration quenching of CaS:Eu^{2+} decrease and the emission intensities of FWHM increase.

Acknowledgements This research was supported by National Natural Science Foundation of China (NSFC) (51504052,51701035), Chongqing Science and Technology Commission (cstc2016jcyjA0157, cstc2014jcyjjq50002), Chongqing Science and technology training program for Young talent (cstc2014kjrc-qnrc50002).

References

1. S.I.I. Oh, Y.K. Jeong, J.G. Kang, Systhesis and luminescence properties of CaS:Eu^{2+}, Si^{4+}, Ga^{3+} for a white LED. Bull. Korean Chem. Soc. **30**, 419–422 (2009)
2. E.F. Schubert, J.K. Kim, H. Luo, J.Q. Xi, Solid-state lighting-a benevolent technology. Rep. Prog. Phys. **69**, 3069–3099 (2006)
3. E.F. Schubert, J.K. Kim, Solid-state light sources getting smart. Science **308**, 1274–1278 (2005)
4. D.Q. Chen, W.D. Xiang, X.J. Liang, J.S. Zhong, H. Yu, M.Y. Ding, H.W. Lu, Z.G. Ji, Advances in transparent glass-ceramic phosphors for white light-emitting diodes—a review. J. Eur. Ceram. Soc. **35**, 859–869 (2015)
5. W. Lehmann, Alkaline earth sulfide phosphors activated by copper, silver, and gold. J. Electrochem. Soc. **117**, 1389–1393 (1970)
6. K.F. Stripp, R. Ward, The effect of activator concentration on the infrared-sensitive phosphor, strontium sulfide—samarium, europium1. J. Am. Chem. Soc. **70**, 401–406 (1948)
7. W. Lehamnn, F.M. Ryan, Cathodoluminescence of CaS:Ce^{3+} and CaS:Eu^{2+} phosphors. J. Electrochem. Soc. **118**, 477–482 (1971)
8. C.F. Guo, D.X. Huang, Q. Su, Methods to improve the fluorescence intensity of CaS:Eu^{2+} red-emitting phosphor for white LED. Mater. Sci. Eng. B **130**, 189–193 (2006)
9. D.D. Jia, X.J. Wang, Alkali earth sulfide phosphors doped with Eu^{2+} and Ce^{3+} for LEDs. Opt. Mater. **30**, 375–379 (2007)
10. D.D. Jia, W.Y. Jia, D.R. Evans, W.M. Dennis, H.M. Liu, J. Zhu, W.M. Yen, Trapping processes in CaS:Eu^{2+}, Tm^{3+}. J. Appl. Phys. **88**, 3402–3407 (2000)
11. D. Jia, Enhancement of long-persistence by Ce Co-doping in CaS:Eu^{2+}, Tm^{3+} red phosphor. J. Electrochem. Soc. **153**, H198–H201 (2006)
12. M. Kim, K.J. Choi, J.K. Park, C.H. Kim, Synthesis characterization and Luminescent properties of CaS:Eu phosphor. J. Am. Ceram. Soc. **89**, 3413–3416 (2006)
13. M. Ando, Y.A. Ono, Temperature effects in the emission characteristics of CaS:Eu thin-film electroluminescent devices. J. Cryst. Growth **117**, 969–974 (1992)
14. N. Yamashita, O. Harada, K. Nakamura, Photoluminescence spectra of Eu^{2+} centers in Ca(S, Se):Eu and Sr(S,Se):Eu. J. Appl. Phys. **34**, 5539–5545 (1995)
15. L. Yang, N. Zhang, R.Y. Zhang, B. Wen, H.L. Li, X.B. Bian, A. CaS, Eu based red-emitting phosphor with significantly improved thermal quenching resistance for LED lighting applications. Mater. Let. **129**, 134–136 (2014)

16. P.F. Smet, I. Moreels, Z. Hens, D. Poelman, Luminescence in sulfides: a rich history and a bright future. Materials **3**, 2834–2883 (2010)
17. L.Y. Zhou, J.S. Wei, L.H. Yi, F.Z. Gong, J.L. Huang, W. Wang, A promising red phosphor MgMoO4:Eu^{3+} for white light emitting diodes. Mater. Res. Bull. **44**, 1411–1414 (2009)
18. P. Dorenbos, Thermal quenching of Eu^{2+} 5d-4f luminescence in inorganic compounds. J. Phys.: Condens. Matter **17**, 8103–8111 (2005)
19. M. Ando, Y.A. Ono, Temperature effects in the emission characteristics of CaS:Eu thin-film electroluminescent devices. Cryst. Growth. **117**, 969–974 (1992)
20. G. Blasse, W.L. Wanmaker, J.W. Ter Vrugt, A. Bril, Fluorescence of Eu^{2+} activated silicates. Philips. Res. Rep. **23**, 189–200 (1968)

Infrared Emission Properties of Cr^{3+}-Doped $NiAl_2O_4$ Spinel Ceramics

Gang He, Jiaxi Liu, Xingxing Jiang and Jiangtao Li

Abstract High infrared emissive $NiAl_{2-x}Cr_xO_4$ (x = 0, 0.1, 0.5, 1.0, 2.0) spinel ceramics were prepared by solid-state reaction method. The phase composition, microstructure, near infrared absorption and infrared emission properties of the prepared samples were investigated. With the addition of Cr^{3+}, the samples maintained a spinel structure, and the grains grow to larger sizes. The absorptivity increased from 0.52 to 0.93 in the near infrared spectral region of 0.75–2.5 μm and the emissivity increased from 0.56 to 0.83 in the infrared spectral region of 3–5 μm with increasing Cr content. The band gap of $NiAl_{2-x}Cr_xO_4$ spinel ceramics decreased with increasing Cr content based on the calculation of the absorbance test results. The first-principles electronic structure calculations for $NiAl_2O_4$ and $NiCr_2O_4$ were performed. The high emissive spinel ceramics shows promising applications for energy-saving in high temperature thermal process equipment field.

Keywords Ceramics · Spinel · Infrared emissivity · Electronic structure

G. He · J. Liu · X. Jiang · J. Li
Key Laboratory of Cryogenics, Technical Institute of Physics and Chemistry, Chinese Academy of Sciences, Beijing 100190, China
e-mail: ipchg@mail.ipc.ac.cn

J. Liu
e-mail: liujiaxi@mail.ipc.ac.cn

X. Jiang
e-mail: xxjiang@mail.ipc.ac.cn

J. Liu · J. Li (✉)
University of Chinese Academy of Sciences, Beijing 100049, China
e-mail: lijiangtao@mail.ipc.ac.cn

Y. Han (ed.), *Advanced Functional Materials*,
https://doi.org/10.1007/978-981-13-0110-0_68

Introduction

Metal aluminate spinel ceramics have attracted significant attention in many areas such as transparent ceramics, magnetic materials, pigments, catalysts and refractory materials, due to its high thermal and chemical stability, hydrophobicity, high mechanical resistance, and low surface acidity [1–3]. The general formula of metal aluminate spinels is denoted as MAl_2O_4 (M=Ni, Mg, Fe, Mn, Zn, Cu, Co, etc.), and tetrahedral sites and octahedral sites are mainly located by divalent M^{2+} cations and trivalent Al^{3+} cation, respectively [1–4]. The overall physicochemical properties of spinel materials mainly depend upon the cation distribution among the tetrahedral and octahedral sites in the crystal lattice. Nickel aluminate ($NiAl_2O_4$) has been extensively investigated as electrocatalysts for the oxidation of organic compounds and nitrous oxide, anodes materials for aluminum electrolysis, and refractory materials [5–7]. Oxide materials with high infrared emissivity have generated significant interest in high temperature thermal process equipment field to save energy by enhancing the radiative heat transfer [8–10]. However, the conventional oxide materials usually possess a relatively low emissivity in the infrared spectral region. It is considered that the doping of transition metal or rare earth ions in spinel structure materials may be a effective way to improve its infrared emission properties [11].

In this study, Cr^{3+}-doped $NiAl_2O_4$ spinel ceramics were prepared by solid-state reaction method. The phase compositions and microstructures of the prepared samples were characterized. The near infrared absorption and infrared emission properties were investigated. To explain the obviously improved infrared emissivity of Cr^{3+}-doped $NiAl_2O_4$ spinels, the electronic structure of the samples were investigated by the first-principles calculations based on density functional theory.

Experimental

In the present study, the Cr^{3+}-doped $NiAl_2O_4$ spinel ceramics were prepared by traditional solid-sate reaction method. High purity NiO, Al_2O_3, Cr_2O_3 powders (99.9%, Sinopharm Chemical Reagent Co., Ltd., Shanghai, China) were used as raw materials. The raw materials were blended together according to the molar ratio of $NiAl_{2-x}Cr_xO_4$ (x = 0, 0.1, 0.5, 1.0, 2.0) by ball-milling for 2 h with alumina balls as milling media. The mixtures were pressed into pellets of 30 mm in and about 3 mm in thickness under a pressure of 120 MPa. The pressed pellets were sintered at 1400 °C for 4 h in air atmosphere.

Phase identification of the sample was characterized by powder X-ray diffraction (XRD; D8Focus, Bruker, Germany) using Cu–Ka radiation with a scanning rate of 6°/min. The microstructure was examined by scanning electron microscopy (SEM; S-4800, Hitachi, Japan). The spectral absorptivity in the wavelength range of 250–2500 nm and the absorption spectrum in the wavelength range of 200–800 nm were recorded by an ultraviolet-visible-near infrared spectrophotometer (UV-VIS-NIR, Cary 5000, Varian, Sunnyvale, CA) equipped with an internal integrating sphere at room temperature. The near infrared absorptivity in the spectral region of 0.75–2.5 μm was calculated by graphical integration of the measured spectral absorptivity. The infrared emissivity in the spectral region of 3–5, 8–14 and 1–22 μm was measured by an IR emissometer (IR-2, Shanghai Institute of Technological Physical, China) at room temperature. Electronic structure of the samples were investigated by first-principles calculations based on density functional theory using the plane-wave pseudopotential approach.

Results and Discussion

The XRD patterns of the prepared $NiAl_{2-x}Cr_xO_4$ (x = 0, 0.1, 0.5, 1.0, 2.0) ceramic materials are shown in Fig. 1. It can be seen that the synthesized samples all possess a cubic spinel structure. The diffraction peak (311) at 2θ = 37.165° decreased to a smaller angle 2θ = 36.031° with increasing Cr content, which corresponds to a increased crystal lattice constant from 0.803 to 0.831 nm, due to a larger size of Cr^{3+} than the Al^{3+}. It can be confirmed from the XRD results that the Cr^{3+} has successfully cooperated into the $NiAl_2O_4$ spinel structure. When the Cr^{3+} doping contents were x = 0 and x = 2.0, the phase compositions of the prepared samples belong to the crystal phase of $NiAl_2O_4$ and $NiCr_2O_4$, respectively.

The SEM images of the prepared $NiAl_{2-x}Cr_xO_4$ (x = 0, 0.1, 0.5, 1.0, 2.0) ceramic materials are shown in Fig. 2. It can be seen that the sample with higher Cr content shows a larger grain size. The grain size of pure $NiAl_2O_4$ sample is about 2–3 μm. While the pure $NiCr_2O_4$ sample shows a lager grain size over 5 μm. With increasing Cr content the morphology of the grains developed to a octahedral spinel crystal shape. The growth of spinel crystals shows an obvious diffusion controlled ledge growth characteristic, as shown in Fig. 2f.

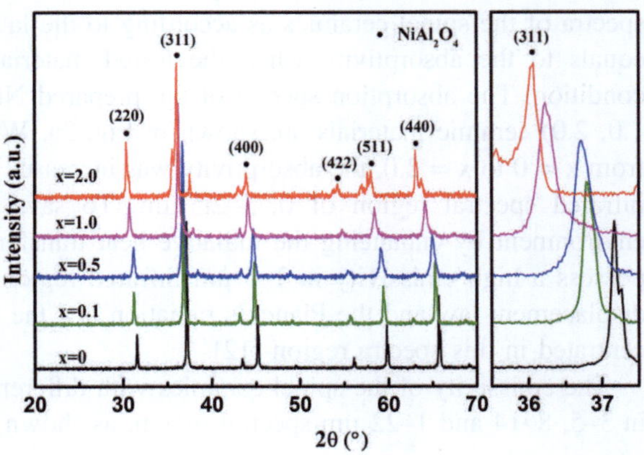

Fig. 1 XRD patterns and the (311) diffraction peaks of $NiAl_{2-x}Cr_xO_4$

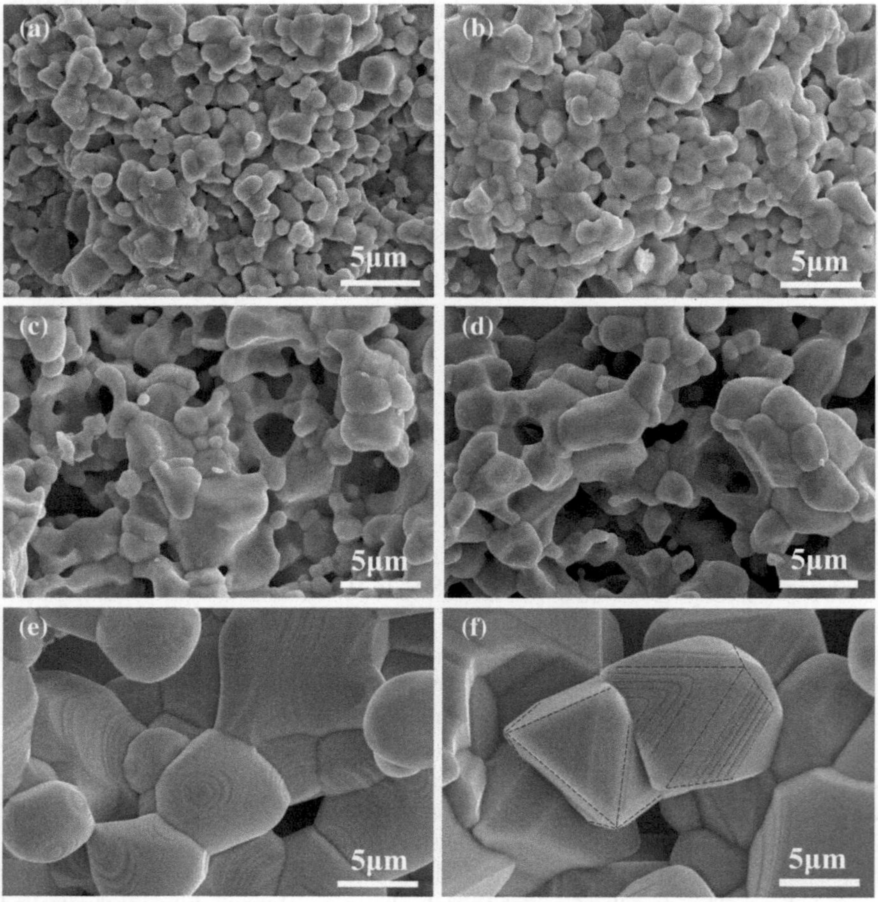

Fig. 2 SEM images of $NiAl_{2-x}Cr_xO_4$ **a** x = 0, **b** x = 0.1, **c** x = 0.5, **d** x = 1.0, **e**, **f** x = 2.0

The near infrared emission property was evaluated by testing the absorption spectra of the spinel ceramics as according to the law of Kirchhoff the emissivity equals to the absorptivity when the tested materials is at thermal equilibrium condition. The absorption spectra of the prepared $NiAl_{2-x}Cr_xO_4$ (x = 0, 0.1, 0.5, 1.0, 2.0) ceramic materials are shown in Fig. 3a. With the Cr content increasing from x = 0 to x = 2.0, the absorptivity was increased from 0.52 to 0.93 in the near infrared spectral region of 0.75–2.5 μm. To save energy in high temperature environment by enhancing the radiative heat transfer, the spinel materials should possess a high emissivity in 1–5 μm infrared region, as according to the Wien's displacement law and the Planck's radiation law the radiation energy mainly concentrated in this spectra region [12].

The emissivity of the spinel ceramics with different Cr contents were measured in 3–5, 8–14 and 1–22 μm spectral region, as shown in Fig. 3b. The emissivity in

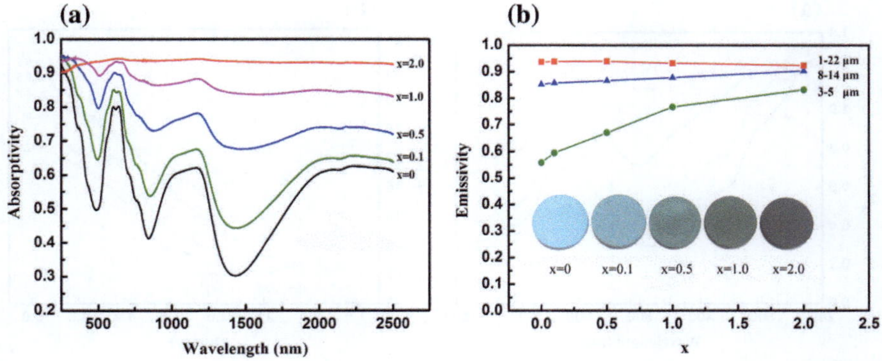

Fig. 3 **a** Absorption spectra **b** infrared emissivity and photographs (insert) of $NiAl_{2-x}Cr_xO_4$

the regions of 8–14 and 1–22 μm increased slightly with increasing Cr content. While the emissivity in the 3–5 μm region shows a significant increasing from 0.56 to 0.83 with increasing Cr content. The color of the prepared spinel ceramics changed from blue (x = 0) to black (x = 2.0), as shown in the insert of Fig. 3b, which is consistent with the results of the absorption spectra in Fig. 3a. The high absorptivity and emissivity in the 0.75–2.5 and 3–5 μm infrared region indicated that the prepared Cr^{3+}-doped $NiAl_2O_4$ spinel ceramics may possess a promising applications for energy-saving in high temperature thermal process equipment field.

Despite the reported results about the synthesis of high emissive $NiCr_2O_4$ powders and coatings [13, 14], the mechanisms for the improved infrared emissivity has not been clarified. To estimate the optical band gaps of the Cr^{3+}-doped $NiAl_2O_4$ the UV-Vis absorption spectra in the range of 200–800 nm were measured, as shown in Fig. 4a. The optical band gaps were calculated based on the equation: $\alpha h\nu = A(h\nu - Eg)^{n/2}$, where α is the absorbance of the sample, h is the Planck's constant, ν is the photo frequency, A is a material dependent constant, and n depends on whether the transition is direct (n = 1) or indirect (n = 4) [15]. As can be seen from Fig. 4b, the band gap of prepared spinel ceramics decreased from 3.34 to 1.80 eV with increasing Cr content. It has been reported that the doping of Cr in the perovskite type and garnet type crystal structure may introduce an impurity energy level into its optical band gaps, which causes a obvious decrease in band gap [8–10]. The greatly improved infrared emissivity may be attributed to the enhancement in electron transition of the impurity energy level and band gap narrowing of the spinels.

To further investigate the physic mechanisms for the improved infrared emissivity of the prepared spinel ceramics the first-principles electronic structures of $NiAl_2O_4$ and $NiCr_2O_4$ were calculated by using the plane-wave pseudopotentials method. The calculated electronic density of states of $NiAl_2O_4$ and $NiCr_2O_4$ are shown in Fig. 5. The band gap of $NiAl_2O_4$ is determined by the O 2p orbitals and the Ni 3d orbitals. The band gap of $NiCr_2O_4$ is determined by the O 2p orbitals and the Cr 3d orbitals, which occupy the top of the valance band and the bottom of the

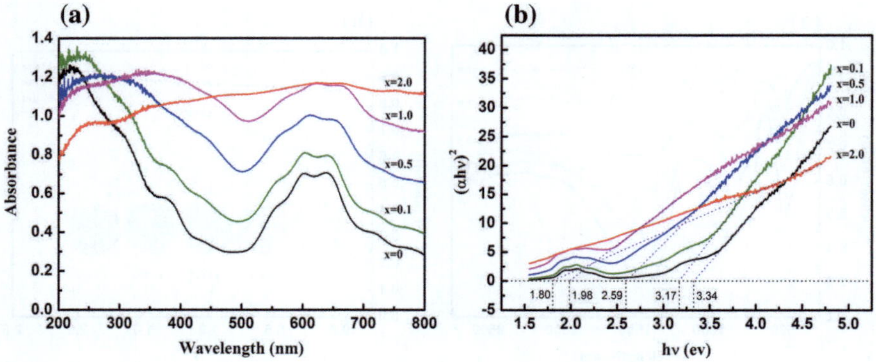

Fig. 4 **a** The ultraviolet absorption spectra of $NiAl_{2-x}CrxO_4$ and **b** the plot of $(\alpha h\nu)_2$ versus hν

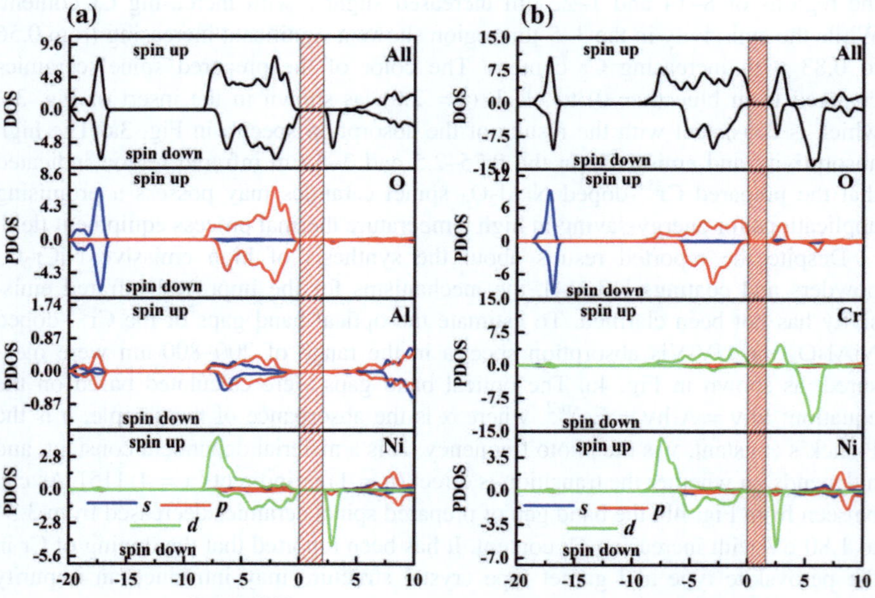

Fig. 5 Electronic density of states of **a** $NiAl_2O_4$ and **b** $NiCr_2O_4$

conduction band, respectively. It can be seen that the band gap of $NiCr_2O_4$ (1.67 eV) is smaller than that of $NiAl_2O_4$ (2.36 eV). Although the band gap values is smaller than the calculated results based on the UV-Vis absorption spectra. But both calculations show the same tendency that the band gap decreases with increasing Cr content. The greatly improved infrared emissivity can be attributed to the changes of crystal and electronic structures of the spinel ceramics.

Summary

High infrared emissive $NiAl_{2-x}Cr_xO_4$ (x = 0, 0.1, 0.5, 1.0, 2.0) spinel ceramics were prepared by solid-state reaction method. The absorptivity increases from 0.52 to 0.93 in 0.75–2.5 μm spectral region and the emissivity increases from 0.56 to 0.83 in 3–5 μm spectral region when the Cr content increases from x = 0 to x = 2.0. The band gap of Cr^{3+}doped-$NiAl_2O_4$ decreases with increasing Cr content as evaluated by the calculations based on the UV-Vis absorption spectra and the first-principles electronic structures. The greatly improved infrared emission property of Cr^{3+}-doped $NiAl_2O_4$ may be caused by the band gap narrowing induced enhancement of electron transition. The high emissive spinel ceramics shows promising applications for energy-saving in high temperature thermal process equipment field.

Acknowledgements This work is supported by The National Key Research and Development Program of China (Nos. 2016YFB0700204) and National Natural Science Foundation of China (Nos. 51432004, 51372255, 51702331) and the Key Laboratory of Cryogenics, Technical Institute of Physics and Chemistry, Chinese Academy of Sciences (No. CRYOQ201704).

References

1. K. Morita, B.N. Kim, H. Yoshida et al., Spark-plasma-sintering condition optimization for producing transparent $MgAl_2O_4$ spinel polycrysta. J. Am. Ceram. Soc. **92**, 1208–1216 (2009)
2. D. Dhak, P. Pramanik, Particle size comparison of soft-chemically prepared transition metal (Co, Ni, Cu, Zn) aluminate spinels. J. Am. Ceram. Soc. **89**, 1014–1021 (2006)
3. C. Ragupathi, J.J. Vijaya, P. Surendhar et al., Comparative investigation of nickel aluminate ($NiAl_2O_4$) nano and microstructures for the structural, optical and catalytic properties. Polyhedron **72**, 1–7 (2014)
4. R.F. Cooley, J.S. Reed, Equilibrium cation distribution in $NiAl_2O_4$, $CuAl_2O_4$, and $ZnAl_2O_4$ spinels. J. Am. Ceram. Soc. **55**, 395–398 (1972)
5. M. Maddahfar, M. Ramezani, M. Sadeghi et al., $NiAl_2O_4$ nanoparticles: synthesis and characterization through modify sol-gel method and its photocatalyst application, J. Mater. Sci. Mater. Electron. **26**, 7745–7750 (2015)
6. S. Jayasree, A. Manikandan, S.A. Antony et al., Magneto-optical and catalytic properties of recyclable spinel $NiAl_2O_4$ nanostructures using facile combustion methods. J. Supercond. Nov. Magn. **29**, 253–263 (2016)
7. C.O. Augustin, K. Hema, L.J. Berchmans et al., Effect of Ce^{4+} substitution on the structural, electrical and dielectric properties of $NiAl_2O_4$ spinel. Phys. Status Solidi A **202**, 1017–1024 (2005)
8. Z. Han, J. Liu, X.W. Li et al., Ca^{2+}-doped $LaCrO_3$: a novel energy-saving material with high infrared emissivity. J. Am. Ceram. Soc. **97**, 2705–2708 (2014)
9. Z. Han, X. Li, J. Ye et al., Significantly enhanced infrared emissivity of $LaAlO_3$ by co-doping with Ca^{2+} and Cr^{3+} for energy-saving applications. J. Am. Ceram. Soc. **98**, 2336–2339 (2015)
10. C.H. Bu, G. He, J.K. Ye et al., High infrared emission property of Ca^{2+}/Cr^{3+} co-doped $Y_3Al_5O_{12}$ ceramics. J. Ceram. Soc. Jap. **124**, 768–773 (2016)
11. Y. Zhang, D. Wen, Infrared emission properties of RE (RE = La, Ce, Pr, Nd, Sm, Eu, Gd, Tb, and Dy) and Mn co-doped $Co_{0.6}Zn_{0.4}Fe_2O_4$ ferrites. Mater. Chem. Phys. **131**, 575–580 (2012)

12. X. He, Y. Li, L. Wang et al., High emissivity coatings for high temperature application: progress and prospect. Thin Solid Films **517**, 5120–5129 (2009)
13. X. Cheng, J. Min, Z. Zhu et al., Preparation of high emissivity $NiCr_2O_4$ powders with a spinel structure by spray drying. Int. J. Miner. Metall. Mater. **19**, 173–178 (2012)
14. Z. Zhu, X. Cheng, W. Ye et al., Synthesis of $NiCr_2O_4$ spinel coatings with high emissivity by plasma spraying. Int. J. Miner. Metall. Mater. **19**, 266–270 (2012)
15. M.A. Butler, Photoelectrolysis and physical properties of the semiconducting electrode WO_2. J. Appl. Phys. **48**, 1914–1920 (1977)

Effect of Curing Temperature on the Properties of Quartz Fiber Felt Reinforced Fused Silica Densified with Simethicone

Ping Zhai, Xiaofeng Duan, Hongbin Li, Hongsheng Wang, Qihong Wei and Changtao Shao

Abstract In this paper, the needle-punched quartz fiber reinforced fused silica composite was densified by impregnating simethicone in vacuum and then solidified at different temperatures. The volume density and mechanical properties were investigated by means of Archimedes method and three point bending method, while the micro-structure was tested by SEM. The results showed that the volume density of composite increased from the non-impregnation of 1.457 g/cm^3 up to the maximum of 1.718 g/cm^3. The compressive strength increased at higher temperatures, up to 84.4 MPa at 500 °C. The tensile strength was weaker with the increase of the temperature, while the bending strength first increased then reduced, and reached the maximum of 43.2 MPa at 160 °C with an increase of 41.64%.

Keywords Simethicone · SiO_{2f}/SiO_2 composites · Needle-punched quartz fiber

Introduction

Fiber reinforced ceramic matrix composites (FRCMC) have been widely used in the field of high temperature structural materials, the advantages of FRCMC lie not only in overcoming the shortcomings of brittle fracture ceramic materials and improving the thermal shock capability, but also in keeping the thermostability, low thermal expansion coefficient, low density and good heat stability of ceramic matrix [1–3]. Yet studies have shown that [4–6] lots of benefits of increasing the dosage of FRCMC can be brought, such as considerable economic benefits and promoting lightweight, miniaturization and high performance of weapon equipment from which national defense benefits are remarkable.

P. Zhai (✉) · H. Li · H. Wang · Q. Wei · C. Shao
Shandong Research & Design Institute of Industrial Ceramics Co., Ltd, Zibo 255031, China
e-mail: zbgty_zhp@126.com

X. Duan
Shandong Jinke Engineering Design Co., Ltd, Zibo 255000, China

Y. Han (ed.), *Advanced Functional Materials*,
https://doi.org/10.1007/978-981-13-0110-0_69

Quartz fiber reinforced fused silica, in which the inorganic silica is gotten from silica gel, has low density, high mechanical strength, in addition, it has high temperature resistance, excellent thermal shock resistance, corrosion resistance and dielectric properties. It has many functions, such as heat protection, load carrying, wave transmitting and so on. Compared with the traditional quartz ceramic materials, the impact properties of quartz fiber reinforced fused silica ceramics are improved obviously. It is the preferred material for flight speed of 5–7 Maher missile in the radome, antenna window and other components [7, 8].

Commonly used quartz fiber preform has different specifications, such as three-dimensional knitting and needle-punched fabric and so on. The quartz fiber reinforced quartz ceramic composite material with three-dimensional braiding structure has the disadvantages of complex process, high preparation cost and long production time [9], as a result, in some areas, needle felt with low-cost and short-time can be used to replace continuous quartz fiber with three-dimensional braided. Compare with continuous quartz fiber reinforced fused silica with three-dimensional braided, needle felt reinforced fused silica lagged behind both in the densification and strength for its high porosity and weak interface of layers. It is necessary to improve the properties of needle-punched quartz fiber reinforced fused silica composites.

Experimental Procedures

Materials. Simethicone and tetraethyl orthosilicate (TEOS, $SiO_2 \geq 28\%$) were used as starting materials for inorganic silica. The organotin reagent ($Bu_2Sn(OAC)_2$, 98%) and 1,1,1,3,3,3-Hexamethyldisilazane ($C_6H_{19}NSi_2$) were used as a catalyst to speed up the chemical reactions of simethicone. Fused silica ($SiO_2 \geq 99.9\%$) was bright in as aggregate to fill larger holes in the needle-punched quartz fiber reinforced fused silica composites with a median particle size between 10 and 30 μm. The needle-punched SiO_{2f}/SiO_2 composites were needled by short length SiO_2 fibers, and then impregnated with slurry and silicon precursor impregnation for several times until the weight of composites was no longer changing.

Preparation. Firstly, according to the mass ratio of 100:0.75:10:0.75:2:4, Simethicone, TEOS, H_2O, $Bu_2Sn(OAC)_2$, $C_6H_{19}NSi_2$, and fused silica powders were stirred evenly after weighting, then heated in water bath of 35–40 °C for 4–6 h. Secondly, take the needle-punched SiO_{2f}/SiO_2 samples to immerse the prepared maceration extract in a vacuum for 30 min, then put the sample and the impregnation solution together into a constant temperature and humidity drying oven for 160 °C to solidify. If used in higher temperature, it is better to excrete the organic compounds from the impregnating solution by higher temperature. In this study, the firing temperature is less than or equal to 500 °C. The flow chart is described in Fig. 1.

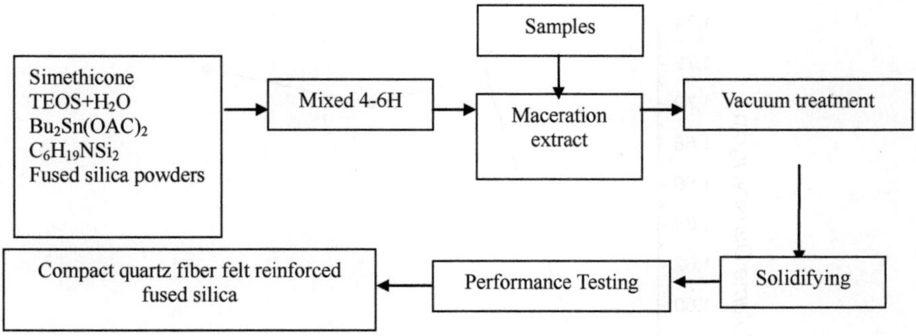

Fig. 1 Experimental process chart of quartz fiber felt reinforced fused silica densified with simethicone

Characterization. The volume density was monitored by using Archimedes drainage method. Scanning electron microscopy (FEI Sirion 200) was used to characterize the micro-structure of the spacemen fractured surfaces. Tensile, compressive and flexural mechanical properties of specimens were determined on a computer controlled universal test machine by three point bending method, and the sample size was 120 mm × 20 mm × 10 mm, 80 mm × 10 mm × 10 mm and 25 mm × 10 mm × 10 mm, respectively, with a loading rate of 0.5 mm/min.

Results and Discussion

Densification. It can be seen from Fig. 2 that with the impregnation of simethicone, the volume density of the samples increase markedly, from the non-impregnation of 1.457 g/cm^3 up to 1.718 g/cm^3, accompanied by an increase of 17.7% (solidified at 160 °C), 16.7% (solidified at 300 °C) and 17.9% (solidified at 500 °C), respectively. These data show that the method of impregnating simethicone in vacuum is of great help to the further densification of needle-punched quartz fiber reinforced fused silica.

Mechanical properties. Figure 3 shows temperature dependence compressive strength of samples. With the increase of solidifying temperature from 160 to 300 °C and to 500 °C, its compressive strength is increased from 34.9 MPa of the non-simethicone sample to 58.1 MPa (solidified at 160 °C), 64.7 MPa (solidified at 300 °C) and 84.4 MPa (solidified at 500 °C) of impregnating simethicone, with a raising proportion of 66.5, 85.4 and 141.8%, respectively, while the compressive modulus is increased from 19.1 GPa of the non-simethicone sample to 20.5 GPa (solidified at 160 °C), 25.6 GPa (solidified at 300 °C) and 36.2 GPa (solidified at 500 °C) of impregnating simethicone, that illustrates the effect of solidifying temperature on the mechanical properties is very significant. The compressive strength and modulus both reach maximum at 500 °C after solidifying, that may be because

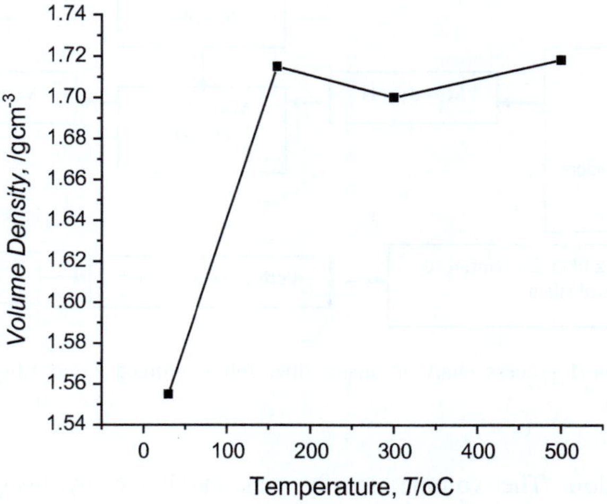

Fig. 2 Effect of firing temperature on the volume density

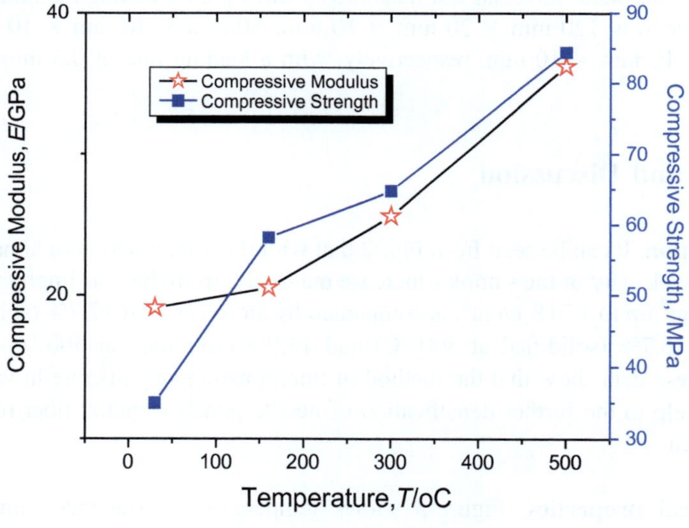

Fig. 3 Effect of temperature on compressive properties

of the fine SiO_2 particles from decomposition of simethicone, which reinforce the interaction between the original matrix particles and fibers, thereby the compressive properties of materials are greatly improved. We can conclude from the results that impregnating simethicone-curing is beneficial to improve the compressive strength of the composites, and the best curing temperature is about 500 °C.

The tensile and flexural properties of samples non-simethicone and with impregnating simethicone cured at 160, 300 and 500 °C are tested as shown in

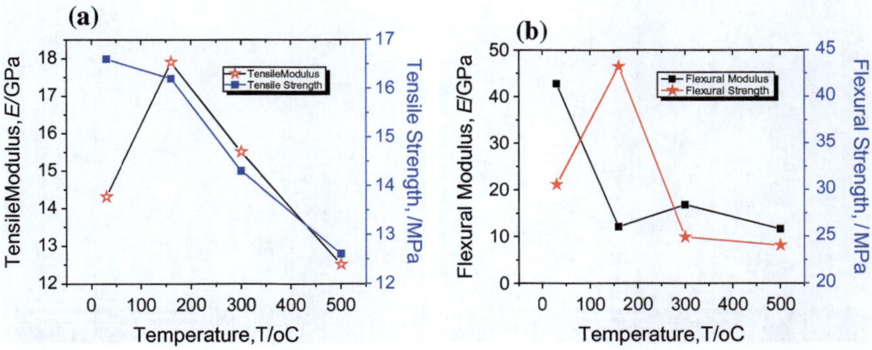

Fig. 4 Effect of temperature on tensile and flexural properties

Fig. 4. From it, we can see there's not much changes between the tensile properties of samples non-simethicone and with impregnating simethicone cured at 160 °C, but when cured at 500 °C, the tensile strength and modulus decrease sharply, drop to 75.9%. That may be the products from the decomposition of simethicone, deposited between or inside fiber bundle or between substrate and fiber bundle, which made the combination of fiber and substrate combined firmly and fibers are not easily pulled out, so the toughening effect of fibers can't occur. Another reason of lower tensile properties may be caused by multiple firing which resulted in lower fiber properties. From Fig. 4b, the flexural strength of the composites with impregnating simethicone cured at 160 °C is 43.2 MPa, 41.64% higher than that of non-simethicone. But when cured at 300 °C, the flexural strength began to decrease markedly by 18.4%, while when fired at 500 °C, the flexural strength changed little and tend to be steady.

Micro-structures. The micro-structure of specimens non-simethicone and impregnating simethicone cured at 160, 300 and 500 °C are shown in Fig. 5. As shown in Fig. 5a, it can be found that without simethicone densification, the inside of material is loose and porous, so the values of density and intensity are low. As illustrated in Fig. 5b–d, the pores between the fibers and the matrix are greatly filled, and the densification of composites is obviously improved. Figure 5b, c can also be clearly seen that there are many fibers extracted from the matrix and the surface of the fibers is relatively smooth, which shows that the damage to the fibers is little. The SiO_{2f}/SiO_2 composite materials were fired at low temperature, so the interaction between the fiber and the matrix is relatively weak, thus the fibers can be more easily pulled out from the matrix of SiO_2 and partial stress spread have been counteracted, which bring the materials better toughness and higher strength. It reveals characteristics of brittle fracture [10–12]. Figure 5d is the SEM picture of micro-structure cross section solidified at 500 °C. It shows that the combination of fibers and matrix is firm, so the fibers do not play the role of strengthening and toughening. That is why the mechanical properties of composites obtained at 500 °C is weak.

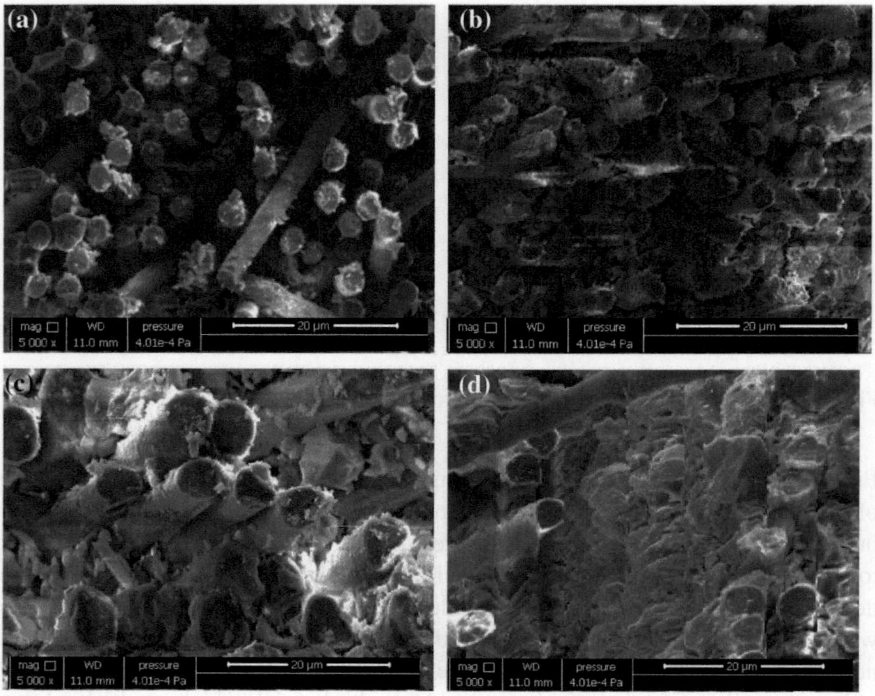

Fig. 5 SEM image of spacemen fired at different temperatures **a** non-simethicone, **b** 160 °C, **c** 300 °C, **d** 500 °C

Summary

Quartz fiber felt reinforced fused silica have been densified by liquid phase penetration-accumulation method, and then solidified at different temperature (less than or equal to 500 °C). The mechanical properties and microstructure have been evaluated. The following conclusions can be drawn.

The method of impregnating simethicone in vacuum markedly improves the density of quartz fiber needled felt reinforced fused silica, increased the non-impregnation of 1.457 g/cm^3 up to the maximum 1.718 g/cm^3 (fired at 500 °C), with an increase of 17.9%. The compressive strength increases with the raise of firing temperature, and reaches maximum 84.4 MPa fired at 500 °C, with an increase of 142%. The tensile strength is the highest at 160 °C, and decreases with the increase of temperature. The flexural strength increases first and then decreases with the increase of temperature, and reaches the highest value at 43.2 MPa at 160 °C, with an increase of 41.64%.

References

1. M. Imuta, J. Gotoh, Development of high temperature materials including CMCs for space application. J. Key Eng. Mater. **164**, 439–444 (1999)
2. X.S. Sun, H.D. Hu, Study on foreign civil aircraft development technology of structural strength of ideas. J. Aerosp. Sci. Technol. **6**, 23–26 (2004)
3. Z.Y. Yang, X. Feng, Z. Su, Meso-structure and elastic properties of 2.5D braided composites. J. Aerosp. Mater. Technol. **40**, 67–71 (2010)
4. L.L. Song, J.L. Li, Y.F. Zhao, L. Chen, Effects of preform structure on thermal conductivity of needle-punched quartz fiber/epoxy composites. J. Acta Mater. Composital Sinica **33**, 955–961 (2016)
5. S.Y. Du, Z.D. Guan, Thinking of China's large aircraft advanced composite technology strategies. J. Compos. Mater. **25**, 1–10 (2008)
6. G.J. Qi, C.R. Zhang, H.F. Hu, Progress in preparation of ceramic matrix composite radome. J. Chin. Ceram. Soc. **33**, 632–638 (2005)
7. J.G. Li, M.S. Cao, Y. Zhang, H.B. Lin, Research progress of high temperature electrical properties of wave transmitting materials abroad. J. Mater. Eng. **2**, 59–62 (2005)
8. L. Chen, Y. Sun, M. Ma, Development of high performance fiber preforms. J. Mater. China **31**, 21–29 (2012)
9. G.J. Qi, C.R. Zhang, H.F. Hu, Preparation of silica reinforced silicon nitride three-dimensional fiber composites using perhydropolysilazane as precursor. J. Mater. Lett. **59**, 3256–3258 (2005)
10. G.J. Qi, C.R. Zhang, H.F. Hu, Crystallization of silica fiber behavior three-dimensional reinforced silicon nitridecomposite. J. Cryst. Growth **284**, 293–296 (2005)
11. H. Gu, Tensile behaviours of quartz, aramid and glass Filaments after NaCl treatment. J. Mater. Des. **30**, 867–870 (2009)
12. N.E. Prasad, S. Kumari, S.V. Kamat, Fracture behaviour of continuous 2D weaved silica fibre reinforced ceramic matrix composites (CFCCs). J. Eng. Fract. Mech. **71**, 2589–2605 (2004)

Preparation of Al_2O_3–Y_2O_3 Composite Coatings on Silica-Based Ceramic Core Surface

Junhao Yan, Huiming Ji, Dingzhong Tang, Jiansheng Yao, Xin Li, Shuxin Niu and Lili Wang

Abstract In order to study the interface reaction between silica-based ceramic core and superalloy during the process to fabricate the directional solidified and single crystal hollow blades, the Al_2O_3–Y_2O_3 composite coatings were prepared by dip-coating method and low pressure air spraying method. The Al_2O_3–Y_2O_3 composite coatings slurry was made through dispersing Al_2O_3 powders and Y_2O_3 powders into yttrium sol or yttrium nitrate solution. The results showed that Al_2O_3 and Y_2O_3 could restrain the interface reaction between ceramic core and superalloy due to their excellent high temperature stability. The coatings had to pre-sinter at 1200 °C so that the coatings themselves as well as the coatings and matrix would have a certain extend bonding strength. The dip-coating method was compared with low pressure air spraying method, as well as yttrium sol binder and yttrium nitrate solution for their influence on coating thickness and crackles. The coatings were characterized by SEM, EDS and XRD. The results indicated that the coating thickness was about 6 μm and easier controlled by low pressure air spraying method than dip-coating method. The coating surface was smoother with fewer microcracks by employing yttrium nitrate solution as coating binder other than yttrium sol. The phases of the coating are mainly Al_2O_3 phase and $YAlO_3$ phase which can restrain the interface reaction.

Keywords Ceramic core · Coatings · Interface reaction · Preparation technology

J. Yan · H. Ji (✉)
Key Laboratory for Advanced Ceramics and Machining Technology of Ministry of Education, School of Materials Science and Engineering, Tianjin University, Tianjin 30072, China
e-mail: jihuiming@tju.edu.cn

J. Yan · D. Tang · J. Yao · X. Li · S. Niu · L. Wang
Science and Technology on Advanced High Temperature Structural Materials Laboratory, Beijing Institute of Aeronautical Materials, Beijing 100095, China

Y. Han (ed.), *Advanced Functional Materials*,
https://doi.org/10.1007/978-981-13-0110-0_70

Introduction

As the properties of aircraft engine and industry gas turbine are continuously improving, the demand of ceramic core acting as a through used for investment casting high efficient cooling single crystal blade is harder than before [1, 2]. At present, silica-based ceramic core is the most commonly used core for blade investment casting [3, 4]. Thermophysical chemical reaction and thermal mechanical penetration will easily happen between ceramic core and molten metal in the process of melting and casting of the aircraft engine blades, since the superalloy contains reactive metal elements such as Ti, Al, Cr, Ta, Hf and Y as well as high reducibility element such as C and the superalloy melt at an extremely high temperature meanwhile has a long time contracting with ceramic core under the circumstance of fine vacuum [5–7]. For example, Chemical scab may generate by low melting eutectic matter; alloy composition may be oxidized; or alloy composition may diffuse to interface, which causes its concentration gradient distribution; and pores may generate by reaction gas, which causes foundry impurities and decreases its mechanical properties.

Preparing coatings on the ceramic core surface can highly improve mechanical properties of the core and restrain the interface reaction. The coatings can not only eliminate the microcracks on the surface of the matrix [8], but also, to some degree, restrain the interface reaction between ceramic core and molten metal for its chemical stability. And then, roughness and properties of the turbine blade internal structure is highly improved as well as lifetime. It has been reported that Halberstadt [9] of Siemens has prepared coatings on the cores for investment casting via high-energy spraying method such as high speed flame spray method and plasma spraying method. After core leaching, the coatings are left on the surface of the foundry internal structure. Coating material is MCrAlY, which considered to be metal coatings including chromium (Cr) and aluminum (Al). Y represents yttrium or a rare earth element. M represents iron (Fe), cobalt (Co) or Nickel (Ni). That coating can prevent foundry inner wall oxided or corroded, which makes the foundry achieve expected lifetime. Beals et al. [10] prepared ceramic coatings on the surface of refractory metal core by chemical vapor deposition (CVD) method. The coatings are used for oxidation protection of the refractory metal core when mould sinter and prevent reaction or dissolution of the refractory metal core. Up to now, there is few reports about preparing coatings on the ceramic core surface.

On the basis of the former researches, this paper prepared Al_2O_3–Y_2O_3 composite coatings by dip-coating method and low pressure air spraying method. The coatings can improve the foundry roughness and restrain the interface reaction between ceramic core and superalloy due to their excellent high temperature stability. The coatings are characterized by SEM, EDS and XRD. The results indicated that the thickness of the coatings is about 6 μm, the surface of the coatings is smooth with few microcracks and the phases are mainly Al_2O_3 and $YAlO_3$ which can restrain the interface reaction.

Experimental

Starting Materials. The starting materials of the coatings were superfine powders including Al_2O_3 powders (D_{50} = 0.2 μm) and Y_2O_3 powders (D_{50} = 50 nm), Hexahydrate yttrium nitrate, citric acid and ammonium hydroxide.

Preparing. First, Al_2O_3 and Y_2O_3 powders were dispersed into yttrium sol or yttrium nitrate solution binder to obtain slurry for coatings. The mass ratio of Al_2O_3 and Y_2O_3 powders was 1:1, the ratio of the total powders (in mass) and the binder (in volume) was 1:5; Then, the Al_2O_3–Y_2O_3 composite coatings were prepared by dip-coating method and low pressure air spraying method on the silica-based ceramic core matrix (test panel (60 × 30 × 6 mm) formed by injection moulding method and composited of 75–80% SiO_2 and 20–25% ZrO_2, all in wt%). Yttrium sol and yttrium nitrate solution acted as coating binders. Yttrium nitrate solution was obtained by dispersing yttrium nitrate into deionized water; while yttrium sol was obtained by adjusting pH via ammonium hydroxide of Yttrium nitrate solution coordinated by citric acid. Model of the dip coater and spray gun are DECO-PBM-V-4L-A and SRIPROL-TE5-10, respectively.

Characterization. The surface and cross-section morphologies were observed by a scanning electron microscope (S4800, Hitachi, Japan), and elemental analysis was characterized by an energy dispersive spectrometer (EDS). Phase composition of the coatings was analyzed by a X-ray diffraction (Bruker D8 advanced). Differential scanning calorimeter (DSC) and thermal gravimetric analyzer (TGA) were used to determine the appropriate pre-sinter temperature.

Results and Discussion

Selection of the Coatings Material. The change of Gibbs free energy is a vital criterion for chemical reaction. It is indispensable to judge the possibilities of the reaction between coatings material with alloy element based on thermodynamics when we choose the coating materials. The Gibbs free energy change of the reaction between coatings material with alloy element is supposed as the following calculation formula:

$$XO(s) + M(l) = MO(s) + X(s) \text{ or } (l) \quad (1)$$

$$\Delta GT = \Delta GT^{\theta} + RT(aMO \cdot aX/aM \cdot aXO) \quad (2)$$

While, XO represents the oxide chosen for coating material; M represents the metal element contained in the superalloy.

Taking reactive metal element Ti, Al, Cr and Hf as example, calculate their Gibbs free energy of the reaction with SiO_2, MgO, Al_2O_3 and Y_2O_3 at 1800 K temperature. The calculation results are is shown in Fig. 1.

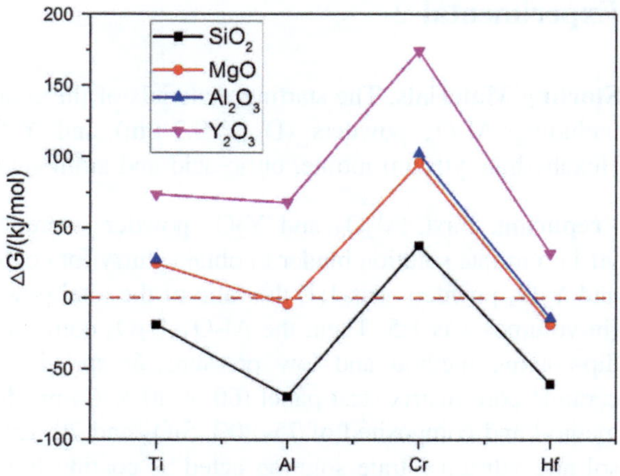

Fig. 1 Gibbs free energy of oxides and metal elements

The results indicate the values of reaction Gibbs free energy between Y_2O_3 with Ti, Al, Cr, Hf are positive; the values of reaction Gibbs free energy between Al_2O_3 with Ti, Cr, are positive, while with Hf is negative; the values of reaction Gibbs free energy between MgO with Ti, Cr, are positive, while with Al, Hf are negative which all the values are lower than the values of Al_2O_3; the values of reaction Gibbs free energy between SiO_2 with these elements are all negative except the value with Cr. Thus, we can consider the order ranked as descending order of the oxides chemical stability is Y_2O_3, Al_2O_3, MgO, SiO_2. Therefore, selecting Y_2O_3 and Al_2O_3 as the coatings material for restraining the interface reaction between ceramic core and superalloy is reasonable.

Given in the cost, this experiment uses Al_2O_3 powders (D_{50} = 0.2 μm) as the aggregate. And fill in the pores produced by Al_2O_3 powders with Y_2O_3 powders, which reinforce and strengthen the framework formed by Al_2O_3 powders. Then we will achieve compact arrangement coatings. Y_2O_3 powders can also have reaction with Al_2O_3 powders and generate $Y_4Al_2O_9$ (YAM), $YAlO_3$ (YAP) and $Y_3Al_5O_{12}$ (YAG), three crystalline phases [11], which are stable at high temperature and don't have reaction with superalloy. Then the coatings are tighter and can reach the aim of restraining interface reaction between ceramic core and superalloy. What's more, using yttrium sol and yttrium nitrate solution as the coating binders. On the one hand, binders can bond Al_2O_3 powders and Y_2O_3 powders together to form tight coatings. On the other hand, binders will have reaction with Al_2O_3 and generate reactant which stable at high temperature.

Determination of Coatings Pre-sinter Temperature. It is indispensable to pre-sinter the coatings at 1200 °C so that the coatings themselves as well as the coatings and matrix will have a certain extend bonding strength. The pre-sintering temperature can't be too high or low. If the pre-sintering temperature is too high, the ceramic core will deform at high temperature and can't be used. On the contrary, if the temperature is too low, the aim of pre-sintering can't be reached. Thus,

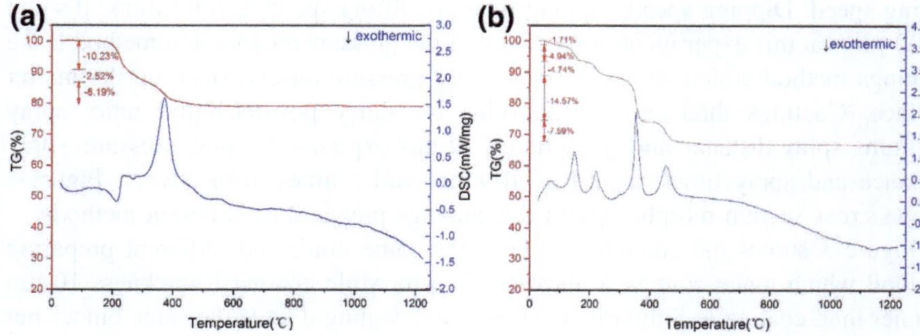

Fig. 2 DSC-TG analysis of the coating slurry (**a** binder yttrium sol, **b** binder yttrium nitrate hexahydrate)

dry coatings slurry to powders, and analysis the powders by DSC and TGA to determine the appropriate pre-sintering temperature, as shown in Fig. 2a, b.

From Fig. 2a, we can see there are two exothermic peaks and one endothermic peak of the coatings slurry powders used yttrium sol as binder. The peak values of the exothermic peaks are 213.1 and 269.4 °C meanwhile weight reduce 10.23 and 2.52%, respectively. It is believed that two exothermic peaks are caused by organic matter contained in yttrium decomposing. The endothermic peak value is 373.6 °C meanwhile weight reduce 8.19%, which is caused by yttrium hydroxide colloid particles dehydration and water decomposing. From about 420 to 1250 °C, the slurry powders continue releasing heat, which we can consider Al_2O_3 powders and Y_2O_3 powders sinter gradually in the process.

Figure 2 shows there are 5 exothermic peaks of the coatings slurry powders used yttrium nitrate solution as binder. Referring to the literature [12], we know that the yttrium nitrate hexahydrate decomposition is a complex process. The first four exothermic peaks are caused by the yttrium nitrate hexahydrate decomposition step-by-step. The last exothermic peak is caused by Y_4O_5 $(NO_3)_2$ deposing to Y_2O_3. From about 450 to 1250 °C, the slurry powders also continue releasing heat, which we can consider Al_2O_3 powders and Y_2O_3 powders sinter gradually in the process, as well.

After 450 °C, the coatings material sinter gradually. Therefore, we select 1200 °C as the coatings sintering temperature for the coatings sintering to maximum extent and core without deformation based on the DSC and TGA results.

Comparison of Dip-coating Method and Low Pressure Air Spraying Method. Preparing coating methods include dip-coating method, low pressure air spraying method, vapor deposition method, sol-gel method, vacuum impregnation method, plasma sputtering method, high energy flame spraying method and so on. At this experiment, we choose dip-coating method and low pressure air spraying method for the coatings preparing based on the coat.

Dip-coating method is one of the coating-preparing methods which binders coerce powders by surface tension and adsorb on the matrix surface. We can control the coating thickness by slurry powder-liquid ratio, dipping speed, dipping time and lifting speed. Dipping speed, dipping time and lifting speed is 1000 μm/s, 0 s and 1000 μm/s at this experiment, respectively. Low pressure air spraying method is the coatings method which slurry is extruded by pressure air and stack on the matrix surface. Coatings thickness is controlled by slurry powder-liquid ratio, spray pressure, spray distance and spray times. At this experiment, spray pressure, spray distance and spray times are 2 bar, 100 mm and 5 times, respectively. Figure 3 shows cross section morphology of the coatings prepared by different methods.

Figure 3 shows the coating a, b have the same binder but different preparing method which make coating a thickness 60 μm while coating b thickness 10 μm thinner than coating a. Similarly, coating c and coating d have the same binder but different preparing method. The result is coating c thickness is 25 μm and coating d thickness is 6 μm thinner than coating c. Thus, thickness of coatings prepared by low pressure air spraying method is far more thinner than the thickness of coatings prepared by dip-coating method. The reason may be the difference of preparing method principle. The principle of dip-coating method depends on the liquor surface tension on the matrix that is the wettability between liquor and matrix. Once the slurry powder-liquid ratio is certain, smaller the wetting angle between liquor and matrix is, higher the surface tension is, which causes coating thickness thicker. The principle of low pressure air spraying method is the slurry extruded by pressure air strikes the matrix surface. In the same way, once the slurry powder-liquid ratio is certain, coatings thickness is determined by spraying pressure, spraying distance and spraying times. Therefore, it is easy to control the coating thickness by using low pressure air spraying method.

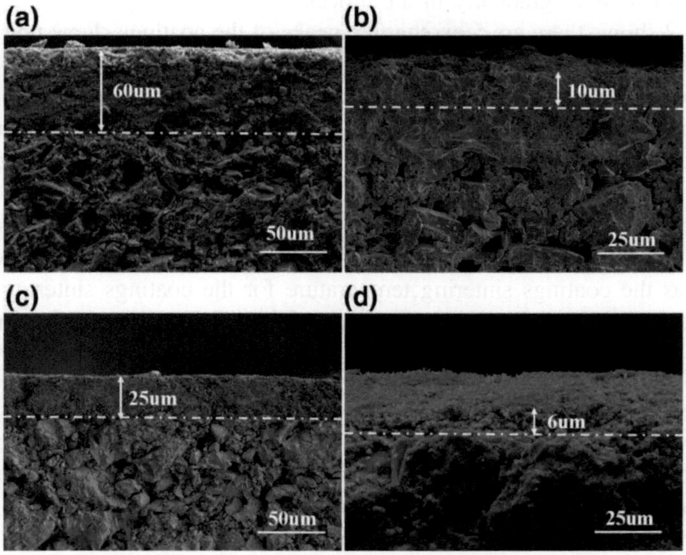

Fig. 3 Coatings cross section morphology (**a–d** 20% powder-liquid ratio. **a, b** yttrium sol binder, **c, d** yttrium nitrate solution binder. **a, c** dip-coating method, **b, d** low pressure air spraying method)

The size of ceramic core acting as through used for investment casting directly decides the size of turbine blade. So we need to control ceramic core shape and size strictly. Thinner coatings prepared by low pressure air method won't change the ceramic core size severely. Low pressure air method takes advantage on ceramic core coatings preparing over dip-coating method.

Influence of Different Binders. This experiment uses yttrium sol and yttrium nitrate solution as binders. Figure 4 shows the surface morphology of the coatings prepared by low pressure air method using the same spray times. We can see that the coating surface is uniform and dense while there exist crackles which distribute dispersively. Coating a, b, c, d thickness is 2, 3, 1 and 1.5 μm, respectively. When the binder is the same, larger the powder-liquid ratio is, thicker the coating surface crackles is. When powder-liquid ratio is the same, thickness of the coating which binder is yttrium nitrate solution is thinner than the thickness of the coating which binder is yttrium sol. The reasons are list as follows.

1. The dense sintering temperature of Al_2O_3 and Y_2O_3 is high, while the coating pre-sintering temperature at this experiment is 1200 °C. So the coatings can't sinter densely, so the coating surface is hard to avoid coming out crackles. And as the powder-liquid ratio rises, the powders content increase and the crackles width enlarges.
2. Yttrium sol acting as binder will turn to gel with shrinkage stress occurring during the drying process, which also enlarges the microcracks width.

If there exist crackles on the coating surface, the molten metal will contact and react with ceramic core in the process of casting, which coatings won't reach the aim of restraining interface reaction. It will also cause many problems such as

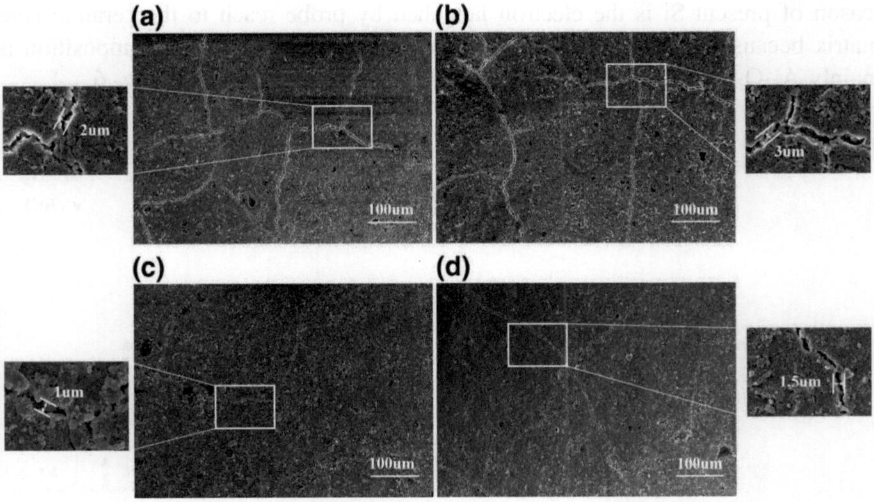

Fig. 4 Coating surface morphology (**a**, **b** yttrium sol binder. **c**, **d** yttrium nitrate solution binder. **a**, **c** 20% powder-liquid ratio. **b**, **d** 30% powder-liquid ratio)

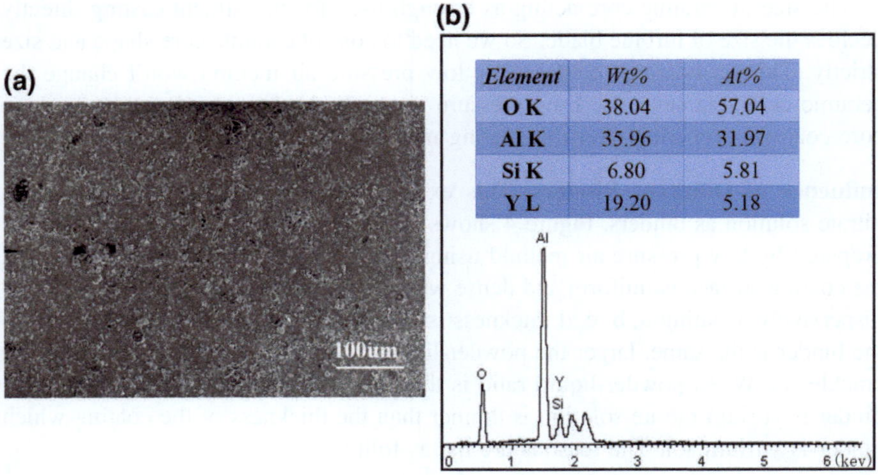

Fig. 5 Coating surface EDS

foundry increasing alloy, foundry inter surface roughness deteriorating, foundry mixing with impurities and so on, which will affect the foundry properties. Hence, the properties of coatings using yttrium nitrate solution as binder are better than the properties of coatings using yttrium sol as binder.

Coatings Constituent Analysis. Thickness and crack width of the coatings prepared by low pressure air spraying method and using yttrium nitrate solution as the binder are minimized to satisfy the operation requirements based on the above analysis. Figure 5 shows the coatings EDS analysis.

Coating surface mainly contains O, Al, Si and Y elements shown in Fig. 5. The reason of present Si is the electron launched by probe reach to the ceramic core matrix because the coatings is too thin. So we believe the coating composition is mainly Al_2O_3, Y_2O_3 and their reactant. XRD analysis is shown in Fig. 6.

Fig. 6 Coating surface XRD

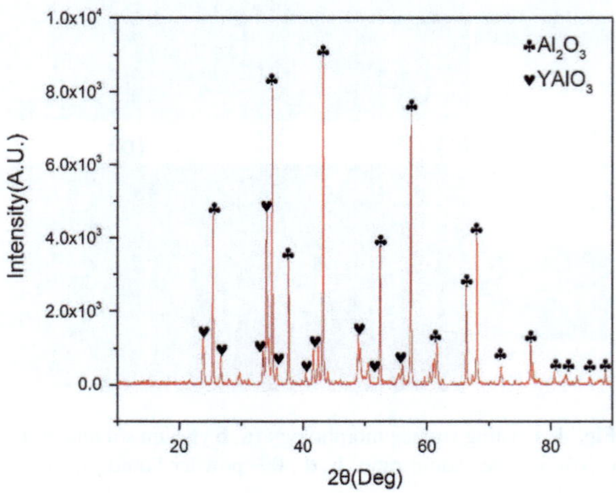

As shown in the Fig. 6, we can see there exist Al_2O_3 phase and $YAlO_3$ phase in the coatings using yttrium nitrate solution as binder. According to the K value method, the mass ratio of Al_2O_3 and $YAlO_3$ is 61.1 and 38.9%, respectively, which indicates all the yttrium nitrate solution have reaction with Al_2O_3 and then generate $YAlO_3$ (YAP). It is consistent with the EDS results. Therefore, we believe the coatings are composed by Al_2O_3 and $YAlO_3$. And SiO_2 doesn't exist in the coatings.

Summary

Form the above analysis, we can draw the conclusion that the coatings themselves as well as the coatings and matrix will have a certain extend bonding strength after pre-sintering at 1200 °C. The coatings using yttrium nitrate solution as binder and prepared by low pressure air spraying method are optimal, which coating thickness is about 6 μm and coating surface is smooth with few microcracks. The Al_2O_3 and Y_2O_3 composite coating mainly composed by Al_2O_3 and $YAlO_3$ have achieved the purpose of restrain interface reaction.

Further investment casting experiments need to be performed to confirm the coating function.

References

1. Y. Qin, W. Pan, Effect of silica sol on the properties of alumina-based ceramic core composites. Mater. Sci. Eng., A **508**(1), 71–75 (2009)
2. A.S. Rizkalla, D.W. Jones, Mechanical properties of commercial high strength ceramic core materials. Dental Mater. Official Publ. Acad. Dent. Mater. **20**(2), 207–212 (2004)
3. M. Gromada, A. Świeca, M. Kostecki et al., Ceramic cores for turbine blades via injection moulding. J. Mater. Process. Technol. **220**, 107–112 (2015)
4. R.C. Reed, *The Superalloys: Fundamentals and Applications* (Cambridge University Press, Cambridge, 2006), p. 122
5. L. Zheng, C. Xiao, G. Zhang et al., Key factors in quality control of a high Cr content cast Ni-base superalloy K4648. Rare Met. **30**(1), 410–413 (2011)
6. L.X. Zhang, J.M. Shi, H.W. Li et al., Interfacial microstructure and mechanical properties of ZrB_2, SiC, C ceramic and GH99 superalloy joints brazed with a Ti-modified FeCoNiCrCu high-entropy alloy. Mater. Des. **97**, 230–238 (2016)
7. L. Jiang, S. Guo, Y. Bian et al., Interfacial behaviors of magnesia partially stabilized zirconia by nickel-based superalloy. Mater. Lett. **181**, 313–316 (2016)
8. F. Valenza, M.L. Muolo, A. Passerone, Wetting and interactions of Ni- and Co-based superalloys with different ceramic materials. J. Mater. Sci. **45**(8), 2071–2079 (2010)
9. K. Halberstadt, Process of making a hollow member having an internal coating: EP, EP1593445 (2007)

10. J.T. Beals, J. Persky, D.M. Shah, et al., Refractory metal core coatings. US (2009)
11. M.C. Wilding, P.F. Mcmillan, A. Navrotsky, Calorimetric study of glasses and liquids in the polyamorphic system Y_2O_3–Al_2O_3. Phys. Chem. Glasses **43**(6), 306–312 (2002)
12. P. Melnikov, V.A. Nascimento, L.Z.Z. Consolo et al., Mechanism of thermal decomposition of yttrium nitrate hexahydrate, $Y(NO_3)_3{\cdot}6H_2O$ and modeling of intermediate oxynitrates. J. Therm. Anal. Calorim. **111**(1), 115–119 (2013)

Preparation of $LaPO_4$ Nanoparticles by Coprecipitation Method

Hongying Dong, Yingchai Shuang, Qinghong Sun, Qi Ren and Wen Ma

Abstract $LaPO_4$ nanopowders were prepared by co-precipitation method using $NH_4H_2PO_4$ and $La(NO_3)_3$ as raw materials. The influence of the composition of raw materials and the heat treatment process on the composition and microstructure of the synthesized powders were investigated by thermogravimetric-differential scanning calorimetry, X-ray diffraction and transmission electron microscopy. The synthesis mechanism of $LaPO_4$ powder was discussed. The results showed that the precipitate prepared using $NH_4H_2PO_4$ and $La(NO_3)_3$ with mole ratio of 1:1 was hexagonal $LaPO_4{\cdot}0.5H_2O$, and it transformed into hexagonal $LaPO_4$ after calcination at 400 °C. The precipitate calcined at 900 °C was not well crystalized monoclinic $LaPO_4$, and it crystalized completely with particle size about 200 nm after calcination at 1000 °C.

Keywords Coprecipitation method · $LaPO_4$ · Nanopowder

H. Dong · Y. Shuang · Q. Sun · Q. Ren
School of Chemical Engineering, Inner Mongolia University of Technology,
Hohhot 010051, China
e-mail: donghongying@imut.edu.cn

Y. Shuang
e-mail: shuangyingchai@qq.com

Q. Sun
e-mail: 1034630937@qq.com

Q. Ren
e-mail: 2628540301@qq.com

W. Ma (✉)
School of Materials Science and Engineering,
Inner Mongolia University of Technology, Hohhot 010051, China
e-mail: wma@imut.edu.cn

Q. Ren · W. Ma
Inner Mongolia Key Laboratory of Thin Film and Coatings Technology,
Hohhot 010051, China

Y. Han (ed.), *Advanced Functional Materials*,
https://doi.org/10.1007/978-981-13-0110-0_71

Introduction

$LaPO_4$ is a kind of high performance structural ceramic materials that can be used in harsh conditions because of its high melting point, good high-temperature phase stability, low thermal conductivity [1–3], as well as good chemical compatibility with Al_2O_3 [4], thus making it a promising thermal barrier coating (TBC) material. The synthesis of $LaPO_4$ ceramic powder is of great significance for study as a TBC material. The power preparation methods are mainly consist of sol-gel method [3, 5], hydrothermal method [6] and chemical precipitation method [6, 7], in which the chemical precipitation method can be used for a large-batch production of ceramic powders with small particle size and relatively uniform chemical composition. Therefore, $LaPO_4$ ceramic powder was prepared by precipitation method in this work. $LaPO_4$ ceramic powder was prepared by precipitation method using two systems of $La(NO_3)_3$ + $NH_4H_2PO_4$ and $La(NO_3)_3$ + Na_3PO_4, respectively, and the difference between the two systems was studied.

Experimental Procedure

The prepared $NH_4H_2PO_4$ or Na_3PO_4 solution was added dropwise to the $La(NO_3)_3$ solution [La:P = 1:1 (in molar ratio)], and precipitation was continued during the dropping and stirring process. The solution was centrifuged and washed with deionized water, followed by drying at 80 °C for 24 h to obtain $LaPO_4$ precursor powder. The chemical equations taken in the process are listed as follows:

$$La(NO_3)_3 + NH_4H_2PO_4 \rightarrow LaPO_4\downarrow + NH_4NO_3 + 2HNO_3 \quad (1)$$

$$La(NO_3)_3 + Na_3PO_4 \rightarrow LaPO_4\downarrow + 3NaNO_3 \quad (2)$$

The $LaPO_4$ precursor powder was then calcined at 400, 600, 900, 1000 and 1100 °C for 2 h, respectively.

The thermogravimetric-differential thermal analysis of the dried $LaPO_4$ precursor powder was carried out using TG-DSC analyzer (STA 449PC, Netzsch, Germany). Phase composition of the powders calcined at different temperatures was characterized by XRD (D/MAX 2200, Rigaku Co. Ltd., Japan). The powders were also characterized by Raman spectroscopy (InVia Microscope Raman spectrometer, Renishaw Co. Ltd., UK). The microstructure of $LaPO_4$ powder was observed by field emission scanning electron microscopy (Quanta FEG 650, FEI Co. Ltd., USA) and transmission electron microscopy (JEM-1200EX, JEOL Co. Ltd., Japan).

Results and Discussion

TG-DSC Analyses of $LaPO_4$ Precursor Powders. Figure 1 shows the TG-DSC curves of the $LaPO_4$ precursor powder prepared by the coprecipitation method using different starting materials. There is a significant weight loss by 11% during the whole heating process for the system of $La(NO_3)_3$ + $NH_4H_2PO_4$ as shown in Fig. 1a. The main mass loss occurred before 300 °C by 10%, and the weight did not change while the temperature increased to 400 °C. Two endothermic peaks are shown at about 90 and 320 °C in DSC curve in Fig. 1a, corresponding to evaporation of adsorbed water and elimination of crystal water in the precursor, respectively. A significant exothermic peak appears at 180 °C with a mass loss of 3% due to the thermal decomposition of NH_4NO_3, which is accompanied by a certain amount of heat loss. Between 450 and 850 °C, DSC curve shows many clutter peaks, corresponding to the process of phase transition from hexagonal $LaPO_4$ to monoclinic $LaPO_4$. A significant exothermic peak at 1000 °C corresponds to the well crystallization of monoclinic $LaPO_4$. In comparison to the system of $La(NO_3)_3$ + Na_3PO_4, the weight loss completed above 600 °C accompanied by an obvious exothermal peak in the system of $La(NO_3)_3$ + Na_3PO_4 (Fig. 1b), which might be attributed to the decomposition of intermediate product.

XRD Analyses of $LaPO_4$ Ceramic Powders. Figure 2 shows the XRD patterns of the as-precipitated $LaPO_4$ powder and the $LaPO_4$ precursor powder after calcination at 400, 600, 900 and 1000 °C for 2 h. In the system of $La(NO_3)_3$ + $NH_4H_2PO_4$ as shown in Fig. 2a, the as-precipitated powder before calcination is hexagonal $LaPO_4{\cdot}0.5H_2O$. After calcination at 400 °C, it is the mixed phases of monoclinic $LaPO_4$ and hexagonal $LaPO_4$. With an increase of calcination temperature up to 600 °C, hexagonal $LaPO_4$ began to transform into monoclinic phase, but the crystallization is not complete. The product after calcination at 900 °C is the monoclinic $LaPO_4$ with incomplete crystallization, and the crystallization of monoclinic $LaPO_4$ completed after calcination at 1000 °C. Therefore, for production of monoclinic $LaPO_4$, the calcination temperature should be higher than 1000 °C.

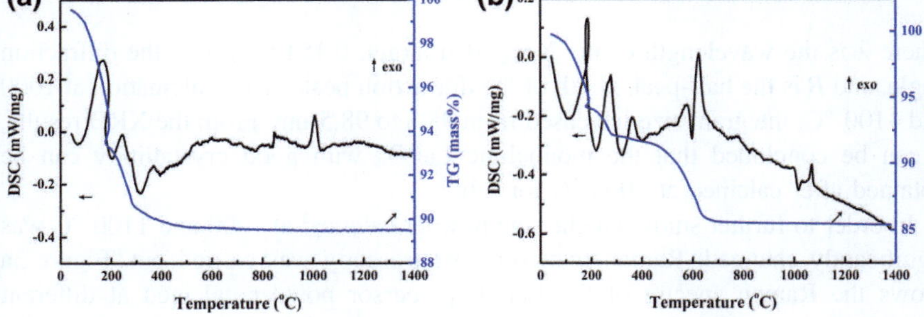

Fig. 1 TG-DSC curves of $LaPO_4$ precursor powders prepared from different starting materials: **a** $La(NO_3)_3$ + $NH_4H_2PO_4$, **b** $La(NO_3)_3$ + Na_3PO_4

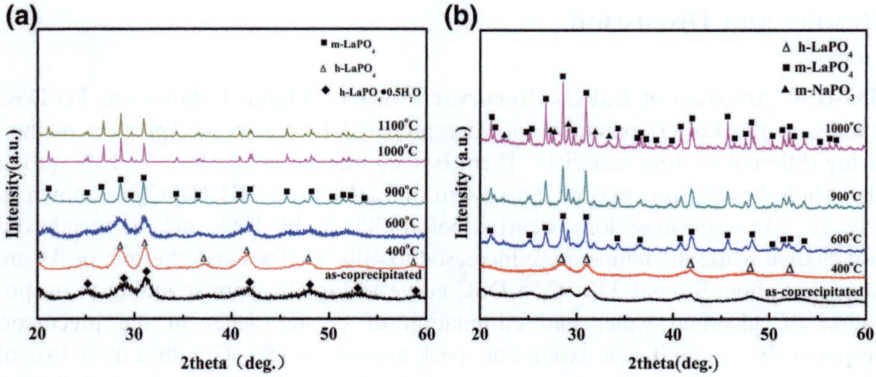

Fig. 2 XRD patterns of the $LaPO_4$ precursor powder calcined at different temperature from different starting materials: **a** $La(NO_3)_3 + NH_4H_2PO_4$, **b** $La(NO_3)_3 + Na_3PO_4$

In the system of $La(NO_3)_3 + Na_3PO_4$ as shown in Fig. 2b, it can be seen that the phase transition temperature of $LaPO_4$ powder also occurs after 600 °C, and after calcination at 900 °C, it is monoclinic $LaPO_4$ with minor amount of monoclinic $NaPO_3$. This is might attributed to the existing of NaH_2PO_4 in Na_3PO_4 raw material, and NaH_2PO_4 is easily hydrolyzed at about 300 °C to produce $NaPO_3$, and The apparent $NaPO_3$ phase can be observed in 600 °C and above. $NaPO_3$ phase is not observed at 400 °C probably because it's covered by the wide diffraction peak of hexagonal $LaPO_4$ with incomplete crystallization.

According to the experimental results as mentioned above, the system of La $(NO_3)_3 + NH_4H_2PO_4$ was chosen to produce $LaPO_4$. The results show that the final exothermic peak appears at about 1030 °C in Fig. 1a, but the XRD results of the powder after calcination at 1100 °C are basically the same as that obtained by calcination at 1000 °C. And when the calcination temperature rises from 1000 to 1100 °C, the monoclinic $LaPO_4$ grain will grow. The grain size of the precursors after calcination was estimated by using the Scherrer formula (Eq. 3).

$$D = \frac{0.9\lambda}{B\cos\theta} \tag{3}$$

where λ is the wavelength of the X-ray (Cu target, 0.154 nm), θ is the diffraction angle, and B is the half-peak width of the diffraction peak. After calcination at 1000 and 1100 °C, the grain size increased from 95.8 to 98.5 nm. From the XRD results, it can be concluded that the monoclinic $LaPO_4$ with good crystallinity can be obtained after calcined at 1000 °C for 2 h.

In order to further study whether the powder calcined at 1000 and 1100 °C was significantly changed, Raman scattering spectroscopy was carried out. Figure 3a shows the Raman spectra of the $LaPO_4$ precursor powder calcined at different temperatures, and Fig. 3b, c show an enlarged view of the Raman spectra in different band regions. In Fig. 3a, the number of characteristic peaks and peak

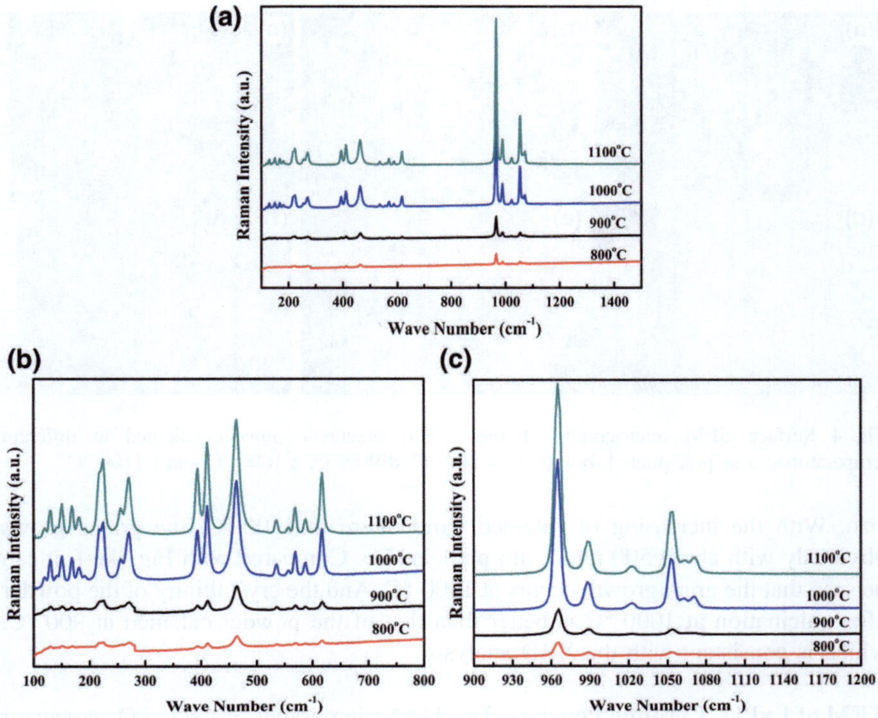

Fig. 3 Raman spectra of $LaPO_4$ precursor powders calcined at different temperatures

intensities after calcination at 1000 and 1100 °C are significantly different than those after calcination at 800 and 900 °C. In Fig. 3b, c, it can be seen that after the calcination at 800 and 900 °C, about 15 characteristic peaks correspond to monoclinic $LaPO_4$ characteristic peaks, whereas there are 23 characteristic peaks of monoclinic $LaPO_4$ in the spectra of 1000 and 1100 °C, and there is no obvious difference between them. From the Raman spectra, it can be seen that the some peaks of monochromatic $LaPO_4$ appeared at 800 °C and above, the spectra of the $LaPO_4$ precursor powder after calcination at 1000 °C are isologous. This is in good agreement with XRD and TG-DSC analyses, indicating that a high purity monoclinic $LaPO_4$ could be synthesized after calcination at 1000 °C for 2 h.

SEM of $LaPO_4$ Ceramic Powders. Figure 4 shows the morphology of the $LaPO_4$ precursor powder after calcination at different temperatures. It can be seen that the grains size of the as-precipitated $LaPO_4$ powder and the powders after calcination at 400 and 600 °C is relatively small, also the powder morphology has no significant change after calcination below 600 °C. However, when the calcination temperature increases to 900 °C, the powder morphology changes significantly with obvious grain growth. With further increasing calcination temperature to 1000 °C, it can be seen that the morphology is quasi-spherical particles with about 200 nm particle

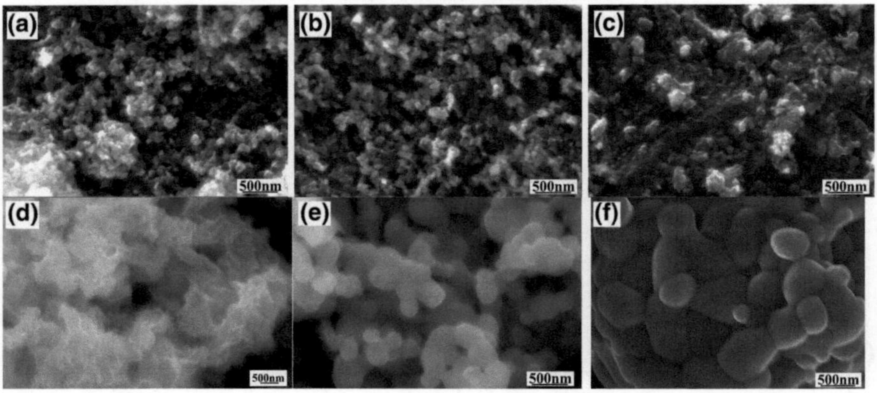

Fig. 4 Surface SEM micrographs of the $LaPO_4$ precursor powder calcined at different temperatures: **a** as-precipitated, **b** 400 °C, **c** 600 °C, **d** 900 °C, **e** 1000 °C and **f** 1100 °C

size. With the increasing of calcined temperature to 1100 °C, the grains grows obviously with about 500 nm–1 μm particle size. Compared with Fig. 4a–f, it can be seen that the grain growth occurs at 1000 °C. And the crystallinity of the powder after calcination at 1000 °C is better than that of the powder calcined at 900 °C, which is consistent with the XRD analysis.

TEM of $LaPO_4$ Ceramic Powders. The TEM micrographs of the $LaPO_4$ precursor powder after calcination at different temperatures are shown in Fig. 5. The morphology of the $LaPO_4$ precursor powder calcined at 400 °C is similar to that of the as-precipitated powder with particle size of about 100 nm. And the rod particles with

Fig. 5 TEM micrographs of the $LaPO_4$ powders after calcination at different temperatures: **a** as-precipitated, **b** 400 °C, **c** 600 °C, **d** 900 °C, **e** 1000 °C and **f** 1100 °C

size of about 250 nm in length developed after calcination at 600 °C, followed by changing into spherical particle in a certain extent after calcination at 900 °C, finally the spherical particles developed after calcination at 1000 °C with particle size of about 200 nm and after calcination at 1100 °C with particle size of about 600 nm. Combined with the XRD results, the main crystal phases are hexagonal $LaPO_4$ and monoclinic $LaPO_4$ calcined below 600 °C and above 600 °C, respectively.

Summary

$LaPO_4$ ceramic powder was prepared by co-precipitation method using La $(NO_3)_3 \cdot 6H_2O$ and $NH_4H_2PO_4$ as raw materials. The as-precipitated powder is hexagonal $LaPO_4 \cdot 0.5H_2O$, and turns into hexagonal $LaPO_4$ after calcination at 400 °C, following by gradual transformation from hexagonal to monoclinic. The monoclinic $LaPO_4$ develops completely after calcination at 1000 °C. The as-precipitated hexagonal $LaPO_4 \cdot 0.5H_2O$ and hexagonal $LaPO_4$ particles calcined at 400 °C have rod morphology with particle size of about 100 nm. After calcination at 1000 °C, monoclinic $LaPO_4$ particles are spherical with particle size of about 200 nm.

Acknowledgements The authors gratefully acknowledge the financial supports of the National Natural Science Foundation of China (51462026, 51672136), and the Inner Mongolia Natural Science Foundation (No. 2017MS0503).

References

1. K. Rajesh, B. Sivakumar, P. Krishna Pillai, Synthesis of nanocrystalline lanthanum phosphate for low temperature densification to monazite ceramics. Mater. Lett. **58**, 1687–1691 (2004)
2. A.Z. Zainurul, M. Rusop, S. Abdullah, Sol-gel synthesis & photoluminescence of multiple layer $LaPO_4$ nanostructure thin films. Adv. Mater. Res. **667**, 68–73 (2013)
3. A.Z. Zainurul, M.N. Asiah, C.H. Rosmani, F.S. Husairi, M. Rusop, S. Abdullah, Preparation of $LaPO_4$ with different morphologies and fluorescence properties by sol-gel spin-coating method. Adv. Mater. Res. **832**, 562–566 (2014)
4. S. Gallini, J.R. Jurado, M.T. Colomer, Combustion synthesis of nanometric powders of $LaPO_4$ and Sr-substituted $LaPO_4$. Chem. Mater. **17**, 4154–4161 (2005)
5. G. Rui, Q. Dong, L. Wei, Sol-gel synthesis and photoluminescence of $LaPO_4$:Eu^{3+} nanorods. Trans. Nonferrous Met. Soc. China **20**, 432–436 (2010)
6. M. Xianfeng, L. Lei, L. Ye, The preparation and properties of $LaPO_4$: Ce, Tb phosphors. Rare Met. Mater. Eng. **37**, 425–427 (2008)
7. R. Wang, W. Pan, M. Jiang, J. Chen et al., Mechanical properties and microstructure of layered $LaPO_4$ ceramics. J. Inorg. Mater. **18**, 109–114 (2003)

Preparation of 3Y-TZP Nanoceramics by a Modified Two-Step Sintering with Ultrahigh Heating and Cooling Rates

Feng Chen and Zhiqiao Yan

Abstract 3 mol% yttria stabilized tetragonal zirconia polycrystal (3Y-TZP) nanoceramics were prepared separately through traditional two-step sintering (TSS) and a modified TSS, namely rapid-treating TSS (RT-TSS) in which ultrahigh heating and cooling rates during the first step were adopted. The comparison studies showed that in RT-TSS, the first-step sintering temperature required for full densification could be obviously lowered and the grain growth be more effectively suppressed than that in TSS. A nearly fully densified (>98%TD) ceramics with an average grain size of 150 nm was obtained through RT-TSS. While, the ceramics prepared by TSS showed the larger grain size of 180 nm. The results also showed that the suppression of grain growth in RT-TSS didn't so strongly rely on the 'frozen' structure. Supersaturated vacancies generated during the cooling process might play a more important role.

Keywords Two-step sintering · Nanoceramics · 3Y-TZP

Introduction

Nanoceramics with grain size below 100 nm have received great attention due to their attractive electrical, optical, mechanical and chemical properties. It is well known that full densification is a pre-requisite for the achievement of nanoceramics with excellent properties. In order to obtain dense nanoceramics, nanocrystalline powder is always used. Although the high specific surface energy provides high

F. Chen
Guangdong Institute of Materials and Processing, Guangzhou 510650, China
e-mail: chenfengcsu@163.com

Z. Yan (✉)
National Engineering Research Center for Powder
Metallurgy of Titanium & Rare Metal, Guangzhou 510650, China
e-mail: zhiqiaoyan@163.com

Y. Han (ed.), *Advanced Functional Materials*,
https://doi.org/10.1007/978-981-13-0110-0_72

driving force and decreases the sintering temperature, grain growth of nanopowder inevitably accompanies with densification during sintering, which seriously deteriorates the advantages of nanoceramics [1–4]. Fabrication of fully densified nanoceramics remains a great challenge in materials science.

In the past decades, some special techniques, such as hot isostatic pressing [5], spark-plasma sintering [6–8], and microwave sintering [9], have been successfully applied to prepare fully densified nanoceramics. Comparing to these techniques, conventional sintering in air pressure is still highly desirable for its low cost and easy operation.

Two-step sintering (TSS), first originated by Chen and Wang [10], is an approach which does not require additional pressure or special equipments to achieve densification without grain growth during the final stage of sintering. In TSS, samples are heated to a high temperature T_1 to conduct a first-step sintering and then cooled to a lower temperature T_2 to conduct a second-step sintering, during which there is only densification without grain growth.

The originators pointed out that a sufficiently high density over 75% obtained during the first step was a necessary condition for a successful TSS. And it is widely agreed in the reported papers up to date. For instance, Wang et al. found that the critical density of the ceramic after first-step sintering was higher than 75% for sintering Y_2O_3 [11], $BaTiO_3$ [12], and NiCuZn [12]. Mazaheri et al. showed the values were 78% and 83% for nanocrystalline ZnO [13] and 3Y-TZP [14], respectively. Li et al. reported a value of 82% for nanometric particles of alumina [15]. For the same submicron alumina powder, Bodisova et al. [16] and Hesabi et al. [17] declared different values of 92 and 75%, respectively. However, it is not easy to achieve such high critical density. Sometimes the treatment temperature or time during the first step has to be increased, which often results in unwanted grain growth [15]. What's more, heating or cooling rate is a very important parameter in sintering. But its influence is scarcely discussed in the reported papers on TSS method.

In the present work, fully densified (>98%) 3 mol% yttria stabilized tetragonal zirconia polycrystal (3Y-TZP) ceramics was prepared through not only TSS but also a modified TSS, in which ultrahigh heating and cooling rates were adopted. It is surprising to find that fully densified nanoceramics can be obtained through this modified TSS even if the relative density after the first step was as low as 54%, which is in sharp contrast with that required by a successful TSS.

Experimental Procedure

Powder Conforming and Sintering. 3Y-TZP powder with an average size of ~30 nm was supplied from Tosoh Co. Ltd (Tokyo, Japan) and compacted in a steel cylindrical die by uniaxial pressing at 350 MPa to produce pellets with 10 mm

Table 1 Sintering cycles used in the present work

Regime	T_1 (°C)	t_1^a (min)	T_2 (°C)	t_2^b (h)
TSS1	1450	3	1350	20
TSS2	1450	3	1300	20
TSS3	1450	3	1250	20
TSS4	1450	3	1200	20
TSS5	1450	3	1150	20
TSS6	1400	3	1150	20
TSS7	1350	3	1150	20
TSS8	1300	3	1150	20
RT-TSS1	1450	3	1150	20
RT-TSS2	1400	3	1150	20
RT-TSS3	1350	3	1150	20
RT-TSS4	1300	3	1150	20

[a]Holding time at T_1
[b]Holding time at T_2

diameter and 2 mm height. The green density was ~50% of the theoretical density (TD) measured by the volumetric method. Subsequent sintering was conducted in air.

(a) Two-step sintering (TSS): Eight different two-step sintering cycles were used. The compacts were heated to T1 at the rate of 5 °C·min^{-1} after delubrication at 600 °C for 1 h. A dwell time of 3 min was given. The samples were cooled to T_2 with a rate of 50 °C·min^{-1}. Dwell time at T_2 was 20 h. The detailed sintering regimes were showed in Table 1 marked as TSSX (X = 1 − 8).
(b) Rapid-treating two-step sintering (RT-TSS): Compacts were heated up to 600 °C at the rate of 5 °C min^{-1} and soaked for 1 h, and then the samples were heated up to T_1 at a rate of ~500 °C·min^{-1} and soaked for 3 min. Subsequently they were cooled quickly at a rate of ~200 °C·min^{-1} to room temperature in air. Finally the samples were put into the furnace with T_2 and soaked for 20 h. The detailed sintering regimes were showed in Table 1 marked as RT-TSSX (X = 1 – 4).

Characterization. The density of the sintered samples was measured by Archimedes method and expressed in percentage of TD of 3Y-TZP (6.10 g cm^{-3}). The microstructure of the sintered samples was studied by field emission scanning electron microscopy (FE-SEM) (Model: JSM 6335F NT). Average grain size was estimated from the FE-SEM micrographs using the linear intercept method (ASTM Standard E112). For each specimen, 15 line segments were considered and Mendelson [18] multiplying factor (1.56) was used.

Results and Discussion

Relative Density and Microstructure of TSS Samples. Figure 1 shows the variation of relative density and average grain size of samples sintered according to the schemes from TSS1 to TSS5 whose T_2 was changed from 1350 to 1150 °C. In order to achieve sufficiently high density after the first step and obtain experimental data with high comparability, T_1 as high as 1450 °C was selected and t_2 was kept constant, 20 h. As shown in Fig. 1, grain growth becomes significant when T_2 is above 1250 °C. But when T_2 is reduced from 1250 to 1150 °C, grain size is almost unchanged. These results suggest that grain-boundary migration is suppressed when T_2 is below 1250 °C. In addition, when T_2 is 1150 or 1200 °C, density of sintered samples is over 99%. With further increasing T_2 from 1200 to 1350 °C, the relative density decreases from 99 to 97% inversely. The decreasing of relative density may be attributed to the unhomogeneous grain growth which is demonstrated by the increasing standard deviation of grain size with increasing T_2 because the nonuniform grain growth would lead to form closed pores easily and hence impede densification. So in further experiments, T_2 was kept at 1150 °C while T_1 was varied, which will establish the lowest T_1 for a successful TSS.

Figure 2 shows the variation of density of samples sintered according to the schemes from TSS5 to TSS8 in which T_1 was changed from 1450 to 1300 °C and T_2 as well as dwell time t_2 was kept constant at 1150 °C and 20 h, respectively. The density decreases with decreasing T_1. And when T_1 is below 1400 °C, relative density of samples after the second step cannot reach 98%. For some T_1, for instance, 1400 °C, T_2 = 1150 °C seems high enough to fabricate a fully dense ceramics. And the grain size of TSS6 (T_1 = 1400 °C, T_2 = 1150 °C) samples is about 180 nm as shown in Fig. 2b. For another T_1, for instance, 1300 °C, the same T_2 becomes too low so that densification in the second step proceeds for a while and then the driving force becomes exhausted. In order to obtain a dense sample at 1150 °C, the sintering temperature during the first step must be over 1400 °C. And

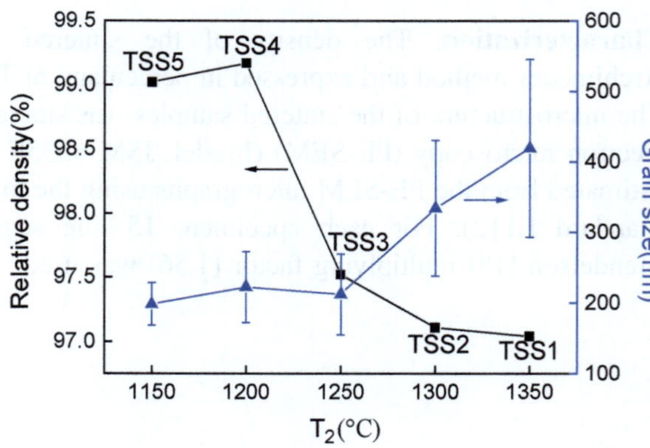

Fig. 1 Relative density and grain size of 3Y-TZP samples sintered at the regimes from TSS1 to TSS5

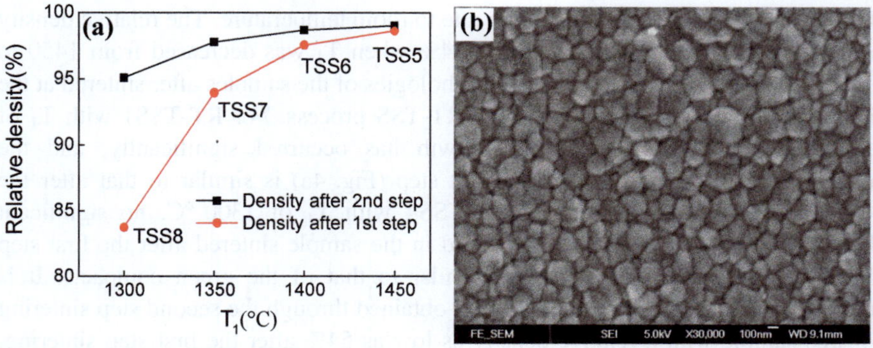

Fig. 2 **a** Relative density of 3Y-TZP samples sintered at the regimes from TSS5 to TSS8; **b** SEM micrograph of the sample sintered by TSS6 process

the relative density after the first step in this experiment is as high as about 97.5%. So it seems that a critical T_1 temperature, always figured as critical density, exists.

Generation and migration of vacancies are considered as the most important and most basic sintering phenomena [19]. And it is well-known that vacancy concentration of equilibrium state increases exponentially with increasing temperature. When samples were cooled rapidly from T_1 to a lower T_2, some extra vacancies would survive and the vacancy concentration in these samples would be higher than that in the samples with T_2 of equilibrium state. Obviously, the higher the difference between T_1 and T_2 is, the larger the residual vacancy concentration becomes, and then the higher the final density reaches. In another consideration, more vacancies would survive with increased cooling rate from T_1 and T_1 may be lowered further than that required in the traditional TSS. So a modified two-step sintering, namely rapid-treating two-step sintering (RT-TSS) was designed.

Relative density, microstructure and grain size of RT-TSS. Figure 3 shows the variation of relative density and average grain size of samples sintered under the schemes from RT-TSS1 to RT-TSS4. The samples were rapidly treated at T_1 and

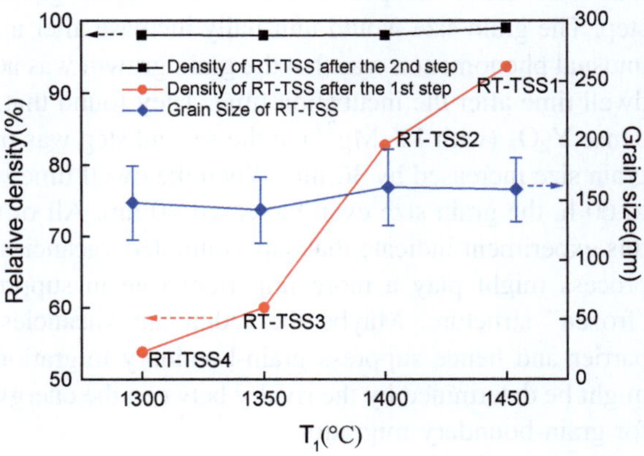

Fig. 3 Relative density and grain size of 3Y-TZP samples sintered at the regimes from RT-TSS1 to RT-TSS4

then cooled with an ultrahigh cooling rate to room temperature. The relative density of samples was decreased from 94 to 54% when T_1 was decreased from 1450 to 1300 °C. Figure 4 shows the SEM morphologies of the samples after sintered at the first-step and the second step through RT-TSS process. For RT-TSS1 with T_1 of 1450 °C, densification and grain growth has occurred significantly, and the microstructure of samples after the first step (Fig. 4a) is similar to that after the second step (Fig. 4b). While for RT-TSS4 with T_1 of 1300 °C, no significant densification or grain growth is observed in the sample sintered after the first step (Fig. 4g) and the microstructure is similar to that of the green compacts. It is interesting that fully dense ceramics was obtained through the second step sintering for this sample with a relative density as low as 54% after the first step sintering. The medium density of samples treated rapidly at 1300 and 1350 °C is 54 and 60%, respectively, and all samples reach nearly full density (>98%) after the second step. While as shown in Fig. 2, for traditional TSS, the samples cannot be fully densified finally when T_1 is 1300 or 1350 °C. In addition, it is also found that the samples cannot be fully densified when T_1 in RT-TSS is further decreased to 1250 °C. What's more, the pore structures after the first-step rapid-treating with various T_1 are obviously different from one another, as shown in Fig. 4. But no matter how much different T_1 is in the first step of RT-TSS, the final grain size is almost the same ($\sim$150 nm). It seems that the final grain size is almost not influenced by the sintering temperature of the first step or the transitional pore structures in RT-TSS. It is mainly determined by the temperature in the second step. And the grain size obtained by RT-TSS ($\sim$150 nm) is smaller than that ($\sim$180 nm) by a successful TSS.

Although it is widely approved that the grain growth could be suppressed through a successful TSS, there exist controversies about the mechanism of suppressing grain growth during TSS. The originators of TSS believed the suppression of grain growth in TSS might be attributed to the pinning effects of pores in the "frozen" structure [10, 11]. However, Binner [4] considered that the sample after heating to T_1 could be treated as a high green density body and the mechanisms taking place during the second step of TSS might be similar to those occurring conventionally but shifted to higher densities. Wang's research [11] demonstrated that fully dense samples couldn't resist grain growth indefinitely in the second step. The grain size would unusually increase after a long incubation time. A more unusual phenomenon was that the grain growth was accelerated with prolonging the dwell time after the incubation time. They found that, when the dwell time of fully dense Y_2O_3 (with 1% Mg^{2+}) in the second step was prolonged from 20 to 40 h, the grain size increased by 30 nm. When the dwell time was further prolonged from 40 to 60 h, the grain size even increased 40 nm. All of these as well as the results of this experiment indicate that supersaturated vacancies generated during the cooling process might play a more important role in suppressing grain growth than the "frozen" structure. Maybe the redundant vacancies will build a higher energy barrier and hence suppress grain-boundary migration. And the suppressing effect might be determined by the rivalry between the energy barrier and activation energy for grain-boundary migration.

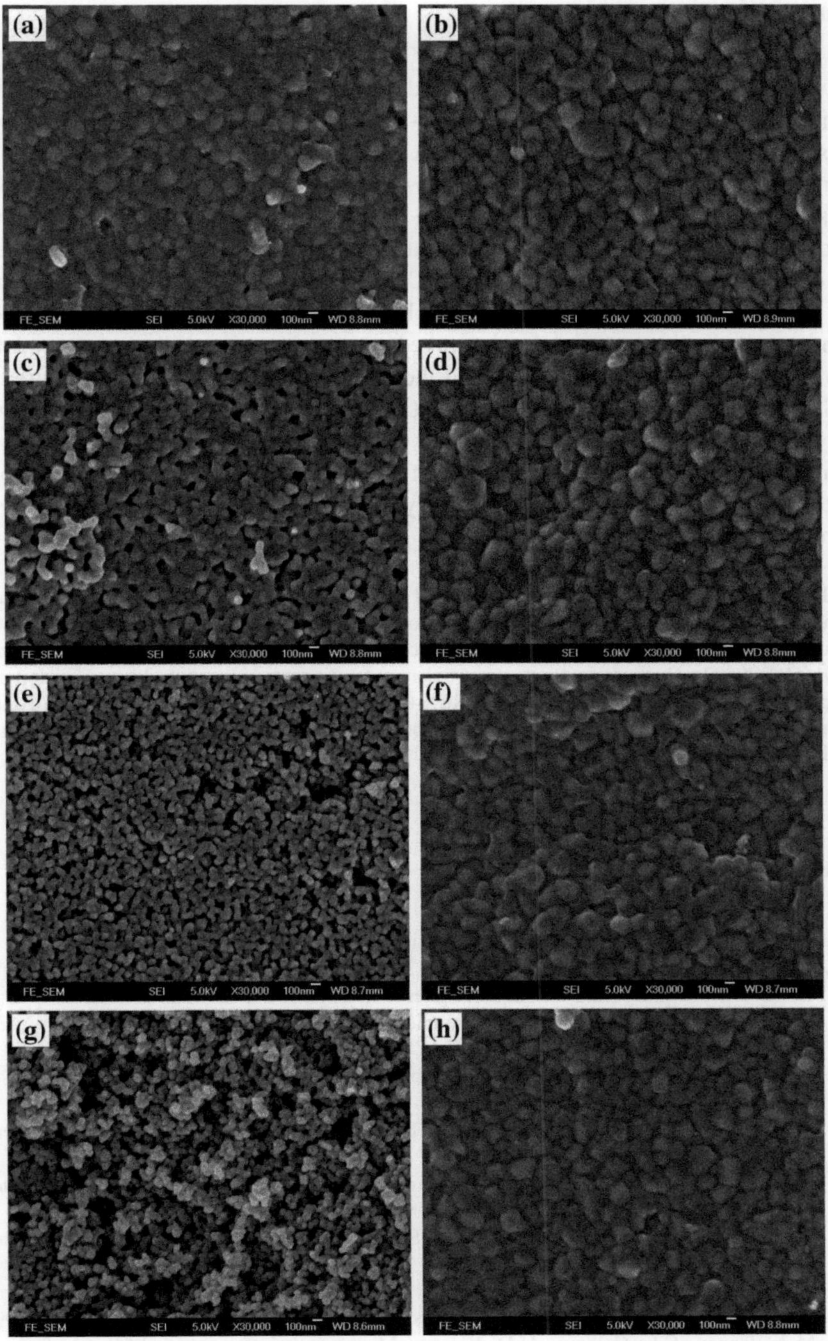

Fig. 4 SEM micrographs of 3Y-PZT samples sintered at RT-TSS regimes. Left, after the first step; Right, after the second step. **a**, **b** RT-TSS1; **c, d** RT-TSS2; **e**, **f** RT-TSS3; **g**, **h** RT-TSS4

Summary

In summary, the influence of heating and cooling rate in the first step of two-step sintering was investigated and a rapid-treating two-step sintering method was designed to sinter 3Y-TZP ceramics. A nearly fully densified (>98%) 3Y-TZP ceramics with an average grain size of ~150 nm was fabricated. The RT-TSS is designed to suppress the grain growth in the first step and suffer a moderate grain growth in the second step. When the heating rate increases dramatically from tens of degree Celsius to hundreds of degree Celsius, the total time duration in the range of high temperatures which will trigger obvious grain growth is sharply decreased. And when cooling rate is also improved, more vacancies will survive and the samples are not necessary to undergo the temperature as high as that required in TSS. The main success of RT-TSS ought to be attributed to suppress grain growth by lowering the sintering temperature or time duration in the first step as well as simultaneously keep densification robust in the second step by achieving super-saturated vacancy concentration in the cycle of heating and cooling process with ultrahigh rate.

Acknowledgements This work was supported by Guangdong Provincial Industrial High-tech Project (No. 2015A010105020), Guangzhou Sci-tech Project (No. 201707010145), Zhongshan-Guangdong Academy of Sciences technology transfer special foundation (2016G1FC0007), Guangdong Academy of Sciences Implements Innovation-driven Development Capacity Building Project (No. 2017 GDASCX-0117), Guangdong Provincial Innovation Ability Construction Project (No. 2016B070701024) and Guangzhou Innovation Platform Construction and Sharing Project (No. 201509010003).

References

1. G. Wilde, *Nanostructured Materials* (Elsevier, Oxford, 2009)
2. V. Viswanathan, T. Laha, K. Balani, A. Agarwal, S. Seal, Challenges and advances in nanocomposite processing techniques. Mater. Sci. Eng., R **54**, 121–285 (2006)
3. D.J. Green, O. Guillon, J. Rodel, Constrained sintering: a delicate balance of scales. J. Eur. Ceram. Soc. **28**, 1451–1466 (2008)
4. J. Binner, B. Vaidhyanathan, Processing of bulk nanostructured ceramics. J. Eur. Ceram. Soc. **28**, 1329–1339 (2008)
5. A. Krell, P. Blank, H. Ma, T. Hutzler, M.P.B. Bruggen, R. Apetz, Transparent sintered corundum with high hardness and strength. J. Am. Ceram. Soc. **86**, 12–18 (2003)
6. M. Omori, Sintering, consolidation, reaction and crystal growth by the spark plasma system (SPS). Mater. Sci. Eng. **287A**, 183–188 (2000)
7. Z. Shen, M. Johnsson, Z. Zhao, M. Nygren, Spark plasma sintering of alumina. J. Am. Ceram. Soc. **85**, 1921–1927 (2002)
8. T. Takeuchi, M. Tabuchi, H. Kageyama, Y. Suyama, Preparation of dense $BaTiO_3$ ceramics with submicrometer grains by spark plasma sintering. J. Am. Ceram. Soc. **82**, 939–943 (1999)
9. M.A. Janney, C.L. Calhoun, H.D. Kimrey, Microwave sintering of solid oxide fuel cell materials: I, Zirconia-8 mol% Yttria. J. Am. Ceram. Soc. **75**, 341–346 (1992)

10. I.W. Chen, X.H. Wang, Sintering dense nanocrystalline ceramics without final-stage grain growth. Nature **404**, 168–171 (2000)
11. X.H. Wang, P.L. Chen, I.W. Chen, Two-step sintering of ceramics with constant grain-size, I. Y_2O_3. J. Am. Ceram. Soc. **89**, 431–437 (2006)
12. X.H. Wang, X.Y. Deng, H.L. Bai, Two-step sintering of ceramics with constant grain-size, II: $BaTiO_3$ and Ni-Cu-Zn Ferrite. J. Am. Ceram. Soc. **89**, 438–443 (2006)
13. M. Mazaheri, A.M. Zahedi, S.K. Sadrnezhaad, Two-step sintering of nanocrystalline ZnO compacts: effect of temperature on densification and grain growth. J. Am. Ceram. Soc. **91**, 56–63 (2008)
14. M. Mazaheri, A. Simchi, F. Golestani-Fard, Densification and grain growth of nanocrystalline 3Y-TZP during two-step sintering. J. Eur. Ceram. Soc. **28**, 2933–2939 (2008)
15. J.G. Li, Y.P. Ye, Densification and grain growth of Al_2O_3 nanoceramics during pressureless sintering. J. Am. Ceram. Soc. **89**, 139–143 (2006)
16. K. Bodisova, P. Sajgalik, D. Galusek, Two-stage sintering of alumina with submicrometer grain size. J. Am. Ceram. Soc. **90**, 330–332 (2007)
17. Z.R. Hesabi, M. Haghighatzadeh, M. Mazaheri, D. Galusek, S.K. Sadrnezhaad, Suppression of grain growth in sub-micrometer alumina via two-step sintering method. J. Eur. Ceram. Soc. **29**, 1371–1377 (2009)
18. M.I. Mendelson, Average grain size in polycrystalline ceramics. J. Am. Ceram. Soc. **52**, 443–446 (1969)
19. S.-J.L. Kang, *Sintering: Densification, Grain Growth & Microstructure* (Elsevier, Oxford, 2005)

Phase Transformation Mechanism of Self-reinforced Y-α-SiAlON Ceramic Tool Material Manipulated by Two-Step Sintering

Jing Li and Ruiming Yin

Abstract In this study, self-reinforced Y-α-SiAlON was prepared by two-step phase transformation using single or dual rare-earth oxide addition. The process was implemented by a two-step sintering technique. The mechanism of two-step phase transformation and microstructure evolution of Y-α-SiAlON with or without Ce_2O_3 addition were investigated in detail. The predominant phase of the sample after the first step (i.e., pressureless sintering at low temperature) comprised elongated β-SiAlON grains. By contrast, the majority phase of sample after the second step (i.e., gas pressure sintering at high temperature) included coarse, elongated α-SiAlON grains. The driving force for β-to-α SiAlON phase transformation and liquid viscosity decreased, whereas the amount of liquid phase increased because of Ce_2O_3 addition, which contributed to anisotropic growth and resulted in more elongated α-SiAlON grains. The sample co-doped with Y and Ce exhibited the best mechanical properties with fracture toughness of 7.8 MPa $m^{1/2}$ and Vickers hardness of 19.81 GPa.

Keywords Self-reinforced Y-α-SiAlON · Two-step sintering · Phase transformation

J. Li (✉) · R. Yin (✉)
School of Metallurgy and Environment, Central South University, Changsha 410083, Hunan, China
e-mail: 1055288859@qq.com

R. Yin
e-mail: yin_ruiming@126.com

R. Yin
College of Metallurgical Engineering, Hunan University of Technology, Zhuzhou 412008, Hunan, China

Y. Han (ed.), *Advanced Functional Materials*,
https://doi.org/10.1007/978-981-13-0110-0_73

Introduction

SiAlON ceramics are ideal ceramic tool materials because of their superior mechanical properties, such as high strength, hardness, wear resistance, and corrosion resistance [1, 2]. The two main crystal structures of SiAlON ceramics, namely, α-SiAlON and β-SiAlON, have been used in ceramic tools [3, 4]. α-SiAlON possesses high hardness and good thermal properties, but its equiaxed crystal structure bestows low fracture toughness. β-SiAlON crystallizes in the form of elongated grains, resulting in high fracture toughness but low hardness. Therefore, development of both hard and tough SiAlON ceramics has been a research focus in ceramic tool material field.

In recent years, extensive work has been focused on improving mechanical properties (hardness, fracture toughness, and wear resistance) of SiAlON ceramics by addition of a second phase to the SiAlON matrix and achieved important findings [5, 6]. However, some defects may be induced in the matrix because of mismatch of thermal expansion coefficient and elastic modulus among different crystal phases, thus limiting application of the material in ceramic tools. At present, α/β-SiAlON ceramics are the most successfully applied material in the ceramic tool market because of the combination of high hardness of equiaxed α-SiAlON grains and relatively high toughness and strength of elongated β-SiAlON grains. When equiaxed α-SiAlON is converted to elongated grains, and the amount of low-hardness β-SiAlON is reduced to the smallest extent, elongated α-SiAlON grains will retain high hardness, wear resistance, and fracture toughness, properties which are highly suitable for application in ceramic tools. According to this viewpoint, a series of studies reported that hard and tough α-SiAlON ceramics can be developed by carefully controlling nucleation and growth processes to achieve self-reinforced α-SiAlON [7, 8]. These discoveries represent a significant breakthrough for α-SiAlON application as ceramic tool material. Chen et al. [7] discovered that tough α-SiAlON with whisker-like microstructure can be achieved by using low-activity β-Si_3N_4 as starting material. High stability of β-Si_3N_4 leads to slow nucleation, resulting in extremely few nuclei available during the initial sintering process. Then, relatively few grains can grow considerably in the (001) direction to form elongated grains. However, densification presents a problem. At present, hot pressing is the most successful fabrication process for preparation of dense α-SiAlON ceramics with elongated microstructure [9, 10]. However, to a certain extent, hot pressing depresses growth of elongated grains. Zhou et al. [10] reported that growth of elongated grains in preparation of dense Y-α-SiAlON was promoted by pressureless sintering using fine α-Si_3N_4 powders, whereas density exceeded 98% within the temperature range of 1750 to 1920 °C. Notably, as starting powders, coarse β-Si_3N_4 restrain phase transformation, resulting in low nucleation rate and density and thereby promoting formation of long rod-like α-SiAlON grains [11]. In recent years, two-step sintering has been explored to control grain growth [12, 13]. Liu et al. [8] manipulated nucleation and grain growth during sintering by two-step hot pressing, and a self-reinforced

Yb-α-SiAlON with fracture toughness of 7.5 MPa $m^{1/2}$ and Vickers hardness of 21.6 GPa was successfully prepared. Values of "m" and "n" significantly influence the microstructure and relative density of the α-SiAlON ceramics prepared by two-step hot-pressing–sintering, and relatively small "n" value and large "m" value benefit preparation of highly dense elongated Y-α-SiAlON grains with small aspect ratio and uniform microstructure [14].

This paper reports a two-step sintering process for preparing self-reinforced Y-α-SiAlON with active α- Si_3N_4 as starting material. The first-step aims to prepare high-density composite ceramics with elongated β-SiAlON grains as the main component by pressureless sintering at low temperature. The objective of the second-step sintering is to achieve β-SiAlON to elongated α-SiAlON phase transformation by gas-pressure sintering at high temperature. In addition, behavioral mechanism of the two-step phase transformation and microstructure development of Y-α-SiAlON with or without Ce_2O_3 addition were investigated in detail.

Experimental Procedure

Material preparation. Based on the viewpoint of Shan et al. [14], composition of Y-α-SiAlON in this study is $Y_{0.5}Si_{9.75}Al_{2.25}O_{0.75}N_{15.25}$ (m = 1.5, n = 0.75). Extra addition of 2 wt% Y_2O_3 or 2 wt% Y_2O_3 and 6 wt% Ce_2O_3 was used to promote sintering densification. Starting composition was prepared using α-Si_3N_4 (SN-E10, containing 1.27% oxygen, UBE Industries, Japan), Al_2O_3 (99.99 wt%, A.R., Beijing F&F Chemical Industrial Co., Ltd., China), AlN (grade C, H.C. Stark, Berlin, Germany), Y_2O_3 (99.99 wt%, HC Starck), and Ce_2O_3 (99.99%, Aladdin) as raw materials. Surface oxygen content of Si_3N_4 (2.38 wt% SiO_2) and AlN (1.83 wt % Al_2O_3) was considered when calculating overall composition.

An α-Si_3N_4 suspension with concentration of 28.8 g/L with absolute ethanol as solvent was prepared. Then, Al_2O_3, AlN, Y_2O_3, and rare-earth additives were added to the suspension. A 5 mL volume of PEG400 was used as dispersant. The slurry was agitated, ultrasonically dispersed for 60 min, and subsequently dried at 500 °C. After drying, composite powders were milled and sieved to 100–150 mesh, pressed uniaxially to a maximum pressure of 25 MPa, and isostatically cold-pressed at 300 MPa to form dense green compacts. Then, the green compact was sintered by two-step sintering. Figure 1 shows in detail heating procedures of the two-step sintering process, where the sample was replaced in the furnace to complete the second sintering process. YS-1 corresponds to the sample with 2 wt% Y_2O_3 addition after first-step sintering; YS-2 refers to the sample after second-step sintering; Ce/YS-1 is the sample with 2 wt% Y_2O_3 and 6 wt% Ce_2O_3 addition after the first step of sintering; and Ce/YS-2 indicates the sample with 2 wt% Y_2O_3 and 6 wt% Ce_2O_3 addition after second-step sintering. Nitrogen was introduced into the furnace at 1150 °C. Heating rate was 10 °C/min, and cooling rate was set at 15 °C/min.

Characterization Techniques. Sintered samples were polished by standard diamond polishing technique, by which a minimum of 4 mm-thick layer was

Fig. 1 Sintering schedules for two-step sintering procedure of samples

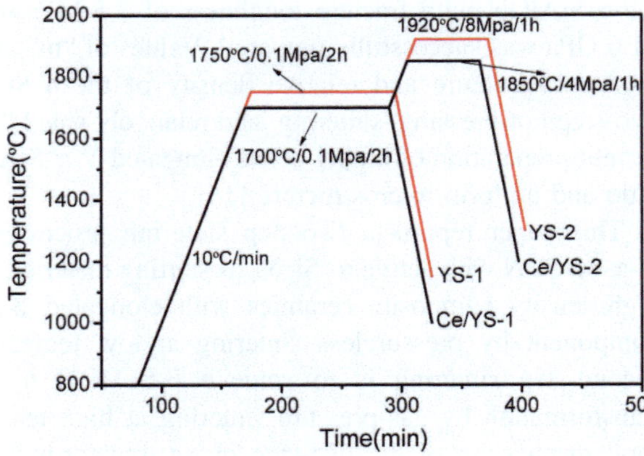

removed. Phase composition was identified by X-ray diffraction (XRD; Rigaku D/MAX-IIIC, Japan) with Cu radiation at 40 kV and 50 mA at a scan rate of 4°/min to record diffraction patterns. Microstructures of polished surface were observed by scanning electron microscopy (SIRION 200, FEI, USA) associated with energy dispersive spectroscopy (INCA X-Act, FEI, USA). Density was measured in deionized water according to Archimedes' principle. Vickers hardness at room temperature was tested by a Vickers diamond indenter (HV-50). The standard procedure was characterized by application of a 10 kg load for 15 s. Means of the single-edge notched beam were used to measure room-temperature fracture toughness in a universal material testing machine (Instron-5569). The test sample was established with dimensions of 2 mm × 4 mm × 20 mm. The span was 16 mm, kerf width measured 0.25 mm, and depth reached 1.6–2.4 mm. Pressure was applied to samples with a loading rate of 0.05 mm/min. Fracture toughness was calculated by the following equation [15]:

$$K_{IC} = \frac{3PL\sqrt{a}}{2bh^2}\left[1.93 - 3.07\left(\frac{a}{h}\right) + 14.53\left(\frac{a}{h}\right)^2 - 25.07\left(\frac{a}{h}\right)^3 + 25.80\left(\frac{a}{h}\right)^4\right] \quad (1)$$

where K_{IC} represents fracture toughness ($MPa{\cdot}m^{1/2}$), P is applied load (N), L refers to span (mm), a is notch depth (mm), b represents the width of specimen (mm), h is height of specimen (mm). The reported fracture toughness is the average of six tests.

Results and Discussion

Phase Transformation. Figure 2 shows XRD patterns of samples designated as $Y_{0.5}Si_{9.75}Al_{2.25}O_{0.75}N_{15.25}$ with 2 wt% Y_2O_3. High amounts of β-SiAlON and α-SiAlON and a small amount of Y_2SiAlO_5N were detected in YS-1 after the first sintering step. Notably, the beta phase to alpha phase ratio (I_β/I_α) reached 1.413. XRD analysis of YS-2 revealed that α-SiAlON was the major phase, whereas minor

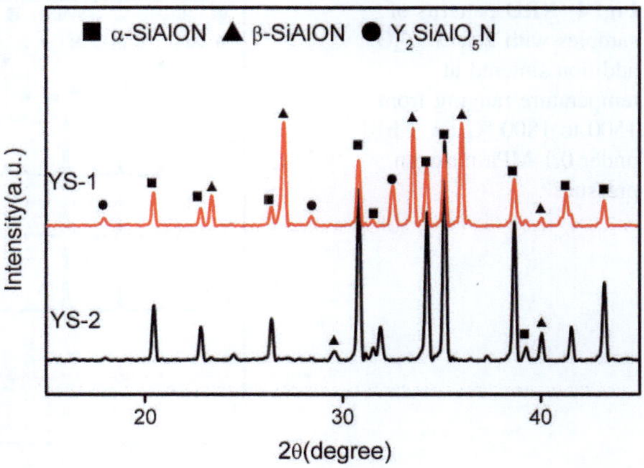

Fig. 2 XRD patterns of samples with 2 wt% Y_2O_3 addition

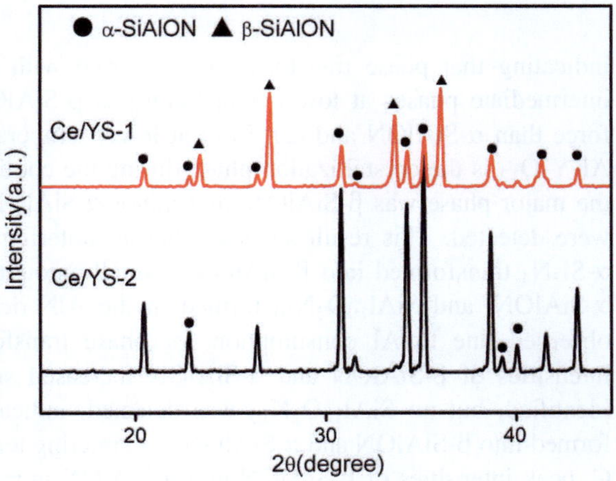

Fig. 3 XRD patterns of samples with 2 wt% Y_2O_3 and 6 wt% Ce_2O_3 addition

β-SiAlON was detected after the second sintering step. Figure 3 shows XRD results of samples with 2 wt% Y_2O_3 and 6 wt% Ce_2O_3. According to XRD analysis, I_β/I_α of Ce/YS-1 totaled 1.826, which is higher than that of YS-1 (1.413). This finding indicates that addition of Ce_2O_3 promotes formation of β-SiAlON because Ce^{3+}, which possesses a large ionic radius, is difficult to incorporate into α-structure, resulting in low reaction kinetics and nucleation of α-SiAlON. This condition encourages formation of β-SiAlON during the first step of pressure sintering at low temperature. XRD analysis of Ce/YS-2 shows that only the α-SiAlON phase was detected, suggesting the existence of a single-phase system after second step.

Two-Step Phase Transformation Mechanism. The first-step phase evolution at low temperature was studied. Figure 4 displays XRD patterns of samples sintered at temperature ranging from 1500 to 1800 °C for 1 h under 0.1 MPa nitrogen pressure. Notably, the sample sintered at 1500 °C predominantly exhibited α-Si_3N_4, $Al_5Y_3O_{12}$, and small amounts of β-SiAlON, Al_2O_3, and AlN polymorphs,

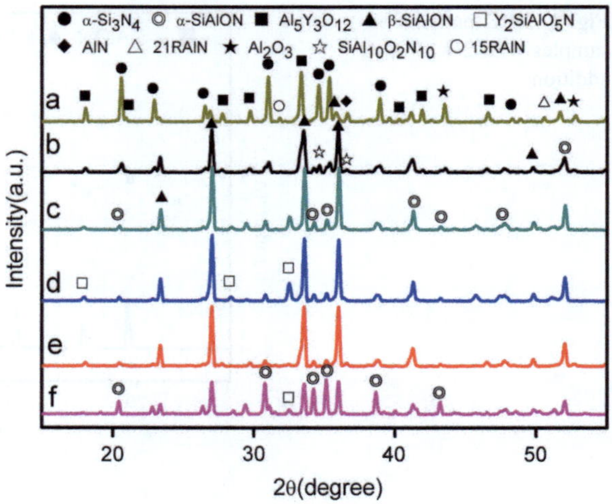

Fig. 4 XRD patterns of samples with 2 wt% Y_2O_3 addition sintered at temperature ranging from 1500 to 1800 °C for 1 h under 0.1 MPa nitrogen pressure

indicating that phase transformation occurred with $Al_5Y_3O_{12}$ and β-SiAlON as intermediate phases at lower temperature as β-SiAlON exhibits a higher driving force than α-SiAlON and can form at lower temperature. According to literature, $Al_5Y_3O_{12}$ is the crystallization phase during the cooling process [16]. At 1600 °C, the major phase was β-SiAlON, and minor α-SiAlON, α-Si_3N_4, and $SiAl_{10}O_2N_{10}$ were detected. This result suggests that as sintering temperature increased, most α-Si_3N_4 transformed into β-SiAlON, a small amount of α-Si_3N_4 transformed into α-SiAlON, and $SiAl_{10}O_2N_{10}$ formed in the AlN-rich region. No $Al_5Y_3O_{12}$ was observed due to Al consumption for phase transformation. At 1650 °C, peak intensities of β-SiAlON and α-SiAlON increased significantly. Y_2SiAlO_5N was identified, but no $SiAl_{10}O_2N_{10}$ was detected, indicating that $SiAl_{10}O_2N_{10}$ transformed into β-SiAlON and α-SiAlON. As sintering temperature increased to 1700 ° C, peak intensities of β-SiAlON and α-SiAlON increased insignificantly, whereas that of Y_2SiAlO_5N was further enhanced. This result can be explained by the increased amount of Si and N because a portion of SiAlON was incorporated into the glass phase, thereby promoting precipitation of Y_2SiAlO_5N. As temperature increased to 1750 °C, I_β/I_α increased to 1.413 with a small amount of Y_2SiAlO_5N. After sintering at 1800 °C for 1 h, peak intensity of α-SiAlON evidently increased, whereas that of β-SiAlON decreased significantly. Meanwhile, β to α-SiAlON phase transformation occurred above 1800 °C. Therefore, 1750 °C is favorable temperature for obtaining the highest I_β/I_α after first step of sintering.

The effect of nitrogen pressure on second-step of phase evolution at high temperature was also researched. As shown in XRD analysis results, SiAlON decomposed under low nitrogen pressure and elevated temperature; thus, nitrogen pressure should be increased. Figure 5 presents nitrogen pressure effect on phase evolution characterized by XRD analysis. As nitrogen pressure was increased within 2–8 MPa, decomposition of SiAlON was restrained, and peak intensity of α-SiAlON decreased significantly. When temperature reached 1920 °C under

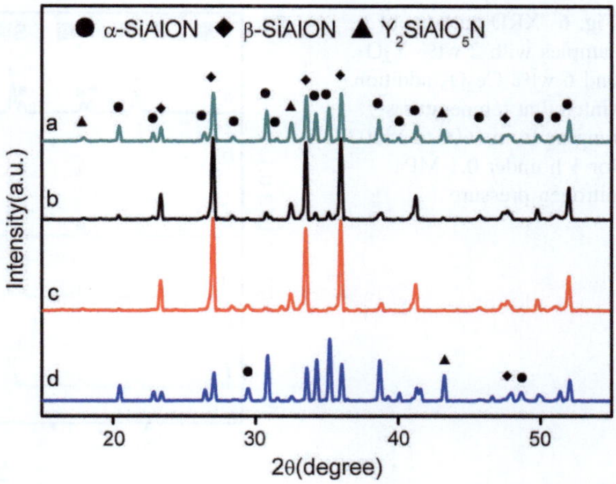

Fig. 5 XRD patterns of samples with 2 wt% Y_2O_3 addition sintered at high temperature under different nitrogen pressures

8 MPa nitrogen pressure, the major phase was α-SiAlON, and Iα/Iβ totaled 5.263. XRD analysis indicated that when nitrogen pressure increased, the driving force for α-SiAlON nucleation decreased, resulting in high temperature required for its formation. Increase in nitrogen pressure promotes N dissolution into silicate glass, resulting in replacement of a portion of O in [SiO4] tetrahedron by N. Thus, network structure and thermal stability of the glass phase are strengthened, and glass crystallization tendency is weakened. Y precipitation from glass phase was inhibited; thus, Y cannot coordinate for the α-SiAlON vacancy bond, leading to low driving force for α-SiAlON formation.

Effect of Ce_2O_3 Addition on Two-Step Phase Transformation. Effect of Ce_2O_3 addition on first-step phase evolution was analyzed. Figure 6 shows XRD results of samples with addition of 2 wt% Y_2O_3 and 6 wt% Ce_2O_3 sintered at temperatures ranging from 1600 to 1850 °C for 1 h under 0.1 MPa nitrogen pressure. According to XRD analysis, peak intensity of α-SiAlON increased significantly compared with the sample without Ce_2O_3 addition at 1600 °C, indicating that addition of Ce_2O_3 promotes formation of α-SiAlON at high temperature. Within the temperature range of 1650 to 1700 °C for 1 h, peak intensity of β-SiAlON further increased, whereas that of α-SiAlON gradually decreased, indicating that phase transformation (α to β-SiAlON) occurred continually. At 1700 °C, I_β/I_α reached 1.529. At temperature ranging from 1750 to 1800 °C for 1 h, peak intensity of β-SiAlON decreased, whereas that of α-SiAlON increased, indicating the initiation of β to α SiAlON phase transformation. I_β/I_α decreased from 1.184 to 0.791. Beyond 1850 °C, peak intensity of α-SiAlON decreased because of α-SiAlON decomposition under low nitrogen pressure. Therefore, to obtain high I_β/I_α value after the first sintering step, sintering temperature for the sample added with Ce_2O_3 should be 1700 °C.

The effect of Ce_2O_3 addition on second step of phase evolution was also studied. Figure 7 shows XRD patterns of samples added with 2 wt% Y_2O_3 and 6 wt% Ce_2O_3 and sintered at 1850 °C for 1 h under 2, 4, and 8 MPa nitrogen pressure.

Fig. 6 XRD patterns of samples with 2 wt% Y_2O_3 and 6 wt% Ce_2O_3 addition sintered at temperatures ranging from 1600 to 1850 °C for 1 h under 0.1 MPa nitrogen pressure

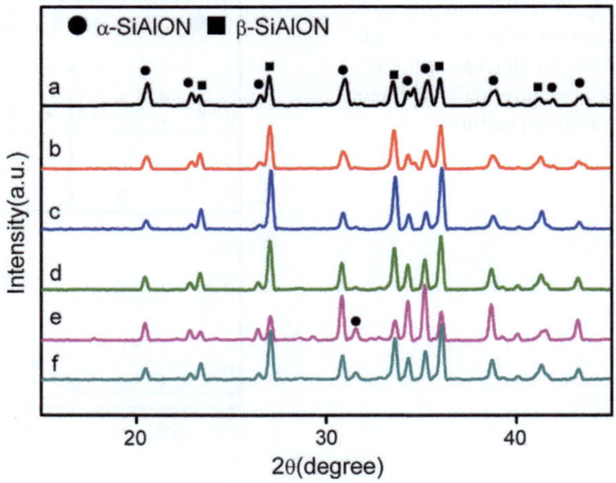

Fig. 7 XRD patterns of samples with 2 wt% Y_2O_3 and 6 wt% Ce_2O_3 addition at 1850 °C for 1 h under different nitrogen pressure

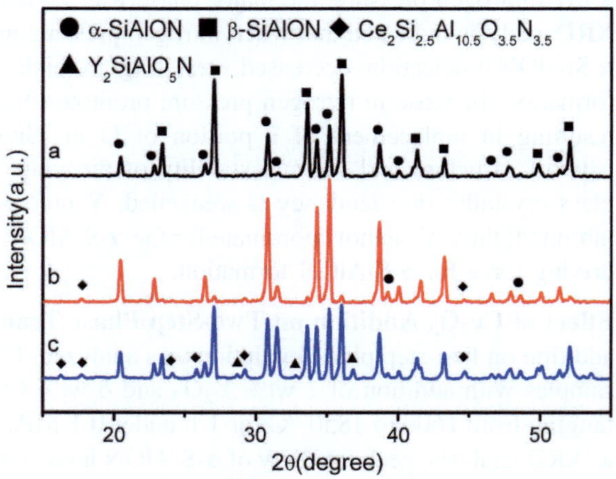

High amounts of β-SiAlON and α-SiAlON were detected in the sample sintered at 1850 °C for 1 h under 2 MPa nitrogen pressure. Iβ/Iα reached 1.826. When nitrogen pressure reached 4 MPa, peak intensity of α-SiAlON was markedly enhanced, and minor β-SiAlON and $Ce_2Si_{2.5}Al_{10.5}O_{3.5}N_{3.5}$ were detected. However, above 8 MPa, peak intensity of α-SiAlON decreased, and low Iβ/Iα of 1.068 was observed. Peak intensity of $Ce_2Si_{2.5}Al_{10.5}O_{3.5}N_{3.5}$ slightly increased, and a small amount of Y_2SiAlO_5N was detected due to increased amount of liquid phase because of Ce_2O_3 addition. As nitrogen pressure increased, the amount of N dissolution into liquid phase increased, leading to precipitation of a certain amount of $Ce_2Si_{2.5}Al_{10.5}O_{3.5}N_{3.5}$. However, the amount of rare earth ions in the glass phase decreased. To supplement the deficiency of rare earth ions in the glass phase, the Y ions running out from α-SiAlON cell into glass phase resulted in decreased α-SiAlON under 8 MPa nitrogen pressure.

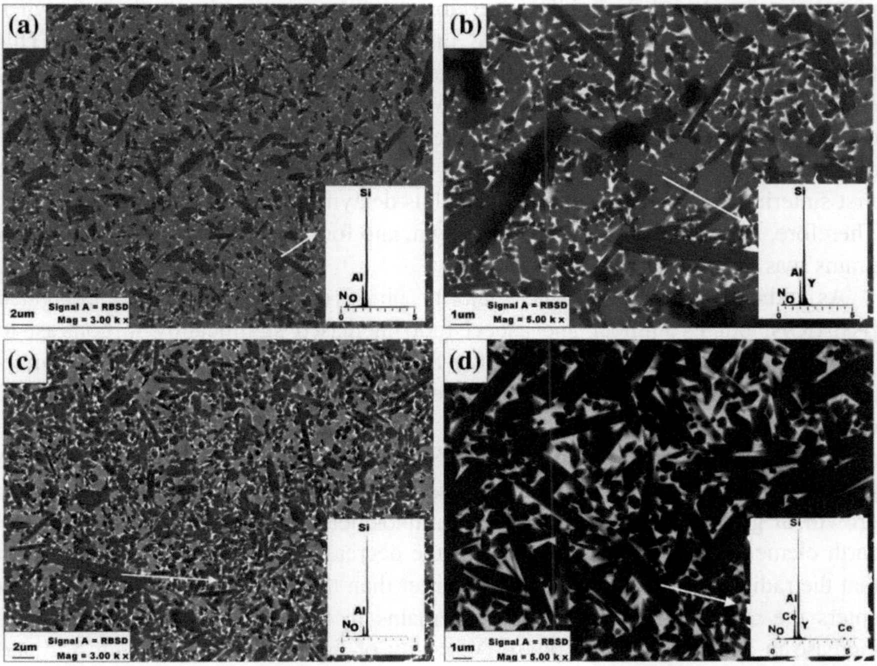

Fig. 8 BSE images of damaged surfaces of YS-1, YS-2, Ce/YS-1, and Ce/YS-2, where EDS for an elongated grain in each sample was included

Microstructure Development and Mechanical Properties. Backscattered electron (BSE) images of damaged surfaces of YS-1, YS-2, Ce/YS-1, and Ce/YS-2 are shown in Fig. 8, where EDS for an elongated grain in each sample was included. BSE imaging was used to differentiate component contrasts in a material. The component with small atomic number produces a strong BSE signal and forms a dark region, whereas the component with large atomic number produces a weak BSE signal and forms a bright region. Therefore, β-SiAlON ($Si_{6-Z}Al_ZO_ZN_{8-Z}$) should be the darkest zone, whereas α-SiAlON should be the brighter zone because of the existence of certain rare earth elements. The glass phase should be the brightest zone because of high amounts of rare earth elements, as identified by EDS analysis. Figure 8a indicates that a large number of elongated β-SiAlON grains were uniformly distributed in a matrix of YS-1, and length of these elongated grains varied from 1 to 4 μm with diameters of less than 1 μm. Several α-SiAlON grains with an equiaxed morphology and particle size of approximately 1–3 μm existed in YS-1. These grains were present because nucleation of α-SiAlON can occur at any site on the surface of α-Si_3N_4 particles and at multiple sites on the surface of one particle, thus leading to absence of preferential orientation and an equiaxed morphology. According to literature [17], β-SiAlON develops into an elongated morphology, whereas α-SiAlON develops into an equiaxed morphology when grains are heterogeneously nucleated on the surface of α-Si_3N_4 particles. This finding is consistent with the experimental result in Fig. 8a. Figure 8b shows that elongated α-SiAlON becomes the majority phase, whereas a small amount of elongated β-SiAlON is distributed in the YS-2 matrix. This result can be due to difficulty in rapidly dissolving relatively large β-SiAlON grains formed after the first sintering step into the liquid phase, thus delaying β to α phase transformation. Therefore, nucleation rate was slowed down, and formation of elongated α-SiAlON grains was promoted.

As indicated in Fig. 8c, the majority phase of Ce/YS-1 is still elongated β-SiAlON, but its number and aspect ratio increased significantly in contrast to YS-1. Figure 8d shows that most of α-SiAlON grains exhibited elongated morphology with a small amount of coarse equiaxed morphology in the Ce/YS-2 matrix. The number and aspect ratio of elongated α-SiAlON grains of Ce/YS-2 evidently increased relative to that of YS-2. Shelby [18] showed that increase in liquid phase and decrease in liquid viscosity all contributed to the anisotropic growth of grains. In addition, for the aluminosilicate liquid phase containing rare earth elements, viscosity increases with the decrease in rare-earth ion size. Given that the radius of Ce^{3+} ion is markedly larger than that of Y^{3+} ion, Ce^{3+} ion, which enters the α structure with difficulty, remains in the liquid phase, implying the increase in amount of liquid phase. Viscosity of liquid phase containing Y^{3+} and Ce^{3+} should be higher than that of liquid phase containing Y^{3+} only. Therefore, the relatively small-radius Y^{3+} ion was promoted to enter the α-structure, and anisotropic growth of α-SiAlON was promoted, eventually leading to the increase in number and aspect ratio of elongated α-SiAlON grains. The presence of a small amount of equiaxed α-SiAlON grains was attributed to homogeneous nucleation and growth of coarse α-SiAlON after the first sintering step.

As shown in Table 1, all samples can be sintered to more than 98.3% of the theoretical density by a two-step sintering procedure. This result is due to the presence of sufficient liquid phase, which attributed to the high m value of designated samples, during sintering, thereby promoting densification [14]. The sample with added with 6 wt% Ce_2O_3 possesses the best mechanical properties. In addition, fracture toughness increased from 7.14 MPa $m^{1/2}$ to 7.8 MPa $m^{1/2}$ with increasing Vickers hardness from 18.65 to 19.81 GPa possibly because of the formation of more elongated α-SiAlON grains in Ce/YS-2.

Table 1 Mechanical properties of samples after two-step sintering procedure

Samples	Mechanical properties		
	Density (%)	H_V (GPa)	K_{IC} (MPa·$m^{1/2}$)
YS1	98.3	15.29	6.6
YS2	99.1	18.65	7.14
Ce/YS1	98.4	17.30	6.8
Ce/YS2	99.6	19.81	7.8

Conclusions

Through addition of single or dual rare-earth oxides, self-reinforced Y-α-SiAlON with elongated grains can be successfully prepared by a two-step sintering method. The sample co-doped with Y and Ce exhibites the best mechanical properties with fracture toughness of 7.8 MPa $m^{1/2}$ and Vickers hardness of 19.81 GPa, which are attributed to the formation of more elongated α-SiAlON grains. The first pressureless sintering step at low temperature is benefitted for the control of phase transformation. Microstructure morphology exhibites a large number of elongated β-SiAlON grains, and a certain amount of equiaxed α-SiAlON uniformly distributes in the ceramic matrix after the first pressureless sintering step. After the second gas-pressure sintering step at high temperature, β to α phase transformation is delayed, thus slowing down nucleation rate. The formation of elongated α-SiAlON grains is promoted, resulting in elongated α-SiAlON as the majority phase. In addition, the driving force for β to α-SiAlON phase transformation decreases, the amount of liquid phase increases, and liquid viscosity decreases due to Ce_2O_3 addition, which is contributed to anisotropic growth and subsequently results in more elongated α-SiAlON grains.

Acknowledgements This work was funded by the National Natural Science Foundation of China (51574120).

References

1. Y. Rouquié, M.I. Jones, Influence of additives and compositions on the nitridation and formation of SiAlONs produced by reaction bonding and silicothermal reduction. J. Asian Ceram. Soc. **1**, 53–64 (2013)
2. G. Tu, S. Wu, J. Liu, Y. Long, B. Wang, Cutting performance and wear mechanisms of SiAlON ceramic cutting tools at high speed dry turning of gray cast iron. Int. J. Refract. Met. Hard Mater. **54**, 330–334 (2016)
3. J. Niu, X. Yi, K. Harada, T. Akiyama, Spark plasma sintering of α/β-SiAlON synthesized by salt-assisted combustion. J. Alloys Compd. **689**, 266–270 (2016)
4. M. Koshiyama, H. Sako, M. Ohno, K. Matsuura, Relationships between spark plasma sintering temperature and mechanical properties of combustion-synthesized α- and β-SiAlON. J. Jpn I. Met. **79**, 191–194 (2015)
5. A. Demir, Effect of Nicalon SiC fibre heat treatment on short fibre reinforced β-SiAlON ceramics. J. Eur. Ceram. Soc. **32**, 1405–1411 (2012)
6. J.C. Garrett, I. Sigalas, M. Herrmann, E.J. Olivier, J.H. O'Connell, cBN reinforced Y-α-SiAlON composites. J. Eur. Ceram. Soc. **33**, 2191–2198 (2013)
7. I.W. Chen, A. Rosenflanz, A tough SiAlON ceramics based on α-Si_3N_4 with a whisker-like microstructure. Nature **38**(9), 701–704 (1997)
8. G. Liu, K. Chen, H. Zhou, K. Ren, X. Ning, C. Pereira, J.M.F. Ferreira, Phase transformation and microstructure development of in situ toughened Yb-α-SiAlON ceramics prepared by two-step hot-pressing. Mater. Chem. Phys. **98**, 159–164 (2006)
9. Z.B. Yu, D.P. Thompson, A.R. Bhatti, Self-reinforcement in Li-α-SiAlON ceramics. J. Mater. Sci. **36**, 3343–3353 (2001)

10. C.R. Zhou, Z.B. Yu, V.D. Krstic, Pressureless sintered self-reinforced Y-α-SiAlON ceramics. J. Eur. Ceram. Soc. **27**, 437–443 (2007)
11. N.C. Acikbas, R. Kumar, F. Kara, H. Mandal, B. Basu, Influence of β-Si_3N_4 particle size and heat treatment on microstructural evolution of α:β-SiAlON ceramics. J. Eur. Ceram. Soc. **31**, 629–635 (2011)
12. I.S. Chi, S.K. Bux, M.M. Bridgewater, K.E. Star, S. Firdosy, V. Ravi, J.P. Fleurial, Mechanically robust SiAlON ceramics with engineered porosity via two-step sintering for applications in extreme environments. Mrs Adv. **1**, 1–7 (2015)
13. G. Magnani, A. Brentari, E. Burresi, G. Raiteri, Ceram. Int. **40**, 1759–1763 (2014)
14. Y. Shan, J. Xu, J. Li, Q. Huang, Effects of composition on microstructure and relative density of two step hot-pressed Y-α-SiAlON ceramics. J. Chin. Ceram. Soc. **36**, 1625–1628 (2008)
15. F. Ye, C. Liu, Y. Zhou, Q. Meng, Microstructure and properties of self-reinforced α-SiAlON ceramics doped with Lu_2O_3. Mater. Sci. Eng., A **496**, 143–149 (2008)
16. F. Ye, M.J. Hoffmann, S. Holzer, Y. Zhou, M. Iwasa, J. Am. Ceram. Soc. **86**, 2136–2142 (2003)
17. M. Zenotchkine, R. Shuba, J.S. Kim, I.W. Chen, Effect of seeding on the microstructure and mechanical properties of α-SiAlON: I, Y-SiAlON. J. Am. Ceram. Soc. **85**, 1254–1259 (2002)
18. J.E. Shelby, J.T. Kohli, Rare-earth aluminosilicate glasses. J. Am. Ceram. Soc. **73**, 39–42 (1989)

Preparation and Characterization of Electro-responsive Graphene Oxide/Hydrolyzed Polyacrylamide/Sodium Alginate Hydrogel Fiber

Li Peng, Jinghua Gong and Jinghong Ma

Abstract Electro-responsive hydrogels have attracted great attentions because the electric stimuli is easy to be applied and controlled. Thus, it is expected to generate electro-responsive hydrogel fiber to improve electro-response rate. In this paper, graphene oxide/hydrolyzed polyacrylamide/sodium alginate (GO/HPAM/SA) hydrogel fibers were prepared by microfluidic spinning. The structure and morphologies of hydrogel fibers were characterized by FTIR and SEM. The electro-response behaviours were also studied. The results indicate that GO/HPAM/SA hydrogel fibers exhibit bending deformation in an electric field. In Na_2SO_4 solution, the hydrogel fibers first bend toward the cathode, and then bend toward the anode. It is suggested the bending of the hydrogel fibers is caused by two major factors, one is the osmotic pressure difference, another is the pH gradient induced by electrochemical reactions.

Keywords Electro-responsive hydrogel fiber · Bending behaviour
Osmotic pressure difference · pH gradient

Introduction

Electro-responsive hydrogels can transform electric energy into mechanical motion by shrinking, swelling, or bending deformation and have great potential application in tissue engineering, drug release, actuator and sensor [1, 2]. However, the

L. Peng · J. Gong · J. Ma (✉)
State Key Laboratory for Modification of Chemical Fibers and Polymer Materials, College of Materials Science and Engineering, Donghua University, Shanghai 201620, China
e-mail: mjh68@dhu.edu.cn

L. Peng
e-mail: berry_pl@163.com

J. Gong
e-mail: gjh@dhu.edu.cn

Y. Han (ed.), *Advanced Functional Materials*,
https://doi.org/10.1007/978-981-13-0110-0_74

deformation is depended on the diffusion or migration of ions, which results in the relatively slow response rate and limits practical applications. Thus, more attentions are focused on the improvement of electro-response rate of hydrogels.

As reported by the literature, the response rate is inversely proportional to the smallest dimension of hydrogels [3]. Thus, forming hydrogel fibers that own smaller diameter than bulk hydrogels can significantly improve electro-response rate. Recently, microfluidic spinning is considered as a promising tool to prepare hydrogel fiber with micrometer diameter [4]. Shin et al. reported the continuous generation of Ca-alginate hydrogel fiber by microfluidics [5]. Lim et al. reported PNIPA/Ca-alginate hydrogel fiber by microfluidics and photo-polymerization [6]. Also, GO/PNIPA/SA hydrogel fiber was reported by microfluidics and free radical polymerization [7, 8]. Thus, it is expected to generate electro-responsive hydrogel fiber by microfluidics.

Besides, graphene oxide (GO) is usually introduced to improve the electro-response of hydrogels. Tai et al. reported the GO/PAA hydrogels and studied its electro-response [9]. The GO/PAA hydrogel took less time to reach the same bending angle compared with the PAA hydrogel. Zhang et al. reported the incorporation of GO improved the electro-response of GO/PAM hydrogel [10]. Moreover, under alkaline conditions, the response rate and bending angle of the hydrogel were further increased, which was attributed to the hydrolyzed of PAM.

In summary, the electro-responsive hydrogels can improve the electro-response by forming hydrogel fiber and adding GO. In our previous study, the GO/PNIPA/SA hydrogel fiber needed 120 s to reach a 90° bending angle. The electro-response of the hydrogel fiber was attributed to the ionizable groups in GO and SA. Thus, the electro-response rate was not fast enough. To further improve the electro-response, the ionizable polymer network was inspired to introduce the hydrogel fiber to replace the non-ionized PNIPA network. However, the microfluidic spinning was depended on the ion-crosslinking between SA and Ca^{2+}, which was influenced by the addition of ionic monomer such as MAA. As a result, GO/HPAM/SA hydrogel fiber was prepared by two steps. Firstly, GO/PAM/SA hydrogel fiber was generated by microfluidics and free radical polymerization. Then, GO/HPAM/SA hydrogel fiber was prepared by hydrolyzing under alkaline condition. Meanwhile, the electro-responsive behaviour of GO/HPAM/SA hydrogel fiber was studied.

Experimental

Materials. Graphene oxide (GO) was obtained from Nanjing XFNANO Materials Tech Co., Ltd. Acrylamide (AM) was purchased from TCI Co., Ltd. N,N′-Methylenebisacrylamide (BIS) was obtained from Sigma Co., Ltd. N,N,N′,N′-tetramethylethylene diamine (TEMED) and sodium alginate (SA) were purchased from Aladdin Co., Ltd. Potassium peroxodisulfate (KPS), calcium chloride ($CaCl_2$), sodium hydroxide (NaOH) and sodium sulphate (Na_2SO_4) were obtained from

Shanghai Chemical Co., Ltd. Deionized water was bubbled with N_2 gas for more than 1 h prior to use.

Preparation of GO/PAM/SA hydrogel fiber. The preparation of GO/PAM/SA hydrogel fiber was reference to our previous reported [8]. GO (0–0.06 g) was dispersed in 9.5 mL deionized water under ultrasound, and the pH value was adjusted to 7.0 by NaOH. Then 0.1 g SA, 1.0 g AM and 0.01 g BIS were added to GO aqueous dispersion and stirred in an ice water bath. Finally, 0.01 g KPS (the KPS was dissolved in 0.5 mL deionized water prior to use) and 10 μL TEMED were added. As shown in the microfluidic spinning device (Fig. 1), the pre-gel solution was injected into the core layer and 200 mM $CaCl_2$ solution was injected into the sheath layer. Due to the ionic crosslinking between SA and Ca^{2+}, the as-spun fiber was generated and coiled on an aluminum tube. Then the as-spun fiber was uncoiled and preserved in the soybean oil dissolved photo-initiator BDK (1%, w/v). After irradiation with UV light for 30 min, the as-spun fiber was transferred to an incubator chamber at 20 °C for free radical polymerization. Finally, the hydrogel fiber was decalcified by saturated Na_2SO_4 solution to prepare GO_m/PAM/SA hydrogel fiber. The number m represents the weight percentage of GO against AM.

Preparation of GO/HPAM/SA hydrogel fiber. The hydrolysis of GO_m/PAM/SA hydrogel fiber was performed in 5 wt% NaOH aqueous solution (the mass ratio of hydrogel fiber and NaOH was 1:5) at 65 °C for 1 h to obtain graphene oxide/hydrolyzed polyacrylamide/sodium alginate (GO_m/HPAM/SA) hydrogel fiber.

Characterization. The compositions of hydrogel fibers were analyzed by FTIR (Nicolet 6700, Thermo Fisher) in a KBr flake. The morphologies of hydrogel fibers were also observed by SEM (SU8010, Hitachi). The electro-responsive behaviours of GO/HPAM/SA hydrogel fibers were investigated in 0.1 M Na_2SO_4 solution. As shown in Fig. 2, the distance of two parallel graphite electrodes was 20 mm and the voltage between electrodes was 20 V. The hydrogel fiber with length in 1.0 cm was put in the centre of electrodes and fixed at one end. The GO_6/HPAM/SA hydrogel fiber was placed on the aluminum foil (Cathode). By applying 5 V, a copper wire (Anode) was touched the hydrogel fiber.

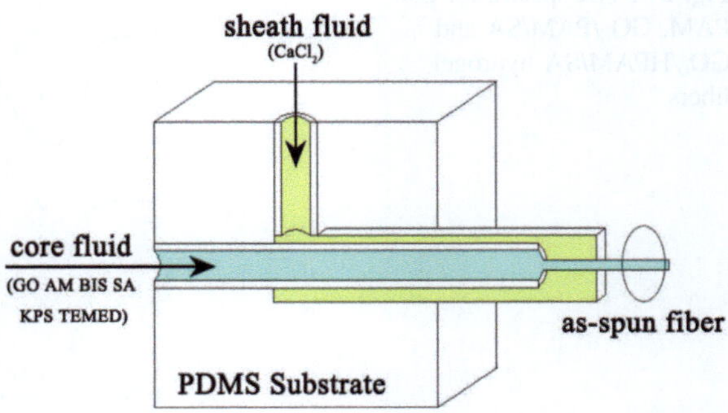

Fig. 1 Schematic of the microfluidic spinning device

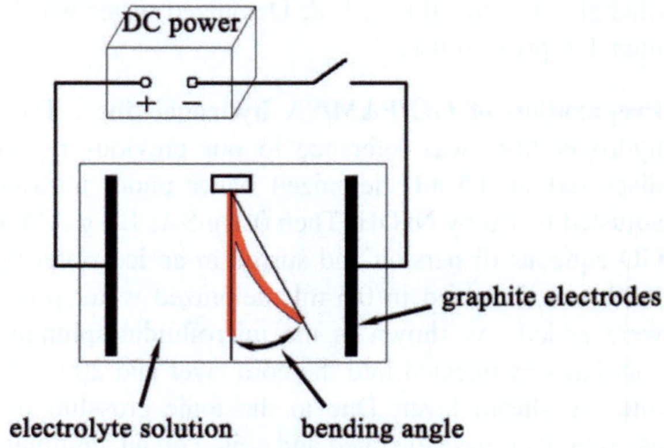

Fig. 2 Schematic for testing the electro-responsive behaviour of GO/HPAM/SA hydrogel fiber. The bending angle is defined positive when the hydrogel fiber bends to the anode side

Result and Discussion

Structure. The hydrogel fibers were analyzed by FTIR. Figure 3 shows the FTIR spectra of GO_0/PAM/SA and GO_0/HPAM/SA hydrogel fibers. In the FTIR spectrum of PAM the characteristic peaks at 1650 and 1602 cm^{-1} are attributed to C=O and N–H in the $-CONH_2$, which can also be observed in the GO_0/PAM/SA hydrogel fiber. This indicates the existence of PAM in the hydrogel network. Similarly, the band at 3434 cm^{-1} for O–H stretching in the SA can be observed in GO_0/PAM/SA hydrogel fiber, implying the existence of SA in the hydrogel network. Compared with that of unhydrolyzed GO_0/PAM/SA hydrogel fiber, a new and strong peak is found at 1550 cm^{-1} in the FTIR spectrum of the hydrolyzed GO_0/HPAM/SA hydrogel fiber. The peak is ascribed to C=O stretching in the $-COO^-$, resulting from the hydrolysis of $-CONH_2$ [11].

Figure 4 shows SEM images of GO/HPAM/SA hydrogel fibers. The hydrogel fibers show porous structure, which indicates the formation of 3D crosslinked

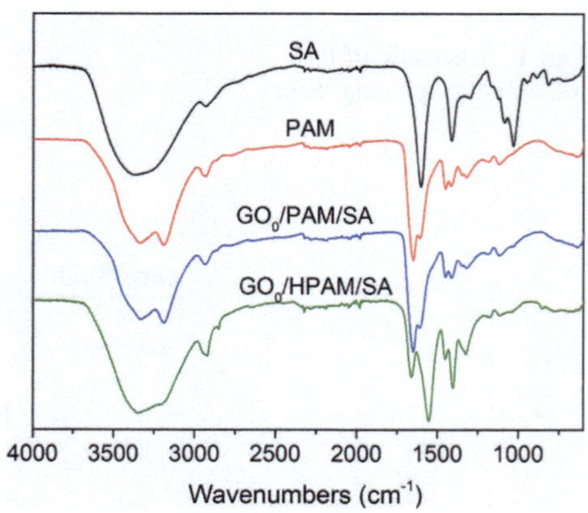

Fig. 3 FTIR spectra for SA, PAM, GO_0/PAM/SA and GO_0/HPAM/SA hydrogel fibers

Fig. 4 SEM images of **a** GO_0/HPAM/SA, **b** GO_2/HPAM/SA, **c** GO_4/HPAM/SA and **d** GO_6/HPAM/SA hydrogel fibers

network. The pore size decreases with the increase of GO content, which is caused by the hydrogen bonding interaction between N–H of HPAM and O–H of GO [12].

Electro-response. Electro-responsive hydrogels can exhibit bending deformation in an electric field, which are attributed to the ionizable groups of hydrogel network. Shiga et al. reported the bending deformation results from the osmotic pressure difference within the hydrogel [2]. In an electric field, the polyions are kept fixed in hydrogel network while the mobile ions move toward their counter electrodes. The osmotic pressure difference are induced by the occurrence of an ionic concentration gradient. As a result, polyanion hydrogels bend toward the cathode while polycation hydrogels bend toward the anode.

Another factor influencing the bending behaviours of polyelectrolyte hydrogels is the pH gradient [13, 14]. In Na_2SO_4 solution, the following electrochemical reactions occur.

$$\text{Anode electrode: } 2H_2O \rightarrow 4H^+ + 4e + O_2$$

$$\text{Cathode electrode: } 4H_2O + 4e \rightarrow 4OH + 2H_2$$

The produced ions (H^+, OH^-) would move toward their counter-electrodes and cause the pH gradient inside the hydrogel, resulting in the bending deformation.

Because the carboxylate groups in GO, HPAM and SA belong to anionic, GO/HPAM/SA hydrogel fibers are response to an electric field. It is interesting that GO/HPAM/SA hydrogel fibers first bend toward the cathode, and then bend toward the anode in 0.1 M Na_2SO_4 solution under 20 V dc voltage (Fig. 5b). This indicates the bending deformation of GO/HPAM/SA hydrogel fibers are caused by the two factors. One is the osmotic pressure difference that induces bending toward the cathode. The other is the pH gradient that induces bending toward the anode [15].

To visualize the pH change inside the hydrogel fiber on the application of electric field, a methyl orange immobilized GO_0/HPAM/SA hydrogel fiber was used in the bending experiment. The colour of the hydrogel fiber first is orange (Fig. 5a). After the electric field is applied for 60 s, the colour of the hydrogel fiber at the anode side becomes red, implying the presence of the H^+. Therefore, the GO/HPAM/SA hydrogel fibers bend toward the anode due to the pH gradient.

Figure 5b shows the influence of GO content on the bending behaviours of GO/HPAM/SA hydrogel fibers. The GO_0/HPAM/SA hydrogel fiber reaches −7° bending angle after 25 s, and then takes 35 s to reach 90° bending angle. Nevertheless, the GO_6/HPAM/SA hydrogel fiber reaches −23° bending angle after 25 s, and then takes 22 s to reach 90° bending angle. Compared with the GO_0/HPAM/SA hydrogel fiber, the electro-response rate of the GO_6/HPAM/SA hydrogel fiber is significantly improved. This is because the increase of GO content leads to an increase of the polyion density of hydrogel network, resulting in larger osmotic pressure difference and also more sensitivity to the pH gradient. Thus, the electro-response rate of GO/HPAM/SA hydrogel fiber increases with increasing GO content.

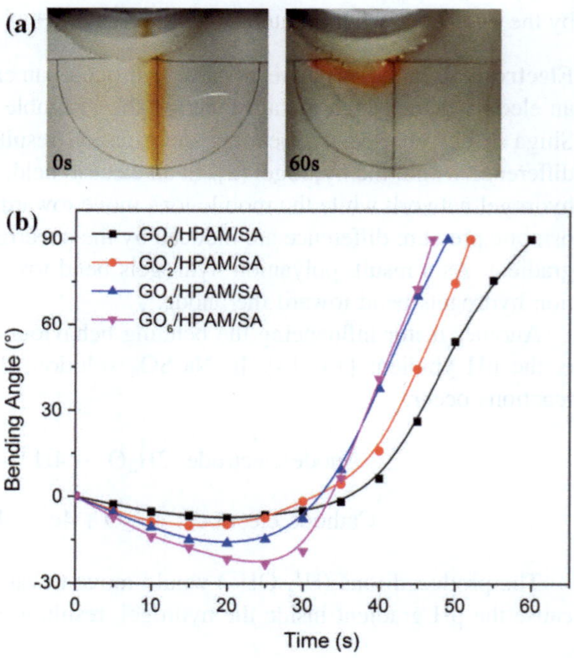

Fig. 5 **a** Photographs of GO_0/HPAM/SA hydrogel fiber at 0 and 60 s (the left side is anode); **b** the bending behaviours of GO/HPAM/SA hydrogel fibers in 0.1 M Na_2SO_4 solution under 20 V dc voltage

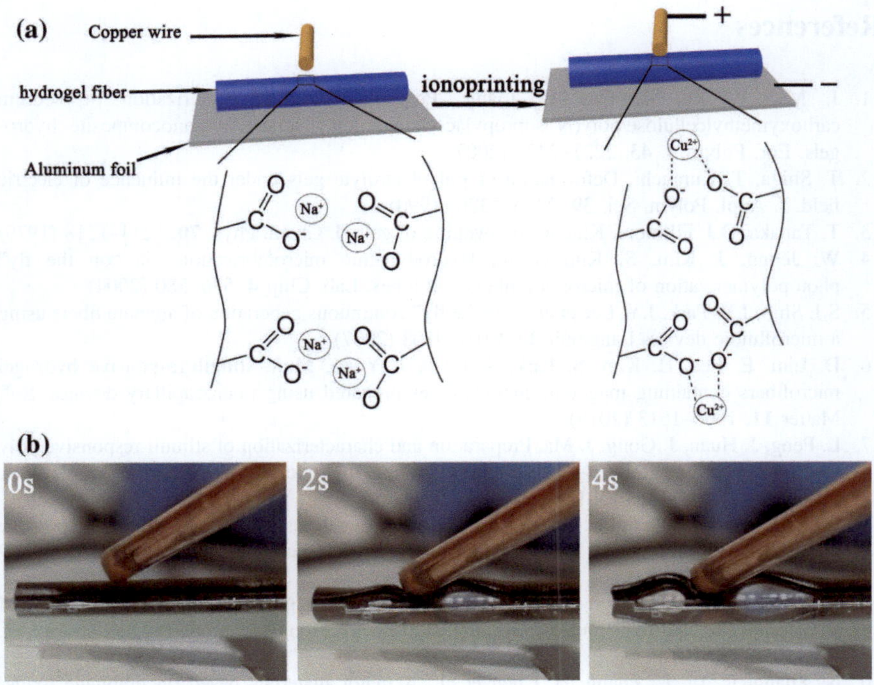

Fig. 6 **a** By applying a voltage to the Cu anode, Cu^{2+} ions was generated and locally associated with GO_6/HPAM/SA hydrogel fiber. **b** The photographs of the actuation of GO_6/HPAM/SA hydrogel fiber at 0, 2 and 4 s

Actuator. The schematic of actuator is shown in Fig. 6a. By applying a 5 V dc voltage to Cu anode, the Cu^{2+} ions were generated at the interface between the copper wire and hydrogel fiber. Then the Cu^{2+} was associated with the COO^- on the polymer backbone and formed localized ionic crosslinks in the hydrogel network, leading to a directed shrinking of the hydrogel fiber [16]. Thus, the hydrogel fiber can be actuated in air. The response rate is very fast, only taking several seconds (Fig. 6b).

Conclusion

GO/HPAM/SA hydrogel fiber is successfully prepared by two steps method. Under a non-contact dc voltage, the hydrogel fiber first bends toward the cathode, and then bends toward the anode. Meanwhile, the electro-response rate of the hydrogel fiber increases with the increase of GO content. Besides, the hydrogel fiber can be actuated in air. Therefore, the electro-responsive GO/HPAM/SA hydrogel fiber has great potential in artificial muscles, drug delivery systems, sensors and actuators.

References

1. J. Ma, Y. Xu, B. Fan, B. Liang, Preparation and characterization of sodium carboxymethylcellulose/poly(N-isopropylacrylamide)/clay semi-IPN nanocomposite hydrogels. Eur. Polym. J. **43**, 2221–2228 (2007)
2. T. Shiga, T. Kurauchi, Deformation of polyelectrolyte gels under the influence of electric field. J. Appl. Polym. Sci. **39**, 2305–2320 (1990)
3. T. Tanaka, D.J. Fillmore, Kinetics of swelling of gels. J. Chem. Phys. **70**, 1214–1218 (1979)
4. W. Jeong, J. Kim, S. Kim et al., Hydrodynamic microfabrication via "on the fly" photopolymerization of microscale fibers and tubes. Lab. Chip **4**, 576–580 (2004)
5. S.J. Shin, J.Y. Park, J.Y. Lee et al., "On the fly" continuous generation of alginate fibers using a microfluidic device. Langmuir **23**, 9104–9108 (2007)
6. D. Lim, E. Lee, H. Kim, S. Park, S. Beak, J. Yoon, Multi stimuli-responsive hydrogel microfibers containing magnetite nanoparticles prepared using microcapillary devices. Soft Matter **11**, 1606–1613 (2015)
7. L. Peng, J. Huan, J. Gong, J. Ma, Preparation and characterization of stimuli-responsive poly (N-isopropylacrylamide)/Ca-alginate hydrogel fiber by microfluidic spinning. Mater. Sci. Forum **898**, 2360–2365 (2017)
8. L. Peng, Y. Liu, J. Gong, K. Zhang, J. Ma, Continous fabrication of multi-stimuli responsive graphene oxide composite hydrogel fibres by microfluidics. RSC Adv. **7**, 19243–19249 (2017)
9. Z. Tai, J. Yang, Y. Qi, X. Yan, Q. Xue, Synthesis of a graphene oxide-polyacrylic acid nanocomposite hydrogel and its swelling and electroresponsive properties. RSC Adv. **3**, 12751–12757 (2013)
10. N. Zhang, R. Li, L. Zhang, H. Chen et al., Actuator materials based on graphene oxide/ polyacrylamide composite hydrogels prepared by in situ polymerization. Soft Matter **7**, 7231–7239 (2011)
11. Q. Zhao, J. Sun, Y. Lin, Q. Zhou, Study of the properties of hydrolyed polyacrylamide hydrogels with various pore structures and rapid pH-sensitivities. React. Funct. Polym. **70**, 602–609 (2010)
12. J. Fan, Z. Shi, M. Lian, H. Li, J. Yin, Mechanically strong graphene oxide/sodium alginate/ polyacrylamide nanocomposite hydrogel with improved dye adsorption capacity. J. Mater. Chem. A. **1**, 7433–7443 (2013)
13. S.Y. Kim, H.S. Lee, Y.M. Lee, C.N. Jeong, Properties of electroresponsive poly(vinylalcohol)/poly(acrylic acid) IPN hydrogels under an electric stimulus. J. Appl. Polym. Sci. **73**, 1675–1683 (1999)
14. S.J. Kim, K.J. Lee, S.I. Kim, Y.M. Lee, T.D. Chung, S.H. Lee, Electrochemical behaviour of an interpenetrating polymer network hydrogel composed of poly(propylene glycol) and poly (acrylic acid). J. Appl. Polym. Sci. **89**, 2301–2305 (2003)
15. H.L. Lim, J.C. Chuang, T. Tran, A. Aung, G. Arya, S. Varghese, Dynamic electromechanical hydrogel matrices for stem cell culture. Adv. Funct. Mater. **21**, 55–63 (2011)
16. E. Palleau, D. Morales, M.D. Dickey, O.D. Velev, Reversible patterning and actuation of hydrogels by electrically assisted ionoprinting. Nat. Commun. **4**, 2257 (2013)

Preparation and Study of Silicone Polyurethane Low Friction Coating

Yongyue Li, Kun Yan, Jinghong Ma and Jinghua Gong

Abstract In this work, polyether functional polydimethylsiloxane (PDPS) were introduced into the backbone of waterborne polyurethane (WPU) as a part of soft segment to synthesize a series of silicone polyurethane (PSU) emulsions by pre-polymerization method. Then the PSU coatings were prepared on the PET cloth by dip coating. The structure and performance of the coatings were characterized by Fourier transform infrared spectroscopy (FTIR), Differential scanning calorimetry (DSC), water dynamic contact angles (DCA), friction coefficient and scanning electron microscopy-energy dispersive spectrometer (SEM-EDS). The effect of the molecular weight and content PDPS on friction property of the PSU coatings was analyzed. The results showed that PDPS size and content had an evident influence on the lubricity of PSU coating, the friction coefficient of the coating was the lowest when the PDPS M_W is 1000 and the content is 28 wt%. Moreover, the prepared PSU emulsion was greatly stable.

Keywords Polydimethylsiloxane · Polyurethane · Friction · Coating

Y. Li · K. Yan · J. Ma (✉) · J. Gong
State Key Laboratory for Modification of Chemical Fibers and Polymer Materials,
College of Materials Science and Engineering, Donghua University,
Shanghai, China
e-mail: mjh68@dhu.edu.cn

Y. Li
e-mail: yueyuexiaochou@163.com

K. Yan
e-mail: yankun69@163.com

J. Gong
e-mail: gjh@dhu.edu.cn

Y. Han (ed.), *Advanced Functional Materials*,
https://doi.org/10.1007/978-981-13-0110-0_75

Introduction

Waterborne polyurethane (WPU) is multiblock copolymer consisting of alternating soft and hard segments. WPU has been widely used in adhesive and coating industries because of the merits of non-toxic, environmentally friendly, wear resistance, good mechanical properties and high elasticity.

Constant need for new and functional materials results in the application of various macromolecule diols during the preparation of WPUs, such as polydimethylsiloxane (PDMS), because of their unique and special properties including high thermal stability, low surface energy, great polymer chain flexibility, low glass transition temperature and oxidation resistance [1, 2]. PDMS segments have a tendency to move toward the surface due to its low surface energy, which is the main reason for achieving beneficial surface morphology of the WPUs. In recent years, the application of polysiloxanes modified waterborne polyurethane is mainly concentrated on its hydrophobic property [3], thermal property and mechanical properties [4]. However, there's little research on its friction property. Sainz-García E et al. showed that the addition of PDMS can reduce the friction coefficient of the coatings and a higher PDMS percent resulted in a lower friction coefficient [5]. What's more, the molecular weight of PDMS segments had a marked effect on the morphology of the obtained polymers [6], the lubricating effect of the styrene-dimethylsiloxane block copolymer films was found to increase with increasing size of the PDMS block [7].

In the present work, polyether functional PDMS (PDPS) were introduced into the backbone of WPU as a part of soft segment to synthesize a series of silicone polyurethane (PSU) emulsions via pre-polymerization method. Then the PSU coatings were prepared on the PET cloth by dipping method. The structure and performance of the coatings were characterized by FTIR, DSC, contact angle, friction coefficient and SEM-EDS. The effect of PDPS on friction property, thermal properties, surface morphology and wettability of PSU coatings was analyzed.

Experimental

The silicone polyurethane (PSU) emulsions containing different contents (0, 14, 21 and 28 wt%) and molecular chain length (Mn = 1000, 4000 g/mol) of PDPS were synthesized by pre-polymerization method by using Methylene-bis (4-cyclohexylisocyanate) (HMDI), polytetramethyene glycol (PTMG, Mn = 1000 g/mol) and 2,2-bis(hydroxymethyl) butyric acid (DMBA) as raw materials. The obtained PSU emulsions were referred as PSUX-Y, X and Y represent molecular weight and content of PDPS, respectively. For example, PSU1-28 means that the molecular weight of PDPS used in the polymerization reaction was 1000, and its additive amount is 28 wt%. The PSU emulsion was diluted with isopropanol to a solution mass fraction of 15%. Then the PSU coatings were prepared on PET film by scraping and on PET cloth by dipping, and then dried in vacuum oven at 120 °C for 10 min.

The diameters of the PSU particles were measured by nano-particle size and potential analyzer (MS 2000). FTIR of PSU coatings were recorded using a Nicolet 8700 Fourier infrared spectrometer. DSC was carried out with a 204F1 instrument (NETZSCH) at a heating rate of 10°C min^{-1}. The friction coefficient of coatings on PET cloth was measured by MXD-02 friction coefficient tester. The morphology of coatings on PET cloth was observed under a SU-8010 (Hitachi) SEM. The surface mapping of PSU1-28 was conducted using the subsidiary EDS. The DCA were measured on an OCA40 Micro automatic video micro-contact angle analyzer.

Results and Discussion

Particle Size Distributions and Stability of PSU Emulsion. There is no sediment in all samples after centrifugal settling, so the stability of PSUs emulsion is good. Particle size distributions of PSU are shown in Fig. 1. It is observed that particle size increases slowly with increasing the content and molecular weight of PDPS. The hydrophilic groups and segments in PSU chains were favor of location on the particle surface, whereas the hydrophobic ingredient PDPS tended to be packaged inside the core of the particles as much as possible. Higher content or molecular weight of PDPS led to the increase of hydrophobicity and form bigger volume in the inner of particles, thus the particle size increases [8].

FTIR Analysis. The composition of PSU samples was confirmed by FTIR spectra shown in Fig. 2. The absorptions at 1260 cm^{-1} (Si–CH_3), 1089 cm^{-1} (Si–O–Si), and 806 cm^{-1} (CH_3–Si) for all samples were detected and ascribed to the PDPS introduced into the PSU backbone [3], indicating that the PDPS was successfully incorporated into the PSU. The disappearance of band around 2265 cm^{-1} in PSU shows that all the NCO groups have reacted completely. The existence of the N–H bands around 3330 cm^{-1} and 1529 cm^{-1} and –C=O band around 1739 cm^{-1}

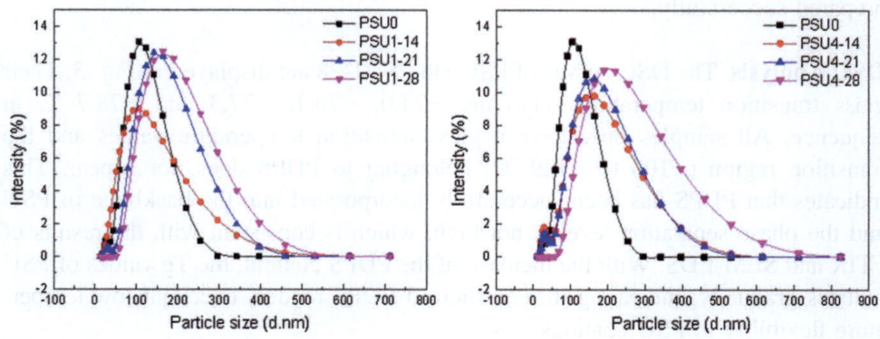

Fig. 1 Particle size distributions of PSU

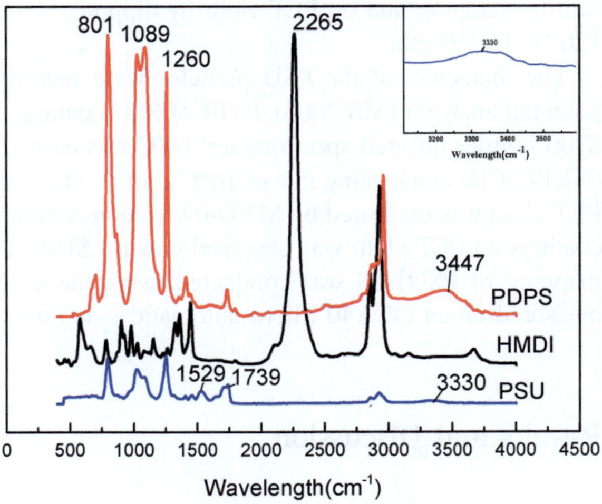

Fig. 2 FTIR spectra of PSU coating, HMDI and PDPS

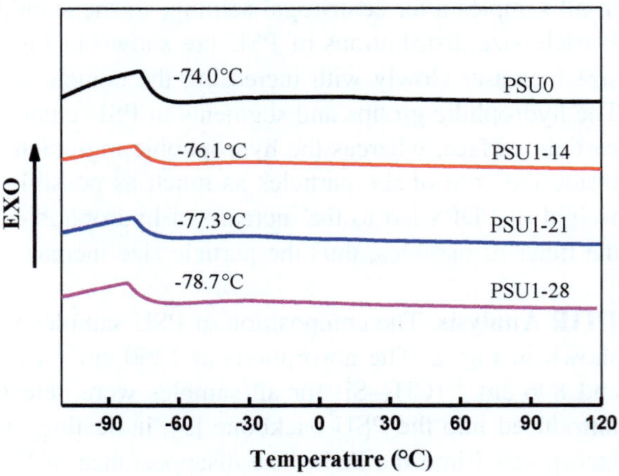

Fig. 3 DSC curves of PSU coatings

demonstrates that the formation of the carbamate group, PSU coatings have the molecular structure of polyurethane. To sum up, silicone polyurethane have been prepared successfully.

DSC Analysis. The DSC curves of PSU0 to PSU1-28 are displayed in Fig. 3. Their glass transition temperatures (Tg) are −74.0, −76.1, −77.3 and −78.7 °C in sequence. All samples only have a glass transition temperature values and the transition region (−100 to −129 °C) belonging to PDPS does not appear. This indicates that PDPS has been successfully incorporated into the backbone of PSU and the phase separation level is not high, which is consistent with the results of FTIR and SEM-EDS. With the increase of the PDPS content, the Tg values of PSU reduces gradually, indicating that plasticized PDPS renders excellent low temperature flexibility to PSU coatings.

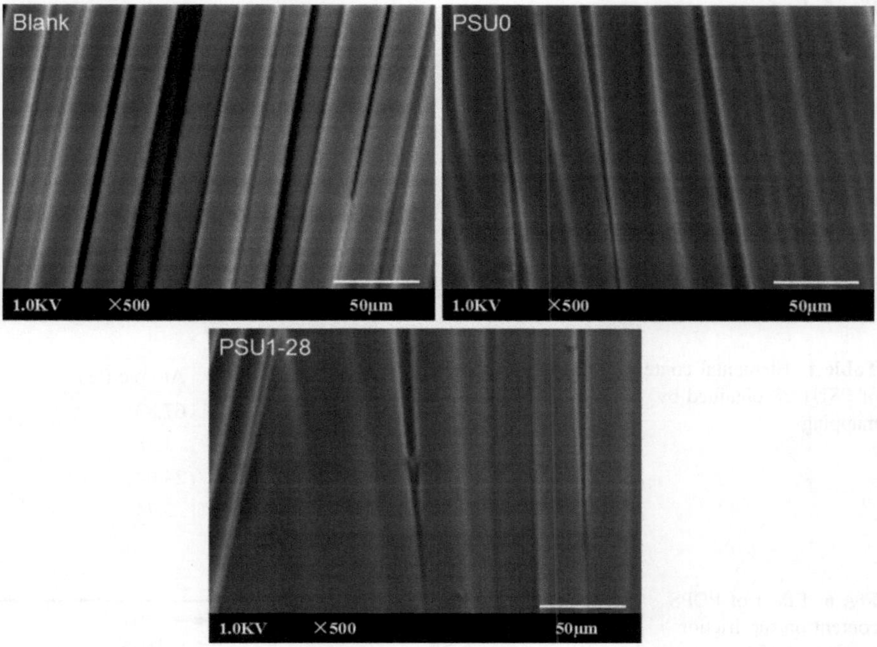

Fig. 4 SEM surface images of PSU coatings

Surface Morphology of Coatings. The SEM surface images of the PET cloth and the PSU coatings are displayed in Fig. 4. The PSU coating fills the gap between the fibers, and makes the cloth surface smoother. And not any clear spots can be found in the image of PSU1-28, preliminary illustrating the uniform distribution of element on its surface [1]. The surface mapping spectrum and the elemental content of PSU1-28 are displayed in Fig. 5 and Table 1, respectively. The Si content of the bulk of PSU1-28 is 3.74%, which reveals the existence of silicon on the surface of coating, the elements distribute uniformly in the bulk of PSU1-28 and no regional enrichment can be found. This shows that PDPS segments have good compatibility with other segments and the obvious phase separation does not occur.

Friction Coefficient Analysis. The dry sliding friction of the polymer elastomer coating is often composed of deformation friction and adhesive friction. Figure 6 illustrates the variation in the friction coefficient with the silicone content for PSU coatings. It can be seen from Fig. 6 that the PSU1 coatings with higher silicone content have lower friction coefficients. This is due to the fact that the low cohesive energy density and high chain flexibility of PDPS segments lead to stronger hydrophobicity on the coating surface. However, for the PSU4 coatings, when the PDPS content increases to 28%, the friction coefficient of the coating increased, which may be due to the deformation of PDPS segments in the process of sliding.

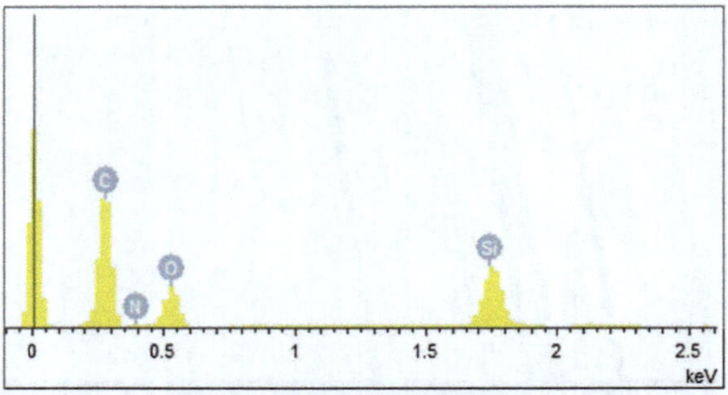

Fig. 5 Surface mapping spectrum of PSU1-28

Table 1 Elemental content of PSU1-28 obtained by mapping

Element	Weight (%)	Atomic (%)
C	59.65	67.83
N	4.49	4.38
O	28.16	24.04
Si	7.69	3.74

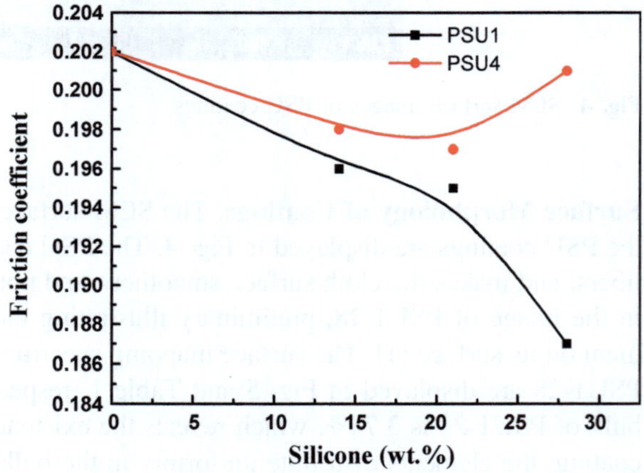

Fig. 6 Effect of PDPS content on the friction coefficient of PSU coatings

Dynamic Water Contact Angle Analysis. The water dynamic contact angle of PSU coatings were presented in Fig. 7. The contact angle increased from 77° to 112° when the PDPS content was increased from 0 to 28%, indicating that introduction of PDPS can greatly improve the surface hydrophobicity of the coatings. This result is expected because PDPS has much lower surface energy than the other components in the PSU copolymers [9]. When PDPS has been used as a soft segment in the copolymer, the siloxane segments tend to migrate toward the air/coating interface to minimize the interfacial energy during the formation of the coating, resulting in forming the siloxane enriched surface [10, 11]. Thus, the surface water contact angle of the coatings increased with the increasing of PDPS content and molecular weight.

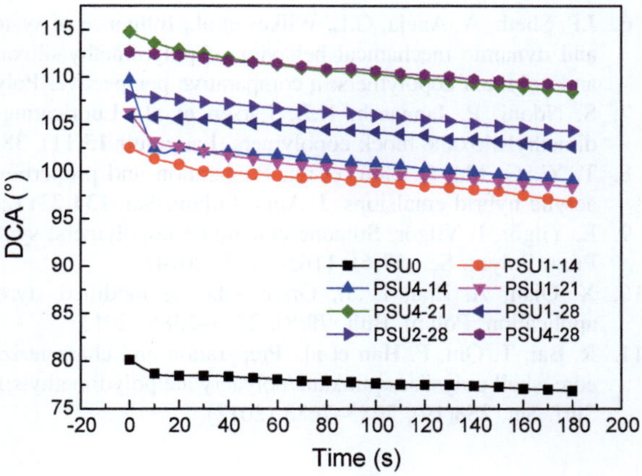

Fig. 7 Water contact angle of PSU coatings

Conclusions

The data of FTIR and DSC showed that a series of silicone polyurethane (PSU) emulsions were prepared successfully. Friction coefficient demonstrates the introduction of PDPS in the PSU can significantly reduce the friction coefficient of the coating. Furthermore, the friction coefficient of the coating is the lowest when the PDMS M_W is 1000 and the content is 28 wt%. Contact angles shows that PDPS segments with low surface tension migrated to the surface, endowing PSU coatings with good hydrophobic property. DSC analysis reveals that PDPS can give PSU coatings good flexibility at low temperature. In combination with FTIR, SEM, EDS and DSC results, there is no obvious phase separation in both bulk and surface of PSU coatings, while the compatibility between PDPS and other segments is good.

References

1. G. Zheng, M. Lu, X. Rui, The effect of polyether functional polydimethylsiloxane on surface and thermal properties of waterborne polyurethane. Appl. Surf. Sci. **399**, 272–281 (2017)
2. R. Hernandez, J. Weksler, A. Padsalgikar et al., Microstructural organization of three-phase polydimethylsiloxane-based segmented polyurethanes. Macromolecules **40**(15), 5441–5449 (2007)
3. Z. Wu, H. Wang, X. Tian et al., The effects of polydimethylsiloxane on transparent and hydrophobic waterborne polyurethane coatings containing polydimethylsiloxane. Phys. Chem. Chem. Phys. **16**(14), 6787–6794 (2014)
4. U.S. Chung, J.H. Min, P.C. Lee et al., Polyurethane matrix incorporating PDMS-based self-healing microcapsules with enhanced mechanical and thermal stability. Colloids Surf. A **518**, 173–180 (2017)
5. E. Sainz-García, F. Alba-Elías, R. Múgica-Vidal et al., Promotion of tribological and hydrophobic properties of a coating on TPE substrates by atmospheric plasma-polymerization. Appl. Surf. Sci. **371**, 50–60 (2016)

6. J.P. Sheth, A. Aneja, G.L. Wilkes et al., Influence of system variables on the morphological and dynamic mechanical behavior of polydimethylsiloxane based segmented polyurethane and polyurea copolymers: a comparative perspective. Polymer **45**(20), 6919–6932 (2004)
7. S. Ndoni, P. Jannasch, N.B. Larsen et al., Lubricating effect of thin films of styrene–dimethylsiloxane block copolymers. Langmuir **15**(11), 3859–3865 (1999)
8. T. Yi, G. Ma, C. Hou, et al., Preparation and properties of poly(siloxane-ether-urethane)-acrylic hybrid emulsions. J. Appl. Polym. Sci. **134**(23) (2017)
9. E. Yilgör, I. Yilgör, Silicone containing copolymers: synthesis, properties and applications. Prog. Polym. Sci. **39**(6), 1165–1195 (2014)
10. X. Qian, A. Zhu, L. Ji, Organosilicone modified styrene-acrylic latex: preparation and application. Polym. Bull. **70**(8), 2373–2385 (2013)
11. R. Bai, T. Qiu, F. Han et al., Preparation and characterization of inorganic–organic trilayer core–shell polysilsesquioxane/polyacrylate/polydimethylsiloxane hybrid latex particles. Appl. Surf. Sci. **258**(19), 7683–7688 (2012)

Preparation of Superhydrophobic PVDF/P(HEMA-MMA-SMA)/SiO_2 Composite Membranes by Electrospinning

Yuanxia Hu, Jinghong Ma and Jinghua Gong

Abstract The hydrophobic random copolymer P(HEMA-MMA-SMA) which contained 2-hydroxyethyl methacrylate (HEMA), methyl methacrylate (MMA) and stearyl methacrylate (SMA) chains was successfully synthesized via radical polymerization. The composites membranes of poly (vinylidene fluoride) (PVDF) and P (HEMA-MMA-SMA) blends and PVDF/P(HEMA-MMA-SMA)/SiO_2 were prepared by electrospinning. Scanning Electron Microscopy (SEM) and Water Contact Angle (WCA) revealed that morphology and hydrophobic properties of the membrane surface. With the condition of 12 wt% SiO_2, the maximum WCA (154°) of the electrospun composite film can be obtained. Superhydrophobic materials have been widely used in the fields, such as oil-water separation and self-cleaning.

Keywords Methacrylate · Poly(vinylidene fluoride) · SiO_2 · Random copolymer Blends · Superhydrophobic · Electrospinning · Membranes

Introduction

Hydrophobicity of material surfaces is a key surface property and the most common interface phenomenon. It usually appears with the contact angle of water on the solid surface. When the WCA is greater than 150° and the water sliding angle (WSA) is less than 10°, this surface is referred to as a superhydrophobic surface. Superhydrophobic phenomenon is ubiquitous in nature, such as the silver ragwort leaf, lotus leaf, Bamboo leaf and mosquitoes [1]. Due to its excellent hydrophobic

Y. Hu
College of Materials Science and Engineering, Donghua University, Shanghai 201620, China
e-mail: 2398429893@qq.com

J. Ma (✉) · J. Gong
The State Key Laboratory for Modification of Chemical Fibers and Polymer Materials, Donghua University, Shanghai 201620, China
e-mail: mjh68@dhu.edu.cn

Y. Han (ed.), *Advanced Functional Materials*,
https://doi.org/10.1007/978-981-13-0110-0_76

properties, superhydrophobic materials have been widely used in the fields of oil-water separation, antifouling, anti-icing and self-cleaning [2].

Today, many techniques have been developed to create rough surface textures, such as plasma etching, self-assembly, phase separation, etc. [3–5]. In addition, electrospinning is an effective method for preparing continuous micron or nanofibers. It is suitable for a wide range of polymers such as polymers, carbon materials, metals, metal oxides, minerals etc. [6–8]. Fiber membranes prepared by electrospinning have the advantages of material diversity, high aspect ratio, high porosity, high specific surface area and uniform structure. In recent years, under the appropriate conditions, by controlling the surface roughness of the preparation of superhydrophobic surfaces [1, 9]. With the development of electrospinning technology [10, 11], many polymers and composites have been widely used in many areas. The structures and morphologies of electrospun fiber membranes will be controlled by polymer type, concentration of solutions, environmental parameters and electrospinning conditions and other factors. Electrospinning is the one of the most promising technologies in the development of polymer with low surface energy of superhydrophobic surface.

Owing to its good performance, such as low surface energy, excellent chemical resistance, high mechanical strength, and very good thermal stability, PVDF is widely used in the production of superhydrophobic membranes in recent years [12]. Recently it has been reported that the nonwoven fibrous membrane has a hydrophobic behavior at a water contact angle of 136°, which composed of purely electrospun PVDF nanofibers [13]. For further improvement of the wettability of those nano-fibrous membranes, hydrophobicity methacrylic acid long chain alcohol ester (such as SMA, lauryl methacryla (LMA)) and inorganic nanoparticles (such as SiO_2, TiO_2 and Al_2O_3) can be added into the PVDF nano-fiber membranes during the electrospinning [14].

In this work, we have synthesized poly(2-Hydroxyethyl methacrylate-co- methyl methacrylate -co-stearyl methacrylate) (P(HEMA-MMA-SMA)) random copolymer. What is more, the MMA to SMA ratio can be controlled, that is, the hydrophilic–hydrophobic balance can be tuned. Also, preliminary experiments showed that homogeneous solution was obtained when blending this hydrophobic copolymer with PVDF in a mixture solvent of N, N-dimethylformamide(DMF)/ ethyl acetate. By adjusting the content of SiO_2 nano-particles, it is possible to control the micro/nano-structures on the electrospinning membranes surface prepared by the protrusions. Through the addition of SiO_2 nano-particles, the electrospun PVDF/P(HEMA-MMA-SMA) composite membranes can be tuned from hydrophobic to superhydrophobic.

Experimental

Materials. PVDF was supplied by Arkema (Shanghai Agency, China). 2-Hydroxyethyl methacrylate (HEMA) and stearyl methacrylate (SMA) were provided by Guangdong Wengjiang Chemical Reagent Co., Ltd. 2,2′-azobisisobutyronitrile (AIBN) was bought from Sinopharm Chemical Reagent Co., Ltd. The hydrophobic SiO_2 particle with diameter range from 7 to 40 nm was provided by Aladdin Reagent Co., Ltd. DMF, methyl methacrylate (MMA), ethyl acetate were supplied by Shanghai Lingfeng Chemical Reagent Co., Ltd. The PVDF powder needs to dry at 80 °C for whole day before use. AIBN and MMA need to purify by re-crystallization and reduced pressure distillation respectively.

Synthesis of P(HEMA-MMA-SMA). A method of obtaining P(HEMA-MMA-SMA) copolymer is developed via conventional solution free radical polymerization. Total weight of solutions was kept constant to 60 g. Solvent weight was always 42 g. MMA weight varied between 13.3 and 9.7 g while SMA was in the range from 2.7 to 6.3 g. The designations of the blend composites generated are listed in Table 1. The mixture of HEMA, MMA, SMA, AIBN, DMF, and ethyl acetate were added into the flask, then kept the flask at 70 °C in N_2 under stirring vigorously for 6 h. The polymer solution was precipitated with methanol, and purified by repeated precipitation of the polymer. Precipitation was filtered, then dried at 45 °C for 12 h to obtain P(HEMA-MMA-SMA).

Preparation of PVDF/P(HEMA-MMA-SMA) and PVDF/P(HEMA-MMA-SMA)/SiO_2 Precursor Solution. Firstly, the PVDF powders were dissolved in a mixture solvent of DMF/ethyl acetate (v:v = 7:3). In order to obtain a transparent homogenous solution, PVDF solution was stirred at 55 °C for 2 h. Secondly, when the PVDF/DMF/ethyl acetate solution reached 25 °C, the random copolymers P (HEMA-MMA-SMA) solution was mixed. The volume ratio of the DMF and ethyl acetate in the resulting PVDF/P(HEMA-MMA-SMA)/DMF/ethyl acetate solution was 7:3. The concentration of PVDF and P(HEMA-MMA-SMA) in the resulting PVDF/P(HEMA-MMA-SMA)/DMF/ethyl acetate solution were 15 and 7.5%, respectively. In the following sections, the polymer blends of PVDF and different SMA concentration of P(HEMA-MMA-SMA) are denoted as DFMAi, where i is the weight portions of SMA concentration of P(HEMA-MMA-SMA). For example, DFMA15 indicates the polymer blends consisting of PVDF and P(HEMA-MMA-SMA15).

Then several groups of precursor solutions with different mass ratios of SiO_2 nano-particles/PVDF (0, 2, 4, 6, 8, 10, 12, 14%) were prepared respectively. In order to disperse SiO_2 nano-particles in DFMA blend solution, the DFMA/SiO_2 solutions were stirred vigorously for 12 h at 25 °C.

Tab.1 Designations and corresponding compositions of the blend composites generated

Sample designation	HEMA (g)	MMA (g)	SMA (g)	AIBN (g)	DMF (g)	Ethyl acetate (g)
P(HEMA-MMA-SMA15)	2	13.3	2.7	0.18	30	12
P(HEMA-MMA-SMA20)	2	12.4	3.6	0.18	30	12
P(HEMA-MMA-SMA25)	2	11.5	4.5	0.18	30	12
P(HEMA-MMA-SMA30)	2	10.6	5.4	0.18	30	12
P(HEMA-MMA-SMA35)	2	9.7	6.3	0.18	30	12

Superhydrophobic Membrane Preparation by Electrospinning. To electrospin fibrous DFMA and DFMA/SiO_2 composites membranes, the blend PVDF/P (HEMA-MMA-SMA)/DMF/ethyl acetate and DFMA/SiO_2/DMF/ethyl acetate precursor solution was injected in the plastic syringe of 5 mL, which connected to a stainless steel needle. And the solution feed rate was set as 2 ml h^{-1}, then 18 kV was forced between the grounded aluminum foil and needle tip 15 cm above. After electrospinning, a nonwoven DFMA and DFMA/SiO_2 composites membrane were prepared onto the aluminum foil, respectively. Then fibrous composite membranes were peeled off from the aluminum foil. In order to remove the influence of the residual solvent, the prepared fibrous composite membranes were dried at 45 °C for 12 h.

Characterization. The contact angle meter measured the WCA of fiber membranes, and used 3 μL distilled water as sample. The same sample at different locations were 5 times the test, and then took the average. To determine the stability of the superhydrophobic of the DFMA/SiO_2 composites membranes, the contact angle change with the drop age of the membrane was recorded by dynamic contact angle (OCA40, Dataphysics, Germany). The morphologies of the prepared membranes were obtained by scanning electron microscopy (SEM, HITACHI S-3000, Japan). Image-pro Plus 5 image analysis software was used to analyze the diameter of nanofiber in SEM. The functional groups of the P(HEMA-MMA-SMA) were characterized by Tensor 27 infrared spectrometer (FT-IR).

Results and Discussion

Chemical Composition of the Membrane. Figure 1 shows FTIR spectrum of P (HEMA-MMA-SMA). In the IR spectrum of random copolymer P (HEMA-MMA-SMA), the characteristic absorption peak of -OH group appears near 3524 cm^{-1}, which belongs to HEMA. Two peaks at 2928 and 2855 cm^{-1} were associated with methyl or methylene groups. The strong peak at 1727 cm^{-1} was ascribed to the stretching vibration of carbonyl group, indicating the presence of methacrylate. Two peaks at 1448 and 1387 cm^{-1} were characteristic peaks of MMA, and another peak at 1242 and 1148 cm^{-1} were attributed to the stretching vibration of C–O–C. The peak at 750 cm^{-1} could be ascribed to –(CH_2)n– ($n \geq 4$) groups from SMA. Meanwhile, the peak at 1640 cm^{-1}, which was attributed to C=C groups from acrylate monomers, disappeared, verifying that MMA, SMA, HEMA have been well copolymerized.

The Effect of P(HEMA-MMA-SMA) on the Wettability and Morphologies of the Electrospun Membranes. Figure 2a, b shows the SEM micrographs and diameter distributions of the pure PVDF and DFMA membranes. With increasing polymers concentration from 15.0 to 22.5 wt%, the increase of fiber mean diameter was obvious from above figure (from 0.51 to 1.67 μm). Therefore, an increase in fiber diameter can be achieved by increasing the polymer concentration in the

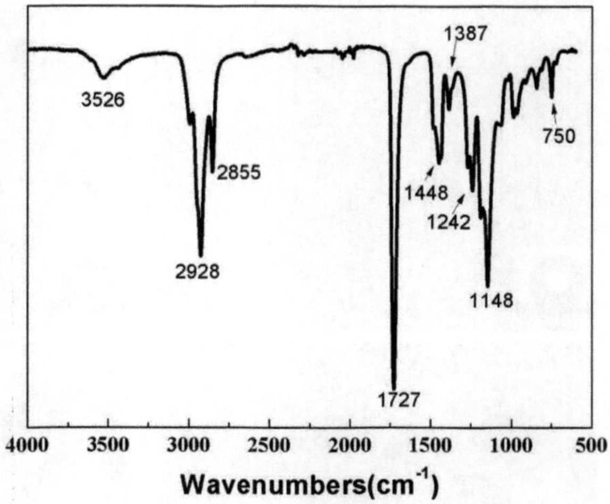

Fig. 1 FTIR spectrum of P (HEMA-MMA-SMA)

electrospinning solution. When the SMA weight increased from 2.7 to 6.3 g, the mean diameters of the DFMA composite fibers reduced from 1.67 to 0.81 μm, and the WCAs of the DFMA composite membranes increased from 138.6° to 143.6°. The increase of hydrophobic SMA results in the WCA variation of the DFMA composite membranes. Therefore, we choose the appropriate P(HEMA-MMA-SMA) (P(HEMA-MMA-SMA35)) to prepare superhydrophobic DFMA35/SiO_2 nanocomposite membranes.

The Influence of SiO_2 on the Hydrophobicity of the Electrospun Membranes. Analyzed from Fig. 3a, the SiO_2 weight influenced the hydrophobic properties of the DFMA35 composite membranes. Increasing the weight ratio of SiO_2 results in the WCA increase of DFMA35/SiO_2 nano-composite membranes. When the content of SiO_2 increased from 0 to 12%, the WCA of the composite nanofiber membranes increased from 143.6° to 154.0°, indicating that the adding of SiO_2 particles significantly enhanced the hydrophobicity of the composite nanofiber membrane. But further increasing SiO_2 weight to 14 wt%, the WCA value of nanofiber membranes decreased to 148°, due to the SiO_2 particles are dispersed unevenly, and agglomerate together. The maximum WCA of 154° can be obtained at SiO_2 weight of 12 wt%. It also can be seen that the dynamic water contact of the DFMA35/SiO_2 (12 wt%) composite membrane almost remain stable during the drop aging. As shown in Fig. 4, electrospun composite nanofiber membranes exist micron and nano-scale roughness, that is, hierarchical rough structure. Due to the micro/nano hierarchical structure constructed on the fiber surface, a large number of papillae are formed on the surface of the fiber, and a large amount of air was stored between the fibers and the papillae of the fiber. When the water drops are dropped on nanofiber membranes, a continuous water-air -solid interface, due to air contact angle up to 180.0°, so as to maintain a stable superhydrophobic state.

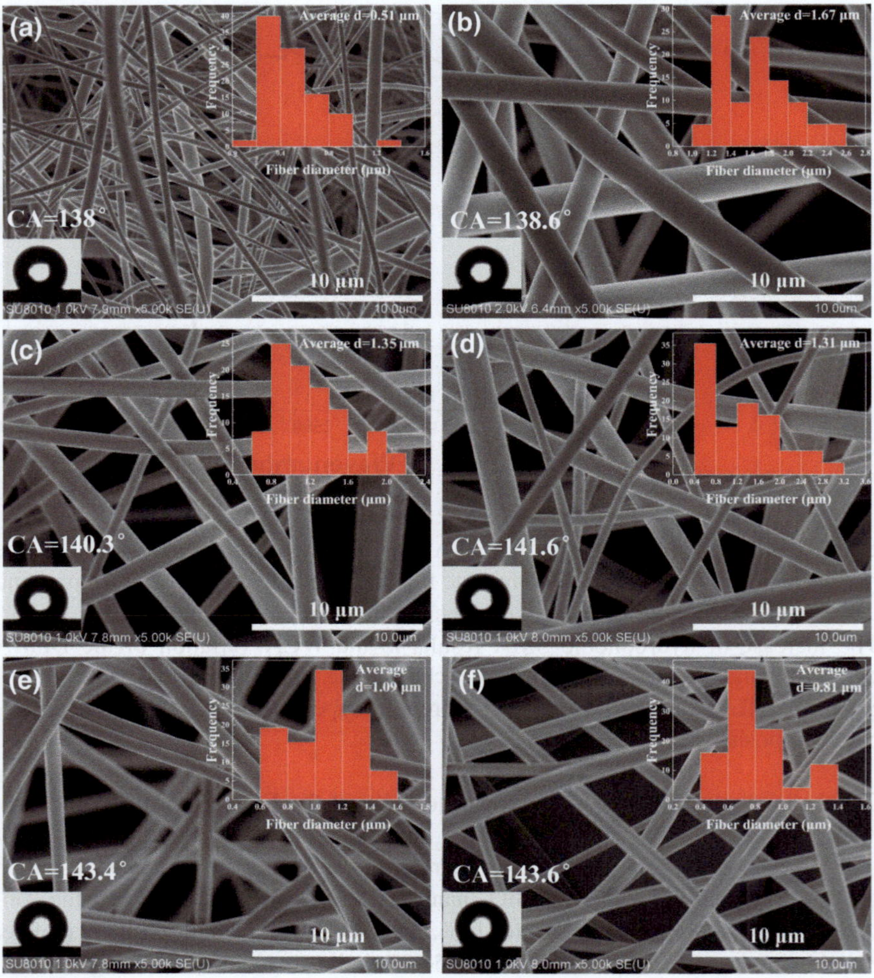

Fig. 2 SEM micrographs of the PVDF electrospun membranes (**a**), the DFMA15 electrospun membranes (**b**), the DFMA20 electrospun membranes (**c**), the DFMA25 electrospun membranes (**d**), the DFMA30 electrospun membranes (**e**), the DFMA35 electrospun membranes (**f**). The two insets in each SEM micrograph are the fiber diameter distribution of an individual DFMA fiber and the water contact measurement, respectively

The Influence of SiO_2 on the Morphologies of the Electrospun Membranes. Figure 4 shows the SEM micrographs and fiber diameter distribution of DFMA35/ SiO_2 composite nano-fibers with different SiO_2 contents. It can be found that the SiO_2 particles are deposited on the nano-fiber surfaces, resulting in a nano-sized rough structure, meanwhile increasing the roughness of the nano-fiber membrane. Observed from Fig. 4a–c, as the content of SiO_2 nano-particles increases, the mean diameter of the fibers gradually increases, the uniformity of the diameter distribution decreases, and the surface roughness of the nano-fibers increases. When the

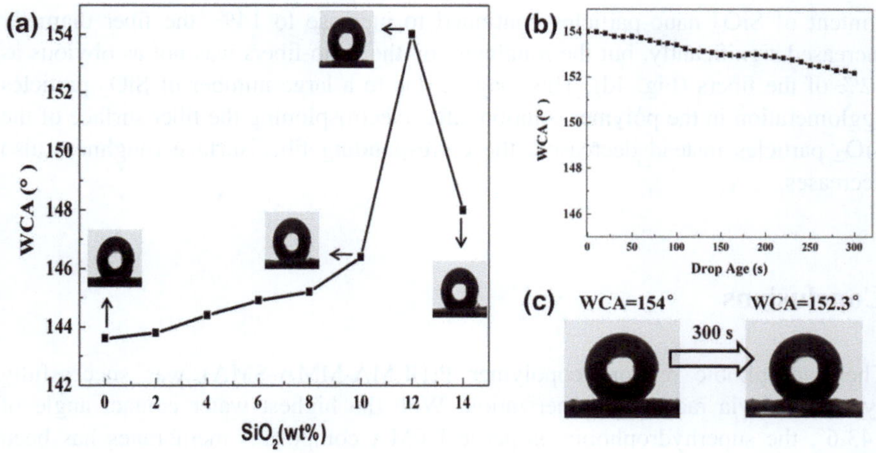

Fig. 3 **a** The effect of SiO_2 weight on the wetting behavior of the DFMA35 composite membranes, **b** dynamic water contact angle of the DFMA35/SiO_2 (12 wt%) composite membrane versus drop age, **c** pictures of water drops on the DFMA35/SiO_2 (12 wt%) composite membrane surfaces

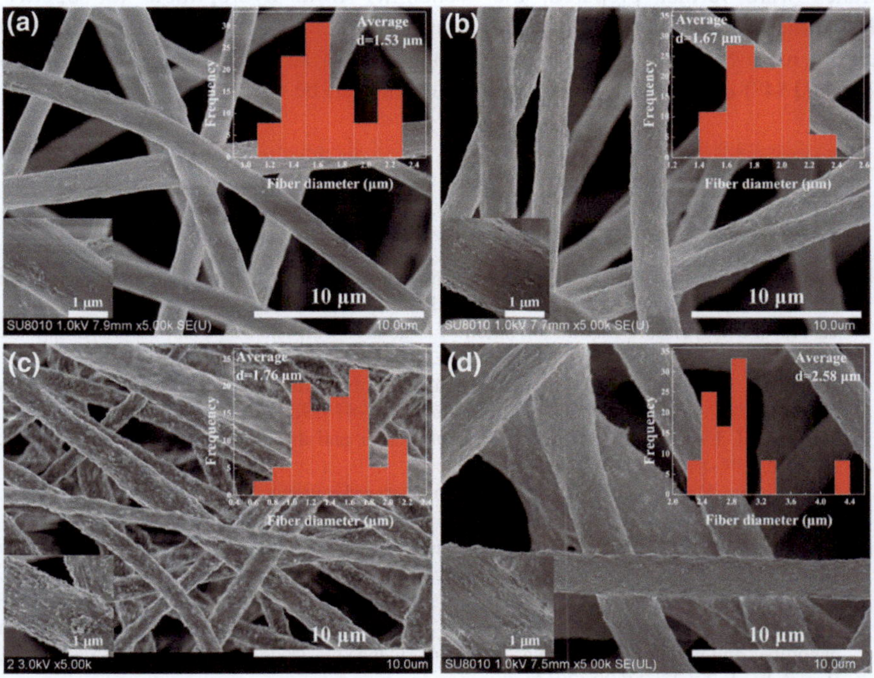

Fig. 4 SEM micrographs of the DFMA35 electrospun membranes with 8% (**a**), 10% (**b**), 12% (**c**), 14% (**d**) SiO_2. The two insets in each SEM micrograph are the fiber diameter distribution of an individual DFMA fiber and the high-magnification images for a single nanofiber, respectively

content of SiO_2 nano-particles continued to increase to 14%, the fiber diameter increased significantly, but the roughness on the nano-fibers was not as obvious as 12% of the fibers (Fig. 4d). This may be due to a large number of SiO_2 particles agglomeration in the polymer solution, after electrospinning the fiber surface of the SiO_2 particles instead decreased, the corresponding fiber surface roughness also decreases.

Conclusions

The hydrophobic random copolymer P(HEMA-MMA-SMA) was successfully synthesized via radical polymerization. With the highest water contact angle of 143.6°, the superhydrophobic propertie DFMA composites membranes has been successfully prepared by electrospinning low-cost means. By rationally controlling the SMA concentration of the P(HEMA-MMA-SMA) copolymer in the electrospinning solution, the diameter and surface morphology of the DFMA fiber can be changed to control the superhydrophobic property of the electrospun fibrous DFMA composite membranes. The electrospun fibrous DFMA/SiO_2 composites membranes were further prepared, the wettability from hydrophobic to superhydrophobic of the electrospun DFMA composites membranes could be obtained by adding nano SiO_2 particles. With the condition of 12 wt% SiO_2, the maximum WCA (154°) of the electrospun composite film can be obtained.

References

1. X. Wang, B. Ding, J. Yu, Engineering biomimetic superhydrophobic surfaces of electrospun nanomaterials. Nano Today **6**, 510–530 (2011)
2. W. Xu, J. Song, J. Sun, Rapid fabrication of large-area, corrosion-resistant superhydrophobic Mg alloy surfaces. Acs. Appl. Mater. Inter. **3**, 4404–4414 (2011)
3. J.M. Lim, G.R. Yi, J.H. Moon, Superhydrophobic films of electrospun fibers with multiple-scale surface morphology. Langmuir **23**, 7981–7989 (2007)
4. Z.Z. Gu, H. Uetsuka, K. Takahashi, Structural color and the lotus effect. Angew. Chemint. Edit. **42**, 894–897 (2003)
5. H.M. Shang, Y. Wang, K. Takahashi, Nanostructured superhydrophobic surfaces. J. Mater. Sci. **40**, 3587–3591 (2005)
6. Z. Zhou, X.F. Wu, High-performance porous electrodes for pseudosupercapacitors based on graphene-beaded carbon nanofibers surface-coated with nanostructured conducting polymers. J. Power Sources **262**, 44–49 (2014)
7. Z. Zhou, X.F. Wu, H. Hou, Electrospun carbon nanofibers surface-grown with carbon nanotubes and polyaniline for use as high-performance electrode materials of supercapacitors. Rsc. Adv. **4**, 23622–23629 (2014)
8. X.F. Wu, A. Rahman, Z. Zhou, Electrospinning core-shell nanofibers for interfacial toughening and self-healing of carbon-fiber/epoxy composites. J. Appl. Polym. Sci. **129**, 1383–1393 (2013)

9. D. Li, Y. Xia, Direct fabrication of composite and ceramic hollow nanofibers by electrospinning. Nano Lett. **4**, 933–938 (2004)
10. J. Lee, C. Boo, W.H. Ryu, Development of omniphobic desalination membranes using a charged electrospun nanofiber scaffold. Acs. Appl. Mater. Inter. **8**, 11154–11161 (2016)
11. Z. Liu, H. Wang, E. Wang, Superhydrophobic poly (vinylidene fluoride) membranes with controllable structure and tunable wettability prepared by one-step electrospinning. Polymer **82**, 105–113 (2016)
12. F. Liu, N.A. Hashim, Y. Liu, Progress in the production and modification of PVDF membranes. J. Membrane. Sci. **375**, 1–27 (2011)
13. M. Wang, D. Fang, N. Wang, Preparation of PVDF/PVP core–shell nanofibers mats via homogeneous electrospinning. Polymer **55**, 2188–2196 (2014)
14. M. Sethupathy, V. Sethuraman, P. Manisankar, Preparation of PVDF/SiO_2 composite nanofiber membrane using electrospinning for polymer electrolyte analysis. Soft Nanoscience Letters. **3**, 37–48 (2013)

Uniformly Polyaniline-decorated Carbon Nanofibers as Active Materials for Improved Supercapacitor Properties

Xiaoxiao Yang, Guang Li and Yang Wang

Abstract As electrode material for supercapacitor, the conducting polyaniline (PANI) and carbon nanofibers (CNFs) formed nanocomposites have also been studied, but how to load stable polyaniline on carbon fiber surface is also difficult. In this paper, aniline and CNFs were prepared in situ emulsion polymerization, using hydrochloric acid (HCl) and sodium dodecyl sulfate (SDS) as dopant, sodium dodecyl sulfate (SDS) can be as an emulsifier at the same time. SDS as an emulsifier is to strong the dispersibility and stability of PANI on the surface of CNFs and further to improve the specific capacitance and cycle life of nanocomposite. The measurements showed that the PANI uniformly coating had grown on the surface of CNFs. The CNFs/PANI nanocomposites and PANI were as electrode materials. Electrochemical analysis illustrated the specific capacitance of the CNFs/PANI nanocomposites was much better than pure materials. The high specific capacitance of 588 F/g at a current density of 0.5 A/g is due to the special structure (much better than the pure PANI) and capacitance retention of 55% after 2000 CD cycles at a current density of 10 A/g (also better than the pure PANI), indicating the electrode material is potential for supercapacitors.

Keywords Emulsion polymerization · Polyaniline · Carbon nanofiber Nanocomposites · Supercapacitors · Specific capacitance · Electrode material

X. Yang · G. Li (✉) · Y. Wang
State Key Laboratory for Modification of Chemical Fibers and Polymer Materials, College of Materials Science and Engineering, Donghua University, Shanghai, China
e-mail: lig@dhu.edu.cn

X. Yang
e-mail: yangxiaoxiao0812@163.com

Y. Han (ed.), *Advanced Functional Materials*,
https://doi.org/10.1007/978-981-13-0110-0_77

Introduction

In the past few years, the global economy has got a rapid development, electronic technology has also improved. Under the pressure of the resource exhaustion and the environmental contamination urgently, developing some sustainable energy storage is extremely urgent, for example fuel cells, supercapacitors and so on [1]. The higher power densities and longer cycle life of supercapacitors is better than secondary batteries and its energy densities is higher than traditional capacitors. Therefore, supercapacitors have got much attention and have been applicated in many aspects [2].

Depend on the difference of charge storage mechanism, supercapacitors have two forms: electrical double-layer capacitors (EDLCs) and pseudo-capacitors [3]. The positive and negative electrodes of are stored energy by forming a double layer between the electrode surface and electrolyte and occurs on the surface of electrode material, materials are generally carbon-based materials. Different from EDLCs, in pseudo-capacitors, energy storage is thorough fast reversible redox reactions of materials and occurs in the interior of material, materials are generally transition metal oxides and conducting polymers [4].

Carbon-based materials, such as carbon fibers, have displayed favorable flexibility and hence been promising as freestanding and flexible capacitors. Especially, electrospun carbon-based nanofibers (such as CNFs) have been widely used in electrochemical fields. Its low electrical resistivity, high porosity, high stability and their high surface area, which facilitate the charge transfer.

However, because of the limited electrical charge separation between electrolyte materials and electrode materials interface, the CNFs have low energy density, which limits their application [5]. Many methods were used to ameliorate its low energy density, such as by controlling diameter to increase the surface areas [6], introduce metallic cation on the surface [7] or dope nonmetallic elements [8], or induce some pseudocapacitive materials [9–11]. Compared these methods, inducing pseudocapacitive materials is more effective, combining high power density of CNFs and high energy density of pseudocapacitive material can obtain higher performance.

PANI (Polyaniline), one conducting polymer, is of high specific capacitance and rate capability, which has been an excellent pseudocapacitive material [12, 13]. PANI-based electrodes, materials have been studied by many researchers. For example, Chaudhari et al. using electrospinning technique to synthesize PANI nanofibers, which showed a specific capacitance of 267 F/g at a current density of 0.35 A/g [14]. However, without polymer nanofibers as supports, the PANI were so brittle that its freestanding electrodes were fit for applications in supercapacitors. Therefore, some researchers combined the advantage properties of CNFs and PANI. To achieve the better specific capacitance, the PANI mass loading should be strictly controlled and the uniformity are especially essential, which was helpful to improve the capacitance of materials [15, 16]. For example, Jyongsik Jang et al. fabricate Polyaniline coated carbon adopted one-step vapor deposition polymerization technology and possessed excellent capacitance performance with a value of 264 F/g [17]. Then Yongliang Cheng et al. successfully prepared the cross-linked carbon nanofibers by situ polymerization, indicating a specific capacitance of 175 F/g with an excellent rate capability and cycling stability [18]. Fujun Miao et al. prepared polyacrylonitrile@polyaniline core-shell nanofibers combined electrospinning technology show 346 F/g at a scan rate of 5 mV/s [19].

In this paper, we used ammonium persulfate as an oxidizing agent and through a simple emulsion polymerization of aniline monomers with the CNFs. It was worth nothing that during situ polymerization we introduced dodecyl sulfate (SDS) as an anionic surfactant form doped macromolecular functional protonic acid on CNFs chain acted as a kind of stable membranes for PANI, the generated polyaniline particles can be stabilized by electrostatic repulsion or embedded in SDS, so inducing PANI to disperse uniformly and to form regular structure on the surface of CNFs. The obtained nanocomposites had relatively stable structure maintained the pristine flexibility of CNFs and displaying improved electrical conductivity and electrochemical performances compared with CNFs. Meanwhile, it had much higher specific capacitance.

Experimental

Materials and Instrumentation. Aniline was purified, ammonium persulfate (APS), hydrochloric (HCl), SDS (sodium dodecyl sulfate) N, N-dimethyl formamide (DMF), Polyacrylonitrile (PAN, Mw = 150,000), distilled water, Acetone.

Preparation of CNFs Networks. Carbon nanofibers(CNFs) were formed by electrospinning technique 10 wt% polyacrylonitrile (PAN) in the N, N-dimethylacetamide (DMF) was as spinning solution, the solution was heated and dissolved at the degree of 50 °C and stirred for several hours. After the solution cooled, poured it into 10 ml plastic syringe. The spinning speed was 0.9 ml/h. The electrospinning voltage and distance respectively were 18 kV and 18 cm. Then the obtained films were placed in a vacuum oven to remove solvent. The temperature of pre-oxidation was 280 °C, continued for 1 h and followed by carbonization at 1400 °C.

Preparation of CNFs-PANI Nanocomposites. In the process of synthesis, 0.034 g of SDS was dissolved in 4.5 ml of 1 M HCl solution and stirred, then add certain amount of CNFs controlling the rate of CNFs/PANI respectively are 1:1, 1:3, 1:5 and 1:7 and followed by sonicating for 1 h. Subsequently, adding 1.2 mmol of aniline (0.27 M) to the CNFs-SDS solution and sonicated for 1 h. In another 20 ml beaker, dissolving 1.8 mmol of ammonium persulfate (0.4 M) in 4.5 ml of 1 M HCl and added APS solution to the CNF-SDS-aniline dispersion, stirred for 8 h in the condition of ice water. Then the obtained solution was deposited in acetone, followed by filtration and lavation with large amounts of water and acetone. Finally, the obtained hybrid materials were dried at 80 °C for 24 h. Then labeling final samples are as CNFs/PANI-1, CNFs/PANI-3, CNFs/PANI-5 and CNFs/PANI-7, respectively.

General Characterization. FT-IR analysis were performed with a Nicolet Nexus FT-IR spectrometer (model: iS10, Minnesota). XRD were obtained on a PHI-5400 X-ray diffractometer (US). Morphological studies were carried out using Hitachi S-4800FE-SEM (Hitachi, Tokyo, Japan), The field-emission transmission electron microscope (FE-TEM) was performed with JEM-2100F (JEOL LTD, Japan). Thermogravimetric analysis (TGA) were performed on TA Instruments TG 209 F1 Iris (Germany) and the heating rate was 20 °C/min, which was under a nitrogen atmosphere. In every general characterization, we selected CNFs/PANI-5 to analyze during all of composites which due to its better electrochemical performance.

Electrochemical Characterization. The electrochemical performances of the materials were measured by symmetrical three-electrode system in 1 M H_2SO_4 aqueous electrolyte. The active materials on the electrolyte was 2 mg/cm^2, where the obtained materials were used as working electrodes, and saturated calomel electrode (SCE) and platinum wire were used as reference and counter electrodes, respectively. The voltage window was −0.2 to 1.0 V. The scanning rate was consecutively 10, 30, 50, 100, 150 mV/s. and measured at room temperature.

Results and Discussion

Structural Morphological and Composition Characterization. FE-SEM and TEM images of pure CNFs and CNFs/PANI-5 hybrid materials are shown in Fig. 1. AS shown in Fig. 1a, the diameters of CNFs were measured to be about 150 nm. Figure 1b, c showed the presence of PANI, whose diameters were about 300–400 nm. And the two parts of PANI and CNFs were marked in Fig. 1c (TEM images of CNFs/PANI-5 nanomaterials). Figure 1d was the EDS of CNFs/PANI-5 and showed that different kinds of elements uniformly had dispersed on the surface of CNFs, which indicated the growth of uniform PANI lamellar on the CNFs. In the process, SDS played an important role during the formation of thick and lamellar morphology of PANI. And the process was shown in Scheme 1. SDS was as the bridge connecting CNFs and PANI and as an anionic surfactant, forming doped macromolecular functional protonic acid on CNFs chain. The generated polyaniline particles can be stabilized by electrostatic repulsion or embedded in SDS. Therefore PANI can exist stably and disperse uniformly on the surface of CNFs [20].

FT-IR spectral analysis (Fig. 2) was introduced to verify the structure of hybrid materials. Spectral analysis indicates that several spectral bands of CNFs/PANI hybrid composites and PANI are matched. 802 cm^{-1} represents 1,4-disubstituted benzene, 1630 cm^{-1} represents –SO_3H group of SDS, the spectral bands of 1130, 1300, and 1580 cm^{-1} are the three kinds of chemical bonds in quinoid ring, respectively represent N=Q=N vibration(quinoid ring), C=N and C=C stretching of

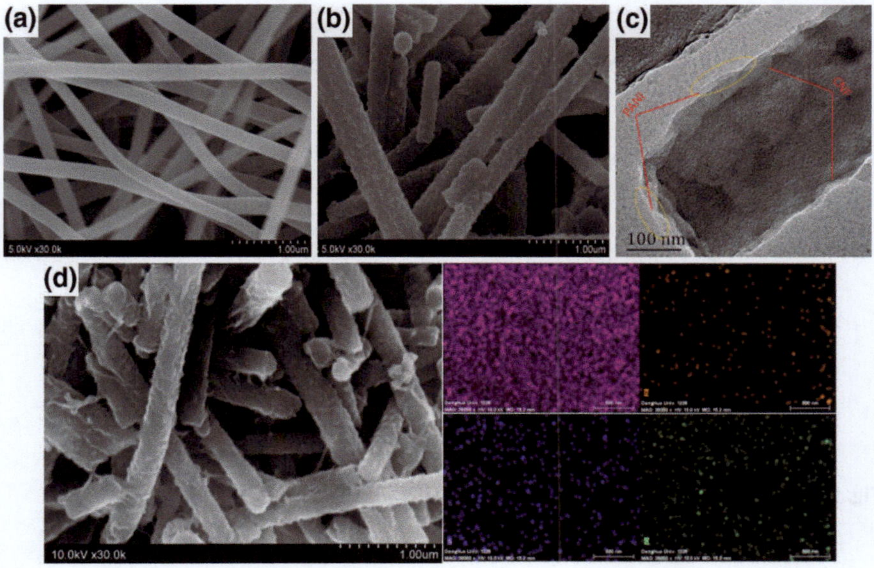

Fig. 1 **a** SEM image of CNFs networks; **b**, **c** TEM images of CNFs/PANI-5; **d** EDS images of CNFs/PANI-5

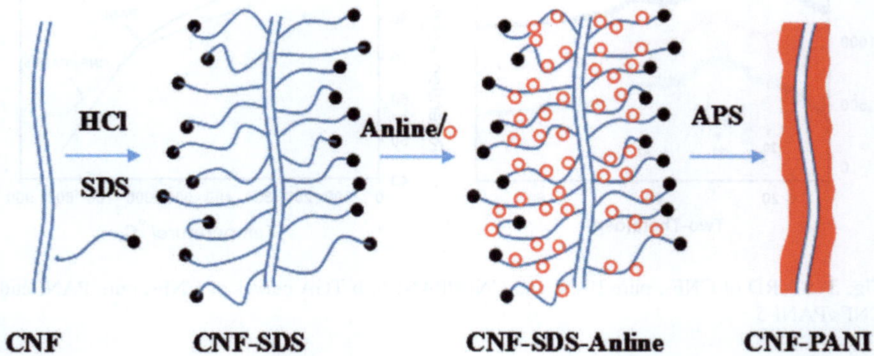

Scheme 1 The process of synthesizing CNFs/PANI nanocomposite in the presence of SDS

quinoid ring 1260 and 1490 cm^{-1} represent C=N and C=C stretching of benzenoid ring, 2920 and 2850 cm^{-1} represent –CH2– stretching of SDS, 3440 cm^{-1} represents N–H stretching. The bands of 1290, 1487 and 1576 cm^{-1} in PANI are successively moved to 1300 1490, 1580 cm^{-1} in CNFs/PANI hybrid composites, at the same time the intensity of peaks occur decrease, suggesting that the aniline had successfully been polymerized and coated on the surface of CNFs by π–π interactions [21]. Compared to pure PANI, the band of CNFs/PANI at 3440 cm^{-1} corresponds to the N–H stretching vibration, meaning PANI layers were well wrapped on CNFs.

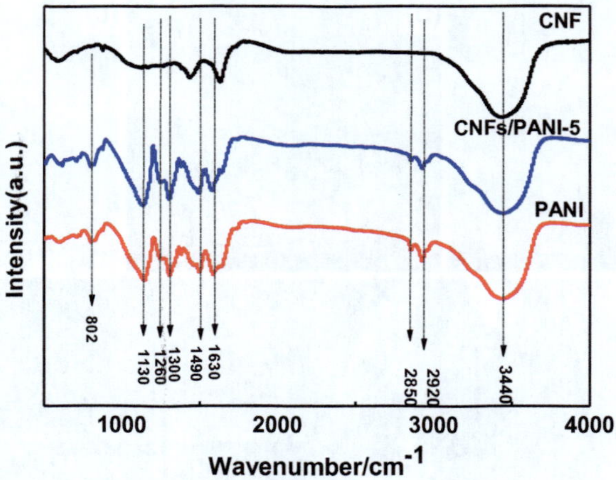

Fig. 2 FT-IR spectra of CNFs, pure PANI and CNFs/PANI-5

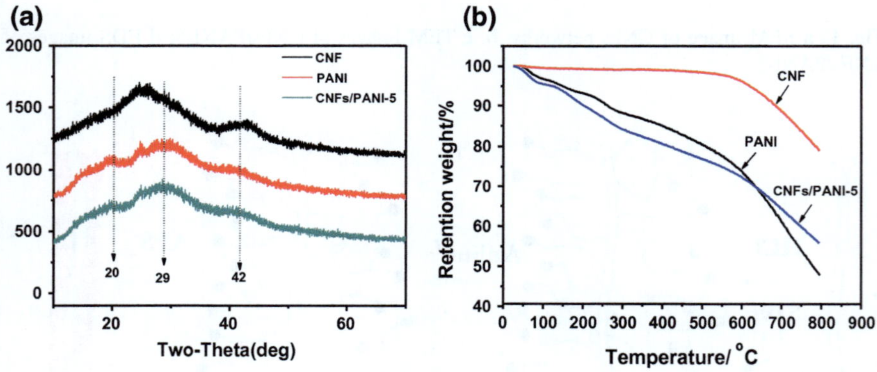

Fig. 3 **a** XRD of CNFs, pure PANI and CNFs/PANI-5; **b** TGA curves of CNFs, pure PANI and CNFs/PANI-5

Typical XRD analysis of pure PANI, CNFs, and the CNFs/PANI-5 nanomaterials are shown in Fig. 3a. The PANI sample shows diffraction peaks at $2\theta = 20°$, 29° and 42° are attributed to the different crystal planes of PANI, respectively [22]. The series of hybrid materials also show diffraction peaks at $2\theta = 20°$, 29° and 42°, which indicate that PANI are coated on the surface. Compared to the CNFs and the CNFs/PANI-5 nanomaterials, the peak of CNFs/PANI-5 peak shift right due to the corporate effect between the PANI and CNFs. Combine the FT-IR spectral analysis, we know they are π–π bonds. The special diffraction peaks of nanocomposites reflect out the excellent corporation between the PANI and CNFs. Therefore, XRD analysis indicate that PANI are coated on the surface of CNFs.

The TGA curves (Fig. 3b) of samples indicate that before 534 °C, the mass of CNFs basically remains unchanged. From 534 to 800 °C, the CNFs suffers 22% weight loss. Pure PANI and CNFs/PANI-5 hybrid materials show three different stages. The first step the samples has got weight loss of about 6% before 150 °C, which is connected with the loss of moisture. From 150 to 600 °C, the dopant molecules and some oligomers are removed lead to the weight loss of 35–45%. Then a fast weight loss is observed for the samples from 600 °C. The TGA curves also show that the addition of CNFs improved the thermal stability of CNFs/PANI-5 nanocomposite, which proves that CNFs is beneficial to strengthen the stability of PANI.

Electrochemical Characterization. The electrochemical performances of PANI and CNFs/PANI are analyzed by CV and GCD tests. For three electrode systems, the specific capacitance of PANI/CNFs nanocomposite were calculated according to Eq. (1) from the CV curves at different scan rates and Eq. (2) from the GCD curves at different current density Where C_{sp} (F/g) represents the specific capacitance, S represents the acreage of CV curves, V represents scan rate, ΔU represents voltage window, M represents the mass of the active electrode materials. I represents the current density, t represents the discharge time.

$$C_{SP} = \frac{S}{2V\Delta UM} \tag{1}$$

$$C_{sp} = \frac{It}{M\Delta U} \tag{2}$$

The scanning rate used in the CV curves (Fig. 4a) of all samples was 50 mV/s. The CV curves show two pairs of redox peaks (0.6/0.35 V) and (0.4/0.05 V), which are due to the interconversion of polyaniline in different valence states. Compared to the area enclosed by the rectangle of other samples, we can know that CNFs/PANI-5 have the largest area, which indicates that the hybrid materials are of excellent capacitance performance. The GCD curves (Fig. 4b) was measured at the current density of 0.5 A/g. Curves exhibit symmetrical and triangular in shape. Owing to combination CNFs and PANI, the discharge curves show two different slopes in the ranges of 1.0–0.6 and 0.6 to –0.2 V [23]. It is known by the discharge time of samples, the discharge time of CNFs/PANI-5 is longest among them, which indicates that the material has excellent specific capacitance. The specific capacitance of all samples were calculated from Eq. (2), CNFs/PANI-5 shows a high specific capacitance of 588 F/g at the current density of 0.5 A/g, however, the specific capacitance of pure PANI is 371 F/g (Fig. 4b).

Another important performance that decides electrodes materials for practical application is the cycle life. Figure 5d shows that, 55% of the initial capacitance is retained after 2000 cycles at the current density of 10 A/g, which is better than pure PANI system (46%). The specific capacitance of CNFs/PANI-5 nanocomposite has improved by 58.5% compared to PANI and good stability of materials depends on

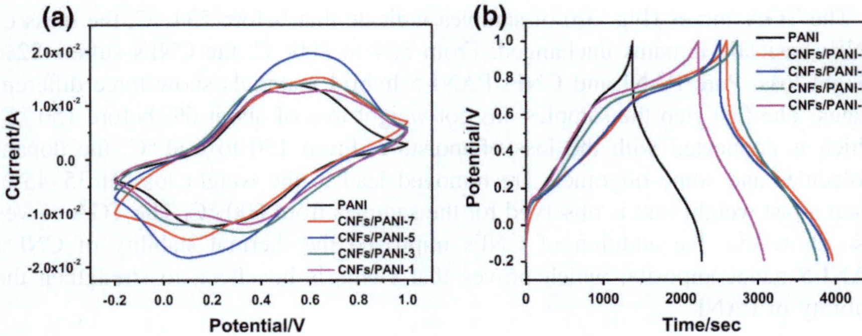

Fig. 4 **a** CV curves of pure PANI, CNFs/PANI-1 to CNFs/PANI-7 measured at a scan rate of 50 mV/s; **b** GCD curves of pure PANI, CNFs/PANI-1 to CNFs/PANI-7 measured at a current density of 0.5 A/g

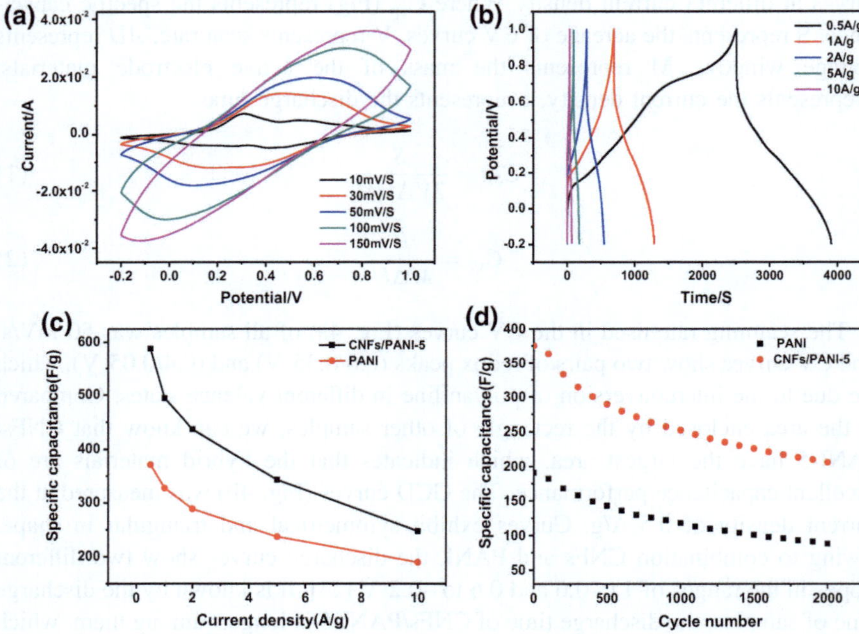

Fig. 5 **a** CV curves of CNFs/PANI-5 composite electrode measured at different scan rates; **b** GCD curves of CNFs/PANI-5 composite electrode measured at different current densities; **c** Plots of specific capacitance for CNFs/PANI-5 and pure PANI at different current densities; **d** Cycle life of CNFs/PANI-5 and pure PANI

its unique package structure, among the structure, the matric material for PANI is CNFs, PANI uniformly distributed on its surface increases the specific surface, which provides higher electroactive area, shorten diffusion lengths and reduced ion-transport barries. The introduction of PANI provides a high conductive pathway, so quicken electrolyte ion penetration. If we can further increase the period stability of this system by coating some materials to prevent the degradation of PANI, therefore CNFs/PANI-5 is one of quite ideal electrode materials in supercapacitors.

Conclusions

CNFs/PANI nanocomposites were fabricated by a simple and efficient approach, in this study, we used electrospinning followed by situ emulsion polymerization method and SDS as an emulsifier. SDS as the bridge between CNFs and PANI can induce PANI films to be wrapped uniformly on the the surface of CNFs, exhibiting excellent capacitive performance, the best special capacitance is 588 F/g and 55% of the initial capacitance is retained after 2000 charge/discharge cycles, which is better than that of the pure PANI. The improved electrode performance is owing to its special structure PANI nanofiber morphology and the synergistic effect between CNFs and PANI. In addition, the period stability of this system also needs to be further increased. These results demonstrate that CNFs/PANI-5 is one of quite ideal electrode materials in supercapacitors.

References

1. G. Wang, L. Zhang, J. Zhang, A review of electrode materials for electrochemical supercapacitors. Chem. Soc. Rev. **41**, 797–828 (2012)
2. H.L. Ferreira, R. Garde, G. Fulli et al., Characterisation of electrical energy storage technologies. Energy **53**, 288–298 (2013)
3. W. Wei, X. Cui, W. Chen, D.G. Ivey, Manganese oxide-based materials as electrochemical supercapacitor electrodes. Chem. Soc. Rev. **40**, 1697–1721 (2011)
4. B.H. Kim, N.N. Bui, K.S. Yang, M.E.D. dela Cruz, J.P. Ferraris, Bull. Korean Chem. Soc. **30**, 1967–1972 (2009)
5. G.H. Yu, X. Xie, L.J. Pan, Z.N. Bao, Y. Cui, Hybrid nanostructured materials for high performance electrochemical capacitors. Nano Energy **2**, 213e34 (2013)
6. P.G. He, L. Liu, W.X. Song, G.P. Xiong, T.S. Fisher, T.F. Chen, Large-scale synthesis and activation of polygonal carbon nanofibers with thin ribbon-like structures for supercapacitor electrodes. RSC Adv. **5**, 31837e44 (2015)
7. X. Chen, B.T. Zhao, Y. Cai, M.O. Tade, Z.P. Shao, Amorphous VeOeC composite nanofibers electrospun from solution precursors as binder- and conductive additive-free electrodes for supercapacitors with outstanding performance. Nanoscale **5**, 12589e97 (2013)
8. L.F. Chen, X.D. Zhang, H.W. Liang, M.G. Kong, Q.F. Guan, P. Chen et al., Synthesis of nitrogene doped porous carbon nanofibers as efficient electrode materials for supercapacitors. ACS Nano **6**, 7092e102 (2012)
9. A.D. Jagadale, V.S. Kumbhar, R.N. Bulakhe, C.D. Lokhande, Influence of electrode position modes on the supercapacitive performance of Co_3O_4 electrodes. Energy **64**, 234e41 (2014)
10. G.P. Xiong, K.P.S.S. Hembram, R.G. Reifenberger, T.S. Fisher, MnO_2-coated graphitic petals for supercapacitor electrodes. J. Power Sources **227**, 254e9 (2013)

11. Z. Shao, H.J. Li, M.J. Li, C.P. Li, C.Q. Qu, B.H. Yang, Fabrication of polyaniline nanowire/ TiO_2 nanotube array electrode for supercapacitors. Energy **87**, 578e85 (2015)
12. D.D. Xu, Q. Xu, K.X. Wang, J. Chen, Z.M. Chen, Fabrication of free-standing hierarchical carbon nanofiber/graphene oxide/polyaniline films for supercapacitors. ACS Appl. Mater. Inter. **6**, 200e9 (2014)
13. S.J. He, X.W. Hu, S.L. Chen, H. Hu, M. Hanif, H.Q. Hou, Needle-like polyaniline nanowires on graphite nanofibers: hierarchical micro/nano-architecture for high performance supercapacitors. J. Mater. Chem. **22**, 5114e20 (2012)
14. S. Chaudhari, Y. Sharma, P.S. Archana, R. Jose, S. Ramakrishna, S. Mhaisalkar, Electrospun polyaniline nanofibers web electrodes. J. Appl. Polym. Sci. **129**, 1660–1668 (2013)
15. J.Y. Wei, J.N. Zhang, Y. Liu, G.H. Xu, Z.M. Chen, Q. Xu, Controlled growth of whisker like polyaniline on carbon nanofibers and their long cycle life for supercapacitors. RSC Adv. **3**, 3957e62 (2013)
16. M. Dirican, M. Yanilmaz, X.G. Zhang, Free-standing polyaniline-porous carbon nanofiber electrodes for symmetric and asymmetric supercapacitors. RSC Adv. **4**, 59427e35 (2014)
17. J. Jang, J. Bae, M. Choi, S.H. Yoon, Fabrication and characterization of polyaniline coated carbon nanofiber for supercapacitor. Carbon **43**, 2730e6 (2005)
18. D.Y. Lei, K. Devarayan, M.K. Seo, Y.G. Kim, B.S. Kim, Flexible polyaniline-decorated carbon fiber nanocomposite mats as supercapacitors. Mater. Lett. **154**, 173e6 (2015)
19. F.J. Miao, C.L. Shao, X.H. Li, N. Lu, K.X. Wang, X. Zhang et al., Flexible solid-state supercapacitors based on freestanding electrodes of electrospun polyacrylonitrile@Polyaniline core-shell nanofibers. Electrochim Acta **176**, 293e300 (2015)
20. B.J. Kim, S.G. Oh, M.G. Han et al., Preparation of polyanline nanoparticles in micellar solutions as polymerization medium. Langmuir **16**(4), 5841–5845 (2000)
21. S.S. Bal, S. Palaniappan, R.Y. Kuk et al., Emulsion polymerization method for polyanline-multiwalled carbon nanotube nanocomposites as supercapacitor materials. J. Solid State Electrochem. **20**, 3447–3457 (2016)
22. F. Miao, C. Shao, X. Li et al., Polyanline-coated electrospun carbon nanofibers with high mass loading and enhanced capacitive performances freestanding electrodes for flexible solid-state supercapcacitors. Energy **95**, 233–241 (2016)
23. L. Liu, Y. Wang, Q. Meng et al., A novel hierarchical graphene/polyaniline hollow microsphere as electrode material for supercapacitor applications. J. Mater. Sci. **52**, 7969 (2017)

Rapid Deposition of Al_2O_3/SiO_2 Composite Coating on Carbon Fiber Fabric

Yuping Zhang, Yonghua Shen, Weiwei Chen, Huanwu Cheng and Lu Wang

Abstract In the paper, a uniform and continuous Al_2O_3/SiO_2 composite coating was prepared on each fiber of carbon fiber fabric via room-temperature aqueous plasma electrolysis technique. The micromorphology and anti-oxidation property of the Al_2O_3/SiO_2-coated carbon fiber fabric were systematically investigated. The dense and continuous Al_2O_3/SiO_2 composite coating was uniformly deposited on each fiber of carbon fiber fabric. The thickness of coating was 120–150 nm. The Al_2O_3/SiO_2 composite coating could protect the carbon fiber fabric from oxidation above 1000 °C. In addition, it could be seen 300 m carbon fiber fabric were coated continuously by high-quality Al_2O_3/SiO_2 composite coating.

Keywords Room-temperature aqueous plasma electrolysis · Al_2o_3/SiO_2 coating Carbon fiber fabric · Oxidation resistance

Introduction

Carbon fiber fabric is a novel textile material, possessing outstanding mechanical properties from its three-dimensional structure, to lay the foundation for its application in industrial, civil and aerospace fields [1–4]. However, the poor oxidation

Y. Zhang · Y. Shen · W. Chen (✉) · H. Cheng · L. Wang
Department of Materials Science and Engineering,
Beijing Institute of Technology, Beijing 100081, China
e-mail: wwchen@bit.edu.cn

Y. Zhang
e-mail: stu_zhangyuping@126.com

Y. Shen
e-mail: shenyh1990@163.com

H. Cheng
e-mail: chenghuanwu@bit.edu.cn

L. Wang
e-mail: wanglu@bit.edu.cn

Y. Han (ed.), *Advanced Functional Materials*,
https://doi.org/10.1007/978-981-13-0110-0_78

resistance of carbon fiber significantly limits the application of the carbon fiber fabric [5, 6]. Therefore, it is an urgent demand to seek an efficient method for the protection of carbon fiber fabric.

A protective coating is most effective in improving the anti-oxidation of carbon fiber. At present, the methods prepared coating on carbon fiber fabric mainly include pyrolysis [7], sol-gel technology [8], dip coating [9], and chemical vapor deposition [10, 11]. Obviously the typical methods provide a technological basis for preparing coating on carbon fiber fabric surface, whereas it seems rather difficult to continuously deposit coating on carbon fiber fabric surface by these methods. In addition, it is also difficult to deposit a coating onto carbon fiber bundle by the above-mentioned method, as thousands of fibers are included in a carbon fiber fabric. Moreover, the traditional methods especially CVD easily destroy the carbon fiber fabric, significantly declining the strength of the carbon fiber fabric. Therefore, there is an urgent demand to develop a new technique for continuously depositing protective coating onto each fiber of carbon fiber fabric.

We have recently developed a novel technique, i.e. room-temperature aqueous plasma electrolysis process for the protection of carbon fiber fabric [12], in order to fast deposit Al_2O_3/SiO_2 composite coating onto each fiber of carbon fiber fabric. The microstructure and anti-oxidation of Al_2O_3/SiO_2-coated carbon fiber fabric were investigated in the present manuscript. It was shown that the Al_2O_3/SiO_2 coating exhibited superior oxidation resistance at elevated temperatures.

Experimental

The room-temperature plasma electrolysis. The carbon fiber fabric with the width of 400 mm and the length of 200 m were used. Figure 1 shows the device diagram of preparing coating on carbon fiber fabric via the room-temperature plasma electrolysis. Carbon fiber fabric was firstly put out from the paying out machine. Then, the carbon fabric passed through cathodic plasma arc zone generated by copper tube, where composite coating was deposited onto carbon fiber fabric surface. The coated carbon fiber fabric was finally coiled onto the reel driven by take-up machine. The electrolyte solution was a mixture of 50 g/L $NaAlO_2$ and 1 g/L $Na_2SiO_3.9H_2O$ aqueous solution. A high frequency switching DC power (TN-KGZ01) was applied with the voltage of 155 V.

Characterization of composite coating. The morphology of Al_2O_3/SiO_2 composite coating was characterized using a scanning electron microscope (SEM). The oxidation resistance of Al_2O_3/SiO_2 composite coating was analyzed using thermo gravimetric analyzer (TGA) under an air atmosphere. The temperature rose from room temperature to 1300 °C at 10 K/min.

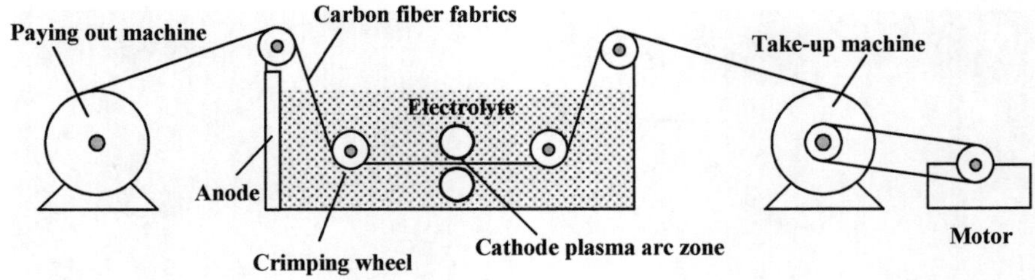

Fig. 1 The device diagram of preparing coating onto carbon fiber fabric by room-temperature aqueous plasma electrolysis

Results and Discussion

Surface and cross-sectional morphologies. Figure 2 shows morphology of Al_2O_3/SiO_2 composite coating. As shown the surface backscattered morphology in Fig. 2a, each fiber of carbon fiber fabric was almost coated by a relatively dense, uniform and continuous composite coating. Figure 2b shows the high magnified microstructure of carbon fiber fabric surface. It can be clearly seen that grooves of carbon fiber surface filled with uniform and thin composite coating.

Figure 3a shows the cross-sectional morphology of carbon fiber fabric. The edge of the cross section was locally magnified with the morphology shown in Fig. 3b, in order to identify the thickness of Al_2O_3/SiO_2 coating. It is clearly observed that the thickness of Al_2O_3/SiO_2 composite coating on carbon fiber fabric was 120–150 nm, as the arrows shown in Fig. 3b.

Oxidation resistance. Figure 4 shows thermogravimetry (TG) and differential thermal analysis (DTA) curves of bare and the Al_2O_3/SiO_2-coated carbon fiber

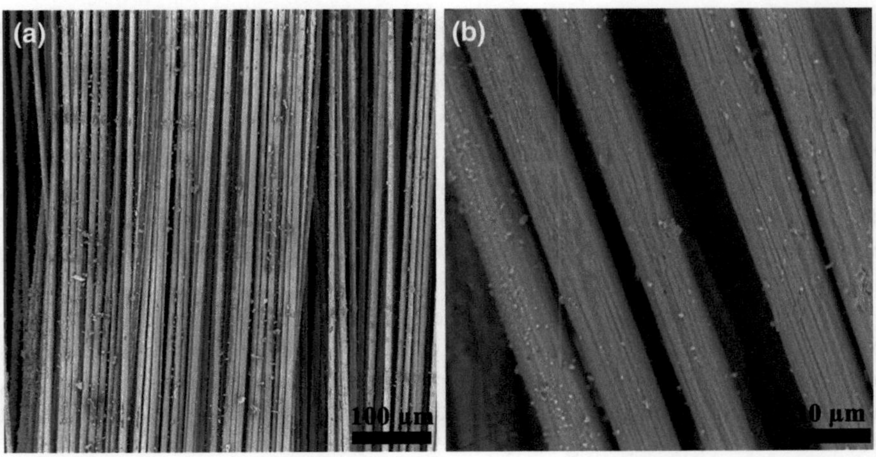

Fig. 2 Surface morphology of the Al_2O_3/SiO_2 composite coating on carbon fiber fabric: **a**, **b** SEM morphologies under different magnification

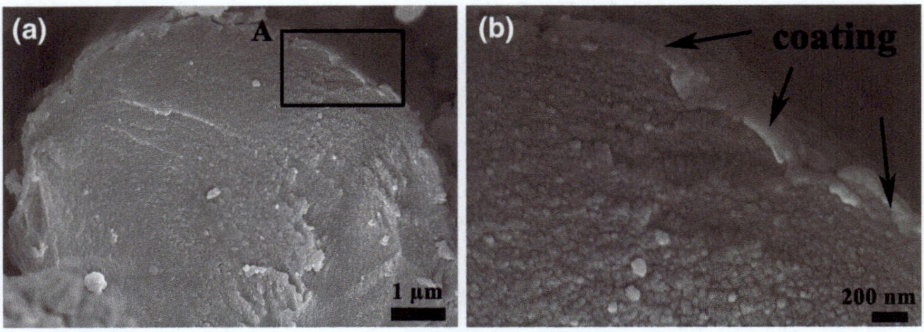

Fig. 3 Cross-sectional morphologies of Al_2O_3/SiO_2 nano-coating on the carbon fiber fabric

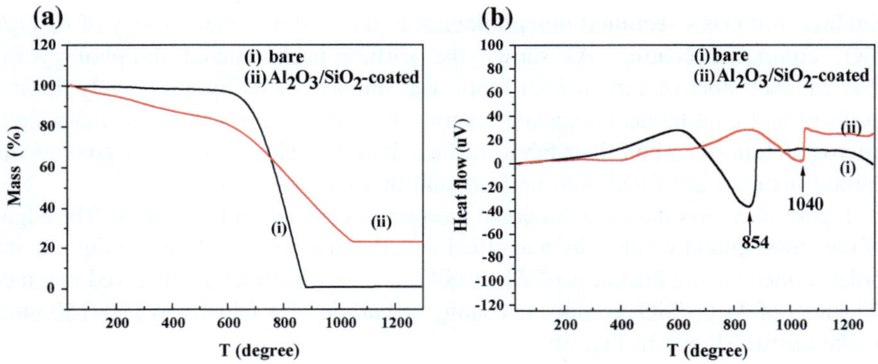

Fig. 4 Mass (**a**), and heat flow (**b**) of bare carbon fiber fabric (i) and the Al_2O_3/SiO_2-coated carbon fiber fabric (ii) at different temperatures

fabric. Followed by the 600 °C for bare sample, a sharp decline was observed from TG curve of the bare sample till 880 °C where carbon fiber was burnout, compared to a bit slower decline for the Al_2O_3/SiO_2-coated carbon fiber fabric till 1050 °C (Fig. 4a). At the elevated temperatures, the protection of the Al_2O_3/SiO_2 coating was much more effective, evidenced by the slower oxidation rate and significant increase of the burn-out temperature from 880 to 1050 °C. Similarly, from Fig. 4b, it could be also obtained that the Al_2O_3/SiO_2 coating hindered the oxidation of the carbon fiber fabric, evidenced by the exothermic peak postponement from 854 to 1040 °C.

However, a slight mass loss was observed for bare carbon fiber fabric below 600 °C, compared to Al_2O_3/SiO_2-coated carbon fiber fabric, which started a mass loss from 50 °C (Fig. 4a). It might be attributed to dehydration of residual H_2SiO_3 and $Al(OH)_3$ in the coating, which was demonstrated by the endothermic peak at 400–600 °C, as shown in Fig. 4b (ii).

The reactions involved in the plasma electrolysis process are suggested as below:

(i) Hydrolysis of Na_2SiO_3

$$Na_2SiO_3 + H_2O \rightarrow H_2SiO_3 + NaOH \quad (1)$$

$$Al(NO_3)_3 + H_2O \rightarrow Al(OH)_3 + HNO_3 \quad (2)$$

(ii) Dehydration of the H_2SiO_3 and $Y(OH)_3$

$$H_2SiO_3 \xrightarrow{\textit{Plasma discharge}} SiO_2 + H_2O \quad (3)$$

$$Al(OH)_3 \xrightarrow{\textit{Plasma discharge}} Al_2O_3 + H_2O. \quad (4)$$

Conclusions

A dense, uniform and continuous Al_2O_3/SiO_2 composite coating deposited onto each fiber of carbon fiber fabric via room-temperature plasma electrolysis. The surface morphology and anti-oxidation were investigated. The thickness of Al_2O_3/SiO_2 composite coating on carbon fiber fabric surface was 120–150 nm. The Al_2O_3/SiO_2 composite coating could protect carbon fiber fabric from oxidation above 1000°C. In addition, it could be seen 300 m carbon fiber fabric were coated by high-quality Al_2O_3/SiO_2 composite coating.

Acknowledgements We thank the Basic Research Program of Beijing Institute of Technology (Grant No: 20150942006) for the financial support. We also thank Dr. Mao yuan Li and Dr. Lin Lu for providing the carbon fiber fabric. In addition, two authors (Yuping Zhang and Yonghua Shen) contributed equally to this work, and should be co-first author.

References

1. H.J. Wang, P.Z. Gao, Z.H. Jin, Preparation and oxidation behavior of three-dimensional braided carbon fiber coated by SiC. Mater. Lett. **59**, 486–490 (2005)
2. H. Li, H. Liang, F. He, Y. Huang, Y.Z. Wan, Air dielectric barrier discharges plasma surface treatment of three-dimensional braided carbon fiber reinforced epoxy composites. Surf. Coat. Tech. **203**, 1317–1321 (2009)
3. A. Khoddamzadeh, R. Liu, M. Liang, Q. Yang, Wear resistant carbon fiber reinforced stellite alloy composites. Mater. Design **56**, 487–494 (2014)
4. L.M. Manocha, A. Warrier, S. Manocha, D.D. Edie, A.A. Ogale, Microstructure of carbon/carbon composites reinforced with pitch-based ribbon-shape carbon fibers. Carbon **41**, 1425–1436 (2003)

5. Y.J. Tong, X.Q. Wang, H. Su, L.H. Xu, Oxidation kinetics of polyacrylonitrile-based carbon fibers in air and the effect on their tensile properties. Corros. Sci. **53**, 2484–2488 (2011)
6. R. Gadiou, S. Serverin, P. Gibot, C. Vix-Guterl, The synthesis of SiC and TiC protective coatings for carbon fibers by the reactive replica process. J. Eur. Ceram. Soc. **28**, 2265–2274 (2008)
7. J.J. Wang, W.S. Lin, X. Wu, Y.Y. Yang, F. Wu, X.Z. Yan, Preparation and oxidation behavior of a double-layer coating for three-dimensional braided carbon fiber. Surf. Coat. Tech. **298**, 58–63 (2016)
8. J.J. Wang, W.S. Lin, X.Z. Yan, X. Wu, F. Wu, Y.Y. Yang, Preparation and microstructure of Al_2O_3-SiO_2-TiO_2 coating on three-dimensional braided carbon fiber by sol-gel technology. Mater. Des. **89**, 928–932 (2016)
9. M. Ghanbarian, E.T. Nassaj, A. Kariminejad, Synthesis of nanostructural turbostratic and hexagonal boron nitride coatings on carbon fiber cloths by dip-coating. Surf. Coat. Tech. **288**, 185–195 (2016)
10. A.Z. Abidin, R. Kozera, M. Höhn, I. Endlera, M. Knautc, A. Boczkowskab, A. Czulakd, P. Malczykd, N. Sobczake, A. Michaelisa, Preparation and characterization of CVD-TiN-coated carbon fibres for applications in metal matrix composites. Thin Solid Films **89**, 479–486 (2015)
11. A. Kariminejad, E.T. Nassaj, M. Ghanbarian, S.A.H. Tabrizi, Effects of PACVD parameters including pulsed direct current and deposition time on nanostructured carbon coating deposited on carbon fiber fabric. Mater. Des. **106**, 184–194 (2016)
12. W.W. Chen, Y.P. Zhang, Y. Meng, H.W. Cheng, L. Wang, A method and device to prepare the coating by copper tubes-assisted cathodic plasma electrolysis, C.N. Patent 201710285417.0 (2017)

Effect of Treatment Temperature on the Structure and Properties of Braiding Reinforced Thermoplastic Polyurethane Medical Hollow Fiber Tube for Invasive Medical Devices

Zhaomin Li, Minglin Qin, Shu Zhu, Muhuo Yu and Yunyun Xue

Abstract To improve mechanical properties of the metal mesh reinforced thermoplastic polyurethane elastomer (TPU) hollow fiber tube, we first discussed the effect of treatment temperatures on microphase separation of the TPU. The chemical microstructures of the TPU were characterized by X-ray diffractometer (XRD). The dynamic mechanical properties of the TPU were measured by DMA. The results exhibited two glass transition temperatures, which showed the distinct microphase separation. In addition, the mechanical property of hollow fiber tube was analyzed by flexural stress and twisting performance, and the mechanical properties of the metal mesh to enhance the hollow fiber tube was increased at different treatment temperature, which indicated that the microphase separation has a great influence on the mechanical properties of the metal mesh to enhance the hollow fiber tube.

Z. Li · M. Qin · S. Zhu · M. Yu (✉)
State Key Laboratory for Modification of Chemical Fibers and Polymer Materials, College of Materials Science and Engineering, Donghua University, Shanghai 201620, China
e-mail: yumuhuo@dhu.edu.cn

Z. Li
e-mail: zmli@microport.com

M. Qin
e-mail: 1139029@mail.dhu.edu.cn

S. Zhu
e-mail: zhushu@dhu.edu.cn

Z. Li · M. Qin
Shanghai Microport Medical (Group) Co., Ltd, Shanghai 201203, China

Y. Xue
School of Materials Science and Engineering, University of Shanghai for Science and Technology, Shanghai 200433, China
e-mail: yyxue@microport.com

Y. Han (ed.), *Advanced Functional Materials*,
https://doi.org/10.1007/978-981-13-0110-0_79

Keywords Treatment temperature · Thermoplastic polyurethane
Mechanical properties

Introduction

Thermoplastic polyurethane (TPU) is a linear multiblock copolymer comprised of alternating hard and soft segments that exhibit a unique combination strength, flexibility and processability due to their phase-separated microstructure [1–4]. It is one of the engineer thermoplastic that has been widely used in many fields such as automobiles, tubes, belts, adhesives and cardiovascular biomaterials [5–9]. Owing to its excellent physicochemical and biocompatibility, a great number of TPU scaffolds were fabricated by solvent casting/particle leaching, foaming, electrospinning, inkjet, salt leaching/freeze drying and thermally induced phase separation. It has been used in engineered tissues such as skin, cardiovascular implants, nerve conduits, cancellous bone graft substitutes, articular cartilage and so on. Meanwhile, TPU has made a great progress in the field of biomedicine due to its excellent comprehensive performances. However, with the rapid development of interventional medical devices, simple TPU tubing has been unable to meet the needs of medical patients, so metal mesh reinforced thermoplastic polyurethane elastomer hollow fiber tube has been involved in medical devices.

The hydrogen bonding between hard segment and its crystallization kinetics plays an important role on TPU phase separation [10]. Numerous methods have been studied for TPU phase separation, such as heat treatment, processing condition, the ratio of HS and SS (soft segment) and so on [11–13]. Heating treatment is one of the most important methods for TPU to change phase separation because TPU can lead to rearrangement of hydrogen crystallization kinetics [14–16]. However, up to now, there is no study on the effect of treatment temperature on phase separation of the mechanical property of the metal mesh reinforced thermoplastic polyurethane elastomer hollow fiber tube. In this communication, we discussed the influence of the treatment temperature on the crystal structure of the TPU. Furthermore, we also discussed the mechanical properties of the metal mesh reinforced thermoplastic polyurethane elastomer hollow fiber tube. The chemical microstructures of the TPU were investigated by X-ray diffractometer (XRD). The results showed that the microphase separation of the optimum treatment temperature were 120 °C. The dynamic mechanical properties of TPU was analyzed by DMA. The results exhibited two glass transition temperatures, which showed the distinct microphase separation. In addition, the mechanical properties of the hollow fiber tube were analyzed by fluxeral strength and twisting performance, the mechanical properties of the metal mesh enhanced hollow fiber tube increased with treatment temperature. This also indicated that the microphase separation has a great influence on the mechanical properties of the metal mesh enhanced hollow fiber tube.

Experimental

Materials. Thermoplastic polyurethane (Pellethane TPU 2363-75D, Shore hardness 75D, Melting point 204–221 °C) was supplied by the Lubrizol Corporation, USA. Metal mesh (304VSS) was purchased from Fort Wayne Corporation, USA.

Preparation of the Metal Mesh to Enhance the Hollow Fiber Tube. It can be seen from Fig. 1, the braid is stretched by a controllable tensioning device and preheated to the Tg and Tm of the coated polymer. After passing through the mandrel into the die, the coated polymer was fully filled in the die with an equal volume and uniform pressure, firmly bonded with the inner layer, leaving the cavity after the air medium to the surface of the annealing and melt relaxation, and then into the cooling water tank to obtain the metal mesh to enhance the hollow fiber tube.

Heat Treatment Method. The polymer particles were pressed into a sheet of 50 × 10 × 1 mm at 210 °C (Labtech Engineering Co., Ltd.), and then heat-treated in a constant-temperature blast dryer (DHG-9053A) at temperatures of 120, 130, 140, 150 °C for 30 min.

Characterizations

XRD Analysis. The crystalline of the TPU was monitored with an X-ray diffracometer (Rigaku Corporation) using Cu–Ka radiation at a generator voltage of 40 kV and current of 40 mA. The data were obtained from 2θ = 10 – 60° at a scanning speed of 2/min.

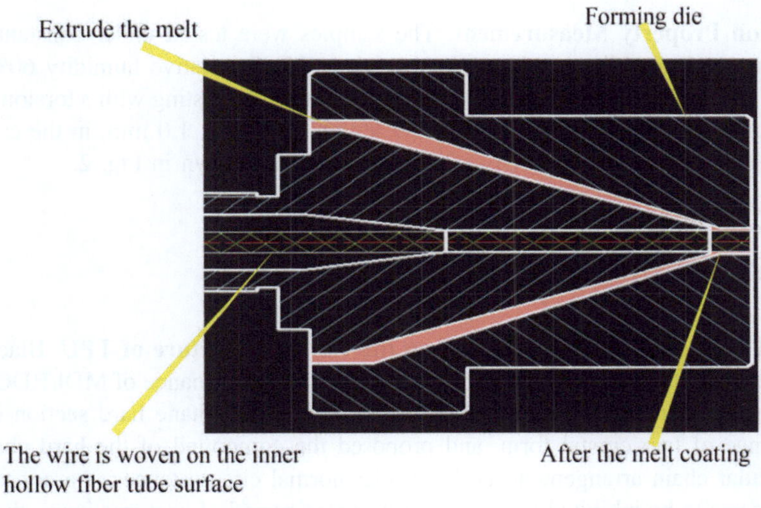

Fig. 1 Preparation of the metal mesh to enhance the hollow fiber tube

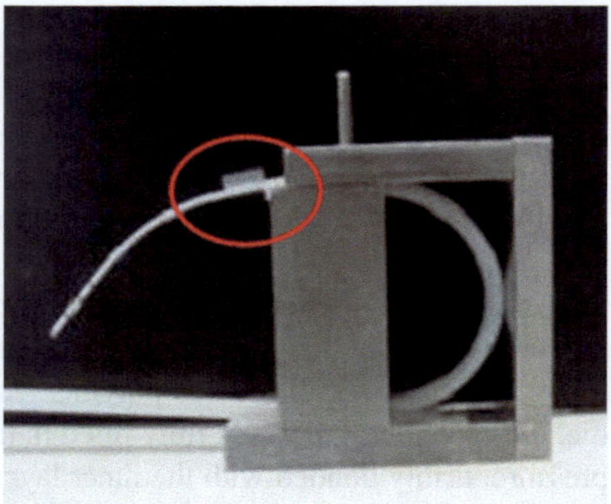

Fig. 2 Schematic diagram of torsion property measurement

Dynamical Mechanical Analysis (DMA). The TPU was cut to 50 × 10 × 1 mm. DMA was carried out using a DMA Q800 (TA Instruments) in the tension mode at a frequency of 1 Hz. The spectra were recorded under nitrogen atmosphere from 150 to 100 °C with a heating rate of 2 °C/min.

Three-point Bending Experiment. The test specimen for three-point bending experiment of mesh reinforced TPU elastomer hollow fiber tube (each length requires 100 ± 1 mm). Then keep these speciments balance for more than 48 h on the constant temperature and humidity box (25 °C, 60%). The test speed is 5 mm/min, and bending displacement is 4 mm. Select the 10 N sensor for the Instron Universal Tensioner (FM-02-003) (replace the 500 N sensor if the upper limit is greater than 10 N). Related equipment: Auxiliary test supportframe (MJ-EP-C-0036); support mold (MJ-EP-C-0040/MJ-EP-C-0063); heating constant temperature water tank (MJ-EP-C-0065). More than five parallel samples were measured for each test.

Torsion Property Measurement. The samples were tested under constant temperature and humidity conditions (temperature 25 °C, relative humidity 60%) for 48 h, and the twist control performance was obtained by testing with a torsion tester (homemade). Adjust the exposed model length of 50.0 ± 1.0 mm, in the catheter distal angle of "0" position paste 3 M tape mark, as shown in Fig. 2.

Result and Discussion

Effect of Treatment Temperature on the Crystal Structure of TPU. Blackwell and Hocker et al. [17, 18] studied the structure and performance of MDI/BDO-type polyurethane, it was found that MDI/BDO-type polyurethane hard section of the existence of two crystal form, and proposed the compound of the hard chain of molecular chain arrangement model. Under normal circumstances, the crystallization seems to be inhibited [3], and after a certain period of heat treatment, the hard region can be suppressed from the state of the crystal "liberation" out to re-form the crystal structure. Figure 3 showed the XRD pattern of TPU under different treatment temperature. It can be observed that the untreated TPU only showed one diffraction peak at $2\theta = 20.3°$ (Fig. 3a), indicating a completely amorphous packing of macromolecules. The reason may be that TPU in the extrusion process was not completely reacted at a higher temperature in a relatively short time, which led to the incomplete separation of soft and hard segments and the incomplete crystallization of the hard segment. From Fig. 3b, c, it can be found that TPU demonstrated an obvious crystal peak at $2\theta = 19.2°$, which is the characteristic peak of the TPU hard domain phase [19]. It is probably because that the immobilization of molecular chain rearrangement movement and the hard segment was released from the crystalline state of inhibition under the heat treatment process, which led to more perfect microphase separation of TPU. The results indicated that treatment temperature has an important influence on the microphase separation of TPU.

Effect of Treatment Temperature on Dynamic Mechanical Properties of TPU. DMA was used to investigate the dynamic mechanical properties of the TPU. Figure 4a showed the storage modulus of the untreated TPU and those treated at different treatment temperatures. It can be seen that the storage modulus of the treated TPU increased compared with untreated one before the glass transition. However, after the glass transition, the storage modulus of TPU decreased rapidly at first and then persisted this state with no change even if the temperature kept rising. Figure 4b exhibited the changes of tan δ of TPU. It can be seen that the untreated TPU sample only showed one glass transition temperature at 53 °C. However, treated TPU at different temperatures exhibited two glass transition temperatures, one at 53 °C and the other at 79 °C, which showed the distinct microphase separation. It is probably because the hydrogen bonding of the TPU was destroyed at higher temperature and the molecular chain was rearranged, which lead a relatively complete soft segment and hard segment structure. Noting that the soft segment

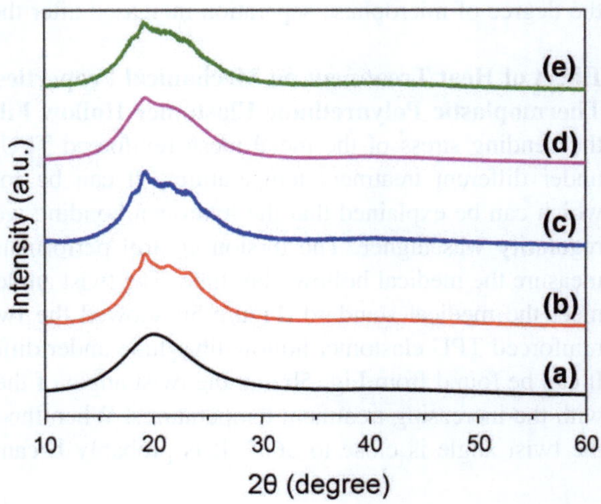

Fig. 3 XRD spectra of the pristine TPU (**a**) and those treated at temperatures of 120, 130, 140, 150 °C (**b**–**e**), respectively

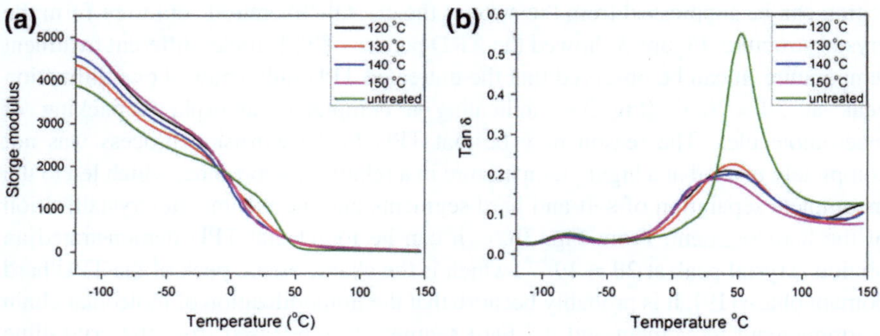

Fig. 4 **a** Storage modulus and (**b**) DMA of the untreated TPU and those treated at temperatures of 120, 130, 140, 150 °C

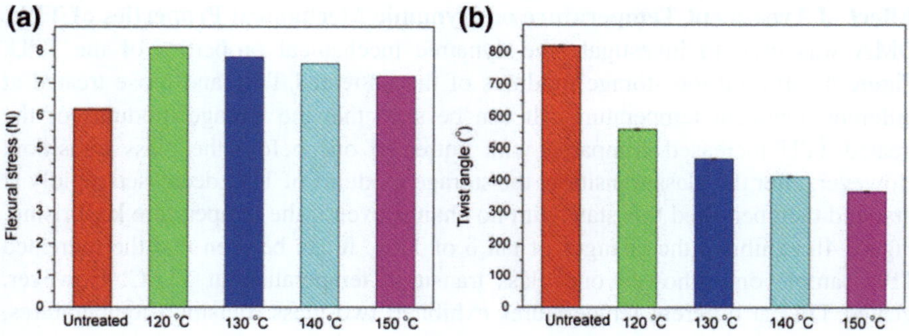

Fig. 5 Mechanical properties of the metal mesh reinforced TPU elastomer hollow fiber tube

melting temperature shifted to the low temperature direction (Fig. 4a) this is because the degree of microphase separation increased after the heat treatment.

Effect of Heat Treatment on Mechanical Properties of Metal Mesh Reinforced Thermoplastic Polyurethane Elastomer Hollow Fiber Tube. Figure 5a showed the bending stress of the metal mesh reinforced TPU elastomer hollow fiber tube under different treatment temperatures. It can be found that the fluxeral stress, which can be explained that the hydrogen bonding were formed and the degree of regularity was higher. The torsion control performance is an important index to measure the medical hollow fiber tube. The twist angle is closer to 360°, which can meet the medical standard. Figure 5b showed the twist angle of the metal mesh reinforced TPU elastomer hollow fiber tube under different treatment temperatures. It can be found from Fig. 5b that the twist angle of the hollow fiber tube decreased with the increasing treatment temperatures. When the temperature reached 150 °C, the twist angle is close to 360°. It is probably because that the crystalline of the polymer tend to be improved, so that the degree of microphase separation of the TPU increased, which led to the decreased twist angle.

Conclusion

The purpose of this article was to investigate the effect of treatment temperature on the mechanical properties of the metal mesh reinforced TPU elastomer hollow fiber tube. The microphase structures of TPU were measured by XRD and DMA, and the result showed the distinct microphase separation after different treatment temperatures. In addition, the mechanical properties of hollow fiber tube were analyzed by fluxeral stress and twisting performance, the flexural stress of the metal mesh to enhance the hollow fiber tube was increased at different treatment temperature, it indicated that microphase separation has a great influence on the mechanical properties of the metal mesh to enhance the hollow fiber tube.

Acknowledgements This work was financially supported by the Nation Basic Research Program (2011 CB606101) of the China 973 Program.

References

1. P.E. Gibson, M.A. Wallace, A.L. Cooper, *Development in Block Copolymer* (Elsevier, London, 1982)
2. N. Hossieny, V. Shaayegan, A. Ameli, M. Saniei, C.B. Park, Polymer **112**, 208–218 (2017)
3. S.L. Cooper, A.V. Tobolsky, J. Appl. Polym. Sci. **10**, 1837–1844 (1966)
4. X.D. Liu, X.Y. Gu, J. Sun, S. Zhang, Carbohyd. Polym. **167**, 356–363 (2017)
5. S. Bourbigot, T. Turf, S. Duquesne, Polymer. Degrad. Stabil. **94**, 1230–12337 (2009)
6. T. Kongkhlang, Y. Kousaka, T. Umemura, D. Nakaya, W. Thuamthong, Y. Pattamamongkolchai, S. Chirachanchai, Polymer **49**, 1676–1684 (2008)
7. S. Nie, C. Peng, S. Yuan, M. Zhang, J. Thermal Anal. Calorim. **113**, 865–871 (2013)
8. X. Chen, C. Ma, C. Jiao, J. Thermal Anal. Calorim. **126**, 633–642 (2016)
9. D.K. Chattopadhyay, D.C. Webster, Prog. Polym. Sci. **34**, 1068–1133 (2009)
10. E. Yilgor, I. Yilgor, Yurtsever, Polymer **43**, 6551–6559 (2002)
11. E. Tocha, H. Janik, H. Debowski, G.J. Vancso, J. Macromol. Sci. Part B Phys. **41**, 1291–1304 (2002)
12. K.B. Wagner, Macromolecules **25**, 5591–5596 (1992)
13. J.A. Miller, S.B. Lin, K.K.S. Hwang, K.S. Wu, P.E. Gibson, S.L. Cooper, Macromolecules **18**, 32–44 (1985)
14. J.T. Koberstein, A.F. Galambos, L.M. Leung, Macromolecules **25**, 6195–6204 (1992)
15. J.T. Koberstein, A.F. Galambos, L.M. Leung, Macromolecules **25**, 56185624 (1992)
16. A. Benegir, S. Michielsen, B. Pourseyhimi, J. Appl. Polym. Sci. **111**, 1246–1256 (2009)
17. L. Leibler, Macromolecules **13**, 1602–1617 (1980)
18. J. Hocker, L. Born, J. Polym. Sci. Polym. Part C: Polym. Lett. Ed. **17**, 723–730 (1979)
19. T. Hosseini-Sianaki, H. Nazockdast, B. Salehnia, E. Nazockdast, Polym. Eng. Sci. **55**, 2163–2173 (2015)

Preparation of Core-Shell $ZrO_2@SiO_2$ Nanoparticles and Its Effect on Properties of Composites

Hao Wang, Xiaoze Jiang, Li Qian, Wenping Chen and Meifang Zhu

Abstract In this study, $ZrO_2@SiO_2$ nanoparticles with core-shell structure were prepared by Stöber method using tetraethyl orthosilicate (TEOS) as the silicon source. The core-shell structure of nanoparticles was characterized via transmission electron microscopy (TEM). The results showed that the surface of ZrO_2 could be uniformly coated with SiO_2 and the thickness of the SiO_2 shell was about 10 nm. The prepared $ZrO_2@SiO_2$ nanoparticles were modified with 3-methacryloxypropyltrimethoxysilane (γ-MPS) and the graft ratios were analyzed by FTIR and TGA. After modified with SiO_2 shell, the graft ratio of γ-MPS on $ZrO_2@SiO_2$ nanoparticles increased to 4.03%. Then the modified $ZrO_2@SiO_2$ nanoparticles were added into the quartz fibers (QFs) reinforced composites to prepare fiber posts with radiopacity. The effects of the nanoparticles on the mechanical properties of fiber posts were investigated by universal testing machine. The flexural strength of the composite material is reduced with the addition of ZrO_2. While with the modified $ZrO_2@SiO_2$ nanoparticles, the flexural strength can be improved even if the content of the inorganic nanoparticles is relatively high.

Keywords ZrO_2 · Core-shell structure · Composites · Surface modification

H. Wang · X. Jiang · L. Qian · W. Chen (✉) · M. Zhu (✉)
State Key Laboratory for Modification of Chemical Fibers and Polymer Materials,
College of Material Science and Engineering, Donghua University,
Shanghai 201620, China
e-mail: chenwp@dhu.edu.cn

M. Zhu
e-mail: zhumf@dhu.edu.cn

H. Wang
e-mail: with28@163.com

Y. Han (ed.), *Advanced Functional Materials*,
https://doi.org/10.1007/978-981-13-0110-0_80

Introduction

Fiber posts for dental restoration were used to support the retention of a core that did not have sufficient coronal dentin and to support occlusal function. Their performances such as mechanical properties, radiopacity, biocompatibility and aesthetics are very important. Radiopacity is particularly important because it makes the posts detectable by X-ray and helps the clinician in establishing a correct diagnostic of technical failures, loss of retention, root fractures, post fracture or periapical lesion [1, 2]. Glass and quartz fiber posts which are usually used do not have enough radiopacity. So the addition of radiopaque agents such as BaO, $BaSO_4$, SrO or ZrO_2 are considered necessary to prepare radiopaque posts.

In recent years, zirconia (ZrO_2) has received considerable attention due to its outstanding features such as good biocompatibility, radiopacity, chemical and thermal stability. The use of nanoparticles as third phase in classical reinforced polymeric composites has also attracted substantial interests [3]. In general, the inclusion of nanoparticles into fiber reinforced composites is confined to a small weight fraction [4–7]. However, in order to get enough radiopacity, a relatively high content of ZrO_2 additions is required, which may result in a decrease in the mechanical properties of the composites [8]. Because of the poor compatibility between inorganic matter and organic matter, the surface of ZrO_2 need to be modified. The surface of inorganic materials are usually modified with a silane coupling agent which allows the particles to be well dispersed in organic matter [9]. Furthermore, some coupling agents with active groups may react with the resin to form chemical bond to reinforce the organic—inorganic interface. However, it is difficult for ZrO_2 to form chemical bonds with coupling agents since there are few active hydroxyl groups on the surface of these heavy metal element oxides [10]. Thus the surface modification becomes more critical and is the key to determine the mechanical properties of composites. Some researchers found that the amount of hydroxyl on the surface of zirconia could be increased by immersion in different corrosive solution, while this kind of treatment may damage the material or even dissolve it [11–13].

The density of hydroxyl on SiO_2 shell surface was high, which could be further modified by functional molecules. The prepared $ZrO_2@SiO_2$ core-shell structure nanoparticles were modified with γ-MPS. In addition, the SiO_2 shell may increase the biointegration of ZrO_2 [14]. The modified nanoparticles were added as radiopaque agent into the quartz fiber reinforced composites. And the properties of the nanoparticles and composites were characterized and analyzed.

Materials and Methods

Materials. Zirconium dioxide (ZrO_2, 50 nm, quartet phase) was obtained from Hefei pure Technology Co., Ltd., Hefei, China. Quartz fiber bundles (B type) with linear density of 72 tex were from Hubei Feilihua Quartz Glass Co., Ltd., China.

Bisphenol A glycerolatedimethacrylate (Bis-GMA) and camphorquinone (CQ, 97%) were purchased from Sigma-Aldrich, USA. Methyl methacrylate (MMA), tetraethyl orthosilicate, polyvinyl pyrrolidone (PVP), ammonia ($NH_3 \cdot H_2O$, 28%), sulphuric acid (H_2SO_4), N-propylamine, cyclohexane and ethanol were obtained from Sinopharm Chemical Reagent Co., Ltd., Shanghai, China. All reagents were used as received without further purification.

Preparation of Core-shell ZrO_2@SiO_2Nanoparticles. The core-shell ZrO_2@SiO_2 nanoparticles were prepared by Stöber method. Typically, 6 g ZrO_2 nanoparticles and 2 g PVP were added in the solution containing 250 mL of water and 350 mL of ethanol. The resultant dispersion was treated by sonication for 0.5 h and stirred for 4 h to adsorb PVP onto the surface of ZrO_2. 5 mL of ammonia aqueous solution (28 wt%) and 5 mL TEOS were dripped into the dispersion under a continuous stirring at room temperature for 8 h. The obtained core-shell ZrO_2@SiO_2 nanoparticles were collected by repetitive centrifugation/washing cycles with water and ethanol to remove the solvents and impurities. The obtained powder was dried in a blast oven at 85 °C for 12 h and then in a vacuum oven at 90 °C for 8 h.

Silane Modification of Nanoparticles. 3 g ZrO_2 or ZrO_2@SiO_2 nanoparticles were dispersed in 300 mL cyclohexane with vigorous stirring for 2 h. Then 0.3 mL γ-MPS and 0.15 mL N-propylamine were added by drops slowly. The obtained dispersion was stirred for 30 min at room temperature and then kept at 60 °C for 2 h. The mixture was evaporated in a rotary evaporator at 80 °C to remove the volatile substances. The obtained powder was dried in a blast oven at 85 °C for 12 h and then in a vacuum oven at 90 °C for 8 h.

Preparation of Fiber Reinforced Posts. To prepare fiber resinforced posts, Bis-GMA, MMA, CQ and 4-EDMAB were firstly mixed with a mass ratio of 1:1:0.004:0.016 at room temperature and then the resin was mixed with different proportions of nanoparticles. The formulation was shown in Table 1. After that, the bundles of modified QFs were immersed in pre-resin and kept in a vacuum oven at 60 °C for 1 h to remove air bubbles. After cooling to room temperature, the QFs bundles together with pre-resin were pulled through a mold and cured with a LED curing light with wavelength of 470 nm for 120 s. FRC post samples with a size of 25 mm × 2 mm × 2 mm were obtained after stripping from the mold.

Characterization

The crystal structure of ZrO_2 and ZrO_2@SiO_2 were characterized with X-ray diffraction (D/Max-2550 PC, Japan) with Cu Kα radiation. Functional groups of ZrO_2 and ZrO_2@SiO_2 before and after silane modification were analyzed with FTIR spectrometer (Nicolet 8700, Germany) together with ATR accessory.

Table 1 Formulation of pre-resin

Number	Nanoparticles contents (%)	Mixed resin (g)	ZrO_2 (g)	$ZrO_2@SiO_2$ (g)
Z0	0	5	0	0
Z1	1	4.95	0.05	0
Z3	3	4.85	0.15	0
Z5	5	4.75	0.25	0
Z7	7	4.65	0.35	0
Z9	9	4.55	0.45	0
Z11	11	4.45	0.55	0
ZS1	1	4.95	0	0.05
ZS3	3	4.85	0	0.15
ZS5	5	4.75	0	0.25
ZS7	7	4.65	0	0.35
ZS9	9	4.55	0	0.45
ZS11	11	4.45	0	0.55

Thermal stability and the graft ratio of silane were analyzed TGA instrument (TG 409 PC, Germany) performed from 50 to 800 °C (heating ratio: 10 °C/min, N_2 atmosphere). Flexural strength and flexural modulus of the composites were measured using a universal mechanical testing machine (Instron 5900, USA). Rectangular bar samples (25 mm × 2 mm × 2 mm) were prepared for the three-point bending test (span 20 mm, crosshead speed 0.75 mm/min). The radiopacity of composites were analyzed using Dental X-ray machine (X'mind D03110). Aluminum ladder (99% high purity aluminum, thickness of each step was 1 mm) was put on the upper part and composite samples (25 mm × 2 mm × 2 mm) were put on the lower part.

Results and Discussion

The morphology of ZrO_2 and the core-shell structure of $ZrO_2@SiO_2$ were characterized by TEM. Figure 1 showed that the diameter of ZrO_2 nanoparticles was in ranges of 30–50 nm. The shell structure was characterized by TEM (Fig. 1b) where the ZrO_2 °C nanoparticles acted as core and the prepared SiO_2 outside the ZrO_2 nanoparticles as shell. The inner ZrO_2 nanoparticles were dark while the continuous SiO_2 shell was relatively bright, and the average thickness of the SiO_2 shell was about 5–10 nm. The results indicated that $ZrO_2@SiO_2$ core-shell structure nanoparticles were successfully synthesized. It can also be observed that there was no obvious change in the morphology between ZrO_2 nanoparticles before and after coating with SiO_2.

It was reported that the quartet phase of ZrO_2 had a better X-ray radiopacity. The XRD patterns of ZrO_2 and $ZrO_2@SiO_2$ nanoparticles were shown in Fig. 2a.

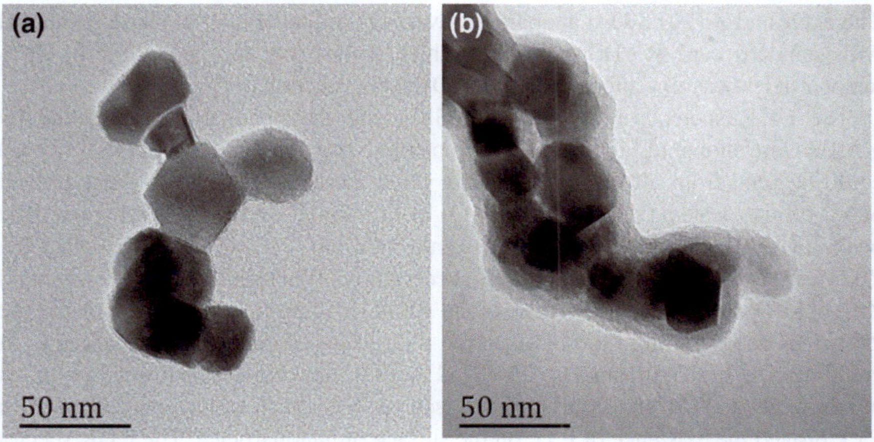

Fig. 1 Transmission electron microscopy (TEM) images of ZrO_2 and $ZrO_2@SiO_2$ nanoparticles: **a** ZrO_2, **b** $ZrO_2@SiO_2$ nanoparticles

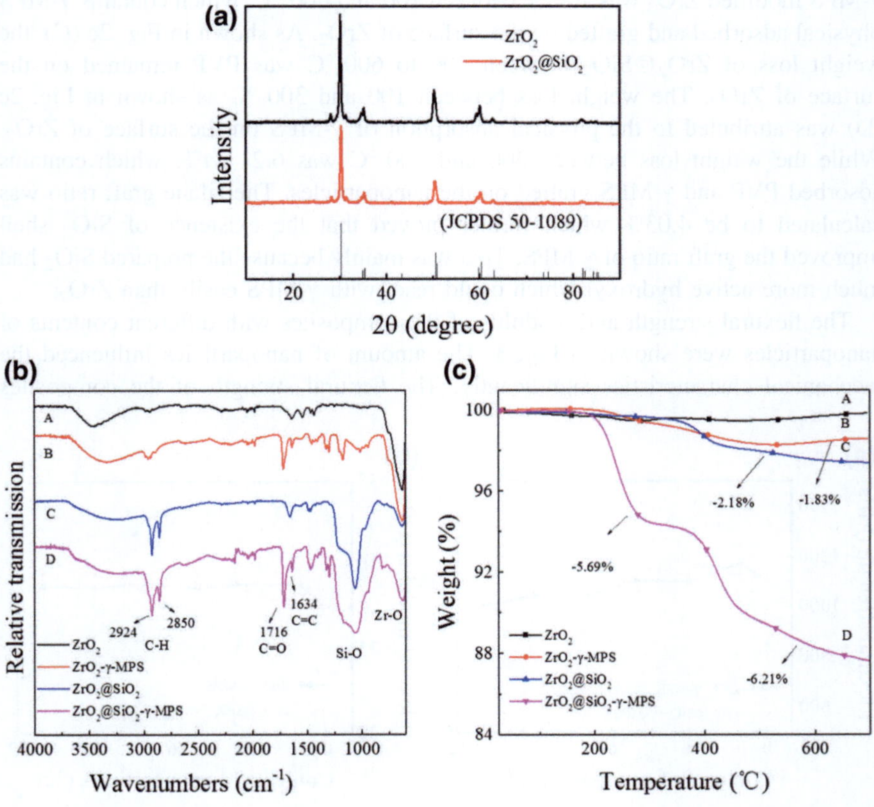

Fig. 2 **a** XRD patterns of ZrO_2 and $ZrO_2@SiO_2$ nanoparticles, **b** FTIR spectra of different nanoparticles, **c** TGA of different nanoparticles

The ZrO_2 and ZrO_2@SiO_2 nanoparticles were proved to be quartet phase compared with standard card JCPDS 50–1089. Which means that the crystalline of ZrO_2 nanoparticles was not affected by the SiO_2 shell structure.

The FTIR spectra of ZrO_2, ZrO_2@SiO_2 and the nanoparticles modified by γ-MPS were shown in Fig. 2b. The characteristic peak of ZrO_2 at 633 cm^{-1} (Zr–O) could be seen from all the four patterns while the existence of SiO_2 was proved based on the peaks at 1063 cm^{-1} (Si–O) in Fig. 2b (C and D). The characteristic peaks of γ-MPS at 1716 cm^{-1} (C=O) and 1634 cm^{-1} (C=C) confirmed that the ZrO_2 and ZrO_2@SiO_2 were silanized successfully. Whereas, the γ-MPS modified ZrO_2@SiO_2 showed stronger peaks than ZrO_2, proving that the SiO_2 outside the ZrO_2 increased grafting ratio of silane on the surface of nanoparticles indirectly. It could also be noticed from Fig. 2b (B, C and D) that characteristic peaks of C–H bonds were at 2924 and 2850 cm^{-1}. Based on Fig. 2b (C), the peaks of C–H in PVP proved the existence of PVP on the surface of nanoparticles. The FTIR spectra confirmed that the nanoparticles were successfully modified by γ-MPS.

The graft ratios of γ-MPS on the surface of different nanoparticles were tested by TG. There was almost no weight loss for unmodified ZrO_2, while the weight loss of γ-MPS modified ZrO_2 was 1.83% between 100 and 600 °C, which contains γ-MPS physical adsorbed and grafted on the surface of ZrO_2. As shown in Fig. 2c (C), the weight loss of ZrO_2@SiO_2 between 300 to 600 °C was PVP remained on the surface of ZrO_2. The weight loss between 100 and 300 °C as shown in Fig. 2c (D) was attributed to the physical absorption of γ-MPS on the surface of ZrO_2. While the weight loss between 300 and 600 °C was 6.21 wt%, which contains adsorbed PVP and γ-MPS grafted on the nanoparticles. The silane graft ratio was calculated to be 4.03% which further proved that the existence of SiO_2 shell improved the graft ratio of γ-MPS. That was mainly because the prepared SiO_2 had much more active hydroxyl which could react with γ-MPS easily than ZrO_2.

The flexural strength and modulus of the composites with different contents of nanoparticles were shown in Fig. 3. The amount of nanoparticles influenced the mechanical characteristics significantly. The flexural strength of the composites

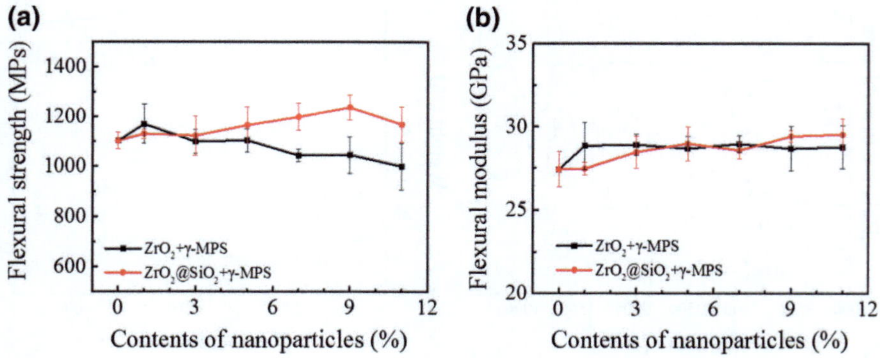

Fig. 3 Mechanical properties of composites with different contents of nanoparticles

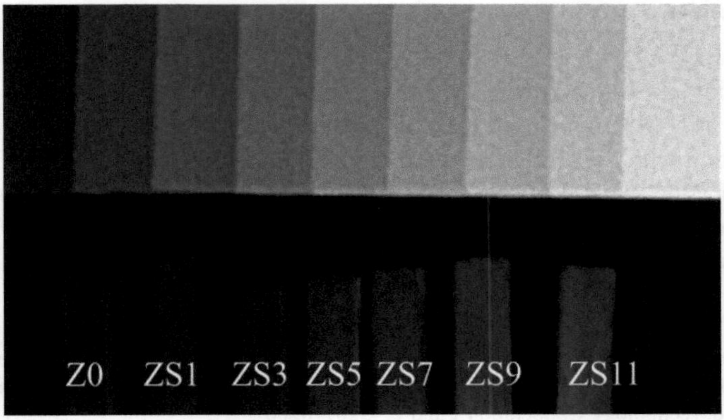

Fig. 4 Radiopacity of composites with different contents of nanoparticles

increased first and then decreased with the increasing of nanoparticles contents. The highest strength of composites with silanized ZrO_2 was 1169.97 ± 79.07 MPs at the content of 1%. With the continuous increase of silanized ZrO_2 amounts, the flexural strength decreased to a lower degree comparing with that of the composites without nanoparticles, which was consistent with the study of Liu [6]. That was because the silane graft ratio of nano-ZrO_2 was low and the compatibility with resin was pool [15]. At the content of 9%, the highest strength of the composite with $ZrO_2@SiO_2$ was 1236.26 ± 50.25 MPa while the strength of the composite with ZrO_2 was 1045.81 ± 73.42 MPa. This means that the presence of the silica shell not only improved the flexural strength of the composites but also increased the amount of the inorganic nanoparticles can be added. The trend of flexural modulus was similar to that of flexural strength. The silane graft ratio increased with the exist of the SiO_2 shell, which resulted in a better compatibility and dispersion in resin. In addition, the double bond of γ-MPS could react with the double bond of the resin, which can improve the interfacial adhesion between organic and inorganic nanoparticles.

As was shown in Fig. 4, the radiopacity of composites with different contents of nanoparticles were tested. The top of Fig. 4 is aluminum ladder and the thickness of each step was 1 mm. The samples at the bottom were (from left to right): Z0, ZS1, ZS3, ZS5, ZS7, ZS9, ZS11. It was obvious to observe that the radiopacity increased with the increase of the nanoparticles content. When the nanoparticles content attained to 7%, the radiopacity was equivalent to 3 mm aluminum ladder, which was better than dentin. The radiopacity at this level allowed the fiber post in the tooth to be recognizable by X-ray detection, meaning that at least 7% ZrO_2 of resin was necessary to obtain enough radiopacity. However, the amount of ZrO_2 at this level would destroy the strength of the composites.

Conclusion

In this study, $ZrO_2@SiO_2$ core-shell structure nanoparticles were prepared and were modified by γ-MPS. The graft ratio of γ-MPS on nanoparticles increased to 4.03% after coating with SiO_2 shell comparing with unmodified ZrO_2 (1.83%). This kind of modification method improved the amount of ZrO_2 nanoparticles added in the composites as well as the flexural strength (12.27%) and modulus (7.27%) of the composites. Finally, the quartz fiber post was prepared with an enough radiopacity (3–4 mm Al) which was higher than that of dentin.

Acknowledgements This work was supported by National Key Research and Development Program of China (2016YFA0201702/2016YFA0201700) and Project of Shanghai International Science and Technology Cooperation Fund (14520710200).

References

1. M. Ghavamnasiri, F. Maleknejad, H. Ameri, M.J. Moghaddas, F. Farzaneh, J.E. Chasteen, J. Conservative Dentistry: JCD **14**, 378–382 (2011)
2. E. Rodrigues, L.M. Pescinini Salzedas, A.C. Botazzo Delbem, D. Pedrini, J. Esthetic Restorative Dentistry **26**, 131–138 (2014)
3. J.-L. Tsai, B.-H. Huang, Y.-L. Cheng, J. Compos. Mater. **43**, 3107–3123 (2009)
4. P. Rosso, L. Ye, K. Friedrich, S. Sprenger, J. Appl. Polym. Sci. **100**, 1849–1855 (2006)
5. I. Isik, U. Yilmazer, G. Bayram, Polymer **44**, 6371–6377 (2003)
6. H. Liu, J. Mo, J. Reinf. Plast. Compos. **29**, 909–920 (2010)
7. A.C. Detomi, R.M. dos Santos, S.L.M. Ribeiro Filho, C.C. Martuscelli, T.H. Panzera, F. Scarpa, Mater. Des. **55**, 463–470 (2014)
8. R.-J. Zhou, T. Burkhart, J. Thermoplast. Compos. Mater. **23**, 487–500 (2010)
9. T. Kang, I. Jang, S.-G. Oh, Colloids Surf. A **501**, 24–31 (2016)
10. E.C. Combe, J. Dent. **1**, 93–97 (1972)
11. H. Fischer, C. Niedhart, N. Kaltenborn, A. Prange, R. Marx, F.U. Niethard, R. Telle, Biomaterials **26**, 6151–6157 (2005)
12. U. Lohbauer, M. Zipperle, K. Rischka, A. Petschelt, F.A. Mueller, J. Biomed. Mater. Res. Part B-Appl. Biomater. **87B**, 461–467 (2008)
13. M. Uchida, H.M. Kim, T. Kokubo, M. Nawa, T. Asano, K. Tanaka, T. Nakamura, J. Biomed. Mater. Res. **60**, 277–282 (2002)
14. C. Caravaca, L. Shi, S. Balvay, P. Rivory, E. Laurenceau, Y. Chevolot, D. Hartmann, L. Gremillard, J. Chevalier, Acta Biomater. **46**, 323–335 (2016)
15. P. Singh, A. Kaushik, Kirandeep, J. Reinf. Plast. Compos. **25**, 119–140 (2006)

Investigation of Tensile Property of Thin Ply Composite Laminate with Open-Hole Assisted by Acoustic Emission Technology

Chunfang Huang, Mingchang He, Yonglyu He, Jiayu Xiao, Jianwei Zhang, Su Ju and Dazhi Jiang

Abstract As the application of carbon fiber reinforced polymer composite (CFRP) laminates in aerospace and aircraft structures, opening hole is inevitable in design, fabricate and assemblage. However, the open-hole procedure has a bad effect on mechanical properties of the composite laminate. In this paper, characters of acoustic emission (AE) and curve of stress-strain of composite laminate with open-hole under tensile test were obtained. It was found that onset damage stress of a thin ply laminate detected using acoustic emission was higher than that of the standard ply laminate with the decreasing of ply thickness, while ultimate open-hole tensile strength of the thin ply laminate was lower than that of the standard ply laminate. More homogeneous microstructure and balanced stacking sequence make the thin ply laminate have higher initial damage stress. AE signals of matrix damage and delamination are presumably classified by the amplitude, which validate superior damage suppression and delamination resistance in the thin ply laminates. Stress concentration around the open-hole cannot be relaxed under tensile test as

C. Huang · M. He · Y. He · J. Xiao · J. Zhang · S. Ju · D. Jiang (✉)
Department of Materials Science and Engineering, National University of Defense Technology, Changsha, Hunan, People's Republic of China
e-mail: jiangdz@nudt.edu.cn

C. Huang
e-mail: hcf8819sk@126.com

M. He
e-mail: hcm_boy@126.com

Y. He
e-mail: yonglyu.he@foxmail.com

J. Xiao
e-mail: jiayuxiao@tom.com

J. Zhang
e-mail: jwzhang.nudt@Gmail.com

S. Ju
e-mail: suju-nudt@nudt.edu.cn

Y. Han (ed.), *Advanced Functional Materials*,
https://doi.org/10.1007/978-981-13-0110-0_81

higher damage suppression and delamination resistance in the thin ply laminate. That can lead the thin ply laminate has lower open-hole tensile strength.

Keywords Thin-ply · Composite laminates · Acoustic emission
Open-hole tension · Damage onset

Introduction

CFRP laminates were widely used in aircraft and aerospace structures because of their excellent specific properties and advantages in saving weight. However, opening hole is inevitable in design, manufacture and assemblage of composite laminates and the open-hole procedure has a bad effect on mechanical properties of the composite laminate. Composite manufacturers are able to obtained a appropriate way to fabricate fabrics almost without of fiber twist and crimp which significantly counteracting fiber performance. Comparing to laminates consisting of standard thickness plies, thin-ply CFRP composites have more excellent mechanical properties and broader design space. Besides, thin ply laminates also demonstrate superior mechanical properties of damage suppression and delamination resistance [1–3].

In the field of thin ply composites, substantial progress has been made in recent years [4, 5]. Thickness of commercially available thin-ply prepreg can be reduced to about 15 μm by tow spreading technology. Using thinner plies is not only to produce thinner and lighter laminates and structures, but to provide higher strength and damage resistance as well because of positive size effects.

Compared with laminates by standard thickness plies, thin ply laminates exhibited superior characteristics in damage suppression according to recently researches. Tomohiro Yokozeki et al. [2, 6] studied damage accumulation behaviors of quasi-isotropic laminates ([[45/0-45/90]ns]) made of thin (0.07 mm) and standard (0.14 mm) plies respectively in transverse indentation test. The standard ply laminates showed delaminations near the back surface of the specimen and inside the laminate, however, delamitation only occurred inside plies and the delamination appeared prior to fiber fractures in the thin-ply laminates. They also investigated compressive strength properties and CAI (compression strength after impact) of the laminates with/without open-hole prepared by different thickness plies. Using thin-ply prepregs resulted in substantially reducing in matrix cracking and delamination, which may result in higher compressive strength properties. Amacher et al. [7] experimentally studied the mechanical properties and size effect of thin ply composites. The results showed that tapering ply thickness will result in dramatic improvements in first-ply failure, fatigue life and ultimate strength. Thin-ply composites exhibit different failure behavior that significantly delayed damage growth and quasi-brittle failure were observed instead of extensive delamination and transverse cracking patterns. Fuller and Wisnom [8] investigated the damage suppression and pseudo-ductility of thin ply (0.03 mm) angle laminates ($[\pm\theta_5]_s$)

with the ply angles varied in range of 15° and 45°. X-ray computed tomography and microscopy have shown that using the thin ply can suppress delamination of the laminates. Sihn [1] also demonstrated that thin ply laminates have superior damage resistance against static, fatigue and impact loading. There are a few other papers also confirming the damage and delamination suppression in thin-ply composite laminates [9, 10]. Additionally, onset and propagation behaviors of the matrix cracking and delaminations were also dependent on ply thickness and stacking sequence [11–13].

Inside the material, sudden release of strain energy derived from damage formation generate the Acoustic Emission (AE) signals. Acoustic Emission (AE) registration can monitor and record damage on-line and in situ for a structure during loading. Different from other non-destructive methods, AE is an efficient method that can continuously monitor the damage of structures [14].

In this paper, quasi-isotropic and cross-ply laminate specimens were prepared with varying ply thickness (0.020, 0.055 and 0.125 mm, respectively). Ultimate tensile strength and failure behavior of laminates with open-hole were investigated. Initial damage stress of the thin ply laminate subjected to tensile loads was detected by acoustic emission.

Experimental

The prepregs used in the specimen preparation were consisted of T300 carbon fiber and epoxy 901. Areal weight of fabric in the prepreg, thicknesses of the prepregs and their codes were shown in Table 1.

To investigate the mechanical behaviors of the thin and standard ply laminates with open-hole under tensile test, the following quasi-isotropic laminates (QILs)and cross-ply laminates (CPLs) were prepared as shown in Table 2. Tapering the ply thickness, there are two available types of laminate thickness scaling strategies: laminate block scaling (for example $[90_m/45_m/0_m/\text{-}45_m]_s$) and sub-laminate scaling (for example $[90/45/0/\text{-}45]_{ms}$). Compared with laminate block scaling, sub-laminate scaling laminates have shown higher properties in tension, open-hole compression and shearing et al. [1, 3]. So, thin ply laminates adopted sub-laminate scaling in this paper.

The specimens were manufactured by compression molding processing with the same laminate thickness and fiber volume fraction. The curing process of T300/901 material system was set to 80 °C/45 min/1 atm, then 115 °C/20 min/2 MPa and 135 °C/2.5 h/2 MPa.

Table 1 T300/EP901 prepregs used in the specimen preparation

Thickness (mm)		Areal weight of fabric (g/m^2)	Ply code
0.020	Thin	20	T-20
0.055		54	T-55
0.125	Standard	125	T-125

Table 2 Stacking of QILs and CPLs

Ply	QILs	CPLs
T-20	$[0/45/90/-45]_{15s}$	$[0/90]_{30s}$
T-55	$[0/45/90/-45]_{6s}$	$[0/90]_{12s}$
T-125	$[0/45/90/-45]_{3s}$	$[0/90]_{5s}$

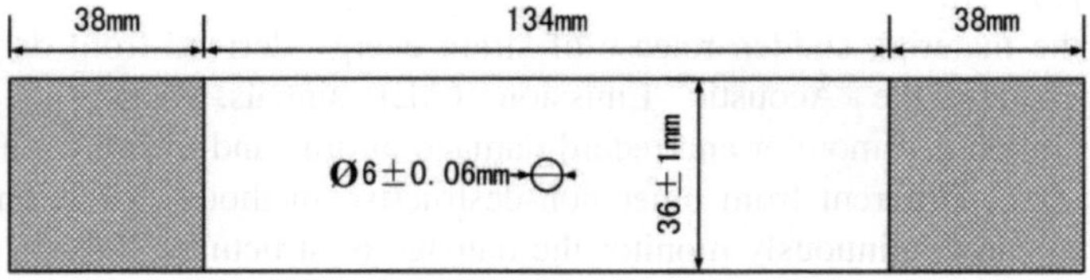

Fig. 1 Dimensions of specimen with open-hole under tension test

Figure 1 showed the specimen dimensions based on ASTM D5766-2002. The specimen width is 36 ± 1 mm and the length is 210 mm. The notch consists of a centrally located hole, 6 ± 0.06 mm in diameter. Loading rate of the test is 1 mm/min. The tensile strength of the laminates with open-hole can be calculated as follows:

$$S = P_{\max}/A \tag{1}$$

where:

S is tensile strength, MPa,
$P_{\max}$ is maximum load prior to failure, N,
A is gross cross-sectional area (disregarding hole), mm^2

All specimens for the tensile test were instrumented with an acoustic emission system consisted of two acoustic emission sensors, a PCI-2 board and AE-Win software (as shown in Fig. 2a). The acoustic emission system was used to register the AE signals of microscopic damage inside the specimen during tensile test under a measured threshold condition of 52 dB induced by environment and test machine. Two AE sensors were located at the middle of the back side of the specimen with a distance of 100 mm apart, to monitor the AE events timely. The parameters in Table 3 are the setting of the AE acquisition system. To calibrate the data acquisition system, a 2B pencil lead break procedure was done for each specimens before test. Figure 2b shows the schematic definition of the features of an AE signal such as rise time, duration, energy, amplitude and count.

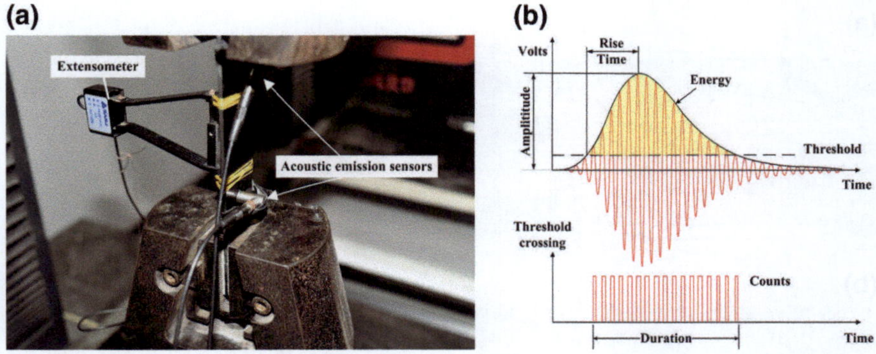

Fig. 2 **a** Acoustic emission system was applied in test **b** Definitions for the AE signal features [14, 15]

Table 3 Setting the AE acquisition system parameters

Parameter	Value and unit
Sampling rate	5 MHz
Preamp gain	40 dB
Threshold	52 dB

Results and Discussion

Optical Micrograph

Optical micrographs were used to check the microstructure of the laminates. Optical micrographs of all specimens are show in Fig. 3. The microstructures of the T-20 and T-55 ply laminates were observed to be more uniform and interfaces between the plies were more homogeneous. Microstructure of the T-125 ply laminate was observed to be in-homogeneous with more rich-resin and poor-resin regions. The difference can be denoted by visible fiber bundles and rich resin regions in the T-125 ply laminate but uniform microstructure for the thin ply laminate. Additionally, the number and size of defects in thin-ply laminate is less and smaller than standard-ply laminate.

Tensile Property of the Laminates with Open-Hole

Ultimate Tensile Strength

The CPLs and QILs with different ply thickness and open-hole were investigated in tensile test. The tensile strengths (calculated by Eq. 1) of CPLs and QILs were shown in Fig. 4. By tapering ply thickness, respectively, lower ultimate tensile

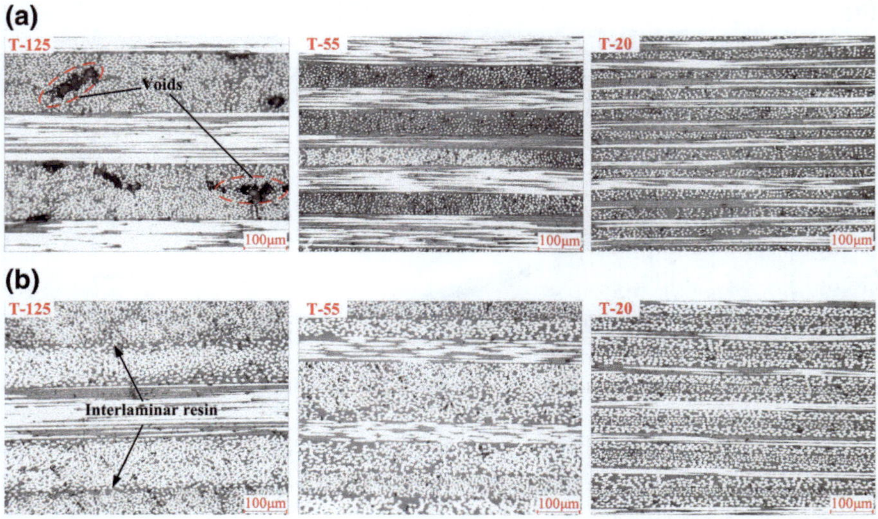

Fig. 3 Optical micrographs of specimens: **a** Cross-ply laminate and **b** Quasi-isotropic laminate

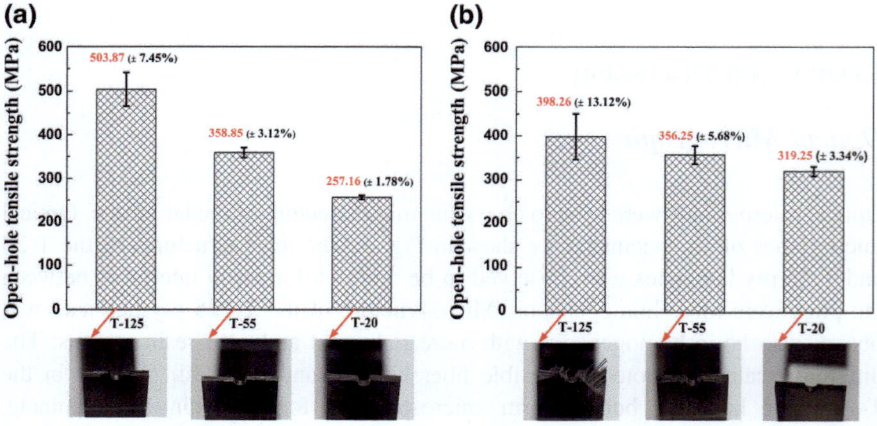

Fig. 4 Average open-hole tensile strength and failure photograph of the **a** cross-ply laminates and **b** quasi-isotropic laminates

strengths were obtained. In quasi-static tensile tests of cross-ply specimens with open-hole, the T-20 ply specimens demonstrated a 28.78% lower ultimate strength (358.85 MPa) than the T-125 ply specimens (503.87 MPa). Especially the ultimate tensile strength of the T-55 ply specimens with open-hole (257.16 MPa) was only a half of that of the T-125 ply specimens. For quasi-isotropic laminates, the T-20 ply specimens demonstrated a 19.84% lower ultimate strength (319.25 MPa) than the T-125 ply specimens (398.26 MPa), and the T-55 ply specimens showed a 10.55% lower ultimate strength (356.25 MPa) than the T-125 ply specimens. Furthermore,

T-55 and T-20 ply specimens demonstrated a smaller standard deviation that means they have more stable mechanical properties.

From the photograph of specimens after test, by tapering ply thickness, a quasi-brittle failure mode was obtained for T-55 and T-20 ply specimens. Especially for T-20 ply laminates, both quasi-isotropic and cross-ply stacking specimens showed a net-section failure mode and the failed section is perpendicular to the loading direction. However, the T-125 ply quasi-isotropic laminate specimens showed a pull-out failure mode with the failed sections at 45° and −45° angels as well as matrix cracking and delamination from the side view. The T-125 ply cross-ply laminate specimens also showed a multi-mode failure with delamination and pull-out. Comparing the failure modes of laminates with different ply thickness, it is clearly that a transition from multi-mode failure in T-125 ply laminates to a quasi-brittle failure mode in the T-20 ply laminates, while this quasi-brittle failure mode often occurs in isotropic materials. This failure mode transition observed in this paper corresponded well to the results given in [1, 7].

Onset Damage Stress

With simultaneous AE monitoring in the tensile test, AE can provide useful information of signal features about damage evolution in the quasi-isotropic and cross-ply open-hole specimens. A typical stress-strain curve and its relationship with the amplitude of AE signals were shown in Figs. 5 and 6. All AE signals were recorded in the test and the significant signal of onset damage of specimen was detected. It can be clearly seen from the Figs. 5 and 6 that less AE signals were detected in thin ply laminate specimens than standard ply laminate specimen. Onset damage stress was detected by AE events and the stress and strain of each specimen at onset damage were shown in Table 4. By tapering ply thickness, respectively, higher onset damage stress and strain are obtained in thin ply laminate specimens with both cross-ply and quasi-isotropic stacking types. Comparing the measured stress of damage onset to the ultimate strength, it can see that the thin ply laminates fail without obvious damage accumulation. In contrast, damage in the standard ply laminates started earlier, but propagated up to a higher load steadily.

A lot of literatures provided enough evidences that the AE amplitude played a important role in the classification of failure modes for laminated composite

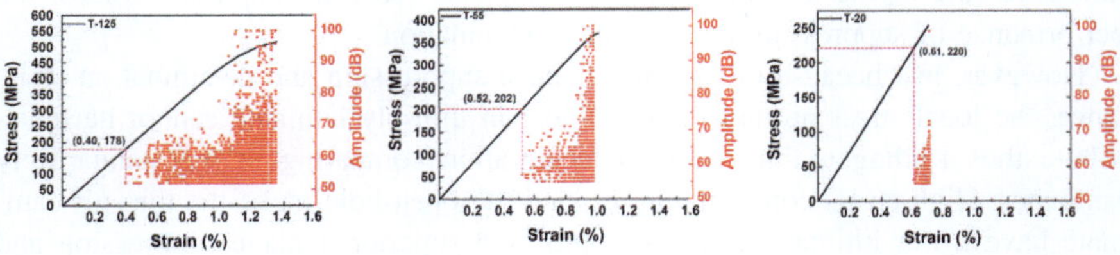

Fig. 5 Onset of damage in open-hole tensile test using acoustic technique for CPLs

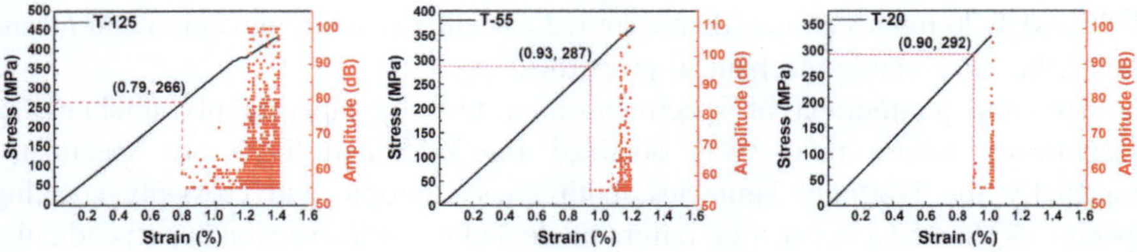

Fig. 6 Onset of damage in open-hole tensile test using acoustic technique for QILs

Table 4 Onset damage stress of specimens detected by AE technique

Stacking type	Ply	Onset damage stress (MPa)	Onset damage strain (%)
CP	T-125	178	0.40
	T-55	202	0.52
	T-20	220	0.61
QI	T-125	266	0.79
	T-55	287	0.93
	T-20	292	0.90

materials [16–20]. According to existing researches, the amplitude ranges of AE signals corresponding to different failure modes can be observed. It was found that lower amplitude was associated with matrix cracking and delamination, and high amplitude with fiber breaking. In this test, AE signals recorded for all specimens can be presumably classified to two classes: (1) signals with 52–75 dB amplitude and (2) signals with 75–100 dB amplitude. Based on the literature data and the classification results in this paper, the signals with 52–75 dB amplitude should be related to the delamination and matrix cracking. Less delamination and matrix cracking related signals appear in the thin-ply laminates, especially in the T-20 ply laminate. Therefore, the improvement of damage suppression and delamination resistance of the thin ply laminates were validated by AE signals data in the tensile test.

Onset damage stress and ultimate strength of quasi-isotropic and cross-ply laminates with open-hole in tension exhibited an opposite trend with reducing of the ply thickness. Increasing of onset damage stress can be owed to uniform microstructure and balance stacking sequence, as they provide the thin-ply laminates superior damage suppression and delamination resistance. In a word, the microstructure of the thin ply laminate composite plays an important role in the performance of suppressing damage and delamination.

However, just because of superior damage suppression and delamination resistance, the local stress around the open-hole in thin-ply laminate cannot be redistribute thus leading to high stress concentration comparing to the standard-ply laminates. High stress concentration around the open-hole make the thin-ply laminate have lower ultimate tensile strength, and superior damage suppression and delamination resistance make thin-ply laminate have less damage propagating paths and exhibit a quasi-brittle failure mode. On the contrary, the damage progression helps the standard ply laminate to improve the tensile strength because the stress concentration around the open-hole can be released in a certain extent.

Conclusions

According to the present results, higher onset damage stress and lower ultimate strength of the specimens with open-hole in tensile test were obtained for CPLs and QILs with thinner ply. Comparing to the standard ply laminates, more uniform microstructure and balance stacking sequence in the thin ply laminates resulted in superior damage and delamination suppression which were validated by AE detecting. Beside, thin ply specimens demonstrated a smaller standard deviation also owed to uniform microstructure and balance stacking sequence. However, superior damage suppression and delamination resistance of the thin ply laminates resulted in higher stress concentration around the hole and the thin ply laminate exhibit lower ultimate tensile strength. What's more, redistribution of the local stress around the hole was suppressed and crack and delamination cannot propagate freely which make the thin-ply laminate specimen exhibit a quasi-brittle failure mode in the tensile test. AE signals data validated the improvement of damage and delamination suppression of the thin ply laminates under the tensile test.

Acknowledgements We gratefully acknowledge the financial support by National Natural Science Foundation of China (11202231, U1537101).

References

1. S. Sihn, R.Y. Kim, K. Kawabe, S.W. Tsai, Experimental studies of thin-ply laminated composites. Compos. Sci. Technol. **67**, 996–1008 (2007)
2. T. Yokozeki, A. Kuroda, A. Yoshimura, Damage characterization in thin-ply composite laminates under out-of-plane transverse loadings. Compos. Struct. **93**, 49–57 (2010)
3. H. Sasayama, K. Kawabe, S. Tomoda, I. Ohsawa, K. Kageyama, N. Ogata, Effect of lamina thickness on first ply failure in multidirectionally laminated composites, in *Proceedings of the 8th Japan International SAMPE Symposium & Exhibition* (JISSE-8), Tokyo, Japan, November 2003
4. J.C. Chen, C.G. Chao, Numerical simulation and experimental investigation for design of a carbon fiber tow pneumatic spreading system. Carbon **43**, 2514–2529 (2005)
5. H.M. EL-Dessouky, C.A. Lawrence, Ultra-lightweight carbon fiber/thermoplastic composite material using spread tow technology. Compos.: Part B. **50**, 91–97 (2013)
6. T. Yokozeki, Y. Aoki, T. Ogasawara, Experimental characterization of strength and damage resistance properties of thin-ply carbon fiber/toughened epoxy laminates. Compos. Struct. **82**, 382–389 (2008)
7. R. Amacher, J. Cugnoni, J. Botsis, Thin ply composites: experimental characterization and modeling of size-effects. Compos. Sci. Technol. **101**, 121–132 (2014)

8. J.D. Fuller, M.R. Wisnom, Pseudo-ductility and damage suppression in thin ply CFRP angle-ply laminates. Compos. A **69**, 64–71 (2015)
9. H. Sasayama, K. Kawabe, S. Tomoda, I. Ohsawa, K. Kageyama, N. Ogata, Effect of lamina thickness on first ply failure in multidirectionally laminated composites, in *Proceedings of the 8th Japan International SAMPE Symposium & Exhibition (JISSE-8)*, Tokyo, Japan 2003
10. K. Yamaguchi, H.T. Hahn, The improved ply cracking resistance of thin-ply laminates, in *Proceedings of 15th International Conference on Composite Materials (ICCM-15)*, Durban, South Africa 2005
11. F.W. Crossman, W.J. Warren, A.S.D. Wang, G.E. Law Jr., Initiation and growth of transverse cracks and edge delamination in composite laminates: part 2. Experimental correlation. J. Compos. Mater. **14**, 88–108 (1980)
12. R.Y. Kim, S.R. Soni, Experimental and analytical studies on the onset of delamination in laminated composites. J. Compos. Mater. **18**, 70–80 (1984)
13. K.A. Lavoie, C. Soutis, J. Morton, Apparent strength scaling in continuous fiber composite laminates. Compos. Sci. Technol. **60**, 283–299 (2000)
14. M. Fotouhi, P. Suwarta, M. Jalalvand, G. Czel, M.R. Wisnom, Detection of fibre fracture and ply fragmentation in thin-ply UD carbon/glass hybrid laminates using acoustic emission. Compos.: Part A **86**, 66–76 (2016)
15. M. Huang, L. Jiang, P.K. Liaw, ChR Brooks, R. Seeley, D.L. Klarstrom, Using acoustic emission in fatigue and fracture materials research. Nondestr. Eval.: Overview **50**, 11 (1998)
16. R.E. Guerjouma, J.C. Baboux, D. Ducret, N. Godin, P. Guy, S. Huguet et al., Nondestructive evaluation of damage and failure of fiber reinforced polymer composites using ultrasonic waves and acoustic emission. Adv. Eng. Mater. **3**, 601–608 (2001)
17. J. Bohse, Acoustic emission characteristics of micro-failure processes in polymer blends and composites. Compos. Sci. Technol. **60**(8), 1213–1226 (2000)
18. S.C. Woo, N.S. Choi, Analysis of fracture process in single-edge-notched laminated composites based on the high amplitude acoustic emission events. Compos. Sci. Technol. **67**, 1451–1458 (2007)
19. M. Fotouhi, M. Ahmadi, Investigation of the mixed-mode delamination in polymer-matrix composites using acoustic emission technique. J. Reinf. Plast. Comp. **33**(19), 1767–1782 (2014)
20. R. Palazzetti, A. Zucchelli, C. Gualandi, M.L. Focarete, L. Donati, G. Minak, Influence of electrospun Nylon 6,6 nanofibrous mats on the interlaminar properties of Gr–epoxy composite laminates. Compos. Struct. **94**(2), 571–579 (2012)

Effects of Purification on the Structure and Properties of Hardwood Kraft Lignin

Erqiang Yin, Zhe Zhou, Shichao Wang and Meifang Zhu

Abstract Poor melt spinnability of lignin hinders the development of lignin-based high-value products, which can attribute to the diversity of its component and structure. In this work, hardwood kraft lignin (HKL), which was obtained from paper pulping process, was filtered using a 0.05 μm ceramic membrane followed by the heat-treatment under vacuum to evaluate the effects of different purification on the chemical structure, components, and thermal properties of HKL. Results showed that the contents of ash and lignin-carbohydrate complexes (LCCs), the molecular weight and its distribution were significantly decreased after purification, while the basic structure of HKL did not change. The thermal stability and carbon content increased a lot after the heat-treatment, while the content of methoxy and β-β or β-1 linkages decreased. However, the melt spinnability of HKL decreased significantly and many defects appeared in the obtained carbon fibers after heat-treatment for a long time.

Keywords Hardwood kraft lignin · Melt-spinning · Purification
Carbon fibers

Introduction

Lignin, an amorphous polymer created through enzymatic dehydrogenative polymerization of coumaryl alcohol, coniferyl alcohol and sinapyl alcohol, is found in most terrestrial plants in the approximate range of 15–40% dry weight [1]. The structure of lignin is relatively complex, including p-hydroxyphenyl (H), guaiacyl

E. Yin · Z. Zhou (✉) · S. Wang · M. Zhu (✉)
State Key Laboratory for Modification of Chemical Fibers
and Ploymer Materials, College of Materials Science and Engineering,
Donghua University, Shanghai 201620, China
e-mail: zzhe@dhu.edu.cn

M. Zhu
e-mail: zhumf@dhu.edu.cn

Y. Han (ed.), *Advanced Functional Materials*,
https://doi.org/10.1007/978-981-13-0110-0_82

(G) and syringyl (S) units, lignin-carbohydrate complexes (LCCs) and many linkages [2–4]. The component and the structure of lignin depend on various factors, such as the plant genus and the growth period. According to the different sources, lignin can be classified into softwood lignin, hardwood lignin and grass lignin. A great deal of work has showed that softwood lignin was almost exclusively of G units, while hardwood lignin was mainly constituted by G and S units, and grass lignin was made up of H, G and S units [2, 3, 5]. Many methods have been used to separate lignin from the main structure, and different kinds of lignin were obtained as a result, including kraft lignin, enzymatic hydrolysis lignin, organic lignin and pyrolytic lignin [6]. Among all the lignin, kraft lignin attracts even more attention due to its large production [7]. Since the carbon content of kraft lignin is relatively high, kraft lignin has been used to prepare carbon fibers to increase its value. The precursor of lignin-based carbon fibers can be prepared using electrospinning [8], solution spinning [9] or melt spinning [10]. Compared with solution spinning and electrospinning methods, melt spinning showed an absolute advantage due to the low production cost, high purity and few defects. Therefore, more interests have arisen in developing lignin into value-added products using melt spinning method.

However, the poor melt spinnability of lignin has limited the further development of lignin-based value-added products due to the impurities and the complex structure [11]. Lots of researches have been focused on the removing of the impurities, including the organosolv purification and ceramic purification. Baker et al. purified hardwood kraft lignin (HKL)using organic solvents and found that after the purification, the glass transition temperature (T_g), softening point and melting point of lignin decreased from 130.2, 172, 202 °C to 88.0, 108, 128 °C, respectively [12]. The purified lignin could be easily processed into fibers by melt spinning method. However, due to the low thermal stability, the obtained fibers were adhered with each other during the thermal stabilization process. Saito et al. [13] also found that the thermal stability of kraft lignin decreased significantly after the purification with methanol. For example, the T_g decreased from 153 to 117 °C. Similar to those studies, Fang et al. investigated the effects of different purification methods on the properties of kraft lignin, including ultrasonic extraction, solvent extraction, dialysis, and hot water treatment. Results showed that the lignin fractions obtained by ultrasonic extraction using ethanol/acetone mixture had a narrow molecular mass distribution, low contents of inorganic compounds and carbohydrates, a lower T_g compared with that of the untreated lignin [14]. Different from the methods mentioned above, Brodin et al. reported that industrial black liquors from kraft pulping were subjected to fractionation by ceramic membranes followed by acidification and isolation of the lignin, which was more suitable for industrial production [15]. The purification process reduced the carbohydrate content and did not change the T_g of lignin significantly. In addition, ion exchange membrane was also used to further decrease the ash content of lignin. Other methods, such as enzymatic hydrolysis and pickling could decrease the carbohydrates content of lignin effectively, while protein contaminants were introduced at the same time [16].

Although many purification methods have been reported, few researches were concerned in details about the change of structure and properties of lignin under different treatments during the purification process. In this paper, HKL obtained from paper pulping process was purified using a 0.05 μm ceramic membrane followed by vacuum heat treatment to evaluate the effects of different treatments on the chemical structure, components, and thermal properties of HKL. Furthermore, melt-spun lignin fibers and their based carbon fibers were prepared, which was used to further investigate the effect of purification process on the morphology of obtained carbon fibers.

Materials and Methods

Materials. Hardwood kraft lignin (HKL, LignoBoost pulping process) was provided by SuzanoPapel e Celulose S.A. Corp., Brazil. Sodium hydroxide (analytical pure). Sulfuric acid (98%) was purchased from Sinopharm Group Chemical Reagent Co., Ltd and used as received without further purification. Deuterated dimethyl sulfoxide (DMSO-d_6) used for NMR analysis was purchased from Sigmae Aldrich.

Purification and Heat Treatment of Lignin. The purification process was performed using a ceramic membrane with a cut-off value of 0.05 μm [17]. HKL was firstly dissolved in sodium hydroxide solution (pH = 12) and fractionated through the ceramic membrane. Next, the pH of penetrated solution was adjusted to 2.0 using sulfuric acid followed by the filtration of the precipitation. The purified HKL was obtained by washing the precipitation to neutral, then filtered and dried under 80 °C for 12 h. HKLs used for melt spinning were prepared after the heat-treatment of purified HKL under vacuum at 200 °C for 0.5, 2 and 4 h, respectively.

Preparation of Lignin Fibers and Their Carbonized Fibers. Purified HKL and HKL treated under vacuum for 0.5, 2 and 4 h were labeled as HKL-0.05 μm, HKL-T30, HKL-T120 and HKL-T240, respectively. All HKL samples were fed into at win screw micro-compounder equipped with a take-up device. The temperatures of melt spinning were 210–240 °C, and the winding rates were 10–100 m/min. HKL melt-spun fibers were adhered to a ceramic ark, then thermally stabilized and carbonized according to our former work [17].

Structural and Properties Characterization. Nuclear magnetic resonance (^{1}H NMR, ^{13}C NMR, COSY and HSQC) spectra were recorded on a Bruker Avance 600 MHz spectrometer at 25 °C with DMSO-d_6 as the solvent according to the reported process [18, 19]. HKL samples (50 mg for ^{1}H NMR, 180 mg for ^{13}C NMR) were dissolved in 1 mL of deuterated dimethyl sulfoxide (DMSO-d_6, 99.8% D). The spectra widths of HSQC were 6600 and 33,500 Hz for ^{1}H and ^{13}C dimensions, respectively [20].

The molecular weights of HKLs were determined by gel permeation chromatography (GPC, Agilent 1200, USA) with a refraction index detector (RID) on a PLaquagel-OH50 column (300 × 7.7 mm, Polymer Laboratories Ltd.), calibrated with PL polystyrene standards (M_W = 1320, 9200, 66,000, 435,500, Polymer Laboratories Ltd.) according to Sun et al. [18].

Elemental analysis was carried out in an elemental analyzer Vario EL III (Elementar, Hanau, Germany). The oxygen content was deduced from the difference with respect to the total sample. The methoxy contents of HKL samples were determined by ^{1}H NMR according to Mousavioun and Doherty [21]. Ash content analysis was carried out in a muffle furnace. The dry HKL of 8 g samples in a crucible was calcined in air at 900 °C for 6 h before the determination of the residual mass. The carbohydrate associated with HKL fractions was determined by high-performance anion exchange chromatography according to Wen et al. [20].

Differential scanning calorimetry (DSC) analysis and thermogravimetric analysis (TGA) were conducted according to Wang et al. [19] using a TA instruments Q20 and Netzsch (TG 209 F1 Iris) thermogravimetric analyzer.

The morphology of HKL-based carbon fibers was characterized using scanning electron microscopy (SEM) (JSM-5600LV, Japan).

Results and Discussion

Chemical Structural Analysis. NMR techniques are powerful methods for the determination of the structural features of lignin. The structures of HKL, HKL-0.05 μm, HKL-T30, HKL-T120 and HKL-T240 were investigated using ^{1}H NMR, ^{13}C NMR, COSY and HSQC NMR spectrometry, as shown in Figs. 1, 2, 3 and 4, respectively.

^{1}H NMR Spectra. As shown in Fig. 1, all the ^{1}H NMR spectra of HKL samples were almost the same. Signals around 6.60 ppm attributed to aromatic protons in

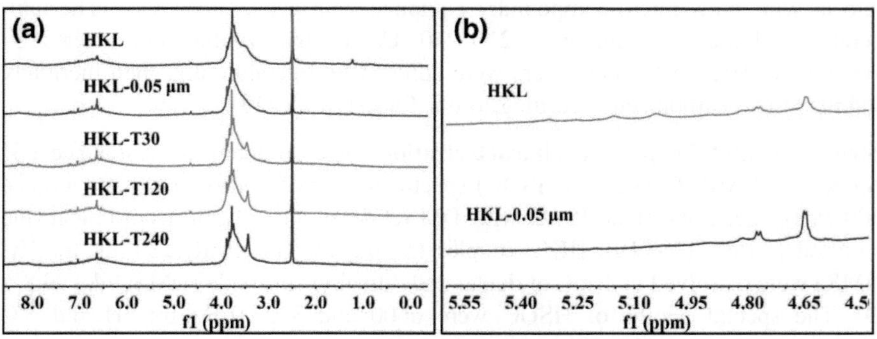

Fig. 1 **a** ^{1}H NMR spectra of HKL, HKL-0.05 μm, HKL-T30, HKL-T120 and HKL-T240; **b** HKL and HKL-0.05 μm with ^{1}H NMR local amplification spectra

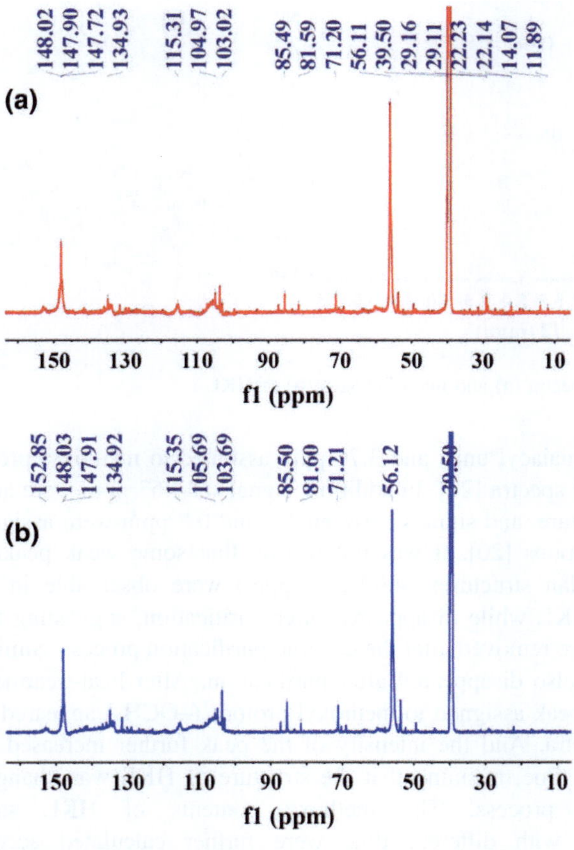

Fig. 2 ^{13}C NMR spectra of HKL **a** before and **b** after purification

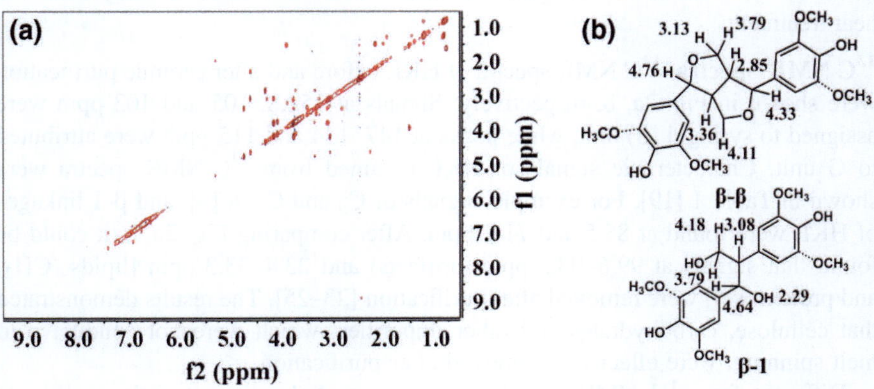

Fig. 3 ^{1}H-^{1}H COSY spectrum (**a**) and main linkages (**b**) of HKL

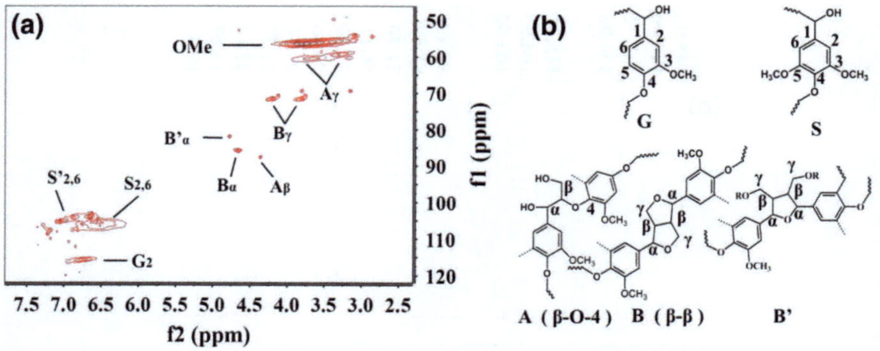

Fig. 4 HSQC spectra (**a**) and main linkages (**b**) of HKL

syringyl and guaiacyl units and 3.76 ppm assigned to methoxyl protons ($-OCH_3$) appeared in all spectra [22]. In addition, signals at 4.63 ppm were attributed to $H_β$ in β-O-4 structure, and signals between 1.5 and 0.8 ppm were assigned to protons in aliphatic groups [20]. It was noteworthy that some weak peaks attributed to protons in xylan structures (4.99–5.31 ppm) were observable in the ^{1}H NMR spectrum of HKL while disappeared after purification, suggesting that some carbohydrates were removed after the ceramic purification process. Similarly, the peak near 3.5 ppm also disappeared after purification. After heat-treatment under vacuum, a sharp peak assigned to methoxyl protons ($-OCH_3$) appeared at 3.4 ppm in ^{1}H NMR spectra. And the intensity of the peak further increased with a longer heat-treatment time, meaning that the structure of HKL was changed during the heat-treatment process. The methoxy contents of HKL samples under heat-treatment with different time were further calculated according to the description of Mousavioun and Doherty [21]. The calculated contents of HKL-T30, HKL-T120 and HKL-T240 were 24.14, 22.95 and 22.56%, respectively, which indicated that the methoxy group contents decreased after a long time for heat-treatment.

^{13}C NMR Spectra. ^{13}C NMR spectra of HKL before and after ceramic purification were shown in Fig. 2a, b, respectively. Signals at 152.8, 105 and 103 ppm were assigned to syringyl (S) unit, while peaks at 147, 134 and 115 ppm were attributed to G unit. Characteristic signals of HKL obtained from ^{13}C-NMR spectra were shown in Table 1 [19]. For example, signals of $C_α$ and $C_γ$ in β-β and β-1 linkages of HKL were found at 85.5 and 71.2 ppm. After comparing Fig. 2a, b, it could be found that signals at 99.6–93.3 ppm (artifacts) and 23.4–33.3 ppm [lipids (CH_2) and protein (sc)] were removed after purification [23–25]. The results demonstrated that cellulose, carbohydrates and other impurities, which were not conducive to melt spinning, were effectively removed after purification.

Different from ^{1}H NMR, no obvious change of the characteristic peaks was found in the ^{13}C NMR spectra of HKL samples. To further investigate the differences among the structures of HKL, the ratio of β-β and β-1 linkages (85.7, 81.8

Table 1 Main chemical shifts (δ, ppm) of ^{13}C NMR spectrum of HKL

δ (ppm)	Assignment	δ (ppm)	Assignment
153.2	C_3/C_5, etherified S	105.6–103.4	C_2/C_6, non-etherified S
148.3	C_3, non-etherified G	85.9	C_α in β-1 units
135.3	C_1, etherified S and etherified G	71.6	C_γ in β-β units
115.7	C_5, non-etherified G	56.4	OCH_3 in S and G

and 71.6 ppm) to methoxy groups (56.4 ppm) in syringyl and guaiacyl units was calculated via the integration of the peaks. Results showed that the ratio of β-β and β-1 linkages to methoxy groups for HKL, HKL-0.05 μm, HKL-T30, HKL-T120 and HKL-T240 were 0.09, 0.069, 0.067, 0.06 and 0.03, respectively. After purification, the ratio decreased a lot due to the removement of impurities. In addition, the content of β-β and β-1 linkages almost kept the same after heat-treatment for 0.5 h, led to a larger break of the linkages presented after a longer treating time.

^{1}H-^{1}H COSY Spectra. ^{1}H–^{1}H COSY spectra provided evidence for the proton coupling of double or triple bond. In general, cross-peak 4.64/6.08 ppm could be attributed to the H_β/H_α couplings in β-O-4 linkage of lignin [26]. As shown in Fig. 3, it was difficult to find the cross-peak of H_β/H_α couplings because most β-O-4 linkages of lignin were broken after kraft pulping process. While the signals of H_β/H_α couplings and H_β/H_γ couplings in β-β linkages were found at 3.36/4.76 ppm and 3.79/4.11 ppm [19], H_α/H_α couplings and H_α/H_β couplings in β-1 linkages also appeared at 4.64/3.08 ppm and 3.08/4.18, 3.79 ppm. Besides, the signals of coupling of H_α and the proton of hydroxy connecting with C_α were found at cross-peak 4.63/2.28 ppm.

In order to investigate the effect of vacuum heat-treatment time on the structure of HKL, the intensity changes of proton couplings in β-β and β-1 linkages were compared under the same test conditions and concentrations. Results showed that the intensity of the relevant cross-peaks in β-β and β-1 linkages increased due to the fact that ceramic membrane filtration process improved the purity of HKL. The intensity of the cross-peak 3.37/4.76 ppm and 4.63/3.07 ppm in β-β or β-1 linkages increased to 5.5 and 2.2 from 2.9 and 1.1, respectively due to the decrease of signals from the impurities. After heat-treatment for 0.5, 2 and 4 h under vacuum, the intensity of the cross-peak 3.37/4.76 ppm and 4.63/3.07 ppm were 4.0, 3.0, 2.0 and 1.7, 1.3, 0.9 respectively, which suggested that vacuum heat treatment destroyed the β-β and β-1 linkages, and the degree of the damage increased with the increase of heat-treatment time.

^{13}C-^{1}H HSQC Spectra. To obtain a more complete structural characterization of kraft lignin, the 2D-HSQC NMR spectra of all samples were analyzed. The C_γ–H_γ correlations in β-O-4 linkages were observed at 59.3/3.3 ppm and 60.3/3.6 ppm in HKL-0.05 μm, and 58.9/3.3 ppm and 59.5/3.7 ppm in HKL. The structure of β-β linkages (B) was observed with C_γ–H_γ at 71.4/3.8,4.2 ppm, and C_α–H_α at 85.7/

4.6 ppm. Signals of 105.5/6.5 ppm and 115.6/6.7 ppm from S and G units could be clearly observed in all of the spectra. As observed from Fig. 4, the main structures of HKL before and after purification were S and G units, and the S/G ratios in HKL and HKL-0.05 μm were about 2.4 and 2.74, respectively. Results suggested that S units slightly increased after purification. After vacuum heat treatment, the ratios of S and G units in HKL-T30, HKL-T120, and HKL-T240 were 6.4, 6.5 and 7.6, respectively, suggesting that long time heat-treatment could break the G units or convert G units into S units.

Component Analysis. The influence of purification and vacuum heat-treatment on the contents of element, carbohydrate and ash, as well as molecular weight and its distribution were investigated. As shown in Table 2, the carbon contents of HKL increased from 59.73 to 61.88% after a 0.05 μm-ceramic membrane filtration, and the value further increased to 65.05, 65.57 and 65.92% after vacuum heat-treatment for 0.5, 2 and 4 h, respectively. While the ash contents of HKL decreased from 1.25 to 0.51% after the purification and further reduced to 0.20% after vacuum heat treatment for 0.5 h. The decrease of ash content was beneficial to the formation of HKL fibers, as well as the mechanical properties of the carbonized fibers.

In addition, the molecular weight and distribution of HKL were decreased after purification, which due to the fact that ceramic membrane filtration effectively removed the lignin-carbohydrate complexes (LCCs) with a large molecular weight. After heat-treatment for 0.5 h, the weight-average molecular weight (M_w) of HKL

Table 2 Component analysis on purification process and heat treatment of HKL

Components	HKL	HKL-0.05 μm	HKL-T30	HKL-T120	HKL-T240
Rhm (%)	0.015	0.011	0.002	0.001	0.000
Ara (%)	0.069	0.018	0.008	0.003	0.003
Gal (%)	0.193	0.061	0.031	0.019	0.025
Glu (%)	0.357	0.062	0.057	0.051	0.053
Xyl (%)	0.856	0.129	0.044	0.014	0.024
Glua (%)	0.013	0.018	0.020	0.019	0.017
Gala (%)	0.011	0.003	0.004	0.001	0.005
Total (%)	1.515	0.301	0.167	0.108	0.127
Ash (%)	1.25	0.51	0.20	0.16	0.13
Elemental analysis					
C (%)	59.73	61.88	65.05	65.57	65.92
H (%)	5.96	5.85	5.61	5.61	5.60
O (%)	34.31	32.27	29.34	28.82	28.48
N (%)	$\leq$0.05	$\leq$0.05	$\leq$0.05	$\leq$0.05	$\leq$0.05
OCH_3 (%)	23.35	23.35	24.14	22.95	22.56
M_w	1738.3	1702.2	3130.5	3676.6	3686.9
M_n	1144.7	1145.0	1414.9	1482.0	1414.6
PDI	1.52	1.49	2.21	2.48	2.60

was 1.8 times higher than that of purified HKL, and the number-average molecular weight (M_n) increased slightly, showing that vacuum heat-treatment removed the low-molecular weight volatile components which led to a condensed structure of HKL.

As shown in Table 2, the effects of ceramic membrane filtration and vacuum heat treatment on carbohydrate contents were investigated. After purification of HKL, the total contents of carbohydrates decreased significantly from 1.515 to 0.301%, especially the galactose, glucose and xylose contents. After treatment under vacuum for a long time heat, the content of carbohydrates further decreased due to the volatilization of carbohydrates and the degradation of LCCs.

Thermal Properties Analysis. The thermal properties of HKLs were characterized using DSC. As shown in Fig. 5a, the glass transition temperatures (T_g) of HKL were almost the same (145 °C) before and after purification, and there was a big exothermic peak between 180 to 205 °C. Afterheat-treatment under vacuum for 0.5, 2 and 4 h, the T_g of HKL increased to 153.5, 153.8 and 157.2 °C, respectively, and the exothermic peak disappeared due to the removement of volatiles.

The effects of purification process and vacuum heat-treatment on the thermal stability of HKLs were also investigated using TGA. As shown in Fig. 5b, the thermal degradation way of HKL was one-step degradation [27]. The initial degradation temperature (T_i, temperature at 3% decomposition) of HKL was essentially constant after the ceramic membrane purification, while after vacuum heat treatment for 0.5 h, an increase of nearly 100 °C appeared. At the same time, the residual carbon weight at 900 °C (W_f) of HKL increased with a longer heat-treatment time, indicating that vacuum heat-treatment increased the degree of the condensed structure of HKL, as well as the thermal stability (Table 3).

Morphology of HKL Fibers and Their Based Carbon Fibers. Figure 6a depicted the melt-spun HKL fibers. It was difficult to spin HKL into fibers without purification due to the existence of impurities. After purification and heat-treatment for 0.5 h, the melt spinnability of HKL improved significantly and continuously

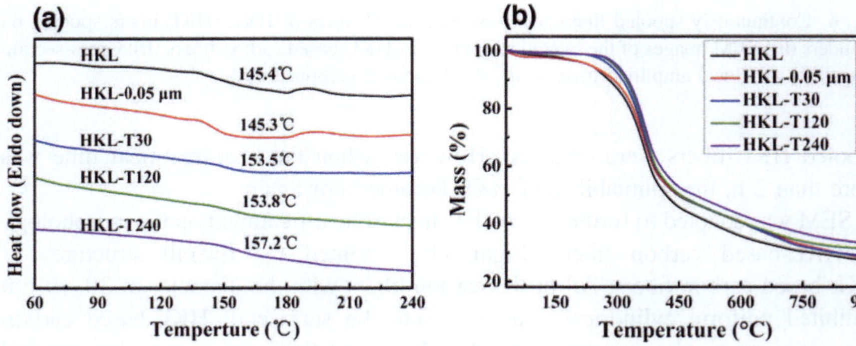

Fig. 5 DSC (**a**) and TG curves (**b**) of HKL, HKL-0.05 μm, HKL-T30, HKL-T120 and HKL-T240

Table 3 Data obtained from the TG curves of HKL

Samples	$T_{3\%}$ (°C)	T_{max} (°C)	Residual carbon (%)
HKL	161.1	370.8	27.1658
HKL-0.05 μm	169.9	366.4	28.1246
HKL-T30	260.5	367.0	28.7571
HKL-T120	274.3	365.6	31.2259
HKL-T240	281.2	365.6	33.1565

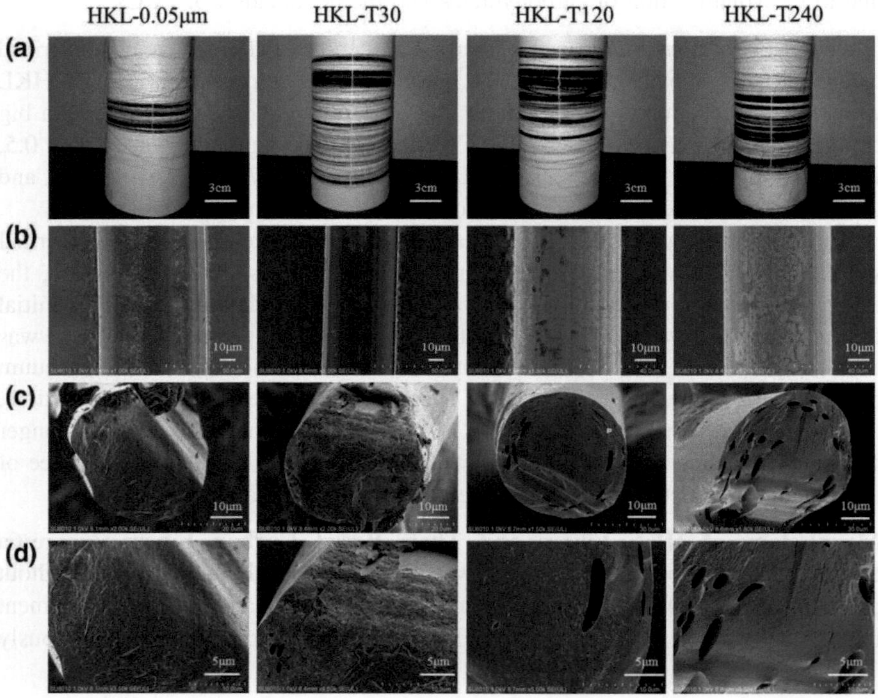

Fig. 6 Continuously spooled fibers and post-carbonized fibers of HKL (HKL fibers spooled on cylinders (**a**); SEM images of the overall structures for HKL-based carbon fibers (**b**); Cross-section images (**c**) and local amplified images (**d**) of HKL-based carbon fibers)

spooled HKL fibers were obtained. However, when the heat-treatment time was more than 2 h, the spinnability of HKL became poor again.

SEM was adapted to further study the effect of heat-treatment on the morphology of HKL-based carbon fibers. Figure 6b presented the overall structures of HKL-based carbon fibers. All of the carbon fibers with the diameter of 30–70 μm exhibited uniform cylindrical structures, and the surface of HKL-based carbon fibers was relatively smooth. Figure 6c, d showed the cross-section images and amplified images of HKL-based carbon fibers, respectively. Some pores caused by the volatilization of LCCs could be seen on the cross-section of HKL-based carbon fibers. After a long time heat-treatment, the degree of condensed structure and the melt viscosity increased significantly. The bubbles existed in the melt were not easily removed during the melt processing process. As a result, many pores appeared in HKL fibers and their carbonized fibers.

Conclusions

This work provided a simple method to purify lignin derived from black liquor by using ceramic purification and vacuum heat-treatment. The contents of ash and LCCs, the molecular weight and its distribution are significantly decreased after purification, while the thermal stability and the basic structure of HKL do not change. After heat treatment under vacuum, volatiles are removed and the purity of HKL is further increased, as well as the thermal stability. The carbon contents are increased to above 65%, while the content of methoxy and β-β or β-1 linkages decreases. In addition, after a long heat-treatment time, the residual carbon is also increased obviously, which is beneficial to the preparation of carbon fibers. However, the melt spinning temperature of HKL and the defects in obtained carbon fibers increase a lot.

Acknowledgements We are grateful for the financial support of this research from Program for Innovative Research Team in University of Ministry of Education of China (IRT_16R13).

References

1. D. Kai, M.J. Tan, P.L. Chee, Y.K. Chua, Y.L. Yap, X.J. Loh, Towards lignin-based functional materials in a sustainable world. Green Chem. **18**(5), 1175–1200 (2016)
2. H. Chung, N.R. Washburn, Extraction and types of lignin, in *Lignin in Polymer Composites*, ed. by F. Omar, S. Mohini (William Andrew, New York, 2016), pp. 13–25
3. S. Haghdan, S. Renneckar, G.D. Smith, Sources of lignin, in *Lignin in Polymer Composites*, ed. by F. Omar, S. Mohini (William Andrew, New York, 2016), pp. 1–11
4. B.C. Zhao, B.Y. Chen, S. Yang, T.Q. Yuan, A. Charlton, R.C. Sun, Structural variation of lignin and lignin–carbohydrate complex in eucalyptus grandis×E. urophylla during its growth process. ACS Sustain. Chem. Eng. **5**(1), 1113–1122 (2017)
5. S. Chatterjee, T. Saito, Lignin-derived advanced carbon materials. Chemsuschem **8**(23), 3941–3958 (2015)
6. A.A. Ogale, M. Zhang, J. Jin, Recent advances in carbon fibers derived from biobased precursors. J. Appl. Polym. Sci. **133**(45), 43794 (2016)
7. H. Mainka, O. Täger, E. Körner, L. Hilfert, S. Busse, F.T. Edelmann, A.S. Herrmann, Lignin-an alternative precursor for sustainable and cost-effective automotive carbon fiber. J. Mater. Res. **4**(3), 283–296 (2015)
8. C. Lai, P. Kolla, Y. Zhao, H. Fong, A.L. Smirnova, Lignin-derived electrospun carbon nanofiber mats with supercritically deposited Ag nanoparticles for oxygen reduction reaction in alkaline fuel cells. Electrochim. Acta **130**, 431–438 (2014)

9. M. Zhang, A.A. Ogale, Carbon fibers from dry-spinning of acetylated softwood kraft lignin. Carbon **69**, 626–629 (2014)
10. S. Kubo, J.F. Kadla, Lignin-based carbon fibers: effect of synthetic polymer blending on fiber properties. J. Polym. Environ. **13**(2), 97–105 (2005)
11. Q. Sun, R. Khunsupat, K. Akato, J. Tao, N. Labbé, N.C. Gallego, J.J. Bozell, T.G. Rials, G.A. Tuskan, T.J. Tschaplinski, A.K. Naskar, Y. Pu, A.J. Ragauskas, A study of poplar organosolv lignin after melt rheology treatment as carbon fiber precursors. Green Chem. **18**(18), 5015–5024 (2016)
12. D.A. Baker, N.C. Gallego, F.S. Baker, On the characterization and spinning of an organic-purified lignin toward the manufacture of low-cost carbon fiber. J. Appl. Polym. Sci. **124**(1), 227–234 (2012)
13. T. Saito, J.H. Perkins, F. Vautard, H.M. Meyer, J.M. Messman, B. Tolnai, A.K. Naskar, Methanol fractionation of softwood kraft lignin: impact on the lignin properties. Chemsuschem **7**(1), 221–228 (2014)
14. W. Fang, M. Alekhina, O. Ershova, S. Heikkinen, H. Sixta, Purification and characterization of kraft lignin. Holzforschung **69**(8), 943 (2015)
15. I. Brodin, E. Sjöholm, G. Gellerstedt, Kraft lignin as feedstock for chemical products: the effects of membrane filtration. Holzforschung **63**(3), 290 (2009)
16. D.S. Argyropoulos, Y.J. Sun, E. Paluš, Isolation of residual kraft lignin in high yield and purity. J. Pulp Pap. Sci. **28**(2), 50–54 (2002)
17. S. Wang, Z. Zhou, H. Xiang, W. Chen, E. Yin, T. Chang, M. Zhu, Reinforcement of lignin-based carbon fibers with functionalized carbon nanotubes. Compos. Sci. Technol. **128**, 116–122 (2016)
18. Y.C. Sun, J.L. Wen, F. Xu, R.C. Sun, Structural and thermal characterization of hemicelluloses isolated by organic solvents and alkaline solutions from Tamarix austromongolica. Bioresour. Technol. **102**(10), 5947–5951 (2011)
19. S. Wang, Y. Li, H. Xiang, Z. Zhou, T. Chang, M. Zhu, Low cost carbon fibers from bio-renewable lignin/poly(lactic acid) (PLA) blends. Compos. Sci. Technol. **119**, 20–25 (2015)
20. J.L. Wen, Z. Sun, Y.C. Sun, S.N. Sun, F. Xu, R.C. Sun, Structural characterization of alkali-extractable lignin fractions from bamboo. J. Biobased Mater. Bioenergy **4**(4), 408–425 (2010)
21. P. Mousavioun, W.O.S. Doherty, Chemical and thermal properties of fractionated bagasse soda lignin. Ind. Crop. Prod. **31**(1), 52–58 (2010)
22. A. Tejado, C. Pena, J. Labidi, J.M. Echeverria, I. Mondragon, Physico-chemical characterization of lignins from different sources for use in phenol-formaldehyde resin synthesis. Bioresour. Technol. **98**(8), 1655–1663 (2007)
23. J.L. Wen, S.L. Sun, B.L. Xue, R.C. Sun, Recent advances in characterization of lignin polymer by solution-state nuclear magnetic resonance (NMR) methodology. Materials **6**(1), 359–391 (2013)
24. H. Kono, S. Yunoki, T. Shikano, M. Fujiwara, T. Erata, M. Takai, CP/MAS ^{13}C NMR study of cellulose and cellulose derivatives. 1. Complete assignment of the CP/MAS^{13}C NMR spectrum of the native cellulose. J. Am. Chem. Soc. **124**(25), 7506–7511 (2002)
25. A. Yamazawa, T. Iikura, A. Shino, Y. Date, J. Kikuchi, Solid-, solution-, and gas-state NMR monitoring of ^{13}C-cellulose degradation in an anaerobic microbial ecosystem. Molecules **18**(8), 9021–9033 (2013)
26. K. Lundquist, Analysis of lignins by ^{1}H NMR spectroscopy. Nord. Pulp Pap. Res. J. **4**(3), 210–213 (1989)
27. M. Brebu, C. Vasile, Thermal degradation of lignin—a review. Cellul. Chem. Technol. **44**(9), 353–363 (2010)

Preparation and Gas Sensing Property of PEDOT/Silica Aerogel Fibers

Si Meng, Xi-yue Huang, Xing-ping Wang, Jun-yan Zhang, Wen-ping Chen and Mei-fang Zhu

Abstract Porous materials, as their high specific surface area, are commonly used in NH_3-sensing. And when NH_3-sensing materials are made into the form with interconnected pore structure, low-dimensional structure or hierarchical structure, their gas-sensitivity will be improved remarkably. Here, firstly silica aerogel fibers were prepared by wet spinning. Those fibers do not only preserve high specific surface area (maximum one is 835.8 m^2/g) and network organization, but also have unique hierarchical structure (macropores in the surface and mesopores in the middle regions) and regular hollow structure. And then PEDOT was synthesized and loaded on the net skeletons of silica aerogel fibers by gas phase polymerization. The specific surface area and pore structure of PEDOT/silica aerogel fibers can be adjusted by change aging bath. Finally, the results of NH_3 sensitivity experiment showed that all fibers have good gas sensitive, especially A0.5-PEDOT/SAFs (resistance variation is over 40%).

Keywords Aerogel fiber · PEDOT · Gas phase polymerization
Gas sensing · Hierarchical

S. Meng · X. Huang · X. Wang · J. Zhang · W. Chen · M. Zhu (✉)
The State Key Laboratory for Modification of Chemical Fibers
and Polymer Materials, College of Materials Science and Engineering,
Donghua University, Shanghai 201620, China
e-mail: zhumf@dhu.edu.cn

S. Meng
e-mail: mengsi_simon@163.com

X. Huang
e-mail: empt_y@qq.com

X. Wang
e-mail: wxp2100486@sina.com

J. Zhang
e-mail: 13817145747@163.com

W. Chen
e-mail: chwp@dhu.edu.cn

Y. Han (ed.), *Advanced Functional Materials*,
https://doi.org/10.1007/978-981-13-0110-0_83

Introduction

Ammonia sensors are heavily used in agriculture and chemical industry. And the NH_3-sensing material is the core component of ammonia sensors [1, 2]. As NH_3-sensing materials, conductive polymer materials can be used at room temperature or even cooler temperatures, so they are perceived to be more universal than inorganic NH_3-sensing materials [3–6]. However, currently all kinds of conductive polymer NH_3-sensing materials have a much lower sensitivity compared to inorganic one. Therefore, it is the key to make conductive polymer NH_3-sensing materials enter practical and commercialize stage that improving the sensitivity of NH_3-sensing materials [7–10].

PEDOT is a typical conductive polymer materials, and it also has a quite low sensitivity. But compared to other conductive polymer materials, it has higher conductivity and better stability. Those properties have important effect on sensors. In addition, its preparation method is simple. Therefore, PEDOT is one of the hottest topics in sensing materials.

In the field of NH_3-sensing, porous materials often show better sensitivity as their high specific surface area [11, 12]. And when the porous NH_3-sensing materials are made into the form with interconnected pore structure [13, 14], low-dimensional structure [15, 16] or hierarchical structure [17, 18], their gas-sensitivity will be improved remarkably. However, PEDOT with those structures is hardly prepared directly. Therefore, some materials, which have such structures, were chosen as substrate to achieve this. At present, most of substrate materials just have one or two of such structures. Although the sensitivity of PEDOT on those substrate had significantly improvement, it is much lower than the sensitivity of polyaniline, let alone inorganic NH_3-sensing materials [19–21]. If a kind of substrate material with all those structures is prepared to load PEDOT, the sensitivity of PEDOT must be improved further.

Herein, a kind of silica aerogel fibers (SAFs) with high specific surface area, hierarchical structure and hollow structure were prepared as substrate by simple wet spinning. The hollow structure could be beneficial to improve sensitivity [22, 23]. Silica aerogel is a kind of porous material with ultra-high specific surface area and interconnected pore structure. And it is prepared from precursor, for example tetraethyl orthosilioate, sodium silicate and water glass, by sol-gel method. In this work, water glass was selected as precursor and spinning dope, and dilute sulfuric acid was selected as catalyzer, reactant and coagulating bath to prepare silica aerogel precursor fibers (SAPFs) by wet spinning. And then the prepared SAPFs transformed into SAFs by aging and drying. The prepared SAFs do not only preserve high specific surface area and network organization, but also have unique hierarchical structure and regular hollow structure. PEDOT was synthesized and loaded on the net skeletons of silica aerogel fibers by gas phase polymerization, thus PEDOT/SAFs was prepared. Benefited from gas phase polymerization, the gas phase 3,4-ethoxylene dioxy thiophene (EDOT) can enter the pores of SAFs and form a thin PEDOT film on the skeletons [24–26]. The morphology, structure and sensing performance of SAFs and PEDOT/SAFs were studied. And the pore sizes and specific surface area were adjusted to optimize their sensing performance by change aging bath.

Introduction

Materials. Water glass (Be° = 50, Na_2O:SiO_2 molar ratio = 1:2.2) was purchased from Tongxiang Xiangyang water glass mill, China. Sulfuric acid was purchased from Sinopharm Chemical Reagent Co., Ltd, China. Ethanol was purchased from Changshu Yangyuan Chemical Co., Ltd, China. Iron(III) p-toluene sulfonate hexahydrate and 3,4-ethoxylene dioxy thiophene (EDOT) were purchased from Energy Chemical Co., Ltd, China. Silver paste was purchased from Beijing ZhongJingkeyi Technology Co., Ltd, China. Copper foil was purchased from Dongguan Xinshi Packing Material Co., Ltd, China.

Preparation of SAFs. A typical procedure for preparing SAFs involves three steps, spinning, aging and drying. Initially, 10 ml water glass as spinning dope was filled in an injector. 0.5 mol/L sulfuric acid as coagulating bath was prepared and stored in a polypropylene sink. The needle (inside diameter is 0.4 mm) of the injector connected to another needle (inside diameter is 0.4 mm too) in coagulating bath by a PTFE tube. Water glass was injected into coagulating bath through PTFE tube and needle by injection pump at a rate of 0.1 ml/min. And glass water was translated into solid fibrous orthosilicic acid, namely silica aerogel precursor fibers (SAPFs), rapidly, when it was injected into coagulating bath. The reaction is shown as Eq. 1. The SAPFs was winded continuously by winder at a rate of 80 mm/s. Then the SAPFs were soaked in aging bath for 3 days. The aging bath is different concentration of diluted sulfuric acid solution (0.5, 0.1, 0.05, 0.01 and 0 mol/L, namely pure water). During 3 days aging, the orthosilicic acid fibers would be slowly translated into silica gel fibers by polycondensation, as shown in Eq. 2. Then the solution in silica gel fibers was replaced by ethanol. Finally, the silica gel fibers were further translated into silica aerogel fibers (SAFs) after drying for 2 day at 80°.

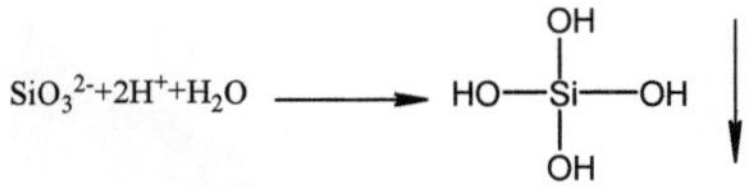

(1)

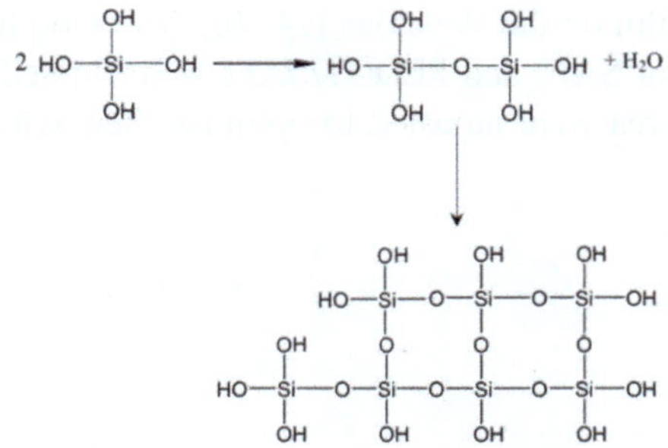

(2)

Preparation of PEDOT/SAFs. 10 mg Iron(III) p-toluene sulfonate hexahydrate dissolved in 100 mg ethanol, namely 10 wt% Iron(III) p-toluene sulfonate hexahydrate solution. And then a strand of SAFs were soaked in Iron(III) p-toluene sulfonate hexahydrate solution for 30 min. Take out SAFs and let them air-dry for 30 min. The colors of SAFs changed from colorless transparent to primrose yellow. Then put it on the copper mesh in a bottle with about 5 ml EDOT. Tighten the lid of bottle and put it in oven at 80° for 5 min. During the 5 min, EDOT) can enter the pores of SAFs, meet Iron(III) p-toluene sulfonate hexahydrate in SAFs, have polymerization and form a thin PEDOT film on the skeletons. And the colors of fibers changed from primrose yellow to black. The unreacted EDOT on fibers was washed with ethanol. In this way the PEDOT/SAFs were prepared.

Fabrication of Sensing Test Samples. 1 mm thick glass plate and PP plate were cut into uniform rectangle pieces, 2.5 cm × 1.5 cm. And a 0.5 cm × 1.5 cm small rectangle piece was removed from each PP rectangle pieces along midline of their short edge. In this way PP U-type baseplates were got. One side of PP U-type baseplates was pasted with glass rectangle piece and another side was pasted with two pieces of copper foil on the two legs of PP U-type baseplates respectively. And then a strand of PEDOT/SAFs was paste with copper foil by silver paste. In this

Fig. 1 Sketch of sensing test sample

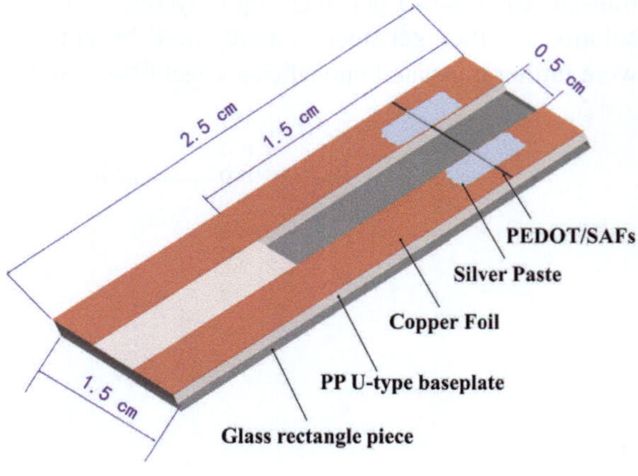

Table 1 Names of SAFs and PEDOT/SAFs

Sample	Aging bath concentration (mol/L)	Loading PEDOT or not
A0.5-SAFs	0.5	Not
A0.1-SAFs	0.1	Not
A0.05-SAFs	0.05	Not
A0.01-SAFs	0.01	Not
A0-SAFs	0	Not
A0.5-PEDOT/SAFs	0.5	Loading
A0.1-PEDOT/SAFs	0.1	Loading
A0.05-PEDOT/SAFs	0.05	Loading
A0.01-PEDOT/SAFs	0.01	Loading
A0-PEDOT/SAFs	0	Loading

way the sensing test samples were fabricated. Its sketch is shown as Fig. 1. And all sample names are listed in Table 1.

Surface Area Analysis. All samples were analyzed for surface area, pore size and pore volume using N_2 adsorption–desorption isotherms (NOVATOUCH LX1, QUANTACHROME INSTRUMENTS, America). The out gas temperature was maintained at 120 °C for 5 h.

Scanning Electron Microscope. Scanning electron microscopy (SEM) images of the samples were all taken to show the morphology with SU8010, Hitachi, Japan.

Fourier-Transform Infra Red (FT-IR) Spectroscopy. Functional groups of EDOT and PEDOT were identified out using fourier-transform infra red (FT-IR) spectrometry (Spectrum BXII, Perkin Elmer, USA).

Sensing Test. The conductivity of the samples was measured using a two-point probe to show the NH_3 sensitivity of PEDOT/SAFs by Keithly 2000, USA. The environment of samples were continuously alternate between air and 1000 ppm NH_3 every 5 min. All sensing measurements were carried out at 25 °C and 60% relative humidity.

Results and Discussions

Pore Structure and Morphology of SAFs. Figure 2a is the N_2 adsorption-desorption isotherm and pore size distribution of A0.5-SAFs. As silica aerogel fibers, they have high specific surface area (760.1 m^2/g), and a lot of mesopores (average pore width is 5.29 nm). Further analysis of the results show that the adsorption isotherm does not have a plateau at high pressure ($p/p_0 > 0.9$). It is suggested that those fibers also have a number of macroporouses, and it was proven by SEM (Fig. 2d–g). The predicted macroporouses can be observed in the outside

Fig. 2 **a** Nitrogen adsorption-desorption isotherms and pore-size distribution curve of A0.5-SAFs; **b**, **c** SEM images of A0.5-SAFs (hollow structure and morphology); **d**, **e** Cross-sectional images near outside surface and inside surface of A0.5-SAFs; **f**, **g** SEM images of outside surface and inside surface of A0.5-SAFs, inserts are images of partial enlargement and the scale bar is 100 nm

surfaces and inside surface. The hollow structure and uniform fiber shape can also be observed by SEM (Fig. 2b, c). Interesting that the outside surface and inside surface are loose, but the middle ring regions between outside surface and inside surface are dense. Those results indicated that the pores in surface are mainly macroporouses, the pores in he middle ring regions are mainly mesopores. Together they form a unique hierarchically macro-meso-macroporous structure.

From Fig. 2, we can find all samples have N_2 adsorption–desorption isotherms hysteretic loop and have not plateau at high pressure. Those results suggest that all samples have hierarchically macro-meso-macroporous structure. Figure 3 and Table 2 show that all samples have high specific surface area, and the specific surface area of samples prepared under high concentration coagulation bath (0.5, 0.1 and 0.05 mol/L) are much higher than under low concentration coagulation bath (0.01 and 0 mol/L). And the mesopore sizes increase first and then decrease with increase of coagulation bath concentration.

FT-IR Spectroscopy. After the polymerization from EDOT to PEDOT, the C-H which adjoined sulfur atom will disappear. Therefore, the corresponding absorption peak (about 890.9 nm^{-1}) of fourier-transform infra red spectroscopy will disappear too. And this phenomenon was observed from Fig. 4.

Sensing Test. The obtained results of sensing tests are given in Fig. 5. What's interesting is that the resistance variations of PEDOT/SAFs based on A0.01-SAFs and A0-SAFs (about 40 and 70% respectively) are much higher others. However, their specific surface area is much lower than others. This phenomenon is contrary to the reports. In order to explore the causes leading to this phenomenon, the pore structure and morphology of PEDOT/SAFs were studied too.

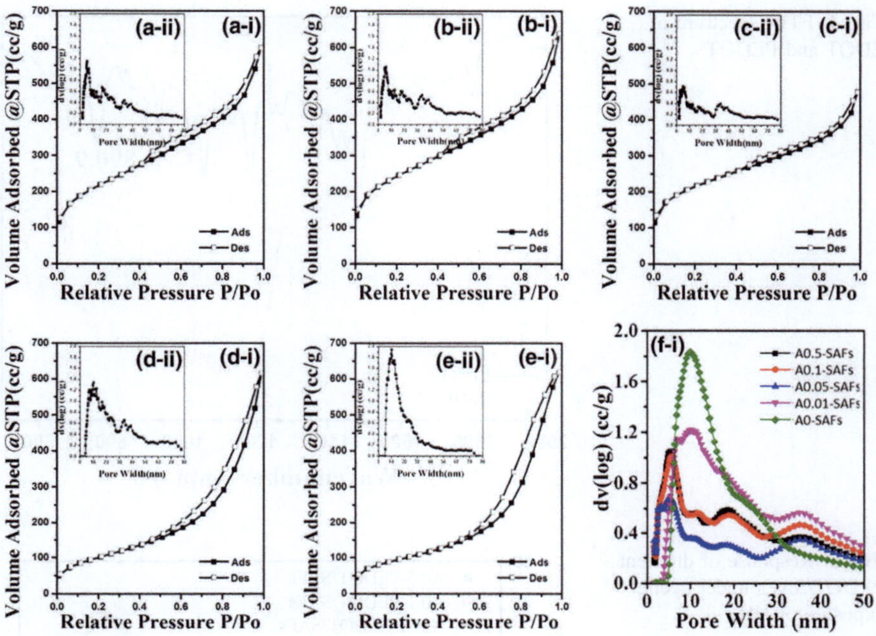

Fig. 3 Nitrogen adsorption-desorption isotherms and pore-size distribution curve of SAFs prepared using different concentrations of sulphuric acid as ageing bath. **a** A0.5- SAFs; **b** A0.1-SAFs; **c** A0.05-SAFs; **d** A0.01-SAFs; **e** A0-SAFs

Table 2 BET data of SAFs prepared using different concentrations of sulphuric acid as ageing bath

Sample	Specific surface area (m^2/g)	Pore volume (cc/g)	Pore width (nm)
A0.5-SAFs	760.1	0.93	5.29
A0.1-SAFs	835.8	0.99	5.88
A0.05-SAFs	731.6	0.74	4.89
A0.01-SAFs	364.8	0.96	10.13
A0-SAFs	333.8	0.96	10.13

Pore Structure and Morphology of PEDOT/SAFs. Firstly, the surface morphologies of PEDOT/SAFs were studied by SEM (Fig. 6). There were only a handful of macropores in the surface of all PEDOT/SAFs except A0-PEDOT/SAFs. And their macropore sizes were smaller than A0.01-PEDOT/SAFs and A0-PEDOT/SAFs.

Their N_2 adsorption-desorption isotherm and pore size distribution shown that all samples still have mesopores and macropores Fig. 7 and Table 3. But their specific surface area was much lower than corresponding SAFs. In addition, the peak heights of dv(log) were all decrease markedly. Those suggest that the pores

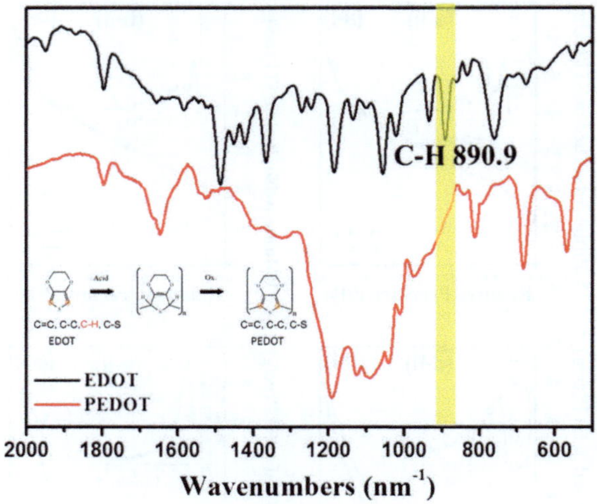

Fig. 4 FTIR spectrum of EDOT and PEDOT

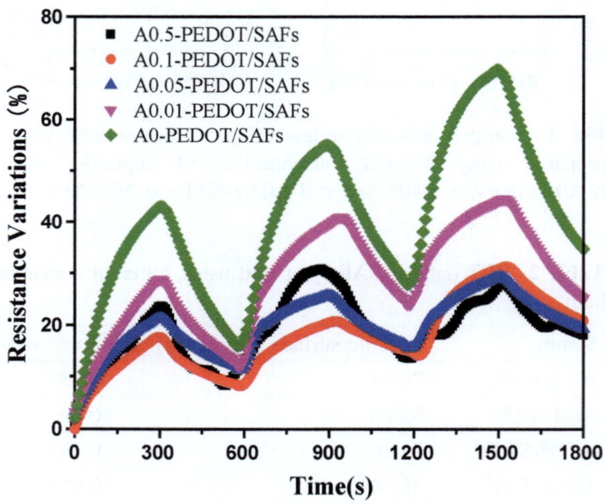

Fig. 5 Response of different PEDOT/SAFs under cyclic exposure of NH_3

narrows down and even were partly blocked by PEDOT, and this limited access for NH_3 gas. Because A0-SAFs have largest mesopore and macropore sizes, their pores have enough port area to pass NH_3 gas. Especially, there were lots of macropores retained in the surface of A0-PEDOT/SAFs. This ensures rapid entrance of NH_3 gas through macropores in the surface to mesopores in the middle ring region. Therefore, A0-PEDOT/SAFs have best gas sensitivity. Based on the analysis, we can expect the gas sensitivity should further improve, if the pore size of substrate is larger or the thickness of PEDOT is thinner. And it will be researched in our next work.

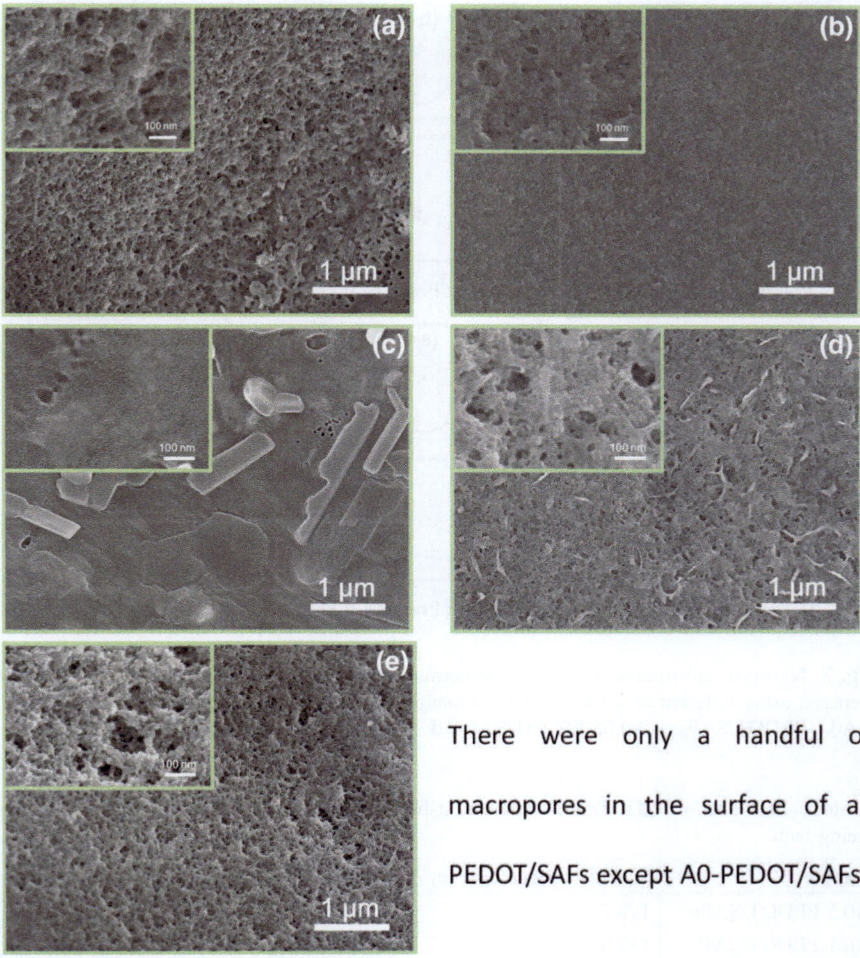

Fig. 6 SEM of surface morphologies of PEDOT/SAFs, **a** A0.5-PEDOT/SAFs; **b** A0.1-PEDOT/SAFs; **c** A0.05-PEDOT/SAFs; **d** A0.01-PEDOT/SAFs; **e** A0-PEDOT/SAFs, inserts are images of partial enlargement and the scale bar is 100 nm

Summary

In conclusion, we prepared a kind of silica aerogel fibers (SAFs) with high specific surface area (maximum one is 835.8 m^2/g), hierarchical structure and hollow structure by wet spinning. Based on SAFs, we further prepared a series of gas sensor materials, A0.5-PEDOT/SAFs, A0.1-PEDOT/SAFs, A0.05-PEDOT/SAFs, A0.01-PEDOT/SAFs and A0-PEDOT/SAFs. And A0-PEDOT/SAFs shows the best gas sensitivity. The reasons were analyzed via pore structure and morphology of PEDOT/SAFs. Finally, further improvement is put forwarded.

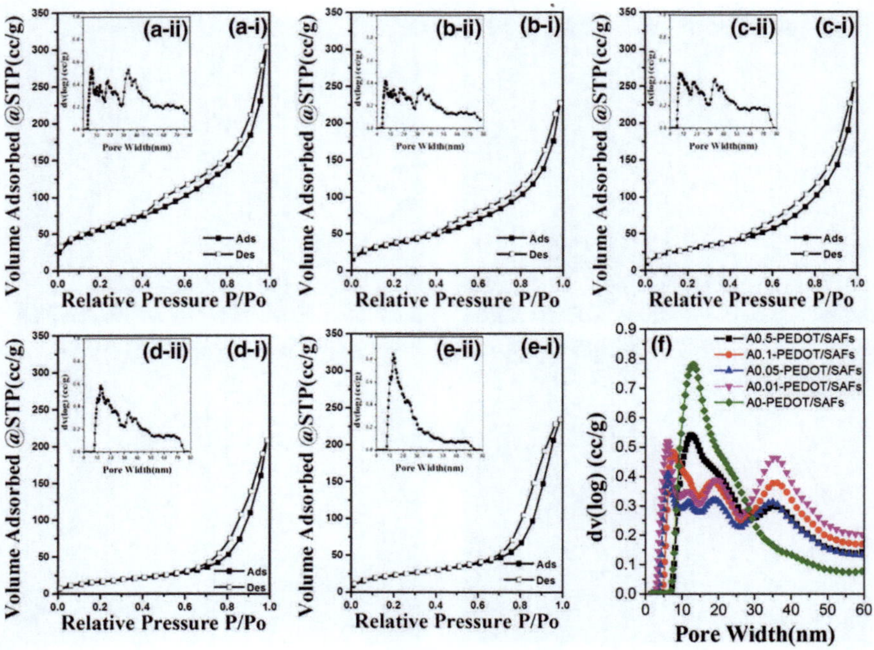

Fig. 7 Nitrogen adsorption-desorption isotherms and pore-size distribution curve of SAFs prepared using different concentrations of sulphuric acid as ageing bath. **a** A0.5-PEDOT/SAFs; **b** A0.1-PEDOT/SAFs; **c** A0.05-PEDOT/SAFs; **d** A0.01-PEDOT/SAFs; **e** A0-PEDOT/SAFs

Table 3 BET data of PEDOT/SAFs prepared using different concentrations of sulphuric acid as ageing bath

Sample	Specific surface area (m^2/g)	Pore volume (cc/g)	Pore width (nm)
A0.5-PEDOT/SAFs	199.7	0.47	5.88
A0.1-PEDOT/SAFs	132.0	0.35	5.88
A0.05-PEDOT/SAFs	105.6	0.40	6.79
A0.01-PEDOT/SAFs	60.1	0.32	12.55
A0-PEDOT/SAFs	82.8	0.36	12.55

Acknowledgements This work was supported by Shanghai Fundamental Research Projects (Project No. 16JC1400701) and Program for Changjiang Scholars and Innovative Research Team in University (IRT16R13).

References

1. J. Zhang, C. Zhang, J. Xia, Q. Li, D. Jiang, X. Zheng, Mixed-potential NH_3 sensor based on Ce0.8Gd0.2O1.9 solid electrolyte. Sens. Actuators B-Chem. **249**, 76–82 (2017)
2. A. Marutaphan, Y. Seekaew, C. Wongchoosuk, Hierarchically porous carbon derived from polymers and biomass: effect of interconnected pores on energy applications. Nanoscale Res. Lett. **12**, 90 (2017)
3. M. Eising, C.E. Cava, R.V. Salvatierra, A.J. Gorgatti Zarbin, L.S. Roman, Doping effect on self-assembled films of polyaniline and carbon nanotube applied as ammonia gas sensor. Sens. Actuators B-Chem. **245**, 25–33 (2017)
4. H. Hoang Thi, G. Ho Truong, H. Van Nguyen, T. Tran, T. Van Chu, Two-dimensional net-like SnO_2/ZnO heteronanostructures for high-performance H_2S gas sensor. Sens. Actuators B-Chem. **249**, 348–356 (2017)
5. X. Wang, S. Meng, W. Ma, J. Pionteck, M. Gnanaseelan, Z. Zhou, B. Sun, Z. Qin, M.F. Zhu, Fabrication and gas sensing behavior of poly (3, 4-ethylenedioxythiophene) coated polypropylene fiber with engineered interface. React. Funct. Polym. **112**, 74–80 (2017)
6. M. Tebyetekerwa, X. Wang, I. Marriam, P. Dan, S.Y. Yang, M.F. Zhu, Green approach to fabricate Polyindole composite nanofibers for energy and sensor applications. Mater. Lett. **209**, 400–403 (2017)
7. U. Lange, N.V. Roznyatouskaya, V.M. Mirsky, Conducting polymers in chemical sensors and arrays. Anal. Chim. Acta **614**, 1–26 (2008)
8. D. Li, J. Huang, R.B. Kaner, Polyaniline nanofibers: a unique polymer nanostructure for versatile applications. Acc. Chem. Res. **42**, 135–145 (2009)
9. S. Virji, J.X. Huang, R.B. Kaner, B.H. Weiller, Polyaniline nanofiber gas sensors: examination of response mechanisms. Nano Lett. **4**, 491–496 (2004)
10. W. Zhou, Y. Guo, H. Zhang, Y. Su, M. Liu, B. Dong, A highly sensitive ammonia sensor based on spinous core-shell PCL-PANI fibers. J. Mater. Sci. **52**, 6554–6566 (2017)
11. Z. Jing, J. Zhan, Fabrication and gas-sensing properties of porous ZnO nanoplates. Adv. Mater. **20**, 4547–4551 (2008)
12. L.E. Kreno, K. Leong, O.K. Farha, M. Allendorf, R.P. Van Duyne, J.T. Hupp, Metal-organic framework materials as chemical sensors. Chem. Rev. **112**, 1105–1125 (2012)
13. C.-J. Chang, C.-K. Lin, C.-C. Chen, C.-Y. Chen, E.-H. Kuo, Gas sensors with porous three-dimensional framework using TiO_2/polymer double-shell hollow microsphere. Thin Solid Films **520**, 1546–1553 (2011)
14. S. Dutta, A. Bhaumik, K.C.W. Wu, Hierarchically porous carbon derived from polymers and biomass: effect of interconnected pores on energy applications. Energy Environ. Sci. **7**, 3574–3592 (2014)
15. D. Fu, C. Zhu, X. Zhang, C. Li, Y. Chen, Two-dimensional net-like SnO_2/ZnO heteronanostructures for high-performance H_2S gas sensor. J. Mater. Chem. A **4**, 1390–1398 (2016)
16. E. Montes, U. Schwingenschlogl, Superior selectivity and sensitivity of blue phosphorus nanotubes in gas sensing applications. J. Mater. Chem. C **5**, 5365–5371 (2017)
17. T. Liu, J. Liu, Q. Liu, R. Li, H. Zhang, X. Jing, J. Wang, Shape-controlled fabrication and enhanced gas sensing properties of uniform sphere-like $ZnFe_2O_4$ hierarchical architectures. Sens. Actuators B-Chem. **250**, 111–120 (2017)
18. N. Wei, H. Cui, X. Wang, X. Xie, M. Wang, L. Zhang, J. Tian, Hierarchical assembly of In_2O_3 nanoparticles on ZnO hollow nanotubes using carbon fibers as templates: enhanced photocatalytic and gas-sensing properties. J. Colloid Interface Sci. **498**, 263–270 (2017)
19. B. Le Ouay, M. Boudot, T. Kitao, T. Yanagida, S. Kitagawa, T. Uemura, Nanostructuration of PEDOT in porous coordination polymers for tunable porosity and conductivity. J. Am. Chem. Soc. **138**, 10088–10091 (2016)

20. Y. Yang, S. Li, W. Yang, W. Yuan, J. Xu, Y. Jiang, In situ polymerization deposition of porous conducting polymer on reduced graphene oxide for gas sensor. ACS Appl. Mater. Interfaces **6**, 13807–13814 (2014)
21. E. Zampetti, S. Pantalei, A. Muzyczuk, A. Bearzotti, F. De Cesare, C. Spinella, A. Macagnano, A high sensitive NO_2 gas sensor based on PEDOT-PSS/TiO_2 nanofibres. Sens. Actuators B: Chem. **176**, 390–398 (2013)
22. J. Tan, M. Dun, L. Li, J. Zhao, W. Tan, Z. Lin, X. Huang, Synthesis of hollow and hollowed-out Co_3O_4 microspheres assembled by porous ultrathin nanosheets for ethanol gas sensors: responding and recovering in one second. Sens. Actuators B-Chem. **249**, 44–52 (2017)
23. W. Tan, J. Tan, L. Li, M. Dun, X. Huang, Nanosheets-assembled hollowed-out hierarchical Co_3O_4 microrods for fast response/recovery gas sensor. Sens. Actuators B-Chem. **249**, 66–75 (2017)
24. B. Winther-Jensen, O. Winther-Jensen, M. Forsyth, D.R. Macfarlane, High rates of oxygen reduction over a vapor phase-polymerized PEDOT electrode. Science **321**, 671–674 (2008)
25. B. Winther-Jensen, K. West, Vapor-phase polymerization of 3,4-ethylenedioxythiophene: a route to highly conducting polymer surface layers. Macromolecules **37**, 4538–4543 (2004)
26. B. Winther-Jensen, J. Chen, K. West, G. Wallace, Vapor phase polymerization of pyrrole and thiophene using iron(III) sulfonates as oxidizing agents. Macromolecules **37**, 5930–5935 (2004)

A Novel Janus Thermosensitive Hydrogel with Bidirectional Shape Changing Property

Yan Li, Kai Hou, Peiling Wei, Zhouqi Meng and Meifang Zhu

Abstract A facile strategy has been successfully established to fabricate janus nanocomposite (NC) hydrogel with anisotropic swelling behavior, which results in inhomogeneous shape change. This janus hydrogel was fabricated via two-step free radical polymerization by using clay as crosslinker and N-isopropylacrylamide (NIPAM) as monomer nut with different molar ratio in two layers, respectively. The free radical polymerization of PNIPAM at the interface provided a homogeneous conjunctive manner to tolerate the inhomogeneous tension of both layer after swelling. Moreover, the novel janus hydrogel could be well designed with bidirectional shape changing under temperature variation. Therefore, diverse thermoresponsive architectures could be built to provide a great potential in the field of cargo transportation, soft robots and actuators.

Keywords Nanocomposite hydrogel · Shape change · N-sopropyl acrylamide Anisotropic swelling

Introduction

Smart hydrogel is an important intelligent material with three dimensional network structures, which possess excellent swelling and deswelling behavior under external stimulus, such as temperature [1], pH [2], and light [3]. However, general smart hydrogel always shows homogeneous swelling behavior, which means that the dimension changing ratio was approximate in all direction because the homogeneous crosslinking structure formed in the fabricated process. Therefore, the promising

Y. Li · K. Hou · P. Wei · Z. Meng · M. Zhu (✉)
State Key Laboratory for Modification of Chemical Fibers and Polymer Materials, College of Materials Science and Engineering, Donghua University, Shanghai 201620, China
e-mail: zhumf@dhu.edu.cn

Y. Li
e-mail: li_yan0305@163.com

Y. Han (ed.), *Advanced Functional Materials*,
https://doi.org/10.1007/978-981-13-0110-0_84

applications of these homogeneous hydrogel were tissue engineering and drug delivery et al. [4, 5]. Naturally, the anisotropic deformation is the foundation to realize the function of body life and manual devices. Based on this phenomenon, more and more studies focus on the fabrication of new type of hydrogel devices with anisotropic deformation on macroscope [6–9]. For example, this hydrogel has been used as smart actuators to biomimic manipulators under external stimuli [10, 11].

To date, there are two main approaches have been utilized for designing anisotropic hydrogel. First, discrepant network structure was constructed in the process of polymerization by external fields interference such as force and temperature. e.g., Luo presented a heterobifunctional crosslinker via hydrothermal synthesis [12] and successfully fabricated elastic anisotropic hydrogel with gradient porous structure because of the pressure effection in hydrothermal reactor. Akashi reported a PNIPAM based hydrogel with asymmetrical pore structure by well-designed asymmetrical silica template electrophoresis and photopolymerization, sequentially [13]. In addition, freezing-coasting method is successfully used to fabricate asymmetry hydrogel [14]. However, such strategies just produce particular gradient structure, which means that it is difficult to realize diversity of deformation to realize the complex application of hydrogel as intelligent device. Second, many researchers take efforts on multilayer construction of hydrogel, in which each layer possess particular function and property by physical or chemistry connected of interface among each layers such as supramolecular [15], electrophoretic [16], host-gust interaction [17] and metal-ligand [18]. But all the method need specially-functional monomer, which confined the adaption seriously. So a simple way was strongly anticipated to construct the anisotropic hydrogel with multilayer. Chu focused on the design of facile hydrogel actuator by rearranging hydrogen bonding between polymers and clay nanosheets in diverse hydrogel subunits. This manner can realize the diversity transformation of hydrogel by assembling different subunits conveniently, but the variation of hydrogel shape was monotonously [19]. Therefore, a simple way to fabricate multilayer hydrogel with bidirectional shape change is crucial.

In this study, based on the thermoresponsive poly(N-isopropylacrylamide) (PNIPAm) hydrogel, we established a two-step free radical polymerization process and successfully constructed a janus hydrogel with reversible multidimensional deformation property due to temperature–triggered swelling-deswelling process, which provided a great potential in the field of cargo transportation, soft robots and actuators.

Materials and Methods

Materials. N- isopropyl acrylamide(NIPAM) (Acros Co. Ltd.) is purified by recrystallization using a toluene/n-hexane mixture. The physical cross-linker nanoclay (Laponite XLS, 92.32 wt% $Mg_{5.34}Li_{0.66}Si_8O_{20}(OH)_4Na_{0.66}$ and 7.68 wt% $Na_4P_2O_7$) with platelets 30 nm in diameter and 1 nm thick, is provided by

Rockwood Ltd. UK. The initiator Potassium persulfate (KPS) and accelerator N, N, N′, N′-tetramethyl ethylene diamine (TEMED), (Sinopharm Chemical Reagent Co. Ltd.,) are used directly, without any further purification. The purified water is obtained by a water purification system (Heal Force Bio-Meditech Holdings Ltd.).

Synthesis of PNIPAM/Clay Janus Nanocomposite Hydrogel. The fabrication process of PNIPAM/clay NC hydrogel was shown in Scheme 1. Typically, nanoclay and purified NIPAM were added into water (18 mL) in sequence, and then dispersed by stirring uniformly. After that, the initiator (KPS, 2 wt%, 2 ml) and accelerator TEMED (30 μL) were added into NIPAM/clay mixture and transferred into a polytetrafluoroethylene (PTFE) mold, in which the in situ free radical polymerization was carried out to form the first layer of janus hydogel. This process was illustrated by the orange part in Scheme 1. Afterwards, when in situ free radical polymerization was carried out for few minutes, another clay/NIPAM dispersion with different molar ratio was transferred onto the first layer and keep it at room temperature until the polymerization process was completed totally. The formed janus hydrogel was coded in terms of $a(C_{m1}N_{x1})/b(C_{m2}N_{x2})$. "*a*" stands for the lower layer of janus hydrogel in the mold, "*b*" stands for the upper layer, "m" and "x" represent the clay and NIPAM quality ratio in the hydrogel, respectively. The janus hydrogel was made by mold of different size. The hydrogel samples for interface mechanical property test were made into diameter 5 mm, height of 20 mm. The samples were $50 \times 5 \times 20$ mm^3 for swelling. In addition to, the janus hydrogel could be cut into various shapes to designing different self-shaping devices.

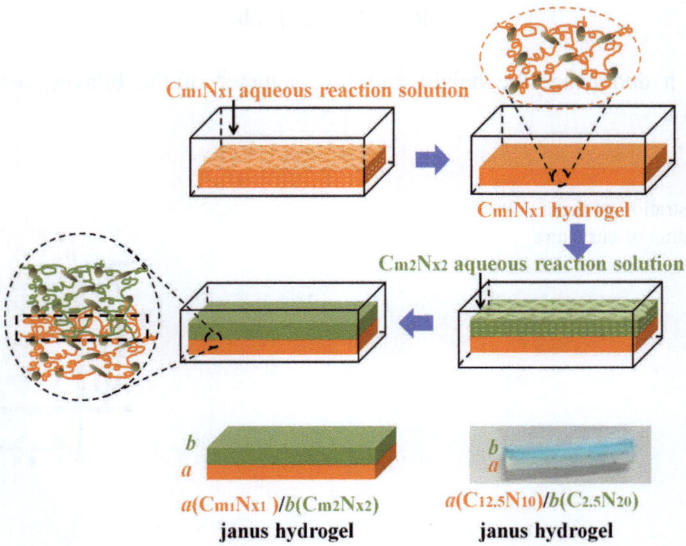

Scheme 1 Schematic diagram of preparing $a(C_{m1}N_{x1})/b(C_{m2}N_{x2})$ bialyer hydrogel

Characterization

1. The morphologies of janus hydrogel was observed by scanning electron microscopy (SEM, JSM-5600LV, JEOL). Before observation, the sample was brittle fractured in liquid nitrogen and then lyophilized by a freeze drier (Lyo Quest, Telstar) at −70 °C for about 48 h.
2. The interface mechanical property of janus hydrogel was measured by tension testing, which were performed on a tensile machine (Instron 5969, Instron company). The cylindrical hydrogel specimens with diameter of 5 mm and height of 20 mm were used for tension tests at a stretch speed of 10 mm/min.
3. The thermosensitivity of PNIPAM hydrogel was determined by turbidity measurement at 610 nm using a PerkinElmer UV-vis Lambda 35 spectrophotometer with a heating rate of 0.5 °C/min from 15 to 45 °C.
4. The swelling behavior was carried out by gravimetric method at different temperature. The swelling ratio (SR) was calculated as:

   $$SR = (W_r - W_d)/W_d. \quad (1)$$

   where W_r is the weight of the swollen hydrogel and W_d is the weight of the dried gel.
5. The $a(C_{12.5}N_{10})/b(C_{2.5}N_{20})$ janus hydrogel was put into deionized water, which was placed on a heating stage (TS62, Instec) to change the temperature from 25 to 45 °C for more than 30 min to reach fully swollen. The radius of curvature (R) was calculated using the equation below [20]

   $$R = \left(h^2 + a^2\right)/2h. \quad (2)$$

   where h and a are the height and the diameter of the bilayer, respectively (Fig. 1).

Fig. 1 Illustration of the a and h in radius of curvature (R) equation

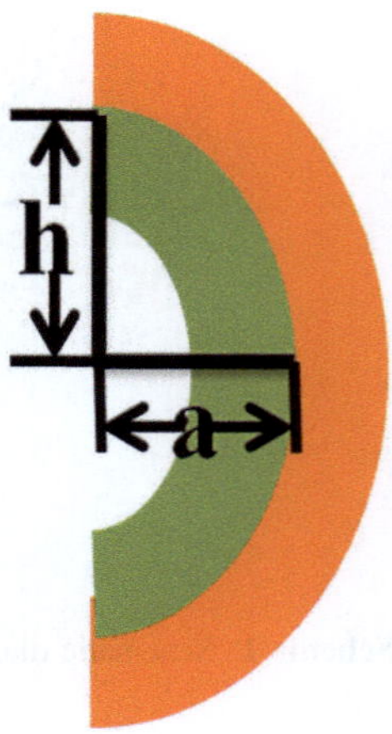

Results and Discussion

Morphology of the Janus Hydrogel. In this study, the $a(C_{12.5}N_{10})/b(C_{2.5}N_{20})$ janus hydrogel was chosen. The inorganic clay was introduced in $a(C_{12.5}N_{10})$ and $b(C_{2.5}N_{20})$ as physical cross-linker, which promised the uniform porous structure of hydrogel as shown in Fig. 2. Because of the different crosslinking density of both side, the pore size was different obviously to form the anisotropic structure of hydrogel, which induced the anisotropic swelling behavior. Moreover, the interface of both side was obvious but not separately. It is because that the interface was formed by free radical polymerization, in which the PNIPAM chain was chemical linked both side. As a result, the interface was well compact to ensure the excellent mechanical property which will be confirmed later.

The Mechanical Property in Interface of $a(C_{12.5}N_{10})/b(C_{2.5}N_{20})$ Janus Hydrogel. In order to determine the relationship between interface property and polymerization process, we measured the interface bonding force via tensile test under the condition that the second layer was initiated after different reaction times (5/10/15/20/25 and 30 min) of the first layer. We can see that, with the reaction times of first layer increased, the strain increased firstly and then decreased, which was shown in Fig. 3. This phenomenon was due to that, when reaction time was short, the first layer was still in solution state, as a result, the two kinds of solution maybe mixed after polymerization. With the reaction time increasing, the first layer started to gelled but not completely, which means that there were numerous residual monomer at the interface and then the residual monomer could be polymerized with the NIPAM monomer belonging to the second layer and result in the better interface adhesion. Further on, when the polymerization of first layer was accomplished, the polymerization between two layers was decreased, and the fracture stress was lower than before.

The relationship between the fracture stress and reaction time was calculated based on the following equation:

$$y = -0.02288x^2 + 0.81205x + 3.48159. \tag{3}$$

where y represents the fracture stress, x stands for time.

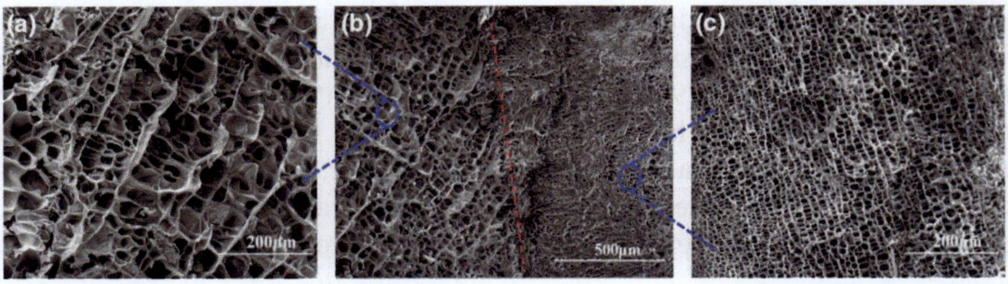

Fig. 2 SEM images of **a** $b(C_{2.5}N_{20})$, **b** $a(C_{12.5}N_{10})/b(C_{2.5}N_{20})$, and **c** $a(C_{12.5}N_{10})$ freeze-dried hydrogel

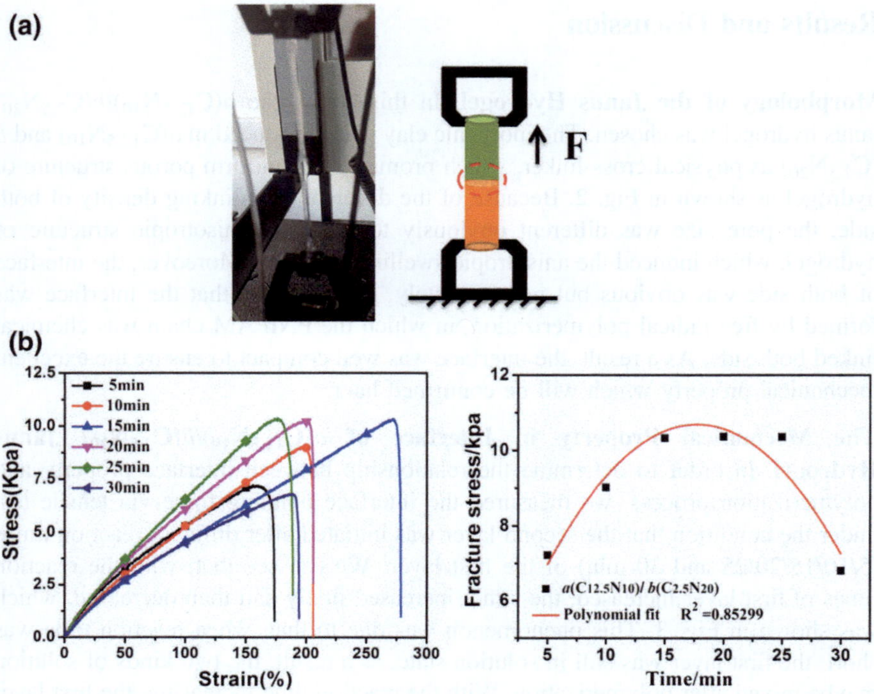

Fig. 3 Tensile tests **a** Schematic tensile test of the $a(C_{12.5}N_{10})/b(C_{2.5}N_{20})$ **b** Fracture stress and strain values of $a(C_{12.5}N_{10})/b(C_{2.5}N_{20})$ at the $a(C_{12.5}N_{10})$ was initiated after different reaction times

And from this equation, we can calculate that the best time to add the second layer was at the first layer reaction about 17 min.

Dynamic Thermoresponsive Anisotropic Swelling Behaviors of Bialyer NC Hydrogel. The hydrogel has janus structure, which leads to bilayer bending to different directions with different temperatures. The responsive bending degree was calculated by curvature value. Accordingly, the higher bending degree of the hydrogel is positive corelated with the possess absolute curvature value. Figure 4 shows the curvature of the bilayer hydrogel, which gradually decreased from positive to negative values with the temperature decreasing from 45 to 23 °C. When $a(C_{12.5}N_{10})/b(C_{2.5}N_{20})$ hydrogel straight in deionized water at 23 °C, the curvature was became negative and stood for bending toward the *a* side. In the process of the temperature increased to 29 °C, the curvature reached a maximum value. With the temperature keep raising, hydrogel unfolded slowly to be straight again and bent toward the *b* even formed a circle at the end. Meanwhile, relative length, which was defined as the ratio of the length of hydrogel strip at certain time to its initial length, deceased along with the temperature changing process. In brief, dynamic

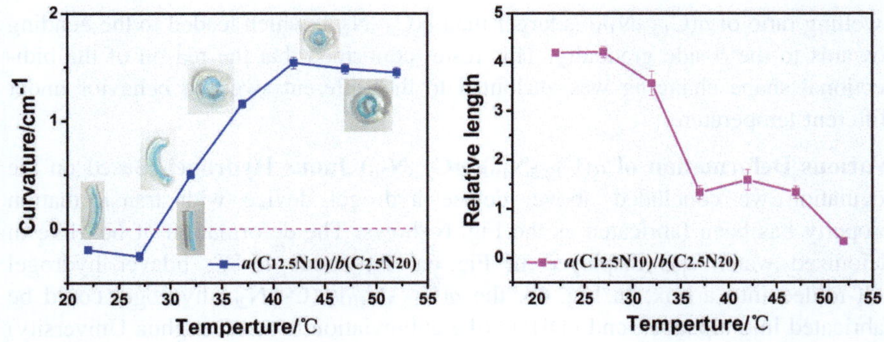

Fig. 4 Curvature value and relative length of $a(C_{12.5}N_{10})/b(C_{2.5}N_{20})$ hydrogel with temperature increasing from 23 to 45 °C

thermoresponsive curvature value of the $a(C_{12.5}N_{10})/b(C_{2.5}N_{20})$ divided into two parts: at first, the janus hydrogel bent towards *a*, and curvature value decreasing with the temperature from 23 to 30 °C; and then, it bent towards *b*, and curvature value upon increasing with the temperature rising from 30 to 45 °C.

In view of the above results, a self-bending deformation of the bilayer hydrogel with the swelling changes of every layer is proposed in Fig. 5. The $b(C_{2.5}N_{20})$ hydrogel swelling ratio was larger than $a(C_{12.5}N_{10})$ hydrogel when temperature lower than 34 °C. It was attributed to the crosslinking density of $b(C_{2.5}N_{20})$ is higher $a(C_{12.5}N_{10})$ hydrogel, which resulted in a higher water retention properties of $b(C_{2.5}N_{20})$. However, hydrogel was featured with different volume phase transition temperatures (VPTTs). As the Uv-vis result shows (Fig. 6), the $b(C_{2.5}N_{20})$ hydrogel got to (VPTTs) first at about 29.7 °C which trigger the deswelling process, therefore the swelling ratio of $b(C_{2.5}N_{20})$ began to approach that of $a(C_{12.5}N_{10})$. With the temperature keep rising to 34 °C, the swelling of $a(C_{12.5}N_{10})$ and deswelling of $b(C_{2.5}N_{20})$ diminished the gap of swelling ratio. After 34 °C, the

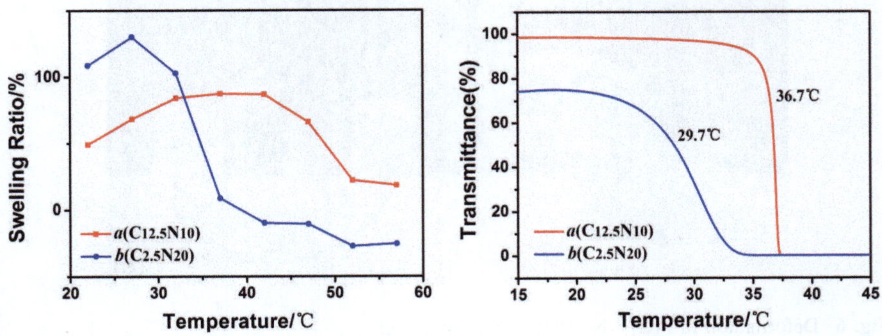

Fig. 5 The swelling ratio and the Transmittance of $a(C_{12.5}N_{10})$ and $b(C_{2.5}N_{20})$ hydrogel with temperature increasing from 23 to 45 °C

swelling ratio of $a(C_{12.5}N_{10})$ is larger than $b(C_{2.5}N_{20})$, which leaded to the bending towards to the b side gradually. This result confirmed that the reason of the bidirectional shape changing was attributed to the different swelling behavior under different temperature.

Various Deformation of $a(C_{12.5}N_{10})/b(C_{2.5}N_{20})$ Janus Hydrogel. Based on the regulation we concluded above, devise hydrogel device with transformation property has been fabricated as the Fig. 6 shows. The deformation of box-like in deionized water was displayed in Fig. 6a. At about 45 °C, bilayer hydrogel self-folded into a box. In Fig. 6b, the $a(C_{12.5}N_{10})/b(C_{2.5}N_{20})$ hydrogel could be fabricated into alphabet, and "DHU" (the abbreviation of we Donghua University) was obtained with temperature from 23 to 45 °C. And also built the shape of "S" and "U" showed in Fig. 6c. These can stand in the table which proved this janus hydrogel has certain mechanical properties to stand external pressure. Overall, the janus hydrogel can be easily designed by varying the number, size and shape of the hydrogel layers, which has the prospect on soft machine and other sensitive acuators.

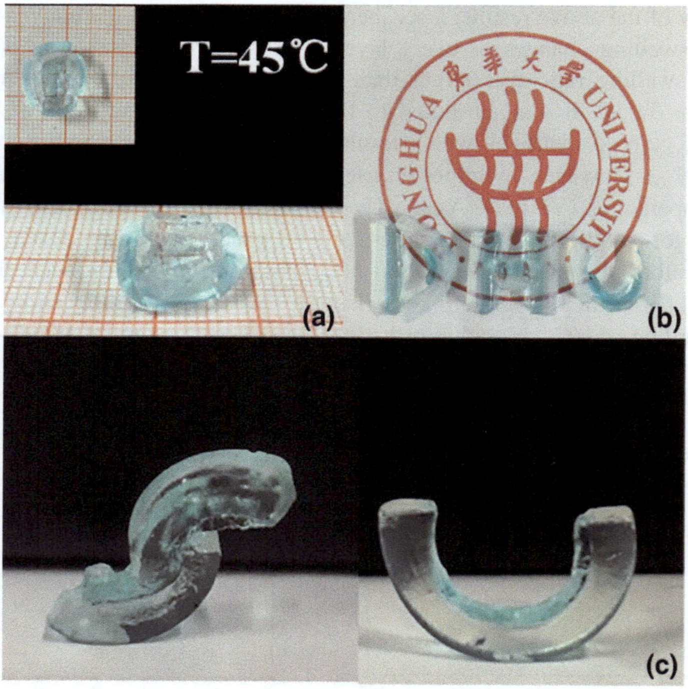

Fig. 6 Deformation of $a(C_{12.5}N_{10})/b(C_{2.5}N_{20})$ hydrogel

Conclusions

In this study, a facile method were developed for fabricating janus hydrogel with inhomogeneous structure, and good mechanical properties were obtained due to the strong hydrogen bonding between every layer. Moreover, the hydrogel can bend towards different directions varying with the change of temperature, and this process can be recycled when the temperature is recycling. Besides, this nanomaterial can also be used under other stimulations just by adjusting the layer's component. So it has highly promising for the development of stimuli responsive hydrogel actuators in the soft robots and actuators fields.

Acknowledgements This research is financially supported by National Key Research and Development Program of China (2016YFA0201702/2016YFA0201700), Project of Shanghai International Science and Technology Cooperation Fund (14520710200), and Key Program of National Natural Science Foundation of China (51733002).

References

1. T.Y. Liu, S.H. Hu, D.M. Liu, Biomedical nanoparticle carriers with combined thermal and magnetic responses. Nano Today **4**, 52–65 (2009)
2. M. Koenig, R. Kumar, C. Hussal, pH-Responsive Aminomethyl functionalized poly (p-xylylene) coatings by chemical vapor deposition polymerization. Macromol. Chem. Phys. **218** (2017)
3. H.J. Cho, M. Chung, M.S. Shim, Engineered photo-responsive materials for near-infrared-triggered drug delivery. J. Ind. and Eng. Chem. **31**, 15–25 (2015)
4. M. Mehrali, A. Thakur, C.P. Pennisi, Nanoreinforced hydrogel for tissue engineering: biomaterials that are compatible with load-bearing and electroactive tissues. Adv. Mater. **29** (2017)
5. H.R. Culver, J.R. Clegg, N.A. Peppas, Analyte-Responsive hydrogels: intelligent materials for biosensing and drug delivery. Accounts. Chem. Res. **50**, 170–178 (2017)
6. C. Ma, X. Le, X. Tang, A Multiresponsive anisotropic hydrogel with macroscopic 3D complex deformations. Adv. Funct. Mater. **26**, 8670–8676 (2016)
7. J. Kim, C. Kim, Y.S. Song, Reversible self-bending soft hydrogel microstructures with mechanically optimized designs. Chem. Eng. J. **321**, 384–393 (2017)
8. S. Wu, F. Yu, H. Dong, A hydrogel actuator with flexible folding deformation and shape programming via using sodium carboxymethyl cellulose and acrylic acid. Carbohydr. Polym. **173**, 526–534 (2017)
9. E. Palleau, D. Morales, M.D. Dickey, Reversible patterning and actuation of hydrogel by electrically assisted ionoprinting. Nat. Commun. **4**, 2257 (2013)
10. C. Yao, Z. Liu, C. Yang, Poly (N-isopropylacrylamide)-Clay nanocomposite hydrogel with responsive bending property as temperature-controlled manipulators. Adv. Funct. Mater. **25**, 2980–2991 (2015)
11. Y. Haraguchi, T. Shimizu, T. Sasagawa, Fabrication of functional three-dimensional tissues by stacking cell sheets in vitro. Nat. Protoc. **7**, 850–858 (2012)
12. R. Luo, J. Wu, N. Dinh, Gradient porous elastic hydrogel with shape-memory property and anisotropic responses for programmable locomotion. Adv. Funct. Mater. **25**, 7272–7279 (2016)

13. T. Asoh, M. Matsusak, T. Kaneko, Fabrication of temperature-responsive bending hydrogel with a nanostructured gradient. Adv. Mater. **20**, 2080–2083 (2008)
14. X. Peng, C. He, J. Liu, Biomimetic jellyfish-like PVA/graphene oxide nanocomposite hydrogel with anisotropic and pH-responsive mechanical properties. J. Mater. Sci. **51**, 5901–5911 (2016)
15. C. Qin, Y. Feng, W. Luo, A supramolecular assembly of cross-linked azobenzene/polymers for a high-performance light-driven actuator. J. Mater. Chem. A. **3**, 16453–16460 (2015)
16. T.A. Asoh, A. Kikuchi, Electrophoretic adhesion of stimuli-responsive hydrogel. Chem. Commun. **46**, 7793–7795 (2010)
17. A. Harada, R. Kobayashi, Y. Takashima, Macroscopic self-assembly through molecular recognition. Nature Chem. **3**, 34–37 (2011)
18. H. Abe, Y. Hara, S. Maeda, Surface modification method for adhesion of gels. Chem. Lett. **43**, 243–245 (2013)
19. C. Yao, Z. Liu, C. Yang, Smart hydrogel with inhomogeneous structures assembled using nanoclay-cross-linked hydrogel subunits as building blocks. ACS Appl. Mat. Interfaces. **8**, 21721–21730 (2016)
20. J. Duan, X. Liang, K. Zhu, Bilayer hydrogel actuators with tight interfacial adhesion fully constructed from natural polysaccharides. Soft Matter **13**, 345–354 (2017)

Surface Modification of Quartz Fiber by Aqueous Plasma Electrolysis

Yang Meng, Weiwei Chen, Huanwu Cheng and Yuping Zhang

Abstract The mechanical properties of quartz and ceramic fiber are easily destroyed at high temperature. There are many process methods for continuous fiber surface treatment, commonly used in a chemical vapor deposition method and sol gel method. In this work, a method of preparing surface coating is proposed to protect the mechanical properties of fiber at high temperature, at the same time to achieve high efficiency, low cost, and to meet the requirements of Engineering production. Treatment of quartz and ceramic fibers in a continuous liquid equal plasma, plasma bombardment of the fibers is carried out, in order to achieve the purpose of preparation of coatings. After coating the surface of the quartz fiber, the ceramic coating is coated by the liquid plasma process, and the breaking strength is tested after the high temperature treatment.

Keywords Fiber · Liquid phase · Plasma · Coating

Introduction

Quartz fiber is a kind of high performance inorganic fiber, and is widely used in ceramics, organic silicon, nitride phosphate matrix material, due to its low dielectric constant, low dielectric loss, and low elastic modulus [1, 2]. However, quartz fiber is brittle and poor wear resistance, significantly limiting its applications in indus-

Y. Meng · W. Chen (✉) · H. Cheng · Y. Zhang
Department of Materials Science and Engineering,
Beijing Institute of Technology, Beijing 100081, China
e-mail: wwchen@bit.edu.cn

Y. Meng
e-mail: johnzhengnan@163.com

H. Cheng
e-mail: chenghuanwu@bit.edu.cn

Y. Zhang
e-mail: stu_zhangyuping@126.com

Y. Han (ed.), *Advanced Functional Materials*,
https://doi.org/10.1007/978-981-13-0110-0_85

tries. Therefore, there is an urgent demand for preparing a protective coating on quartz fiber.

Ceramic coatings, such as Y_2O_3, ZrO_2 and Al_2O_3, attract a wide attention, with their high hardness, high temperature resistance, wear resistance and corrosion resistance [3, 4]. Meanwhile, Al_2O_3 ceramic has the advantages of low cost, high hardness, low friction coefficient, high temperature resistance, oxidation resistance, corrosion resistance and abrasion resistance. Therefore, it is widely used as thermal insulation coating, and superconducting insulating coating [5, 6].

At present, the methods prepared the coating on the quartz fiber mainly include sol gel method: prepared by gel and after drying, sintering and preparation of ceramic coating [7, 8]. However, the method is easy to cause damage for the fiber, leading to the decline of fiber strength. Therefore, how to rapidly prepare the ceramic coating on the fiber surface without reducing the strength of the quartz fiber is an urgent problem to be solved [9].

Aqueous plasma electrolysis is a new technique for the Al_2O_3 coating deposition on the quartz fiber. The microstructure, oxidation resistance and tensile strength of the Al_2O_3-coated quartz fiber were investigated in the present manuscript. It was shown that the Al_2O_3 coating exhibited superior protection for the quartz fiber at elevated temperatures [10–12].

Experimental

The novel aqueous plasma electrolysis. The schematic diagram of continuously depositing coating on quartz fiber by the aqueous plasma electrolysis is shown in Fig. 1. The 50 g/L $AlCl_3$ (Chemical Reagent, Sinopharm Chemical Reagent Co., Ltd) is added to deionized water to prepare a suitable electrolyte solution, and the prepared electrolyte solution is placed in the reaction vessel. The cathode and anode electrode are introduced, and the electrode is connected with a high-frequency direct current power supply. The voltage is applied to the cathode and the anode to form a uniform plasma arc region. The quartz fiber continuously passed through the

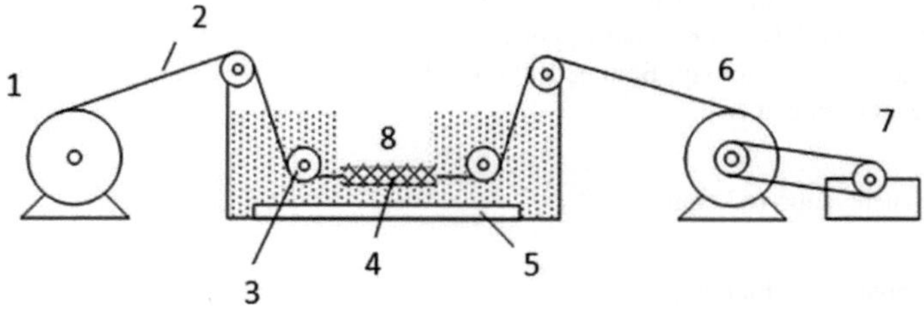

Fig. 1 The schematic diagram of continuously depositing coating on quartz fiber by the aqueous plasma electrolysis. 1-Paying out machine; 2-Quartz fiber; 3-Crimping wheel; 4-Cathode plasma arc region; 5-Anode; 6-Take-up machine; 7-Motor; 8- Electrolyte solution

plasma arc zone in a certain speed, where the ceramic coating was deposited on the quartz fiber, finally the coated fiber was coiled by the take-up machine.

Characterization of coatings. The morphologies of the coating were analyzed using a scanning electron microscope (SEM). The thermal stability of the coating was explored using a thermo gravimetric analyzer (TGA, STA 7300, Hitachi) from ambient temperature to 1300 °C under an air atmosphere at a heating rate of 10 K/min. In order to identify the high-temperature strength of the fiber, the fiber was annealed at 600 °C for 30 min in air, and then the tensile strength was tested according to GB/T 7689.5-2013.

Results and Discussion

Energy spectrum. Figure 2 shows the element composition and morphology of fibers prepared by a fixed winding machine at constant arc voltage. the SEM shows that an uneven coating has been deposited on the bright surface of the original quartz fibers. From the energy spectrum, it can be observed that the content of Al element is 1 wt%. The element of Si and O belong to the quartz fibers, while C element belongs to the epoxy resins covered on the fiber surface.

However, the deposition degree of the coating is negatively related to the speed of the take-up wheel, i.e., the slower the speed of the take-up wheel is, the more the deposition of the coating. And from the morphological analysis, different growth rates, there are different growth morphology, as a whole step like growth. It is obvious that the growth direction of the step is perpendicular to the direction of movement of the quartz fiber. For the specific mechanism, a lot of people have put forward the corresponding model, but they are difficult to fully explain.

Surface and cross-sectional morphologies. The Al_2O_3 ceramic coating was characterized as shown in Fig. 3a. A relatively dense, continuous and uniform

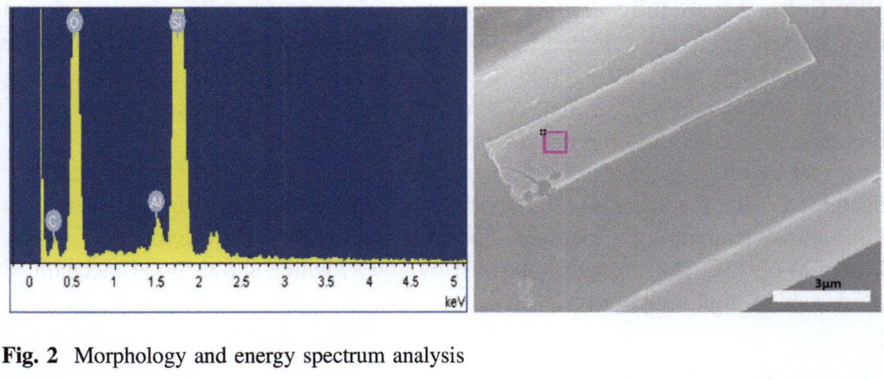

Fig. 2 Morphology and energy spectrum analysis

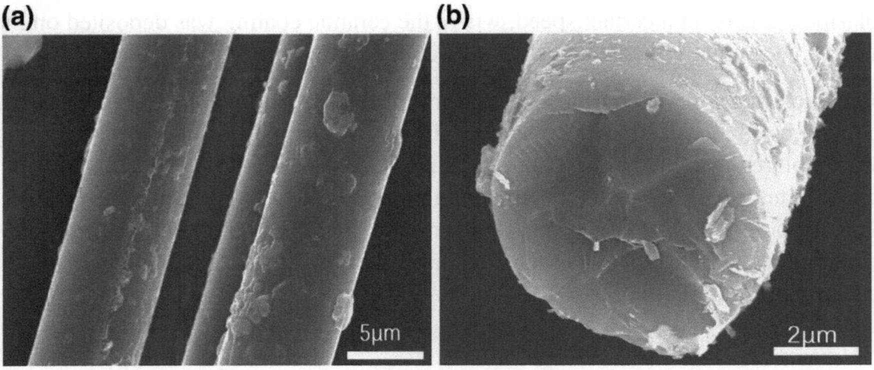

Fig. 3 SEM surface and cross-sectional morphologies of the Al_2O_3 ceramic coating on the quartz fiber, **a** ×4500 magnification, **b** ×10,000 magnification

coating was almost formed on each fiber within the fiber bundle as the surface backscattered morphology shown in Fig. 3a, b shows the cross-sectional morphologies of the Al_2O_3 coating. It could be shown that Al_2O_3 coating were successfully deposited on the quartz fiber.

Thermo-gravimetric analysis. The quartz fiber processing by liquid phase plasma, with epoxy resin have been removed from the surface, it can be regard as a random sample in the experiment. While the fiber had been heated in the air, the weight loss is analyzed, as shown in Fig. 4.

It is shown in Fig. 4 that the weight lose of untreated quartz fiber is about 1%. This phenomenon can be attribute to the decomposition of epoxy resin, which is usually used as a protective coating for quartz fiber. The TG curve obtained by liquid phase plasma shows a large mass lose, contributing by the decomposition of $Al(OH)_3$. The decomposition products of $Al(OH)_3$ are Al_2O_3 and H_2O, which also confirms that the composition of coatings prepared by liquid phase plasma is Al $(OH)_3$.

After processing coatings in specific parameter, samples were kept in dry oven at 100 °C for 2 h to fully dry the solution on fiber surface. The processed and unprocessed fibers were treated together in high temperature SX-G12123 furnace for 5 min at 450, 500, 550 and 600 °C, respectively. Then the samples were cooling at room temperature for 30 min. A tensile text was used to investigate the fibers' mechanical strength and the results were shown in Fig. 5.

It can be observed from Fig. 5 that the mechanical strength of fibers gradually decreased with the increase of temperature. Compared with original fibers, the mechanical strength of processed fibers does not decrease significantly. After heated at 600 °C, the strength improved by 164%, which shows good mechanical properties in high temperature.

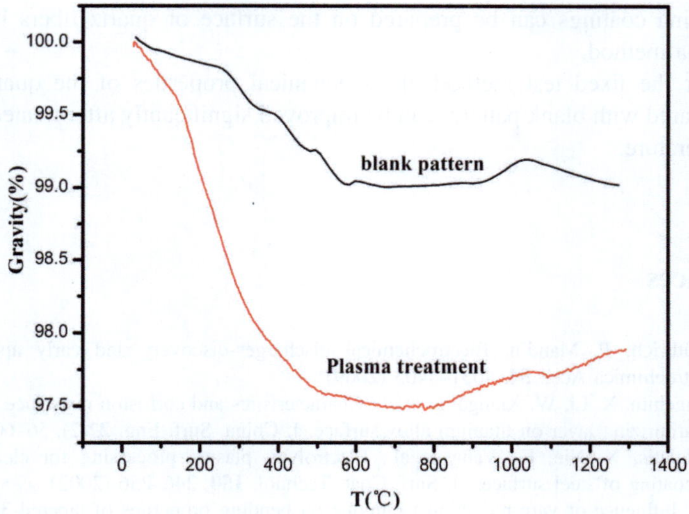

Fig. 4 TG contrast of quartz fiber treated by liquid plasma and blank quartz fiber

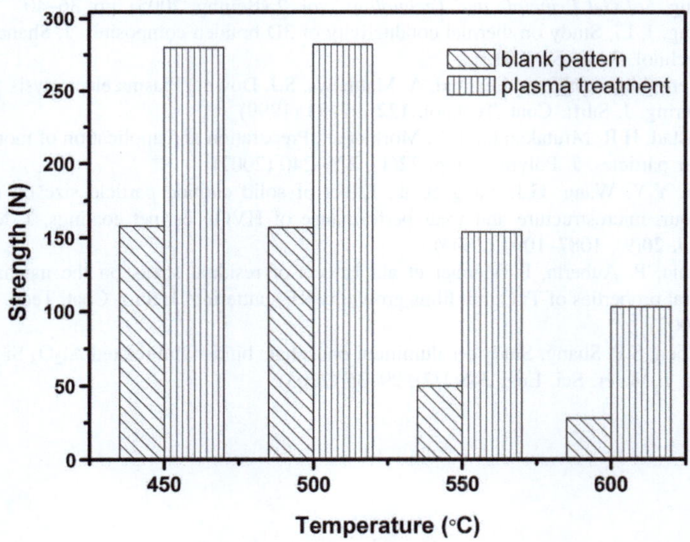

Fig. 5 Comparison of mechanical strength of fiber after treatment at high temperature

Conclusions

In this article, a method of preparing coatings on fiber surface by aqueous plasma electrolysis was analyzed, and the following conclusions were drawn:

(1) Alumina coatings can be prepared on the surface of quartz fibers by liquid plasma method.
(2) Under the fixed test method, the mechanical properties of the quartz fiber, compared with blank pattern, can be improved significantly after treated in high temperature.

References

1. R. Wüthrich, P. Mandin, Electrochemical discharges-discovery and early applications. J. Electrochimica Acta. **54**, 4031–4403 (2009)
2. H. Zongchun, X. Li, W. Xiangqing et al., Characteristics and corrosion resistance of Plasma Nitrocarburizing layer on titanium alloy surface. J. China, Surf. Eng. **22**(2), 56–60 (2009)
3. E.I. Meletis, X. Nie, F. Wang et al., Electrolytic plasma processing for cleaning and metal-coating of steel surfaces. J. Surf. Coat. Technol. **150**, 246–256 (2002)
4. Z. Liu, Influence of yarn-reduction technique on bending properties of tapered 3D braided composites. J. Mater. Rev. **28**(2), 85–90 (2014)
5. L. Chen, Y. Sun, M. Ma, Development of high performance fiber preforms. J. Mater. China **31**(10), 21–29 (2012)
6. J. Huang, *Sol Gel Principle and Technology*, vol. 2 (Beijing, 2005), pp. 36–40
7. L. Zhang, J. Li, Study on thermal conductivity of 3D braided composites. J. Shandong Text. Sci. Technol. **5**, 50–53 (2015)
8. A.L. Yerokhin, X. Nie, A. Leyland, A. Matthews, S.J. Dowey, Plasma electrolysis for surface engineering. J. Surft. Coat. Technol. **122**, 97–93 (1999)
9. J. Ugelstad, H.R. Mfutakamba, P.C. Mork et al., Preparation and application of monodisperse polymer particles. J. Polym. Symp. **72**(1), 225–240 (2007)
10. C.J. Li, Y.Y. Wang, G.J. Yang et al., Effect of solid carbide particle size on deposition behaviour, microstructure and wear performance of HVOF cermet coatings. J. Mater. Sci. Technol. **20**(9), 1087–1090 (2004)
11. A. Mania, P. Auberta, F. Mercier et al., Effects of residual stress on the mechanical and structural properties of Ti C thin films grown by RF sputtering. J. Surf. Coat. Tech. **194**, 190–195 (2005)
12. D.C. Chen, S.P. Shang, Study on aluminum phosphate binder and related Al_2O_3 Si C ceramic coating. J. Mater. Sci. Eng. **348**(1/2), 29–35 (2003)

GO/DNA/Au/PANi Nanocomposite with High Photoconductive Responses to Visible Light and NIR

Mingjun Gao, Shisheng Lv, Jianxun Qiu, Wei Du, Xintao Zhang, Xiaochun He, Xiangming Li, Xingfa Ma and Guang Li

Abstract In synthesis of nanomaterials with bottom-up approach, DNA (deoxyribonucleic acid) chain segment took an important role in self-assembly of nano/micro-structured materials. DNA chain segment not only has good recognition characteristics, but also holds excellent transporting properties of delivery for drugs, protein or gene. To functionalize and smart the nanocomposites for applications in several devices, in this paper, single-stranded DNA chain and graphene oxide (GO) were hybridized, and Au nanoparticles were formed in situ with chloroauric acid and aniline as reducing agent. The GO/DNA/Au/PANi (polyaniline) nanocomposite were characterized by several approaches. The photoconductive

M. Gao · S. Lv · J. Qiu · W. Du · X. Zhang · X. He · X. Li · X. Ma (✉)
School of Environmental and Material Engineering, Center of Advanced Functional Materials, Yantai University, Yantai 264005, China
e-mail: xingfazju@163.com; xingfamazju@aliyun.com

M. Gao
e-mail: gmj2188@126.com

S. Lv
e-mail: lvss19920119@qq.com

J. Qiu
e-mail: qjxun@163.com

W. Du
e-mail: dw311@yeah.net

X. Zhang
e-mail: zxtyt@163.com

X. He
e-mail: hexiaochun76@126.com

X. Li
e-mail: li_xiangming1981@126.com

G. Li
National Laboratory of Industrial Control Technology, Institute of Cyber-Systems and Control, Zhejiang University, Hangzhou 310027, China
e-mail: guangli@zju.edu.cn

Y. Han (ed.), *Advanced Functional Materials*,
https://doi.org/10.1007/978-981-13-0110-0_86

responses to weak visible light and 808 nm NIR (near infrared) were studied based on interdigital electrodes of Au on flexible polymer film substrate with casting method. The results indicate that the resulting nanocomposite exhibits good photo-switching behaviors, the ratio of On/Off is 1–2 orders of magnitudes, while the response and recovery are very rapidly. It can be used in developing an intelligent nano-carriers with external stimuli responses, light detector to NIR, nano-machine controlled by light, biomimetic materials, and so on.

Keywords DNA · Graphene oxide · Au · Photo-induced charge generation External stimuli response · Photoconductive properties

Introduction

It is well known that deoxyribonucleic acid (DNA) is one of the most important biological macromolecule since its ability to carry genetic information via hydrogen-bonded nucleobase pairs. DNA chain segment with different length can assembled many nanostructure with external stimuli response properties. Functional DNA nanostructure or nanocomposite would led to the creation of supramolecular systems with applications in artificial molecular machines, including DNA based walkers [1, 2], motors, tweezers, nanorobots, et al. The applications on relative DNA are very popularly, such as discrimination of metal ions [3], molecular recognition [4], cell imaging [5], biosensors [6–8], remote controlling DNA hydrogel [9–12], drug delivery [13], nanocomposites with metal oxides, sulfides, noble metals, carbon nanomaterials [14–29], and so on.

Currently, graphene or graphene oxides and their nanocomposites were received considerable, due to their excellent physical and chemical properties, high specific surface area, excellent biocompatibility. Graphene based nanocomposites are widely used in various applications, including drug and protein delivery [30, 31], biosensors [32–38], chemical sensors [39–45], supercapacitor [46–49], lithium-ion batteries [50–52], solar cells [53, 54], adsorption of metal ions [55, 56], infrared imaging photodetectors [57], et al. DNA/GO hybridization can construct some artificial molecular machines or smart nanocomposites and devices with responsiveness to various stimuli. DNA/GO nanocomposite would also been designed sensors with high sensitive for the detection of biomolecules, metal ions, and small molecules.

Nanostructured Au have excellent optical properties ideal for chemical and biosensor construction due to its surface plasmon resonance strong depending on their size and distance [58–65]. Otherwise, Au nanomaterials have good biocompatibility and low toxicity. The addition of a small amount of Au in DNA/GO nanocomposite would enhance the multi-functionalities of materials, and widen their applications. Synthesis of Au nanostructure is very simple, generally chloroauric acid was reduced with appropriate reducing agents, and then the preparation of conductive polymer was often used oxidizing agents. Therefore, the mixture of chloroauric acid and aniline can led to the formation of Au/PANi nanocomposite automatically.

PANi as a representative conductive polymer has also been extensively studied and widely applied in various fields due to their outstanding optical properties and electrical properties. These studies including biomedical applications [66–69], flexible device [70], new energy devices [71–73], supercapacitors [74–79], sensors [80], photodetector [81], photocatalysts [82, 83], electrochromic device [84], and so on.

For many years, the authors are very interested in low-dimensional conductive polymers or organic-inorganic functional nanocomposites and their properties studies [85–89]. To develop smart nanocomposites, in this paper, single-stranded DNA oligonucleotides and graphene oxide (GO) was hybridized, and Au nanoparticle were formed in situ with chloroauric acid and aniline as reducing agent. Generally, external stimuli response involves in pH, temperature, electric field, magnetic fields, light, et al. Regarding the properties of functional materials, visible light activity of materials for demands of energy and environmental is important, and NIR activity for biomedical is favorable. And for biomedical applications, two wavelength windows are needed. One is 808 nm, the other is 1064 nm. In this study, the emphasis was put on the photoconductive responses to weak visible light and 808 nm NIR to develop nanocomposites or devices controlled by light. Some good results were obtained.

Experimental Details

Materials. $HAuCl_4$ (AR), $HPtCl_4$ (AR), ammonium persulfate (AR), hydrochloric acid (AR), aniline (AR), Potassium permanganate (AR), sulfuric acid (Chemical Reagent, i.e. CP), nitric acid (CP), hydrochloric acid (AR), sodium hydroxide (AR) were purchased from Sinopharm Chemical Reagent Co., Ltd., et al. DNA chain segment with different length were provided by Shihua Chen (associate professor) of College of Life Science, Yantai University. Imported graphite (AR), 325 mesh (purity 99.8%). Multiwalled carbon nanotube (MWCNT) was purchased from Shenzhen Nanotech Port Co., Ltd.

Synthesis of Graphene Oxide. The preparation of grapheme oxide was using chemical oxidation method, which is according to the previous report utilizing graphite, concentrated sulfuric acid, concentrated nitric acid, potassium permanganate, et al. [90]. 1 g graphite was transferred about 500 mL GO solution.

Otherwise, the preparation of grapheme oxide was using chemical method, which is according to the previous report utilizing carbon nanotube, concentrated sulfuric acid, concentrated nitric acid, potassium permanganate, et al. 1 g CNTs was transferred about 1000 mL GO solution.

Hybrid of Graphene Oxide/DNA. In the experiment, 40 μL, 80 μL, 320 μL DNA with different chain segment (the concentration of DNA is 2198, 6134, 2917 ng/μL) were added in 300–500 mL glass vessel, then heated 95 °C about 15–20 min for formation of single-stranded DNA chains. 30 mL GO solution above-prepared were added rapidly, with ultrasound about 15–20 min. Standing 24 h for later use.

Preparation of DNA/GO/Au/PANi Nanocomposites. 50 mL $HAuCl_4$ (or $HPtCI_4$) solution (the concentration is 0.001 g/mL) was added above-prepared DNA/GO hybrid sample respectively, 200 mL H_2O was added, and ultrasounded for 15–20 min. 0.02 M aniline solution 50 mL was added slowly in the mixture, keeping for 24 h at room temperature. The resulted products were washed with deionised water repeatedly several times.

Morphology Observation with SEM. The scanning electron microscopy (SEM) observation was carried out with Hitachi S-4800. The obtained sample was washed with deionized water, deposited on an aluminum foil substrate, dried at room temperature and then sputtered with a thin layer of Pt on the surface for the SEM observation.

Morphology Observation with TEM. The TEM observation was carried out with JEM-1011. The nanocomposite suspension were deposited on cupper mesh coated with carbon film, dried at room temperature.

Measurement of UV-Vis Spectrum. The UV-Vis was determined by a TU-1810 spectrophotometer and the samples were using suspension.

Measurement of FTIR Spectra. FTIR (Fourier transform infrared spectroscopy) spectra were taken with KBr, and recorded on IR Prestige-21 Fourier transform infrared spectrometer. Sample and KBr powders were mixed and pressed into a small slice, and then dried at room temperature for determination.

Measurement of Raman Spectra. Raman spectra were recorded on a LabRAM HR800 (HORLBA JY) using powers sample.

Photoresponses of Nanocomposite to Visible Light and NIR. The nanocomposite suspension was casted on the interdigital electrodes of Au on flexible PET (polyethylene terephthalate) substrate, after drying, the photoconductive responses to visible light (20–25 W) and 808 nm NIR with low-power were determined with LK2000A Electrochemical Work Station from LANLIKE Chemistry and Electron High Technology Co., Ltd (China).

Results and Discussion

It is well known that graphene oxide holds abundant chemical groups on its surface, such as –COOH, –OH, epoxide group, and single-stranded DNA chains also has some active chemical groups, hybrid of GO and single-stranded DNA chains is very easy. The properties of nanocomposites strongly depending on the morphologies

Fig. 1 SEM (**a**) and TEM (**b**) images of GO/DNA/Au/PANi nanocomposite

were considered, the SEM and TEM of nanocomposite produced were observed carefully. The representative TEM images of GO/DNA/Au/PANi nanocomposite is shown in Fig. 1a. The representative SEM images of GO/DNA/Au/PANi nanocomposite is shown in Fig. 1b.

Figure 1a showed that a very thin layer of polymer and Au nanoparticle were deposited on the slice of graphene oxide. Au nanoparticle was more clearly, the size is about 1–3 nm or so, and the layer of graphene oxide (GO) was also observed, and the size of GO sheets is not uniform. Some holes were appeared in the big slice of graphene oxide. This is the results of overoxidation in the preparation of graphene oxide with carbon nanotube as raw materials.

As shown from Fig. 1b, some polymer were coated on the surfaces of Au nanoparticles and GO. These polymers should be DNA and some PANi.

To confirm the presence of GO and Au nanoparticles of nanocomposite, Raman spectra and XRD pattern were examined. The results are shown in Fig. 2a, b.

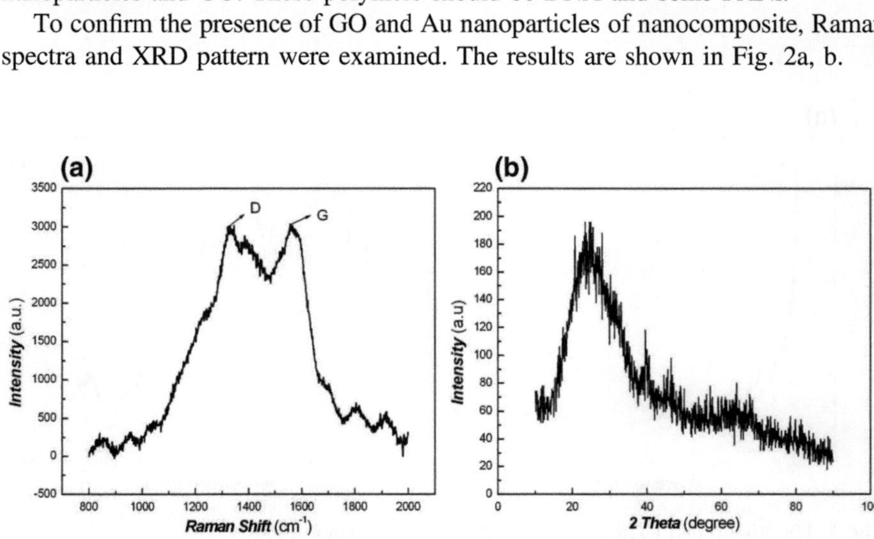

Fig. 2 Raman spectra (**a**) and XRD pattern (**b**) of GO/DNA/Au/PANi nanocomposite

As shown from Fig. 2a, D band (defect of C) and G band (C sp^2) were clearly observed in the GO/DNA/Au/PANi nanocomposite. This illustrated that the nanocomposite contained some GO.

Figure 2b exhibited that there are some diffraction peaks of Au nanoparticles, such as 2θ in 20.22, 40.69, 44.39, 45.55, et al. These peaks belong to face-centered of Au. Because the amount of Au is small, the diffraction peaks of Au nanoparticle are not very strong.

The UV-Vis and FTIR of GO/DNA/Au/PANi nanocomposite were also carried out. The results are shown in Fig. 3a, b.

The UV-Vis of GO/DNA/Au/PANi nanocomposite indicated that it is very similar curve compared with pure PANi. It illustrated the nanocomposite synthesized contained PANi. PANi not only immobilized Au nanoparticles, but also was favor to catch some biomolecules for biomedical applications. Otherwise, the adsorption of GO/DNA/Au/PANi nanocomposite was almost covered the whole region of visible light and extended to NIR (near-infrared). It would be utilized more efficiently the visible light and NIR for environmental fields and biomedical applications.

FTIR of GO/DNA/Au/PANi nanocomposite showed that some characteristic bands of DNA (1103.84 cm^{-1} is stretching vibration of bond of phosphodiester, 1400 cm^{-1} belongs to stretching vibration of glucosidic bond, 1622 cm^{-1} is attributed to the stretching vibration of cytosine), 1146 cm^{-1} is attributed to the bending vibration of benzene ring. 1372 cm^{-1} is the absorption band of C–N of PANi, et al. Therefore, the nanocomposite contained DNA and PANi.

Figure 3a showed that the nanocomposite synthesized had good adsorption in visible light region and NIR. Therefore, the photoconductive response properties of nanocomposite to weak visible light and 808 nm NIR were examined. The representative results are shown in Fig. 4a, b.

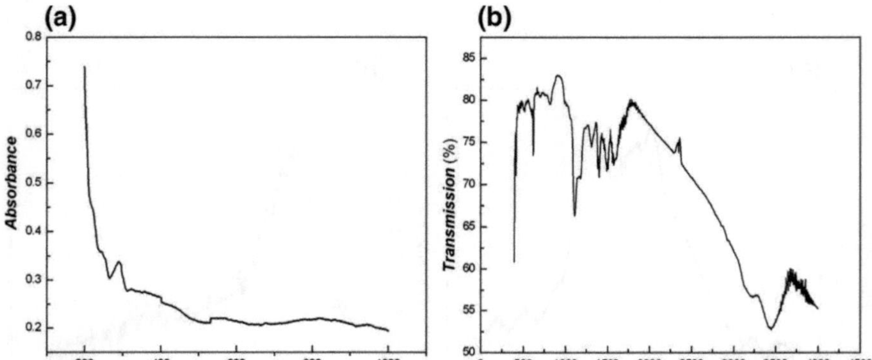

Fig. 3 UV-Vis (**a**) and FTIR (**b**) of GO/DNA/Au/PANi nanocomposite

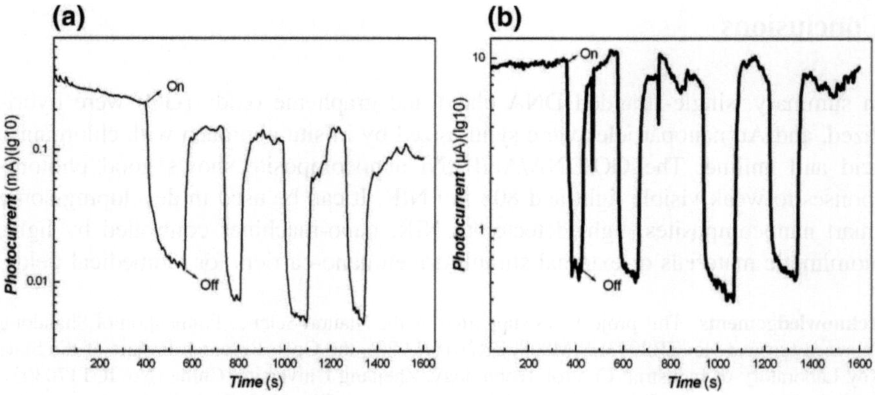

Fig. 4 Photoresponses of GO/DNA/Au/PANi nanocomposite to weak visible light (**a**) and 808 nm NIR of 200 mW (**b**)

Fig. 5 The photoresponse of GO/DNA/Pt/PANi nanocomposite to weak visible light

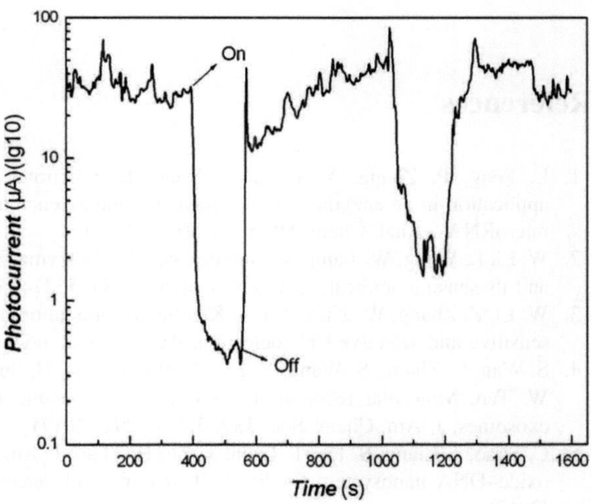

As shown from Fig. 4, the GO/DNA/Au/PANi nanocomposite showed good photo-switching behaviors to weak visible light and 808 nm NIR. The ratio of On/Off is about two orders. It would be developed light detectors, NIR driven nano-carriers, photo-remote nanocomposites, et al.

In the experiments, Au was instead of Pt and some similar results were also obtained. The representative results can also be seen in Fig. 5.

Conclusions

In summary, single-stranded DNA chain and grapheme oxide (GO) were hybridized, and Au nanoparticles were synthesized by in situ approach with chloroauric acid and aniline. The GO/DNA/Au/PANi nanocomposite shows good photoresponses to weak visible light and 808 nm NIR. It can be used in developing some smart nanocomposites, light detector to NIR, nano-machines controlled by light, biomimetic materials or external stimuli driven nano-carriers for biomedical fields.

Acknowledgements This project was supported by the Natural Science Foundation of Shandong Province (project No. ZR2013EMM008, ZR2016JL020), the Open Research Project of the State Key Laboratory of Industrial Control Technology, Zhejiang University, China (No. ICT170303), Zhejiang University, the Science Research Project of Yantai City (2013ZH349). Thanks were due to associate professor Shihua Chen of College of Life Science, Yantai University, providing a large number of DNA chain segment with different length, his student (Yingchang Lu) doing some part of experiments.

We declare that we have no conflict of interest.

References

1. L. Peng, P. Zhang, Y. Chai, R. Yuan, Bi-directional DNA walking machine and its application in an enzyme-free electrochemiluminescence biosensor for sensitive detection of microRNAs. Anal. Chem. **89**, 5036–5042 (2017)
2. W. Li, L. Wang, W. Jiang, A catalytic assembled enzyme-free three-dimensional DNA walker and its sensing application. Chem. Commun. **53**, 5527–5530 (2017)
3. W. Li, Z. Zhang, W. Zhou, J. Liu, Kinetic discrimination of metal ions using DNA for highly sensitive and selective Cr^{3+} detection. ACS Sens. **2**, 663–669 (2017)
4. S. Wan, L. Zhang, S. Wang, Y. Liu, C. Wu, C. Cui, H. Sun, M. Shi, Y. Jiang, L. Li, L. Qiu, W. Tan, Molecular recognition-based DNA nanoassemblies on the surfaces of nanosized exosomes. J. Am. Chem. Soc. **139**, 5289–5292 (2017)
5. C. Shao, J. Liang, S. He, T. Luan, J. Yu, H. Zhao, J. Xu, L. Tian, pH-Responsive graphene oxide–DNA nanosystem for live cell imaging and detection. Anal. Chem. **89**, 5445–5452 (2017)
6. A.H. Loo, C.K. Chua, M. Pumera, DNA biosensing with 3D printing technology. Analyst **142**, 279–283 (2017)
7. P. Zhang, Z. Li, H. Wang, Y. Zhuo, R. Yuan, Y. Chai, DNA nanomachine-based regenerated sensing platform: a novel electrochemiluminescence resonance energy transfer strategy for ultra-high sensitive detection of microRNA from cancer cells. Nanoscale **9**, 2310–2316 (2017)
8. X. Lin, A.P. Ivanov, J.B. Edel, Selective single molecule nanopore sensing of proteins using DNA aptamer-functionalised gold nanoparticles. Chem. Sci. **8**, 3905–3912 (2017)
9. X. Ma, Z. Yang, Y. Wang, G. Zhang, Y. Shao, H. Jia, T. Cao, R. Wang, D. Liu, Remote controlling DNA hydrogel by magnetic field. ACS Appl. Mater. Interfaces. **9**, 1995–2000 (2017)
10. Y. Vyborna, M. Vybornyi, Robert Häner, Functional DNA-grafted supramolecular polymers – chirality, cargo binding and hierarchical organization. Chem. Commun. **53**, 5179–5181 (2017)

11. W. Liao, S. Lilienthal, J.S. Kahn, M. Riutin, Y.S. Sohn, R. Nechushtai, I. Willner, pH- and ligand-induced release of loads from DNA–acrylamide hydrogel microcapsules. Chem. Sci. **8**, 3362–3373 (2017)
12. D. Wang, Y. Hu, P. Liu, D. Luo, Bioresponsive DNA hydrogels: beyond the conventional stimuli responsiveness. Acc. Chem. Res. **50**, 733–739 (2017)
13. B.L. Li, M.I. Setyawati, L. Chen, J. Xie, K. Ariga, C. Lim, S. Garaj, D.T. Leong, Directing assembly and disassembly of 2D MoS_2 nanosheets with DNA for drug delivery. ACS Appl. Mater. Interfaces. **9**, 15286–15296 (2017)
14. S.S. Hinman, K.S. McKeating, Q. Cheng, DNA linkers and diluents for ultrastable gold nanoparticle bioconjugates in multiplexed assay development. Anal. Chem. **89**, 4272–4279 (2017)
15. P. Miao, Y. Tang, L. Wang, DNA modified Fe_3O_4@Au magnetic nanoparticles as selective probes for simultaneous detection of heavy metal ions. ACS Appl. Mater. Interfaces. **9**, 3940–3947 (2017)
16. C. Heck, J. Prinz, A. Dathe, V. Merk, O. Stranik, W. Fritzsche, J. Kneipp, I. Bald, Gold nanolenses self-assembled by DNA origami. ACS Photonics **4**, 1123–1130 (2017)
17. K.L. Jiménez-Monroy, N. Renaud, J. Drijkoningen, D. Cortens, K. Schouteden, C. Haesendonck, W.J. Guedens, J.V. Manca, L.D.A. Siebbeles, F.C. Grozema, P.H. Wagner, High electronic conductance through double-helix DNA molecules with fullerene anchoring groups. J. Phys. Chem. A **121**, 1182–1188 (2017)
18. P. Zhan, P.K. Dutta, P. Wang, G. Song, M. Dai, S. Zhao, Z. Wang, P. Yin, W. Zhang, B. Ding, Y. Ke, Reconfigurable three-dimensional gold nanorod plasmonic nanostructures organized on DNA origami tripod. ACS Nano **11**, 1172–1179 (2017)
19. C. Qiao, J. Wu, Z. Huang, X. Cao, J. Liu, B. Xiong, Y. He, E.S. Yeung, Sequence-modulated interactions between single multivalent DNA conjugated gold nanoparticles. Anal. Chem. **89**, 5592–5597 (2017)
20. A. Noorbakhsh, M. Khakpoor, M. Rafieniya, E. Sharifi, M. Mehrasa, Highly sensitive electrochemical hydrogen peroxide sensor based on iron oxide-reduced graphene oxide-chitosan modified with DNA-celestine blue. Electroanalysis **29**, 1113–1123 (2017)
21. Y. Eguchi, T. Kato, T. Tanaka, T. Maruyama, A DNA–gold nanoparticle hybrid hydrogel network prepared by enzymatic reaction. Chem. Commun. **53**, 5802–5805 (2017)
22. N. Yang, T. You, X. Liang, C. Zhang, L. Jiang, P. Yin, An ultrasensitive near-infrared satellite SERS sensor: DNA self-assembled gold nanorod/nanospheres structure. RSC Adv. **7**, 9321–9327 (2017)
23. R. Wang, I. Bowling, W. Liu, Cost effective surface functionalization of gold nanoparticles with a mixed DNA and PEG monolayer for nanotechnology applications. RSC Adv. **7**, 3676–3679 (2017)
24. N. Li, Y. Zhu, T. Liu, S. Liu, S. Lin, Y. Shi, H. Luo, N. Li, Turn-on fluorescence detection of pyrophosphate anion based on DNA-attached cobalt oxyhydroxide. New J. Chem. **41**, 1993–1996 (2017)
25. J. McMorrow, M. Freeley, M. Palma, DNA-wrapped single-walled carbon nanotube assemblies. Ind. Eng. Chem. Res. **56**, 5302–5308 (2017)
26. A.M. Maier, W. Bae, D. Schiffels, J.F. Emmerig, M. Schiff, T. Liedl, Self-assembled DNA tubes forming helices of controlled diameter and chirality. ACS Nano **11**, 1301–1306 (2017)
27. T.D. Jorgenson, A.M. Mohammed, D.K. Agrawal, R. Schulman, Self-assembly of hierarchical DNA nanotube architectures with well-defined geometries. ACS Nano **11**, 1927–1936 (2017)
28. S.K. Albert, H.V.P. Thelu, M. Golla, N. Krishnan, R. Varghese, Modular synthesis of supramolecular DNA amphiphiles through host–guest interactions and their self-assembly into DNA-decorated nanovesicles. Nanoscale **7**, 5425–5432 (2017)
29. L. Yan, H. Shi, X. Sui, Z. Deng, L. Gao, MoS_2-DNA and MoS_2 based sensors. RSC Adv. **7**, 23573–23582 (2017)

30. X. Su, C. Chan, J. Shi, M. Tsang, Y. Pan, C. Cheng, O. Gerile, M. Yang, A graphene quantum dot@Fe_3O_4@SiO_2 based nanoprobe for drug delivery sensing and dual-modal fluorescence and MRI imaging in cancer cells. Biosens. Bioelectron. **92**, 489–495 (2017)
31. C. He, Z. Shi, C. Cheng, C. Nie, M. Zhou, L. Wang, C. Zhao, Highly swellable and biocompatible graphene/heparin-analogue hydrogels for implantable drug and protein delivery. RSC Adv. **6**, 71893–71904 (2016)
32. M. Mehrzad-Samarin, F. Faridbod, A.S. Dezfuli, M.R. Ganjali, A novel metronidazole fluorescent nanosensor based on graphene quantum dots embedded silica molecularly imprinted polymer. Biosens. Bioelectron. **92**, 618–623 (2017)
33. H. Liu, W. Na, Z. Liu, X. Chen, X. Su, A novel turn-on fluorescent strategy for sensing ascorbic acid using graphene quantum dots as fluorescent probe. Biosens. Bioelectron. **92**, 229–233 (2017)
34. J. Yoon, T. Lee, J. Jo, B. Oh, J. Choi, Electrochemical H_2O_2 biosensor composed of myoglobin on MoS_2 nanoparticle-graphene oxide hybrid structure. Biosens. Bioelectron. **93**, 14–20 (2017)
35. S. Dong, Q. Bi, C. Qiao, Y. Sun, X. Zhang, X. Lu, L. Zhao, Electrochemical sensor for discrimination tyrosine enantiomers using graphene quantum dots and β-cyclodextrins composites. Talanta **173**, 94–100 (2017)
36. W. Liu, Y. Ma, G. Sun, S. Wang, J. Deng, H. Wei, Molecularly imprinted polymers on graphene oxide surface for EIS sensing of testosterone. Biosens. Bioelectron. **92**, 305–312 (2017)
37. H. Kanso, M.B.G. García, L.F. Llano, S. Ma, R. Ludwig, P.F. Bolado, D.H. Santos, Novel thin layer flow-cell screen-printed graphene electrode for enzymatic sensors. Biosens. Bioelectron. **93**, 298–304 (2017)
38. P. Liang, Q. Li, Z. Wu, J. Jiang, R. Yu, Graphene oxide–peptide nanoassembly as a general approach for monitoring the activity of histone deacetylases. Analyst. **141**, 3989–3992 (2016)
39. C. Zhang, R. Hao, B. Zhao, Y. Hao, Y. Liu, A ternary functional Ag@GO@Au sandwiched hybrid as anultrasensitive and stable surface enhanced Raman scattering platform. Appl. Surf. Sci. **409**, 306–313 (2017)
40. T. Zhan, Z. Tan, X. Tian, W. Hou, Ionic liquid functionalized graphene oxide-Au nanoparticles assembly for fabrication of electrochemical 2,4-dichlorophenol sensor. Sens. Actuators, B **246**, 638–646 (2017)
41. P.N. Joshi, S. Agawane, M.C. Athalye, V. Jadhav, D. Sarkar, R. Prakash, Multifunctional inulin tethered silver-graphene quantum dots nanotheranostic module for pancreatic cancer therapy. Mater. Sci. Eng., C **78**, 1203–1211 (2017)
42. L. Liu, D. Zhang, Q. Zhang, X. Chen, G. Xu, Y. Lu, Q. Liu, Smartphone-based sensing system using ZnO and graphene modified electrodes for VOCs detection. Biosens. Bioelectron. **93**, 94–101 (2017)
43. Y. Song, J. Lü, B. Liu, C. Lu, Temperature responsive polymer brushes grafted from graphene oxide: an efficient fluorescent sensing platform for 2,4,6-trinitrophenol. J. Mater. Chem. C. **4**, 7083–7092 (2016)
44. D. Huang, Z. Yang, X. Li, L. Zhang, J. Hu, Y. Su, N. Hu, G. Yin, D. He, Y. Zhang, Three-dimensional conductive networks based on stacked SiO_2@graphene frameworks for enhanced gas sensing. Nanoscale **9**, 109–118 (2017)
45. B. An, Y. Ma, W. Li, M. Su, F. Li, Y. Song, Three-dimensional multi-recognition flexible wearable sensor via graphene aerogel printing. Chem. Commun. **52**, 10948–10951 (2016)
46. J. Li, Y. Sun, D. Li, H. Yang, X. Zhang, B. Lin, Novel ternary composites reduced-graphene oxide/zine oxide/poly(pphenylenediamine) for supercapacitor: synthesis and properties. J. Alloy. Compd. **708**, 787–795 (2017)
47. G.G. Jang, B. Song, K. Moon, C. Wong, J.K. Keum, M.Z. Hu, Particle size effect in porous film electrodes of ligand-modified graphene for enhanced supercapacitor performance. Carbon **119**, 296–304 (2017)

48. T. Qin, B. Liu, Y. Wen, Z. Wang, X. Jiang, Z. Wan, S. Peng, G. Cao, D. He, Freestanding flexible grapheme foams@polypyrrole@MnO_2 electrodes for highperformance supercapacitors. J. Mater. Chem. A. **4**, 9196–9203 (2016)
49. S. Li, Z. Wang, H. Jiang, L. Zhang, J. Ren, M. Zheng, L. Dong, L. Sun, Plasma-induced highly efficient synthesis of boron doped reduced graphene oxide for supercapacitors. Chem. Commun. **52**, 10988–10991 (2016)
50. C. Xu, Y. Jing, J. He, K. Zhou, Y. Chen, Q. Li, J. Lin, W. Zhang, Self-assembled interwoven CoS_2/CNTs/graphene architecture as anode for high-performance lithium ion batteries. J. Alloy. Compd. **708**, 1178–1183 (2017)
51. R. Tian, W. Wang, Y. Huang, H. Duan, Y. Guo, H. Kang, H. Li, H. Liu, 3D composites of layered MoS_2 and grapheme nanoribbons for high performance lithium-ion battery anodes. J. Mater. Chem. A. **4**, 13148–13154 (2016)
52. M. Shi, T. Wu, X. Song, J. Liu, L. Zhao, P. Zhang, L. Gao, Active Fe_2O_3 nanoparticles encapsulated in porous g-C3N4/graphene sandwich-type nanosheets as a superior anode for high-performance lithium-ion batteries. J. Mater. Chem. A. **4**, 10666–10672 (2016)
53. W. Wei, K. Sun, Y. Hu, Direct conversion of CO_2 to 3D graphene and its excellent performance for dye-sensitized solar cells with 10% efficiency. J. Mater. Chem. A. **4**, 12054–12057 (2016)
54. X. Gan, R. Lv, H. Zhu, L. Ma, X. Wang, Z. Zhang, Z. Huang, H. Zhu, W. Ren, M. Terronesdfg, F. Kang, Polymer-coated graphene films as anti-reflective transparent electrodes for Schottky junction solar cells. J. Mater. Chem. A. **4**, 13795–13802 (2016)
55. R.P. Medina, E.T. Nadres, F.C. Ballesteros Jr., D.F. Rodrigues, Incorporation of graphene oxide into a chitosan–poly (acrylic acid) porous polymer nanocomposite for enhanced lead adsorption. Environ. Sci. Nano. **3**, 638–646 (2016)
56. Y. Shen, X. Zhu, B. Chen, Size effects of graphene oxide nanosheets on the construction of three-dimensional graphene-based macrostructures as adsorbents. J. Mater. Chem. A. **4**, 12106–12118 (2016)
57. N. Guo, W. Hu, T. Jiang, F. Gong, W. Luo, W. Qiu, P. Wang, L. Liu, S. Wu, L. Liao, X. Chena, W. Lu, High-quality infrared imaging with grapheme photodetectors at room temperature. Nanoscale **8**, 16065–16072 (2016)
58. S. Xu, T. Gao, X. Feng, X. Fan, G. Liu, Y. Mao, X. Yu, J. Lin, X. Luo, Near infrared fluorescent dual ligand functionalized Au NCs based multidimensional sensor array for pattern recognition of multiple proteins and serum discrimination. Biosens. Bioelectron. **97**, 203–207 (2017)
59. Q. Huang, X. Lin, J. Zhu, Q. Tong, Pd-Au@carbon dots nanocomposite: Facile synthesis and application as an ultrasensitive electrochemical biosensor for determination of colitoxin DNA in human serum. Biosens. Bioelectron. **94**, 507–512 (2017)
60. H. Zhang, L. Zhang, Y. Han, Y. Yu, M. Xu, X. Zhang, L. Huang, S. Dong, RGO/Au NPs/N-doped CNTs supported on nickel foam as an anode for enzymatic biofuel cells. Biosens. Bioelectron. **97**, 34–40 (2017)
61. M. Amoli-Diva, R. Sadighi-Bonabi, K. Pourghazi, Switchable on/off drug release from gold nanoparticles-grafted dual light- and temperature-responsive hydrogel for controlled drug delivery. Mater. Sci. Eng., C **76**, 242–248 (2017)
62. Y. Wei, X. Li, X. Sun, H. Ma, Y. Zhang, Q. Wei, Dual-responsive electrochemical immunosensor for prostate specific antigen detection based on Au-CoS/graphene and CeO_2/ionic liquids doped with carboxymethyl chitosan complex. Biosens. Bioelectron. **94**, 141–147 (2017)
63. C. Liu, H. Tai, P. Zhang, Z. Ye, Y. Su, Y. Jiang, Enhanced ammonia-sensing properties of PANI-TiO_2-Au ternaryself-assembly nanocomposite thin film at room temperature. Sens. Actuators, B **246**, 85–95 (2017)
64. I. Schmidt, A. Gad, G. Scholz, H. Boht, M. Martens, M. Schilling, H.S. Wasisto, A. Waag, U. Schröder, Gold-modified indium tin oxide as a transparent window in optoelectronic diagnostics of electrochemically active biofilms. Biosens. Bioelectron. **94**, 74–80 (2017)

65. Y. Zhao, X. Li, Y. Liu, L. Zhang, F. Wang, Y. Lu, High performance surface-enhanced Raman scattering sensing basedon Au nanoparticle-monolayer graphene-Ag nanostar array hybrid system. Sens. Actuators, B **247**, 850–857 (2017)
66. A.J. Hackett, J. Malmström, J. Travas-Sejdic, Functionalization of conducting polymers for biointerface applications. Prog. Polym. Sci. **70**, 18–33 (2017)
67. H. Rao, M. Chen, H. Ge, Z. Lu, X. Liu, P. Zou, X. Wang, H. He, X. Zeng, Y. Wang, A novel electrochemical sensor based on Au@PANI composites film modified glassy carbon electrode binding molecular imprinting technique for the determination of melamine. Biosens. Bioelectron. **87**, 1029–1035 (2017)
68. X. Zhu, J. Zhao, C. Wang, Acid and base dual-controlled cargo molecule release from polyaniline gated-hollow mesoporous silica nanoparticles. Polym. Chem. **7**, 6467–6474 (2016)
69. X. Gong, G. Fei, W. Fu, M. Fang, X. Gao, B. Zhong, L. Zhang, Flexible strain sensor with high performance based on PANI/PDMS films. Org. Electron. **47**, 51–56 (2017)
70. Y. Luan, J. Noh, S.H. Kim, Facile control of stretchability and electrical resistance of elastomer/polyaniline composites for stretchable conductors. Mater. Chem. Phys. **190**, 68–73 (2017)
71. E. Alkuam, M. Mohammed, T. Chen, Fabrication of CdS nanorods and nanoparticles with PANI for (DSSCs) dye-sensitized solar cells. Sol. Energy **150**, 317–324 (2017)
72. R. Song, K. Yan, Z. Lin, J.S.C. Loo, L. Pan, Q. Zhang, J. Zhang, J. Zhu, Inkjet-printed porous polyaniline gel as an efficient anode for microbial fuel cells. J. Mater. Chem. A. **4**, 14555–14559 (2016)
73. D. Hursa'n, A. Korma'nyos, K. Rajeshwar, C. Jana'ky, Polyaniline films photoelectrochemically reduce CO_2 to alcohols. Chem. Commun. **52**, 8858–8861 (2016)
74. K.Z. Htut, M. Kim, E. Lee, G. Lee, S.H. Baeck, S.E. Shim, Biodegradable polymer-modified graphene/polyaniline electrodes for supercapacitors. Synth. Met. **227**, 61–70 (2017)
75. D. Gu, C. Ding, Y. Qin, H. Jiang, L. Wang, L. Shen, Behavior of electrical charge storage/release in polyaniline electrodes of symmetric supercapacitor. Electrochim. Acta **245**, 146–155 (2017)
76. C. Chang, Z. Hu, T. Lee, Y. Huang, W. Ji, W. Liu, J. Yeh, Y. Wei, Biotemplated hierarchical polyaniline composite electrodes with high performance for flexible supercapacitors. J. Mater. Chem. A. **4**, 9133–9145 (2016)
77. Q. Liu, S. Jing, S. Wang, H. Zhuo, L. Zhong, X. Peng, R. Sun, Flexible nanocomposites with ultrahigh specific areal capacitance and tunable properties based on a cellulose derived nanofiber-carbon sheet framework coated with polyaniline. J. Mater. Chem. A. **4**, 13352–13362 (2016)
78. J. Chao, J. Deng, W. Zhou, J. Liu, R. Hu, L. Yang, M. Zhu, O.G. Schmidt, Hierarchical nanoflowers assembled from MoS_2/polyaniline sandwiched nanosheets for high-performance supercapacitors. Electrochim. Acta **243**, 98–104 (2017)
79. R. Wang, Q. Wu, X. Zhang, Z. Yang, L. Gao, J. Ni, O.K.C. Tsu, Flexible supercapacitors based on a polyaniline nanowire-infilled 10 nm-diameter carbon nanotube porous membrane by in situ electrochemical polymerization. J. Mater. Chem. A. **4**, 12602–12608 (2016)
80. Y. Guo, T. Wang, F. Chen, X. Sun, X. Li, Z. Yu, P. Wan, X. Chen, Hierarchical graphene–polyaniline nanocomposite films for high-performance flexible electronic gas sensors. Nanoscale **8**, 12073–12080 (2016)
81. X. Yang, Y. Liu, H. Lei, B. Li, An organic–inorganic broadband photodetector based on a single polyaniline nanowire doped with quantum dots. Nanoscale **8**, 15529–15537 (2016)
82. T. Pham, B. Lee, V.N. Nguyen, V. Dao, Novel photocatalytic activity of vanadium-doped tantalum nitride sensitized/protected by polyaniline for efficient visible light water splitting. J. Catal. **352**, 13–21 (2017)
83. M. Radoičić, G. Cirić-Marjanović, V. Spasojević, P. Ahrenkiel, M. Mitrića, T. Novakovićd, Z. Saponjića, Superior photocatalytic properties of carbonized PANI/TiO_2 nanocomposites. Appl. Catal. B: Environ. **213**, 155–166 (2017)

84. Y. Tian, X. Zhang, S. Dou, L. Zhang, H. Zhang, H. Lv, L. Wang, J. Zhao, Y. Li, A comprehensive study of electrochromic device with variable infrared emissivity based on polyaniline conducting polymer. Sol. Energy Mater. Sol. Cells **170**, 120–126 (2017)
85. X. Ma, G. Li, M. Wang, Y. Cheng, R. Bai, H. Chen, Preparation of nano-wire structured polyaniline composite and its gas-sensitivity studies. Chem.-A Eur. J. **12**, 3254–3260 (2006)
86. X. Ma, M. Wang, G. Li, H. Chen, R. Bai, Preparation of polyaniline-TiO_2 composite film with in-situ polymerization approach and its gas-sensitivity at room temperature. Mater. Chem. Phys. **98**, 241–247 (2006)
87. X. Ma, A. Liu, H. Xu, G. Li, Growth of one-dimensional ZnO with ramification-like in the presence of polyaniline. Colloid Polym. Sci. **285**, 1631–1635 (2007)
88. X. Ma, H. Xu, G. Li, M. Wang, H. Chen, S. Chen, Gas-response studies of polyaniline composite film containing zeolite to chemical vapors. Macromol. Mater. Eng. **291**, 1539–1546 (2006)
89. X. Ma, M. Gao, X. He, G. Li, Morphology tailoring of nano/micro-structured conductive polymers, composites and their applications in chemical sensors. Recent Pat. Nanotechnol. **4**, 150–163 (2010)
90. J. Zheng, X. Ma, X. He, M. Gao, G. Li, Praparation, characterizations, and its potential applications of PANi/graphene oxide nanocomposite. Procedia Eng. **27**, 1478–1487 (2012)

Effect of Boron Content on Structure and High Thermal Stability of Polyborosilazane Precursor

Chenyu Zhang, Yong Liu, Keqing Han, Xuefeng Chang and Muhuo Yu

Abstract In this paper, a novel polyborosilazane (PBSZ) for preparing SiBNC ceramics was successfully prepared via co-polymerization of propylamine ($C_3H_7NH_2$), trichlorosilane ($HSiCl_3$) with different amounts of boron trichloride (BCl_3). Pyrolysis of these precursors yields amorphous SiBNC ceramics, which with different boron content whereas similar Si/C/N ratio. The FTIR, NMR, EA and TGA were used to analyze the polymerization mechanism, composition, structure and ceramic yield of PBSZ samples with different boron content. The results indicated that these PBSZ contained the major framework of –Si–N–B– and six-membered boron-nitrogen rings. As boron content increased, the residual ceramic weight of PBSZ at 900 °C increased. When the molar ratio of BCl_3:$HSiCl_3$ was 1:1, the ceramic yield of PBSZ was 71.6%. However, the ceramic yield decreased when the boron content was further increased.

Keywords Polyborosilazane · SiBNC ceramic · High temperature resistance Ceramic yield

Introduction

The outstanding performances such as high temperature stability, high oxidation stability, high resistance against crystallization, high strength and high modulus of amorphous SiBNC ceramics have drawn considerable scientific attentions in recent years [1–6]. Due to its specific properties, SiBNC ceramics have been candidate materials for application under extreme conditions like in heat engines, as well as in aerospace and aviation [7, 8].

C. Zhang · Y. Liu · K. Han · X. Chang · M. Yu (✉)
State Key Laboratory for Modification of Chemical Fibers and Polymer Materials, Key Laboratory of High-Performance Fiber and Product, Ministry of Education, College of Materials Science and Engineering, Donghua University, Shanghai 201620, China
e-mail: yumuhuo@dhu.edu.cn

Y. Han (ed.), *Advanced Functional Materials*,
https://doi.org/10.1007/978-981-13-0110-0_87

The polymer-derived ceramic (PDC) route is suitable for the preparation of SiBNC ceramics [9]. The main advantage of this route lies in it allow for the generation of these materials with controlled composition and a feasibility of tailoring polymers with desired properties of the precursors and the ceramic materials [10, 11]. The molecular structure of the polymeric precursor and the atomic ratios of the elements determine the thermal properties of these materials [12, 13]. In order to obtain high-performance SiBNC ceramics, many types of polyborosilazane (PBSZ) have been prepared by researchers. Jansen [14, 15] synthesized N-methyl PBSZs by replaced partial silicon-bonded chlorine with methyl groups and introduced two more kinds of precursor $Cl_2MeSi–NH–BCl_2$ (MADB) and $ClMe_2Si–NH–BCl_2$ (DADB). Sneddon [16] prepared PBSZs by chemical modification of hydridopolysilazanes with various monofunctional borane. Tang [17] prepared preceramic polymer with high yield by condensation reaction of dichoromethylsilane, BCl_3, and hexamethyldisilazane. Mou [18] synthesized precursor for SiBNC ceramic fibers by co-polymerization of trichlorosilane ($SiHCl_3$), boron trichloride (BCl_3) and methyl amine ($MeNH_2$). Most of these synthesis are completed in one step, the polymerization step proceeds more quickly and provides polymers with a high ceramic yield. Compared to the materials derived from Si–N–C precursors, the introduction of boron in the Si–B–N–C system generally improves the high temperature stability of the respective ceramics [19]. However, the reasons for this remarkable characteristic are not yet understood in detail and far less work has been reported for the correlation of boron content doping on structure and high thermal stability of polyborosilazane precursor [13].

Therefore, a novel polymeric precursor for SiBNC ceramic materials was successfully prepared via co-polymerization of propylamine ($C_3H_7NH_2$), trichlorosilane ($HSiCl_3$) and different amounts of boron trichloride (BCl_3) in one step. The aim of this research is to study the effect of boron content doping on structure and high thermal stability of polyborosilazane precursor.

Experimental

Materials. Solution of boron trichloride in n-hexane (BCl_3, BTC, 1 mol/L) (Beijing Multi Technology Co.), trichlorosilane ($HSiCl_3$, 99.99% purity) (Bai Lingwei Chemical Co. Ltd), propylamine ($C_3H_7NH_2$, 98.50% purity) (Guoyao chemical Co. Ltd) and n-hexane (Sinopharm Chemical Reagent Co. Ltd). n-hexane was purified by distillation under nitrogen from sodium. Nitrogen and ammonia with 99.99% purity were used during preparation.

Polymer Precursors Synthesis. Polymer precursors were synthesized via copolymerization method using BCl_3, $SiHCl_3$ and $C_3H_7NH_2$. The molar ratios between monomers (BCl_3:$SiHCl_3$: $C_3H_7NH_2$) were set at 1:2:18, 1:1.5:15, 1:1:12 and 1:0.5:9. For example of 1:2:18, under nitrogen atmosphere, n-hexane (2 L) was cooled to −15 °C in a 3 L glass-lined reactor equipped with a brine-ice cooled reflux condenser. The mixture of BCl_3 (0.15 mol, 150 ml) and $HSiCl_3$ (0.3 mol, 30 ml) in n-hexane was added dropwise to the $C_3H_7NH_2$ (2.7 mol, 222 ml)/ n-hexane solution under vigorous stirring. During the addition of the mixture, the temperature of the solution was kept below −10 °C. After the addition was completed, the resulting mixture was stirred at room temperature overnight. Then the polymer solution was separated and the precipitated product was removed with purified n-hexane. After the oligomer of polyborosilizane (150 ml) was heated at 270 °C under dry nitrogen atmosphere, the polymer precursor was collected as a yellow powders.

Characterization. Fourier transform infrared (FTIR) spectra of the samples were obtained using KBr pellets on a Nicolet 8700 spectrometer in the 4000–4000 cm^{-1} frequency range. ^{11}B-nuclear magnetic resonance (NMR) spectra was performed in $CDCl_3$ with a Bruker Avance 400 spectrometer operating at 128.37 MHz. Tetramethylsilane (TMS) was used as a reference. ^{29}Si solid-state NMR spectra was obtained at 79.49 MHz under MAS conditions. Thermogravimetric analysis (TGA) of the polymer to ceramic conversion was conducted with a Netzsch 209 FIIris instrument. The samples were heated in flowing nitrogen atmosphere from 50 to 900 °C at a heating rate of 10 °C/min in alumina crucible. For the element composition analysis (EA) of the samples pyrolysized at different temperatures, a N analysis (Leco TC-436, Leco Corporation, USA) was used to record the nitrogen content of the ceramic samples, carbon analysis (Leco TC-200, Leco Corporation, USA) was applied to determine the carbon content and the analysis of other elements was carried out with an Elementar Vario III Aanlyser and a Leeman Prodigy ICP-OES.

Results and Discussion

Synthesis and Elements Analysis. Polymeric precursor was synthesized according to Eq. 1 and the molecular ratios setting was according to Table 1.

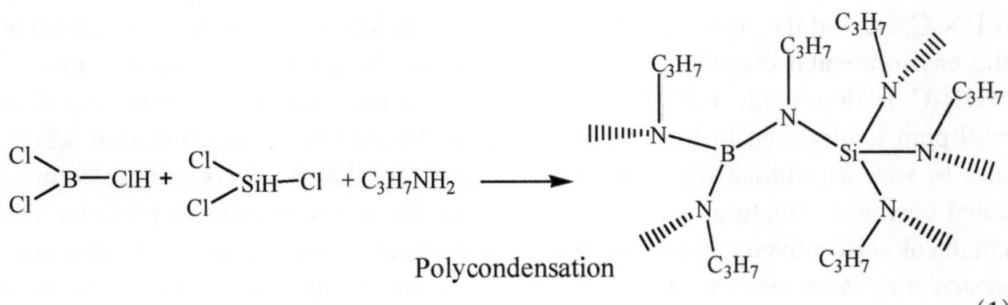

(1)

The BCl_3 and $SiHCl_3$ was dissolved in n-hexane and added into $C_3H_7NH_2$ drop-wise. As shown in Table 1, the molar ratios were 1:2:18 (compound 1),

Table 1 Composition of polymeric precursors synthesized with different BCl_3:$SiHCl_3$: $C_3H_7NH_2$ molar ratios

BCl_3:$SiHCl_3$:$C_3H_7NH_2$ (mmol)	Contents (wt%)					
	B	Si	C	N	H	O
1:2:18	5.2	26.4	51.2	10.1	6.4	0.7
1:1.5:15	6.1	23.3	49.8	11.4	8.3	1.1
1:1:12	7.6	19.8	52.3	10.6	8.7	1.0
1:0.5:9	8.7	11.9	59.3	10.4	8.9	0.8

1:1.5:15 (compound 2), 1:1:12 (compound 3) and 1:0.5:9 (compound 4). The element contents were also illustrated and a little amount of oxygen element is discovered in the system. This is because both the polymeric intermediates and polymer precursors are sensitive to moisture. However, during the synthesis processes and the transition of byproducts, materials contacted with the oxygen is inevitable.

Structure Analysis. In order to investigate the effects of boron content on the structure of polymeric precursor, polymeric precursors with different boron content were synthesized. The FTIR spectrum of PBSZ samples with different boron content is illustrated in Fig. 1. The peak at 3454 cm^{-1} is attributed to N-H bond. Three intense absorptions at 2953, 2891 and 2811 cm^{-1} are caused by C–H stretching vibrations. B–H bonds can be observed at 2477 cm^{-1} and Si-H bonds can be observed at 2110 cm^{-1}. The characteristic peak of B–N is visible at 1475 and 1357 cm^{-1}, and the signal at 909 cm^{-1} is caused by the stretching vibration of the Si–N linkage, which suggesting the presence of a Si–N–B bridge. As shown in the Fig. 1, the structures of PBSZ samples are similar except that the intensity absorption band of Si–H decreased with the increasing of boron content. The fragmentation Si–H bonds is attributed to more electron withdrawing groups are substituted by less ones.

For further determine the structure differences of the PBSZ samples with different boron content, the ^{29}Si-NMR and ^{11}B-NMR spectra were recorded in Figs. 2 and 3, respectively. As shown in Fig. 2a, b, the signal at ca.-14 ppm attributed to $SiHN_2C$ site and the signal at −35 ppm assigned to SiN_3C site, which indicated that the environmental chemistry of silicon is similar. Moreover, the structure unit of $SiHN_2C$ is dominant. With the increasing of boron content, a broad signal at −19 ppm is observed in Fig. 2c, d, which can be ascribed to a tetrahedral $SiHN_3$ unit as well as a broad resonance signal, indicating that it covers a more complicated chemical structure as SiN_4, Si–SiN_3 and SiN in six-membered ring etc. The chemical shift moved towards high field at elevated boron content. It is also suggested that more electron withdrawing groups are substituted by less ones. As a result, Si–H bonds fragmented and a new Si–Si bond formed, which agreed with FT-IR results well.

In the ^{11}B-NMR spectra (Fig. 3), a signal located at 25–27 ppm is attributed to the BN_3 units. It covers the BN_3 units in the linear structure of Si–N–B and the BN

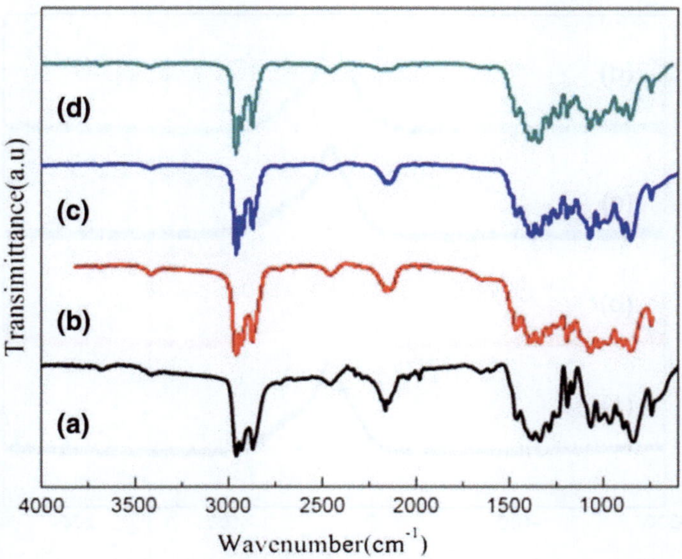

Fig. 1 FTIR spectra of PBSZ samples with different boron content. The molar ratios of BCl_3:$HSiCl_3$: $C_3H_7NH_2$ are, **a** 1:2:18, **b** 1:1.5:15, **c** 1:1:12, **d** 1:0.5:9

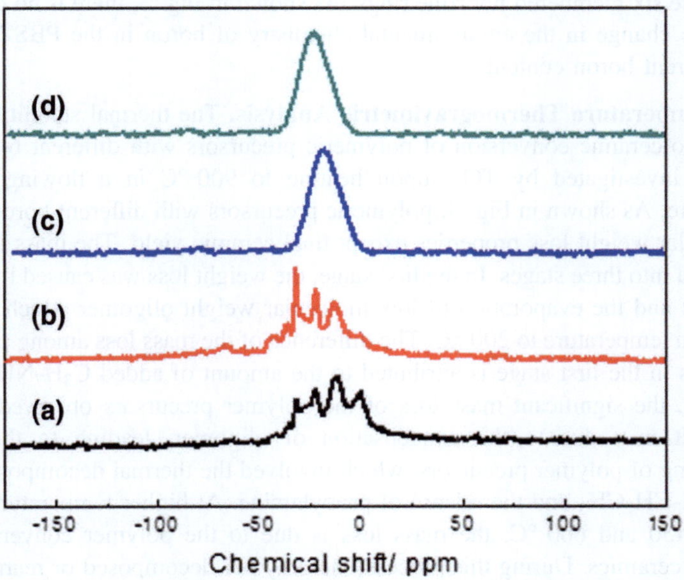

Fig. 2 ^{29}Si-NMR spectra of PBSZ samples with different boron content. The molar ratios of BCl_3:$HSiCl_3$: $C_3H_7NH_2$ are, **a** 1:2:18, **b** 1:1.5:15, **c** 1:1:12, **d** 1:0.5:9

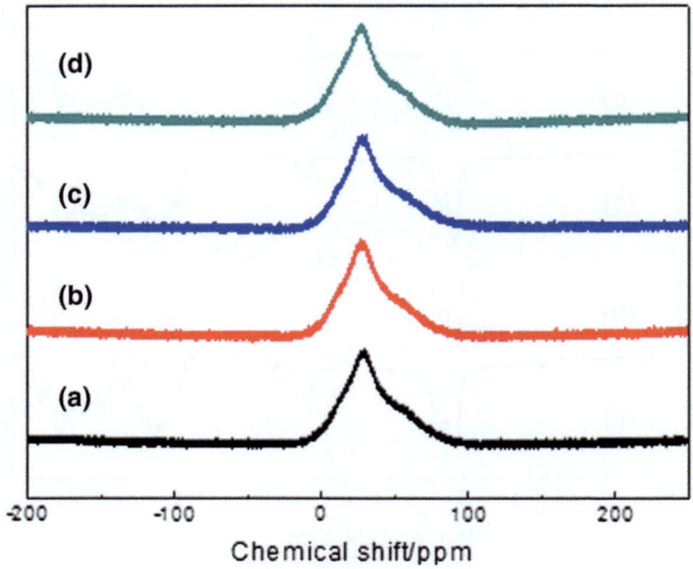

Fig. 3 ^{11}B-NMR spectra of PBSZ samples with different boron content. The molar ratios of BCl_3: $HSiCl_3$: $C_3H_7NH_2$ are, **a** 1:2:18, **b** 1:1.5:15, **c** 1:1:12, **d** 1:0.5:9

units in the six membered borazine rings. As shown in Fig. 3, there is no change or almost no change in the environmental chemistry of boron in the PBSZ samples with different boron content.

High Temperature Thermogravimetric Analysis. The thermal stability induced polymer-to-ceramic conversion of polymeric precursors with different boron content was investigated by TGA upon heating to 900 °C in a flowing nitrogen atmosphere. As shown in Fig. 4, polymeric precursors with different boron content have similar weight loss properties except final ceramic yield. The mass losses are all divided into three stages. In the first stage, the weight loss was caused by the loss of solvent and the evaporation of low molecular weight oligomer which occurred from room temperature to 200 °C. The difference of the mass loss among polymeric precursors in the first stage is attributed to the amount of added $C_3H_7NH_2$. At the stage two, the significant mass loss of the polymer precursors observed at 200–450 °C. It may due to the condensation of oligomer, leading to the further crosslinking of polymer precursors, which involved the thermal decomposition end group $CH_3CH_2CH_3$ and the release of propylamine. At higher temperatures, range between 450 and 600 °C, the mass loss is due to the polymer conversion into inorganic ceramics. During this process, the polymer decomposed or rearranged to transform the polymeric structure to the ceramic phase, the breaking of N–C bonds occurred and the propylamine released. In addition, from 600 to 800 °C, a tiny weight loss could be observed.

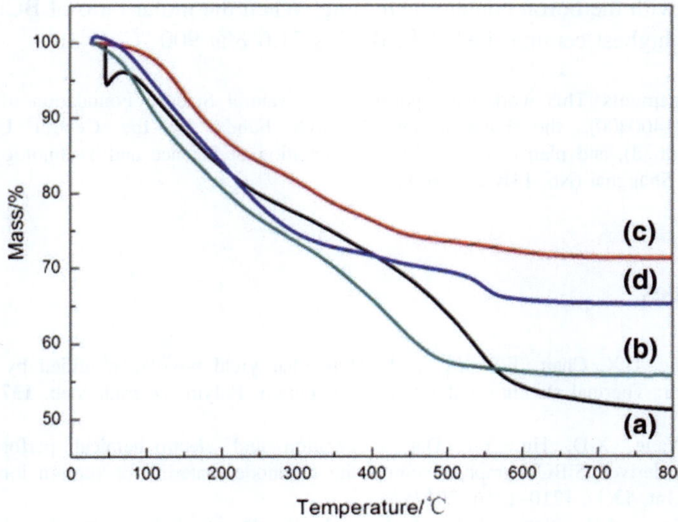

Fig. 4 TG of PBSZ samples with different boron content. The molar ratios of BCl_3:$HSiCl_3$:$C_3H_7NH_2$ are, **a** 1:2:18, **b** 1:1.5:15, **c** 1:1:12, **d** 1:0.5:9

Table 2 Ceramic yield of compounds 1–4 at 900 °C

Polymeric precursor	1	2	3	4
Ceramic yield (%)	50.9	56.3	71.6	64.6

The final ceramic yield of polymeric precursors with different boron content were shown in Table 2. As we can see, the ceramic yield increased as the boron content increased. However, the ceramic yield decreased when the molar ratios of BCl_3:$HSiCl_3$ over 1:1. This result is agreement with the previous research, which indicated that boron content in the region from 5 to 17% is able to provide remarkably high temperature stability. [19, 20] When the molar ratios of BCl_3:$HSiCl_3$ is 1:1, an overall 71.6% ceramic yield was obtained at 900 °C.

Conclusion

A polymeric precursor for producing SiBNC ceramics was synthesized via copolymerization of BCl_3, $HSiCl_3$ and $C_3H_7NH_2$. FT-IR and NMR results revealed that a network structure containing Si–N, B–N, N–H, N–$CH_2CH_2CH_3$ bonds within the polymeric precursor was observed. Boron content nearly has no effect on the structure of polyborosilazane precursor. High thermal stability of polyborosilazane precursor is dependent on the boron content. Ceramic yield first increased then decreased with the boron content increasing. When the molar ratio of BCl_3:$HSiCl_3$ is 1:1, the highest ceramic yield of PBSZ is 71.6% at 900 °C.

Acknowledgements This work was sponsored by Natural Science Foundation of Shanghai (No. 15ZR1400400), the Fundamental Research Funds for the Central Universities (No. 15D110638), and plan of action for the innovation of Science and Technology in Basic Research of Shanghai (No. 14JC1400100).

References

1. Y. Wang, L.X. Chen, T.T. Xu et al., High char yield novolac modified by Si-B-N-C precursor: Thermal stability and structural evolution. Polym. Degrad. Stab. **137**, 184–196 (2017)
2. S.H. Wang, X.D. Hu, Y.J. Dai, Preparation and electrochemical performance of polymer-derived SiBCN-graphene composite as anode material for lithium ion batteries. Ceram. Int. **43**(1), 1210–1216 (2017)
3. Q. Zhang, Z.H. Yang, D.C. Jia et al., Synthesis and structural evolution of dual-boron-source-modified polysilazane derived SiBCN ceramics. New J. Chem. **40**(8), 7034–7042 (2016)
4. L.J. Yang, Y. Zhang, X. Cheng, Performance and Structure of Polymer Derived SiBCN Ceramics. Prog. Chem. **28**(2–3), 308–316 (2016)
5. A. Viard, P. Miele, S. Bernard, Polymer-derived ceramics route toward SiCN and SiBCN fibers: from chemistry of polycarbosilazanes to the design and characterization of ceramic fibers. J. Ceram. Soc. Jpn. **124**(10), 967–980 (2016)
6. X. Sun, H.T. Liu, J.S. Li et al., Effects of CVD SiBCN interphases on mechanical and dielectric properties of SiCf/SiC composites fabricated via a PIP process. Ceram. Int. **42**(1), 82–89 (2016)
7. M. Weinmann, T.W. Kamphowe, J. Schuhmacher et al., Design of polymeric Si-BCN ceramic precursors for application in fiber-reinforced composite materials. Chem. Mater. **12** (8), 2112–2122 (2000)
8. M. Peuckert, T. Vaahs, M. Brück, Ceramics from organometallic polymers. Adv. Mater. **2**(9), 398–404 (1990)
9. Y. Tang, J. Wang, X. Li et al., Polymer-derived SiBN fiber for high-temperature structural/functional applications. Chem.–A Eur. J. **16**(22), 6458–6462 (2010)
10. Y. Tang, J. Wang, X. Li et al., Thermal stability of polymer derived SiBNC ceramics. Ceram. Int. **35**(7), 2871–2876 (2009)
11. U. Müller, M. Weinmann, M. Jansen, Cl 2 MeSi–NH–BCl 2 and ClMe 2 Si–NH–BCl 2: novel processable single source precursors of amorphous Si/C/B/N ceramics. J. Mater. Chem. **18**(31), 3671–3679 (2008)
12. A. Müller, P. Gerstel, M. Weinmann et al., Correlation of boron content and high temperature stability in Si–B–C–N ceramics II. J. Eur. Ceram. Soc. **21**(12), 2171–2177 (2001)
13. A. Müller, P. Gerstel, M. Weinmann et al., Correlation of boron content and high temperature stability in Si–B–C–N ceramics. J. Eur. Ceram. Soc. **20**(14), 2655–2659 (2000)
14. J. Wilfert, M. Jansen, One-pot synthesis of DMTA-analogous SiBNC preceramic polymer. J. Mater. Chem. **22**(19), 9782–9786 (2012)
15. T. Jäschke, M. Jansen, Improved durability of Si/B/N/C random inorganic networks. J. Eur. Ceram. Soc. **25**(2), 211–220 (2005)
16. J. Lee, D.P. Butt, R.H. Baney et al., Synthesis and pyrolysis of novel polysilazane to SiBCN ceramic. J. Non-Cryst. Solids **351**(37), 2995–3005 (2005)

17. Y. Tang, J. Wang, X. Li et al., One-pot synthesis of novel polyborosilazane to SiBNC fibres. Inorg. Chem. Commun. **12**(7), 602–604 (2009)
18. S. Mou, Y. Liu, K. Han et al., Synthesis and characterization of amorphous SiBNC ceramic fibers. Ceram. Int. **41**(9), 11550–11554 (2015)
19. S. Bernard, M. Weinmann, P. Gerstel et al., Boron-modified polysilazane as a novel single-source precursor for SiBCN ceramic fibers: synthesis, melt-spinning, curing and ceramic conversion. J. Mater. Chem. **15**(2), 289–299 (2005)
20. Y. Peng, K. Han, X. Zhao et al., Large-scale preparation of SiBN ceramic fibres from a single source precursor. Ceram. Int. **40**(3), 4797–4804 (2014)

Dispersion of Graphene Oxide in Polyvinylidene Difluoride and Its Improvement of Photoresponse Properties of Nanocomposite

Xintao Zhang, Shizhen Wang, Jianxun Qiu, Wei Du, Xiaochun He, Mingjun Gao, Xiangming Li, Xingfa Ma and Guang Li

Abstract Graphene oxide (GO) and its nanocomposites are widely used in energy storage devices, membrane separation, multi-functional materials due to their good charge transferring property and excellent mechanical, physical, chemical properties. The performances of materials and devices are strongly dependent on the dispersion effects of GO in matrix. In the applications of polymer/GO nanocomposites, the re-dispersion of dried GO in matrix with good effects is still having some difficulties. Good dispersion of GO solution with polymer aqueous would be obtained easily. However, the dispersion of GO solution in non-aqueous polymers is still a great challenge. In this paper, attempts of the dispersion of GO solution in non-aqueous polymers were carried out with oil/water interface mixed approach. In

X. Zhang · S. Wang · J. Qiu · W. Du · X. He · M. Gao · X. Li · X. Ma (✉)
School of Environmental and Material Engineering, Center of Advanced Functional Materials, Yantai University, Yantai 264005, China
e-mail: xingfazju@163.com; xingfamazju@aliyun.com

X. Zhang
e-mail: zxtyt@163.com

S. Wang
e-mail: 1355664505@qq.com

J. Qiu
e-mail: qjxun@163.com

W. Du
e-mail: dw311@yeah.net

X. He
e-mail: hexiaochun76@126.com

M. Gao
e-mail: gmj2188@126.com

X. Li
e-mail: li_xiangming1981@126.com

G. Li
National Laboratory of Industrial Control Technology, Institute of Cyber-Systems and Control, Zhejiang University, Hangzhou 310027, China
e-mail: guangli@zju.edu.cn

Y. Han (ed.), *Advanced Functional Materials*,
https://doi.org/10.1007/978-981-13-0110-0_88

the study, PVDF [poly-(vinylidene difluoride)] selected as representative polymer is based on the following reason. PVDF is a well-known organic multi-functional polymer, which is widely used in several devices due to its outstanding piezoelectric, pyroelectric, and dielectric performance. PVDF/GO nanocomposites were obtained with interface blend, and a series of characterizations were performed. The photoconductive responses to weak visible light and 808 nm NIR (near infrared) were studied based on interdigital electrodes of Au on flexible polymer film substrate with casting method. The results indicate that the GO has good dispersion in PVDF matrix and no obvious aggregation of GO is observed. The resulting nanocomposite exhibits good photo-switching behavior to visible light, while the response and recovery are very rapidly. The response to weak 808 nm NIR is relatively weak, this may be the results of PVDF possessing some absorbance to NIR. It can be seen that an intelligent composite film with external stimuli responses is developed and can be used in many fields, such as electronic skin, sensors and actuators, information storage, artificial muscle, biomimetic films, membrane separation controlled with external stimuli.

Keywords Graphene oxide · Polyvinylidene difluoride · Interface blend
Nanocomposite film · Photoconductivity improvement

Introduction

Recently, many advanced functional materials and several devices have been obtained excellent progresses. However, to integrate their multi-functionalities or smart properties and widen new applications are still great challenges. Graphene and its nanocomposites as dazzling star were developed rapidly due to their excellent physical properties. Graphene based nanocomposites involved in graphene/inorganic functional materials, graphene/polymers, graphene/noble metals, graphene/organic small molecules, et al. And they are extensively used in energy, environmental fields [1–15]. The remarkable property of graphene is excellent charge-transfer characteristics, and the graphene oxide also has abundant chemical groups on its surface, such as, −COOH, −OH, epoxide group. Therefore, graphene oxide can be constructed polymer based nanocomposites utilizing interactions of chemical groups and π–π stacking.

Recent several years, there are still a large number of reports on GO/polymers nannocomposites. For example, Mina et al. [16] constructed fluorescent nanosensor based on graphene quantum dots embedded silica molecularly imprinted polymer. Dong et al. [17] fabricated an electrochemical sensor using graphene quantum dots and β-cyclodextrins composites. Liu et al. [18] reported molecularly imprinted polymers on graphene oxide surface for sensor. Song et al. [19] introduced an fluorescent sensing platform for determination of 2,4,6-trinitrophenol using temperature responsive polymer brushes grafted graphene oxide. Liang et al. [20] assembled graphene oxide/peptide nanocomposite. Li et al. [21] synthesized reduced-graphene oxide/zine oxide/poly(pphenylenediamine) for supercapacitor. Qin et al. [22] prepared freestanding flexible graphene foams@polypyrrole@MnO_2 electrodes. Gan et al. [23] used polymer/graphene films as anti-reflective transparent electrodes for solar cells. Medina et al. [24] reported graphene oxide/chitosan/pol(acrylic acid) porous polymer nanocomposite for enhanced lead adsorption, and so on. And recently, carbon nanomaterial-based multilayer polymeric actuators controlled by light, electrochemical, and electric, etc., have also attracted more and more attention. Therefore, to develop carbon based smart materials is one of the directions of multi-functional nanocomposites.

Smart responsive materials can respond to external stimuli, such as temperature, pH, pressure, heat, electric, light, magnetic fields and other external stimuli, and it also can convert these different energy forms to mechanical energy or physical signals. The flexible smart devices are mainly depending on some functional polymers or polymer based nanocomposites. Compared with inorganic functional materials, multi-functional polymers are limited. Some electroactive polymers (EAPs) include dielectric elastomers, conducting polymers, piezoelectric polymers, electrostrictive materials, ionic polymers are suitable to applications of flexible smart devices. Therefore, PVDF is a good candidate for developing flexible smart multi-functional materials and devices since PVDF is an interesting fluoropolymer with remarkable properties such as thermal stability, chemical inertness to solvents and acids as well as outstanding piezo-, pyro- and ferroelectric properties. For many years, modified PVDF or PVDF based composites and their applications are received considerable world attentions. For instance, Gao et al. [25] developed ultrathin multi-functional graphene/PVDF layers for flexible displays. Xiao et al. [26] constructed a fish-like robot based on high performance graphene/PVDF bimorph actuation materials. Sui et al. [27] enhanced the dielectric and ferroelectric properties of PVDF composite flexible films through doping agent. Anithakumari et al. [28] prepared PVDF/$BaFe_{12}O_{19}$ composites with enhancement of dielectric, ferroelectric and magneto-dielectric properties. Karan et al. [29] reported Fe-rGO incorporated PVDF based flexible nanocomposite film with enhanced thermal conductivity and dielectric property for energy storage applications. Abdelhamid et al. [30] prepared multiferroic PVDF/Fe_3O_4 hybrid films with reduced graphene oxide and ZnO nanofillers. Meng et al. [31] introduced PVDF/PVDF-TrFE blended films with enhanced dielectric and ferroelectric properties. Yu et al. [32] constructed highly stretchable and conductive nanofibrous PANi/PVDF strain sensors. Zabek et al. [33] examined graphene ink laminate structures on PVDF for pyroelectric thermal energy harvesting and waste heat recovery, et al.

Otherwise, in environmental fields, membrane separation technology is regarded as one of the most efficient ways to handle a variety of wastewaters because of its low energy and time consumption, high separation efficiency, and simple operation. PVDF is one of the most extensively studied and attractive polymer matrices for membrane separation owing to its unique mechanical properties, electro-active properties, chemical and thermal stability. Recently, many studies have described treatment of oil-contaminated wastewater, microfiltration hollow fiber membranes, hybrid ultrafiltration membrane with enhanced hydrophilicity of a thermo-responsive properties, porous membrane for selective adsorption and molecular filtration, and hollow fiber membrane for CO_2 capture using modified PVDF or PVDF composites films [34–44]. In new energy and environmental fields, studies on modification and enhancements of physical properties of PVDF by new synthesis method or inorganic/organic nanocomposites are appeared continuously [45–57]. For example, synthesis of PVDF based dendrimers is also good ideas [45].

For many years, the authors are very interested in organic functional materials or organic-inorganic functional nanocomposites and their properties [58–62]. In this paper, to improve the dispersion of GO in non-aqueous polymers matrix, attempts of the dispersion of GO solution in non-aqueous polymers were carried out with oil/water interface mixed approach. The properties study was put on the external stimuli responses of GO/PVDF nanocomposite, mainly referring to weak visible light and 808 nm NIR for applications of some flexible devices controlled by light, some good results were obtained.

Experimental Details

Materials. Imported graphite (AR), 325 mesh (purity 99.8%). Potassium permanganate (AR), sulfuric acid (Chemical Reagent, i.e. CP), nitric acid (CP), hydrochloric acid (AR), sodium hydroxide (AR), DMF (Dimethyl Formamide) were purchased from Sinopharm Chemical Reagent Co., Ltd., et al. PVDF (FR902) was purchased from Shanghai 3F New Material Co., Ltd.

Synthesis of Graphene Oxide. The preparation of graphene oxide was using chemical method, which is according to the previous report utilizing graphite, concentrated sulfuric acid, concentrated nitric acid, potassium permanganate, et al. [63]. About 1 g graphite was transferred 500 mL GO solution.

Blend of Graphene Oxide Solution and PVDF Solution at Interface of Oil/Water. In the experiment, about 20–30 mL graphene oxide was added in 250 mL glass vessel, 10–20 mL FVDF solution (concentration is about 0.05 g/mL) were added, precipitation of PVDF was appeared, filtrated, then 30 mL DMF was added, the GO/PVDF solution was formed for later use.

Morphology Observation with SEM. The field emission scanning electron microscopy (SEM) observation was carried out with JSM—6700 (JEOL). The GO/PVDF solution was casted on an aluminum foil substrate, dried at room temperature and then sputtered with a thin layer of Pt on the surface for the SEM observation.

Morphology Observation with TEM. The TEM observation was carried out with JEM-1011. The GO/PVDF solution was casted on cupper mesh coated with carbon film, dried at room temperature.

Measurement of UV-Vis Spectrum. The UV-Vis was determined by a TU-1810 spectrophotometer and the samples were using GO/PVDF solution.

Measurement of FTIR Spectra. FTIR (Fourier transform infrared spectroscopy) spectra were taken with KBr, and recorded on IR Prestige-21 Fourier transform infrared spectrometer. Sample and KBr powders were mixed and pressed into a small slice, and then dried at room temperature for measurement.

Measurement of Raman Spectra. Raman spectra were recorded on a LabRAM HR800 (HORLBA JY) using GO/PVDF powers sample.

TGA (Thermo Gravimetric Analyzer). TGA was carried out with SDT Q600 Simultaneous DSC-TGA Instrument produced by The United States of America TA Instrument Company. The heating rate is 10 °C/min in range of room temperature to 600 °C under high purity N_2 protection.

Photoresponses of Nanocomposite to Visible Light and NIR. The GO/PVDF solution was casted on the interdigital electrodes of Au on flexible PET substrate, after drying, the photoconductive responses to weak visible light (20–25 W) and 808 nm NIR with low-power were determined with LK2000A Electrochemical Work Station from LANLIKE Chemistry and Electron High Technology Co., Ltd (China).

Results and Discussion

It is well known that GO has good dispersion effects in water-soluble polymers, but in the oil-soluble polymers, it is difficult to reach satisfying results. PVDF is a polar polymer, which can be dissolve in DMF. DMF belongs to polar solvent, which can be mixed with water. Therefore, GO solution (water) and PVDF solution (DMF) can mix at the interface of water and DMF. Since PVDF did not dissolve in water, precipitation of PVDF at the interface of water and DMF was appeared. GO with good dispersion was entered in PVDF matrix. The representative TEM images of GO/PVDF nanocomposite is shown in Fig. 1a. The representative SEM images of GO/PVDF nanocomposite is shown in Fig. 1b.

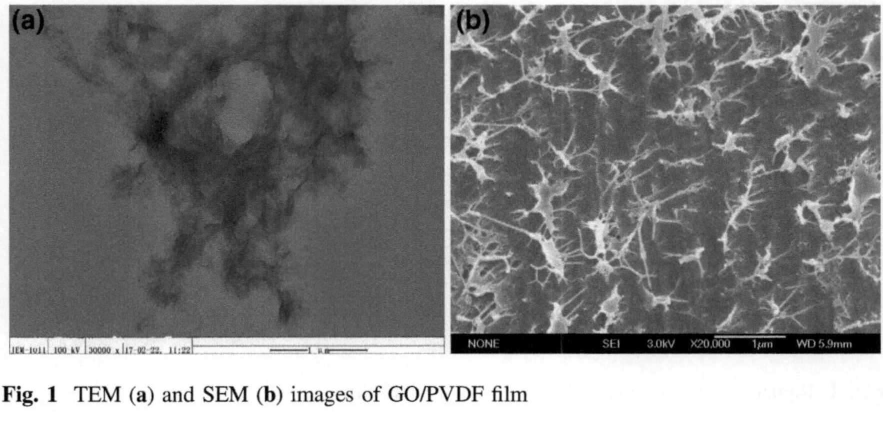

Fig. 1 TEM (**a**) and SEM (**b**) images of GO/PVDF film

As shown form Fig. 1a, the GO had good dispersion in PVDF matrix, no obvious aggregation of GO were observed. It illustrated that the interface blend is an effective way for obtaining GO/PVDF nanocomposite.

Figure 1b showed that the graphene oxide (GO)/PVDF film has polyporous structure. Therefore, it would be many potential applications in environmental fields, especially for membrane separation with multi-functionalities.

To confirm the presence of GO and PVDF, Raman spectra and FTIR were carried out. The results are shown in Fig. 2a, b.

Figure 2a showed that D band (defect of C) and G band (C sp^2) were clearly obversed in GO/PVDF nanocomposite. This illustrated that the GO/PVDF nanocomposite contained some GO.

As shown from Fig. 2b, 3446.9 cm^{-1} belongs to stretching vibration of –OH of GO, 2975 cm^{-1} is attributed to the stretching vibration of C–H, 1645 cm^{-1} is C=C, 1072 cm^{-1} is C–F, 1713 cm^{-1} is the stretching vibration of –COOH, et al. Therefore, GO/PVDF nanocomposite also contained some PVDF.

The UV-Vis and TGA of GO/PVDF nanocomposite were examined. The results are shown in Fig. 3a, b.

Figure 3a showed that the adsorption of GO/PVDF covered the whole region of visible light and extended to NIR (near-infrared). The edge of band is near to 900 nm. Therefore, the obtained nanocomposite would be utilized more efficiently the visible light and NIR.

Figure 3b indicated that the weight loss of GO/PVDF nanocomposite is much smaller than that of pure PVDF below 465 °C. Over 465 °C, the weight loss of GO/PVDF and PVDF is almost same. The main reason is as followed: graphene oxide has abundant chemical groups on its surface, such as, −COOH, −OH, epoxide group. These groups have good adsorption properties, including physical adsorption and chemical adsorption. Below 465 °C, the weight loss of GO/PVDF nanocomposite is attributed to the removal of physical adsorption and chemical adsorption and some degradation of PVDF. Over 465 °C, the weight loss of GO/PVDF nanocomposite is mainly results of degradation and carbonization of PVDF.

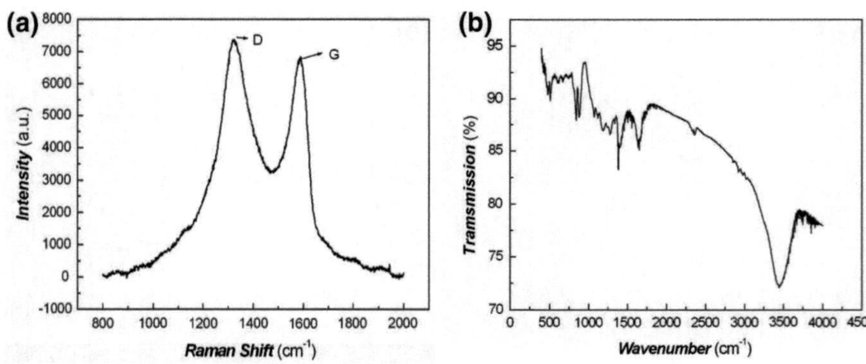

Fig. 2 Raman spectra (**a**) and FTIR (**b**) of GO/PVDF film

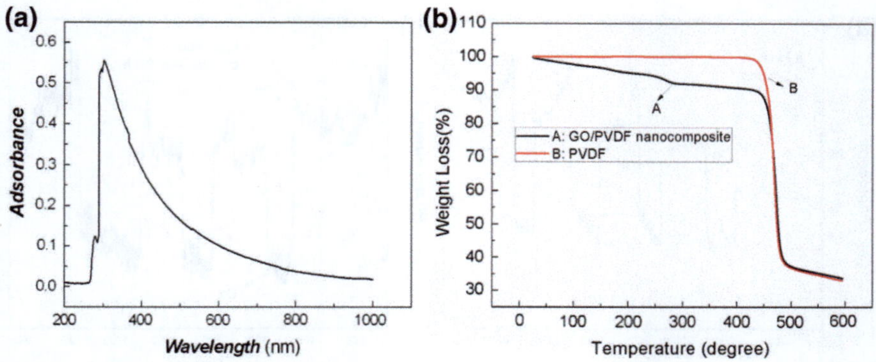

Fig. 3 UV-Vis (**a**) and TGA (**b**) of GO/PVDF film

Figure 3a showed that the GO/PVDF nanocomposite had good adsorption in visible light and NIR. Therefore, the photoconductive properties of nanocomposite to weak visible light and 808 nm NIR were examined. The representative results are shown in Fig. 4a, b.

Figure 4 showed that GO/PVDF nanocomposite to weak visible light has good photoresponse behavior, to 808 nm NIR is not very high. This is the results of PVDF having some adsorption to NIR.

Repeatability of photoresponse to visible light was also examined. Good results were obtained. The representative results are shown in Fig. 5.

As shown from Fig. 5, although some similar results were obtained in the cycle tests, an interesting results were appeared. The current of baseline is increased with the times of tests, and the ratio of on/off is also enhanced with the times of tests. This illustrated that the photo-induced charges in GO/PVDF nanocomposite film were produced very easily to weak light. Nanocomposite prolonged the lifetime photo-induced charges, avoided the recombination of carrier, and the GO/PVDF

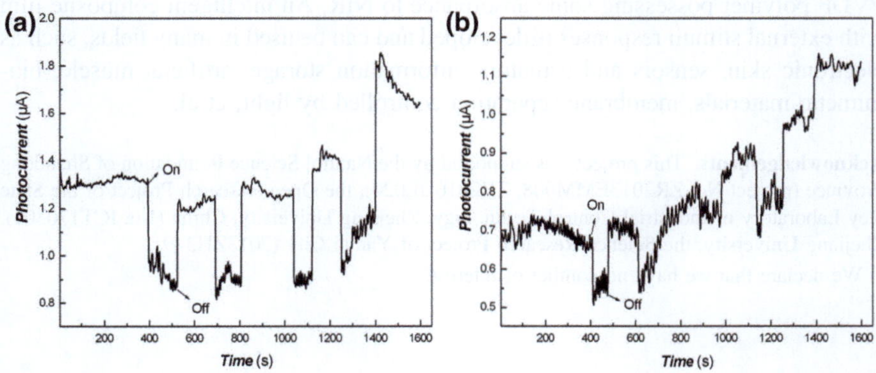

Fig. 4 Photoresponses of GO/PVDF to weak visible light (**a**) and 808 nm NIR of 200 mW (**b**)

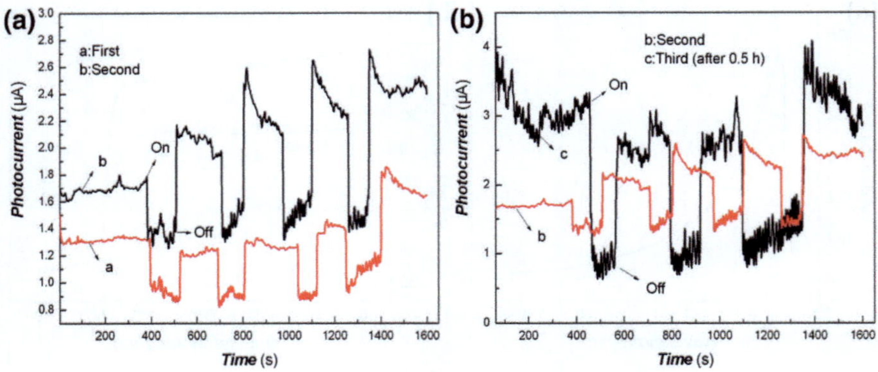

Fig. 5 Repeatability of photoresponses of GO/PVDF to weak visible light (**a** first test; **b** second test; **c** third test)

nanocomposite showed good activities to visible light and NIR (near-infrared). It would be developing light detector, intelligent composite film with external stimuli responses, electronic skin, sensors and actuators, information storage, artificial muscle, biomimetic materials, membrane separation with photocatalyst, multi-functional nanocomposite, et al.

Conclusions

In summary, dispersion of GO solution in non-aqueous polymers was realized with oil/water interface mixed approach. No obvious aggregation of GO is observed in PVDF matrix, and the GO/PVDF nanocomposite exhibits good photo-switching behavior to weak visible light, while the response and recovery are very rapidly. The response to weak 808 nm laser is relatively weak, this may be the results of PVDF polymer possessing some absorbance to NIR. An intelligent composite film with external stimuli responses is developed and can be used in many fields, such as electronic skin, sensors and actuators, information storage, artificial muscle, biomimetic materials, membrane separation controlled by light, et al.

Acknowledgements This project was supported by the Natural Science Foundation of Shandong Province (project No. ZR2013EMM008, ZR2016 JL020), the Open Research Project of the State Key Laboratory of Industrial Control Technology, Zhejiang University, China (No. ICT170303), Zhejiang University, the Science Research Project of Yantai City (2013ZH349).

We declare that we have no conflict of interest.

References

1. C. Zhang, R. Hao, B. Zhao, Y. Hao, Y. Liu, Appl. Surf. Sci. **409**, 306–313 (2017)
2. T. Zhan, Z. Tan, X. Tian, W. Hou, Ionic liquid functionalized graphene oxide-Au nanoparticles assembly for fabrication of electrochemical 2,4-dichlorophenol sensor. Sens. Actuators, B **246**, 638–646 (2017)
3. P.N. Joshi, S. Agawane, M.C. Athalye, V. Jadhav, D. Sarkar, R. Prakash, Mater. Sci. Eng., C **78**, 1203–1211 (2017)
4. L. Liu, D. Zhang, Q. Zhang, X. Chen, G. Xu, Y. Lu, Q. Liu, Biosens. Bioelectron. **93**, 94–101 (2017)
5. B. An, Y. Ma, W. Li, M. Su, F. Li, Y. Song, Chem. Commun. **52**, 10948–10951 (2016)
6. G.G. Jang, B. Song, K. Moon, C. Wong, J.K. Keum, M.Z. Hu, Carbon **119**, 296–304 (2017)
7. C. Xu, Y. Jing, J. He, K. Zhou, Y. Chen, Q. Li, J. Lin, W. Zhang, J. Alloy. Compd. **708**, 1178–1183 (2017)
8. R. Tian, W. Wang, Y. Huang, H. Duan, Y. Guo, H. Kang, H. Li, H. Liu, J. Mater. Chem. A. **4**, 13148–13154 (2016)
9. M. Shi, T. Wu, X. Song, J. Liu, L. Zhao, P. Zhang, L. Gao, J. Mater. Chem. A **4**, 10666–10672 (2016)
10. W. Wei, K. Sun, Y. Hu, J. Mater. Chem. A **4**, 12054–12057 (2016)
11. Y. Shen, X. Zhu, B. Chen, J. Mater. Chem. A **4**, 12106–12118 (2016)
12. X. Sua, C. Chana, J. Shia, M. Tsangb, Y. Panc, C. Cheng, O. Gerilea, M. Yang, Biosens. Bioelectron. **92**, 489–495 (2017)
13. H. Liu, W. Na, Z. Liu, X. Chen, X. Su, A novel turn-on fluorescent strategy for sensing ascorbic acid using graphene quantum dots as fluorescent probe. Biosens. Bioelectron. **92**, 229–233 (2017)
14. J. Yoon, T. Lee, J. Jo, B. Oh, J. Choi, Electrochemical H_2O_2 biosensor composed of myoglobin on MoS_2 nanoparticle-graphene oxide hybrid structure. Biosens. Bioelectron. **93**, 14–20 (2017)
15. H. Kanso, M.B.G. García, L.F. Llano, S. Ma, R. Ludwig, P.F. Bolado, D.H. Santos, Novel thin layer flow-cell screen-printed graphene electrode for enzymatic sensors. Biosens. Bioelectron. **93**, 298–304 (2017)
16. M. Mehrzad-Samarin, F. Faridbod, A.S. Dezfuli, M.R. Ganjali, A novel metronidazole fluorescent nanosensor based on graphene quantum dots embedded silica molecularly imprinted polymer. Biosens. Bioelectron. **92**, 618–623 (2017)
17. S. Dong, Q. Bi, C. Qiao, Y. Sun, X. Zhang, X. Lu, L. Zhao, Electrochemical sensor for discrimination tyrosine enantiomers using graphene quantum dots and β-cyclodextrins composites. Talanta **173**, 94–100 (2017)
18. W. Liu, Y. Ma, G. Sun, S. Wang, J. Deng, H. Wei, Molecularly imprinted polymers on graphene oxide surface for EIS sensing of testosterone. Biosens. Bioelectron. **92**, 305–312 (2017)
19. Y. Song, J. Lü, B. Liu, C. Lu, J. Mater. Chem. C. **4**, 7083–7092 (2016)
20. P. Liang, Q. Li, Z. Wu, J. Jiang, R. Yu, Analyst **141**, 3989–3992 (2016)
21. J. Li, Y. Sun, D. Li, H. Yang, X. Zhang, B. Lin, J. Alloy. Compd. **708**, 787–795 (2017)
22. T. Qin, B. Liu, Y. Wen, Z. Wang, X. Jiang, Z. Wan, S. Peng, G. Cao, D. He, J. Mater. Chem. A **4**, 9196–9203 (2016)
23. X. Gan, R. Lv, H. Zhu, L. Ma, X. Wang, Z. Zhang, Z. Huang, H. Zhu, W. Ren, M. Terronesdfg, F. Kang, J. Mater. Chem. A **4**, 13795–13802 (2016)
24. R.P. Medina, E.T. Nadres, F.C. Ballesteros Jr., D.F. Rodrigues, Environ. Sci.: NANO **3**, 638–646 (2016)
25. S. Gao, X. Wu, H. Ma, J. Robertson, A. Nathan, Ultrathin multi-functional graphene-PVDF layers for multidimensional touch interactivity for flexible displays. ACS Appl. Mater. Interfaces. **9**, 18410–18416 (2017)

26. P. Xiao, N. Yi, T. Zhang, Y. Huang, H. Chang, Y. Yang, Y. Zhou, Y. Chen, Construction of a fish-like robot based on high performance graphene/PVDF bimorph actuation materials. Adv. Sci. **3**, 1500438 (2016)
27. Y. Sui, W. Chen, J. Ma, R. Hu, D. Liu, Enhanced dielectric and ferroelectric properties in PVDF composite flexible films through doping with diisopropylammonium bromide. RSC Adv. **6**, 7364–7369 (2016)
28. P. Anithakumari, B.P. Mandal, E. Abdelhamid, R. Naikb, A.K. Tyagi, Enhancement of dielectric, ferroelectric and magneto-dielectric properties in PVDF–$BaFe_{12}O_{19}$ composites: a step towards miniaturizated electronic devices. RSC Adv. **6**, 16073–16080 (2016)
29. S.K. Karan, A.K. Das, R. Bera, S. Paria, A. Maitra, N.K. Shrivastava, B.B. Khatua, Effect of g-PVDF on enhanced thermal conductivity and dielectric property of Fe-rGO incorporated PVDF based flexible nanocomposite film for efficient thermal management and energy storage applications. RSC Adv. **6**, 37773–37783 (2016)
30. E.H. Abdelhamid, O.D. Jayakumar, V. Kotari, B.P. Mandal, R. Rao, V.M. Naik, R. Naik, A. K. Tyagi, Multiferroic PVDF–Fe_3O_4 hybrid films with reduced graphene oxide and ZnO nanofillers. RSC Adv. **6**, 20089–20094 (2016)
31. N. Meng, X. Zhu, R. Mao, M.J. Reece, E. Bilotti, Nanoscale interfacial electroactivity in PVDF/PVDF-TrFE blended films with enhanced dielectric and ferroelectric properties. J. Mater. Chem. C **5**, 3296–3305 (2017)
32. G. Yu, X. Yan, M. Yu, M. Jia, W. Pan, X. He, W. Han, Z. Zhang, L. Yu, Y. Long, Patterned, highly stretchable and conductive nanofibrous PANI/PVDF strain sensors based on electrospinning and in situ polymerization. Nanoscale **8**, 2944–2950 (2016)
33. D. Zabek, K. Seunarine, C. Spacie, C. Bowen, Graphene ink laminate structures on poly (vinylidene difluoride) (PVDF) for pyroelectric thermal energy harvesting and waste heat recovery. ACS Appl. Mater. Interfaces. **9**, 9161–9167 (2017)
34. C. Luo, Q. Liu, Oxidant-induced high-efficient mussel-inspired modification on PVDF membrane with superhydrophilicity and underwater superoleophobicity characteristics for oil/ water separation. ACS Appl. Mater. Interfaces. **9**, 8297–8307 (2017)
35. Y. Zhu, W. Xie, F. Zhang, T. Xing, J. Jin, Superhydrophilic in-situ-cross-linked zwitterionic polyelectrolyte/ PVDF-blend membrane for highly efficient oil/water emulsion separation. ACS Appl. Mater. Interfaces. **9**, 9603–9613 (2017)
36. Q. Zhou, Z. Wang, H. Shen, Z. Zhu, L. Liu, L. Yang, L. Cheng, Morphology and performance of PVDF TIPS microfiltration hollow fiber membranes prepared from PVDF/ DBP/DOP systems for industrial application. J. Chem. Technol. Biotechnol. **91**, 1697–1708 (2016)
37. N.A. Samsure, N.A. Hashim, N.M.N. Sulaiman, C.Y. Chee, Alkaline etching treatment of PVDF membrane for water filtration. RSC Adv. **6**, 22153–22160 (2016)
38. J. Cai, S. Zhou, Y. Zhao, A. Xue, Y. Zhang, M. Li, W. Xing, Enhanced hydrophilicity of a thermo-responsive PVDF/palygorskite-g-PNIPAAM hybrid ultrafiltration membrane via surface segregation induced by temperature. RSC Adv. **6**, 62186–62192 (2016)
39. K. Ji, X. Ma, H. Xu, J. Yin, X. Jiang, Hyperbranched poly (ether amine)@poly (vinylidene fluoride) porous membrane (hPEA@PVDF) for selective adsorption and molecular filtration of hydrophilic dyes. J. Mater. Chem. A **5**, 10470–10479 (2017)
40. X. Zhao, Y. Chen, H. Xuan, C. He, Investigation of one-dimensional multi-functional zwitterionic Ag nanowires as a novel modifier for PVDF ultrafiltration membranes. New J. Chem. **40**, 441–446 (2016)
41. G. Lopez, M. Guerre, B. Ameduri, J. Habas, V. Ladmiral, Photocrosslinked PVDF-based star polymer coatings: an all-in-one alternative to PVDF/PMMA blends for outdoor applications. Polym. Chem. **8**, 3045–3049 (2017)
42. L. Peng, W. Lei, P. Yu, Y. Luo, Polyvinylidene fluoride (PVDF)/hydrophobic nanosilica (H-SiO2) coated superhydrophobic porous materials for water/oil separation. RSC Adv. **6**, 10365–10371 (2016)

43. K. Venkatesh, G. Arthanareeswaran, A. Chandra Bose, PVDF mixed matrix nano-filtration membranes integrated with 1D-PANI/TiO_2 NFs for oil–water emulsion separation. RSC Adv. **6**, 18899–18908 (2016)
44. Z. Zhang, X. Wu, L. Wang, B. Zhao, J. Li, H. Zhang, RSC Adv. **7**, 13451–13457 (2017)
45. E. Folgado, M. Guerre, C. Bijani, V. Ladmiral, A. Caminade, B. Ameduri, A. Ouali, Polym. Chem. **7**, 5625–5629 (2016)
46. D. Guo, K. Cai, Y. Wang, J. Mater. Chem. C **5**, 2531–2541 (2017)
47. Y. Li, J. Sun, L. Liu, F. Yang, Environ. Sci. NANO **4**, 335–345 (2017)
48. K. Zhang, Z. Cui, G. Xing, Y. Feng, S. Meng, RSC Adv. **6**, 100079–100089 (2016)
49. A.K. Solarajan, V. Murugadoss, S. Angaiah, J. Appl. Polym. Sci. **134**, 45177 (2017)
50. I. Chae, S. Ahmed, H.B. Atitallah, J. Luo, Q. Wang, Z. Ounaies, S.H. Kim, Macromolecules **50**, 2838–2844 (2017)
51. S.D.I. Bhagavathula, V. Kokkarachedu, D.Q. Acuna, R. Koduri, S. Veluri, V. Reddy, J. Appl. Polym. Sci. **134**, 44983 (2017)
52. J. Ma, X. Nan, J. Liu, Polym. Adv. Technol. **28**, 166–173 (2017)
53. A.A. Issa, M. Al-Maadeed, A.S. Luyt, M. Mrlik, M.K. Hassan, J. Appl. Polym. Sci. **133**, 43594 (2016)
54. A.S. Motamedi, H. Mirzadeh, F. Hajiesmaeilbaigi, S. Bagheri-Khoulenjani, M.A. Shokrgozar, J. Biomed. Mater. Res., Part A **105**, 1984–1993 (2017)
55. S.K. Karan, R. Bera, S. Paria, A.K. Das, S. Maiti, A. Maitra, B.B. Khatua, Adv. Energy Mater. **6**, 1601016 (2016)
56. X. Chen, X. Han, Q. Shen, PVDF-based ferroelectric polymers in modern flexible electronics. Adv. Electron. Mater. 1600460 (2017)
57. F.A. Sanchez, J. González-Benito, Polym. Compos. **38**, 227–235 (2017)
58. X. Ma, B. Zhang, Q. Cong, X. He, M. Gao, G. Li, Organic/inorganic nanocomposites of ZnO/CuO/chitosan with improved properties. Mater. Chem. Phys. **178**, 88–97 (2016)
59. Q. Cong, X. He, M. Gao, X. Ma, G. Li, ZnO/CuS heterostructured nanocomposite and its organic functionalization. Mater. Res. Innovations **18**, 740–746 (2014)
60. Q. Cong, H. Geng, X. He, M. Gao, X. Ma, G. Li, Surface modification of ZnO nanosheets with Au/polyaniline and their properties. Mater. Res. Innovations **18**, 30–36 (2014)
61. X. Ma, M. Wang, G. Li, H. Chen, R. Bai, Mater. Chem. Phys. **98**, 241–247 (2006)
62. X. Ma, B. Zhang, Q. Cong, X. He, M. Gao, G. Li, Highly-enhanced performance of TiO_2 nanotube attached CdS quantum dots. Curr. Nanosci. **12**, 500–507 (2016)
63. J. Zheng, X. Ma, X. He, M. Gao, G. Li, Praparation, characterizations, and its potential applications of PANi/graphene oxide nanocomposite. Procedia Eng. **27**, 1478–1487 (2012)

Antimony Doped Tin Oxide Infrared Shielding Films for Cooling Silicon Solar Cells

Ruiqianling Gao, Hao Xiong, Ran Li, Chengyi Hou, Yaogang Li, Qinghong Zhang and Hongzhi Wang

Abstract The operating temperature of silicon solar cells is strongly impacting on their photoelectrical conversion efficiency (PCE). The increasing temperature of the silicon solar cells can degrade device performance significantly. It is well known that antimony-doped tin oxide (ATO) shows the very high transmittivity in visible light region and heat-insulating properties. In this work, ATO nanoparticles with a crystallite size of 8 nm via a peroxo-route and hydrothermal process were prepared. Then the films were deposited by spin-coating method, furtherly studying the effect of varying Sb dopant concentration and film thickness on solar-heat shielding performance and device performance. To evaluate the IR-shielding and silicon solar cells properties, the ATO films on glass were used as the cover glass on the top of silicon solar cells under solar simulator. The experimental results showed that in presence of ATO films, PCE was effectively kept, which has proved that the properties of the ATO films could affect the cell performance after being irradiated for 30 min. Among them, 6-layer ATO film with 7 mol% doped concentration was the champion for silicon solar cells to keep its efficiency. After being irradiated for 30 min, the efficiency of solar cell was maintained 97.21% as the initial, which was much higher than that of in presence of bare glass.

Keywords Antimony doped tin oxide · Infrared shielding films
Cooling · Silicon solar cells

R. Gao · H. Xiong · R. Li · C. Hou · H. Wang
State Key Laboratory for Modification of Chemical Fibers
and Polymer Materials, College of Materials Science and Engineering,
Donghua University, Shanghai 201620, China

Y. Li · Q. Zhang (✉)
Engineering Research Center of Advanced Glasses
Manufacturing Technology, MOE, College of Materials Science and Engineering,
Donghua University, Shanghai 201620, China
e-mail: zhangqh@dhu.edu.cn

Y. Han (ed.), *Advanced Functional Materials*,
https://doi.org/10.1007/978-981-13-0110-0_89

Introduction

Silicon solar cells are truly mainstream in the solar industry currently, and its highest efficiency has even exceeded 26%, which has been certified by the National Renewable Energy Laboratory [1]. However, the silicon solar cells' surface temperature increase sharply under the sun light illuminating continuously, which causes a rapid drop in the photoelectric transformation efficiency and irreversible damage to the silicon materials [2].

When an electron–hole pair with energy greater than the band gap of silicon ($E = 1.12$ eV, $\lambda = 1100$ nm) is created, this part of the photon's energy is fully converted into heat to increase the cell temperature [3]. The increase in operating temperature of the crystalline silicon solar cells results in decrease of conversion efficiency of about 0.45%/K [4]. In practice, the operating temperature of a solar cell increases by $\sim$20 °C compared to ambient temperature in summer [5]. Therefore, decreasing the surface temperature in solar cells is crucial to keep device performance.

In traditional, to address the problem mentioned above, heat exchange is the most common and efficient method, including heat pipe passive cooling [6], liquid immersion cooling [7–9], active cooling by flow of water [10–12] and air-cooling [13]. But their performance mainly depends on the heat transfer area and wind velocity, in some case water and additional power needed for pumping. Those factors must be taken into consideration while designing the system, thus heat exchange is a more complicated project.

In order to reduce the device temperature more effectively, researchers developed radiative cooling through the photonic crystal structure. Very recently, Zhu et al. [14, 15] etched periodical ultra-small patterns on ultra-thin quartz glass via e-beam lithography techniques to form a transparent overlay. The result demonstrated that the overlay let visible light pass through to the solar cells while cooling down 13 K outside. Hossain et al. [16] designed an array microstructure shaped conical metamaterial (CMM) pillars leading to selectively matching the thermal emission (absorption). The structures have the ability to cool down 9 K below the ambient temperature during day-time operation. However, microstructure preparation is very expensive and difficult in the mass production, which greatly increases the cost of manufacturing the solar cell.

Herein, antimony doped tin oxide (ATO) on glass was introduced onto the silicon solar cells. ATO is a kind of transparent conductive oxides (TCOs) with promising applications for energy savings because of its high optical transparency and low transmittance in the near-infrared [17–19]. Moreover, ATO has good chemical stability and low cost compared to indium tin oxide. In this work, ATO hydrosol was prepared via hydrothermal method using peroxo complex as precursors. The sol was deposited onto a glass substrate obtaining an oxide thin film after subsequently dried and sintered. The effect of the Sb doping concentration and film thickness on the IR shielding performance and solar cell performance were further studied.

Experimental Section

All chemicals and reagents were purchased in the highest grade commercially available and used without further purification. Tin tetrachloride ($SnCl_4 \cdot 5H_2O$) and antimony chloride ($SbCl_3$) were acquired from Sinopharm Chemical Reagent. Ultrapure water (UPW, 18.2 MΩ cm) was obtained from a Milli-Q water purification system (Millipore Corp., Bedford, MA, USA). Polycrystalline silicon solar cell plate in area of 20 mm × 20 mm was obtained from Shanghai Solar Energy Research Center.

Synthesis of the ATO Sol. $SnCl_4 \cdot 5H_2O$ was dissolved in ultrapure water to a 0.25 M solution. Then, $NH_3 \cdot H_2O$ solution (1.0 M) was slowly added into the $SnCl_4$ solution with continuous stirring at 70 °C until a pH value of 6 was reached. The prepared precipitates were centrifuged and washed several times with ultrapure water to remove chlorides. 60 mL of H_2O_2 (30%) was added to the precipitates as peptizing agent, and stood for over 12 h at room temperature, obtaining a dispersed sol. Next, varying concentration of $SbCl_3$ was added into the dispersed sol with stirring for 1 h. Then the resulting mixture is transferred to a 100 mL Teflon-lined autoclave and heated at 180 °C for 12 h. After cooling down naturally, the resultant precipitate is collected and rinsed with ultrapure water several times to remove impurities and dispersed in ultrapure water.

Preparation of ATO Film. Pre-cleaning of glass slide substrates was further cleaned using solution of surfactant, deionized (DI) water, ethyl alcohol, and acetone for 10 min by ultrasonicator respectively, then treated in O_2 plasma (DT-01, Suzhou Omega Machinery Electronic Technology Co., Ltd). ATO film was formed through spin-coating at 3000 rpm for 35 s and dried at 150 °C for 25 min. The procedure was repeated several times for increasing the thickness of films. Then, the film was annealed at 500 °C for 30 min.

Characterization. The ATO sol (dried at 60 °C and ground into a fine powder) and the obtained films were characterized using a X-ray diffraction (XRD) equipment (Model D/max 2550 V, Rigaku Co. Tokyo, Japan) under Cu Ka (l = 1.5406 Å) irradiation. The morphologies of particles in the sol were obtained by transmission electron microscopy (TEM) (Model JEM-2100F JEOL, Tokyo, Japan). The surface morphologies and cross-sectional microstructure of the resultant ATO film was observed by field-emission scanning electron microscopy (FESEM, Model S-4800, Hitachi, Japan). Dopant concentration of Sb in ATO nanoparticles was characterized using FESEM attached energy dispersive X-ray (EDS) mode. The optical transmission/absorption was measured using a UV/vis/NIR spectrophotometer (Shimadzu UV-3600, Tokyo, Japan) in the range of 300 − 2500 nm. The photocurrent density–voltage (J–V) curves of the solar cells were recorded using a Keithley model 2400 source measure unit. A solar simulator (Model 96160 Newport Co., USA) equipped with a 300 W Xenon lamp was used as a light source, where light intensity was adjusted using an NREL calibrated Si solar cell with KG-1 filter for approximating AM 1.5G one sun light intensity. The cell performance parameters, including short-circuit current density (Jsc), open-circuit voltage (Voc), fill factor (FF = Pmax/(Jsc × Voc)), and photoelectron conversion efficiency (ŋ(%) = Jsc × Voc × FF/total incident energy × 100), were measured and calculated from the J–V characteristics. The incident-photon-to-current conversion efficiency (IPCE) spectra were measured as a function of wavelength using a specially designed IPCE system (Newport Co., USA).

Result and Discussion

The Synthesis of ATO Sol. The mechanism of the formation of ATO nanoparticles is described as follows. The procedure is divided into two steps: synthesis of the SnO_2 precursor sol and hydrothermal synthesis of ATO sol. In first step, SnO_2 precursor sol was synthesized by a peroxo-route. $NH_3{\cdot}H_2O$ was added to neutralize the tin tetrachloride solution, then obtaining SnO_2 precipitate. Next, the SnO_2 precipitate reacted with H_2O_2 to form a peroxy stannate acid sol, in similar with the formation of peroxotitanium acid [20]. Although, the peroxy stannate acid is metastable and easily dehydrates to generate SnO_2 at room temperature. H_2O_2 could be adsorbed onto SnO_2 crystal particle, which prevents nanometer particles from heavily agglomeration. After that, $SbCl_3$ was added into the dispersed sol and was stirred until well mixed. At the elevated temperature, $SbCl_3$ could be easily hydrolyzed in aqueous solution to form Sb_2O_3 clusters, and the H_2O_2 in the mixture also is absorbed on the surface of Sb_2O_3 and possibly form peroxy-Sb complex. Both Sb_2O_3 clusters and other Sb species are highly activated, and in the following hydrothermal process, Sb atoms substitute Sn atoms in tin oxide lattice.

Figure 1a shows the digital images of SnO_2 precursor sol and ATO sol (5 mol% Sb), the dramatic color change can be observed from the digital images. After the hydrothermal reaction, the sample exhibits the characteristic bluish color of ATO, which can be explained by the coexistence of two different oxidation states of antimony [21]. As shown in Fig. 1b, nanoparticles in SnO_2 precursor sol were highly agglomerated of ultrafine crystallites and the inter-granular boundary was not obvious, which implies that SnO_2 nanoparticles had poor crystallinity and some amount of peroxy-stannate acid might be also occurred. The XRD pattern of the SnO_2 precursor (Fig. 1c) shows weaker diffraction peaks, which indicates both peroxy-stannate acid and SnO_2 have been formed during the peroxo-route. After the hydrothermal reaction with $SbCl_3$ (5 mol%) at 180 °C for 12 h, the TEM image shows that the ATO nanoparticles are highly crystallized (Fig. 3d) and well-dispersed (Fig. 3e), and the average particles size is of approximately 8 nm.

Influence of Sb Concentration. Figure 2 shows the XRD spectrum of dried ATO powder doped with 3, 5, 7 and 9 mol% Sb prepared by hydrothermal reaction, respectively. All XRD patterns show the main diffraction peaks at 26.61°, 33.89°, and 51.78°, corresponding to the (110), (101), and (211) planes of rutile SnO_2 (PDF card No.41-1445) without detective level of other phases, such as Sb_2O_3 or Sb_2O_5,

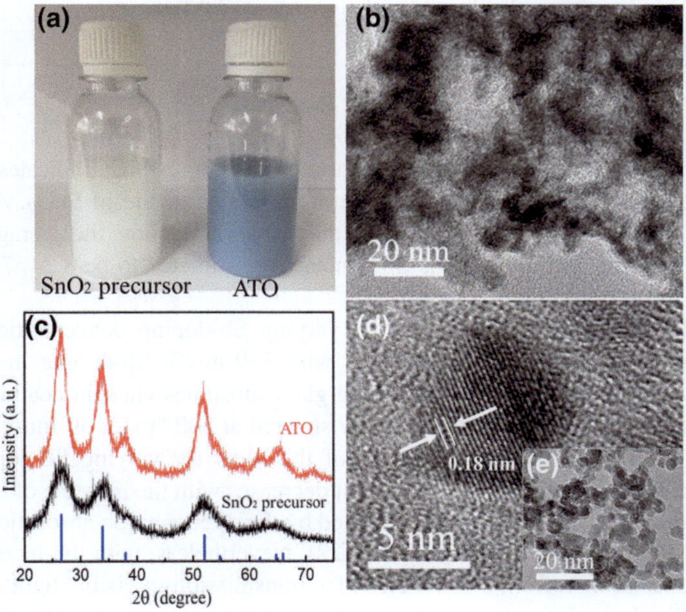

Fig. 1 **a** Digital images of SnO_2 precursor sol and ATO nanoparticles dispersed in sol, **b** TEM images of SnO_2 precursor sol, **c** XRD patterns of dried SnO_2 precursor sol and ATO sol, **d** HRTEM and **e** TEM images of ATO after hydrothermal reaction

Fig. 2 XRD patterns of dried ATO sol after hydrothermal reaction at varying nominal Sb

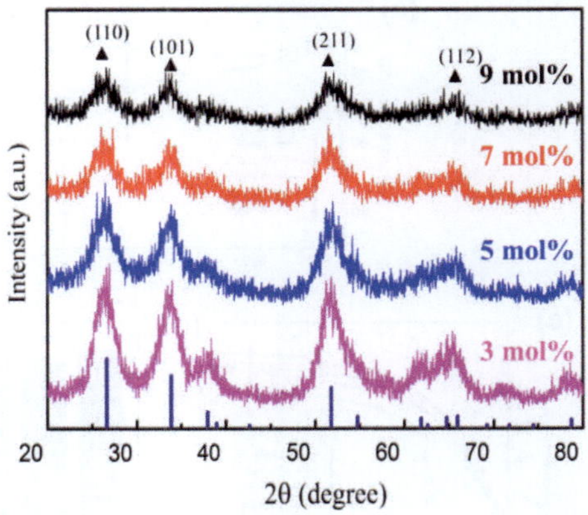

implies that the dopant Sb may substitute Sn in SnO_2 lattice. From HRTEM shown in Fig. 1d, the lattice spacing for the ATO crystals is about 0.18 nm, which corresponds to (211) plane distance, consistent with the observation from XRD. The EDS analysis (as shown in Table 1) confirms the actual doping concentration of the Sb in the as-synthesized ATO nanoparticles is close to the prescribed Sb doping amount. The full width at half maximum (FWHM) of ATO broadens

Table 1 Summary of normalized mol% of Sb in synthesized ATO

Prescribed Sb (mol%)	3	5	7	9
Actual Sb detected by EDS (mol%)	1.3	4.5	6.3	8.4

significantly with Sb doping concentration increasing, which indicates that the excessive Sb doped SnO_2 might decrease the crystal structure of SnO_2. According to the (211) diffraction peak of ATO with the Scherrer equation, the average particle size of ATO with 3, 5, 7, and 9% Sb content are estimated to be 8.24, 7.44, 6.53, and 5.81 nm respectively.

In order to explore the influence of varying Sb-doping concentration on the properties of ATO thin films, the films with 3–9 mol% Sb-doping in a similar thickness (6 layers) were deposited on the glass substrates via spin-coating method at a speed of 3000 rpm, and subsequently sintered at 500 °C for 30 min. Figure 3a shows the optical transmission spectra of the films for varying Sb-doping concentration, the transmission in visible light decreases with the increase of Sb doping concentration overall. This can be explained by the direct photon absorption and the increased light scattering from particles [22]. Nevertheless, with the increasing Sb doping concentration from 3 to 7 mol%, the transmission in visible light decreased slightly. Similarly, from the Fig. 3a one can notice transmission in the IR range is also closely related to the Sb doping concentration. IR shielding properties of film improve with the increasing of Sb from 3 to 9 mol%, caused by the free electron

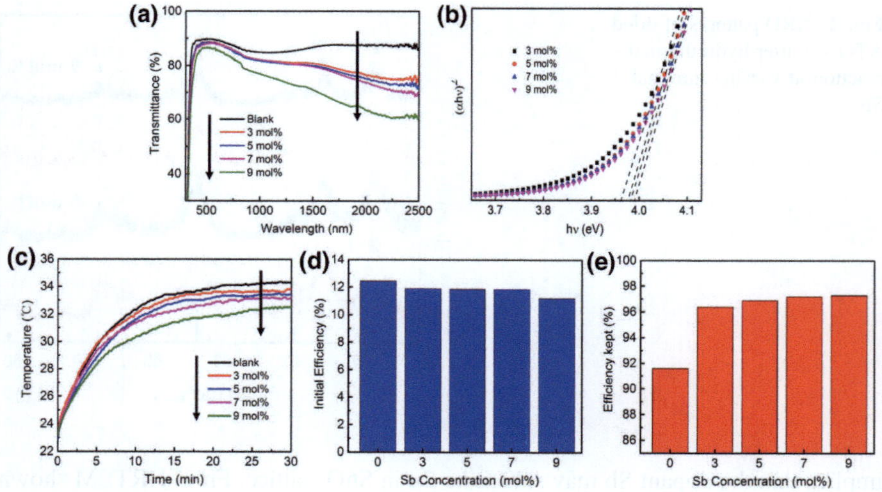

Fig. 3 Effect of Sb contents on optical properties of ATO thin film and silicon solar cell performance **a** optical transmission spectra of ATO films at different nominal Sb amounts, **b** Tauc plots of the $(\alpha h\nu)^2$ versus hν curves of ATO films with different antimony content, **c** the temperature-time curves of silicon solar cell at different antimony content, **d** the initial efficiency of silicon solar cell at different antimony content, and **e** the efficiency kept of silicon solar cell at different antimony content after being irradiated by solar simulator for 30 min

absorption in the near infrared region. The increase in Sb content results in the improvement of the free electron concentration, which lead to an enhancement of NIR absorption [22].

According to the transmittance (T) of thin film, film thickness (*d*), the relationship between photon energy (hν) and absorption coefficient (α), optical band gap can be evaluated by Tauc's relation [23]:

$$\alpha h\nu = A(h\nu - Eg)^n \quad (1)$$

$$\alpha = \frac{\ln 1/T}{d}. \quad (2)$$

Where A is a constant, n is a constant (direct bandgap semiconductor n = 1/2, indirect bandgap semiconductor n = 2), and Eg is the direct band gap energy. It is known that ATO thin film is a direct band gap semiconductor. The extrapolations of the linear region of curve at zero absorption gives the values of optical band gap. The optical band gaps are calculated shown in Fig. 3b, Tauc plots of the ATO thin films exhibit a blue shift towards higher band gap (3.96–3.99 eV) with increasing dopant concentration.

To test the IR shielding performance of ATO film with varying Sb amounts, the experiments of the silicon cells with ATO thin films were carried out using the solar simulators at room temperature. The apparatus contains the silicon solar cell plate and a home-made mask covered by glass. A glass with plated ATO film layer was placed on the top of mask as a cover, and exposure under the sunlight simulator continuously for 30 min. The back surfaces of the solar cell are instrumented with an infrared thermometer (optris LSLT) connected to a computer for recording temperature changes.

The temperature-time curves of solar cell are showed in Fig. 3c. From the result it's found that the increase of Sb content results in a decrease of temperature of solar cells, owing to low transmission in IR range. The temperature of solar cell with ATO thin film decreases maximally by 1.8 °C, compared with the control one, this is supported by the optical transmission spectra of the films for varying Sb-doping concentration (see Fig. 3a). The less light transmission in IR region, the less infrared (thermal) energy.

On the other hand, the solar cells absorb most incident irradiance, and the light at wavelength equal or less to the absorption band edge excited hole-electron pairs is converted into the electric power, the rest is converted into heat, which results in high temperature of the solar cell and low efficiency [2]. Meanwhile, the decrease in transmittance of thin film at wavelength less than 1100 nm results in decrease of the initial efficiency of solar cells. Hence, both transmission in visible region and IR shielding properties can influence the photoelectric conversion efficiency of solar cell after being irradiated by solar simulator for 30 min. Figure 3d shows the influence of different antimony content on initial cell efficiency. The solar cell covered with 9 mol% Sb-doped ATO film had poor initial cell efficiency, which is due to the low transmittance in visible light and near infrared region. It should be pointed out that the solar cell covered with 3, 5, 7 mol% Sb-doped ATO film had about the same initial efficiency, while the efficiency difference decreased with the increase of Sb concentration (Fig. 3e). Hence, efficiency difference depends on the IR-shielding performance when the initial efficiency was similar cases, the better IR-shielding performance, the high efficiency kept (efficiency kept = final efficiency/initial efficiency). The result indicates the device covered with 7 mol% Sb-doped ATO film has the better performance. More details about efficiency values of silicon solar cells in 30 min are shown in Table 2.

Influence of Film Thickness. To investigate the structure and morphologies of ATO films with variable layers, the ATO thin films (7 mol%) were prepared via multiple spin coating method at a speed of 3000 rpm, and all films were sintered at 500 °C for 30 min. The morphologies of ATO films with variable layers are examined by FESEM (shown in Fig. 4). It was shown that the surface morphologies of the films are rough, because the thin films were formed by the aggregated ATO nanoparticles on the glass slide. The cross section FESEM images of the corresponding ATO films were also exhibited in the Fig. 4. It is noted that the thickness of the films increases from the cross section FESEM images increasing. The thickness of the films is about 151.5, 273.8, 464.8 and 570.6 nm for spin coating with 2, 4, 6, 8 layers, respectively. On the basis of these results, it can conclude that the thickness of each layer was about ~75 nm. The total film thickness increases by roughly the same amount as the original film thickness, suggesting that the thickness of the films could be controlled.

Crystallinity and structural characteristics of the ATO thin films (7 mol%) with varying layers ware characterized by XRD. As shown in Fig. 5, the intensity of crystal surface diffraction was enhanced with the layers of the thin films increasing, and the full width at half maximum (FWHM) became sharper, due to more crystals coated on the glass.

Table 2 Efficiency values of silicon solar cells covered with variable Sb doping content ATO films in 30 min

Time (min)	Efficiency (%) (Blank)	Efficiency (%) (3 mol%)	Efficiency (%) (5 mol%)	Efficiency (%) (7 mol%)	Efficiency (%) (9 mol%)
0	12.44	11.90	11.87	11.85	11.18
3	11.95	11.76	11.74	11.77	11.14
6	11.78	11.68	11.69	11.71	11.08
9	11.68	11.65	11.64	11.64	11.00
12	11.61	11.61	11.59	11.60	10.98
15	11.57	11.58	11.55	11.58	10.96
18	11.53	11.54	11.54	11.56	10.93
21	11.48	11.52	11.53	11.55	10.90
24	11.45	11.50	11.52	11.54	10.89
27	11.42	11.49	11.51	11.53	10.88
30	11.39	11.47	11.50	11.52	10.87

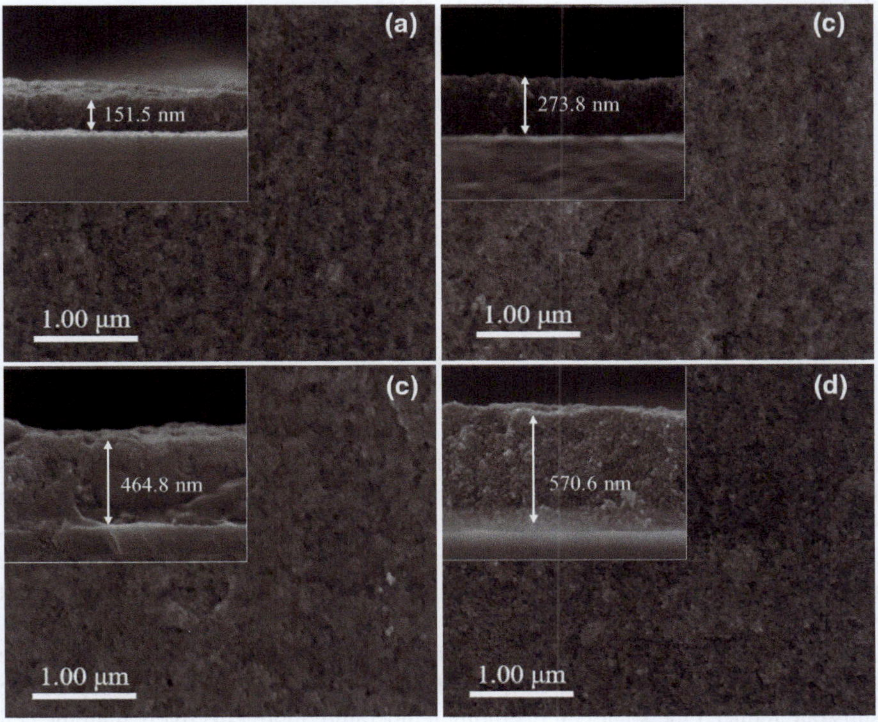

Fig. 4 FESEM images of **a** 2 layers, **b** 4 layers, **c** 6 layers, and **d** 8 layers 7 mol% ATO thin films

Fig. 5 XRD patterns of variable layers ATO thin films

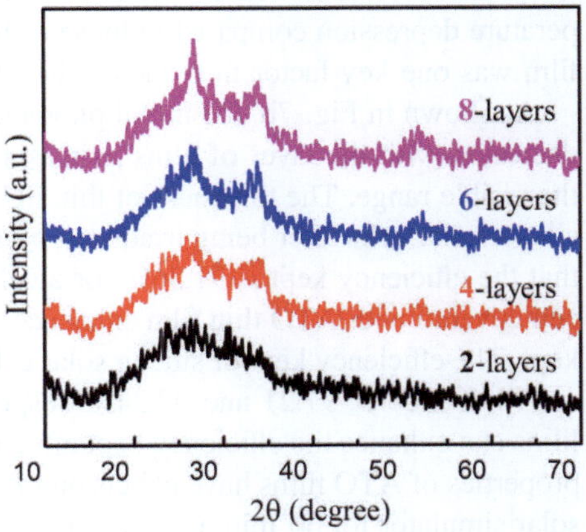

From the corresponding transmission spectra (Fig. 6a) we have obtained, one can conclude that the transmittance in the IR region decreases with thickness of ATO film increasing. The free-carrier absorption results in the transmission decrease [17]. The increase in free electron number will lead to the enhanced NIR

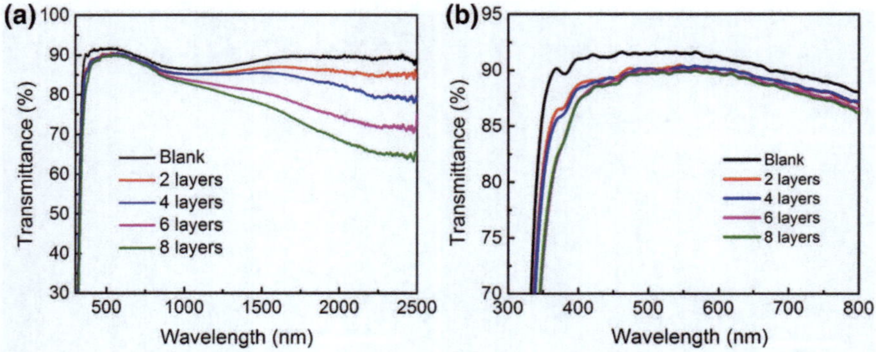

Fig. 6 Influence of thickness on the optical properties of ATO (7 mol%) thin film

absorption [24]. Moreover, the transmittance of the glass substrates can reach as high as 90% in visible light region, while that of the ATO films gradually decreases with increasing thickness of film, but still all exceeds 87%, as shown in Fig. 6b. This can be explained by light scattering effects of tiny ATO particles. The surface roughness had great influence on the transmission of the films, this is consistent with the observation from SEM.

The temperature-time curves of solar cells are showed in Fig. 7a, it can be concluded that the ATO thin film which attached at the base of the top of the substrate reduced the solar cell operating temperature more effectively than the cell without ATO film. The increase in thickness of thin films results in the decrease of the solar cell temperature. After being irradiated by solar simulator for 30 min, the temperature of solar cell covered with ATO film shows further 0.5–1.5 °C temperature depression compared to those of blank glasses. Therefore, the thickness of film was one key factor in reducing the solar cell temperature.

As shown in Fig. 7b, the initial photoelectric conversion efficiency of solar cells decreases with the layer of films increasing, due to the decreased transmittance in the visible range. The thickness of thin films has a huge impact on the efficiency of silicon solar cells after being irradiated by solar simulator for 30 min. It is analyzed that the efficiency kept is 91.56% for a cell without ATO thin film, while that of a cell covered with ATO thin film enhances, and the thicker films, the high efficiency kept. The efficiency kept of silicon solar cells cover with 2, 4, 6, 8 layers ATO films are 94.40, 95.75, 97.21 and 97.34%, respectively (Fig. 7c) which prove that ATO films can enhance the efficiency kept effectively. Both visible light and IR-shielding properties of ATO films have effects on the cells' efficiency after being irradiated by solar simulator for 30 min. Considering two factors synthetically, when 6 layers of ATO thin film is employed as cover glass, solar cell has the best performance after 30 min. More about the effect of irradiation time on the silicon solar cell efficiency is shown in Table 3.

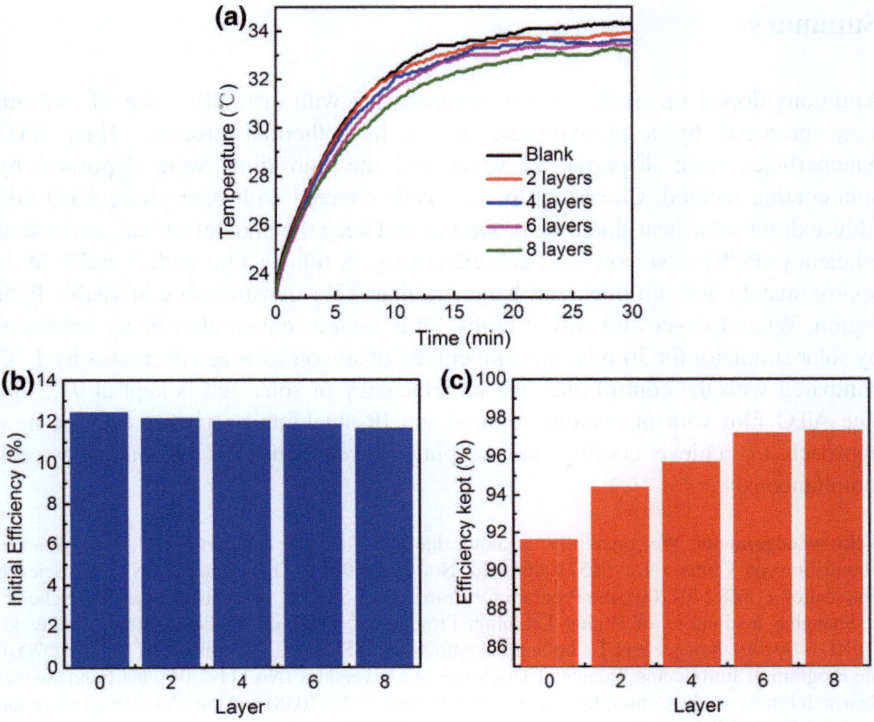

Fig. 7 Influence of thickness on silicon solar cell performance, **a** the temperature-time curves of silicon solar cell at different layers, **b** the initial efficiency of silicon solar cell at different layers, and **c** the efficiency kept of silicon solar cell at different layers after being irradiated by solar simulator for 30 min

Table 3 Efficiency values of silicon solar cells covered with different layers ATO films in 30 min

Time (min)	Efficiency (%) (Blank)	Efficiency (%) (2 layers)	Efficiency (%) (4 layers)	Efficiency (%) (6 layers)	Efficiency (%) (8 layers)
0	12.44	12.15	12.00	11.85	11.65
3	11.95	11.99	11.89	11.77	11.58
6	11.78	11.90	11.79	11.71	11.53
9	11.68	11.82	11.75	11.64	11.48
12	11.61	11.77	11.71	11.60	11.45
15	11.57	11.70	11.67	11.58	11.41
18	11.53	11.64	11.64	11.56	11.39
21	11.48	11.58	11.60	11.55	11.37
24	11.45	11.53	11.56	11.54	11.36
27	11.42	11.48	11.52	11.53	11.35
30	11.39	11.47	11.49	11.52	11.34

Summary

Antimony-doped tin oxide (ATO) nanoparticles with crystallite size of ~8 nm were prepared by a peroxo-route and a hydrothermal process. Then ATO nanoparticles were dispersed in water and the thin films were deposited by spin-coating method. Compared to solar cells covered with bare glass, ATO thin films exhibit solar-heat shielding properties and keep the photoelectrical conversion efficiency (PCE) of silicon solar cell effectively. A 6-layer film with 7 mol% Sb is approximately 460 nm thick and has more than 87% transmittance in visible light region. When 6-layer film with 7 mol% Sb is used as cover, after being irradiated by solar simulator for 30 min, the temperature of silicon solar cell decreases by 1 °C compared with the control one, and the efficiency of solar cell is kept at 97.21%. The ATO film with outstanding optical and IR-shielding properties offers a new approach to achieve cooling and keeping the efficiency of silicon solar cells simultaneously.

Acknowledgements We gratefully acknowledge the financial support by Natural Science Foundation of China (No. 51572046, and No. 51590902), The Shanghai Natural Science Foundation (15ZR1401200), the Program for Professor of Special Appointment (Eastern Scholar) at Shanghai Institutions of Higher Learning, Program of Shanghai Academic Research Leader (16XD1400100), Science and Technology Commission of Shanghai Municipality (16JC1400700), the Program of Introducing Talents of Discipline to Universities (No.111-2-04), the Fundamental Research Funds for the Central Universities (CUSF-DH-D-2017038) and the Major Program of the National Natural Science Foundation of China (51590902).

References

1. M.A. Green et al., Solar cell efficiency tables (Version 45). Prog. Photovolt: Res. Appl. **23**, 1–9 (2015)
2. S. Sargunanathan, A. Elango, S.T. Mohideen, Performance enhancement of solar photovoltaic cells using effective cooling methods: a review. Renew. Sustain. Energy Rev. **64**, 382–393 (2016)
3. C. Strümpel et al., Modifying the solar spectrum to enhance silicon solar cell efficiency—an overview of available materials. Sol. Energy Mater. Sol. Cells **91**, 238–249 (2007)
4. E. Skoplaki, J. Palyvos, Sol. Energy **83**, 614–624 (2009)
5. M. Chandrasekar, T. Senthilkumar, Experimental demonstration of enhanced solar energy utilization in flat PV (photovoltaic) modules cooled by heat spreaders in conjunction with cotton wick structures. Energy **90**, 1401–1410 (2015)
6. X. Tang, Z.H. Quan, Y.H. Zhao, Experimental investigation of solar panel cooling by a novel micro heat pipe array. Energy Power Eng. **2**, 171–174 (2010)
7. X.Y. Han, Y.P. Wang, L. Zhu, Electrical and thermal performance of silicon concentrator solar cells immersed in dielectric liquids. Appl. Energy **88**, 4481–4489 (2011)
8. Y.A. Abrahamyan et al., The efficiency of solar cells immersed in liquid dielectrics. Sol. Energy Mater. Sol. Cells **73**, 367–375 (2002)
9. Y. Wang et al., The performance of silicon solar cells operated in liquids. Appl. Energy **86**, 1037–1042 (2009)

10. M. Abdolzadeh, M. Ameri, Improving the effectiveness of a photovoltaic water pumping system by spraying water over the front of photovoltaic cells. Renew. Energy **34**, 91–96 (2009)
11. S. Krauter, Increased electrical yield via water flow over the front of photovoltaic panels. Sol. Energy Mater. Sol. Cells **82**, 131–137 (2004)
12. A. Kordzadeh, The effects of nominal power of array and system head on the operation of photovoltaic water pumping set with array surface covered by a film of water. Renew. Energy **35**, 1098–1102 (2010)
13. J.K. Tonui, Y. Tripanagnostopoulos, Air-cooled PV/T solar collectors with low cost performance improvements. Sol. Energy **8**, 498–511 (2007)
14. L. Zhu, A. Raman, K.X. Wang, M.A. Anoma, S. Fan, Radiative cooling of solar cells. Optica **1**, 32–38 (2014)
15. L. Zhu, A. Raman, S. Fan, Radiative cooling of solar absorbers using a visibly transparent photonic crystal thermal blackbody. PNAS **112**, 12282–12287 (2015)
16. M.M. Hossain, B. Jia, M. Gu, A metamaterial emitter for highly efficient radiative cooling. Adv. Opt. Mater. **3**, 1047–1051 (2015)
17. T. Abendroth et al., Optical and thermal properties of transparent infrared blocking antimony doped tin oxide thin films. Thin Solid Films **624**, 152–159 (2017)
18. Y.S. Li, J. Liu, J. Liang, X.B. Yu, D.J. Li, A.C.S. Appl, Mater. Interfaces **7**, 6574–6583 (2015)
19. J. Montero, C. Guillén, J. Herrero, Nanocrystalline antimony doped tin oxide (ATO) thin films: a thermal restructuring study. Surf. Coat. Tech. **211**, 37–40 (2012)
20. D.F. Qian, Y.G. Li, Q.H. Zhang, G.Y. Shi, H.Z. Wang, J. Alloy. Compd. **509**, 10121–10126 (2011)
21. T. Nütz, U. Felde, M. Haase, J. Chem. Phys. **110**, 12142–12150 (1999)
22. J.M. Xu et al., Influence of Sb doping on the structural and optical properties of tin oxide nanocrystals. Cryst. Eng. Commun. **15**, 3296–3300 (2013)
23. H.R. Fallah, M. Ghasemi, A. Hassanzadeh, H. Steki, Mater. Res. Bull. **42**, 487–496 (2007)
24. T.R. Giraldi et al., Effect of thickness on the electrical and optical properties of Sb doped SnO_2 (ATO) thin films. J. Electroceram. **13**, 159–165 (2004)

Preparation and Properties of Cool-Feeling PA6 Fiber

Yiren Li, Xiaomin Zhao, Lian Tang, Peng Ji, Chaosheng Wang and Huaping Wang

Abstract In order to increase comfort of fabric, the development of fiber with the function of cool-feeling has been become the hotspot of chemical fiber industry. The cool-feeling fibers were prepared in this paper. The nano bowlder powder, aluminum nitride (AlN) powder, silicon carbide (SiC) powder were used as the cooling function powder, and the powders were modified by silane coupling agent 3-Aminopropyltriethoxysilane(KH-550). Nylon 6 (PA6) fibers with cooling function were prepared by melt blending method and melt spinning method. Further more, the properties of cool-feeling PA6 chip, fiber and fabric were characterized. The results show that the particle size of the powders become smaller and the distribution become narrower, the scale mainly distribute from 400 to 600 nm after the surface modification by KH-550. The results of thermal conductivity test and fabric contact cold test show that, the thermal conductivity of PA6 composite fiber reaches maximum with the value of 1.73 W/(m/K) when modified by AlN powder and bowlder powder. And the corresponding Q-max value of the fabric up to 0.162 J/cm^2 s demonstrates that the cooling effect of composite fiber reach optimum.

Y. Li · X. Zhao · L. Tang · P. Ji · C. Wang (✉) · H. Wang
State Key Laboratory for Modification of Chemical Fibers and Polymer Materials,
College of Materials Science and Engineering, Donghua University, Shanghai, China
e-mail: cswang@dhu.edu.cn

Y. Li
e-mail: renyilzx@163.com

X. Zhao
e-mail: 13916221072@163.com

L. Tang
e-mail: marstanglian@sina.com

P. Ji
e-mail: jipeng@dhu.edu.cn

H. Wang
e-mail: wanghp@dhu.edu.cn

Y. Han (ed.), *Advanced Functional Materials*,
https://doi.org/10.1007/978-981-13-0110-0_90

Keywords PA6 · Cooling function · Silane coupling agent · Nano powder
Surface modification

Introduction

The cool-feeling fiber is an attractive fiber in the textile industry [1–3]. The ideal cool-feeing fiber can provide human body temperature comfort at all times. There are many kinds of cool-feeing fiber, and the mechanism of cool-feeling is different. For example, CoolMaxfiber [4], developed by DuPont Co, is a hollow fiber with a cross sectional shape of four tubular and breathable walls. Because of the unique physical structure, CoolMaxfiber has the properties of good water absorption, perspiration and breathability, and keeps the skin dry and comfortable. At present, it has been used in sports clothes. There is another type of fiber that improves fiber cooling properties by increasing the moisture absorption and the evaporation moisture of fibers. For example, the Japanese company Li Cola's "Sophista" fiber [5], which is a seed core composite fiber. Besides that, there is also a way to enhance the thermal conductivity of fibers to achieve a cool-feeling effect by compound matrix with high thermal conductivity filter, such as bowlder powder, mica powder, borneol. This method has been widely used due to its quick effect, low cost and remarkable effect. At present, cool-feeling fiber mainly consists of two kinds: polyester and polyamide. Cool-feeling polyester fiber technology is relatively mature and low cost. However, cool-feeling polyamide fiber, with its unique advantages of moisture absorption, cannot been replaced by polyester in some fields, such as underdress and so on. Thus, the development of cool-feeling polyamide fiber is necessary.

Experimental Section

Surface Modification of Inorganic Powders. Inorganic powder (Bowlder powder, AlN, SiC) 50 g and 250 mL ethanol were added into three flasks, the pH value of the system was adjusted to 3–5 with formic acid, stirring one hour under water bath at 65 °C(solution A); While, adding 0.75 g KH-550 to 20 mL ethanol and ultrasonic dispersion one hour(solution B); A and B were mixed at 65 °C for one hour, and modified inorganic powders were obtained after treatment filtering and vacuum drying.

Preparation of Cool PA6 Fiber and Fabric. The composites were prepared through melt blending method, Then, cool-feeling PA6 fiber was obtained by melt spinning. Finally, the cool-feeling PA6 fibers were weaved to form knitted fabric on circular machine. Spinning process parameters: the spinning temperature is around 270 °C; the winding speed is 800 m/min; the heat setting temperature is 120 °C, the hot roller temperature is 60 °C and the draw ratio is 3 (Tables 1 and 2).

Table 1 Process parameters of twin screw extruder

Section	Zone 1	Zone 2	Zone 3	Zone 4	Zone 5	Zone 6
Temperature (°C)	220	245	260	270	260	245

Table 2 Formulation of cool-feeling PA6

Formula	Addition amount (wt%)	Resin code	Fiber code	Fabric code
Bowlder:AlN = 1:1	0.5	RYA-0.5	FYA-0.5	TYA-0.5
Bowlder:AlN = 1:1	1.0	RYA-1	FYA-1	TYA-1
Bowlder:AlN = 1:1	1.5	RYA-1.5	FYA-1.5	TYA-1.5
Bowlder:AlN = 1:1 (Unmodified)	1.0	RYAW-1	FYAW-1	TYAW-1
Bowlder:SiC = 1:1	1.0	RYS-1	FYS-1	TYS-1
Bowlder	1.0	RY-1	FY-1	TY-1
AlN	1.0	RA-1	FA-1	TA-1
SiC	1.0	RS-1	FS-1	TS-1

Characterization. The morphologies of membranes were observed using S-4800 field emission scanning electron microscope (FE-SEM) after covered with gold. The crystallization and melt behavior of cool-feeling PA6 composites were measured by Q-20 DSC. The mechanical properties of fibers were investigated by using XQ-1A monofilament strength machine.

Results and Discussion

Reflectance Analysis of Inorganic Powders on Visible and Near Infrared. Visible light and high infrared reflectivity material can reflect the sun's rays, thereby inhibiting the material surface temperature. Therefore, the effect of the powder on the visible light and infrared radiation has a great influence on the cooling performance of the modified PA6 fiber. As shown in Table 3, the reflectance of SiC powder to visible and near infrared bands is more lower than other powders. Therefore, the main research objects of this experiment are bowlder powder and AlN powder.

Table 3 The reflectivity of the cool function powder on the visible and near infrared

Sample	Code (Unmodified)	Code (modified)	Visible reflectance (%)	Near infrared reflectance (%)
Bowlder	PW-Y	P-Y	72.6	64.6
AlN	PW-A	P-A	57.9	76.6
SiC	PW-S	P-S	38.5	31.6

Analysis of Particle Size. After one hour of ultrasonic oscillation of the powder in ethanol, the particle size of the powder was tested at 25 °C by Laser Nano Particle Size Analyzer. Figure 1 shows the particle size distribution of the Bowlder (a), AlN (b) and SiC (c) before and after silane coupling agent modification. As shown in the Fig. 1, the particle size of the powder is reduced and the distribution is narrow after silane coupling agent modification. The particle size of PW-Y was reduced from 700–1100 nm to 400–600 nm, PW-A was reduced from 500–700 nm to 450–550 nm, and PW-S was reduced from 500–700 nm to 450–600 nm, demonstrating that the powder surface treatment effect is successful. This is due to the Silane coupling agent acts as a bridge between the inorganic phase and the organic phase [6]. The surface energy of the powder is high without surface modification, which can been easy to attract each other and agglomerate. The silane coupling agent can reduce the surface energy of powders, and make the powder dispersed more evenly.

Melt and Crystallization Properties of Polyamide Composites. DSC conditions: Ramp 20 °C/min to 300 °C, isothermal for 5 min, and then ramp 10 °C/min to room temperature, finally, Ramp 20 °C/min to 300 °C again. As shown in Fig. 2 and Table 4. The crystallization temperature of cool-feeling PA6 resins are

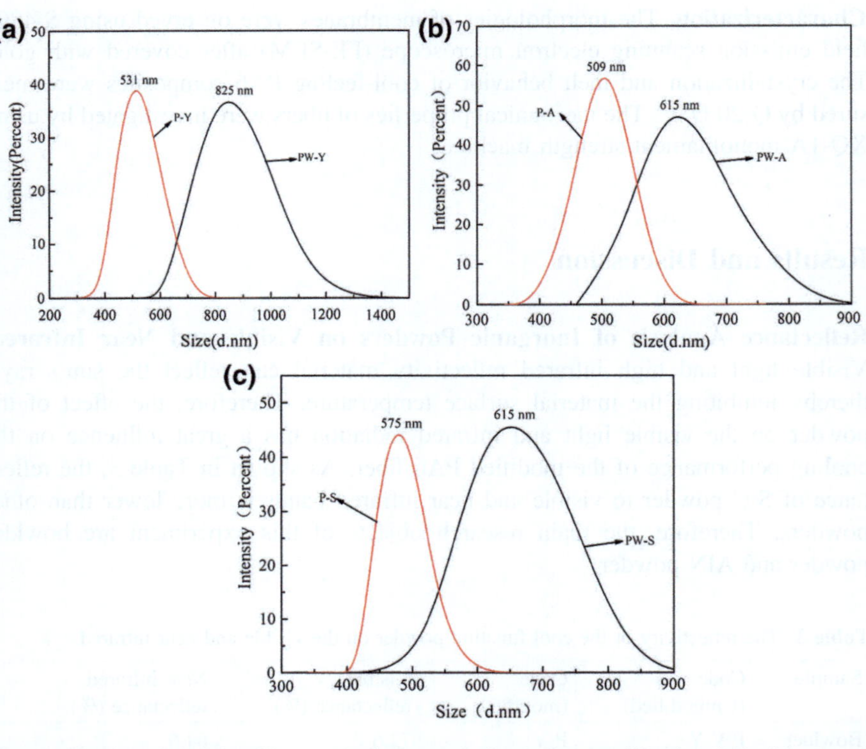

Fig. 1 **a** Particle size of Bowlder powder before and after surface modification; **b** Particle size of AlN before and after surface modification; **c** Particle size of SiC before and after surface modification

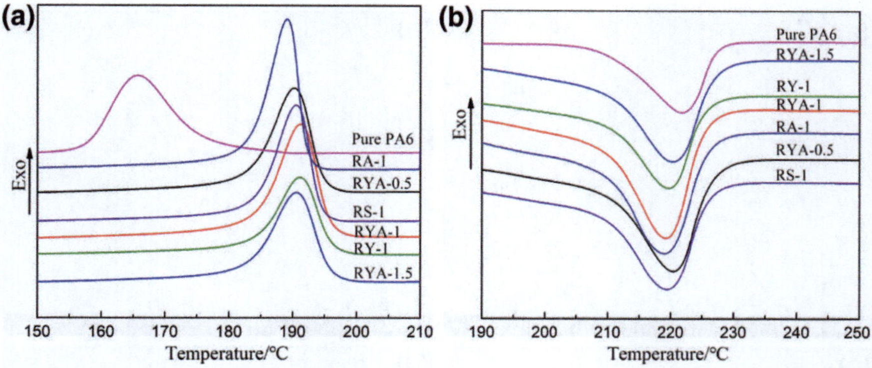

Fig. 2 Cooling crystallization curve and melting curve of DSC, **a** Cooling crystallization curve **b** melting curve

Table 4 Thermal properties of cold sensitive polyamide resin

Resin Code	Tm (°C)	ΔHm (J/g)	Tc (°C)	Xc (%)
Pure PA6	221.5	66.93	165.8	29.10
RYA-0.5	220.4	62.59	190.2	27.21
RYA-1	219.1	65.01	191.2	28.27
RYA-1.5	220.1	63.57	190.5	27.64
RY-1	219.6	63.28	189.2	27.51
RA-1	218.7	57.5	190.6	25.00
RS-1	219.6	64.97	191.1	28.25

significantly improved due to the inorganic powder as the role of nucleating agent in the cooling process, reducing interfacial free nuclei in the crystallization of polymer and the barrier of nucleation process [7]. However, the T_m of polyamide slice decreased slightly after adding cooling functional powder, this is because the inorganic powder weakens the force of polymer chain segment.

The melting enthalpy and crystallinity of the samples are reduced. It infers that the inorganic particles destroy the regularity of the macromolecular chains and limit the crystallization of polyamide, which leads to the decrease of crystallinity.

Characterization of Fiber Surface and Fracture Appearance. The surface morphology of FYA-0.5, FYA-1, FYA-1.5, FYAW-1 are shown in Fig. 3a–d, respectively. It can be seen that the inorganic powder unmodified by silane coupling agent is not uniformly mixed with the polyamide matrix, and the agglomeration is obvious in Fig. 3d. In contrast, the inorganic powder modified by silane coupling agent is more homogeneous with polyamide matrix. Thus, Silane coupling agent can obviously improve the compatibility of inorganic powder and polymer matrix.

There were no obvious dispersed particles in the fiber section in Fig. 4a, a small amount of inorganic particles dispersed evenly in Fig. 4b, the distribution of inorganic particles is more dense, but still dispersed evenly and without obvious agglomeration, as shown in Fig. 4c.

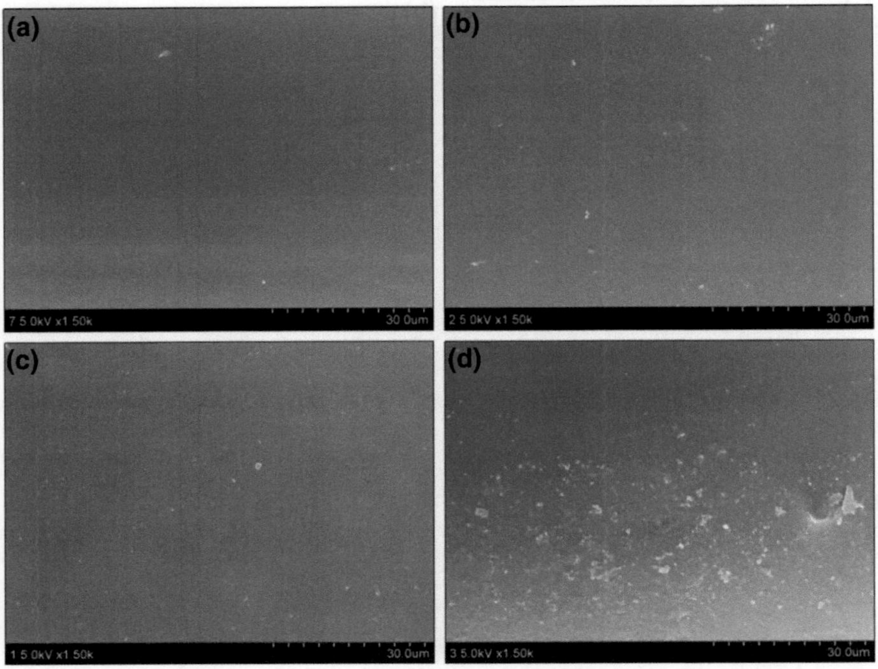

Fig. 3 Surface morphology of the cool PA6 fiber with different contents: **a** FYA-0.5; **b** FYA-1; **c** FYA-1.5; **d** FYAW-1

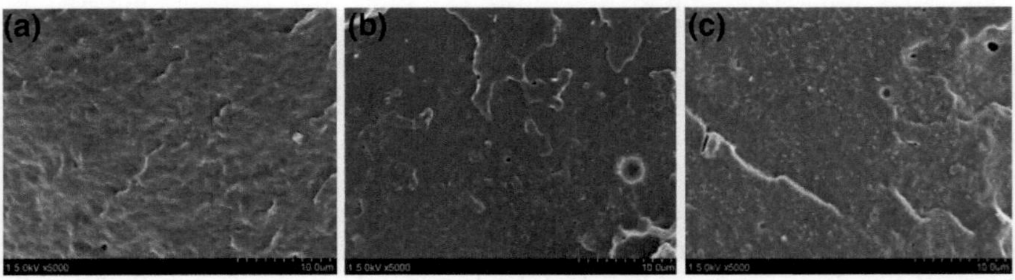

Fig. 4 Fracture appearance of cool-feeling PA6 fiber with different composite powder contents: **a** FYA-0.5; **b** FYA-1; **c** FYA-1.5

Mechanical Properties of Composite Fiber. Test conditions: pre tension: 0.05 cN/dtex; clamping distance: 25 mm; stretching speed: 50 mm/min. As shown in Table 5. The fineness of fiber increased slightly with the increasing of the amount of powder. The strength and elongation of composite fiber decreased due to the increase in the amount of powder affecting the crystallization properties of fibers.

The strength, elongation and modulus of the fiber (FYAW-1) that addition powder without silane coupling agent modified were significantly decreased

Table 5 The mechanical properties of cool-feeling PA6 fiber with different powder content

Fiber code	Titre (dtex/f)	Strength (cN/dtex)	Elongation (%)
FYA-0.5	190/36	2.6	76.0
FYA-1	192/36	2.6	82.7
FYA-1.5	199/36	2.5	81.2
FYAW-1	195/36	2.4	77.0

compared with that (FYA-1) addition the powder modified by silane coupling agent. It is due to the stress concentration caused by the agglomeration of the powder.

Thermal Conductivity of Fibers. As shown in Table 6, the thermal conductivity of fibers FY-1 is the worst, which is caused by the lowest thermal conductivity (unit: W/(m/K) Bowlder: 25, AlN: 150) [8] of Bowlder powder. The thermal conductivity of fibers FYA-1 is the highest. Although the thermal conductivity of AlN powder is the highest, the density of AlN powder is large, and the space between particles is larger after compounding with polyamide. Therefore, it is difficult to form a good heat transfer path, which cannot exert its advantages of high thermal conductivity. However, the combination of the two powders can be combined with the advantages of two powders, forming a good heat conduction path, and improving the thermal conductivity of the fiber. Therefore, the thermal conductivity of the PA6 fiber FYA-1 increases obviously.

Fabric Contact Creeping Chill Characterization of Fabric. Fabric Contact Creeping chill [9] is a method to evaluate cool-feeling fabric. Q-max is proposed by Kawabata [10] to characterize Contact Creeping Chill. The larger of the Q-max value is, the faster the heat flow in the contact with the skin is, the more cold it feels. This value is related to fiber type, fabric structure and surface filoplume [11]. The Q-max value of the fabric was measured by the Contact Creeping Chill Tester (KES-F7). Fabric size: 20 cm × 20 cm. The Q-max value result shows TYA-1 was the highest, up to 0.162 J/cm^2 s, which also indicate the effect of compound modification is better than that of single powder (Table 7).

Table 6 The thermal conductivity of cool-feeling PA6 fiber

Fiber Code	FYA-1	FY-1	FA-1	Pure PA6
Thermal conductivity/(m/K)	0.173	0.162	0.168	0.1325

Table 7 Q-max value of cool-feeling PA6 fabric

Fabric code	Pure PA6	TYA-0.5	TYA-1	TYA-1.5	TYAW-1	TY-1	TA-1	TS-1
Q-max (J/cm^2 s)	0.145	0.150	0.162	0.158	0.153	0.155	0.161	0.159

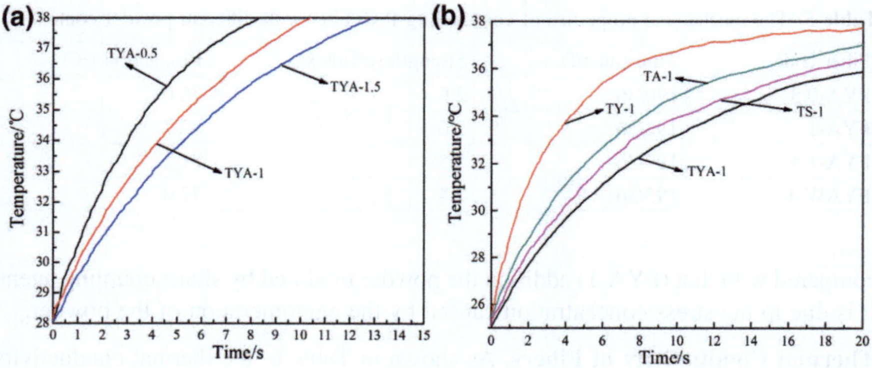

Fig. 5 Temperature rise curve of fabric: **a** Fabric with different content of composite powder **b** Fabric with composite powder and single powder

Characterization of Infrared Thermal Imaging. The temperature change of different cool feeling polyamide fabric was measured by infrared thermal imager (FLIR ThermaCAM A40 M). Test method: Placing dry fabric samples on a heating table at 38 °C, and recorded the process of heating curve, then, removing the fiber sample from the heater, recorded the process of cooling curve. With the increase of the amount of cold sensitive powder added, the heat dissipation rate of fabric faster, the temperature rise rate of fabric is slowed down in (a). The heating rate of TYA-1 is slower than TY-1, TA-1, TS-1of (b) in Fig. 5, which shows that the effect of adding bowlder powder and AlN composite powder is better than that adding single powder.

As shown in Fig. 6, TYA-0.5 has the slowest cooling rate, which indicates that adding functional powder have great influence on the thermal conductivity. In addition, the cooling rate of TYA-1 is faster than TY-1, TA-1, TS-1,which shows

Fig. 6 Cooling curve of fabric in standard atmospheric condition after heating to 36 °C

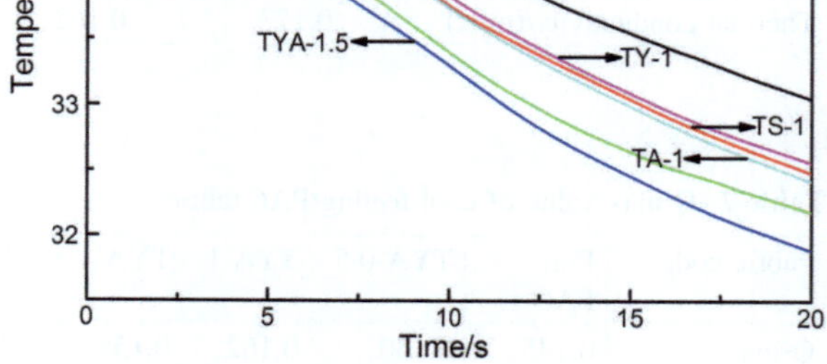

that the effect of adding bowlder powder and AlN powder is better than that adding single powder. furthermore, the more the amount of powder added, the faster the cooling rate of fabrics, the higher thermal conductivity of fabric.

Conclusions

In summary, the cool-feeling PA6 composites were prepared through melt blending method combining inorganic powder and PA6 chip, and the cool-feeling PA6 fibers were obtained through melt spinning method. The results show that the compatibility of inorganic powder and PA6 matrix is improved obviously after adding KH550. The melting points of composites are around 220 °C similar to the pure one. Crystallization analysis suggests that the powders can improve the speed of crystallization rate of PA6. And additive of powders leads to the strength of fiber decrease, but the strength is still greater than 2.4 cN/dtex. The thermal conductivity of PA6 fibers reaches the highest (1.73 W/(m/K)) when modified by AlN powder and bowlder powder. And the Q-max value of PA6 fabric reaches the highest (0.162 J/cm^2 s) when AlN powder and bowlder powder are added to the composite fibers. The fabrication route developed herein also fits to the industry requirements, paving the way toward cool-feeling PA6 fibers and fabric applications at large-scales.

Acknowledgements This work was financially supported by the National Natural Science Foundation of China (61134009), National Basic Research Program of China (2013BAE01B02), and the National Key Research and Development Program of China (2016YFB0302702).

References

1. A. Mangat, L. Hes, V. Bajzik, Effect of biopolishing on warm-cool feeling of knitted fabric: a subjective and an objective evaluations. Autex. Res. J. **2**, 95–102 (2017)
2. Y.C. Sun, Fabric transient cool feeling and effect factor analyses. Cotton. Text. Tech. **10**, 594–597 (2009)
3. X. Cao. Y. Duan, Z. Qian, J. Zhao, Production process of the functional and cool polyester fiber. J. Text. Res. **28**, 12–13 (2007)
4. F. Wang, S. Annaheim, M. Morrissey, R.M. Rossi, Real evaporative cooling efficiency of one-layer tight-fitting sportswear in a hot environment. Scand. J. Med. Sci. Spo. **24**, 129–139 (2014)
5. E. Akiba, Cooling effect of "Sophista". Sen-i. Gakkaishi **06**, 177–178 (2006)
6. A.J. Kinloch, *Adhesion and adhesives science and technology* (Chapman and Hall, London, 1987)
7. J.M. García-Martínez, O. Laguna, S. Areso, E.P. Collar, A thermal and mechanical study under dynamical conditions of polypropylene/mica composites containing atactic polypropylene with succinil-fluoresceine grafted groups as interfacial modifier from the matrix side. J. Polym. Sci., Part B: Polym. Phys. **11**, 1564–1574 (2000)
8. M.J. Laubitz, Thermal conductivity of powders. Can. J. Phys. **7**, 798–808 (2011)

9. M.V. Vivekanadan, S. Raj, S. Sreenivasan, R.P. Nachane, Parameters affecting warm-cool feeling in cotton denim fabrics. Indian J. Fibre. Tex. **2**, 117–121 (2011)
10. M. Yoneda, S. Kawabata, Analysis of transient heat conduction and its applications. J. Text. Mach. Soc. Jpn. **4**, 73–83 (1983)
11. M.J. Pac, M.A. Bueno, M. Renner, S.E. Kasmi, Warm-cool feeling relative to tribological properties of fabrics. Text. Res. J. **9**, 806–812 (2001)

Dispersion of Single-Walled Carbon Nanotubes in Organic Solvents DMAC

Baihua Liu, Jingwei Zhang, Cong Wang, Cuiqing Teng, Hui Zhang and Muhuo Yu

Abstract The effective use of single walled carbon nanotubes (SWCNT) for improving mechanical properties of polymers is the key to the efficient diffusion of carbon nanotubes into organic solvents to form a stable single solution. The single walled carbon nanotubes in dimethylacetamide (DMAc) with different dispersion methods (including homogenizer, ball mill, micro jet machine) were studied in this paper, adding different organic compounds (dispersant, polyacrylonitrile) and different chemical treatment methods (acid purification, etc.) affect the dispersion of single wall carbon nanotubes were also discussed. The best dispersing time, the effects of different conditions on the basic properties of Single-walled carbons and on the dispersed state of carbon nanotubes were determined by using optical microscopy, scanning electron microscopy, Raman and elemental analysis, in order to get the optimal dispersion conditions of single-walled carbon nanotubes in DMAc.

Keywords Single wall carbon nanotubes · Organic solvent DMAc
Dispersion

Introduction

Carbon nanotubes (CNTs) is a new type of carbon structure, which was found by the Japanese NEC company in 1991 [1], it is formed by the carbon atoms in the graphene sheet rolled into a seamless, hollow tube, where the walls of the tubes are

B. Liu · J. Zhang · C. Wang · C. Teng · H. Zhang (✉) · M. Yu (✉)
State Key Laboratory for Modification of Chemical Fiber and Polymer Materials, Donghua University, Shanghai 201620, China
e-mail: zhanghui@dhu.edu.cn

M. Yu
e-mail: yumuhuo@dhu.edu.cn

B. Liu
e-mail: bhliu@mail.dhu.edu.cn

Y. Han (ed.), *Advanced Functional Materials*,
https://doi.org/10.1007/978-981-13-0110-0_91

hexagonal carbon (graphite structure) and often capped at each end [2]. Unlike graphene, diamonds, and C60, carbon nanotubes are typical one-dimensional nanomaterials [3] with diameters of nanometers and aspect ratios greater than 1000, and their carbon and carbon molecules are bonded by SP2 hybrid bonds. This binding effect is more solid than the SP3 hybrid bond in diamond, providing exhibit exceptional material properties that are a consequence of their symmetric structure [4, 5]. Carbon nanotubes have very high strength, toughness and elastic modulus.

The tensile strength of carbon nanotubes reaches 50–200 GPa, 100 times higher than that of the same volume steel, and the flexural strength can reach 14 GPa, and the density is about 1.33 g/cm^3, only 1/6 to 1/7 of the latter. In addition, theoretical and experimental results have shown extremely high elastic modulus, greater than 1 TPa. Indeed, if the reported mechanical properties are accurate, carbon nanotubes may result in an entire new class of advanced materials. Due to its special mechanical properties and one-dimensional cylindrical geometry (nanoscale diameter and very high aspect ratio), it is considered as an ideal filling material for polymer fibers. Due to the large van Edward force attraction between carbon nanotubes, it exists in the entangled aggregate state in general, which affects the homogeneous dispersion in solution or composite [6]. At the same time, the surface is intact and smooth, the suspension keys are few, and it is difficult to bond with the matrix. Therefore, the composite can not achieve the desired performance. To unlock the potential of carbon nanotubes of carbon nanotubes as well as the interactions at the nanotube/ matrix interface. Although this requirement is no different from that for conventional fiber-reinforced composites [7]. Carbon nanotube dispersed [8] is the process of carbon nanotubes particles separated and dispersed in the liquid medium and uniformly distributed throughout the liquid phase, including wetting, aggregation and stabilization of dispersed particles. The commonly used mechanical dispersion methods include ultrasonic treatment, high energy ball milling, grinding, stirring, adding surfactant, strong acid, alkali washing and in-situ growth method [9–14], etc. In addition, Wang, etc. [15] have improved the dispersion of CNTs by chemical treatment of CNTs. Organic solvent DMAC is a solvent for many high-performance fibers. Studying the dispersion of single walled carbon nanotubes in organic solvent DMAC is a prerequisite for the study of carbon nanotubes to enhance high performance fibers.

In the past twenty years, carbon nanotubes have been extensively studied in polymer materials [16–21]. At present, the carbon nanotube dispersion method has been studied depthly, but the dispersion of single walled carbon nanotubes in organic solvent DMAc has not been systematically studied, and how to prepare fully uniform and dispersed carbon nanotubes is still a difficult problem. In this paper, the effects of various dispersion methods and treatments on the dispersion of single-walled carbon nanotubes in organic solvent DMAc are systematically compared.

Experimental Part

Raw Materials and Reagents. Polyacrylonitrile copolymer consist of acrylonitrile (AN)/methyl acrylate (MMA) and itaconic acid (IA), Shanghai Jinshan Petrochemical Industries Co; Single walled carbon nanotubes: Sigma, 805033-25; Dimethylacetamide (DMAC), AR, were purchased from Yong Hua Chemical Co. Ltd.; Dispersing agent were purchased from China Academy of Sciences Chengdu organic chemicals co., which is a kind of modifier C_2H_4O; Concentrated nitric acid: AR, used as received from Ping Hu Chemical Reagent Factory; Concentrated sulfuric acid: AR, used as received from Ping Hu Chemical Reagent Factory; Deionized water: laboratory homemade.

Dispersion Experiments with Different Methods Of Treatment. Method A: The dense solution of SWCNT/DMAc circulated at a pressure of 2400 bar recycled 10 times under the action of micro jet (ATS engineering limited, AMF-3); Method B: The SWCNT were milled 3 h at the rotational speed of 200r/min by planetary ball milling (Changsha Deco equipment limited, DECO-PBM-V-0.4L); Method C: Ultrasonic treatment of carbon nanotubes by probe ultrasonic cracker (Ningbo scientz Biotechnology Co. Ltd., Scientz-950E) 2 h in ice bath with ultrasonic pulse mode (5, 5 s, 50% power). Method D: The SWCNT/DMAc solution circulated at a pressure of 1200 bar recycled 10 times with homogenizer (ATS engineering limited, AH-2010). All the samples were stirred 12 h before processing and after the treatment samples were diluted to 1 mg/100 ml in small sample bottle to be observe the ultrasonic dispersion condition.

Dispersion Experiments Adding Different Polymers. Single walled carbon nanotubes were prepared into small sample bottle at a certain concentration. After magnetic stirring 12 h, the solution was diluted to 1 mg/100 mL. After ultrasonic 15 h by cleaning machine ultrasonic under cold water bath, the solution was divided into 3 parts: "a" without adding, "b" adding 50 wt% of CNTs dispersing agent, and "c" adding PAN/DMAC solution in critical overlap viscosity to observe the dispersion process of ultrasonic cleaning machine.

Dispersion Experiments with Different Chemical Treatments

Nitric Acid Treatment. Take 0.2 g carbon nanotubes were added to 82.5 ml HNO_3 (3 M) and treated with ultrasonic cleaning machine in cold water bath for 2 h and condensation reflux 12 h at 120 degrees centigrade. Filtering with a large amount of deionized water on 0.22 μm filter. The treated samples were vacuum dried for 12 h. The sample was prepared into 1 mg/100 ml SWCNT-1/DMAc, and its dispersion process in ultrasonic water cleaning machine was observed.

Mixed Acid Treatment. Take 0.2 g carbon nanotubes and 30 ml mixed acid (concentrated sulfuric acid/concentrated nitric acid = 3:1) mixed. Ultrasonic treat in a cold bath for 2 h, and filter with 0.22 μm filter using a large amount of deionized water. The treated samples were vacuum dried and for 12 h. The sample was prepared into 1 mg/100 ml SWCNT-2/ DMAc, and its dispersion process in ultrasonic water cleaning machine was observed.

Test and Characterization Method. Optical microscope photos were taken by the BX53 M optical microscope produced by Suzhou star Braun mechanical and Electrical Equipment Co., Ltd. Elemental analysis were measured by using the 5000C ESCA element analysis instrument manufactured by American PHI company. Field emission scanning electron microscopy (FE-SEM, SU8010, Hitachi) was employed to analyze the surface morphology of the samples. Raman spectroscopy (Renishaw, UK) operating under the excitation light source of 633 nm, and the characteristic information peaks of single wall carbon nanotubes were obtained.

Results and Discussion

Effect of Different Methods of Treatment on Dispersion of Carbon Nanotubes in DMAC Solvent

Naked Eye Observation. We can observe the dispersion of the samples at different ultrasonic time and record in Fig. 1. Four samples from left to right is "A" micro jet treated samples, "B" ball milling treated samples, "C" ultrasonic treated samples, "D" homogeneous treated samples. There were a lot of black flocculent suspension before ultrasound. After 10 h, "A" is basically clear; "B" has a lot of particles floating; "C" and "D" have a very small amount of black particles floating. At the moment of 15 h sonicating, all the samples are clear and bright except "B".

It was found that the ultrasonic dispersion of micro jet, ultrasonic breaking and homogenizing treatment is good. However, the dispersed black particles were still observed after ultrasonic treatment with a long time.

Optical Microscope. The samples were dispersed by ultrasonic for 15 h and observed under 50× light microscope. The results were shown in Fig. 2.

Fig. 1 Ultrasonic observation of sample after mechanical treatment

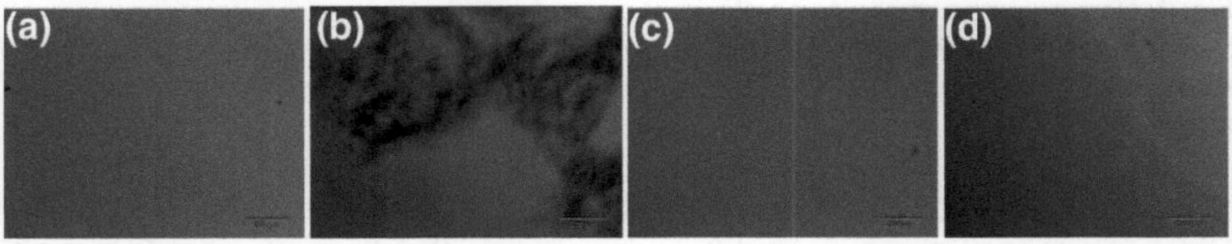

Fig. 2 Light microscope diagram of mechanical treatment

Optical micrographs show respectively by micro jet, ultrasonic and homogenization of samples processed by ultrasound in 47 h system, only a very small amount of carbon black particles; the same time under ball milling the sample after the dispersion effect is poor, there is a large floating floc. The experimental results show that the effect of micro jet milling, ultrasonic crushing and homogenizing is better.

Scanning Electron Microscope. The morphology of the sample after mechanical treatment and ultrasonic dispersion was observed, and the results were shown in Fig. 3.

Raman Spectra. According to SEM analysis, the carbon nanotubes dispersed by micro jet, ultrasonic cracker and homogenizing treatment are better, and the carbon nanotubes after ball milling still have winding phenomenon. Raman spectra of four

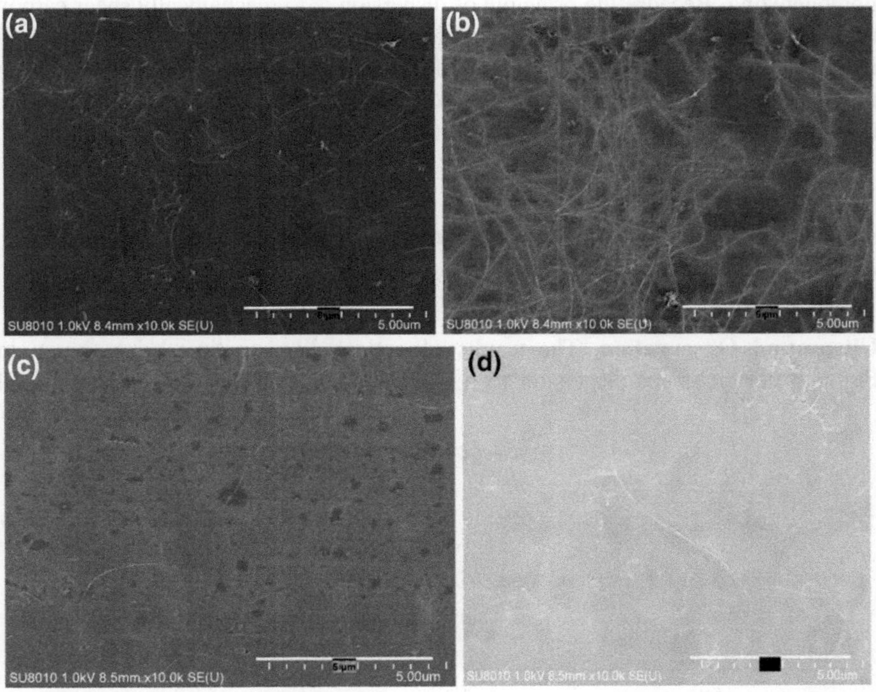

Fig. 3 SEM after mechanical treatment. **a** Micro jet, **b** ball milling, **c** ultrasonic cracker, **d** homogenizing

Fig. 4 Raman spectra of carbon nanotubes/DMAC system after mechanical treatment

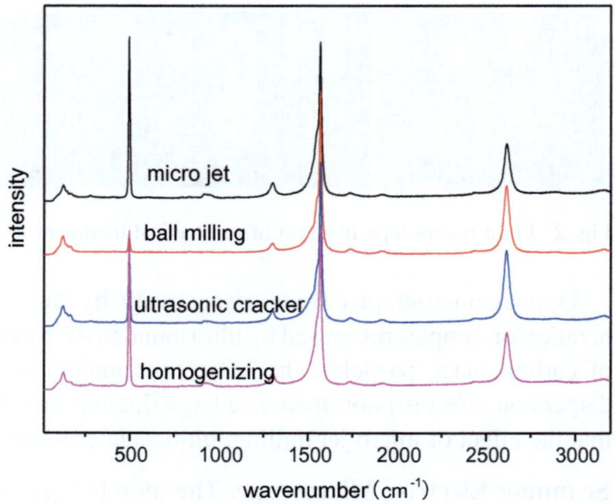

kinds of samples were obtained after ultrasonic treatment, and the results were shown in Fig. 4.

In Fig. 4, G mode at 1570 cm^{-1} of mechanical treatment of carbon nanotubes, normalized. The 470 cm^{-1} peak is the intrinsic information peak of the silicon chip. IG/ID is 8.49/14.52/10.70/10.07 for micro jet, ball milling, ultrasonic cracker, homogenizing. Because the principle of dispersion is to mechanically shear carbon nanotubes, they break and resolve them at defects, Corresponding to the photograph in front, this illuminates that the ball mill does not completely disperse the carbon nanotubes. The micro jet has the best dispersion effect, and ultrasonic cracker, homogenizing has almost the same effect of dispersant.

As Effect of Dispersant and Polyacrylonitrile on Dispersion of Carbon Nanotubes in DMAC Solvent

Ultrasound Observation. The results of optical disperse time were shown in Fig. 5. After ultrasonic dispersion 15 h, there still some agglomerates standing still

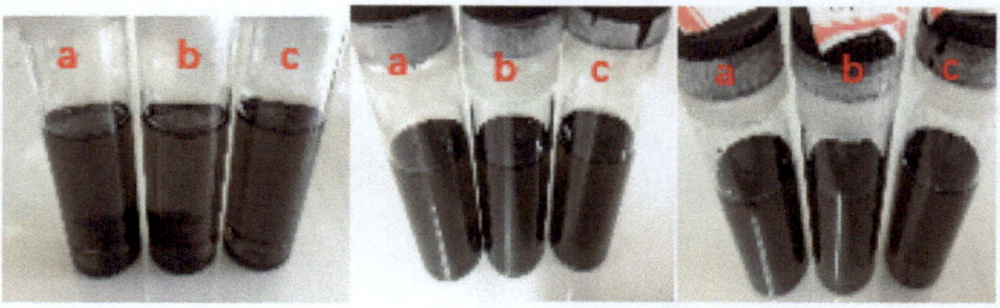

Fig. 5 Plots at different ultrasonic dispersion times

for a while. Three groups go on sonicating, after 29 h sonicating "a" without adding still have some agglomerated carbon nanotubes, but, "b" adding 50% dispersant and "c" with adding PAN/DMAC solution is clear. With sonicating for 34 h, all the CNTs have been dispersed.

As can be seen from Fig. 5, untreated carbon nanotube system is essentially clear in cold water after ultrasonic cleaning using an ultrasonic cleaning machine 15 h. The process of dispersing samples with dispersants and PAN is significantly faster than that of the untreated ones.

Optical Microscope. After the ultrasonic dispersion of 34 h, the samples were placed for 2 days for 3 days (Fig. 6).

Optical microscopy showed that the "c" group showed an obvious reunion phenomenon. It shows that the addition of PAN will promote the dispersion of carbon nanotubes, but the carbon nanotubes can be easily reunited after being placed.

Effect of Dispersant Dosage on Dispersion of Carbon Nanotubes in DMAC Solvent. According to the experiment, dispersant can obviously promote the dispersion of carbon nanotubes in DMAC solvent. Therefore, by setting up the concentration gradient of dispersant, the optimum concentration range of carbon nanotubes can be further explored.

Optical Microscope. SWCNT/DMAC solution was prepared by adding different concentration of dispersant. After ultrasonicating for 35 h, the dispersion was observed under light microscope. The result was shown in Fig. 7.

By observing under optical microscope, the effect of dispersing more than 200% dispersant at the same ultrasonic time is better. A follow-up observation was taken, after a period of standing, low concentration of dispersants occurred in different degrees of agglomeration. But, more than 200% of the dispersant samples remained basically clear, without obvious reunion phenomenon.

Scanning Electron Microscope. The samples after ultrasonic dispersion of 35 h were observed under scanning electron microscope, and the results were shown in Fig. 8.

It can be seen from Fig. 8 that the dispersant is coated on the surface of carbon nanotubes, so that carbon nanotubes have certain organic solvent characteristics, and promote the dispersion of carbon nanotubes in the DMAC system. Carbon nanotubes have no obvious beam forming phenomenon, which is consistent with that observed by the naked eye.

Fig. 6 Optical microscope of the sample. **a** untreated **b** added dispersant **c** added to PAN

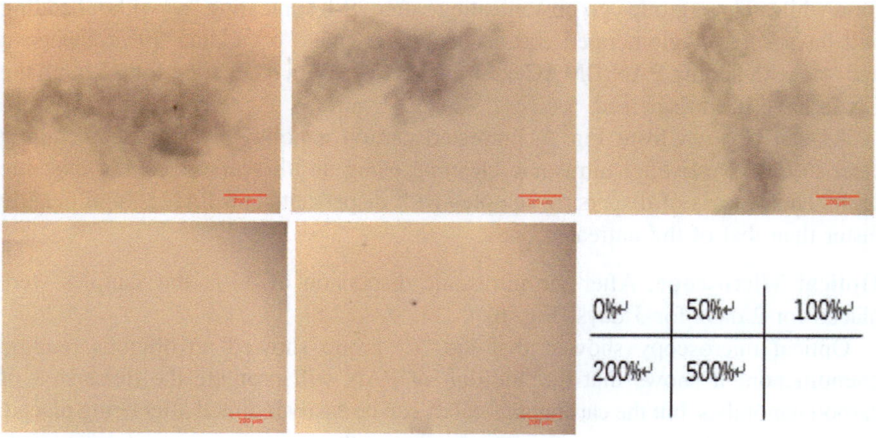

Fig. 7 Optical microscope of carbon nanotubes with different concentration dispersants

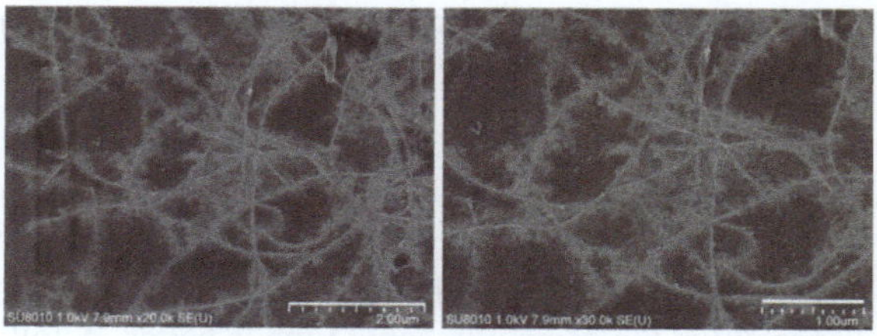

Fig. 8 SEM diagram of SWCNT with dispersant

Effect of Different Chemical Treatment of Carbon Nanotubes on It's Dispersion in DMAC Solvent

Elemental Analysis. Elemental analysis of the solid sample after acidification and drying was carried out as shown in Table 1.

The results of elemental analysis showed that the content of carbon decreased obviously, and the content of H and N increased obviously. It could be concluded that defects on the surface of carbon nanotubes had been occurred and the formation of organic functional groupse. Therefore, acidified carbon nanotubes are easier to disperse in organic solvents

Raman Spectra. Raman spectra of solid carbon nanotubes treated by acidification and drying are shown in Fig. 9.

Table 1 Elemental analysis of carbon nanotubes after acidification

Treatment	C (%)	H (%)	N (%)	Remaining (%)
Untreated	77.58	1.18	0.08	21.16
Nitric acid	75.60	2.11	0.64	21.65
Mixed acid	72.77	2.10	0.15	24.98

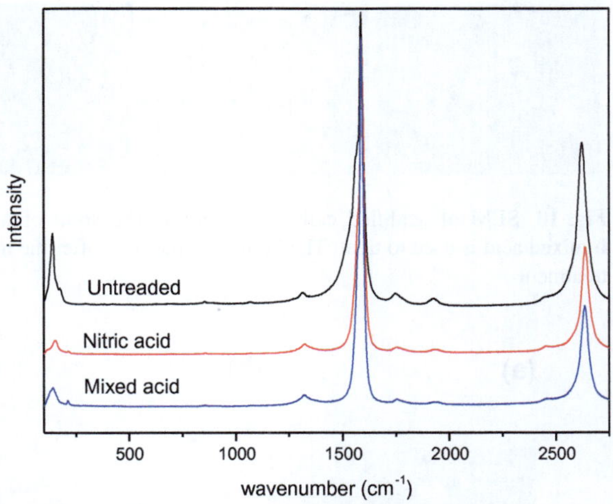

Fig. 9 Raman spectra of carbon nanotubes after acid treatment

Single wall carbon nanotube has a unique radial breathing mode at 150–300 cm^{-1} (RMB), RMB peak can be used to determine the presence of single-walled carbon nanotubes, there is a certain relationship between RMB wave number and the diameter of the single-walled carbon nanotubes, see the formula [20] (1):

$$W_{RMB} = 214/d + 6 \quad (1)$$

By Formula (1), the diameter of the carbon nanotube is inversely related to the wavenumber of RMB. As can be seen from Fig. 9, the RMB shifted to the right and the diameter of the carbon nanotubes decreased, indicating that the larger diameter nanotubes were easily oxidized during acidification. Table 2 shows that after acidification, the ID/IG ratio is less than that of the untreated one, indicating that the amorphous carbon is oxidized in the acidification process, and the purity of SWNTs is increased with fewer defects.

Scanning Electron Microscope. The samples were prepared by acidification and drying of carbon nanotubes solid, and the results were shown in Fig. 10.

According to Fig. 10, after acidification, the acidification of the carbon nanotubes resulted in fracture, reunion, and dispersion of carbon nanotubes at the weak sites. The impurities remained in the samples after treatment with nitric acid. The samples treated with mixed acid had little impurity, which indicated that the oxidation of mixed acids was strong.

Table 2 ID/IG value of Raman spectra of carbon nanotubes after acidizing treatment

Treatment	I_D/I_G
Untreated	0.0584
Nitric acid	0.0407
Mixed acid	0.0392

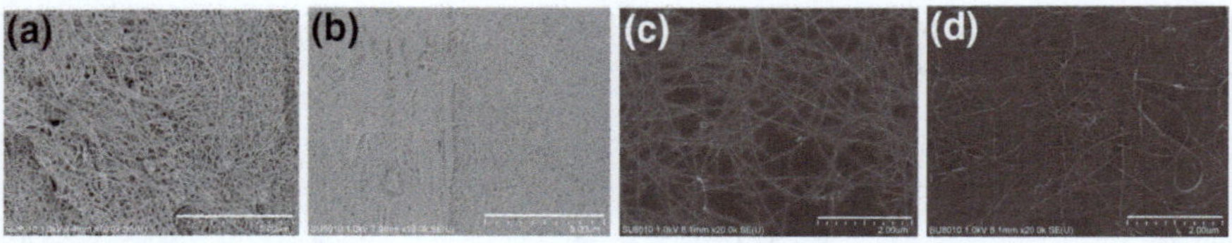

Fig. 10 SEM of acidified carbon nanotubes. The solid of **a** after the treatment of nitric acid **b** mixed acid is used to treat; The liquid samples of **c** after the treatment of nitric acid **d** mixed acid treatment

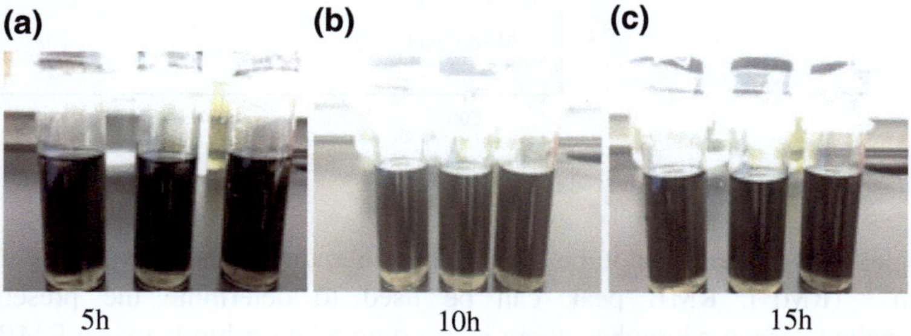

Fig. 11 Ultrasonic observations of untreated (**a**), nitric acid (**b**) and mixed acid treated (**c**) samples

Ultrasonic Observation. After the acidification of carbon nanotubes, the surface grafted with some organic functional groups, then the ultrasound time will be shortened? Ultrasonic observation results as shown in Fig. 11.

Ultrasonic for 5 h, the solution become black. Nitric acid treatment sample (a) and sample treat with mixed acid solution only have very small fluffy solid; As we can see, there still been a small amount of black particles in untreated CNT solution. By observation, it was found that the sample treated by nitric acid and mixed acid is basically clear after ultrasonic dispersion of 10 h, and the time of ultrasonic dispersion is shortened obviously. It shows that acidification treatment can promote the dispersion of carbon nanotubes in DMAC solvent.

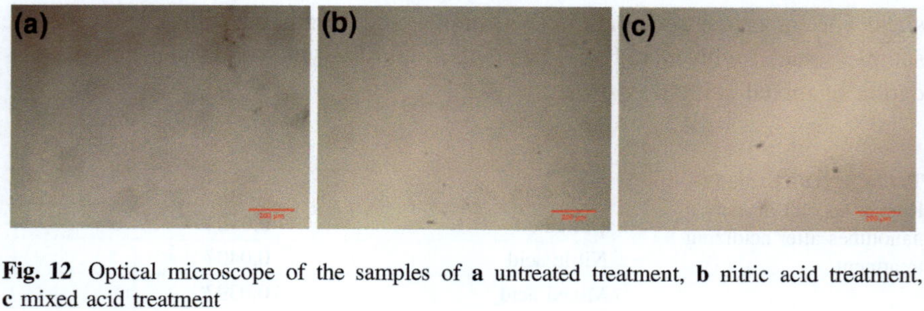

Fig. 12 Optical microscope of the samples of **a** untreated treatment, **b** nitric acid treatment, **c** mixed acid treatment

Optical Microscope. The acidified samples are better dispersed than untreated ones, which were treated with 15 h ultrasound. According to light microscopy, only a small amount of black particles were present in acidified samples. By comparison, a small amount of black floc was found in the untreated samples (Fig. 12).

Conclusions

The experimental results show that the micro jet, ultrasonic and homogenization treatment can shorten the time of carbon nanotubes uniformly dispersed in DMAC, but the milling effect is not obvious, which is concerned with the choice of research method of ball milling liquid system. To sum up, the treatment effect of ultrasonic fragmentation is better. Adding dispersant agent in carbon nanotube system is better than polyacrylonitrile for processing. Because of the addition of polyacrylonitrile solution, the solution of carbon nanotubes is unstable and easy to reunite, although both can make carbon nanotubes be well dispersed.

The effect of dispersant concentration was investigated on the basis of finding and the results show that the dispersant played an important role in the dispersion of carbon nanotubes. It is concluded by experiment that when carbon nanotubes are added more than 200% dispersants, the carbon nanotubes can achieve good dispersion effect, and the time required is short. After a long time, the phenomenon of precipitate aggregation is not observed.

The carbon nanotubes were treated with nitric acid and mixed acid respectively. The characterization results show that the acidification treatment can disperse carbon nanotubes in DMAC uniformly, shorten the time used, and promote the dispersion of carbon nanotubes in DMAC. Because of its strong oxidizing ability, the effect of impurity removal is more obvious.

References

1. S. Iijima, Helical microtubules of graphitic carbon. Nature **354**(6348), 56 (1991)
2. D.S. Bethune, C.H. Klang, M.S. De Vries, G. Gorman, R. Savoy, J. Vazquez, R. Beyers, Cobalt-catalysed growth of carbon nanotubes with single-atomic-layer walls. Nature **363** (6430), 605–607 (1993)
3. E.T. Thostenson, Z. Ren, T.W. Chou, Advances in the science and technology of carbon nanotubes and their composites: a review. Compos. Sci. Technol. **61**(13), 1899–1912 (2001)
4. J.P. Salvetat, J.M. Bonard, N.H. Thomson, A.J. Kulik, L. Forro, W. Benoit, L. Zuppiroli, Mechanical properties of carbon nanotubes. Appl. Phys. A **69**(3), 255–260 (1999)
5. D. Qian, G.J. Wagner, W.K. Liu, M.F. Yu, R.S. Ruoff, Mechanics of carbon nanotubes. Appl. Mech. Rev. **55**(6), 495–533 (2002)
6. L. Vaisman, H.D. Wagner, G. Marom, The role of surfactants in dispersion of carbon nanotubes. Adv. Coll. Interface. Sci. **128**(128–130), 37–46 (2006)
7. T.W. Chou, *Microstructural design of fiber composites* (Cambridge University Press, Cambridge, 2005)

8. J. Hilding, A. Eric, Z. Grulke, Z. George Zhang, F. Lockwood, Dispersion of carbon nanotubes in liquids. J. Dispersion Sci. Technol. **24**(1), 1–41 (2003)
9. H. Sato, M. Sano, Characteristics of ultrasonic dispersion of carbon nanotubes aided by antifoam. Colloids Surf., A **322**(1), 103–107 (2008)
10. N. Pierard, A. Fonseca, Z. Konya, I. Willems, G. Van Tendeloo, J.B. Nagy, Production of short carbon nanotubes with open tips by ball milling. Chem. Phys. Lett. **335**(1), 1–8 (2001)
11. H.G. Chae, Y.H. Choi, M.L. Minus, S. Kumar, Carbon nanotube reinforced small diameter polyacrylonitrile based carbon fiber. Compos. Sci. Technol. **69**(3), 406–413 (2009)
12. J.M. Cao, *The research on dispersion property of carbon nanotubes in water* (Guangzhou Chemistry, 2005)
13. P.K. Rai, A.N.G. Parra-Vasquez, C. Jayanta, R.A. Pinnick, L. Feng, A.K. Sadana et al., Dispersions of functionalized single-walled carbon nanotubes in strong acids: solubility and rheology. J. Nanosci. Nanotechnol. **7**(10), 3378–3385 (2007)
14. C. Park, Z. Ounaies, K.A. Watson, R.E. Crooks, J.S. Smith Jr., S.E. Lowther et al., Dispersion of single wall carbon nanotubes by in situ polymerization under sonication. Chem. Phys. Lett. **364**(3), 303–308 (2002)
15. V.C. Moore, M.S. Strano, E.H. Haroz, R.H. Hauge, R.E. Smalley, J. Schmidt, Y. Talmon, Individually suspended single-walled carbon nanotubes in various surfactants. Nano Lett. **3** (10), 1379–1382 (2003)
16. M.M. And, K.I. Winey, Polymer nanocomposites containing carbon nanotubes. Macromolecules **39**(16), 543–545 (2006)
17. O. Probst, E.M. Moore, D.E. Resasco, B.P. Grady, Nucleation of polyvinyl alcohol crystallization by single-walled carbon nanotubes. Polymer **45**(13), 4437–4443 (2004)
18. J.K.W. Sandler, S. Pegel, M. Cadek, F. Gojny, M.V. Es, J. Lohmar et al., A comparative study of melt spun polyamide-12 fibres reinforced with carbon nanotubes and nanofibres. Polymer **45**(6), 2001–2015 (2004)
19. R. Haggenmueller, J.E.F. And, K.I. Winey, Single wall carbon nanotube/polyethylene nanocomposites: nucleating and templating polyethylene crystallites. Macromolecules **39**(8), 2575 (2006)
20. M.L. Minus, H.G. Chae, S. Kumar, Observations on solution crystallization of poly(vinyl alcohol) in the presence of single-wall carbon nanotubes. Macromol. Rapid Commun. **31**(3), 310 (2010)
21. R. Jain, M.L. Minus, G.C. Han, S. Kumar, Processing, structure, and properties of pan/mwnt composite fibers. Macromol. Mater. Eng. **295**(8), 742–749 (2010)

Preparation and Characterization of TPS/PBAT/Tourmaline Biodegradable Composite Films

Yingmo Hu and Quan Liu

Abstract To improve the compatibility between starch and poly(butylene adipate-co-terephthalate) (PBAT), thermoplastic starch (TPS) was prepared via treatment starch with KH-560, and then the melt blended with PBAT to get degradable TPS/PBAT blend membranes. The compatibility and thermal stability of TPS/PBAT blend membranes were effective improved, and its water absorption was reduced. And the degradable TPS/PBAT/tourmaline composite materials that had the functions of negative ion releasing and far-infrared radiation were prepared by adjunction the surface organic modified tourmaline in the blend procedure of TPS and PBAT. The dispersion stability of tourmaline, negative ion releasing amounts, far-infrared emissivity and mechanical property of TPS/PBAT/tourmaline composite were characterized. The experimental results demonstrated that the TPS/PBAT/tourmaline functional composite materials had excellent negative ion releasing, far-infrared radiation and mechanical properties.

Keywords PBAT · Tourmaline · Composite film · Negative ions
Far-infrared radiation

Introduction

In order to eliminate the environmental pollution from plastic wastes, the researchers have been motivated to concern environmentally friendly materials, for example the degradable polymers [1, 2]. The degradable polymers could be decomposed into water, carbon dioxide, or other small molecules ultimately [3] with the action of the

Y. Hu (✉) · Q. Liu
Beijing Key Laboratory of Materials Utilization of Nonmetallic Minerals and Solid Wastes, National Laboratory of Mineral Materials, School of Materials Science and Technology, China University of Geosciences, Beijing, China
e-mail: huyingmo@cugb.edu.cn

Q. Liu
e-mail: liuquan215@qq.com

Y. Han (ed.), *Advanced Functional Materials*,
https://doi.org/10.1007/978-981-13-0110-0_92

microorganisms in nature such as bacteria, fungi and algae or light [4]. As it was well known, the starch is a renewable resource that can be obtained from kinds of plants [5] and has various favored characteristics including low cost, high processibility, degradability and abundance [6], however, starch was fragility and difficult to make useful material. To conquer the fragility of starch and enhance their compatibility and processibility, many workers had studied the modified process of starch and tried to prepare starch-based composite materials [7, 8].

Poly (butylene adipateco-terephthalate) (PBAT) was widespread concerned recently for its excellent biological degradability and mechanical properties [9, 10], but the high cost limited its wide application [11]. If starch was combined with PBAT, the high performance and degradable starch/PBAT composite could be useful in wide range applications.

On the other hand, the degradable functional materials have become the research hotspot of material field. As a kind of functional mineral, Tourmaline [12] was caused broad attention for its distinctive function of negative ions releasing and far-infrared radiation [13, 14], and was admired as outstanding healthcare material at 21th century.

In order to probe the novel negative ions releasing degradable materials, TPS/PBAT/tourmaline composite was prepared with melt blending procedures, and studied its dispersion stability, intersolubility, negative ion releasing amounts, far-infrared emissivity and mechanical property.

Experimental

Materials

Potato starch was obtained from Beijing Gusong Economics & Trade Co. Ltd; The poly (butylene adipate-co-terephthalate) (PBAT) was provided by the BASF company with the commercial name of Ecoflex-F-Blend-C1200; Silane coupling agent KH-560 (CP grade) was purchased from Sinopharm Chemical Reagent Co., Ltd; Tourmaline powder with diameters of d_{50} = 1.75 μm and d_{97} = 5.23 μm (purity: 98%) was purchased from Yanxin Mineral Co. Ltd., Hebei, China.

The Preparation of Thermoplastic Starch (TPS). The commercial starch was firstly dried at 80 °C for overnight in a ventilated oven and then fully mixed with KH-560 solution in a high-speed mixer, the mixture was placed at 80 °C for 4 h, the TPS was obtained.

The Preparation of TPS/PBAT Compsite Films. The TPS (70 wt%) and PBAT (30 wt%) were mixed homogeneously in a high speed mixer, the mixtures were processed in a double roller open mill at 140 °C for 20 min, then TPS/PBAT composite was heat pressed to films with flat vulcanizer at 140 °C for 5 min with a pressure of 10 MPa.

The Preparation of TPS/PBAT/Tourmaline Composite Films. In a 100 ml four-neck flask, 10 g tourmaline powder and 0.3 g span-60 were added and stirred in 50 ml of toluene at 60 °C for 1 h. The samples were filtrated under reduced pressure, and washed for three times with anhydrous ethanol, and finally dried and placed in a sealed bag. Then TPS and PBAT (70 wt% TPS and 30 wt% PBAT) with 1, 3 or 5 wt% of modified tourmaline were mixed homogeneously and processed with the same method mentioned above to prepare TPS/PBAT/tourmaline composite films.

Mechanical Properties Measurements. The tensile strength and elastic modulus were determined according to GB/T 1040 using a Universal Testing Machine CTM-2500 (Shanghai Association-Instrument Manufacturing Co, Ltd.) at a strain rate of 50 mm/min at room temperature with five replicates. The specimens were cut to dimensions of 100 mm in length and 10 mm in width, and the thickness of the specimens were determined using a manual micrometer caliper.

SEM Analysis. A scanning electron microscope (Japan's Hitachi LTD., Japan) was used to observe the cryogenic fractured surfaces of the samples, operated at an acceleration voltage of 20 kV. The fractured surface was sputtered with gold before observation.

Water Absorption. The water absorption of the composites was investigated according to Chinese standard GB1034-86 on three samples ($10 \times 10 \times 1$ mm^3) to obtain an average value. The specimens were dried at 70 °C for 24 h, then cooled and weighed ($W1$) quickly. The dried specimens were immersed in water at 23 ± 2 °C for 24 h. Then, the specimens were taken out from the distilled water and wiped with a dry filter paper to dry the water on the specimen surfaces, and then promptly tested the wet weight ($W2$). The water absorption of the composites was calculated using the following equation:

$$Water\ absorption = \frac{\mathrm{W2} - W1}{\mathrm{W1}} \quad (1)$$

Thermo-Gravimetric Analysis (TGA). Thermo-gravimetric analysis (TGA) was performed on 10 mg samples using a thermo-gravimeter (STA 449 F5) at 10 °C/min from 30 to 700 °C at a heating rate of 10 °C/min under nitrogen atmosphere.

Mensuration of the Negative Ions. In a closed environment, a sample with the square of 10 cm × 10 cm was put on the sample stage near the instrument to test the negative ion release, using an air ion measuring instrument (DL AIC-1000, AlphaLab).

Far-infrared Emissivity. The sample was prepared into a wafer with a diameter of 2 cm and a thickness of 1 mm to examine the far-infrared emissivity in the wavelength range of 8–25 μm at 100 °C using an Fourier transform infrared spectroscopy (FTIR, Bruker-80 V, Germany).

Results and Discussion

Characterization of TPS/PBAT Film

The Mechanical Properties and Water Absorption. The mechanical properties and water absorption of starch/PBAT films and TPS/PBAT films were shown in Table 1. A higher tensile strength of 14.3 MPa and elastic modulus of 408 MPa were obtained in the TPS/PBAT film with KH-560, increased 34.5% and 6.0% respectively compared with the starch/PBAT film without KH-560. The tensile strength and elastic modulus of samples were attributed to their microstructure and compatibility, in the starch/PBAT film, the starch and PBAT were simply physically blended and had a phase separation phenomena, as a contrast, there brought cross-linked between TPS and PBAT when KH-560 was added [15, 16], and compatibility was improved, so the mechanical properties of TPS/PBAT composite was enhanced.

The water absorption of the starch/PBAT film and TPS/PBAT film were also shown in Table 1. Compared to the water absorption of starch/PBAT film, the water absorption of TPS/PBAT film was decreased 30%. Those also confirmed that the polar functional groups in TPS/PBAT film were reduced and restricted the diffusion of water molecular into the films with addition of KH-560 to enhance interfacial adhesion between TPS and PBAT [17].

SEM Images. The SEM images of the starch/PBAT and TPS/PBAT composites were shown in Fig. 1. In the starch/PBAT film (Fig. 1a), the phase separation between starch and PBAT matrix was observed obviously, and the surface of starch/PBAT composite was rough, it indicated that the weak interfacial adhesion and poor compatibility between PBAT and starch matrix without KH-560. As a contrast, the surface of TPS and PBAT became smooth and no phase separation (Fig. 1b) with the action of KH-560 to improve the compatibility of TPS and PBAT [18].

Thermal Stabilities Properties. The thermo gravimetric analysis of starch and TPS/PBAT film was shown in Fig. 2. The weight-loss curves of starch and TPS/PBAT film presented no obvious distinctions before 300 °C, indicated that only the low molecular and moisture were lost in this stage, with the temperature was raising, weight-loss curves of starch and TPS/PBAT film appeared clear distinctions that the decomposition temperature of starch and TPS/PBAT film were 321 and

Table 1 The physical properties of Starch/PBAT film and TPS/PBAT film

Items	Starch/PBAT film	TPS/PBAT film	Change values	Extent change (%)
Tensile strength (MPa)	10.6	14.3	+3.7	+34.5
Elastic modulus (MPa)	385	408	+23	+6.0
Water absorption (%)	21.91	15.34	−6.57	−30.0

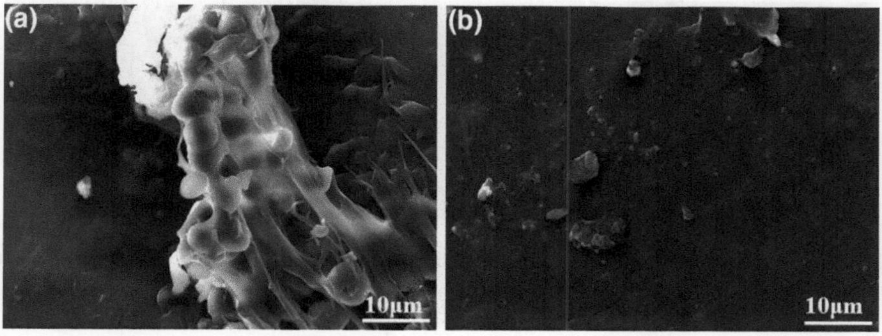

Fig. 1 SEM images of films: **a** starch/PBAT film, **b** TPS/PBAT film

Fig. 2 TG analysis curves of starch and starch/PBAT film

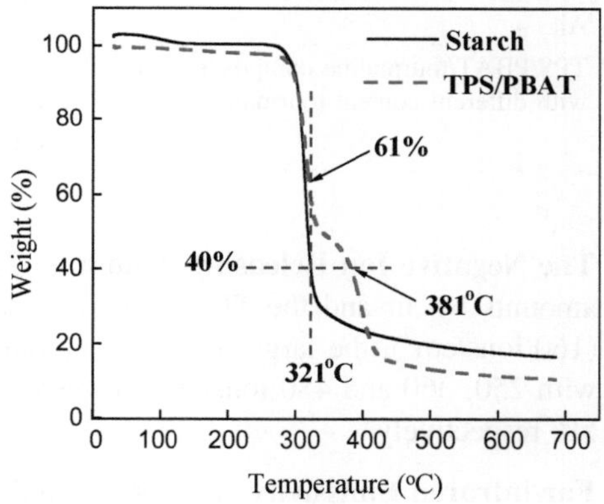

381 °C respectively at 40% residual weight, and the residual weight of starch and TPS/PBAT film were 40 and 61% respectively [19] at 321 °C. These analytic results revealed that the thermal stability of the TPS/PBAT film was significantly enhanced.

Characterization of TPS/PBAT/Tourmaline Film

SEM Images. The TPS/PBAT/tourmaline composite films were prepared with 1, 3, 5% modified tourmaline powders. In order to observe the dispersity of tourmaline in the matrix, the surface morphology of the films were investigated as showing in Fig. 3, when the additive amount of modified tourmaline powder was 1 and 3%, the composite film presented a homogenous microstructure with well dispersity of tourmaline powder in the matrix (Fig. 3a, b), and when the additive amount of modified tourmaline was over 5% (Fig. 3c), the clustering phenomena of tourmaline was obvious.

Fig. 3 SEM images of the dispersion of modified tourmaline in composite films with different dosage: **a** 1%; **b** 3%; **c** 5%

Table 2 Amount of negative ions releasing of air and TPS/PBAT/tourmaline composite films with different content of modified tourmaline

Sample		Amount of Negative Ions (ions/cm^3)
Air		160
TPS/PBAT/tourmaline composite films with different content tourmaline	1%	250
	3%	360
	5%	480

The Negative Ion Releasing Amounts. Table 2 listed the negative ions releasing amounts by air and the TPS/PBAT/tourmaline composite films. Compared to air (160 ions/cm^3), the larger releasing amounts was detected in the composite films with 250, 360 and 480 ions/cm^3 corresponding the tourmaline content of 1, 3 and 5% respectively.

Far-infrared Emissivity. The far-infrared emissivity of the TPS/PBAT/tourmaline composite films were measured and presented in Table 3. It was showed that the far-infrared emissivity of the TPS/PBAT/tourmaline composite films were all above 0.95 with rather high level. Combination with the negative ions releasing amounts, the functional composite materials with negative ions releasing and far-infrared radiation was prepared in this process.

Mechanical Properties. The mechanical properties of the samples were showed in Table 4. The tensile strength and elastic modulus were all better with 3% tourmaline dosage than that of 1 and 5%, and SEM analysis also showed well dispersity of tourmaline at 3% tourmaline dosage (Fig. 3b), 3% tourmaline dosage was the optimal in TPS/PBAT/tourmaline composite.

Table 3 Far-infrared radiation of TPS/PBAT/tourmaline composite films with different content of modified tourmaline

Sample		Far-infrared emissivity
TPS/PBAT/tourmaline composite films with different content tourmaline	1%	0.954
	2%	0.951
	3%	0.951

Table 4 Mechanical properties of TPS/PBAT/tourmaline composite films with different content of modified tourmaline

Sample		Tensile strength (MPa)	Elastic modulus (MPa)
Composite films with different tourmaline content	1%	12.8	324.4
	3%	13.22	370
	5%	11.9	296.6

Conclusion

TPS/PBAT composite film was prepared by melt blending of starch modified with KH-560 and PBAT, as a compatibiliser, the KH-560 was efficient for cross-linking between starch and PBAT and improved mechanical and thermal properties, alleviated water absorption of TPS/PBAT composite films. And the TPS/PBAT/tourmaline composite materials were prepared by adjunction the surface organic modified tourmaline in the blend procedure of TPS and PBAT. The experimental results exhibit that the amounts of negative ions releasing of TPS/PBAT/tourmaline functional composite materials reach to 480 ions/cm^3, the far-infrared emissivity is over 0.95, its tensile strength and elastic modulus are 13.22 MPa and 370 MPa respectively. This work offers a novel procedure for wide application of starch and preparation of degradable functional composite materials.

Acknowledgements This study is supported by the National Natural Science Foundation of China (No. 51372233).

References

1. G. Guo, C. Zhang, Z. Du et al., Structure and property of biodegradable soy protein isolate/PBAT blends. Ind. Crops Prod. **74**, 731–736 (2015)
2. J. Xie, K. Zhang, J. Wu et al., Bio-nanocomposite films reinforced with organo-modified layered double hydroxides: preparation, morphology and properties. Appl. Clay Sci. **126**, 72–80 (2016)
3. H. Wang, D. Wei, A. Zheng et al., Soil burial biodegradation of antimicrobial biodegradable PBAT films. Polym. Degrad. Stab. **116**, 14–22 (2015)
4. D. Raghavan, Characterization of biodegradable plastics. Polym.-Plast. Technol. Eng. **34**, 41–63 (1995)
5. D. Liu, H. Tian, Soy protein nanocomposites: properties and applications, in *Natural Polymers*, vol II, ed. by M.J. John (Royal Society of Chemistry, 2012), pp. 91–112
6. E. Schwach, L. Averous, Starch-based biodegradable blends: morphology and interface properties. Polym. Int. **53**, 2115–2124 (2004)
7. P.S. Garcia, M.V.E. Grossmann, F. Yamashita et al., Citric acid as multifunctional agent in blowing films of starch/PBAT. Quim. Nova **34**, 1507–1510 (2011)
8. S. Nattakan, T. Chuleeporn, S. Orawan, A review: starch-based composite foams. Compos. A Appl. Sci. Manuf. **78**, 246–263 (2015)
9. D. Wei, H. Wang, H. Xiao et al., Morphology and mechanical properties of poly (butylene adipate-co-terephthalate)/potato starch blends in the presence of synthesized reactive

compatibilizer or modified poly (butylene adipate-co-terephthalate). Carbohyd. Polym. **123**, 275–282 (2015)
10. Y. Nabar, J.M. Raquez, P. Dubois et al., Production of starch foams by twin-screw extrusion: effect of maleated poly (butylene adipate-co-terephthalate) as a compatibilizer. Biomacromol **6**, 807–817 (2005)
11. R.P.H. Brandelero, M.V. Grossmann, F. Yamashita, Films of starch and poly (butylene adipate co-terephthalate) added of soybean oil (SO) and Tween 80. Carbohyd. Polym. **90**, 1452–1460 (2012)
12. Y. Hu, X. Yang, The surface organic modification of tourmaline powder by span-60 and its composite. Appl. Surf. Sci. **258**, 7540–7545 (2012)
13. Y. Wang, J.T. Yeh, T. Yue et al., Surface modification of superfine tourmaline powder with titanate coupling agent. Colloid Polym. Sci. **284**, 1465–1470 (2006)
14. K. Krambrock, M.V.B. Pinheiro, S.M. Medeiros et al., Investigation of radiation-induced yellow color in tourmaline by magnetic resonance. Nucl. Instrum. Methods Phys. Res., Sect. B **191**, 241–245 (2002)
15. H. Lv, S. Song, S. Sun et al., Enhanced properties of poly (lactic acid) with silica nanoparticles. Polym. Adv. Technol. **27**, 1156–1163 (2016)
16. J. Balaji, M.G. Sethuraman, Studies on the effects of thiourea and its derivatives doped-Hybrid/zirconium nanocomposite based sol-gel coating for the corrosion behaviour of aluminum metal. Prog. Org. Coat. **99**, 463–473 (2016)
17. S. Lv, J. Gu, H. Tan et al., Modification of wood flour/PLA composites by reactive extrusion with maleic anhydride. J. Appl. Polym. Sci. **133**, 1–9 (2016)
18. Y. Hu, Q. Wang, M. Tang, Preparation and properties of Starch-g-PLA/poly (vinyl alcohol) composite film. Carbohyd. Polym. **96**, 384–388 (2013)
19. J.H. Kim, J.C. Lee, G.H. Kim, Study on poly (butylene adipate-co-terephthalate)/starch composites with polymeric methylenediphenyl diisocyanate. J. Appl. Polym. Sci. **132**, 1–7 (2015)

Fire Performance Analysis of PVC and Cabtyre Cables Based upon the Ignition Characteristics and Fire Growth Indexes

Bosi Zhang, Jiaqing Zhang, Hui Xie, Qiang Li, Liufang Wang and Minghao Fan

Abstract The ignition characteristics and fire growth indexes of the PVC cable (Cable 1), the flame retardant PVC cable (Cable 2), the cabtyre cable (Cable 3), and the cabtyre cable with fireproof coating (Cable 4), were investigated experimentally. Twenty experiments were conducted by cone calorimeter, and the ignition time and fire growth index under different external radiation heat fluxes were focused primarily. Results showed that the ignition time of Cable 2 was longer than that of Cable 1, and the ignition time of Cable 4 was longer than that of Cable 3. The critical heat flux, under which the cable could be ignited for any length of time, was derived through data fitting. In the present study, the critical radiation heat fluxes for Cable 1–4 are 3.5, 9.5, 13.4 and 18.5 kW/m^2. The ratio of peak heat release rate (HRR) to the time reached the peak HRR is defined as fire growth index. The fire growth index of Cable 1 is larger compared with that of Cable 2, and the fire growth index of Cable 3 is larger than that of Cable 4. The fire growth index of cables has well linear relationship with the external radiation heat flux in the present study.

Keywords PVC cables · Cabtyre cables · Ignition characteristics
Fire growth indexes

B. Zhang (✉)
Department of Safety Engineering,
China University of Labor Relations, Beijing 100048, China
e-mail: bosi_zhang@163.com

J. Zhang · H. Xie · L. Wang · M. Fan
State Grid Anhui Electric Power Research Institute, Hefei 230601, China

Q. Li
Chinese People's Armed Police Forced Academy, Langfang 065000, China

Y. Han (ed.), *Advanced Functional Materials*,
https://doi.org/10.1007/978-981-13-0110-0_93

Introduction

Cable fires pose serious threats and cause huge damages to the electrical power. Therefore, it is necessary to find out the fire hazard of cables. The ignition characteristic and fire growth index are two key parameters for assessing fire performance of cables, which represented the ignition hazard and the speed of fire development, respectively.

The fire hazard of cables has earned much attention [1–4]. There are many studies devoted to investigate the fire performance of cables by cone calorimeter. Niu et al. [5] explored the flame retardation of flame retarded PET cable based upon the parameters derived by cone calorimeter, including HRR, total heat release and effective heat of combustion. The effects of enteral radiation heat flux on fire performance of PVC cables were investigated by Wang et al. [6]. Their study was also conducted by cone calorimeter, and the HRR, mass loss rate and smoke release rate were concerned primarily. By using the cone calorimeter, the effects of waterborne flame retardant coating and solvent flame retardant coating on fire performance of cables were investigated experimentally by Zhang et al. [7]. Zhang et al. [8] also explored the fire parameters of Level-A and Level-C flame retardant cables by using cone calorimeter. In above two studies, HRR and toxic gas release rate were mainly concerned.

Based upon the above studies, the cone calorimeter is an effective method to investigate the fire performance of cables. However, most of the studies conducted by cone calorimeter were focused on the parameters of heat hazard and toxic gas release rate of cables. The discussions on ignition hazard and the speed of fire development of cables were not sufficient. Therefore, the critical radiation heat flux of ignition and fire growth index of cables were investigated, and the effects of cable retardant and fireproof coating on ignition hazard and the speed of fire development were also explored in the present study.

Materials and Methods

The ignition characteristic and fire growth index of cables, including the PVC cable (Cable 1), the flame retardant PVC cable (Cable 2), the cabtyre cable (Cable 3), and the cabtyre cable with fireproof coating (Cable 4), were investigated. The ignition characteristic and fire growth index of cables represented the ignition hazard and the fire development speed, respectively. The effects of cable retardant and fireproof coating on relieving the fire hazard of cables were also discussed.

Experiments on fire performance of cables were conducted by cone calorimeter. Before the experiments, the exhaust ventilation rate and radiation heat flux of cone calorimeter were calibrated first. The cables was cut into 10 cm and placed in the specimen tray. In the experiments, the spicemens of cable were ignited by using the spark igniter. Once the spicemen was ignited, the spark igniter removed immediately.

For each kind of cable, five levels of external heat radiation: 25, 35, 50, 65 and 75 kW/m^2, were included. There are twenty experiments conducted in the present study. By using the cone calorimeter, fire parameters, such as ignition time, HRR, mass loss rate, CO release rate, were derived. The ignition characteristics and fire growth indexes were discussed mainly based upon the ignition time and HRR curve.

Results and Discussions

Ignition Characteristics of Cables. The ignition times of Cable 1–4 under different external heat radiations are listed in Table 1. With the increase of external heat radiation, the ignition time of cables became shorter. Under the same external heat radiation, the ignition time of Cable 2 was larger than that of Cable 1, and the ignition time of Cable 4 was larger than that of Cable 3. It indicates that cable retardant and fire proof coating are benefit to decrease the ignition hazard of cables in fire. However, those improvements became much less with the increase of external heat radiation. From the view of enlarge ignition time of cables, fire proof coating had better effect than cable retardant in the present study.

In the present study, the tested cables were ignited under all external radiation heat fluxes. The maximum radiation heat flux, under which the cable cannot be ignited for any length of time, is defined as the critical radiation heat flux of ignition. The critical radiation heat flux represented the threshold value of heat environment, in which the cable had no risk for being ignited.

The critical radiation heat flux of ignition can be calculated based upon the ignition times under different external radiation heat fluxes according to the studies of Janssens [9, 10]. The transformed ignition time of materials, $(1/t_{ig})^n$ is positively related to the external radiation heat fluxes. Janssens [9, 10] pointed out that the values of parameter n are 0.55 and 1 for the thermal thick and thermal thin materials, respectively. The material is thermal thick if the material thickness is larger than the heat penetration thickness, otherwise it's thermal thin material. It is hard to distinguish the type of materials through experiments directly. According to Janssens, the transformed ignition time, $(1/t_{ig})^n$ versus external radiation heat fluxes should be linear fitted with n = 0.55 and n = 1, respectively, and get the correlation coefficients, R^2. The material is thermal thick material if the R^2 with n = 0.55 is larger than the R^2 with n = 1, otherwise the material is thermal thin material. After

Table 1 Ignition times of Cable 1–4

Radiation heat flux		25 kW/m^2	35 kW/m^2	50 kW/m^2	65 kW/m^2	75 kW/m^2
Ignition time (s)	Cable 1	55	26	11	7	6
	Cable 2	83	46	17	10	7
	Cable 3	175	70	31	14	8
	Cable 4	500	297	52	23	20

the appropriate linear fitting is derived, the critical radiation heat flux can be obtained according to the intercept [11, 12]. In this study, the critical radiation heat fluxes of Cable 1–4 were discussed by the above method. The relationships between the transformed ignition time, $(1/t_{ig})^n$ and external radiation heat fluxes, $\dot{q}_e''$ of Cable 1–4, as well as the linear fittings are shown in Figs. 1 and 2.

From Fig. 1, the linear fitting results between the transformed ignition time, $(1/t_{ig})^n$ and the external radiation heat flux for Cable 1 and Cable 2 are shown in Eqs. 1 and 2,

$$(1/t_{ig})^{0.55} = 0.0054\dot{q}_e'' - 0.0190, \quad R^2 = 0.99 \quad ; \quad 1/t_{ig} = 0.0031\dot{q}_e'' - 0.0638, \quad R^2 = 0.99 \tag{1}$$

$$(1/t_{ig})^{0.55} = 0.0052\dot{q}_e'' - 0.0491, \quad R^2 = 0.99 \quad ; \quad 1/t_{ig} = 0.0026\dot{q}_e'' - 0.0636, \quad R^2 = 0.96 \tag{2}$$

The correlation coefficients, R^2 of Cable 2 with n = 0.55 and n = 0.1 were 0.99 and 0.96, respectively. According to the criterion discussed above, the Cable 2 was thermal thick material for analyzing the ignition characteristics. Moreover, the critical radiation heat flux of Cable 2 was 25 kW/m^2 with n = 1, under which the Cable 2 was ignited at 83 s in the experiments. Hence, the critical radiation heat flux of Cable 2 was 9.5 kW/m^2. The correlation coefficients, R^2 of linear fitting for the data of Cable 1 with n = 0.55 and n = 0.1 were almost the same. When n = 0.55, the critical radiation heat flux of Cable 1 was 3.5 kW/m^2. When n = 1, the critical radiation heat flux was 20.1 kW/m^2. According to the experimental results, the ignition time of Cable 1 under 25 kW/m^2 was only 55 s. Therefore, the critical radiation heat flux of Cable 1 was unlikely as large as 20.1 kW/m^2. In

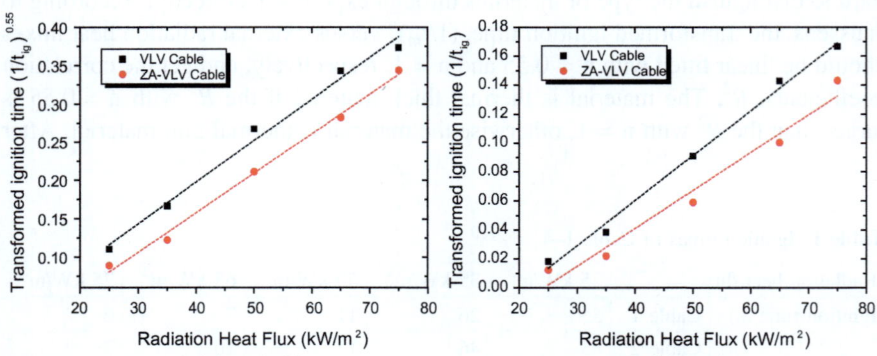

(a) n=0.55 (b) n=1

Fig. 1 Transformed ignition time of PVC cables

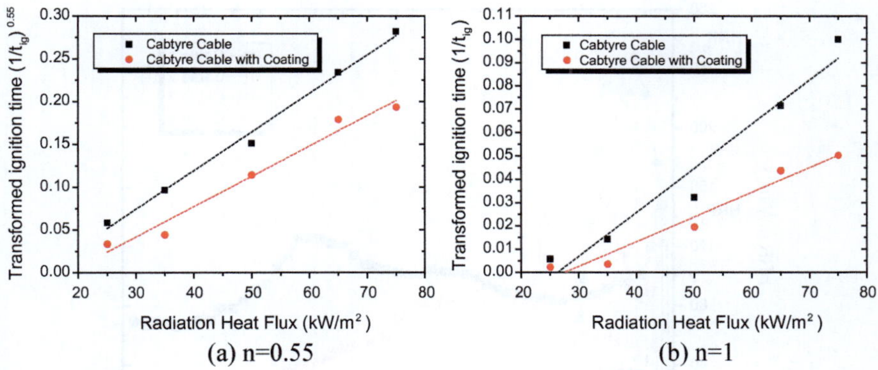

Fig. 2 Transformed ignition time of cabtyre cables

addition, the Cable 2 had higher flame retardant level, so the critical radiation heat flux of Cable 2 should be lower than that of Cable 1. From the above discussion, the Cable 1 should also be regarded as thermal thick material (n = 0.55) for analyzing the ignition characteristics, and the critical radiation heat flux of Cable 1 was 3.5 kW/m^2.

From Fig. 2, the linear fitting results between the transformed ignition time of materials, $(1/t_{ig})^n$ and the external radiation heat flux for Cable 3 and Cable 4 are shown in Eqs. 3 and 4.

$$(1/t_{ig})^{0.55} = 0.0045\dot{q}_e'' - 0.0605, \quad R^2 = 0.99 \quad ; \quad 1/t_{ig} = 0.0019\dot{q}_e'' - 0.0294, \quad R^2 = 0.94 \tag{3}$$

$$(1/t_{ig})^{0.55} = 0.0035\dot{q}_e'' - 0.0646, \quad R^2 = 0.97 \quad ; \quad 1/t_{ig} = 0.0011\dot{q}_e'' - 0.0294, \quad R^2 = 0.95 \tag{4}$$

From Eqs. (3) and (4), the correlation coefficients, R^2 with n = 0.55 were larger than those with n = 1 for Cable 3 and Cable 4. It means that Cable 3 and Cable 4 should be regarded as thermal thick material for discussing the ignition characteristics. The critical radiation heat fluxes of Cable 3 and Cable 4 were 13.4 and 18.5 kW/m^2, respectively.

To sum up, the cabtyre cable had better performance than the PVC cable from the perspective of critical radiation heat flux of ignition in the present study. The critical radiation heat flux of Cable 2 was larger than that of Cable 1, and the critical radiation heat flux of Cable 4 is larger than that of Cable 3. It implies that cable retardant and fireproof coating both have positive effects on decreasing the ignition hazard of cables.

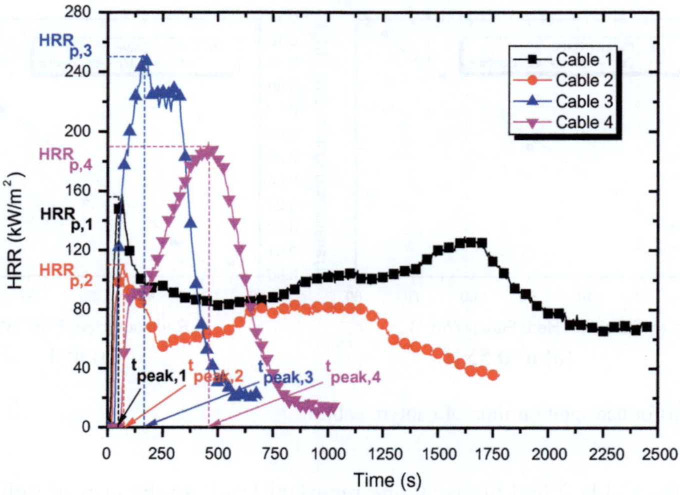

Fig. 3 HRR curve of Cable 1–4 under the external radiation heat flux of 50 kW/m^2

Fire Growth Indexes of Cables. The fire growth index was defined as the ratio of peak HRR to the time reached the peak HRR. The HRR versus time curves of Cable 1–4 under the external heat radiation of 50 kW/m^2 are shown in Fig. 3. It is observed that the HRR curves of Cable 1 and Cable 2 had two peaks. The first peak was higher, and the second peak was not apparent. The first peak was selected for calculating the fire growth index in the present study.

Fire growth indexes with different external radiation heat fluxes of Cable 1–4 are shown in Figs. 4 and 5. The fire growth index increased with the external heat radiation flux. It means that the cable fire develops fast under high external heat radiation flux. In above discussions, the cabtyre cable had better performance on ignition characteristics than the PVC cable. However, this advantage disappeared from the perspective of fire growth index. The comparison among fire growth indexes of Cable 1–4 showed the similar rules with those draw from the ignition time. Cable retardant and fireproof coating were benefit for delaying fire development of cables, and the latter one showed better performance.

As presented in the previous studies [13], the fire growth index had linear relationship with the external radiation heat flux. The linear relationships between the fire growth index and external radiation flux for Cable 1–4 are shown in Eqs. 5–8. The FGI_i (i = 1, 2, 3 and 4) represented the fire growth indexes of Cable 1, 2, 3 and 4, respectively.

$$\mathrm{FGI}_1 = 0.038\dot{q}_e'' + 0.367, \quad R^2 = 0.97 \tag{5}$$

$$\mathrm{FGI}_2 = 0.028\dot{q}_e'' + 0.147, \quad R^2 = 0.95 \tag{6}$$

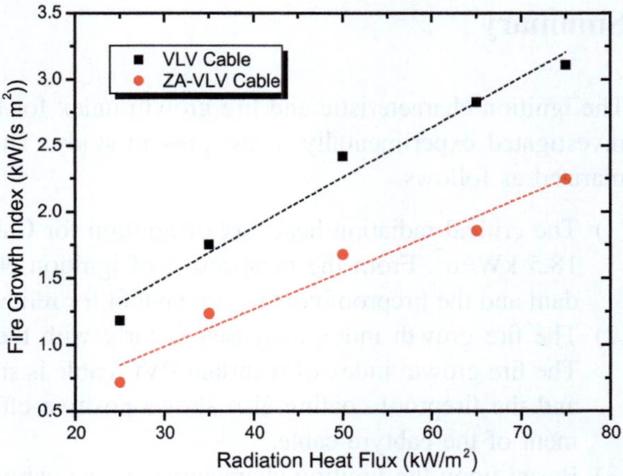

Fig. 4 Fire growth index of PVC cables

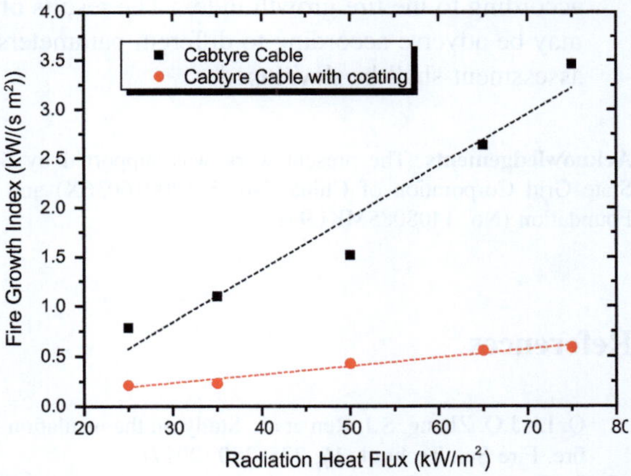

Fig. 5 Fire growth index of cabtyre cables

$$FGI_3 = 0.053\dot{q}_e'' - 0.749, \quad R^2 = 0.94 \tag{7}$$

$$FGI_4 = 0.008\dot{q}_e'' - 0.025, \quad R^2 = 0.95 \tag{8}$$

The linearity between the fire growth index and the external radiation heat flux was well for Cable 1–4. The slope in Eq. 6 was smaller than that in Eq. 5, and the slope in Eq. 8 was smaller than that in Eq. 7. It indicates that cable retardant and fireproof coating had more obvious effect on delaying fire development speed under high external radiation fluxes.

Summary

The ignition characteristic and fire growth index for PVC and cabtyre cables were investigated experimentally in the present study. The main conclusions are summarized as follows.

(1) The critical radiation heat flux of ignition for Cable 1-4 are 3.5, 9.5, 13.4 and 18.5 kW/m^2. From the perspective of ignition characteristics, the cable retardant and the fireproof coating are benefit for relieving ignition hazard of cables.
(2) The fire growth index increases linearly with the external radiation heat flux. The fire growth index of retardant PVC cable is smaller than that of PVC cable, and the fireproof coating also shows positive effect on delaying fire development of the cabtyre cable.
(3) Based upon the ignition characteristics, the cabtyre cable shows better performance than the PVC cable, while the PVC cable has a little advantage according to the fire growth index. The results of fire risk assessment on cables may be adverse according to different parameters, therefore the comprehensive assessment shall be developed.

Acknowledgements The present work was supported by Science and Technology Project of State Grid Corporation of China (No. 52120016036X) and Anhui Provincial Natural Science Foundation (No. 1408085MKL94).

References

1. Q. Li, J.Q. Zhang, S.J. Ren et al., Study on the insulation failure prediction model of cable in fire. Fire Sci. Technol. **36**, 275–280 (2017)
2. L. Courty, J.P. Garo, External heating of electrical cables and auto-ignition investigation. J. Hazard. Mater. **321**, 528–536 (2017)
3. J.Q. Zhang, M.H. Fan, W. Li, et al., Fire Safety of Cables in Power Grid: Tracking Combustion Test Standards of Cables and New Insights on Test Framework, CIGRE (International Council on Large Electric systems). (2016) B1-214
4. L. Li, X.J. Huang, K. Bi et al., An enhanced fire hazard assessment model and validation experiments for vertical cable trays. Nucl. Eng. Des. **301**, 32–38 (2016)
5. M.J. Niu, X.F. Li, J.Z. Chen et al., Study on the flame retardation and smoke emission of flame retarded polyethyleneterephthalate with cone. Polym. Mater. Sci. Eng. **18**, 202–204 (2002)
6. W. Wang, H.P. Zhang, Y.T. Wan, A study on the burning behavior of PVC cables in the conic calorimeter test. J. Saf. Environ. **8**, 117–120 (2008)
7. J.Q. Zhang, M.H. Fan, W. Li et al., Study on combustion and toxic gas release characteristics of flame retardant cables. Insulating Mater. **49**, 72–77 (2016)
8. J.Q. Zhang, B.S. Zhang, M.H. Fan et al., Effects of external heat radiation on combustion and toxic gas release of flame retardant cables. Mater. Sci. Forum **898**, 2392–2398 (2017)
9. M. Janssens, *Improved method of analysis for the LIFT apparatus, Part 1: ignition* (Intersci. Commun., London, 1993)
10. M. Janssens, Piloted ignition of wood: a review. Fire Mater. **15**, 151–167 (1991)

11. J. Quintiere, A theoretical basis for flammability properties. Fire Mater. **30**, 175–214 (2006)
12. J. Luche, T. Rogamue, F. Richard et al., Characterization of thermal properties and analysis of combustion behavior of PMMA in a cone calorimeter. Fire Saf. J. **46**, 451–461 (2011)
13. R. Chen, Thermal decomposition and combustion characteristics of typical metro train interior materials, Doctoral Thesis, University of Science and Technology of China, Hefei, 2016

Loading and Controlled Releasing of Anti-cancer Drug Bortezomib by Glucose-Containing Diblock Copolymer

Xiao-Ting Zhang, Hai-Liang Dong, Zhong-Li Niu, Jia-Ming Xu, Dan-Yue Wang, Han Tong, Xiao-Ze Jiang and Mei-Fang Zhu

Abstract A glucose-containing diblock copolymer was employed as nanocarrier in this study for delivery of the anticancer drug bortezomib (BTZ). Our system was based on pH-induced dynamical conjugation of boronic acid on BTZ to cis-diols on glucose-containing polymer. Diblock copolymer poly(ethylene glycol)-b-poly (gluconamidoethyl methacrylate) (PEG-PGAMA), was firstly synthesized via atom transfer radical polymerization(ATRP) by successive polymerization of monomer gluconamidoethyl methacrylate (GAMA) using a PEG-based ATRP macroinitiator. BTZ was then loaded in glucose-containing copolymer as chemical conjugation occurred of boronic acid to glucose groups and the drug-released behavior of this system was simulated in vitro. The results demonstrated that PEG-PGAMA copolymer had strong ability to bind BTZ at physiological pH of 7.4; it could also effectively release BTZ at acid pH of 5.5(close to environment of cancer tissue or the subcellular endosome) in a pH-dependent manner. In our study, a facile and interesting nanocarrier system for anti-cancer drug bortezomib (BTZ) was provided with a kind of glucose-containing block copolymer without any need of chemical modification, which only utilized dynamic chemical complexation to reach effective drug-loading and controlled release of BTZ upon responsiveness to external pH.

X.-T. Zhang · H.-L. Dong · Z.-L. Niu · J.-M. Xu · D.-Y. Wang · H. Tong
X.-Z. Jiang (✉) · M.-F. Zhu (✉)
Shanghai National Key Laboratory for Fiber Material Modification,
Department of Materials Science and Engineering,
Donghua University, Shanghai 201620, China
e-mail: xiaozejiang@dhu.edu.cn

M.-F. Zhu
e-mail: zmf@dhu.edu.cn

X.-T. Zhang
e-mail: 18817325774@163.com

H.-L. Dong
e-mail: lddonghailiang@163.com

Z.-L. Niu
e-mail: 18817327600@163.com

Y. Han (ed.), *Advanced Functional Materials*,
https://doi.org/10.1007/978-981-13-0110-0_94

Keywords Dynamic chemical complexation · Glucose · BTZ · Diblock copolymer · pH-responsive

Introduction

Stimuli-responsive polymeric nanocarriers used for drug delivery are getting considerable attention in the field of cancer treatment due to their ability to preferentially gather in the cancer tissue by the enhanced permeation and retention effect (EPR) [1, 2], and realize controlled drug release in response to biological stimuli such as pH [3–5], temperature [6, 7], redox reactions [8, 9], and enzymes [10]. Following much efforts on unsolved problems such as premature release and drug loss during storage, there has been increasing interest in introducing dynamic bond into design of polymer carrier [11–15]. Boronic acid can bind 1,2- or 1,3-diols reversibly forming boronate esters with fast dual responsiveness to external pH [14, 16], which was extensively utilized for carbohydrate sensors [17–19], nano-reactors [20], drug delivery systems [21, 22], and self-healing materials [23]. Based on this pH-induced dynamic interaction, anticancer drug bortezomib (BTZ) containing boronic acids can be delivered by diols-containing polymer through chemical attachment. This kind of drug delivery system is expected to load BTZ at physiological pH of 7.4 and release payload at acid pH of 5.5 (tumor extra-cellular pH 6.5–7.2, endosomal/lysosomal pH 4.5–6 [24, 25]).

Bortezomib [26, 27] (BTZ) is a kind of protease inhibitor with boronic acid groups, which has been approved to be used for the treatment of multiple myeloma and cell lymphoma. However, considering the pharmacokinetics, BTZ would be quickly removed in the blood circulation due to the combination with non-specific protein and dose-dependent toxicity, and these shortcomings limit its effect on tumor inhibition. Base on this, employing polymer micelles system as drug delivery of BTZ obviously increased the drug loading capacity with many desirable feature such as significant targeting, sustained release and reducing the adverse effects of drugs on the body. BTZ can conjugate with diols-containing polymer to be passively targeted and preferentially accumulated in the lesion and released under the stimuli of slight acid pH at tumor tissues.

Zhang's group [28] used catechol-containting polymeric micelles to load BTZ in the outer shell by complexation of BTZ with polydopamine and this drug delivery system was demonstrated to be able to realize controlled release upon pH-trigger. In our study, glucose was employed to be component of polymer providing considerable diols conjugated with BTZ. Glucose has excellent hydrophilicity and biocompatibility and glucose-containing block copolymer could be synthesized via atom transfer radical polymerization (ATRP) [29, 30] which has been widely used to prepare well-defined block copolymers [31–35]. Herein we synthesized glucose-containing diblock copolymer poly(ethylene glycol)-b-poly(gluconamidoethyl methacrylate)(PEG- PGAMA) via ATRP by successive polymerization of monomer gluconamidoethyl methacrylate (GAMA) using a PEG-based ATRP macroinitiator. We demonstrated the dynamic complexation of BTZ with glucose by a colorimetric assay based on the indicator of alizarin red s (ARS). As BTZ was loaded in PEG-PGAMA, the drug releasing behavior of this system was simulated in vitro and the result showed appreciable effect of drug-loading at physiological pH of 7.4 and controlled release at slight acid pH of 5.5.

Experimental Section

Materials. Monohydroxy-capped poly(ethylene glycol) (PEG_{113}-OH) with a mean degree of polymerization, DP, of 113 (Mn = 5000 g/mol, Mw/Mn = 1.02) was purchased from Aldrich and dried at 30 °C under vacuum overnight prior to use. 2, 2-bipyridine (bpy), cuprous bromide (CuBr), 2-bromoisobutyryl bromide, 2-aminoethyl methacrylic acid hydrochloride, *D*-glucono-lactone, bortezomib (BTZ) were all purchased from J&K. Methanol, toluene, triethylamine (TEA), 1-methyl-2-pyrrolidinone (NMP) were distilled from CaH_2 under reduced pressure. Dichloromethane, dimethyl sulfoxideand (DMSO) and other organic solvents were purchased from Sinopharm Group Chemical Reagent Co., Ltd., and directly used without further purification.

Preparation of GAMA Monomer. D-glucono-lactone(10.0 g, 56.2 mmol) was dissolved in 100 mL of dry methanol in a 250 mL round bottom flask which was stirred and heated for 2 h in an oil bath at 40 °C. After azeotropic distillation of 50 mL of methanol at reduced pressure to remove traces of water with 50 mL of fresh methanol added, the solution of 2-aminoethyl methacrylic acid hydrochloride (11.0 g, 67 mmol) in methanol (50 mL) was added with well stirred, then excess triethylamine was added and the reaction mixture was stirred at 20 °C overnight. The stirred solution was filtered, concentrated and precipitated into dichloromethane. After filtered, the product was washed several times with dichloromethane and isopropanol and dried under vacuum. 11.46 g of white solid was obtained in yield of 66.67%.

Preparation of PEG_{113}-Br Macroinitiator. PEG_{113}-OH (10 g, 2 mmol) and triethylamine (0.34 mL, 2.5 mmol) were dissolved in 90 mL of toluene in a 250 mL round bottom flask. Then 2-bromoisobutyryl bromide (0.3 mL, 2.5 mmol) was added dropwise via constant pressure dropping funnel for 1 h and the reaction mixture was stirred at room temperature for 48 h. The stirred solution was filtered to remove the triethylamine salt, evaporated to remove toluene and dissolved in dichloromethane to pass through alumina column (200–300 mesh). The resulting solution was washed with saturated sodium bicarbonate solution and saturated sodium chloride, then the organic layers were collected and dried over $MgSO_4$. The solution dried was concentrated under reduced pressure, precipitated into a 10-fold excess of ether. After filtered, the product was dried under vacuum overnight to remove the solvent. 7.1 g of white solid was obtained in yield of 71.1%.

Preparation of PEG-PGAMA Diblock Copolymer. A reaction eggplant flask was charged with PEG_{113}-Br macroinitiator (0.25 g, 0.048 mmol), GAMA monomer (0.30 g, 0.98 mmol) and NMP(2.0 mL) which were stirred under a N_2 atmosphere. Then 2, 2'-bipyridine(bpy) and $Cu^{I}Br$ were added to start the polymerization at room temperature for 4 h. The reaction mixture was dialyzed and then freeze-dried under vacuum to give 0.45 g of white solid in yield of 81.8%.

Preparation of Aqueous Solution of Drug-loaded PEG-PGAMA Diblock Copolymer. According to the method reported [36], 4 mg of BTZ and 40 mg of the glucose-containing block copolymer were dissolved in 2 mL of DMSO, then 10 mL of NaOHaq(pH 9) was slowly added dropwise under stirring. The solution was dialyzed using a dialysis bag (MWCO 3500) for 24 h with changing water once every 1 h. The prepared drug-loaded micelles were freeze-dried for 48 h.

Characterization. All ^{1}H NMR spectra were recorded in D_2O or d-chloroform or DMSO using a Bruker 400 MHz spectrometer.

Molecular weight distributions were assessed by gel permeation chromatography (GPC) in DMF containing 0.5 g/L LiBr at a flow rate of 1.0 mL min^{-1} using a series of two linear Styragel columns (HT2, HT4 and HT5), a Waters 1515 pump and waters 2414 differential refractive index detector (set at 30 °C) and an oven temperature of 25 °C. Calibration was performed using a series of near-monodisperse polystyrene standards.

The forming of dynamic chemical bonds between boronic acid and glucose were analysized by PerkinElmer Lambda Fluorescence Spectrometer with ARS(alizarin red s) as the color indicator and 460 nm of excitation wavelength to record emission spectrums of samples in the range from 500 to 600 nm.

Standard UV ray absorbtion curve of BTZ, entrapment efficiency of drug by nanocarrier and drug cumulative drug-releasing morphological curves were measured by UV-Vis spectrophotometer (Beijing General Analytical Spectrometer).

Results and Discussion

Synthesis of Glucose-containing Diblock Copolymer. Both PEG and PGAMA blocks are well water-soluble and PEG-GAMA could be synthesized via ATRP using PEG-Br as macroinitiator at room-temperature due to highly reactive GAMA monomer. The general synthesis routines of three glucose-containing block copolymers are shown in Scheme 1. Well defined PEG-GAMA diblock copolymer was obtained with the DP was 20 from ^{1}H NMR analysis (as shown Fig. 1). The DP of the PGAMA block was calculated by comparing the integral ratio of resonance at 4.0 ppm (c + e + f) and 3.6 ppm(a'), characteristic of PGAMA and PEG blocks, respectively. THF GPC traces (as shown in Fig. 2) revealed a narrow and symmetric peak for both macroinitiator PEG-Br and diblock copolymer PEG-GAMA with an Mn of 3100 and 6150 g/mol, an Mw/Mn of 1.02 and 1.20, thus denoted as PEG-b-$PGAMA_{20}$.

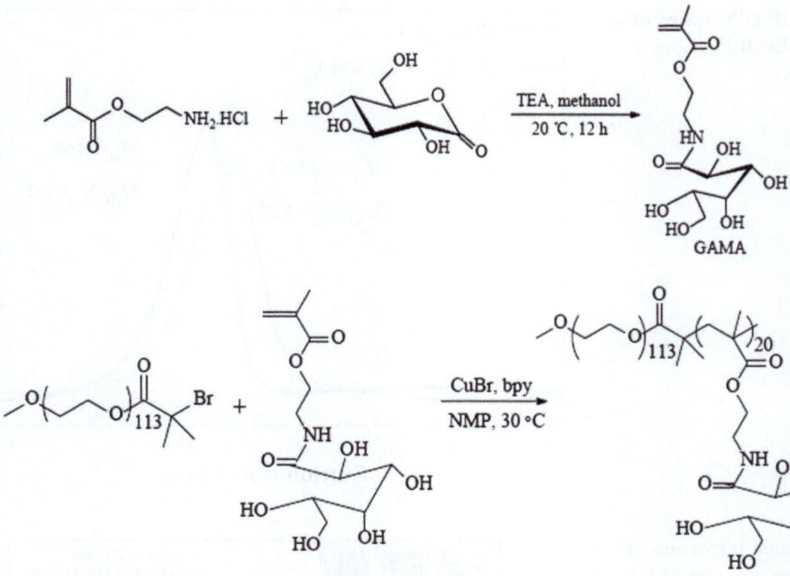

Scheme 1 Synthetic route for the preparation of diblock copolymer PEG-b-PGAMA

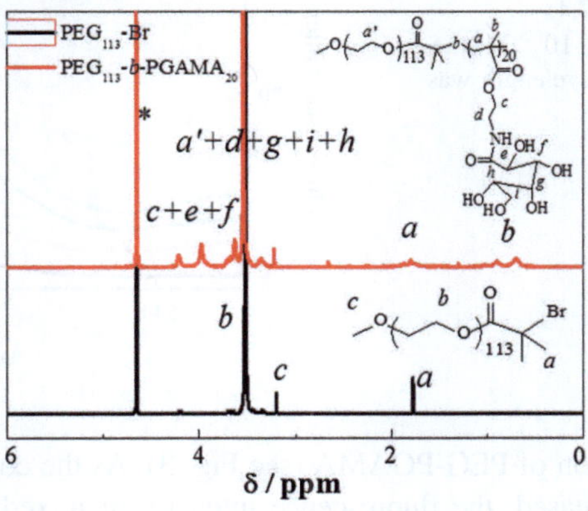

Fig. 1 ^{1}H NMR spectra recorded for **a** PEG_{113}-Br, **b** PEG_{113}-b-$PGAMA_{20}$

Bortezomib Loaded Glucose-Containing Driblock Copolymer. ARS is a catechol dye that will exhibit dramatic changes in color upon binding to boronic acid [37]. Bortezomib significantly changed the color of PBS solution of ARS (pH 7.4) from burgundy to yellow. The color returned when PEG_{113}-$PGAMA_{20}$ diblock copolymer was added, indicating that there is interaction between bortezomib and PEG-PGAMA. ARS also exhibits dramatic changes in fluorescence intensity after forming esters with boronic acids [37] and the fluorescence spectra were recorded

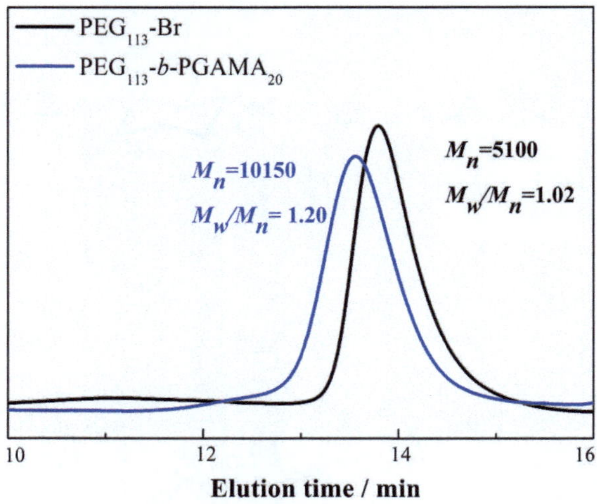

Fig. 2 THF GPC traces of **a** PEG_{113}-Br, **b** PEG_{113}-b-$PGAMA_{20}$

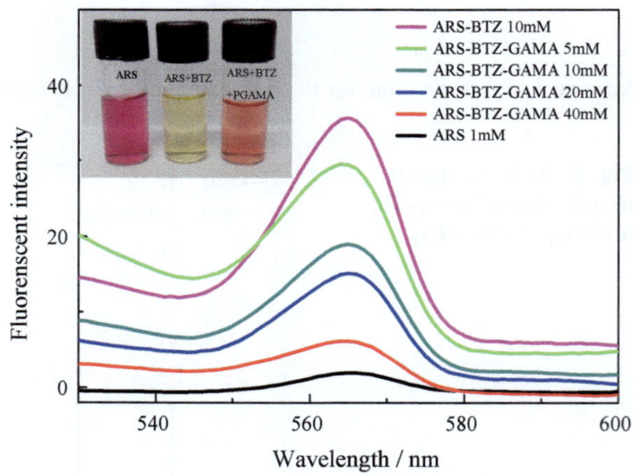

Fig. 3 Visual inspection of diol binding and release for glypolymer corresponding fluorescence spectra using ARS and BTZ in PBS buffer (0.1 M, pH 7.4) ($C_{ARS} = 1 \times 10^{-5}$ M; the excitation wavelength was 460 nm)

with addition of PEG-PGAMA (see Fig. 3). As the concentration of PEG-PGAMA added increased, the fluorescence intensity of mixed solution of ARS and bortezomib obviously decreased which demonstrated the formation of boronic esters between boronic acid containing bortezomib and glucose groups as ARS was deterred from complexation with bortezomib.

Standard UV Ray Absorbtion Curve of BTZ. There is ultraviolet absorption for bortezomib in the UV-visible region [38] and the wavelength of 260 nm was commonly chose for quantitative analysis of bortezomib. A series of solution of bortezomibine with different concentrations (c) in DMSO were prepared and then their absorbance (A) at 260 nm wavelength was measured to give the standard curve of bortezomib in DMSO (Fig. 4):

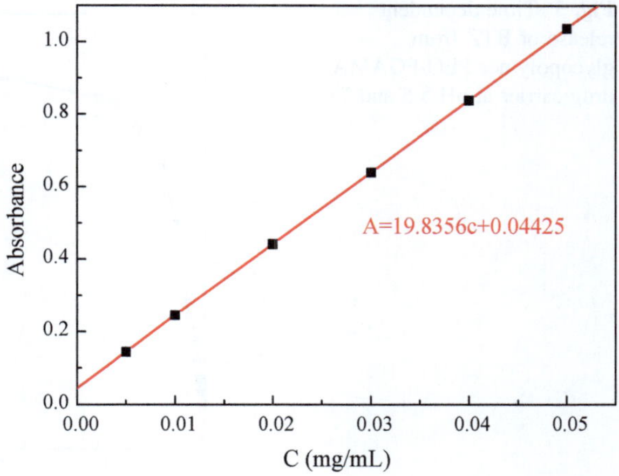

Fig. 4 The absorbance curve of BTZ in DMSO

$$A = 19.8356c + 0.04425 \left(R^2 = 0.99986\right) \quad (1)$$

In the same way, the standard curves of bortezomib were obtained in PBS solution at pH 5.5 and pH 7.4, respectively:

$$A = 14.7178c + 0.02825 \left(R^2 = 0.99986\right) \quad (2)$$

$$A = 27.1760c + 0.00230\left(R^2 = 0.99990\right) \quad (3)$$

The standard curves of bortezomib can be used to calculate the amount of loaded and released bortezomib. The drug loaded capacity (DLC) of PEG113-PGAMA20 is 12.4%, and the drug loaded encapsulation efficiency (DLE) is 84.5%.

Bortezomib Controlled Releasing Behavior of Glucose-Containing Block Copolymer. The drug-released behavior simulated in vitro was provided in Fig. 5. Drug-releasing rate of PEG_{113}-$PGAMA_{20}$ was slow at pH 7.4 with drug cumulative release is 35% within 10 h and 60% within 60 h, respectively. The results suggested that PEG_{113}-$PGAMA_{20}$ copolymer is able to chemically complex stably with bortezomib which can effectively waken burst release of bortezomib at 7.4. Nearly half of all drugs loaded in PEG_{113}-$PGAMA_{20}$ copolymer was still released after 60 h at pH 7.4 that might because part of drugs was loaded in copolymer not by dynamical chemical comoplexation but by physical interaction to attach to polymer-drug carrier surface and the latter is not stable as the former. As pH-induced acid-sensitive boric acid ester bonds were broken between cis-diols and BTZ at pH 5.5, a large amount of drugs was released with drug cumulative release is 80% within 10 h and 90% within 60 h, respectively and that showed excellent effect of controlled BTZ release by PEG_{113}-$PGAMA_{20}$ copolymer.

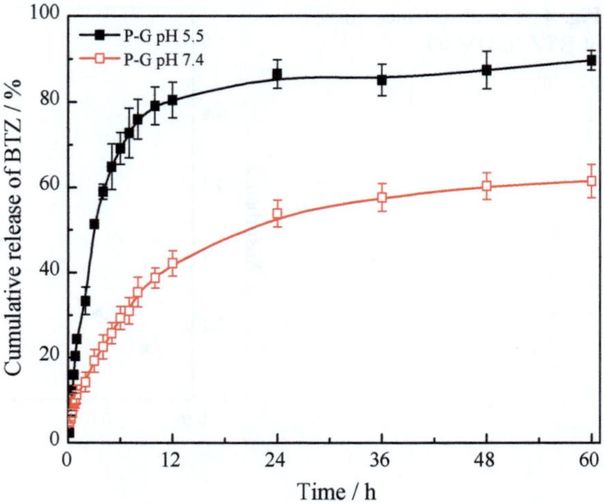

Fig. 5 Time-dependent release of BTZ from glycopolymer PEG-PGAMA drug carrier at pH 5.5 and 7.4

Conclusions

We have synthesized glucose-containing triblock copolymer that PEG-PGAMA which was demonstrated to be able to load anticancer drug bortezomib(BTZ) at physiological pH 7.4. This kind of BTZ delivery system shows great potential to heighten the drug-loading capacity by pH-induced dynamic covalent conjugation along with BTZ is endowed with some water-solubility at alkaline pH. Besides, the result of drug-releasing behavior simulated in vitro indicates that PEG-PGAMA can release considerable payload with appreciable controlled release effect at pH 5.5 that is close to slight acid environment in tumor issue. Our study provides a very facile and interesting pH-responsive delivery system for anticancer drug bortezomib with further great design flexibility in structure of containing-glucose polymer nanocarriers to minimize the premature release of drug and accomplish accurate controlled release in vivo.

Acknowledgements The authors gratefully acknowledged the financial support of the National Natural Science Foundation of China (No. 21204010, No. 51473035), and the Research Program of Shanghai Science and Technology Commission (No. 13NM1400102).

References

1. H. Maeda, L.W. Seymour, Y. Miyamoto, Conjugates of anticancer agents and polymers: advantages of macromolecular therapeutics in vivo. Cheminform **24**, 351–362 (1993)
2. R.K. Jain, Delivery of molecular medicine to solid tumors: lessons from in vivo imaging of gene expression and function. Science **271**, 1079–1080 (1996)

3. K. Zhou, Y. Wang, X. Huang, K. Lubyphelps, B.D. Sumer, J. Gao, Tunable, ultrasensitive pH-responsive nanoparticles targeting specific endocytic organelles in living cells. Angew. Chem. **50**, 6109–6114 (2011)
4. Y. Lee, T. Ishii, H. Cabral, H.J. Kim, J.H. Seo, N. Nishiyama, H. Oshima, K. Osada, K. Kataoka, Inside cover: charge-conversional polyionic complex micelles & mdash; efficient nanocarriers for protein delivery into cytoplasm (Angew. Chem. Int. Ed. 29/2009). Angew. Chem. **48**, 5309–5312 (2010)
5. E.S. Lee, K. Na, Y.H. Bae, Super pH-sensitive multifunctional polymeric micelle. Nano Lett. **5**, 325 (2005)
6. B. Law, Protease-sensitive fluorescent nanofibers. Bioconjug. Chem. **18**, 1701–1704 (2007)
7. K. Wookhyun, T. Julie, I. Emmanuel, L. Sébastien, L.C. Elliot, Self-assembly of thermally responsive amphiphilic diblock copolypeptides into spherical micellar nanoparticles. Angew. Chem. **49**, 4257–4260 (2010)
8. R.A. Petros, P.A. Ropp, J.M. Desimone, Reductively labile PRINT particles for the delivery of doxorubicin to HeLa cells. J. Am. Chem. Soc. **130**, 5008 (2008)
9. Y. Li, K. Xiao, J. Luo, W. Xiao, J.S. Lee, A.M. Gonik, J. Kato, T. Dong, K.S. Lam, Well-defined. Reversible disulfide cross-linked micelles for on-demand paclitaxel delivery, biomaterials **32**, 6633–6645 (2011)
10. B. Law, C.H. Tung, Proteolysis: a biological process adapted in drug delivery, therapy, and imaging. Bioconjug. Chem. **20**, 1683 (2009)
11. Y. Liu, Y. Zhang, Y. Guan, New polymerized crystalline colloidal array for glucose sensing. Chem. Commun. **1**, 1867–1869 (2009)
12. G. Pasparakis, A. Cockayne, C. Alexander, Control of bacterial aggregation by thermoresponsive glycopolymers. J. Am. Chem. Soc. **129**, 11014 (2007)
13. M.C. Roberts, M.C. Hanson, A.P. Massey, E.A. Karren, P.F. Kiser, Dynamically restructuring hydrogel networks formed with reversible covalent crosslinks. Adv. Mater. **19**, 2503–2507 (2010)
14. D. Roy, J.N. Cambre, B.S. Sumerlin, Sugar-responsive block copolymers by direct RAFT polymerization of unprotected boronic acid monomers. Chem. Commun. **21**, 2477–2479 (2008)
15. Y. Zhang, Y. Guan, S. Zhou, Permeability control of glucose-sensitive nanoshells. Biomacromol **8**, 3842–3847 (2007)
16. Y. Qin, V. Sukul, D. Pagakos, A. Chengzhong Cui, F. Jäkle, Preparation of organoboron block copolymers via ATRP of silicon and boron-functionalized monomers. Macromolecules **38**, 800–804 (2005)
17. S. Li, E.N. Davis, J. Anderson, Q. Lin, Q. Wang, Development of boronic acid grafted random copolymer sensing fluid for continuous glucose monitoring. Biomacromol **10**, 113–118 (2009)
18. C. Cannizzo, S. Amigoni-Gerbier, C. Larpent, Boronic acid-functionalized nanoparticles: synthesis by microemulsion polymerization and application as a re-usable optical nanosensor for carbohydrates. Polymer **46**, 1269–1276 (2005)
19. K.T. Kim, J.J.L.M. Cornelissen, R.J.M. Nolte, J.C.M.V. Hest, Polymeric monosaccharide receptors responsive at neutral pH. J. Am. Chem. Soc. **131**, 13908 (2009)
20. K.T. Kim, J.J.L.M. Cornelissen, R.J.M. Nolte, J.C.M.V. Hest, A polymersome nanoreactor with controllable permeability induced by stimuli-responsive block copolymers. Adv. Mater. **21**, 2787–2791 (2010)
21. L. Zhang, Y. Lin, J. Wang, W. Yao, W. Wu, X. Jiang, A facile strategy for constructing boron-rich polymer nanoparticles via a boronic acid-related reaction. Macromol. Rapid Commun. **32**, 534–539 (2011)
22. B. Wang, R. Ma, G. Liu, Y. Li, X. Liu, Y. An, L. Shi, Langmuir ACS J. Surf. Colloids **25**, 12522 (2009)
23. L. He, D.E. Fullenkamp, J.G. Rivera, P.B. Messersmith, pH responsive self-healing hydrogels formed by boronate-catechol complexation. Chem. Commun. **47**, 7497–7499 (2012)

24. S. Liu, J.V.M. Weaver, A. Maud Save, S.P. Armes, Synthesis of pH-responsive shell cross-linked micelles and their use as nanoreactors for the preparation of gold nanoparticles. Langmuir **18**, 1347–1353 (2002)
25. S. Farhat, I. Marek, From vinyl sulfides, sulfoxides, and sulfones to vinyl transition metal complexes, Cheminform **33** (2010)
26. K. Xiao, Y.P. Li, C. Wang, S. Ahmad, M. Vu, K. Kuma, Y.Q. Cheng, K.S. Lam, Biomaterials **67**, 183–193 (2015)
27. U. Hasegawa, M. Moriyama, H. Uyama, A.J.V.D. Vlies, Antioxidant micelles for bortezomib delivery. Colloid Polym. Sci. **293**, 1887–1892 (2015)
28. R. Zhang, S. Su, K. Hu, L. Shao, X. Deng, W. Sheng, Y. Wu, Nanoscale **7**, 19722–19731 (2015)
29. J.S. Wang, K. Matyjaszewski, Controlled/"living" radical polymerization. halogen atom transfer radical polymerization promoted by a Cu(I)/Cu(II) redox process. Macromolecules **23**, 7901 (1995)
30. K. Matyjaszewski, J. Xia, Atom transfer radical polymerization. J. Chem. Educ. **87**, 2921–2990 (2001)
31. Q. Ma, E.E. Remsen, T. Kowalewski, K.L. Wooley, Two-dimensional, shell-cross-linked nanoparticle arrays. J. Am. Chem. Soc. **123**, 4627 (2001)
32. Q. Ma, E.E. Remsen, T. Kowalewski, A.J. Schaefer, K.L. Wooley, Environmentally-responsive, entirely hydrophilic, shell cross-linked (SCK) nanoparticles. Nano Lett. **1**, 651–655 (2001)
33. X.S. Wang, R.A.J. And, S.P. Armes, Facile synthesis of acidic copolymers via atom transfer radical polymerization in aqueous media at ambient temperature. Chem. Commun. **18**, 1817–1818 (2000)
34. X.S. Wang, R.A. Jackson, S.P. Armes, Facile synthesis of acidic copolymers via atom transfer radical polymerization in aqueous media at ambient temperature. Macromolecules **33**, 255–257 (2000)
35. X.S.W. And, S.P. Armes, Facile atom transfer radical polymerization of methoxy-capped oligo(ethylene glycol) methacrylate in aqueous media at ambient temperature. Macromolecules **33**, 6640–6647 (2000)
36. P. Huang, W. Wang, J. Zhou, F. Zhao, Y. Zhang, J. Liu, J. Liu, A. Dong, D. Kong, J. Zhang, ACS Appl. Mater. Interfaces. **7**, 6340 (2015)
37. G. Springsteen, B. Wang, ChemInform abstract: Alizarin Red S. as a general optical reporter for studying the binding of boronic acids with carbohydrates. Chem. Commun. **17**, 1608–1609 (2001)
38. J.Z. Du, X.J. Du, C.Q. Mao, J. Wang, Tailor-made dual pH-sensitive polymer-doxorubicin nanoparticles for efficient anticancer drug delivery. J. Am. Chem. Soc. **133**, 17560 (2011)

Trimethylisophthtalodinitrile Oxide as a Room-Temperature Curing Agent for Curing HTPB

Xiaochuan Wang, Yuanjie Shu, Yonglin Lei, Jichuan Huo, Xianming Lu, Qian Zhang, Yingying Lu and Wei Wang

Abstract In order to cure hydroxyl-terminated polybutadiene (HTPB) at room temperature, Trimethylisophthtalodinitrile oxide (TINO) was synthesized in four steps from mesitylene with total yield of 42.0%. Its structure was characterized by FT-IR ^{1}H NMR and ^{13}C NMR. Make curing agent TINO/HTPB equals 12, 14, 16, 18 and 20% to prepare the dumbbell-shaped cross-linked polymers. The FT-IR, tensile strength, DSC, TG and contact angle of the cross-linked polymers were tested and analyzed. With the increasing of TINO, tensile strength of cross-linked polymers increased from 0.077 to 0.300 MPa and the elongation at break decreased from about 437 to 104%. The Tg of the crosslink polymers are −75.01, −73.92, −73.42, −70.09 and −71.78 °C, respectively. The contact angles are 116.91°, 100.66°, 93.68°, 86.25° and 87.35°. The results showed that TINO can cure HTPB under room temperature and the cross-linked polymers are expected to be used in the composite solid propellant.

Keywords Trimethylisophthtalodinitrile oxide · HTPB · Room-temperature Curing agent

Introduction

Hydroxyl-terminated polybutadiene (HTPB) is a low molecular-weight liquid rubber. It has been used in composite solid propellant for a very long time for binder the whole propellant grains [1–3]. As a telechelic polymer, HTPB has a lot

X. Wang · Y. Shu (✉) · X. Lu · Q. Zhang · Y. Lu · W. Wang
Xi'an Modern Chemistry Research Institute, Xi'an 710065, China
e-mail: 379806759@qq.com

X. Wang
e-mail: 365200401@163.com

Y. Lei · J. Huo
Southwest University of Science and Technology, Mianyang 621000, China
e-mail: 645830274@qq.com

Y. Han (ed.), *Advanced Functional Materials*,
https://doi.org/10.1007/978-981-13-0110-0_95

of advantages, such as low glass transition temperature, good transparency, low viscosity, and so on. HTPB are usually cross-linked by the curing agent of isocyanate to form HTPB-based polyurethanes [4–7]. The polyurethane three-dimensional network system shows excellent mechanical properties. However, the HTPB-isocyanate systems are easily suffered from the humidity sensitivity of curing process and the inherent incompatibility of isocyanate with new oxidizers [8–11]. Furthermore, HTPB-isocyanate system needs a higher temperature (60–80 °C) to cure.

For there are a lot of C=C bonds in the HTPB molecule, the 1,3-dipolar cycloaddition reaction between nitrile oxide and C=C groups can be used to cure HTPB. Nitrile oxides are organic compounds which contain –CNO bound directly to carbon atom [12–14]. There are two –CNO groups in each trimethylisophthtalodinitrile oxide (TINO) molecule. TINO are stable definitely under room temperature. The –CNO group reacts with C=C bond to form the isoxazoline. Therefore, TINO can be used as a room-temperature curing agent for curing unsaturated polymers [15, 16].

In this study, the cross-linked polymers based on trimethylisophthtalodinitrile oxide (TINO) and HTPB were investigated. The cross-linked polymers were tested and the results were analyzed.

Materials and Methods

Materials. Hydroxyl-terminated polybutadiene (average $M_n \sim$ 3000 by VPO) were obtained from the Liming Research Institute of Chemical Industry (Luoyang, Henan province, China). Hydrobromic acid (33 wt% in acetic acid), 2-nitropropaneand mesitylene were obtained from the Energy Chemical (Shanghai, China). Analytical pure reagents including acetic acid, dichloromethane, isopropane, and alcohol were obtained from the Xi'an Chemical Reagents Factory (Xi'an, Shanxi province, China). Analytical pure reagents including potassium hydroxide, sodium hydroxide, hydroxylamine hydrochloride, sodium hypochlorite, and paraformaldehyde were obtained from the Chengdu Kelong Chemical Reagent Company (Chengdu, Sichuan province, China).

Synthesis of Trimethylisophthtalodinitrile Oxide [17]. The TINO was synthesized from mesitylene in four steps with total yield of 42.0%. Its structure was characterized by FT-IR, ^{1}H NMR, and ^{13}C NMR. IR(KBr), wave number(cm^{-1}): 2985(–CH_3), 2296(C≡N–, st), 1324 cm^{-1} (–N=O, st), 892(Ar–H). ^{1}H NMR (500 MHz, $CDCl_3$), δ 7.06 (s, 1H), 2.60 (s, 3H), 2.46 (s, 6H); ^{13}C NMR (125 MHz, $CDCl_3$), δ 145.0, 144.3, 129.6, 113.5, 21.2, 20.3. The synthetic scheme was shown as the following formula (Fig. 1).

Preparation of the Cross-linked Polymer. Hydroxyl-terminated polybutadiene (HTPB) was used to investigate the potential of TINO as a crosslinking agent. For the preparation of the cross-linked polymer, TINO, HTPB and dichloromethane

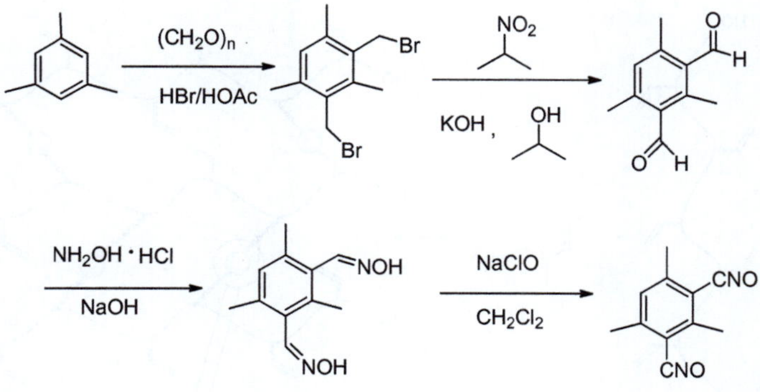

Fig. 1 Synthesis of TINO

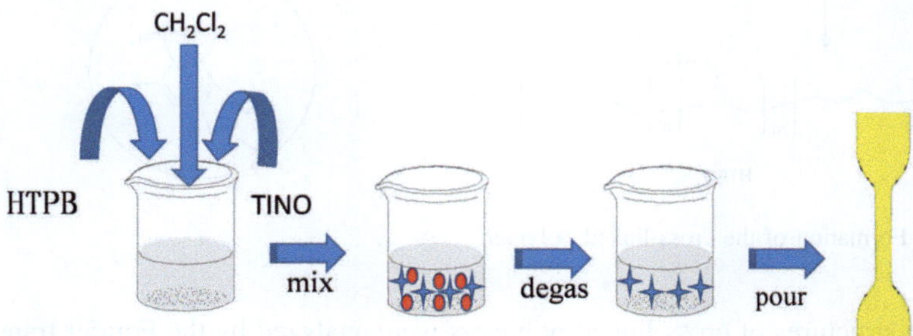

Fig. 2 Preparation flow chart of cross-linked polymer

were mixed. Make curing agent TINO/HTPB equals 12, 14, 16, 18, and 20% to prepare the dumbbell-shaped cross-linked polymers. Take 12% as example, 1.44 g TINO was mixed with 12 g HTPB and degassed for 30 min. The mixture was poured into a dumbbell-shaped Teflon mold and cured at 25 °C for 3 days. All the cross-linked polymers were sent to measure the mechanical properties using a testing machine at a speed of 100 mm/min at 25 °C. The preparation flow chart of cross-linked polymer is shown in Fig. 2.

TINO reacts with unsaturated bonds of HTPB to form isoxazoline which makes the polymer cross-linked. Schematic diagram for the formation of the cross-linked polymer is shown in Fig. 3.

Characterization

According to GB/T 528-2009, the mixture was poured into a dumbbell-shaped Teflon mold with a thickness of 4.0 mm for tensile test. The test was carried out on a C45.504 computer-controlled electronic testing machine at a rate of 100 mm/min.

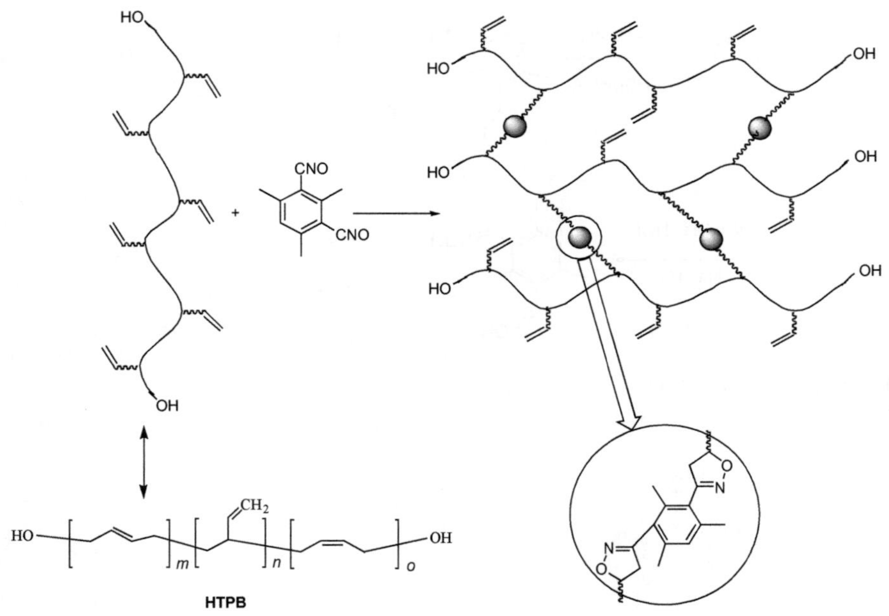

Fig. 3 Formation of the cross-linked polymer

The structures of cross-linked polymers were analyzed by the Fourier transform infrared Spectroscopy (FT-IR). The FT-IR were recorded on a Nicolet 5700 FT-IR spectrophotometer by the KBr-pellet method and scanned from 4000 to 400 cm^{-1}, with a resolution superior to 0.5 cm^{-1}.

Static contact angle measurements were carried out on a DSA30 contact angle measuring instrument (Kruss, Germany) by deposing a water drop on the surface of cross-linked polymer.

The thermal properties of cross-linked polymers were characterized by differential scanning calorimetry (DSC, Q2000, TA company, America)and thermogravimetric analysis (TG, SDT Q160, TA company, America) in a purified nitrogen atmosphere with a flow of 100 mL/min.

Results and Discussion

Mechanical Properties of Cross-linked Polymers. The tensile strength and elongation at break of the cured polymers (different content of curing agent TINO) are shown in Fig. 4. With the increasing of curing agent TINO, the tensile strength increased and the elongation at break decreased. The tensile strength of the cured polymers increased by about 290% (from about 0.077 to 0.300 MPa), while the elongation at break decreased from about 437 to 104%. This is because there are

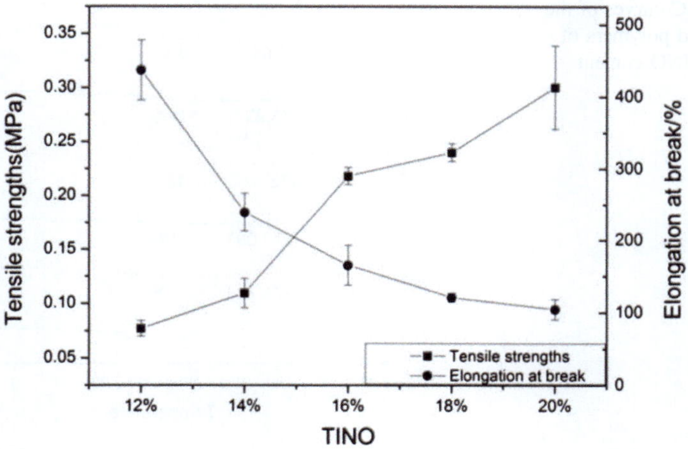

Fig. 4 Tensile strength and elongation at break of different content of TINO

two –CNO groups in each TINO molecule and there are about fifty-five C=C bounds in each HTPB. The –CNO group reacts with C=C bounds to form the isoxazoline. With the increasing of –CNO group, more isoxazoline in the cross-linked polymer are formed. As a result, the tensile strength increased and the elongation at break decreased.

FT-IR Spectra of Cross-linked Polyners. The FT-IR spectra of the cross-linked polymers of different TINO content are shown in Fig. 5. As it shows, an absorption peak of C = C (1654 cm^{-1}) exists, which means the cross-linked polymers still contains some C=C bonds. Compare the spectra of 12, 14, 16, and 18%, the strong absorption peak of –CNO (2300 cm^{-1}) [18] disappeared, which means 1,3-dipolar cycloaddition reaction of –CNO groups and C=C bounds completed. As can be

Fig. 5 FT-IR spectra of different content of TINO

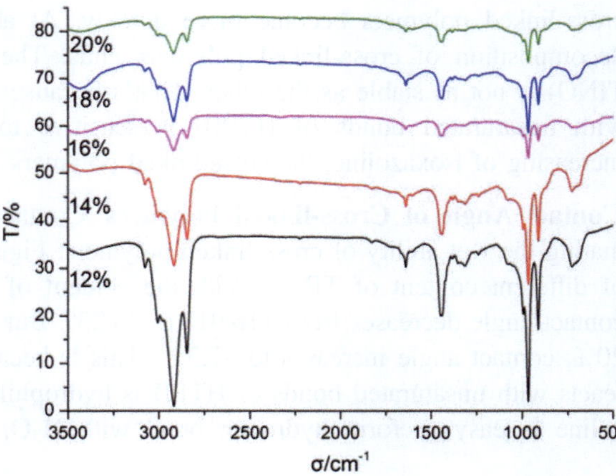

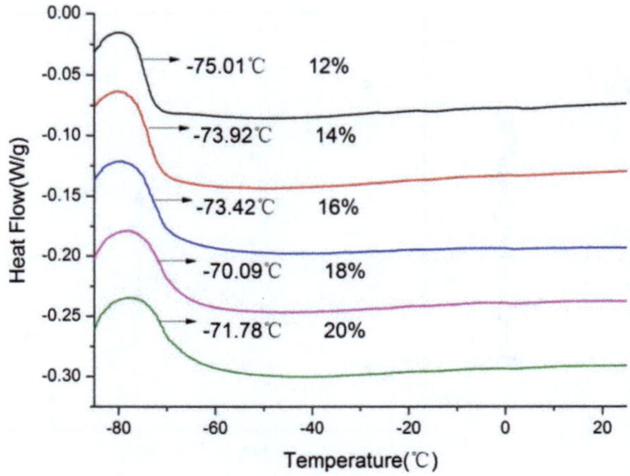

Fig. 6 DSC curves of the cross-linked polymers of different TINO content

seen in the spectrum of 20%, there still have the absorption peak of –CNO (2300 cm^{-1}) indicating the incompletion of 1,3-dipolar cycloaddition reaction.

Thermal Properties of Cross-linked Polymers. Figure 6 exhibits T_g results of the cross-linked polymers of different TINO content. DSC curves show different content of TINO, T_g at −75.01, −73.92, −73.42, −70.09, −71.78 °C, respectively. T_g is determined by the motion of polymer molecule segments and it is the lowest temperature for rubber using. When the cross-linked polymer under T_g, though there is not thermal effects, the specific heat capacity has a big change. There is a sidestep in the DSC curve. With the increasing of TINO, T_g increase from −75.01 to −70.09 °C. It indicates the cross-linked net became better. However, when the TINO is 20%, the T_g is lower than 18%. That's because part –CNO groups of TINO were not reacted with C=C bounds of HTPB.

Figure 7 shows the TG curves of the cross-linked polymers of different content of TINO. Through thermal gravity analysis, the thermal decomposition temperature of the cross-linked polymers can be seen in Fig. 7. The cross-linked polymers begin to decompose at about 250 °C. At about 350 °C, the decomposition of the cross-linked polymers become more quickly. At about 500 °C, the process of decomposition of cross-linked polymers ends. The cross-linked polymer (20% TINO) is not as stable as the other. That's because the isoxazoline (TINO reacts with unsaturated bonds of HTPB) is sensitive to the temperature. With the increasing of isoxazoline, the cross-linked polymers become unstable.

Contact Angle of Cross-linked Polymers. Contact angle is important for estimating the wet ability of cross-linked polymers. Figure 8 shows the contact angle of different content of TINO. With the content of curing agent increasing, the contact angle decreases from 116.91° to 86.25°. But when the content of TINO is 20%, contact angle increases to 87.35°. This is because the isoxazoline of TINO reacts with unsaturated bonds of HTPB is hydrophilic. The O atom in the isoxazoline is easy to form hydrogen bond with H_2O. With the content of TINO

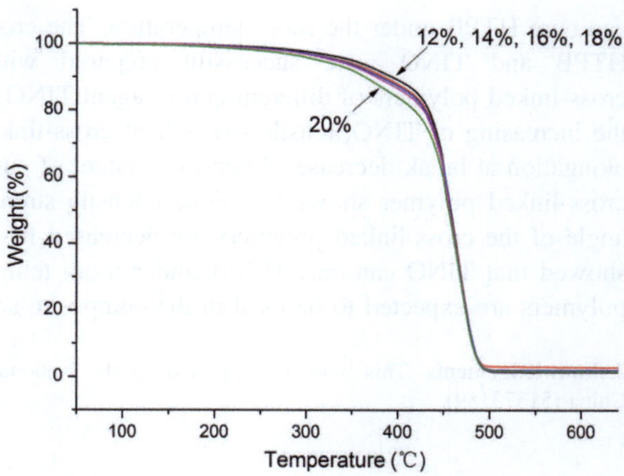

Fig. 7 TG curves of different content of TINO

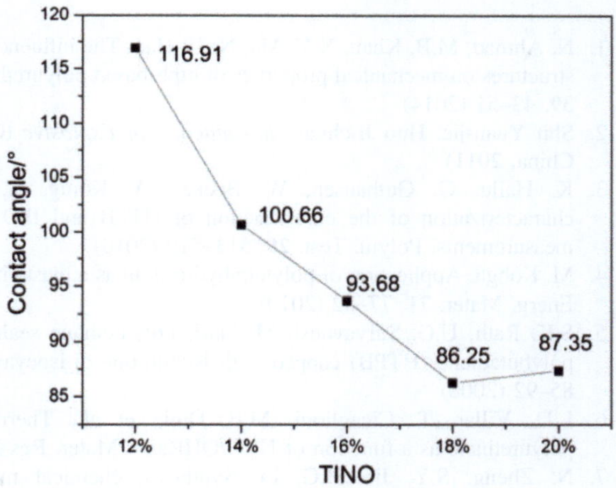

Fig. 8 Contact angle of different content of TINO

increasing, the isoxazoline of cross-linked polymer increase. There will form more hydrogen bonds; the wet ability of cross-linked polymer became better. Therefore the contact angles of cross-linked polymers decrease step by step. But when the content of TINO is 20%, there are still some –CNO groups weren't taken part in the reaction, therefore the contact angle increases.

Conclusions

Trimethylisophthtalodinitrile oxide (TINO) is an effective room-temperature curing agent for curing HTPB. It was synthesized from mesitylene with total yield of 42.0% and its structure was characterized by FT-IR ^{1}H NMR and ^{13}C NMR. TINO can cure HTPB under the room temperature. The cross-linked polymers based on HTPB and TINO were successful prepared with rubbery properties. The cross-linked polymers of different curing agent TINO content were prepared. With the increasing of TINO, tensile strength of cross-linked polymer increase and the elongation at break decrease. When the content of curing agent TINO is 20%, the cross-linked polymer shows the highest tensile strength. The T_g and the contact angle of the cross-linked polymers are decreased first then increased. The results showed that TINO can cure HTPB under room temperature and the cross-linked polymers are expected to be used in the composite solid propellant.

Acknowledgements This work is supported by the National Natural Science Foundation of China (51373159).

References

1. N. Ahmad, M.B. Khan, X.Y. Ma, N. Ul-Haq, The influence of cross-linking/chain extension structures on mechanical properties of htpb-based polyurethane elastomers. Arab. J. Sci. Eng. **39**, 43–51 (2014)
2. Shu Yuan-jie, Huo Ji-chuan, *Introduction of Explosive* (Chemical Industry Press, Beijing, China, 2011)
3. K. Hailu, G. Guthausen, W. Becker, A. Konig, A. Bendfeld, E. Geissler, In-situ characterization of the cure reaction of HTPB and IPDI by simultaneous NMR and IR measurements. Polym. Test. **29**, 513–519 (2010)
4. M. Kohga, Application of polytetrahydrofuran as a plasticizer of HTPB binder. Sci. Technol. Energ. Mater. **71**, 77–82 (2010)
5. S.K. Rath, U.G. Suryavansi, M. Patri, Polyurethane sealant based on hydroxy terminated polybutadiene (HTPB) capped with isophorone di-isocyanate (IPDI). J. Polym. Mater. **25**, 85–92 (2008)
6. L.D. Villar, T. Cicaglioni, M.F. Diniz et al., Thermal aging of HTPB/IPDI-based polyurethane as a function of NCO/OHRatio. Mater. Res-Ibero-Am. J. **14**, 372–375 (2011)
7. N. Zheng, S.Y. Jie, B.G. Li, Synthesis, chemical modifications and applications of hydroxyl-terminated polybutadiene. Prog. Chem. **28**, 665–672 (2016)
8. D.H. Lee, K.T. Kim, Y. Jang et al., 1,2,3-Triazole cross-linked polymers as binders for solid rocket propellants. J. Appl. Polym. Sci. **131**, 40594 (2014)
9. T. Keicher, W. Kuglstatter, S. Eisele et al., Isocyanate-free curing of glycidyl azide polymer (GAP) with bis-propargyl-succinate (II). Propell. Explos. Pyrot. **34**, 210–217 (2009)
10. K. Menke, T. Heintz, W. Schweikert et al., Formulation and properties of ADN/GAP propellants. Propell. Explos. Pyrot. **34**, 218–230 (2009)
11. H. Li, F.Q. Zhao, Q.Q. Yu et al., A comparison of triazole cross-linked polymers based on poly-AMMO and GAP: mechanical properties and curing kinetics. J. Appl. Polym. Sci. **133**, 43341 (2016)
12. L. Yudina, C. Eldsäter, P. Goede, et al., Synthesis, analysis and evaluation of 2,4,6-trimethylisophthtalodinitrile-N-oxide as a potential new curing agent for AND-based formulations//International Annual Conference of ICT (2005), 36th (Energetic Materials): 123/1-123/9
13. B.S. Huffman, R.A. Schultz, P.J. Schlom, Novel reagents for heat-activated polymer crosslinking. Polym. Bull. **47**, 159–166 (2001)
14. S.A. Kotel'nikov, V.S. Sukhinin, A.S. Eemilov, Kinetics of formation and curing of oligoether urethane allyl ester. Russ. J. Appl. Chem. **75**, 477–479 (2002)

15. D.V. Tsyganov, A.P. Yakubov, L.I. Belenkii, et al., Synthesis and properties of stable aromatic bis(nitrile oxides). Russ. Chem. B+ **40**, 1238–1243 (1991)
16. H. Feuer, *Nitrile oxides, nitrones, and nitronates in organic synthesis* (Wiley, Hoboken, 2008)
17. E. Hess, J. W. Mcfarland, U.S. Patent 3,258,397 (1966)
18. X.C. Wang, Y.J. Shu, X.M. Lu et al., Application of 1,3-dipolar cycloaddition reaction in curing energetic binders. Propell. Polym. Mater. **15**, 32–37 (2017)

Effect of Coagulants on Phase Morphology and Property of Cellulose/Silk Fibroin Blend Fiber

Yongbo Yao, Zhiyong Yan, Zhe Li, Honglei Yi, Yumei Zhang and Huaping Wang

Abstract In order to study the effect of coagulants on phase morphology and property of cellulose/silk fibroin blend fiber, coagulants of water, ethanol, ethanol/1-butyl-3-methylimidazolium chloride ([BMIM]Cl) were chosen for coagulation bath. Ethanol is effective coagulant for cellulose and silk fibroin. The uniform dispersion of silk fibroin along the radial distribution can be found when [BMIM]Cl was added in ethanol. The tensile strength of the blend fiber can be improved by adjusting coagulation bath composition.

Keywords Cellulose · Silk fibroin · Coagulant · Phase morphology Blend fiber

Introduction

Silk, one kind of natural protein fibers, is a high-grade garment material. However, the application of such fiber is limited due to insufficient supply of silk worms. In order to make full use of silk resource, waste silk can be used to prepare regenerated silk fiber. But the molecular chain of silk fibroin breaks during dissolution process, and the mechanical property of such fiber is weak. Comprehensive utilization of silk can be achieved by preparation of blend fiber. Cellulose is the most abundant natural polymer in the world, which can be used to produce regenerated cellulose/silk fibroin fibers by wet spinning [1]. Marsano et al. fabricated cellulose/silk fibroin blend fibers using N-methylmorpholine N-oxide hydrate ($NMMO/H_2O$) and N, N-dimethylacetamide/LiCl as co-solvent [2, 3], it is found that the miscibility of

Y. Yao (✉) · Z. Li · H. Yi
Jiaxing University, Jiaxing, Zhejiang 314001, China
e-mail: yaoyongbo@foxmail.com

Z. Yan · Y. Zhang · H. Wang
State Key Laboratory for Modification of Chemical Fibers and Polymer Materials, Donghua University, Shanghai 201620, China
e-mail: yzyong77@163.com

Y. Han (ed.), *Advanced Functional Materials*,
https://doi.org/10.1007/978-981-13-0110-0_96

cellulose and silk fibroin can be improved by co-dissolving. In our previous research, ionic liquid 1-butyl-3-methylimidazolium chloride ([BMIM]Cl) was chosen as co-solvent, the "fibril-like" phase morphology of silk fibroin in the blend fiber was detected, and the tensile strength of the blend fiber meets the demand of textile [4].

In the process of wet spinning, coagulation conditions are related to the structure and property of the fibers [5, 6]. So far, the effect of coagulation conditions on the phase morphology of cellulose/silk fibroin blend fibers has been rarely studied. It has been proved that alcohols (methanol, ethanol, et al.) can be used to coagulate cellulose and silk fibroin [1, 7]. Water, the common coagulant for cellulose, cannot deduce the random coil conformation state of silk fibroin to crystalize structure. But it is useful to study the state of silk fibroin phase during coagulation process when water is coagulant. Otherwise, the phase morphology of cellulose and silk fibroin can be adjusted by changing coagulation methods. According to Zhou's study, the mechanical property of cellulose/silk fibroin blend film can be enhanced by vapored methanol and cold pressing during vacuum dry [8].

In present work, two-stage coagulation baths were adopted in blend fiber preparation. Water, ethanol, ethanol/[BMIM]Cl were used as coagulants in the first stage, and ethanol was chosen as coagulant in the later stage. The interaction of cellulose and silk fibroin, phase morphology as well as mechanical property of the blend fibers were measured. The mechanism of coagulation conditions on the phase morphology and property of the blend fiber was discussed.

Experiment

Materials. Cellulose (degree of polymerization (DP) = 500) was provided from Shandong Silver Eagle Chemical Fiber Co. Ltd, China. Silk fibroin was degummed from silk cocoon in boiling 0.5% (w/w) Na_2CO_3 solution with a bath ratio of 1:50 and then washed with deionized water to remove silk sericin [9].

[BMIM]Cl was synthesized and purified in our laboratory as previously described [10].

Spinning Solution Preparation. Cellulose and silk fibroin were dried under vacuum at 80 °C for 5 h. The 6 wt% concentration of cellulose/silk fibroin/[BMIM] Cl solution were prepared by kneading at 90 °C, and the weight ratio of cellulose/ silk fibroin is 8/2. After that, the blend solution was kept under vacuum at 90 °C to remove air bubble.

Spinning Trials. The cellulose/silk fibroin blend fiber was spun through a laboratory-scale dry-jet wet spinning equipment. The blend solution was extruded through the spinneret orifices of 0.2 mm in diameter at 85 °C. The air gap distance is 50 mm and the air temperature is 25 °C. The draw ratios of the first and second coagulation baths are 3 and 1.1 separately, and the coagulation temperature of first and second coagulation baths are 20 and 25 °C respectively. The coagulators in the

Table 1 Coagulants in the first coagulation bath during spinning process

Sample	CS-W	CS-E	CS-EI
Coagulants	Water	Ethanol	Ethanol/[BMIM]Cl (9/1w/w)

first coagulation bath is listed in Table 1, and the coagulator in the second coagulation bath is ethanol. The as-spun fiber was further washed in ethanol, then dried and taken up.

Measurements

The protein content in the blend fibers was obtained by testing nitrogen (N) element content with element analyzer (Elementar Vario EL III, German). The protein content was calculated according to the Eq. 1 [11]:

$$P_{SF}\% = 5.97N\% \tag{1}$$

Where P_{SF} is protein content, N is the nitrogen content in blend fibers.

Fourier transform infrared spectra (FTIR) was measured by a Nicolet spectrometer (Nexus 670, Thermo Fisher, USA) at wavenumber of 700–4000 cm^{-1} with a resolution of 0.09 cm^{-1}. A bundle of blend fibers were pinched under total internal reflection accessory.

The cross-section morphology of the fibers was measured on environment scanning electron microscope (SEM) (Quanta-250, FEI, USA). The fibers were cryogenically broken under liquid nitrogen and sputter-coated with a thin layer of gold.

The intrinsic fluorescence of silk fibroin in the blend fibers was captured by a laser scanning confocal microscope (LSCM) (TCS SP5, Leica, German), the excitation wavelength is 488 nm and the emission wavelength range is 495–550 nm.

The mechanical property of the fibers was tested on a fiber mechanical strength tester (XQ-2, China). The stretching rate is 20 mm/min with a 20 mm gauge length. The diameter of the fiber was measured using microscope (XSZ–360AP, China). The tensile strength, initial modulus and elongation were calculated as the average of at least 20 measurements from stress-strain.

The thermal property of the fibers was measured by a thermogravimetric analysis (Netzsch TG 209F1). The measurements were carried out under nitrogen atmosphere with a temperature range from room temperature to 600 °C at a heating rate of 10 °C/min.

Results and Discussion

Protein Content of Cellulose/Silk Fibroin Blend Fibers. The silk fibroin content of the blend fiber was calculated by nitrogen element content analysis. When water acts as coagulant, the protein content of the blend fiber is only 3.9 wt%, and the weight loss of protein is 80.5%. The viscosity of silk fibroin/[BMIM]Cl is very low [12, 13], it moves to the outermost of the fiber and diffused into the coagulation bath. However, when ethanol is used as coagulant, the weight loss of silk fibroin plunged to 32.5 wt%. It means that silk fibroin can be coagulated by ethanol. When choosing ethanol/[BMIM]Cl as coagulant, the weight loss further declined to 21.0 wt%. This is because the diffusion rate of silk fibroin/[BMIM]Cl is slower by adding [BMIM]Cl, and the movement of silk fibroin along the fiber radial is reduced. So, ethanol, not water, can be used to fabricate cellulose/silk fibroin blend fiber (Table 2).

Interaction of Cellulose and Silk Fibroin in the Blend Fibers. FTIR is opted for studying the interaction between cellulose and silk fibroin. The FTIR of cellulose/ silk fibroin blend fibers is shown in Fig. 1. The blend fiber coagulated by water shows characteristic peaks at 1658, 1626 cm^{-1}, which are related to random coil or silk I structure [14], and it does not transfer to β-Sheet or silk II structure in ethanol

Table 2 Protein content of cellulose/silk fibroin blend fibers

Sample	Protein content in the solution (wt%)	Protein content in the blend fiber (wt%)	Weight loss of protein (wt%)
CS-W	20	3.9	80.5
CS-E	20	13.5	32.5
CS-EI	20	15.8	21.0

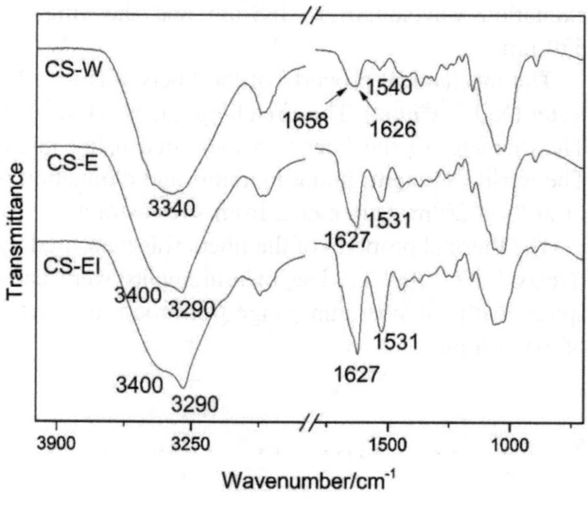

Fig. 1 FTIR of cellulose/silk fibroin blend fibers

bath. When ethanol was used as coagulant, the characteristic peaks of β-Sheet or silk II structure at 1637, 1531 cm^{-1} bands can be found [12]. Water coagulated cellulose/silk fibroin blend fiber has a broad band at 3340 cm^{-1}, which is caused by the intra- and intermolecular hydrogen bonding of cellulose molecules. The peaks become narrow and shift to 3290 cm^{-1} in ethanol coagulated fiber, owing to the intermolecular hydrogen bonds between OH group of cellulose and NH group of silk fibroin [15].

Phase Morphology of Silk Fibroin in the Blend Fiber. The morphology of cellulose/silk fibroin blend fibers was characterized by SEM and LSCM microscopy. As shown in Fig. 2, the blend fiber coagulated with water has dense cross-section structure, because little silk fibroin exists in the blend fiber. The outer layer of the blend fiber coagulated with ethanol is rough, which is caused by the enrichment of silk fibroin. The cross-section of the blend fiber coagulated with ethanol/[BMIM]Cl is more uniform, the dispersion of the silk fibroin along the radial direction becomes uniform owing to relative slower diffusion process.

In our earlier study, the phase morphology of silk fibroin in the blend fiber was detected by LSCM. The silk fibroin exists as "fibril-like" in the blend fiber, and the radial dimension is about 0.5–1.0 μm [4]. As shown in Fig. 3, there is few scattered green region (silk fibroin) in the LSCM image of the blend fiber coagulated with water. In contrast, silk fibroin disperses as "fibril-like" in the blend fiber with ethanol as coagulation bath. When ethanol/[BMIM]Cl is coagulant, the distribution

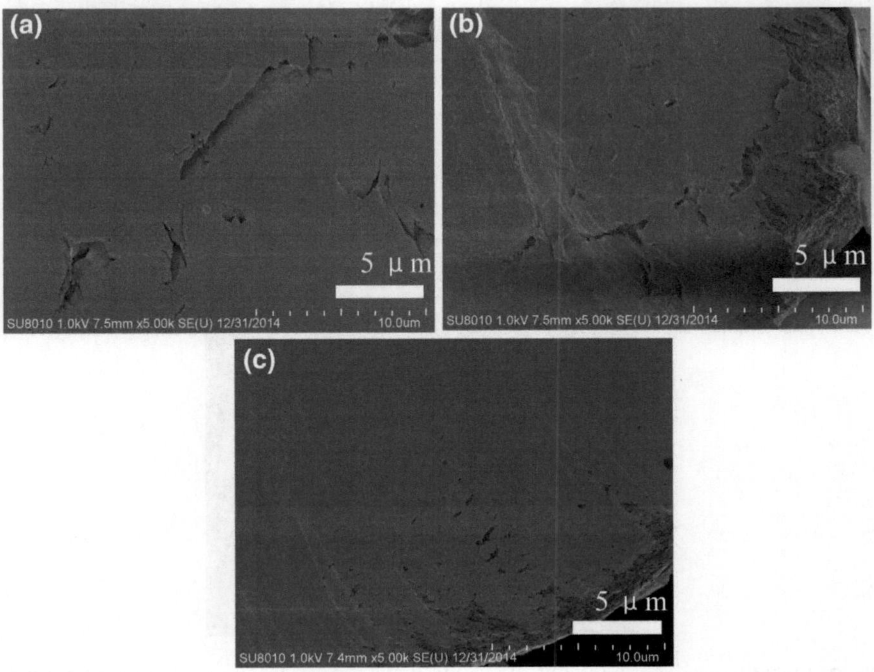

Fig. 2 SEM of the cross-section of cellulose/silk fibroin blend fibers: **a** CS-W, **b** CS-E, **c** CS-EI

of silk fibroin along the fiber axes is more continuous, which can also be speculated from the information of silk fibroin along the radial direction by SEM images.

The effect of coagulation condition on the phase morphology of silk fibroin in the blend fiber is shown in Scheme 1. If coagulant is water, silk fibroin cannot be solidified, almost all of the silk fibroin comes into the bath. Then when water is replaced by ethanol as coagulant, the low viscosity of silk fibroin phase migrates to the high extensional stress region which is located in the surface of solution flow [16]. Furthermore, the diffusion rate of [BMIM]Cl decreases by adding [BMIM]Cl in the coagulation bath, thus the uniform dispersion of silk fibroin in the blend fiber is formed.

Mechanical Property of Cellulose/Silk Fibroin Blend Fibers. The mechanical properties of cellulose/silk fibroin blend fibers are shown in Table 3. The diameters of the blend fiber vary from 60.6 to 75.1 μm. The tensile strength of the blend fiber in water coagulation bath is poor, which is caused by random coil or silk I structure of silk fibroin. The mechanical property of cellulose/silk fibroin blend fiber with

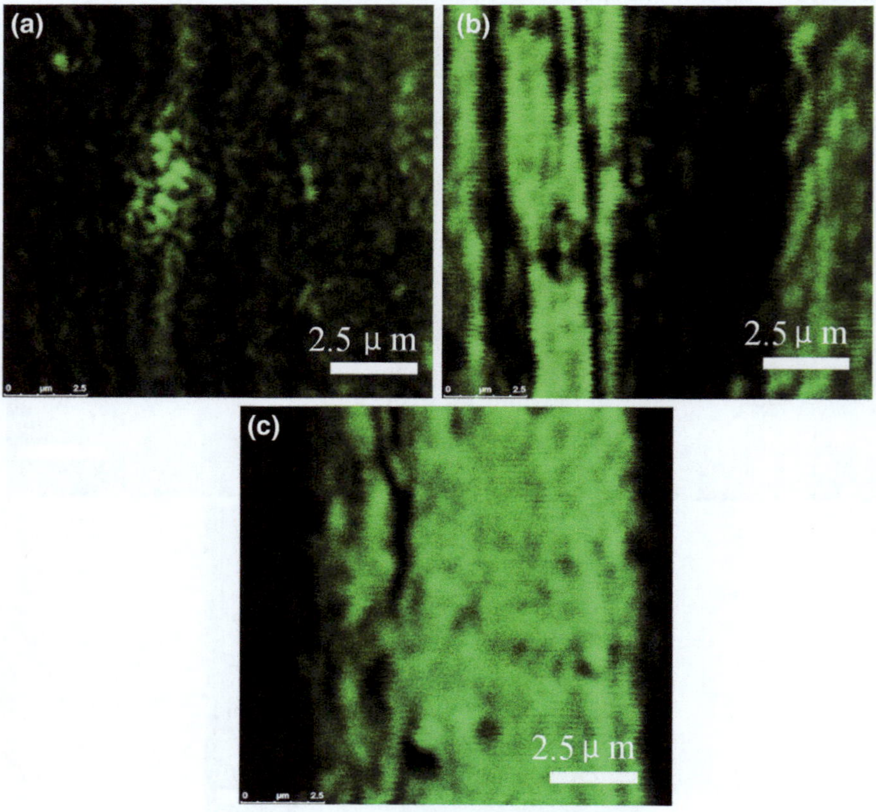

Fig. 3 LSCM images of cellulose/silk fibroin blend fibers along the fiber axes: **a** CS-W, **b** CS-E, **c** CS-EI

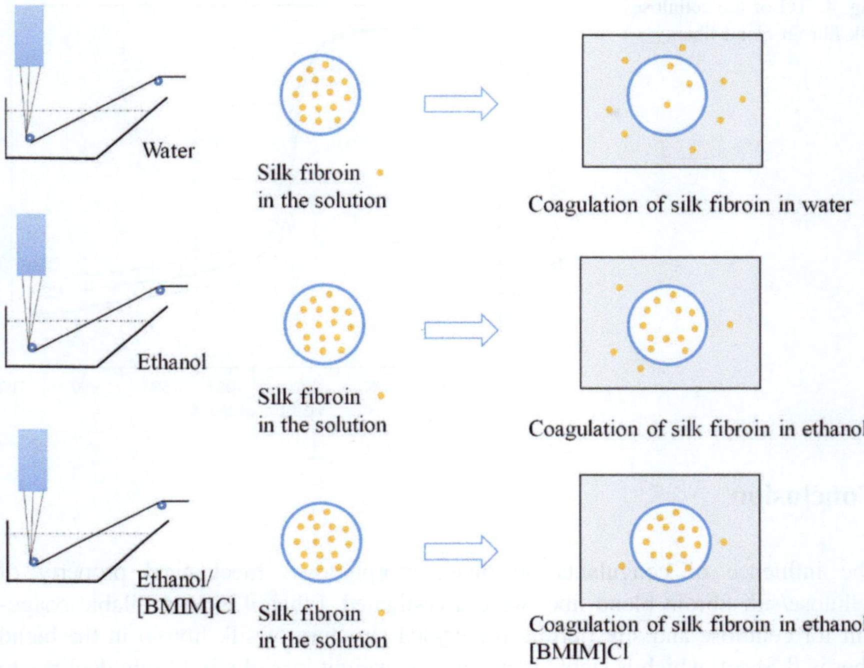

Scheme 1 Schematic of the coagulation process of silk fibroin

Table 3 Mechanical properties of cellulose and cellulose/silk fibroin blend fibers

Sample	Diameter (μm)	E (GPa)	σ_b (MPa)	Elongation (%)
CS-W	66.1	0.77	73.9	13.8
CS-E	60.6	3.89	108.4	15.8
CS-EI	75.1	3.17	120.4	18.5

ethanol/[BMIM]Cl as coagulant is better than that of ethanol alone, owing to the uniform distribution of silk fibroin in the blend fiber.

Thermal Stability of Cellulose/Silk Fibroin Blend Fibers. The thermal property of cellulose/silk fibroin blend fiber was studied through analysis of thermogravimetric (TG). The thermal degradation of cellulose/silk fibroin fibers is one-step degradation process. The onset temperatures of decomposition (T_{onset}) of the three samples are almost same, and the residual weight increased with the increasing of silk fibroin in the blend fiber [8, 17] (Fig. 4).

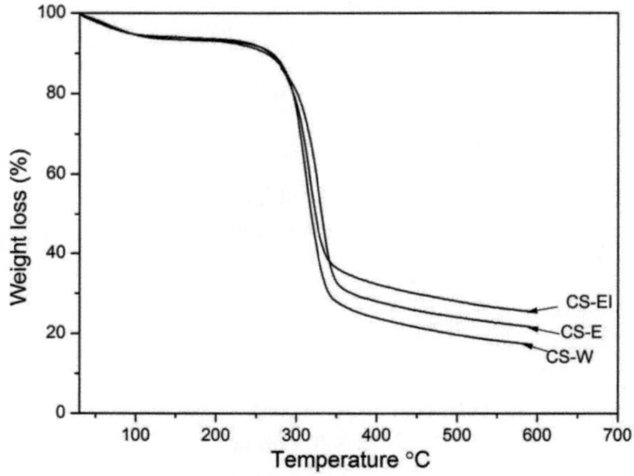

Fig. 4 TG of the cellulose/silk fibroin blend fibers

Conclusion

The influence of coagulants on phase morphology, mechanical property of cellulose/silk fibroin blend fiber were investigated. Ethanol is an available coagulant for cellulose and silk fibroin, the crystal structure of silk fibroin in the blend fiber is β-Sheet, which is stable in water. The weight loss of silk fibroin declines to 21.0 wt% when [BMIM]Cl is added in the ethanol coagulation bath, and the dispersion of silk fibroin along the fiber axes is also more continuous. The mechanical property of the blend fiber can be regulated by changing coagulation bath composition, and the tensile strength of the blend fiber can be up to 120.4 MPa.

Acknowledgements This work was financially supported by Zhejiang Provincial Natural Science Foundation of China (LQ17E030002) and National Natural Science Foundation of China (21704034, 51273041).

References

1. H.P. Fink, P. Weigel, H.J. Purz, J. Ganster, Structure formation of regenerated cellulose materials from NMMO-solutions. Prog. Polym. Sci. **26**, 1473–1524 (2001)
2. E. Marsano, M. Canetti, G. Conio, P. Corsini, G. Freddi, Fibers based on cellulose-silk fibroin blend. J. Appl. Polym. Sci. **104**, 2187–2196 (2007)
3. E. Marsano, P. Corsini, M. Canetti, G. Freddi, Regenerated cellulose-silk fibroin blends fibers. Int. J. Biol. Macromol. **43**, 106–114 (2008)
4. Y. Yao, E. Zhang, X. Xia, J. Yu, K. Wu, Y. Zhang, Morphology and properties of cellulose/silk fibroin blend fiber prepared with 1-butyl-3-methylimidazolium chloride as solvent. Cellulose **22**, 625–635 (2015)
5. J.W. Gooch, *Wet Spinning* (Springer, New York, 2011), p. 810
6. Y.X. Wang, C.G. Wang, M.J. Yu, Effects of different coagulation conditions on polyacrylonitrile fibers wet spun in a system of dimethylsulphoxide and water. J. Appl. Polym. Sci. **104**, 3723–3729 (2007)

7. M. Tsukada, Y. Gotoh, M. Nagura, N. Minoura, N. Kasai, G. Freddi, Structural changes of silk fibroin membranes induced by immersion in methanol aqueous solutions. J. Polym. Sci. Poly. Phys. **32**, 961–968 (1994)
8. L. Zhou, Q. Wang, J. Wen, X. Chen, Z. Shao, Preparation and characterization of transparent silk fibroin/cellulose blend films. Polymer **54**, 5035–5042 (2013)
9. H. Cao, X. Chen, L. Huang, Z. Shao, Electrospinning of reconstituted silk fiber from aqueous silk fibroin solution. Mat. Sci. Eng C-Mater. **29**, 2270–2274 (2009)
10. S. Csihony, C. Fischmeister, C. Bruneau, I.T. Horvath, P.H. Dixneuf, First ring-opening metathesis polymerization in an ionic liquid. Efficient recycling of a catalyst generated from a cationic ruthenium allenylidene complex. New J. Chem. **26**, 1667–1670 (2002)
11. G. Yang, L.N. Zhang, Y.G. Liu, Structure and microporous formation of cellulose/silk fibroin blend membranes I. Effect of coagulants. J. Membrane Sci. **177**, 153–161 (2000)
12. Y. Yao, K.S. Mukuze, Y. Zhang, H. Wang, Rheological behavior of cellulose/silk fibroin blend solutions with ionic liquid as solvent. Cellulose **21**, 675–684 (2014)
13. Y. Yao, X. Xia, K.S. Mukuze, Y. Zhang, H. Wang, Study on the temperature-induced sol-gel transition of cellulose/silk fibroin blends in 1-butyl-3-methylimidazolium chloride via rheological behavior. Cellulose **21**, 3737–3743 (2014)
14. L.F. Drummy, D.M. Phillips, M.O. Stone, B.L. Farmer, R.R. Naik, Thermally induced alpha-helix to beta-sheet transition in regenerated silk fibers and films. Biomacromol **6**, 3328–3333 (2005)
15. T. Kondo, C. Sawatari, R.S. Manley, D.G. Gray, Characterization of hydrogen-bonding in cellulose synthetic-polymer blend systems with regioselectively substituted methylcellulose. Macromolecules **27**, 210–215 (1994)
16. Z. Pan, Y. Chen, M. Zhu, C. Jiang, Z. Xu, W. Lu, J. Pionteck, The non-uniform phase structure in blend fiber. II. The migration phenomenon in melt spinning. Fibers. Polym. **11**, 625–631 (2010)
17. S. Shang, L. Zhu, J. Fan, Physical properties of silk fibroin/cellulose blend films regenerated from the hydrophilic ionic liquid. Carbohyd. Polym. **86**, 462–468 (2011)

Insulation Failure of Class 1E Cable for Nuclear Power Station Under Fire Heat Radiation

Jinmei Li, Jiaqing Zhang, Qiang Li, Yichen Yang and Minghao Fan

Abstract In order to discover the regularity of insulation failure of typical cable for nuclear power plant in fire, the experimental research method was used to study the insulation failure of Class 1E cable for nuclear power plant under different simulated fire conditions by using cable thermal radiation test furnace in this paper. The results show that the insulation failure time and the breakdown time of the cable decrease exponentially with the increase of the ambient heat radiation flux, and the insulation failure temperature and the insulation breakdown temperature of the cable are basically the same.

Keywords Insulation failure · Class 1E cable · Thermal radiation
Fire

Introduction

Fire is one of the most direct and major threats to safe operation of nuclear power plants [1]. As the flow of nuclear power plant equipment, "arteries" and "nerve", in the event of a fire in the cable due to environmental high temperature and strong thermal radiation lead to insulation failure possible, and then lose the power transmission, transmission of information, resulting in further expansion of the disaster. Previous studies on the insulation failure of cables under fire conditions have been studied extensively. Studies have shown that the insulation failure of the

J. Li · Q. Li (✉)
Chinese People's Armed Police Force Academy, Xichang Road 220, Langfang 065000, China
e-mail: sterops@foxmail.com

J. Zhang · M. Fan
State Grid Anhui Electric Power Research Institute, Jinzhai Road 73, Hefei 230026, China

Y. Yang
Cangzhou Fire Detachment, South Jianshe Street 20, Cangzhou 061000, China

Y. Han (ed.), *Advanced Functional Materials*,
https://doi.org/10.1007/978-981-13-0110-0_97

cable under fire conditions is related to the ambient heat flux and temperature, regardless of the number of cable cores [2, 3]. Multi-core cable insulation failure usually occurs in close to the sheath layer (or shield) within the location [4]. When the cable insulation fails, the temperature at the insulation failure point (cable insulation failure temperature) is related to the thermal characteristics of the cable material. The insulation failure temperature of the same material cable is in a certain temperature range [5, 6]. As a potential nuclear power in China, there are still some deficiencies in the study of the insulation failure of nuclear cable fire. Therefore, this paper selects the Class 1E Category K3 cable for nuclear power plants, and studies the insulation failure rules of the cables by simulating the thermal radiation environment provided by the SDR-1 cable heat radiation experimental furnace.

Experimental Setup

SDR-1 Type Plastic Cable Radiation Furnace. The simulation of the fire environment temperature and radiation environment is made by the self-developed SDR-1 cable heat radiation experimental furnace, as shown in Fig. 1. The maximum rated output power of the experimental furnace is 60 kW, and different fire scenarios can be simulated by setting the output power added by the furnace.

Design of Experimental Conditions. The experimental cables are made of Class 1E Category K3 300/500 V copper core conductor, halogen free low smoke flame retardant XLPE insulated sheath PE sheathed cable (WDZ-HK3-PYJP3YJP3).

Referring to the condition setting [6] of the CAROLFIRE project, the range of the heat radiation flux in the experiment is limited to 3–30 kW/m^2. In order to study the failure of cables under different scales of fire, the internal radiation flux of

Fig. 1 SDR-1 cable thermal radiation experimental furnace

SDR-1 type plastic cable radiant furnace was set to 3, 6, 9, 15, 21 and 30 kW/m^2. In the experiment, it is found that under the condition of 3 kW/m^2, the cable will not be ignited and the outer surface of the sheath is basically intact. Under the condition of 6 kW/m^2, the combustion of the cable is serious. In order to further study the effect of heat flux on the ignition of cables, three working conditions such as 4.2, 4.8 and 5.4 kW/m^2 were added.

Result and Analysis

Cable Temperature and Ignition Time. The variation of the surface temperature of the cable with time under different heat flux is shown in Fig. 2. The lateral temperature of the cable increases logically with time. The highest value of the surface temperature of the cable increases with the increase of the heat flux, and the heating rate increases with the increase of the heat radiation flux.

By taking the surface temperature before ignition as a measure of fire conditions, the relationship between fire temperature and heat radiation flux can be obtained

$$T_{fire} = 196.7\ \ln(\dot{Q}) + 89.7. \quad (1)$$

Where T_{fire} is the temperature of a fire condition, °C, $\dot{Q}$ is the heat radiation flux of the cable, kW/m^2.

Similarly, Fig. 3 shows the variation of the internal temperature of the cable with time under different heat radiation flux. It can be seen that the heating rate of the cable increases with the increase of the heat flux, but when the heat flux is 30 kW/m^2, the internal temperature rise rate of the cable is reduced. This can be explained by the fact

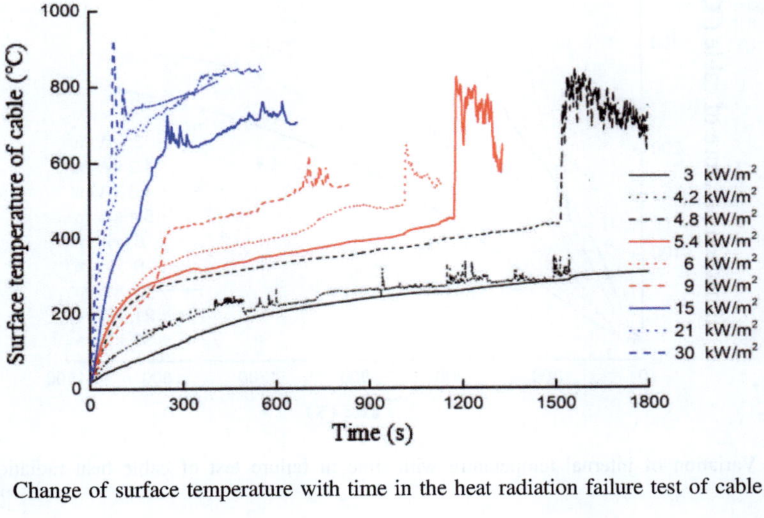

Fig. 2 Change of surface temperature with time in the heat radiation failure test of cable

that when the heat flux is small, the thermal insulation performance of the cable jacket layer is relatively stable, and the heat of conduction increases with the increase of the radiation flux. Therefore, the internal heating rate increases with the increase of the heat flux Increase, when the external heat flux is large, the cable jacket is quickly carbonized, the thermal conductivity will be reduced, resulting in lower internal heating rate.

It is found that when the ambient heat flux is less than or equal to 4.2 kW/m^2, the WDZ-HK3-PYJP3YJP3 type cable will not be ignited, and the cable will be ignited at a heat flux greater than or equal to 4.8 kW/m^2. The relationship between the time of ignition and the heat flux is shown in Fig. 4.

The relation between the fitting time and the heat flux of the cable can be obtained

$$t_{ig} = 2148.8e^{-0.131\dot{Q}}. \tag{2}$$

Where t_{ig} is the ignition time of cable, s; $\dot{Q}$ is the heat flux of cable, kW/m^2.

Failure Time and Temperature of Cable Insulation. Since the ignition time of the WDZ-HK3-PYJP3YJP3 cable is less than its failure time when the heat radiation flux is greater than or equal to 21 kW/m^2, in the discussion of heat flux effect on failure time, without considering the radiation heat flux is greater than or equal to 21 kW/m^2. As can be seen from Fig. 5, the failure time of the cable is shortened with the increase of the heat flux released.

Using the exponential function to fit the cable failure time and the relationship between heat radiation flux, can be

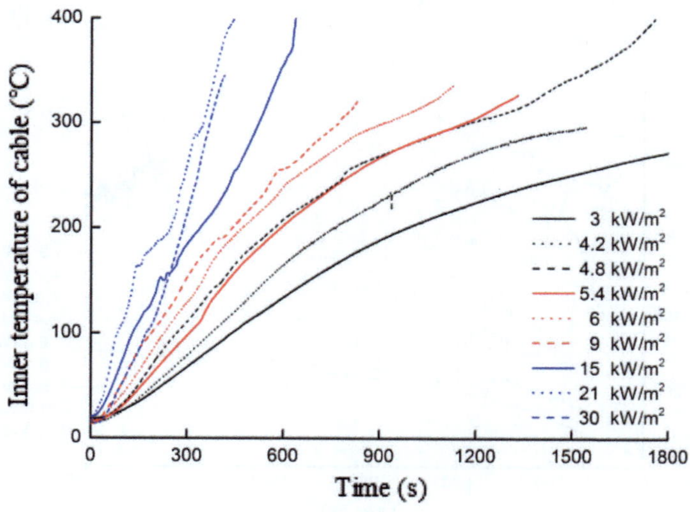

Fig. 3 Variation of internal temperature with time in failure test of cable heat radiation test furnace

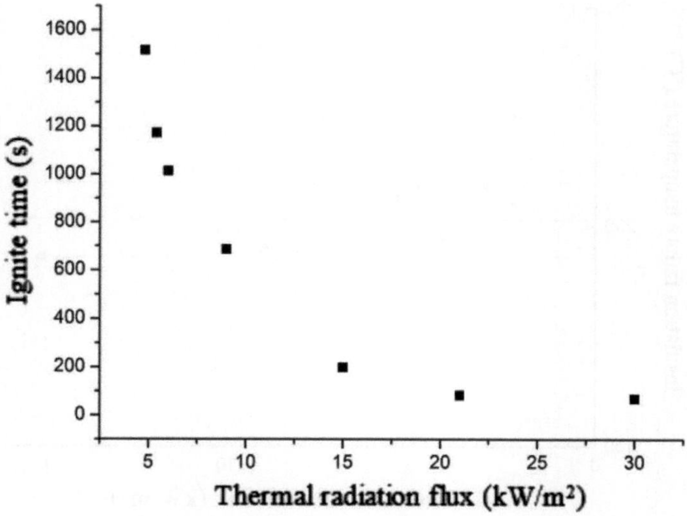

Fig. 4 Relation between cable ignition time and heat radiation flux

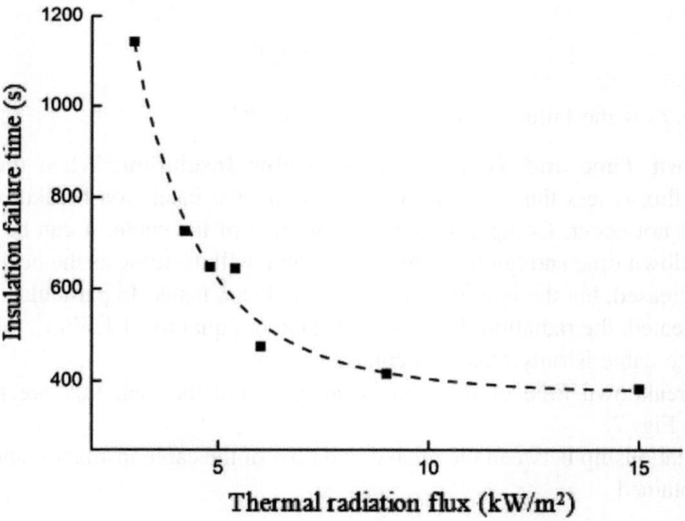

Fig. 5 Relation between cable failure time and heat radiation flux

$$t_f = 388.02 + 4298e^{-0.583\dot{Q}} \tag{3}$$

Where t_f is the cable failure time, s; $\dot{Q}$ is the heat flux of cable, kW/m^2.

As shown in Fig. 6, the failure temperature fluctuates in a small range as the heat radiation flux of the cable increases.

The failure temperature is fitted to obtain

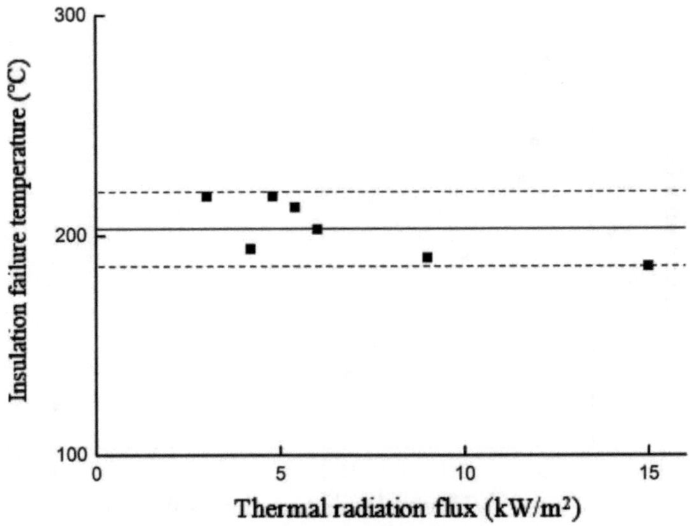

Fig. 6 Relationship between cable insulation failure temperature and heat radiation flux

$$T_f = 203.15 \pm 8.4\%. \tag{4}$$

Where T_f is the failure temperature of cable, °C.

Breakdown Time and Temperature of Cable Insulation. When the thermal radiation flux is less than or equal to 4.2 kW/m^2, the insulation breakdown of the cable will not occur. Comparing the ignition time of the cable, it can be seen that the breakdown time and ignition time of the cable will decrease as the heat radiation flux is increased, but the ignition time will be reduced faster. In particular, when the cable is heated, the radiation flux is greater than or equal to 21 kW/m^2, the ignition time of the cable is only tens of seconds.

The breakdown time of the cable insulation and the heat flux are plotted as shown in Fig. 7.

The relationship between the breakdown time of the cable insulation and the heat flux is obtained

$$t_j = 376.7 + 2271\mathrm{e}^{-0.1968\dot{Q}} \tag{5}$$

Where t_j is the cable breakdown time, s; $\dot{Q}$ is the heat flux of cable, kW/m^2.

When the cable thermal radiation flux is strong, occurrence of cable insulation breakdown, the sheath has violent combustion, the combustion of the internal temperature distribution of cable will also have a great impact, and observed in experiments, violent combustion will damage the cable skin, the arrangement of the internal thermocouple displacement occurs, unable to accurately measure the temperature.

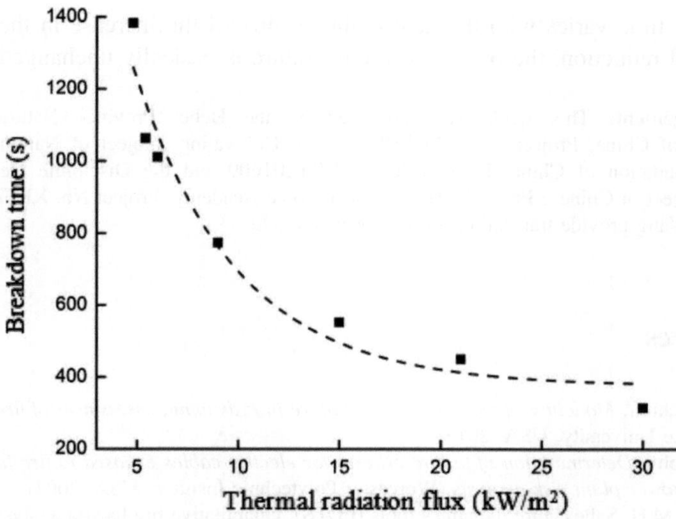

Fig. 7 Relation between cable breakdown time and thermal radiation flux

When the heat radiation flux of the cable is less than or equal to 15 kW/m^2, the breakdown temperature of the cable is less affected by the combustion, and the relationship between the breakdown temperature and the heat radiation flux is obtained.

$$T_j = 308.8 \pm 6.7\% \tag{6}$$

Where T_j is the cable breakdown temperature, °C.

Conclusion

The integrity of cable maintenance function plays an important role in the safe operation of nuclear safety related equipment in fire. In order to master the fire law of typical cables in nuclear power plants, the failure law of WDZ-HK3-PYJP3YJP3 cables under different simulated fire conditions was studied by using a cable heat radiation test furnace. The results show that when the heat radiation flux of the cable is greater than or equal to 21 kW/m^2, the ignition time is less than the insulation failure time. At this time, the cable does not need to be subjected to failure analysis. When the heat radiation flux of the cable is less than or equal to 15 kW/m^2, the insulation failure time decreases exponentially with the increase of the heat radiation flux, and the failure temperature is almost constant. When the heat radiation flux of the cable is less than or equal to 15 kW/m^2, the insulation breakdown time varies with the heat radiation flux of the increase in the form of exponential reduction, the breakdown temperature is basically unchanged.

Acknowledgements This work was sponsored by the Hebei Province Natural Science Foundation of China, Project No. E2017507012, the Cultivating Subject of National Natural Science Foundation of China, Project No. ZKJJPY201609 and the Discipline Development Support Subject of Chinese People's Armed Police Force Academy, Project No. XKFZ2017004. Yonggang Wang provide translation service for this article.

References

1. M.D. Bucknor, *Modeling of electrical cable failure in a dynamic assessment of fire risk* (The Ohio State University, USA, 2013)
2. J.E. Murphy, *Determination of failure criteria for electric cables exposed to fire for use in a nuclear power plant risk analysis* (Worcester Polytechnic Institute, USA, 2004)
3. N. Iqbal, M.H. Salley, Fire dynamics tools (FDTS): Quantitative fire hazard analysis methods for the U.S. nuclear regulatory commission fire protection inspection program, NUREG-1805. U.S. Nuclear Regulatory Commission, Washington DC (2004)
4. K. McGrattan, Cable response to live fire (CAROLFIRE) Volume 3: Thermally-induced electrical failure (THIEF) model, NUREG/CR-6931, Vol. 3. National Institute of Standards and Technology, Maryland, U.S. Nuclear Regulatory Commission, Washington DC (2007)
5. S.P. Nowlen, F.J. Wyant, Cable response to live fire (CAROLFIRE) Volume 1: Test descriptions and analysis of circuit response data, NUREG/CR-6931, Vol. 1. U.S. Nuclear Regulatory Commission, Washington DC, 2008
6. S.P. Nowlen, F.J. Wyant, Cable response to live fire (CAROLFIRE) Volume 2: Cable fire response data for fire model improvement, NUREG/CR-6931, Vol. 2. U.S. Nuclear Regulatory Commission, Washington DC (2008)

Insulation Failure Mechanism of Cable in Fire Environment

Qiang Li, Jiaqing Zhang, Jinmei Li, Yichen Yang and Minghao Fan

Abstract In order to reveal the insulation failure mechanism of the cable under the fire condition, the conductive mechanism and aging mechanism of the cable insulation material were analyzed theoretically, and the relationship between the thermal decomposition weight loss curve and the insulation resistance and temperature of the cable insulation material was constructed. The results show that the reason for the failure of the cable in the fire is that the insulation performance of the insulating material decreases after heating. For different thermal aging degree of the cable, the decomposition process of insulating materials are divided into two stages. When the failure occurs, the insulating material is not decomposed; when the breakdown occurs, the insulating material is in the first stage of thermal loss. Therefore, the failure temperature of the cable can be considered mainly depends on the characteristics of its insulating material, regardless of the structure, shape and size of the cable.

Keywords Cable · Insulation failure · Fire environment · Mechanism study

Introduction

The electrical conductivity of the material is caused by the movement of the carriers present inside the material. The concentration and mobility of carriers are the microscopic physical quantities that characterize the conductivity of the material [1]. For a cable insulation material, its conductivity is related to the number of

Q. Li · J. Li (✉)
Chinese People's Armed Police Force Academy, Xichang Road 220,
Langfang 065000, China
e-mail: lijinmei119@163.com

J. Zhang · M. Fan
State Grid Anhui Electric Power Research Institute, Jinzhai Road 73,
Hefei 230026, China

Y. Yang
Cangzhou Fire Detachment, South Jianshe Street 20, Cangzhou 061000, China

Y. Han (ed.), *Advanced Functional Materials*,
https://doi.org/10.1007/978-981-13-0110-0_98

carriers contained in the insulating material and its moving speed. XLPE—common insulating material for cable, its conductivity process is mainly impurity ion conductivity and intrinsic electron conductivity. When a fire occurs, the impurities inside the material are ionized, its impurity ionic conductance dominates and are the main carriers of their conductivity [2].

In addition, in the process of manufacture, laying and using, the cable will be affected by many aging factors. These factors include thermal aging, mechanical aging, water aging, electrical aging, radiation aging, overload, etc. [3]. The aging of cable insulation is a complex process. In the aging process, the electrical properties of insulating materials, physical properties will change. For XLPE, in the initiation of the aging of the single factor or multi-factor role, there is a threshold, when the insulation material to bear the environmental factors do not exceed a certain value, the material will not occur aging [4]. For the cable insulation material, it usually bear the aging of the multi-factor interaction, in which case the aging state of the cable is not a single factor superimposed [5]. Normally, multi-factor simultaneous action will lead to a new internal aging mechanism of insulating materials. The results show that temperature is the main reason leading to deterioration of cable insulation. The essence of thermal aging is the chemical reaction of insulating materials in high temperature environment. Thermal aging can be divided into two types: thermal degradation and thermal oxygen degradation [6]. Thermal cracking is the cracking reaction of the insulating material under thermal conditions and anaerobic conditions. Most of the cable laying environment is the atmospheric environment, so the simple pyrolysis reaction is not much. Thermal oxygen degradation refers to the double reaction of the molecular fracture and XLPE of the insulating material under aerobic conditions. It produces peroxy radicals or other oxygen-containing groups. Thermal oxygen degradation is a chain reaction that the initiation phase activation energy is higher, but the free chain growth stage of the activation energy is low, the reaction is easy to carry out and relatively rapid. The chemical reaction leads to the change of the molecular structure of the insulating material, which leads to the change of its physical properties, and the mechanical properties and electrical properties can be observed macroscopically.

When the cable insulation material is heated in the fire, it will occur pyrolysis. Because of its different composition and processing process, the pyrolysis process will show different characteristics. The pyrolysis of the cable insulation will have an effect on its electrical properties, and that process can react to the extent that the cable insulation is reduced to some extent. Therefore, this paper analyzes the mechanism of cable failure by studying the pyrolysis performance of cable insulation.

Thermogravimetric Experiment

Experimental Materials and Sample Preparation. The experimental cable is made of 1E grade 300/500 V copper core cable for nuclear power plant (WDZ-HK3-PYJP3YJP3). Experiments were carried out before and aged, 19 kinds of working conditions of the cable that it's insulating material stripped, dried and cut into granular.

Experimental Parameter Setting. TGA/SDTA 851 Thermogravimetric/Synchronous differential thermal analyzer, the heating rate was 10 °C/min, air, the flow rate was 25 ml/min, the temperature range was 25–600 °C.

Experimental Results and Discussion

TG-DTG Analysis of Non-aged Cable Insulation Materials. As shown in Fig. 1, under the experimental conditions, the thermal loss of XLPE mainly has two stages. In the first stage, the temperature range is about 270–420 °C, and the weight loss rate of the material at this stage is 15%. The main weight loss reason may be the decomposition of small molecules such as additives and by-products during the production of XLPE. In the third stage, the temperature range is about 420–520 °C, and the weight loss rate of the material at this stage is 40%. The main weight loss reason is the cross-linked polyethylene molecular chain fracture, the formation of small molecular products.

Combined with the failure temperature and breakdown temperature of the cable, the failure temperature of the cable is before the weight loss of the cross-linked polyethylene. The breakdown temperature of the cable is in the first stage temperature range of the weight loss of the cross-linked polyethylene. From the analysis of the conductive mechanism of the cable insulation material, it can be seen that the nature of the cable insulation material does not change before the failure, so the cable insulation resistance and thermodynamic temperature ratio of the natural logarithm and the thermodynamic temperature is proportional to the reciprocal. As the

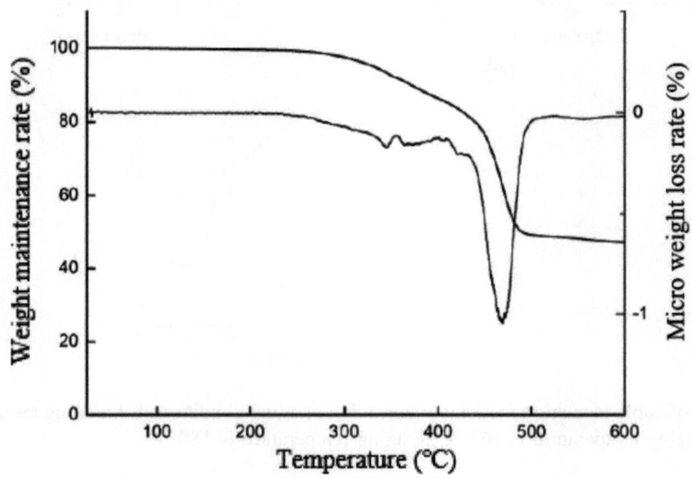

Fig. 1 TG-DTG of non-aged cable insulation

temperature increases, the insulation resistance decreases, resulting in visible failure. With the further increase of temperature, the insulating material is thermally decomposed, and the polar substance with conductivity is generated. With the accumulation of the number of carriers and the increase in energy, a macroscopic observation of the dielectric breakdown is formed under the action of the applied electric field.

TG-DTG Analysis of Aged Cable Insulation Materials. Figure 2 is the TG curve of cable insulation materials with different aging time under different aging environment temperature. It can be seen that the weight loss of XLPE aged is also divided into two stages. The failure temperature of the cable aged is before the weight loss of the cross-linked polyethylene. The breakdown temperature of the cable is in the range of the first stage of the weight loss of the XLPE, which is the same as that of a non-aged cable.

In order to further study the influence of aging on the performance of the cable insulation material, the decomposition temperature of the sample is taken as the tangent, and the intersection point of the two tangent lines is the initial critical temperature for the decomposition of the cable insulation material. The initial critical temperature of the cable insulation material under various aging conditions

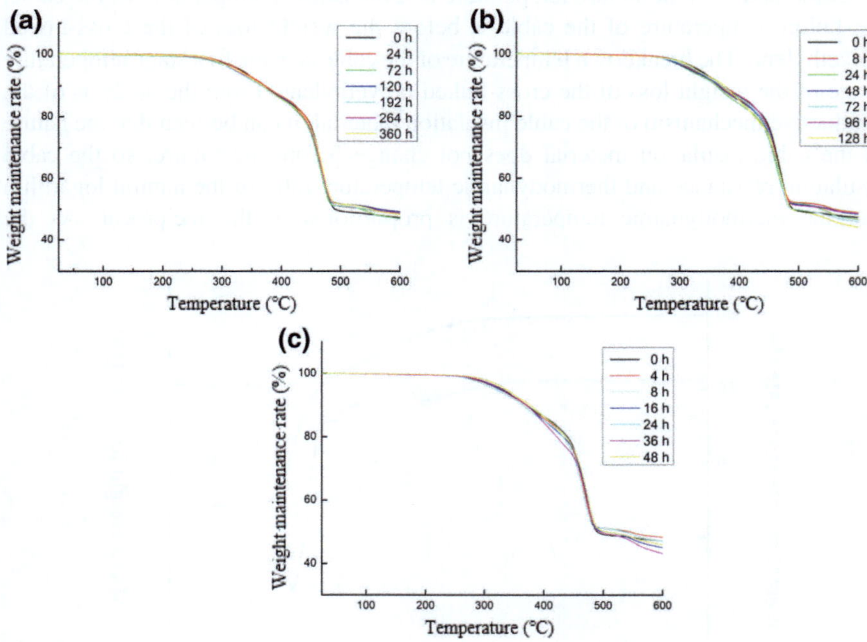

Fig. 2 TG of cable insulation material under different aging conditions for **a** aging temperature is 150 °C; **b** aging temperature is 165 °C; **c** aging temperature is 180 °C

Table 1 Decomposition initial temperature of cable insulation materials under different aging conditions

Aging temperature (°C)	Aging time (h)	Decomposition of the initial critical temperature (°C)
150	0	277.1
	24	271.6
	72	265.6
	120	262.4
	192	261.8
	264	260.1
	360	257.9
165	0	277.1
	8	260.8
	24	259.7
	48	256.4
	72	253.2
	96	252.1
	128	250.1
180	0	277.1
	4	271.9
	8	269.7
	16	268.9
	24	267.2
	36	254.2
	48	248.8

is shown in Table 1. It can be seen that the initial decomposition critical temperature of the cable insulation material will decrease with the increase of aging time at different aging temperature, which indicates that the stability of XLPE has decreased after aging. Due to the decrease of thermal stability of the insulating material, the cable samples are more likely to decompose during the failure test. These decomposition products enhance the conductivity of the material, resulting in the decrease of cable insulation performance, so the aging cable is more prone to insulation failure. This is consistent with the experimental phenomena observed in the previous experiments.

As can be seen from Fig. 2, the breakdown temperature of the cable is in the first stage of weightlessness. This is due to the occurrence of breakdown needs the accumulation of the heat and carriers, so the change of initial decomposition of the critical temperature have little influence on the breakdown time and temperature, which is consistent with the basically unchanged breakdown temperature measured.

Analysis of Relationship Between Cable Insulation Resistance and Temperature

Relationship between Cable Insulation Resistance and Temperature in Different Fire Conditions. In a certain range of temperature, the relationship between the natural logarithm value of the cable insulation resistance and the temperature and the reciprocal of the thermodynamic temperature of the insulation is linear. The experimental results of cable failure under different heat radiation flux is considered as X axis and as y axis plotted, as shown in Fig. 3.

It can be seen from Fig. 3, in different heat radiation flux, for the XLPE cable, $\ln(R_i/T)$ and $1/T$ in the 90–250 °C temperature range is consistent with the linear law. Combined with the thermal decomposition characteristics of the cable, it can be found that the thermal decomposition of the cable has not occurred at all in this temperature range, which is consistent with the theoretical deduction. Meanwhile, by comparing the cable failure temperature, it can be seen that the cable failure temperature is in this linear change within the temperature range. Under the different heat radiation flux, the law of cable $\ln(R_i/T)$ change with $1/T$ is basically the same and keeps fluctuation in a small range to maintain a small range of fluctuations. This is consistent with the previous measurements of the failure temperature to maintain a certain value.

Relationship between Cable Insulation Resistance and Temperature under Different Aging Degree. The experimental results of the cable failure under different aging degree, $\ln(R_i/T)$ are considered ac X axis and $1/T$ as Y axis plotted, as shown in Fig. 4.

As can be seen from Fig. 4, the $\ln(R_i/T)$ and $1/T$ of cables with different aging degrees are basically linear in the range of 90–250 °C degrees of temperature, but their initial values and slopes decrease after aging. With the increase of aging temperature, the initial value of the $\ln(R_i/T)$ decreases, and its relationship with the aging temperature is obtained

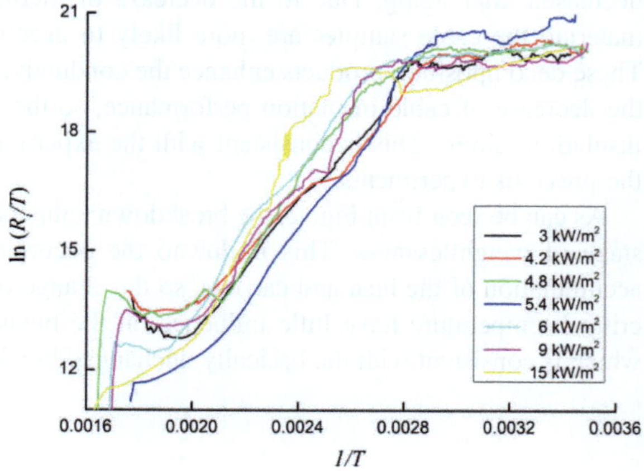

Fig. 3 $\ln(R_i/T) - 1/T$ of unaged cable

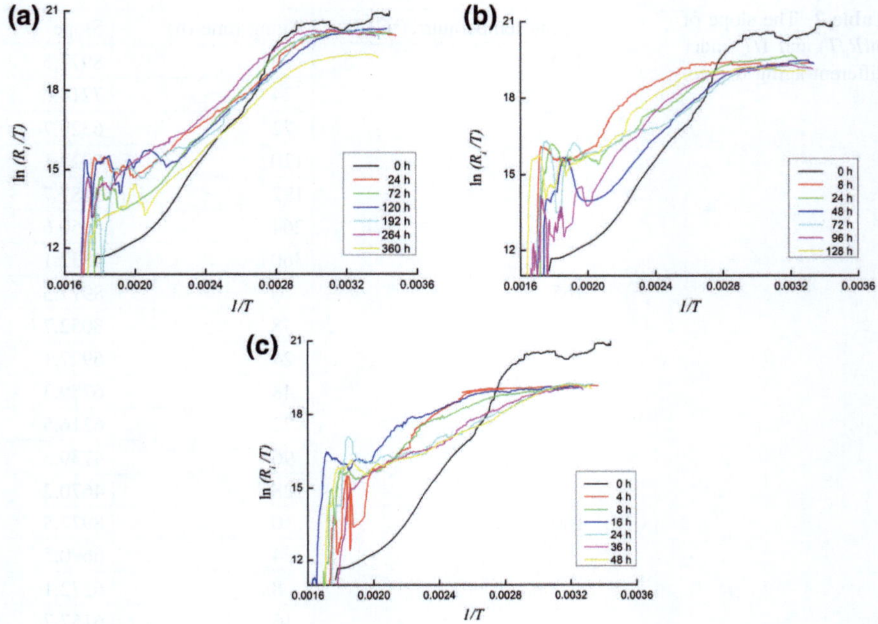

Fig. 4 ln(R_i/T) change with $1/T$ of cable under different aging conditions for **a** aging temperature is 150 °C; **b** aging temperature is 165 °C; **c** aging temperature is 180 °C

$$y = -0.0283t_h + 29.9 \tag{1}$$

Where y is the initial value of ln(R_i/T), t_h is aging temperature, °C.

Using Fig. 4, calculate the slope of the ln(R_i/T) and 1/T of cable at 90–250 °C for different aging conditions, as shown in Table 2.

The calculated slope in Table 2 and the corresponding aging years are plotted, as shown in Fig. 5.

The relation between the fitted slope and the corresponding aging years can be obtained

$$s = -337.4t_a + 7580 \tag{2}$$

Where s is the slope of the ln(R_i/T) to the $1/T$, t_a is the aging years, and the unit is the year.

Through analysis, we can see that the slope of the ln(R_i/T) to the $1/T$, is E_a/k, in the formula, E_a is the conductance activation energy. It can be seen that the conductance activation energy of the cable insulation materials decreases with the increase of aging years. This indicates that the conductivity of XLPE is enhanced with the increase of aging age.

Table 2 The slope of $\ln(R_i/T)$ and $1/T$ under different aging degree

Aging temperature (°C)	Aging time (h)	Slope
150	0	8977.5
	24	7214.6
	72	6329.7
	120	6023.1
	192	5287.7
	264	5050.6
	360	4837.1
165	0	8977.5
	8	8052.7
	24	6927.1
	48	6789.3
	72	6216.5
	96	4739.3
	128	4670.2
180	0	8977.5
	4	6690.5
	8	6272.1
	16	6157.7
	24	4201.7
	36	3213.7
	48	3176.7
	36	3213.7
	48	3176.7

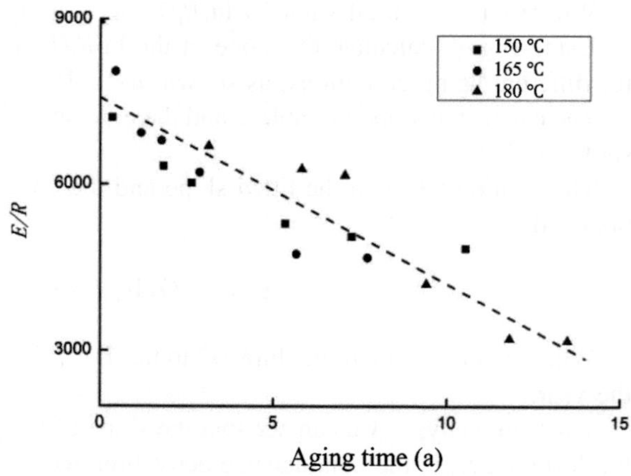

Fig. 5 Relationship between calculated slope and aging age

The failure temperature and failure time of the cable after aging are shortened, which is due to the decrease of the failure temperature of the cable after aging due to the change of the conductivity of the XLPE, the failure time decreases with the decrease of the failure temperature.

Conclusion

Through the combination of theoretical analysis and experimental research, the failure mechanism of cable fire is discussed, and the main conclusions are as follows:

(1) Through the thermogravimetric analysis of the cable insulation material, the XLPE did not thermal decomposition at the time of cable failure, and the XLPE was in the first weight loss stage at the time of cable breakdown; with the increase of the aging degree, the initial decomposition temperature of XLPE is reduced.
(2) Through the analysis of the cable curve $\ln(R_i/T)$ with $1/T$, the linear relationship is good in the range of 90–250 °C; with the increase of the aging degree, the initial value of the cable insulation resistance decreases, and the conduction activation energy of the cable insulation material decreases.
(3) According to the test results, it is considered that the failure of the cable in the fire is that the insulation performance of the insulating material decreases after heating. Before the failure temperature, the relationship between the natural logarithm value of the cable insulation resistance and the temperature and the reciprocal of the thermodynamic temperature of the insulation is linear. Combined with the results of theoretical analysis, it can be concluded that the failure temperature of the cable mainly depends on the characteristics of the insulating material, which is independent of the structure, shape and size of the cable. Therefore, it is feasible to establish the cable internal heat conduction numerical model to predict the failure time of the cable.

Acknowledgements This work was sponsored by the Hebei Province Natural Science Foundation of China, Project No. E2017507012, the Cultivating Subject of National Natural Science Foundation of China, Project No. ZKJJPY201609 and the Discipline Development Support Subject of Chinese People's Armed Police Force Academy, Project No. XKFZ2017004. Zhengbing Dong provide translation service for this article.

References

1. J. Tian, Y. Zhou, Carrier mobility of polyethylene and mechanism of space charge packet formation. High Voltage Eng. **12**, 2882–2888 (2010)
2. Z. Chen, J. Zhao, B. Ouyang, J. Li, H. Li, S. Wang, Difference in insulation characteristics and mechanism between AC and DC XLPE cable's material. High Voltage Eng. **09**, 2644–2652 (2014)
3. C. Zhou, The Study of Estimating Degradation Of Rubber Insulation Low-Voltage Cable Based on Frequency Dielectric Spectroscopy. Daliang University and Technology, 2012
4. F. Liu, Evaluation of the insulation aging of 35 kV and below XLPE power cables. Shanghai Jiao Tong University, 2014
5. Y. Liang, The Research of Aging Assessment for Cross-linked Polyethylene Insulation Effected with Multi-factors. Chongqing University, 2015
6. R. Huo, Research on Thermal Aging Monitoring and Fast Prediction on the Life of XLPE. Shanghai Jiao Tong University, 2014

Microstructure Evolution of Nanoporous Copper Fabricated by Electrochemical Dealloying Mn–Cu Alloy

Jinyi Wang, Yi Yang, Mingrui Zhang and Sen Yang

Abstract The electrochemical dealloying behavior of single-phase Mn-Cu alloy in hydrochloric acid solution by applying different dealloyed voltages was investigated. The concentrations of HCl solution are also explored as a parameter for modulating the nanoporous microstructure. Phase constitution and microscopic morphology of precursor alloy and dealloyed samples were characterized using X-ray diffraction and scanning electron microscopy. The precursor alloy was single-phase $CuMn_4$ solid solution. After dealloying, only face-centered cubic copper phase can be identified. The results show that different applied voltages make a small difference on nanoporous structure. The concentration of hydrochloric acid solution has a remarkable influence on the dealloying process and nanoporous microstructure. The dealloying in a 0.5 M HCl solution results in nanoporous copper with ligament size range of 43.76–160.66 nm. The high concentration of dealloying solution leads to the formation of coarsened nano structure.

Keywords Corrosion · Copper · Porous · Electrochemical

J. Wang · Y. Yang · M. Zhang · S. Yang (✉)
School of Materials Science and Engineering, Nanjing University of Science and Technology, No. 200 Xiaolingwei, Nanjing 210094, China
e-mail: yangsen@njust.edu.cn

J. Wang
e-mail: blackblack00@163.com

Y. Yang
e-mail: 359519529@qq.com

M. Zhang
e-mail: 824128576@qq.com

Y. Han (ed.), *Advanced Functional Materials*,
https://doi.org/10.1007/978-981-13-0110-0_99

Introduction

Nanoporous metals (NPMs), as novel functional materials, have attracted considerable attention in a wide variety of potential applications including biosensors [1, 2], actuators [3], photocatalysis [4], surface-enhanced Raman scattering (SERS) [5, 6], fuel cells [7, 8], and so on. Lang et al. [9] have demonstrated that hybrid structure of NPG and MnO_2 have enhanced conductivity resulting in a specific capacitance. Yu et al. [10] reported nanoporous metal-supported tin foil as an alternative anode for rechargeable Lithium ion batteries with improved electrochemical performance.

Recently, dealloying, as a simple and efficient way, has been growing into the most important method to fabricate nanoporous metals. Dealloying is a selective dissolution process. In most situations, it occurs in a solution to selectively remove the less noble elements and the more noble elements retain and rearrange to form a three-dimensional nano pore-ligament structure [11]. The properties of NPMs can be optimized through adjusting pore and ligament size and distribution. Typically, fabrication methods result in controlled porosity by kinds of strategies including design of precursors [12–14], controlling over the dealloying parameters [15–19] and so on. In general, the composition of less noble elements is usually rich and the more noble elements are below its parting limit [20]. Xu et al. [21] obtained NPG with ligament size less than 6 nm by electrochemical dealloying, under similar free corrosion conditions, free corrosion dealloying requires more time and results in ligaments of ~ 30 nm.

To obtain biocontinuous porous structure by dealloying binary alloys, the precursor alloy must have complete single-phase solid solubility. The best fits these requirements is Ag–Au system. The evolution of the dealloyed morphology has been examined in detail [11, 20]. However, gold and silver are both noble metals, besides their high price, the global reserve is getting less. Because of low density and low cost advantages, nanoporous copper has attracted extensive interest. The precursor systems of Al–Cu [22], Mg–Cu [23], Zn–Cu [24] and Zr–Cu [25] have been used. Hayes et al. [26] have studied the free corrosion dealloying behavior of Mn–Cu alloy. Mn–Cu alloy is suggested to be a suitable precursor due to the single phase solid solubility and the high difference of the standard reversible potential between Mn/Mn^{2+} and Cu/Cu^{2+}. The microstructure of NPC with a rapid growth of pore sizes as the Cu composition (15–45%) of precursor alloy decreased [27]. Both inorganic [26] and organic [28] acid as dealloying medium have been used to fabricate nanoporous structure. In most cases, free corrosion dealloying was conducted. It usually took a few days to obtain the nanoporous metals. In the present work, we have explored the effective electrochemical dealloying behavior of low-cost $Mn_{72}Cu_{28}$ precursor alloy in hydrochloric acid solutions.

Experimental

Mn–Cu alloy with nominal composition of 28 at.% Cu was prepared from pure Mn (99.99%) and pure Cu (99.999%). Induction melting was employed to melt the raw materials in a copper crucible in an argon atmosphere. After cooling down in situ, the ingot was cut into slices of 0.5 mm in thickness by wire cut electrical discharge machine (WEDM). The slices were grinded and cleaned in an ultrasonic cleaner. The electrochemical dealloying experiments were performed in 0.1 M HCl solution at 25 °C in a thermostatic water bath. After dealloying, the samples were rinsed with deionized water for several times.

The phase constitution was characterized using X-ray diffraction (XRD, Bruker-AXS D8 Advance, Cu Kα source radiation). The microstructure was observed using a scanning electron microscopy (SEM, FEI Quanta 250 FEG). The characteristic scale length of ligaments and pores were measured more than 100 sits by using the single chord length method, and then summarized statistically. The electrochemical experiments were performed in a three-electrode system, platinum plate was used as the counter electrode, samples as the working electrode, saturated calomel electrode as the reference electrode and 0.1 M HCl solution as electrolyte.

Results and Discussion

Figure 1a shows the XRD patterns of $Mn_{72}Cu_{28}$ precursor alloy, it is composed of a single-phase $CuMn_4$ (Powder Diffraction File (PDF) No. 65-5589). The initial sample had a polycrystalline structure. After electrochemical dealloyed on the condition of applied voltage −500, −300 and −100 mV for 1800s, the XRD patterns were also polycrystalline structure. They are composed of a single

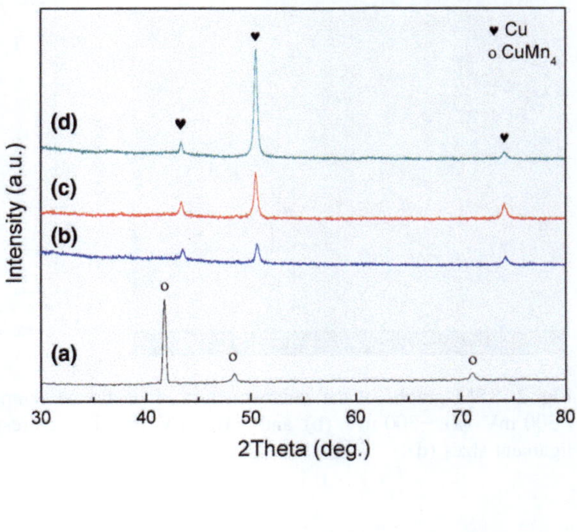

Fig. 1 XRD patterns of Mn–Cu precursor alloy (**a**) and electrochemical dealloyed samples at applied voltage of −500 mV (**b**), −300 mV (**c**) and −100 mV (**d**) in 0.1 M HCl solutions for 1800 s at 25 °C

face-centered cubic (fcc.) Cu phase (Fig. 1b–d). The diffraction peaks of all dealloyed samples attributed to (111), (200), (220), and appeared at almost the same diffraction angles. That indicated only single copper phase existed.

Nanoporous copper was synthesized by electrochemical dealloying of Mn from $Mn_{72}Cu_{28}$ (at.%) alloys in 0.1 M HCl solution for 1800 s at 25 °C. The resulting nanostructure was shown in Fig. 2. For the as-dealloyed samples at applied voltage of −500 mV (2a), −300 mV (2b) and −100 mV (2c), all the microscopic morphologies are similar. Nanoporous structure with an open-cell three-dimensional structure and nanosized ligaments were clearly observed. Applying a voltage to dealloying process can accelerate the dissolution of Mn and improve the uniformity of resulted nanoporous structure. The presence of cracks was showed in Fig. 2b, c. The surface of crack area was smooth and bicontinuous. They were most likely formed during the dealloying process. There are several cracking mechanisms: residual stress, capillary forces or coherency stresses [26].

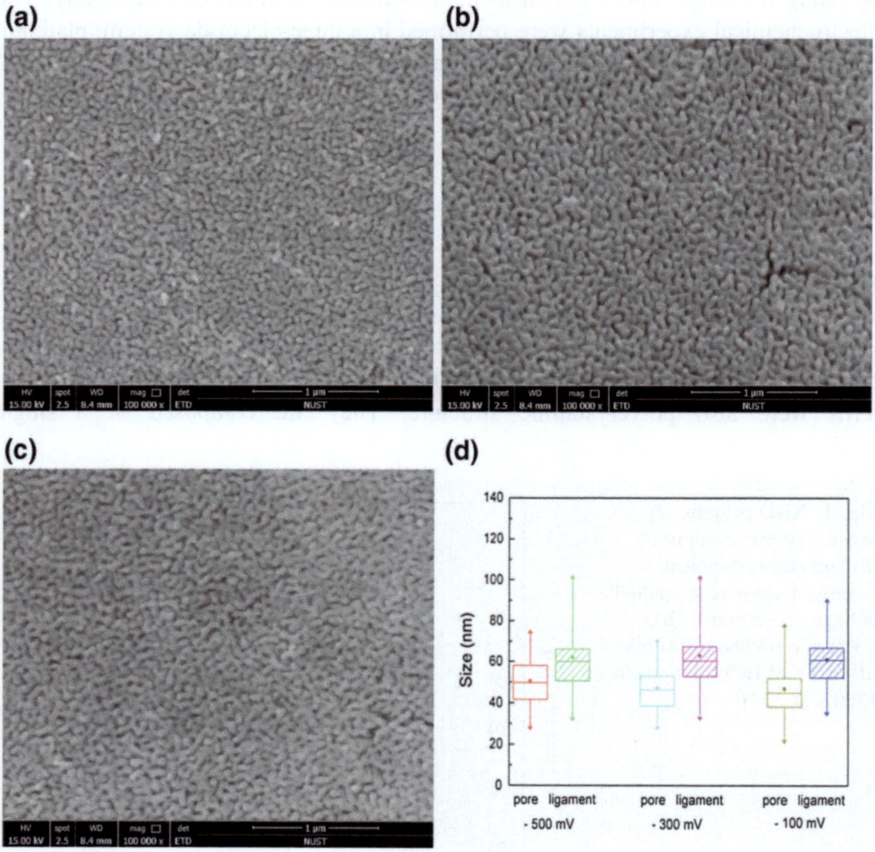

Fig. 2 SEM graphs of the microstructure of dealloyed samples upon different applied voltage −500 mV (**a**) −300 mV (**b**) and −100 mV (**c**). The corresponding distribution of pore and ligament sizes (**d**)

Figure 2d shows the distribution of nanoporous pore and ligament size. The average pore size of dealloyed $Mn_{72}Cu_{28}$ ribbons was 49.68 nm under the electrochemical dealloying condition of applied −500 mV. That of dealloyed by −300 and −100 mV were reduced to 46.27 and 44.86 nm, respectively. With increasing applied voltage, there is a minor reduction of obtained pore size. On the contrary, on the condition of higher voltage, the average ligament size got a small coarsened. It has been found that different applied voltages did not significantly affect the size of nanoporous structure. This was likely due to the single-phase precursor rather than multiphase. There are no different critical potential values in homogeneous phase precursor alloy. Although three different voltages were applied, complete dealloying occurred and unique nanoporous structure were produced.

Figure 3a–c shows SEM images of electrochemical dealloyed $Mn_{72}Cu_{28}$ alloy under applied voltage of −300 mV. The precursors were immersed in different

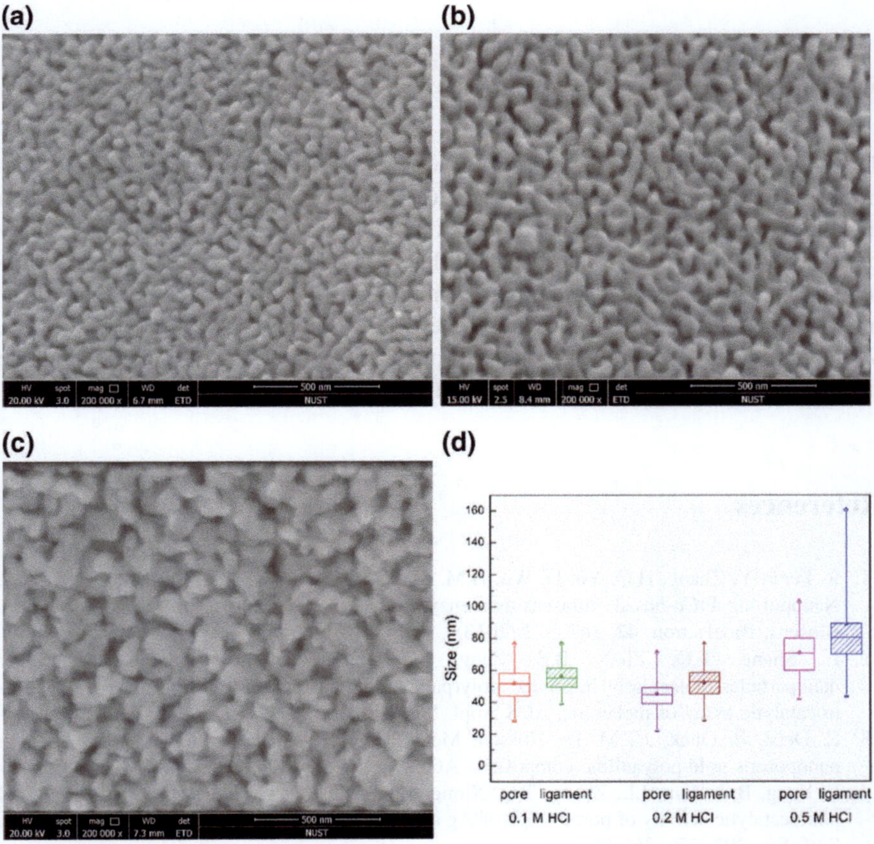

Fig. 3 SEM graphs of the microstructure of dealloyed samples upon different concentrations of HCl solution, 0.1 M (**a**), 0.2 M (**b**) and 0.5 M (**c**). The corresponding distribution of pore and ligament sizes (**d**)

concentration of HCl solutions. Figure 3d shows the corresponding porosity size distribution. For the 0.1 M HCl solution, three-dimensional bicontinuous nanoporous copper can be prepared, and the average pore size and ligament size were about 51.72 and 55.33 nm. In the case of 0.2 M HCl solution, the porous structure was smoother and its size was almost equal to that of 0.1 M HCl solution. As the concentration to 0.5 M, the pore and ligament was wider and the average ligament was coarsened to 89.31 nm. The microstructure exhibited island-like ligament instead of three-dimensional bicontinuous nanoporous structure.

The dealloying process can be viewed as a balance between roughening and smoothing. Dissolution of less-noble element caused surface roughening while diffusion and rearrangement of more-noble element induced smoothing. The ligament size depends on the competition of dissolution rate and the diffusion rate [11]. During dealloying process under high concentration of HCl solution condition, adsorption of Cl^- can accelerate the surface diffusion of copper atoms and induce the significant coarsening.

Summary

Three-dimensional bicontinuous nanoporous copper has been fabricated by electrochemical dealloying single-phase $Mn_{72}Cu_{28}$ precursor alloy. Applied a voltage to dealloying process can accelerate the dissolution of Mn and improve the uniformity of resulted nanoporous structure. The different applied voltages make a small difference on nanoporous structure. Concentration of dealloying solution has a significant influence on the dealloying process and formation of nanoporous copper. With higher solution concentration, adsorption of Cl^- can increase the diffusion rate of copper, resulting in coarsened nanoporous structure with larger length scale.

References

1. R. Feng, Y. Zhang, H.Q. Yu, D. Wu, H.M. Ma, B.C. Zhu, C.X. Xu, H. Li, B. Du, Q. Wei, Nanoporous PtCo-based ultrasensitive enzyme-free immunosensor for zeranol detection. Biosens. Bioelectron. **42**, 367–372 (2013)
2. L. Shang, F.Q. Zhao, B.Z. Zeng, 3D porous graphene-porous PdCu alloy nanoparticles-molecularly imprinted poly(para-aminobenzoic acid) composite for the electrocatalytic assay of melamine. ACS Appl. Mater. Inter. **6**(21), 18721–18727 (2014)
3. E. Detsi, P. Onck, J.T.M. De Hosson, Metallic muscles at work: high rate actuation in nanoporous gold/polyaniline composites. ACS Nano **7**(5), 4299–4306 (2013)
4. F. Yang, B.Z. Tian, J.L. Zhang, T.Q. Xiong, T.T. Wang, Preparation, characterization, and photocatalytic activity of porous AgBr@Ag and AgBr I@Ag plasmonic photocatalysts. Appl. Surf. Sci. **292**, 256–261 (2014)
5. S.O. Kucheyev, J.R. Hayes, J. Biener, T. Huser, C.E. Talley, A.V. Hamza, Surface-enhanced Raman scattering on nanoporous Au. Appl. Phys. Lett. **89**(5), 053102 1-3 (2006)

6. J.B. Zeng, F.S. Zhao, M. Li, C.H. Li, T.R. Lee, W.C. Shih, Morphological control and plasmonic tuning of nanoporous gold disks by surface modifications. J. Mater. Chem. C. **3**(2), 247–252 (2015)
7. R.Y. Wang, C. Wang, W.B. Cai, Y. Ding, Ultralow-platinum-loading high-performance nanoporous electrocatalysts with nanoengineered surface structures. Adv. Mater. **22**(16), 1845–1848 (2010)
8. J. Snyder, K. Livi, J. Erlebacher, Oxygen reduction reaction performance of [MTBD][beti]-encapsulated nanoporous NiPt alloy nanoparticles. Adv. Func. Mater. **23**(44), 5494–5501 (2013)
9. X.Y. Lang, A. Hirata, T. Fujita, M.W. Chen, Nanoporous metal/oxide hybrid electrodes for electrochemical supercapacitors. Nat. Nanotechnol. **6**(4), 232–236 (2011)
10. Y. Yu, L. Gu, X.Y. Lang, C.B. Zhu, T. Fujita, M.W. Chen, J. Maier, Li storage in 3D nanoporous Au-supported nanocrystalline tin. Adv. Mater. **23**(21), 2443–2447 (2011)
11. J. Erlebacher, M.J. Aziz, A. Karma, N. Dimitrov, K. Sieradzki, Evolution of nanoporosity in dealloying. Nature **410**(6827), 450–453 (2001)
12. W. Wang, Y. Yang, Y.Q. Liu, F.X. Wang, Y.M. Kang, Z.Q. Lei, Dealloyed different atom ratios Pd-x(FeCo)(10-x), nanoparticle: promising electrocatalyst towards ethylene glycol oxidation. Int. J. Hydrogen Energ. **41**(1), 300–306 (2016)
13. M.Y. Niu, W.C. Xu, S.L. Zhu, Y.Q. Liang, Z.D. Cui, X.J. Yang, A. Inoue, Synthesis of nanoporous CuO/TiO_2/Pd-NiO composite catalysts by chemical dealloying and their performance for methanol and ethanol electro-oxidation. J. Power Sour. **362**, 10–19 (2017)
14. W.Y. Zhang, Y.N. Li, G.Z. Li, Q.B. Wang, H.P. Tang, Z.P. Xi, M. Fang, Fabrication of nanoporous structure silver film by dealloying of Ag-Al alloy and control of pore size. Porous Met. Met. Foams: Metfoam **2007**, 325 (2008)
15. Y.M. Sabri, K.M.M. Kabir, A.E. Kandjani, S.K. Bhargava, Studying the effect of dealloying Cu-Au nanostructures on their mercury sensing performance. Sens. Actuators B-Chem. **245**, 273–281 (2017)
16. Z.H. Zhang, Y. Wang, Z. Qi, C. Somsen, X.G. Wang, C.C. Zhao, Fabrication and characterization of nanoporous gold composites through chemical dealloying of two phase Al-Au alloys. J. Mater. Chem. **19**(33), 6042–6050 (2009)
17. J. Snyder, K. Livi, J. Erlebacher, Dealloying silver/gold alloys in neutral silver nitrate solution: Porosity evolution, surface composition, and surface oxides. J. Electrochem. Soc. **155**(8), C464–C473 (2008)
18. S. Sattayasamitsathit, Y.E. Gu, K. Kaufmann, S. Minteer, R. Polsky, J. Wang, Tunable hierarchical macro/mesoporous gold microwires fabricated by dual-templating and dealloying processes. Nanoscale **5**(17), 7849–7854 (2013)
19. Z.Q. Li, X.M. Wang, X. Lu, Refinement of nanoporous silver by adding surfactant to the electrolyte. ECS Electrochem. Lett. **3**(4), C13–C16 (2014)
20. J. Erlebacher, An atomistic description of dealloying—porosity evolution, the critical potential, and rate-limiting behavior. J. Electrochem. Soc. **151**(10), C614–C626 (2004)
21. C.X. Xu, J.X. Su, X.H. Xu, P.P. Liu, H.J. Zhao, F. Tian, Y. Ding, Low temperature CO oxidation over unsupported nanoporous gold. J. Am. Chem. Soc. **129**(1), 42–43 (2007)
22. Z. Qi, C.C. Zhao, X.G. Wang, J.K. Lin, W. Shao, Z.H. Zhang, X.F. Bian, Formation and characterization of monolithic nanoporous copper by chemical dealloying of Al-Cu alloys. J. Phys. Chem. C **113**(16), 6694–6698 (2009)
23. C.C. Zhao, Z. Qi, X.G. Wang, Z.H. Zhang, Fabrication and characterization of monolithic nanoporous copper through chemical dealloying of Mg-Cu alloys. Corros. Sci. **51**(9), 2120–2125 (2009)
24. Q. Yang, S.H. Liang, B.B. Han, J. Wang, R. Mao, Preparation and properties of enhanced bulk nanoporous coppers. Mater. Lett. **73**, 136–138 (2012)
25. H.B. Lu, Y. Li, F.H. Wang, Dealloying behaviour of Cu-20Zr alloy in hydrochloric acid solution. Corros. Sci. **48**(8), 2106–2119 (2006)

26. J.R. Hayes, A.M. Hodge, J. Biener, A.V. Hamza, K. Sieradzki, Monolithic nanoporous copper by dealloying Mn-Cu. J. Mater. Res. **21**(10), 2611–2616 (2006)
27. J.W. Zheng, S.C. Zhang, W.B. Liu, Y.L. Xing, Influence of alloy composition on nanoporous structure by dealloying Mn-Cu ribbons. Rare Met. **30**, 370–374 (2011)
28. S. An, S. Zhang, W. Liu, H. Fang, M. Zhang, Y. Yu, Dealloying behavior of Mn–30Cu alloy in acetic acid solution. Corros. Sci. **75**, 256–261 (2013)

CeO_2 Nanorod-Supported Transition Metal Catalysts Prepared by Dealloying for CO Oxidation

Dong Duan, Haiyang Wang, Xiaolong Zhang, Caihua Wei and Zhanbo Sun

Abstract In this study, Al–Ce, Al–Ce–Ni, and Al–Ce–Mn precursor alloys were prepared by melt-spinning, and then dealloyed in 20% NaOH aqueous solution and calcined in air. The catalytic activities of the as-prepared precursors were evaluated for the oxidation of CO at atmospheric pressure. The results revealed that a novel CeO_2 nanorod-supported skeleton structure could be obtained by dealloying and calcining the as-quenched Al–Ce ribbons. The CO conversion temperatures for the CeO_2/NiO and CeO_2/MnO_2 composites were reduced to about 147 and 136 °C, respectively, which indicated the superior activity of these composites as compared to pure CeO_2 (Zhang et al. in Nanotechnology 28, 45602, 2017) [11]. The reason for the increased catalytic performance can be attributed to the large number of active sites provided by the composite materials. The synergistic effects at the NiO/MnO_2 and CeO_2 interface as well as the increase in the concentration of oxygen vacancies in the composites also play important roles in the enhancement of the catalytic performance.

Keywords Al–Ce precursor alloy · Dealloying · Support-skeleton structure Composite · CO oxidation

Introduction

As an important rare-earth oxide, CeO_2 has unique oxidation and reduction properties owing to its ability to rapidly switch between the two oxidation states represented by Ce^{4+} and Ce^{3+}. This characteristic makes CeO_2 a prospective candidate for many applications such as catalysts [1, 2] (especially, for the catalytic oxidation of harmful gases such as CO), battery materials [3, 4], surface acoustic wave sensors [5], and lactates of enzyme biosensors [6]. However, the catalytic activity of pure CeO_2 for CO oxidation is still unsatisfactory because CeO_2 is prone to

D. Duan · H. Wang · X. Zhang · C. Wei · Z. Sun (✉)
Xi'an Jiaotong University, Xi'an 710049, China
e-mail: szb@mail.xjtu.edu.cn

Y. Han (ed.), *Advanced Functional Materials*,
https://doi.org/10.1007/978-981-13-0110-0_100

sintering at high temperatures. Therefore, it is usually not used in the pure form, but in combination with other oxides or active metals as a structural p-electron promoter or as a co-catalyst. Typically, the main effects of CeO_2 as a catalyst additive in CO oxidation are as follows [7]: (1) improved dispersion of the active ingredients, (2) increased stability of the carrier, so that a greater number of CO active sites can be produced at the interface, (3) increased CO oxidation rate because of the presence of oxygen vacancies, and (4) rapid oxygen storage according to the changes in process conditions, thereby promoting the redox reaction.

In the past few years, Ni–Ce–O and Mn–Ce–O alloy systems have been found to be effective catalysts for the low-temperature oxidation of CO, and have therefore been extensively studied. Lu et al. prepared CeO_2/NiO composites by a simple sintering method; the Ce^{3+} and Ce^{4+} ions in the molten solution were polymerized and precipitated by oxidization and hydrolysis, respectively. A nanosized mesoporous NiO/CeO_2 composite was obtained by subsequent sintering and used in a water-gas reforming reaction [8]. Liu et al. prepared a heterogeneous (1−x) CeO_2·xNiO composite by a combination of the hydrothermal method and calcination. According to their report, Ni^{2+} can be doped into a CeO_2 lattice and dispersed on the surface of the CeO_2 nanoparticles by changing the sintering temperature between 400 and 700 °C; such a technique is different from the traditional sol-gel and coprecipitation methods [9]. Wei et al. prepared a three-dimensional CeO_2/MnO_2 composite by a KIT-6 template method and evaluated the diffusion of the molecules using molecular diffusion method. The nanoporous CeO_2/MnO_2 composite has a large adsorption-specific surface and a suitable pore size, which make it favorable for diffusion in the selective catalytic reduction process of NH_3 to NO [10]. However, the above manufacturing processes are too complicated; their high production costs and environmental unfriendliness make it impossible to achieve large-scale industrial production, which limits their scope of application.

The baize-like monolithic CeO_2 and 7NiO/10CeO_2 (Ni:Ce at. ratio = 7:10) nanorod catalysts were successfully prepared by a combined dealloying and calcination process in our previous work, and the catalytic activities were evaluated using the CO catalytic oxidation process [11]. However, the catalytic activities of the composite materials were not studied for other proportions of Ni: Ce. In this work, a series of NiO/CeO_2 composites were prepared by adjusting the proportions of Ni and Ce in the precursor alloys. The results obtained for the 7NiO/10CeO_2 [11] and $Al_{85}Ni_{15}$ [11] material systems are re-displayed in this paper for a clear comparison with the results from the present study. Further, a series of MnO_2/CeO_2 composites were prepared by similar methods. A significantly higher catalytic activity than that obtained for pure CeO_2 was achieved. The results of our study are expected to provide a new direction for the synthesis of nanocomposites.

Experimental Methods and Characterization

$Al_{90}Ni_{10}$ [11], $Al_{90-x}Ni_xCe_{10}$ (x = 0 [11], 5, 7 [11], 10 at.%), and $Al_{88}Mn_{12-x}Ce_x$ (x = 0, 2, 4, 6, 8, 10 at.%) were prepared from the corresponding pure elements (Al: 99.9%, Ni: 99.9%, Ce: 99.9%, and Mn: 99.9%) according to the ingredients. The accurately weighed pure metals were arc-melted in an argon atmosphere to yield ingots. Subsequently, the pre-alloyed ingots were re-melted in a quartz tube under high-frequency induction heating. The metal solution was blown onto the surface of a melt-spinning single roller at a speed of 33 m s^{-1}. The resulting ribbons were typically 30 μm in thickness and 3 mm in width. The prepared melt-spun ribbons were dealloyed in a 20 wt% NaOH aqueous solution at room temperature for 3 h and then placed in a water bath at 80 °C for 10 h to remove residual Al. The corroded alloy ribbons were rinsed several times in deionized water and dried in an oven at 50 °C for 6 h. Subsequently, the dealloyed ribbons were calcined at 350 and 400 °C for 1 h, respectively.

The phase constitutions of the prepared samples were analyzed using a Bruker D8 Advance X-ray diffractometer (XRD). The microstructures and compositions were characterized using a JEM-2100 high-resolution transmission electron microscope (TEM) and a JSM-7000F field-emission scanning electron microscope (SEM). The catalyst (100 mg) was placed inside a quartz tube with an inner diameter of 6 mm to be measured. An asbestos column with a length of about 1 cm was used to fix the catalyst in the tube; the quartz tube was then placed inside a variable-temperature reaction furnace.

A reaction gas mixture consisting of 1%CO, 20%O_2, and 79%N_2 (volume) was fed into the reactor by using a mass flow meter (Brooks 5850E) at a total flow rate of 100 ml min^{-1} (60,000 h^{-1}). The concentrations of the produced CO_2 and unreacted CO gases were analyzed online using a GC-7900 gas chromatograph equipped with a flame ionization detector (FID). The CO conversion rate (%) was determined by the changes in the inlet and outlet concentrations of CO, and is given by the formula Eq. 1:

$$CO_{conversion} = \frac{C_{in} - C_{out}}{C_{in}} \times 100\% \quad (1)$$

where C_{in} represents the inlet concentration of CO and C_{out} is the outlet CO concentration.

Results and Discussion

Figure 1 shows the XRD patterns of the melt-spun $Al_{85}Ni_{15}$ [11], $Al_{90-x}Ni_xCe_{10}$ (x = 0 [11], 5, 7 [11], 10 at.%), and $Al_{88}Mn_xCe_{12-x}$ (x = 2, 4, 6, 8, and 10 at.%) ribbons after dealloying and calcinating. Only the definite peaks of CeO_2 appear in the diffraction pattern after the Al–Ce–Ni systems were dealloyed. However, the NiO peaks were too weak to be distinguished when the Ni content in the precursor alloy was between 0 and 10%, as shown in Fig. 1a. The XRD results of the Al–Ce–Mn alloy are similar to those of Al–Ce–Ni, and are shown in Fig. 1b.

The microstructures of the dealloyed melt-spun $Al_{90}Ce_{10}$ [11] ribbons are shown in Fig. 2 for after calcination at 400 °C. The appearance and growth of weeds can be confirmed in the plane-section image of the sample whose morphology is shown in Fig. 2a. The corresponding cross-sectional image exhibiting the nanorod morphology can be seen in Fig. 2b.

Figure 3 shows the plane and cross-section images of dealloyed $Al_{90-x}Ni_xCe_{10}$ (x = 5, 7 [11], 10 at.%) after calcination at 400 °C. The plane morphology of

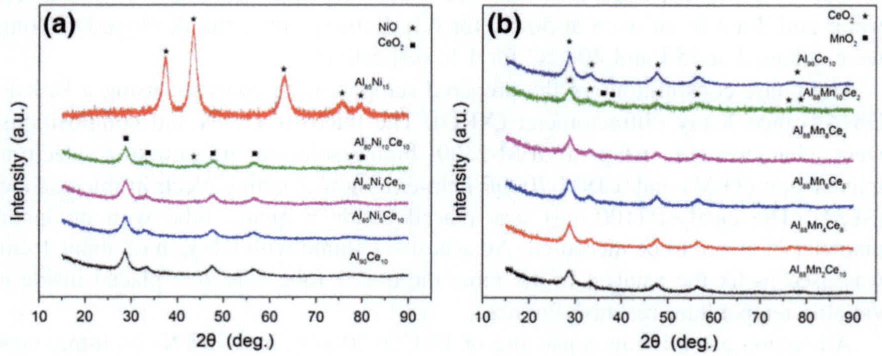

Fig. 1 XRD patterns of dealloyed **a** $Al_{90-x}Ni_xCe_{10}$ (x = 0, 5, 7 [11], 10 at.%) and $Al_{85}Ni_{15}$ [11], and **b** $Al_{88}Mn_xCe_{12-x}$ (x = 2, 4, 6, 8, and 10 at.%), and $Al_{90}Ce_{10}$ ribbons. The dealloyed $Al_{90-x}Ni_xCe_{10}$ and $Al_{85}Ni_{15}$ ribbons were calcined at 400 °C, while the dealloyed $Al_{88}Mn_xCe_{12-x}$ ribbons were calcined at 350 °C

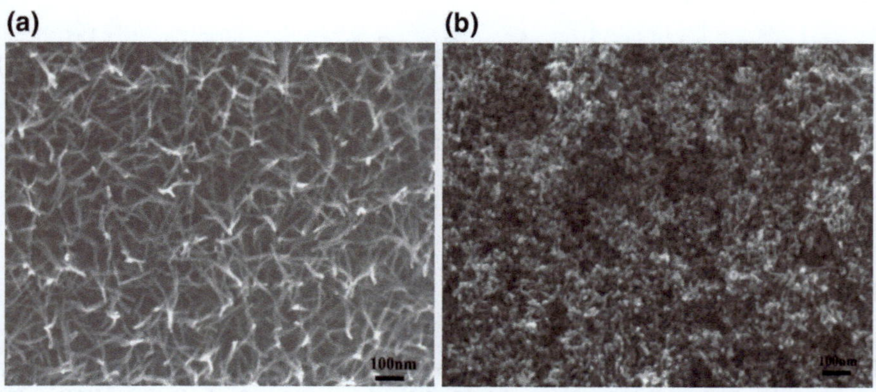

Fig. 2 Microstructures of **a** the plane section and **b** cross-section of dealloyed $Al_{90}Ce_{10}$ [11] after calcination at 400 °C

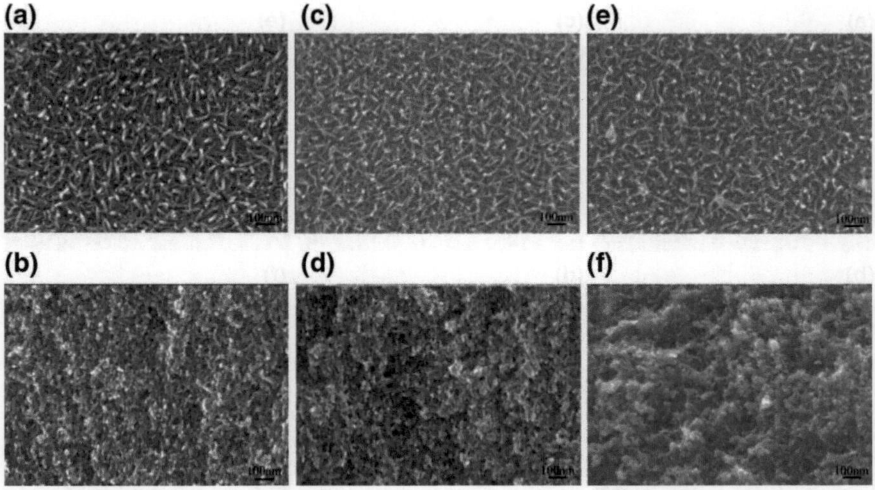

Fig. 3 Plane-section (**a**, **c**, **e**) and cross-section (**b**, **d**, **f**) SEM images of **a**, **b** $Al_{85}Ni_5Ce_{10}$, **c**, **d** $Al_{83}Ni_7Ce_{10}$ [11], and **e**, **f** $Al_{80}Ni_{10}Ce_{10}$ after dealloying and calcination at 400 °C

Table 1 Atomic ratios of dealloyed $Al_{85}Ni_{15}$ [11] and $Al_{90-x}Ni_xCe_{10}$ (x = 0, 5, 7 [11], 10 at.%) samples

Dealloyed sample	Ce (at.%)	Ni (at.%)	Al (at.%)
$Al_{85}Ni_{15}$	0	92.28	7.72
$Al_{85}Ce_{10}Ni_5$	63.42	29.04	7.54
$Al_{83}Ce_{10}Ni_7$	55.93	33.80	5.26
$Al_{80}Ce_{10}Ni_{10}$	46.18	46.88	6.94

dealloyed Al–Ni–Ce for different Ni contents is similar to that of Al–Ce, as seen in Figs. 3a, c, e. All the cross-sectional images reveal a nanorod morphology, similar to the case of the dealloyed Al–Ce, and can be seen in Figs. 3b, d, f.

The atomic ratios of Ce and Ni in the dealloyed $Al_{85}Ni_{15}$ [11] and $Al_{90-x}Ni_xCe_{10}$ (x = 0 [11], 5, 7 [11],10 at.%) sample were close to the nominal values found in the precursor alloys, as shown in Table 1, indicating that the loss of Ce and Ni during dealloying was small. Consequently, most of the Ni was retained in the sample.

Figure 4 shows the plane and cross-section scanning images of the dealloyed $Al_{88}Mn_{12-x}Ce_x$ (x = 2, 4, 6, 8, and 10 at.%) samples after calcination at 350 °C. When the Mn content was 2 at.% in the precursor alloy, the microstructure of the corresponding dealloyed sample closely resembled that of Al–Ce, as shown in Fig. 4a, b. When the Mn content increased to 6 at.%, the nanorods in the samples were significantly coarsened, which led to a decrease in porosity, as seen in Fig. 4c, d. The plane-section morphology of the dealloyed $Al_{88}Mn_8Ce_4$ sample was composed of irregular nanosheets with diameters of about 100–300 nm, as observed in Fig. 4e. The cross-sectional scanning image in Fig. 4f reveals the thickness of nanosheets to be about 5 nm.

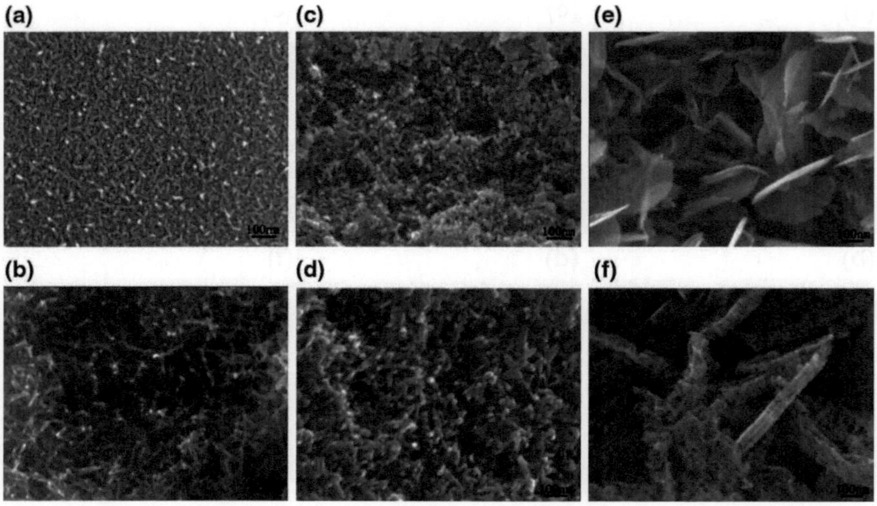

Fig. 4 Plane-section (**a**, **c**, **e**) and cross-section (**b**, **d**, **f**) SEM images of **a**, **b** $Al_{88}Mn_2Ce_{10}$, **c**, **d** $Al_{88}Mn_6Ce_6$, and **e**, **f** $Al_{88}Mn_8Ce_4$ after dealloying

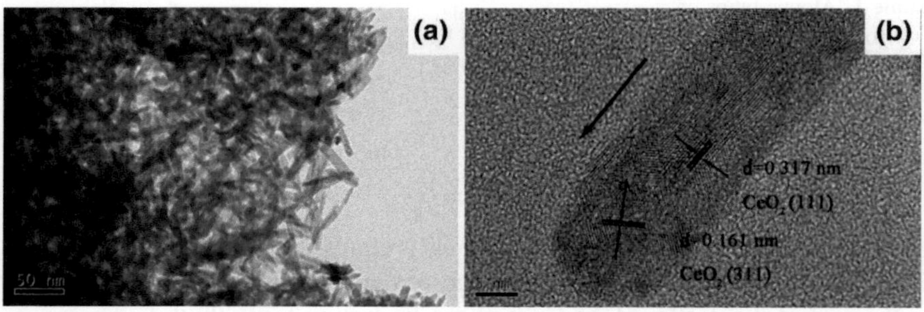

Fig. 5 **a** TEM and **b** HRTEM images of the dealloyed and calcined at 400 °C for the $Al_{90}Ce_{10}$ [11] ribbons

Figure 5 displays a TEM image of the dealloyed $Al_{90}Ce_{10}$ [11] ribbons. It is clearly observed that the nanorods appear in the samples, and their diameter and length are about 7–13 nm and 30–100 nm respectively. HRTEM analysis indicates that the nanorods are composed of CeO_2 nanoparticles (Fig. 5b). Further analysis reveals that the CeO_2 nanorods support each other to form a structure with nanoscale pores.

The above mentioned results show that Al can removed from the Al–Ce alloys by using aqueous NaOH aqueous, which also causes release of Ce atoms. It has been widely recognized that the Ce atoms can react quickly with hot water to generate $Ce(OH)_3$ nuclei; this hydroxide can develop into nanorods due to the anisotropic growth of different nuclei under basic hydrothermal conditions. As a result, CeO_2 nanorods are obtained through the dehydration and oxidation of the

$Ce(OH)_3$ nanorods during their calcination in air. Based on these results, it can be concluded that the CeO_2 nanorods can be obtained by the dealloying and calcination of the $Al_{90}Ce_{10}$ precursor alloy. These nanorods show a unique three-dimensional pore support structure, which can be called the nanorod support skeleton. This structure is significantly influenced by the presence of a third component and its concentration in the alloy, as can be seen in Fig. 4 for different Mn contents.

Figure 6 shows the TEM (Fig. a, c, e) and HRTEM images (Fig. b, d, f) of $Al_{83}Ni_7Ce_{10}$ (Fig. a, b) [11], $Al_{85}Ni_5Ce_{10}$ (Fig. c, d), and $Al_{80}Ni_{10}Ce_{10}$ (Fig. e, f) after dealloying and calcination at 400 °C. The TEM images of the dealloyed and calcined samples are similar to those of CeO_2. It is noteworthy that a large amount of NiO is not clearly observed under the TEM; only a few NiO nanoparticles could be observed by HRTEM analysis (Fig. 6b). However, no NiO nanoparticles are found in the $Al_{85}Ni_5Ce_{10}$ and $Al_{80}Ni_{10}Ce_{10}$ samples shown in Fig. 6d, f.

According to Figs. 1a, 3, 5, and 6b, and Table 1, most of the NiO nanoparticles are highly dispersed on the surface of CeO_2; however, the characteristic lattice images are not observed by HRTEM.

Figure 7 shows the TEM and HRTEM images of the sample obtained by dealloying $Al_{88}Mn_8Ce_4$ ribbons and calcining them at 350 °C. Many nanorods with diameters of 10 nm and lengths of 100 nm could be seen in Fig. 7a. As shown in Fig. 7b, the surface of the nanorods was loaded with many particles having diameters of about 5 nm and interplanar spacings of 2.11 Å. This value of interplanar spacing was found to be consistent with that of the MnO_2 (101) plane, thus confirming that the particles were composed of MnO_2. The support skeleton having an interplanar spacing of 3.11 Å was composed of the CeO_2 particles.

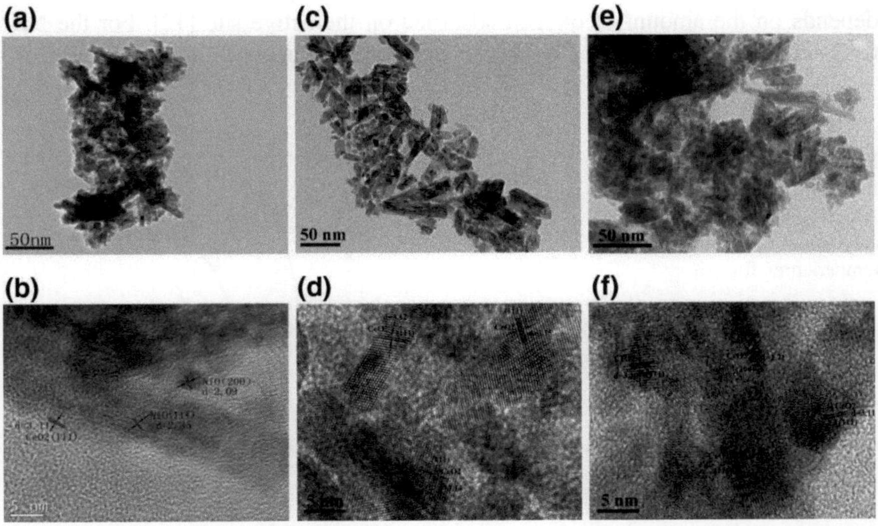

Fig. 6 TEM (**a**, **c**, **e**) and HRTEM (**b**, **d**, **f**) images of the samples **a**, **b** $Al_{83}Ni_7Ce_{10}$ [11], **c**, **d** $Al_{85}Ni_5Ce_{10}$, and **e**, **f** $Al_{80}Ni_{10}Ce_{10}$ after dealloying and calcining at 400 °C

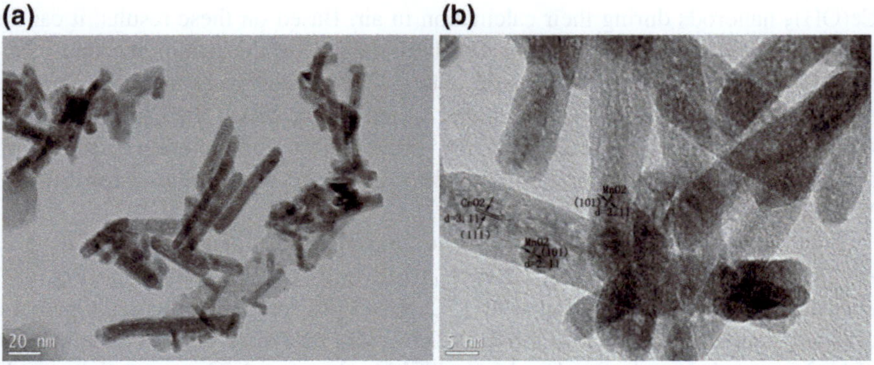

Fig. 7 **a** TEM and **b** HRTEM images of $Al_{88}Mn_8Ce_4$ after dealloying and calcination at 350 °C

From the previous work [11] and Fig. 8, it can be seen that the best calcination temperatures for the CeO_2/NiO and CeO_2/MnO_2 composites are 400 and 350 °C, respectively.

Figure 9 shows the CO conversion as a function of temperature for different oxide systems. It can be seen from Fig. 9a that when CO conversion is 50%, the corresponding temperature for the catalytic oxidation of CO is 250 °C; the complete conversion temperature is 360 °C for pure CeO_2. The 50% CO conversion and the complete CO conversion temperatures of the CeO_2/NiO composite are greatly reduced with increasing amounts of NiO added. When the atomic ratio of Ni:Ce is 7:10 [11], the catalytic oxidation of CO is 50% at 147 °C and the complete conversion temperature is 200 °C, as seen in Fig. 9a. This phenomenon is attributed to the redox reaction occurring at the active site, and the rate of the redox reaction depends on the amount of oxygen adsorbed on the active site [12]. For the NiO/CeO_2 composite, the system is oxygen-rich and conducive to the occurrence of a

Fig. 8 CO conversion as a function of temperature for $Al_{88}Mn_8Ce_4$ dealloyed and calcined at different temperatures for 1 h

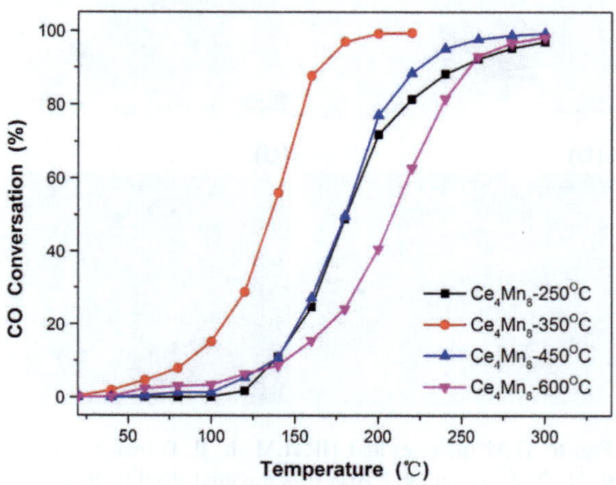

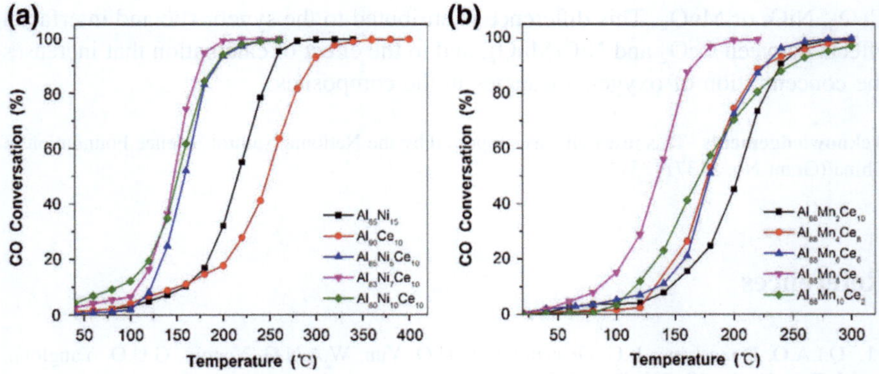

Fig. 9 CO conversion as a function of temperature for **a** $Al_{85}Ni_{15}$ [11] and $Al_{90-x}Ni_xCe_{10}$ (x = 0 [11], 5, 7 [11], 10 at.%) dealloyed and calcined at 400 °C for 1 h, and **b** $Al_{88}Mn_{12-x}Ce_x$ (x = 2, 4, 6, 8, and 10 at.%) dealloyed and calcined at 350 °C for 1 h

redox reaction. After the formation of the weak nanorods that support the skeleton, the number of unsaturated sites on the surface of the composite material is increased, which is beneficial for the adsorption and synergistic effects of oxygen. These two points contribute to the promotion of the catalytic performance.

It can be seen from Fig. 9b that the dealloyed and calcined (at 350 °C) $Al_{88}Mn_8Ce_4$ produces the CeO_2/MnO_2 composite with the best catalytic performance. When the reaction temperature is 136 °C, CO conversion is 50%, and when the reaction temperature is 200 °C, the conversion is 100%. When the Mn content is greater or less than 8 at.%, the catalytic performance is greatly reduced. The results show that the catalytic performance of the CeO_2/MnO_2 composite can be further controlled by regulating the original alloy composition. In addition, the catalytic performance of CeO_2/MnO_2 nanocomposites is significantly higher than that of pure MnO_2; this is attributed to not only the surface effects but also the redox reaction occurring at the interface between CeO_2 and MnO_2 [13]. In essence, most of the active sites are present at the interface between CeO_2 and MnO_2, and the concentration of active sites varies with the CeO_2 and MnO_2 contents because the oxygen atoms will migrate between the two oxides. The interaction between CeO_2 and MnO_2 results in the formation of a Mn_3O_4 phase, which indicates a synergistic effect between them.

Conclusion

In this work, CeO_2, NiO/CeO_2, and MnO_2/CeO_2 nanorod support-skeleton composites were successfully prepared by dealloying melt-spun Al–Ce, Al–Ce–Ni, and Al–Ce–Mn alloys in 20% NaOH aqueous solution. The NiO/CeO_2 and MnO_2/CeO_2 composites exhibit higher catalytic activity for the oxidation of CO than that of pure

CeO_2, NiO, or MnO_2. This difference is attributed to the synergistic and interfacial effects between CeO_2 and NiO/MnO_2, and to the effect of calcination that increases the concentration of oxygen vacancies in the composites.

Acknowledgements This research was supported by the National Natural Science Foundation of China (Grant No. 51371135).

References

1. Q.I.A.O. Dongsheng, L.U. Guanzhong, G.U.O. Yun, W.A.N.G. Yanqin, G.U.O. Yanglong, J. Rare Earths **28**, 742–746 (2010)
2. T.K. Nevanperä, S. Ojala, N. Bion, F. Epron, R.L. Keiski, Appl. Catal. B **182**, 611–625 (2016)
3. T. Saravanan, M. Shanmugam, P. Anandan, M. Azhagurajan, K. Pazhanivel, M. Arivanandhan, Y. Hayakawa, R. Jayavel, Dalton Trans. **44**, 9901–9908 (2015)
4. M.A. Guo-Qiang, Z.-Y. Wen, Q.-S. Wang, J. Jin, X.-W. Wu, J.-C. Zhang, J. Inorg. Mater. 913–918 (2015)
5. Y. Liu, H. Huang, L. Wang, D. Cai, B. Liu, D. Wang, Q. Li, T. Wang, Sens. Actuators, B Chem. **223**, 730–737 (2016)
6. A. Uzunoglu, L.A. Stanciu, Anal. Chim. Acta **909**, 121 (2016)
7. J. Kašpar, P. Fornasiero, M. Graziani, Catal. Today **50**, 285–298 (1999)
8. B. Lu, K. Kawamoto, Mater. Res. Bull. **53**, 70–78 (2014)
9. X. Liu, Y. Zuo, L. Li, X. Huang, G. Li, RSC Adv. **4**, 6397–6406 (2014)
10. Y. Wei, Y. Sun, W. Su, J. Liu, RSC Adv. **5**, 26231–26235 (2015)
11. X. Zhang, K. Li, W. Shi, C. Wei, X. Song, S. Yang, Z. Sun, Nanotechnology **28**, 45602 (2017)
12. J. Wan, L. Zhou, H. Deng, F. Zhan, R. Zhang, J. Mol. Catal. A: Chem. **407**, 67–74 (2015)
13. M. Sadeghinia, M. Rezaei, E. Amini, Korean J. Chem. Eng. **30**, 2012–2016 (2013)

Effect of Corrosion Potential on the Structure and Properties of Nanoporous Nickel

Jinxiong Wu, Siyue Dai, Xiankang Shan, Yu Zhang and Xiuling Yan

Abstract Taking the Ni–Mn alloy with 25% Ni as the precursor, nanoporous nickel (NPN) was fabricated by electrochemical dealloying, and the effect of corrosion potential on the material microstructure and electrochemical performance was studied in the article. Morphology, element species, content and phase of the sample were characterized and analyzed by scanning electron microscope (SEM), energy dispersive spectrometer (EDS) and X-ray diffractometer (XRD). Different from the previously report that the material surface presented the structure of "mud crack", materials prepared in this article remained the porous skeleton structure, but also reduced the large cracks and increased the hairline cracks. After a deep research, it is found that the corrosion potential determines the morphology and structure of porous material. This was due to the fact that corrosion rate became quicker when the corrosion voltage was higher and the degree of corrosion increased accordingly; precursor alloy defects increased as the dissolution of Mn, and the internal stress increased as a result; in addition, the diffusion rate of Ni was so low that it couldn't refill the vacancy of Mn timely, and then more cracks occurred, therefore there were significant differences for the number and size of cracks. In order to further understand the electrochemical performance of the electrode material prepared under a proper potential, cyclic voltammetry and charge-discharge technology were used to characterize the electrochemical performance of NPN. The results show that the specific capacitance of NPN was

J. Wu · S. Dai · X. Shan · Y. Zhang · X. Yan (✉)
School of Chemistry and Environmental Science, Yili Normal University,
Xinjiang 835000, China
e-mail: xlyan1212@163.com

J. Wu
e-mail: 18403433951@163.com

S. Dai
e-mail: 18699966487@163.com

X. Shan
e-mail: m13401754998@163.com

Y. Zhang
e-mail: 15261032863@163.com

Y. Han (ed.), *Advanced Functional Materials*,
https://doi.org/10.1007/978-981-13-0110-0_101

1.69 F cm^{-2} which is 3 times of that of the nanoporous nickel martial reported previously. Therefore the prepared nanocomposite nickel in this article is a high-quality electrode capacitor material.

Keywords Nanoporous Ni · Corrosion potential · Ratio between capacitance and area · Electrode capacitor

Introduction

Nanoporous metals are one of emerging materials in recent years, which combine the excellent features of function materials and structural materials, manifesting with a high specific surface area and particular interfacial characteristics, and have a wide application prospect in catalyzing, sensing and fuel cell etc. [1–4]. Electrochemical dealloying is an effective way to prepare nanoporous metals with three-dimensional open pore and bicontinuous ligament-channel structure and also is the research focus in recent years. The common binary dealloy systems are mainly focusing on the solid solutions, such as Au–Ag, Au–Zn, Cu–Mn, Ni–Al and Ni–Mn etc. [5–12]. As one of them, Ni–Mn alloy is inexpensive and easy to obtain, and have great advantage of preparing nanoporous Ni. Such as Hakamada et al. [13] obtained nanoporous nickel with a skeleton size of 10–20 nm, by electrochemical dealloying with rolled Ni–Mn alloy as raw materials; Dan et al. [14] carried out the dealloying treatment of $Ni_{38}Mn_{62}$ alloy in 0.17 mol L^{-1} $NH_4H_2PO_4$ solution, and found that when the open circuit potential was changed, the as-prepared nanopore type changed correspondingly. When the open circuit potential was −0.45 V, it formed the nanopore with several hundred nanometers, but only a few tenths of nanometers when it's −0.4 V. While the current researches on nanoporous metals have made significant progress, there are also shows some problems. The preparation technology of nanoporous metals is still under developing, there are no sufficient researches about influence on applied potential's dynamics of the dealloying, as well as researches on the electrochemical performance of the prepared nanoporous metal material [15–22].

Therefore it's necessary to conduct the research on the action mechanism of applied potential and the electrochemical performance of nanoporous metal materials. Focusing on the influencing mechanism of applied potential on the porous structure, the electrochemical properties of nanoporous nickel electrode materials were investigated, which prepared by electrochemical dealloying and possessed large specific surface area and three-dimensional continuous structure. It is expected that the research results can provide new ideas for the application of nanoporous nickel in electrochemical capacitors.

Experiment

Materials and Reagents. Pure Ni and Mn cube (Purity > 99.99%) were chosen as primary materials.

NaOH, $(NH_4)_2SO_4$, acetone and alcohol were analytic reagent (AR) and were bought from Tianjin Damao Chemical Rreagent Company. All purchased reagents were used directly without further processing. All solutions were prepared by deionized water.

Preparation and Characterization of Nanoporous Ni. Ni–Mn alloy were made by reining pure Ni and Mn (99.99%) in arc-furnace, followed by melting-spinning. The atomic content of Ni is controlled at 25%. For the $Ni_{25}Mn_{75}$ precursor alloy material, its grain size, morphology, compact degree, and distribution characteristics were characterized by XJP-3 optical metalloscope (OM). Dealloying was performed to prepare nanoporous nickel in at three-electrode system with a platinum foil as counter electrode and saturated calomel electrode (SCE) as reference electrode, and Ni-Mn alloy as working electrode. All electrochemical measurements were carried out using a CHI 660D potentiost (CHI instrument, Shanghai, China). The electrolyte was purged with pure nitrogen for 20 min prior to measuring. Nanoporous nickel was fabricated by electrochemical dealloying Ni–Mn alloy foils in NaOH solution at a suitable potential which selected by means of Tafel curve, linear sweep voltammetry (LSV) and other methods.

The morphology and composition of nanoporous nickel were characterized by scanning electron microscope (JSM-7500, JEOL) equipped with Energy Dispersive X-ray Spectroscopy. The crystallographic information was obtained with X-ray diffractometer (D8-ADVANCE, Bruker AXS).

Result and Analysis

Alloy Phase Diagram. Ni–Mn alloy is the continuous solid solution with γ-MnNi formed by Ni and γ-Mn in a scope of 353–1455 °C, the standard electrode potential of Ni and Mn are −0.25 and −1.182 V respectively, which meet the requirements of dealloying, that is the requirements of larger standard electrode potential difference and critical threshold [23]. Therefore Ni–Mn alloy with Ni 25 (wt%) was selected as precursor alloy, and the annealing temperature was 900 °C.

After long time storage, the segregation of the structure and composition of alloy material might happen, and the continuous corrosion channels wouldn't be guaranteed during the process of dealloying. A and B in Fig. 1 were constitution diagrams before and after the annealing of $Ni_{25}Mn_{75}$ alloy respectively, and it's obvious from the diagrams that the material grains after annealing were finer and more uniform so that they could provide more grain boundaries and diffusion channels for the dealloying and also promote the diffusion of inert atoms to obtain

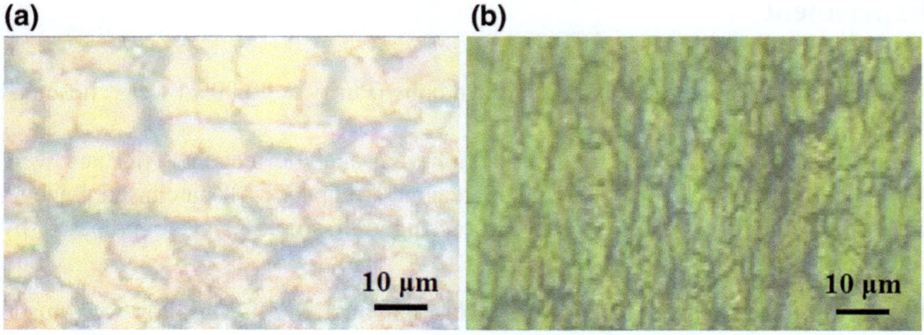

Fig. 1 $Ni_{25}Mn_{75}$ alloy phase diagram before (**a**) and after annealing (**b**)

multi-layer nanoporous Ni with uniform pores. That's the reason why the sample after annealing treatment was selected as the precursor to prepare nanoporous Ni.

Determination of Corrosion Potential. Corrosion potential has great impact on the morphology, structure and electrochemical performance of electrode material. So the determination of corrosion potential is very useful for preparing electrode material with good electrochemical performance.

In order to determine the scope of corrosion potentials, Tafel polarization curve test was conducted in $Ni_{25}Mn_{75}$ alloy electrode material and then corrosion potential was determined by measuring the anodic polarization curve; for the purpose of studying the impact of corrosion solution concentration on the material internal corrosion potential, Tafel polarization curve and anodic polarization curve tests were conducted indifferent concentration $(NH_4)_2SO_4$ solutions. Figure 2a is the Tafel polarization curve of $Ni_{25}Mn_{75}$ measured in 0.1, 1 and 3 mol L^{-1} $(NH_4)_2SO_4$ solutions, and it could be seen that corresponding corrosion potential of different concentration solutions was −0.88, −0.98 and −1.06 V respectively. Compared with the change trend of material corrosion potentials in different concentration corrosion solutions, it could be found that the critical potential of alloy decreased along with the increasing of $(NH_4)_2SO_4$ solution concentration, which is accordant with the result in literatures [24]. It was mainly due to that the corrosion current and corrosion rate increased with the increasing concentration of etchant solution, and the critical potential reduced accordingly. Taking fully into account the corrosion rate, value of critical potential as well as the uniformity and size of the pore after corrosion, selected 0.1 mol L^{-1} $(NH_4)_2SO_4$ solution as corrosion solution and −900 mV as corrosion potential according to the anodic polarization curves (see Fig. 2b), the electric current density keeps constant value (less than 1 μA cm^{-2} [25]) for about 1 h without decay, which means the base metal (Mn) of Ni–Mn alloy can be seen as removed completely and the dealloying process is finished basically.

The crystal structure and phase constitution of nanoporous nickel were characterized by X-ray diffraction (XRD), as shown in Fig. 3.

The nanoporous nickel (a) shows broad diffraction peaks with 2θ values at around 44.3°, 51.6° and 76.0°, which can be assigned to (111), (200) and

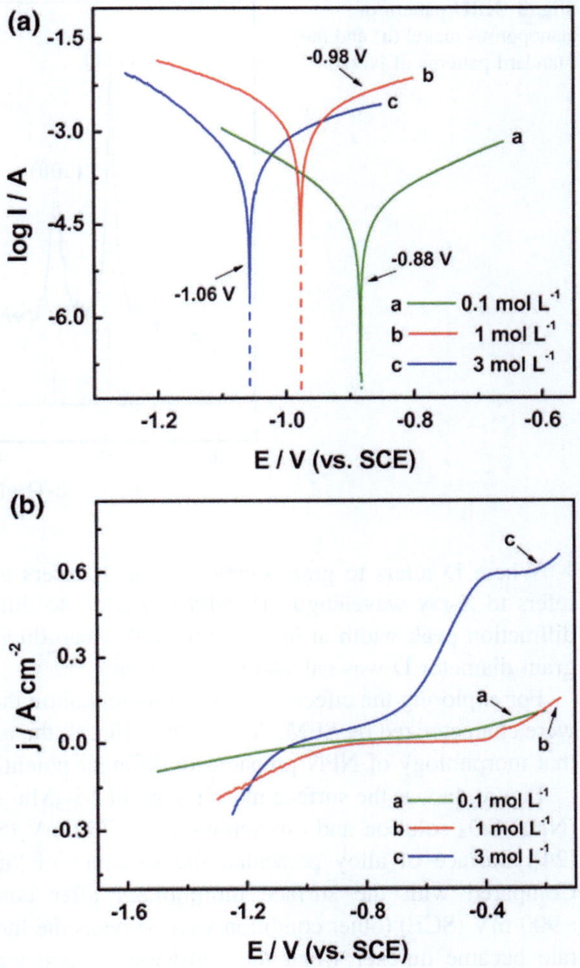

Fig. 2 Tafel polarization curves (**a**) and anodic polarization curve (**b**) in $(NH_4)_2SO_4$ solutions at different concentrations

(220) diffractions for a cubic crystals respectively. Compared with the standard XRD pattern of Ni (b), the diffraction peaks of materials are at 2θ = 44.5°, 51.8°, 76.4°, which indicates that the nickel in the materials are still in the form of elemental nickel, and its lattice constant a, b and c all equaled 3.524. It approved that Mn in the prepared NPN material had been corroded completely, nano-pores with uniform size were formed and the elementary substance Ni established its skeleton. According to the Scherrer formula [26]:

$$D = \frac{K\gamma}{B\cos\theta} \quad (1)$$

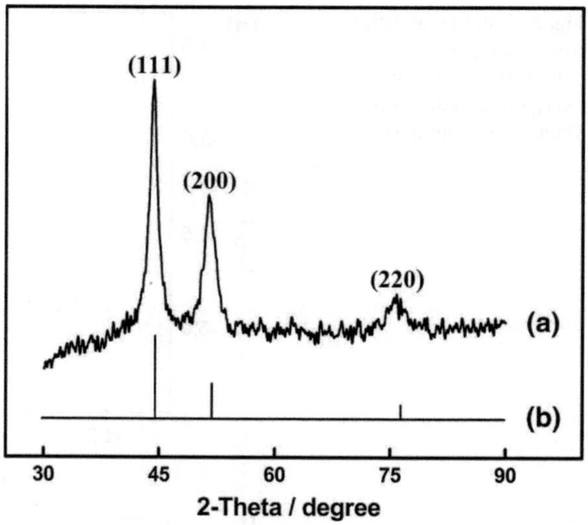

Fig. 3 XRD pattern of nanoporous nickel (**a**) and the standard patterns of Ni (**b**)

Where D refers to grain diameter, nm; K refers to Scherrer constant (0.89); γ refers to X-ray wavelength (0.15406); θ refers to diffraction angle (°); B refers to diffraction peak width at half height (rad). According to the Scherrer formula, the grain diameter D was calculated as 17.6 nm.

For exploring the effect of corrosion potential on the microstructure, the samples were characterized by SEM. As shown in Fig. 4, the results of SEM analysis shows that morphology of NPN prepared in different potential is quite different.

Figure 4a was the surface morphology of $Ni_{25}Mn_{75}$ alloy corroded in 1 mol L^{-1} $(NH_4)_2SO_4$ solution and on condition of −750 mV (SEC) for 1 h in the literature [24]. Surface of alloy presented the structure of "mud crack" after dealloying. Compared with the surface morphology after corrosion in the condition of −900 mV (SCE) (other condition were same as the literature), due to that corrosion rate became quicker when the corrosion voltage was higher and the degree of corrosion increased accordingly, and the ligament size and pore size were 23 and 20.9 nm, respectively; disfigurement of precursor alloy increased as the dissolution of Mn, and the internal stress increased as a result; in addition, the diffusion rate of Ni was so low that it couldn't refill the vacancy of Mn timely, and then more cracks occurred, therefore there was a significant difference for the number and size of cracks. When material was in condition of −900 mV (SCE), the amount of Ni was higher than that of material in −750 mV (SCE) mentioned in the literature (see Fig. 4c), Ni was reserved in the alloy as the supported structure to cover the vacancy caused by the selective corrosion of Mn [27], and then pores were formed so that the frame of alloy was supported, so the cracks of the material surface were reduced and became shallow. Combined with the energy spectrum analysis (Fig. 4c), it's found that the surface residual Mn of material mentioned in the literature was 19.35% after dealloying in the condition of −750 mV (SCE) and it was 7.81% in the condition of −900 mV (SCE). It indicates that increasing the

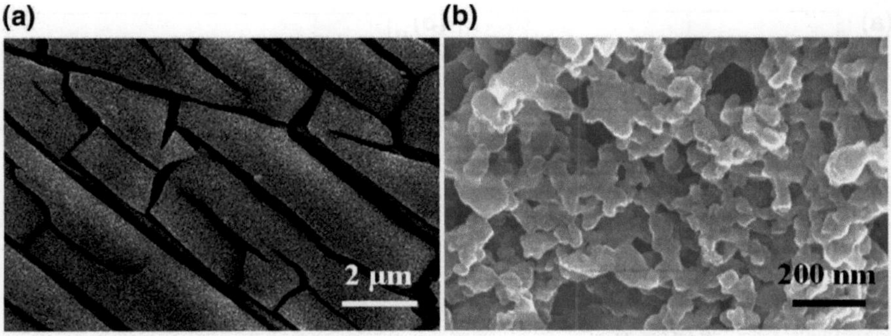

(c)

	Element	Mass[%]
A	Ni K	80.65
	Mn K	19.35
B	Ni K	92.82
	Mn K	7.81

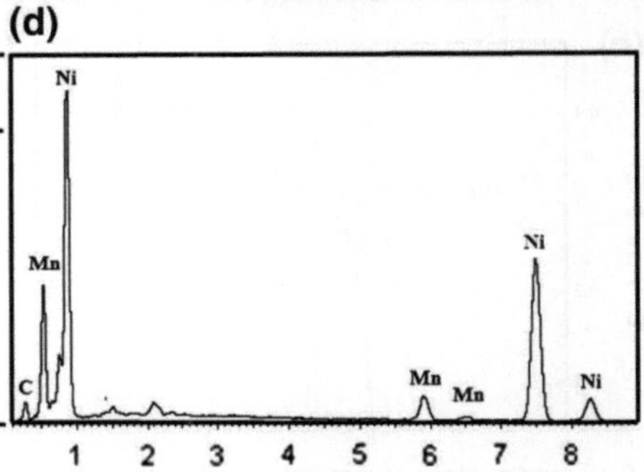

Fig. 4 Surface morphology of $Ni_{25}Mn_{75}$ alloy corroded in 1 mol L^{-1} $(NH_4)_2SO_4$ solution for 1 h at −750 mV (**a**) (in literatures) and at −900 mV (**b**); corresponding composition table of A and B materials (**c**); corresponding EDS spectra of material A (**d**)

corrosion potential appropriately could reduce the residual amount of the base component (Mn) and it would contribute to the complete dealloying of the precursor alloy and also provide favorable condition for material to form the porous structure. It is also needs to be noticed that appropriate residue could have a positive influence on the catalyst activity, and if all residual Mn were removed, the catalyst activity would not increase but decreased instead [28].

The electrochemical performance of NPN was characterized by cyclic voltammetry and galvanostatic charge-discharge technology. Through the analysis of cyclic voltammetry curve of electrode (Fig. 5a), many features could be found like the oxidation-reduction potential, reversibility of electrode reaction and its difficulty etc. From Fig. 5a, it could be found that when the scanning rate of electrode material prepared from $Ni_{25}Mn_{75}$ was 50 mV s^{-1}, its oxidation peak and reduction peak were 0.39 and 0.15 V respectively. However, due to the scanning rate increased, polarization would increase accordingly, when scanning rate was increased from 5 to 100 mV s^{-1}, the oxidation peak potential of Ni moved 0.1 V to the positive direction, peak current of oxidation peak increased by 9.4 times, while its corresponding reduction peak potential moved 0.1 V to the negative direction, and peak current of reduction peak increased by 6.5 times. In addition, along with the increasing of scanning rate, response current increased, and it meant that

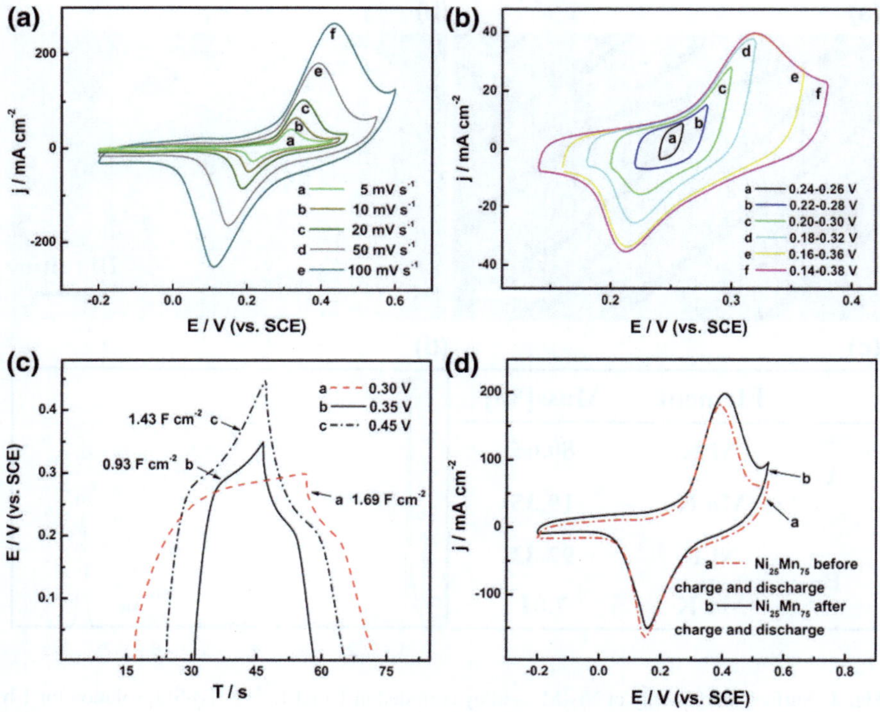

Fig. 5 Cyclic voltammetry curves of NPN at different scanning rates (**a**) and in different potentials range (**b**); charge-discharge curve of NPN at different potentials (**c**); comparison graph of cyclic voltammetry curves before and after charge-discharge performance test of NPN (**d**)

electrode had good electrochemical reversibility and showed that catalytic kinetics of material mainly was controlled by the diffusion [29].

Figure 5b was the cyclic voltammetry curve in different potential scopes of electrode material prepared from $Ni_{25}Mn_{75}$ after electrochemical dealloying. From Fig. 5b, it could be found that when potential range at 0.26–0.28 V (vs. SCE), response current would not change with the change of potentials; when potential range at 0.36–0.38 V (vs. SCE), response current would also not change with the change of potentials; and the 0.3–0.32 V (vs. SCE) is the transition range of the first two potential ranges; when the potential was over 0.3 V (vs. SCE), the stable response current would change sharply; then 0.3 V was selected as the high potential parameter value when the testing of chronopotentiometry, and in order to verify the validity of the selected high potential value, 0.35 and 0.45 V were selected as high potentials to draw the curve C. It could be seen from Fig. 5c that when the voltage was over 0.3 V, the voltage increased significantly, when the current density was 0.028 A cm^{-2}, the specific capacitance was 1.69, 0.93 and 1.43 F cm^{-2} respectively when the high potential was 0.3, 0.35 and 0.4 V. It proved obviously the validity of the selected high potential value, and the high potential value would have an impact on the measured capacitance value of material to some extent [30]. To study the impact of several charge-discharge curve tests on the cyclic voltammetry performance of material, cyclic voltammetry testing was conducted to the material after the charge-discharge testing. As shown in Fig. 5d, potential of redox peak moved 0.01 V to the positive after charge and discharge, and the peak pattern became shaper, which explained that charge-discharge testing wouldn't make material loss but make it more uniform and finer.

Conclusion

1. The corrosion speed is faster when concentration of corrosion solution is larger and corrosion voltage is higher, in the meantime degree of corrosion will increase accordingly. The defects of precursor alloy increases with the dissolution of Mn, and the internal stress increases as a result. In addition, the diffusion rate of Ni is too low to refill the vacancy of Mn timely, and then much more cracks occur, therefore there is a significant difference for the number and size of cracks compared with those in literature.
2. When the scanning speed of the prepared electrode material prepared from $Ni_{25}Mn_{75}$ is 50 mV s^{-1}, its oxidation peak and reduction peak are 0.39 and 0.15 V respectively. When the high potential is 0.3 V and the current density is 0.028 A cm^{-2}, ratio between capacitance and area of material is 1.69 F cm^{-2} which is 3 times of that of the nanoporous nickel martial reported previously. Therefore the prepared nano-composite nickel is a kind of higher performance capacitor material.

Acknowledgements Financial supports from the Nature Science Foundation of China (No. 51561030), "13th Five-Year" Key Discipline of Xinjiang Uygur Autonomous Region, Graduate Innovative Project of Yili Normal University (No. 2016YSY037), and Key Disciplines of Xinjiang Uygur Autonomous Region of Yili Normal University (No. 2017GIP001).

References

1. A. Wittstock, V. Zielasek, J. Biener, C.M. Friend, M. Bäumer, Nanoporous gold catalysts for selective gas-phase oxidative coupling of methanol at low temperature. Science **327**, 319–322 (2010)
2. R. Wang, C. Wang, W.B. Cai, Y. Ding, Adv. Mater. **22**, 1845–1848 (2010)
3. K. Bonroy, J.M. Friedt, F. Frederix, W. Laureyn, S. Langerock, A. Campitelli, M. Sára, G. Borghs, B. Goddeeris, P. Declerck, Anayticall Chem. **76**, 4299–4306 (2004)
4. D. Kramer, R.N. Viswanath, J. Weissmüller, Surface-stress induced macroscopic bending of nanoporous gold cantilevers. Nano Lett. **4**, 793–796 (2004)
5. M. Raney, Catalysts from alloys. Ind. Eng. Chem. **32**, 1199–1203 (1940)
6. H.X. Zhang, Z.F. Wang, M.Z. Yang, Q.B. Deng, The effect of an external magnetic field on the dealloying process of the Ni-Al alloy in alkaline solution **19**, 18167–18171 (2017)

7. Y. Ding, Y.J. Kim, J. Erlebacher, Nanoporous gold leaf: "ancient technology". Adv. Mater. **16**, 1897–1900 (2004)
8. Y. Ding, J. Erlebacher, Nanoporous metals with controlled multimodal pore size distribution. J. Am. Chem. Soc. **125**, 7772–7773 (2003)
9. Y.N. Li, H.P. Tang, J.Y. Wang, G.Z. Li, X.T. Kang, W.Y. Zhang, Y. Ge, B.J. Yang, Research progress of nanoporous nickel and nickel alloy by dealloying. Mater. China **30**, 49–53 (2011)
10. J.F. Huang, I.W. Sun, Adv. Func. Mater. **15**, 989–994 (2005)
11. J.R. Hayes, A.M. Hodge, J. Biener, A.V. Hamza, K. Sieradzki, Monolithic nanoporous copper by dealloying Mn-Cu. J. Mater. Res. **21**, 2611–2616 (2006)
12. M. Hakamada, M. Mabuchi, Fabrication of nanoporous palladium by dealloying and its thermal coarsening. J. Alloy. Compd. **479**, 326–329 (2009)
13. M. Hakamada, M. Mabuchi, Preparation of nanoporous Ni and Ni-Cu by dealloying of rolled Ni-Mn and Ni-Cu-Mn alloys. J. Alloy. Compd. **485**, 583–587 (2009)
14. Z.H. Dan, F.X. Qin, S. Yu, M. Izumi, N. Hara, Intermetallics **31**, 157–164 (2012)
15. Y. Yang, Preparation research on porous metallic materials. Technol. Dev. Chem. Ind. **40**, 21–23 (2011)
16. W.Y. Zhang, Z.P. Xi, M. Fang, Y.N. Li, G.Z. Li, L. Zhang, Research progresses on metallic nanoporous materials. Rare Metal Mater. Eng. **37**, 1129–1133 (2008)
17. L. Sun, C.L. Chien, P.C. Searson, Fabrication of nanoporous nickel by electrochemical dealloying **16**, 3125–3129 (2004)
18. J. Jayaraj, B.J. Park, D.H. Kim, W.T. Kim, E. Fleury, Nanometer-sized porous Ti-based metallic glass. Scripta Mater. **55**, 1063–1066 (2006)
19. X.L. Tan, Y.J. Tang, Y. Liu, J.S. Luo, K. Li, X.B. Liu, Progress in research on preparations of nanoporours metals by dealloying. Mater. Rev. **23**, 68–71 (2009)
20. Y.D. Kan, W.J. Liu, M.L. Zhong, M.X. Ma, Development of nanoporous metals prepared by dealloying. Heat Treat. Met. **33**, 43–46 (2008)
21. K.L. Guo, L. Wang, Q. Zhang, Some research developments on mechanical property of nanoporous metals. Mater. Rev. **31**, 97–102 (2017)
22. J. Chen, W.C. Hu, K. Du, D. Dong, S.Y. Zhang, L. Zhang, Research progress in fabrication of nanoporous metal materials. Mater. Rev. **24**, 24–27 (2010)
23. I.C. Oppenheim, D. Trevor, C.D. Chidsey, K. Sieradzki, In situ scanning tunneling microscopy of corrosion of silver-gold alloys. Science **254**, 687–689 (1991)
24. Z. Sun, M.E. Dissertation, Dalian Jiaotong University (2012)
25. Q. Zheng, Z.H. Zhang, H.L. Jia, Y.J. Qu, G.D. Liu, X.F. Bian, J. Alloy. Compd. **472**, 71–78 (2009)
26. M.X. Wang, Z.Q. Wei, N.C. Li, X.F. Zhao, P.X. Yan, Investigation of particle size of Ni nanopowders. Mater. Rev. **20**, 166–168 (2006)
27. Y. Ding, Nanoporous metals: a new class of nanostructured energy materials. J. Shandong Univ. (Nat. Sci. Ed.) **46**, 121–133 (2011)
28. H. Lei, Doctoral Dissertation, Dalian Institute of Chemical Physics, Chinese Academy of Sciences, 2003
29. Z.Z. Gao, Mester's Dissertation, Yanshan University, 2012
30. W. Zhang, K.Y. Liu, Y. Zhang, H. Zhen, G. Su, Studies of charge-discharge process in MnO_2 supercapacitor. Chemistry **70**, 217–221 (2007)

Effect of Two-Step Anodization on Structure of TiO_2 Nanotube Arrays

Siyu Chen, Shuhui Zhang, Zuojun Tan and Shu Zhang

Abstract A two-step anodic oxidation was used to prepare TiO_2 nanotube arrays with controllable structure and reproducibility by the isobarically anodization of pure Ti foil. The morphology and structure of TiO_2 nanotube arrays were characterized by high resolution scanning electron microscopy (SEM). Moreover, the formation mechanism of TiO_2 nanotube arrays was discussed. While the effects of anodic oxidation voltage and oxidation time on the diameter and length of nanotubes were further studied. The "field-assisted dissolution" model can perfectly explain the four stages of the first oxidation process, combined with the theory of oxygen bubble mould, to produce an electron flow during the growth of the oxide film, and oxygen bubbles hinder the upward growth of the barrier oxide film. The formation of pores resulted by bubbles evolution can be a good explanation of the TiO_2 nanotubes which show a large tube containing the structure of small tubes. The results demonstrate that with the increasing of oxidation time and oxidation voltage, the tube diameter, tube length and tube number of large tube have a tendency to increase obviously. The preparation of TiO_2 nanotubes by anodic oxidation is simple and easy to operate. It can efficiently and easily prepare TiO_2 nanotubes with excellent morphology and broaden its application range.

Keywords Anodic TiO_2 nanotubes · Two-step anodization · Formation mechanism

S. Chen · Z. Tan · S. Zhang (✉)
College of Science, Huazhong Agricultural University, Wuhan 430070, China
e-mail: zhangshu@mail.hzau.edu.cn

S. Zhang
College of Engineering, Huazhong Agricultural University, Wuhan 430070, China

Z. Tan · S. Zhang
Institute of Applied Physics, Huazhong Agricultural University, Wuhan 430070, China

Y. Han (ed.), *Advanced Functional Materials*,
https://doi.org/10.1007/978-981-13-0110-0_102

Introduction

Nano-materials have been favored by researches since it first appeared because of its unique physical and chemical properties. Especially TiO_2 nanomaterials prepared now are mostly nano-particles, TiO_2 thin films and TiO_2 nanotubes. The preparation process of the first two materials is quite mature, but due to their small electron transport rates and considering that the electron hole is easy to form composite, so they are largely limited in practical applications. TiO_2 nanotubes overcome these shortcomings because of its good photoelectric properties and a large specific surface area, so TiO_2 nanotubes have widely acknowledged in the photocatalytic, dye-sensitized solar cells, gas sensors [1–4] in recent years.

At present, there are many kinds of preparation methods of TiO_2 nanotube arrays, including major ones like template method, hydrothermal synthesis method and electrochemical anodic oxidation method. Among them, electrochemical anodic oxidation method is simple, low-cost, easy to control the reaction rate along with high order of nanotubes and neat arrangement, so it is commonly used. In 2001, the Grimes research team [5] reported for the first time that high-purity Ti-based films were used to fabricate uniformly ordered TiO_2 nanotube arrays in HF electrolytes by electrochemical anodic oxidation. After over ten years of development, the researchers optimized the anodic oxidation technology by adjusting the oxidation voltage, electrolyte concentration, electrolyte composition and reaction time to prepare a more excellent TiO_2 nanotube array [6–9]. At present, most of the preparation of TiO_2 nanotube arrays adopts an anodic oxidation method. Though this method could prepare the orderly array of TiO_2 nanotubes, the regularity of nanotubes appears to be not very good [10]. Since Wang et al. [11] first proposed a quadratic anodic oxidation process in 2009 and found that the TiO_2 nanotubes prepared by two-step anodic oxidation overcome the problem that the surface of TiO_2 nanotubes prepared by primary oxidation is rough and easy to be detached from the titanium matrix [10]. So far, the optimum conditions for the synthesis of TiO_2 nanotubes by anodic oxidation are not clear yet, especially in the two-step anodic oxidation. In order to investigate the law of controlling the growth of TiO_2 nanotubes by anodic oxidation method to broaden the application range, it is very important to efficiently and easily prepare TiO_2 nanotubes with excellent morphology. In this study, the order of TiO_2 nanotubes was prepared by secondary anodic oxidation method, followed with a further investigation into the effects of oxidation voltage and oxidation time on the microstructure of TiO_2 nanotubes in the second oxidization and the exploration of controllable preparation method of TiO_2 nanotubes.

Experimental Details

In this experiment, purity titanium foil (99.9%, 0.2 mm-thick, 20 mm $\times$ 50 mm) was used as substrate. TiO_2 nanotubes were prepared on titanium foil with two-step anodic oxidation method. The titanium foil was placed into the acetone, anhydrous ethanol and deionized water in order with 15–20 min ultrasonic shock for each, to remove the oil and other debris from sample's surface. The two-electrode system was used to carry out the anodizing where the titanium plate was placed as anode and the graphite rod as cathode, with the electrode spacing of 6 cm. The electrolyte was 0.30 wt% NH_4F, 2% volume H_2O (DI) and the ethylene glycol system. The first anodic oxidation was carried out with 60 V DC, and the oxidation time was 8 h. Then the sample was ultrasonically cleaned in the deionized water for several minutes, and the oxide layer was peeled off from the titanium substrate. The treated titanium sheet was re-immersed in the electrolyte, and the positive electrode was charged for the second oxidation (see Table 1 for oxidation). Then samples were rinsed with a large amount of deionized water, and dried in an oven to obtain highly ordered TiO_2 nanotubes. Real-time acquisition of current-time curves was completed via the Keithley 2000 multimeter. The microscopic morphology of the samples was observed by Hitachi S-5500 (Japan) high resolution scanning electron microscopy (SEM), of which the resolution could reach nanometer level.

Results and Discussions

Mechanism Analysis of Two-step Anodic Oxidation. Figure 1 is the scanning electron micrographs of sample S3. We first oxidized the titanium plate at 60 V for 8 h. On the titanium surface grew a layer of thick TiO_2 nanotube film. The combination of the film and the titanium substrate was not very solid and easily to be separated from the substrate with ultrasonic cleaning. After the separation of the first oxidation of TiO_2 nanotubes, an orderly arranged pit pattern was left on the surface of the titanium substrate, and then the second anodic oxidation was carried out on substrate. The nanotubes continued to grow towards the inside of the titanium plate along the direction of the pits. After a short period of time, a highly ordered array of TiO_2 nanotubes was obtained. The tube after the first oxidation remained very clear, covering the top of the second oxidized nanotubes, and a large

Table 1 Secondary oxidization conditions

Sample name	Experiment voltage (V)	Experiment duration (min)
S1	30	20
S2	30	30
S3	30	50
S4	50	20
S5	50	30
S6	50	50

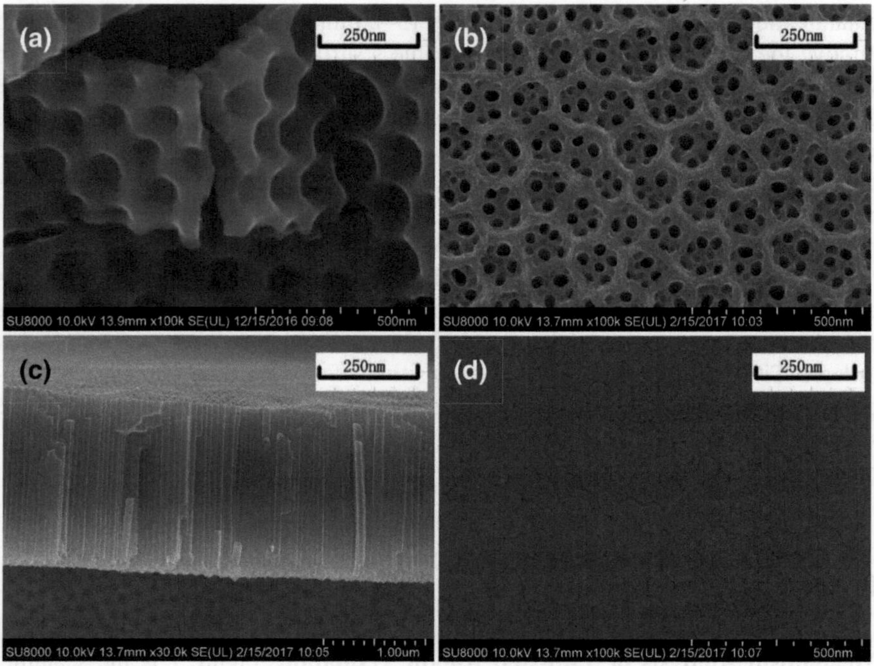

Fig. 1 A pit substrate left after the first oxidation (**a**) and a front view (**b**), a side view (**c**), a bottom view (**d**) of the sample S3

tube often contained several "tubes". With regard to the formation of porous oxidative nanotubes, the traditional theory seems to be unable to explain the experimental phenomenon of two-step oxidation. In 1992, Parkhutik and Shershulsky [12] modeled the "field-assisted solubilization" pore formation process. They analyzed the entire pipeline generation process into four stages according to the current-time curve of the anodic oxidation process at constant voltage: (I) The current on the current-time curve sharply decreases and this stage is the formation period of composite oxide film; (II) the current decline slows down and the cracks appear on the surface of the oxide film in this stage; (III) when small holes grow to a certain level, the current will be on the rise again. At this time, there will be no more etching reaction at the tube mouth and the electrolyte enters the pores to further widen a deepen then pore's bottom; (IV) the current is turning stable, and the growth rate and the etching rate of the barrier layer become equal with the holes being no longer deepened. Considering the four stages of I, II, III and IV of the first oxidized current-time curve we obtained in the experiment shown in Fig. 2, the experimental phenomena occurred in the first oxidation well reasonably explained. However, the "field-assisted dissolution" can not explain the current-time curve after the second oxidation. Until 2008, Zhu et al. [13] put

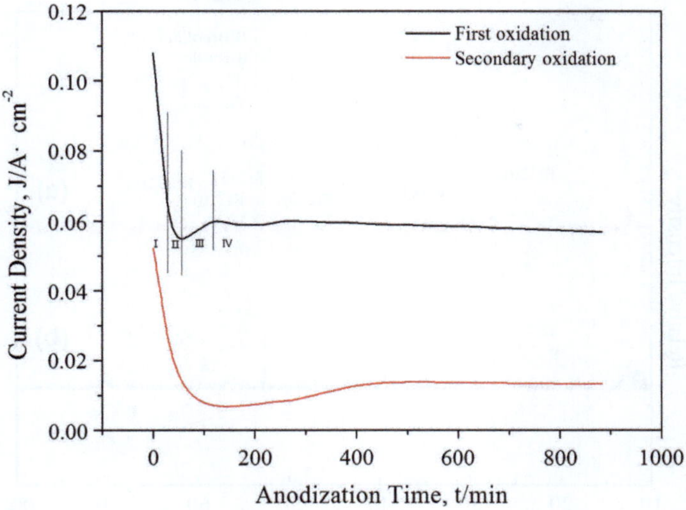

Fig. 2 Current-time curve

forward the "oxygen bubble mould" growth model. They believed that there were two currents in the growth process of the oxide film. The migration of the anions and cations forms an ionic current, and an oxide film is formed at the two interfaces of the barrier layer. When the oxide film of the barrier layer grows to a certain thickness, O^{2-} or OH^{-} discharges and produces electrons to release oxygen. The presence of oxygen bubbles hinders the upward growth of the oxide film of the barrier layer, and these small bubbles play a mold effect, forming pores in the process of bubble precipitation, as shown in Fig. 1 The big pores on the surface of the porous nanotubes that we prepared are the pits left after the first oxidation. The second oxidation continued to form on this basis. However, through the comparison of the bottom view and the top view in Fig. 1, the so-called "tube" may be a closed tube or a stop tube, and in fact there is only one real major tube among these two or three small tubes and the outlet of the major tube is smaller than the major tube outlet of the first oxide film. Other small holes are only very shallow part of the surface of the results. The reason behind this shape is caused by O_2 splitting into smaller bubbles during spilling over. The TiO_2 nanotubes without annealing only had titanium base peak as shown in Fig. 3b. The results indicated that, the samples had amorphous structures during the anodic oxidation and it began to transform to rutile and brookite at 600 °C. The heat treatment caused crystallization of the amorphous titania nanotube arrays, it contribute to bonding strength with the substrate of titanium.

Effect of Secondary Oxidation Conditions for Formation of TiO_2 Nanotubes. In the experiment, the oxidation voltage and the growth time are the parameters that

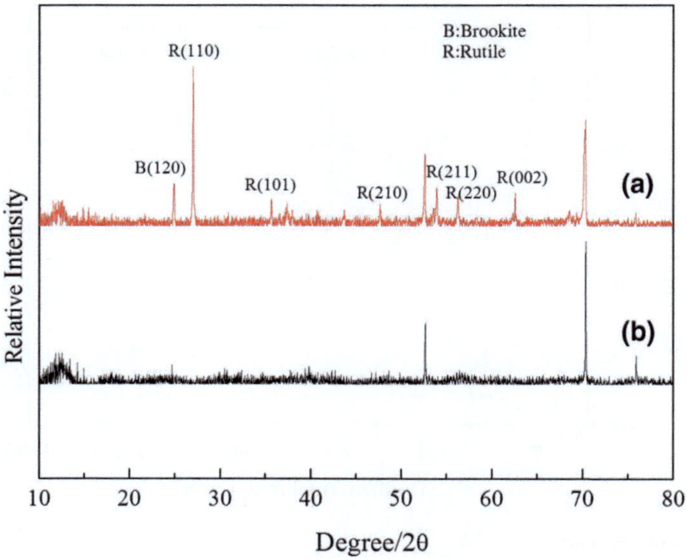

Fig. 3 XRD patterns of **a** annealing at 600 °C and **b** unannealing of TiO_2 nanotubes

can be adjusted. The second oxidation can change the two parameters, and the TiO_2 nanotube arrays with different morphologies can be obtained. Figures 4 and 5 show the top and side views of the TiO_2 nanotube array under different preparation conditions. Along with Table 2, the change in the length of the titanium dioxide nanotube with the preparation conditions could be clearly seen. From the top view, the first oxidation of the pit is clearly visible, covering the second oxidation of the formation of the top of the nanotubes, and the diameter of the big tube is much larger than the diameter of the small tube as a large tube often "contains" a few small tube. Comparing S1, S2 with S3, the average diameter of the big tube increases from 100 to 120 nm with the increase of the oxidation time under the oxidation condition of 30 V. The tube wall is also thinned, and the average pore diameter of the tube increases from 25 to 40 nm, as the number of small tubes is also increasing. Comparing S4, S5 with S6, the oxidation time increased at 50 V, and the average diameter of the tube increased from 95 to 110 nm. The tube wall was thinned, and the number of tubules increased by 3 to 4, with the tube length also continuing to increase. The morphology of TiO_2 nanotube arrays has been significantly improved to obtain a plain surface. Comparing S3 with S6, the diameter and length of the nanotubes increase with the increase of the voltage in the range of 30–50 V, which is explained by that the increased voltage accelerated the etching of the titanium dioxide on the electrolyte. The increase of the two-step oxidation time makes the etching time of the electrolyte on the titanium dioxide longer, both of which will increase the depth of etching, that is, increase the length

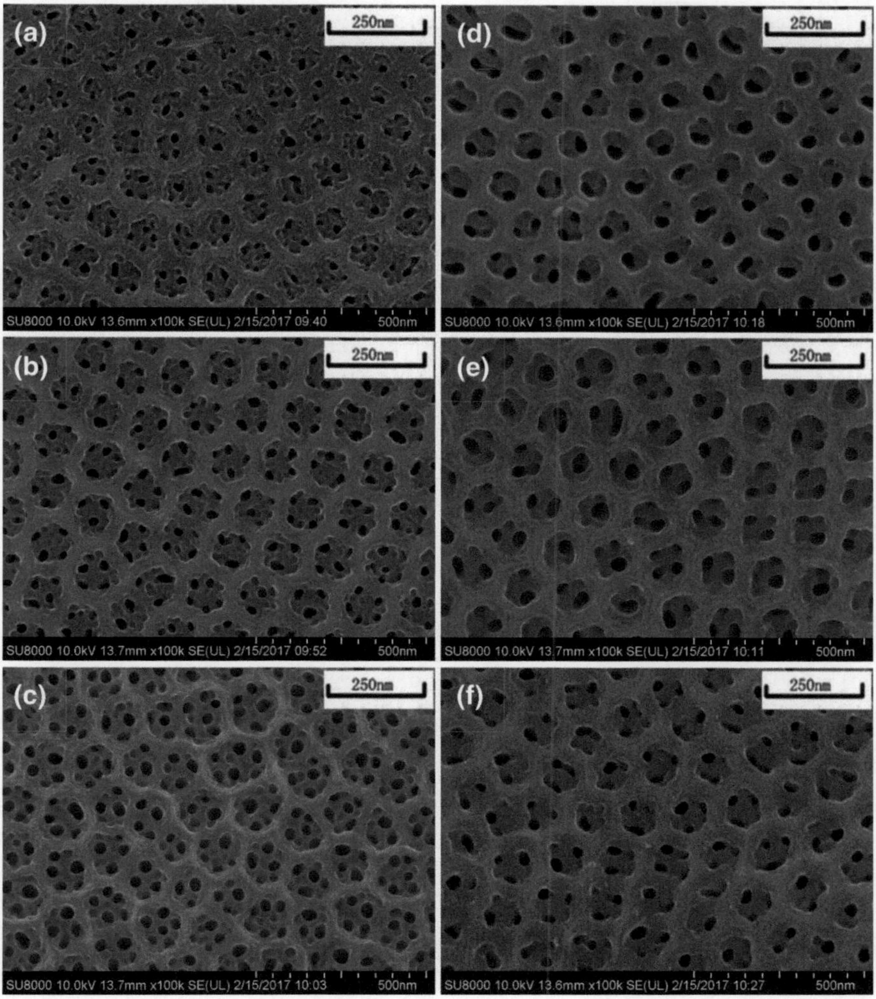

Fig. 4 Top view of samples S1, S2, S3, S4, S5 and S6

of titanium dioxide nanotubes. Considering the principle of the preceding link, we believe that the reason for this result is that the pores are formed by two-step oxidation, whether it is increasing the voltage or increasing the oxidation time, and its essence is to increase the electrolyte to titanium dioxide. Therefore, with the increase of oxidation time and oxidation voltage, the diameter of tube, pipe length and tube number of tubule are obviously increasing. In the process of actual production, the expected tube diameter and tube length through the control of both these two.

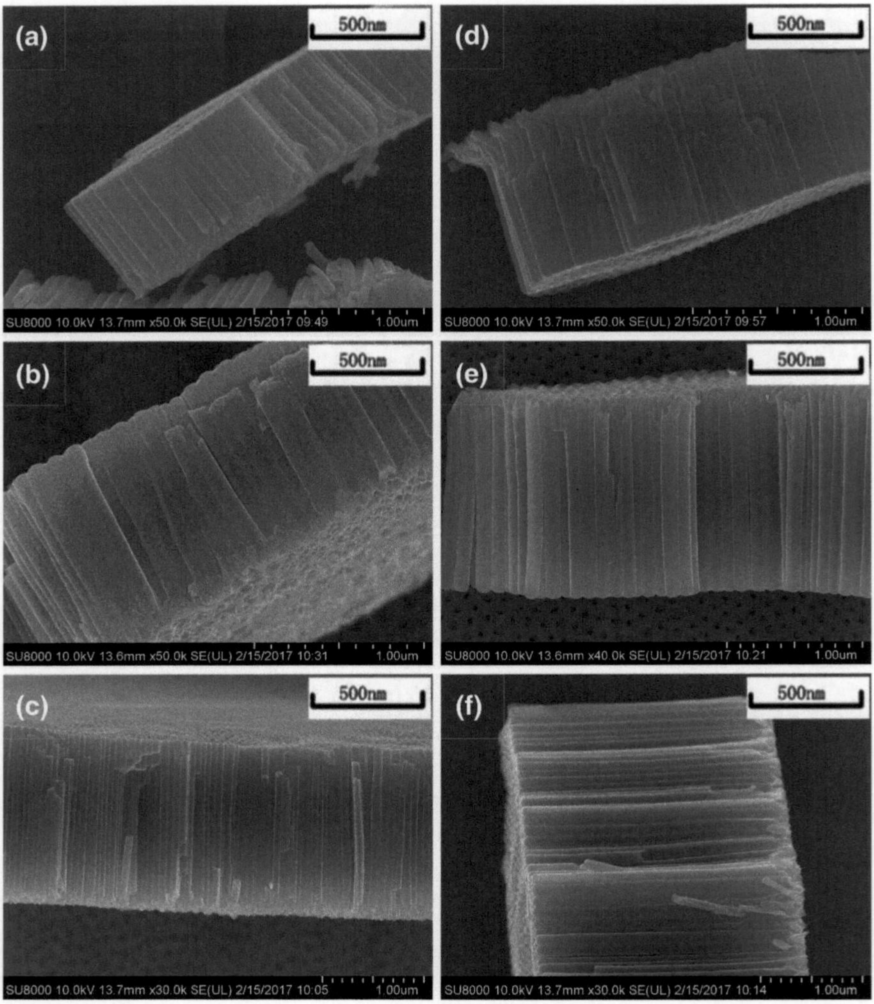

Fig. 5 Side view of samples of S1, S2, S3, S4, S5 and S6

Table 2 Morphology parameters of TiO_2 nanotubes

Sample name	Average tube length (nm)	Average tube diameter of big tube (nm)	Average tube diameter of small tube (nm)	Average tube amount of small tube (nm)
S1	682.00	100.00	33.20	3–4
S2	880.00	123.00	35.82	4–5
S3	1700.00	126.70	40.00	5–6
S4	1100.00	106.00	39.67	2–3
S5	1500.00	113.00	43.25	3
S6	2600.00	110.00	45.47	3–4

Conclusions

In this paper, TiO_2 nanotubes were prepared on the titanium foil with two-step anodic oxidation method. The formation mechanism of TiO_2 nanotube arrays was discussed. Combined with field-assisted dissolution and oxygen bubble mould, the structure of TiO_2 nanotubes big tubes containing the small tubes could be well explained. The effects of anodic oxidation voltage and oxidation time on the diameter and length of nanotubes were also studied. The results showed that with the increase of oxidation time and oxidation voltage, the tube diameter, tube length and tube number of large tube were all increasing. In the experimental process, since we only took two groups of voltage and three groups of oxidation time and only discussed the two-step oxidation conditions, so the relationship between the nature and preparation conditions of titanium dioxide nanotubes was only explored qualitatively. If a further quantitative exploration is expected to dive into the relationship, the voltage and oxidation time of the change span need to be increased on one hand, and the voltage and oxidation time point density need to be elevated on the other hand. In this case, the relationship between the morphology of the tube and the preparation conditions could be clearly and precisely understood, so that the desired titanium dioxide nanotubes can be eventually obtained through controlling the preparation conditions.

References

1. GK. Mor et al., A room-temperature TiO_2-nanotube hydrogen sensor able to self-clean photoactively from environmental contamination. Mater. Res. **19**(2), 628–634 (2004)
2. B. O'Regan, M. Grätzel, A low-cost, high-efficiency solar cell based on dye- sensitized colloidal TiO_2 films. Nature **353**(6346), 737–740 (1991)
3. M. Paulose et al., Backside illuminated dye-sensitized solar cells based on titania nanotube array electrodes. Nanotechnology **17**(5), 1446–1448 (2006)
4. G.K. Mor et al., A self-cleaning, room-temperature titania-nanotube hydrogen gas sensor. Sens. Lett. **1**(1), 42–46 (2003)
5. Qing Gong, et al. J. Central S. Univ. **42**(11), 3270–3276 (2011)
6. G. Butail et al., Branched titania nanotubes through anodization voltage control. Thin Solid Films **520**(1), 235–238 (2011)
7. Z. Xu et al., Mater. Sci. Technol. **28**(10), 865–870 (2012)
8. S. Yoriya, C.A. Grimes, Self-assembled anodic TiO_2 nanotube arrays: electrolyte properties and their effect on resulting morphologies. J. Mater. Chem. **21**(1), 102–108 (2010)
9. N.M. Deyab, et al. J. Power Source. **280**(1), 339–346 (2015)
10. S. Li et al., Anodization fabrication of highly ordered TiO_2 nanotubes. J. Phys. Chem. C. **113** (29), 12759–12765 (2009)
11. D. Wang et al., A novel protocol toward perfect alignment of anodized TiO_2 nanotubes. Adv. Mater. **21**(19), 1964–1967 (2009)
12. V.P. Parkhutik, V.I. Shershulsky, Theoretical modelling of porous oxide growth on aluminium. J. Phys. D Appl. Phys. **25**(8), 1258 (1992)
13. X. Zhu et al., Oxygen bubble mould effect: serrated nanopore formation and porous alumina growth. Monatsh. Chem. Chem. Monthly. **139**(9), 999–1003 (2008)

The Influence Factor of Silicon Carbide Ceramic Sintering

Chunlei Duan, Xiaochen Li, Qiang Zhao, Wenlin Zhao, Lihua Fu, Guangrong Zhang and Yuhong Chen

Abstract Sintering is the key process to a pressureless sintering silicon carbide ceramic. From the perspective of industrial production, this paper discusses the influence factors of big size silicon carbide ceramic sintering from three aspects: sintering curves, the density of green body and particle size of raw material. The results were as follows: (1) Under the conditions that heating rate from 1000 to 1500 °C and from 2100 to 2150 °C is 3 °C per minute and 0.5 °C per minute respectively, a big size product of densification of the ceramics has been obtained. (2) Using dry pressing at 50 MPa, dwell time 2 min, and then with CIP at 180 MPa, dwell time 5 min, the green body densities will more uniform from center to edge. (3) Applying raw materials particle size below 0.6 μm the big size product will get more dense material. Combining with the above three methods. The final product can achieve densification. The sintered sample can been achieved with density of 3.12 g/cm^3, Vickers hardness of 26 GPa, and three-point flexure strength 420 MPa.

Keywords Pressureless sintering · Silicon carbide

C. Duan (✉) · X. Li · Q. Zhao · W. Zhao · L. Fu · G. Zhang
Ningxia Machinery Research Institute Co., Ltd, Yinchuan 750001, China
e-mail: dcl163@163.com

X. Li
e-mail: lixiaochen@nxjixie.com

Q. Zhao
e-mail: zhaoqiang@nxjixie.com

W. Zhao
e-mail: zhaowenlin@nxjixie.com

L. Fu
e-mail: fulihua@nxjixie.com

G. Zhang
e-mail: zhangguangrong@nxjixie.com

Y. Chen
School of Materials Science and Engineering,
Beifang University of Nationalities, Yinchuan 750001, China
e-mail: lyhchen@163.com

Y. Han (ed.), *Advanced Functional Materials*,
https://doi.org/10.1007/978-981-13-0110-0_103

Introduction

As a common structural ceramics, silicon carbide ceramics have high hardness, high strength and low density, which are widely used in various fields. Silicon carbide is a covalent compound, so it is very difficult to sintering. Pressureless sintering silicon carbide was invented by Prochazka [1], Prochazka suggested that adding carbon and boron could improve the sintering properties of silicon carbide. In this process, carbon can react with silicon dioxide on the surface of silicon carbide to improve the surface energy [2]. And the same time, Boron can reduce the grain boundary energy [2–4].

Sintering is the key process to manufacturing a pressureless sintering silicon carbide ceramic. The sintering process of silicon carbide is a very difficult process, which usually requires a temperature of 2150°–2180° [5, 6]. During the sintering process, the high temperature will cause the exaggerated grain growth of silicon carbide grain, thus reducing the mechanical properties of the material. To a big size silicon carbide ceramics, some time the surface grain has been abnormal grain growth, while the internal density is very low. So there are many factors can influence sintering performance.

Experimental Procedure

Commercially available α-SiC (Ningxia Machinery Research Institute CO., Ltd., China) was used as raw powder. The properties of the powder are shown in Table 1.

Thermogravimetric analysis of binder was tested at temperature from 30 to 800 °C, with heating rate 5 °C/min (test equipment: Netzsch-STA449C).

Rectangle sample with length of 150 mm, width of 100 mm and thickness of 100 mm were prepared by dry press or use dry pressing followed by cold isostatic pressing (Details will be discussed in the paper).

The equipment used in sintering is high temperature sintering furnace, and argon gas is used as the protective gas in sintering process. Density is detected by Archimedes method. (ASTMC373). Polished samples were observed using scanning

Table 1 Properties of the Submicron silicon carbide powder

Properties	Units	Values
Particle size d50	μm	≤0.7
Purity	wt%	≥99.0
T. Fe	wt%	≤0.03
F. C	wt%	≤0.3
F. Si	wt%	≤0.15
SiO_2	wt%	≤0.15

electron microscopy (Shimadzu- SSX-550).The flexural strength is tested by three point bending method. Ten samples as bars of 3 mm × 4 mm × 36 mm were tested in according to the standard GB/T 6569-2006 (cross head speed 0.5 mm/min).

Results and Discussion

The Influence of Sintering Curve

Most time there are 3 ways to improve sintering performance. First, reduce grain boundary. Second, enhance surface energy. Third, use ultrafine silicon carbide powder [1].

To big size sample, because the size of the product is large, when it sintering, the internal and external temperatures are different. In this experiment, it will adjust sintering curve. Normally there are three stages, which are very important for sintering. In initial stage, binder will remove from green body, in intermediate stage, carbon will react with dioxide silicon. In final stage, the pore will remove and grain will grow [7].

From the results of Fig. 1, before 500 °C binders will removal completely. And the same time, between 1000 and 1500 °C carbon will react with dioxide silicon. Based on these, the experiment design two type of sintering curve, Type A is normal sintering curve, it usually to sintering small size products, Type B sintering curve prolong before 500°, the heating rate 3 °C per minute from 1000 to 1500 °C, the heating rate 0.5 °C per minute from 2100 to 2150°C (Fig. 2).

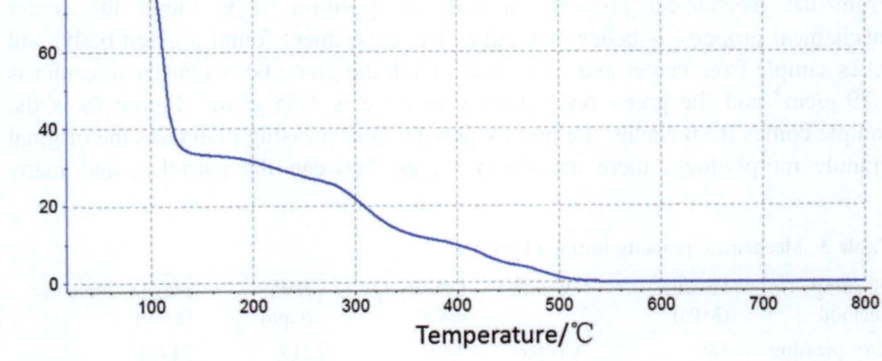

Fig. 1 The thermal analysis of binder

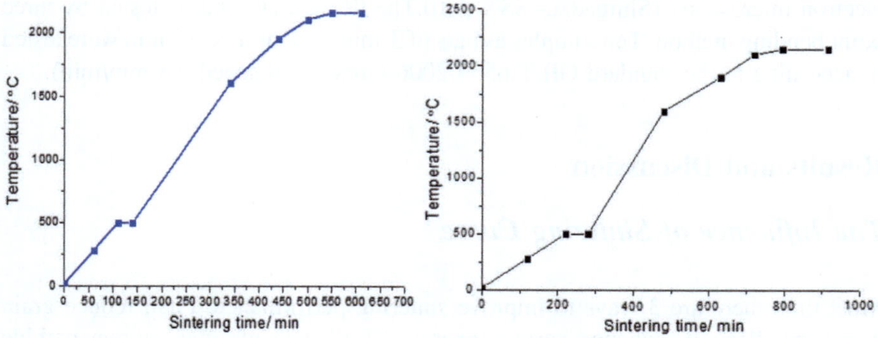

Fig. 2 Sintering curve

Table 2 Sintering results

Shaping method	Pressure (MPa)	Sintering curve	Density (g/cm^3)
Dry pressing	180	Type A	3.04
Dry pressing	180	Type B	3.08

The results of two different sintering curves were shown in Table 2. From the results, the density of the sample which is sintered by type B sintering curve is better, but the results also not very well, the density reached 3.08 g/cm^3. In order to find what is the real reason the experiment grinding the sample, take the sample from center and edge. And then test mechanical property, the results as shown in Table 3.

It is shown that the density and other property is big different from center and edge.

The Influence of Pressure and Green Body Density

From the mechanical property of different position, it is found the center mechanical property is better than edge. The experiment found a green body, and takes sample from center and edge. It is found, the green body density in center is 1.79 g/cm^3 and the green body density in edge is 1.85 g/cm^3. Figure 3a is the sample comes from center, the premix powder after pressing maintains the original granule morphology, there are obvious gaps between the particles, and many

Table 3 Mechanical property testing (Type B)

Shaping method	Pressure (MPa)	Position	Density (g/cm^3)	HV5 (Gpa)	Flexure strength (MPa)
Dry pressing	180	Center	3.06	21.6	317.6
Dry pressing	180	Edge	3.13	24.2	402.2

Fig. 3 Microstructure for green body

Table 4 Forming experiment

N	Shaping method	Pressure (MPa)	Holding time (min)	Shaping method	Pressure (Mpa)	Holding time (min)
1	Dry pressing	180	5	CIP	–	–
2	Dry pressing	180	10	CIP	–	–
3	Dry pressing	50	2	CIP	140	5
4	Dry pressing	50	2	CIP	140	10
5	Dry pressing	50	2	CIP	180	5
6	Dry pressing	50	2	CIP	180	10

CIP Cold Isostatic Pressing

defects in the interior. On the contrary, the sample comes from edge (Fig. 3b), it is almost dense. Some studies have suggested that the green body density will impact on the sintered density [8]. It is why sintered density in different position is different, so it needs to make the green body density more uniform. An experiment was designed in the next step as schemed in the Table 4.

It is found from Fig. 4, the dark colour is the green body density of center sample, and light colour is the green body density of edge sample. Compared with only use dry pressing, Use dry pressing at 50 MPa, dwell time 2 min, and then with CIP at 180 MPa, dwell time 10 min. The product will get best property performance (Table 5).

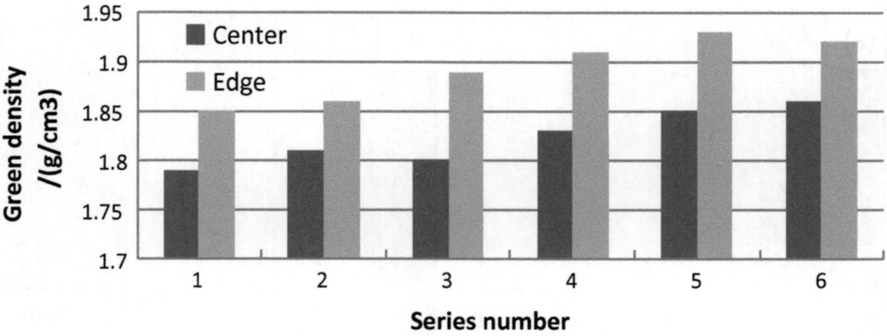

Fig. 4 Green body density result

Table 5 Mechanical property testing

N	Position	Density (g/cm^3)	HV5 (GPa)	Flexure strength (MPa)
6	Center	3.11	23.3	395.9
6	Edge	3.13	25.9	420.0

The Influence of Particle Size

This experiment select 2 different particle sizes, and then performed by dry press with CIP, with pressure of 180 MPa, sintering temperature 2150 °C, dwell time 1.5 h. And then properties were tested.

The mechanical properties (density, hardness, flexural strength) of sintered by different particle size are summarized in Table 6. Use fine particle size raw material exhibited high hardness and flexure strength values. Because fine particle size can enhance surface energy, it can make material denser. And the same time, Fig. 5 includes two pictures of 3000 times SEM pictures. Picture 1 is sintered by 0.65 μm raw materials, it is shown that the grain shape is elongated grain, grain size from 5-10 μm, many pores can be seen from the picture. Picture 2 is sintered by 0.51 μm raw materials, it is shown that the grain size is about 0.5 μm, the material almost dense, only a few pore can find from the picture.

Table 6 Mechanical property testing

N	Grinding time (h)	Particle size (μm)	Density (g/cm^3)	HV5 (GPa)	Flexure strength (MPa)
1	54	0.650	3.11	23.3	395
2	60	0.510	3.12	26.0	420

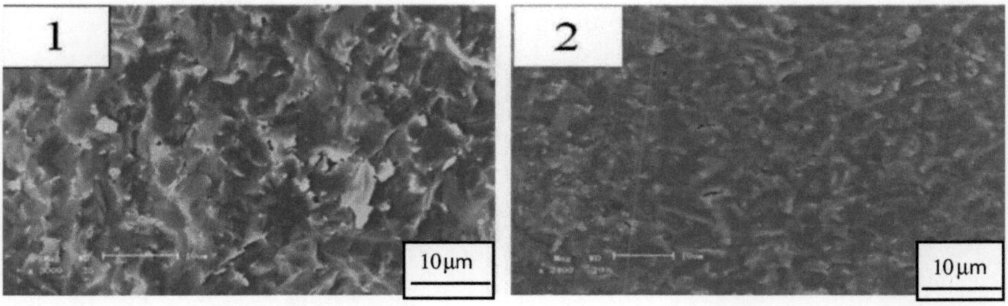

Fig. 5 SEM pictures of different particle size

Conclusion

(1) Under conditions of the heating rate 3 °C per minute from 1000 to 1500 °C, the heating rate 0.5 °C per minute from 2100 to 2150 °C, a big size product of densification of the ceramics has been obtained.
(2) Use dry pressing at 50 MPa, dwell time 2 min, and then with CIP at 180 MPa, dwell time 5 min, the green body densities will be more uniform from center to edge.
(3) For a big size product, applying raw materials particle size below 0.6 μm will get more dense material.

Combining with the above three methods. The final product can achieve densification. The sintered sample can be achieved with density of 3.12 g/cm^3, Vickers hardness of 26 GPa, three-point flexure strength of 420 MPa.

References

1. S. Prochazka, Hot pressed silicon carbide. U.S. Patent 3853566, 1974
2. E.R. Maddrell, Pressureless sintering of silicon carbide. J. Mater. Sci. Lett. **6**, 486–488 (1987)
3. W. Van Rijswjk, D.J. Shanafeld, Effects of carbon as a sintering aid in silicon carbide. J. Am. Ceram. Soc. **73**, 148–149 (1990)
4. R. Hamminger, Carbon in clusions in sintered silicon carbide. J. Am. Ceram. Soc. **72**, 1741–1744 (1989)
5. Giuseppe Magnanin, Alida Brentari, Pressureless sintered silicon carbide with enhanced mechanical properties obtained by the two-step sintering method. Ceram. Int. **40**, 1759–1763 (2014)
6. V. Lankau, Preparation and thermoelectric characterization of SiC-B_4C composites. J. Electron. Mater. **39**, 1809–1813 (2010)
7. A.M. Merlo, The contribution of surface engineering to the product performance in the automotive industry. Surf. Coat. Technol. **174**, 21–26 (2003)
8. Duan Chunlei, Influence of forming method on pressureless sintered silicon carbide. China Ceram. **49**, 61–63 (2013)

Effect of Functional Additives on Performance of Internal Combustion Engine Lubricating Oil

Jingqiu Mu, Jinlong Wu, Fenglan Han, Xueyan Li and Yaguang Wang

Abstract In order to improve the quality of lubricants and prolong the life of oil products, functional additives should be added to the base oil. The functional additives of lubricating oil mainly include detergent, dispersant, antioxidant and rust inhibitor. The purpose of present research is to investigate the effect of additives and the dosage of them on the performance of internal combustion engine lubricants. Certain amount of additives were added to the base oil for testing. Detergent T106A and T115B, dispersant T154B, antioxidant T203 and rust inhibitor T705 were selected as functional additives in the experiment. The detergency of the oil was evaluated by coking test. The dispersibility of lubricating oil was evaluated by spot dispersion experiment at low temperature and high. The kinematic viscosity of the lubricating oil was determined according to the method of GB/T265. The infrared spectra of the lubricating oil were used to determine the oxidation value and nitrification value and the flash point of the lubricant oil was measured by flash point meter. The results showed that both detergent T106A and T115B had a good clean ability, and T106A showed better effect than T115B does. In combination with the standard of oil use in internal combustion engine, economic benefit and the reference amount provided by the manufacturer, the optimal dosage of detergent T106A is 2.0 wt%, and the optimal dosage of detergent T115B is 3.0 wt%. The dispersant T154B had good sensitivity at high temperature and also at low temperature, and the optimal dosage was 3.5 wt%. Antioxidant T203 had a good anti-corrosion function, and the optimal dosage was 3.0 wt%. The addition of the rust inhibitor T705 resulted in an overall reduction in the flash point and viscosity of the lubricating oil, and the optimal dosage of rust inhibitor T705 was 4.0 wt%.

Keywords Lubricants · Functional additives

J. Mu · J. Wu · F. Han (✉) · X. Li · Y. Wang
North Minzu University, Yinchuan 750021, China
e-mail: 625477897@qq.com

Y. Han (ed.), *Advanced Functional Materials*,
https://doi.org/10.1007/978-981-13-0110-0_104

Introduction

In general, the commonly used lubricating oil is made up of two parts, the basic oil and the additive. Base oil accounted for 70–90 wt% of the entire lubricating oil, which determined the basic physical and chemical properties of lubricating oil [1]. But the simple base oil can not meet the needs of the use, it is necessary to add some additives, such as detergents, dispersants, antioxidants, anti-wear agent to improve the quality and prolonged life of lubricating oil [2]. Although the proportion of additive is small, it has a great impact on performance. Adding additives to the base oil can improve the quality of lubricating oil and expand the variety of lubricating oil, and it is also an important means to improve the performance of lubricating oil and reduce environmental pollution.

Various additives have their respective functions. Most of the detergent is alkaline. The main function of detergent is to neutralize the acid which is produced by the oxidation of the lubricating oil during the use, so as to prevent the corrosion of the device and prolong the life of lubricant. At the same time, the detergent can form a protective film on the metal surface to prevent the deposition of the glial and other particles on the metal surface [3]. Dispersant is mainly used to disperse solid particles such as carbon, which can be dissolved in oil to prevent formation of sludge and dissolve acid oxidation products at the same time [4]. The biggest difference between dispersant and detergent is that dispersant do not produce ash when it is burned. So dispersant is called non-ash dispersant [5]. Among many additives, the amount of antioxidants is second only to the detergent and dispersant. ZDDP is still mainly used at home and abroad. With the development of science, other auxiliary antioxidants have also been developed. Water enters into the oil becomes an inevitable phenomenon because the device is not tightly sealed during operation. The combination of oil, water and oxygen often causes the mechanical device to rust [6]. Therefore, it is necessary to add a certain amount of anti-rust agent to the lubricating oil to solve this problem. Antioxidants can form a thin film on the metal surface, and it blocks water and oxygen and metal contact, thus preventing rust.

Engine lubricant is a kind of lubricant which is directly used in internal combustion engine and can reduce the friction. Choosing the right engine oil can not only prolong the life of the internal combustion engine, but also can save fuel and improve the efficiency of work [7]. With the increasing maturity of the internal combustion engine and the increasing awareness of environmental protection, there is an increasing expectation and demand for the engine oil [8]. The development of the mechanical industry and the expansion of the automobile market have greatly stimulated the internal combustion engine oil industry. The operation of the automobile engine produces a lot of noise and vibration, and the great friction can affect the life of the engine. In order to overcome this problem, in the process of improving the manufacturing technology of the internal combustion engine, the lubricating oil has been updated and gradually moved to high-end. In the face of more and more serious emissions, the country has put forward higher emission standards, and the quality of internal combustion engine lubricants has been improved [9].

In this study, the internal combustion engine commonly used additives (detergents T106A and T115B, dispersant T154B, antioxidant T203, rust inhibitor T705) as experimental functional additives. The effects of different amounts on the cleanliness, dispersibility, kinematic viscosity, oxidation value, nitrification value and flash point of lubricating oil were investigated.

Experimental Methods

Major Reagents. Base oil 400SN, base oil 150BS, base oil 250 N were obtained from Shenzhen Zhongrun Tong Chemical Co. Ltd; detergent T106A, detergent T115B, dispersant T154B, antioxidant T203, rust inhibitor T705 were obtained from Jinzhou Baxter Chemical Co. Ltd.

Functional Additives Single Agent Experiment. According to the reference dosage range provided by the manufacturer, the above functional additives were added to the base oil of the internal combustion engine by a certain quality score gradient. The specific performance of lubricating oil was determined by the following specific detection methods, and it determined the optimum range of application of the single additive.

Assessment of Clean Performance. Different concentrations of detergents were added in base oil which had been prepared, and made it fully mixing. The cleaning effect of lubricating oil was determined by coking experiment. Coking experiment was used to evaluate the thermal stability of lubricating oil, it was judged by the coke content of lubricating oil heating and high temperature aluminum generated short contact.

Evaluation Method of Dispersion Performance. Dispersant with certain mass fraction was added to base oil, and made the oil and dispersant fully mixed. Spot dispersion test was adopted to determine the dispersion of oil. Speckle dispersion test was one of the most convenient methods for measuring the dispersibility of oils. Test oil was dropped in the center of a flat filter paper. After 24 h of static, the diameter of the diffusion ring and the oil ring were measured. The percentage of the two was the diffusion coefficient. The experiment was divided into low temperature (100 ± 5 °C) and high temperature (250 ± 10 °C). The dispersity was evaluated by measuring the dispersion coefficient at different temperatures.

Evaluation of Antioxidant Properties. The oil was further oxidized by laboratory simulated oxidation experiments. At a constant temperature of 160 °C, the oil was blown out of the air for 12 h and stirred at the same time. The oxygen resistance of lubricating oil is characterized by the kinetic viscosity of the oil at 100 °C and the oxidation value and nitration value:

Viscosity. Viscosity is an important indicator of the assessment of oil, and the viscosity of the oil was tested by using the Ubbelohde viscometer.

The Oxidation Value and Nitrification Value. The oxidation value and nitration value were determined by infrared spectroscopy. Infrared spectroscopy is an important tool to study the decline of oil. In this experiment, the specific absorption peaks were analyzed by WQF-520A Fourier Transform Infrared Spectrometer. The degree of oxidation is determined by obtaining the relative amount of oxidation products and nitration products produced in the process of oxidation.

Test Method for Rust Resistance. Add a certain percentage of rust inhibitor to the base oil, and mix evenly. The influence of rust inhibitor T705 on the viscosity and flash point of lubricating oil was determined.

Results and Discussion

Detergents. Different concentrations of lubricating oil with detergent T106A or T115B were prepared. Coking experiment results are shown in Fig. 1.

As it can be seen from Fig. 1, the amount of coke is decreasing linearly with the increase of the amount of detergent T106A, or T115B, which indicates that the cleaning effect is improved with the increase of T106A or T115B addition. The results showed that the cleaning agent T106A and T115B had good cleaning effect, and the effect of T106A is stronger than T115B.

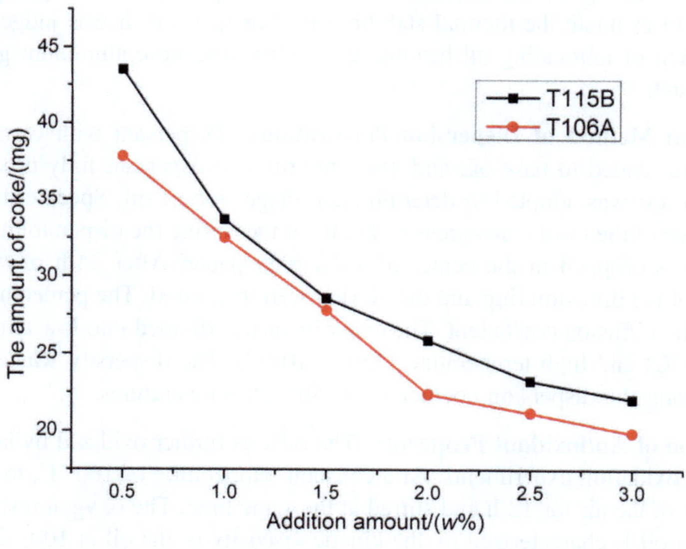

Fig. 1 Detergent T106A, T115B single-agent susceptibility test

In combination with the standard of oil use in internal combustion engine, economic benefit and the reference amount provided by the manufacturer, the optimal dosage of detergent T106A is 2.0 wt%, and the optimal dosage of detergent T115B is 3.0 wt%.

Dispersant. Different concentrations of lubricating oil with dispersant T154B were prepared. The results of the dispersion coefficient of the spot dispersion test at low temperature (100 °C) and high temperature (250 °C) are shown in Fig. 2.

As it can be seen from Fig. 2, the dispersion coefficient at low temperature and high temperature of lubricating oil gradually increased with the amount of dispersant T154B increased, the dispersion performance gradually increased. According to the dispersion experiment of low temperature and high temperature spots, T154B had good dispersion properties for internal combustion engine oil.

According to the standard of lubricants used in internal combustion engine, the manufacturers provide reference amount and economic benefit, the optimal dosage of dispersant T154B was 3.5 wt%.

Antioxidants. Different concentrations of lubricating oil with antioxidant T203 were prepared. The results of antioxidant T203 single agent on the kinematic viscosity of the susceptibility test are shown in Fig. 3, and the results of effect of additive amount of antioxidant T203 on oxidation value and nitration value are shown in Fig. 4.

As shown in Fig. 3, antioxidant T203 had influence on the kinematic viscosity (at 100 °C) of the lubricating oil, and the viscosity increased gradually with the increasing amount of additive. When the addition amount was less than 1.5*w*%, the

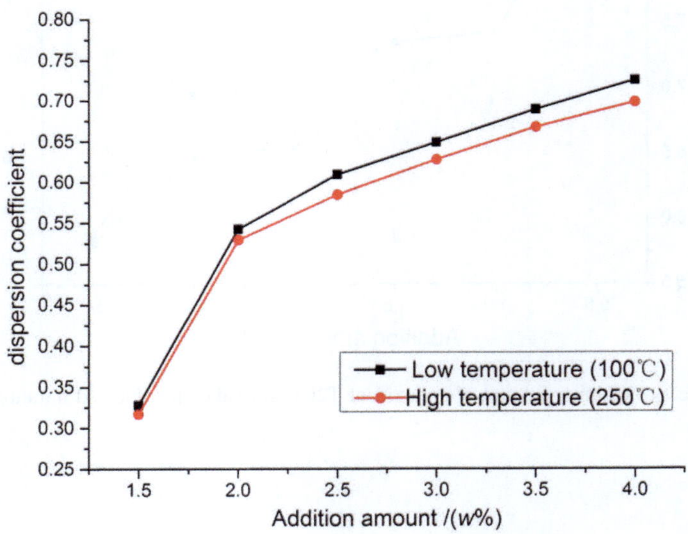

Fig. 2 Results of the dispersion coefficient of the spot dispersion test at low temperature (100 °C) and high temperature (250 °C)

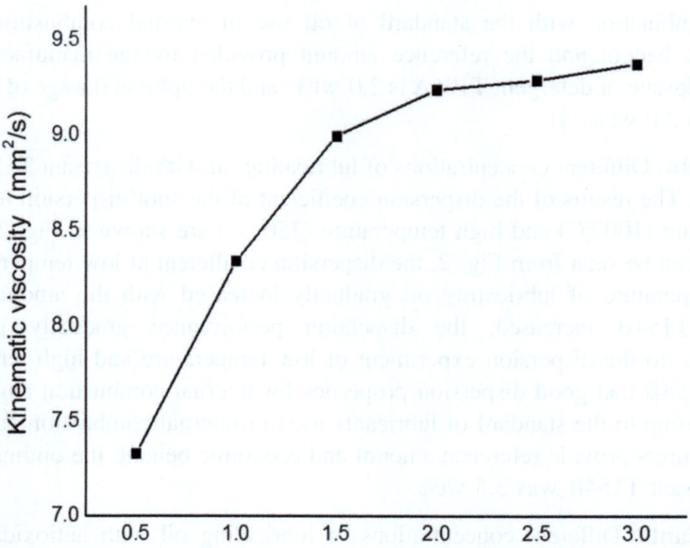

Fig. 3 Antioxidant T203 single agent on the Kinematic viscosity of the susceptibility test

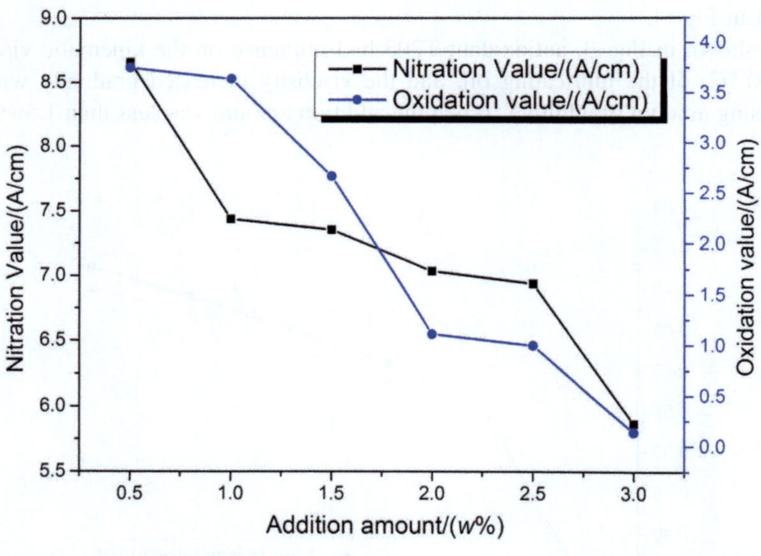

Fig. 4 Effect of additive amount of antioxidant T203 on oxidation value and nitration value

viscosity increased obviously. The curve was still increasing after adding more than 1.5 wt%, but the trend was to be gentle and the viscosity did not change too much.

As shown in Fig. 4, both the oxidation value and the nitrification value showed a decreasing trend with the increase of T203. The results showed that Antioxidant T203 had a good anti corrosion function. In combination with the standard of oil use in internal combustion engine, economic benefit and the reference amount provided by the manufacturer, and the optimal dosage was 3.0 wt%.

Rust Inhibitor. Different concentrations of lubricating oil with rust inhibitor T705 were prepared. The results of antioxidant T203 single agent on the viscosity of the susceptibility test are shown in Fig. 5, and the results of sensitivity test of kinematic viscosity of rust inhibitor T705 are shown in Fig. 6.

As can be seen from Fig. 5, the flash point of the oil dropping regularly after adding different amounts of rust inhibitor T705. It is possible that the flash point of the rust inhibitor is lower than that of base oil, so the flash point of oil is changed after adding rust inhibitor, but the over all influence is very small. It is also possible that the poor thermal stability of the rust inhibitor, a part of volatilization at high temperature results the reduction of flash point. The experiment showed that the addition of T705 would reduce the flash point of base oil.

It can be seen from Fig. 6, the overall presentation of oil viscosity decreased with the increase of the amount of T705. The addition of the rust inhibitor T705 resulted in an overall reduction in the flash point and viscosity of the lubricating oil, and the optimal dosage of rust inhibitor T705 was 4.0 wt%.

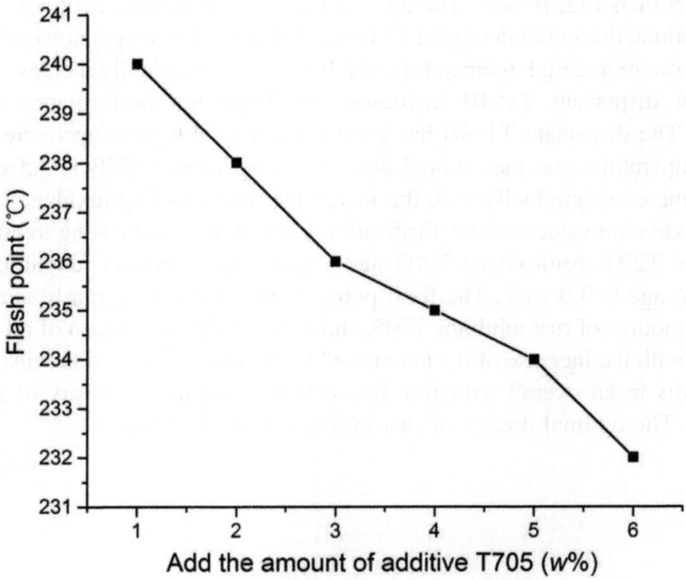

Fig. 5 Susceptibility test of rust inhibitor T705 to flash point

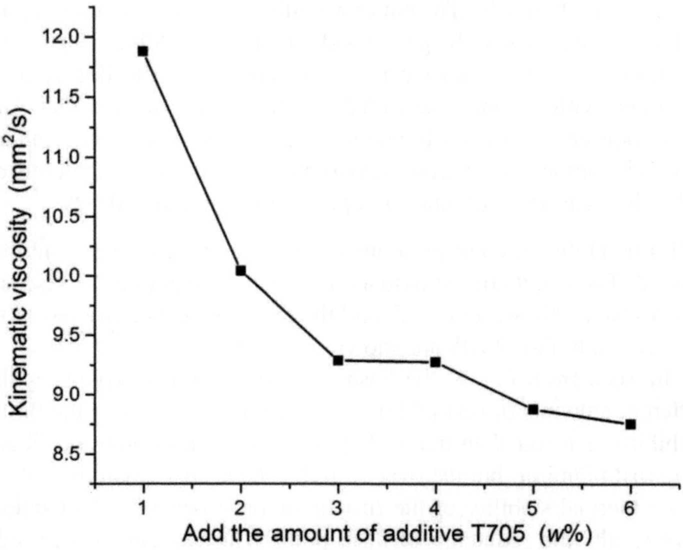

Fig. 6 Sensitivity test of kinematic viscosity of rust inhibitor T705

Conclusion

The cleaning effect is improved with the increase of T106A or T115B addition. Both detergent T106A and T115B have a good clean ability, and T106A shows the better effect than T115B does. The optimal dosage of detergent T106A is 2.0 wt%, and the optimal dosage of detergent T115B is 3.0 wt%. The dispersion coefficient at low temperature and high temperature of lubricating oil gradually increases with the amount of dispersant T154B increases, the dispersion performance gradually increases. The dispersant T154B has good sensitivity at high temperature and also at low temperature, and the optimal dosage of dispersant T154B is 3.5 wt%. The viscosity increases gradually with the increasing amount of antioxidant T203, and both the oxidation value and the nitrification value show a decreasing trend with the increase of T203. Antioxidant T203 has a good anti-corrosion function, and the optimal dosage is 3.0 wt%. The flash point of the oil drops regularly after adding different amounts of rust inhibitor T705, and the overall presentation of oil viscosity decreases with the increase of the amount of T705. The addition of the rust inhibitor T705 results in an overall reduction in the flash point and viscosity of the lubricating oil. The optimal dosage of rust inhibitor T705 is 4.0 wt%.

References

1. W-q Yao, Y-l Wu, J-y Liu, Base oil and additives for modern gasoline engine lube. J. Chang'an Univ. (Nat. Sci. Ed.) **03**, 71–74 (2003)
2. S. Liang, J. Fan, J. Zhang, L. Ding, Research progress in lubricating oil detergent metalation process. Chem. Ind. Eng. Prog. **08**, 1451–1456 (2010)
3. L. Huang, Lecture 4: the mechanism of lubricating oil detergent, the main varieties and performance. Petrol. Technol. **01**, 88–95 (2016)
4. Wang Hui, Deng Yuanli, Development of lubricant ashless dispersant. Lubricating Oil **05**, 39–43 (2007)
5. W. Huang, Chapter 5: Lubricant dispersant varieties and performance. Petrol. Technol. **02**, 90–96 (2016)
6. J. Li, J. Wang, C. Li, W. Zhang, J. Wang, Research on water-resistance rust inhibitor of turbine oils. Lubr. Eng. **08**, 126–128 (2016)
7. Z.H.A.O. Feng-tian, L.I.U. Yao-feng, F.A.N.G. Di, Key points of internal combustion engine lubricating oil. Light Veh. Technol., Z **2**, 76–78 (2015)
8. F-g Jiang, The development status and trend of lubricating oil in automobile internal combustion engine. China Petrol. Chem. Ind. Stan. Qual. **08**, 101 (2011)
9. H.-x. Zhao, Application and development of lubricating oil in internal combustion engine. China Automotive Engineering Society Automotive Fuel and Lubricating Oil Branch. China Automotive Engineering Society Fuel and Lubricant Branch of the 13th Annual Conference Proceedings. China Society of Automotive Engineers With Lubricants Branch (2008), p. 5

MIX
Papier aus verantwortungsvollen Quellen
Paper from responsible sources
FSC® C105338
FSC
www.fsc.org

If you have any concerns about our products,
you can contact us on
ProductSafety@springernature.com

In case Publisher is established outside the EU,
the EU authorized representative is:
Springer Nature Customer Service Center GmbH
Europaplatz 3, 69115 Heidelberg, Germany

Printed by Libri Plureos GmbH
in Hamburg, Germany